STRUCTURAL LOAD DETERMINATION

STRUCTURAL LOAD DETERMINATION

2024 IBC® and ASCE/SEI 7-22

David A. Fanella, Ph.D., S.E., P.E., F.ASCE, F.SEI, F.ACI

New York Chicago San Francisco Athens London
Madrid Mexico City Milan New Delhi
Singapore Sydney Toronto

Structural Load Determination: 2024 IBC® and ASCE/SEI 7-22

1 2 3 4 5 6 7 8 9 LKV 28 27 26 25 24 23

Library of Congress Control Number: 2023946036

ISBN 978-1-264-96170-2
MHID 1-264-96170-7

Sponsoring Editor
Ania Levinson

Editorial Supervisor
Janet Walden

Project Manager
Radhika Jolly,
KnowledgeWorks Global Ltd.

Acquisitions Coordinator
Olivia Higgins

Copy Editor
Nupur Mehra

Proofreader
Rupnarayan

Indexer
Michael Ferreira

Production Supervisor
Lynn M. Messina

Composition
KnowledgeWorks Global Ltd.

Illustration
KnowledgeWorks Global Ltd.

Art Director, Cover
Jeff Weeks

ICC Staff

Executive VP and Director of Business Development
Mark Johnson

Senior VP, Product Development
Hamid Naderi

VP and Technical Director
Doug Thornburg

Product Development Manager
Mary Lou Luif

About the Author

David A. Fanella, Ph.D., S.E., P.E., F.ASCE, F.SEI, F.ACI, is the Senior Director of Engineering at the Concrete Reinforcing Steel Institute. He has 30 years of experience in the design of a wide variety of buildings and other structures. Dr. Fanella has authored many technical publications, including the second edition of a textbook on reinforced concrete design for McGraw Hill. He is a Member of a number of ACI Committees and is a Fellow of ASCE, SEI, and ACI. He is a Voting Member of ASCE/SEI 7 Committee, Minimum Design Loads and Associated Criteria for Buildings and Other Structures. Dr. Fanella is a licensed Structural Engineer and a licensed Professional Engineer in Illinois and is a past board member and a past president of the Structural Engineers Association of Illinois.

About the International Code Council

The International Code Council is the leading global source of model codes and standards and building safety solutions that include product evaluation, accreditation, technology, codification, training, and certification. The Code Council's codes, standards, and solutions are used to ensure safe, affordable, and sustainable communities and buildings worldwide.

The International Code Council (ICC) family of solutions includes the ICC Evaluation Service (ICC-ES), S. K. Ghosh Associates, the International Accreditation Service (IAS), the General Code, ICC NTA, ICC Community Development Solutions, Alliance for National & Community Resilience (ANCR), ICC Consulting Services, and American Legal Publishing, which are dedicated to the construction of safe, sustainable, affordable, and resilient structures.

Washington DC Headquarters:

200 Massachusetts Ave, NW, Suite 250, Washington, DC 20001

Regional Offices:

Eastern Regional Office (BIR)
Central Regional Office (CH)
Western Regional Office (LA)
Distribution Center (Lenexa, KS)
888-ICC-SAFE (888-422-7233); www.iccsafe.org

Family of Solutions:

About the National Council of Structural Engineers Associations

The National Council of Structural Engineers Associations (NCSEA) was founded in 1993 as an autonomous federation of state Structural Engineers Associations (SEAs) from throughout the United States. NCSEA advances the practice of structural engineering by representing and strengthening these 44 state-based SEA Member Organizations. In addition, NCSEA provides industry-leading education through its webinars, Annual Summit, publications, and magazines. Based in Chicago, NCSEA's governing body is its Board of Directors.

20 N. Wacker Drive, Suite 750, Chicago, IL 60606

1-312-649-4600; www.NCSEA.com

About S. K. Ghosh Associates

S. K. Ghosh Associates LLC is a member of the ICC Family of Solutions. The company provides seismic and code-related consulting services to engineers, businesses, trade associations, code-writing bodies, and governmental agencies involved in the design and construction of buildings and other structures that are impacted by the provisions of building codes and structural design standards. Technical support is provided through publications, webinars, seminars, peer reviews, research projects, computer programs, code interpretations and comparisons, a website, and other means.

334 East Colfax Street, Unit E, Palatine, IL 60067

1-847-991-2700; www.skghoshassociates.com

About the Structural Engineering Institute

The Structural Engineering Institute (SEI) was created on October 1, 1996. SEI is a full-service, discipline-oriented, and semi-autonomous institute within the American Society of Civil Engineers (ASCE). SEI involves all facets of the structural engineering community including practicing engineers, research scientists, academicians, technologists, material suppliers, contractors, and owners. By facilitating coalitions or as an independent activity, SEI is committed to advancing the structural engineering profession and rapidly responding to the emerging needs of the broad structural engineering community.

1801 Alexander Bell Drive, Reston, VA 20191

1-800-548-2723; www.asce.org/sei

Contents

Preface

This edition updates this publication to the 2024 *International Building Code*® (IBC®) and the 2022 edition of *Minimum Design Loads and Associated Criteria for Buildings and Other Structures* (ASCE/SEI 7-22).

Like previous editions, this edition is an essential resource for civil and structural engineers, architects, plan check engineers, and students who need an efficient and practical approach to load determination under the 2024 IBC and ASCE/SEI 7-22 standard. It illustrates the application of code provisions and methodology for determining structural loads through the use of numerous flowcharts and practical design examples. Included are the following major topics:

- Load combinations for allowable stress design, load and resistance factor (strength) design, seismic load combinations with vertical load effect and special seismic load combinations, and
- Dead loads, live loads (including live load reduction), rain loads, snow loads, ice loads, wind and tornado loads, earthquake load effects, flood loads, and tsunami loads.

Practical example problems are included at the end of most of the chapters. Solutions to these problems, which are available in a free downloadable companion document to this publication, further illustrate the proper application of the code provisions.

Numerous changes occurred in the 2022 edition of ASCE/SEI 7. Some of the major changes are as follows:

- Digital data are available for all hazards in the ASCE 7 Hazard Tool.
- Updated wind speeds along the U.S. hurricane coastline.
- Revised provisions for main force wind-resisting systems and components and cladding of elevated buildings, including removal of the simplified methods for wind load determination.
- A new chapter on tornado loads and effects, including tornado wind speed maps.
- New long return period hazard maps for wind and tornados.
- Updated tsunami data for Hawaii and many populated locations in California, which were coordinated with the state agencies.
- New tsunami provisions for above-ground horizontal pipelines.
- Updated ground snow load data to reflect more recent snow load data and reliability-targeted values.
- A revised method for estimating snow drifts, which includes a new wind parameter.
- A revised method of determining rain loads, which explicitly includes ponding.
- New risk-targeted atmospheric ice load data for the continental United States and Alaska.
- New multi-period response spectrum data that streamlines the determination of earthquake load effects.
- New seismic force-resisting systems, including reinforced concrete ductile coupled shear walls.

- New provisions for the determination of design seismic diaphragm forces for buildings with rigid shear walls and flexible diaphragms.
- New seismic provisions for supported and interconnected (coupled) nonbuilding structures.

Two major revisions also occurred to this edition of the publication:

- Many new flowcharts and completely worked-out examples have been included to enhance the overall learning experience.
- Both inch-pound and S.I. units are used throughout this edition, including in the equations, figures, tables, flowcharts, and examples.

Structural Load Determination: *2024 IBC® and ASCE/SEI 7-22* is a multipurpose resource for civil and structural engineers, architects, building officials, plan checkers, individuals studying for licensing exams and university professors and students because it can be used as a self-learning guide as well as a reference manual.

Acknowledgments

The author is deeply grateful to John "Buddy" Showalter, P.E., M.ASCE, Senior Staff Engineer, Product Development, at the International Code Council for his thorough review of this manuscript. His helpful comments and suggestions are sincerely appreciated.

Enhance Your Study Experience

The Solutions Manual to Structural Load Determination is a free bonus learning tool just right for you.

Structural Load Determination: 2024 IBC® and ASCE/SEI 7-22 includes a downloadable companion Solutions Manual to help enhance understanding of how to solve structural load problems. Included in the book at the end of most chapters are many practical and real-situation problems, and the Solutions Manual provides their solutions; these will be of value to practitioners, students, candidates for engineering license exams, building officials, and other building industry professionals. The Solutions Manual covers Chapters 2 through 7 and includes the following:

- Chapter 2—Load combinations
- Chapter 3—Dead, live, rain, and soil lateral loads
- Chapter 4—Snow and ice loads
- Chapter 5—Wind and tornado loads
- Chapter 6—Earthquake loads
- Chapter 7—Flood and tsunami loads

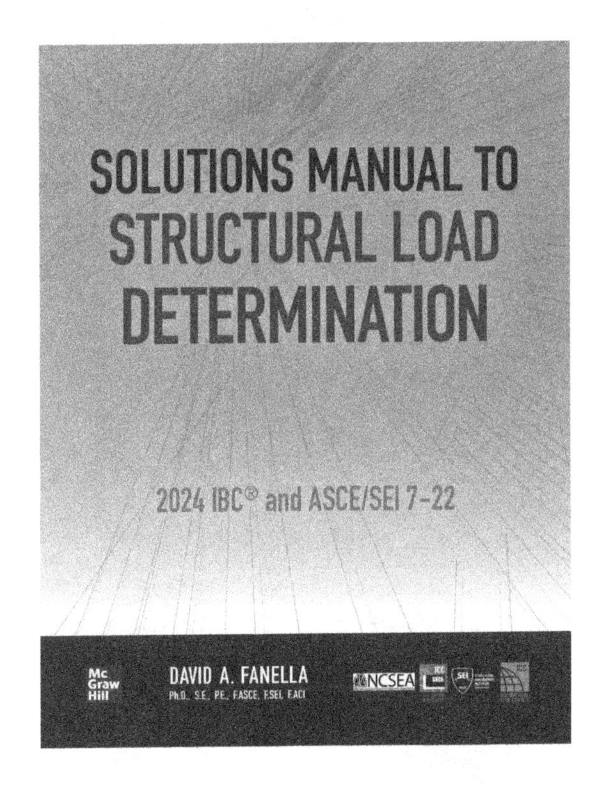

To download your free bonus Solutions Manual to Structural Load Determination, visit:

www.iccsafe.org/solutionsmanual2024
or
https://www.mhprofessional.com/StructuralLoad2024

STRUCTURAL LOAD DETERMINATION

CHAPTER 1
Introduction

1.1 Overview

The purpose of this book is to assist in the proper determination of structural loads in accordance with the 2024 edition of the *International Building Code®* (IBC®) (Reference 1.1) and the 2022 edition of ASCE/SEI 7 *Minimum Design Loads and Associated Criteria for Buildings and Other Structures* (Reference 1.2). Chapter 16 of the IBC, Structural Design, prescribes minimum structural loading requirements for buildings, other structures, and their nonstructural components. The design strengths or allowable stress limits in the material design specifications must be greater than or equal to the effects due to these minimum loads, which are combined in accordance with the applicable load combinations in IBC Section 1605.

The IBC references the load provisions in ASCE/SEI 7, which is one of several codes and standards referenced in Chapter 35 of the IBC. These referenced standards are considered part of the requirements of the IBC to the prescribed extent of each reference (see Section 101.4 of the IBC).

The seismic requirements of the 2024 IBC and ASCE/SEI 7-22 are based primarily on those in Reference 1.3. The National Earthquake Hazards Reduction Program (NEHRP) recommended seismic provisions, which has been updated every 3 to 5 years since the first edition in 1985, contains state-of-the-art criteria for the design and construction of buildings subject to the effects of earthquake ground motion located anywhere in the United States and its territories. Life safety is the primary goal of the provisions. The requirements are also intended to enhance the performance of high-occupancy buildings and to improve the capability of essential facilities to function during and after a design-basis earthquake.

In addition to minimum design load requirements, IBC Chapter 16 includes important criteria that have a direct impact on the design of buildings and other structures. For example, provisions for structural integrity are given in Section 1616, which are applicable to high-rise buildings assigned to risk category III or IV with structural systems consisting of bearing walls or frames. "High-rise buildings" are defined in Section 202 as a building with an occupied floor or occupiable roof located more than 75 ft (22.9 m) above the lowest level of fire department vehicle access. Definitions of bearing wall structures and frame structures are also given in Section 202.

Risk categories are defined in IBC Table 1604.5. Buildings and other structures must be assigned to one of the risk categories in that table based on the risk to human life, health, and welfare associated with their damage or failure by nature of their occupancy or use.

Risk category I buildings and structures are usually unoccupied and, as such, result in negligible risk to the public should they fail. Included are agricultural facilities (such as barns), certain temporary facilities, and minor storage facilities.

Many buildings and structures, including most residential, commercial, and industrial facilities, fall under risk category II. According to IBC Table 1604.5, any building or structure not listed in risk category I, III, or IV is assigned to risk category II.

Places of public assembly, educational occupancies, institutional facilities, and structures associated with utilities required to protect the health and safety of a community, such as power-generating stations and water and sewage treatment facilities, are included in risk category III. Also included are buildings and structures containing certain amounts of toxic or explosive substances exceeding a threshold quantity established by the building official or that are sufficient to pose a threat to the public if released.

Essential structures, including hospitals, fire stations, police stations, rescue facilities, and designated emergency shelters are some of the types of structures included in risk category IV. Power-generating stations and other public utility facilities required as emergency backup facilities and buildings or structures that house quantities of highly toxic materials exceeding the amounts specified in IBC Table 1604.5 also fall in this risk category.

For buildings and other structures located within a tsunami design zone defined in Chapter 6 of ASCE/SEI 7, the required tsunami risk categories are the same risk categories given in ASCE/SEI 1.5 (IBC Table 1604.5) subject to the modifications in ASCE/SEI 6.4.

Importance factors are directly related to the risk category of a building or structure and are referenced in ASCE/SEI 7 Chapter 11 (seismic) for the four risk categories noted above (see ASCE/SEI Table 1.5-2). In general, larger importance factors are assigned in situations where the consequence of failure due to a design seismic event may be severe.

1.2 Scope

The load requirements of the IBC and ASCE/SEI 7 are presented in this book in a straightforward manner with emphasis placed on the proper application of the provisions in everyday practice.

Code provisions have been organized in comprehensive flowcharts, which provide a road map that guides the reader through the requirements. Included in the flowcharts are the applicable section numbers and equation numbers from the IBC and ASCE/SEI 7 that pertain to the specific requirements.

Numerous completely worked-out design examples are included in the chapters that illustrate the proper application of the code requirements. These examples follow the steps provided in the referenced flowcharts. Problems are included at the end of each chapter (except for Chapter 8), and solutions to these problems are available in a companion document to this book.

Throughout this book, section numbers from the IBC are referenced as illustrated by the following: Section 1613 of the 2024 IBC is denoted as IBC 1613. Similarly, Section 11.4 from ASCE/SEI 7-22 is referenced as ASCE/SEI 11.4.

Chapter 2 covers the required load combinations that must be considered when designing a building or its members for a variety of load effects. Load combinations using strength design or allowable stress design are both included. Examples are provided that illustrate the strength design and allowable stress design load combinations for different types of members subject to a variety of load effects.

Dead, live, and rain loads are discussed in Chapter 3. The general method and an alternate method of live load reduction are covered, and flowcharts and examples illustrate both methods. Numerous examples demonstrate the calculation of design rain loads for roofs.

Design provisions for snow loads are given in Chapter 4. A series of flowcharts highlight the requirements, and examples show the determination of flat roof snow loads, sloped roof snow loads, unbalanced roof snow loads, and snow drift loads on a variety of flat and sloped roofs, including gable roofs, monoslope roofs, sawtooth roofs, and curved roofs. Examples are also given that illustrate the design snow loads for parapets and rooftop units. Included in Chapter 4 are design provisions for atmospheric ice loads.

Chapter 5 presents the design requirements for wind and tornado loads. Flowcharts give step-by-step procedures on how to determine design wind and tornado pressures on main wind force-resisting systems and components and cladding of enclosed, partially enclosed, and open buildings using the provisions outlined in ASCE/SEI 7. Worked-out examples illustrate the design requirements for a variety of buildings and structures.

Earthquake loads are presented in Chapter 6. Information on how to determine the design ground accelerations, site class, and seismic design category (SDC) of a building or structure is included, as are the various methods of analysis and their applicability for regular and irregular buildings and structures. Flowcharts and examples are provided that cover seismic design criteria and seismic design requirements for building structures, nonstructural components, and non-building structures.

Chapter 7 contains the requirements for flood and tsunami loads. Included is information on flood and tsunami hazard areas and hazard zones. Methods are provided for the determination of hydrostatic loads, hydrodynamic loads, wave loads, and impact loads. Examples illustrate load calculations for a variety of different building and load types.

Load paths are covered in Chapter 8. Gravity and lateral load paths are presented for various types of structures. The role of diaphragms and collectors are discussed for wind and seismic loads. Included are details that illustrate common load paths for a variety of situations.

Both inch-pound and S.I. units are used throughout this book, including in the equations, figures, tables, flowcharts, examples, and problems. In the examples, calculations are performed independently using both sets of units; in other words, the calculations are not performed in one set of units and then converted to the other. Thus, in some cases, the numerical results in inch-pound units do not "exactly" convert to the corresponding numerical results in S.I. units or vice versa.

1.3 References

1.1. International Code Council (ICC). 2023. 2024 *International Building Code*. Washington, DC.

1.2. Structural Engineering Institute of the American Society of Civil Engineers (ASCE). 2022. *Minimum Design Loads and Associated Criteria for Buildings and Other Structures*. ASCE/SEI 7-22, Reston, VA.

1.3. Building Seismic Safety Council (BSSC). 2020. *NEHRP Recommended Seismic Provisions for New Buildings and Other Structures*. FEMA-P-2082, Washington, DC.

CHAPTER 2

Load Combinations

2.1 Introduction

Structural members of buildings and other structures must be designed to resist the strength load combinations of ASCE/SEI 2.3, the allowable stress design load combinations of ASCE/SEI 2.4, or the alternative allowable stress design load combinations of IBC 1605.2 (IBC 1605.1).

Four exceptions are given related to the provisions of IBC 1605.1:

1. Modifications to the load combinations of ASCE/SEI 2.3, ASCE/SEI 2.4, and IBC 1605.2 in ASCE/SEI Chapters 18 and 19 must also apply.
2. Where the allowable stress load combinations in ASCE/SEI 2.4 are used, the effects due to flat roof snow loads of 45 lb/ft^2 (2.15 kN/m^2) and roof live loads of 30 lb/ft^2 (1.44 kN/m^2) or less need not be combined with the effects due to seismic loads. Where flat roof snow loads exceed 45 lb/ft^2 (2.15 kN/m^2), 15 percent of the effects due to the flat roof snow load must be combined with the effects due to seismic loads.
3. Where the allowable stress design load combinations in ASCE/SEI 2.4 are used, the effects due to crane hook loads need not be combined with the effects due to roof live loads or with more than three-fourths the snow loads or one-half the wind loads.
4. Where design for tornado loads is required, the alternative allowable stress design load combinations in IBC 1605.2 are not permitted where tornado loads govern the design.

The load combinations in ASCE/SEI 2.3, ASCE/SEI 2.4, or IBC 1605.2 are permitted to be used to check the overall structural stability of a structure, including stability against overturning, sliding, or buoyancy (IBC 1605.1.1). Where the load combinations in ASCE/SEI 2.3 are used, strength reduction factors related to soil resistance must be provided by the registered design professional. Stability of retaining walls must be verified in accordance with IBC 1807.2.3.

Load combinations must be investigated with one or more of the nonpermanent loads set equal to zero (ASCE/SEI 2.3.1). It is possible that the most critical load effects on a member occur when one or more loads are not present.

Prior to examining the various load combinations, a brief introduction on load effects is given in Section 2.2.

2.2 Load Effects

The load effects included in the IBC and ASCE/SEI 7 load combinations are given in Table 2.1. More details on these load effects can be found in those documents, as well as in subsequent chapters of this book (see the References column in Table 2.1, which gives specific locations where more information can be found on the various load effects).

Table 2.1 Load Effects in the IBC and ASCE/SEI 7

Load Effect		References
A_k	Load or load effect due to an extraordinary event, A	ASCE/SEI 2.5 and Section 2.5 of this book
D	Dead load	IBC 1606, ASCE/SEI Chapter 3 and Chapter 3 of this book
D_i	Weight of ice	IBC 1614, ASCE/SEI Chapter 10 and Chapter 4 of this book
E	Earthquake load	IBC 1613, ASCE/SEI 12.4 and Chapter 6 of this book
F	Load caused by fluids with well-defined pressures and maximum heights other than those caused by groundwater pressure	ASCE/SEI Chapter 2
F_a	Flood load	IBC 1612, ASCE/SEI Chapter 5 and Chapter 7 of this book
H	Load due to lateral earth pressure (including lateral earth pressure from fixed or moving surcharge loads), ground water pressure, or pressure of bulk materials	IBC 1610, ASCE/SEI Chapter 3 and Chapter 3 of this book
L	Live load	IBC 1607, ASCE/SEI Chapter 4 and Chapter 3 of this book
L_r	Roof live load	IBC 1607, ASCE/SEI Chapter 4 and Chapter 3 of this book
N	Notional load for structural integrity	ASCE/SEI Chapter 1 and Section 2.6 of this book
R	Rain load	IBC 1611, ASCE/SEI Chapter 8 and Chapter 3 of this book
S	Snow load	IBC 1608, ASCE/SEI Chapter 7 and Chapter 4 of this book
T	Cumulative effect of self-straining forces and effects due to contraction or expansion resulting from environmental or operational temperature changes, shrinkage, moistures changes, creep in component materials, movement caused by differential settlement or combinations thereof	ASCE/SEI Chapter 2
W	Wind load	IBC 1609, ASCE/SEI Chapters 26 to 31 and Chapter 5 of this book
W_i	Wind-on-ice	IBC 1614, ASCE/SEI Chapter 10 and Chapter 4 of this book
W_T	Tornado load	IBC 1609, ASCE/SEI Chapter 32 and Chapter 5 of this book

2.3 Load Combinations for Strength Design

2.3.1 Basic Combinations

Basic load combinations for strength design are given in ASCE/SEI 2.3.1 (see Table 2.2). These equations, which apply only to strength limit states, establish the minimum required strength that needs to be provided in the members of a building or structure. Serviceability limit states for deflection, vibration, drift, camber, expansion and contraction, and durability are given in ASCE/SEI Appendix C.

Table 2.2 Basic Strength Design Load Combinations

ASCE/SEI Equation No.	Load Combination	Principal Load
1a	$1.4D$	D
2a	$1.2D+1.6L+(0.5L_r$ or $0.3S$ or $0.5R)$	L
3a	$1.2D+(1.6L_r$ or $1.0S$ or $1.6R)+(L$ or $0.5W)$	L_r or S or R
4a	$1.2D+1.0(W$ or $W_T)+L+(0.5L_r$ or $0.3S$ or $0.5R)$	W or W_T
5a	$0.9D+1.0(W$ or $W_T)$	W or W_T

The following exceptions in ASCE/SEI 2.3.1 apply:

1. The load factor on live load, L, in combinations 3a and 4a is permitted to equal 0.5 for all occupancies where the minimum uniformly distributed live load, L_o, in ASCE/SEI Table 4.3-1 is less than or equal to 100 lb/ft^2 (4.78 kN/m^2). This exception does not apply to garages or areas occupied as places of public assembly.
2. Snow load, S, in combinations 2a and 4a must be taken as either the flat roof snow load, p_f, determined in accordance with ASCE/SEI 7.3 or the sloped roof snow load, p_s, determined in accordance with ASCE/SEI 7.4.
3. Where tornado load, W_T, is used in combination 4a, ($0.5L_r$ or $0.3S$ or $0.5R$) is permitted to be replaced with $0.5(L_r$ or $R)$.

The load combinations in Table 2.2 are applicable to structures assigned to SDC A. Wind loads act in more than one direction on a building or structure, and the appropriate sign of the wind load must be considered in the load combinations.

The load factors were developed using a first-order probabilistic analysis and a broad survey of the reliabilities inherent in contemporary design practice. The load combinations in Table 2.2 are meant to be used in the design of any structural member regardless of material in conjunction with the appropriate nominal resistance factors set forth in the individual material specifications. References 2.1 and 2.2 provide information on the development of these load factors along with additional background material.

Load combinations are constructed by adding to the dead (permanent) load one of the principal loads at its maximum lifetime value. Also included in the combinations are other variable loads with load factors less than 1.0; these are companion loads that represent arbitrary point-in-time values for those loads.

The third exception in ASCE/SEI 2.3.1 permits snow load effects, S, to be taken as zero in cases where tornado loads are used because tornados generally occur in warm weather.

Where fluid loads, F, are present, their effects must be included with the same load factor as dead load, D, in combinations 1a through 4a (see Table 2.3). Fluid load effects occur in tanks and other storage containers due to stored liquid products, which are generally considered to have characteristics of both a dead load and a live load. It is not a purely permanent load because the tank or storage container can go through cycles of being emptied and refilled. The fluid load effect must be included in load combinations 1a through 4a where it adds to the effects from the other loads. Because the wind load effects, W, can be present when the tank is either full or empty, F is not incorporated in load combination 5a; that is, the maximum effects occur when F is set equal to zero.

Where loads, H, are present, their effects must be included as follows:

- Where the effect of H adds to the principal load effect, the load factor on H must be 1.6.
- Where the effect of H resists the principal load effect, the load factor on H must be 0.9 where the load H is permanent; for all other conditions, the load factor must be 0.

Table 2.3 Strength Design Load Combinations Including Fluid Loads, F

ASCE/SEI Equation No.	Load Combination
1a	$1.4(D+F)$
2a	$1.2(D+F)+1.6L+(0.5L_r$ or $0.3S$ or $0.5R)$
3a	$1.2(D+F)+(1.6L_r$ or $1.0S$ or $1.6R)+(L$ or $0.5W)$
4a	$1.2(D+F)+1.0(W$ or $W_T)+L+(0.5L_r$ or $0.3S$ or $0.5R)$
5a	$0.9D+1.0(W$ or $W_T)$

2.3.2 Load Combinations Including Flood Loads

Structures located in a designated flood zone (see Chapter 7 of this book) must be designed for the load combinations in Table 2.4 in addition to the applicable basic load combinations in Table 2.2 (ASCE/SEI 2.3.2).

Table 2.4 Strength Design Load Combinations Including Flood Loads, F_a

ASCE/SEI Equation No.	Load Combination
V Zones or Coastal A Zones*	
4b	$1.2D+1.0W+2.0F_a+L+(0.5L_r$ or $0.3S$ or $0.5R)$
5b	$0.9D+1.0W+2.0F_a$
Noncoastal A Zones*	
4b	$1.2D+0.5W+1.0F_a+L+(0.5L_r$ or $0.3S$ or $0.5R)$
5b	$0.9D+0.5W+1.0F_a$

*See ASCE/SEI 5.2 for definitions of V zones, coastal A zones, and noncoastal A zones (also see Chapter 7 of this book).

2.3.3 Load Combinations Including Atmospheric Ice and Wind-on-Ice Loads

Structures subjected to atmospheric ice loads, D_i, and wind-on-ice loads, W_i, must be designed for the load combinations in Table 2.5 (ASCE/SEI 2.3.3).

Table 2.5 Strength Design Load Combinations Including Atmospheric Ice Loads, D_i, and Wind-on-Ice Loads, W_i

ASCE/SEI Equation No.	Load Combination
2b	$1.2D+1.6L+0.2D_i+0.3S$
4c	$1.2D+L+D_i+W_i+0.3S$
4d	$1.2D+D_i$
5c	$0.9D+D_i+W_i$

2.3.4 Load Combinations Including Self-Straining Forces and Effects

In cases where self-straining loads, T, must be considered, their effects in combination with other loads are to be determined by ASCE/SEI 2.3.4. Instead of calculating self-straining effects based on upper bound values like other variable load effects, the most probable effect expected at any arbitrary point in time should be used. The load factor on T must not be taken less than 1.0. The following strength design load combinations in ASCE/SEI C2.3.4 should be considered when checking the capacity of a structure or structural element for effects of self-straining forces and effects:

$$1.2D + 1.2T + 0.5L \tag{2.1}$$

$$1.2D + 1.6L + 1.0T \tag{2.2}$$

2.3.5 Load Combinations for Nonspecified Loads

A methodology on how to develop strength design load criteria where no information on loads or load combinations is given in ASCE/SEI 7 or where performance-based design in accordance with ASCE/SEI 1.3.1.3 is being utilized is given in ASCE/SEI 2.3.5. Detailed information on how to develop such load criteria that is consistent with the methodology used in ASCE/SEI 7 can be found in ASCE/SEI C2.3.5.

2.3.6 Basic Combinations with Seismic Load Effects

Structures subjected to the seismic load effects, E, defined in ASCE/SEI 12.4.2 must be designed for the load combinations in Table 2.6 in addition to the load combinations in Table 2.2 where $E = E_h \pm E_v$ (ASCE/SEI 2.3.6).

Table 2.6 Strength Design Load Combinations with Seismic Load Effects, E_h and E_v

ASCE/SEI Equation No.	Load Combination
6	$1.2D + E_v + E_h + L + 0.15S$
7	$0.9D - E_v + E_h$

The terms E_h and E_v are defined in ASCE/SEI 12.4.2.1 and 12.4.2.2, respectively:

- E_h = horizontal seismic load effect $= \rho Q_E$
- E_v = vertical seismic load effect $= 0.2S_{DS}D$

where ρ is the redundancy factor determined in accordance with ASCE/SEI 12.3.4, Q_E is the effects of the horizontal seismic forces due to the base shear, V, or the seismic force acting on a component of a structure, F_p, and S_{DS} is the design, 5 percent damped spectral response acceleration parameter at short periods determined in accordance with ASCE/SEI 11.4.4.

Substituting E_h and E_v into load combinations 6 and 7 results in the following:

Load combination 6: $(1.2 + 0.2S_{DS})D + \rho Q_E + L + 0.15S$
Load combination 7: $(0.9 - 0.2S_{DS})D + \rho Q_E$

According to the first exception in ASCE/SEI 12.4.2.2, E_v must be determined by ASCE/SEI Equation (12.4-4b) where the effects of vertical seismic ground motion are required to be determined in accordance with ASCE/SEI 11.9:

$$E_v = 0.3S_{av}D \tag{2.3}$$

In this equation, S_{av} is the design vertical response spectral acceleration, which is equal to two-thirds the value of the MCE_R vertical response acceleration, S_{aMv}, determined in accordance with ASCE/SEI 11.9.1 or 11.9.2.

In the second exception in ASCE/SEI 12.4.2.2, E_v is permitted to be taken as zero for either of the following conditions:

- In ASCE/SEI Equations (12.4-1), (12.4-2), (12.4-5), and (12.4-6) for structures assigned to SDC B
- In ASCE/SEI Equation (12.4-2) when determining demands on the soil-structure interface of foundations

Strength design load combinations for structures assigned to SDC B are given in Table 2.7. In accordance with the second exception in ASCE/SEI 12.4.2.2, the vertical seismic load effect, E_v, is set equal to zero and in accordance with ASCE/SEI 12.3.4.1, the redundancy factor, ρ, is set equal to 1.0. Seismic loads act in more than one direction on a building or structure, and the appropriate sign of the seismic load must be considered in the load combinations.

Table 2.7 Strength Design Load Combinations for Structures Assigned to SDC B

ASCE/SEI Equation No.	Load Combination	Principal Load
1a	$1.4D$	D
2a	$1.2D + 1.6L + (0.5L_r$ or $0.3S$ or $0.5R)$	L
3a	$1.2D + (1.6L_r$ or $1.0S$ or $1.6R) + (L$ or $0.5W)$	L_r or S or R
4a	$1.2D + 1.0(W$ or $W_T) + L + (0.5L_r$ or $0.3S$ or $0.5R)$	W or W_T
5a	$0.9D + 1.0(W$ or $W_T)$	W or W_T
6	$1.2D + Q_E + L + 0.15S$	E
7	$0.9D + Q_E$	E

Strength design load combinations for structures assigned to SDC C are given in Table 2.8. In accordance with ASCE/SEI 12.3.4.1, the redundancy factor, ρ, is set equal to 1.0.

Table 2.8 Strength Design Load Combinations for Structures Assigned to SDC C

ASCE/SEI Equation No.	Load Combination	Principal Load
1a	$1.4D$	D
2a	$1.2D + 1.6L + (0.5L_r$ or $0.3S$ or $0.5R)$	L
3a	$1.2D + (1.6L_r$ or $1.0S$ or $1.6R) + (L$ or $0.5W)$	L_r or S or R
4a	$1.2D + 1.0(W$ or $W_T) + L + (0.5L_r$ or $0.3S$ or $0.5R)$	W or W_T
5a	$0.9D + 1.0(W$ or $W_T)$	W or W_T
6	$(1.2 + 0.2S_{DS})D + Q_E + L + 0.15S$	E
7	$(0.9 - 0.2S_{DS})D + Q_E$	E

Strength design load combinations for structures assigned to SDC D, E, or F are given in Table 2.9.

Table 2.9 Strength Design Load Combinations for Structures Assigned to SDC D, E, or F

ASCE/SEI Equation No.	Load Combination	Principal Load
1a	$1.4D$	D
2a	$1.2D+1.6L+(0.5L_r$ or $0.3S$ or $0.5R)$	L
3a	$1.2D+(1.6L_r$ or $1.0S$ or $1.6R)+(L$ or $0.5W)$	L_r or S or R
4a	$1.2D+1.0(W$ or $W_T)+L+(0.5L_r$ or $0.3S$ or $0.5R)$	W or W_T
5a	$0.9D+1.0(W$ or $W_T)$	W or W_T
6	$(1.2+0.2S_{DS})D+\rho Q_E+L+0.15S$	E
7	$(0.9-0.2S_{DS})D+\rho Q_E$	E

Structures required to resist seismic load effects with overstrength, E_m, must be designed for the load combinations in Table 2.10 in addition to the load combinations in Table 2.2 where $E_m = E_{mh} \pm E_v$ (ASCE/SEI 2.3.6).

Table 2.10 Strength Design Load Combinations with Seismic Load Effects Including Overstrength, E_{mh} and E_v

ASCE/SEI Equation No.	Load Combination
6	$1.2D+E_v+E_{mh}+L+0.15S$
7	$0.9D-E_v+E_{mh}$

The term E_{mh} is defined in ASCE/SEI 12.4.3.1 as follows:

E_{mh} = horizontal seismic load effect including overstrength $= \Omega_0 Q_E$

where Ω_0 is the overstrength factor given in ASCE/SEI Table 12.2-1 based on the seismic force-resisting system (SFRS).

Substituting E_{mh} and E_v into load combinations 6 and 7 results in the following:

Load combination 6: $(1.2+0.2S_{DS})D+\Omega_0 Q_E+L+0.15S$

Load combination 7: $(0.9-0.2S_{DS})D+\Omega_0 Q_E$

The horizontal seismic load effect including overstrength, E_{mh}, need not be taken larger than the capacity-limited horizontal seismic load effect, E_{cl}, which is equal to the maximum effect that can develop in a member as determined by a rational, plastic mechanism analysis. Additional information on how to determine E_{cl} is given in ASCE/SEI C12.4.3.2.

Two exceptions are given in ASCE/SEI 2.3.6 pertaining to the basic load combinations with seismic load effects. In the first exception, the load factor on L in load combination 6 is permitted to equal 0.5 for all occupancies where the minimum uniformly distributed live load, L_o, in ASCE/SEI Table 4.3-1 is less than or equal to 100 lb/ft^2 (4.78 kN/m^2). This exception does not apply to garages or areas occupied as places of public assembly. In the second exception, snow load, S, must be taken as either the flat roof snow load, p_f, determined in accordance with ASCE/SEI 7.3 or the sloped roof snow load, p_s, determined in accordance with ASCE/SEI 7.4.

Where fluid loads, F, are present, their effects must be included with the same load factor as dead load, D, in load combinations 6 and 7.

Where loads, H, are present, their effects must be included as follows:

- Where the effect of H adds to the principal load effect, the load factor on H must be 1.6.
- Where the effect of H resists the principal load effect, the load factor on H must be 0.9 where the load H is permanent; for all other conditions, the load factor must be 0.

The most unfavorable effects from seismic loads must be investigated where appropriate, but they need not be considered to act simultaneously with wind or tornado loads.

2.3.7 Alternative Method for Loads from Water in Soil

The provisions in ASCE/SEI 2.3.7 are permitted to be used to combine loads from soil and water in soil instead of the requirements in ASCE/SEI 2.3.1. This alternative method separates the calculation of lateral loads due to soil pressure from ground water pressure in soil (see Section 3.4.4 of this book). Lateral soil pressures are calculated in accordance with ASCE/SEI 3.2.1 (see Section 3.4.2 of this book) and are designated H_{eb}. Load due to ground water pressure in soil is designated H_w and must be based on the maximum ground water elevation.

Loads H_{eb} and H_w must be included in the basic load combinations in ASCE/SEI 2.3.1 as follows:

1. Where the effect of H_{eb} adds to the principal load effect, determine H_{eb} based on the maximum ground water elevation and include H_{eb} with a load factor of 1.6.
2. Where the effect of H_{eb} resists the principal load effect, determine H_{eb} based on the minimum ground water elevation and include H_{eb} with a load factor of 0.9 where the load H_{eb} is permanent or a load factor of 0 for all other conditions.
3. Where the effect of H_w adds to the principal load effect, include H_w based on the maximum ground water with a load factor of 1.0.
4. Where the effect of H_w resists the principal load effect and the soil is permanent, determine H_w based on the minimum ground water elevation and include H_w with a load factor of 1.0, otherwise assign a load factor of 0 to H_w.

2.4 Load Combinations for Allowable Stress Design

2.4.1 Basic Combinations

The basic load combinations where allowable design is used are given in ASCE/SEI 2.4.1 (see Table 2.11). Design of structural members must be based on the load combinations causing the most unfavorable effects on the member, which could occur when one or more loads are not acting.

Table 2.11 Basic Load Combinations for Allowable Stress Design

ASCE/SEI Equation No.	Load Combination	Principal Load
1a	D	D
2a	$D+L$	L
3a	$D+(L_r$ or $0.7S$ or $R)$	L_r or S or R
4a	$D+0.75L+0.75(L_r$ or $0.7S$ or $R)$	L
5a	$D+0.6(W$ or $W_T)$	W or W_T
6a	$D+0.75L+0.75[0.6(W$ or $W_T)]$ $+0.75(L_r$ or $0.7S$ or $R)$	W or W_T
7a	$0.6D+0.6(W$ or $W_T)$	W or W_T

The following exceptions in ASCE/SEI 2.4.1 apply:

1. Snow load, S, in load combinations 4a and 6a must be taken as either the flat roof snow load, p_f, determined in accordance with ASCE/SEI 7.3 or the sloped roof snow load, p_s, determined in accordance with ASCE/SEI 7.4.
2. For nonbuilding structures where the wind or tornado load is determined using force coefficients, C_f, from ASCE/SEI Figures 29.4-1, 29.4-2, and 29.4-3 and the projected area contributing wind or tornado force to a foundation element is greater than 1,000 ft^2 (92.9 m^2) on either a vertical or horizontal plane, it is permitted to replace $(W \text{ or } W_T)$ with $0.9(W \text{ or } W_T)$ in load combination 7a for the design of the foundation, excluding anchorage of the structure to the foundation.
3. It is permitted to replace $0.75(L_r \text{ or } 0.7S \text{ or } R)$ with $0.75(L_r \text{ or } R)$ in load combination 6a where W_T is used in this combination.

Wind loads act in more than one direction on a building or structure, and the appropriate sign of the wind load must be considered in the load combinations.

The 0.75 factor in some of the load combinations accounts for the unlikelihood that two or more variable loads acting at the same time will all attain their maximum value simultaneously.

Load combination 7a addresses cases where the effects of lateral or uplift forces counteract the effects of gravity loads. A factor of 0.6 is applied to the dead load, D, which is meant to limit the dead load that resists horizontal loads to approximately two-thirds of its actual value. The legacy building codes specified that the overturning moment and sliding due to wind load must not exceed two-thirds of the dead load stabilizing moment. This provision was not typically applied to all members in the building. The load combinations in Table 2.11 apply to the design of all members in a structure and provide for overall stability of a structure.

Many material standards and building codes in the past permitted a one-third increase in allowable stress for load combinations including wind or seismic effects. This increase is not permitted when the basic allowable stress design load combinations in ASCE/SEI 2.4.1 are used unless it can be demonstrated that such an increase is justified by structural behavior caused by rate or duration of loading (e.g., in the case of some soils or wood, it is permitted to adjust the allowable stresses based on increased strength under short-term loading).

In the design of tanks and other industrial structures, the wind force coefficients, C_f, in ASCE/SEI 7 do not account for area averaging in the determination of wind forces, unlike enclosed structures where area averaging is accounted for in the pressures coefficients, C_p. The second exception permits a 10 percent reduction in wind or tornado loads used in the design of nonbuilding structure foundations and self-anchored ground-supported tanks.

The third exception permits snow load effects, S, to be taken as zero in cases where tornado loads are used because tornados generally occur in warm weather.

Where fluid loads, F, are present, their effects must be included with the same load factor as dead load, D, in load combinations 1a through 6a (see Table 2.12). The fluid load effect must be included in load combinations 1a through 6a where it adds to the effects from the other loads. Because the wind load effects, W, can be present when the tank is either full or empty, F is not incorporated in load combination 7a; that is, the maximum effects occur when F is set equal to zero.

Where loads, H, are present, their effects must be included as follows:

- Where the effect of H adds to the principal load effect, the load factor on H must be 1.0.
- Where the effect of H resists the principal load effect, the load factor on H must be 0.6 where the load H is permanent; for all other conditions, the load factor must be 0.

Table 2.12 Allowable Stress Design Load Combinations Including Fluid Loads, F

ASCE/SEI Equation No.	Load Combination
1a	$D+F$
2a	$D+F+L$
3a	$D+F+(L_r \text{ or } 0.7S \text{ or } R)$
4a	$D+F+0.75L+0.75(L_r \text{ or } 0.7S \text{ or } R)$
5a	$D+F+0.6(W \text{ or } W_T)$
6a	$D+F+0.75L+0.75[0.6(W \text{ or } W_T)]+0.75(L_r \text{ or } 0.7S \text{ or } R)$
7a	$0.6D+0.6(W \text{ or } W_T)$

2.4.2 Load Combinations Including Flood Loads

Structures located in a designated flood zone (see Chapter 7 of this book) must be designed for the load combinations in Table 2.13 in addition to the applicable basic combinations in Table 2.11 (ASCE/SEI 2.4.2).

Table 2.13 Allowable Stress Design Load Combinations Including Flood Loads, F_a

ASCE/SEI Equation No.	Load Combination
V Zones or Coastal A Zones*	
5b	$D+0.6W+1.5F_a$
6b	$D+0.75L+0.75(0.6W)+0.75(L_r \text{ or } 0.7S \text{ or } R)+1.5F_a$
7b	$0.6D+0.6W+1.5F_a$
Noncoastal A Zones*	
5b	$D+0.6W+0.75F_a$
6b	$D+0.75L+0.75(0.6W)+0.75(L_r \text{ or } 0.7S \text{ or } R)+0.75F_a$
7b	$0.6D+0.6W+0.75F_a$

*See ASCE/SEI 5.2 for definitions of V zones, coastal A zones, and noncoastal A zones (also see Chapter 7 of this book).

2.4.3 Load Combinations Including Atmospheric Ice and Wind-on-Ice Loads

Structures subjected to atmospheric ice loads, D_i, and wind-on-ice loads, W_i, must be designed for the load combinations in Table 2.14 (ASCE/SEI 2.4.3).

Table 2.14 Allowable Stress Design Load Combinations Including Atmospheric Ice Loads, D_i, and Wind-on-Ice Loads, W_i

ASCE/SEI Equation No.	Load Combination
1b	$D+0.7D_i$
2b	$D+L+0.7D_i$
3b	$D+0.7D_i+0.7W_i+0.7S$
7c	$0.6D+0.7D_i+0.7W_i$

2.4.4 Load Combinations Including Self-Straining Forces and Effects

In cases where self-straining loads, T, must be considered, their effects in combination with other loads are to be determined by ASCE/SEI 2.4.4. The load factor on T must not be taken less than 0.75. The following strength design load combinations in ASCE/SEI C2.4.4 should be considered:

$$1.0D + 1.0T \tag{2.4}$$

$$1.0D + 0.75(L + T) \tag{2.5}$$

2.4.5 Basic Combinations with Seismic Load Effects

Structures subjected to the seismic load effects, E, defined in ASCE/SEI 12.4.2 must be designed for the load combinations in Table 2.15 in addition to the load combinations in Table 2.11 where $E = E_h \pm E_v$ (ASCE/SEI 2.4.5).

Table 2.15 Allowable Stress Design Load Combinations with Seismic Load Effects, E_h and E_v

ASCE/SEI Equation No.	Load Combination
8	$D + 0.7E_v + 0.7E_h$
9	$D + 0.525E_v + 0.525E_h + 0.75L + 0.1S$
10	$0.6D - 0.7E_v + 0.7E_h$

The terms E_h and E_v are defined in ASCE/SEI 12.4.2.1 and 12.4.2.2, respectively:

- E_h = horizontal seismic load effect = ρQ_E
- E_v = vertical seismic load effect = $0.2S_{DS}D$

where ρ is the redundancy factor determined in accordance with ASCE/SEI 12.3.4, Q_E is the effects of the horizontal seismic forces due to the base shear, V, or the seismic force acting on a component of a structure, F_p, and S_{DS} is the design, 5 percent damped spectral response acceleration parameter at short periods determined in accordance with ASCE/SEI 11.4.4.

Substituting E_h and E_v into load combinations 8, 9, and 10 results in the following:

Load combination 8: $(1.0 + 0.14S_{DS})D + 0.7\rho Q_E$

Load combination 9: $(1.0 + 0.105S_{DS})D + 0.525\rho Q_E + 0.75L + 0.1S$

Load combination 10: $(0.6 - 0.14S_{DS})D + 0.7\rho Q_E$

According to the first exception in ASCE/SEI 12.4.2.2, E_v must be determined by ASCE/SEI Equation (12.4-4b) where the effects of vertical seismic ground motion are required to be determined in accordance with ASCE/SEI 11.9.

In the second exception in ASCE/SEI 12.4.2.2, E_v is permitted to be taken as zero for either of the following conditions:

- In ASCE/SEI Equations (12.4-1), (12.4-2), (12.4-5), and (12.4-6) for structures assigned to SDC B
- In ASCE/SEI Equation (12.4-2) when determining demands on the soil-structure interface of foundations

The allowable stress design load combinations for structures assigned to SDC B are given in Table 2.16. In accordance with the second exception in ASCE/SEI 12.4.2.2, the vertical seismic load effect, E_v, is set equal to zero and in accordance with ASCE/SEI 12.3.4.1, the

Table 2.16 Allowable Stress Design Load Combinations for Structures Assigned to SDC B

ASCE/SEI Equation No.	Load Combination	Principal Load
1a	D	D
2a	$D+L$	L
3a	$D+(L_r$ or $0.7S$ or $R)$	L_r or S or R
4a	$D+0.75L+0.75(L_r$ or $0.7S$ or $R)$	L
5a	$D+0.6(W$ or $W_T)$	W or W_T
6a	$D+0.75L+0.75[0.6(W$ or $W_T)]+0.75(L_r$ or $0.7S$ or $R)$	W or W_T
7a	$0.6D+0.6(W$ or $W_T)$	W or W_T
8	$D+0.7Q_E$	E
9	$D+0.525Q_E+0.75L+0.1S$	E
10	$0.6D+0.7Q_E$	E

redundancy factor, ρ, is set equal to 1.0. Seismic loads act in more than one direction on a building or structure, and the appropriate sign of the seismic load must be considered in the load combinations.

The allowable stress design load combinations for structures assigned to SDC C are given in Table 2.17. In accordance with ASCE/SEI 12.3.4.1, the redundancy factor, ρ, is set equal to 1.0.

Table 2.17 Allowable Stress Design Load Combinations for Structures Assigned to SDC C

ASCE/SEI Equation No.	Load Combination	Principal Load
1a	D	D
2a	$D+L$	L
3a	$D+(L_r$ or $0.7S$ or $R)$	L_r or S or R
4a	$D+0.75L+0.75(L_r$ or $0.7S$ or $R)$	L
5a	$D+0.6(W$ or $W_T)$	W or W_T
6a	$D+0.75L+0.75[0.6(W$ or $W_T)]+0.75(L_r$ or $0.7S$ or $R)$	W or W_T
7a	$0.6D+0.6(W$ or $W_T)$	W or W_T
8	$(1.0+0.14S_{DS})D+0.7Q_E$	E
9	$(1.0+0.105S_{DS})D+0.525Q_E+0.75L+0.1S$	E
10	$(0.6-0.14S_{DS})D+0.7Q_E$	E

The allowable stress design load combinations for structures assigned to SDC D, E, or F are given in Table 2.18.

Structures required to resist seismic load effects with overstrength, E_m, must be designed for the load combinations in Table 2.19 in addition to the load combinations in Table 2.11, where $E_m = E_{mh} \pm E_v$ (ASCE/SEI 2.4.5).

Table 2.18 Allowable Stress Design Load Combinations for Structures Assigned to SDC D, E, or F

ASCE/SEI Equation No.	Load Combination	Principal Load
1a	D	D
2a	$D+L$	L
3a	$D+(L_r$ or $0.7S$ or $R)$	L_r or S or R
4a	$D+0.75L+0.75(L_r$ or $0.7S$ or $R)$	L
5a	$D+0.6(W$ or $W_T)$	W or W_T
6a	$D+0.75L+0.75[0.6(W$ or $W_T)]+0.75(L_r$ or $0.7S$ or $R)$	W or W_T
7a	$0.6D+0.6(W$ or $W_T)$	W or W_T
8	$(1.0+0.14S_{DS})D+0.7\rho Q_E$	E
9	$(1.0+0.105S_{DS})D+0.525\rho Q_E+0.75L+0.1S$	E
10	$(0.6-0.14S_{DS})D+0.7\rho Q_E$	E

Table 2.19 Allowable Stress Design Load Combinations with Seismic Load Effects Including Overstrength, E_{mh} and E_v

ASCE/SEI Equation No.	Load Combination
8	$D+0.7E_v+0.7E_{mh}$
9	$D+0.525E_v+0.525E_{mh}+0.75L+0.1S$
10	$0.6D-0.7E_v+0.7E_{mh}$

The term E_{mh} is defined in ASCE/SEI 12.4.3.1 as follows:

$$E_{mh} = \text{horizontal seismic load effect including overstrength} = \Omega_0 Q_E$$

where Ω_0 is the overstrength factor given in ASCE/SEI Table 12.2-1 based on the SFRS.

Substituting E_{mh} and E_v into load combinations 8, 9, and 10 results in the following:

Load combination 8: $(1.0+0.14S_{DS})D+0.7\Omega_0 Q_E$

Load combination 9: $(1.0+0.105S_{DS})D+0.525\Omega_0 Q_E+0.75L+0.1S$

Load combination 10: $(0.6-0.14S_{DS})D+0.7\Omega_0 Q_E$

The horizontal seismic load effect including overstrength, E_{mh}, need not be taken larger than the capacity-limited horizontal seismic load effect, E_{cl}, which is equal to the maximum effect that can develop in a member as determined by a rational, plastic mechanism analysis. Additional information on how to determine E_{cl} is given in ASCE/SEI C12.4.3.2.

Two exceptions are given in ASCE/SEI 2.4.5 pertaining to the basic load combinations with seismic load effects. In the first exception, snow load, S, in load combination 9 must be taken as either the flat roof snow load, p_f, determined in accordance with ASCE/SEI 7.3 or the sloped roof snow load, p_s, determined in accordance with ASCE/SEI 7.4. In the second exception, it is permitted to replace $0.6D$ with $0.9D$ in load combination 10 for the design of special reinforced masonry shear walls where the walls satisfy the requirement in ASCE/SEI 14.4.2.

Where fluid loads, F, are present, their effects must be included with the same load factor as dead load, D, in load combinations 8, 9, and 10.

Where loads, H, are present, their effects must be included as follows:

- Where the effect of H adds to the primary variable load effect, the load factor on H must be 1.0.
- Where the effect of H resists the primary variable load effect, the load factor on H must be 0.6 where the load H is permanent; for all other conditions, the load factor must be 0.

The most unfavorable effects from seismic loads must be investigated where appropriate but they need not be considered to act simultaneously with wind or tornado loads.

2.4.6 Alternative Allowable Stress Design Load Combinations

The alternative allowable stress design load combinations in IBC 1605.2 are permitted to be used instead of the allowable stress design load combinations in ASCE/SEI 2.4 (see Table 2.20).

Table 2.20 Alternative Allowable Stress Design Load Combinations in IBC 1605.2

IBC Equation No.	Load Combination
16-1	$D+L+(L_r$ or $0.7S$ or $R)$
16-2	$D+L+0.6W$
16-3	$D+L+0.6W+0.7S/2$
16-4	$D+L+0.7S+0.6(W/2)$
16-5	$D+L+0.7S+E/1.4$
16-6	$0.9D+E/1.4$

Two exceptions are given in IBC 1605.2:

1. Crane hook loads need not be combined with roof live loads or with more than three-fourth of the snow load or one-half of the wind load.
2. Flat roof snow loads of 45 lb/ft^2 (2.15 kN/m^2) or less and roof live loads of 30 lb/ft^2 (1.44 kN/m^2) or less need not be combined with seismic loads. Where flat roof snow loads exceed 45 lb/ft^2 (2.15 kN/m^2), 15 percent of the flat roof snow load must be combined with the seismic loads.

The following requirements apply to the alternative allowable stress design load combinations:

1. For load combinations that include wind or seismic loads, allowable stresses are permitted to be increased or load combinations reduced where permitted by the material chapter of the IBC or the referenced standards.
2. For load combinations that include the counteracting effects of dead and wind load, two-thirds of the minimum dead load likely to be in place during a design wind event is permitted to be used in the load combinations.
3. Where sliding, overturning, and soil bearing are evaluated at the soil-structure interface, the reduction of foundation overturning permitted by ASCE/SEI 12.13.4 is not permitted.
4. Where foundations are proportioned for loads which include seismic loads, the vertical seismic load effect, E_v, in ASCE/SEI 12.4.2.2 is permitted to be taken as zero.
5. Where required in ASCE/SEI Chapters 12, 13, and 15, load combinations including overstrength in ASCE/SEI 2.4.5 must be used.

2.5 Load Combinations for Extraordinary Events

2.5.1 Overview

Requirements for extraordinary loads and events are given in ASCE/SEI 1.4.5, which references ASCE/SEI 2.5. Minimum requirements for strength and stability of a structure are provided in ASCE/SEI 2.5 where the owner or applicable code requires that the structure be able to withstand the effects from extraordinary events without disproportionate collapse.

Extraordinary events arise from service or environmental conditions not traditionally considered in the design of ordinary buildings because their probability of occurrence is low and their duration is short. Fires, explosions, and vehicular impact are examples of such events. The purpose of these requirements is to help ensure that buildings and structures have sufficient strength and ductility and are adequately tied together so that damage caused by the extraordinary event is relatively small.

2.5.2 Load Combinations

ASCE/SEI Equation (2.5-1) must be used to check the capacity of a structure or structural element to withstand the effects of an extraordinary event:

$$(0.9 \text{ or } 1.2)D + A_k + 0.5L + 0.15S \tag{2.6}$$

where A_k is the load or load effect resulting from the extraordinary event, A.

A factor of 0.9 is used on the dead load effect, D, where the dead load has a stabilizing effect; otherwise, a load factor of 1.2 must be used. The companion actions $0.5L$ and $0.15S$ correspond approximately to the mean of the yearly maximum live and snow loads, respectively. Roof live loads, L_r, and rain loads, R, are not included in this load combination because they have short durations in comparison to S, and thus, the probability of them occurring with A_k is negligible.

ASCE/SEI Equation (2.5-2) must be used to check the residual load-carrying capacity of a structure or structural element following the occurrence of a damaging event. Selected load-bearing members are to be removed from the structure assuming they have been critically damaged (i.e., they have essentially no load-carrying capacity) and the capacity of the damaged structure is evaluated by the following load combination:

$$(0.9 \text{ or } 1.2)D + 0.5L + 0.2(L_r \text{ or } 0.7S \text{ or } R) \tag{2.7}$$

2.5.3 Stability Requirements

The stability of the entire structure and each of its members must be checked after an extraordinary event. Any method that includes second-order effects is permitted. ASCE/SEI C2.5 provides a rational method for meeting this requirement.

2.6 Load Combinations for General Structural Integrity Loads

Provisions for structural integrity are given in IBC 1616 and are applicable to buildings classified as high-rise buildings in accordance with IBC 403 and assigned to risk category III or IV with frame structures or bearing wall structures. A high-rise building is defined in IBC 202 as a building with an occupied floor or occupiable roof located more than 75 ft (22.9 m) above the lowest level of fire department vehicle access. Risk categories III and IV are defined in IBC Table 1604.5. Specific load combinations are not included in these prescriptive requirements; rather, the requirements are meant to improve the redundancy and ductility of these types of framing systems in the event of damage due to an abnormal loading event. General design and detailing requirements are provided for frame structures and bearing wall structures.

General structural integrity requirements applicable to all structures are given in ASCE/SEI 1.4:

- **A continuous load path in accordance with ASCE/SEI 1.4.1 must be provided.** A continuous path to the lateral force-resisting system is essential to ensure that the loads are transmitted properly. The members and connections in that path must be designed to resist the applicable load combinations, and any smaller part of the structure must be tied to the remainder of the structure with elements that can resist a minimum force equal to 5 percent of its weight.
- **A complete lateral force-resisting system with adequate strength to resist the forces indicated in ASCE/SEI 1.4.2 must be provided.** In each of two orthogonal directions, a structure must be able to resist lateral forces at each floor level that are equal to 1 percent of the total dead load assigned to that level [see ASCE/SEI Equation (1.4-1)]. These forces must be applied at each floor level simultaneously in the direction of analysis and can be applied independently in the two orthogonal directions. Any structure that has been explicitly designed for stability, which includes the necessary second-order effects, automatically complies with this requirement.
- **Members of the structural system must be connected to their supporting members in accordance with ASCE/SEI 1.4.3.** A positive connection must be provided to resist horizontal forces acting parallel to the member. Each beam, girder or truss must have adequate connections to its supporting elements or to slabs acting as diaphragms. The connection must have the strength to resist a minimum force equal to 5 percent of the unfactored dead load plus live load reaction imposed by the supported member on the supporting member. In cases where the supported element is connected to a diaphragm, the supporting member also must be connected to the diaphragm.
- **Structural walls must be anchored to diaphragms and supports in accordance with ASCE/SEI 1.4.4.** Load-bearing walls and shear walls (i.e., walls that provide lateral shear resistance) must be adequately anchored to the floor and roof members that provide lateral support of the wall or that are supported by the wall. A direct connection must be provided that can resist a strength-level horizontal force perpendicular to the plane of the wall of at least 0.2 times the weight of the wall that is tributary to the connection or 5 lb/ft^2 (0.24 kN/m^2), whichever is greater.
- **Extraordinary loads and events must be considered where applicable.** Design for resistance to extraordinary loads and events must be in accordance with ASCE/SEI 2.5 (see Section 2.5 of this book).

The above minimum strength criteria help ensure structural integrity is maintained for anticipated and minor unanticipated loading events that have a reasonable chance of occurring during the life of the structure. Guidelines for providing general structural integrity are given in ASCE/SEI C1.4.

The load combinations for general structural integrity loads are given in ASCE/SEI 2.6 for strength design and allowable stress design:

- Strength design notional load combinations
  1. $1.2D+N+L+0.15S$
  2. $0.9D+N$
- Allowable stress design notional load combinations
  1. $D+0.7N$
  2. $D+0.75(0.7N)+0.75L+0.75(L_r \text{ or } 0.7S \text{ or } R)$
  3. $0.6D+0.7N$

Notional loads, N, are specified in ASCE/SEI 1.4 for structural integrity.

Information on general collapse and limited local collapse is given in ASCE/SEI C1.4 along with examples of general collapse.

2.7 Examples

2.7.1 Example 2.1—Strength Design Load Combinations for a Column Subjected to Axial Loads

Determine the strength design load combinations for a column in a multistory office building given the following axial loads:

- Dead load, $D = 78$ kips (347 kN)
- Live load, $L = 38$ kips (169 kN)
- Roof live load, $L_r = 13$ kips (58 kN)
- Balanced snow load, $S = 30$ kips (133 kN)

The live load on the floors is less than 100 lb/ft^2 (4.79 kN/m^2), and the building has a gable roof.

Solution

The basic strength design load combinations in ASCE/SEI 2.3.1 are given in Table 2.21 for this column. Only the applicable load effects from the design data are included.

Because the floor live load is less than 100 lb/ft^2 (4.79 kN/m^2), the load factor on L in load combinations 3a and 4a is taken as 0.5 in accordance with the first exception in ASCE/SEI 2.3.1. When one or more of the variable loads (live, roof live, and snow) are taken equal to zero, the resulting factored loads are less than those in Table 2.21.

Table 2.21 Basic Strength Design Load Combinations for the Column in Example 2.1

ASCE/SEI Equation No.	Load Combination	Axial Load, kips (kN)
1a	$1.4D$	109 (486)
2a	$1.2D + 1.6L + 0.5L_r$	161 (716)
	$1.2D + 1.6L + 0.3S$	163 (727)
3a	$1.2D + 1.6L_r + 0.5L$	133 (594)
	$1.2D + 1.0S + 0.5L$	143 (634)
4a	$1.2D + 0.5L + 0.5L_r$	119 (530)
	$1.2D + 0.5L + 0.3S$	122 (541)
5a	$0.9D$	70 (312)

2.7.2 Example 2.2—Strength Design Load Combinations for a Column Subjected to Combined Axial Loads and Bending Moments

Determine the strength design load combinations for a column in a multistory office building given the design data in Table 2.22. The live load on the floors is less than 100 lb/ft^2 (4.79 kN/m^2), and the building has a flat roof. The term SSR refers to sidesway right and SSL refers to sidesway left. The adopted sign convention is given in Figure 2.1.

Table 2.22 Design Data for Example 2.2

Load		Axial Load, kips (kN)	Bending Moment at Bottom, ft-kips (kN-m)
Dead load, *D*		78 (347)	−15 (−20)
Live load, *L*		38 (169)	−5 (−7)
Roof live load, L_r		13 (58)	0
Balanced snow load, *S*		30 (134)	0
Wind load, *W*	SSR	−32 (−142)	−75 (−102)
	SSL	32 (142)	75 (102)

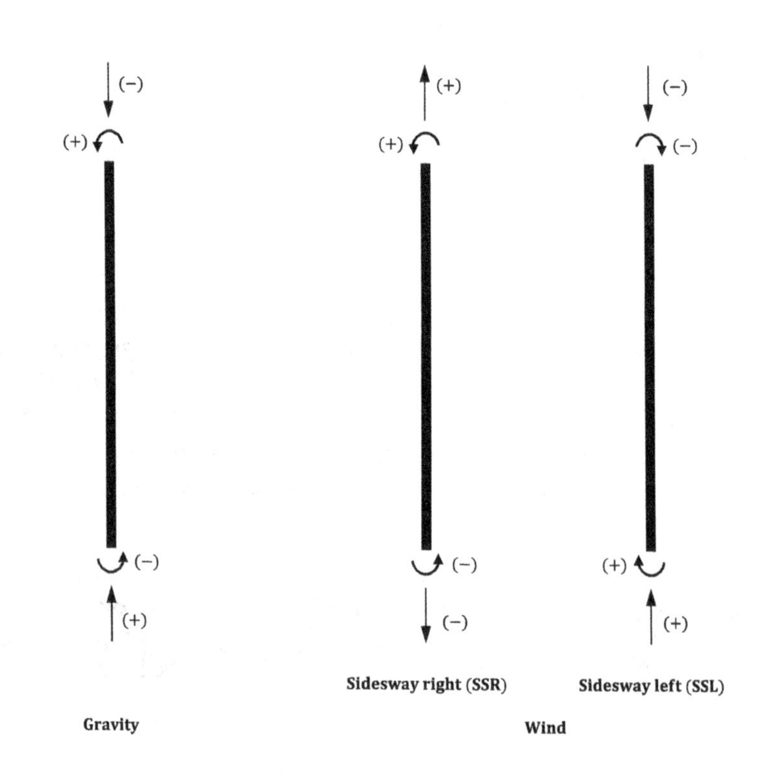

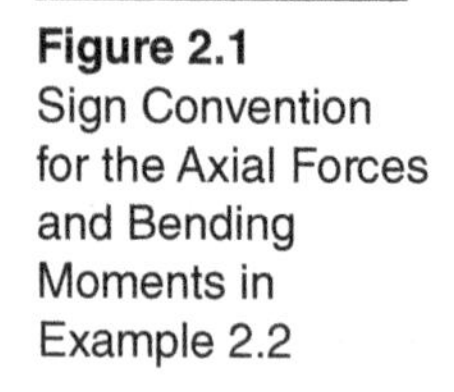

Figure 2.1 Sign Convention for the Axial Forces and Bending Moments in Example 2.2

Solution

The basic strength design load combinations in ASCE/SEI 2.3.1 are given in Table 2.23 for this column. Only the applicable load effects from the design data are included.

Because the floor live load is less than 100 lb/ft^2 (4.79 kN/m^2), the load factor on *L* in load combinations 3a and 4a is taken as 0.5 in accordance with the first exception in ASCE/SEI 2.3.1.

Because the structure can sway to the right (SSR) and to the left (SSL) due to the wind loads, load combinations must be investigated for both cases.

In general, all the load combinations in Table 2.23 must be investigated when designing a column subjected to axial loads and bending moments; it may not be obvious which of the load combinations is the most critical in such cases.

2.7.3 Example 2.3—Strength Design Load Combinations for a Beam Subjected to Bending Moments and Shear Forces

Determine the strength design load combinations for a beam in a university building given the design data in Table 2.24. The occupancy of the building is classified as a place of public assembly. The adopted sign convention is given in Figure 2.2.

Table 2.23 Basic Strength Design Load Combinations for the Column in Example 2.2

ASCE/SEI Equation No.	Load Combination		Axial Load, kips (kN)	Bending Moment, ft-kips (kN-m)
1a	$1.4D$		109 (486)	−21 (−28)
2a	$1.2D+1.6L+0.5L_r$		161 (716)	−26 (−28)
	$1.2D+1.6L+0.3S$		163 (727)	−26 (−28)
3a	$1.2D+1.6L_r+0.5L$		133 (594)	−21 (−28)
	$1.2D+1.6L_r+0.5W$	SSR	98 (438)	−56 (−75)
		SSL	130 (580)	20 (27)
	$1.2D+1.0S+0.5L$		143 (635)	−21 (−28)
	$1.2D+1.0S+0.5W$	SSR	108 (479)	−56 (−75)
		SSL	140 (621)	20 (27)
4a	$1.2D+1.0W+0.5L+0.5L_r$	SSR	87 (388)	−96 (−130)
		SSL	151 (672)	55 (75)
	$1.2D+1.0W+0.5L+0.3S$	SSR	90 (399)	−96 (−130)
		SSL	154 (683)	55 (75)
5a	$0.9D+1.0W$	SSR	38 (170)	−89 (−120)
		SSL	102 (454)	62 (84)

Table 2.24 Design Data for Example 2.3

Load		Bending Moment, ft-kips (kN-m)			Shear Force, kips (kN)	
		Left Support	Near Midspan	Right Support	Left Support	Right Support
Dead load, D		−80 (−109)	52 (71)	−250 (−339)	33 (147)	50 (222)
Live load, L		−16 (−22)	35 (48)	−50 (−68)	12 (53)	15 (67)
Wind load, W	SSR	160 (217)	—	−160 (−217)	−16 (−71)	16 (71)
	SSL	−160 (−217)	—	160 (217)	16 (71)	−16 (−71)
Seismic, Q_E	SSR	50 (68)	—	−50 (−68)	−5 (−22)	5 (22)
	SSL	−50 (−68)	—	50 (68)	5 (22)	−5 (−22)

The seismic design data are as follows (more information on seismic design can be found in Chapter 6 of this book):

- Redundancy factor, $\rho = 1.0$
- Design spectral response acceleration parameter at short periods, $S_{DS} = 0.5$

Solution

The strength design load combinations in ASCE/SEI 2.3.1 and 2.3.6 are given in Table 2.25 for this beam. Only the applicable load effects from the design data are included.

The term, Q_E, is the effect of code-prescribed horizontal seismic forces on the beam determined from a structural analysis.

Because the occupancy of the building is classified as a place of public assembly, the load factor on L in load combinations 3a, 4a, and 6 must be taken as 1.0 in accordance with the first exceptions in ASCE/SEI 2.3.1 and 2.3.6.

Figure 2.2 Sign Convention for the Bending Moments and Shear Forces in Example 2.3

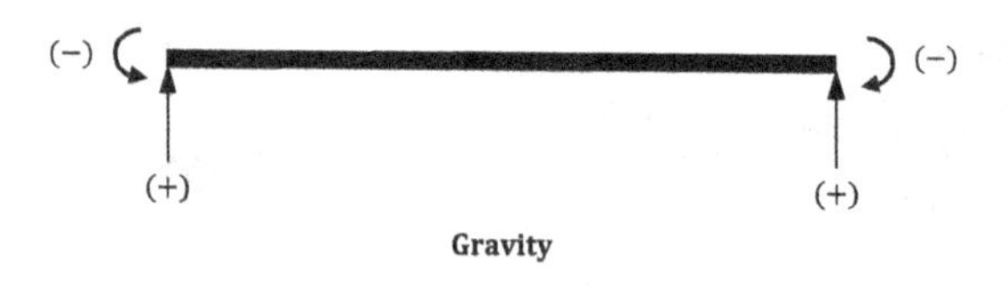

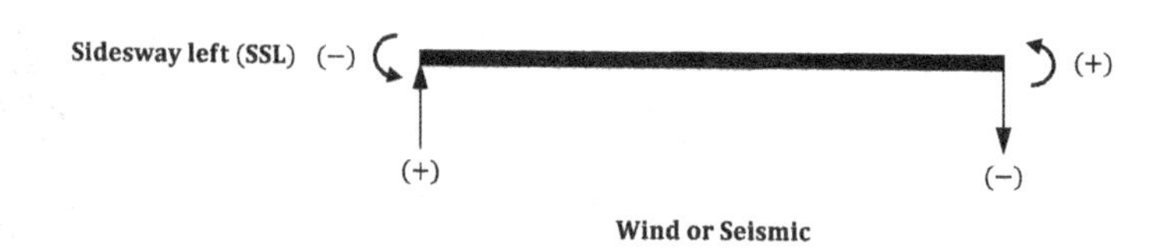

Table 2.25 Strength Design Load Combinations for the Beam in Example 2.3

ASCE/SEI Equation No.	Load Combination		Bending Moment, ft-kips (kN-m)			Shear Force, kips (kN)	
			Left Support	Near Midspan	Right Support	Left Support	Right Support
1a	$1.4D$		−112 (−153)	73 (99)	−350 (−475)	46 (206)	70 (311)
2a	$1.2D+1.6L$		−122 (−166)	118 (162)	−380 (−516)	59 (261)	84 (374)
3a	$1.2D+L$		−112 (−153)	97 (133)	−350 (−475)	52 (229)	75 (333)
	$1.2D+0.5W$	SSR	−16 (−22)	62 (85)	−380 (−515)	32 (141)	68 (302)
		SSL	−176 (−239)	62 (85)	−220 (−298)	48 (212)	52 (231)
4a	$1.2D+W+L$	SSR	48 (64)	97 (133)	−510 (−692)	36 (158)	91 (404)
		SSL	−272 (−370)	97 (133)	−190 (−258)	68 (300)	59 (262)
5a	$0.9D+W$	SSR	88 (119)	47 (64)	−385 (−522)	14 (61)	61 (271)
		SSL	−232 (−315)	47 (64)	−65 (−88)	46 (203)	29 (129)
6	$1.3D+Q_E+L$	SSR	−70 (−96)	103 (140)	−425 (−577)	50 (222)	85 (378)
		SSL	−170 (−232)	103 (140)	−325 (−441)	60 (266)	75 (334)
7	$0.8D+Q_E$	SSR	−14 (−19)	42 (57)	−250 (−339)	21 (96)	45 (200)
		SSL	−114 (−155)	42 (57)	−150 (−203)	31 (140)	35 (156)

The seismic load effect, E, is defined as follows (ASCE/SEI 12.4.2):

- For use in load combination 6:
  $E = E_h + E_v = \rho Q_E + 0.2S_{DS}D = Q_E + (0.2 \times 0.5)D = Q_E + 0.1D$
- For use in load combination 7:
  $E = E_h - E_v = \rho Q_E - 0.2S_{DS}D = Q_E - (0.2 \times 0.5)D = Q_E - 0.1D$

Load combinations 6 and 7 can be rewritten as follows:

Load combination 6: $1.2D + E_h + E_v + L = (1.2 + 0.1)D + Q_E + L = 1.3D + Q_E + L$

Load combination 7: $0.9D + E_h - E_v = (0.9 - 0.1)D + Q_E = 0.8D + Q_E$

Because the structure can sway to the right (SSR) and to the left (SSL) due to the wind and seismic loads, load combinations must be investigated for both cases.

The beam must be designed for the maximum negative and positive bending moments and maximum shear forces in Table 2.25.

2.7.4 Example 2.4—Allowable Stress Design Load Combinations for a Beam Subjected to Bending Moments and Shear Forces

Determine the allowable stress design load combinations for the beam in Example 2.3. The adopted sign convention is given in Figure 2.2.

SOLUTION

The allowable stress design load combinations in ASCE/SEI 2.4.1 and 2.4.5 are given in Table 2.26 for this beam. Only the applicable load effects from the design data are included.

The term, Q_E, is the effect of code-prescribed horizontal seismic forces on the beam determined from a structural analysis.

The seismic load effect, E, is defined as follows (ASCE/SEI 12.4.2):

- For use in load combinations 8 and 9:
  $E = E_h + E_v = \rho Q_E + 0.2S_{DS}D = Q_E + (0.2 \times 0.5)D = Q_E + 0.1D$
- For use in load combination 10:
  $E = E_h - E_v = \rho Q_E - 0.2S_{DS}D = Q_E - (0.2 \times 0.5)D = Q_E - 0.1D$

Load combinations can be rewritten as follows:

Load combination 8: $D + 0.7E_h + 0.7E_v = [1 + (0.7 \times 0.1)]D + 0.7Q_E = 1.07D + 0.7Q_E$

Load combination 9: $D + 0.525E_h + 0.525E_v + 0.75L = [1 + (0.525 \times 0.1)]D + 0.525Q_E + 0.75L =$
$1.05D + 0.525Q_E + 0.75L$

Load combination 10: $0.6D + 0.7E_h - 0.7E_v = [0.6 - (0.7 \times 0.1)]D + 0.7Q_E = 0.53D + 0.7Q_E$

Because the structure can sway to the right (SSR) and to the left (SSL) due to the wind and seismic loads, load combinations must be investigated for both cases.

The beam must be designed for the maximum negative and positive bending moments and maximum shear forces in Table 2.26.

2.7.5 Example 2.5—Alternative Allowable Stress Design Load Combinations for a Beam Subjected to Bending Moments and Shear Forces

Determine the alternative allowable stress design load combinations for the beam in Example 2.3. The adopted sign convention is given in Figure 2.2.

Table 2.26 Allowable Stress Design Load Combinations for the Beam in Example 2.4

ASCE/SEI Equation No.	Load Combination		Bending Moment, ft-kips (kN-m)			Shear Force, kips (kN)	
			Left Support	Near Midspan	Right Support	Left Support	Right Support
1a	D		−80 (−109)	52 (71)	−250 (−339)	33 (147)	50 (222)
2a	$D+L$		−96 (−131)	87 (119)	−300 (−407)	45 (200)	65 (289)
4a	$D+0.75L$		−92 (−126)	78 (107)	−288 (−390)	42 (187)	61 (272)
5a	$D+0.6W$	SSR	16 (21)	52 (71)	−346 (−469)	23 (104)	60 (265)
		SSL	−176 (−239)	52 (71)	−154 (−209)	43 (190)	40 (179)
6a	$D+0.75L+0.45W$	SSR	−20 (−28)	78 (107)	−360 (−488)	35 (155)	68 (304)
		SSL	−164 (−224)	78 (107)	−216 (−292)	49 (219)	54 (240)
7a	$0.6D+0.6W$	SSR	48 (65)	31 (43)	−246 (−334)	10 (46)	40 (176)
		SSL	−144 (−196)	31 (43)	−54 (−73)	29 (131)	20 (91)
8	$1.07D+0.7Q_E$	SSR	−51 (−69)	56 (76)	−303 (−410)	32 (142)	57 (253)
		SSL	−121 (−164)	56 (76)	−233 (−315)	39 (173)	50 (222)
9	$1.05D+0.525Q_E+0.75L$	SSR	−70 (−95)	81 (111)	−326 (−443)	41 (183)	66 (295)
		SSL	−122 (−167)	81 (111)	−274 (−371)	46 (206)	61 (272)
10	$0.53D+0.7Q_E$	SSR	−7 (−10)	28 (38)	−168 (−227)	14 (63)	30 (133)
		SSL	−77 (−105)	28 (38)	−98 (−132)	21 (93)	23 (102)

Solution

The alternative allowable stress design load combinations in IBC 1605.2 are given in Table 2.27 for this beam. Only the applicable load effects from the design data are included.

The term, Q_E, is the effect of code-prescribed horizontal seismic forces on the beam determined from a structural analysis.

The seismic load effect, E, is defined as follows (ASCE/SEI 12.4.2):

- For use in IBC Equation 16-5:

$$E = E_h + E_v = \rho Q_E + 0.2S_{DS}D = Q_E + (0.2\times 0.5)D = Q_E + 0.1D$$

Table 2.27 Alternative Allowable Stress Design Load Combinations for the Beam in Example 2.5

IBC Equation No.	Load Combination		Bending Moment, ft-kips (kN-m)			Shear Force, kips (kN)	
			Left Support	Near Midspan	Right Support	Left Support	Right Support
16-1	$D+L$		−96 (−131)	87 (119)	−300 (−407)	45 (200)	65 (289)
16-2	(D or 0.67D) $+L+0.6W$*	SSR	26 (35)	87 (119)	−396 (−537)	25 (109)	75 (332)
		SSL	−192 (−224)	87 (119)	−122 (−165)	55 (243)	39 (173)
16-4	(D or 0.67D) $+L+0.3W$*	SSR	−22 (−30)	87 (119)	−348 (−472)	29 (130)	70 (310)
		SSL	−144 (−196)	87 (119)	−170 (−230)	50 (221)	44 (194)
16-5	$1.07D+L$ $+Q_E/1.4$	SSR	−66 (−90)	91 (124)	−353 (−479)	44 (195)	72 (320)
		SSL	−137 (−187)	91 (124)	−282 (−382)	51 (226)	65 (289)
16-6	$0.83D$ $+Q_E/1.4$	SSR	−31 (−42)	43 (59)	−243 (−330)	24 (106)	45 (200)
		SSL	−102 (−139)	43 (59)	−172 (−233)	31 (138)	38 (169)

*Two-thirds of the dead load must be used where the effects of dead loads and wind loads counteract (IBC 1605.2).

- For use in IBC Equation 16-6:
  $E=E_h-E_v=\rho Q_E-0.2S_{DS}D=Q_E-(0.2\times 0.5)D=Q_E-0.1D$

IBC Equations 16-5 and 16-6 can be rewritten as follows:

IBC Equation 16-5: $D+L+E/1.4=D+L+Q_E/1.4+0.1D/1.4=1.07D+L+Q_E/1.4$

IBC Equation 16-6: $0.9D+E/1.4=0.9D+Q_E/1.4-0.1D/1.4=0.83D+Q_E/1.4$

Because the structure can sway to the right (SSR) and to the left (SSL) due to the wind and seismic loads, load combinations must be investigated for both cases. The sign convention adopted in this case is depicted in Figure 2.2.

The beam must be designed for the maximum negative and positive bending moments and maximum shear forces in Table 2.27.

2.7.6 Example 2.6—Strength Design Load Combinations Including Overstrength for a Collector Beam Subjected to Axial Loads, Bending Moments and Shear Forces

Determine the strength design load combinations for a simply supported collector beam in a residential building using the design data in Table 2.28. The live load on the floors is less than 100 lb/ft^2 (4.79 kN/m^2). Seismic design data are as follows (see Chapter 6 of this book for information on seismic design):

- System overstrength factor, $\Omega_0=2.5$
- Design spectral response acceleration parameter at short periods, $S_{DS}=1.0$
- Seismic design category: D

Table 2.28 Design Data for the Collector Beam in Example 2.6

Load	Axial Load, kips (kN)	Bending Moment, ft-kips (kN-m)	Shear Force, kips (kN)
Dead, D	0	703 (953)	56 (249)
Live, L	0	235 (319)	19 (85)
Seismic, Q_E	±50 (±222)	0	0

Solution

- Load combination 1a: $1.4D$
  Bending moment: $1.4 \times 703 = 984$ ft-kips
  Shear force: $1.4 \times 56 = 78$ kips
- Load combination 2a: $1.2D + 1.6L$
  Bending moment: $(1.2 \times 703) + (1.6 \times 235) = 1{,}220$ ft-kips
  Shear force: $(1.2 \times 56) + (1.6 \times 19) = 98$ kips

Because the beam is a collector beam in a building assigned to SDC D, the beam must be designed for the strength design load combinations including overstrength (see ASCE/SEI 12.10.2.1 and Chapter 6 of this book).

- Load combination 6: $(1.2 + 0.2S_{DS})D + \Omega_0 Q_E + 0.5L$
  Axial load: $0 + (2.5 \times 50) + 0 = 125$ kips tension or compression
  Bending moment: $(1.4 \times 703) + 0 + (0.5 \times 235) = 1{,}102$ ft-kips
  Shear force: $(1.4 \times 56) + 0 + (0.5 \times 19) = 88$ kips

The load factor on L is permitted to equal 0.5 in accordance with exception 1 in ASCE/SEI 2.3.6.

- Load combination 7: $(0.9 - 0.2S_{DS})D + \Omega_0 Q_E$
  Axial load: $0 + (2.5 \times 50) = 125$ kips tension or compression
  Bending moment: $= (0.7 \times 703) + 0 = 492$ ft-kips
  Shear force: $(0.7 \times 56) + 0 = 39$ kips

In S.I.:

- Load combination 1a: $1.4D$
  Bending moment: $1.4 \times 953 = 1{,}334$ kN-m
  Shear force: $1.4 \times 249 = 349$ kN
- Load combination 2a: $1.2D + 1.6L$
  Bending moment: $(1.2 \times 953) + (1.6 \times 319) = 1{,}654$ kN-m
  Shear force: $(1.2 \times 249) + (1.6 \times 85) = 435$ kN

Because the beam is a collector beam in a building assigned to SDC D, the beam must be designed for the strength design load combinations including overstrength (see ASCE/SEI 12.10.2.1 and Chapter 6 of this book).

- Load combination 6: $(1.2 + 0.2S_{DS})D + \Omega_0 Q_E + 0.5L$
  Axial load: $0 + (2.5 \times 222) = 555$ kN tension or compression
  Bending moment: $(1.4 \times 953) + 0 + (0.5 \times 319) = 1{,}494$ kN-m
  Shear force: $(1.4 \times 249) + 0 + (0.5 \times 85) = 391$ kN

The load factor on L is permitted to equal 0.5 in accordance with exception 1 in ASCE/SEI 2.3.6.

- Load combination 7: $(0.9-0.2S_{DS})D+\Omega_0 Q_E$
  Axial load: $0+(2.5\times 222)=555$ kN tension or compression
  Bending moment: $(0.7\times 953)+0=667$ kN-m
  Shear force: $(0.7\times 249)+0=174$ kN

The collector beam and its connections must be designed to resist the combined effects from the following:

- Flexure and axial tension
- Flexure and axial compression
- Shear

2.7.7 Example 2.7—Allowable Stress Design Load Combinations Including Overstrength for a Collector Beam Subjected to Axial Loads, Bending Moments and Shear Forces

Determine the allowable stress design load combinations for the simply supported collector beam in Example 2.6.

SOLUTION

- Load combination 1a: D
  Bending moment: 703 ft-kips
  Shear force: 56 kips
- Load combination 1b: $D+L$
  Bending moment: $703+235=938$ ft-kips
  Shear force: $56+19=75$ kips

Because the beam is a collector beam in a building assigned to SDC D, the beam must be designed for the allowable stress design load combinations including overstrength (see ASCE/SEI 12.10.2.1 and Chapter 6 of this book).

- Load combination 8: $(1.0+0.14S_{DS})D+0.7\Omega_0 Q_E$
  Axial load: $0+(0.7\times 2.5\times 50)=88$ kips tension or compression
  Bending moment: $(1.14\times 703)+0=801$ ft-kips
  Shear force: $(1.14\times 56)+0=64$ kips
- Load combination 9: $(1.0+0.105S_{DS})D+0.525\Omega_0 Q_E+0.75L$
  Axial load: $0+(0.525\times 2.5\times 50)+0=66$ kips tension or compression
  Bending moment: $(1.105\times 703)+0+(0.75\times 235)=953$ ft-kips
  Shear force: $(1.105\times 56)+0+(0.75\times 19)=76$ kips
- Load combination 10: $(0.6-0.14S_{DS})D+0.7\Omega_0 Q_E$
  Axial load: $0+(0.7\times 2.5\times 50)=88$ kips tension or compression
  Bending moment: $(0.46\times 703)+0=323$ ft-kips
  Shear force: $(0.46\times 56)+0=26$ kips

In S.I.:

- Load combination 1a: D
  Bending moment: 953 kN-m
  Shear force: 249 kN

- Load combination 1b: $D+L$
  Bending moment: $953+319=1,272$ kN-m
  Shear force: $249+85=334$ kN

Because the beam is a collector beam in a building assigned to SDC D, the beam must be designed for the allowable stress design load combinations including overstrength (see ASCE/SEI 12.10.2.1 and Chapter 6 of this book).

- Load combination 8: $(1.0+0.14S_{DS})D+0.7\Omega_0 Q_E$
  Axial load: $0+(0.7\times 2.5\times 222)=389$ kN tension or compression
  Bending moment: $(1.14\times 953)+0=1,086$ kN-m
  Shear force: $(1.14\times 249)+0=284$ kN
- Load combination 9: $(1.0+0.105S_{DS})D+0.525\Omega_0 Q_E+0.75L$
  Axial load: $0+(0.525\times 2.5\times 222)+0=291$ kN tension or compression
  Bending moment: $(1.105\times 953)+0+(0.75\times 319)=1,292$ kN-m
  Shear force: $(1.105\times 249)+0+(0.75\times 85)=339$ kN
- Load combination 10: $(0.6-0.14S_{DS})D+0.7\Omega_0 Q_E$
  Axial load: $0+(0.7\times 2.5\times 222)=389$ kN tension or compression
  Bending moment: $(0.46\times 953)+0=439$ kN-m
  Shear force: $(0.46\times 249)+0=115$ kN

The collector beam and its connections must be designed to resist the combined effects from the following:

- Flexure and axial tension
- Flexure and axial compression
- Shear

2.7.8 Example 2.8—Alternative Allowable Stress Design Load Combinations Including Overstrength for a Collector Beam Subjected to Axial Loads, Bending Moments, and Shear Forces

Determine the alternative allowable stress design load combinations for the simply supported collector beam in Example 2.6.

SOLUTION

- IBC Equation 16-1: $D+L$
  Bending moment: $703+235=938$ ft-kips
  Shear force: $56+19=75$ kips

Because the beam is a collector beam in a building assigned to SDC D, the beam must be designed for the alternative allowable stress design load combinations including overstrength (see ASCE/SEI 12.10.2.1 and Chapter 6 of this book).

- IBC Equation 16-5: $\left(1.0+\frac{0.2S_{DS}}{1.4}\right)D+L+\frac{\Omega_0 Q_E}{1.4}$

  Axial load: $0+0+\frac{2.5\times 50}{1.4}=89$ kips tension or compression

  Bending moment: $\left[\left(1.0+\frac{0.2\times 1.0}{1.4}\right)\times 703\right]+235+0=1,038$ ft-kips

  Shear force: $\left[\left(1.0+\frac{0.2\times 1.0}{1.4}\right)\times 56\right]+19+0=83$ kips

- IBC Equation 16-6: $\left(0.9-\frac{0.2S_{DS}}{1.4}\right)D+\frac{\Omega_0 Q_E}{1.4}$

  Axial load: $0+\frac{2.5\times 50}{1.4}=89$ kips tension or compression

  Bending moment: $\left[\left(0.9-\frac{0.2\times 1.0}{1.4}\right)\times 703\right]+0=532$ ft-kip

  Shear force: $\left[\left(0.9-\frac{0.2\times 1.0}{1.4}\right)\times 56\right]+0=42$ kips

In S.I.:

- IBC Equation 16-1: $D+L$
  Bending moment: $953+319=1{,}272$ kN-m
  Shear force: $249+85=334$ kN

Because the beam is a collector beam in a building assigned to SDC D, the beam must be designed for the alternative allowable stress design load combinations including overstrength (see ASCE/SEI 12.10.2.1 and Chapter 6 of this book).

- IBC Equation 16-5: $\left(1.0+\frac{0.2S_{DS}}{1.4}\right)D+L+\frac{\Omega_0 Q_E}{1.4}$

  Axial load: $0+0+\frac{2.5\times 222}{1.4}=396$ kN tension or compression

  Bending moment: $\left[\left(1.0+\frac{0.2\times 1.0}{1.4}\right)\times 953\right]+319+0=1{,}408$ kN-m

  Shear force: $\left[\left(1.0+\frac{0.2\times 1.0}{1.4}\right)\times 249\right]+85+0=370$ kN

- IBC Equation 16-6: $\left(0.9-\frac{0.2S_{DS}}{1.4}\right)D+\frac{\Omega_0 Q_E}{1.4}$

  Axial load: $0+\frac{2.5\times 222}{1.4}=396$ kN tension or compression

  Bending moment: $\left[\left(0.9-\frac{0.2\times 1.0}{1.4}\right)\times 953\right]+0=722$ kN-m

  Shear force: $\left[\left(0.9-\frac{0.2\times 1.0}{1.4}\right)\times 249\right]+0=189$ kN

The collector beam and its connections must be designed to resist the combined effects from the following:

- Flexure and axial tension
- Flexure and axial compression
- Shear

2.7.9 Example 2.9—Allowable Stress Design Load Combinations for a Pile Subjected to Axial Loads

Determine the allowable stress design load combinations for a pile supporting a residential building using the design data in Table 2.29. The building is in a coastal A zone.

Solution

The allowable stress design load combinations in ASCE/SEI 2.4.1 and 2.4.2 are given in Table 2.30 for this pile. Only the applicable load effects from the design data are included.

Table 2.29 Design Data for the Pile in Example 2.9

Load	Axial Load, kips (kN)
Dead, D	8 (36)
Live, L	6 (27)
Roof live load, L_r	4 (18)
Wind, W	±26 (±116)
Flood, F_a	±2 (±9)

Table 2.30 Allowable Stress Design Load Combinations for the Pile in Example 2.9

ASCE/SEI Equation No.	Load Combination	Axial Load, kips (kN)
1a	D	8 (36)
2a	$D+L$	14 (63)
3a	$D+L_r$	12 (54)
4a	$D+0.75L+0.75L_r$	16 (70)
5a	$D+0.6W$	24 (106)
	$D-0.6W$	−8 (−34)
5b	$D+0.6W+1.5F_a$	27 (119)
	$D-0.6W-1.5F_a$	−11 (−47)
6a	$D+0.75L+0.45W+0.75L_r$	27 (122)
	$D+0.75L-0.45W+0.75L_r$	4 (18)
6b	$D+0.75L+0.45W+0.75L_r+1.5F_a$	30 (136)
	$D+0.75L-0.45W+0.75L_r-1.5F_a$	1 (4)
7a	$0.6D+0.6W$	20 (91)
	$0.6D-0.6W$	−11 (−47)
7b	$0.6D+0.6W+1.5F_a$	23 (105)
	$0.6D-0.6W-1.5F_a$	−14 (−62)

The pile must be designed for the axial compression and tension loads in Table 2.21 in combination with bending moments caused by the wind and flood loads. Shear forces and deflection at the tip of the pile must also be checked. Finally, the embedment length of the pile must be sufficient to resist the maximum net tension load.

More information on flood loads can be found in Chapter 7 of this book.

2.8 References

2.1. Ellingwood, B. 1981. "Wind and Snow Load Statistics for Probabilistic Design." *Journal of the Structural Division*, 107(7):1345–1350.

2.2. Galambos, T.V., Ellingwood, B., MacGregor, J.G., and Cornell, C.A. 1982. "Probability-Based Load Criteria: Assessment of Current Design Practice." *Journal of the Structural Division*, 108(5):959–977.

2.9 Problems

2.1. Determine the strength design load combinations for a beam on a typical floor of a multistory residential building using the nominal bending moments in Table 2.31. Assume the live load on the floor is less than 100 lb/ft^2 (4.79 kN/m^2).

Table 2.31 Design Data for Problem 2.1

Load	Bending Moment, ft-kips (kN-m)		
	Exterior Negative	Positive	Interior Negative
Dead, *D*	−13.3 (−18.0)	43.9 (59.5)	−53.2 (−72.1)
Live, *L*	−12.9 (−17.5)	42.5 (57.6)	−51.6 (−70.0)

2.2. Determine the strength design load combinations for a beam that is part of an ordinary moment frame in an office building using the bending moments and shear forces in Table 2.32. Assume the live load on the floor is less than 100 lb/ft^2 (4.79 kN/m^2).

Table 2.32 Design Data for Problem 2.2

Load	Bending Moment, ft-kips (kN-m)		Shear Force, kips (kN)
	Left Support	Midspan	
Dead, *D*	−57.6 (−78.1)	41.1 (55.7)	11.8 (52.5)
Live, *L*	−22.5 (−30.5)	16.2 (22.0)	4.6 (20.5)
Wind, *W*	±54.0 (±73.2)	—	±4.8 (±21.4)

2.3. Given the design data in Problem 2.2, determine the allowable stress design load combinations.

2.4. Given the design data in Problem 2.2, determine the alternative allowable stress design load combinations.

2.5. Determine the strength design load combinations for a column in an office building using the nominal axial loads, bending moments, and shear forces in Table 2.33. Assume the live loads on the floors are equal to 100 lb/ft^2 (4.79 kN/m^2), $\rho = 1.0$ and $S_{DS} = 0.41$.

Table 2.33 Design Data for Problem 2.5

Load	Axial Load, kips (kN)	Bending Moment at Bottom, ft-kips (kN-m)	Shear Force at Bottom, kips (kN)
Dead, *D*	167.9 (746.9)	21.3 (28.9)	2.3 (10.2)
Live, *L*	41.5 (184.6)	21.0 (28.5)	2.2 (9.8)
Roof live, L_r	14.9 (66.3)	—	—
Wind, *W*	±13.6 (±60.5)	±121.0 (±164.1)	±11.1 (±49.4)
Seismic, Q_E	±36.4 (±161.9)	±432.1 (±585.9)	±42.2 (±187.7)

2.6. Determine the strength design load combinations for a shear wall in a parking garage using the nominal axial forces, bending moments, and shear forces in Table 2.34. Assume $\rho = 1.0$ and $S_{DS} = 1.0$.

Table 2.34 Design Data for Problem 2.6

Load	Axial Load, kips (kN)	Bending Moment at Bottom, ft-kips (kN-m)	Shear Force at Bottom, kips (kN)
Dead, D	645 (2,869)	0	0
Live, L	149 (663)	0	0
Seismic, Q_E	0	±4,280 (±5,803)	±143 (±636)

2.7. Determine the strength design load combinations and the combinations for strength design with overstrength for a collector beam in an assembly building using the axial loads, bending moments, and shear forces in Table 2.35. Assume $S_{DS} = 0.9$ and $\Omega_0 = 2.0$.

Table 2.35 Design Data for Problem 2.7

Load	Axial Load, kips (kN)	Bending Moment, ft-kips (kN-m)		Shear Force, kips (kN)
		Negative	Positive	
Dead, D	0	−80.6 (−109.3)	53.7 (72.8)	29.7 (132.1)
Live, L	0	−42.1 (−57.1)	30.4 (41.2)	19.0 (84.5)
Seismic, Q_E	±241 (±1,072)	0	0	0

2.8. Given the design data in Problem 2.7, determine the allowable stress design load combinations.

2.9. Determine the strength design load combinations for a beam in a commercial building with a curved roof using the following uniformly distributed loads:

- Dead, $D = 75$ lb/ft (1,095 N/m)
- Rain, $R = 200$ lb/ft (2,919 N/m)
- Roof live, $L_r = 100$ lb/ft (1,459 N/m)
- Balanced snow, $S = 125$ lb/ft (1,824 N/m)

2.10. Given the design data in Problem 2.9, determine the allowable stress design load combinations.

CHAPTER 3

Dead, Live, Rain, and Soil Lateral Loads

3.1 Dead Loads

Dead loads, *D*, are the weights of construction materials and fixed service equipment attached to or supported by the building or structure. Various types of such loads are listed in IBC 202 under "Dead Load." The effects of dead loads must be considered in the design of structural members in buildings and structures (IBC 1606.1).

Dead loads are permanent loads, that is, loads in which variations over time are rare or of small magnitude. Variable loads, such as live loads and wind loads, are not permanent. It is important to know the distinction between permanent and variable loads when applying the provisions for load combinations (see IBC 1605, ASCE/SEI Chapter 2 and Chapter 2 of this book for information on load combinations).

The actual weights of materials of construction and equipment must be considered as dead load for purposes of design. However, such weights are not usually known during the design phase, so estimated weights are often used. Where the effect of gravity loads and lateral loads are additive, estimated dead loads are assumed to be greater than anticipated so that the design of certain structural elements is conservative. This is not acceptable when considering load combinations where gravity loads and lateral loads counteract. For example, it would be unconservative to design for uplift on a structure using an overestimated dead load.

Minimum design dead loads for various types of common construction components, including ceilings, roof and wall coverings, floor fill, floors and floor finishes, frame partitions and frame walls are given in ASCE/SEI Table C3.1-1a in pounds per square foot (lb/ft^2) and in ASCE/SEI Table C3.1-1b in kilonewtons per square meter (kN/m^2). Minimum densities for common construction materials are given in ASCE/SEI Table C3.1-2 in pounds per cubic foot (lb/ft^3) and kilonewtons per cubic meter (kN/m^3). The weights and densities in these tables can be used as a guide when estimating dead loads. Actual weights of construction materials and equipment can be greater than tabulated values, so it is always prudent to verify weights with manufacturers or other similar resources prior to design. In cases where information on dead load is unavailable, values of dead loads used in design must be approved by the building official (IBC 1606.2).

The weight of fixed service equipment must be considered as dead load for design purposes (IBC 1606.3). This includes plumbing stacks and risers; electrical feeders; heating, ventilation, and air conditioning (HVAC) systems; elevators and elevator machinery; fire protection systems; and process equipment such as vessels, tanks, piping, and cable trays.

The weight of photovoltaic panel systems and their supports must be considered as dead load (IBC 1606.4). Loads from ballasted systems not permanently attached to the structure must be included.

For vegetative and landscaped roofs, the weight of landscaping materials (such as soil, plants, and drainage layer materials) and hardscaping materials (such as walkways, fences, and walls) must be considered as dead load (IBC 1606.5). When determining the maximum effects on the structure, the following load cases must be considered:

- The weight of fully saturated soil and drainage layer materials where this weight is additive to other loads.
- The weight of fully dry soil and drainage layer materials where this weight acts to counteract uplift forces.

3.2 Live Loads

3.2.1 Overview

Live loads, which are determined in accordance with IBC 1607, are produced by the use and occupancy of a building or structure and do not include construction loads, environmental loads (such as wind loads, snow loads, rain loads, earthquake loads, and flood loads), or dead loads (see the definition of "Live Load" in IBC 202).

Live loads are transient in nature and vary in magnitude over the life of a structure. Studies have shown that building live loads consist of both a sustained portion and a variable portion. The sustained portion is based on general day-to-day use of the facilities and normally varies during the life of the structure due to tenant modifications and changes in occupancy, for example. The variable portion of the live load is typically created by events such as temporary storage and similar unusual events.

Uniformly distributed and concentrated live loads are given in IBC Table 1607.1 as a function of occupancy or use. The occupancy category listed in the table is not necessarily group-specific (occupancy groups are defined in IBC Chapter 3). For example, an office building with a Business Group B classification may also have storage areas with live loads of 125 lb/ft^2 (6.00 kN/m^2) or 250 lb/ft^2 (11.97 kN/m^2) depending on the type of storage.

The uniform and concentrated live loads in IBC Table 1607.1 are minimum loads. The structure may have to be designed for live loads greater than the tabulated values, but in no case is the structure to be designed for live loads less than the minimum values. For occupancies not listed in the table, live loads used in design must be approved by the building official (IBC 1607.2).

Concurrent application of the uniform and concentrated live loads in IBC Table 1607.1 is not required. Structural members are designed based on the maximum effects due to the application of either a uniform live load or a concentrated live load and need not be designed for the effects due to both loads applied at the same time.

The uniformly distributed live loads in ASCE/SEI Table C4.3-1 can be used as a guide for common occupancies, and live load statistics for some typical occupancies are given in ASCE/SEI Table C4.3-2.

3.2.2 Uniform Live Loads

Overview

Buildings must be designed for the minimum uniform live loads in IBC Table 1607.1 based on occupancy or use (IBC 1607.3). The possibility of later changes of occupancy involving loads heavier than originally anticipated should be considered. Uniform loads greater than those in IBC Table 1607.1 may be required, but in no case is the uniform load to be taken less than the minimum value in the table.

Live loads acting on a sloping surface are assumed to act vertically on a horizontal projection of that surface (IBC 1607.3).

Partial Loading of Floors

Uniform live loads applied over selected spans of a structure or a member may produce greater effects than when applied over the full length. Thus, partial live loading must be considered in the design of structural members (IBC 1607.3.1). In general, full dead loads must be applied on all spans in combination with floor live loads on spans selected to produce the greatest load effect at each location under consideration. It is permitted to use reduced uniform live loads in accordance with IBC 1607.13 when investigating partial floor loading.

Checkerboard live load patterns applied to continuous members produce higher positive moments than live loads applied to all spans, and live loads on either side of a support produce greater negative moments. Cantilevers must not rely on live load on the back span for equilibrium.

Illustrated in Figure 3.1 are the four loading patterns that must be investigated for a three-span continuous system subject to uniform dead and live loads. Similar load patterns can be derived for systems with other span conditions.

Figure 3.1 Distribution of Floor Loads for a Three-span Continuous System

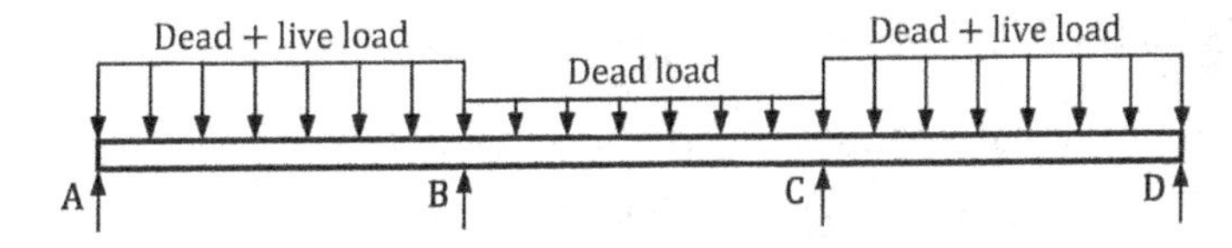

Loading pattern for maximum negative moment at support A or D and maximum positive moment in span AB or CD

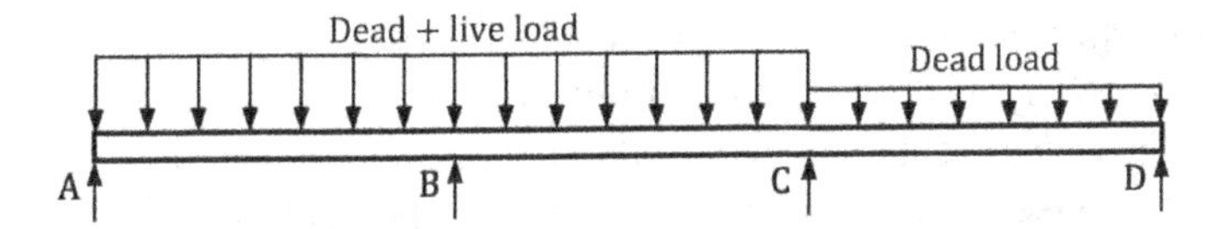

Loading pattern for maximum negative moment at support B

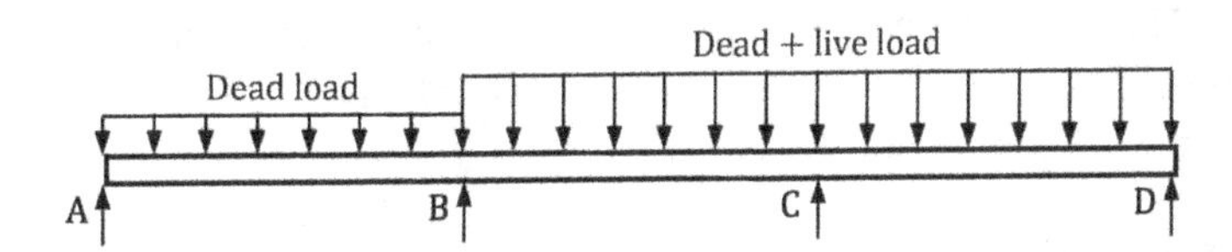

Loading pattern for maximum negative moment at support C

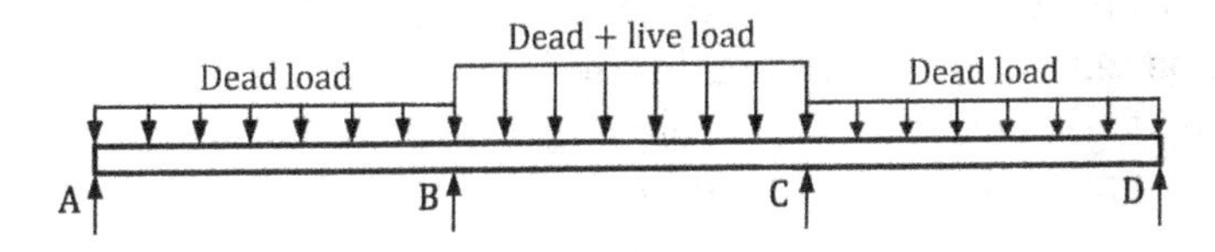

Loading pattern for maximum positive moment in span BC

Partial Roof Loading

In cases where the uniform roof loads in IBC Table 1607.1 are permitted to be reduced to less than 20 lb/ft^2 (0.96 kN/m^2) in accordance with IBC 1607.14, the reduced roof live load must be applied to adjacent spans or to alternate spans to produce the greatest effects on the supporting structure, similar to partial loading for floor systems.

Partial or pattern loading must be considered regardless of the magnitude of the uniform load where a uniform roof live load is caused by occupancy.

3.2.3 Concentrated Live Loads

Minimum concentrated live loads are given in IBC Table 1607.1 (IBC 1607.4). Unless specified otherwise, concentrated loads are to be uniformly distributed over an area of 2.5 ft by 2.5 ft (762 mm by 762 mm) and are to be located to produce the maximum load effects in the supporting structural members.

3.2.4 Partition Loads

Requirements for partition loads are given in IBC 1607.5. Partitions that can be relocated (i.e., those types that are not permanently attached to the structure) must be considered as live loads in offices and other buildings because they are variable by nature. The weight of built-in partitions that cannot be moved is considered a dead load in accordance with IBC 1602.

A uniformly distributed live load of at least 15 lb/ft^2 (0.72 kN/m^2) must be included for moveable partitions unless the minimum specified uniform live load is greater than or equal to 80 lb/ft^2 (3.83 kN/m^2). The minimum partition load is based on the following assumptions:

- A typical partition consists of a 10-foot (3.05-m) high wood or steel stud wall.
- 0.5 in. (13 mm) gypsum board is attached on each side of the stud wall.
- The wall assembly weighs 8 lb/ft^2 (0.38 kN/m^2).

Based on these assumptions, the wall load on the floor is 80 lb per linear foot (1.17 kN/m). If the partitions are spaced on a 10-foot (3.05-m) grid on center (which is an extremely dense spacing over a whole bay), the average distributed load is 16 lb/ft^2 (0.77 kN/m^2). Because the partitions are not likely to be spaced this closely over large areas, a minimum value of 15 lb/ft^2 (0.72 kN/m^2) is reasonable for design. A larger design load for partitions should be considered if taller wall heights, heavier walls, or a higher density of partitions is anticipated.

Because partition loads are live loads, they are not permitted to resist uplift loads and are subject to the partial floor loading requirements in IBC 1607.3.1 (see Section 3.2.2 of this book).

The distribution of partition loads is fundamentally different from the distribution of typical office live loads, so partition loads are not permitted to be reduced in accordance with IBC 1607.13.

Unlike other live loads, partition loads are both semipermanent and firmly attached to a structure. Thus, partition weight is required to be part of the effective seismic weight when determining earthquake forces on a structure.

3.2.5 Helipads

Minimum uniform and concentrated live loads to be used in the design of helipads are given in IBC 1607.6 (see Table 3.1). The concentrated loads are not required to act concurrently with other uniform or concentrated live loads, and the uniformly distributed loads must not be reduced.

Helipads must be marked to indicate the maximum take-off weight: the take-off weight limitation must be indicated in units of thousands of pounds and placed in a box that is in the bottom right corner of the landing area as viewed from the primary approach path. The box must be a minimum of 5 ft (1,524 mm) in height.

3.2.6 Passenger Vehicle Garages

Live load requirements for floors in garages and portions of a building used for the storage of motor vehicles are given in IBC 1607.7. The supporting structural members must be designed for either a uniform load of 40 lb/ft^2 (1.92 kN/m^2) or the following concentrated loads:

- 3,000 lb (13.35 kN) acting over an area of 4.5 in. by 4.5 in. (114 mm by 114 mm) for garages restricted to passenger vehicles accommodating not more than nine passengers.
- 2,250 lb (10.00 kN) per wheel for mechanical parking structures without slab or deck used for storing passenger vehicles only.

Table 3.1 Live Loads for Helipads

Load Type	Minimum Load
Uniform	• 40 lb/ft^2 (1.92 kN/m^2) for a helicopter with a maximum take-off weight of 3,000 lb (13.35 kN) or less. • 60 lb/ft^2 (2.87 kN/m^2) for a helicopter with a maximum take-off weight greater than 3,000 lb (13.35 kN).
One concentrated load	A concentrated load equal to 3,000 lb (13.35 kN) applied over an area of 4.5 in. by 4.5 in. (114 mm by 114 mm) and located to produce the maximum load effects on the structural elements under consideration.
Two single concentrated loads	• Each concentrated load is equal to 0.75 times the maximum take-off weight of the helicopter. • The concentrated loads must be applied on the landing pad 8 ft (2,438 mm) apart to produce the maximum load effects on the structural elements under consideration. • The concentrated loads must be applied over an area of 8 in. by 8 in. (203 mm by 203 mm) on the supporting structure.

3.2.7 Heavy Vehicle Loads

Loads

Live load requirements for heavy vehicle loads [i.e., vehicle loads greater than 10,000 lb (44.48 kN)] are given in IBC 1607.8. In general, portions of structures where such vehicles can have access must be designed for live loads—including impact and fatigue—determined in accordance with codes and specifications required by the jurisdiction having authority for the design and construction of roadways and bridges in the same location of the structure (IBC 1607.8.1). The American Association of State Highway and Transportation Officials (AASHTO) design specifications (Reference 3.1) have been adopted by many jurisdictions throughout the United States and contain provisions on how to determine these live loads.

Fire Truck and Emergency Vehicles

Portions of a structure accessible to fire trucks and emergency vehicles must be designed for the greater of the following loads:

- Actual operational loads, which include the total vehicle load, the individual wheel loads, and the outrigger reactions, where applicable. These operational loads do not act concurrently.
- The live load specified in IBC 1607.8.1 (see above).

Emergency vehicle loads need not be assumed to act concurrently with other uniform live loads.

Heavy Vehicle Garages

Garages accommodating vehicle loads exceeding 10,000 lb (44.48 kN) must be designed for the live loads in IBC 1607.8, except the design need not include the effects due to impact or fatigue. It is permitted to design garage floors for actual vehicle weights provided the following are satisfied:

- The loads and their placement are based on a rational analysis.
- The loads are greater than or equal to 50 lb/ft^2 (2.39 kN/m^2) and are not reduced.
- The loads are approved by the building official.

ASCE/SEI 4.10.2 makes specific reference to Reference 3.1 for the design of garages accommodating trucks and buses.

Forklifts and Moveable Equipment

Structural members supporting forklifts and similar moveable equipment must be designed for a minimum live load corresponding to the total vehicle or equipment load and the individual wheel loads. These loads must be posted. Impact and fatigue loads must also be considered in design. It is permitted to account for impact by increasing the vehicle and wheel loads by 30 percent.

Posting

Similar to heavy live loads for floors, the live load for heavy vehicles must be posted in accordance with IBC 106.1.

3.2.8 Handrails, Guards, Grab Bars, and Seats

Handrails and Guards

Handrails and guards for stairs, balconies, and similar elements must be designed for the live loads in ASCE/SEI 4.5.1 (IBC 1607.9). The following live loads must be considered:

1. A single concentrated load of 200 lb (0.89 kN) applied in any direction at any point on the handrail or top rail to produce the maximum load effect on the element being considered and to transfer this load through the supports to the structure (load condition 1 in Figure 3.2).
2. A uniform load of 50 lb/ft (0.73 kN/m) applied in any direction along the handrail or top rail and to transfer this load through the supports to the structure (load condition 2 in Figure 3.2).

Figure 3.2 Handrail Concentrated and Uniform Loads

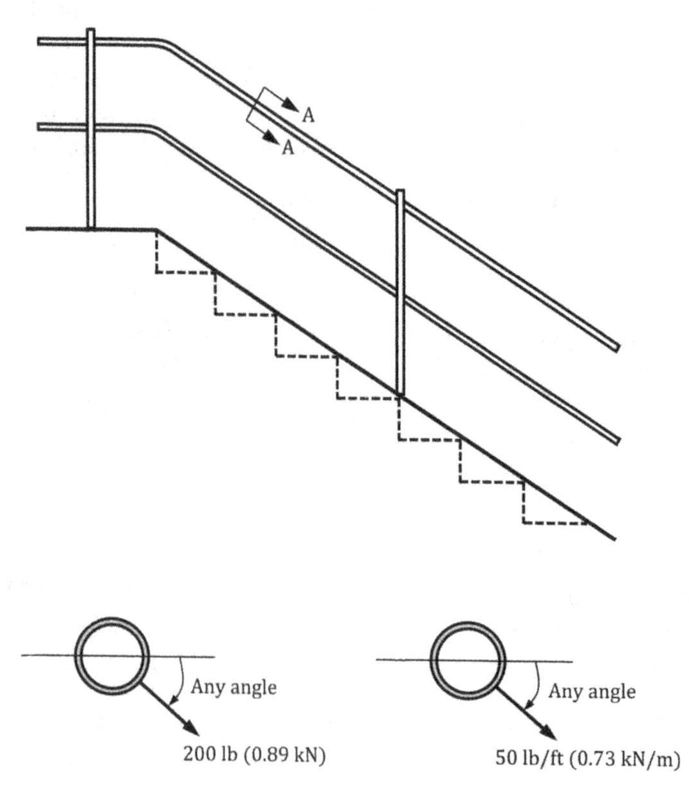

3. A load of 50 lb (0.22 kN) distributed normal to a 12 in. by 12 in. (305 mm by 305 mm) area on balusters, panel fillers, and guard system infill components, including all rails, (i.e., on all elements except the handrail or top rail) located to produce the maximum load effects (see Figure 3.3).

Figure 3.3
Rail Load

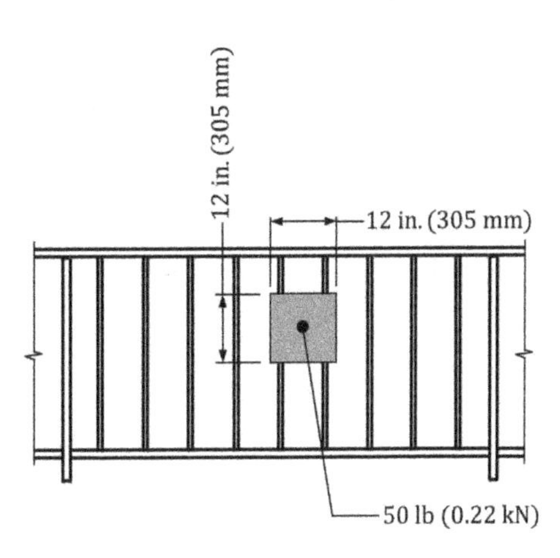

The concentrated load in load condition 1 is meant to simulate the maximum anticipated load from a person grabbing or falling into the handrail or guard while the uniform load in load condition 2 is the maximum anticipated load on a handrail or guard from use by a crowd of people on a stairway. These loads can occur in any direction and need not be applied concurrently (see Figure 3.2).

Reactions due to the guard system component load in Figure 3.3 need not be superimposed with the concentrated and uniform loads in Figure 3.2.

Exceptions to these requirements recognize the circumstances in occupancies where the handrail or guard is inaccessible to the public. In particular, only the single concentrated load (load condition 1) must be considered in one- and two-family dwellings. In Group I-3, F, H, and S occupancies in areas not accessible to the public with an occupant load of 50 or less, the minimum load is 20 lb/ft (0.29 kN/m) instead of 50 lb/ft (0.73 kN/m).

The second exception differs from that in ASCE/SEI 4.5.1 in three ways. First, Group I-3 institutional occupancies (correctional centers, jails, prisons, and the like) are included in the IBC exception and not in the ASCE/SEI 7 exception. Second, the exception is applicable to occupant loads less than 50 in the IBC, while in ASCE/SEI 7, it is applicable to 50 or less. Finally, the IBC requires a minimum line load of 20 lb/ft (0.29 kN/m) in such occupancies; in ASCE/SEI 7, a minimum line load is not required.

The area load is a localized load for guard members and components; the balusters in Figure 3.3 resisting this load are those within the one-square-foot (0.093 m^2) area in the plane of the guard. This area load need not be superimposed with any other loads.

IBC 1607.9 also stipulates that glass handrail assemblies and guards comply with IBC 2407, which includes minimum requirements related to material properties and support conditions.

Grab Bars, Shower Seats, and Accessible Benches

Live load requirements for grab bars, shower seats and accessible benches are given in IBC 1607.9 (these components are typically required in providing accessibility in accordance with IBC Chapter 11). Such elements must be designed for a single concentrated load of 250 lb (1.11 kN) applied in any direction and at any point of the grab bar, seat, or bench to produce maximum load effects. This loading is consistent with *ADA Standards for Public Accommodations and Commercial Facilities*.

The same requirement is given in ASCE/SEI 4.5.2 but for grab bar systems and shower seats only.

3.2.9 Vehicle Barrier Systems

Vehicle barriers are defined in IBC 202 as a component or a system of components positioned at open sides of a parking garage floor or ramp or at building walls that act as restraints for vehicles. In other words, these barriers provide a passive restraint system at locations where vehicles could fall to a lower level.

Vehicle barrier systems for passenger vehicles must be designed to resist a single concentrated load of 6,000 lb (26.70 kN) applied horizontally over an area not to exceed 12 in. by 12 in. (305 mm by 305 mm) in any direction to the barrier system at heights between 1 ft-6 in. (460 mm) and 2 ft-3 in. (686 mm) above the floor or ramp surface, which accounts for varying bumper heights, barrier configurations, and anchorage methods (see IBC 1607.11, ASCE/SEI 4.5.3, and Figure 3.4). This load, which includes the effects from impact, must be located to produce the maximum load effects, and it need not act concurrently with any handrail or guard loading (see Section 3.2.8 of this book).

Figure 3.4 Vehicle barrier system load requirements for passenger vehicles

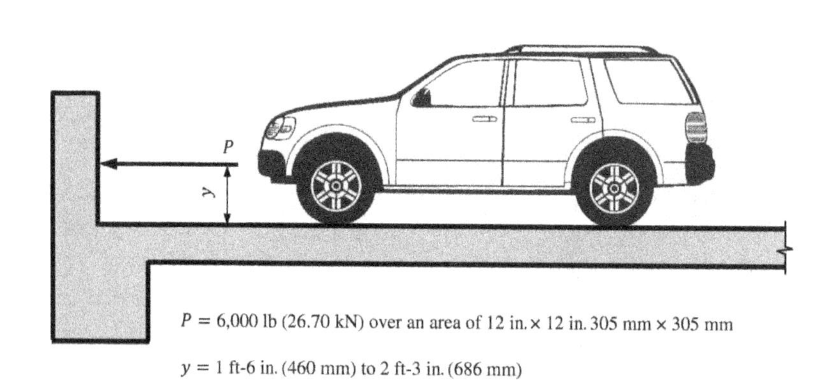

For garages and buildings accommodating trucks, buses, and similar vehicles, the vehicle barrier systems must be designed in accordance with Reference 3.1.

3.2.10 Impact Loads

Overview

The live loads in IBC 1607.3 through 1607.11 include an allowance for impact that is normally attributed to such loads (IBC 1607.12). Impact loads with unusual vibration and impact forces, such as those from elevators and machinery, must be accounted for in design because such elements produce additional forces and deflections in the supporting structural systems.

Elevators

The static load of an elevator must be increased to account for motion effects. The loads on the members supporting an elevator are significantly higher than the weight of the elevator and its occupants when the elevator comes to a stop. The rate of acceleration and deceleration has a significant impact on this effect. Elements subjected to dynamic loads imparted by elevators must be designed for the impact loads and deflection limits specified in ASME A17.1 *Safety Code for Elevators and Escalators* (IBC 1607.12 and ASCE/SEI 4.6.2). An impact factor equal to 100 percent is specified in ASME A17.1, which means the weight must be increased by a factor of 2 to account for impact. Deflection limits on the applicable elements are also specified. A dynamic analysis is generally not required; use of an equivalent static load is usually sufficient.

Machinery

The weight of machinery and moving loads must be increased as follows to account for impact:

- Shaft- or motor-driven light machinery: 20 percent
- Reciprocating machinery or power-driven units: 50 percent

These impact factors include the effects due to vibration; thus, a larger impact factor is assigned to reciprocating machinery and power-driven units because this type of equipment vibrates more than shaft- or motor-driven light machinery. It is always good practice to acquire impact factors for specific pieces of equipment from manufacturers because such factors can be greater than those specified in IBC 1607.12 and ASCE/SEI 4.6.3.

Elements Supporting Hoists for Façade Access and Building Maintenance Equipment

Structural elements (such as roof anchors, wall-mounted anchors, and davits) supporting hoists used for accessing the façade and building maintenance equipment must be designed for a live load equal to the greater of the following (IBC 1607.12 and ASCE/SEI 4.6.4):

- 2.5 times the rated load of the hoist
- The stall load of the hoist

Where load combinations for strength design are used, these members must be designed for a live load equal to $1.6 \times 2.5 = 4.0$ times the rated load plus other applicable factored loads; this matches the Occupational Safety and Health Administration (OSHA) requirements in Standard 1910.66, *Powered Platforms for Building Maintenance* for the design of such elements. Where allowable stress design is used, 2.5 times the rated load results in a comparable design when a safety factor of 1.6 is used in determining the allowable stresses.

According to Code of Federal Regulations Section 1926.451 of OSHA *Safety Standards for Scaffolds used in the Construction Industry*, supporting elements must also be designed for 1.5 times the stall load of the hoist. Where strength design in accordance with IBC 1605.1 is used, the total factored live load is equal to 1.6 times the stall load, which is slightly more conservative than the 1.5 times the stall load requirement by OSHA. The same requirements are given in ASCE/SEI 4.6.4.

Fall Arrest, Lifeline, and Rope Descent System Anchorages

Live load requirements for fall arrest, lifeline, and rope descent system anchorages are given in IBC 1607.12 and ASCE/SEI 4.6.5. Such anchorages and any other pertinent structural elements supporting these anchorages must be designed for a live load not less than 3,100 lb (13.8 kN) for each attached line in any direction the load can be applied.

In the case of strength design, the total factored live load is equal to $1.6 \times 3{,}100 = 4{,}960$ lb (22.1 kN), which is essentially equal to the OSHA requirement of 5,000 lb (22.1 kN) for the design of these anchorages. For allowable stress design, a design live load of 3,100 lb results in a comparable design when a safety factor of 1.6 is used. The requirements in ASCE/SEI 4.6.5 are the same as those in IBC 1607.12.

Anchorages of horizontal lifelines and the structural elements supporting these anchorages must be designed for the maximum tension force in the horizontal lifeline due to the prescribed live loads. Much larger loads can develop at the end anchorages when a fall arrest load is applied perpendicular to the length of a horizontal lifeline due to the geometry of the horizontal cable. These increased forces must be considered in design.

Illustrated in Figure 3.5 is a hoist used to access the exterior of a building. The hoist is supported by a davit system, which is anchored to the roof structure of the building. The structural member(s) supporting the davit anchorage must be designed for the live loads specified in IBC 1607.12 in combination with all other applicable loads. The structural members supporting the wall-mounted fall arrest/lifeline anchors must be designed in accordance with IBC 1607.12.

3.2.11 Reduction in Uniform Live Loads

Overview

The minimum uniformly distributed live loads, L_o, in IBC Table 1607.1 are permitted to be reduced in accordance with the requirements of IBC 1607.13. Both methods are discussed below.

Figure 3.5 Elements Supporting Hoists and Fall Arrest/Lifeline Anchorages

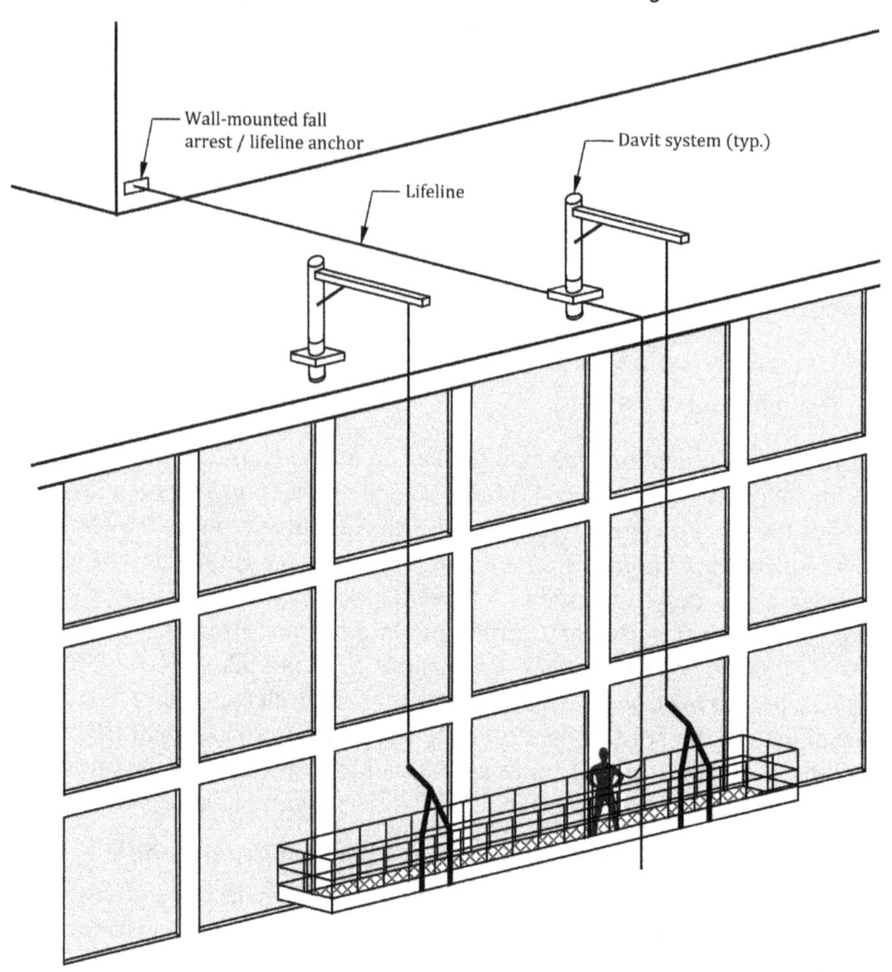

Roof live loads are permitted to be reduced in accordance with IBC 1607.14 (see Section 3.2.12 of this book).

Basic Uniform Live Load Reduction

Structural members for which $K_{LL}A_T \geq 400$ square feet (37.16 m^2) are permitted to be designed for a reduced uniform live load, L, determined by IBC Equation 16-7, subject to the limitations of IBC 1607.13.1.1 through 1607.13.1.3:

$$L = L_o\left(0.25 + \frac{15}{\sqrt{K_{LL}A_T}}\right) = L_o(\text{RM})\ (\text{lb/ft}^2) \tag{3.1}$$

$$L = L_o\left(0.25 + \frac{4.57}{\sqrt{K_{LL}A_T}}\right) = L_o(\text{RM})\ (\text{kN/m}^2) \tag{3.2}$$

where RM is the reduction multiplier.

For members supporting one floor, L must not be taken less than $0.50L_o$, and for members supporting two or more floors, L must not be taken less than $0.40L_o$.

In these equations, K_{LL} is the live load element factor in IBC Table 1607.13.1 and A_T is the tributary area in square feet (m^2).

The live load element factor, K_{LL}, converts the tributary area of a structural member, A_T, to an influence area, which is considered to be the adjacent floor area from which the member derives its load. The following equation can be used to determine K_{LL}:

$$K_{LL} = \frac{\text{Influence area}}{\text{Tributary area}} \tag{3.3}$$

The influence area for interior column C2 in Figure 3.6 is equal to the area of the four bays adjacent to the column:

$$\text{Influence area} = (\ell_A + \ell_B)(\ell_2 + \ell_3)$$

Figure 3.6 Influence Areas and Tributary Areas for a Column and Beam

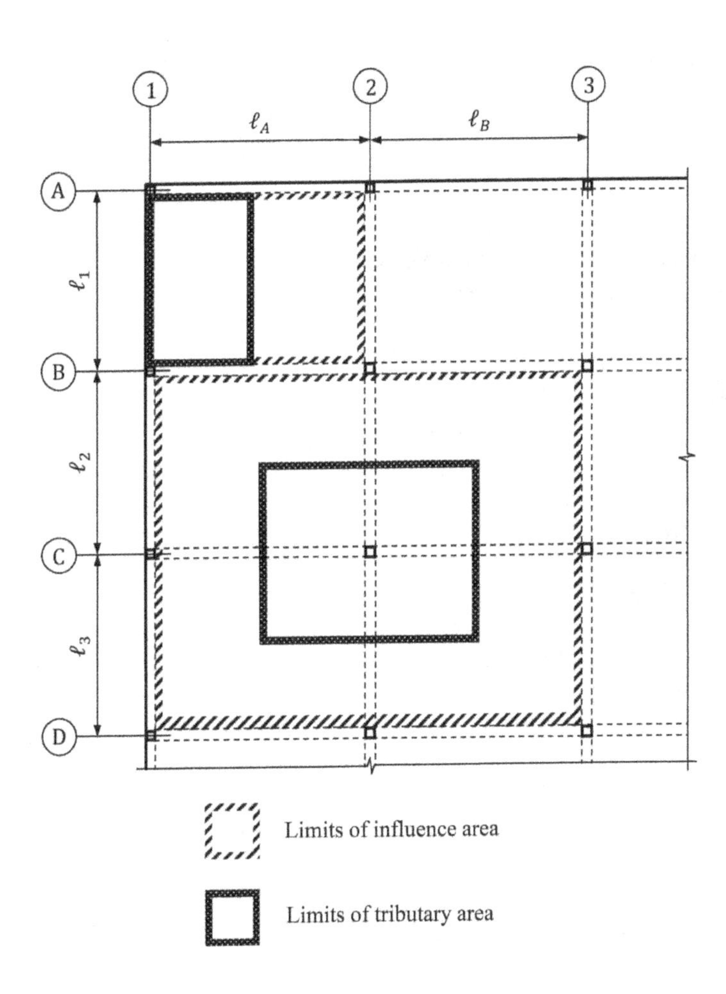

The tributary area, A_T, supported by this column is equal to the product of the tributary widths in both directions (a tributary width in this case is equal to the sum of one-half of the span lengths on both sides of the column):

$$A_T = \left(\frac{\ell_A}{2} + \frac{\ell_B}{2}\right)\left(\frac{\ell_2}{2} + \frac{\ell_3}{2}\right)$$

Using Equation (3.3), K_{LL} is equal to the following for this interior column:

$$K_{LL} = \frac{(\ell_A + \ell_B)(\ell_2 + \ell_3)}{\frac{1}{4} \times (\ell_A + \ell_B)(\ell_2 + \ell_3)} = 4$$

This matches the value given in IBC Table 1607.13.1 for an interior column. It can also be determined that K_{LL} is equal to 4 for an exterior column (other than a corner column) without cantilever slabs.

The influence area for the beam on column line 1 between A and B is equal to the area of the bay adjacent to the beam:

$$\text{Influence area} = \ell_A \ell_1$$

The tributary area, A_T, supported by this beam is equal to the tributary width times the length of the beam:

$$A_T = \left(\frac{\ell_A}{2}\right)\ell_1$$

Thus, K_{LL} is equal to the following for this edge beam without a cantilever slab:

$$K_{LL} = \frac{\ell_A \ell_1}{\frac{1}{2} \times \ell_A \ell_1} = 2$$

This matches the value given in IBC Table 1607.13.1.

Values of K_{LL} for other elements can be derived in a similar fashion. Typical influence areas, tributary areas, and corresponding values of K_{LL} for a variety of elements are given in ASCE/SEI Figure C4.7-1.

The variation of the reduction multiplier (RM) with respect to the influence area, $K_{LL}A_T$, is given in Figure 3.7. The minimum influence area of 400 square feet (37.16 m^2) and the limits of 0.5 and 0.4 (which are the maximum permitted reductions for members supporting one floor and two or more floors, respectively) are indicated in the figure.

The following limitations on live load reduction by this method are given in IBC 1607.13.1.1 through 1607.13.1.3:

- One-way slabs: Live load reduction on one-way slabs is permitted provided the tributary area, A_T, does not exceed an area equal to the slab span times a width normal to the span of 1.5 times the slab span (IBC 1607.13.1.1). The live load will be somewhat higher for a one-way slab with an aspect ratio of 1.5 than for a two-way slab with the same aspect ratio. This

Figure 3.7 Reduction Multiplier (RM) for Live Load in Accordance with IBC 1607.13.1

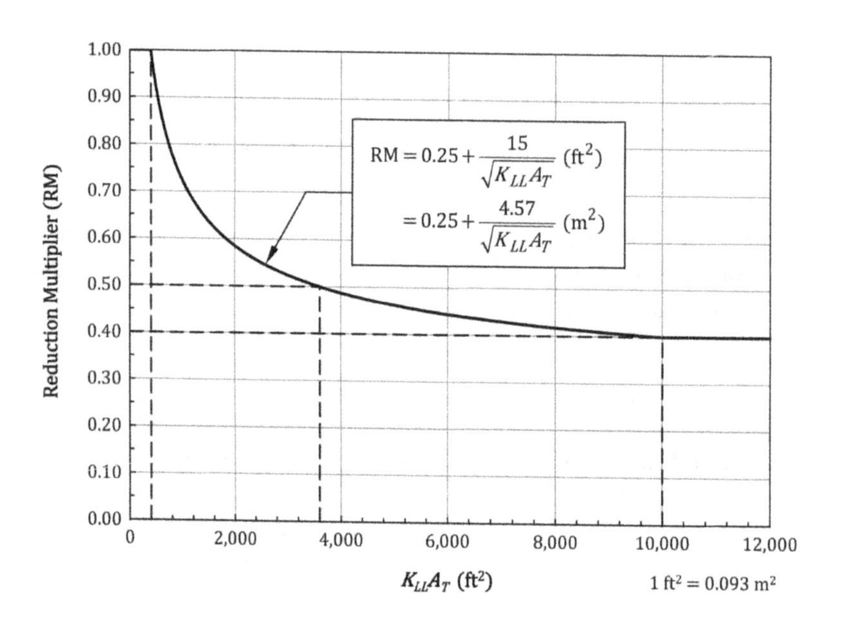

recognizes the benefits of higher redundancy that results from two-way action. The same requirement is in ASCE/SEI 4.7.6.

- Heavy live loads: Live loads greater than 100 lb/ft^2 (4.79 kN/m^2) are not permitted to be reduced except for members supporting two or more floors where a maximum 20 percent reduction is permitted (IBC 1607.13.1.2). The reduced live load in such cases must not be taken less than L determined by IBC 1607.13.1. In buildings supporting relatively large live loads, such as storage buildings, several adjacent bays may be fully loaded; therefore, live loads must not be reduced in those situations. Data in actual buildings indicate that the floor in any story is seldom loaded with more than 80 percent of the nominal live load. Thus, a maximum live load reduction of 20 percent is permitted for members supporting two or more floors, such as columns and walls. The same requirement is in ASCE/SEI 4.7.3.
- Passenger vehicle garages: The live load in passenger vehicle garages is not permitted to be reduced, except for members supporting two or more floors; in such cases, the maximum reduction is 20 percent, but the reduced live load must not be taken less than L determined by IBC 1607.13.1 (IBC 1607.13.1.3). Parking garage live loads are unlike live loads in office and residential occupancies: vehicles are parked in regular patterns and the garages are often full compared to office or residential space where the live loads are most of the time spatially random. Thus, live load reduction is not permitted except for members that support two or more floors, such as columns or walls. The same requirement is in ASCE/SEI 4.7.4.

In addition to the above limitations, live load reduction is not permitted in assembly areas. Due to the nature of assembly occupancies, there is a high probability that the entire floor will be subjected to full uniform live load. According to Footnote a in IBC Table 1607.1, live load reduction is not permitted for the following in assembly areas:

- Fixed seats fastened to the floor
- Lobbies
- Moveable seats
- Assembly platforms
- Bleachers, folding and telescopic seating and grandstands
- Stadiums and arenas with fixed seats fastened to the floor
- Other assembly areas

Live load reduction in accordance with IBC 1607.13.1.2 is permitted for members supporting two or more stage floors in an assembly area (Footnote b in IBC Table 1607.1). Live load reduction in an assembly occupancy is permitted only for follow spot, projection, and control rooms.

Flowchart 3.1 in Section 3.5 of this book can be used to determine basic uniform live load reduction in accordance with IBC 1607.13.1.

Alternative Uniform Live Load Reduction

An alternative method of uniform live load reduction, which is based on provisions in the 1997 Uniform Building Code (Reference 3.2), is given in IBC 1607.13.2. This method is not in ASCE/SEI 7. IBC Equation 16-8 can be used to obtain a reduction factor, R, in percent for members supporting an area greater than or equal to 150 square feet (13.94 m^2) where the live load is less than or equal to 100 lb/ft^2 (4.79 kN/m^2):

$$R = 0.08(A - 150)\ (\text{ft}^2) \tag{3.4}$$

$$R = 0.861(A - 13.94)\ (\text{m}^2) \tag{3.5}$$

In these equations, A is the area supported by the member in square feet (m^2).

Values of R determined by these equations must not exceed the least of the following:

- 40 percent for members supporting one floor.
- 60 percent for members supporting two or more floors.
- The value of R determined by IBC Equation 16-9: $R = 23.1(1 + D/L_o)$ where D and L_o are the dead load and unreduced live load per square foot (m^2) of area supported, respectively.

It is evident that the maximum live load reduction is proportional to the dead to live load ratio. For heavy framing systems, the reduction is permitted to be greater than that for lighter framing systems. The reasoning behind this is the system with the heavier dead load is overstressed proportionally less than one with a lighter dead load.

The reduced live load, L, is determined by the following equation:

$$L = L_o\left(1 - \frac{R}{100}\right) \tag{3.6}$$

Similar to the basic method of uniform live load reduction, live loads are not permitted to be reduced in the following cases:

- In assembly occupancies.
- Where the live load exceeds 100 lb/ft^2 (4.79 kN/m^2) except for the following:
  1. Members supporting two or more floors, in which case the live load may be reduced by a maximum of 20 percent.
  2. In occupancies other than storage where it can be shown by a registered design professional that such a reduction is warranted.
- In passenger vehicle parking garages except for members supporting two or more floors, in which case the live load may be reduced by a maximum of 20 percent.

Reduction of live load on one-way slab systems is permitted by this method where the area, A, is limited to the product of slab span and a width normal to the span of 0.5 times the slab span.

Flowchart 3.2 in Section 3.5 of this book can be used to determine alternative uniform live load reduction in accordance with IBC 1607.13.2.

3.2.12 Reduction in Uniform Roof Live Loads

Overview

Roofs must be designed to resist dead, live, wind, and where applicable, rain, snow, and earthquake loads. Minimum roof live loads are given in IBC Table 1607.1 for various occupancies and uses.

The minimum uniformly distributed live loads of roofs and marquees, L_o, in IBC Table 1607.1 are permitted to be reduced in accordance with IBC 1607.14.

Ordinary Roofs, Awnings, and Canopies

Ordinary flat, pitched, and curved roofs, and awnings and canopies other than those of fabric construction supported by a skeletal structure are permitted to be designed for the following reduced uniform roof live load, L_r, per square foot (m^2) of horizontal projection supported by a member (IBC 1607.14):

$$L_r = L_o R_1 R_2 \tag{3.7}$$

where L_o = unreduced roof live load per square foot (m^2) of horizontal roof projection supported by the member determined in accordance with IBC Table 1607.1

$$R_1 = \begin{cases} 1.0 \text{ for } A_t \leq 200 \text{ square feet} \\ 1.2 - 0.001A_t \text{ for } 200 \text{ square feet} < A_t < 600 \text{ square feet} \\ 0.6 \text{ for } A_t \geq 600 \text{ square feet} \end{cases}$$

$$\text{In S.I.: } R_1 = \begin{cases} 1.0 \text{ for } A_t \leq 18.58 \text{ m}^2 \\ 1.2 - 0.011A_t \text{ for } 18.58 \text{ m}^2 < A_t < 55.74 \text{ m}^2 \\ 0.6 \text{ for } A_t \geq 55.74 \text{ m}^2 \end{cases}$$

$$R_2 = \begin{cases} 1.0 \text{ for } F \leq 4 \\ 1.2 - 0.05F \text{ for } 4 < F < 12 \\ 0.6 \text{ for } F \geq 12 \end{cases}$$

A_t = tributary area in square feet (m^2) supported by the member

F = number of in. of rise per foot (in S.I., $F = 0.12 \times$ slope, where the slope is expressed as a percentage) for a sloped roof.

= rise-to-span ratio multiplied by 32 for an arch or dome.

Roof live load reduction is based on the tributary area, A_t, of the member being considered and the slope of the roof. No live load reduction is permitted for members supporting less than or equal to 200 square feet (18.58 m^2) as well as for roof slopes less than or equal to 4:12; it is less likely that live loads on a roof member will reach maximum values as the slope of the roof increases. In no case is the reduced roof live load to be taken less than 12 lb/ft^2 (0.58 kN/m^2). This minimum load accounts for occasional loading due to the presence of workers and materials during repair operations.

In structures such as greenhouses, where special scaffolding is used as a work surface for workers and materials during maintenance and repair operations, a uniform roof live load less than that determined by IBC 1607.14 is not permitted unless approved by the building official.

Occupiable Roofs

Live loads on areas of roofs designated as occupiable (such as vegetative roofs, roof gardens, landscaped roofs or for assembly) and marquees, are live loads normally associated with floors rather than roofs. Therefore, such live loads are permitted to be reduced in accordance with the requirements in IBC 1607.13 for uniform live loads on floors (IBC 1607.14.2).

3.2.13 Awnings and Canopies

Awnings and canopies must be designed for the uniform live loads in IBC Table 1607.1 and all other applicable loads.

3.2.14 Photovoltaic Panel Systems

Overview

Roof structural members providing support for photovoltaic panel must be designed in accordance with IBC 1607.14.3.

Roof Live Load

Roof structures supporting photovoltaic panel systems must be designed for the following two conditions:

1. The dead load of the photovoltaic panel system in combination with applicable uniform and concentrated roof live loads from IBC Table 1607.1.

   The load combinations in IBC 1605 are to be used to design roof members supporting these systems. According to the exception in this section, roof live loads need not be applied to the areas of the roof structure where the clear space between the panels and the roof surface is 24 in. (610 mm) or less (see Figure 3.8). It is assumed that human traffic is unlikely in areas beneath a panel system with a clear space less than or equal to 24 in. (610 mm). Only roof live loads are permitted to be omitted from these areas of the roof;

Figure 3.8
Area of Roof Beneath a Photovoltaic Panel System Where Roof Live Load Need Not Be Applied

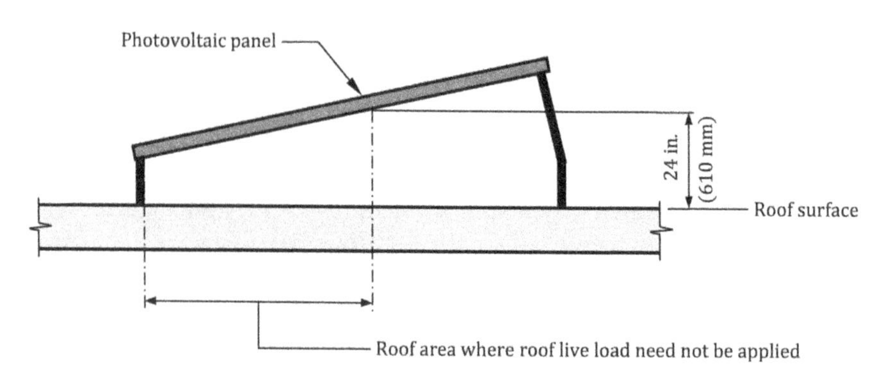

other applicable roof loads, such as rain and snow, must be included. Snow drift loads created by the panels or modules must also be included, where applicable.

2. Applicable uniform and concentrated roof loads from IBC Table 1607.1 without the photovoltaic panel system present.

   This condition considers the possibility that the panels may be removed during the service life of the building; in such cases, the roof structure must be able to support all applicable loads over the entire roof area.

The above conditions are the same as those in ASCE/SEI 4.16.1.

Photovoltaic Panels or Modules

Roof members supporting photovoltaic panels or modules must be designed for the full dead load of the panels and modules and the dead load from the ballast (IBC 1607.14.3.2). Concentrated loads from the support frames must be combined with the load conditions in IBC 1607.14.3.1 and other applicable loads. Snow drift loads created by the panels or modules must also be included, where applicable.

Photovoltaic Panels Installed on Open Grid Roof Structures

Structures with open grid framing and no roof deck or sheathing (such as carports and shade structures) supporting photovoltaic panel systems must be designed to support the uniform and concentrated roof live loads specified in IBC 1607.14.3 except the uniform roof live load is permitted to be reduced to 12 lb/ft^2 (0.57 kN/m^2). The full concentrated roof live load must be applied to the structure as specified.

The above requirements are the same in ASCE/SEI 4.16.3.

Ground-Mounted Photovoltaic Systems

Roof photovoltaic live loads need not be considered for ground-mounted photovoltaic panel systems with independent structures without accessible/occupied space underneath. All other applicable loads and combinations in accordance with IBC 1605 must be considered.

Ballasted Photovoltaic Panel Systems

Roof members supporting ballasted photovoltaic panel systems must satisfy the following:

- The members must be analyzed and designed in accordance with the general design requirements in IBC 1604.4.
- The members must be checked for deflection in accordance with IBC 1604.3.6.
- The ponding requirements in IBC 1611 must be checked.

3.2.15 Crane Loads

Overview

Requirements for crane loads are given in IBC 1607.15. Crane live loads are required to be the rated capacity of the crane. The design loads for runway beams, including connections and support brackets, of moving bridge cranes and monorail cranes must be determined in accordance with ASCE/SEI 4.9. Design loads must include the maximum wheel loads of the crane and the vertical impact, lateral, and longitudinal forces induced by the moving crane.

A typical top-running bridge crane is depicted in Figure 3.9. The trolley and hoist move along the crane bridge, which is supported by the runway beams and support columns. The entire crane assemblage can also move along the length of the runway beams.

Figure 3.9 Top-Running Bridge Crane

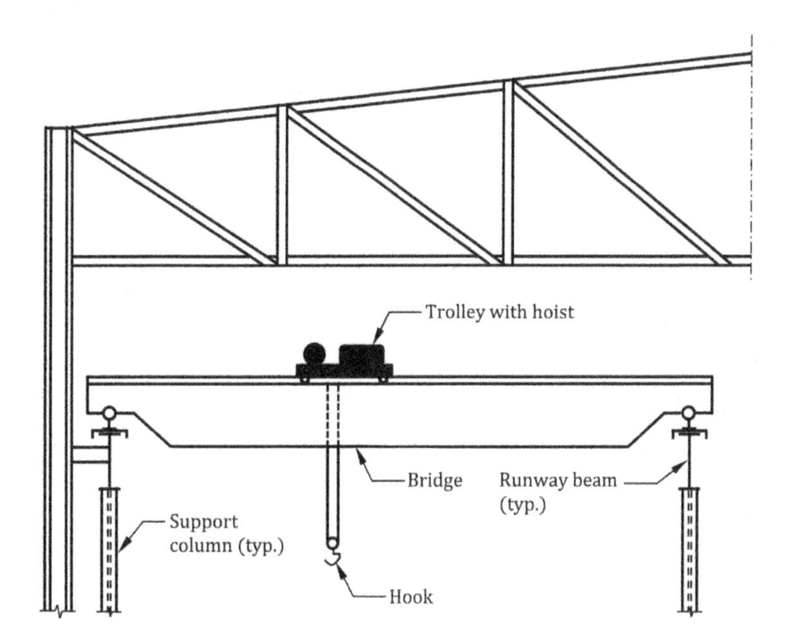

Maximum Wheel Loads

The maximum wheel loads to be used in the design of the supporting members are equal to the weight of the bridge plus the sum of the rated capacity and the weight of the trolley (ASCE/SEI 4.9.2). The trolley is to be positioned on its runway at the location where the resulting load effect is maximum; generally, this occurs when the trolley is moved as close to the supporting members as possible.

Vertical Impact Force

To account for the vertical impact force or vibration caused by the starting and stopping movement of the suspended weight from the crane and by the movement of the crane along the rails, the maximum wheel loads must be increased by the percentages in ASCE/SEI 4.9.3.

The percent increase depends on the assigned bridge crane service class, which are defined in ASCE/SEI 4.9.3.1. The descriptions of these classes are based on those in *Specifications for Top Running Bridge and Gantry Type Multiple Girder Electric Overhead Traveling Cranes* by the Crane Manufacturers Association of America.

Lateral Force

A lateral force acting perpendicular to the crane runway beams is generated by the transverse movement of the crane, that is, by movement that occurs perpendicular to the runway beam (see Figure 3.10). According to ASCE/SEI 4.9.4, the magnitude of this load on crane runway beams

Figure 3.10
Crane Loads on a Runway Beam

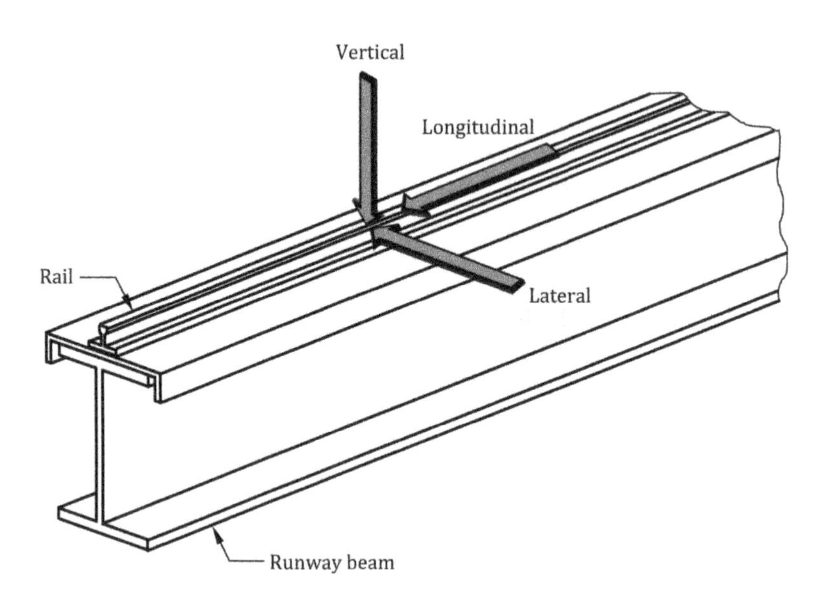

with electronically powered trolleys must be 20 percent of the sum of the rated capacity of the crane and the weight of the hoist and trolley. It is assumed this load acts horizontally at the traction surface of the runway beam and is distributed to the runway beam and supporting structure (such as columns) based on the lateral stiffness of the members.

Longitudinal Force

A longitudinal force is generated on a crane runway beam by acceleration, deceleration, and braking of the crane bridge beam (see Figure 3.10). This load must be 10 percent of the maximum wheel loads of the crane and is assumed to act horizontally at the traction surface of the runway beam in either direction parallel to the beam (ASCE/SEI 4.9.5). Bridge cranes with hand-geared bridges are exempt from this requirement.

3.2.16 Interior Walls and Partitions

Overview

Interior walls and partitions are subjected to lateral loads from occupants, occasional impact from moving furniture or equipment, and from pressurization from heating, ventilating, and air-conditioning (HVAC) systems. To account for this, a 5 lb/ft^2 (0.240 kN/m^2) horizontal load must be applied to interior walls and partitions exceeding 6 ft (1,829 mm) in height (IBC 1607.16).

Fabric Partitions

Requirements for fabric partitions that exceed 6 ft (1,829 mm) in height are given in IBC 1607.16.1. A horizontal load equal to 5 lb/ft^2 (0.240 kN/m^2) must be applied to the partition framing only. The total area used to determine the distributed load is the area of the fabric face between the framing members to which the fabric is attached. The distributed load on these framing members must be uniform and is based on the length of the member. In addition, a 40 lb (0.176 kN) load must be applied over an 8-in. (203 mm) diameter area of the fabric face at a height of 54 in. (1,372 mm) above the floor. This condition is meant to simulate the load caused by a person leaning against the fabric using their hand as the point of contact.

Fire Walls

Fire walls and their supports must be designed to withstand a minimum horizontal allowable stress load of 5 lb/ft^2 (0.240 kN/m^2) [IBC 1607.16.2]. This load requirement meets the structural stability requirements of IBC 706.2 where the structure on either side of the fire wall has collapsed.

3.2.17 Fixed Ladders

Live load requirements for fixed ladders with rungs are given in IBC 1607.10. A single concentrated load of 300 lb (1.33 kN) must be applied at any point to produce the maximum load effect on the element under consideration. Additional 300-lb (1.33-kN) loads are required for every 10 ft (3.05 m) of ladder height (ASCE/SEI 4.5.4).

A concentrated load of 100 lb (0.455 kN) is required to be applied to the top of each rail extension of a fixed ladder that extends above a floor or platform at the top of the ladder. This load is to be applied in any direction and at any height up to the top of the side rail extension.

Ship's ladders with treads instead of rungs must be designed to resist the stair loads in IBC Table 1607.1.

3.2.18 Library Stack Rooms

Live load requirements for library stack rooms are given in IBC 1607.17. Minimum uniform and concentrated live loads are given in IBC Table 1607.1:

- Minimum uniform load: 150 lb/ft^2 (7.18 kN/m^2)
- Minimum concentrated load: 1,000 lb (4.45 kN)

The minimum live loads are based on the following limitations:

- Nominal book stack height of no more than 90 in. (2,290 mm).
- Nominal shelf depth of no more than 12 in. (305 mm) for each face.
- Parallel rows of double-faced book stacks separated by aisles not less than 36 in. (914 mm) in width.

The design of the structural members supporting library stacks must account for the actual conditions where library shelving installation does not fall within the parameter limits noted above. Some examples of heavy loading are given in ASCE/SEI C4.13.

Live load reduction is not permitted for members supporting library stack rooms except for members supporting two or more floors where the live loads are permitted to be reduced by a maximum of 20 percent, but not less than the reduced live load determined in accordance with IBC 1607.13.

3.2.19 Seating for Assembly Uses

According to IBC 1607.18, bleachers, folding and telescopic seating, and grandstands must be designed for the loads in *Standard for Bleachers, Folding and Telescopic Seating, and Grandstands* (ICC 300).

The following horizontal sway loads must be applied to each row of seats in stadiums and arenas with fixed seats in addition to the minimum loads in IBC Table 1607.1:

- 24 lb per linear foot (0.35 kN/m) of seat applied in a direction parallel to each row of seats.
- 10 lb per linear foot (0.15 kN/m) of seat applied in a direction perpendicular to each row of seats.

The parallel and perpendicular loads need not be applied simultaneously.

It is recommended in ASCE/SEI C4.14 that gymnasium balconies with stepped floors for seating be treated as arenas and be designed for the horizontal sway forces given above.

3.2.20 Sidewalks, Vehicular Driveways, and Yards Subject to Trucking

Sidewalks, vehicular driveways, and yards may be subject to loads from trucks driving on these surfaces. In such cases, the supporting structure must be designed for either a uniform load of 250 lb/ft^2 (11.97 kN/m^2) or a concentrated load equal to 8,000 lb (35.60 kN) acting over an area of 4.5 in. by 4.5 in. (114 mm by 114 mm) [see IBC 1607.19 and IBC Table 1607.1].

Where appropriate, other uniform loads determined in accordance with an approved method containing provisions for truck loading, like those in Reference 3.1, must be considered.

3.2.21 Stair Treads

According to IBC 1607.20, stair treads must be designed for a concentrated load of 300 lb (1.34 kN) applied on an area of 2 in. by 2 in. (51 mm by 51 mm). This concentrated load need not be applied simultaneously with the minimum uniform live loads of 40 lb/ft^2 (1.92 kN/m^2) for one- and two-family dwellings and 100 lb/ft^2 (4.79 kN/m^2) for all other cases in IBC Table 1607.1.

3.2.22 Residential Attics

Overview

Three separate attic live loads could exist in one- and two-family dwellings, and descriptions of each are given in the following sections (IBC 1607.21).

Uninhabitable Attics Without Storage

An uninhabitable attic area without storage in a residential structure is defined as follows:

- For rafter construction, the maximum clear height between the joists and rafters is less than 42 in. (1,067 mm).
- For truss construction, there are not two or more adjacent trusses with web configurations capable of accommodating a rectangle that is 42 in. (1,067 mm) in height by at least 24 in. (610 mm) in width within the plane of the trusses.

Such areas must be designed for a live load of 10 lb/ft^2 (0.48 kN/m^2) in accordance with IBC Table 1607.1. This live load need not act concurrently with any other live load.

Uninhabitable Attics with Storage

An uninhabitable attic area with storage in a residential structure is defined as follows (see Figure 3.11):

- For rafter construction, the maximum clear height between the joists and rafters is greater than or equal to 42 in. (1,067 mm).
- For truss construction, there are two or more adjacent trusses with web configurations capable of accommodating a rectangle that is 42 in. (1,067 mm) in height by at least 24 in. (610 mm) in width within the plane of the trusses.

A ceiling joist in raft construction or the bottom chord of a truss in an uninhabitable attic area with storage must be designed for a live load of 20 lb/ft^2 (0.96 kN/m^2) only where both of the following conditions are satisfied (see IBC Table 1607.1):

- The attic area is accessed from an opening not less than 20 in. (508 mm) in width by 30 in. (762 mm) in length located where the clear height in the attic is greater than or equal to 30 in. (762 mm).
- The slope of the ceiling joist or the bottom chord of the truss is less than or equal to 2 units vertical in 12 units horizontal.

In other areas of the attic, the ceiling joists and truss bottom chords must be designed for a live load of 10 lb/ft^2 (0.48 kN/m^2).

Attics Served by Stairs

Attic spaces accessible by stairs other than pull-down stairs are considered to be habitable and must be designed for a live load of 30 lb/ft^2 (1.44 kN/m^2) in accordance with IBC Table 1607.1 for habitable attics and sleeping areas.

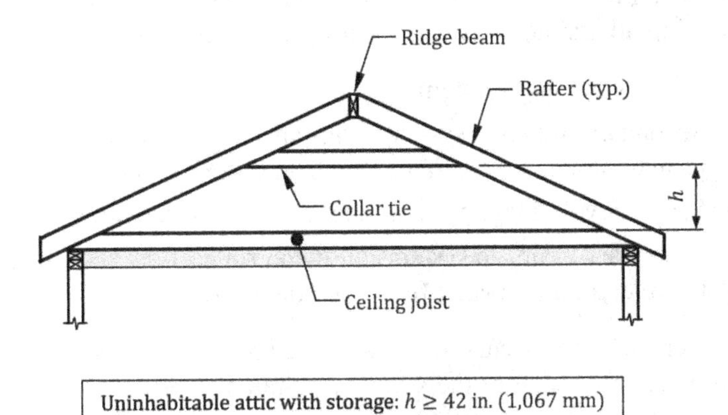

Rafter Construction

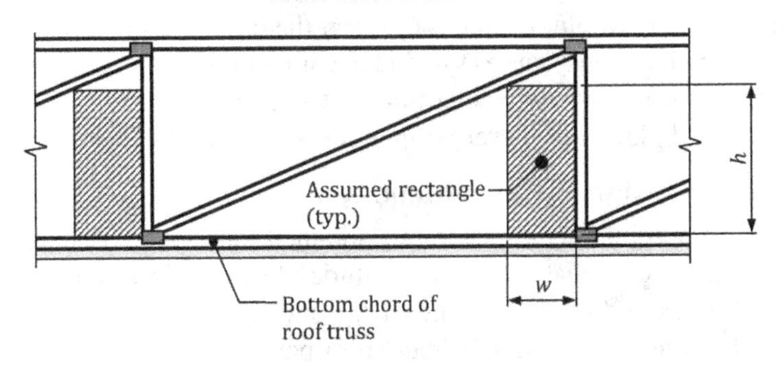

Uninhabitable attic with storage: A rectangular box with $h \geq 42$ in. (1,067 mm) and $w \geq 24$ in. (610 mm) can be accommodated between the webs of two adjacent trusses

Truss Construction

Figure 3.11 Uninhabitable Attic with Storage

3.3 Rain Loads

3.3.1 Overview

Design rain loads must be determined in accordance with ASCE/SEI Chapter 8 (IBC 1611.1). The design procedure in Table 3.2 can be used to determine R based on these requirements. Additional information needed to calculate R is given in the sections of this book indicated in the table.

Table 3.2 Design Procedure to Determine Design Rain Load, *R*

Step	Procedure	Section Number
1	Determine the rainfall intensity, i, for the Risk Category given in IBC Table 1611.1.	3.3.2
2	Determine the flow rate, Q.	3.3.3
3	Obtain the static head, d_s.	3.3.4
4	Determine the hydraulic head, d_h.	3.3.5
5	Determine the ponding head, d_p.	3.3.6
6	Determine the design rain load, R.	3.3.7

Design rain load, R, is determined based on the amount of water that can accumulate on a roof assuming all drainage systems meeting any of the following criteria are blocked:

- Primary drainage system.
- Secondary drainage systems with an inlet that is vertically separated from the inlet to the primary drainage system by less than 2 in. (51 mm).
- Secondary drainage systems sharing drain lines with the primary system.
- Secondary drainage systems with controlled flow roof drains, which limit the rate or delay the release of rainwater flow from the roof.

When blockage occurs, water will rise above the primary roof drain until it reaches the elevation of the roof edge or the secondary drainage system. The depth of water above the primary drain at the design rainfall intensity is based on the flow rate of the secondary system, which varies widely depending on the type of secondary system that is used.

The secondary drainage system for structural loads (SDSL) is the roof drainage system that drains the rainwater off the roof when the criteria listed above are blocked or not working. The type and location of the SDSL and the amount of rainwater above their inlets under design conditions must be known to determine R. Coordination among the design team (architectural, structural, and plumbing) is very important when establishing design rain loads.

3.3.2 Design Rainfall Intensity, *i*

Design rainfall intensity, i, can be obtained for a given Risk Category in IBC 1604.5 by entering an address or the latitude and longitude of the site in References 3.3 and 3.4.

The design flow rate of the SDSL and the resulting hydraulic head, d_h, must be based on a rainfall intensity in in. per hour (mm per hour) equal to or greater than a 15-minute duration storm with a return period given in IBC Table 1611.1 for a given risk category.

3.3.3 Flow Rate, *Q*

The flow rate, Q, of rainwater through a single drainage system can be determined by ASCE/SEI Equations (C8.2-1) and (C8.2-1.SI):

$$Q = 0.0104Ai \text{ (gal/min)} \tag{3.8}$$

$$Q = (0.278 \times 10^{-6})Ai \text{ (m}^3\text{/s)} \tag{3.9}$$

where the tributary area, A, is in square feet (m^2) and design rainfall intensity, i, is in in. per hour (mm per hour).

The constants in these equations are obtained based on the units associated with the variables in the equations:

In Eq. (3.8):

$$\text{Constant} = \text{ft}^2 \times \frac{\text{in.}}{\text{h}} \times \frac{1\text{ ft}}{12\text{ in.}} \times \frac{7.48\text{ gal}}{\text{ft}^3} \times \frac{1\text{ h}}{60\text{ min}} = 0.0104$$

In Eq. (3.9):

$$\text{Constant} = \text{m}^2 \times \frac{\text{mm}}{\text{h}} \times \frac{1\text{ m}}{1{,}000\text{ mm}} \times \frac{1\text{ h}}{3{,}600\text{ s}} = 0.278 \times 10^{-6}$$

The tributary area, A, is equal to the tributary roof area plus one-half the wall area that diverts rainwater onto the roof (where applicable) serviced by a single drain outlet in the secondary drainage system (SDSL). Relatively large walls adjacent to roofs have the potential to divert substantial wind-driven rain flow down the wall to the roof.

3.3.4 Static Head, d_s

The static head, d_s, is the depth of water on the undeflected roof up to the inlet of the SDSL and is determined in the design of the combined drainage system (see Figures 3.12 through 3.14). It is usually specified in the range of 2 in. to 4 in. (51 mm to 102 mm) in depth.

3.3.5 Hydraulic Head, d_h

The hydraulic head, d_h, is equal to the depth of water on the undeflected roof above the inlet of the SDSL required to achieve the design flow. It is related to Q and the type and size of the secondary drainage system. Methods to determine d_h are given in the following figures for three types of SDSL:

- Figure 3.12 for roof edge overflow (see ASCE/SEI C8.2)
- Figure 3.13 for roof drains (see ASCE/SEI C8.2)
- Figure 3.14 for scuppers (see ASCE/SEI C8.2 and Reference 3.5)

3.3.6 Ponding Head, d_p

Overview

Ponding head, d_p, is equal to the depth of water due to the deflections of the roof subjected to the unfactored rain load, R, and the unfactored dead load. Several methods are available to determine d_p, two of which are discussed below.

Iterative Method

The following iterative method can be used to determine d_p:

1. Calculate the design rain load, R, neglecting the ponding contribution, that is, assuming $d_p = 0$ (see Section 3.3.7 of this book on how to calculate R).
2. Determine the maximum deflection due to R calculated in step 1 and the unfactored dead loads from a structural analysis and set that maximum deflection equal to d_p.
3. Recalculate R using d_p from step 2.
4. Determine the maximum deflection due to R calculated in step 3 and the unfactored dead loads from a structural analysis and set that maximum deflection equal to the updated d_p.
5. Perform successive analyses, each with rain loads based on the updated deflected shape of the roof, until d_p converges to the correct value or diverges, which indicates ponding instability.

Figure 3.12 Determination of d_h—Roof Edge Overflow

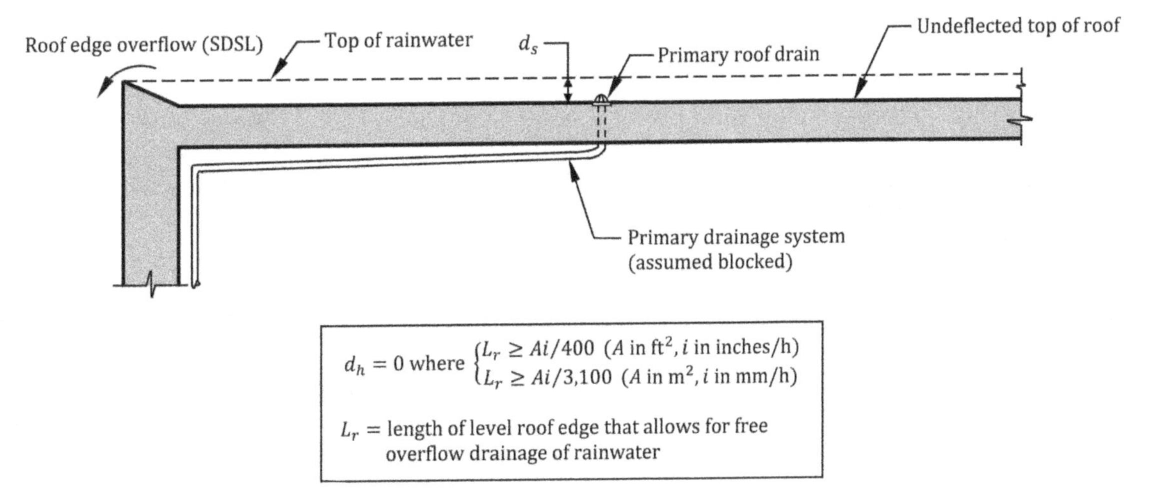

Figure 3.13 Determination of d_h—Roof Drains

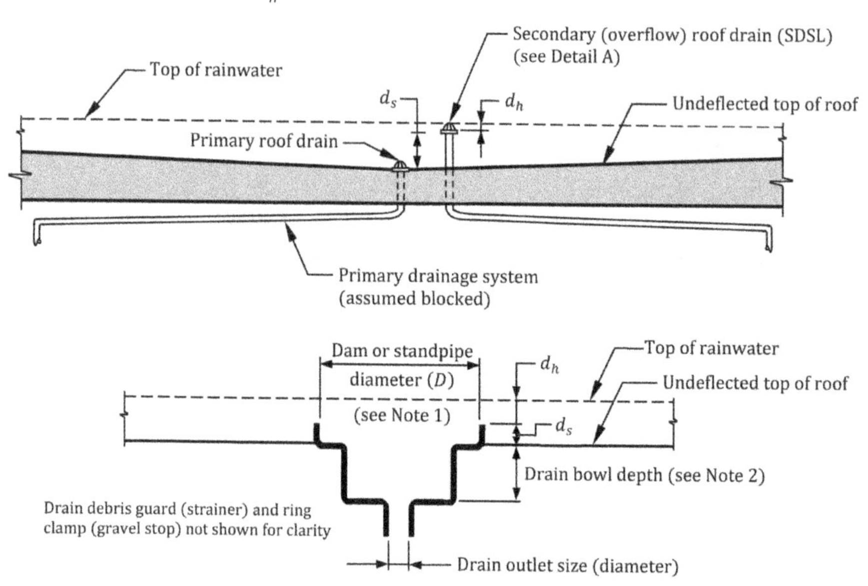

Detail A - Dam or Standpipe System

Determine d_h from ASCE Tables C8.2-1 (in.) and C8.2-2 (mm) for a given drainage system and Q.

Notes

1. For weir flow and transition flow regime designations (cells that are not shaded) in ASCE/SEI Tables C8.2-1 and C8.2-2:

   - Where the specified secondary (overflow) drain dam or standpipe diameter differs from what is provided in ASCE Tables C8.2-1 and C8.2-2, the hydraulic head can be adjusted by ASCE/SEI Eq. (C8.2-3) for a given Q:

   $d_{h2} = (D_1/D_2)^{0.67} d_{h1} \geq 0.8 d_{h1}$

2. For orifice flow regime designations for roof drains, as shown in the shaded cells in ASCE/SEI Tables C8.2-1 or C8.2-2:

   - Where the depth of the specified drain bowl is less than the depth of the tested drain bowl (indicated in the tables), the difference in drain bowl depth should be added to d_h from the tables to determine the design hydraulic head and total head.

   - Where the depth of the specified drain bowl is greater than that indicated in the tables, the difference in drain bowl depth can be subtracted from d_h in the tables to determine the design hydraulic head and total head. It is advisable not to use an adjusted design hydraulic head less than 80 percent of the d_h provided in the tables for a given flow rate, Q.

Approximate Amplified First-Order Analysis

The approximate amplified first-order analysis in ASCE/SEI C8.2 can also be used to determine R, which then can be used to determine d_p, if needed. This approximate method is applicable to rectangular bays consisting of primary members supporting equally spaced secondary members (see Figure 3.15).

The following steps can be used to determine R and d_p:

1. Determine the total load, which is equal to the dead load plus the rain load, R, excluding the ponding contribution (i.e., R is calculated assuming $d_p = 0$; see Section 3.3.7 of this book).
2. Calculate the flexibility coefficient, C_p, for the primary members by ASCE/SEI Eq. (C8.2-5):

$$C_p = \frac{\gamma L_s L_p^4}{\pi^4 E I_p}$$

Figure 3.14 Determination of d_h—Scuppers

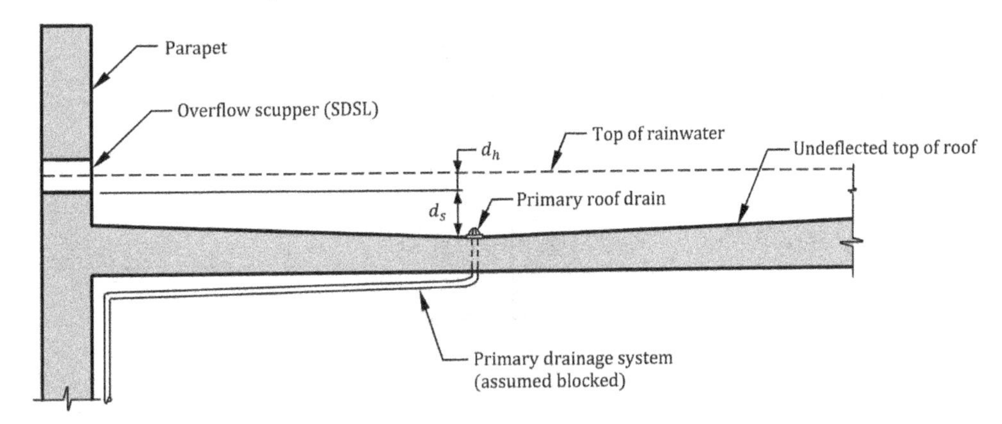

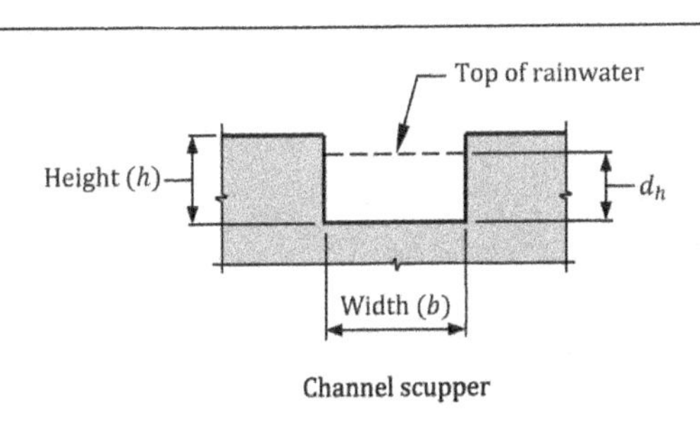

Determine d_h by one of the following methods:

1. ASCE/SEI Tables C8.2-3 (in.) and C8.2-4 (mm) for a given channel scupper system and Q.
2. For channel scuppers:

$$d_h = \left(\frac{Q}{2.9b}\right)^{2/3} \text{ (in.) for } d_h < h$$

$$d_h = \left[\frac{(17.65 \times 10^6)Q}{b}\right]^{2/3} \text{ (mm) for } d_h < h$$

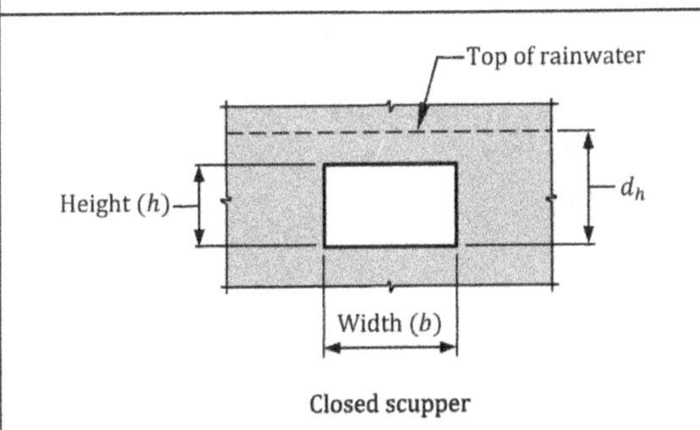

Determine d_h by one of the following methods:

1. ASCE/SEI Tables C8.2-3 (in.) and C8.2-4 (mm) for a given closed scupper system and Q.
2. For closed scuppers:

$$d_h = \left(\frac{Q}{4.3bh}\right)^{2} + 0.5h \text{ (in.) for } d_h \geq h$$

$$d_h = \left[\frac{(12.00 \times 10^6)Q}{bh}\right]^{2} + 0.5h \text{ (mm) for } d_h \geq h$$

Where $d_h < h$, use the equations for channel scuppers to determine d_h.

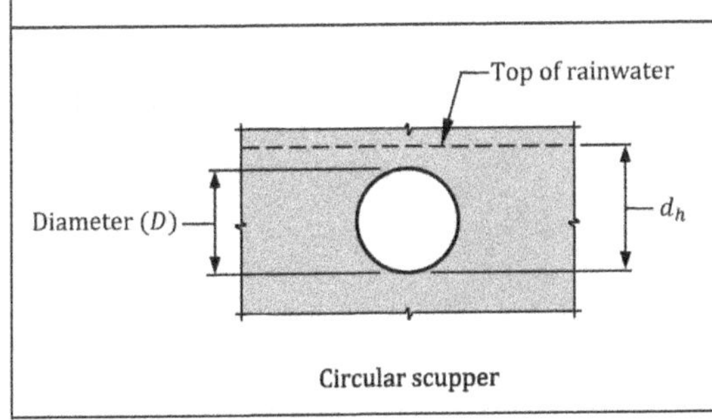

Determine d_h from ASCE/SEI Tables C8.2-5 (in.) and C8.2-6 (mm) for a given scupper diameter and Q.

where γ = unit weight of water = 62.4 lb/ft³ (9.80 kN/m³)

L_s = span of secondary members

L_p = span of primary members

E = modulus of elasticity of the members

I_p = moment of inertia of the primary members

3. Calculate the flexibility coefficient, C_s, for the secondary members by ASCE/SEI Eq. (C8.2-6):

$$C_s = \frac{\gamma S L_s^4}{\pi^4 E I_s}$$

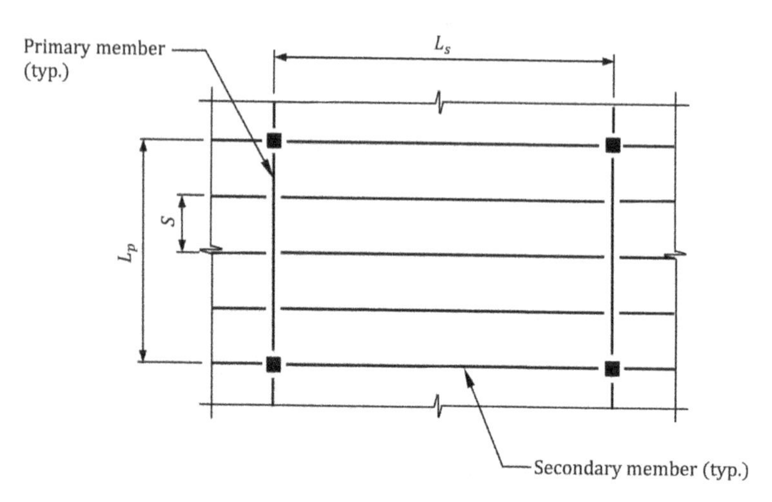

Figure 3.15 Rectangular Bay Used in the Determination of an Approximate Ponding Head, d_p

where S = spacing of secondary members
L_s = span of secondary members
I_s = moment of inertia of the secondary members

4. Calculate the ponding amplification factor, B_p, by ASCE/SEI Eq. (C8.2-4):

$$B_p = \frac{1}{1 - 1.15C_p - C_s}$$

5. Calculate the total load (dead load plus rain load including the effects of ponding) by multiplying the total load in step 1 by the ponding amplification factor, B_p, calculated in step 4.
6. Determine the design rain load, R, by subtracting the dead load from the total load calculated in step 5.
7. Determine d_p from ASCE/SEI Eq. (8.2-1) or ASCE/SEI Eq. (8.2-1.SI).

3.3.7 Design Rain Load, *R*

The design rain load, R, is determined by ASCE/SEI Eq. (8.2-1) or ASCE/SEI Eq. (8.2-1.SI):

$$R = 5.2(d_s + d_h + d_p) \text{ (lb/ft}^2\text{)} \tag{3.10}$$

$$R = 0.0098(d_s + d_h + d_p) \text{ (kN/m}^2\text{)} \tag{3.11}$$

The constants in these equations are equal to the unit load of rainwater, which is the density per unit depth of rainwater:

$$\text{In Eq. (3.10)}: \frac{62.4 \text{ lb/ft}^3}{12 \text{ in./ft}} = 5.2 \text{ lb/ft}^2\text{/in.}$$

$$\text{In Eq. (3.11)}: \frac{9.8 \text{ kN/m}^3}{1{,}000 \text{ mm/m}} = (0.0098 \text{ kN/m}^2\text{)/mm}$$

3.3.8 Bays with Low Slope

Where roofs do not have adequate slope or have insufficient and/or blocked drains to remove water due to rain (or melting snow), water will tend to pond in low areas, which will cause the roof structure to deflect. These low areas will subsequently attract even more water, leading to additional deflection. The structural members supporting the roof must be stiff enough so that deflections will not continually increase until instability occurs, resulting in localized failure (i.e., ponding instability).

The leftmost bay adjacent to the parapet and the center bays of the roof structure in Figure 3.16 must be designed for rain loads in accordance with ASCE/SEI 8.2 because these bays can impound rainwater during the design event. Ponding is addressed for these bays in the equation for the design rain load, R, which includes the ponding head, d_p (see Equations [3.10] and [3.11]). Even

Figure 3.16 Examples of Bays Subject to Rain Loads and Requiring Additional Investigation for Ponding Loads

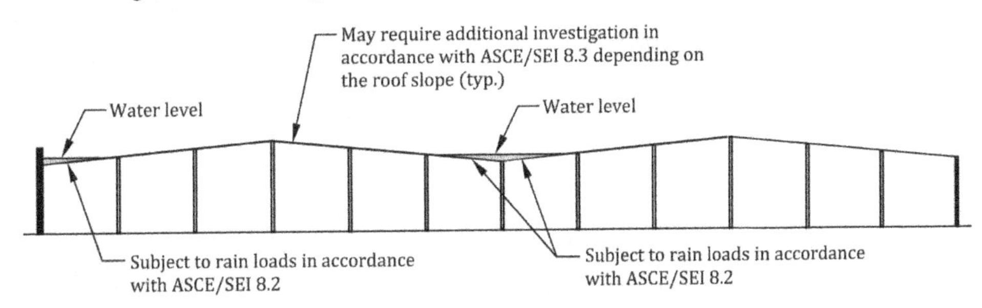

though rain loads need not be considered for the other bays, an investigation in accordance with ASCE/SEI 8.3 needs to be performed to determine whether ponding loads must be considered.

It is possible for isolated ponds to form in low points in the deflected shape of a roof in cases where the roof slope is low. Free-draining bays and internal bays may be susceptible to ponding loads even though they are not required to be subjected to rain loads in accordance with ASCE/SEI 8.2. An additional investigation for potential ponding loads must be performed where either of the following conditions are met (ASCE/SEI 8.3):

- The roof slope is less than ¼ in. per foot (1.19 degrees)
- The bay is adjacent to a free-draining edge with secondary members parallel to the free-draining edge and the roof slope is less than β where $\beta = [\pi + (L_s/S)]/20$ (in./foot). The terms L_s and S are the span and spacing of the secondary members, respectively.

These limits were derived assuming rectangular bays with sinusoidal deflected shapes, a maximum deflection to span ratio of 1/240, and a rigid sidewall (or primary roof member) at the free draining edge (i.e., the vertical deflection of the sidewall or primary roof member is negligible compared to the deflection of the secondary and other primary members). Where secondary members are perpendicular and adjacent to a free edge, it is assumed no water is impounded in a bay (i.e., no local minimum is achieved in the deflected roof shape) where β (in in.) for a run of 1-foot (i.e., a β on 12 roof slope) is greater than or equal to the value determined by ASCE/SEI Eq. (C8.3-1):

$$\beta = \frac{1}{20}\left(\pi + \frac{L_p}{L_s}\right) \tag{3.12}$$

In this equation, L_p and L_s are the spans of the primary and secondary members, respectively (see Figure 3.17). Values of β determined by Eq. (3.12) are given in Figure 3.18 for primary and secondary spans ranging from 20 to 60 ft (6.10 to 18.3 m).

A roof bay with secondary members parallel and adjacent to the free draining edge of the roof is depicted in Figure 3.19. If the roof was rigid, rainwater would flow off the free edge. Ponding can occur if the deflection of the first secondary member upslope from the free draining edge exceeds the initial difference in elevation. It is assumed no local minimum is achieved in the deflected roof shape where β is greater than or equal to the value determined by ASCE/SEI Eq. (C8.3-2), which is the same equation in the second condition in ASCE/SEI 8.3:

$$\beta = \frac{1}{20}\left(\pi + \frac{L_s}{S}\right) \tag{3.13}$$

where S is the center-to-center spacing of the secondary members. Values of β determined by Eq. (3.13) are given in Figure 3.20 for secondary beam spans and spacings ranging from 20 to 60 ft (6.10 to 18.3 m) and 3 to 12 ft (0.91 to 3.7 m), respectively.

For interior bays, it is assumed no water is impounded in a bay where $\beta \geq \pi/20 = 0.16$ in. (4 mm) regardless of the orientation of the secondary members. However, a 0.25-in. (6-mm) limit is specified in the first condition in ASCE/SEI 8.3.

Figure 3.17 Roof Bay Where Secondary Members Are Perpendicular and Adjacent to the Free Draining Edge of the Roof

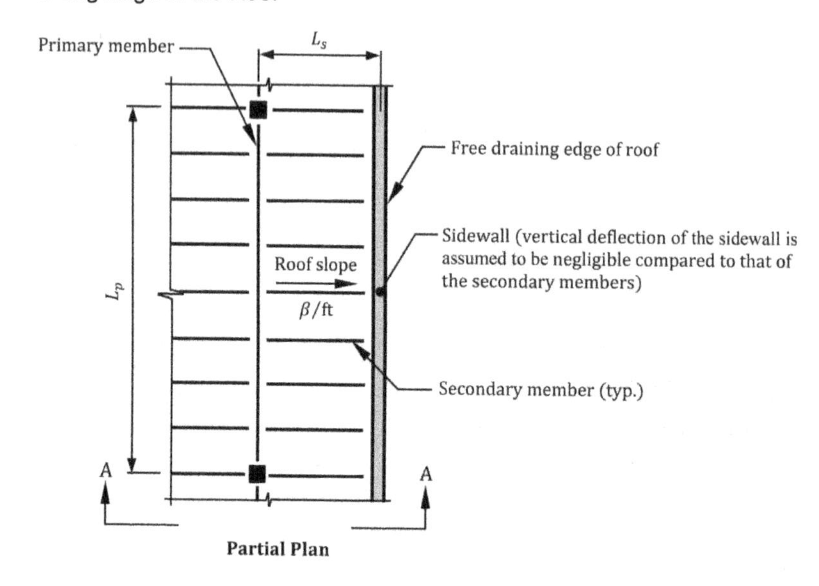

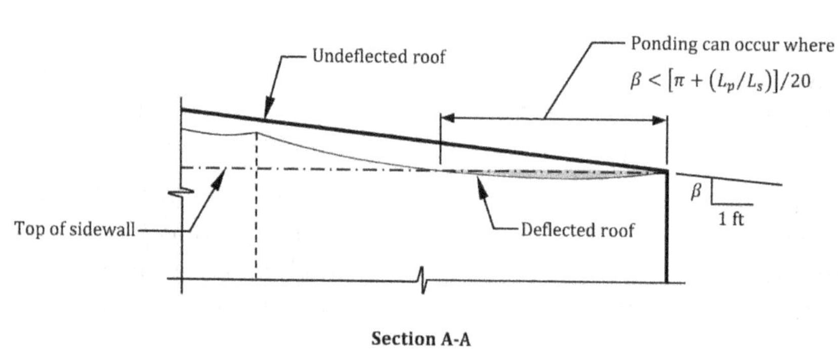

Figure 3.18 Roof Rise, β, for Roof Bays Where Secondary Members Are Perpendicular and Adjacent to the Free Draining Edge of the Roof

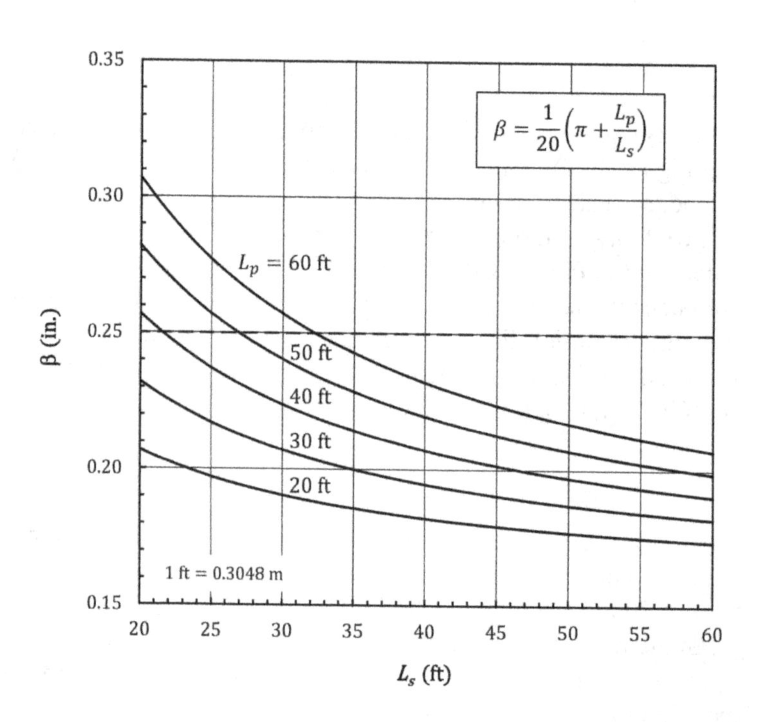

Figure 3.19 Roof Bay Where Secondary Members Are Parallel and Adjacent to the Free Draining Edge of the Roof

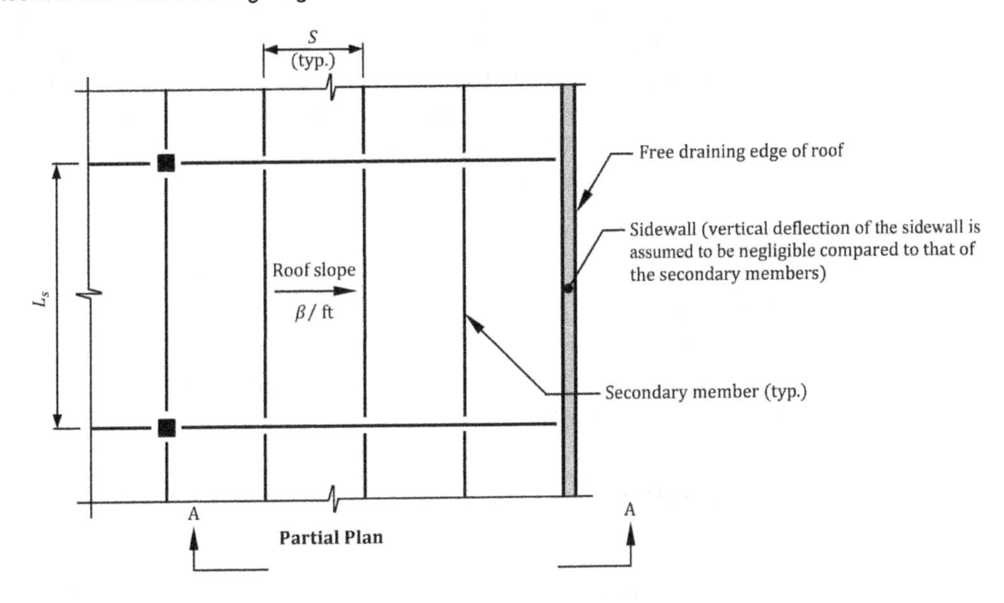

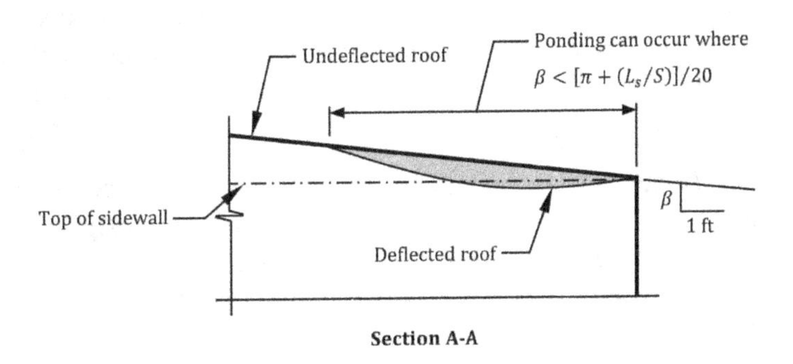

Figure 3.20 Roof Rise, β, for Roof Bays Where Secondary Members Are Parallel and Adjacent to the Free Draining Edge of the Roof

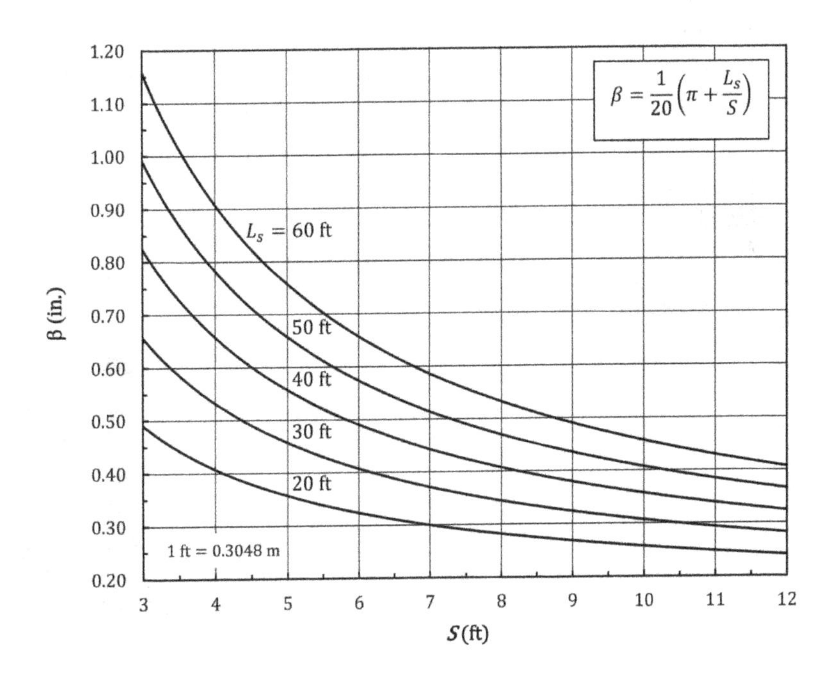

Not all bays requiring additional investigation in accordance with ASCE/SEI 8.3 will be required to be designed for rain loads in accordance with ASCE/SEI 8.2.

In all cases, the dead load, D, and the full design rain load, R, must be used when performing a ponding analysis. A design method for roof structures considering loads from ponding is given in Reference 3.6; this method accounts for the flexural rigidities of the primary and secondary structural members. If it is determined by analysis that ponds can form, Equations (3.10) and (3.11) can be used to determine R where the water level would be taken as the elevation of the free-draining edge plus a nominal hydraulic head, d_h, such as 0.25 in. (6 mm).

3.3.9 Drainage to Existing Roofs

According to ASCE/SEI 8.4, drainage systems for new construction are not permitted to discharge water onto existing roofs unless the roof is evaluated for the additional load determined by the requirements in ASCE/SEI Chapter 8. Where the structure of the existing roof is unable to support the additional load and strengthening the existing structural members is cost prohibitive, it may be possible to install overflow drainage systems, such as scuppers, to help mitigate the additional load.

3.4 Soil Loads and Hydrostatic Pressure

3.4.1 Overview

Structures below grade must be designed to resist the lateral loads caused by adjacent soil. A geotechnical investigation is usually undertaken to determine the magnitude of the soil pressure. In cases where the results of such an investigation are not available, the lateral soil loads in IBC Table 1610.1 are to be used (similar design lateral loads are provided in ASCE/SEI Table 3.2-1).

3.4.2 Design Lateral Soil Load, *H*

The design lateral soil load, H, depends on the type of soil and the boundary conditions at the top of the wall. Walls restricted from horizontal movement at the top are to be designed for the at-rest pressures in IBC Table 1610.1, while walls free to deflect and rotate at the top are to be designed for the active pressures. The distribution of at-rest soil pressure over the height of a reinforced concrete foundation wall restrained by the reinforced concrete slab is illustrated in Figure 3.21.

Foundation walls not extending more than 8 ft (2.44 m) below grade and laterally supported at the top by flexible diaphragms are permitted to be designed for the active pressure values given in IBC Table 1610.1.

Lateral soil pressures are not provided for the expansive soils identified by Note b in IBC Table 1610.1 because these soils have unpredictable characteristics. These soils absorb water and tend to shrink and swell to a higher degree than other soils. As these soils swell, relatively large

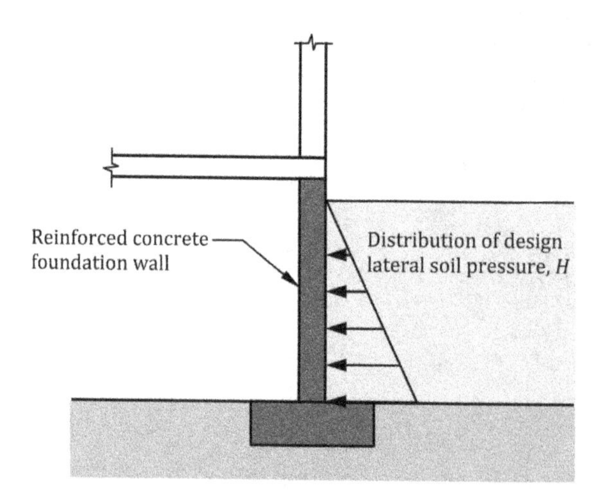

Figure 3.21 Distribution of At-Rest Soil Pressure on a Foundation Wall

forces can be exerted on the soil-retaining structure. As such, expansive soils are not suitable as backfill (see Note b).

3.4.3 Hydrostatic Pressure

In addition to lateral pressures from soil, walls must be designed to resist the effects of hydrostatic pressure due to undrained backfill (unless a drainage system is installed) and to any surcharge loads resulting from sloping backfills or from driveways or parking spaces in close proximity to a wall. Submerged or saturated soil pressures include the weight of the buoyant soil plus the hydrostatic pressure.

Requirements for the design of horizontal elements supported directly on soil, such as slabs on grade and basement slabs, are given in ASCE/SEI 3.2.2. Full hydrostatic pressure must be applied over the entire area of such elements where applicable. Elements supported by expansive soils must be designed to tolerate the movement or resist the upward loads caused by the expansive soil, or the expansive soil is to be removed or stabilized around and beneath the structure.

3.4.4 Alternative Method for Loads from Water in Soil

The provisions in ASCE/SEI 3.3 are permitted to be used instead of the requirements in ASCE/SEI 3.2 for the determination of loads from water in soil.

This alternative method separates the calculation of lateral loads due to soil pressure from ground water pressure in soil. Lateral loads from soil pressure are calculated in accordance with ASCE/SEI 3.2.1 (see Section 3.4.2 of this book) and are designated H_{eb} (see ASCE/SEI 2.3.7). Load due to ground water pressure in soil is designated H_w and must be based on the maximum ground water elevation.

The maximum ground water elevation must be established such that the annual probability of exceedance does not exceed the following values:

- Risk category I: 0.0024 (mean recurrence interval of 400 years)
- Risk category II: 0.0012 (mean recurrence interval of 800 years)
- Risk category III: 0.0006 (mean recurrence interval of 1,700 years)
- Risk category IV: 0.0003 (mean recurrence interval of 3,300 years)

Site specific information is needed to establish maximum ground water elevations. Where such information is not available, it is conservative to assume the maximum ground water elevation occurs at the ground surface for structures not located in a design flood zone defined in ASCE/SEI Chapter 5.

In cases where lateral pressures are used to resist nonpermanent loads and the soil is considered to be permanent, H_w must be based on the minimum ground water elevation, which in this case, must be established so that the annual probability of exceedance does not exceed the same values specified for the maximum ground water elevation. The minimum ground water elevation need not be taken below the lowest portion of the structure.

If this alternative method is used, the load combinations in ASCE/SEI 2.3.7 must be used (see Section 2.3.7 of this book).

3.5 Flowcharts

A summary of the flowcharts provided in this chapter is given in Table 3.3.

Table 3.3 Summary of Flowcharts Provided in Chapter 3

Flowchart	Title
3.1	Basic Uniform Live Load Reduction (IBC 1607.13.1)
3.2	Alternate Uniform Live Load Reduction (IBC 1607.13.2)

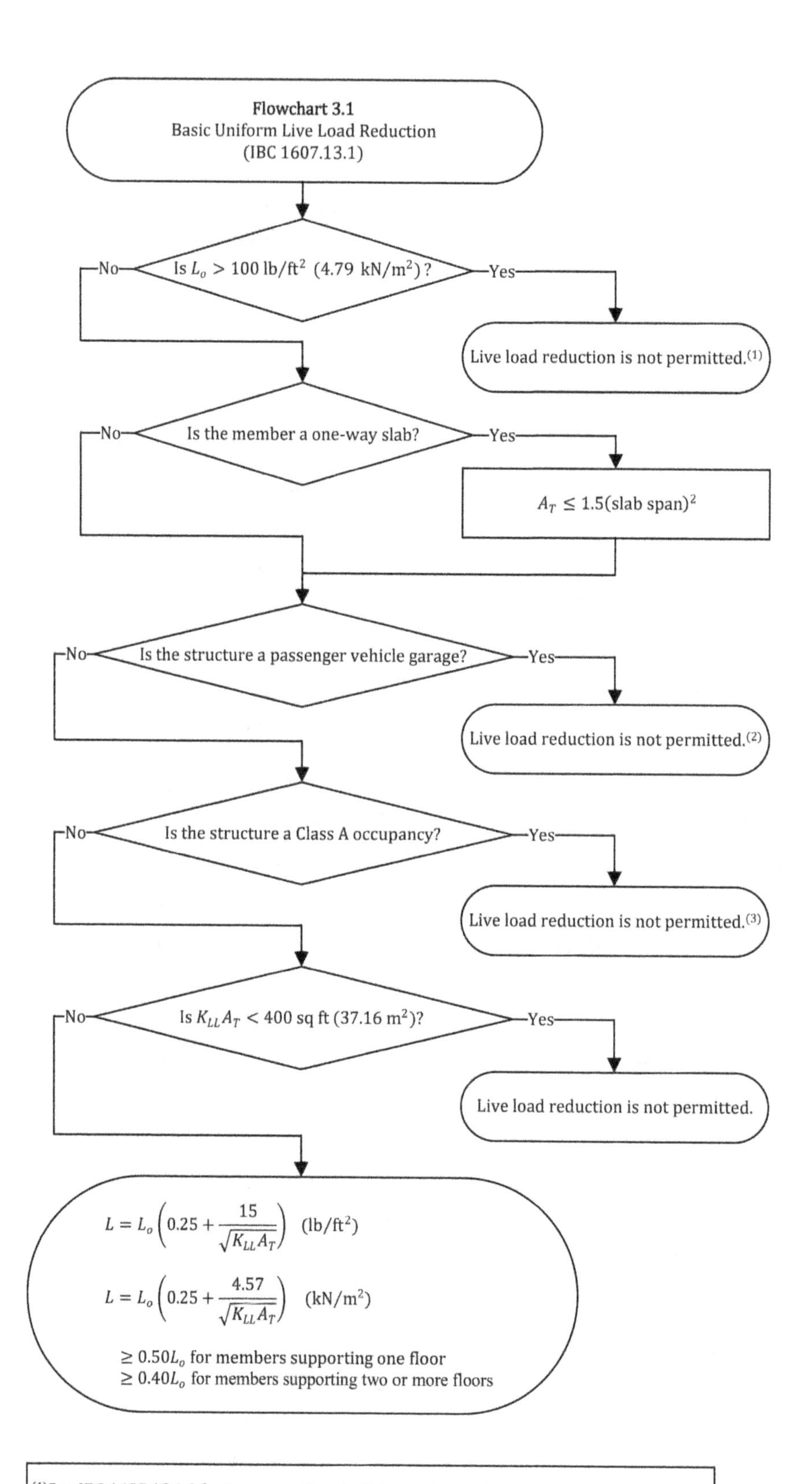

(1)See IBC 1607.13.1.2 for two exceptions to this requirement.
(2)Live loads for members supporting two or more floors are permitted to be reduced by a maximum of 20 percent (IBC 1607.13.1.3).
(3)Live loads in follow spot, projections and control rooms are permitted to be reduced (see IBC Table 1607.1). Stage floors are permitted to be reduced only in accordance with the two exceptions in IBC 1607.13.1.2.

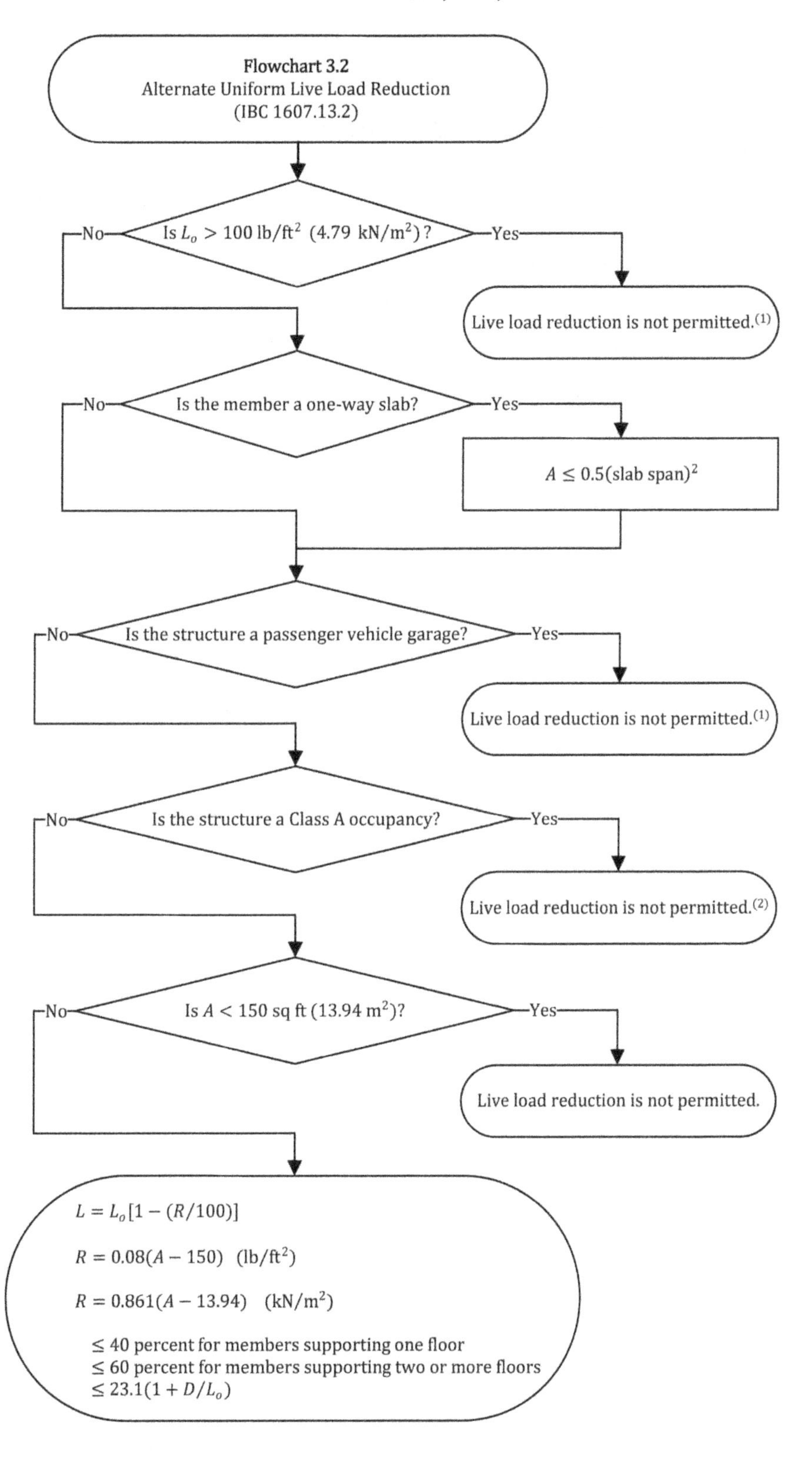

(1)Live loads for members supporting two or more floors are permitted to be reduced by a maximum of 20 percent (IBC 1607.13.1.3).
(2)Live loads in follow spot, projections and control rooms are permitted to be reduced (see IBC Table 1607.1). Stage floors are permitted to be reduced only in accordance with Item 1 in IBC 1607.13.2.

3.6 Examples

3.6.1 Example 3.1—Basic Uniform Live Load Reduction, Interior Column (IBC 1607.13.1)

The typical floor and roof plan of a 10-story office building is illustrated in Figure 3.22. Determine the total live loads in column B3 in each story considering basic uniform live load reduction in accordance with IBC 1607.13.1.

Figure 3.22 Typical Floor and Roof Plan of the 10-Story Office Building in Example 3.1

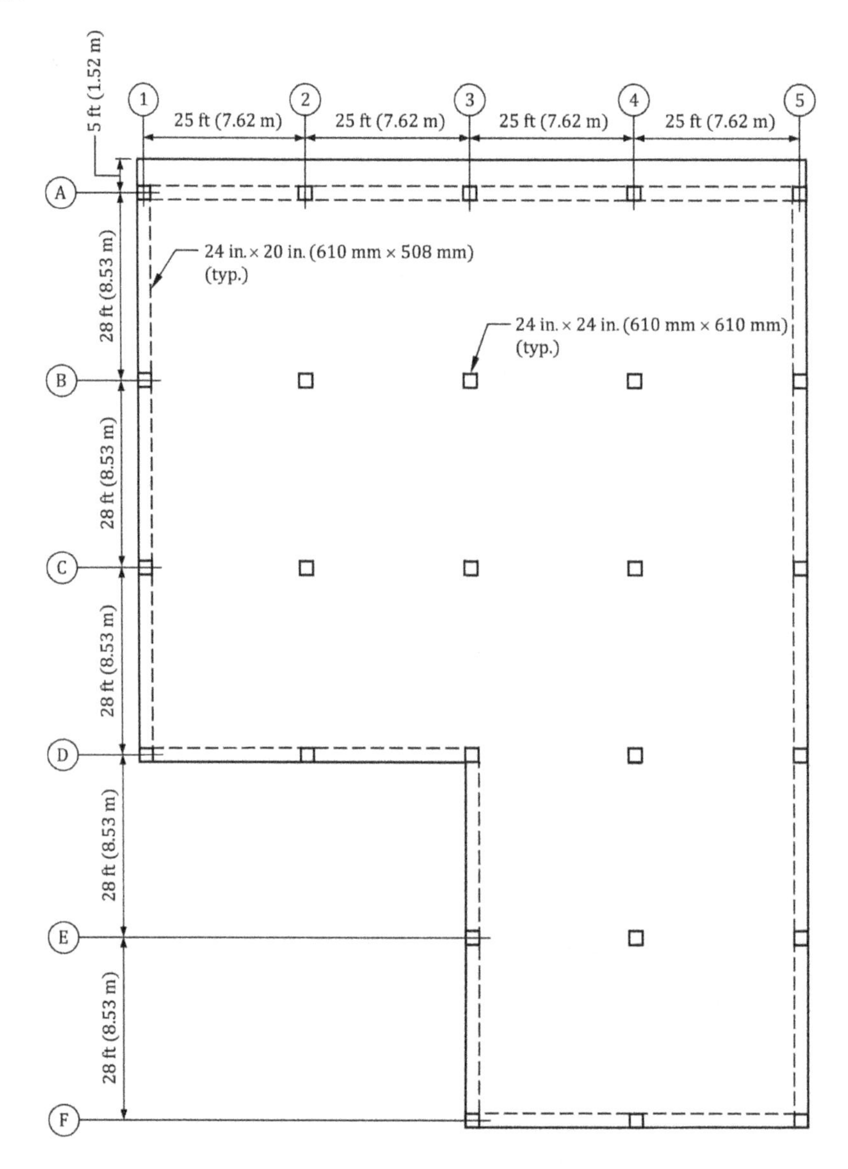

The roof level is an ordinary flat roof (slope of 2.39 degrees) that is not occupiable. For purposes of this example, neglect rain and snow loads.

The ninth floor is designated as a file storage floor with a live load of 200 lb/ft^2 (9.58 kN/m^2). All other floors are typical office floors with moveable partitions. Neglect live loads in lobbies and corridors on all floors for purposes of this example.

Solution

Step 1—Determine the axial live load in column B3 in the tenth story

The reduced uniform roof live load, L_r, is determined by IBC Eq. 16-10:

$$L_r = L_o R_1 R_2$$

The minimum roof live load, L_o, for an ordinary flat roof that is not occupiable is equal to 20 lb/ft² (0.96 kN/m²) in accordance with IBC Table 1607.1.

Tributary area, A_t, of column B3:

$$A_t = 25 \times 28 = 700 \text{ ft}^2$$

$$A_t = 7.62 \times 8.53 = 65.0 \text{ m}^2$$

Because $A_t > 600$ ft² (55.7 m²), $R_1 = 0.6$ (IBC Eq. 16-13)

Because $F = 1/2 < 4$, $R_2 = 1.0$ (IBC Eq. 16-14)

Therefore, $L_r = 20 \times 0.6 \times 1.0 = 12$ lb/ft², which is equal to the permitted minimum reduced roof live load in IBC 1607.14.1.

Axial load $= 12 \times 700/1{,}000 = 8.4$ kips

In S.I.:

$$L_r = 0.96 \times 0.6 \times 1.0 = 0.58 \text{ kN/m}^2$$

Axial load $= 0.58 \times 65.0 = 37.7$ kN

Step 2—Determine the axial live load in column B3 in the ninth story

Because the ninth floor is designated a file storage floor with a live load of 200 lb/ft² (9.58 kN/m²), which exceeds 100 lb/ft² (4.79 kN/m²), and because the column at this level does not support two or more floors of occupancy other than roof occupancy, the live load is not permitted to be reduced (IBC 1607.13.1.2).

$$\text{Axial load} = 200 \times 700/1{,}000 = 140.0 \text{ kips}$$

$$\text{Axial load} = 9.58 \times 65.0 = 622.7 \text{ kN}$$

Step 3—Determine the axial live load in column B3 in a typical story

The reduced uniform live load, L, is determined by IBC Eq. 16-7:

$$L = L_o\left(0.25 + \frac{15}{\sqrt{K_{LL}A_T}}\right) \text{(lb/ft}^2\text{)}$$

$$L = L_o\left(0.25 + \frac{4.57}{\sqrt{K_{LL}A_T}}\right) \text{(kN/m}^2\text{)}$$

For an office occupancy, the minimum reducible live load, L_o, is equal to 50 lb/ft² (2.39 kN/m²) in accordance with IBC Table 1607.1.

The minimum partition load is equal to 15 lb/ft² (0.72 kN/m²) because the specified live load is less than 80 lb/ft² (3.83 kN/m²); this uniform live load is not reducible (IBC 1607.5).

For an interior column, the live load element factor, K_{LL}, is equal to 4 from IBC Table 1607.13.1.

The tributary area, A_T, to be used in IBC Eq. 16-7 in the determination of L at a particular story is equal to the sum of the tributary areas above that story where the uniform floor live load is permitted to be reduced.

For the determination of L in the eighth story, the tributary area of the roof cannot be included because it is not a floor level and the tributary area of the ninth floor cannot be included because the live load is not permitted to be reduced at that level. Therefore, $A_T = 1 \times 700 = 700$ ft² (65.0 m²), and L is equal to the following:

$$L = L_o\left(0.25 + \frac{15}{\sqrt{K_{LL}A_T}}\right) = 50 \times \left(0.25 + \frac{15}{\sqrt{4 \times 700}}\right) = 0.534 \times 50 = 26.7 \text{ lb/ft}^2 > 0.50L_o$$

Axial load due to L:

$$\Sigma P_R = L(K_{LL}A_T)/K_{LL} = [26.7 \times (4 \times 700)]/(4 \times 1{,}000) = 18.7 \text{ kips}$$

Axial load due to partitions:

$$P_N = 15 \times 700/1{,}000 = 10.5 \text{ kips}$$

In S.I.:

$$L = L_o\left(0.25 + \frac{4.57}{\sqrt{K_{LL}A_T}}\right)$$

$$= 2.39 \times \left(0.25 + \frac{4.57}{\sqrt{4 \times 65.0}}\right) = 0.533 \times 2.39 = 1.28 \text{ kN/m}^2 > 0.50L_o$$

Axial load due to L:

$$\Sigma P_R = L(K_{LL}A_T)/K_{LL} = [1.28 \times (4 \times 65.0)]/4 = 83.2 \text{ kN}$$

Axial load due to partitions:

$$P_N = 0.72 \times 64.6 = 46.5 \text{ kN}$$

Similarly, in the first story, $A_T = 8 \times 700 = 5{,}600$ ft² (520.0 m²), and L is equal to the following:

$$L = L_o\left(0.25 + \frac{15}{\sqrt{K_{LL}A_T}}\right) = 50 \times \left(0.25 + \frac{15}{\sqrt{4 \times 5{,}600}}\right) = 0.35 \times 50 = 17.5 \text{ lb/ft}^2 < 0.40L_o$$

Thus, $L = 0.40 \times 50 = 20.0$ lb/ft²

Axial load due to L:

$$\Sigma P_R = L(K_{LL}A_T)/K_{LL} = [20.0 \times (4 \times 5{,}600)/(4 \times 1{,}000) = 112.0 \text{ kips}$$

Axial load due to partitions:

$$P_N = 15 \times 700/1{,}000 = 10.5 \text{ kips}$$

In S.I.:

$$L = L_o\left(0.25 + \frac{4.57}{\sqrt{K_{LL}A_T}}\right)$$

$$= 2.39 \times \left(0.25 + \frac{4.57}{\sqrt{4 \times 520.0}}\right) = 0.35 \times 2.39 = 0.84 \text{ kN/m}^2 < 0.40L_o$$

Thus, $L = 0.40 \times 2.39 = 0.96$ kN/m²

Axial load due to L:

$$\Sigma P_R = L(K_{LL}A_T)/K_{LL} = [0.96 \times (4 \times 520.0)]/4 = 499.2 \text{ kN}$$

Axial load due to partitions:

$$P_N = 0.72 \times 64.6 = 46.5 \text{ kN}$$

A summary of the axial live loads for column B3 is given in Table 3.4.

Table 3.4 Summary of Axial Live Loads for Column B3

Story	L_o, lb/ft² (kN/m²)		$K_{LL}A_T$, ft² (m²)	RM	L, lb/ft² (kN/m²)	P_N, kips (kN)	ΣP_N, kips (kN)	ΣP_R, kips (kN)	$\Sigma P_N + \Sigma P_R$, kips (kN)
	N	R							
10	0	20 (0.96)	—	—	12.0 (0.58)	0	0	8.4 (37.7)	8.4 (37.7)
9	200 (9.58)	0	—	—	—	140.0 (622.7)	140.0 (622.7)	0	148.4 (660.4)
8	15 (0.72)	50 (2.39)	2,800 (260)	0.53	26.7 (1.28)	10.5 (46.8)	150.5 (669.5)	18.7 (83.2)	177.6 (790.4)
7	15 (0.72)	50 (2.39)	5,600 (520)	0.45	22.5 (1.08)	10.5 (46.8)	161.0 (716.3)	31.5 (140.4)	200.9 (894.4)
6	15 (0.72)	50 (2.39)	8,400 (780)	0.41	20.7 (0.99)	10.5 (46.8)	171.5 (762.9)	43.5 (193.3)	223.4 (993.9)
5	15 (0.72)	50 (2.39)	11,200 (1,040)	0.40	20.0 (0.96)	10.5 (46.8)	182.0 (809.7)	56.0 (249.6)	246.4 (1,097.0)
4	15 (0.72)	50 (2.39)	14,000 (1,300)	0.40	20.0 (0.96)	10.5 (46.8)	192.5 (856.5)	70.0 (312.0)	270.9 (1,206.2)
3	15 (0.72)	50 (2.39)	16,800 (1,560)	0.40	20.0 (0.96)	10.5 (46.8)	203.0 (903.3)	84.0 (374.4)	295.4 (1,315.4)
2	15 (0.72)	50 (2.39)	19,600 (1,820)	0.40	20.0 (0.96)	10.5 (46.8)	213.5 (950.1)	98.0 (436.8)	319.9 (1,424.6)
1	15 (0.72)	50 (2.39)	22,400 (2,080)	0.40	20.0 (0.96)	10.5 (46.8)	224.0 (996.9)	112.0 (499.2)	344.4 (1,533.8)

N = nonreducible uniform live load, lb/ft² (kN/m²)
R = reducible uniform live load, lb/ft² (kN/m²)
RM = reduction multiplier = $0.25 + (15/\sqrt{K_{LL}A_T})$ (ft²) = $0.25 + (4.57/\sqrt{K_{LL}A_T})$ (m²)
P_N = axial load due to the nonreducible uniform live load, kips (kN)
ΣP_N = cumulative axial load due to the nonreducible live load, kips (kN)
ΣP_R = cumulative axial load due to the reducible live load, kips (kN)

In the first story, the total axial live load is equal to the axial load in the tenth story plus the cumulative axial load due to the nonreducible live load, $\sum P_N$, plus the cumulative axial load due to the reducible live load, $\sum P_R$:

Total axial live load in the first story $= 8.4 + 224.0 + 112.0 = 344.4$ kips

Total axial live load in the first story $= 37.7 + 996.9 + 499.2 = 1{,}533.8$ kN

3.6.2 Example 3.2—Basic Uniform Live Load Reduction, Edge Column with a Cantilever (IBC 1607.13.1)

For the 10-story office building in Figure 3.22, determine the total live loads in column A3 in each story considering basic uniform live load reduction in accordance with IBC 1607.13.1. Use the design data in Example 3.1.

Solution

Step 1—Determine the axial live load in column A3 in the tenth story

The reduced uniform roof live load, L_r, is determined by IBC Eq. 16-10:

$$L_r = L_o R_1 R_2$$

The minimum roof live load, L_o, for an ordinary flat roof that is not occupiable is equal to 20 lb/ft^2 (0.96 kN/m^2) in accordance with IBC Table 1607.1.

Tributary area, A_t, of column A3:

$$A_t = 25 \times \left(\frac{28}{2} + 5\right) = 475 \text{ ft}^2$$

$$A_t = 7.62 \times \left(\frac{8.53}{2} + 1.52\right) = 44.1 \text{ m}^2$$

Because 200 ft$^2 < A_t < 600$ ft^2, $R_1 = 1.2 - 0.001A_t = 0.73$ (IBC Eq. 16-12)

In S.I.:

$$R_1 = 1.2 - 0.011A_t = 0.72$$

Because $F = 1/2 < 4$, $R_2 = 1.0$ (IBC Eq. 16-14)

Therefore,

$$L_r = 20 \times 0.73 \times 1.0 = 14.6 \text{ lb/ft}^2$$

Axial load $= 14.6 \times 475/1{,}000 = 6.9$ kips

In S.I.:

$$L_r = 0.96 \times 0.72 \times 1.0 = 0.69 \text{ kN/m}^2$$

Axial load $= 0.69 \times 44.1 = 30.4$ kN

Step 2—Determine the axial live load in column A3 in the ninth story

Because the ninth floor is designated a file storage floor with a live load of 200 lb/ft^2 (9.58 kN/m^2), which exceeds 100 lb/ft^2 (4.79 kN/m^2), and because the column at this level does not support two or more floors of occupancy other than roof occupancy, the live load is not permitted to be reduced (IBC 1607.13.1.2).

Axial load $= 200 \times 475/1{,}000 = 95.0$ kips

Axial load $= 9.58 \times 44.1 = 422.5$ kN

Step 3—Determine the axial live load in column A3 in a typical story

The reduced uniform live load, L, is determined by IBC Eq. 16-7:

$$L = L_o\left(0.25 + \frac{15}{\sqrt{K_{LL}A_T}}\right)\text{(lb/ft}^2\text{)}$$

$$L = L_o\left(0.25 + \frac{4.57}{\sqrt{K_{LL}A_T}}\right)\text{(kN/m}^2\text{)}$$

For an office occupancy, the minimum reducible live load, L_o, is equal to 50 lb/ft^2 (2.39 kN/m^2) in accordance with IBC Table 1607.1.

The minimum partition load is equal to 15 lb/ft^2 (0.72 kN/m^2) because the specified live load is less than 80 lb/ft^2 (3.83 kN/m^2); this uniform live load is not reducible (IBC 1607.5).

For an edge column with a cantilever, the live load element factor, K_{LL}, is equal to 3 from IBC Table 1607.13.1.

For comparison purposes, calculate K_{LL} using the influence and tributary areas:

$$K_{LL} = \frac{\text{Influence area}}{\text{Tributary area}} = \frac{50\times(28+5)}{25\times\left(\frac{28}{2}+5\right)} = 3.5$$

In S.I.:

$$K_{LL} = \frac{\text{Influence area}}{\text{Tributary area}} = \frac{15.24\times(8.53+1.52)}{7.62\times\left(\frac{8.53}{2}+1.52\right)} = 3.5$$

$K_{LL} = 3$ is used in the remainder of this example.

The tributary area, A_T, to be used in IBC Eq. 16-7 in the determination of L at a particular story is equal to the sum of the tributary areas above that story where the uniform floor live load is permitted to be reduced.

For the determination of L in the eighth story, the tributary area of the roof cannot be included because it is not a floor level and the tributary area of the ninth floor cannot be included because the live load is not permitted to be reduced at that level. Therefore, $A_T = 1\times 475 = 475$ ft^2 (44.1 m^2), and L is equal to the following:

$$L = L_o\left(0.25 + \frac{15}{\sqrt{K_{LL}A_T}}\right) = 50\times\left(0.25 + \frac{15}{\sqrt{3\times 475}}\right) = 0.65\times 50 = 32.5\text{ lb/ft}^2 > 0.50L_o$$

Axial load due to L:

$$\Sigma P_R = L(K_{LL}A_T)/K_{LL} = [32.5\times(3\times 475)]/(3\times 1{,}000) = 15.4\text{ kips}$$

Axial load due to partitions:

$$P_N = 15\times 475/1{,}000 = 7.1\text{ kips}$$

In S.I.:

$$L = L_o\left(0.25 + \frac{4.57}{\sqrt{K_{LL}A_T}}\right)$$

$$= 2.39\times\left(0.25 + \frac{4.57}{\sqrt{3\times 44.1}}\right) = 0.65\times 2.39 = 1.55\text{ kN/m}^2 > 0.50L_o$$

Axial load due to L:

$$\sum P_R = L(K_{LL}A_T)/K_{LL} = [1.55 \times (3 \times 44.1)]/3 = 68.4 \text{ kN}$$

Axial load due to partitions:

$$P_N = 0.72 \times 44.1 = 31.8 \text{ kN}$$

Similarly, in the first story, $A_T = 8 \times 475 = 3{,}800$ ft² (352.8 m²), and L is equal to the following:

$$L = L_o\left(0.25 + \frac{15}{\sqrt{K_{LL}A_T}}\right)$$

$$= 50 \times \left(0.25 + \frac{15}{\sqrt{3 \times 3{,}800}}\right) = 0.391 \times 50 = 19.5 \text{ lb/ft}^2 < 0.40L_o$$

Thus, $L = 0.40 \times 50 = 20.0$ lb/ft²

Axial load due to L:

$$\sum P_R = L(K_{LL}A_T)/K_{LL} = [20.0 \times (3 \times 3{,}800)/(3 \times 1{,}000) = 76.0 \text{ kips}$$

Axial load due to partitions:

$$P_N = 15 \times 475/1{,}000 = 7.1 \text{ kips}$$

In S.I.:

$$L = L_o\left(0.25 + \frac{4.57}{\sqrt{K_{LL}A_T}}\right)$$

$$= 2.39 \times \left(0.25 + \frac{4.57}{\sqrt{3 \times 352.8}}\right) = 0.391 \times 2.39 = 0.93 \text{ kN/m}^2 < 0.40L_o$$

Thus, $L = 0.40 \times 2.39 = 0.96$ kN/m²

Axial load due to L:

$$\sum P_R = L(K_{LL}A_T)/K_{LL} = [0.96 \times (3 \times 352.8)]/3 = 338.7 \text{ kN}$$

Axial load due to partitions:

$$P_N = 0.72 \times 44.1 = 31.8 \text{ kN}$$

A summary of the axial live loads for column A3 is given in Table 3.5.

In the first story, the total axial live load is equal to the axial load in the tenth story plus the cumulative axial load due to the nonreducible live load, $\sum P_N$, plus the cumulative axial load due to the reducible live load, $\sum P_R$:

Total axial live load in the first story = 6.9 + 151.8 + 76.0 = 234.7 kips

Total axial live load in the first story = 30.4 + 676.9 + 338.7 = 1,046.0 kN

Table 3.5 Summary of Axial Live Loads for Column A3

Story	L_o, lb/ft² (kN/m²)		$K_{LL}A_T$, ft² (m²)	RM	L, lb/ft² (kN/m²)	P_N, kips (kN)	ΣP_N, kips (kN)	ΣP_R, kips (kN)	$\Sigma P_N + \Sigma P_R$, kips (kN)
	N	R							
10	0	20 (0.96)	—	—	14.6 (0.69)	0	0	6.9 (30.4)	6.9 (30.4)
9	200 (9.58)	0	—	—	—	95.0 (422.5)	95.0 (422.5)	0	101.9 (452.9)
8	15 (0.72)	50 (2.39)	1,425 (132.3)	0.65	32.5 (1.55)	7.1 (31.8)	102.1 (454.3)	15.4 (68.4)	124.4 (553.1)
7	15 (0.72)	50 (2.39)	2,850 (264.6)	0.53	26.6 (1.27)	7.1 (31.8)	109.2 (486.1)	25.3 (112.0)	141.4 (628.5)
6	15 (0.72)	50 (2.39)	4,275 (396.9)	0.48	24.0 (1.15)	7.1 (31.8)	116.3 (517.9)	34.2 (152.2)	157.4 (700.5)
5	15 (0.72)	50 (2.39)	5,700 (529.2)	0.45	22.5 (1.08)	7.1 (31.8)	123.4 (549.7)	42.8 (190.5)	173.1 (770.6)
4	15 (0.72)	50 (2.39)	7,125 (661.5)	0.43	21.5 (1.03)	7.1 (31.8)	130.5 (581.5)	51.1 (227.1)	188.5 (839.0)
3	15 (0.72)	50 (2.39)	8,550 (793.8)	0.41	20.6 (0.99)	7.1 (31.8)	137.6 (613.3)	58.7 (262.0)	203.2 (905.7)
2	15 (0.72)	50 (2.39)	9,975 (926.1)	0.40	20.0 (0.96)	7.1 (31.8)	144.7 (645.1)	66.5 (296.4)	218.1 (971.9)
1	15 (0.72)	50 (2.39)	11,400 (1,058.4)	0.40	20.0 (0.96)	7.1 (31.8)	151.8 (676.9)	76.0 (338.7)	234.7 (1,046.0)

N = nonreducible uniform live load, lb/ft² (kN/m²)
R = reducible uniform live load, lb/ft² (kN/m²)
RM = reduction multiplier = $0.25 + (15/\sqrt{K_{LL}A_T})$ (ft²) = $0.25 + (4.57/\sqrt{K_{LL}A_T})$ (m²)
P_N = axial load due to the nonreducible uniform live load, kips (kN)
ΣP_N = cumulative axial load due to the nonreducible live load, kips (kN)
ΣP_R = cumulative axial load due to the reducible live load, kips (kN)

3.6.3 Example 3.3—Basic Uniform Live Load Reduction, Two-Way Slab (IBC 1607.13.1)

Determine the total live load for the two-way slab bound by column lines BC23 on a typical floor of the 10-story office building in Figure 3.22 considering basic uniform live load reduction in accordance with IBC 1607.13.1. Use the design data in Example 3.1.

SOLUTION

The reduced uniform live load, L, is determined by IBC Eq. 16-7:

$$L = L_o\left(0.25 + \frac{15}{\sqrt{K_{LL}A_T}}\right) \text{(lb/ft}^2\text{)}$$

$$L = L_o\left(0.25 + \frac{4.57}{\sqrt{K_{LL}A_T}}\right) \text{(kN/m}^2\text{)}$$

For an office occupancy, the minimum reducible live load, L_o, is equal to 50 lb/ft² (2.39 kN/m²) in accordance with IBC Table 1607.1.

For a two-way slab, the live load element factor, K_{LL}, is equal to 1 from IBC Table 1607.13.1, and the tributary area, A_T, is equal to $25 \times 28 = 700$ ft² (65.0 m²).

Therefore,

$$L = L_o\left(0.25 + \frac{15}{\sqrt{K_{LL}A_T}}\right) = 50 \times \left(0.25 + \frac{15}{\sqrt{1 \times 700}}\right) = 0.82 \times 50 = 41.0 \text{ lb/ft}^2$$

In S.I.:

$$L = L_o\left(0.25 + \frac{4.57}{\sqrt{K_{LL}A_T}}\right) = 2.39 \times \left(0.25 + \frac{4.57}{\sqrt{1 \times 65.0}}\right) = 0.82 \times 2.39 = 1.96 \text{ kN/m}^2$$

The minimum partition load is equal to 15 lb/ft² (0.72 kN/m²) because the specified live load is less than 80 lb/ft² (3.83 kN/m²); this uniform live load is not reducible (IBC 1607.5).

Total live load = 41.0 + 15.0 = 56.0 lb/ft²

Total live load = 1.96 + 0.72 = 2.68 kN/m²

3.6.4 Example 3.4—Basic Uniform Live Load Reduction, Beams, and One-Way Slab (IBC 1607.13.1)

Determine the total live loads in pounds per linear foot (kN per meter) for the following members on the typical floor of the classroom depicted in Figure 3.23 considering basic uniform live load reduction in accordance with IBC 1607.13.1: (a) secondary beam, (b) primary beam on line 2, and (c) one-way slab. Partitions are not utilized in the classrooms.

SOLUTION

(a) Secondary beam

The reduced uniform live load, L, is determined by IBC Eq. 16-7:

$$L = L_o\left(0.25 + \frac{15}{\sqrt{K_{LL}A_T}}\right) \text{(lb/ft}^2\text{)}$$

$$L = L_o\left(0.25 + \frac{4.57}{\sqrt{K_{LL}A_T}}\right) \text{(kN/m}^2\text{)}$$

Figure 3.23 Typical Floor Plan of the Classroom in Example 3.4

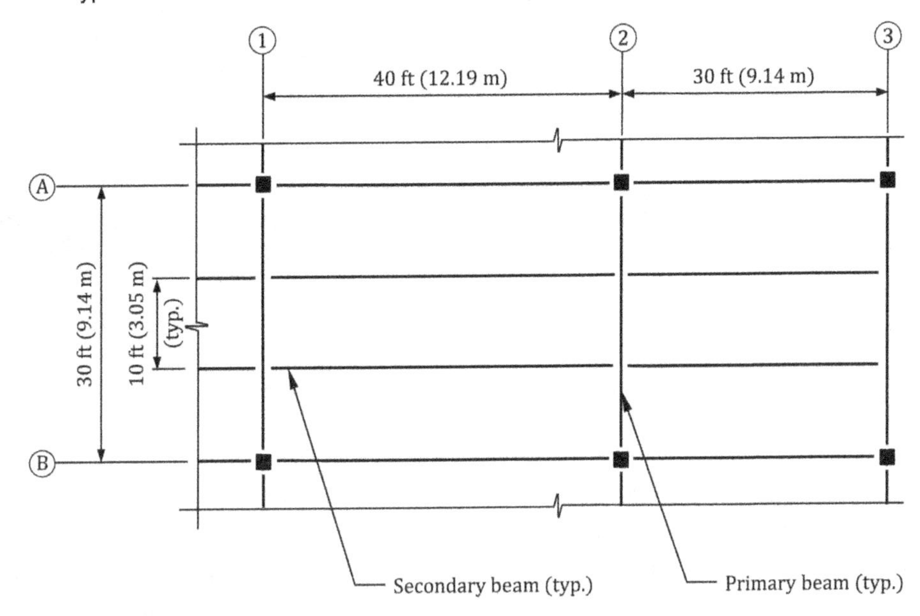

For a classroom occupancy, the minimum reducible live load, L_o, is equal to 40 lb/ft^2 (1.92 kN/m^2) in accordance with IBC Table 1607.1.

For an interior beam, the live load element factor, K_{LL}, is equal to 2 from IBC Table 1607.13.1, and the tributary area, A_T, is equal to $10 \times 40 = 400$ ft^2 (37.18 m^2).

Therefore,

$$L = L_o\left(0.25 + \frac{15}{\sqrt{K_{LL}A_T}}\right) = 40 \times \left(0.25 + \frac{15}{\sqrt{2 \times 400}}\right) = 0.78 \times 40 = 31.2 \text{ lb/ft}^2$$

Total load $= 31.2 \times 10 = 312$ lb/ft

In S.I.:

$$L = L_o\left(0.25 + \frac{4.57}{\sqrt{K_{LL}A_T}}\right) = 1.92 \times \left(0.25 + \frac{4.57}{\sqrt{2 \times 37.18}}\right) = 0.78 \times 1.92 = 1.50 \text{ kN/m}^2$$

Total load $= 1.50 \times 3.05 = 4.58$ kN/m

(b) Primary beam on line 2

For the primary beam on line 2 (interior beam), the live load element factor, K_{LL}, is equal to 2 from IBC Table 1607.13.1, and the tributary area, A_T, is equal to $[0.5 \times (40 + 30)] \times 30 = 1{,}050$ ft^2 (97.48 m^2).

Therefore,

$$L = L_o\left(0.25 + \frac{15}{\sqrt{K_{LL}A_T}}\right) = 40 \times \left(0.25 + \frac{15}{\sqrt{2 \times 1{,}050}}\right) = 0.58 \times 40 = 23.2 \text{ lb/ft}^2$$

Total load $= 23.2 \times 35 = 812$ lb/ft

In S.I.:

$$L = L_o\left(0.25+\frac{4.57}{\sqrt{K_{LL}A_T}}\right) = 1.92\times\left(0.25+\frac{4.57}{\sqrt{2\times 97.48}}\right) = 0.58\times 1.92 = 1.11\ \text{kN/m}^2$$

Total load $= 1.11\times 10.67 = 11.84$ kN/m

(c) One-way slab

For a one-way slab, the live load element factor, K_{LL}, is equal to 1 from IBC Table 1607.13.1, and the tributary area, A_T, is equal to $10\times(1.5\times 10) = 150$ ft^2 (13.95 m^2) in accordance with IBC 1607.13.1.1.

$K_{LL}A_T = 1\times 150 = 150$ ft^2 (13.95 m^2), which is less than 400 ft^2 (37.16 m^2), so live load reduction is not permitted (IBC 1607.13.1).

Total load $= 40.0 \times 1 = 40$ lb/ft

Total load $= 19.2 \times 1 = 19.2$ kN/m

3.6.5 Example 3.5—Alternate Uniform Live Load Reduction, Interior Column (IBC 1607.13.2)

Determine the total live loads in column B3 in each story of the 10-story office building illustrated in Figure 3.22 considering alternate uniform live load reduction in accordance with IBC 1607.13.2. Use the design data in Example 3.1. Assume a dead to live load ratio, D/L_o, equal to 2.0.

Solution

Step 1—Determine the axial live load in column B3 in the tenth story

The reduced uniform roof live load, L_r, is determined by IBC Eq. 16-10:

$$L_r = L_o R_1 R_2$$

The minimum roof live load, L_o, for an ordinary flat roof that is not occupiable is equal to 20 lb/ft^2 (0.96 kN/m^2) in accordance with IBC Table 1607.1.

Tributary area, A_t, of column B3:

$$A_t = 25\times 28 = 700\ \text{ft}^2$$

$$A_t = 7.62\times 8.53 = 65.0\ \text{m}^2$$

Because $A_t > 600$ ft^2 (55.7 m^2), $R_1 = 0.6$ (IBC Eq. 16-13)

Because $F = 1/2 < 4$, $R_2 = 1.0$ (IBC Eq. 16-14)

Therefore, $L_r = 20\times 0.6\times 1.0 = 12$ lb/ft^2, which is equal to the permitted minimum reduced roof live load in IBC 1607.14.1.

Axial load $= 12\times 700/1{,}000 = 8.4$ kips

In S.I.:

$$L_r = 0.96\times 0.6\times 1.0 = 0.58\ \text{kN/m}^2$$

$$\text{Axial load} = 0.58\times 65.0 = 37.7\ \text{kN}$$

Step 2—Determine the axial live load in column B3 in the ninth story

Because the ninth floor is designated a file storage floor with a live load of 200 lb/ft^2 (9.58 kN/m^2), which exceeds 100 lb/ft^2 (4.79 kN/m^2), and because the column at this level does not support two or more floors of occupancy other than roof occupancy, the live load is not permitted to be reduced (IBC 1607.13.1.2).

Axial load $= 200 \times 700/1{,}000 = 140.0$ kips

Axial load $= 9.58 \times 65.0 = 622.7$ kN

Step 3—Determine the axial live load in column B3 in a typical story

The reduced uniform live load, L, is determined by the following equation:

$$L = L_o\left(1 - \frac{R}{100}\right)$$

The reduction factor, R, is determined by IBC Eq. 16-8:

$$R = 0.08(A - 150)\ (\text{ft}^2)$$

$$R = 0.861(A - 13.94)\ (\text{m}^2)$$

where A is the tributary area in square feet (m²).

R must not exceed the least of the following:

- 40 percent for members supporting one floor
- 60 percent for members supporting two or more floors
- $R = 23.1(1 + D/L_o) = 23.1(1 + 2.0) = 69.3$ percent

For an office occupancy, the minimum reducible live load, L_o, is equal to 50 lb/ft² (2.39 kN/m²) in accordance with IBC Table 1607.1.

The minimum partition load is equal to 15 lb/ft² (0.72 kN/m²) because the specified live load is less than 80 lb/ft² (3.83 kN/m²); this uniform live load is not reducible (IBC 1607.5).

For an interior column, the live load element factor, K_{LL}, is equal to 4 from IBC Table 1607.13.1.

The tributary area, A, to be used in IBC Eq. 16-8 in the determination of L at a particular story is equal to the sum of the tributary areas above that story where the uniform floor live load is permitted to be reduced.

For the determination of L in the eighth story, the tributary area of the roof cannot be included because it is not a floor level and the tributary area of the ninth floor cannot be included because the live load is not permitted to be reduced at that level. Therefore, $A = 1 \times 700 = 700$ ft² (65.0 m²), and R and L are equal to the following:

$$R = 0.08(A - 150) = 0.08 \times (700 - 150) = 44 \text{ percent} > 40 \text{ percent, use } 40 \text{ percent}$$

$$L = L_o\left(1 - \frac{R}{100}\right) = 50 \times \left(1 - \frac{40}{100}\right) = 30.0\ \text{lb/ft}^2$$

Axial load due to L:

$$\sum P_R = 30.0 \times 700/1{,}000 = 21.0\ \text{kips}$$

Axial load due to partitions:

$$P_N = 15 \times 700/1{,}000 = 10.5\ \text{kips}$$

In S.I.:

$$R = 0.861(A - 13.94) = 0.861 \times (65.0 - 13.94) = 44 \text{ percent} > 40 \text{ percent, use } 40 \text{ percent}$$

$$L = L_o\left(1 - \frac{R}{100}\right) = 2.39 \times \left(1 - \frac{40}{100}\right) = 1.43\ \text{kN/m}^2$$

Axial load due to L:

$$\Sigma P_R = 1.43 \times 65.0 = 93.0 \text{ kN}$$

Axial load due to partitions:

$$P_N = 0.72 \times 64.6 = 46.5 \text{ kN}$$

Similarly, in the first story, $A_T = 8 \times 700 = 5{,}600$ ft^2 (520.0 m^2), and R and L are equal to the following:

$R = 0.08(5{,}600 - 150) = 0.08 \times (700 - 150) = 436$ percent > 60 percent, use 60 percent

$$L = L_o\left(1 - \frac{R}{100}\right) = 50 \times \left(1 - \frac{60}{100}\right) = 20.0 \text{ lb/ft}^2$$

Axial load due to L:

$$\Sigma P_R = 20.0 \times 5{,}600/1{,}000 = 112.0 \text{ kips}$$

Axial load due to partitions:

$$P_N = 15 \times 700/1{,}000 = 10.5 \text{ kips}$$

In S.I.:

$R = 0.861(A - 13.94) = 0.861 \times (520.0 - 13.94) = 436$ percent > 60 percent, use 60 percent

$$L = L_o\left(1 - \frac{R}{100}\right) = 2.39 \times \left(1 - \frac{60}{100}\right) = 0.96 \text{ kN/m}^2$$

Axial load due to L:

$$\Sigma P_R = 0.96 \times 520.0 = 499.2 \text{ kN}$$

Axial load due to partitions:

$$P_N = 0.72 \times 64.6 = 46.5 \text{ kN}$$

A summary of the axial live loads for column B3 is given in Table 3.6.
In the first story, the total axial live load is equal to the axial load in the tenth story plus the cumulative axial load due to the nonreducible live load, ΣP_N, plus the cumulative axial load due to the reducible live load, ΣP_R:

Total axial live load in the first story $= 8.4 + 224.0 + 112.0 = 344.4$ kips

Total axial live load in the first story $= 37.7 + 996.9 + 499.2 = 1{,}533.8$ kN

3.6.6 Example 3.6—Calculation of Design Rain Load for a Roof with Edge Overflow

Determine the design rain load, R, on the roof in Figure 3.24 given the following design data:

- Location: Overland Park, KS (Latitude = 38.92°, Longitude = −94.66°)
- Risk Category: II (IBC Table 1604.5)
- Secondary drainage system for structural load (SDSL): rainwater overflow on two edges
- Roof slope: ½ in./foot (2.39 degrees)

Table 3.6 Summary of Axial Live Loads for Column B3

Story	L_o, lb/ft² (kN/m²)		A, ft² (m²)	R (%)	L, lb/ft² (kN/m²)	P_N, kips (kN)	ΣP_N, kips (kN)	ΣP_R, kips (kN)	$\Sigma P_N + \Sigma P_R$, kips (kN)
	N	R							
10	0	20 (0.96)	—	—	12.0 (0.58)	0	0	8.4 (37.7)	8.4 (37.7)
9	200 (9.58)	0	—	—	—	140.0 (622.7)	140.0 (622.7)	0	148.4 (660.4)
8	15 (0.72)	50 (2.39)	700 (65)	40	30.0 (1.43)	10.5 (46.8)	150.5 (669.5)	21.0 (93.0)	179.9 (800.2)
7	15 (0.72)	50 (2.39)	1,400 (130)	60	20.0 (0.96)	10.5 (46.8)	161.0 (716.3)	28.0 (124.8)	197.4 (878.8)
6	15 (0.72)	50 (2.39)	2,100 (195)	60	20.0 (0.96)	10.5 (46.8)	171.5 (762.9)	42.0 (187.2)	221.9 (987.8)
5	15 (0.72)	50 (2.39)	2,800 (260)	60	20.0 (0.96)	10.5 (46.8)	182.0 (809.7)	56.0 (249.6)	246.4 (1,097.0)
4	15 (0.72)	50 (2.39)	3,500 (325)	60	20.0 (0.96)	10.5 (46.8)	192.5 (856.5)	70.0 (312.0)	270.9 (1,206.2)
3	15 (0.72)	50 (2.39)	4,200 (390)	60	20.0 (0.96)	10.5 (46.8)	203.0 (903.3)	84.0 (374.4)	295.4 (1,315.4)
2	15 (0.72)	50 (2.39)	4,900 (455)	60	20.0 (0.96)	10.5 (46.8)	213.5 (950.1)	98.0 (436.8)	319.9 (1,424.6)
1	15 (0.72)	50 (2.39)	5,600 (520)	60	20.0 (0.96)	10.5 (46.8)	224.0 (996.9)	112.0 (499.2)	344.4 (1,533.8)

N = nonreducible uniform live load, lb/ft² (kN/m²)
R = reducible uniform live load, lb/ft² (kN/m²)
R = reduction multiplier = $0.08(A - 150)$ (ft²) = $0.861(A - 13.94)$ (m²)
P_N = axial load due to the nonreducible uniform live load, kips (kN)
ΣP_N = cumulative axial load due to the nonreducible live load, kips (kN)
ΣP_R = cumulative axial load due to the reducible live load, kips (kN)

Figure 3.24 Roof Plan, Example 3.6

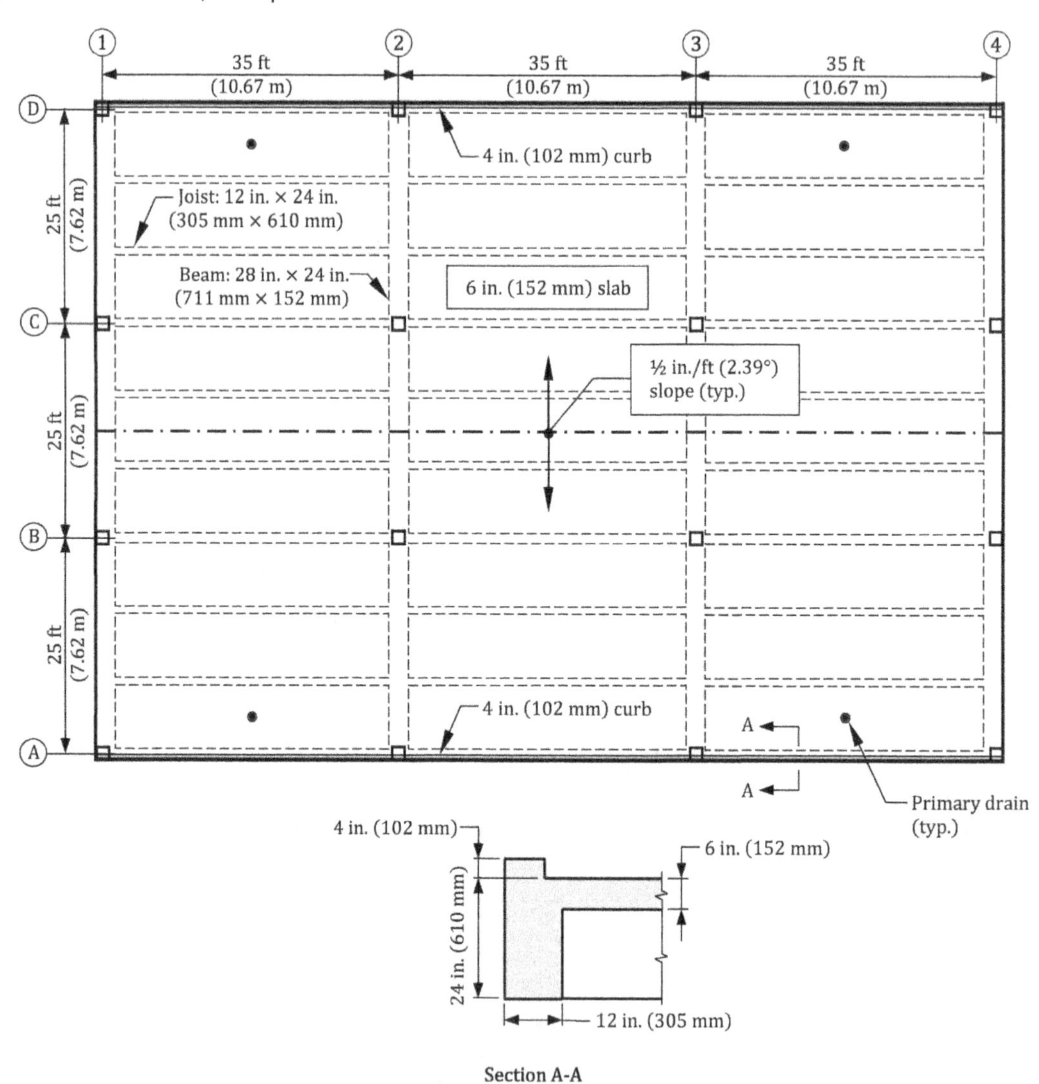

Solution

The design procedure in Table 3.2 of this book is used to determine R.

Step 1—Determine the rainfall intensity, i, for the given risk category

The design storm return period for Risk Category II is 100 years (IBC Table 1611.1). Using Reference 3.3, the 15-minute rainfall intensity, i, is equal to 7.58 in./h (192.43 mm/h).

Step 2—Determine the flow rate, Q

Flow rate, Q, is determined by Equations (3.8) and (3.9) of this book:

$$Q = 0.0104Ai \text{ (gal/min)}$$

$$Q = (0.278 \times 10^{-6})Ai \text{ (m}^3\text{/s)}$$

Tributary area, A, to a roof drain:

$$A = (105 \times 75)/4 = 1{,}969 \text{ ft}^2$$

$$Q = 0.0104Ai = 0.0104 \times 1{,}969 \times 7.58 = 155.2 \text{ gal/min}$$

In S.I.:

$$A = (32.00 \times 22.86)/4 = 182.9 \text{ m}^2$$

$$Q = (0.278 \times 10^{-6})Ai = (0.278 \times 10^{-6}) \times 182.9 \times 192.43 = 0.0098 \text{ m}^3/\text{s}$$

Step 3—Determine the static head, d_s

From the roof section in Figure 3.24, $d_s = 4$ in. (102 mm)

Step 4—Determine the hydraulic head, d_h

Check if the length of the roof edge, L_r, is greater than the limiting values in ASCE/SEI Equations (C8.2-2) and (C8.2-2.SI):

$$L_r = 105/2 = 52.5 \text{ ft} > Ai/400 = 1{,}969 \times 7.58/400 = 37.3 \text{ ft}$$

$$L_r = 32.0/2 = 16.0 \text{ m} > Ai/3{,}100 = 182.9 \times 192.43/3{,}100 = 11.35 \text{ m}$$

Because L_r is greater than the limiting value, d_h can be taken as 0.25 in. (6 mm; see ASCE/SEI C8.2).

Step 5—Determine the ponding head, d_p

The iterative method is used to determine d_p.

Step 5a—Calculate the design rain load, R, assuming $d_p = 0$

$$R = 5.2(d_s + d_h + d_p) = 5.2 \times (4.0 + 0.25 + 0) = 22.1 \text{ lb/ft}^2$$

$$R = 0.0098(d_s + d_h + d_p) = 0.0098 \times (102 + 6 + 0) = 1.06 \text{ kN/m}^2$$

Step 5b—Determine the maximum deflection due to R calculated in Step 5a and the unfactored dead loads and set that maximum deflection equal to d_p.

Reference 3.7 was used to analyze the structure. The maximum deflection due to R and the unfactored dead loads (i.e., the unfactored self-weight of the reinforced concrete framing system in Figure 3.24 assuming normalweight concrete with a density of 150 lb/ft^3 [23.56 kN/m^3] plus the unfactored superimposed dead load of 15 lb/ft^2 [0.72 kN/m^2]) is equal to 0.52 in. (13 mm). Thus, $d_p = 0.52$ in. (13 mm).

Step 5c—Recalculate R using d_p from Step 5b

$$R = 5.2(d_s + d_h + d_p) = 5.2 \times (4.0 + 0.25 + 0.52) = 24.8 \text{ lb/ft}^2$$

$$R = 0.0098(d_s + d_h + d_p) = 0.0098 \times (102 + 6 + 13) = 1.19 \text{ kN/m}^2$$

Step 5d—Determine the maximum deflection due to R calculated in Step 5c and the unfactored dead loads and set that maximum deflection equal to d_p

From the analysis, the maximum deflection due to R and the unfactored dead loads is equal to 0.53 in. (14 mm).

Step 5e—Recalculate R using d_p from Step 5d

$$R = 5.2(d_s + d_h + d_p) = 5.2 \times (4.0 + 0.25 + 0.53) = 24.9 \text{ lb/ft}^2$$

$$R = 0.0098(d_s + d_h + d_p) = 0.0098 \times (102 + 6 + 14) = 1.20 \text{ kN/m}^2$$

Step 5f—Determine the maximum deflection due to R calculated in Step 5e and the unfactored dead loads and set that maximum deflection equal to d_p

From the analysis, the maximum deflection due to R and the unfactored dead loads is equal to 0.53 in. (14 mm), which is the same maximum deflection obtained from the analysis in Step 5d.

Therefore, the analysis converges to $d_p = 0.53$ in. (14 mm) and the design rain load, R, is equal to 24.9 lb/ft^2 (1.20 kN/m^2).

For comparison purposes, R is determined using the approximate amplified first-order analysis in ASCE/SEI C8.2. The following steps can be used to determine R:

1. Determine the total load, which is equal to the dead load plus the rain load, R, excluding the ponding contribution (i.e., R is calculated assuming $d_p = 0$; see Section 3.3.7 of this book).

   Based on the members sizes in Figure 3.24 and assuming normal-weight concrete with a density of 150 lb/ft³ (23.56 kN/m³), the self-weight of the structural system is equal to 141 lb/ft² (6.75 kN/m²). The superimposed dead load is equal to 15 lb/ft² (0.72 kN/m²).

   The design rain load, R, assuming is equal to the following:

$$R = 5.2(d_s + d_h + d_p) = 5.2 \times (4.0 + 0.25 + 0) = 22.1 \text{ lb/ft}^2$$

$$R = 0.0098(d_s + d_h + d_p) = 0.0098 \times (102 + 6 + 0) = 1.06 \text{ kN/m}^2$$

   Therefore, the total load is equal to 141 + 15 + 22.1 = 178.1 lb/ft²
   In S.I.:

$$\text{Total load} = 6.75 + 0.72 + 1.06 = 8.53 \text{ kN/m}^2$$

2. Calculate the flexibility coefficient, C_p, for the primary members by ASCE/SEI Eq. (C8.2-5):

$$C_p = \frac{\gamma L_s L_p^4}{\pi^4 E I_p}$$

   where γ = unit weight of water = 62.4 lb/ft³ (9.80 kN/m³)
   L_s = span of secondary members = 35 ft (10.67 m)
   L_p = span of primary members = 25 ft (7.62 m)
   E = modulus of elasticity of the members = 3,605 kips/in.² (2.49 × 10⁷ kN/m²)
   I_p = moment of inertia of the primary members

$$= \frac{28 \times 24^3}{12} = 32{,}256 \text{ in.}^4$$

$$= \frac{0.71 \times 0.61^3}{12} = 0.0134 \text{ m}^4$$

   Therefore,

$$C_p = \frac{\gamma L_s L_p^4}{\pi^4 E I_p} = \frac{\frac{62.4}{1{,}000} \times \left(\frac{1}{12}\right)^3 \times (35 \times 12) \times (25 \times 12)^4}{\pi^4 \times 3{,}605 \times 32{,}256} = 0.0108$$

   In S.I.:

$$C_p = \frac{\gamma L_s L_p^4}{\pi^4 E I_p} = \frac{9.80 \times 10.67 \times (7.62)^4}{\pi^4 \times (2.49 \times 10^7) \times 0.0134} = 0.0108$$

3. Calculate the flexibility coefficient, C_s, for the secondary members by ASCE/SEI Eq. (C8.2-6):

$$C_s = \frac{\gamma S L_s^4}{\pi^4 E I_s}$$

   where S = spacing of secondary members = 8.33 ft (2.54 m)
   L_s = span of secondary members = 35 ft (10.67 m)

I_s = moment of inertia of the secondary members

$$= \frac{12 \times 24^3}{12} = 13{,}824 \text{ in.}^4$$

$$= \frac{0.31 \times 0.61^3}{12} = 0.0059 \text{ m}^4$$

Therefore,

$$C_s = \frac{\gamma S L_s^4}{\pi^4 E I_s} = \frac{\frac{62.4}{1{,}000} \times \left(\frac{1}{12}\right)^3 \times (8.33 \times 12) \times (35 \times 12)^4}{\pi^4 \times 3{,}605 \times 13{,}824} = 0.0231$$

In S.I.:

$$C_s = \frac{\gamma S L_s^4}{\pi^4 E I_s} = \frac{9.80 \times 2.54 \times (10.67)^4}{\pi^4 \times (2.49 \times 10^7) \times 0.0059} = 0.0225$$

4. Calculate the ponding amplification factor, B_p, by ASCE/SEI Eq. (C8.2-4):

$$B_p = \frac{1}{1 - 1.15C_p - C_s} = \frac{1}{1 - (1.15 \times 0.0108) - 0.0231} = 1.04$$

In S.I.:

$$B_p = \frac{1}{1 - 1.15C_p - C_s} = \frac{1}{1 - (1.15 \times 0.0108) - 0.0225} = 1.04$$

5. Calculate the total load (dead load plus rain load including the effects of ponding) by multiplying the total load in step 1 by the ponding amplification factor, B_p, calculated in step 4.

$$\text{Total load} = 1.04 \times 178.1 = 185.2 \text{ lb/ft}^2$$

$$\text{Total load} = 1.04 \times 8.53 = 8.87 \text{ kN/m}^2$$

6. Determine the design rain load, R, by subtracting the dead load from the total load calculated in step 5.

$$R = 185.2 - (141.0 + 15.0) = 29.2 \text{ lb/ft}^2$$

$$R = 8.87 - (6.75 + 0.72) = 1.40 \text{ kN/m}^2$$

The design rain load calculated by this approximate method is conservative compared to the design rain load determined by the iterative approach.

3.6.7 Example 3.7—Calculation of Design Rain Load for a Roof with Drains—Standpipe System

Determine the design rain load, R, on the roof in Figure 3.25 given the following design data:

- Location: Chicago, IL (Latitude = 41.87°, Longitude = −87.65°)
- Risk Category: II (IBC Table 1604.5)
- Secondary drainage system for structural load (SDSL): roof drains consisting of overflow standpipes with a 6-in. (152-mm) diameter (see Figure 3.13)

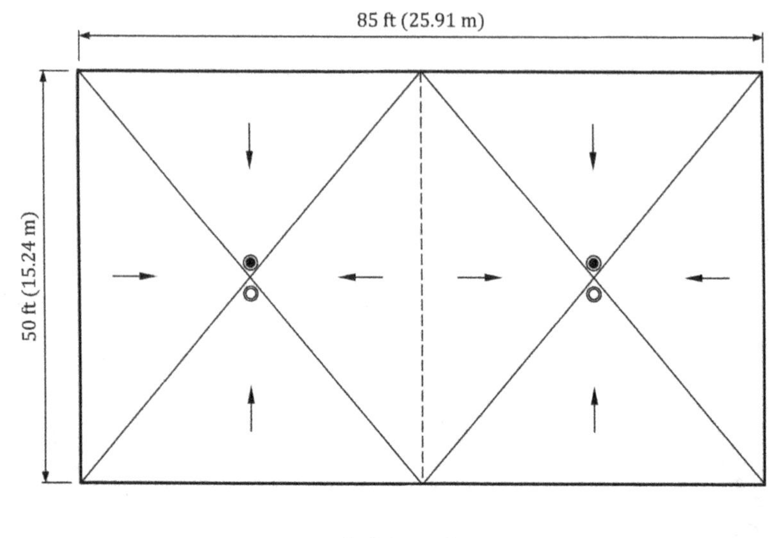

Figure 3.25 Roof Plan, Example 3.7

- Static head, d_s: inlet of the overflow standpipe is set 2 in. (51 mm) above the roof surface
- Ponding head, d_p: $d_p = 0.8$ in. (20 mm) from an iterative analysis
- Roof slope: ½ in./foot (2.39 degrees)

Solution

The design procedure in Table 3.2 of this book is used to determine R.

Step 1—Determine the rainfall intensity, i, for the given risk category
The design storm return period for Risk Category II is 100 years (IBC Table 1611.1). Using Reference 3.3, the 15-minute rainfall intensity, i, is equal to 6.35 in./h (161.24 mm/h).

Step 2—Determine the flow rate, Q
Flow rate, Q, is determined by Equations (3.8) and (3.9) of this book:

$$Q = 0.0104Ai \text{ (gal/min)}$$

$$Q = (0.278 \times 10^{-6})Ai \text{ (m}^3\text{/s)}$$

Tributary area, A, to each roof drain:

$$A = (85 \times 50)/2 = 2{,}125 \text{ ft}^2$$

$$Q = 0.0104Ai = 0.0104 \times 2{,}125 \times 6.35 = 140.3 \text{ gal/min}$$

In S.I.:

$$A = (25.91 \times 15.24)/2 = 197.4 \text{ m}^2$$

$$Q = (0.278 \times 10^{-6})Ai = (0.278 \times 10^{-6}) \times 197.4 \times 161.24 = 0.0088 \text{ m}^3\text{/s}$$

Step 3—Determine the static head, d_s
From the design data, $d_s = 2$ in. (51 mm)

Step 4—Determine the hydraulic head, d_h
From ASCE/SEI Table C8.2-1 for a 6-in. diameter standpipe: $d_h = 2.5$ in. for $Q = 150$ gal/min > 140.3 gal/min.

From ASCE/SEI Table C8.2-2 for a 152-mm diameter standpipe: $d_h = 64$ mm for $Q = 0.0095$ m²/s > 0.0088 m²/s.

Step 5—Determine the ponding head, d_p

The iterative method was used to determine d_p, which is equal to 0.8 in. (20 mm) in the design data.

Step 6—Calculate the design rain load, R

$$R = 5.2(d_s + d_h + d_p) = 5.2 \times (2.0 + 2.5 + 0.8) = 27.6 \text{ lb/ft}^2$$

$$R = 0.0098(d_s + d_h + d_p) = 0.0098 \times (51 + 64 + 20) = 1.32 \text{ kN/m}^2$$

3.6.8 Example 3.8—Calculation of Design Rain Load for a Roof with Drains—Overflow Dam System

Determine the design rain load, R, on the roof in Figure 3.26 given the following design data:

- Location: Miami, FL (Latitude = 25.78°, Longitude = −80.21°)
- Risk Category: II (IBC Table 1604.5)
- Secondary drainage system for structural load (SDSL): roof drains consisting of the following: (1) overflow dam diameter = 12.75 in. (329 mm), (2) drain outlet size = 6 in. (152 mm), and (3) drain bowl depth = 2 in. (51 mm; see Figure 3.13)
- Static head, d_s: overflow dam height = 2 in. (51 mm)
- Ponding head, d_p: $d_p = 1.2$ in. (31 mm) from an iterative analysis
- Roof slope: ½ in./foot (2.39 degrees)

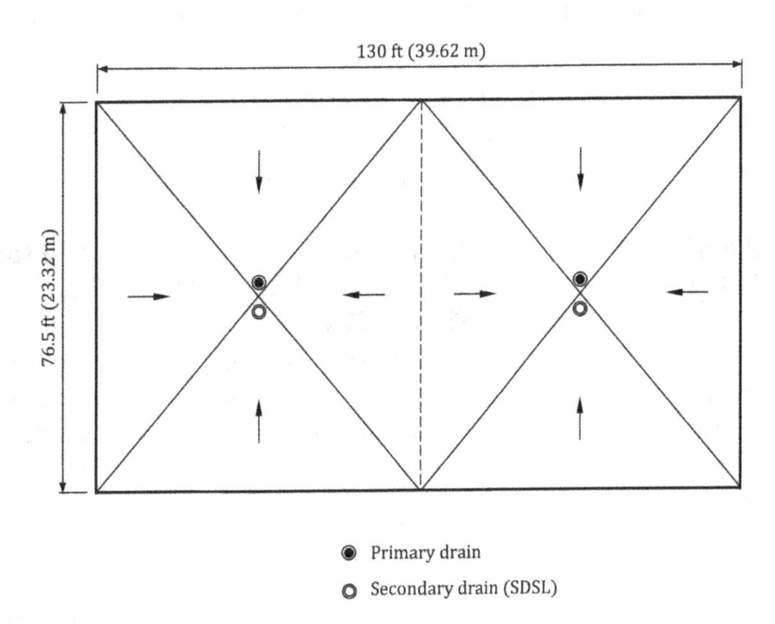

Figure 3.26 Roof Plan, Example 3.8

SOLUTION

The design procedure in Table 3.2 of this book is used to determine R.

Step 1—Determine the rainfall intensity, i, for the given risk category

The design storm return period for Risk Category II is 100 years (IBC Table 1611.1). Using Reference 3.3, the 15-minute rainfall intensity, i, is equal to 9.42 in./h (239.37 mm/h).

Step 2—Determine the flow rate, Q

Flow rate, Q, is determined by Equations (3.8) and (3.9) of this book:

$$Q = 0.0104Ai \text{ (gal/min)}$$

$$Q = (0.278 \times 10^{-6})Ai \text{ (m}^3\text{/s)}$$

Tributary area, A, to each roof drain:

$$A = (130 \times 76.5)/2 = 4{,}973 \text{ ft}^2$$

$$Q = 0.0104Ai = 0.0104 \times 4{,}973 \times 9.42 = 487.2 \text{ gal/min}$$

In S.I.:

$$A = (39.62 \times 23.32)/2 = 462.0 \text{ m}^2$$

$$Q = (0.278 \times 10^{-6})Ai = (0.278 \times 10^{-6}) \times 462.0 \times 239.37 = 0.0307 \text{ m}^3\text{/s}$$

Step 3—Determine the static head, d_s

From the design data, $d_s = 2$ in. (51 mm)

Step 4—Determine the hydraulic head, d_h

The hydraulic head is determined by linear interpolation in ASCE/SEI Tables C8.2-1 and C8.2-2 for the given overflow dam diameter, drain outlet size, drain bowl depth, and Q:

$$d_h = 3.0 + \frac{(3.5 - 3.0) \times (487.2 - 450)}{500 - 450} = 3.4 \text{ in. at } Q = 487.2 \text{ gal/min}$$

$$d_h = 76 + \frac{(89 - 76) \times (0.0307 - 0.0284)}{0.0315 - 0.0284} = 86 \text{ mm at } Q = 0.0307 \text{ m}^3\text{/s}$$

Step 5—Determine the ponding head, d_p

The iterative method was used to determine d_p, which is equal to 1.2 in. (31 mm) in the design data.

Step 6—Calculate the design rain load, R

$$R = 5.2(d_s + d_h + d_p) = 5.2 \times (2.0 + 3.4 + 1.2) = 34.3 \text{ lb/ft}^2$$

$$R = 0.0098(d_s + d_h + d_p) = 0.0098 \times (51 + 86 + 31) = 1.65 \text{ kN/m}^2$$

3.6.9 Example 3.9—Calculation of Design Rain Load for a Roof with Drains—Overflow Dam System with an Overflow Diameter Not Given in ASCE/SEI Tables C8.2-1 and C8.2-2

Determine the design rain load, R, on the roof in Figure 3.26 given the design data in Example 3.8 where a 10-in. (254-mm) overflow dam diameter is specified instead of the 12.75-in. (329 mm) overflow diameter given in Example 3.8. All other data are the same.

SOLUTION

The design procedure in Table 3.2 of this book is used to determine R.

Step 1—Determine the rainfall intensity, i, for the given risk category

The design storm return period for Risk Category II is 100 years (IBC Table 1611.1). Using Reference 3.3, the 15-minute rainfall intensity, i, is equal to 9.42 in./h (239.37 mm/h).

Step 2—Determine the flow rate, Q

Flow rate, Q, is determined by Equations (3.8) and (3.9) of this book:

$$Q = 0.0104Ai \text{ (gal/min)}$$

$$Q = (0.278 \times 10^{-6})Ai \text{ (m}^3\text{/s)}$$

Tributary area, A, to each roof drain:

$$A = (130 \times 76.5)/2 = 4{,}973 \text{ ft}^2$$

$$Q = 0.0104Ai = 0.0104 \times 4{,}973 \times 9.42 = 487.2 \text{ gal/min}$$

In S.I.:

$$A = (39.62 \times 23.32)/2 = 462.0 \text{ m}^2$$

$$Q = (0.278 \times 10^{-6})Ai = (0.278 \times 10^{-6}) \times 462.0 \times 239.37 = 0.0307 \text{ m}^3/\text{s}$$

Step 3—Determine the static head, d_s
From the design data, $d_s = 2$ in. (51 mm)

Step 4—Determine the hydraulic head, d_{h2}, for the specified secondary drain
For weir flow and transition flow regime designations (cells not shaded) in ASCE/SEI Tables C8.2-1 and C8.2-2, calculate d_{h2} by ASCE/SEI Eq. (C8.2-3) (see also Figure 3.13, Note 1):

$$d_{h2} = (D_1/D_2)^{0.67} d_{h1}$$

From Step 4 in Example 3.8, $d_{h1} = 3.4$ in. (86 mm) for $Q = 487.2$ gal/min (0.0307 m^3/s) and $D_1 = 12.75$ in. (329 mm).

$$D_2 = 10.0 \text{ in. (254 mm)}$$

Therefore,

$$d_{h2} = (12.75/10.0)^{0.67} \times 3.4 = 4.0 \text{ in.}$$

$$d_{h2} = (329/254)^{0.67} \times 86 = 102 \text{ mm}$$

Step 5—Determine the ponding head, d_p
The iterative method was used to determine d_p, which is equal to 1.2 in. (31 mm) in the design data.

Step 6—Calculate the design rain load, R

$$R = 5.2(d_s + d_h + d_p) = 5.2 \times (2.0 + 4.0 + 1.2) = 37.4 \text{ lb/ft}^2$$

$$R = 0.0098(d_s + d_h + d_p) = 0.0098 \times (51 + 102 + 31) = 1.80 \text{ kN/m}^2$$

3.6.10 Example 3.10—Calculation of Design Rain Load for a Roof with Drains—Overflow Dam System and an Adjacent Wall Diverting Rainwater onto the Roof

Determine the design rain load, R, on the lower roof in Figure 3.27 given the following design data:

- Location: Sacramento, CA (Latitude = 38.58°, Longitude = −121.49°)
- Risk Category: II (IBC Table 1604.5)
- Secondary drainage system for structural load (SDSL): roof drains consisting of the following: (1) overflow dam diameter = 8 in. (203 mm), (2) drain outlet size = 4 in. (102 mm), and (3) drain bowl depth = 2 in. (51 mm; see Figure 3.13)
- Static head, d_s: overflow dam height = 2 in. (51 mm)
- Ponding head, d_p: $d_p = 0.5$ in. (13 mm) from an iterative analysis
- Roof slope: 1/4 in./foot (1.19 degrees)

Figure 3.27 Building in Example 3.10

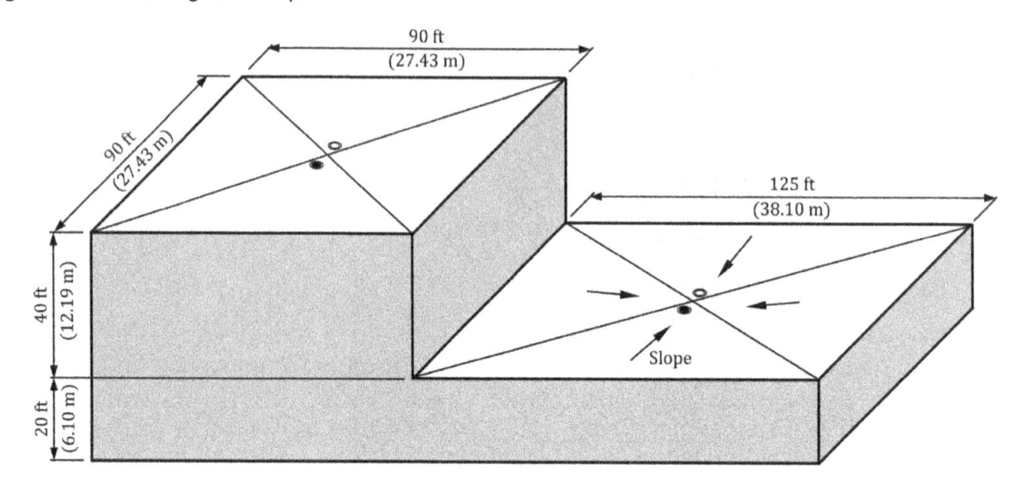

Solution

The design procedure in Table 3.2 of this book is used to determine *R*.

Step 1—Determine the rainfall intensity, i, for the given risk category
The design storm return period for Risk Category II is 100 years (IBC Table 1611.1). Using Reference 3.3, the 15-minute rainfall intensity, i, is equal to 2.46 in./h (62.59 mm/h).

Step 2—Determine the flow rate, Q
Flow rate, Q, is determined by Equations (3.8) and (3.9) of this book:

$$Q = 0.0104Ai \text{ (gal/min)}$$

$$Q = (0.278 \times 10^{-6})Ai \text{ (m}^3\text{/s)}$$

The adjacent wall diverts rainwater onto the lower roof, so A is equal to the area of the lower roof plus one-half of the wall area:

$$A = (125 \times 90) + (0.5 \times 40 \times 90) = 13{,}050 \text{ ft}^2$$

$$Q = 0.0104Ai = 0.0104 \times 13{,}050 \times 2.46 = 333.9 \text{ gal/min}$$

In S.I.:

$$A = (38.10 \times 27.43) + (0.5 \times 12.19 \times 27.43) = 1{,}212 \text{ m}^2$$

$$Q = (0.278 \times 10^{-6})Ai = (0.278 \times 10^{-6}) \times 1{,}212 \times 62.59 = 0.0211 \text{ m}^3\text{/s}$$

Step 3—Determine the static head, d_s
From the design data, $d_s = 2$ in. (51 mm)

Step 4—Determine the hydraulic head, d_h
The hydraulic head is determined by linear interpolation in ASCE/SEI Tables C8.2-1 and C8.2-2 for the given overflow dam diameter, drain outlet size, drain bowl depth, and Q:

$$d_h = 3.0 + \frac{(3.5 - 3.0) \times (333.9 - 300)}{350 - 300} = 3.3 \text{ in. at } Q = 333.9 \text{ gal/min}$$

$$d_h = 76 + \frac{(89 - 76) \times (0.0211 - 0.0189)}{0.0221 - 0.0189} = 85 \text{ mm at } Q = 0.0211 \text{ m}^3\text{/s}$$

Step 5—Determine the ponding head, d_p
The iterative method was used to determine d_p, which is equal to 0.5 in. (13 mm) in the design data.

Step 6—Calculate the design rain load, R

$$R = 5.2(d_s + d_h + d_p) = 5.2 \times (2.0 + 3.3 + 0.5) = 30.2 \text{ lb/ft}^2$$

$$R = 0.0098(d_s + d_h + d_p) = 0.0098 \times (51 + 85 + 13) = 1.46 \text{ kN/m}^2$$

3.6.11 Example 3.11—Calculation of Design Rain Load for a Roof with Drains—Overflow Dam System with Drain Bowl Depth Not Given in ASCE/SEI Tables C8.2-1 and C8.2-2 and an Adjacent Wall Diverting Rainwater onto the Roof

Determine the design rain load, R, on the roof in Figure 3.27 given the design data in Example 3.10 where a 1.5-in. (38-mm) drain bowl depth is specified instead of the 2.0-in. (51-mm) drain bowl depth given in Example 3.10. All other data are the same.

SOLUTION

The design procedure in Table 3.2 of this book is used to determine R.

Step 1—Determine the rainfall intensity, i, for the given risk category
The design storm return period for Risk Category II is 100 years (IBC Table 1611.1). Using Reference 3.3, the 15-minute rainfall intensity, i, is equal to 2.46 in./h (62.59 mm/h).

Step 2—Determine the flow rate, Q
Flow rate, Q, is determined by Equations (3.8) and (3.9) of this book:

$$Q = 0.0104Ai \text{ (gal/min)}$$

$$Q = (0.278 \times 10^{-6})Ai \text{ (m}^3\text{/s)}$$

The adjacent wall diverts rainwater onto the lower roof, so A is equal to the area of the lower roof plus one-half of the wall area:

$$A = (125 \times 90) + (0.5 \times 40 \times 90) = 13{,}050 \text{ ft}^2$$

$$Q = 0.0104Ai = 0.0104 \times 13{,}050 \times 2.46 = 333.9 \text{ gal/min}$$

In S.I.:

$$A = (38.10 \times 27.43) + (0.5 \times 12.19 \times 27.43) = 1{,}212 \text{ m}^2$$

$$Q = (0.278 \times 10^{-6})Ai = (0.278 \times 10^{-6}) \times 1{,}212 \times 62.59 = 0.0211 \text{ m}^3\text{/s}$$

Step 3—Determine the static head, d_s
From the design data, $d_s = 2$ in. (51 mm)

Step 4—Adjust the hydraulic head, d_h, for the specified drain bowl depth
The specified drain bowl depth is less than the depth of the tested drain bowl and the flow regime is orifice flow (shaded portions of ASCE/SEI Tables C8.2-1 and C8.2-2). Therefore, the adjusted hydraulic head is equal to d_h from Example 3.10 plus the difference in drain bowl depths (see Figure 3.13, Note 2]:

$$d_h = 3.3 + (2.0 - 1.5) = 3.8 \text{ in.}$$

$$d_h = 85 + (51 - 38) = 98 \text{ mm}$$

Step 5—Determine the ponding head, d_p
The iterative method was used to determine d_p, which is equal to 0.5 in. (13 mm) in the design data.

Step 6—Calculate the design rain load, R

$$R = 5.2(d_s + d_h + d_p) = 5.2 \times (2.0 + 3.8 + 0.5) = 32.8 \text{ lb/ft}^2$$

$$R = 0.0098(d_s + d_h + d_p) = 0.0098 \times (51 + 98 + 13) = 1.59 \text{ kN/m}^2$$

3.6.12 Example 3.12—Calculation of Design Rain Load for a Roof with Channel Scuppers

Determine the design rain load, R, on the roof in Figure 3.28 given the following design data:

- Location: New Orleans, LA (Latitude = 29.94°, Longitude = −90.09°)
- Risk Category: II (IBC Table 1604.5)
- Secondary drainage system for structural load (SDSL): channels scuppers with a width, b, equal to 24 in. (610 mm) and a height, h, equal to 12 in. (305 mm; see Figure 3.14)
- Static head, d_s: inlet of the scuppers set 2 in. (51 mm) above the roof surface
- Ponding head, d_p: $d_p = 1.5$ in. (38 mm) from an iterative analysis
- Roof slope: 1/2 in./foot (2.39 degrees)

Figure 3.28
Roof Plan, Example 3.12

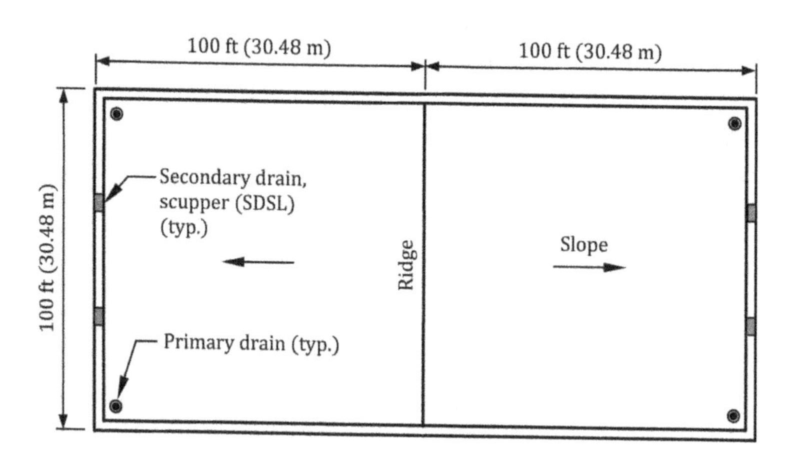

SOLUTION

The design procedure in Table 3.2 of this book is used to determine R.

Step 1—Determine the rainfall intensity, i, for the given risk category

The design storm return period for Risk Category II is 100 years (IBC Table 1611.1). Using Reference 3.3, the 15-minute rainfall intensity, i, is equal to 9.34 in./h (237.13 mm/h).

Step 2—Determine the flow rate, Q

Flow rate, Q, is determined by Equations (3.8) and (3.9) of this book:

$$Q = 0.0104Ai \text{ (gal/min)}$$

$$Q = (0.278 \times 10^{-6})Ai \text{ (m}^3\text{/s)}$$

Tributary area, A, to each scupper:

$$A = (200 \times 100)/4 = 5{,}000 \text{ ft}^2$$

$$Q = 0.0104Ai = 0.0104 \times 5{,}000 \times 9.34 = 485.7 \text{ gal/min}$$

In S.I.:

$$A = (60.96 \times 30.48)/4 = 464.5 \text{ m}^2$$

$$Q = (0.278 \times 10^{-6})Ai = (0.278 \times 10^{-6}) \times 464.5 \times 237.13 = 0.0306 \text{ m}^3/\text{s}$$

Step 3—Determine the static head, d_s

From the design data, $d_s = 2$ in. (51 mm)

Step 4—Determine the hydraulic head, d_h

For channel scuppers where $d_h < h$ (see Figure 3.14):

$$d_h = \left(\frac{Q}{2.9b}\right)^{2/3} = \left(\frac{485.7}{2.9 \times 24.0}\right)^{2/3} = 3.7 \text{ in.} < h = 12.0 \text{ in.}$$

$$d_h = \left[\frac{(17.65 \times 10^6)Q}{b}\right]^{2/3} = \left[\frac{(17.65 \times 10^6) \times 0.0306}{610}\right]^{2/3} = 92 \text{ mm} < h = 305 \text{ mm}$$

Alternatively, determine d_h by linear interpolation in ASCE/SEI Tables C8.2-3 and C8.2-4 for the given scupper width and Q:

$$d_h = 3.0 + \frac{(4-3) \times (485.7 - 360)}{560 - 360} = 3.6 \text{ in. at } Q = 485.7 \text{ gal/min}$$

$$d_h = 76 + \frac{(102 - 76) \times (0.0306 - 0.0227)}{0.0353 - 0.0227} = 92 \text{ mm at } Q = 0.0306 \text{ m}^3/\text{s}$$

Step 5—Determine the ponding head, d_p

The iterative method was used to determine d_p, which is equal to 1.5 in. (38 mm) in the design data.

Step 6—Calculate the design rain load, R

$$R = 5.2(d_s + d_h + d_p) = 5.2 \times (2.0 + 3.7 + 1.5) = 37.4 \text{ lb/ft}^2$$

$$R = 0.0098(d_s + d_h + d_p) = 0.0098 \times (51 + 92 + 38) = 1.77 \text{ kN/m}^2$$

3.6.13 Example 3.13—Calculation of Design Rain Load for a Roof with Closed Scuppers

Determine the design rain load, R, on the roof in Figure 3.28 given the design data in Example 3.12 using two 24-in. (610-mm) wide by 4-in. (102-mm) high closed scuppers instead of the four channel scuppers. All other data are the same.

SOLUTION

The design procedure in Table 3.2 of this book is used to determine R.

Step 1—Determine the rainfall intensity, i, for the given risk category

The design storm return period for Risk Category II is 100 years (IBC Table 1611.1). Using Reference 3.3, the 15-minute rainfall intensity, i, is equal to 9.34 in./h (237.13 mm/h).

Step 2—Determine the flow rate, Q

Flow rate, Q, is determined by Equations (3.8) and (3.9) of this book:

$$Q = 0.0104Ai \text{ (gal/min)}$$

$$Q = (0.278 \times 10^{-6})Ai \text{ (m}^3/\text{s)}$$

Tributary area, A, to each scupper:

$$A = (200 \times 100)/2 = 10{,}000 \text{ ft}^2$$

$$Q = 0.0104Ai = 0.0104 \times 10{,}000 \times 9.34 = 971.4 \text{ gal/min}$$

In S.I.:

$$A = (60.96 \times 30.48)/2 = 929.0 \text{ m}^2$$

$$Q = (0.278 \times 10^{-6})Ai = (0.278 \times 10^{-6}) \times 929.0 \times 237.13 = 0.0612 \text{ m}^3/\text{s}$$

Step 3—Determine the static head, d_s

From the design data, $d_s = 2$ in. (51 mm)

Step 4—Determine the hydraulic head, d_h

For closed scuppers where $d_h > h$ (see Figure 3.14):

$$d_h = \left(\frac{Q}{4.3bh}\right)^2 + 0.5h = \left(\frac{971.4}{4.3 \times 24.0 \times 4.0}\right)^2 + (0.5 \times 4.0) = 7.5 \text{ in.} > h = 4.0 \text{ in.}$$

$$d_h = \left[\frac{(12.00 \times 10^6)Q}{bh}\right]^2 + 0.5h = \left[\frac{(12.00 \times 10^6) \times 0.0612}{610 \times 102}\right]^2 + (0.5 \times 102) = 190 \text{ mm} > h = 102 \text{ mm}$$

Alternatively, determine d_h by linear interpolation in ASCE/SEI Tables C8.2-3 and C8.2-4 for the given scupper size and Q:

$$d_h = 7.0 + \frac{(8-7) \times (971.4 - 924)}{1{,}012 - 924} = 7.5 \text{ in. at } Q = 971.4 \text{ gal/min}$$

$$d_h = 178 + \frac{(203 - 178) \times (0.0612 - 0.0583)}{0.0638 - 0.0583} = 191 \text{ mm at } Q = 0.0612 \text{ m}^3/\text{s}$$

Step 5—Determine the ponding head, d_p

The iterative method was used to determine d_p, which is equal to 1.5 in. (38 mm) in the design data.

Step 6—Calculate the design rain load, R

$$R = 5.2(d_s + d_h + d_p) = 5.2 \times (2.0 + 7.5 + 1.5) = 57.2 \text{ lb/ft}^2$$

$$R = 0.0098(d_s + d_h + d_p) = 0.0098 \times (51 + 191 + 38) = 2.74 \text{ kN/m}^2$$

3.6.14 Example 3.14—Calculation of Design Rain Load for a Roof with Circular Scuppers

Determine the design rain load, R, on the roof in Figure 3.28 given the design data in Example 3.12 using four 12-in. (305-mm) diameter circular scuppers instead of the four channel scuppers. All other data are the same.

SOLUTION

The design procedure in Table 3.2 of this book is used to determine R.

Step 1—Determine the rainfall intensity, i, for the given risk category

The design storm return period for Risk Category II is 100 years (IBC Table 1611.1). Using Reference 3.3, the 15-minute rainfall intensity, i, is equal to 9.34 in./h (237.13 mm/h).

Step 2—Determine the flow rate, Q

Flow rate, Q, is determined by Equations (3.8) and (3.9) of this book:

$$Q = 0.0104Ai \text{ (gal/min)}$$

$$Q = (0.278 \times 10^{-6})Ai \text{ (m}^3\text{/s)}$$

Tributary area, A, to each scupper:

$$A = (200 \times 100)/4 = 5{,}000 \text{ ft}^2$$

$$Q = 0.0104Ai = 0.0104 \times 5{,}000 \times 9.34 = 485.7 \text{ gal/min}$$

In S.I.:

$$A = (60.96 \times 30.48)/4 = 464.5 \text{ m}^2$$

$$Q = (0.278 \times 10^{-6})Ai = (0.278 \times 10^{-6}) \times 464.5 \times 237.13 = 0.0306 \text{ m}^3/\text{s}$$

Step 3—Determine the static head, d_s
From the design data, $d_s = 2$ in. (51 mm)

Step 4—Determine the hydraulic head, d_h
The hydraulic head is determined by linear interpolation in ASCE/SEI Tables C8.2-5 and C8.2-6 for the given circular scupper diameter and Q:

$$d_h = 7.0 + \frac{(8-7) \times (485.7 - 410)}{510 - 410} = 7.8 \text{ in. at } Q = 485.7 \text{ gal/min}$$

$$d_h = 178 + \frac{(203 - 178) \times (0.0306 - 0.0259)}{0.0322 - 0.0259} = 197 \text{ mm at } Q = 0.0309 \text{ m}^3/\text{s}$$

Step 5—Determine the ponding head, d_p
The iterative method was used to determine d_p, which is equal to 1.5 in. (38 mm) in the design data.

Step 6—Calculate the design rain load, R

$$R = 5.2(d_s + d_h + d_p) = 5.2 \times (2.0 + 7.8 + 1.5) = 58.8 \text{ lb/ft}^2$$

$$R = 0.0098(d_s + d_h + d_p) = 0.0098 \times (51 + 197 + 38) = 2.80 \text{ kN/m}^2$$

3.6.15 Example 3.15—Bays with Low Slope

For the roof in Figure 3.24, check if the roof is susceptible to ponding loads assuming there are no curbs at the two edges indicated in the figure.

SOLUTION

Without curbs, rainwater will flow off the free-draining edges of the roof and all bays will not be subject to the design rain loads in ASCE/SEI 8.2. However, if the deflection of the first secondary members (joists) upslope from the free-draining edges exceeds the initial difference in elevation, a pond can form.

Determine the minimum rise, β, to achieve no local minimum by ASCE/SEI Eq. (C8.3-2), which is applicable to secondary members parallel and adjacent to the free-draining edges:

$$\beta = \frac{1}{20}\left(\pi + \frac{L_s}{S}\right) = \frac{1}{20}\left(\pi + \frac{35}{8.33}\right) = 0.4 \text{ in.}$$

The provided β is equal to 0.5 in., which is greater than 0.4 in., so a pond is unlikely to form and the bay will not be subjected to rain loads (ASCE/SEI 8.3, item 2). This is confirmed by analysis where positive slopes to the free-draining edges are maintained for the deflected roof where the deflection is calculated using only the dead loads.

3.7 References

3.1. American Association of State Highway and Transportation Officials. 2020. *AASHTO LRFD Bridge Design Specifications*, 9th ed. Washington, DC.

3.2. International Conference of Building Officials. 1997. *Uniform Building Code*. Whittier, CA.

3.3. American Society of Civil Engineers (ASCE). 2022. ASCE 7 Hazard Tool. https://asce7hazardtool.online/.

3.4. National Oceanic and Atmospheric Administration (NOAA). Precipitation Frequency Data Server (PFDS). https://hdsc.nws.noaa.gov/hdsc/pfds/.

3.5. Factory Mutual Insurance Company. 2016. *Roof Loads for New Construction*, FM Global Property Loss Prevention Data Sheet 1-54. Johnston, RI.

3.6. van Herwijnen, F., Snijder, H.H., and Fijneman, H.J. 2006. "Structural Design for Ponding of Rainwater on Roof Structures." *HERON*, 51(2/3): 115–150.

3.7. Computers and Structures, Inc. (CSI). 2021. ETABS – Building Analysis and Design, Version 20.1.0, Walnut Creek, CA.

3.8 Problems

3.1. Given the 10-story office building in Figure 3.22, determine the total live loads in column A1 using the design data in Example 3.1. Use the basic uniform live load reduction provisions of IBC 1607.13.1.

3.2. Given the 10-story office building in Figure 3.22, determine the total live loads in column D3 using the design data in Example 3.1. Use the basic uniform live load reduction provisions of IBC 1607.13.1.

3.3. Given the 10-story office building in Figure 3.22, determine the total live loads in column B3 using the design data in Example 3.1 except the fifth floor is designated as a file storage floor instead of the ninth floor. Use the basic uniform live load reduction provisions of IBC 1607.13.1.

3.4. Given the 10-story office building in Figure 3.22, determine the total live loads in column B3 using the design data in Example 3.1 except the fifth floor is designated as a file storage floor instead of the ninth floor. Use the alternate uniform live load reduction provisions of IBC 1607.13.2 with a dead-to-live load ratio equal to 2.

3.5. Given the 10-story office building in Figure 3.22, determine the total live loads in column A1 using the design data in Example 3.1. Use the alternate uniform live load reduction provisions of IBC 1607.13.2 with a dead-to-live load ratio equal to 2.

3.6. A transfer beam supports an interior column that supports a flat roof and four floors of office space. The tributary area of the column at each level is 600 square feet (55.74 m^2). The beam also supports a tributary floor area of 300 square feet (27.87 m^2) of office space at the first elevated level. Using the basic uniform live load reduction provisions of IBC 1607.13.1, determine the total live loads on the beam.

3.7. Given the information in Problem 3.6, determine the total live loads on the transfer beam using the alternative uniform live load reduction provisions of IBC 1607.13.2 with a dead-to-live load ratio of 0.75.

3.8. Given the information in Problem 3.6, determine the total live loads using the basic uniform live load reduction provisions of IBC 1607.13.1 assuming all the floors support patient rooms in a hospital.

3.9. Determine the reduced live load at each floor level of an edge column in an 8-story parking garage with a tributary area of 1,080 square feet (100.33 m^2) at each level using the basic uniform live load reduction provisions of IBC 1607.13.1.

3.10. Determine the design rain load, *R*, on a flat roof given the following design data:
- Tributary area of primary roof drain = 3,000 square feet (278.70 m^2)
- Rainfall rate $i = 3.75$ in./hour (95.25 mm/h)
- Channel scupper, 6 in. (152 mm) wide by 4 in. (102 mm) deep
- Vertical distance from primary roof drain to inlet of scupper (static head distance, d_s) = 3 in. (76 mm)
- Ponding head, d_p: $d_p = 2.0$ in. (51 mm) from an iterative analysis

3.11. Given the information in Problem 3.10, determine the design rain load, *R*, assuming 4-in. (102-mm) diameter drains with an 8-in. (203-mm) diameter overflow dam and a 2-in. (51-mm) drain bowl depth are used as the secondary drainage system for structural load (SDSL).

3.12. Determine the design rain load, *R*, on the roof in Figure 3.29 given the following design data:
- Location: Boulder, CO (Latitude = 40.02°, Longitude = −105.26°)
- Risk Category: II (IBC Table 1604.5)
- Secondary drainage system for structural load (SDSL): channels scuppers with a width, *b*, equal to 24 in. (610 mm) and a height, *h*, equal to 12 in. (305 mm; see Figure 3.14)
- Static head, d_s: inlet of the scuppers set 2 in. (51 mm) above the roof surface
- Ponding head, d_p: $d_p = 3.0$ in. (76 mm) from an iterative analysis

Figure 3.29 Roof in Problem 3.12

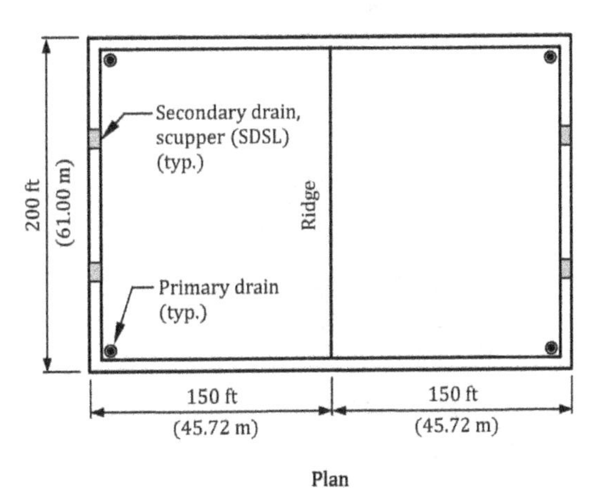

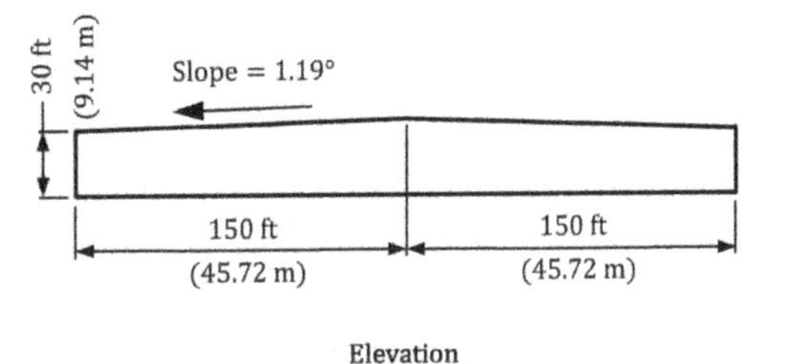

3.13. Given the information in Problem 3.12, determine the design rain load, R, assuming closed scuppers with a width equal to 24 in. (610 mm) and a height, h, equal to 4 in. (102 mm).

3.14. Determine the design rain load, R, on the lower roof in Figure 3.30 given the following design data:

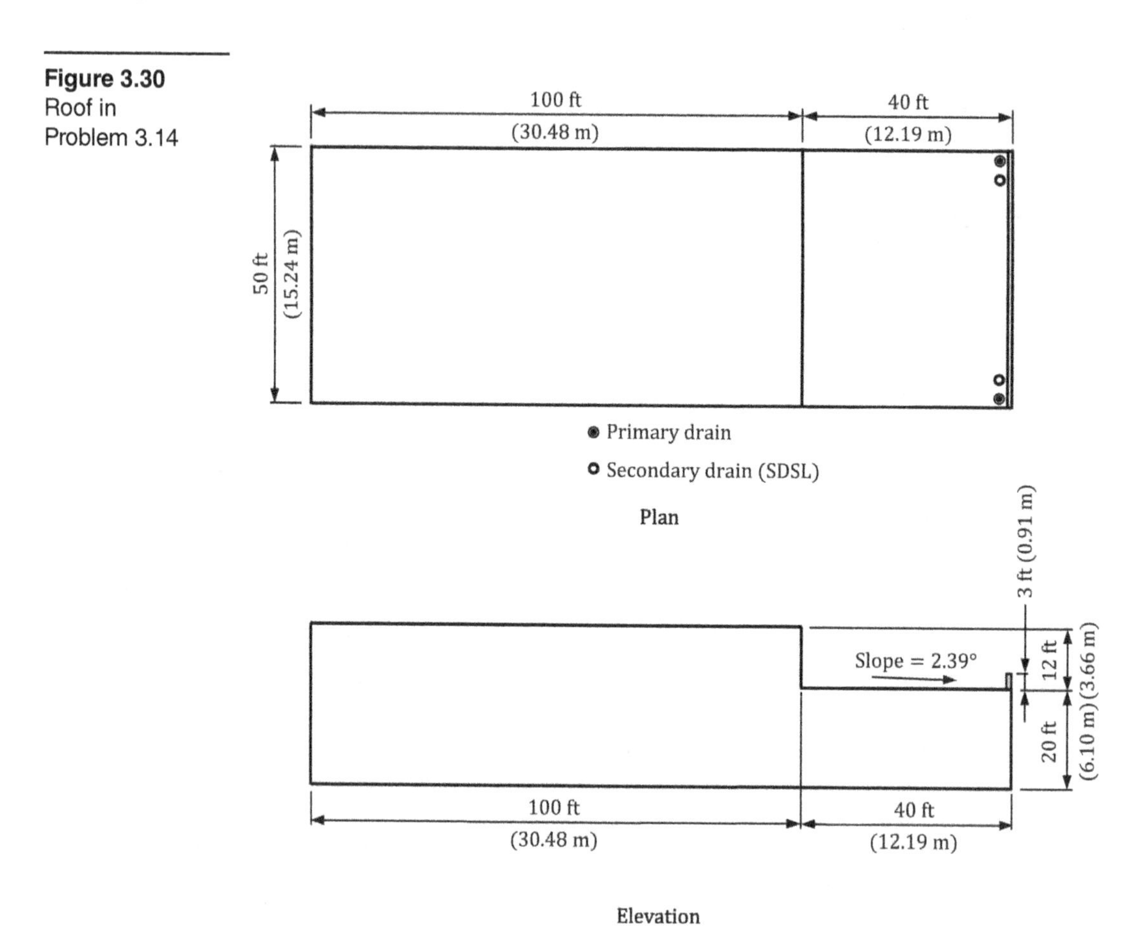

Figure 3.30 Roof in Problem 3.14

- Location: Amherst, NH (Latitude = 42.86°, Longitude = −71.63°)
- Risk Category: III (IBC Table 1604.5)
- Secondary drainage system for structural load (SDSL): (1) overflow dam diameter = 12.75 in. (324 mm), (2) drain outlet size = 8 in. (203 mm), and (3) drain bowl depth = 3.25 in. (83 mm; see Figure 3.13)
- Static head, d_s: inlet of the scuppers set 2 in. (51 mm) above the roof surface
- Ponding head, d_p: $d_p = 1.5$ in. (38 mm) from an iterative analysis

CHAPTER 4

Snow and Ice Loads

4.1 Introduction

Structural members of roofs, balconies, canopies, and similar structures exposed to the environment must be designed for the effects of snow loads in geographic areas where snowfall can occur.

Loads on buildings and other structures due to snow are determined based on the anticipated ground snow load, the occupancy of the building, the exposure, the thermal resistance of the roof structure, and the shape and slope of the roof. Partial loading, unbalanced snow loads due to roof configuration, drift loads on lower or adjacent roofs and on projections such as parapets and mechanical equipment, sliding snow loads, and rain-on-snow loads must also be considered when designing for the effects from snow.

In certain parts of the United States, atmospheric ice loads must be considered in the design of structures and structural members exposed to the elements. Requirements on how to determine ice loads due to freezing rain on a variety of structural shapes, objects, and configurations as a function of the design ice thickness are presented at the end of this chapter.

4.2 Snow Loads

4.2.1 Overview

Design snow loads, S, must be determined in accordance with Chapter 7 of ASCE/SEI 7, but the design roof load must not be taken less than that determined by IBC 1607 (IBC 1608.1). The design procedure in Table 4.1 can be used to determine S based on these requirements. Additional information needed to calculate S is given in the sections of this book indicated in the table.

In lieu of the requirements in ASCE/SEI 7.3 through 7.13, design snow loads are permitted to be determined by thermal performance studies and scale-model studies in wind tunnels or water flumes conducted in accordance with Reference 4.2. Additional information on these alternate procedures is given in ASCE/SEI 7.14.

4.2.2 Ground Snow Loads, p_g

According to IBC 1608.2, ground snow loads, p_g, to be used in the determination of design snow loads must be determined using the ASCE Design Ground Snow Load Geodatabase (see ASCE/SEI 7.2 and Reference 4.1) or IBC Figure 1608.2(1) through 1608.2(4) for the contiguous United States and IBC Table 1608.2 for Alaska. Geocoded values of the risk-targeted design ground snow load can be obtained by entering an address or the latitude and longitude coordinates of the site along with the risk category of the structure (IBC Table 1604.5). The values in the Geodatabase have been determined based on a reliability analysis consistent with the target values in ASCE/SEI Table 1.3-1 corresponding to "failure that is not sudden and does not lead to widespread progression of damage." The adoption of reliability-based design ground snow loads is a significant change from previous requirements for design ground snow loads, which were based on a 50-year mean recurrence interval. Almost 30 years of additional snow load data have also been included in the Geodatabase.

Graphical representations of the data in the Geodatabase are given in ASCE/SEI Figures 7.2-1A through 7.2-1D and IBC Figures 1608.2(1) through 1608.2(4) for the conterminous United States

Table 4.1 Design Procedure to Determine Design Snow Load, S

Step	Procedure	Section Number
1	Determine the ground snow load, p_g, for the site (Reference 4.1).	4.2.2
2	Determine a flat roof snow load, p_f, considering roof exposure and roof thermal condition.	4.2.3
3	Determine the minimum snow load for low-slope roofs, p_m.	4.2.3
4	Determine the sloped roof snow load, p_s.	4.2.4
5	Consider partial snow loading, where applicable.	4.2.5
6	Consider unbalanced roof snow loads, where applicable.	4.2.6
7	Consider snow drifts on lower roofs, where applicable.	4.2.7
8	Consider snow drifts adjacent to roof projections and parapets, where applicable.	4.2.8
9	Consider sliding snow, where applicable.	4.2.9
10	Consider rain-on-snow surcharge load, where applicable.	4.2.10
11	Consider ponding loads, where applicable.	4.2.11
12	Consider snow loads on existing roofs, where applicable.	4.2.12
13	Consider snow loads on open-frame equipment structures, where applicable.	4.2.13

based on risk category. Values of p_g are given in ASCE/SEI Table 7.2-1 and IBC Table 1608.2 for various locations in Alaska. Snow loads are permitted to be taken as zero for Hawaii except in mountainous regions as approved by the building official.

A small number of locations in the Geodatabase indicate a case study must be completed to determine the ground snow load. These apply only to locations higher than any locally available snow measurement locations. A ground snow load value is obtained from the Geodatabase in such cases with a warning that the provided ground snow load lies outside the range of elevations of surrounding measurement locations. It is prudent to check with the local building official to confirm the ground snow load.

Tabulated ground snow loads for Alaska must be used for the specific locations noted in ASCE/SEI Table 7.2-1 and IBC Table 1608.2 based on risk category; these values do not necessarily represent the values relevant at nearby locations. The wide variability of snow loads in Alaska, which is evident from the tabulated values, precludes statewide mapping of ground snow loads. Local records, experience, and the local building official should also be utilized in such cases. For example, Reference 4.3 can be used to acquire ground snow load values for various locations in Alaska.

The snow load provisions in ASCE/SEI Chapter 7 need not be considered for roofs with no potential for drift accumulation or unbalanced snow loading where p_g is less than the factored roof live load. In all other cases, the requirements in ASCE/SEI Chapter 7 need not be considered in either of the following cases:

- $p_g \leq 10$ lb/ft^2 (0.48 kN/m^2) and the length of the roof upwind from any potential drifting location, ℓ_u, is less than or equal to 100 ft (30.48 m).
- $p_g \leq 5$ lb/ft^2 (0.24 kN/m^2) and the length of the roof upwind from any potential drifting location, ℓ_u, is less than or equal to 300 ft (91.44 m).

For decks, balconies, and other near-ground level surfaces or roofs of subterranean spaces whose height above the ground is less than the anticipated depth of the ground snow, h_g (which is equal to p_g/γ where γ is the snow density), the structural members must be designed for a balanced snow load equal to p_g. This requirement is applicable where elements can be buried within the ground snow layer. No reductions due to roof thermal effects or snow blowing off a roof are appropriate in these cases.

In cases where allowable stress design ground snow loads, $p_{g(asd)}$, are required, the ground snow loads, p_g, in IBC Figures 1608.2(1) through 1608.2(4) can be converted to $p_{g(asd)}$ using the IBC Equation 16-17 in IBC 1608.2.1: $p_{g(asd)} = 0.7p_g$.

4.2.3 Flat Roof Snow Loads, p_f

Overview

Once the ground snow load, p_g, has been established, the flat roof snow load, p_f, in lb/ft^2 (kN/m^2) is determined by ASCE/SEI Equation (7.3-1):

$$p_f = 0.7C_eC_tp_g \tag{4.1}$$

Research has shown that the snow load on a roof is usually less than that on the ground in cases where drifting is not prevalent. The 0.7 factor in Equation (4.1) is a conservative ground-to-roof conversion factor used to account for this phenomenon.

The other factors in this equation account for the thermal, aerodynamic, geometric, and occupancy characteristics of a structure at its particular site and are discussed in the following sections. Minimum snow loads, p_m, for low-slope roofs are also covered.

Exposure Factor, C_e

The exposure factor, C_e, accounts for the wind at the site and is related to the type of terrain and the exposure of the roof. Values of C_e are given in ASCE/SEI Table 7.3-1 as a function of the surface roughness category and the type of roof exposure. In general, unabated wind is more likely to blow snow off of a roof than wind impeded in some way; as such, snow loads are likely to be less.

Surface roughness categories B, C, and D are defined in ASCE/SEI 26.7 for wind design (see Table 4.2). A surface roughness category for a specific site should be selected that represents the anticipated conditions during the life of the structure. Buildings located in surface roughness category D are more likely to have smaller snow loads on the roof than those located in surface roughness category B because the roofs in the former are sheltered much less by the surrounding terrain than the latter. This is evident in ASCE/SEI Table 7.3-1 because the value of C_e decreases going from surface roughness category B to D for a given roof exposure.

Table 4.2 Surface Roughness Categories

Surface Roughness Category	Description
B	Urban and suburban areas, wooded areas, or other terrain with numerous, closely spaced obstructions having the size of single-family dwellings or larger
C	Open terrain with scattered obstructions having heights generally less than 30 ft (9.14 m); this category includes flat, open country, and grasslands
D	Flat, unobstructed areas, and water surfaces; this category includes smooth mud flats, salt flats, and unbroken ice

The surface roughness category identified in ASCE/SEI Table 7.3-1 as "Above the tree line in windswept mountainous areas" is to be used at appropriate locations other than high mountain valleys that receive little wind.

Roof exposures are defined as fully exposed, partially exposed, and sheltered (see footnote a in ASCE/SEI Table 7.3-1). A fully exposed condition exists where a roof is exposed on all sides with no shelter provided by adjoining terrain, higher structures, or trees. Roofs with large mechanical equipment, parapets that extend above the height of the balanced snow load or other similar obstructions are not considered to be fully exposed because such conditions can provide some shelter to the wind.

Obstructions are defined as providing shelter when located within a distance of $10h_o$ from the roof, where h_o is the height of the obstruction above the roof (see footnote b in ASCE/SEI Table 7.3-1). The conifers depicted in Figure 4.1 provide shelter to the building roof where the distance, x, between the centerline of the building and the tree line is less than or equal to $10h_o$. In cases where deciduous trees that are leafless in winter surround the site, the fully exposed category is applicable.

Figure 4.1 Sheltered Roof Exposures for Snow Loads

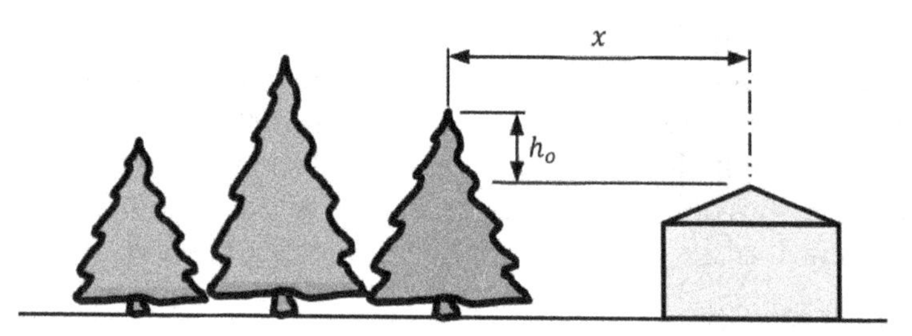

Partially exposed roof exposures are to be used where fully exposed and sheltered conditions do not apply. This is generally the most common roof exposure.

Similar to surface roughness categories, a roof exposure condition must represent the conditions expected during the life of a structure.

Although a single surface roughness category is specified at a particular site, buildings with multiple roofs may have different exposure conditions. For example, the upper roof of a building may have a fully exposed condition while the lower roof, which is sheltered by the higher portion of the building, may be partially exposed.

It is evident from ASCE/SEI Table 7.3-1 that for a given surface roughness category, the value of C_e and, thus, the corresponding snow load, increases going from a fully exposed roof exposure to a sheltered roof exposure.

Thermal Factor, C_t

The thermal factor, C_t, accounts for the amount of heat loss through the roof. In general, more snow will be present on cold roofs than warm roofs. Values of C_t are given in ASCE/SEI Table 7.3-2 as a function of the thermal condition. Like terrain categories and exposure conditions, the selected thermal condition must represent the anticipated conditions during winters for the life of a structure.

The term "structure kept just above freezing" in ASCE/SEI Table 7.3-2 represents a structure where the internal ambient temperature is kept at approximately 40 to 50°F (4 to 10°C) to keep contents from freezing but is usually not intended for human occupancy. Also, the term "cold, ventilated roof" refers to a roof where an airspace is provided between the building envelope and the roof surface through which air is intended to flow to relieve heat buildup under the roof. The term "cold" represents an attic space located outside the insulation envelope that will not be heated or cooled.

ASCE/SEI Table 7.3-2 refers to ASCE/SEI Table 7.3-3 for values of C_t for all structures not included in ASCE/SEI Table 7.3-2, namely, heated structures with unventilated roofs. Values of C_t are given based on the ground snow load, p_g, and the measure of resistance to heat flow through the roof component or assembly per unit area, R_{roof} (or, equivalently, the measure of heat transmission through the roof component or assembly, U_{roof}; note that $U_{\text{roof}} = 1/R_{\text{roof}}$). Linear interpolation is permitted to be used to determine C_t for values of p_g and R_{roof} that fall between those shown in the table (see footnote a in ASCE/SEI Table 7.3-3). Also, for values of $R_{\text{roof}} > 50$ h ft^2°F/Btu (8.80 m^{2}K/W), $C_t = 1.2$ (see footnote b in ASCE/SEI Table 7.3-3).

Minimum Snow Loads for Low-Slope Roofs, p_m

Low-slope roofs are defined in ASCE/SEI 7.3.3 for the following types of roof configurations (see Figure 4.2):

- Monoslope roofs with slopes less than 15 degrees
- Hip and gable roofs with slopes less than 15 degrees
- Curved roofs where the vertical angle from the eaves to the crown is less than 10 degrees

Minimum roof snow loads, p_m, to be applied to low-slope roofs depend on the ground snow load, p_g, and the minimum snow load upper limit, $p_{m,\max}$, which is given in ASCE/SEI Table 7.3-4 as a function of the risk category:

- Case 1: $p_g \leq p_{m,\max} : p_m = p_g$
- Case 2: $p_g > p_{m,\max} : p_m = p_{m,\max}$

The purpose of the minimum snow loads is to account for important situations that may develop on roofs that are relatively flat. For example, in regions where $p_g < p_{m,\max}$, a single storm event can result in loading where the ground-to-roof conversion factor of 0.7 and factors C_e and C_t are not applicable, resulting in a roof load equal to at most p_g.

Figure 4.2 Low-Slope Roofs

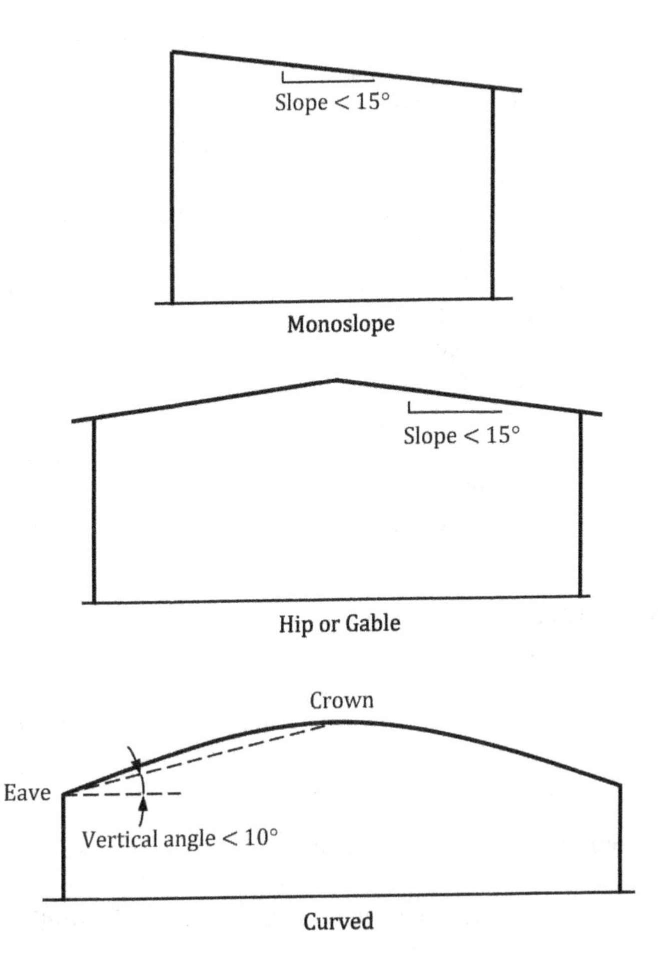

The roof slope limits for consideration of minimum loads are related to the roof slope limits for consideration of unbalanced loads. Unbalanced roof snow loads need to be considered for roofs based on the slopes in ASCE/SEI 7.6.1 (see Section 4.2.6 of this book). In general, if a roof slope is steep enough so that unbalanced loads must be considered, it is unlikely that minimum snow loads would control.

The minimum snow load is a uniform load case to be considered separately from any of the other applicable load cases. It need not be used in determining or in combination with drifting, sliding, unbalanced, or partial snow loads.

Flowchart 4.1 in Section 4.4 of this book can be used to determine the flat roof snow load, p_f.

4.2.4 Sloped Roof Snow Loads, p_s

Overview

Design snow loads for all structures are based on the sloped roof snow load, p_s, which is determined by ASCE/SEI Equation (7.4-1):

$$p_s = C_s p_f \tag{4.2}$$

This snow load, which is also referred to as the balanced snow load, is assumed to act on the horizontal projection of a sloping roof surface. In general, snow loads decrease as the slope of a roof increases: snow is more likely to slide and be blown off a sloping roof compared to a flat roof. In the case of flat roofs, $p_s = p_f$.

The slope factor, C_s, depends on the slope and temperature of the roof, the presence or absence of obstructions, and the degree of slipperiness of the roof surface. Graphs for determining C_s are given in ASCE/SEI Figure 7.4-1 for warm and cold roof conditions, and equations for C_s are given in ASCE/SEI C7.4.

The thermal factor, C_t, is used to determine whether a roof is warm or cold (see Section 4.2.3 of this book). A roof is defined as warm where $C_t < 1.2$ and is defined as cold where $C_t \geq 1.2$. Warm roofs are more likely to shed snow than colder ones. Values of C_s in ASCE/SEI Figures 7.4-1c for cold roofs are greater than or equal to values of C_s for warm roofs for a given roof angle and surface condition.

The ability of a sloped roof to shed snow also depends on the presence of obstructions on the roof and the degree of slipperiness of the roof surface. An obstruction can be considered as anything impeding snow from sliding off a roof. Large vent pipes, snow guards, parapet walls and large rooftop equipment are a few common examples of obstructions that could prevent snow from sliding off a roof. Ice dams and icicles along eaves can also inhibit snow from sliding off warm roofs (see ASCE/SEI 7.4.4 and the discussion below). An obstruction may also occur at the lower portions of sloping roofs near the ground; snow loads can concentrate near the lower portion of the roof because the snow may not be able to completely slide off of the roof due to the proximity of the ground.

Examples of roof surfaces considered to be slippery and not slippery are given in Table 4.3 (see ASCE/SEI 7.4).

Curved Roofs

For portions of curved roofs with a slope exceeding 70 degrees, $C_s = 0$, which means $p_s = 0$ (ASCE/SEI 7.4.2). Balanced snow loads for curved roofs are determined from the loading diagrams in ASCE/SEI Figure 7.4-2 with C_s determined from the appropriate curve in ASCE/SEI Figure 7.4-1.

Additional information on the determination of balanced and unbalanced snow loads for curved roofs is given in Section 4.2.6 of this book.

Multiple Folded Plate, Sawtooth, and Barrel Vault Roofs

Multiple folded plate, sawtooth, and barrel vault roofs are to be designed using $C_s = 1.0$; thus, $p_s = p_f$ for these roof geometries (ASCE/SEI 7.4.3). Additional snow is collected in the valleys

Table 4.3 Slippery and Not Slippery Roof Surfaces

Slippery roof surfaces	Metal
	Slate
	Glass
	Bituminous membranes*
	Rubber membranes*
	Plastic membranes*
Not slippery roof surfaces	Asphalt shingles
	Wood shingles
	Shakes

*For a membrane roof system to be considered slippery, the membrane must be smooth. Membranes with an embedded aggregate or a mineral granule surface are not considered to be smooth.

of these roofs by wind drifting and snow sliding, so no reduction in snow load based on roof slope is permitted.

Determination of balanced and unbalanced snow loads for sawtooth roofs is given in Section 4.2.6 of this book.

Ice Dams and Icicles along Eaves

Relatively heavy loads due to ice accumulation can occur on the cold, overhanging portions of unventilated roofs that drain water over their eaves (ASCE/SEI 7.4.4). It is assumed ice dams can form where the thermal factor, C_t, is less than or equal to 1.1 in accordance with ASCE/SEI 7.3.2. To account for this phenomenon, a uniformly distributed load equal to at least $2p_{f(\text{heated})}$ must be applied to the overhang where $p_{f(\text{heated})}$ is the flat roof snow load for the heated portion of the roof upslope of the exterior wall (see Figure 4.3). This uniform load is applied over the entire roof overhang where the horizontal extent of the overhang is less than or equal to 5 ft (1.52 m). No other loads except dead loads need to be considered when $2p_{f(\text{heated})}$ is applied to the overhang.

The applicable load case where the horizontal extent of the overhang is greater than 5 ft (1.52 m) is depicted in Figure 4.4. The uniform load $2p_{f(\text{heated})}$ need only be applied over a distance of 5 ft (1.52 m) from the eave; the remainder of the overhang is subjected to the uniform flat roof snow load for the unheated portion of the roof, $p_{f(\text{unheated})}$.

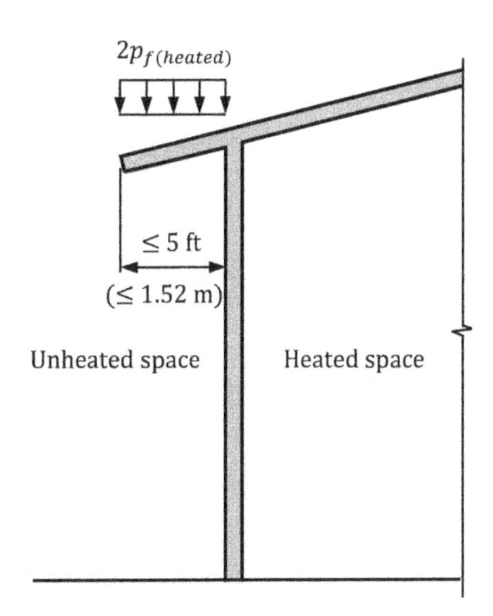

Figure 4.3 Load Case for Ice Dams Where the Horizontal Extent of the Overhang Is Less Than or Equal to 5 ft (1.52 m)

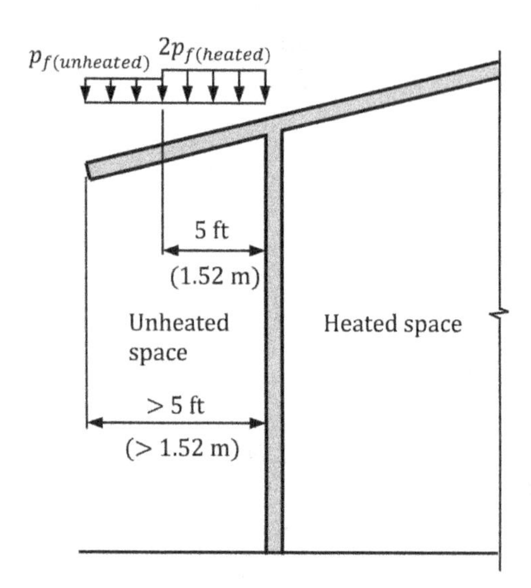

Figure 4.4 Load Case for Ice Dams Where the Horizontal Extent of the Overhang Is Greater Than 5 ft (1.52 m)

Air-Supported Structures

For structures with air-supported roofs with vinyl-coated exterior fabric (which is considered to be slippery), C_s varies linearly from 0.6 at the location where the roof slope is 30 degrees to 1.0 at the location where the slope is 5 degrees (see ASCE/SEI Figure 7.4-3). Values of C_s in ASCE/SEI Figure 7.4-3 match the values in ASCE/SEI Figure 7.4-1a up to a roof slope of 30 degrees for unobstructed slippery surfaces with $C_t = 1.0$. Snow loads are assumed to be zero for roof slopes greater than 30 degrees. A snow load diagram for structures with air-supported roofs is given in Figure 4.5.

Flowchart 4.2 in Section 4.4 of this book can be used to determine the sloped roof snow load, p_s.

Figure 4.5 Snow Load Diagram for Structures with Air-Supported Roofs

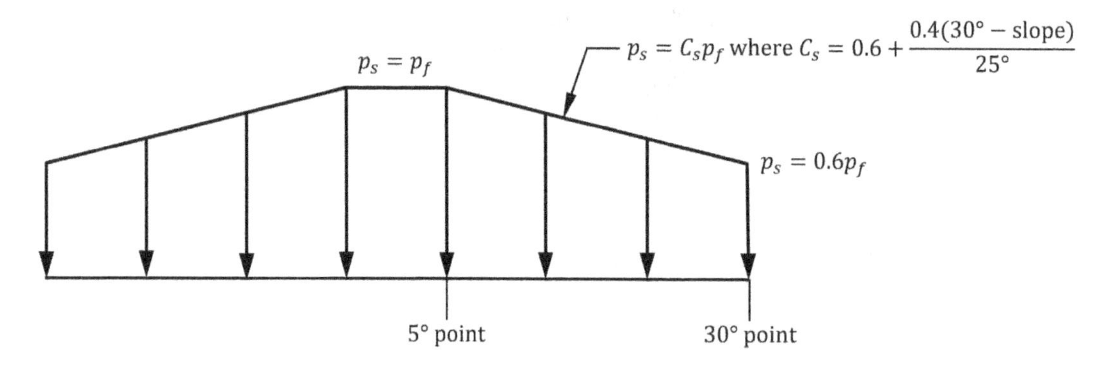

4.2.5 Partial Loading

The partial loading requirements of ASCE/SEI 7.5 must be satisfied for roof framing systems with continuous beams and for other roof systems where removal of snow on one span causes an increase in stress or deflection in an adjacent span.

Only the three load cases depicted in Figure 4.6 need to be investigated for continuous beam systems with and without cantilevered end spans; comprehensive alternate span (checkerboard) loading analyses are not required (see ASCE/SEI Figure 7.5-1):

- Load Case 1: Full balanced snow load, p_s, on either of the exterior spans and $p_s/2$ on all other spans.
- Load Case 2: $p_s/2$ on either exterior span and p_s on all other spans.

Figure 4.6 Partial Loading Diagrams for Continuous Beams with or without Cantilevered End Spans

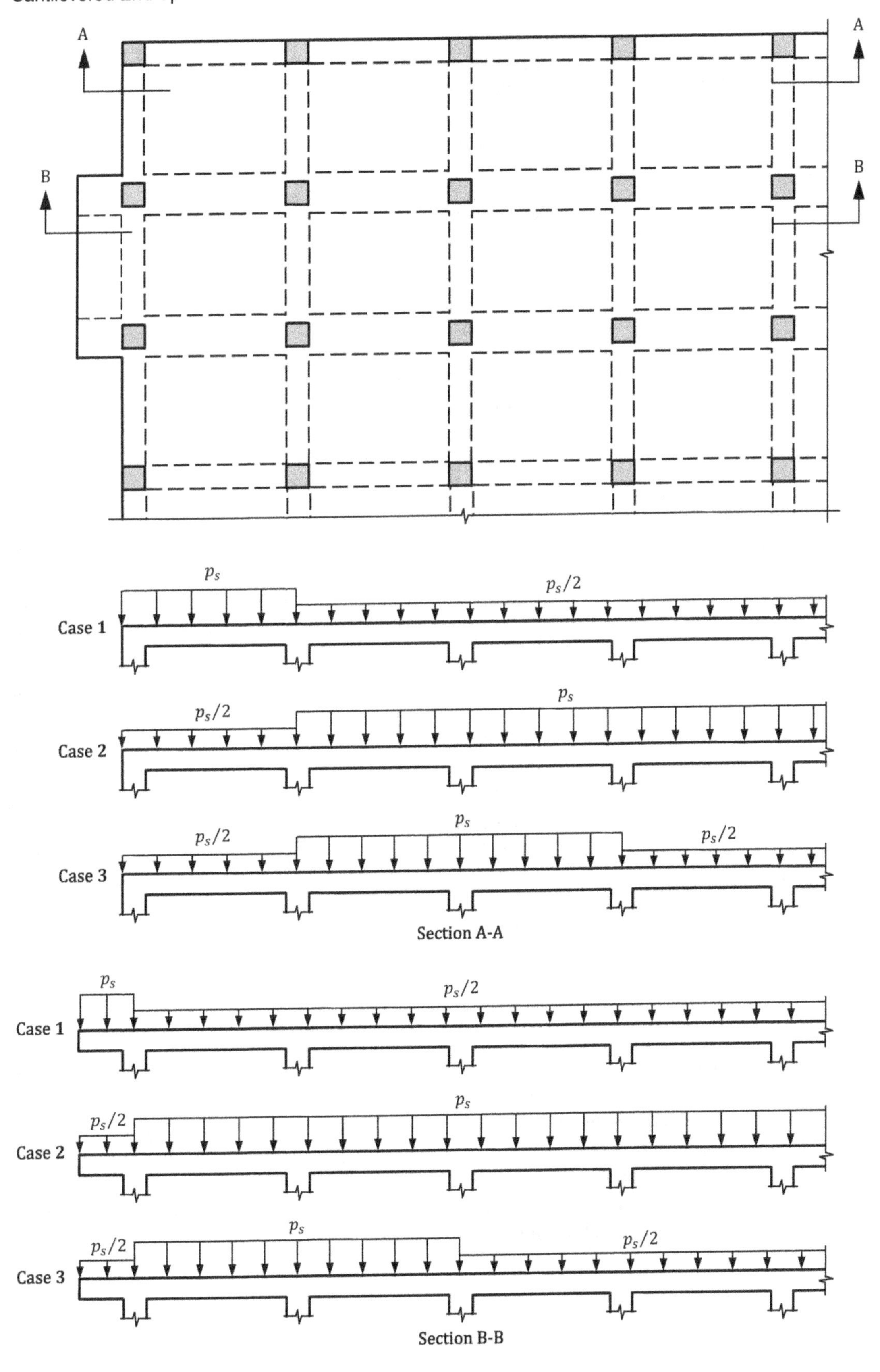

- Load Case 3: All possible combinations of p_s on any two adjacent spans and $p_s/2$ on all other spans. There are $(n-1)$ possible combinations for this case where n is equal to the number of spans in the continuous beam system.

Partial loading need not be applied to structural members spanning perpendicular to the ridgeline in gable roofs with slopes between (1) ½ on 12 (2.39 degrees) and (2) 7 on 12 (30.3 degrees).

4.2.6 Unbalanced Roof Snow Loads

Overview

Unbalanced snow loads occur on sloped roofs from wind (which reduces the snow load on the windward portion of the roof and increases it on the leeward portion) and from sunlight (which melts snow on the portions of the roof exposed to sunlight). Unbalanced loads can be considered drift loads. Requirements are given for the following roof types: (1) hip and gable roofs, (2) curved roofs, (3) multiple folded plate, sawtooth, and barrel vault roofs, and (4) dome roofs.

Wind from all directions must be considered when determining unbalanced snow loads. Roofs must be analyzed for balanced and unbalanced loads separately and the structural members are designed for the critical effects from these two cases.

Hip and Gable Roofs

Provisions for unbalanced snow loads on hip and gable roofs are given in ASCE/SEI 7.6.1. Unbalanced snow loads must be considered for roofs with slopes of ½ on 12 (2.39 degrees) through 7 on 12 (30.3 degrees). Snow drifts typically do not form on roofs with slopes less than or greater than these limiting values.

Two unbalanced load conditions are identified in ASCE/SEI 7.6.1. The first of the two conditions is applicable to roofs with an eave to ridge distance, W, less than or equal to 20 ft (6.10 m) where simply supported prismatic members span from the ridge to the eave. The load on the windward portion of the roof is equal to zero and the load on the leeward portion is equal to p_g, which is uniformly distributed over the entire width. Where the moment and shear capacities of a structural member (such as a roof truss) varies along its length, it is not prismatic, which means the unbalanced load condition is not applicable in such cases.

The second unbalanced load condition is applicable to all other hip and gable roofs. A uniform load equal to $0.3p_s$ is applied to the windward portion of the roof. The load on the leeward portion consists of two parts: (1) the balanced snow load, p_s, uniformly distributed over the entire width and (2) a drift load equal to $h_d\gamma/\sqrt{S}$ uniformly distributed over the length $8h_d\sqrt{S}/3$ measured horizontally from the ridge. The drift height, h_d, in this case is determined by ASCE/SEI Equation (7.6-1):

$$h_d = 1.5\sqrt{\frac{(p_g)^{0.74}(\ell_u)^{0.70}(W_2)^{1.7}}{\gamma}} \tag{4.3}$$

In this equation, ℓ_u is equal to the eave to ridge distance for the windward portion of the roof, W; the term W_2 is the winter wind parameter for the site, which is the percent time the wind speed is above 10 miles per hour (4.5 m/s) in the winter months of October through April (values of W_2 are given in ASCE/SEI Figure 7.6-1 for the conterminous United States and ASCE/SEI Table 7.2-1 for Alaska); and γ is the snow density. Instead of the aforementioned figure and table, values of W_2 can also be obtained from Reference 4.1. The snow density, γ, is determined by ASCE/SEI Equations (7.7-1) and (7.7-1.SI):

$$\gamma = 0.13p_g + 14 \leq 30 \text{ lb/ft}^3 \tag{4.4}$$

$$\gamma = 0.426p_g + 2.2 \leq 4.7 \text{ kN/m}^3 \tag{4.5}$$

Balanced and unbalanced snow loads for hip and gable roofs with an eave to ridge distance, W, of 20 ft (6.10 m) or less and simply supported prismatic members spanning from ridge to eave are given in Figure 4.7. In all other cases, the balanced and unbalanced snow loads are given in Figure 4.8.

Flowchart 4.3 in Section 4.4 of this book can be used to determine the balanced and unbalanced roof snow loads for hip and gable roofs in accordance with ASCE/SEI 7.6.1.

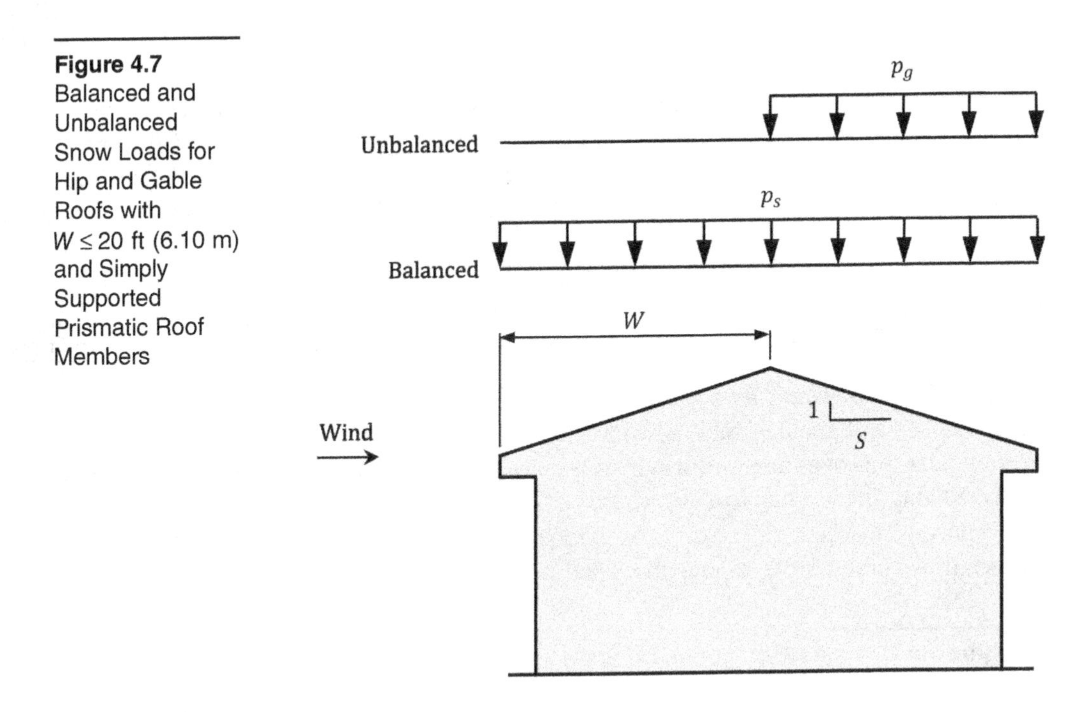

Figure 4.7 Balanced and Unbalanced Snow Loads for Hip and Gable Roofs with $W \leq 20$ ft (6.10 m) and Simply Supported Prismatic Roof Members

Figure 4.8 Balanced and Unbalanced Snow Loads for Hip and Gable Roofs Other Than Those with $W \leq 20$ ft (6.10 m) and Simply Supported Prismatic Roof Members

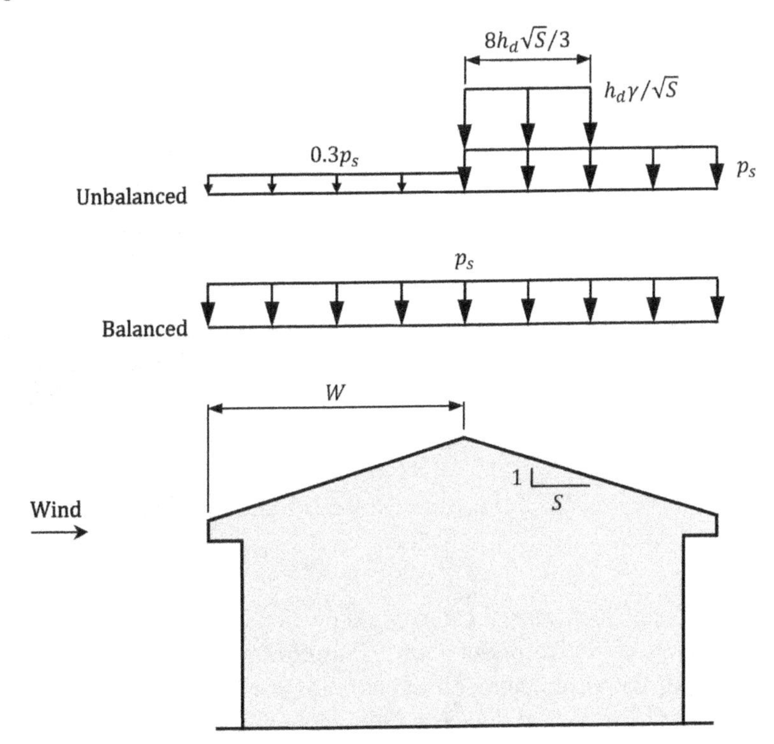

Curved Roofs

Overview

Provisions for unbalanced snow loads on curved roofs are given in ASCE/SEI 7.6.2. Any portion of a curved roof where the slope exceeds 70 degrees is considered free of snow loads; in other words, it is assumed snow is unable to accumulate on such steep portions of a roof. The roof slope is measured from the horizontal to the tangent of the curved roof at that point. In cases where the roof slope exceeds 70 degrees, the point on the roof at a slope of 70 degrees is considered to be the eave (see discussion below).

Balanced and unbalanced load cases are given in ASCE/SEI Figure 7.4-2 as a function of the slope of the roof at the eave. The three cases given in that figure are discussed in the following sections.

Unbalanced snow loads need not be considered where the slope of a straight line from the eaves or from a point on the roof where the tangent slope is equal to 70 degrees to the crown is less than 10 degrees or is greater than 60 degrees.

The provisions in ASCE/SEI 7.6.2 are applicable to curved roofs that are concave downward. For other roof geometries, such as a concave upward curved roof or for complicated site conditions, alternate procedures should be used to establish design snow loads (see ASCE/SEI C7.14).

Case 1—Slope at eaves less than 30 degrees

The balanced and unbalanced loads for Case 1 are given in Figure 4.9. The balanced load is trapezoidal near the eaves and uniform over a segment centered on the crown. The magnitude of the balanced load, p_s, is equal to $C_s p_f$ where p_f is determined in accordance with ASCE/SEI 7.3 (see Flowchart 4.2). At the eaves, $C_{s|eave}$ is determined using the slope of the roof at those locations. For shallow curved roofs, C_s may be equal to 1.0 over the entire roof.

Figure 4.9 Balanced and Unbalanced Snow Loads for Curved Roofs—Slope at Eaves Less Than 30 Degrees

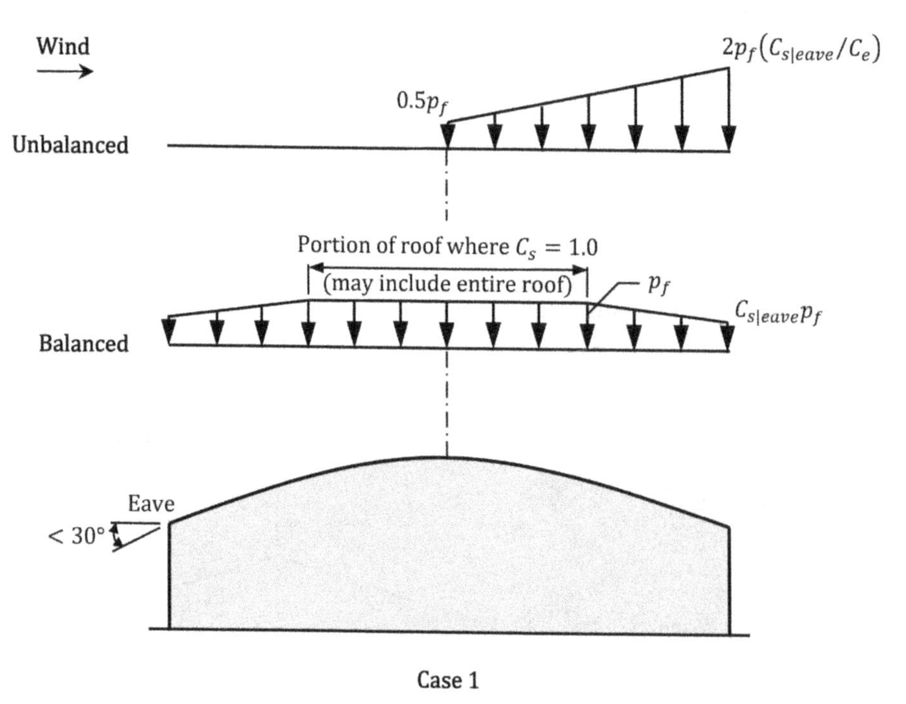

In the unbalanced load case, the windward portion of the roof between the windward eave and the crown is assumed to be free of snow. A trapezoidal load varying from $0.5p_f$ at the crown to $2p_f(C_{s|eave}/C_e)$ at the eave occurs on the leeward portion of the roof where $C_{s|eave}$ is determined using the slope of the roof at the eave and the exposure factor, C_e, is determined in accordance with ASCE/SEI 7.3.1 (see Flowchart 4.1).

Case 2—Slope at eaves 30 degrees to 70 degrees

The balanced load in this case is equal to p_f over the segment of the roof centered at the crown with two sets of trapezoidal loads on each side of this segment (see Figure 4.10). Where the roof slope is equal to 30 degrees, the magnitude of the load is $C_{s|30}p_f$ with the roof slope factor $C_{s|30}$ determined in accordance with ASCE/SEI 7.4. Similarly, the load at the eaves is equal to $C_{s|eave}p_f$ where $C_{s|eave}$ is determined in accordance with ASCE/SEI 7.4 based on the roof angle at that location.

Figure 4.10 Balanced and Unbalanced Snow Loads for Curved Roofs—Slope at Eaves 30 Degrees to 70 Degrees

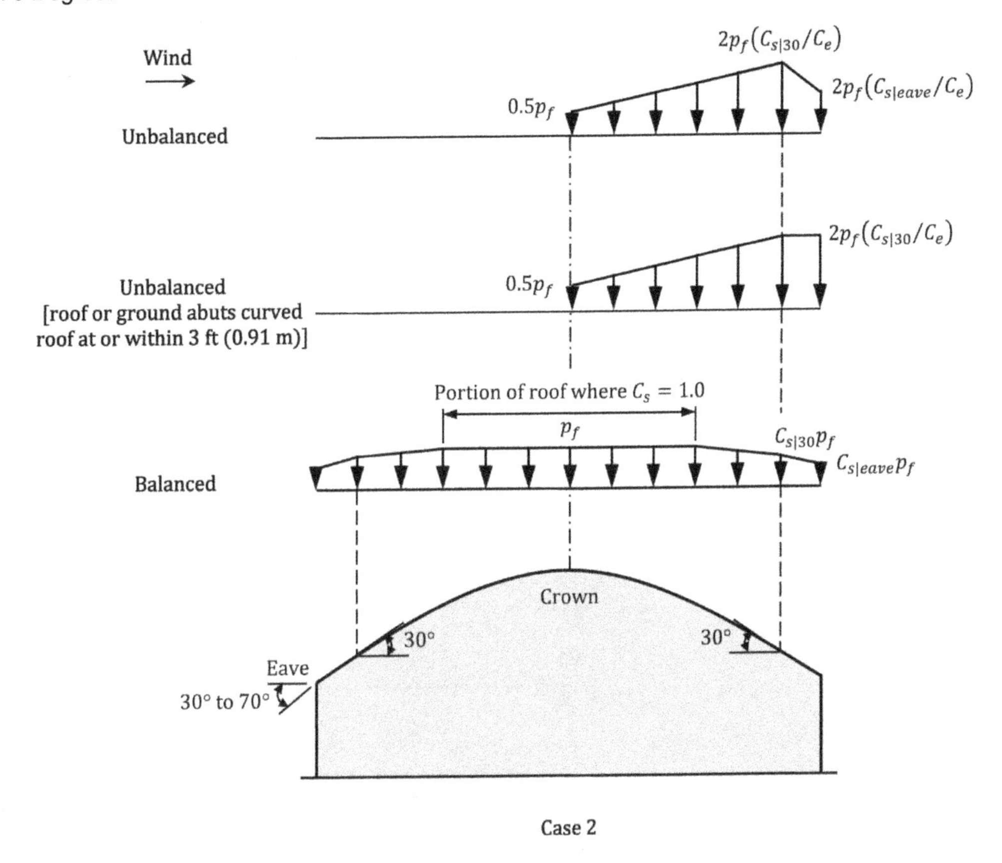

Like in Case 1, the windward portion of the roof between the windward eave and the crown is assumed to be free of snow in the unbalanced load case. Two cases must be considered for the leeward portion. Where the ground or another roof abuts the curved roof at or within 3 ft (0.91 m) of its eaves, the unbalanced load is $0.5p_f$ at the crown and $2p_f(C_{s|30}/C_e)$ at the point where the roof slope is equal to 30 degrees. From that point to the eaves, the load is a constant $2p_f(C_{s|30}/C_e)$. In all other cases, the unbalanced loading is the same as that described above except for the segment of the roof between the point where the slope is 30 degrees and the eaves; in that segment, the load at the 30-degree point is equal to $2p_f(C_{s|30}/C_e)$, which decreases linearly to $2p_f(C_{s|eave}/C_e)$.

Case 3—Slope at eaves greater than 70 degrees

The balanced load is equal to p_f over the segment of the roof centered at the crown with trapezoidal and triangular loads on each side of this segment (see Figure 4.11). The trapezoidal loads extend from the points where $C_s = 1.0$ to the points where the roof slope is 30 degrees. At the 30-degree points, the load is $C_{s|30}p_f$. The triangular loads extend from the points where the roof slope is 30 degrees to the points where it is 70 degrees. The load is zero in the segments between the points where the roof slope is 70 degrees and the eaves.

The unbalanced load for this case is very similar to that for Case 2: The windward portion of the roof between the windward eave and the crown is assumed to be free of snow, and two cases

Figure 4.11 Balanced and Unbalanced Snow Loads for Curved Roofs—Slope at Eaves Greater Than 70 Degrees

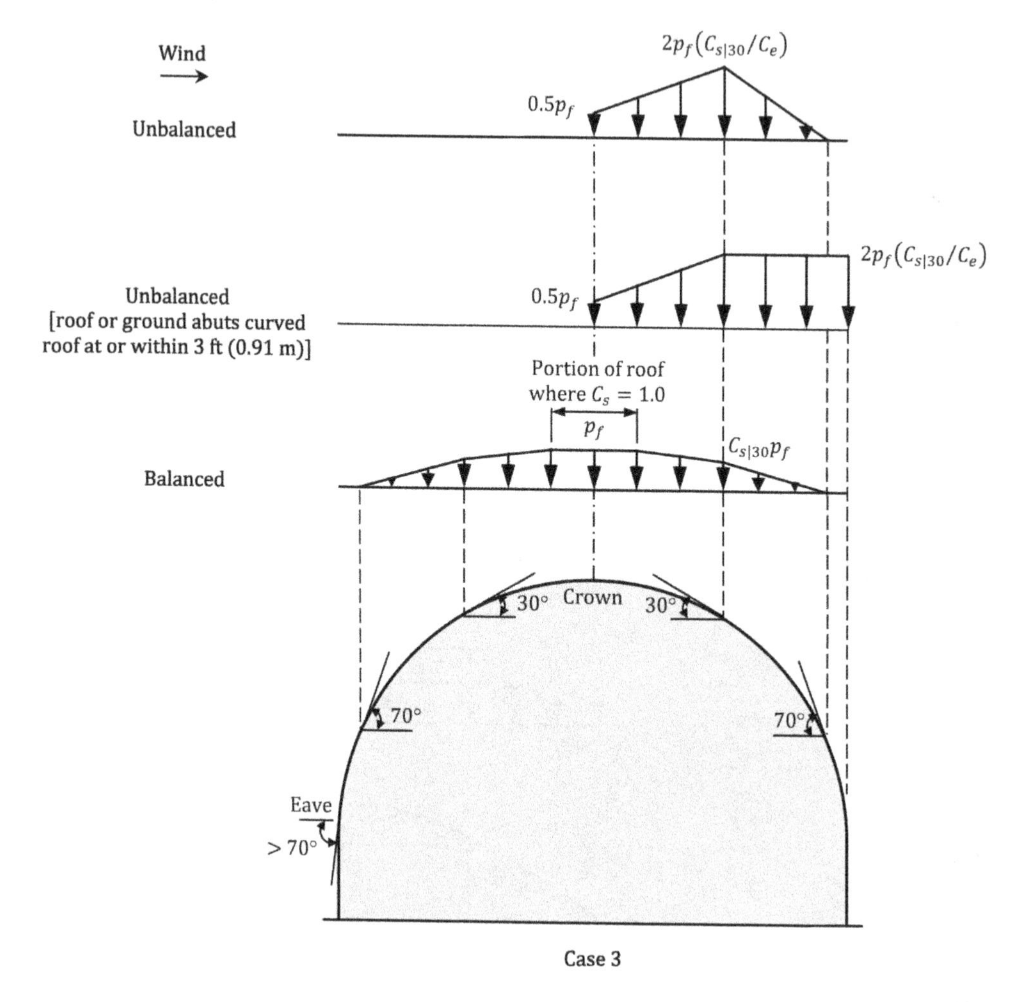

must be considered for the leeward portion depending on whether the ground or another roof abuts the curved roof within 3 ft (0.91 m) of its eaves.

Flowchart 4.4 in Section 4.4 of this book can be used to determine the balanced and unbalanced roof snow loads for curved roofs in accordance with ASCE/SEI 7.6.2.

Multiple Folded Plate, Sawtooth, and Barrel Vault Roofs

Provisions for unbalanced snow loads on multiple folded plate, sawtooth, and barrel vault roofs are given in ASCE/SEI 7.6.3. Unbalanced snow loads must be applied where the slope of the roof exceeds 1.79 degrees. In accordance with ASCE/SEI 7.4.3, $C_s = 1.0$ for these types of roofs, so the balanced snow load, p_s, is equal to the flat roof snow load, p_f (see Figure 4.12 for a sawtooth roof).

Like curved roofs, the unbalanced load is $0.5p_f$ at the crown or ridge of the roof and $2p_f/C_e$ at the valley. The snow load at a valley is limited by the space available for snow accumulation (i.e., it is limited by the depth of the valley). Assume the vertical distance from a valley to the roof ridge is h_r. The maximum snow load occurring at a valley is equal to the load at the ridge, $0.5p_f$, plus the load corresponding to a snow depth equal to h_r, which is equal to γh_r where the snow density, γ, is determined by Equation (4.4) or (4.5). Therefore, the load at the valley is equal to the following:

- If $2p_f/C_e < 0.5p_f + \gamma h_r$, the load at the valley $= 2p_f/C_e$
- If $2p_f/C_e \geq 0.5p_f + \gamma h_r$, the load at the valley $= 0.5p_f + \gamma h_r$

Figure 4.12 Balanced and Unbalanced Roof Loads for a Sawtooth Roof

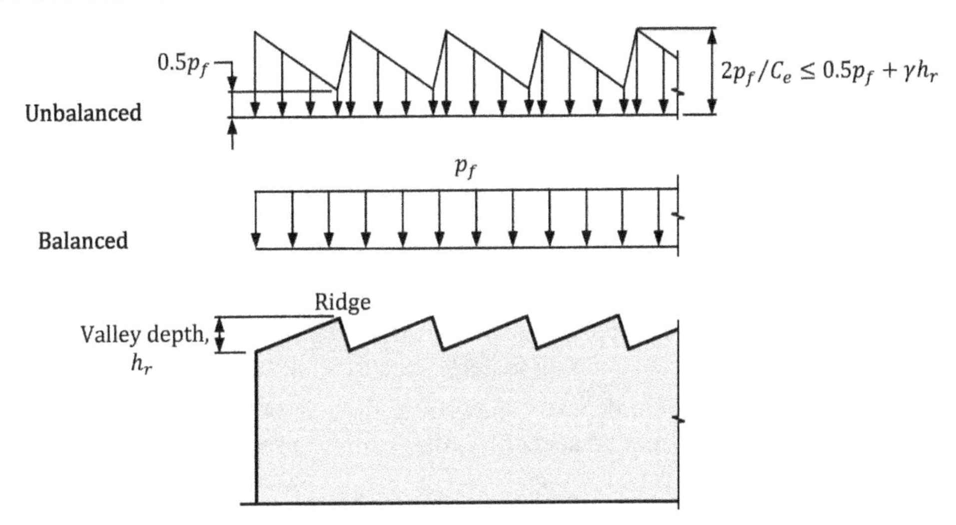

Flowchart 4.5 in Section 4.4 of this book can be used to determine the balanced and unbalanced roof snow loads for multiple folded plate, sawtooth, and barrel vault roofs in accordance with ASCE/SEI 7.6.3.

Dome Roofs

According to ASCE/SEI 7.6.4, unbalanced snow loads on dome roofs are determined in the same manner as for curved roofs. Unbalanced loads must be applied to the downwind 90-degree sector of the dome in plan (see Figure 4.13). The load decreases linearly to zero over 22.5-degree sectors on each side of the 90-degree sector. No snow load is taken on the remaining 225-degree upwind sector.

Figure 4.13 Unbalanced Snow Loads for a Dome Roof

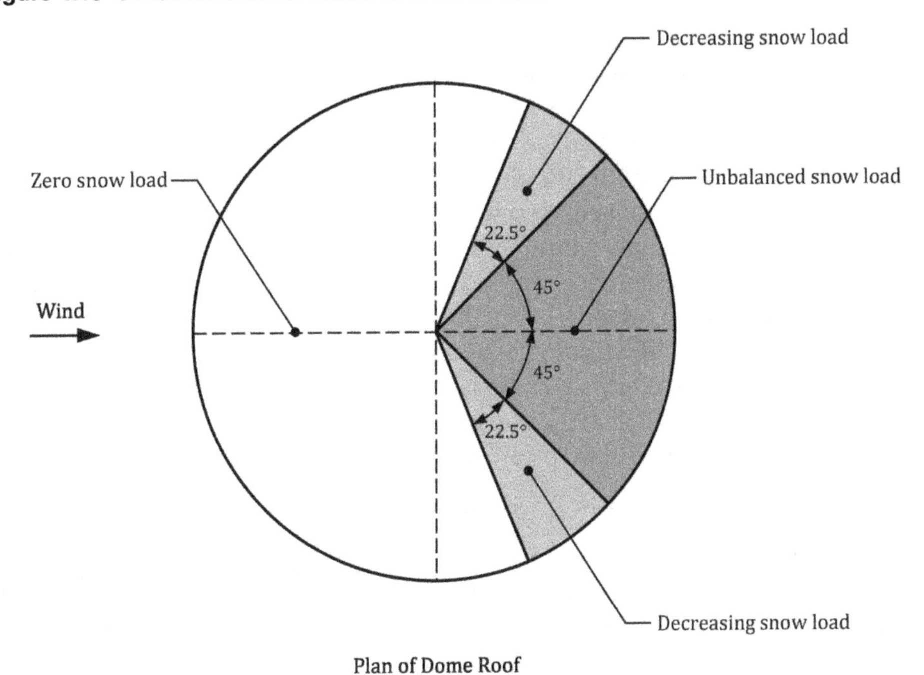

The balanced and unbalanced load distributions for a dome roof can be determined using Flowchart 4.4. In the unbalanced load case, the snow load at the eave or at the location where the roof slope is equal to 70 degrees decreases linearly to zero over the 22.5-degree sector on each side of the 90-degree downwind sector of the roof.

4.2.7 Drifts on Lower Roofs

Overview

Provisions for snow drifts occurring on lower roofs of a building are given in ASCE/SEI 7.7. Two types of drifts are addressed (see ASCE/SEI Figure 7.7-1 and Figure 4.14):

- Leeward drifts: Wind deposits snow from (a) a higher portion of the same building or an adjacent building or (b) a terrain feature (such as a hill) to a lower roof.
- Windward drifts: Wind deposits snow from the windward portion of a lower roof to the portion of the lower roof adjacent to a taller part of the building.

Figure 4.14 Windward and Leeward Snow Drifts

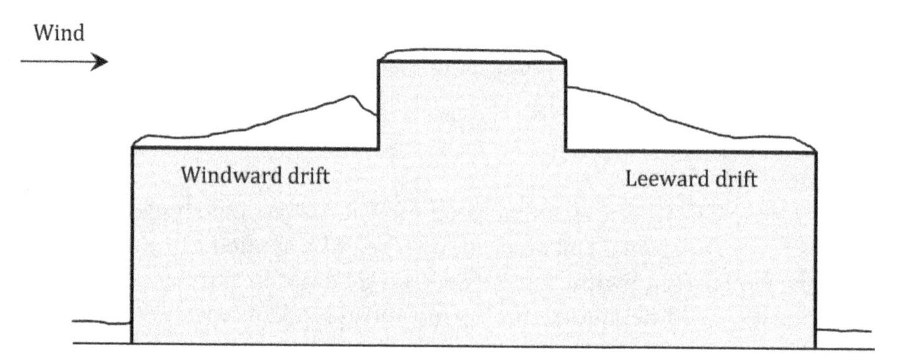

Both types of drifts are covered in the following sections. Also covered are provisions for drifting on lower structures adjacent to taller ones and intersecting drifts at lower roofs.

Lower Roof of a Structure

Wind deposits snow from one area of a roof to another. In cases where there is a change in elevation between roofs, snow will tend to accumulate at this location (see Figure 4.14). Depending on wind direction, either a windward or leeward drift will form. In the case of a windward drift, the wind blows across the length of the lower roof and deposits the snow on the lower roof adjacent to the wall. A leeward drift is formed when the wind blows the snow off of the upper roof on to the lower roof. Because wind can blow in any direction, both types of drifts must be investigated.

Leeward drifts are generally triangular in shape. Windward drifts usually have more complex shapes than leeward ones depending on the height of the wall (or, roof step). For simplicity, a triangular shape is used to characterize windward drifts as well.

The configuration of snow drifts on lower roofs is given in ASCE/SEI 7.7.1 (see ASCE/SEI Figure 7.7-2 and Figure 4.15).

Drift loads need not be considered where the following equation is satisfied:

$$h_c/h_b = (h_{\text{step}} - h_b)/h_b = [h_{\text{step}}/(p_s/\gamma)] - 1 < 0.2 \tag{4.6}$$

In this equation, h_c is the clear height from the top of the balanced snow load to the closet point on the adjacent upper roof (see Figure 4.15). This length can be obtained by subtracting the height of the balanced snow load, $h_b = p_s/\gamma$, from h_{step}, which is the vertical distance from the lower roof to the upper roof.

Figure 4.15 Drift Configuration on a Lower Roof

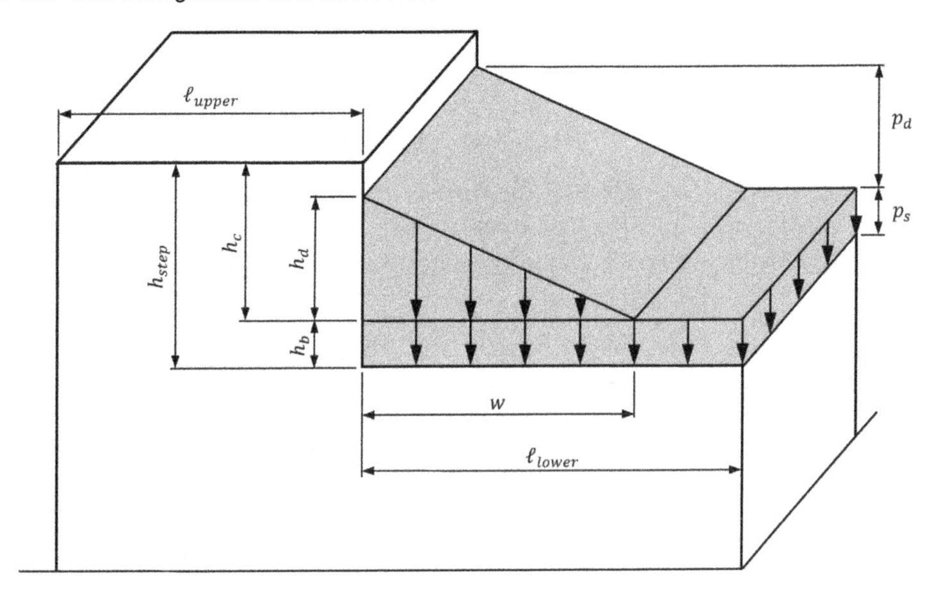

Where drift loads must be considered, the total snow load at the step, p_{total}, is equal to the drift load, p_d, which is assumed to be triangular, plus the uniform balanced snow load, p_s. The maximum drift load is determined by the following equation:

$$p_d = \gamma h_d \qquad (\text{lb/ft}^2,\ \text{kN/m}^2) \tag{4.7}$$

where the snow density, γ, is determined by Equations (4.4) and (4.5). The drift height, h_d, to use in Equation (4.7) is the larger of the leeward and windward drift heights, $h_{d,\text{leeward}}$ and $h_{d,\text{windward}}$, determined in accordance with ASCE/SEI Equation (7.6-1) [see Equation (4.3)]:

- For leeward drifts:

$$h_{d,\text{leeward}} = 1.5\sqrt{\frac{(p_g)^{0.74}(\ell_{\text{upper}})^{0.70}(W_2)^{1.7}}{\gamma}} \le 0.6\ell_{\text{lower}} \tag{4.8}$$

- For windward drifts:

$$h_{d,\text{windward}} = 1.125\sqrt{\frac{(p_g)^{0.74}(\ell_{\text{lower}})^{0.70}(W_2)^{1.7}}{\gamma}} \tag{4.9}$$

In these equations, ℓ_{upper} and ℓ_{lower} are the lengths of the upper and lower roofs, respectively.

The maximum drift load, p_d, the total load, p_{total}, and the width of the drift on the lower roof, w, depends on the clear height, h_c, and the calculated value of h_d:

- Where $h_{d,\text{leeward}} > h_{d,\text{windward}}$ and $h_{d,\text{leeward}} \le h_c = h_{\text{step}} - h_b$:

$$p_d = \gamma h_{d,\text{leeward}}$$

$$w = 4h_{d,\text{leeward}} \le 8h_c$$

$$p_{\text{total}} = p_s + p_d = \gamma(h_b + h_{d,\text{leeward}})$$

- Where $h_{d,\text{leeward}} > h_{d,\text{windward}}$ and $h_{d,\text{leeward}} > h_c = h_{\text{step}} - h_b$:

$$p_d = \gamma h_c$$

$$w = 4h_d^2/h_c \le 8h_c$$

$$p_{\text{total}} = p_s + p_d = \gamma(h_b + h_c) = \gamma h_{\text{step}}$$

- Where $h_{d,\text{windward}} > h_{d,\text{leeward}}$:

$$p_d = \gamma h_{d,\text{windward}}$$

$$w = 8h_{d,\text{windward}}$$

$$p_{\text{total}} = p_s + p_d = \gamma(h_b + h_{d,\text{windward}})$$

In cases where w of either the windward or leeward drift exceeds the width of the lower roof, ℓ_{lower}, the drift load is to taper linearly to zero at the far end of the lower roof (see Figure 4.16). This provision may be applicable to canopies over an entranceway to a building, for example.

Figure 4.16 Load Configuration where the Drift Width Is Greater Than the Length of the Lower Roof

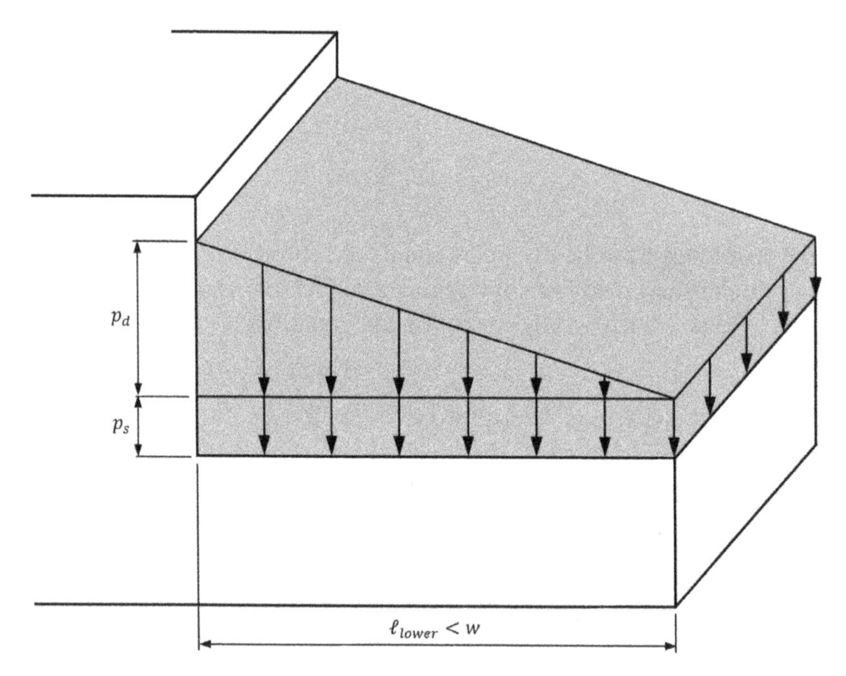

Windward and leeward drifts must be checked independently to determine which controls the design of the supporting members.

Flowchart 4.6 in Section 4.4 of this book can be used to determine the drift loads and widths on a lower roof of a structure.

Adjacent Structures

Leeward and windward drifts can form on the roof of a structure adjacent to a structure with a higher roof. Leeward drifts can form on lower roofs when the horizontal separation distance, s, between the adjacent structures is less than 20 ft (6.10 m) and less than $6h$ where h is the vertical separation distance between the two structures (see ASCE/SEI 7.7.2 and Figure 4.17). The drift load, p_d, in such cases is determined by the requirements for leeward drifts in ASCE/SEI 7.7.1 where the drift height, $h_{d,\text{leeward}}$, is equal to the lesser of the following:

- $1.5\sqrt{\dfrac{(p_g)^{0.74}(\ell_{\text{upper}})^{0.70}(W_2)^{1.7}}{\gamma}}$
- $(6h - s)/6$

In the first of these equations, ℓ_{upper} is the roof length of the adjacent higher structure.

The drift width, w, on the lower roof is equal to six times the lesser of the drift heights, that is, the lesser of $6h_{d,\text{leeward}}$ and $(6h - s)$.

Figure 4.17 Leeward Drift Loads on an Adjacent Roof

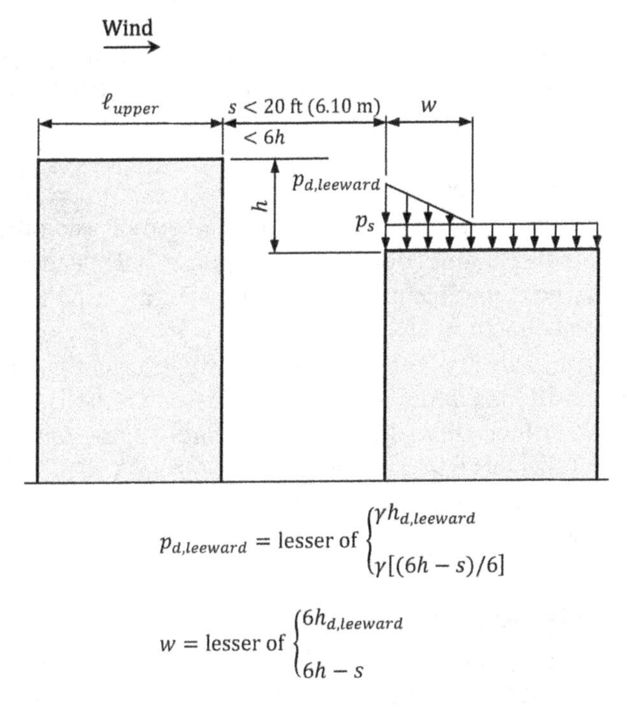

Windward drifts on the lower roof are also determined by the requirements of ASCE/SEI 7.7.1. It is assumed that $h_{d,\text{windward}}$ occurs at the face of the adjacent higher structure and is calculated using the roof length of the lower structure, ℓ_{lower} (see Figure 4.18). Within the separation zone between the two structures, the drift is truncated, and the height of the drift at the edge of the

Figure 4.18 Windward Drift Loads on an Adjacent Roof

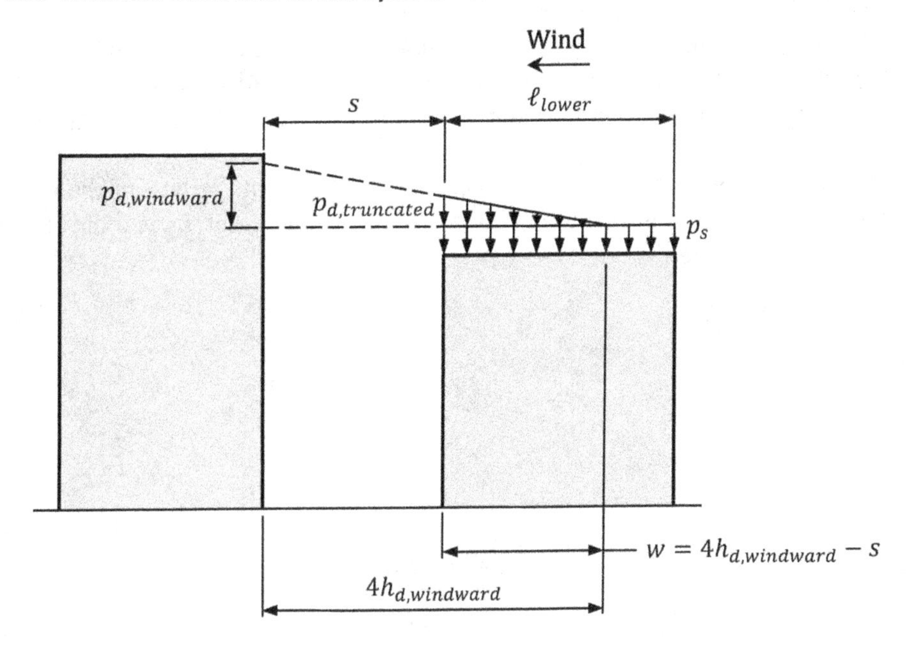

$$p_{d,windward} = \gamma h_{d,windward}$$

$$p_{d,truncated} = \left(1 - \frac{s}{4h_{d,windward}}\right) p_{d,windward}$$

lower structure can be determined from geometry. The total drift width in this case is equal to $4h_{d,\text{windward}}$, which extends from the face of the structure with the higher roof to the point on the lower roof where $p_d = 0$.

Flowchart 4.7 in Section 4.4 of this book can be used to determine leeward and windward drifts on an adjacent lower roof in accordance with ASCE/SEI 7.7.2.

Intersecting Drifts

Intersecting drifts can form at locations such as reentrant corners, parapet wall corners, intersections of a gable roof with the roof step wall of a taller roof, and other similar geometries due to wind acting in multiple directions. Provisions for intersecting snow drifts are given in ASCE/SEI 7.7.3. The requirements in ASCE/SEI 7.7.1 are used to determine the individual drift geometries for each direction.

Intersecting drift loads are considered to occur concurrently and are combined as shown in Figure 4.19. The drift load at the intersection is based on the larger drift height and not on the addition of the two drift heights. Also shown in Figure 4.19 are equations to determine the leeward and windward drift heights based on the requirements in ASCE/SEI 7.7.1.

Additional information on three-dimensional roof snowdrifts, including design examples, is given in Reference 4.4.

4.2.8 Roof Projections and Parapets

Drift loads on roof projections (including rooftop equipment) and parapet walls are determined by the provisions in ASCE/SEI 7.8 and are based on the requirements in ASCE/SEI 7.7.1. Windward drifts are formed on the side of parapet walls. Both windward and leeward drifts can form adjacent to rooftop units and other projections; however, the leeward drift is typically insignificant. So for simplicity, only windward drifts are considered.

The height of the drift, h_d, is equal to three-quarters of the value determined by ASCE/SEI Equation (7.6-1). For parapet walls, ℓ_u is equal to the length of the roof upwind of the parapet wall. For roof projections, ℓ_u is equal to the greater of the length of the roof upwind or downwind of the projection in that direction.

Drift loads are not required where the side of a projection is less than 15 ft (4.57 m) long or where the clear distance between the height of the balanced snow load, $h_b = p_s/\gamma$, and the bottom of the projection (including horizontal supports) is greater than or equal to 2 ft (0.61 m) [see Figure 4.20].

The maximum drift load, p_d, and the total load, p_{total}, at the face of the projection or parapet wall and the width of the drift, w, depends on the clear height, h_c, and the calculated value of h_d (see Figures 4.21 and 4.22):

- Where $h_d \le h_c = h_{\text{step}} - h_b$:

$$p_d = \gamma h_d \text{ and } w = 8h_d$$

$$p_{\text{total}} = p_s + p_d = \gamma(h_b + h_d)$$

- Where $h_d > h_c = h_{\text{step}} - h_b$:

$$p_d = \gamma h_c \text{ and } w = 8h_d$$

$$p_{\text{total}} = p_s + p_d = \gamma(h_b + h_c) = \gamma h_{\text{step}}$$

In these equations, h_{step} is the height of the projection or the height of the parapet wall.

Flowchart 4.8 in Section 4.4 of this book can be used to determine drift loads on roof projections and parapets.

Figure 4.19 Intersecting Drifts at Low Roofs

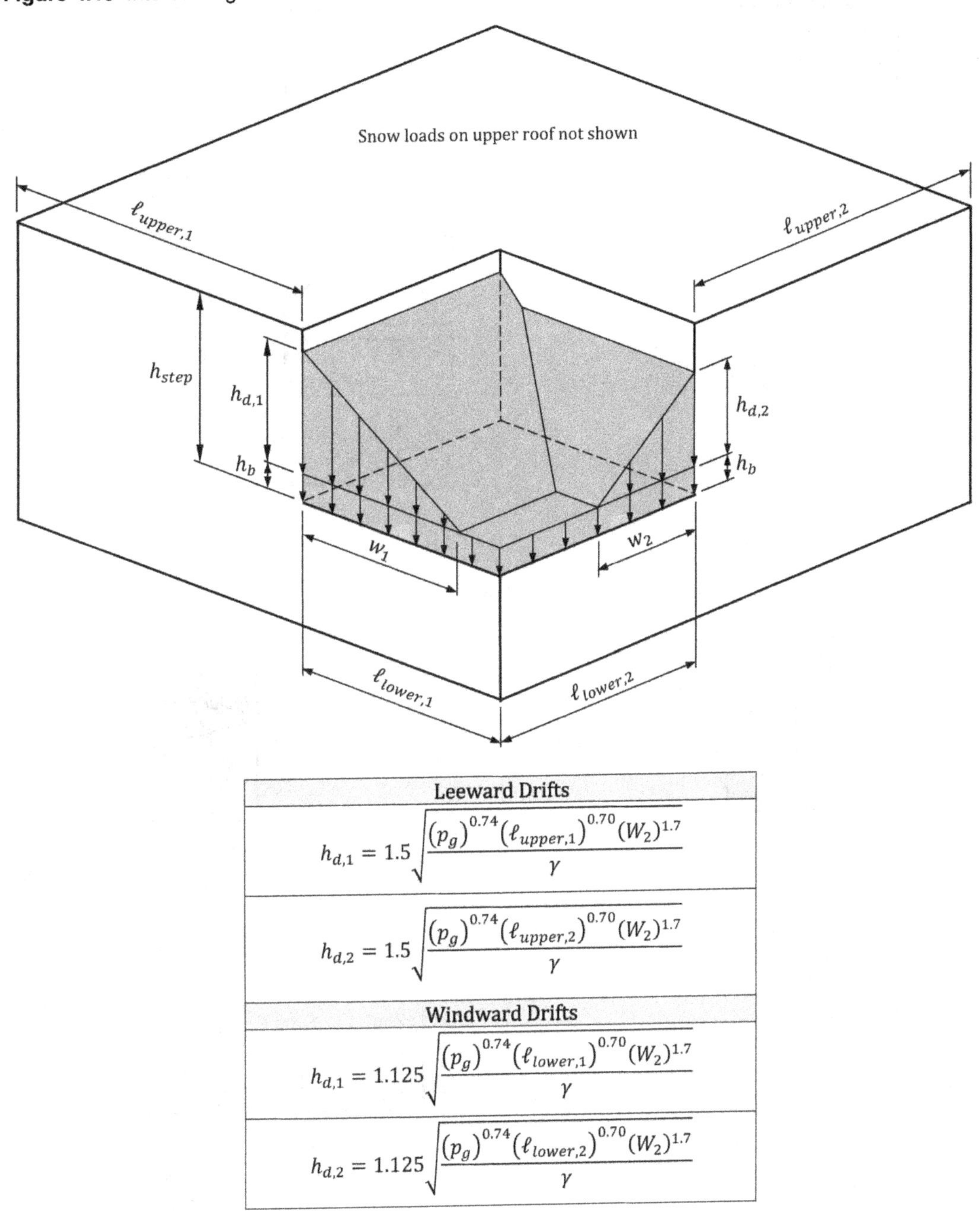

Leeward Drifts
$h_{d,1} = 1.5\sqrt{\frac{(p_g)^{0.74}(\ell_{upper,1})^{0.70}(W_2)^{1.7}}{\gamma}}$
$h_{d,2} = 1.5\sqrt{\frac{(p_g)^{0.74}(\ell_{upper,2})^{0.70}(W_2)^{1.7}}{\gamma}}$
Windward Drifts
$h_{d,1} = 1.125\sqrt{\frac{(p_g)^{0.74}(\ell_{lower,1})^{0.70}(W_2)^{1.7}}{\gamma}}$
$h_{d,2} = 1.125\sqrt{\frac{(p_g)^{0.74}(\ell_{lower,2})^{0.70}(W_2)^{1.7}}{\gamma}}$

4.2.9 Sliding Snow

Provisions for the load caused by snow sliding off a sloped roof onto a lower roof are given in ASCE/SEI 7.9. The sliding snow load is superimposed on the balanced snow load on the lower roof and need not be used in combination with drift, unbalanced, partial, or rain-on-snow loads.

Sliding snow loads are assumed to occur on lower roofs adjacent to slippery upper roofs with slopes greater than ¼ on 12 (1.19 degrees) and nonslippery roofs with slopes greater than 2 on 12 (9.46 degrees).

For adjacent structures with no horizontal separation between the two, the sliding load per length of eave is equal to $0.4p_fW$, where p_f and W are the flat roof snow load and the horizontal distance from the eave to the ridge of the upper roof, respectively (see Figure 4.23). The length

Figure 4.20 Conditions Where Drift Loads Need Not Be Considered

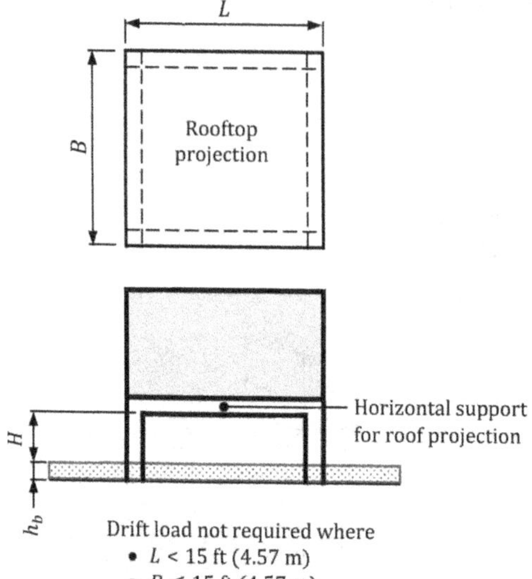

Figure 4.21 Drift Loads for Roof Projections

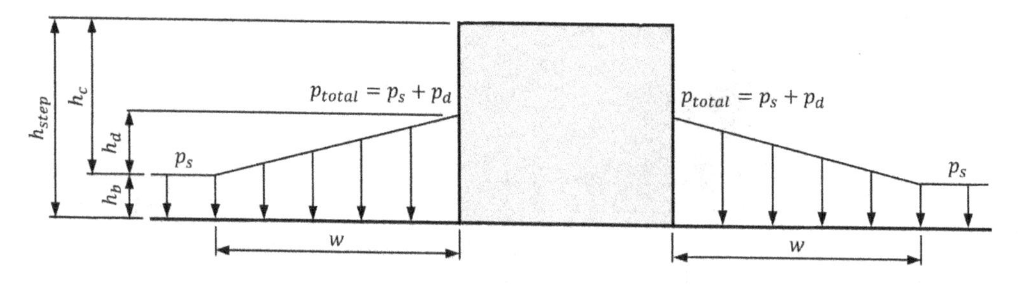

Figure 4.22 Drift Loads for Parapets

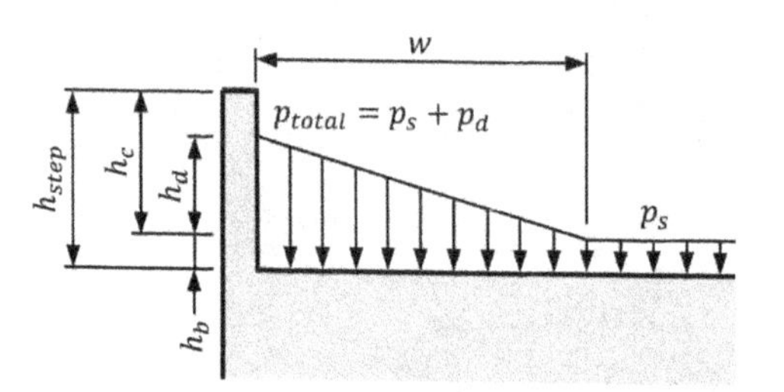

Figure 4.23 Sliding Snow Loads on a Lower Roof with No Horizontal Separation between the Upper and Lower Roofs

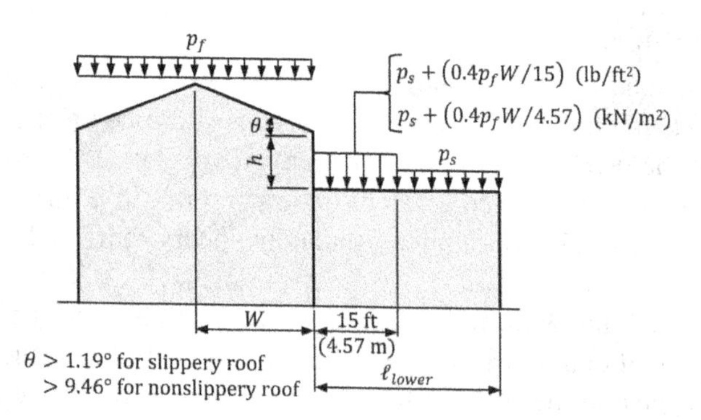

over which the sliding snow acts on the lower roof is 15 ft (4.57 m), which is measured horizontally from the eave of the upper roof. In cases where the length of the lower roof is less than 15 ft (4.57 m), the sliding snow load is permitted to be reduced proportionally, that is, the sliding snow can be reduced to $0.4p_f W(\ell_{\text{lower}}/15)$ [in S.I.: $0.4p_f W(\ell_{\text{lower}}/4.57)$] where ℓ_{lower} is the length of the lower roof.

The total snow load in the area subjected to sliding snow can be determined by the following equations for the case where $\ell_{\text{lower}} \geq 15$ ft (4.57 m):

$$p_{\text{total}} = p_s + (0.4p_f W/15) \text{ (lb/ft}^2\text{)} \tag{4.10}$$

$$p_{\text{total}} = p_s + (0.4p_f W/4.57) \text{ (kN/m}^2\text{)} \tag{4.11}$$

The corresponding total depth of snow is equal to the following:

$$h_{\text{total}} = [p_s + (0.4p_f W/15)]/\gamma \text{ (ft)} \tag{4.12}$$

$$h_{\text{total}} = [p_s + (0.4p_f W/4.57)]/\gamma \text{ (m)} \tag{4.13}$$

If h_{total} on the lower roof is greater than the vertical distance h from the top of the lower roof to the eave of the upper roof, sliding snow is blocked and a portion of the sliding snow remains on the upper roof. In such cases, the total load on the lower roof is equal to γh, which is uniformly distributed over 15 ft (4.57 m) or ℓ_{lower}, whichever is less.

Sliding snow loads on a lower roof horizontally separated by a distance s from an adjacent structure with a higher roof must be considered where $s < 15$ ft ($s < 4.57$ m) and $s < h$ (see Figure 4.24). The sliding snow load in this case is equal to the following:

$$p_{\text{sliding}} = 0.4p_f W(15 - s)/15 \text{ (lb/ft)} \tag{4.14}$$

$$p_{\text{sliding}} = 0.4p_f W(4.57 - s)/4.57 \text{ (kN/m)} \tag{4.15}$$

Figure 4.24 Sliding Snow Loads on a Lower Roof with Horizontal Separation between the Upper and Lower Roofs

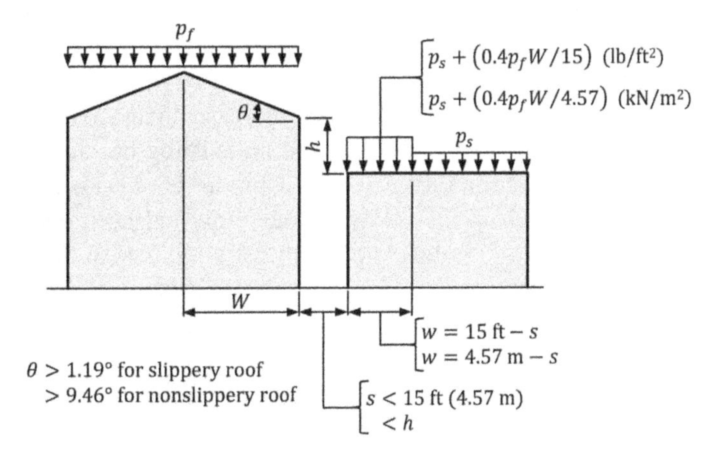

The sliding snow load is distributed over a length equal to (15 ft $- s$) [(4.57 m $- s$)] on the lower roof, and the total snow load is determined by Equations (4.10) and (4.11).

Flowchart 4.9 in Section 4.4 of this book can be used to determine sliding snow loads in accordance with ASCE/SEI 7.9.

4.2.10 Rain-on-Snow Surcharge Load

The snow load provisions in the discussions above consider load effects due to light rain on snow; effects due to heavy rain are not directly considered in the ground snow loads. At locations where p_g is greater than $p_{m,\max}$, it is assumed that because of the relatively deep snowpack, heavy rains have a less likely chance of permeating through the snowpack and draining away; as such, rain-on-snow load effects have been captured in the ground snow load measurements and an additional surcharge to account for this is not required.

A rain-on-snow surcharge load of 8 lb/ft^2 (0.38 kN/m^2) must be added on all roofs meeting the following two conditions (ASCE/SEI 7.10):

- The building is located where $0 < p_g \leq p_{m,\max}$.
- The slope of the roof in degrees is less than $W/50$ where W is the horizontal distance in feet from the ridge to the eave [in S.I.: $W/15.2$ where W is in meters].

This surcharge load applies only to the sloped (balanced) load case and need not be used in combination with drift, sliding, unbalanced, minimum, or partial loads.

4.2.11 Ponding Instability

In cases where roofs do not have adequate slope or have insufficient and/or blocked drains to remove water due to rain or melting snow, water will tend to pond in low areas, which can cause the roof structure to deflect. These low areas will subsequently attract even more water, leading to additional deflection. Sufficient stiffness must be provided so that deflections will not continually increase until instability occurs, resulting in localized failure.

Provisions for ponding instability and progressive deflection of roofs are given in ASCE/SEI 7.11 and ASCE/SEI 8.3 (see Section 3.3.8 of this book). Susceptible bays must be analyzed for the effects from the greater of the snow load or the rain load. The roof structure in these bays must be designed with adequate strength and stiffness to preclude ponding instability.

4.2.12 Existing Roofs

Requirements for increased snow loads on existing roofs due to additions and alterations are given in ASCE/SEI 7.12. Where a new structure with a higher roof is constructed within 20 ft (6.10 m) of an existing structure with a lower roof, both drift and sliding snow loads must be considered on the lower roof. The exposure of the existing roof must also be examined. For example, if the existing roof were fully exposed prior to the new building, it is likely it will be partially exposed or sheltered if the new building is taller, resulting in an increase in snow loads on the existing roof.

Another example of where snow loads may increase on an existing roof is where a new building with a gable roof is constructed alongside an existing building with a gable roof and both roofs are at the same elevation (see ASCE/SEI Figure C7.12-1). The new roof configuration is essentially a folded plate, and the valley created between the roofs will be subjected to a drift load, which did not have to be considered in the original design of the existing roof. In general, the effects of new structures or alterations on existing roofs must be carefully checked and existing roof structural members must be strengthened where required.

4.2.13 Snow on Open-Frame Equipment Structures

Snow loads determined in accordance with the provisions in ASCE/SEI 7.13 must be considered for all levels of open-frame equipment where applicable. In particular, snow accumulations must be considered for the following elements:

- Flooring
- Pipes and cable trays
- Equipment and equipment platforms

The loads to be applied to each type of element are given in Table 4.4.

Table 4.4 Snow Loads on Elements of Open-Frame Equipment Structures

Element		Snow Loads	ASCE/SEI Fig. No.
Flooring	At the top level	• Flat roof snow load, p_f, is applied over the entire level.[1],[2] • Windward drift loads are applied over the applicable widths where applicable.[2],[3]	7.13-1
	Below the top level[4]	Flat roof snow load, p_f, is applied over a portion of the flooring near any open edge with a horizontal width equal to the vertical difference in elevation between the level in question and the next floor above.	
Pipes and cable trays	Pipe diameter or cable tray width $\leq 0.73p_f/\gamma$[2]	Triangular snow load of width D and height $1.37D$ with a maximum intensity equal to $1.37D\gamma$ at $(D/2)$ is applied to the pipe or cable tray where D is the pipe diameter or cable tray width.[5],[6]	7.13-2a
	Pipe diameter or cable tray width $> 0.73p_f/\gamma$[2]	Trapezoidal snow load of width D with a maximum intensity equal to p_f is applied to the pipe or cable tray.[5],[6]	7.13-2b
	Clear spacing, S_p, between multiple adjacent pipes or cable trays $< (p_f/\gamma)$	Uniform cornice load of p_f is applied in the spaces between the pipes or cable trays in addition to the snow load on each individual pipe or cable tray.[6]	7.13-3
Equipment and equipment platforms		Snow loads on the structure must include snow loads on any equipment or equipment platforms supported by the structure.[6]	—

Notes

(1) Snow loads are applicable to flooring (grating, checkered plates, etc.) or elements that can retain snow. It is assumed open-frame members with a width greater than 8 in. (200 mm) can retain snow.

(2) Use $C_t = 1.2$ in the determination of p_f for unheated open-frame equipment structures (see Flowchart 4.1 for the determination of p_f).

(3) The top level of the structure must be designed for windward drift loads in accordance with ASCE/SEI 7.7 and 7.9 where there are wind walls or equivalent obstructions (see Flowchart 4.8 for the determination of drift loads and widths on roof projections).

(4) In cases where the top level of a structure does not have any snow-retaining surfaces, the level below the top level must be designed as the top level.

(5) The diameter or width, D, must include the thickness of insulation, where applicable.

(6) Snow accumulation on pipes, equipment, and equipment platforms need not be considered where the wintertime external surface temperature is greater than 45°F (7.2°C).

4.3 Ice Loads

4.3.1 Overview

Requirements for ice loads due to atmospheric icing are given in ASCE/SEI Chapter 10. An ice-sensitive structure is defined as one in which the effects due to atmospheric ice loading governs the design of part or all of the structure (IBC 202 and ASCE/SEI 10.2). Examples include but are not limited to the following structures: (1) lattice structures, (2) guyed masts, (3) overhead electric and communication lines, (4) light suspension and cable-stayed bridges, (5) aerial cable systems (e.g., for ski lifts and logging operations), (6) amusement rides, (7) open catwalks and platforms, (8) flagpoles, and (9) signs.

Freezing rain is rain or drizzle that falls into a layer of subfreezing air in the earth's surface and freezes on contact with the ground or any other exposed surface to form glaze (clear, high-density ice). Compared to in-cloud icing (which occurs when a supercooled cloud or fog droplets carried by the wind freeze on impact with objects) and snow, freezing rain is considered the cause of the most severe ice loads in most of the contiguous United States. Because values of ice thickness for in-cloud icing and snow are not currently available in a form suitable for inclusion in ASCE/SEI 7, only data for freezing rain are given in ASCE/SEI Chapter 10.

Site-specific studies must be used to determine ice thickness or load for the applicable risk category, concurrent wind speed, and concurrent temperature in the following cases (ASCE/SEI 10.1.1):

- In areas where it has been determined by records or experience that in-cloud icing or snow produce larger loads than freezing rain.
- In special icing regions as indicated in ASCE/SEI Figures 10.4-2 through 10.4-5.
- In regions with complex terrain where examination indicates unusual icing conditions exist.

Additional information and references on site-specific studies can be found in ASCE/SEI C10.1.1.

Design for dynamic load effects resulting from galloping, ice shedding, and aeolian vibrations, to name a few, are not covered in ASCE/SEI Chapter 10 (ASCE/SEI 10.1.2). Such effects must be considered in certain types of ice-sensitive structures.

The provisions of ASCE/SEI Chapter 10 do not apply to structures covered by national standards (e.g., electrical supply and communication systems and communication towers and masts). In such cases, the standards and documents referenced in ASCE/SEI 10.1.3 must be used where applicable.

The design procedure in Table 4.5 can be used to determine design atmospheric ice loads based on the requirements in ASCE/SEI Chapter 10. Additional information needed to calculate the ice loads is given in the sections of this book indicated in the table.

Table 4.5 Design Procedure to Determine Design Atmospheric Ice Loads

Step	Procedure	Section Number
1	Determine the nominal ice thickness, t, the concurrent gust speed, V_c, and the concurrent temperature for the site from the following: • ASCE/SEI Figures 10.4-2 through 10.4-5, 10.5-1, 10.5-2, 10.6-1 and 10.6-2 • Reference 4.1 • Site-specific study	4.3.2
2	Determine the height factor, f_z.	4.3.2
3	Determine the topographic factor for the site, K_{zt}.	4.3.2
4	Determine the design ice thickness, t_d.	4.3.2
5	Determine the ice load based on the cross-sectional area or volume of glaze ice.	4.3.2
6	Determine the wind velocity pressure, q_z, for concurrent gust speed, V_c.	4.3.3
7	Determine the fundamental natural frequency of the structure, n_1, and the corresponding gust-effect factor, G.	4.3.3
8	Determine the wind force coefficients, C_f.	4.3.3
9	Determine the wind-on-ice load, W_i.	4.3.3

4.3.2 Ice Loads Caused by Freezing Rain

Overview

Ice loads caused by freezing rain are determined using the ice weight formed on all exposed surfaces of structural members, guys, components, appurtenances, and cable systems. The following sections provide information needed to calculate ice weight.

Nominal Ice Thickness

Equivalent uniform radial thickness, t, of ice due to freezing rain at a height of 33 ft (10 m) above the ground is given for the contiguous 48 states (ASCE/SEI Figures 10.4-2A through 10.4-2D), Alaska (ASCE/SEI Figures 10.4-3A through 10.4-3D), the Columbia Gorge region (ASCE/SEI Figures 10.4-4A through 10.4-4D), and the Lake Superior region (ASCE/SEI Figures 10.4-5A through 10.4-5D) for Risk Categories I, II, III, and IV (IBC Table 1604.5). The concurrent gust speeds, V_c, are given in ASCE/SEI Figures 10.5-1 and 10.5-2 for the contiguous 48 states and Alaska, respectively. These speeds correspond to the winds occurring during the freezing rainstorm and those occurring between the time the freezing rain stops and the temperature rises to above freezing. Values of t and V_c for an address or latitude and longitude of a site can be obtained from Reference 4.1. Ice thicknesses for Hawaii must be obtained from local meteorological studies.

The data given in these figures are based on studies using the US Army's Cold Regions Research and Engineering Laboratory (CRREL) and simple ice accretion models with historical data from 540 National Weather Service (NWS), military, Federal Aviation Administration (FAA) and Environment Canada weather stations (which are indicated in ASCE/SEI Figure C10.4-1 for the 48 contiguous states and Canada and in ASCE/SEI Figures 10.4-3A through 10.4-3D for Alaska). The models utilize the measured weather and precipitation data to simulate the accretion of ice on horizontal cylinders located 33 ft (10 m) above the ground and oriented perpendicular to the direction of wind in freezing rainstorms. It is assumed the ice remains on the cylinder during the duration of the storm and remains there until after the temperature increases to at least 32°F (0°C).

Special icing regions are also identified on the maps (gray shaded areas) and occur in the western mountainous regions, in the Appalachian Mountains, and in Alaska. In the Cascades of Oregon and Washington, ice thicknesses may exceed the mapped value in foothills and passes, while in the Appalachian Mountains, ice thicknesses may vary significantly over short distances because of local variations in elevation, topography, and exposure. The thicknesses given in ASCE/SEI Figures 10.4-2 through 10.4-5 should be adjusted based on local historical records and experience. Local building officials should be consulted when making such adjustments.

Height Factor

The height factor, f_z, adjusts the mapped values of t in ASCE/SEI Figures 10.4-2 through 10.4-5, which are based on a height of 33 ft (10 m) above ground, to any height z above ground [ASCE/SEI Equations (10.4-4) and (10.4-4.SI)]:

$$f_z = \begin{cases} \left(\dfrac{z}{33}\right)^{0.10} & \text{for } 0 \text{ ft} < z \leq 900 \text{ ft} \\ 1.4 & \text{for } z > 900 \text{ ft} \end{cases} \tag{4.16}$$

$$f_z = \begin{cases} \left(\dfrac{z}{10}\right)^{0.10} & \text{for } 0 \text{ m} < z \leq 275 \text{ m} \\ 1.4 & \text{for } z > 275 \text{ m} \end{cases} \tag{4.17}$$

This factor is similar to the velocity pressure exposure coefficient, K_z, which modifies wind velocity with respect to exposure and height above ground (see ASCE/SEI Chapter 26 and Chapter 5 of this book).

Topographic Factor

Because of wind speed-up effects, t and V_c are greater for buildings and structures situated on hills, ridges, and escarpments than those located on level terrain. To account for these effects, t is modified by $(K_{zt})^{0.35}$ where K_{zt} is the topographic factor determined by ASCE/SEI Equation (26.8-1):

$$K_{zt} = (1 + K_1 K_2 K_3)^2 \tag{4.18}$$

The topographic multipliers K_1, K_2, and K_3 are determined from ASCE/SEI Figure 26.8-1.

Not every hill, ridge, or escarpment requires an increase in t or V_c; these quantities must be increased only when the site conditions and structure locations meet all the conditions in ASCE/SEI 26.8.1. Otherwise, $K_{zt} = 1.0$.

Design Ice Thickness for Freezing Rain

The design ice thickness, t_d, is equal to the nominal ice thickness, t, multiplied by the modification factors above [see ASCE/SEI Equation (10.4-5)]:

$$t_d = t f_z (K_{zt})^{0.35} \quad \text{(in. and mm)} \tag{4.19}$$

This thickness is used in calculating ice loads caused by freezing rain in accordance with ASCE/SEI 10.4.1 and wind loads on ice-covered structures in accordance with ASCE/SEI 10.5 (see below).

Ice Loads Caused by Freezing Rain

Ice loads caused by freezing rain are determined using the volume or cross-sectional area of glaze ice formed on all exposed surfaces of structural members, guys, components, appurtenances, and cable systems (ASCE/SEI 10.4.1).

The cross-sectional area of ice, A_i, formed on structural shapes, prismatic members, and other similar shapes is determined by ASCE/SEI Equations (10.4-1) and (10.4-1.SI):

$$A_i = \frac{\pi t_d}{12}\left(D_c + \frac{t_d}{12}\right) \quad (\text{ft}^2) \tag{4.20}$$

$$A_i = \frac{\pi t_d}{1{,}000}\left(D_c + \frac{t_d}{1{,}000}\right) \quad (\text{m}^2) \tag{4.21}$$

where t_d is in inches in Equation (4.20) and is in millimeters in Equation (4.21).

As noted previously, the equivalent radial ice thickness, t, due to freezing rain and the corresponding design ice thickness, t_d, have been established using a horizontal cylinder oriented perpendicular to the wind, and, thus, are not directly applicable to structural shapes, prismatic members, or other shapes that are not round. However, the ice area determined from Equations (4.20) and (4.21) are the same for all shapes for which the circumscribed circles have equal diameters. The diameter of a cylinder circumscribing a shape or object, D_c, is given in ASCE/SEI Figure 10.4-1 for different cross-sectional shapes (see Figure 4.25). It is assumed in Equations (4.20) and (4.21) that the maximum dimension of the cross-section is perpendicular to the path of the raindrops.

The volume of ice, V_i, formed on flat plates and large three-dimensional objects such as domes and spheres is determined by ASCE/SEI Equations (10.4-2) and (10.4-2.SI):

$$V_i = \frac{\pi t_d}{12} A_s \quad (\text{ft}^3) \tag{4.22}$$

$$V_i = \frac{\pi t_d}{1{,}000} A_s \quad (\text{m}^3) \tag{4.23}$$

Figure 4.25 Dimension D_c for Different Cross-Sectional Shapes

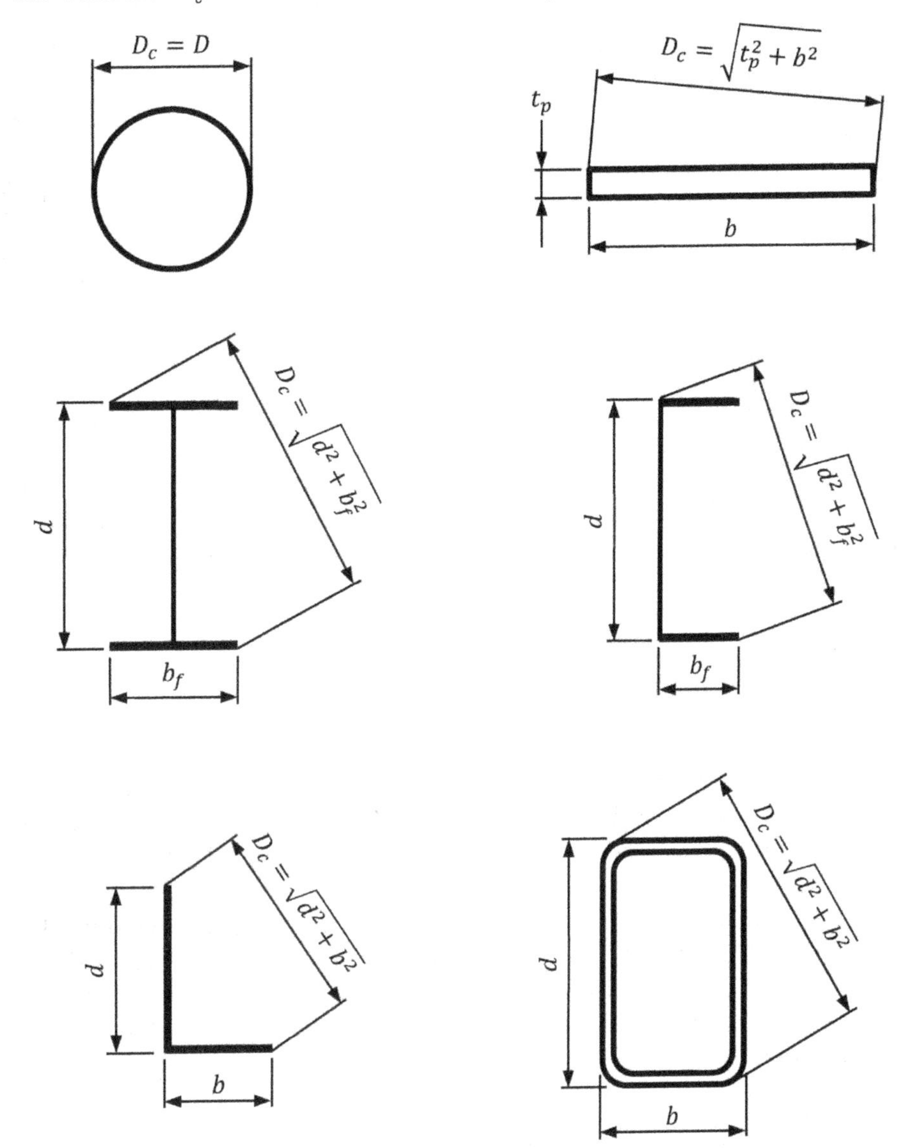

In these equations, A_s is the surface area of one side of a plate or the projected area of a complex shape in square feet (square meters).

The ice volume given by Equations (4.22) and (4.23) is for a flat plate or projected surface oriented perpendicular to the path of the raindrops. For vertical plates, A_s is permitted to be taken equal to 0.8 times the area of one side of the plate. Similarly, A_s is equal to 0.6 times the area of one side of the plate for horizontal plates. For domes and spheres, $A_s = \pi r^2$ [ASCE/SEI Equation (10.4-3)] where r is the radius of the maximum cross-section of a dome or radius of a sphere in feet (meters).

Once A_i or V_i has been determined, the ice load can be calculated by multiplying A_i or V_i by the density of ice, which must not be taken less than 56 lb/ft^3 (900 kg/m^3) [ASCE/SEI 10.4.1].

The equations in Table 4.6 can be used to determine ice loads for the shapes and objects noted above.

Table 4.6 Ice Loads on Shapes and Objects Caused by Freezing Rain

Shape / Object	Ice Load*
Structural shapes, prismatic members, and other similar shapes	$14.66t_d\left(D_c+\frac{t_d}{12}\right)$ (lb/ft) $27.71t_d\left(D_c+\frac{t_d}{1{,}000}\right)$ (N/m)
Vertical plate	$11.73t_dbh$ (lb) $22.17t_dbh$ (N)
Horizontal plate	$8.80t_dbh$ (lb) $16.63t_dbh$ (N)
Domes and spheres	$46.06t_dr^2$ (lb) $87.05t_dr^2$ (N)

* t_d is in in. (mm), D_c is in ft (m), and b (plate width), h (plate height), and r are in ft (m).

4.3.3 Wind on Ice-Covered Structures

Overview

Ice that has formed on structural members, components, and appurtenances increases the projected area exposed to wind and changes the structure's wind drag coefficients. Ice accretions tend to round sharp edges thereby reducing drag coefficients for members like angles and rectangular bars.

Ice-sensitive structures must be designed for the wind loads determined by the provisions in ASCE/SEI Chapters 26 through 31 using increased projected area and the modifications in ASCE/SEI 10.5.1 through 10.5.5, which are discussed next. The loads determined in this fashion are defined as the wind-on-ice loads, W_i (see Chapter 2 of this book). It is assumed in the following discussions that the reader is familiar with the wind load provisions in Chapter 5 of this book.

Wind Velocity Pressure, q_z

Wind loads on ice-covered structures are determined using the wind velocity pressure, q_z, which is determined by ASCE/SEI Equations (26.10-1) and (26.10-1.SI):

$$q_z = 0.00256K_zK_{zt}K_eV_c^2 \quad (\text{lb}/\text{ft}^2) \tag{4.24}$$

$$q_z = 0.613K_zK_{zt}K_eV_c^2 \quad (\text{N/m}^2) \tag{4.25}$$

The concurrent gust speed, V_c, is determined from ASCE/SEI Figures 10.5-1 and 10.5-2 or from Reference 4.1 in miles per hour (mi/h) [meters per second (m/s)]. Flowchart 4.10 in Section 4.4 of this book can be used to determine q_z.

Wind-On-Ice Loads, W_i

Wind-on-ice loads for the structures in ASCE/SEI 10.5.1 through 10.5.5 are given in Table 4.7. The wind directionality factor, K_d, is determined in accordance with ASCE/SEI 26.6. The gust-effect factor, G, is determined in accordance with ASCE/SEI 26.11 based on the fundamental natural frequency of the structure, n_1, which can be determined by any rational method. The gust-effect factor is permitted to be taken as 0.85 for a rigid building or other structure, that is, for structures with $n_1 \geq 1$ Hz. Flexible buildings or other structures are defined as those with $n_1 < 1$ Hz.

Table 4.7 Wind-On-Ice Loads, W_i

Structure	Force Coefficient, C_f	Wind-On-Ice Load, W_i (lb and N)	Notes
Chimneys, tanks, and similar structures (ASCE/SEI 10.5.1)	• For structures with square, hexagonal, and octagonal cross-sections, determine C_f from ASCE/SEI Figure 29.4-1. • For structures with round cross-sections, determine C_f from ASCE/SEI Figure 29.4-1 for round cross-sections with $D\sqrt{q_z} \leq 2.5$ [in S.I.: $D\sqrt{q_z} \leq 5.3$] for all ice thicknesses, gust speeds, and structure diameters.	$q_z K_d G C_f A_f$ [ASCE/SEI Equation (29.4-1) and (29.4-1.SI)]	W_i is calculated based on the area of the structure, including ice, A_f, projected on a vertical plane normal to the wind direction. W_i is assumed to act parallel to the wind direction.
Solid freestanding walls and solid signs (ASCE/SEI 10.5.2)	Determine C_f from ASCE/SEI Figure 29.3-1 for Cases A, B, and C based on the dimensions of the wall or sign, including ice.	$q_h K_d G C_f A_s$ [ASCE/SEI Equation (29.3-1) and (29.3-1.SI)]	W_i is calculated for Cases A, B, and C in ASCE/SEI Figure 29.3-1. A_s is the gross area of the solid freestanding wall or freestanding solid sign, including ice.
Open signs and lattice frameworks* (ASCE/SEI 10.5.3)	• For flat members, determine C_f from ASCE/SEI Figure 29.4-2. • For rounded members and for the additional projected area caused by ice on both flat and rounded members, determine C_f from ASCE/SEI. Figure 29.4-2 for rounded members with $D\sqrt{q_z} \leq 2.5$ [in S.I.: $D\sqrt{q_z} \leq 5.3$] for all ice thicknesses, gust speeds, and member diameters.	$q_z K_d G C_f A_f$ [ASCE/SEI Equation (29.4-1) and (29.4-1.SI)]	W_i is calculated based on the area of all exposed members and elements, including ice, projected on a plane normal to the wind direction. W_i is assumed to act parallel to the wind direction. A_f is the solid area, including ice, projected normal to the wind direction.
Trussed towers* (ASCE/SEI 10.5.4)	Determine C_f from ASCE/SEI Figure 29.4-3. It is acceptable to reduce C_f for the additional projected area caused by ice on both round and flat members by the factor for rounded members in Note 3 of ASCE/SEI Figure 29.4-3.	$q_z K_d G C_f A_f$ [ASCE/SEI Equation (29.4-1) and (29.4-1.SI)]	W_i is to be applied in the directions resulting in maximum member forces and reactions. For all wind directions considered, A_f is the solid area of a tower face, including ice, projected on the plane of that face for the tower segment under consideration.
Guys and cables (ASCE/SEI 10.5.5)	Use $C_f = 1.2$.	$q_z K_d G C_f A_f$ [ASCE/SEI Equation (29.4-1) and (29.4-1.SI)]	See ASCE/SEI C10.5.5.

*The solidity ratio (ratio of solid area to gross area), ε, must be based on the projected area, including ice.

The areas used in the determination of W_i must be increased by adding the design ice thickness, t_d, determined by Equation (4.19) to all free edges of the projected area.

4.3.4 Design Temperatures for Freezing Rain

Some ice-sensitive structures can also be sensitive to changes in temperature. While maximum load effects usually occur at the lowest temperature when the structure is loaded with ice, it is possible for some types of structures, such as overhead cable systems, to experience maximum load effects at or around the melting point of ice, which is 32°F (0°C).

Temperatures concurrent with ice thickness due to freezing rain are given in ASCE/SEI Figures 10.6-1 and 10.6-2 for the contiguous 48 states and Alaska, respectively. The design temperature for ice and wind-on-ice is that from ASCE/SEI Figures 10.6-1 and 10.6-2 or 32°F (0°C), whichever gives the maximum load effect. The design temperature for Hawaii is 32°F (0°C).

4.3.5 Partial Loading

Variations in ice thickness due to freezing rain at a given elevation are usually small over distances of about 1,000 ft (305 m). Thus, partial loading from freezing rain does not usually produce maximum load effects except in certain types of structures. Additional information on this topic can be found in ASCE/SEI C10.7.

4.4 Flowcharts

A summary of the flowcharts provided in this chapter is given in Table 4.8.

Table 4.8 Summary of Flowcharts Provided in Chapter 4

Flowchart	Title
4.1	Flat Roof Snow Load, p_f (ASCE/SEI 7.3)
4.2	Sloped Roof Snow Load, p_s (ASCE/SEI 7.4)
4.3	Unbalanced Snow Loads—Hip and Gable Roofs (ASCE/SEI 7.6.1)
4.4	Unbalanced Snow Loads—Curved Roofs (ASCE/SEI 7.6.2)
4.5	Unbalanced Snow Loads—Sawtooth Roofs (ASCE/SEI 7.6.3)
4.6	Drifts on Lower Roof of a Structure (ASCE/SEI 7.7.1)
4.7	Drifts on Lower Roof of an Adjacent Structure (ASCE/SEI 7.7.2)
4.8	Drifts on Roof Projections and Parapets (ASCE/SEI 7.8)
4.9	Sliding Snow (ASCE/SEI 7.9)
4.10	Wind Velocity Pressure, q_z (ASCE/SEI 26.10)

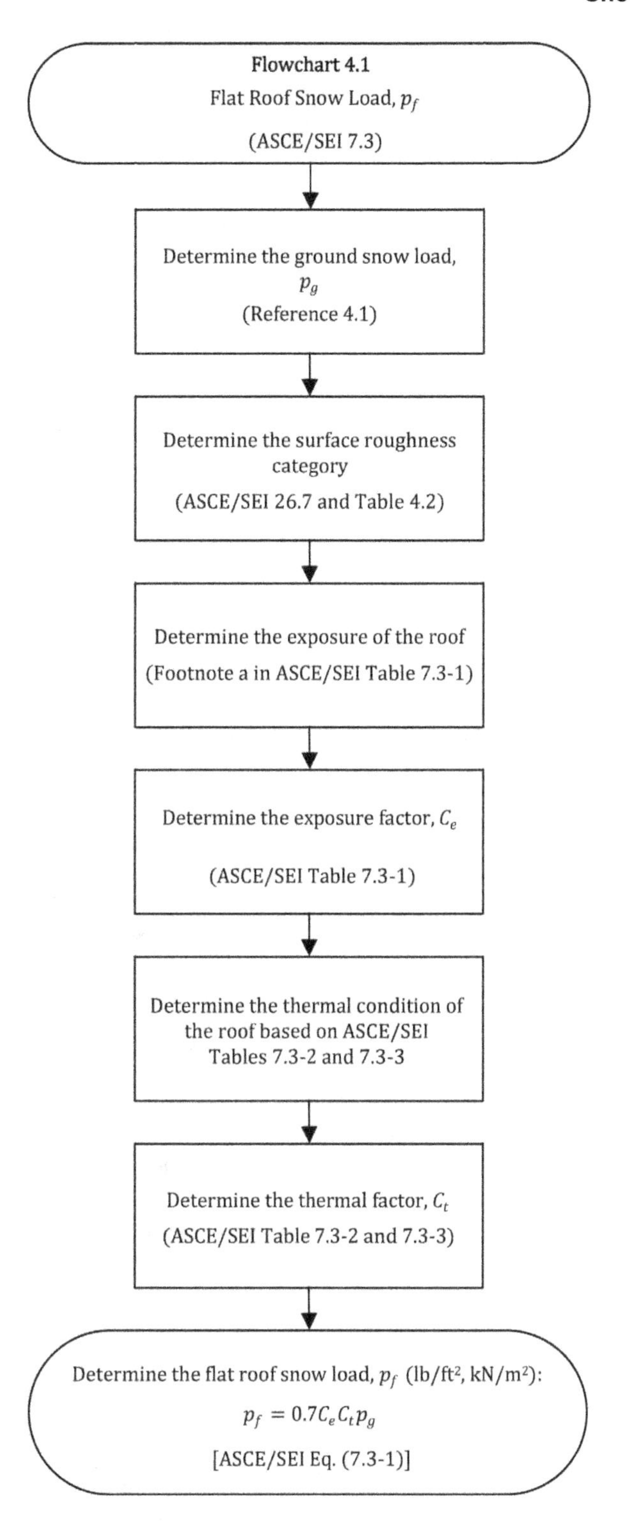

Flowchart 4.1
Flat Roof Snow Load, p_f
(ASCE/SEI 7.3)
Determine the ground snow load, p_g
(Reference 4.1)
Determine the surface roughness category
(ASCE/SEI 26.7 and Table 4.2)
Determine the exposure of the roof
(Footnote a in ASCE/SEI Table 7.3-1)
Determine the exposure factor, C_e
(ASCE/SEI Table 7.3-1)
Determine the thermal condition of the roof based on ASCE/SEI Tables 7.3-2 and 7.3-3
Determine the thermal factor, C_t
(ASCE/SEI Table 7.3-2 and 7.3-3)
Determine the flat roof snow load, p_f (lb/ft^2, kN/m^2):
$p_f = 0.7C_eC_tp_g$
[ASCE/SEI Eq. (7.3-1)]

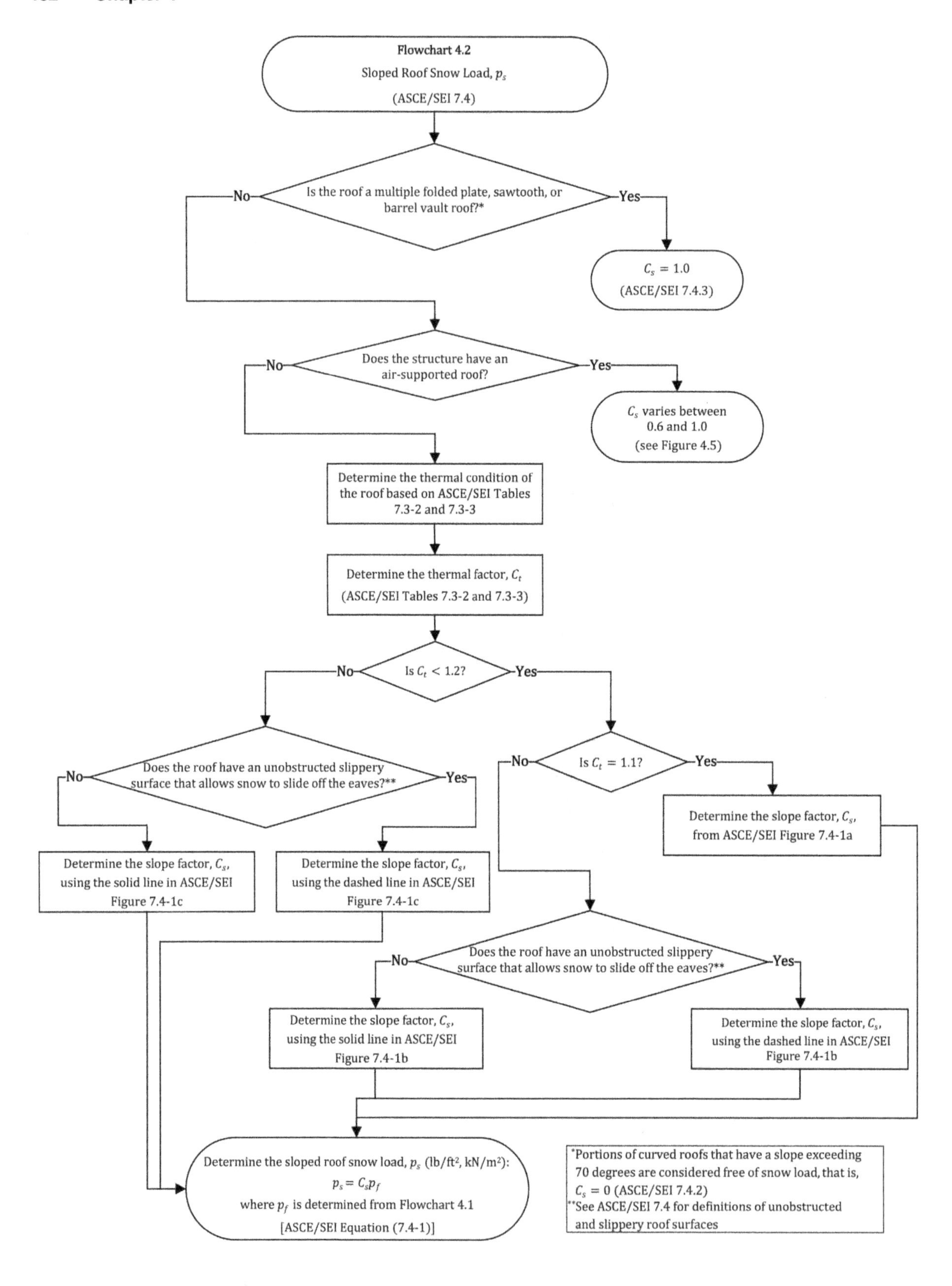
Flowchart 4.2
Sloped Roof Snow Load, p_s
(ASCE/SEI 7.4)
Is the roof a multiple folded plate, sawtooth, or barrel vault roof?*
No
Yes
$C_s = 1.0$
(ASCE/SEI 7.4.3)
Does the structure have an air-supported roof?
No
Yes
C_s varies between 0.6 and 1.0
(see Figure 4.5)
Determine the thermal condition of the roof based on ASCE/SEI Tables 7.3-2 and 7.3-3
Determine the thermal factor, C_t
(ASCE/SEI Tables 7.3-2 and 7.3-3)
Is $C_t < 1.2$?
No
Yes
Does the roof have an unobstructed slippery surface that allows snow to slide off the eaves?**
No
Yes
Is $C_t = 1.1$?
No
Yes
Determine the slope factor, C_s, from ASCE/SEI Figure 7.4-1a
Determine the slope factor, C_s, using the solid line in ASCE/SEI Figure 7.4-1c
Determine the slope factor, C_s, using the dashed line in ASCE/SEI Figure 7.4-1c
Does the roof have an unobstructed slippery surface that allows snow to slide off the eaves?**
No
Yes
Determine the slope factor, C_s, using the solid line in ASCE/SEI Figure 7.4-1b
Determine the slope factor, C_s, using the dashed line in ASCE/SEI Figure 7.4-1b
Determine the sloped roof snow load, p_s (lb/ft², kN/m²):
$p_s = C_s p_f$
where p_f is determined from Flowchart 4.1
[ASCE/SEI Equation (7.4-1)]
*Portions of curved roofs that have a slope exceeding 70 degrees are considered free of snow load, that is, $C_s = 0$ (ASCE/SEI 7.4.2)
**See ASCE/SEI 7.4 for definitions of unobstructed and slippery roof surfaces

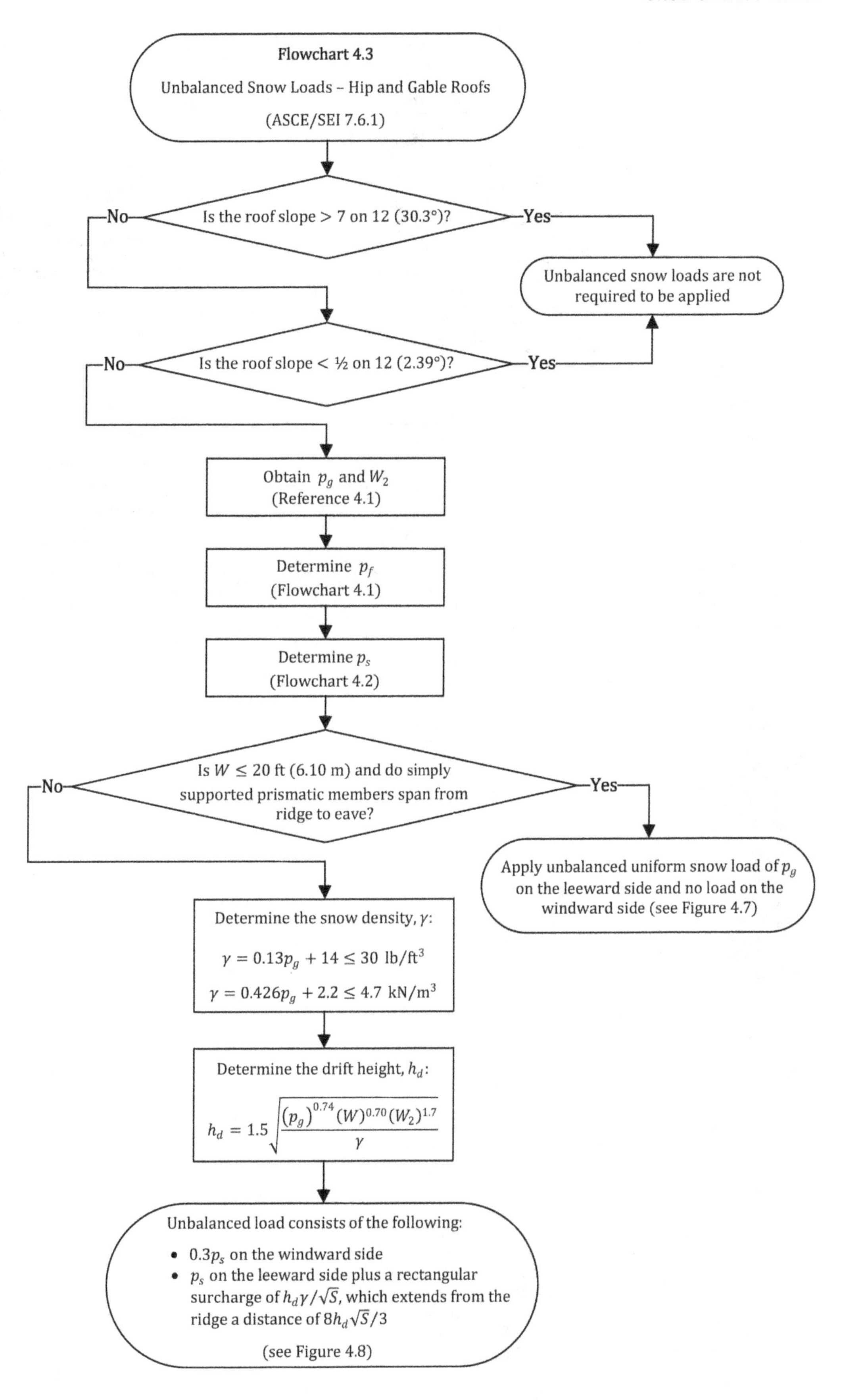
Flowchart 4.3
Unbalanced Snow Loads – Hip and Gable Roofs
(ASCE/SEI 7.6.1)
Is the roof slope > 7 on 12 (30.3°)?
No
Yes
Unbalanced snow loads are not required to be applied
Is the roof slope < ½ on 12 (2.39°)?
No
Yes
Obtain p_g and W_2
(Reference 4.1)
Determine p_f
(Flowchart 4.1)
Determine p_s
(Flowchart 4.2)
Is $W \leq 20$ ft (6.10 m) and do simply supported prismatic members span from ridge to eave?
No
Yes
Apply unbalanced uniform snow load of p_g on the leeward side and no load on the windward side (see Figure 4.7)
Determine the snow density, γ:
$\gamma = 0.13p_g + 14 \leq 30$ lb/ft^3
$\gamma = 0.426p_g + 2.2 \leq 4.7$ kN/m^3
Determine the drift height, h_d:
$h_d = 1.5\sqrt{\frac{(p_g)^{0.74}(W)^{0.70}(W_2)^{1.7}}{\gamma}}$
Unbalanced load consists of the following:
• $0.3p_s$ on the windward side
• p_s on the leeward side plus a rectangular surcharge of $h_d\gamma/\sqrt{S}$, which extends from the ridge a distance of $8h_d\sqrt{S}/3$
(see Figure 4.8)

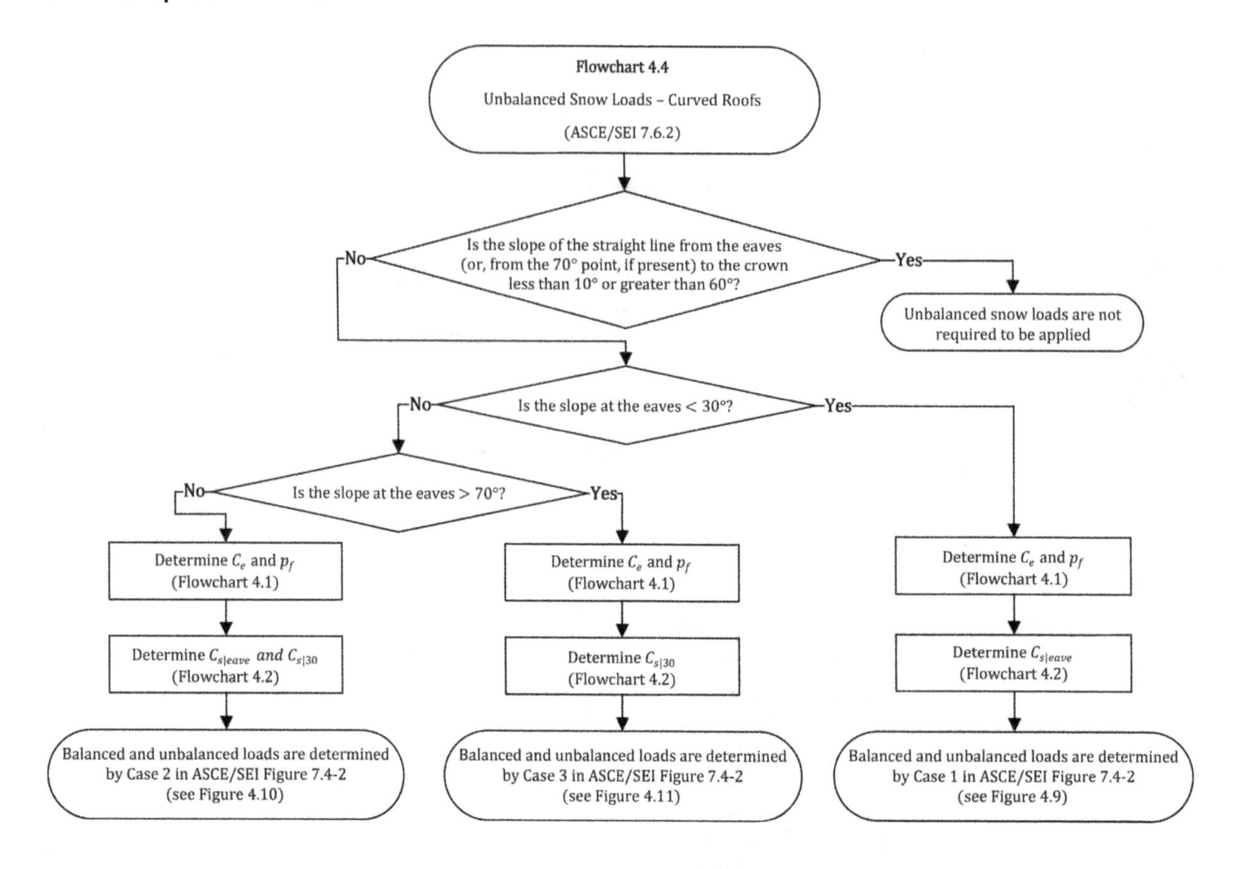

Flowchart 4.4
Unbalanced Snow Loads – Curved Roofs
(ASCE/SEI 7.6.2)
Is the slope of the straight line from the eaves (or, from the 70° point, if present) to the crown less than 10° or greater than 60°?
No
Yes
Unbalanced snow loads are not required to be applied
Is the slope at the eaves < 30°?
No
Yes
Is the slope at the eaves > 70°?
No
Yes
Determine C_e and p_f (Flowchart 4.1)
Determine $C_{s|eave}$ and $C_{s|30}$ (Flowchart 4.2)
Balanced and unbalanced loads are determined by Case 2 in ASCE/SEI Figure 7.4-2 (see Figure 4.10)
Determine C_e and p_f (Flowchart 4.1)
Determine $C_{s|30}$ (Flowchart 4.2)
Balanced and unbalanced loads are determined by Case 3 in ASCE/SEI Figure 7.4-2 (see Figure 4.11)
Determine C_e and p_f (Flowchart 4.1)
Determine $C_{s|eave}$ (Flowchart 4.2)
Balanced and unbalanced loads are determined by Case 1 in ASCE/SEI Figure 7.4-2 (see Figure 4.9)

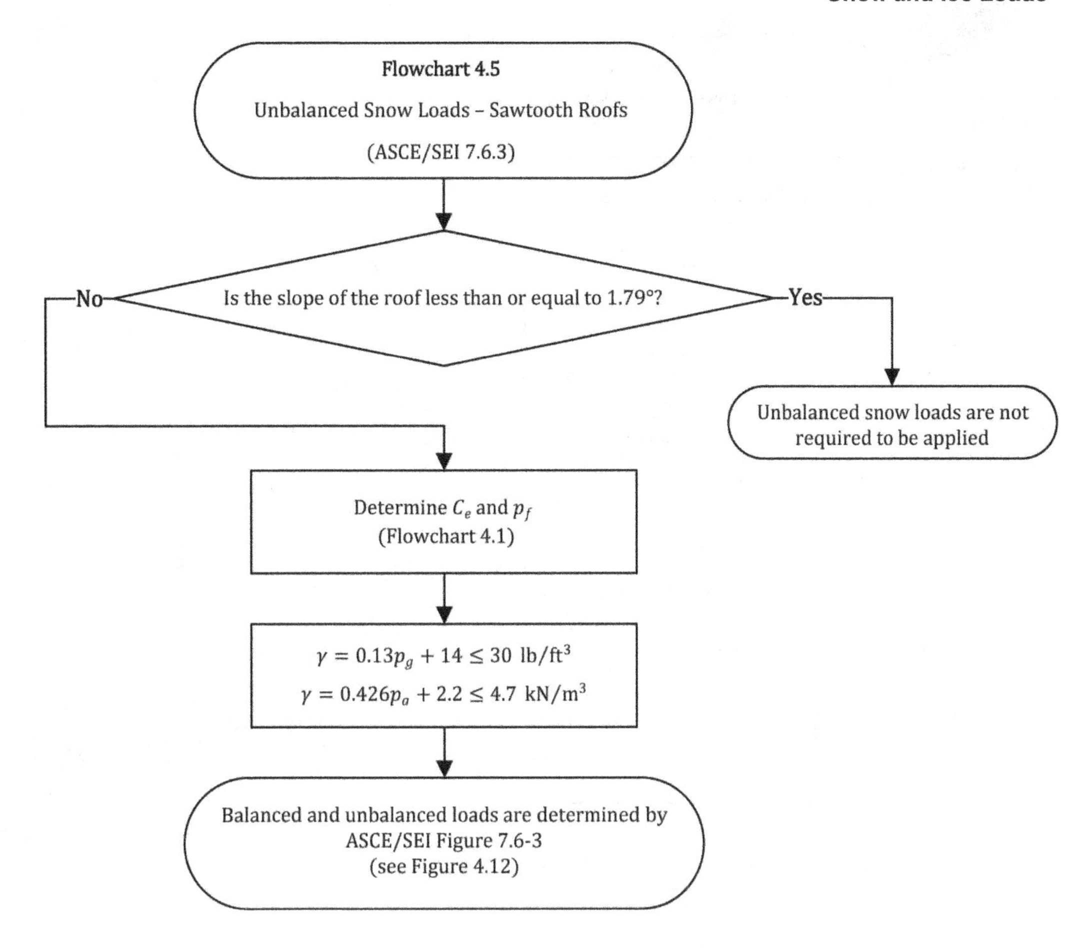
Flowchart 4.5
Unbalanced Snow Loads – Sawtooth Roofs
(ASCE/SEI 7.6.3)
Is the slope of the roof less than or equal to 1.79°?
No
Yes
Unbalanced snow loads are not required to be applied
Determine C_e and p_f (Flowchart 4.1)
$\gamma = 0.13p_g + 14 \leq 30$ lb/ft^3
$\gamma = 0.426p_g + 2.2 \leq 4.7$ kN/m^3
Balanced and unbalanced loads are determined by ASCE/SEI Figure 7.6-3 (see Figure 4.12)

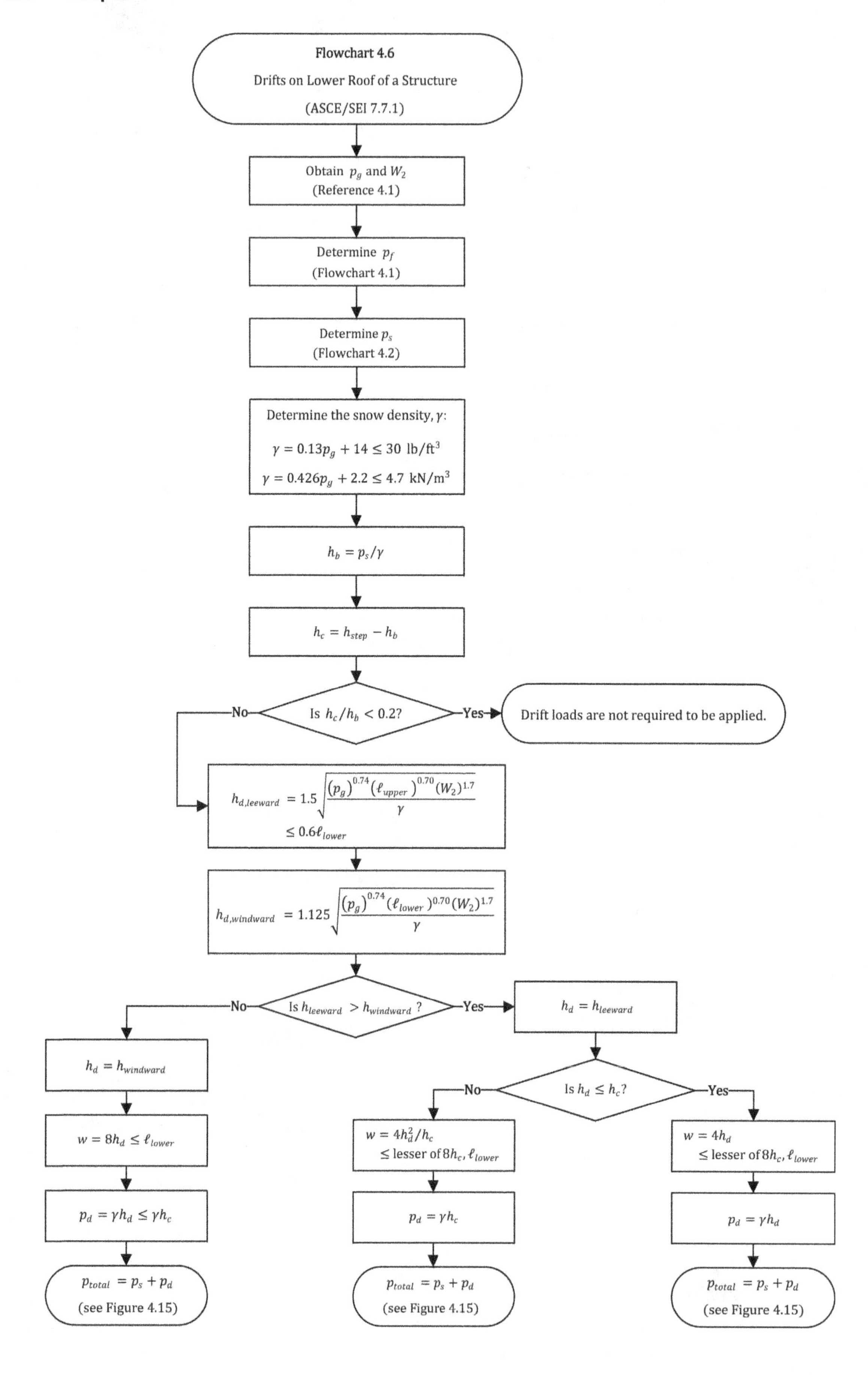

Flowchart 4.6
Drifts on Lower Roof of a Structure
(ASCE/SEI 7.7.1)
Obtain p_g and W_2
(Reference 4.1)
Determine p_f
(Flowchart 4.1)
Determine p_s
(Flowchart 4.2)
Determine the snow density, γ:
$\gamma = 0.13p_g + 14 \leq 30$ lb/ft^3
$\gamma = 0.426p_g + 2.2 \leq 4.7$ kN/m^3
$h_b = p_s/\gamma$
$h_c = h_{step} - h_b$
Is $h_c/h_b < 0.2$?
No
Yes
Drift loads are not required to be applied.
$h_{d,leeward} = 1.5\sqrt{\frac{(p_g)^{0.74}(\ell_{upper})^{0.70}(W_2)^{1.7}}{\gamma}} \leq 0.6\ell_{lower}$
$h_{d,windward} = 1.125\sqrt{\frac{(p_g)^{0.74}(\ell_{lower})^{0.70}(W_2)^{1.7}}{\gamma}}$
Is $h_{leeward} > h_{windward}$?
No
Yes
$h_d = h_{leeward}$
$h_d = h_{windward}$
Is $h_d \leq h_c$?
No
Yes
$w = 8h_d \leq \ell_{lower}$
$w = 4h_d^2/h_c \leq$ lesser of $8h_c, \ell_{lower}$
$w = 4h_d \leq$ lesser of $8h_c, \ell_{lower}$
$p_d = \gamma h_d \leq \gamma h_c$
$p_d = \gamma h_c$
$p_d = \gamma h_d$
$p_{total} = p_s + p_d$
(see Figure 4.15)
$p_{total} = p_s + p_d$
(see Figure 4.15)
$p_{total} = p_s + p_d$
(see Figure 4.15)

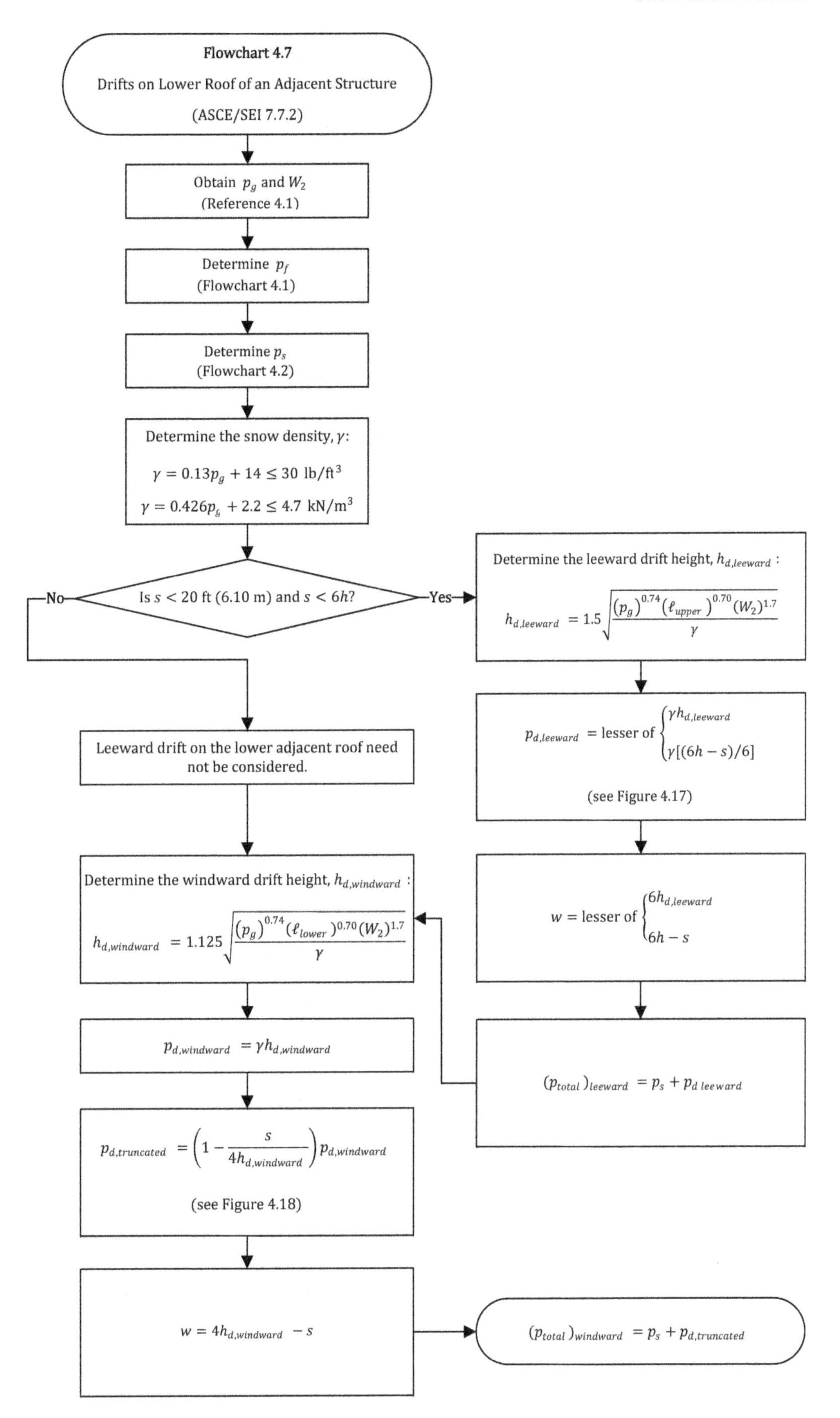

Flowchart 4.7
Drifts on Lower Roof of an Adjacent Structure
(ASCE/SEI 7.7.2)
Obtain p_g and W_2
(Reference 4.1)
Determine p_f
(Flowchart 4.1)
Determine p_s
(Flowchart 4.2)
Determine the snow density, γ:
$\gamma = 0.13p_g + 14 \leq 30$ lb/ft^3
$\gamma = 0.426p_g + 2.2 \leq 4.7$ kN/m^3
Is $s < 20$ ft (6.10 m) and $s < 6h$?
No
Yes
Determine the leeward drift height, $h_{d,leeward}$:
$h_{d,leeward} = 1.5\sqrt{\frac{(p_g)^{0.74}(\ell_{upper})^{0.70}(W_2)^{1.7}}{\gamma}}$
Leeward drift on the lower adjacent roof need not be considered.
$p_{d,leeward} = \text{lesser of} \begin{cases} \gamma h_{d,leeward} \\ \gamma[(6h - s)/6] \end{cases}$
(see Figure 4.17)
Determine the windward drift height, $h_{d,windward}$:
$h_{d,windward} = 1.125\sqrt{\frac{(p_g)^{0.74}(\ell_{lower})^{0.70}(W_2)^{1.7}}{\gamma}}$
$w = \text{lesser of} \begin{cases} 6h_{d,leeward} \\ 6h - s \end{cases}$
$p_{d,windward} = \gamma h_{d,windward}$
$(p_{total})_{leeward} = p_s + p_{d\ leeward}$
$p_{d,truncated} = \left(1 - \frac{s}{4h_{d,windward}}\right)p_{d,windward}$
(see Figure 4.18)
$w = 4h_{d,windward} - s$
$(p_{total})_{windward} = p_s + p_{d,truncated}$

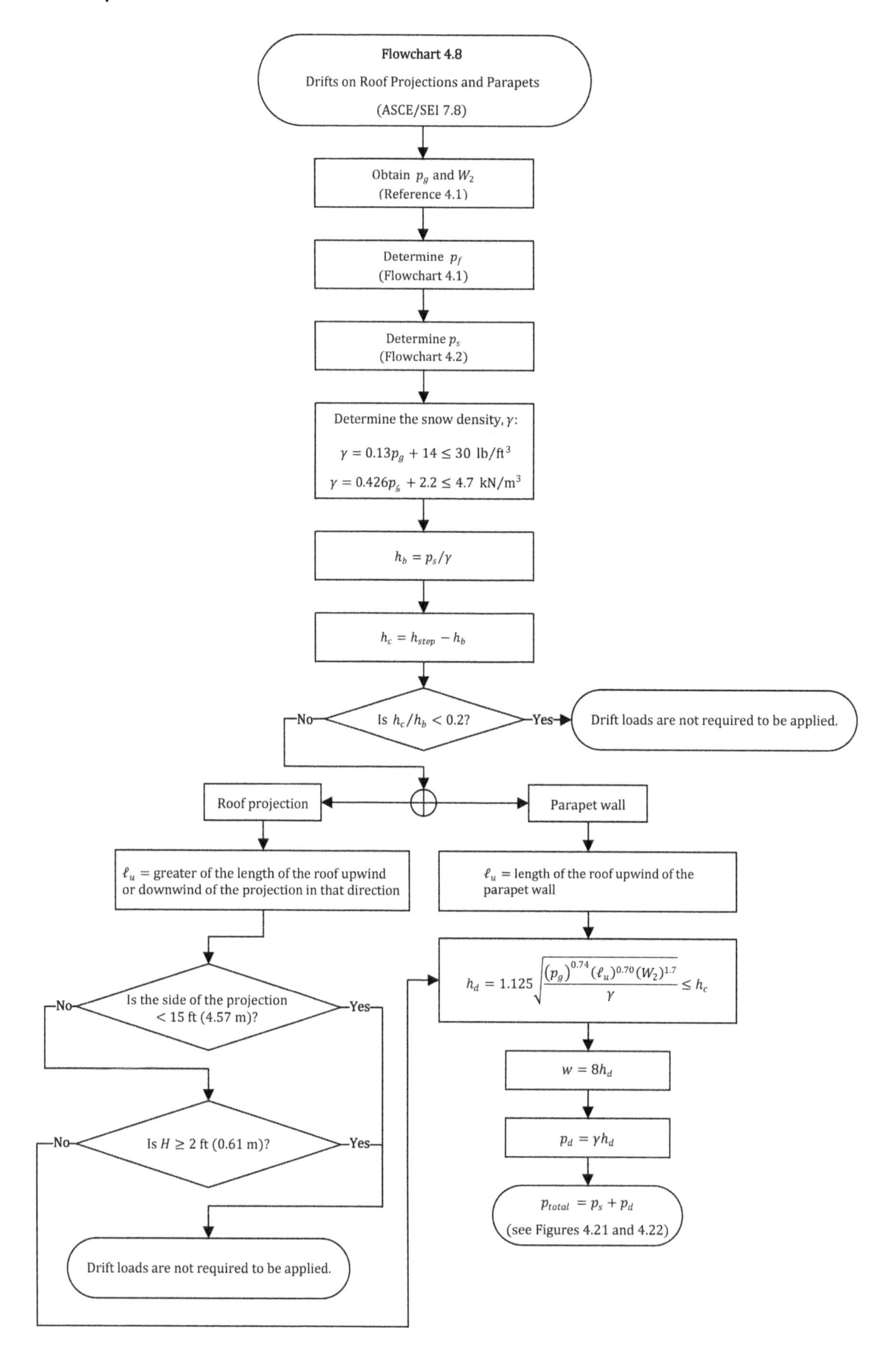

Flowchart 4.8
Drifts on Roof Projections and Parapets
(ASCE/SEI 7.8)
Obtain p_g and W_2
(Reference 4.1)
Determine p_f
(Flowchart 4.1)
Determine p_s
(Flowchart 4.2)
Determine the snow density, γ:
$\gamma = 0.13p_g + 14 \leq 30$ lb/ft^3
$\gamma = 0.426p_g + 2.2 \leq 4.7$ kN/m^3
$h_b = p_s/\gamma$
$h_c = h_{step} - h_b$
Is $h_c/h_b < 0.2$?
No
Yes
Drift loads are not required to be applied.
Roof projection
Parapet wall
ℓ_u = greater of the length of the roof upwind or downwind of the projection in that direction
ℓ_u = length of the roof upwind of the parapet wall
$h_d = 1.125\sqrt{\frac{(p_g)^{0.74}(\ell_u)^{0.70}(W_2)^{1.7}}{\gamma}} \leq h_c$
Is the side of the projection < 15 ft (4.57 m)?
No
Yes
Is $H \geq 2$ ft (0.61 m)?
No
Yes
$w = 8h_d$
$p_d = \gamma h_d$
$p_{total} = p_s + p_d$
(see Figures 4.21 and 4.22)
Drift loads are not required to be applied.

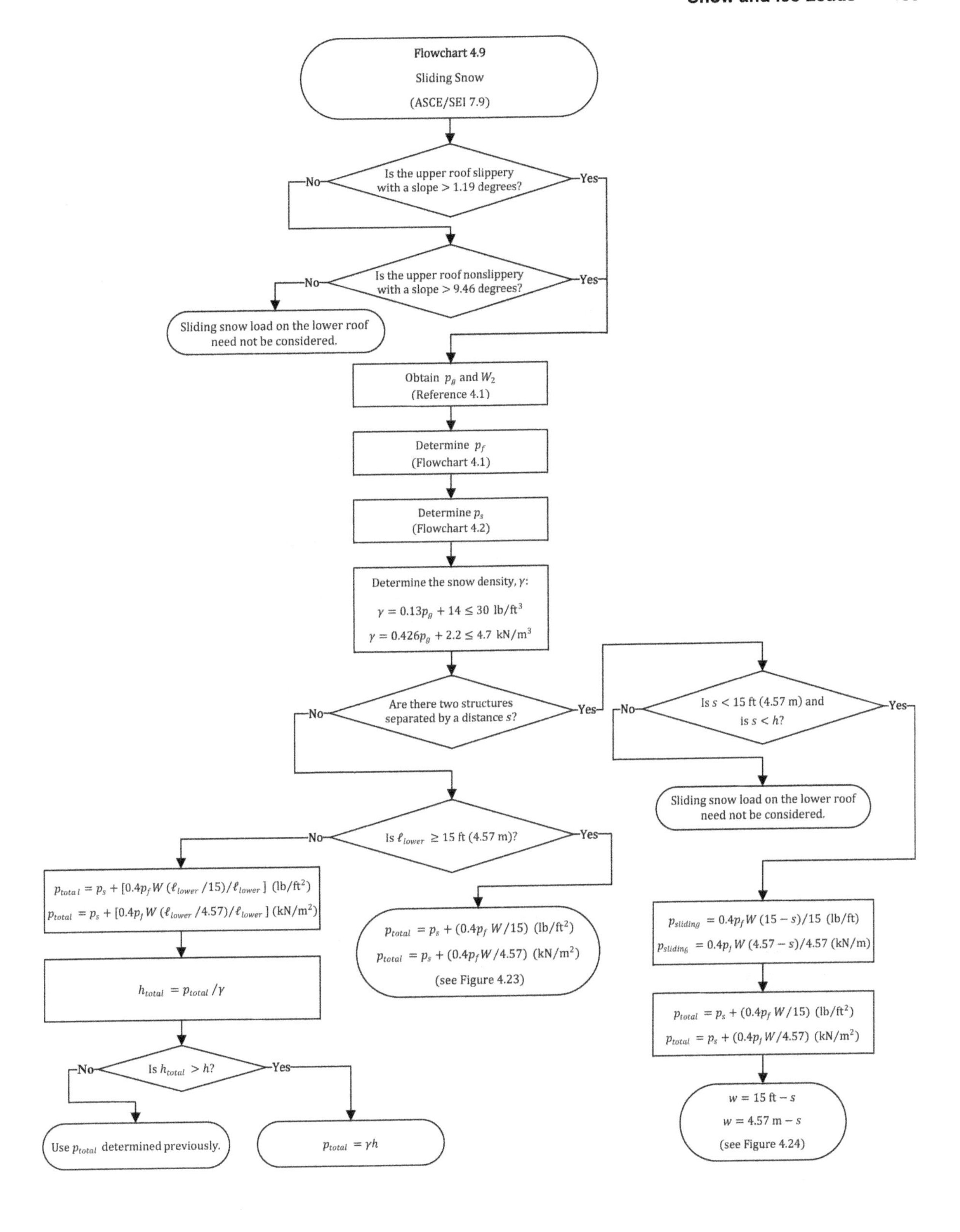

Flowchart 4.9
Sliding Snow
(ASCE/SEI 7.9)
Is the upper roof slippery with a slope > 1.19 degrees?
No
Yes
Is the upper roof nonslippery with a slope > 9.46 degrees?
No
Yes
Sliding snow load on the lower roof need not be considered.
Obtain p_g and W_2 (Reference 4.1)
Determine p_f (Flowchart 4.1)
Determine p_s (Flowchart 4.2)
Determine the snow density, γ:
$\gamma = 0.13p_g + 14 \le 30$ lb/ft³
$\gamma = 0.426p_g + 2.2 \le 4.7$ kN/m³
Are there two structures separated by a distance s?
No
Yes
Is $s < 15$ ft (4.57 m) and is $s < h$?
No
Yes
Sliding snow load on the lower roof need not be considered.
Is $\ell_{lower} \ge 15$ ft (4.57 m)?
No
Yes
$p_{total} = p_s + [0.4p_f W(\ell_{lower}/15)/\ell_{lower}]$ (lb/ft²)
$p_{total} = p_s + [0.4p_f W(\ell_{lower}/4.57)/\ell_{lower}]$ (kN/m²)
$p_{total} = p_s + (0.4p_f W/15)$ (lb/ft²)
$p_{total} = p_s + (0.4p_f W/4.57)$ (kN/m²)
(see Figure 4.23)
$p_{sliding} = 0.4p_f W(15 - s)/15$ (lb/ft)
$p_{sliding} = 0.4p_f W(4.57 - s)/4.57$ (kN/m)
$h_{total} = p_{total}/\gamma$
$p_{total} = p_s + (0.4p_f W/15)$ (lb/ft²)
$p_{total} = p_s + (0.4p_f W/4.57)$ (kN/m²)
Is $h_{total} > h$?
No
Yes
$w = 15$ ft $- s$
$w = 4.57$ m $- s$
(see Figure 4.24)
Use p_{total} determined previously.
$p_{total} = \gamma h$

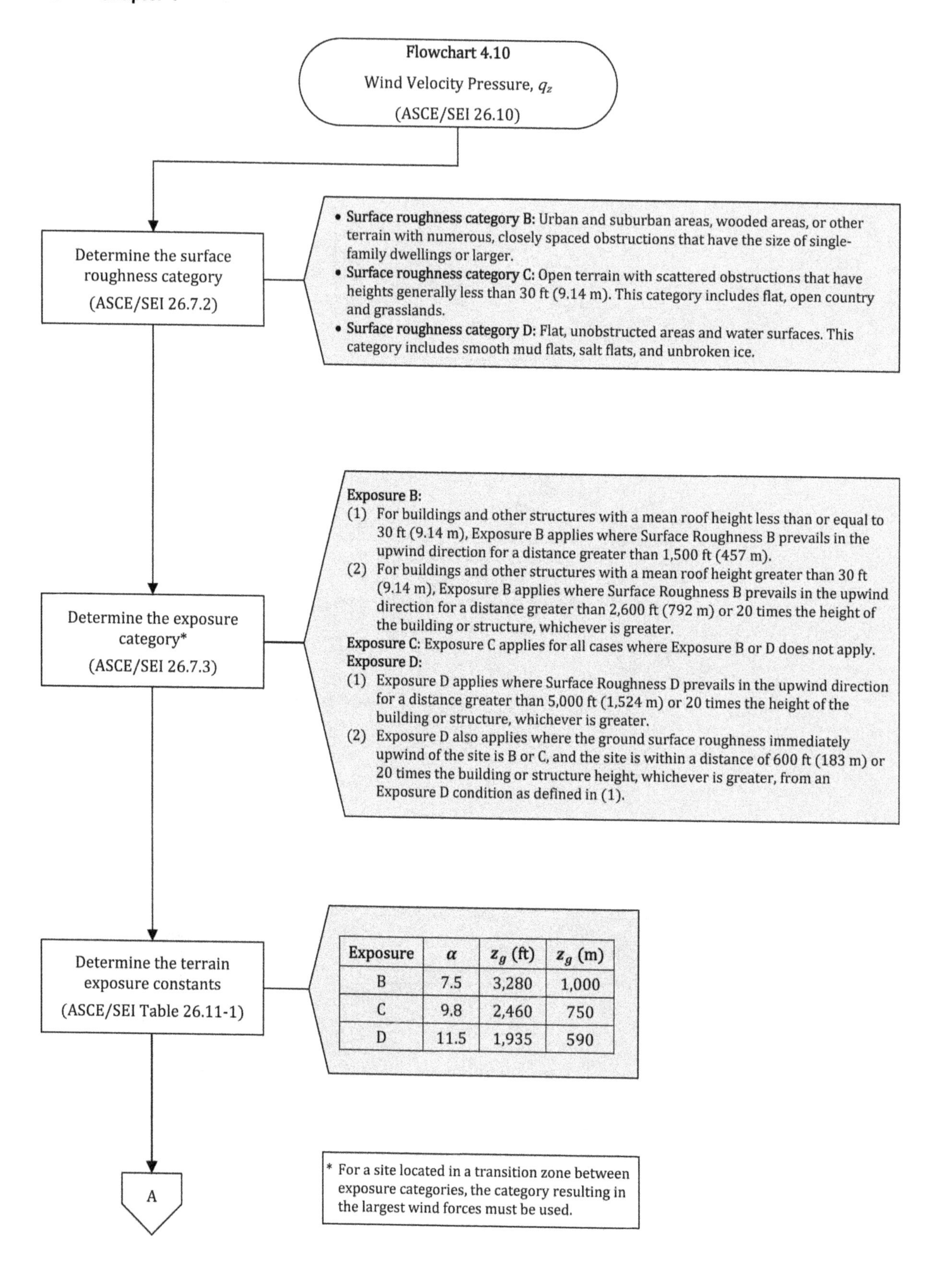
Flowchart 4.10
Wind Velocity Pressure, q_z
(ASCE/SEI 26.10)
Determine the surface roughness category
(ASCE/SEI 26.7.2)
• Surface roughness category B: Urban and suburban areas, wooded areas, or other terrain with numerous, closely spaced obstructions that have the size of single-family dwellings or larger.
• Surface roughness category C: Open terrain with scattered obstructions that have heights generally less than 30 ft (9.14 m). This category includes flat, open country and grasslands.
• Surface roughness category D: Flat, unobstructed areas and water surfaces. This category includes smooth mud flats, salt flats, and unbroken ice.
Determine the exposure category*
(ASCE/SEI 26.7.3)
Exposure B:
(1) For buildings and other structures with a mean roof height less than or equal to 30 ft (9.14 m), Exposure B applies where Surface Roughness B prevails in the upwind direction for a distance greater than 1,500 ft (457 m).
(2) For buildings and other structures with a mean roof height greater than 30 ft (9.14 m), Exposure B applies where Surface Roughness B prevails in the upwind direction for a distance greater than 2,600 ft (792 m) or 20 times the height of the building or structure, whichever is greater.
Exposure C: Exposure C applies for all cases where Exposure B or D does not apply.
Exposure D:
(1) Exposure D applies where Surface Roughness D prevails in the upwind direction for a distance greater than 5,000 ft (1,524 m) or 20 times the height of the building or structure, whichever is greater.
(2) Exposure D also applies where the ground surface roughness immediately upwind of the site is B or C, and the site is within a distance of 600 ft (183 m) or 20 times the building or structure height, whichever is greater, from an Exposure D condition as defined in (1).
Determine the terrain exposure constants
(ASCE/SEI Table 26.11-1)
A
* For a site located in a transition zone between exposure categories, the category resulting in the largest wind forces must be used.

Exposure	α	z_g (ft)	z_g (m)
B	7.5	3,280	1,000
C	9.8	2,460	750
D	11.5	1,935	590

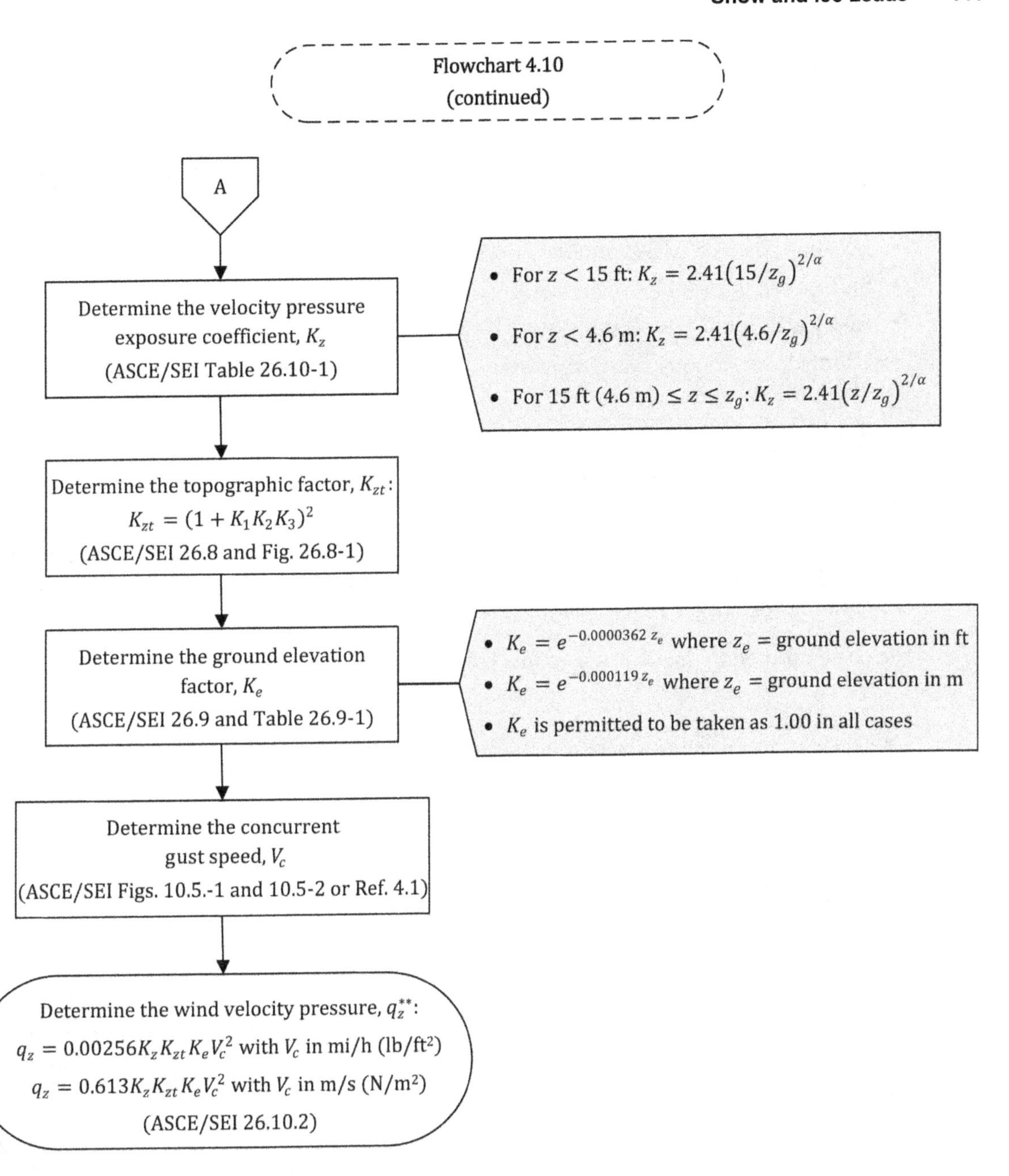

Flowchart 4.10
(continued)
A
Determine the velocity pressure exposure coefficient, K_z
(ASCE/SEI Table 26.10-1)
• For $z < 15$ ft: $K_z = 2.41(15/z_g)^{2/\alpha}$
• For $z < 4.6$ m: $K_z = 2.41(4.6/z_g)^{2/\alpha}$
• For 15 ft (4.6 m) $\leq z \leq z_g$: $K_z = 2.41(z/z_g)^{2/\alpha}$
Determine the topographic factor, K_{zt}:
$K_{zt} = (1 + K_1 K_2 K_3)^2$
(ASCE/SEI 26.8 and Fig. 26.8-1)
Determine the ground elevation factor, K_e
(ASCE/SEI 26.9 and Table 26.9-1)
• $K_e = e^{-0.0000362\, z_e}$ where z_e = ground elevation in ft
• $K_e = e^{-0.000119\, z_e}$ where z_e = ground elevation in m
• K_e is permitted to be taken as 1.00 in all cases
Determine the concurrent gust speed, V_c
(ASCE/SEI Figs. 10.5.-1 and 10.5-2 or Ref. 4.1)
Determine the wind velocity pressure, q_z^{**}:
$q_z = 0.00256 K_z K_{zt} K_e V_c^2$ with V_c in mi/h (lb/ft²)
$q_z = 0.613 K_z K_{zt} K_e V_c^2$ with V_c in m/s (N/m²)
(ASCE/SEI 26.10.2)

** Wind velocity pressure at the mean roof height, q_h, is computed as $q_h = q_z$ using $K_z = K_h$ at the mean roof height, h.

4.5 Examples

4.5.1 Example 4.1—Design Snow Loads on a Monoslope Roof

Determine the design snow loads on the monoslope roof in Figure 4.26 given the following design data:

- Location: Bangor, ME (Latitude = 44.80°, Longitude = −68.77°)
- Surface roughness: C
- Risk Category: II (IBC Table 1604.5)
- Thermal condition: Heated structure with an unventilated roof and $R_{roof} = 40$ h ft²°F/Btu (7.04 m²K/W)
- Roof exposure: Partially exposed
- Roof surface: Smooth bituminous membrane
- Roof obstructions: None
- Roof framing: Primary members spaced 20 ft (6.10 m) on center overhanging a wall and simply supported secondary members spaced at 5 ft (1.52 m) on center parallel to the free-draining edge of the roof

The design procedure in Table 4.1 of this book is used to determine the design snow loads.

Figure 4.26 Elevation of the Building in Example 4.1

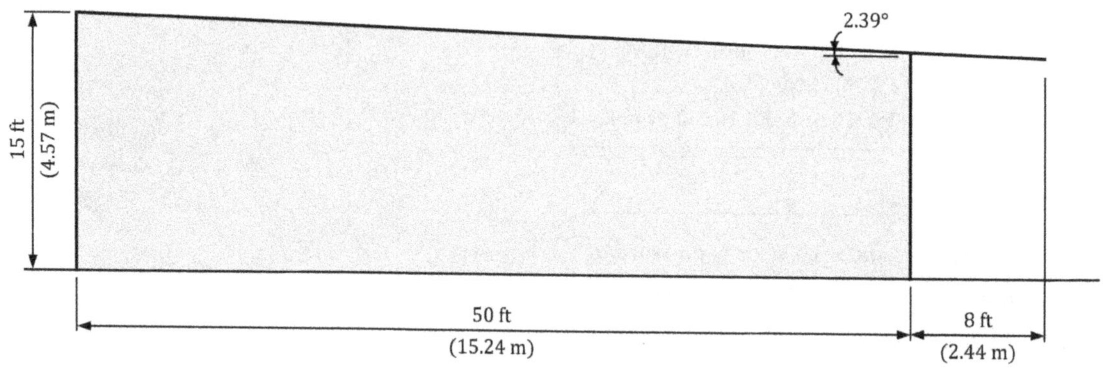

SOLUTION

Step 1—Determine the ground snow load, p_g

Using Reference 4.1, the ground snow load is equal to 106 lb/ft² (5.08 kN/m²) for Risk Category II.

Step 2—Determine the flat roof snow load, p_f

Flowchart 4.1 is used to determine p_f.

Step 2a—Determine the surface roughness category

The surface roughness category is given in the design data as C.

Step 2b—Determine the exposure of the roof

The roof exposure is given in the design data as partially exposed.

Step 2c—Determine the exposure factor, C_e

Given a surface roughness category of C and a partially exposed roof exposure, $C_e = 1.0$ from ASCE/SEI Table 7.3-1.

Step 2d—Determine the thermal condition of the roof

From the design data, it is a heated structure with an unventilated roof with $R_{roof} = 40$ h ft²°F/Btu (7.04 m²K/W).

Step 2e—Determine the thermal factor, C_t

For $R_{roof} = 40$ h ft²°F/Btu (7.04 m²K/W) and $p_g = 106$ lb/ft² (5.08 kN/m²) > 70 lb/ft² (3.36 kN/m²), $C_t = 1.15$ (ASCE/SEI Table 7.3-3).

Step 2f—Determine the flat roof snow load

$$p_f = 0.7C_eC_tp_g = 0.7\times1.0\times1.15\times106 = 85.3 \text{ lb/ft}^2$$

$$p_f = 0.7C_eC_tp_g = 0.7\times1.0\times1.15\times5.08 = 4.09 \text{ kN/m}^2$$

Step 3—Determine the minimum snow load for low-slope roofs, p_m

A minimum snow load, p_m, applies to monoslope roofs with slopes less than 15 degrees (see Figure 4.2). Because the roof slope in this example is equal to 2.39 degrees, minimum snow loads must be considered.

For Risk Category II, the minimum snow load upper limit, $p_{m,\max}$, is equal to 30 lb/ft² (1.44 kN/m²) from ASCE/SEI Table 7.3-4. Because $p_g = 106$ lb/ft² (5.08 kN/m²) > $p_{m,\max}$ = 30 lb/ft² (1.44 kN/m²), $p_m = p_{m,\max} = 30$ lb/ft² (1.44 kN/m²) (ASCE/SEI 7.3.3).

Step 4—Determine the sloped roof (balanced) snow load, p_s

Flowchart 4.2 is used to determine p_s.

Step 4a—Determine the slope factor, C_s

From Step 2e, $C_t = 1.15$, which means the roof is warm.

From the design data, there are no obstructions inhibiting the snow from sliding off the roof. Also, because $C_t > 1.1$, ice dams on all overhanging portions need not be considered for this unventilated roof (if an ice dam can form, it is considered to be an obstruction; see ASCE/SEI 7.4.4).

The roof surface is a smooth bituminous membrane from the design data. According to ASCE/SEI 7.4, smooth bituminous membranes are considered to be slippery surfaces.

Because the roof is unobstructed and slippery with $1.1 < C_t = 1.15 < 1.2$, use the dashed line in ASCE/SEI Figure 7.4-1b to determine the slope factor, C_s. For a roof slope of 2.39 degrees, which is less than 10 degrees, $C_s = 1.0$.

Step 4b—Determine the sloped roof (balanced) snow load, p_s

$$p_s = C_sp_f = 1.0\times85.3 = 85.3 \text{ lb/ft}^2$$

$$p_s = C_sp_f = 1.0\times4.09 = 4.09 \text{ kN/m}^2$$

Step 5—Consider partial loading

From the design data, the secondary members are simply supported, so partial loading need not be considered for these members (ASCE/SEI 7.5).

The primary framing members are continuous over the wall, which means partial loading must be considered for these cantilevered members. The balanced snow load determined in Step 4b is used in partial loading cases; the minimum snow load is not applicable in such cases (ASCE/SEI 7.3.3). With a center-to-center spacing of 20 ft (6.10 m), the partial loads on a typical primary member are determined as follows (see Figure 4.6):

- Case 1:

  Balanced load on cantilevers = $85.3\times20.0 = 1,706$ lb/ft

  Partial load on main span = one-half of balanced load = $0.5\times1,706 = 853$ lb/ft

- Case 2:
  Partial load on cantilevers = one-half of balanced load $= 0.5 \times 1,706 = 853$ lb/ft
  Balanced load on main span $= 1,706$ lb/ft

In S.I.:

- Case 1:
  Balanced load on cantilevers $= 4.09 \times 6.10 = 24.95$ kN/m
  Partial load on main span = one-half of balanced load $= 0.5 \times 24.95 = 12.48$ kN/m
- Case 2:
  Partial load on cantilevers = one-half of balanced load $= 0.5 \times 24.95 = 12.48$ kN/m
  Balanced load on main span $= 24.95$ kN/m

Step 6—Consider unbalanced snow loads
Because this roof is monoslope, unbalanced snow loads need not be considered (ASCE/SEI 7.6).

Step 7—Consider drift loads on lower roofs
Not applicable.

Step 8—Consider drifts on roof projections and parapets
Not applicable.

Step 9—Consider sliding snow
Not applicable.

Step 10—Consider rain-on-snow surcharge loads
According to ASCE/SEI 7.10, a rain-on-snow surcharge load is required for locations where $p_g \le p_{m,\max}$ and $p_g > 0$ on roofs with slopes less than $W/50$ (in S.I.: $W/15.2$). In this example, $p_g = 106$ lb/ft^2(5.08 kN/m^2) $> p_{m,\max} = 30$ lb/ft^2 (1.44 kN/m^2), so a rain-on-snow surcharge load need not be added to the balanced snow load.

Step 11—Consider ponding loads
A ponding analysis considering roof deflections caused by full snow loads in combination with dead and superimposed dead loads must be performed for this roof based on the stiffness of the roof structural members.

Step 12—Consider snow loads on existing roofs
Not applicable.

Step 13—Consider snow loads on open-frame equipment structures
Not applicable.

Design snow loads on the secondary and primary roof members are determined as follows:

- Secondary roof members
  Balanced snow load:
  Uniform design snow load $= 85.3 \times 5.0 = 426.5$ lb/ft
  Uniform design snow load $= 4.09 \times 1.52 = 6.22$ kN/m
  Minimum snow load:
  Uniform design snow load $= 30.0 \times 5.0 = 150.0$ lb/ft
  Uniform design snow load $= 1.44 \times 1.52 = 2.19$ kN/m
- Primary roof members
  Balanced snow load:
  Uniform design snow load $= 85.3 \times 20.0 = 1,706.0$ lb/ft
  Uniform design snow load $= 4.09 \times 6.10 = 24.95$ kN/m
  Minimum snow load:
  Uniform design snow load $= 30.0 \times 20.0 = 600.0$ lb/ft
  Uniform design snow load $= 1.44 \times 6.10 = 8.78$ kN/m
  Partial loads on the primary members are determined in Step 5.

The balanced, minimum, and partial loads for the primary roof members are given in Figure 4.27. These members must be designed for these design snow load cases in combination with other applicable design loads using the appropriate load combinations.

Figure 4.27 Design Snow Load Cases for the Primary Roof Members Supporting the Monoslope Roof in Example 4.1

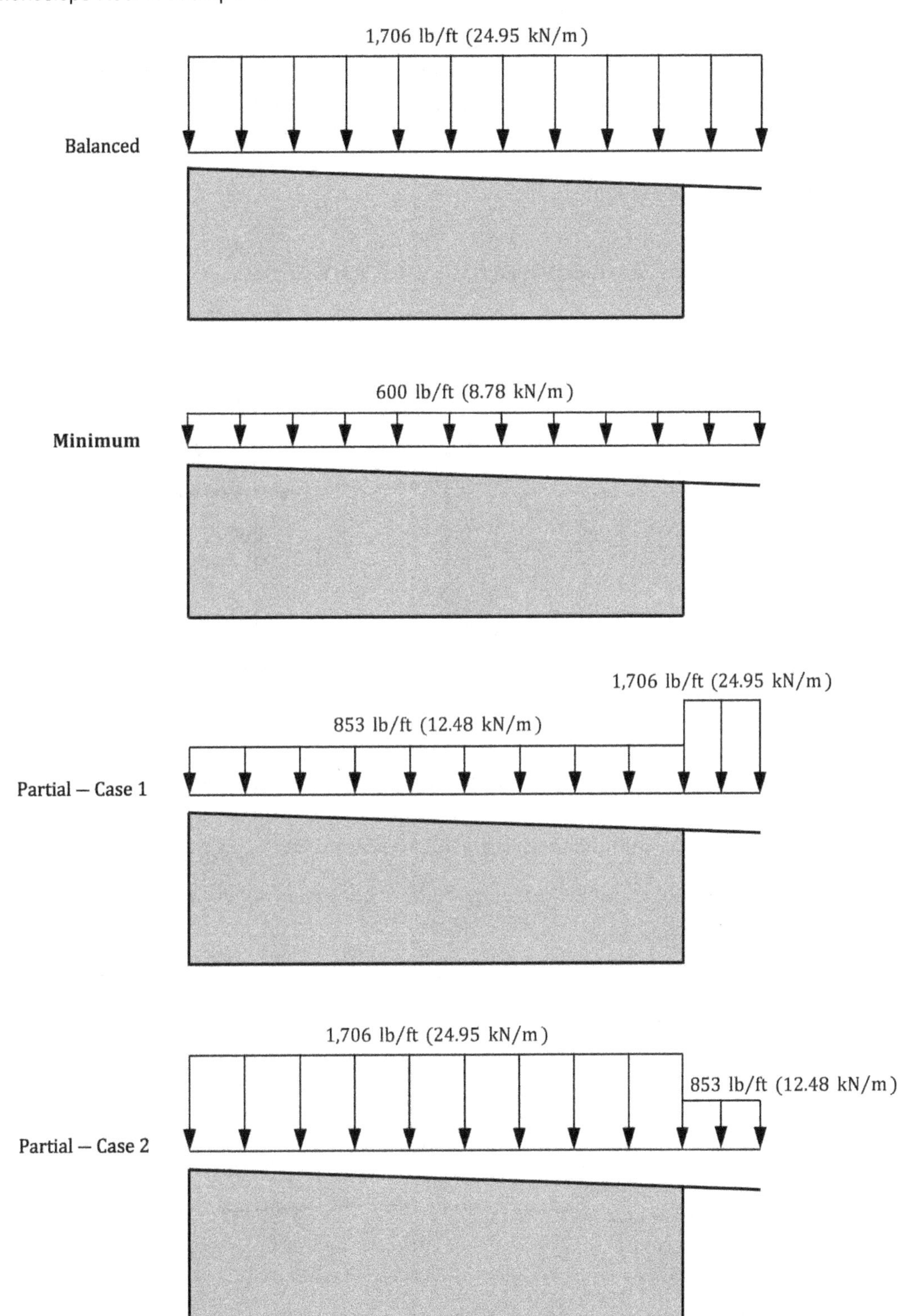

Comments

If the thermal condition in this example was a heated structure with an unventilated roof with $R_{roof} = 30$ h ft²°F/Btu (5.28 m²K/W), then $C_t = 1.10$ in accordance with ASCE/SEI Table 7.3-3, which means an ice dam could form on the roof overhang (ASCE/SEI 7.4.4). Thus, $C_s = 1.0$ for a roof slope of 2.39 degrees from ASCE/SEI Figure 7.4-1a, and the balanced snow load is equal to $1.0 \times 0.7 \times 1.0 \times 1.10 \times 106 = 81.6$ lb/ft² (3.91 kN/m²).

The additional load case in ASCE/SEI 7.4.4 must be considered on the overhang. For a distance of 5 ft (1.52 m) from the face of the structure, the uniformly distributed load attributed to the ice dam is equal to $2p_{f(\text{heated})} = 2 \times 81.6 = 163.2$ lb/ft² (in S.I.: $2p_{f(\text{heated})} = 2 \times 3.91 = 7.82$ kN/m²), which is based on the heated space (see Figure 4.4). For the remaining 3 ft (0.91 m), the uniformly distributed load is based on the unheated condition. Assuming $C_t = 1.2$ from ASCE/SEI Table 7.3-2, $p_{f(\text{unheated})} = 0.7C_eC_tp_g = 0.7 \times 1.0 \times 1.2 \times 106 = 89.0$ lb/ft² (in S.I.: $p_{f(\text{unheated})} = 0.7 \times 1.0 \times 1.2 \times 5.08 = 4.27$ kN/m²). This load case is applied to the overhang with no other loads except dead loads (see Figure 4.28).

Figure 4.28 Ice Dam Load Case Where $C_t = 1.10$ for the Structure in Example 4.1

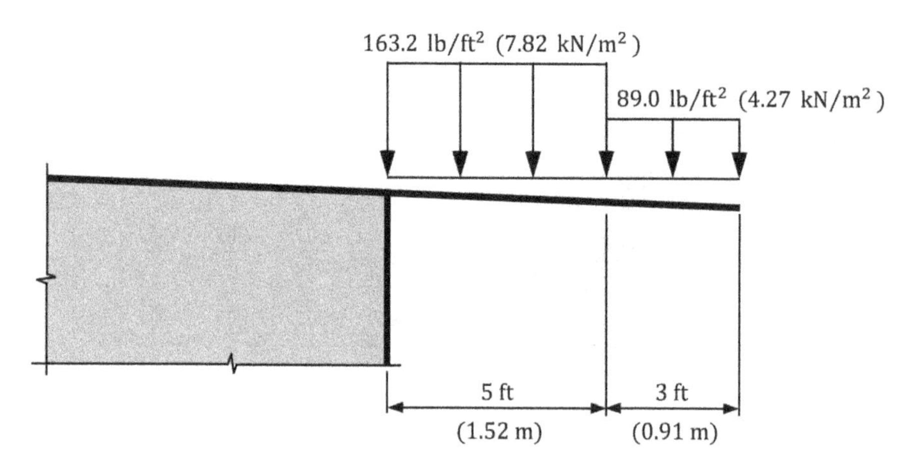

4.5.2 Example 4.2—Design Snow Loads on a Gable Roof (Roof Slope = 2.39 degrees)

Determine the design snow loads on the gable roof in Figure 4.29 given the following design data:

- Location: St. Louis, MO (Latitude = 38.62°, Longitude = −90.23°)
- Surface roughness: C

Figure 4.29 Elevation of the Building in Example 4.2

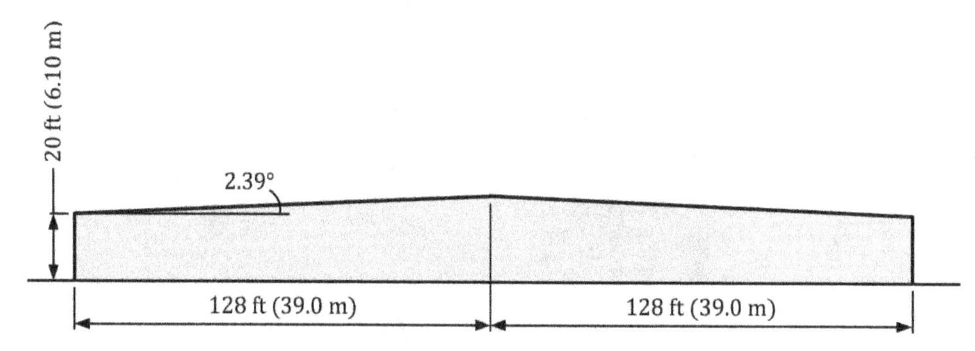

- Risk Category: II (IBC Table 1604.5)
- Thermal condition: Structure is kept just above freezing
- Roof exposure: Partially exposed
- Roof surface: Rubber membrane
- Roof obstructions: None
- Roof framing: All members are simply supported

The design procedure in Table 4.1 of this book is used to determine the design snow loads.

Solution

Step 1—Determine the ground snow load, p_g

Using Reference 4.1, the ground snow load is equal to 23 lb/ft^2 (1.10 kN/m^2) for Risk Category II.

Step 2—Determine the flat roof snow load, p_f

Flowchart 4.1 is used to determine p_f.

Step 2a—Determine the surface roughness category

The surface roughness category is given in the design data as C.

Step 2b—Determine the exposure of the roof

The roof exposure is given in the design data as partially exposed.

Step 2c—Determine the exposure factor, C_e

Given a surface roughness category of C and a partially exposed roof exposure, $C_e = 1.0$ from ASCE/SEI Table 7.3-1.

Step 2d—Determine the thermal condition of the roof

From the design data, the structure is kept just above freezing during the winter.

Step 2e—Determine the thermal factor, C_t

For a structure kept just above freezing, $C_t = 1.2$ (ASCE/SEI Table 7.3-2).

Step 2f—Determine the flat roof snow load

$$p_f = 0.7C_eC_tp_g = 0.7 \times 1.0 \times 1.2 \times 23 = 19.3 \text{ lb/ft}^2$$

$$p_f = 0.7C_eC_tp_g = 0.7 \times 1.0 \times 1.2 \times 1.10 = 0.92 \text{ kN/m}^2$$

Step 3—Determine the minimum snow load for low-slope roofs, p_m

A minimum snow load, p_m, applies to gable roofs with slopes less than 15 degrees (see Figure 4.2). Because the roof slope in this example is equal to 2.39 degrees, minimum snow loads must be considered.

For Risk Category II, the minimum snow load upper limit, $p_{m,\max}$, is equal to 30 lb/ft^2 (1.44 kN/m^2) from ASCE/SEI Table 7.3-4. Because $p_g = 23$ lb/ft^2 (1.10 kN/m^2) $< p_{m,\max} =$ 30 lb/ft^2 (1.44 kN/m^2), $p_m = p_g = 23$ lb/ft^2 (1.10 kN/m^2) (ASCE/SEI 7.3.3).

Step 4—Determine the sloped roof (balanced) snow load, p_s

Flowchart 4.2 is used to determine p_s.

Step 4a—Determine the slope factor, C_s

From Step 2e, $C_t = 1.2$, which means the roof is cold.

From the design data, there are no obstructions inhibiting the snow from sliding off the roof. Also, because $C_t > 1.1$, ice dams on all overhanging portions need not be considered for this roof (if an ice dam can form, it is considered to be an obstruction; see ASCE/SEI 7.4.4).

The roof surface is a rubber membrane from the design data. According to ASCE/SEI 7.4, rubber membranes are considered to be slippery surfaces. Because the roof is unobstructed and slippery with $C_t = 1.2$, use the dashed line in ASCE/SEI Figure 7.4-1c to determine the slope factor, C_s. For a roof slope of 2.39 degrees, which is less than 15 degrees, $C_s = 1.0$.

Step 4b—Determine the sloped roof (balanced) snow load, p_s

$$p_s = C_s p_f = 1.0 \times 19.3 = 19.3 \text{ lb/ft}^2$$

$$p_s = C_s p_f = 1.0 \times 0.92 = 0.92 \text{ kN/m}^2$$

Step 5—Consider partial loading

Because all the roof members are simply supported, partial loading need not be considered (ASCE/SEI 7.5).

Step 6—Consider unbalanced snow loads

Flowchart 4.3 is used to determine if unbalanced snow loads must be considered.

Unbalanced snow loads need not be considered for gable roofs with a slope exceeding 30.3 degrees or with a slope less than 2.39 degrees (ASCE/SEI 7.6.1). The roof slope in this example is equal to 2.39 degrees, so unbalanced loads must be considered.

Because $W = 128$ ft (39.0 m) > 20 ft (6.10 m), the unbalanced load consists of the following (see Figure 4.8):

- Windward side:
  Unbalanced load $= 0.3p_s = 0.3 \times 19.3 = 5.8$ lb/ft^2 applied over the entire length of the windward side
- Leeward side:
  From Reference 4.1, winter wind parameter, $W_2 = 0.45$
  Snow density, $\gamma = 0.13p_g + 14 = (0.13 \times 23) + 14 = 17$ lb/ft^3 < 30 lb/ft^3

$$h_d = 1.5\sqrt{\frac{(p_g)^{0.74}(W)^{0.70}(W_2)^{1.7}}{\gamma}} = 1.5 \times \sqrt{\frac{(23)^{0.74} \times (128)^{0.70} \times (0.45)^{1.7}}{17}} = 3.2 \text{ ft}$$

S = roof slope run for a rise of one $= 1/\tan 2.39° = 24$

$p_s = 19.3$ lb/ft^2 is applied over the entire length of the leeward side

Rectangular surcharge $= h_d\gamma/\sqrt{S} = (3.2 \times 17)/\sqrt{24} = 11.1$ lb/ft^2 is applied on the leeward side a distance equal to $8h_d\sqrt{S}/3 = (8 \times 3.2 \times \sqrt{24})/3 = 41.8$ ft from the ridge

In S.I.:

- Windward side:
  Unbalanced load $= 0.3p_s = 0.3 \times 0.92 = 0.28$ kN/m^2 applied over the entire length of the windward side
- Leeward side:
  From Reference 4.1, winter wind parameter, $W_2 = 0.45$
  Snow density, $\gamma = 0.426p_g + 2.2 = (0.426 \times 1.10) + 2.2 = 2.67$ kN/m^3 < 4.7 kN/m^3

$$h_d = 1.5\sqrt{\frac{(p_g)^{0.74}(W)^{0.70}(W_2)^{1.7}}{\gamma}} = 1.5 \times \sqrt{\frac{(23)^{0.74} \times (128)^{0.70} \times (0.45)^{1.7}}{17}}$$

$$= 3.2 \text{ ft} \times 0.3048 = 0.98 \text{ m}$$

S = roof slope run for a rise of one $= 1/\tan 2.39° = 24$

$p_s = 0.92$ kN/m^2 is applied over the entire length of the leeward side

Rectangular surcharge $= h_d\gamma/\sqrt{S} = (0.98 \times 2.67)/\sqrt{24} = 0.53$ kN/m^2 is applied on the leeward side a distance equal to $8h_d\sqrt{S}/3 = (8 \times 0.98 \times \sqrt{24})/3 = 12.8$ m from the ridge

Step 7—Consider drift loads on lower roofs
Not applicable.

Step 8—Consider drifts on roof projections and parapets
Not applicable.

Step 9—Consider sliding snow
Not applicable.

Step 10—Consider rain-on-snow surcharge loads
According to ASCE/SEI 7.10, a rain-on-snow surcharge load is required for locations where $p_g \leq p_{m,\max}$ and $p_g > 0$ on roofs with slopes less than $W/50$ (in S.I.: $W/15.2$). In this example, $p_g = 23$ lb/ft²(1.10 kN/m²) < $p_{m,\max} = 30$ lb/ft² (1.44 kN/m²) and $W/50 = 128/50 = 2.56$ degrees > 2.39 degrees (in S.I.: $W/15.2 = 39.0/15.2 = 2.57$ degrees > 2.39 degrees). Therefore, an additional 8 lb/ft² (0.38 kN/m²) rain-on-snow surcharge load must be added to the balanced snow load.

Step 11—Consider ponding loads
A ponding analysis considering roof deflections caused by full snow loads in combination with dead and superimposed dead loads must be performed for this roof based on the stiffness of the roof structural members.

Step 12—Consider snow loads on existing roofs
Not applicable.

Step 13—Consider snow loads on open-frame equipment structures
Not applicable.

Design snow loads on the roof for the minimum, balanced, and unbalanced load cases are given in Figure 4.30.

Figure 4.30 Design Snow Load Cases for the Gable Roof in Example 4.2

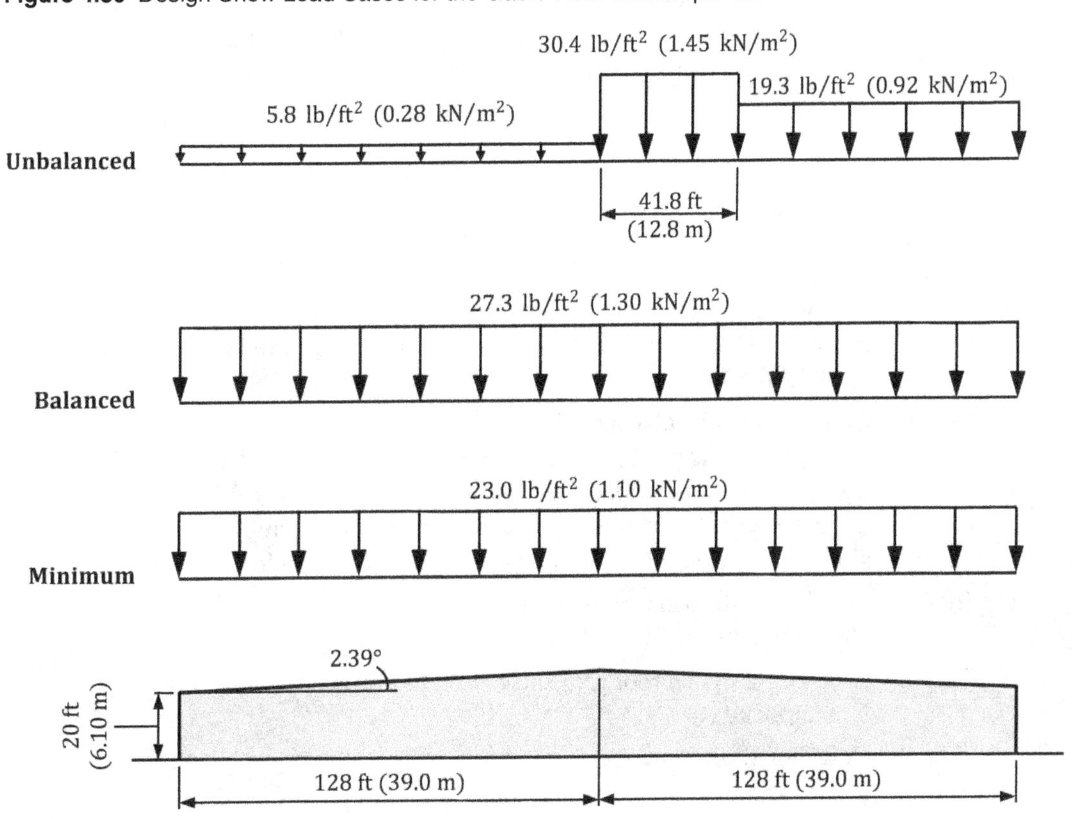

4.5.3 Example 4.3—Design Snow Loads on a Gable Roof (Roof Slope = 1.19 degrees)

Determine the design snow loads on the gable roof in Figure 4.29 for a roof slope of 1.19 degrees given the design data in Example 4.2.

The design procedure in Table 4.1 of this book is used to determine the design snow loads.

SOLUTION

Step 1—Determine the ground snow load, p_g

Using Reference 4.1, the ground snow load is equal to 23 lb/ft² (1.10 kN/m²).

Step 2—Determine the flat roof snow load, p_f

From Example 4.2, $p_f = 19.3$ lb/ft² (0.92 kN/m²).

A minimum snow load, p_m, applies to gable roofs with slopes less than 15 degrees (see Figure 4.2). Because the roof slope in this example is equal to 1.19 degrees, minimum snow loads must be considered.

For Risk Category II, the minimum snow load upper limit, $p_{m,\max}$, is equal to 30 lb/ft² (1.44 kN/m²) from ASCE/SEI Table 7.3-4. Because $p_g = 23$ lb/ft² (1.10 kN/m²) $< p_{m,\max} =$ 30 lb/ft² (1.44 kN/m²), $p_m = p_g = 23$ lb/ft² (1.10 kN/m²) (ASCE/SEI 7.3.3).

Step 4—Determine the sloped roof (balanced) snow load, p_s

From Example 4.2, $C_t = 1.2$.

From the design data, there are no obstructions inhibiting the snow from sliding off the roof. Also, because $C_t > 1.1$, ice dams on all overhanging portions need not be considered for this roof (if an ice dam can form, it is considered to be an obstruction; see ASCE/SEI 7.4.4).

The roof surface is a rubber membrane from the design data. According to ASCE/SEI 7.4, rubber membranes are considered to be slippery surfaces.

Because the roof is unobstructed and slippery with $C_t = 1.2$, use the dashed line in ASCE/SEI Figure 7.4-1c to determine the slope factor, C_s. For a roof slope of 1.19 degrees, which is less than 15 degrees, $C_s = 1.0$.

Therefore,

$$p_s = C_s p_f = 1.0 \times 19.3 = 19.3 \text{ lb/ft}^2$$

$$p_s = C_s p_f = 1.0 \times 0.92 = 0.92 \text{ kN/m}^2$$

Step 5—Consider partial loading

Because all the roof members are simply supported, partial loading need not be considered (ASCE/SEI 7.5).

Step 6—Consider unbalanced snow loads

Flowchart 4.3 is used to determine if unbalanced snow loads must be considered.

Unbalanced snow loads need not be considered for gable roofs with a slope exceeding 30.3 degrees or with a slope less than 2.39 degrees (ASCE/SEI 7.6.1). The roof slope in this example is 1.19 degrees, so unbalanced loads need not be considered.

Step 7—Consider drift loads on lower roofs

Not applicable.

Step 8—Consider drifts on roof projections and parapets

Not applicable.

Step 9—Consider sliding snow

Not applicable.

Step 10—Consider rain-on-snow surcharge loads

According to ASCE/SEI 7.10, a rain-on-snow surcharge load is required for locations where $p_g \le p_{m,\max}$ and $p_g > 0$ on roofs with slopes less than $W/50$ (in S.I.: $W/15.2$). In this example, $p_g = 23$ lb/ft^2(1.10 kN/m^2) $< p_{m,\max} = 30$ lb/ft^2 (1.44 kN/m^2) and $W/50 = 128/50 = 2.56$ degrees > 1.19 degrees (in S.I.: $W/15.2 = 39.0/15.2 = 2.57$ degrees $>$ 1.19 degrees). Therefore, an additional 8 lb/ft^2 (0.38 kN/m^2) rain-on-snow surcharge load must be added to the balanced snow load.

Step 11—Consider ponding loads

A ponding analysis considering roof deflections caused by full snow loads in combination with dead and superimposed dead loads must be performed for this roof based on the stiffness of the roof structural members.

Step 12—Consider snow loads on existing roofs

Not applicable.

Step 13—Consider snow loads on open-frame equipment structures

Not applicable.

Design snow loads on the roof for the minimum and balanced load cases are given in Figure 4.31.

Figure 4.31 Design Snow Load Cases for the Gable Roof in Example 4.3

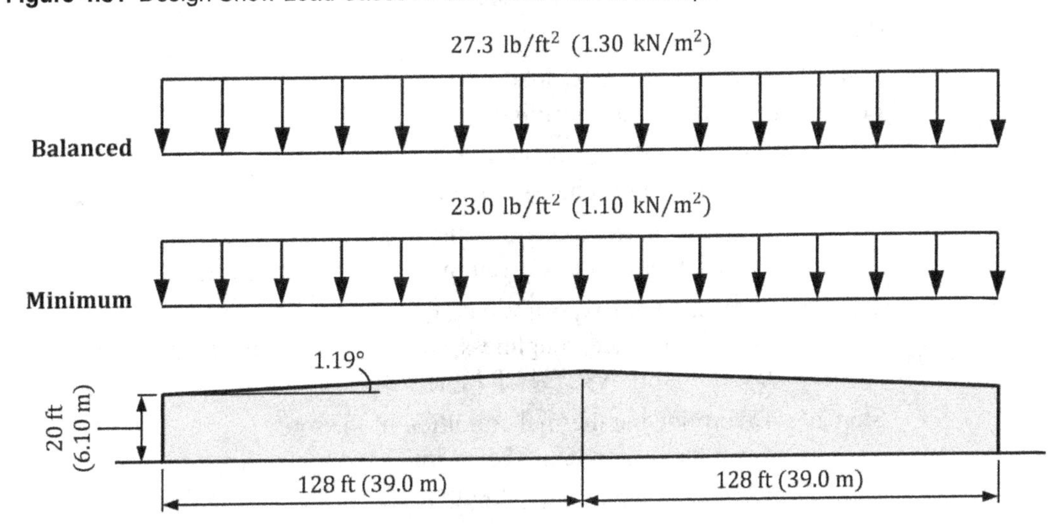

4.5.4 Example 4.4—Design Snow Loads on a Gable Roof (Roof Slope = 16.7 degrees)

Determine the design snow loads on the gable roof in Figure 4.32 given the following design data:

- Location: Bloomington, IL (Latitude = 40.47°, Longitude = −89.01°)
- Surface roughness: C
- Risk Category: I (IBC Table 1604.5)
- Thermal condition: Open-air structure (no walls)
- Roof exposure: Sheltered
- Roof surface: Wood shingles
- Roof obstructions: None
- Roof framing: Trusses spaced 5 ft (1.52 m) on center

The design procedure in Table 4.1 of this book is used to determine the design snow loads.

Figure 4.32 Elevation of the Building in Example 4.4

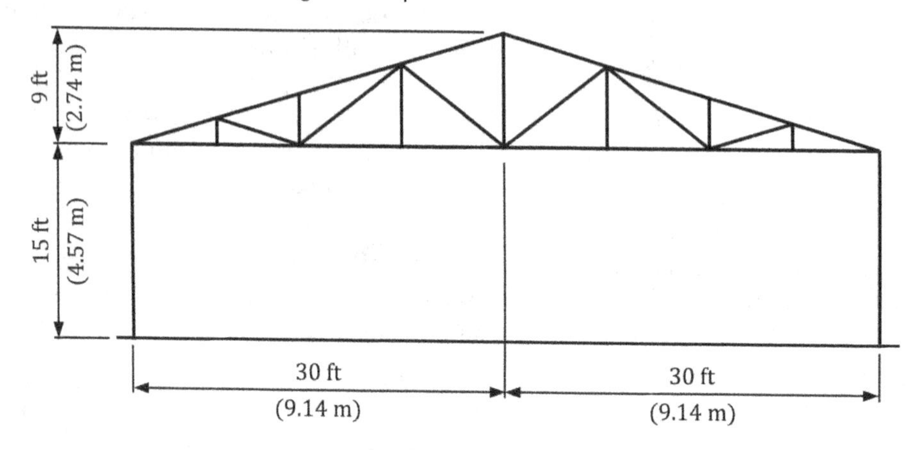

Solution

Step 1—Determine the ground snow load, p_g

Using Reference 4.1, the ground snow load is equal to 27 lb/ft² (1.29 kN/m²) for Risk Category I.

Step 2—Determine the flat roof snow load, p_f

Flowchart 4.1 is used to determine p_f.

Step 2a—Determine the surface roughness category

The surface roughness category is given in the design data as C.

Step 2b—Determine the exposure of the roof

The roof exposure is given in the design data as sheltered.

Step 2c—Determine the exposure factor, C_e

Given a surface roughness category of C and a sheltered roof exposure, $C_e = 1.1$ from ASCE/SEI Table 7.3-1.

Step 2d—Determine the thermal condition of the roof

From the design data, the building is an open-air structure.

Step 2e—Determine the thermal factor, C_t

For an open-air structure, $C_t = 1.2$ (ASCE/SEI Table 7.3-2).

Step 2f—Determine the flat roof snow load

$$p_f = 0.7C_eC_tp_g = 0.7\times1.1\times1.2\times27.0 = 25.0 \text{ lb/ft}^2$$

$$p_f = 0.7C_eC_tp_g = 0.7\times1.1\times1.2\times1.29 = 1.19 \text{ kN/m}^2$$

Step 3—Determine the minimum snow load for low-slope roofs, p_m

A minimum snow load, p_m, applies to gable roofs with slopes less than 15 degrees (see Figure 4.2). Because the roof slope in this example is equal to $\tan^{-1}(9.0/30.0) =$ 16.7 degrees (in S.I.: $\tan^{-1}(2.74/9.14) = 16.7$ degrees), minimum snow loads need not be considered.

Step 4—Determine the sloped roof (balanced) snow load, p_s

Flowchart 4.2 is used to determine p_s.

Step 4a—Determine the slope factor, C_s

From Step 2e, $C_t = 1.2$, which means the roof is cold.

From the design data, there are no obstructions inhibiting the snow from sliding off the roof. Also, because $C_t > 1.1$, ice dams on all overhanging portions need not be considered for this roof (if an ice dam can form, it is considered to be an obstruction; see ASCE/SEI 7.4.4).

The roof surface has wood shingles from the design data. According to ASCE/SEI 7.4, wood shingles are not considered to be slippery.

Because the roof is unobstructed and not slippery with $C_t = 1.2$, use the solid line in ASCE/SEI Figure 7.4-1c to determine the slope factor, C_s. For a roof slope of 16.7 degrees, which is less than 45 degrees, $C_s = 1.0$.

Step 4b—Determine the sloped roof (balanced) snow load, p_s

$$p_s = C_s p_f = 1.0 \times 25.0 = 25.0 \text{ lb/ft}^2$$

$$p_s = C_s p_f = 1.0 \times 1.19 = 1.19 \text{ kN/m}^2$$

Step 5—Consider partial loading

Partial loads need not be applied to structural members spanning perpendicular to the ridgeline in gable roofs with slopes between 2.39 degrees and 30.3 degrees (ASCE/SEI 7.5). Because the slope of this roof is between these limits, partial loading need not be considered (note: partial loads on individual members of roof trusses are usually not considered).

Step 6—Consider unbalanced snow loads

Flowchart 4.3 is used to determine if unbalanced snow loads must be considered.

Unbalanced snow loads need not be considered for gable roofs with a slope exceeding 30.3 degrees or with a slope less than 2.39 degrees (ASCE/SEI 7.6.1). The roof slope in this example is equal to 16.7 degrees, so unbalanced loads must be considered.

Because $W = 30$ ft $(9.14 \text{ m}) > 20$ ft (6.10 m), the unbalanced load consists of the following (see Figure 4.8):

- Windward side:
  Unbalanced load $= 0.3p_s = 0.3 \times 25.0 = 7.5 \text{ lb/ft}^2$ applied over the entire length of the windward side
- Leeward side:
  From Reference 4.1, winter wind parameter, $W_2 = 0.55$
  Snow density, $\gamma = 0.13p_g + 14 = (0.13 \times 27) + 14 = 17.5 \text{ lb/ft}^3 < 30 \text{ lb/ft}^3$

$$h_d = 1.5\sqrt{\frac{(p_g)^{0.74}(W)^{0.70}(W_2)^{1.7}}{\gamma}} = 1.5 \times \sqrt{\frac{(27.0)^{0.74} \times (30)^{0.70} \times (0.55)^{1.7}}{17.5}} = 2.4 \text{ ft}$$

S = roof slope run for a rise of one $= 1/\tan 16.7° = 3.3$

$p_s = 25.0 \text{ lb/ft}^2$ is applied over the entire length of the leeward side

Rectangular surcharge $= h_d\gamma/\sqrt{S} = (2.4 \times 17.5)/\sqrt{3.3} = 23.1 \text{ lb/ft}^2$ is applied on the leeward side a distance equal to $8h_d\sqrt{S}/3 = (8 \times 2.4 \times \sqrt{3.3})/3 = 11.6$ ft from the ridge

Unbalanced load on a roof truss over the 11.6-ft length $= (25.0 + 23.1) \times 5.0 = 241$ lb/ft

Unbalanced load on a roof truss over the remaining length $= 25.0 \times 5.0 = 125$ lb/ft

In S.I.:

- Windward side:
  Unbalanced load $= 0.3p_s = 0.3 \times 1.19 = 0.36 \text{ kN/m}^2$ applied over the entire length of the windward side
- Leeward side:
  From Reference 4.1, winter wind parameter, $W_2 = 0.55$

Snow density, $\gamma = 0.426p_g + 2.2 = (0.426 \times 1.29) + 2.2 = 2.75$ kN/m³ < 4.7 kN/m³

$$h_d = 1.5\sqrt{\frac{(p_g)^{0.74}(W)^{0.70}(W_2)^{1.7}}{\gamma}} = 1.5 \times \sqrt{\frac{(27.0)^{0.74} \times (30)^{0.70} \times (0.55)^{1.7}}{17.5}}$$

$= 2.4 \text{ ft} \times 0.3048 = 0.73 \text{ m}$

S = roof slope run for a rise of one $= 1/\tan 16.7° = 3.3$

$p_s = 1.19$ kN/m² is applied over the entire length of the leeward side

Rectangular surcharge $= h_d\gamma/\sqrt{S} = (0.73 \times 2.75)/\sqrt{3.3} = 1.11$ kN/m² is applied on the leeward side a distance equal to $8h_d\sqrt{S}/3 = (8 \times 0.73 \times \sqrt{3.3})/3 = 3.54$ m from the ridge

Unbalanced load on a roof truss over the 3.54-m length $= (1.19 + 1.11) \times 1.52 = 3.50$ kN/m

Unbalanced load on a roof truss over the remaining length $= 1.19 \times 1.52 = 1.81$ kN/m

Step 7—Consider drift loads on lower roofs

Not applicable.

Step 8—Consider drifts on roof projections and parapets

Not applicable.

Step 9—Consider sliding snow

Not applicable.

Step 10—Consider rain-on-snow surcharge loads

According to ASCE/SEI 7.10, a rain-on-snow surcharge load is required for locations where $p_g \leq p_{m,\max}$ and $p_g > 0$ on roofs with slopes less than $W/50$ (in S.I.: $W/15.2$). In this example, $p_g = 27$ lb/ft²(1.29 kN/m²) $> p_{m,\max} = 25.0$ lb/ft²(1.20 kN/m²) from ASCE/SEI Table 7.3-4 for Risk Category I. Therefore, an additional 8 lb/ft² (0.38 kN/m²) rain-on-snow surcharge load need not be added to the balanced snow load.

Step 11—Consider ponding loads

A ponding analysis considering roof deflections caused by full snow loads in combination with dead and superimposed dead loads must be performed for this roof based on the stiffness of the roof structural members.

Step 12—Consider snow loads on existing roofs

Not applicable.

Step 13—Consider snow loads on open-frame equipment structures

Not applicable.

Design snow loads on the roof for the balanced and unbalanced load cases are given in Figure 4.33.

4.5.5 Example 4.5—Design Snow Loads on a Curved Roof

Determine the design snow loads on the curved roof in Figure 4.34 given the following design data:

- Location: Milwaukee, WI (Latitude = 43.01°, Longitude = −87.91°)
- Surface roughness: D
- Risk Category: IV (IBC Table 1604.5)
- Thermal condition: Unheated structure
- Roof exposure: Fully exposed
- Roof surface: Rubber membrane
- Roof obstructions: None
- Roof framing: All structural members are simply supported

The design procedure in Table 4.1 of this book is used to determine the design snow loads.

Figure 4.33 Design Snow Load Cases for the Gable Roof in Example 4.4

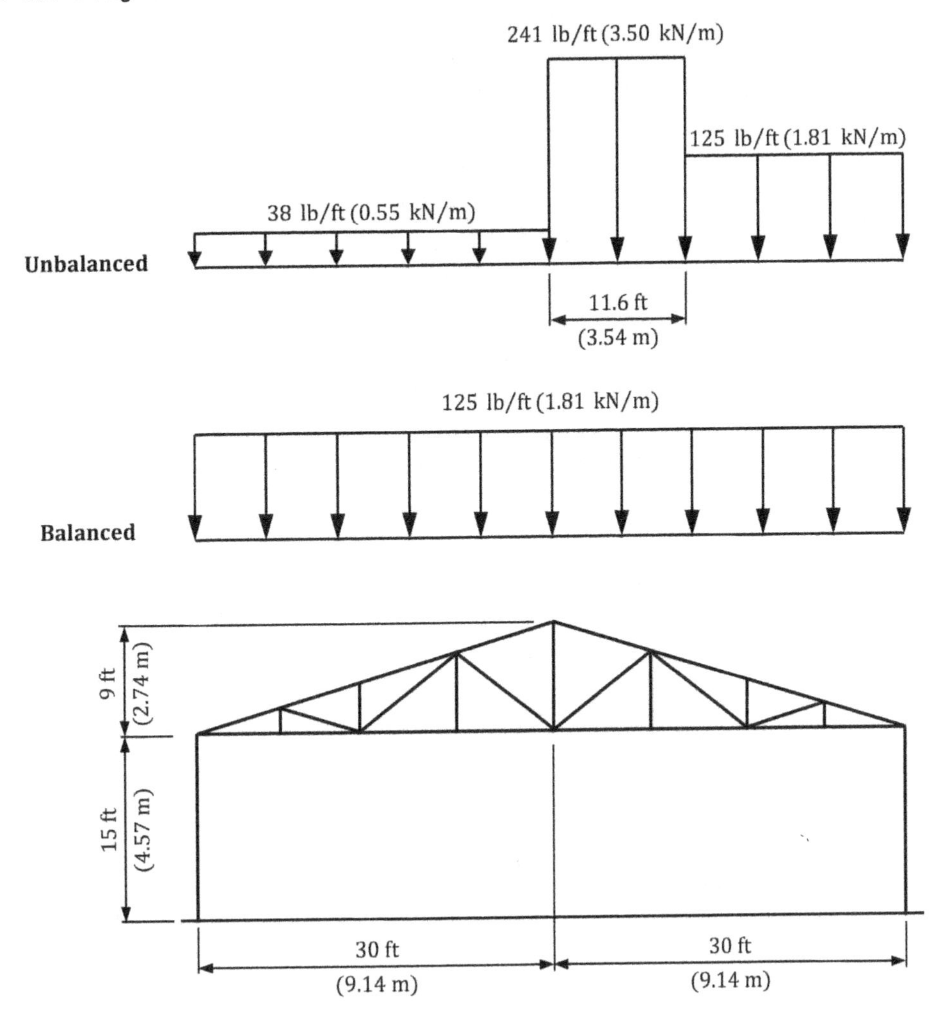

Figure 4.34 Elevation of the Building in Example 4.5

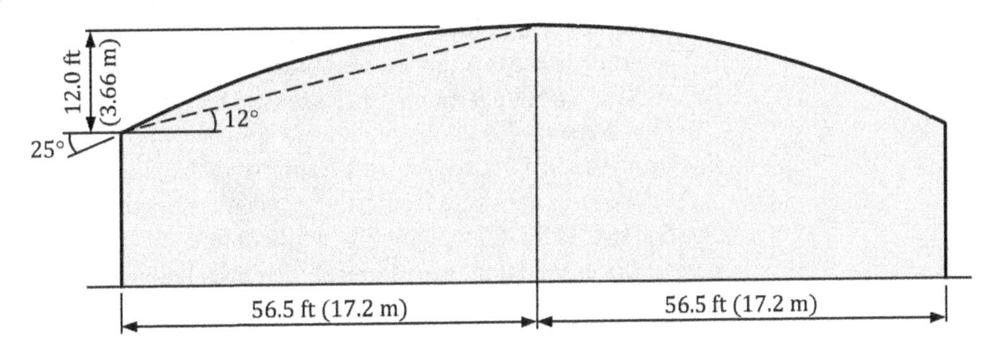

Solution

Step 1—Determine the ground snow load, p_g

Using Reference 4.1, the ground snow load is equal to 77 lb/ft^2 (3.69 kN/m^2) for Risk Category IV.

Step 2—Determine the flat roof snow load, p_f

Flowchart 4.1 is used to determine p_f.

Step 2a—Determine the surface roughness category

The surface roughness category is given in the design data as D.

Step 2b—Determine the exposure of the roof

The roof exposure is given in the design data as fully exposed.

Step 2c—Determine the exposure factor, C_e

Given a surface roughness category of D and a fully exposed roof exposure, $C_e = 0.8$ from ASCE/SEI Table 7.3-1.

Step 2d—Determine the thermal condition of the roof

From the design data, the building is unheated.

Step 2e—Determine the thermal factor, C_t

For an unheated structure, $C_t = 1.2$ (ASCE/SEI Table 7.3-2).

Step 2f—Determine the flat roof snow load

$$p_f = 0.7C_eC_tp_g = 0.7 \times 0.8 \times 1.2 \times 77.0 = 51.7 \text{ lb/ft}^2$$

$$p_f = 0.7C_eC_tp_g = 0.7 \times 0.8 \times 1.2 \times 3.69 = 2.48 \text{ kN/m}^2$$

Step 3—Determine the minimum snow load for low-slope roofs, p_m

A minimum snow load, p_m, applies to curved roofs where the vertical angle from the eaves to the crown is less than 10 degrees (ASCE/SEI 7.3.3). Because the slope for this roof is equal to 12 degrees, minimum roof snow loads need not be considered.

Step 4—Determine the sloped roof (balanced) snow load, p_s

Flowchart 4.2 is used to determine p_s.

Step 4a—Determine the slope factor, C_s

From Step 2e, $C_t = 1.2$, which means the roof is cold.

From the design data, there are no obstructions inhibiting the snow from sliding off the roof. Also, because $C_t > 1.1$, ice dams on all overhanging portions need not be considered for this roof (if an ice dam can form, it is considered to be an obstruction; see ASCE/SEI 7.4.4).

The roof surface is a rubber membrane from the design data. According to ASCE/SEI 7.4, rubber membranes are considered to be slippery.

Because the roof is unobstructed and slippery with $C_t = 1.2$, use the dashed line in ASCE/SEI Figure 7.4-1c to determine the slope factor, C_s.

At the eaves, the slope factor, C_{sleave}, is determined by the following equation in ASCE/SEI C7.4, which is applicable to $C_t \geq 1.2$, unobstructed slippery surfaces, and a tangent slope at the eaves between 15 and 75 degrees:

$$C_{sleave} = 1.0 - \frac{\text{slope} - 15°}{55°} = 1.0 - \frac{25° - 15°}{55°} = 0.82$$

Away from the eaves, C_s is equal to 1.0 where the tangent roof slope is less than or equal to 15 degrees (see the dashed line in ASCE/SEI Figure 7.4-1c). This occurs at approximately 20.7 ft (6.31 m) from the eaves at both ends of the roof.

Step 4b—Determine the sloped roof (balanced) snow load, p_s

At the eaves:

$$p_s = C_{sleave} p_f = 0.82 \times 51.7 = 42.4 \text{ lb/ft}^2$$

$$p_s = C_{sleave} p_f = 0.82 \times 2.48 = 2.03 \text{ kN/m}^2$$

Within the center portion of the roof:

$$p_s = 1.0 \times 51.7 = 51.7 \text{ lb/ft}^2$$

$$p_s = 1.0 \times 2.48 = 2.48 \text{ kN/m}^2$$

Step 5—Consider partial loading

From the design data, the roof structural members are simply supported, so partial loading need not be considered (ASCE/SEI 7.5).

Step 6—Consider unbalanced snow loads

Flowchart 4.4 is used to determine if unbalanced snow loads must be considered.

Unbalanced snow loads on curved roofs need not be considered where the slope of the straight line from the eaves (or, from the 70-degree point, if present) to the crown is less than 10 degrees or greater than 60 degrees (ASCE/SEI 7.6.2). Because the slope of the straight line from the eaves to the crown is equal to 12 degrees in this example, unbalanced snow loads must be considered.

Unbalanced snow loads for this roof are given in Case 1 of ASCE/SEI Figure 7.4-2 where the slope at the eaves is less than 30 degrees (see Figure 4.9):

- Windward side

  The unbalanced load is equal to zero.
- Leeward side

  At the eaves:

  Unbalanced load $= 2p_f(C_{sleave}/C_e) = 2 \times 51.7 \times (0.82/0.80) = 106.0$ lb/ft^2

  Unbalanced load $= 2p_f(C_{sleave}/C_e) = 2 \times 2.48 \times (0.82/0.80) = 5.08$ kN/m^2

  At the crown:

  Unbalanced load $= 0.5p_f = 0.5 \times 51.7 = 25.9$ lb/ft^2

  Unbalanced load $= 0.5p_f = 0.5 \times 2.48 = 1.24$ kN/m^2

Step 7—Consider drift loads on lower roofs

Not applicable.

Step 8—Consider drifts on roof projections and parapets

Not applicable.

Step 9—Consider sliding snow

Not applicable.

Step 10—Consider rain-on-snow surcharge loads

According to ASCE/SEI 7.10, a rain-on-snow surcharge load is required for locations where $p_g \leq p_{m,\max}$ and $p_g > 0$ on roofs with slopes less than $W/50$ (in S.I.: $W/15.2$). In this example, $p_g = 77$ lb/ft^2(3.69 kN/m^2) $> p_{m,\max} = 40.0$ lb/ft^2(1.92 kN/m^2) from ASCE/SEI Table 7.3-4 for Risk Category IV. Therefore, an additional 8 lb/ft^2 (0.38 kN/m^2) rain-on-snow surcharge load need not be added to the balanced snow load.

Step 11—Consider ponding loads

A ponding analysis considering roof deflections caused by full snow loads in combination with dead and superimposed dead loads must be performed for this roof based on the stiffness of the roof structural members.

Step 12—Consider snow loads on existing roofs
Not applicable.

Step 13—Consider snow loads on open-frame equipment structures
Not applicable.

Design snow loads on the roof for the balanced and unbalanced load cases are given in Figure 4.35.

Figure 4.35 Design Snow Load Cases for the Curved Roof in Example 4.5

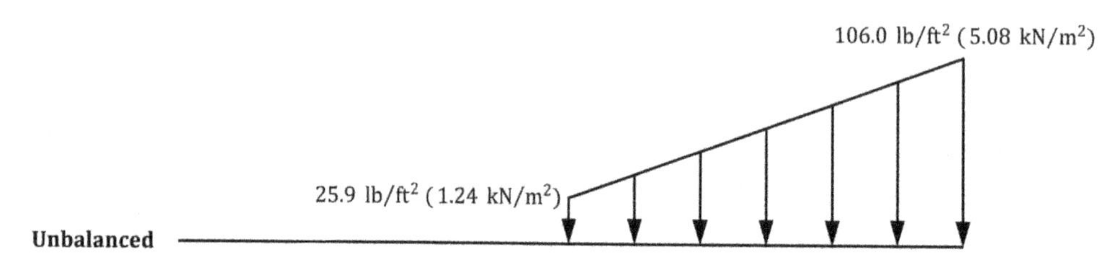

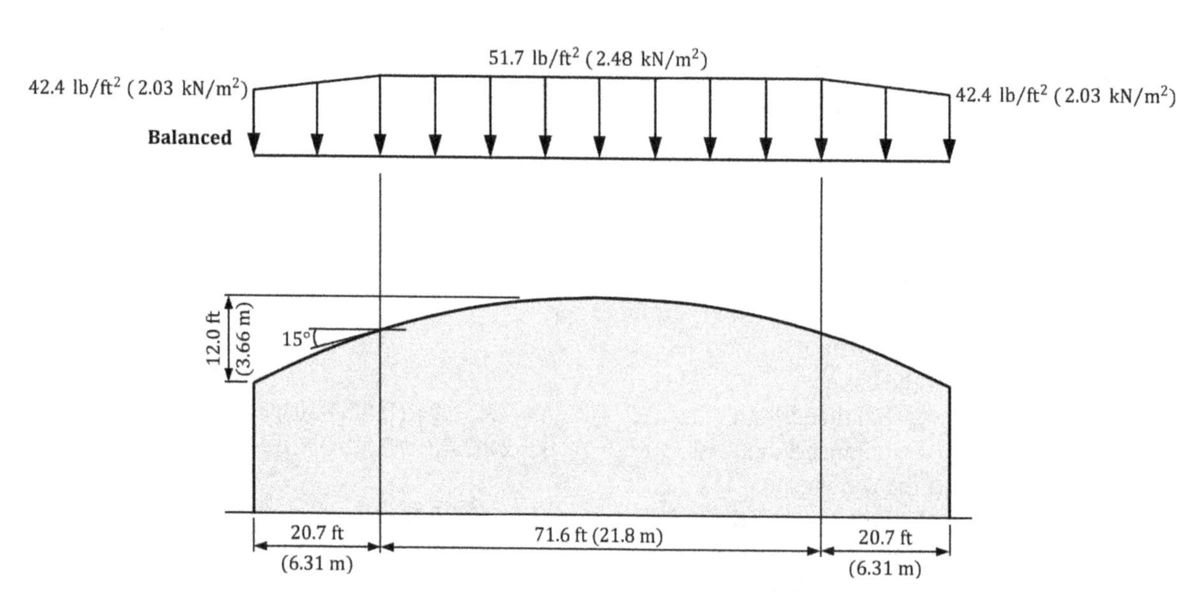

4.5.6 Example 4.6—Design Snow Loads on a Sawtooth Roof

Determine the design snow loads on the sawtooth roof in Figure 4.36 given the following design data:

- Location: Birmingham, AL (Latitude = 33.53°, Longitude = –86.77°)
- Surface roughness: C
- Risk Category: III (IBC Table 1604.5)
- Thermal condition: Cold, ventilated roof
- Roof exposure: Partially exposed
- Roof surface: Glass
- Roof obstructions: None
- Roof framing: All structural members are simply supported

The design procedure in Table 4.1 of this book is used to determine the design snow loads.

Figure 4.36 Elevation of the Building in Example 4.6.

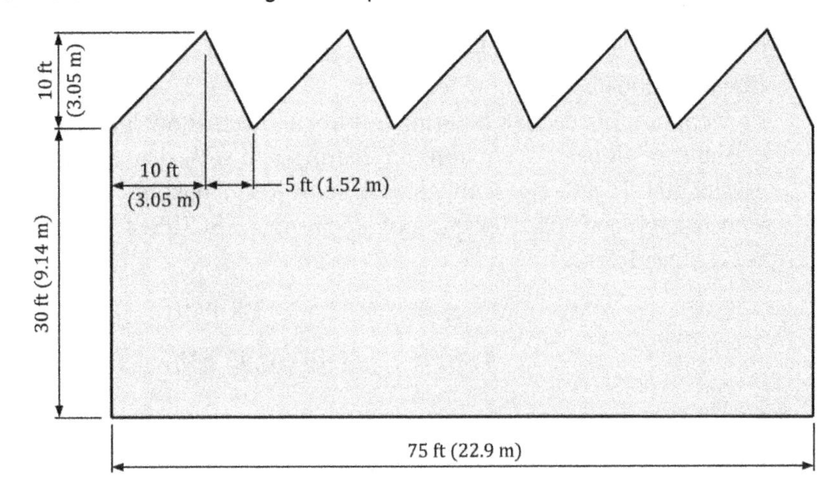

Solution

Step 1—Determine the ground snow load, p_g

Using Reference 4.1, the ground snow load is equal to 28 lb/ft^2 (1.34 kN/m^2) for Risk Category III.

Step 2—Determine the flat roof snow load, p_f

Flowchart 4.1 is used to determine p_f.

Step 2a—Determine the surface roughness category

The surface roughness category is given in the design data as C.

Step 2b—Determine the exposure of the roof

The roof exposure is given in the design data as partially exposed.

Step 2c—Determine the exposure factor, C_e

Given a surface roughness category of C and a partially exposed roof exposure, $C_e = 1.0$ from ASCE/SEI Table 7.3-1.

Step 2d—Determine the thermal condition of the roof

From the design data, the building has a cold, ventilated roof.

Step 2e—Determine the thermal factor, C_t

For a building with a cold, ventilated roof, $C_t = 1.2$ (ASCE/SEI Table 7.3-2).

Step 2f—Determine the flat roof snow load

$$p_f = 0.7C_eC_tp_g = 0.7 \times 1.0 \times 1.2 \times 28.0 = 23.5 \text{ lb/ft}^2$$

$$p_f = 0.7C_eC_tp_g = 0.7 \times 1.0 \times 1.2 \times 1.34 = 1.13 \text{ kN/m}^2$$

Step 3—Determine the minimum snow load for low-slope roofs, p_m

Minimum snow load requirements are not applicable to sawtooth roofs.

Step 4—Determine the sloped roof (balanced) snow load, p_s

Flowchart 4.2 is used to determine p_s.

Step 4a—Determine the slope factor, C_s

In accordance with ASCE/SEI 7.4.3, $C_s = 1.0$ for sawtooth roofs.

Step 4b—Determine the sloped roof (balanced) snow load, p_s

$$p_s = C_sp_f = 1.0 \times 23.5 = 23.5 \text{ lb/ft}^2$$

$$p_s = C_sp_f = 1.0 \times 1.13 = 1.13 \text{ kN/m}^2$$

Step 5—Consider partial loading

From the design data, the roof structural members are simply supported, so partial loading need not be considered (ASCE/SEI 7.5).

Step 6—Consider unbalanced snow loads

Flowchart 4.5 is used to determine if unbalanced snow loads must be considered.

The roof slopes are equal to 45 degrees and 63.4 degrees. Because these slopes are greater than 1.79 degrees, unbalanced snow loads must be considered (ASCE/SEI 7.6.3). Unbalanced snow loads for this roof are given in ASCE/SEI Figure 7.6-3 (see Figure 4.12).

- At the ridge:

$$0.5p_f = 0.5 \times 23.5 = 11.8 \text{ lb/ft}^2$$

$$0.5p_f = 0.5 \times 1.13 = 0.57 \text{ kN/m}^2$$

- At the valley:

$$2p_f/C_e = (2 \times 23.5)/1.0 = 47.0 \text{ lb/ft}^2$$

The load at the valley is limited by the space available for snow accumulation.

$$\gamma = 0.13p_g + 14 = (0.13 \times 28.0) + 14 = 17.6 \text{ lb/ft}^3 < 30.0 \text{ lb/ft}^3$$

Ridge height, $h_r = 10.0$ ft
Maximum load at the valley:

$$0.5p_f + \gamma h_r = (0.5 \times 23.5) + (17.6 \times 10.0) = 187.8 \text{ lb/ft}^2$$

Because the unbalanced load of 47.0 lb/ft² at the valley is less than 187.8 lb/ft², use 47.0 lb/ft² as the load at the valley.

In S.I.:

$$2p_f/C_e = (2 \times 1.13)/1.0 = 2.26 \text{ kN/m}^2$$

The load at the valley is limited by the space available for snow accumulation.

$$\gamma = 0.426p_g + 2.2 = (0.426 \times 1.34) + 2.2 = 2.77 \text{ kN/m}^3 < 4.7 \text{ kN/m}^3$$

Ridge height, $h_r = 3.05$ m
Maximum load at the valley:

$$0.5p_f + \gamma h_r = (0.5 \times 1.13) + (2.77 \times 3.05) = 9.0 \text{ kN/m}^2$$

Because the unbalanced load of 2.26 kN/m² at the valley is less than 9.0 kN/m², use 2.26 kN/m² as the load at the valley.

Step 7—Consider drift loads on lower roofs

Not applicable.

Step 8—Consider drifts on roof projections and parapets

Not applicable.

Step 9—Consider sliding snow

Not applicable.

Step 10—Consider rain-on-snow surcharge loads

According to ASCE/SEI 7.10, a rain-on-snow surcharge load is required for locations where $p_g \leq p_{m,\max}$ and $p_g > 0$ on roofs with slopes less than $W/50$ (in S.I.: $W/15.2$). In this example, $p_g = 28$ lb/ft^2(1.34 kN/m^2) $< p_{m,\max} = 35.0$ lb/ft^2(1.68 kN/m^2) from ASCE/SEI Table 7.3-4 for Risk Category III. Also, the roof slopes are greater than

$W/50 = 10/50 = 0.2$ degrees and $5/50 = 0.1$ degrees (In S.I.: $W/15.2 = 3.05/15.2 = 0.2$ degrees and $1.52/15.2 = 0.1$ degrees). Therefore, an additional 8 lb/ft² (0.38 kN/m²) rain-on-snow surcharge load need not be added to the balanced snow load.

Step 11—Consider ponding loads
A ponding analysis considering roof deflections caused by full snow loads in combination with dead and superimposed dead loads must be performed for this roof based on the stiffness of the roof structural members.

Step 12—Consider snow loads on existing roofs
Not applicable.

Step 13—Consider snow loads on open-frame equipment structures
Not applicable.

Design snow loads on the roof for the balanced and unbalanced load cases are given in Figure 4.37.

Figure 4.37 Design Snow Load Cases for the Sawtooth Roof in Example 4.6

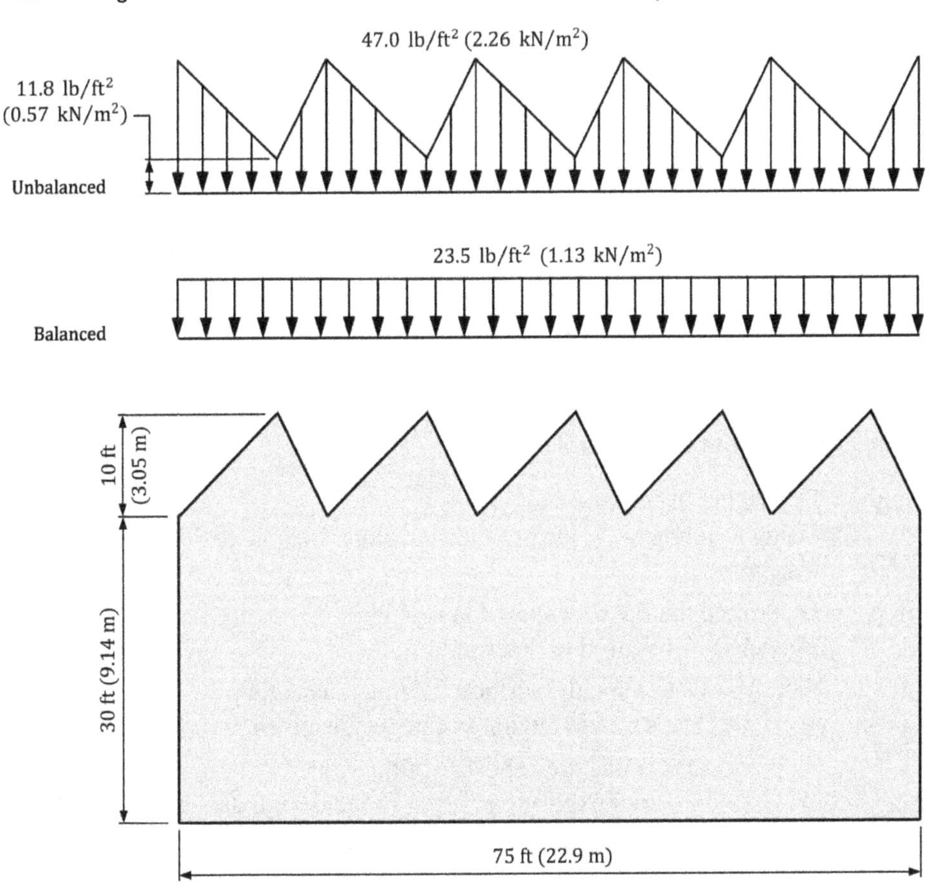

4.5.7 Example 4.7—Design Snow Loads on an Upper Roof and a Lower Roof Including Sliding Snow

Determine the design snow loads on the upper roof and lower roof of the building in Figure 4.38 given the following design data:

- Location: Fort Collins, CO (Latitude = 40.59°, Longitude = −105.08°)
- Surface roughness: B

Figure 4.38 Elevation of the Building in Example 4.7

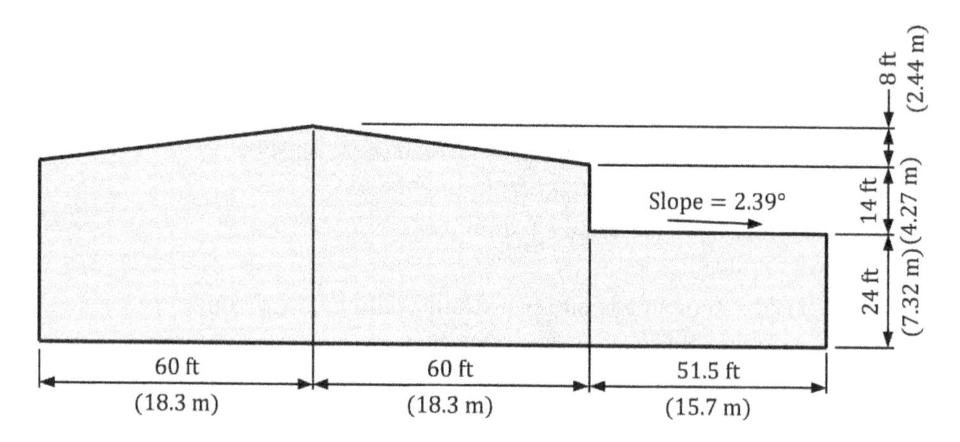

- Risk Category: II (IBC Table 1604.5)
- Thermal condition: Heated structure with an unventilated roof and $R_{roof} = 50$ h ft^2°F/Btu (8.80 m^{2}K/W)
- Roof exposure: Sheltered
- Roof surface: Metal
- Roof obstructions: None
- Roof framing: All structural members are simply supported for both roofs

The design procedure in Table 4.1 of this book is used to determine the design snow loads.

Solution

Snow Loads on the Upper Roof

Step 1—Determine the ground snow load, p_g

Using Reference 4.1, the ground snow load is equal to 39.0 lb/ft^2 (1.87 kN/m^2) for Risk Category II.

Step 2—Determine the flat roof snow load, p_f

Flowchart 4.1 is used to determine p_f.

Step 2a—Determine the surface roughness category

The surface roughness category is given in the design data as B.

Step 2b—Determine the exposure of the roof

The roof exposure is given in the design data as sheltered.

Step 2c—Determine the exposure factor, C_e

Given a surface roughness category of B and a sheltered roof exposure, $C_e = 1.2$ from ASCE/SEI Table 7.3-1.

Step 2d—Determine the thermal condition of the roof

From the design data, the building has an unventilated roof with $R_{roof} =$ 50 h ft^2°F/Btu (8.80 m^{2}K/W).

Step 2e—Determine the thermal factor, C_t

For a building with an unventilated roof with $R_{roof} = 50$ h ft^2°F/Btu (8.80 m^{2}K/W), $C_t = 1.19$ for p_g = 39.0 lb/ft^2 (1.87 kN/m^2) (ASCE/SEI Table 7.3-3).

Step 2f—Determine the flat roof snow load

$$p_f = 0.7C_eC_tp_g = 0.7\times1.2\times1.19\times39.0 = 39.0 \text{ lb/ft}^2$$

$$p_f = 0.7C_eC_tp_g = 0.7\times1.2\times1.19\times1.87 = 1.87 \text{ kN/m}^2$$

Step 3—Determine the minimum snow load for low-slope roofs, p_m

A minimum snow load, p_m, applies to gable roofs with slopes less than 15 degrees (see Figure 4.2). Because the upper roof slope in this example is equal to $\tan^{-1}(8/60) =$ 7.6 degrees (in S.I.: $\tan^{-1}(2.44/18.3) = 7.6$ degrees), minimum snow loads must be considered.

For Risk Category II, the minimum snow load upper limit, $p_{m,\max}$, is equal to 30 lb/ft² (1.44 kN/m²) from ASCE/SEI Table 7.3-4. Because $p_g = 39$ lb/ft² (1.87 kN/m²) $> p_{m,\max} =$ 30 lb/ft² (1.44 kN/m²), $p_m = p_{m,\max} = 30$ lb/ft² (1.44 kN/m²) (ASCE/SEI 7.3.3).

Step 4—Determine the sloped roof (balanced) snow load, p_s

Flowchart 4.2 is used to determine p_s.

Step 4a—Determine the slope factor, C_s

From Step 2e, $C_t = 1.19$, which means the roof is warm.

From the design data, there are no obstructions inhibiting the snow from sliding off the roof. Also, because $C_t > 1.1$, ice dams on all overhanging portions need not be considered for this unventilated roof (if an ice dam can form, it is considered to be an obstruction; see ASCE/SEI 7.4.4).

The roof surface is metal from the design data. According to ASCE/SEI 7.4, a metal roof is considered to be slippery.

Because the roof is unobstructed and slippery with $1.1 < C_t = 1.19 < 1.2$, use the dashed line in ASCE/SEI Figure 7.4-1b to determine the slope factor, C_s. For a roof slope of 7.6 degrees, which is less than 10 degrees, $C_s = 1.0$.

Step 4b—Determine the sloped roof (balanced) snow load, p_s

$$p_s = C_sp_f = 1.0\times39.0 = 39.0 \text{ lb/ft}^2$$

$$p_s = C_sp_f = 1.0\times1.87 = 1.87 \text{ kN/m}^2$$

Step 5—Consider partial loading

From the design data, the upper roof structural members are simply supported, so partial loading need not be considered (ASCE/SEI 7.5).

Step 6—Consider unbalanced snow loads

Flowchart 4.3 is used to determine if unbalanced snow loads must be considered.

Unbalanced snow loads need not be considered for gable roofs with a slope exceeding 30.3 degrees or with a slope less than 2.39 degrees (ASCE/SEI 7.6.1). The roof slope in this example is 7.6 degrees, so unbalanced loads must be considered.

Because $W = 60$ ft (18.3 m) > 20 ft (6.10 m), the unbalanced load consists of the following (see Figure 4.8):

- Windward side:
  Unbalanced load $= 0.3p_s = 0.3\times39.0 = 11.7$ lb/ft² applied over the entire length of the windward side
- Leeward side:
  From Reference 4.1, winter wind parameter, $W_2 = 0.45$
  Snow density, $\gamma = 0.13p_g + 14 = (0.13\times39) + 14 = 19.1$ lb/ft³ < 30 lb/ft³

$$h_d = 1.5\sqrt{\frac{(p_g)^{0.74}(W)^{0.70}(W_2)^{1.7}}{\gamma}} = 1.5\times\sqrt{\frac{(39.0)^{0.74}\times(60.0)^{0.70}\times(0.45)^{1.7}}{19.1}} = 2.8 \text{ ft}$$

S = roof slope run for a rise of one $= 1/\tan 7.6° = 7.5$
$p_s = 39.0$ lb/ft^2 is applied over the entire length of the leeward side
Rectangular surcharge $= h_d\gamma/\sqrt{S} = (2.8\times19.1)/\sqrt{7.5} = 19.5$ lb/ft^2 is applied on the leeward side a distance equal to $8h_d\sqrt{S}/3 = (8\times2.8\times\sqrt{7.5})/3 = 20.5$ ft from the ridge

In S.I.:

- Windward side:
  Unbalanced load $= 0.3p_s = 0.3\times1.87 = 0.56$ kN/m^2 applied over the entire length of the windward side
- Leeward side:
  From Reference 4.1, winter wind parameter, $W_2 = 0.45$
  Snow density, $\gamma = 0.426p_g + 2.2 = (0.426\times1.87) + 2.2 = 3.00$ kN/m^3 < 4.7 kN/m^3

$$h_d = 1.5\sqrt{\frac{(p_g)^{0.74}(W)^{0.70}(W_2)^{1.7}}{\gamma}} = 1.5\times\sqrt{\frac{(39.0)^{0.74}\times(60.0)^{0.70}\times(0.45)^{1.7}}{19.1}}$$

$$= 2.8 \text{ ft}\times0.3048 = 0.85 \text{ m}$$

S = roof slope run for a rise of one $= 1/\tan 7.6° = 7.5$
$p_s = 1.87$ kN/m^2 is applied over the entire length of the leeward side
Rectangular surcharge $= h_d\gamma/\sqrt{S} = (0.85\times3.00)/\sqrt{7.5} = 0.93$ kN/m^2 is applied on the leeward side a distance equal to $8h_d\sqrt{S}/3 = (8\times0.85\times\sqrt{7.5})/3 = 6.21$ m from the ridge

Step 7—Consider drift loads on lower roofs
See the section of this solution below for snow loads on the lower roof.

Step 8—Consider drifts on roof projections and parapets
Not applicable.

Step 9—Consider sliding snow
See the section of this solution below for snow loads on the lower roof.

Step 10—Consider rain-on-snow surcharge loads
According to ASCE/SEI 7.10, a rain-on-snow surcharge load is required for locations where $p_g \leq p_{m,\max}$ and $p_g > 0$ on roofs with slopes less than $W/50$ (in S.I.: $W/15.2$). In this example, $p_g = 39.0$ lb/ft^2(1.87 kN/m^2) $> p_{m,\max} = 30.0$ lb/ft^2(1.44 kN/m^2) from ASCE/SEI Table 7.3-4 for Risk Category II. Therefore, an additional 8 lb/ft^2 (0.38 kN/m^2) rain-on-snow surcharge load need not be added to the balanced snow load.

Step 11—Consider ponding loads
A ponding analysis considering roof deflections caused by full snow loads in combination with dead and superimposed dead loads must be performed for this roof based on the stiffness of the roof structural members.

Step 12—Consider snow loads on existing roofs
Not applicable.

Step 13—Consider snow loads on open-frame equipment structures
Not applicable.

Design snow loads on the roof for the minimum, balanced, and unbalanced load cases are given in Figure 4.39.

Snow Loads on the Lower Roof

Steps 1 through 5—Results are the same as those for the upper roof.

$$p_g = 39.0 \text{ lb/ft}^2(1.87 \text{ kN/m}^2)$$

$$C_e = 1.2$$

Figure 4.39 Design Snow Load Cases for the Upper Gable Roof in Example 4.7

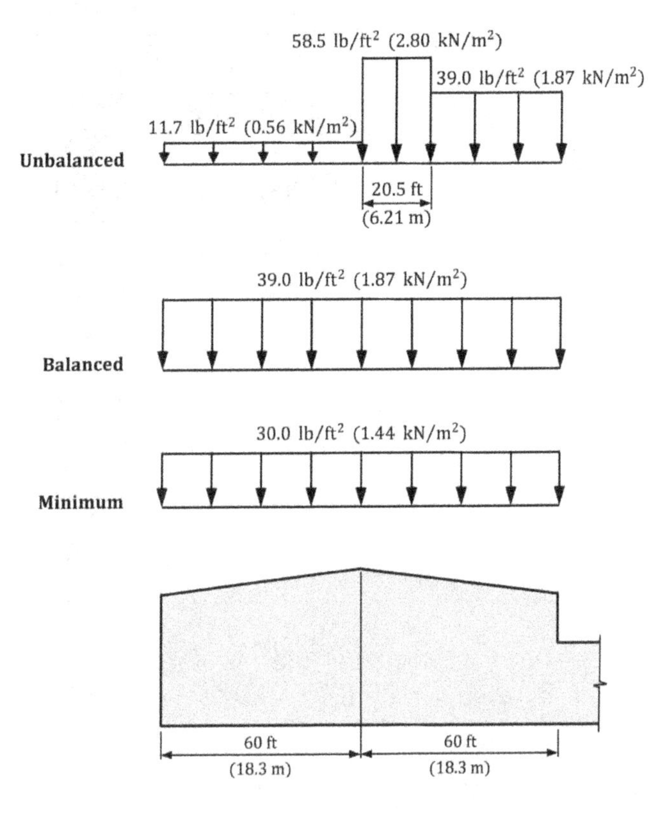

$$C_t = 1.19$$

$$p_f = p_s = 39.0 \text{ lb/ft}^2 (1.87 \text{ kN/m}^2)$$

$$p_m = 30.0 \text{ lb/ft}^2 (1.44 \text{ kN/m}^2)$$

$$\gamma = 19.1 \text{ lb/ft}^3 (3.00 \text{ kN/m}^3)$$

Ice dam loads and partial loading need not be considered.

Step 6—Consider unbalanced snow loads

Because this is a monoslope roof, unbalanced snow loads need not be considered.

Step 7—Consider drift loads on lower roofs

Use Flowchart 4.6 to determine the leeward and windward drifts on the lower roof.

Step 7a—Determine the height of the balanced snow load, h_b

$$h_b = p_s / \gamma = 39.0/19.1 = 2.0 \text{ ft}$$

Step 7b—Determine the clear height from the top of the balanced snow load to the upper roof, h_c

$$h_c = h_{\text{step}} - h_b = [14.0 - (51.5 \times \tan 2.39°)] - 2.0 = 9.9 \text{ ft}$$

Step 7c—Determine if drift loads must be considered

Drift loads must be considered where $h_c/h_b \geq 0.2$.

$$h_c/h_b = 9.9/2.0 = 4.95 > 0.2$$

Therefore, drift loads must be considered.

Step 7d—Determine the leeward drift height, $h_{d,\text{leeward}}$

$$h_{d,\text{leeward}} = 1.5\sqrt{\frac{(p_g)^{0.74}(\ell_{\text{upper}})^{0.70}(W_2)^{1.7}}{\gamma}}$$

$$= 1.5\times\sqrt{\frac{(39.0)^{0.74}\times(120.0)^{0.70}\times(0.45)^{1.7}}{19.1}}$$

$$= 3.6 \text{ ft} < 0.6\ell_{\text{lower}} = 0.6\times 51.5 = 30.9 \text{ ft}$$

Step 7e—Determine the windward drift height, $h_{d,\text{windward}}$

$$h_{d,\text{windward}} = 1.125\sqrt{\frac{(p_g)^{0.74}(\ell_{\text{lower}})^{0.70}(W_2)^{1.7}}{\gamma}}$$

$$= 1.125\times\sqrt{\frac{(39.0)^{0.74}\times(51.5)^{0.70}\times(0.45)^{1.7}}{19.1}} = 2.0 \text{ ft}$$

Step 7f—Determine the drift height, h_d

$$h_d = \text{greater of}\begin{cases} h_{d,\text{leeward}} = 3.6 \text{ ft} \\ h_{d,\text{windward}} = 2.0 \text{ ft} \end{cases}$$

Step 7g—Determine the drift width, w

Because $h_d = 3.6$ ft $< h_c = 9.9$ ft, $w = 4h_d = 4\times 3.6 = 14.4$ ft $< 8h_c = 79.2$ ft and $w < \ell_{\text{lower}} = 51.5$ ft.

Step 7h—Determine the drift load, p_d

$$p_d = \gamma h_d = 19.1\times 3.6 = 68.8 \text{ lb/ft}^2$$

Step 7i—Determine the total snow load at the step

$$p_{\text{total}} = p_s + p_d = 39.0 + 68.8 = 107.8 \text{ lb/ft}^2$$

In S.I.:

Step 7a—Determine the height of the balanced snow load, h_b

$$h_b = p_s/\gamma = 1.87/3.00 = 0.62 \text{ m}$$

Step 7b—Determine the clear height from the top of the balanced snow load to the upper roof, h_c

$$h_c = h_{\text{step}} - h_b = [4.27 - (15.7\times\tan 2.39°)] - 0.62 = 3.00 \text{ m}$$

Step 7c—Determine if drift loads must be considered

Drift loads must be considered where $h_c/h_b \geq 0.2$.

$$h_c/h_b = 3.00/0.62 = 4.8 > 0.2$$

Therefore, drift loads must be considered.

Step 7d—Determine the leeward drift height, $h_{d,\text{leeward}}$

$$h_{d,\text{leeward}} = 1.5\sqrt{\frac{(p_g)^{0.74}(\ell_{\text{upper}})^{0.70}(W_2)^{1.7}}{\gamma}}$$

$$= 1.5\times\sqrt{\frac{(39.0)^{0.74}\times(120.0)^{0.70}\times(0.45)^{1.7}}{19.1}}$$

$$= 3.6 \text{ ft}\times 0.3048 = 1.10 \text{ m} < 0.6\ell_{\text{lower}} = 0.6\times 15.7 = 9.42 \text{ m}$$

Step 7e—Determine the windward drift height, $h_{d,\text{windward}}$

$$h_{d,\text{windward}} = 1.125\sqrt{\frac{(p_g)^{0.74}(\ell_{\text{lower}})^{0.70}(W_2)^{1.7}}{\gamma}}$$

$$= 1.125 \times \sqrt{\frac{(39.0)^{0.74} \times (51.5)^{0.70} \times (0.45)^{1.7}}{19.1}} = 2.0 \text{ ft} \times 0.3048 = 0.61 \text{ m}$$

Step 7f—Determine the drift height, h_d

$$h_d = \text{greater of} \begin{cases} h_{d,\text{leeward}} = 1.10 \text{ m} \\ h_{d,\text{windward}} = 0.61 \text{ m} \end{cases}$$

Step 7g—Determine the drift width, w

Because $h_d = 1.10\,\text{m} < h_c = 3.00\,\text{m}$, $w = 4h_d = 4 \times 1.10 = 4.40\,\text{m} < 8h_c = 24.0\,\text{m}$ and $w < \ell_{\text{lower}} = 15.7$ m.

Step 7h—Determine the drift load, p_d

$$p_d = \gamma h_d = 3.00 \times 1.10 = 3.30 \text{ kN/m}^2$$

Step 7i—Determine the total snow load at the step

$$p_{\text{total}} = p_s + p_d = 1.87 + 3.30 = 5.17 \text{ kN/m}^2$$

Step 8—Consider drifts on roof projections and parapets

Not applicable.

Step 9—Consider sliding snow

Use Flowchart 4.9 to determine the sliding snow load on the lower roof.

Loads caused by snow sliding must be considered because the upper roof is slippery with a slope greater than ¼ on 12 (1.19 degrees) [ASCE/SEI 7.9].

The sliding snow load is equal to the following where $\ell_{\text{lower}} > 15$ ft (4.57 m) [see Figure 4.23]:

$$p_{\text{sliding}} = 0.4 p_f W/15 = (0.4 \times 39.0 \times 60.0)/15 = 62.4 \text{ lb/ft}^2$$

$$p_{\text{sliding}} = 0.4 p_f W/4.57 = (0.4 \times 1.87 \times 18.3)/4.57 = 3.00 \text{ kN/m}^2$$

The total snow load over the 15-ft (4.57-m) width is equal to the sliding snow load plus the balanced snow load:

$$p_{\text{total}} = p_s + (0.4 p_f W/15) = 39.0 + 62.4 = 101.4 \text{ lb/ft}^2$$

$$p_{\text{total}} = p_s + (0.4 p_f W/4.57) = 1.87 + 3.00 = 4.87 \text{ kN/m}^2)$$

The depth of snow for the total snow load $= p_{\text{total}}/\gamma = 101.4/19.1 = 5.3$ ft [in S.I.: $p_{\text{total}}/\gamma = 4.87/3.00 = 1.62$ m]. This depth is less than the distance from the eave of the upper roof to the top of the lower roof at the interface, which is equal to $h_{\text{step}} = 14.0 - (51.5 \times \tan 2.39°) = 11.9$ ft [in S.I.: $h_{\text{step}} = 4.27 - (15.7 \times \tan 2.39°) = 3.62$ m]. Therefore, sliding snow is not blocked and the full load can be developed over the 15-ft (4.57-m) width.

Step 10—Consider rain-on-snow surcharge loads

According to ASCE/SEI 7.10, a rain-on-snow surcharge load is required for locations where $p_g \le p_{m,\max}$ and $p_g > 0$ on roofs with slopes less than $W/50$ (in S.I.: $W/15.2$). In this example, $p_g = 39.0 \text{ lb/ft}^2 (1.87 \text{ kN/m}^2) > p_{m,\max} = 30.0 \text{ lb/ft}^2 (1.44 \text{ kN/m}^2)$ from

ASCE/SEI Table 7.3-4 for Risk Category II. Therefore, an additional 8 lb/ft^2 (0.38 kN/m^2) rain-on-snow surcharge load need not be added to the balanced snow load.

Step 11—Consider ponding loads
A ponding analysis considering roof deflections caused by full snow loads in combination with dead and superimposed dead loads must be performed for this roof based on the stiffness of the roof structural members.

Step 12—Consider snow loads on existing roofs
Not applicable.

Step 13—Consider snow loads on open-frame equipment structures
Not applicable.

Design snow loads on the roof for the minimum, balanced, drift, and sliding load cases are given in Figure 4.40.

Figure 4.40 Design Snow Load Cases for the Lower Monoslope Roof in Example 4.7

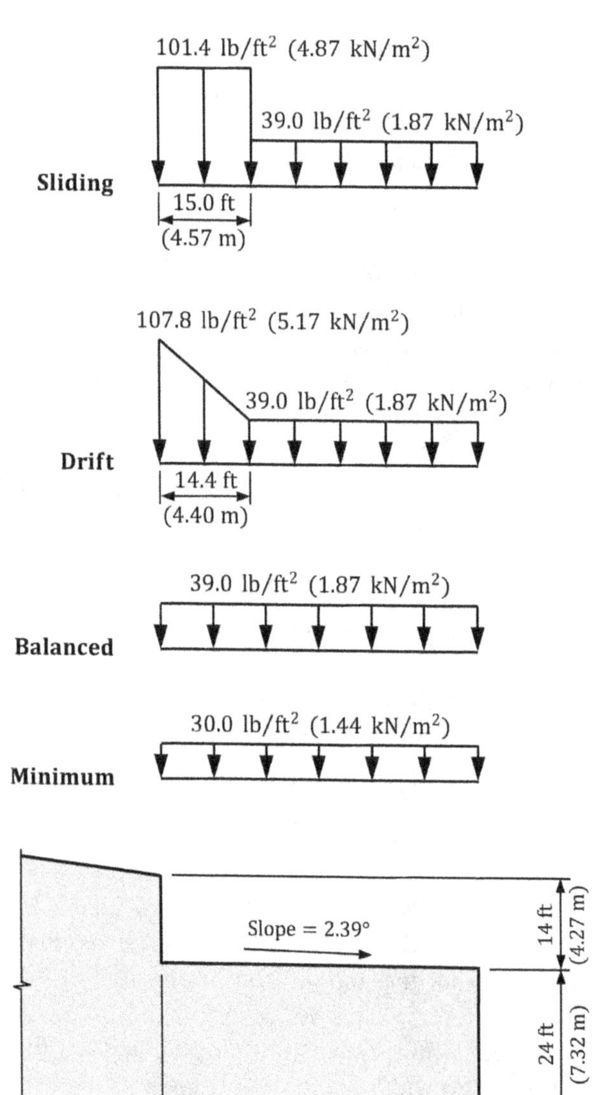

4.5.8 Example 4.8—Design Snow Loads on an Adjacent Lower Roof Including Sliding Snow and Horizontal Separation

Determine the design snow loads on the lower roof of the building in Figure 4.41 given the design data in Example 4.7.

Figure 4.41 Elevation of the Buildings in Example 4.8.

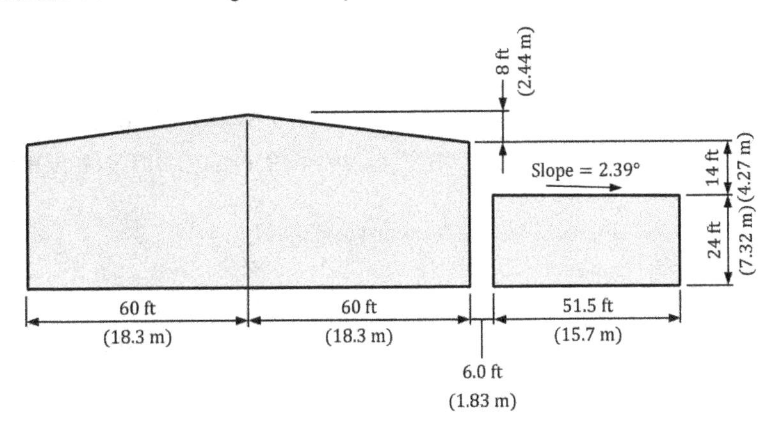

SOLUTION

Steps 1 through 5—Results are the same as in Example 4.7 for the lower roof.

$$p_g = 39.0 \text{ lb/ft}^2 (1.87 \text{ kN/m}^2)$$

$$C_e = 1.2$$

$$C_t = 1.19$$

$$p_f = p_s = 39.0 \text{ lb/ft}^2 (1.87 \text{ kN/m}^2)$$

$$p_m = 30.0 \text{ lb/ft}^2 (1.44 \text{ kN/m}^2)$$

$$\gamma = 19.1 \text{ lb/ft}^3 (3.00 \text{ kN/m}^3)$$

Ice dam loads and partial loading need not be considered.

Step 6—Consider unbalanced snow loads

Because this is a monoslope roof, unbalanced snow loads need not be considered.

Step 7—Consider drift loads on lower roofs

Use Flowchart 4.7 to determine the leeward and windward drifts on the lower roof.

Step 7a—Determine if a leeward drift on the lower adjacent roof must be considered

Leeward drifts must be considered where $s < 20$ ft and $s < 6h$ (ASCE/SEI 7.7.2).

$$h = 14.0 - (51.5 \times \tan 2.39°) = 11.9 \text{ ft}$$

$$s = 6.0 \text{ ft} < 20 \text{ ft and } s < 6h = 6 \times 11.9 = 71.4 \text{ ft}$$

Therefore, a leeward drift on the lower roof must be considered.

Step 7b—Determine the leeward drift height, $h_{d,\text{leeward}}$, based on the length of the adjacent higher structure

$$h_{d,\text{leeward}} = 1.5\sqrt{\frac{(p_g)^{0.74}(\ell_{\text{upper}})^{0.70}(W_2)^{1.7}}{\gamma}}$$

$$= 1.5 \times \sqrt{\frac{(39.0)^{0.74} \times (120.0)^{0.70} \times (0.45)^{1.7}}{19.1}} = 3.6 \text{ ft}$$

Step 7c—Determine the leeward drift load, $p_{d,\text{leeward}}$

$$p_{d,\text{leeward}} = \text{lesser of} \begin{cases} \gamma h_{d,\text{leeward}} = 19.1 \times 3.6 = 68.8 \text{ lb/ft}^2 \\ \gamma[(6h - s)/6] = 19.1 \times \{[(6 \times 11.9) - 6.0]/6\} = 208.2 \text{ lb/ft}^2 \end{cases}$$

Step 7d—Determine the leeward drift width, w

$$w = \text{lesser of} \begin{cases} 6h_{d,\text{leeward}} = 6 \times 3.6 = 21.6 \text{ ft} \\ 6h - s = (6 \times 11.9) - 6.0 = 65.4 \text{ ft} \end{cases}$$

Step 7e—Determine the total snow load at the lower roof edge based on a leeward drift

$$(p_{\text{total}})_{\text{leeward}} = p_s + p_{d,\text{leeward}} = 39.0 + 68.8 = 107.8 \text{ lb/ft}^2$$

Step 7f—Determine the windward drift height, $h_{d,\text{windward}}$, based on the length of the lower structure

$$h_{d,\text{windward}} = 1.125\sqrt{\frac{(p_g)^{0.74}(\ell_{\text{lower}})^{0.70}(W_2)^{1.7}}{\gamma}}$$

$$= 1.125 \times \sqrt{\frac{(39.0)^{0.74} \times (51.5)^{0.70} \times (0.45)^{1.7}}{19.1}} = 2.0 \text{ ft}$$

Step 7g—Determine the windward drift load, $p_{d,\text{windward}}$

$$p_{d,\text{windward}} = \gamma h_{d,\text{windward}} = 19.1 \times 2.0 = 38.2 \text{ lb/ft}^2$$

Step 7h—Determine the truncated windward drift load at the edge of the lower roof, $p_{d,\text{truncated}}$

$$p_{d,\text{truncated}} = \left(1 - \frac{s}{4h_{d,\text{windward}}}\right) p_{d,\text{windward}} = \left(1 - \frac{6.0}{4 \times 2.0}\right) \times 38.2 = 9.6 \text{ lb/ft}^2$$

Step 7i—Determine the windward drift width, w

$$w = 4h_{d,\text{windward}} - s = (4 \times 2.0) - 6.0 = 2.0 \text{ ft}$$

Step 7j—Determine the total snow load at the lower roof edge based on a windward drift

$$(p_{\text{total}})_{\text{windward}} = p_s + p_{d,\text{truncated}} = 39.0 + 9.6 = 48.6 \text{ lb/ft}^2$$

In S.I.:

Step 7a—Determine if a leeward drift on the lower adjacent roof must be considered
Leeward drifts must be considered where $s < 6.10$ m and $s < 6h$ (ASCE/SEI 7.7.2).

$$h = 4.27 - (15.7 \times \tan 2.39°) = 3.62 \text{ m}$$

$$s = 1.83 \text{ m} < 6.10 \text{ m and } s < 6h = 6 \times 3.62 = 21.7 \text{ m}$$

Therefore, a leeward drift must be considered.

Step 7b—Determine the leeward drift height, $h_{d,\text{leeward}}$, based on the length of the adjacent higher structure

$$h_{d,\text{leeward}} = 1.5\sqrt{\frac{(p_g)^{0.74}(\ell_{\text{upper}})^{0.70}(W_2)^{1.7}}{\gamma}}$$

$$= 1.5 \times \sqrt{\frac{(39.0)^{0.74} \times (120.0)^{0.70} \times (0.45)^{1.7}}{19.1}} = 3.6 \text{ ft} \times 0.3048 = 1.10 \text{ m}$$

Step 7c—Determine the leeward drift load, $p_{d,\text{leeward}}$

$$p_{d,\text{leeward}} = \text{lesser of} \begin{cases} \gamma h_{d,\text{leeward}} = 3.00 \times 1.10 = 3.30 \text{ kN/m}^2 \\ \gamma[(6h - s)/6] = 3.00 \times \{[(6 \times 3.62) - 1.83]/6\} = 9.95 \text{ kN/m}^2 \end{cases}$$

Step 7d—Determine the leeward drift width, w

$$w = \text{lesser of} \begin{cases} 6h_{d,\text{leeward}} = 6 \times 1.10 = 6.60 \text{ m} \\ 6h - s = (6 \times 3.62) - 1.83 = 19.9 \text{ m} \end{cases}$$

Step 7e—Determine the total snow load at the lower roof edge based on a leeward drift

$$(p_{\text{total}})_{\text{leeward}} = p_s + p_{d,\text{leeward}} = 1.87 + 3.30 = 5.17 \text{ kN/m}^2$$

Step 7f—Determine the windward drift height, $h_{d,\text{windward}}$, based on the length of the lower structure

$$h_{d,\text{windward}} = 1.125\sqrt{\frac{(p_g)^{0.74}(\ell_{\text{lower}})^{0.70}(W_2)^{1.7}}{\gamma}}$$

$$= 1.125 \times \sqrt{\frac{(39.0)^{0.74} \times (51.5)^{0.70} \times (0.45)^{1.7}}{19.1}} = 2.0 \text{ ft} \times 0.3048 = 0.61 \text{ m}$$

Step 7g—Determine the windward drift load, $p_{d,\text{windward}}$

$$p_{d,\text{windward}} = \gamma h_{d,\text{windward}} = 3.00 \times 0.61 = 1.83 \text{ kN/m}^2$$

Step 7h—Determine the truncated windward drift load at the edge of the lower roof, $p_{d,\text{truncated}}$

$$p_{d,\text{truncated}} = \left(1 - \frac{s}{4h_{d,\text{windward}}}\right) p_{d,\text{windward}} = \left(1 - \frac{1.83}{4 \times 0.61}\right) \times 1.83 = 0.46 \text{ kN/m}^2$$

Step 7i—Determine the windward drift width, w

$$w = 4h_{d,\text{windward}} - s = (4 \times 0.61) - 1.83 = 0.61 \text{ m}$$

Step 7j—Determine the total snow load at the lower roof edge based on a windward drift

$$(p_{\text{total}})_{\text{windward}} = p_s + p_{d,\text{truncated}} = 1.87 + 0.46 = 2.33 \text{ kN/m}^2$$

Step 8—Consider drifts on roof projections and parapets
Not applicable.

Step 9—Consider sliding snow
Use Flowchart 4.9 to determine the sliding snow load on the lower roof.

Loads caused by snow sliding must be considered because the upper roof is slippery with a slope greater than ¼ on 12 (1.19 degrees) and for separated structures, $h/s = 11.9/6 = 1.98 > 1$ and $s = 6.0$ ft < 15 ft (in S.I.: $h/s = 3.62/1.83 = 1.98 > 1$ and $s = 1.83$ m < 4.57 m).

The sliding snow load is equal to the following (see Figure 4.24):

$$p_{\text{sliding}} = 0.4 p_f W/15 = (0.4 \times 39.0 \times 60.0)/15 = 62.4 \text{ lb/ft}^2$$

$$p_{\text{sliding}} = 0.4 p_f W/4.6 = (0.4 \times 1.87 \times 18.3)/4.57 = 3.00 \text{ kN/m}^2$$

The total snow load is equal to the sliding snow load plus the balanced snow load:

$$p_{\text{total}} = p_s + (0.4 p_f W/15) = 39.0 + 62.4 = 101.4 \text{ lb/ft}^2$$

$$p_{\text{total}} = p_s + (0.4 p_f W/4.6) = 1.87 + 3.00 = 4.87 \text{ kN/m}^2$$

The sliding snow loads acts over a width equal to $15.0 - 6.0 = 9.0$ ft $(2.74$ m$)$.

Step 10—Consider rain-on-snow surcharge loads
According to ASCE/SEI 7.10, a rain-on-snow surcharge load is required for locations where $p_g \le p_{m,\max}$ and $p_g > 0$ on roofs with slopes less than $W/50$ (in S.I.: $W/15.2$). In this example, $p_g = 39.0$ lb/ft^2(1.87 kN/m^2) $> p_{m,\max} = 30.0$ lb/ft^2(1.44 kN/m^2) from ASCE/SEI Table 7.3-4 for Risk Category II. Therefore, an additional 8 lb/ft^2 (0.38 kN/m^2) rain-on-snow surcharge load need not be added to the balanced snow load.

Step 11—Consider ponding loads
A ponding analysis considering roof deflections caused by full snow loads in combination with dead and superimposed dead loads must be performed for this roof based on the stiffness of the roof structural members.

Step 12—Consider snow loads on existing roofs
Not applicable.

Step 13—Consider snow loads on open-frame equipment structures
Not applicable.

Design snow loads on the lower roof for the minimum, balanced, leeward drift, and sliding load cases are given in Figure 4.42.

Figure 4.42 Design Snow Load Cases for the Lower Monoslope Roof in Example 4.8

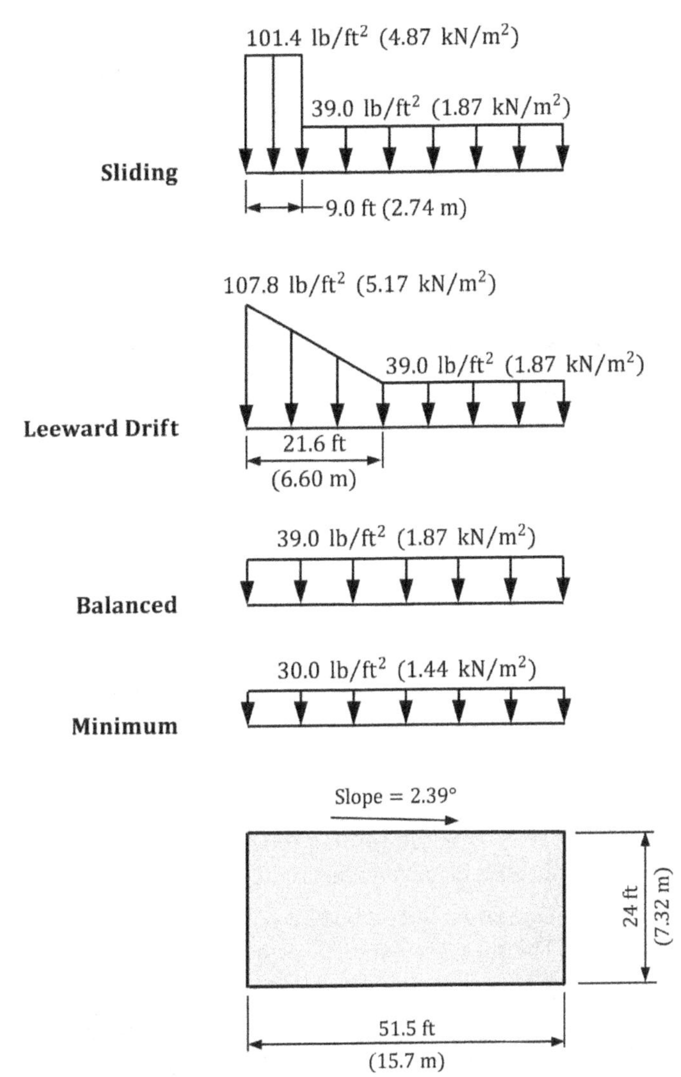

4.5.9 Example 4.9—Design Snow Loads on a Building with a Canopy

Determine the design snow loads on the roof and canopy of the building in Figure 4.43 given the following design data:

- Location: Tacoma, WA (Latitude = 47.23°, Longitude = −122.48°)
- Surface roughness: C
- Risk Category: II (IBC Table 1604.5)
- Thermal condition: Heated structure with an unventilated roof and $R_{roof} = 30$ h ft^2°F/Btu (5.28 m^{2}K/W)
- Roof exposure: Partially exposed
- Roof surface: Smooth bituminous
- Roof obstructions: None
- Roof framing: All structural members are simply supported

Figure 4.43 Elevation of the Building in Example 4.9

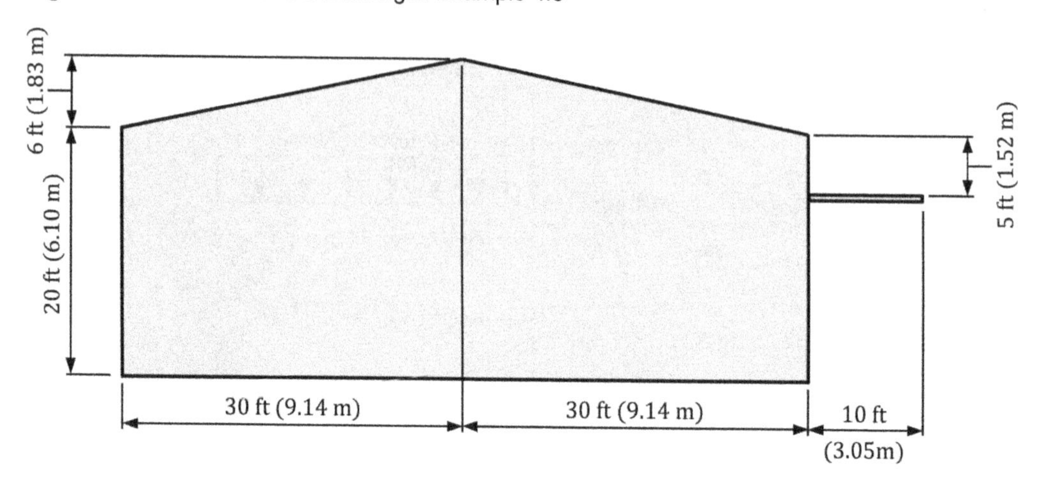

SOLUTION

Snow Loads on the Roof

Step 1—Determine the ground snow load, p_g

Using Reference 4.1, the ground snow load is equal to 41.0 lb/ft^2 (1.96 kN/m^2) for Risk Category II.

Step 2—Determine the flat roof snow load, p_f

Flowchart 4.1 is used to determine p_f.

Step 2a—Determine the surface roughness category

The surface roughness category is given in the design data as C.

Step 2b—Determine the exposure of the roof

The roof exposure is given in the design data as partially exposed.

Step 2c—Determine the exposure factor, C_e

Given a surface roughness category of C and a partially exposed roof exposure, $C_e = 1.0$ from ASCE/SEI Table 7.3-1.

Step 2d—Determine the thermal condition of the roof

From the design data, the building has an unventilated roof with $R_{\text{roof}} =$ 30 h ft^2°F/Btu (5.28 m^{2}K/W).

Step 2e—Determine the thermal factor, C_t

For a building with an unventilated roof with $R_{\text{roof}} = 30$ h ft^2°F/Btu (5.28 m^{2}K/W), $C_t = 1.13$ for $p_g = 41.0$ lb/ft^2 (1.96 kN/m^2) (ASCE/SEI Table 7.3-3).

Step 2f—Determine the flat roof snow load

$$p_f = 0.7C_eC_tp_g = 0.7 \times 1.0 \times 1.13 \times 41.0 = 32.4 \text{ lb/ft}^2$$

$$p_f = 0.7C_eC_tp_g = 0.7 \times 1.0 \times 1.13 \times 1.96 = 1.55 \text{ kN/m}^2$$

Step 3—Determine the minimum snow load for low-slope roofs, p_m

A minimum snow load, p_m, applies to gable roofs with slopes less than 15 degrees (see Figure 4.2). Because the upper roof slope in this example is equal to $\tan^{-1}(6/30) =$ 11.3 degrees (in S.I.: $\tan^{-1}(1.83/9.14) = 11.3$ degrees), minimum snow loads must be considered.

For Risk Category II, the minimum snow load upper limit, $p_{m,\max}$, is equal to 30.0 lb/ft² (1.44 kN/m²) from ASCE/SEI Table 7.3-4. Because $p_g = 41.0$ lb/ft²(1.96 kN/m²) > $p_{m,\max} = 30.0$ lb/ft² (1.44 kN/m²), $p_m = p_{m,\max} = 30.0$ lb/ft² (1.44 kN/m²)(ASCE/SEI 7.3.3).

Step 4—Determine the sloped roof (balanced) snow load, p_s

Flowchart 4.2 is used to determine p_s.

Step 4a—Determine the slope factor, C_s

From Step 2e, $C_t = 1.13$, which means the roof is warm.

From the design data, there are no obstructions inhibiting the snow from sliding off the roof. Also, because $C_t > 1.1$, ice dams on all overhanging portions need not be considered for this unventilated roof (if an ice dam can form, it is considered to be an obstruction; see ASCE/SEI 7.4.4).

The roof surface is smooth bituminous from the design data. According to ASCE/SEI 7.4, a smooth bituminous roof is considered to be slippery.

Because the roof is unobstructed and slippery with $1.1 < C_t = 1.13 < 1.2$, use the dashed line in ASCE/SEI Figure 7.4-1b to determine the slope factor, C_s. In lieu of using the graph in the figure, use the following equation given in ASCE/SEI C7.4, which is the equation for the dashed line in ASCE/SEI Figure 7.4-1b for roof slopes between 10 degrees and 70 degrees, inclusive:

$$C_s = 1.0 - \frac{\text{slope} - 10°}{60°} = 1.0 - \frac{11.3° - 10°}{60°} = 0.98$$

Step 4b—Determine the sloped roof (balanced) snow load, p_s

$$p_s = C_s p_f = 0.98 \times 32.4 = 31.8 \text{ lb/ft}^2$$

$$p_s = C_s p_f = 0.98 \times 1.55 = 1.52 \text{ kN/m}^2$$

Step 5—Consider partial loading

From the design data, the upper roof structural members are simply supported, so partial loading need not be considered (ASCE/SEI 7.5).

Step 6—Consider unbalanced snow loads

Flowchart 4.3 is used to determine if unbalanced snow loads must be considered.

Unbalanced snow loads need not be considered for gable roofs with a slope exceeding 30.3 degrees or with a slope less than 2.39 degrees (ASCE/SEI 7.6.1). The roof slope in this example is 11.3 degrees, so unbalanced loads must be considered.

Because $W = 30$ ft (9.14 m) > 20 ft (6.10 m), the unbalanced load consists of the following (see Figure 4.8):

- Windward side:
  Unbalanced load $= 0.3p_s = 0.3 \times 31.8 = 9.5$ lb/ft² applied over the entire length of the windward side
- Leeward side:
  From Reference 4.1, winter wind parameter, $W_2 = 0.35$
  Snow density, $\gamma = 0.13p_g + 14 = (0.13 \times 41) + 14 = 19.3$ lb/ft³ < 30 lb/ft³

$$h_d = 1.5\sqrt{\frac{(p_g)^{0.74}(W)^{0.70}(W_2)^{1.7}}{\gamma}} = 1.5 \times \sqrt{\frac{(41.0)^{0.74} \times (30.0)^{0.70} \times (0.35)^{1.7}}{19.3}} = 1.8 \text{ ft}$$

S = roof slope run for a rise of one $= 1/\tan 11.3° = 5$

$p_s = 31.8$ lb/ft² is applied over the entire length of the leeward side

Rectangular surcharge $= h_d\gamma/\sqrt{S} = (1.8 \times 19.3)/\sqrt{5} = 15.5$ lb/ft² is applied on the leeward side a distance equal to $8h_d\sqrt{S}/3 = (8 \times 1.8 \times \sqrt{5})/3 = 10.7$ ft from the ridge

In S.I.:

- Windward side:
  Unbalanced load $= 0.3p_s = 0.3 \times 1.52 = 0.46$ kN/m² applied over the entire length of the windward side
- Leeward side:
  From Reference 4.1, winter wind parameter, $W_2 = 0.35$
  Snow density, $\gamma = 0.426p_g + 2.2 = (0.426 \times 1.96) + 2.2 = 3.04$ kN/m³ < 4.7 kN/m³

$$h_d = 1.5\sqrt{\frac{(p_g)^{0.74}(W)^{0.70}(W_2)^{1.7}}{\gamma}} = 1.5 \times \sqrt{\frac{(41.0)^{0.74} \times (30.0)^{0.70} \times (0.35)^{1.7}}{19.3}}$$

$$= 1.8 \text{ ft} \times 0.3048 = 0.55 \text{ m}$$

S = roof slope run for a rise of one $= 1/\tan 11.3° = 5$

$p_s = 1.52$ kN/m² is applied over the entire length of the leeward side

Rectangular surcharge $= h_d\gamma/\sqrt{S} = (0.55 \times 3.04)/\sqrt{5} = 0.75$ kN/m² is applied on the leeward side a distance equal to $8h_d\sqrt{S}/3 = (8 \times 0.55 \times \sqrt{5})/3 = 3.28$ m from the ridge

Step 7—Consider drift loads on lower roofs
See the section of this solution below for snow loads on the canopy.

Step 8—Consider drifts on roof projections and parapets
Not applicable.

Step 9—Consider sliding snow.
See the section of this solution below for snow loads on the canopy

Step 10—Consider rain-on-snow surcharge loads
According to ASCE/SEI 7.10, a rain-on-snow surcharge load is required for locations where $p_g \le p_{m,\max}$ and $p_g > 0$ on roofs with slopes less than $W/50$ (in S.I.: $W/15.2$). In this example, $p_g = 41.0$ lb/ft²(1.96 kN/m²) $> p_{m,\max} = 30.0$ lb/ft²(1.44 kN/m²) from ASCE/SEI Table 7.3-4 for Risk Category II. Therefore, an additional 8 lb/ft² (0.38 kN/m²) rain-on-snow surcharge load need not be added to the balanced snow load.

Step 11—Consider ponding loads
A ponding analysis considering roof deflections caused by full snow loads in combination with dead and superimposed dead loads must be performed for this roof based on the stiffness of the roof structural members.

Step 12—Consider snow loads on existing roofs
Not applicable.

Step 13—Consider snow loads on open-frame equipment structures
Not applicable.

Design snow loads on the roof for the minimum, balanced, and unbalanced load cases are given in Figure 4.44.

Snow Loads on the Canopy

Step 1—Determine the ground snow load, p_g
Using Reference 4.1, the ground snow load is equal to 41.0 lb/ft² (1.96 kN/m²) for Risk Category II.

Step 2—Determine the flat roof snow load, p_f
Flowchart 4.1 is used to determine p_f.

Step 2a—Determine the surface roughness category
The surface roughness category is given in the design data as C.

Figure 4.44 Design Snow Load Cases for the Gable Roof in Example 4.9

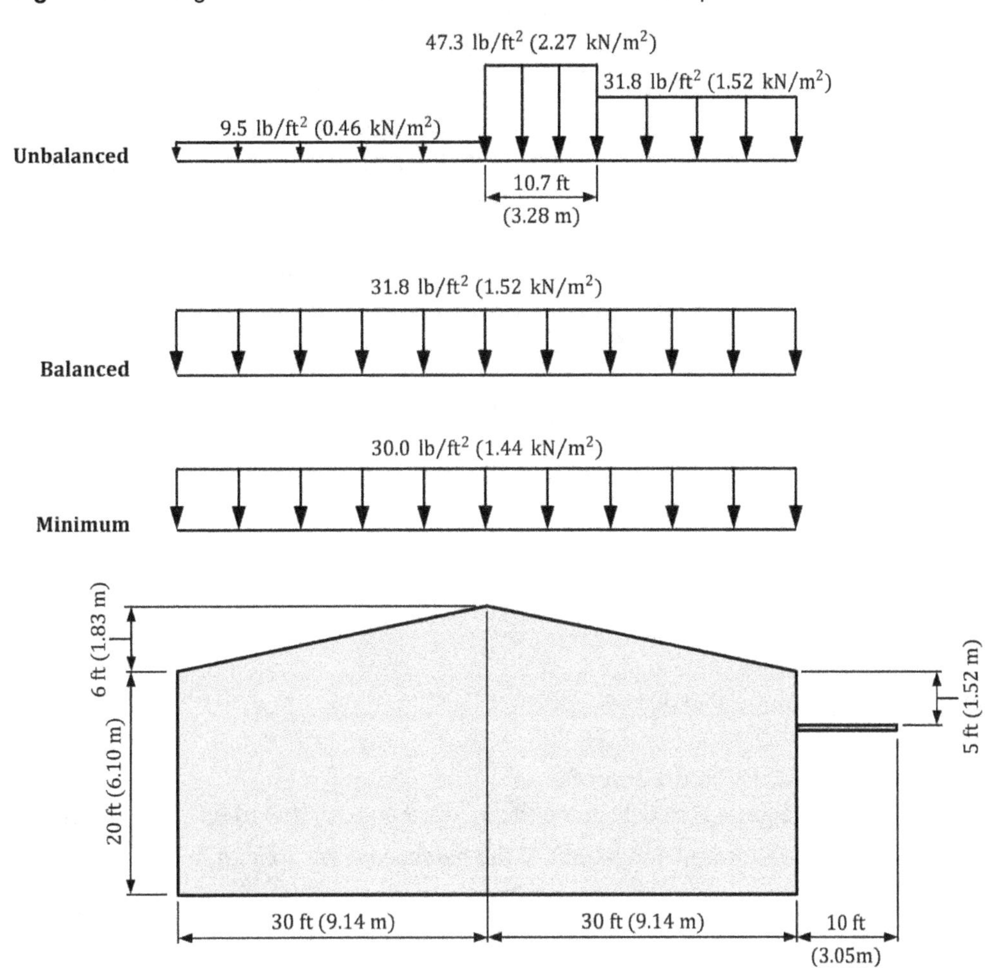

Step 2b—Determine the exposure of the roof

Assume the canopy is partially exposed due to the presence of the building.

Step 2c—Determine the exposure factor, C_e

Given a surface roughness category of C and a partially exposed roof exposure, $C_e = 1.0$ from ASCE/SEI Table 7.3-1.

Step 2d—Determine the thermal condition of the roof

The canopy is an open-air structure.

Step 2e—Determine the thermal factor, C_t

For an open-air structure, $C_t = 1.2$ from ASCE/SEI Table 7.3-2.

Step 2f—Determine the flat roof snow load

$$p_f = 0.7C_eC_tp_g = 0.7 \times 1.0 \times 1.2 \times 41.0 = 34.4 \text{ lb/ft}^2$$

$$p_f = 0.7C_eC_tp_g = 0.7 \times 1.0 \times 1.2 \times 1.96 = 1.65 \text{ kN/m}^2$$

Step 3—Determine the minimum snow load for low-slope roofs, p_m

Because the canopy is essentially flat, a minimum snow load, p_m, must be considered.

For Risk Category II, the minimum snow load upper limit, $p_{m,\max}$, is equal to 30.0 lb/ft² (1.44 kN/m²) from ASCE/SEI Table 7.3-4. Because $p_g = 41.0$ lb/ft² (1.96 kN/m²) > $p_{m,\max} = 30.0$ lb/ft² (1.44 kN/m²), $p_m = p_{m,\max} = 30.0$ lb/ft² (1.44 kN/m²) (ASCE/SEI 7.3.3).

Step 4—Determine the sloped roof (balanced) snow load, p_s

Flowchart 4.2 is used to determine p_s.

Step 4a—Determine the slope factor, C_s

From Step 2e, $C_t = 1.2$, which means the canopy is cold.

Assume there are no obstructions inhibiting the snow from sliding off the canopy. Also, ice dams are not applicable (if an ice dam can form, it is considered to be an obstruction; see ASCE/SEI 7.4.4).

Assume the surface of the canopy is smooth bituminous. According to ASCE/SEI 7.4, a smooth bituminous surface is considered to be slippery.

Because the canopy is unobstructed and slippery with $C_t = 1.2$, use the dashed line in ASCE/SEI Figure 7.4-1c to determine the slope factor, C_s. For a flat canopy, $C_s = 1.0$.

Step 4b—Determine the sloped roof (balanced) snow load, p_s

$$p_s = C_s p_f = 1.0 \times 34.4 = 34.4 \text{ lb/ft}^2$$

$$p_s = C_s p_f = 1.0 \times 1.65 = 1.65 \text{ kN/m}^2$$

Step 5—Consider partial loading

Not applicable.

Step 6—Consider unbalanced snow loads

Not applicable.

Step 7—Consider drifts on lower roofs

Use Flowchart 4.6 to determine the leeward and windward drifts on the canopy.

Step 7a—Determine the height of the balanced snow load, h_b

$$h_b = p_s/\gamma = 34.4/19.3 = 1.8 \text{ ft}$$

Step 7b—Determine the clear height from the top of the balanced snow load to the upper roof, h_c

$$h_c = h_{\text{step}} - h_b = 5.0 - 1.8 = 3.2 \text{ ft}$$

Step 7c—Determine if drift loads must be considered

Drift loads must be considered where $h_c/h_b \geq 0.2$.

$$h_c/h_b = 3.2/1.8 = 1.8 > 0.2$$

Therefore, drift loads must be considered.

Step 7d—Determine the leeward drift height, $h_{d,\text{leeward}}$

From Reference 4.1, winter wind parameter, $W_2 = 0.35$

$$h_{d,\text{leeward}} = 1.5\sqrt{\frac{(p_g)^{0.74}(\ell_{\text{upper}})^{0.70}(W_2)^{1.7}}{\gamma}}$$

$$= 1.5 \times \sqrt{\frac{(41.0)^{0.74} \times (60.0)^{0.70} \times (0.35)^{1.7}}{19.3}}$$

$$= 2.3 \text{ ft} < 0.6\ell_{\text{lower}} = 0.6 \times 10.0 = 6.0 \text{ ft}$$

Step 7e—Determine the windward drift height, $h_{d,\text{windward}}$

$$h_{d,\text{windward}} = 1.125\sqrt{\frac{(p_g)^{0.74}(\ell_{\text{lower}})^{0.70}(W_2)^{1.7}}{\gamma}}$$

$$= 1.125 \times \sqrt{\frac{(41.0)^{0.74} \times (10.0)^{0.70} \times (0.35)^{1.7}}{19.3}} = 0.9 \text{ ft}$$

Step 7f—Determine the drift height, h_d

$$h_d = \text{greater of} \begin{cases} h_{d,\text{leeward}} = 2.3 \text{ ft} \\ h_{d,\text{windward}} = 0.9 \text{ ft} \end{cases}$$

Step 7g—Determine the drift width, w
Because $h_d = 2.3 \text{ ft} < h_c = 3.2 \text{ ft}$, $w = 4h_d = 4 \times 2.3 = 9.2 \text{ ft} < 8h_c = 25.6 \text{ ft}$ and $w < \ell_{\text{lower}} = 10.0 \text{ ft}$.

Step 7h—Determine the drift load, p_d

$$p_d = \gamma h_d = 19.3 \times 2.3 = 44.4 \text{ lb/ft}^2$$

Step 7i—Determine the total snow load at the step

$$p_{\text{total}} = p_s + p_d = 34.4 + 44.4 = 78.8 \text{ lb/ft}^2$$

In S.I.:

Step 7a—Determine the height of the balanced snow load, h_b

$$h_b = p_s/\gamma = 1.65/3.04 = 0.54 \text{ m}$$

Step 7b—Determine the clear height from the top of the balanced snow load to the upper roof, h_c

$$h_c = h_{\text{step}} - h_b = 1.52 - 0.54 = 0.98 \text{ m}$$

Step 7c—Determine if drift loads must be considered
Drift loads must be considered where $h_c/h_b \geq 0.2$.

$$h_c/h_b = 0.98/0.54 = 1.8 > 0.2$$

Therefore, drift loads must be considered.

Step 7d—Determine the leeward drift height, $h_{d,\text{leeward}}$
From Reference 4.1, winter wind parameter, $W_2 = 0.35$

$$h_{d,\text{leeward}} = 1.5\sqrt{\frac{(p_g)^{0.74}(\ell_{\text{upper}})^{0.70}(W_2)^{1.7}}{\gamma}}$$

$$= 1.5 \times \sqrt{\frac{(41.0)^{0.74} \times (60.0)^{0.70} \times (0.35)^{1.7}}{19.3}}$$

$$= 2.3 \text{ ft} \times 0.3048 = 0.70 \text{ m} < 0.6\ell_{\text{lower}} = 0.6 \times 3.05 = 1.83 \text{ m}$$

Step 7e—Determine the windward drift height, $h_{d,\text{windward}}$

$$h_{d,\text{windward}} = 1.125\sqrt{\frac{(p_g)^{0.74}(\ell_{\text{lower}})^{0.70}(W_2)^{1.7}}{\gamma}}$$

$$= 1.125 \times \sqrt{\frac{(41.0)^{0.74} \times (10.0)^{0.70} \times (0.35)^{1.7}}{19.3}} = 0.9 \text{ ft} \times 0.3048 = 0.27 \text{ m}$$

Step 7f—Determine the drift height, h_d

$$h_d = \text{greater of} \begin{cases} h_{d,\text{leeward}} = 0.70 \text{ m} \\ h_{d,\text{windward}} = 0.27 \text{ m} \end{cases}$$

Step 7g—Determine the drift width, w

Because $h_d = 0.70 \text{ m} < h_c = 0.98$ m, $w = 4h_d = 4 \times 0.70 = 2.80 \text{ m} < 8h_c = 7.84$ m and $w < \ell_{\text{lower}} = 3.05$ m.

Step 7h—Determine the drift load, p_d

$$p_d = \gamma h_d = 3.04 \times 0.70 = 2.13 \text{ kN/m}^2$$

Step 7i—Determine the total snow load at the step

$$p_{\text{total}} = p_s + p_d = 1.65 + 2.13 = 3.78 \text{ kN/m}^2$$

Step 8—Consider drifts on roof projections and parapets

Not applicable.

Step 9—Consider sliding snow

Use Flowchart 4.9 to determine the sliding snow load on the canopy.

Loads caused by snow sliding must be considered because the roof is slippery with a slope greater than ¼ on 12 (1.19 degrees).

Because $\ell_{\text{lower}} = 10$ ft (3.05 m) $<$ 15 ft (4.57 m), the sliding snow is permitted to be reduced proportionally (ASCE/SEI 7.9):

$$p_{\text{sliding}} = 0.4 p_{f,\text{roof}} W(\ell_{\text{lower}}/15)/\ell_{\text{lower}}$$

$$= 0.4 \times 32.4 \times 30.0 \times (10/15)/10 = 25.9 \text{ lb/ft}^2$$

$$p_{\text{sliding}} = 0.4 p_{f,\text{roof}} W(\ell_{\text{lower}}/4.57)/\ell_{\text{lower}}$$

$$= 0.4 \times 1.55 \times 9.14 \times (3.05/4.57)/3.05 = 1.24 \text{ kN/m}^2$$

The total snow load over the 10-ft (3.05-m) width is equal to the balanced snow load plus the sliding snow load:

$$p_{\text{total}} = p_s + 0.4 p_{f,\text{roof}} W(\ell_{\text{lower}}/15)/\ell_{\text{lower}} = 34.4 + 25.9 = 60.3 \text{ lb/ft}^2$$

$$p_{\text{total}} = p_s + 0.4 p_{f,\text{roof}} W(\ell_{\text{lower}}/4.57)/\ell_{\text{lower}} = 1.65 + 1.24 = 2.89 \text{ kN/m}^2$$

The depth of snow for the total snow load $= p_{\text{total}}/\gamma = 60.3/19.3 = 3.1$ ft (in S.I.: $p_{\text{total}}/\gamma = 2.89/3.04 = 0.95$ m). This depth is less than the distance from the eave of the roof to the top of the canopy, which is equal to 5 ft (1.52 m). Therefore, sliding snow is not blocked and the full load can be developed over the 10-ft (3.05-m) width.

Step 10—Consider rain-on-snow surcharge loads

According to ASCE/SEI 7.10, a rain-on-snow surcharge load is required for locations where $p_g \le p_{m,\max}$ and $p_g > 0$ on roofs with slopes less than $W/50$ (in S.I.: $W/15.2$). In this example, $p_g = 41.0$ lb/ft^2(1.96 kN/m^2) $> p_{m,\max} = 30.0$ lb/ft^2(1.44 kN/m^2) from ASCE/SEI Table 7.3-4 for Risk Category II. Therefore, an additional 8 lb/ft^2 (0.38 kN/m^2) rain-on-snow surcharge load need not be added to the balanced snow load.

Step 11—Consider ponding loads

Not applicable.

Step 12—Consider snow loads on existing roofs

Not applicable.

Step 13—Consider snow loads on open-frame equipment structures

Not applicable.

Design snow loads on the canopy for the minimum, balanced, drift, and sliding load cases are given in Figure 4.45.

Figure 4.45 Design Snow Load Cases for the Canopy in Example 4.9

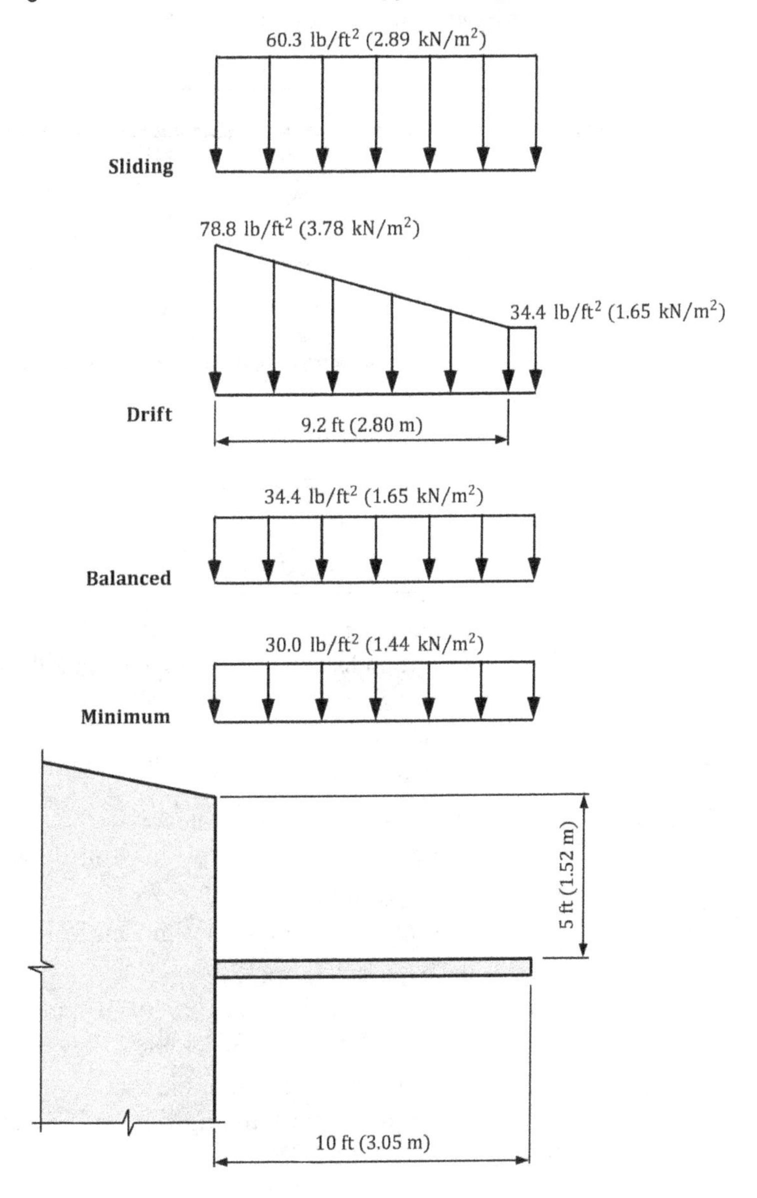

4.5.10 Example 4.10—Design Snow Loads on a Building with a Parapet

Determine the design snow loads on the roof of the building in Figure 4.46 given the following design data:

- Location: Lincoln, NE (Latitude = 40.81°, Longitude = –96.70°)
- Surface roughness: B
- Risk Category: II (IBC Table 1604.5)
- Thermal condition: Heated structure with an unventilated roof and $R_{roof} = 35$ h ft²°F/Btu (6.16 m²K/W)
- Roof exposure: Partially exposed
- Roof surface: Concrete slab with waterproofing
- Roof obstructions: Parapet on four sides of the building
- Roof framing: All structural members are simply supported

The roof is flat except for sloped localized areas around roof drains to facilitate drainage.

Figure 4.46 Roof Plan of the Building in Example 4.10

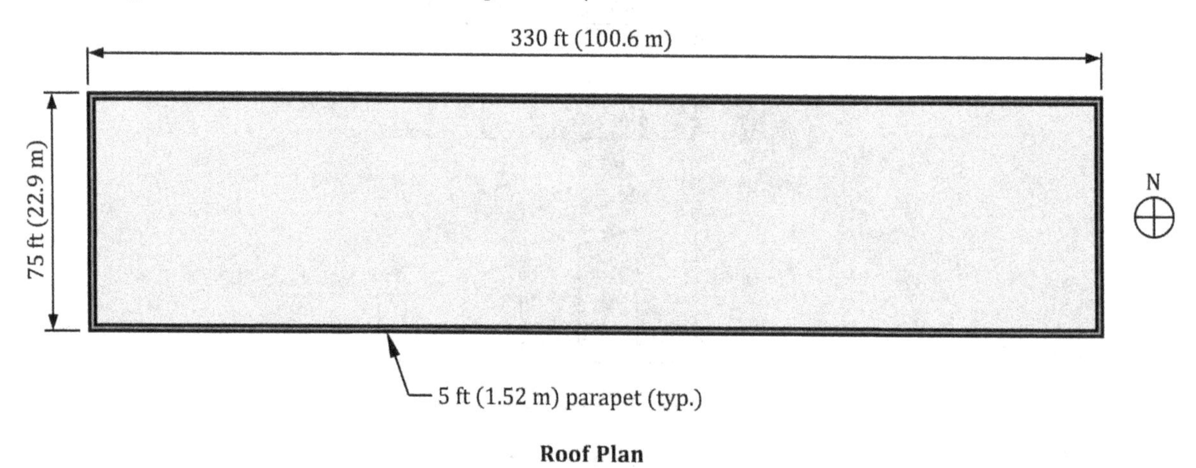

SOLUTION

Step 1—Determine the ground snow load, p_g

Using Reference 4.1, the ground snow load is equal to 45.0 lb/ft² (2.15 kN/m²) for Risk Category II.

Step 2—Determine the flat roof snow load, p_f

Flowchart 4.1 is used to determine p_f.

Step 2a—Determine the surface roughness category

The surface roughness category is given in the design data as B.

Step 2b—Determine the exposure of the roof

The roof exposure is given in the design data as partially exposed.

Step 2c—Determine the exposure factor, C_e

Given a surface roughness category of B and a partially exposed roof exposure, $C_e = 1.0$ from ASCE/SEI Table 7.3-1.

Step 2d—Determine the thermal condition of the roof

From the design data, the building has an unventilated roof with $R_{roof} =$ 35 h ft²°F/Btu (6.16 m²K/W).

Step 2e—Determine the thermal factor, C_t

For a building with an unventilated roof with $R_{\text{roof}} = 35$ h ft^2°F/Btu (6.16 m^{2}K/W), $C_t = 1.14$ by double linear interpolation from ASCE/SEI Table 7.3-3 for $p_g = 45.0$ lb/ft^2 (2.15 kN/m^2).

Step 2f—Determine the flat roof snow load

$$p_f = 0.7C_eC_tp_g = 0.7 \times 1.0 \times 1.14 \times 45.0 = 35.9 \text{ lb/ft}^2$$

$$p_f = 0.7C_eC_tp_g = 0.7 \times 1.0 \times 1.14 \times 2.15 = 1.72 \text{ kN/m}^2$$

Step 3—Determine the minimum snow load for low-slope roofs, p_m

A minimum snow load, p_m, must be considered in this example because the roof is essentially flat.

For Risk Category II, the minimum snow load upper limit, $p_{m,\max}$, is equal to 30.0 lb/ft^2 (1.44 kN/m^2) from ASCE/SEI Table 7.3-4. Because $p_g = 45.0$ lb/ft^2(2.15 kN/m^2) > $p_{m,\max} = 30.0$ lb/ft^2(1.44 kN/m^2), $p_m = p_{m,\max} = 30.0$ lb/ft^2(1.44 kN/m^2) (ASCE/SEI 7.3.3).

Step 4—Determine the sloped roof (balanced) snow load, p_s

Flowchart 4.2 is used to determine p_s.

Because the roof is flat, $C_s = 1.0$, and the sloped roof (balanced) snow load, p_s, is equal to the following:

$$p_s = C_sp_f = 1.0 \times 35.9 = 35.9 \text{ lb/ft}^2$$

$$p_s = C_sp_f = 1.0 \times 1.72 = 1.72 \text{ kN/m}^2$$

Step 5—Consider partial loading

From the design data, the roof structural members are simply supported, so partial loading need not be considered (ASCE/SEI 7.5).

Step 6—Consider unbalanced snow loads

Not applicable.

Step 7—Consider drift loads on lower roofs

Not applicable.

Step 8—Consider drifts on roof projections and parapets

According to ASCE/SEI 7.8, drift loads at parapet walls and other roof projections are determined using the provisions of ASCE/SEI 7.7.1 for windward drifts.

Windward drifts occur at parapet walls and Flowchart 4.8 is used to determine the windward drift load.

Step 8a—Determine the snow density, γ

$$\gamma = 0.13p_g + 14 = (0.13 \times 45) + 14 = 19.9 \text{ lb/ft}^3 < 30 \text{ lb/ft}^3$$

Step 8b—Determine the height of the balanced snow load, h_b

$$h_b = p_s/\gamma = 35.9/19.9 = 1.8 \text{ ft}$$

Step 8c—Determine the clear height from the top of the balanced snow load to the top of the parapet wall, h_c

$$h_c = h_{\text{step}} - h_b = 5.0 - 1.8 = 3.2 \text{ ft}$$

Step 8d—Determine if drift loads must be considered

Drift loads must be considered where $h_c/h_b \geq 0.2$.

$$h_c/h_b = 3.2/1.8 = 1.8 > 0.2$$

Therefore, drift loads must be considered.

Step 8e—Determine the drift height, h_d

- Wind in the north-south direction

  ℓ_u = length of the roof upwind of the wall = 75.0 ft

  From Reference 4.1, winter wind parameter, $W_2 = 0.55$

$$h_d = 1.125\sqrt{\frac{(p_g)^{0.74}(\ell_u)^{0.70}(W_2)^{1.7}}{\gamma}}$$

$$= 1.125 \times \sqrt{\frac{(45.0)^{0.74} \times (75.0)^{0.70} \times (0.55)^{1.7}}{19.9}} = 2.8 \text{ ft} < h_c = 3.2 \text{ ft}$$

- Wind in the east-west direction

  ℓ_u = length of the roof upwind of the wall = 330.0 ft

$$h_d = 1.125\sqrt{\frac{(p_g)^{0.74}(\ell_u)^{0.70}(W_2)^{1.7}}{\gamma}}$$

$$= 1.125 \times \sqrt{\frac{(45.0)^{0.74} \times (330.0)^{0.70} \times (0.55)^{1.7}}{19.9}} = 4.7 \text{ ft} > h_c = 3.2 \text{ ft}$$

Step 8f—Determine the drift width, w

- Wind in the north-south direction

$$w = 8h_d = 8 \times 2.8 = 22.4 \text{ ft}$$

- Wind in the east-west direction

$$w = 8h_d = 8 \times 4.7 = 37.6 \text{ ft}$$

Step 8g—Determine the drift load, p_d

- Wind in the north-south direction

$$p_d = \gamma h_d = 19.9 \times 2.8 = 55.7 \text{ lb/ft}^2$$

- Wind in the east-west direction

$$p_d = \gamma h_c = 19.9 \times 3.2 = 63.7 \text{ lb/ft}^2$$

Step 8h—Determine the total snow load at the face of the parapet

- Wind in the north-south direction

$$p_{\text{total}} = p_s + p_d = 35.9 + 55.7 = 91.6 \text{ lb/ft}^2$$

- Wind in the east-west direction

$$p_{\text{total}} = p_s + p_d = 35.9 + 63.7 = 99.6 \text{ lb/ft}^2$$

In S.I.:

Step 8a—Determine the snow density, γ

$$\gamma = 0.426p_g + 2.2 = (0.426 \times 2.15) + 2.2 = 3.12 \text{ kN/m}^3 < 4.7 \text{ kN/m}^3$$

Step 8b—Determine the height of the balanced snow load, h_b

$$h_b = p_s/\gamma = 1.72/3.12 = 0.55 \text{ m}$$

Step 8c—Determine the clear height from the top of the balanced snow load to the top of the parapet wall, h_c

$$h_c = h_{\text{step}} - h_b = 1.52 - 0.55 = 0.97 \text{ m}$$

Step 8d—Determine if drift loads must be considered

Drift loads must be considered where $h_c / h_b \geq 0.2$.

$$h_c / h_b = 0.97 / 0.55 = 1.8 > 0.2$$

Therefore, drift loads must be considered.

Step 8e—Determine the drift height, h_d

- Wind in the north-south direction

  ℓ_u = length of the roof upwind of the wall = 22.9 m

  From Reference 4.1, winter wind parameter, $W_2 = 0.55$

$$h_d = 1.125\sqrt{\frac{(p_g)^{0.74}(\ell_u)^{0.70}(W_2)^{1.7}}{\gamma}}$$

$$= 1.125 \times \sqrt{\frac{(45.0)^{0.74} \times (75.0)^{0.70} \times (0.55)^{1.7}}{19.9}}$$

$$= 2.8 \text{ ft} \times 0.3048 = 0.85 \text{ m} < h_c = 0.97 \text{ m}$$

- Wind in the east-west direction

  ℓ_u = length of the roof upwind of the wall = 100.6 m

$$h_d = 1.125\sqrt{\frac{(p_g)^{0.74}(\ell_u)^{0.70}(W_2)^{1.7}}{\gamma}}$$

$$= 1.125 \times \sqrt{\frac{(45.0)^{0.74} \times (330.0)^{0.70} \times (0.55)^{1.7}}{19.9}}$$

$$= 4.7 \text{ ft} \times 0.3048 = 1.43 \text{ m} > h_c = 0.97 \text{ m}$$

Step 8f—Determine the drift width, w

- Wind in the north-south direction

$$w = 8h_d = 8 \times 0.85 = 6.80 \text{ m}$$

- Wind in the east-west direction

$$w = 8h_d = 8 \times 1.43 = 11.4 \text{ m}$$

Step 8g—Determine the drift load, p_d

- Wind in the north-south direction

$$p_d = \gamma h_d = 3.12 \times 0.85 = 2.65 \text{ kN/m}^2$$

- Wind in the east-west direction

$$p_d = \gamma h_c = 3.12 \times 0.97 = 3.03 \text{ kN/m}^2$$

Step 8h—Determine the total snow load at the face of the parapet

- Wind in the north-south direction

$$p_{\text{total}} = p_s + p_d = 1.72 + 2.65 = 4.37 \text{ kN/m}^2$$

- Wind in the east-west direction

$$p_{\text{total}} = p_s + p_d = 1.72 + 3.03 = 4.75 \text{ kN/m}^2$$

Step 9—Consider sliding snow.
Not applicable.

Step 10—Consider rain-on-snow surcharge loads
According to ASCE/SEI 7.10, a rain-on-snow surcharge load is required for locations where $p_g \leq p_{m,\max}$ and $p_g > 0$ on roofs with slopes less than $W/50$ (in S.I.: $W/15.2$). In this example, $p_g = 45.0$ lb/ft^2(2.15 kN/m^2)$>$ $p_{m,\max} = 30.0$ lb/ft^2(1.44 kN/m^2) from ASCE/SEI Table 7.3-4 for Risk Category II. Therefore, an additional 8 lb/ft^2 (0.38 kN/m^2) rain-on-snow surcharge load need not be added to the balanced snow load.

Step 11—Consider ponding loads
A ponding analysis considering roof deflections caused by full snow loads in combination with dead and superimposed dead loads must be performed for this roof based on the stiffness of the roof structural members.

Step 12—Consider snow loads on existing roofs
Not applicable.

Step 13—Consider snow loads on open-frame equipment structures
Not applicable.

Design balanced and drift snow loads are given in Figure 4.47.

Figure 4.47 Design Snow Load Cases for the Roof in Example 4.10

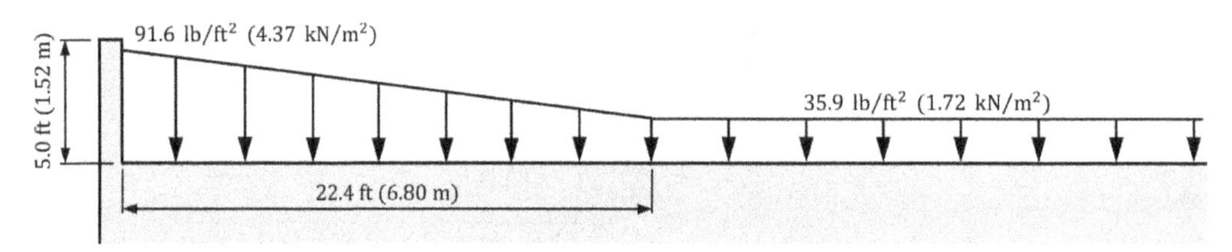

Parapet Walls at North and South Faces

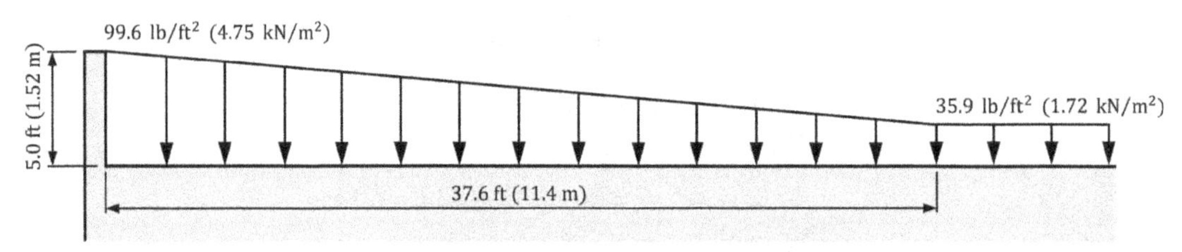

Parapet Walls at East and West Faces

4.5.11 Example 4.11—Design Snow Loads on a Building with a Rooftop Unit

Determine the design snow loads on the roof of the building in Figure 4.48 using the design data in Example 4.10. The roof is flat except for sloped localized areas around roof drains to facilitate drainage and in this example, the roof has no parapet walls.

SOLUTION

Steps 1 through 7—Results are the same as in Example 4.10.

$$p_g = 45.0 \text{ lb/ft}^2 \ (2.15 \text{ kN/m}^2)$$
$$C_e = 1.0$$
$$C_t = 1.14$$
$$p_f = p_s = 35.9 \text{ lb/ft}^2 \ (1.72 \text{ kN/m}^2)$$
$$p_m = 30.0 \text{ lb/ft}^2 \ (1.44 \text{ kN/m}^2)$$

Figure 4.48 Roof Plan of the Building and Elevation of the Rooftop Unit in Example 4.11

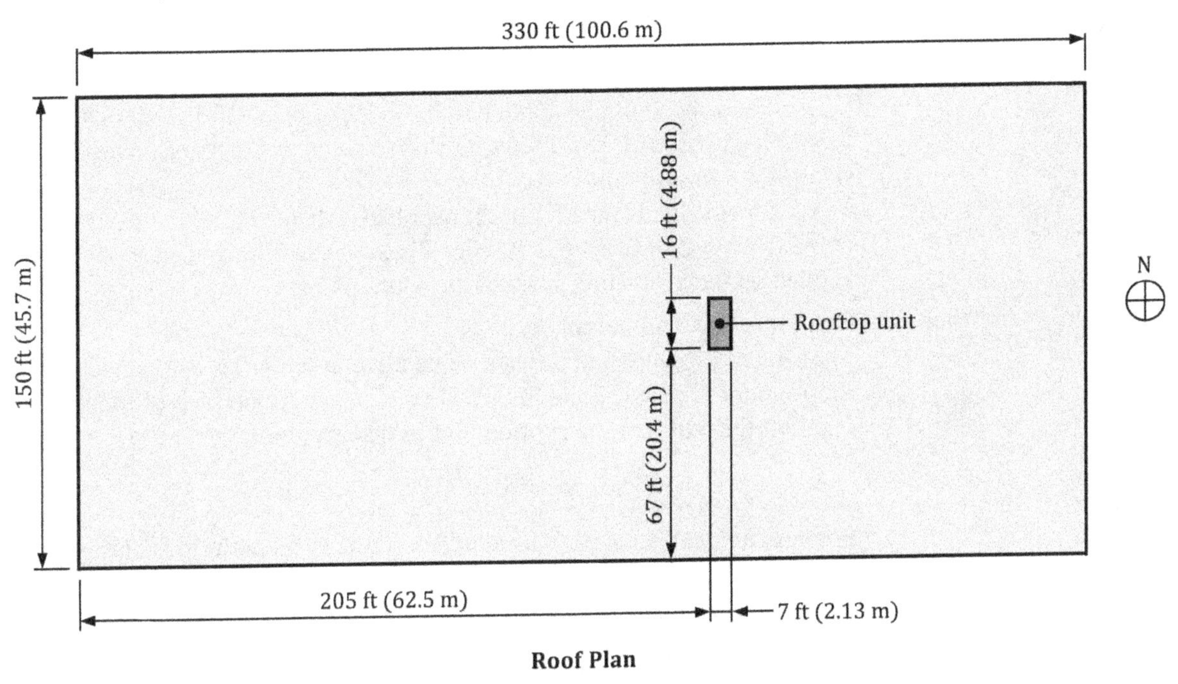

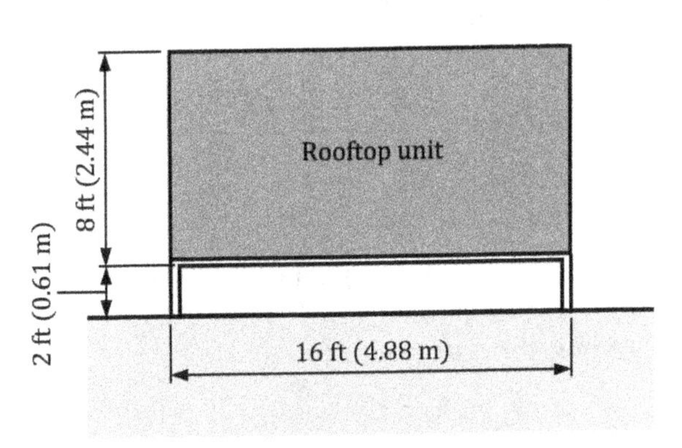

Step 8—Consider drifts on roof projections and parapets

According to ASCE/SEI 7.8, drift loads at parapet walls and other roof projections are determined using the provisions of ASCE/SEI 7.7.1 for windward drifts.

Flowchart 4.8 is used to determine the windward drift load.

Step 8a—Determine the snow density, γ

$$\gamma = 0.13p_g + 14 = (0.13 \times 45) + 14 = 19.9 \text{ lb/ft}^3 < 30 \text{ lb/ft}^3$$

Step 8b—Determine the height of the balanced snow load, h_b

$$h_b = p_s / \gamma = 35.9/19.9 = 1.8 \text{ ft}$$

Step 8c—Determine the clear height from the top of the balanced snow load to the top of the rooftop unit, h_c

$$h_c = h_{\text{step}} - h_b = 10.0 - 1.8 = 8.2 \text{ ft}$$

Step 8d—Determine if drift loads must be considered

Drift loads must be considered where $h_c/h_b \geq 0.2$.

$$h_c/h_b = 8.2/1.8 = 4.6 > 0.2$$

Therefore, drift loads must be considered.

Drift loads need not be applied to the sides of the rooftop unit that are less than 15 ft (ASCE/SEI 7.8). Therefore, drift loads are not required on the 7-ft-long sides of the rooftop unit.

Also, drift loads need not be applied where $H \geq 2$ ft. In this case, $H = 2.0 - h_b = 2.0 - 1.8 = 0.2$ ft, which is less than 2 ft, so drift loads must be applied on the 16-ft-long sides of the rooftop unit.

Step 8e—Determine the drift height, h_d

The larger of the upwind fetches perpendicular to the 16-ft-long sides of the rooftop unit is equal to the longer clear distance to the face of the unit plus one-half the width of the rooftop unit in that direction:

$$\ell_u = 205.0 + (7.0/2) = 208.5 \text{ ft}$$

For simplicity, this length is used to determine h_d on both sides of the rooftop unit.

From Reference 4.1, winter wind parameter, $W_2 = 0.55$

$$h_d = 1.125\sqrt{\frac{(p_g)^{0.74}(\ell_u)^{0.70}(W_2)^{1.7}}{\gamma}}$$

$$= 1.125 \times \sqrt{\frac{(45.0)^{0.74} \times (208.5)^{0.70} \times (0.55)^{1.7}}{19.9}} = 4.0 \text{ ft} < h_c = 8.2 \text{ ft}$$

Step 8f—Determine the drift width, w

$$w = 8h_d = 8 \times 4.0 = 32.0 \text{ ft}$$

Step 8g—Determine the drift load, p_d

$$p_d = \gamma h_d = 19.9 \times 4.0 = 79.6 \text{ lb/ft}^2$$

Step 8h—Determine the total snow load at the faces of the rooftop unit

$$p_{\text{total}} = p_s + p_d = 35.9 + 79.6 = 115.5 \text{ lb/ft}^2$$

In S.I.:

Step 8a—Determine the snow density, γ

$$\gamma = 0.426p_g + 2.2 = (0.426 \times 2.15) + 2.2 = 3.12 \text{ kN/m}^3 < 4.70 \text{ kN/m}^3$$

Step 8b—Determine the height of the balanced snow load, h_b

$$h_b = p_s/\gamma = 1.72/3.12 = 0.55 \text{ m}$$

Step 8c—Determine the clear height from the top of the balanced snow load to the top of the rooftop unit, h_c

$$h_c = h_{\text{step}} - h_b = 3.05 - 0.55 = 2.50 \text{ m}$$

Step 8d—Determine if drift loads must be considered

Drift loads must be considered where $h_c/h_b \geq 0.2$.

$$h_c/h_b = 2.50/0.55 = 4.6 > 0.2$$

Therefore, drift loads must be considered.

Drift loads need not be applied to the sides of the rooftop unit that are less than 4.57 m (ASCE/SEI 7.8). Therefore, drift loads are not required on the 2.13-m-long sides of the rooftop unit.

Also, drift loads need not be applied where $H \geq 0.61$ m. In this case, $H = 0.61 - h_b = 0.61 - 0.55 = 0.06$ m, which is less than 0.61 m, so drift loads must be applied on the 4.88-m-long sides of the rooftop unit.

Step 8e—Determine the drift height, h_d

The larger of the upwind fetches perpendicular to the 4.88-m-long sides of the rooftop unit is equal to the longer clear distance to the face of the unit plus one-half the width of the rooftop unit in that direction:

$$\ell_u = 62.5 + (2.13/2) = 63.6 \text{ m}$$

For simplicity, this length is used to determine h_d on both sides of the rooftop unit.

From Reference 4.1, winter wind parameter, $W_2 = 0.55$

$$h_d = 1.125\sqrt{\frac{(p_g)^{0.74}(\ell_u)^{0.70}(W_2)^{1.7}}{\gamma}}$$

$$= 1.125 \times \sqrt{\frac{(45.0)^{0.74} \times (208.5)^{0.70} \times (0.55)^{1.7}}{19.9}}$$

$$= 4.0 \text{ ft} \times 0.3048 = 1.22 \text{ m} < h_c = 2.50 \text{ m}$$

Step 8f—Determine the drift width, w

$$w = 8h_d = 8 \times 1.22 = 9.76 \text{ m}$$

Step 8g—Determine the drift load, p_d

$$p_d = \gamma h_d = 3.12 \times 1.22 = 3.81 \text{ kN/m}^2$$

Step 8h—Determine the total snow load at the faces of the rooftop unit

$$p_{\text{total}} = p_s + p_d = 1.72 + 3.81 = 5.53 \text{ kN/m}^2$$

Step 9—Consider sliding snow.

Not applicable.

Step 10—Consider rain-on-snow surcharge loads

According to ASCE/SEI 7.10, a rain-on-snow surcharge load is required for locations where $p_g \leq p_{m,\max}$ and $p_g > 0$ on roofs with slopes less than $W/50$ (in S.I.: $W/15.2$). In this example, $p_g = 45.0 \text{ lb/ft}^2 (2.15 \text{ kN/m}^2) > p_{m,\max} = 30.0 \text{ lb/ft}^2 (1.44 \text{ kN/m}^2)$ from ASCE/SEI Table 7.3-4 for Risk Category II. Therefore, an additional 8 lb/ft^2 (0.38 kN/m^2) rain-on-snow surcharge load need not be added to the balanced snow load.

Step 11—Consider ponding loads

A ponding analysis considering roof deflections caused by full snow loads in combination with dead and superimposed dead loads must be performed for this roof based on the stiffness of the roof structural members.

Step 12—Consider snow loads on existing roofs

Not applicable.

Step 13—Consider snow loads on open-frame equipment structures

Not applicable.

Design balanced and drift snow loads are given in Figure 4.49.

Figure 4.49 Design Snow Load Cases for the Roof in Example 4.11

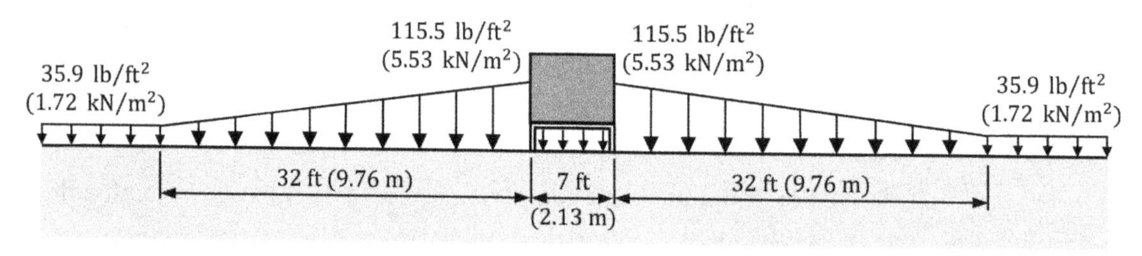

4.5.12 Example 4.12—Design Snow Loads on an Existing Building Adjacent to a New Building

Determine the design snow loads on the roof of the existing building in Figure 4.50 before and after the construction of the new adjacent building given the following design data:

- Location: Anchorage, AK (Latitude = 61.22°, Longitude = −149.89°)
- Surface roughness: B
- Risk Category for both buildings: II (IBC Table 1604.5)
- Thermal condition:
  Existing building: Unheated
  New building: Structure kept just above freezing
- Roof exposure for both buildings: Partially exposed
- Roof surface for both buildings: Metal

Figure 4.50 Elevations of the Existing and New Buildings in Example 4.12

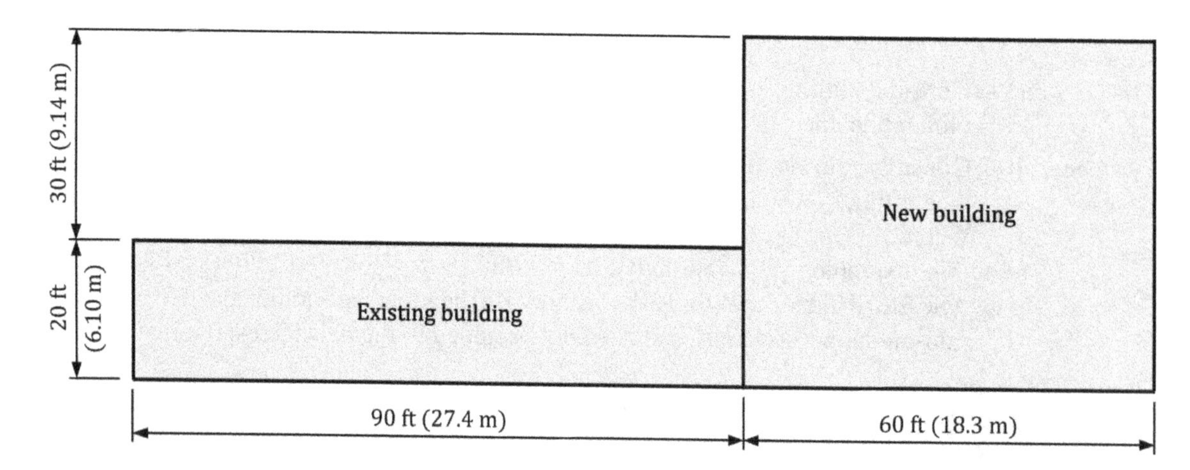

- Roof obstructions for both buildings: None
- Roof framing: All structural members are simply supported

Both roofs are flat except for sloped localized areas around roof drains to facilitate drainage.

SOLUTION

Snow Loads Prior to Construction of the New Building

Step 1—Determine the ground snow load, p_g

Using ASCE/SEI Table 7.2-1 or Reference 4.1, the ground snow load is equal to 80.0 lb/ft^2 (3.83 kN/m^2) for Risk Category II.

Step 2—Determine the flat roof snow load, p_f

Flowchart 4.1 is used to determine p_f.

Step 2a—Determine the surface roughness category

The surface roughness category is given in the design data as B.

Step 2b—Determine the exposure of the roof

The roof exposure is given in the design data as partially exposed.

Step 2c—Determine the exposure factor, C_e

Given a surface roughness category of B and a partially exposed roof exposure, $C_e = 1.0$ from ASCE/SEI Table 7.3-1.

Step 2d—Determine the thermal condition of the roof

From the design data, the building is unheated.

Step 2e—Determine the thermal factor, C_t

For an unheated building, $C_t = 1.2$ from ASCE/SEI Table 7.3-2.

Step 2f—Determine the flat roof snow load

$$p_f = 0.7C_eC_tp_g = 0.7 \times 1.0 \times 1.2 \times 80.0 = 67.2 \text{ lb/ft}^2$$

$$p_f = 0.7C_eC_tp_g = 0.7 \times 1.0 \times 1.2 \times 3.83 = 3.22 \text{ kN/m}^2$$

Step 3—Determine the minimum snow load for low-slope roofs, p_m

A minimum snow load, p_m, must be considered in this example because the roof is essentially flat.

For Risk Category II, the minimum snow load upper limit, $p_{m,\max}$, is equal to 30.0 lb/ft^2 (1.44 kN/m^2) from ASCE/SEI Table 7.3-4. Because $p_g = 80.0$ lb/ft^2(3.83 kN/m^2) > $p_{m,\max} = 30.0$ lb/ft^2 (1.44 kN/m^2), $p_m = p_{m,\max} = 30.0$ lb/ft^2 (1.44 kN/m^2)(ASCE/SEI 7.3.3).

Step 4—Determine the sloped roof (balanced) snow load, p_s

Flowchart 4.2 is used to determine p_s.

Because the roof is flat, $C_s = 1.0$, and the sloped roof (balanced) snow load, p_s, is equal to the following:

$$p_s = C_sp_f = 1.0 \times 67.2 = 67.2 \text{ lb/ft}^2$$

$$p_s = C_sp_f = 1.0 \times 1.72 = 3.22 \text{ kN/m}^2$$

Step 5—Consider partial loading

From the design data, the roof structural members are simply supported, so partial loading need not be considered (ASCE/SEI 7.5).

Step 6—Consider unbalanced snow loads

Not applicable.

Step 7—Consider drift loads on lower roofs
Not applicable.

Step 8—Consider drifts on roof projections and parapets
Not applicable.

Step 9—Consider sliding snow.
Not applicable.

Step 10—Consider rain-on-snow surcharge loads
According to ASCE/SEI 7.10, a rain-on-snow surcharge load is required for locations where $p_g \leq p_{m,\max}$ and $p_g > 0$ on roofs with slopes less than $W/50$ (in S.I.: $W/15.2$). In this example, $p_g = 80.0$ lb/ft^2 (3.83 kN/m^2) $> p_{m,\max} = 30.0$ lb/ft^2 (1.44 kN/m^2) from ASCE/SEI Table 7.3-4 for Risk Category II. Therefore, an additional 8 lb/ft^2 (0.38 kN/m^2) rain-on-snow surcharge load need not be added to the balanced snow load.

Step 11—Consider ponding loads
A ponding analysis considering roof deflections caused by full snow loads in combination with dead and superimposed dead loads must be performed for this roof based on the stiffness of the roof structural members.

Step 12—Consider snow loads on existing roofs
Not applicable.

Step 13—Consider snow loads on open-frame equipment structures
Not applicable.

Design minimum and balanced snow loads on the roof of the existing building prior to the construction of the new building are given in Figure 4.51.

Figure 4.51 Design Snow Load Cases on the Roof of the Existing Building Prior to the Construction of the New Building in Example 4.12

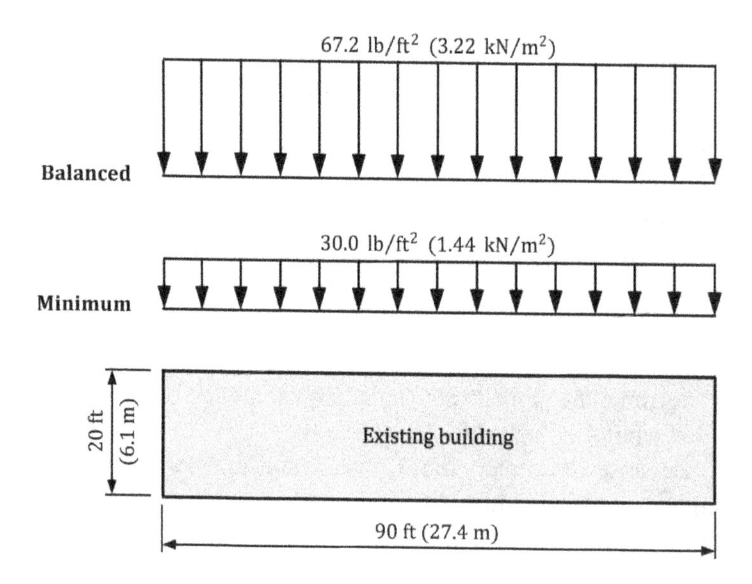

Snow Loads after Construction of the New Building

The snow loads on the existing roof prior to construction of the new building are also applicable after construction of the new building. After construction, drift loads on the existing roof must also be considered. Sliding snow loads need not be considered because the upper roof is essentially flat (see ASCE/SEI 7.9).

Step 7—Consider drift loads on lower roofs

Use Flowchart 4.6 to determine the leeward and windward drifts on the lower roof.

Step 7a—Determine the height of the balanced snow load, h_b

Snow density, $\gamma = 0.13p_g + 14 = (0.13 \times 80) + 14 = 24.4 \text{ lb/ft}^3 < 30 \text{ lb/ft}^3$

$$h_b = p_s/\gamma = 67.2/24.4 = 2.8 \text{ ft}$$

Step 7b—Determine the clear height from the top of the balanced snow load to the upper roof, h_c

$$h_c = h_{\text{step}} - h_b = 30.0 - 2.8 = 27.2 \text{ ft}$$

Step 7c—Determine if drift loads must be considered

Drift loads must be considered where $h_c/h_b \geq 0.2$.

$$h_c/h_b = 27.2/2.8 = 9.7 > 0.2$$

Therefore, drift loads must be considered.

Step 7d—Determine the leeward drift height, $h_{d,\text{leeward}}$

From Reference 4.1, winter wind parameter, $W_2 = 0.2$

$$h_{d,\text{leeward}} = 1.5\sqrt{\frac{(p_g)^{0.74}(\ell_{\text{upper}})^{0.70}(W_2)^{1.7}}{\gamma}}$$

$$= 1.5 \times \sqrt{\frac{(80.0)^{0.74} \times (60.0)^{0.70} \times (0.2)^{1.7}}{24.4}}$$

$$= 1.6 \text{ ft} < 0.6\ell_{\text{lower}} = 0.6 \times 90.0 = 54.0 \text{ ft}$$

Step 7e—Determine the windward drift height, $h_{d,\text{windward}}$

$$h_{d,\text{windward}} = 1.125\sqrt{\frac{(p_g)^{0.74}(\ell_{\text{lower}})^{0.70}(W_2)^{1.7}}{\gamma}}$$

$$= 1.125 \times \sqrt{\frac{(80.0)^{0.74} \times (90.0)^{0.70} \times (0.2)^{1.7}}{24.4}} = 1.4 \text{ ft}$$

Step 7f—Determine the drift height, h_d

$$h_d = \text{greater of} \begin{cases} h_{d,\text{leeward}} = 1.6 \text{ ft} \\ h_{d,\text{windward}} = 1.4 \text{ ft} \end{cases}$$

Step 7g—Determine the drift width, w

Because $h_d = 1.6 \text{ ft} < h_c = 27.2 \text{ ft}$, $w = 4h_d = 4 \times 1.6 = 6.4 \text{ ft} < 8h_c = 217.6 \text{ ft}$ and $w < \ell_{\text{lower}} = 90.0 \text{ ft}$

Step 7h—Determine the drift load, p_d

$$p_d = \gamma h_d = 24.4 \times 1.6 = 39.0 \text{ lb/ft}^2$$

Step 7i—Determine the total snow load at the step

$$p_{\text{total}} = p_s + p_d = 67.2 + 39.0 = 106.2 \text{ lb/ft}^2$$

In S.I.:

Step 7a—Determine the height of the balanced snow load, h_b

Snow density, $\gamma = 0.426 p_g + 2.2 = (0.426 \times 3.83) + 2.2$

$$= 3.83 \text{ kN/m}^3 < 4.70 \text{ kN/m}^3$$

$$h_b = p_s / \gamma = 3.22/3.83 = 0.84 \text{ m}$$

Step 7b—Determine the clear height from the top of the balanced snow load to the upper roof, h_c

$$h_c = h_{\text{step}} - h_b = 9.14 - 0.84 = 8.30 \text{ m}$$

Step 7c—Determine if drift loads must be considered

Drift loads must be considered where $h_c / h_b \geq 0.2$.

$$h_c / h_b = 8.30/0.84 = 9.9 > 0.2$$

Therefore, drift loads must be considered.

From Reference 4.1, winter wind parameter, $W_2 = 0.2$

$$h_{d,\text{leeward}} = 1.5 \sqrt{\frac{(p_g)^{0.74} (\ell_{\text{upper}})^{0.70} (W_2)^{1.7}}{\gamma}}$$

$$= 1.5 \times \sqrt{\frac{(80.0)^{0.74} \times (60.0)^{0.70} \times (0.2)^{1.7}}{24.4}}$$

$$= 1.6 \text{ ft} \times 0.3048 = 0.49 \text{ m} < 0.6 \ell_{\text{lower}} = 0.6 \times 27.4 = 16.4 \text{ m}$$

Step 7e—Determine the windward drift height, $h_{d,\text{windward}}$

$$h_{d,\text{windward}} = 1.125 \sqrt{\frac{(p_g)^{0.74} (\ell_{\text{lower}})^{0.70} (W_2)^{1.7}}{\gamma}}$$

$$= 1.125 \times \sqrt{\frac{(80.0)^{0.74} \times (90.0)^{0.70} \times (0.2)^{1.7}}{24.4}} = 1.4 \text{ ft} \times 0.3048 = 0.43 \text{ m}$$

Step 7f—Determine the drift height, h_d

$$h_d = \text{greater of} \begin{cases} h_{d,\text{leeward}} = 0.49 \text{ m} \\ h_{d,\text{windward}} = 0.43 \text{ m} \end{cases}$$

Step 7g—Determine the drift width, w

Because $h_d = 0.49 \text{ m} < h_c = 8.30 \text{ m}, w = 4h_d = 4 \times 0.49 = 1.96 \text{ m} < 8h_c = 66.4 \text{ m}$ and $w < \ell_{\text{lower}} = 27.4 \text{ m}$

Step 7h—Determine the drift load, p_d

$$p_d = \gamma h_d = 3.83 \times 0.49 = 1.88 \text{ kN/m}^2$$

Step 7i—Determine the total snow load at the step

$$p_{\text{total}} = p_s + p_d = 3.22 + 1.88 = 5.10 \text{ kN/m}^2$$

Design balanced and drift snow loads on the roof of the existing building after construction of the new building are given in Figure 4.52.

Figure 4.52 Design Snow Load Cases on the Roof of the Existing Building after Construction of the New Building in Example 4.12

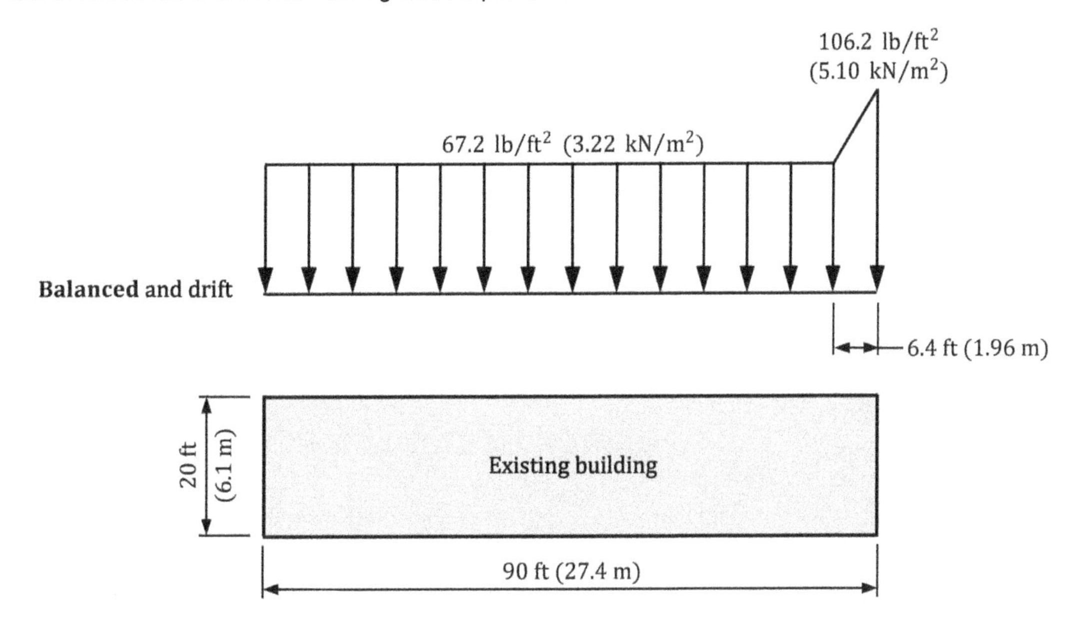

4.5.13 Example 4.13—Design Snow Loads on Elements in an Open-Frame Equipment Structure

Determine the design snow loads on the following elements in an open-frame equipment structure: (a) a 12-in. (305-mm) diameter pipe (which includes the insulation on the pipe) and (b) the adjacent cable trays in Figure 4.53. Assume $p_g = 30 \text{ lb/ft}^2$ (1.44 kN/m²), $C_e = 1.0$, and Risk Category II.

Figure 4.53 Adjacent Cable Trays in Example 4.13

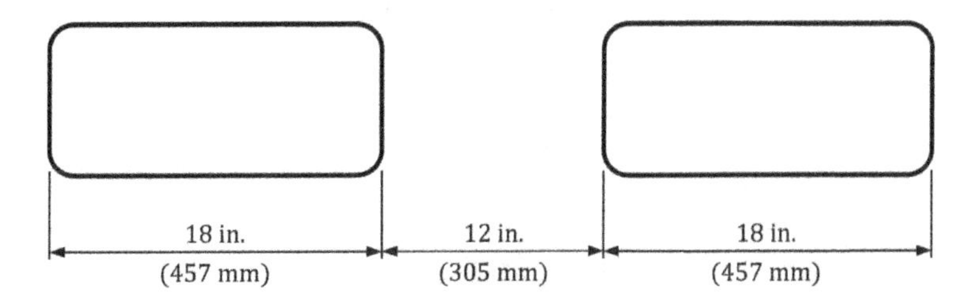

Solution

Part (a)—Snow loads on the 12-in. (305-mm) diameter pipe

Step 1—Determine the flat roof snow load, p_f

Step 1a—Determine the exposure factor, C_e

From the design data, $C_e = 1.0$.

Step 1b—Determine the thermal factor, C_t

For open-frame equipment structures, $C_t = 1.2$ (ASCE/SEI 7.13).

Step 1c—Determine the flat roof snow load, p_f

$$p_f = 0.7C_eC_tp_g = 0.7\times1.0\times1.2\times30.0 = 25.2 \text{ lb/ft}^2$$

$$p_f = 0.7C_eC_tp_g = 0.7\times1.0\times1.2\times1.44 = 1.21 \text{ kN/m}^2$$

Step 2—Determine the snow density, γ

$$\gamma = 0.13p_g + 14 = (0.13\times30.0) + 14 = 17.9 \text{ lb/ft}^3 < 30.0 \text{ lb/ft}^3$$

$$\gamma = 0.426p_g + 2.2 = (0.426\times1.44) + 2.2 = 2.81 \text{ kN/m}^3 < 4.7 \text{ kN/m}^3$$

Step 3—Determine the depth of the flat roof snow load

$$\text{Depth} = p_f/\gamma = 25.2/17.9 = 1.4 \text{ ft} = 16.8 \text{ in.}$$

$$\text{Depth} = p_f/\gamma = 1.21/2.81 = 0.43 \text{ m} = 430 \text{ mm}$$

Step 4—Determine the snow load on the pipe

Pipe diameter $= 12$ in. $< 0.73p_f/\gamma = (0.73\times25.2\times12)/17.9 = 12.3$ in.

Because the pipe diameter is less than $0.73p_f/\gamma$, the snow load on the pipe is triangular [see ASCE/SEI 7.13.3 and ASCE/SEI Figure 7.13-2(a)].

Height of the snow on the pipe $= 1.37D = 1.37\times12.0 = 16.4$ in.

Maximum snow load $= 1.37D\gamma = 1.37\times(12.0/12)\times17.9 = 24.5$ lb/ft^2

With an assumed angle of repose = 70 degrees, area of snow, $A = (12.0\times16.4)/2 = 98.4$ in.^{2}

Snow load per length of pipe $= \gamma A = 17.9\times(98.4/144) = 12.2$ lb/ft

In S.I.:

Pipe diameter $= 305$ mm $< 0.73p_f/\gamma = (0.73\times1.21)/2.81 = 0.314$ m $= 314$ mm

Because the pipe diameter is less than $0.73p_f/\gamma$, the snow load on the pipe is triangular [see ASCE/SEI 7.13.3 and ASCE/SEI Figure 7.13-2(a)].

Height of the snow on the pipe $= 1.37D = 1.37\times305 = 418$ mm

Maximum snow load $= 1.37D\gamma = 1.37\times0.305\times2.81 = 1.17$ kN/m^2

With an assumed angle of repose = 70 degrees, area of snow $A = (305\times418)/2 = 63{,}745$ mm^2 $= 0.064$ m^2

Snow load per length of pipe $= \gamma A = 2.81\times0.064 = 0.18$ kN/m

The design snow load on the pipe is given in Figure 4.54.

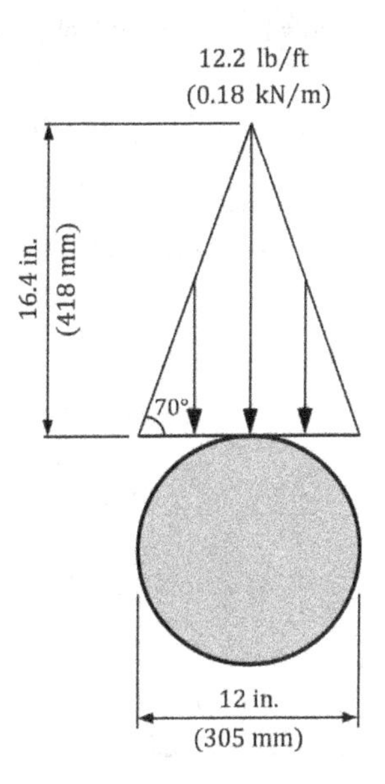

Figure 4.54 Design Snow Load on the Pipe in Example 4.13

Part (b)—Snow loads on the adjacent cable trays

The 18-in. width of each cable tray is greater than $0.73p_f/\gamma = 12.3$ in. Therefore, the snow load on each cable tray is trapezoidal with a maximum pressure of $p_f = 25.2$ lb/ft^2 [see ASCE/SEI 7.13.3 and ASCE/SEI Figure 7.13-2(b)].

Height of the snow on the cable tray, $h = p_f/\gamma = (25.2 \times 12)/17.9 = 16.9$ in.
Clear space between the adjacent cable trays $S_p = 12.0$ in. $< h = 16.9$ in.

Therefore, an additional uniform cornice load of p_f must be applied in the space between the adjacent cable trays (see ASCE/SEI 7.13.3 and ASCE/SEI Figure 7.13-3).

With an assumed angle of repose = 70 degrees, the area of snow tributary to each cable tray is equal to the following:

$$A = 16.9 \times \left[18.0 + \frac{12.0}{2} - \left(\frac{1}{2} \times \frac{16.9}{\tan 70^\circ}\right)\right] = 353.6 \text{ in.}^2$$

Snow load per cable tray $= \gamma A = 17.9 \times (353.6/144) = 44.0$ lb/ft

In S.I.:

The 457-mm width of each cable tray is greater than $0.73p_f/\gamma = 314$ mm. Therefore, the snow load on each cable tray is trapezoidal with a maximum pressure of $p_f = 1.21$ kN/m^2 [see ASCE/SEI 7.13.3 and ASCE/SEI Fig. 7.13-2(b)].

Height of the snow on the cable tray $h = p_f/\gamma = 1.21/2.81 = 0.43$ m $= 430$ mm
Clear space between the adjacent cable trays $S_p = 305$ mm $< h = 430$ mm

Therefore, an additional uniform cornice load of p_f must be applied in the space between the adjacent cable trays (see ASCE/SEI 7.13.3 and ASCE/SEI Figure 7.13-3).

With an assumed angle of repose = 70 degrees, the area of snow tributary to each cable tray is equal to the following:

$$A = 430 \times \left[457 + \frac{305}{2} - \left(\frac{1}{2} \times \frac{430}{\tan 70^\circ}\right)\right] = 228{,}436 \text{ mm}^2 = 0.228 \text{ m}^2$$

Snow load per cable tray $= \gamma A = 2.81 \times 0.228 = 0.64$ kN/m
The design snow load on the cable trays is given in Figure 4.55.

Figure 4.55 Design Snow Load on the Cable Trays in Example 4.13

4.5.14 Example 4.14—Atmospheric Ice Loads on a Tank

Determine the atmospheric ice loads for the circular reinforced concrete tank in Figure 4.56. The tank, which is located on a relatively flat Exposure B site in Grand Rapids, MI (latitude = 43.02°, longitude = −85.65°), does not contain hazardous, toxic, or explosive materials. Assume the natural fundamental frequency of the tank, n_1, is greater than 1 Hz. The design procedure in Table 4.5 of this book is used to determine the atmospheric ice loads.

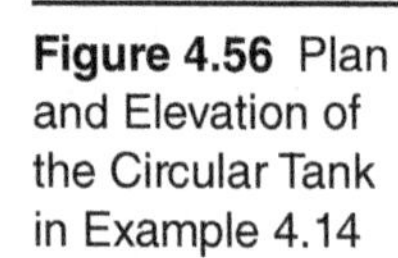

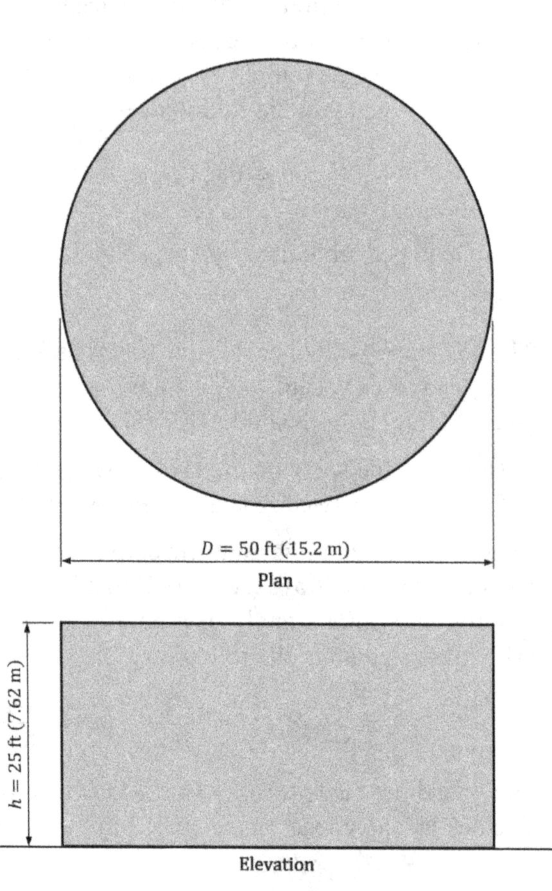

Figure 4.56 Plan and Elevation of the Circular Tank in Example 4.14

Solution

Step 1—Determine the nominal ice thickness, t, and the concurrent gust speed, V_c

In accordance with IBC Table 1604.5, the tank is classified under Risk Category II. Using Reference 4.1, $t = 0.96$ in. (24 mm) and $V_c = 47$ mi/h (21 m/s).

Step 2—Determine the height factor, f_z

$$\text{For } z = 25 \text{ ft}: \quad f_z = \left(\frac{z}{33}\right)^{0.10} = \left(\frac{25}{33}\right)^{0.10} = 0.97$$

$$\text{For } z = 7.62 \text{ m}: \quad f_z = \left(\frac{z}{10}\right)^{0.10} = \left(\frac{7.62}{10}\right)^{0.10} = 0.97$$

Step 3—Determine the topographic factor, K_{zt}

Because the tank is located on a relatively flat site, $K_{zt} = 1.0$ (ASCE/SEI 26.8.2).

Step 4—Determine the design ice thickness, t_d

$$t_d = tf_z(K_{zt})^{0.35} = 0.96 \times 0.97 \times (1.0)^{0.35} = 0.93 \text{ in.}$$

$$t_d = tf_z(K_{zt})^{0.35} = 24 \times 0.97 \times (1.0)^{0.35} = 23 \text{ mm}$$

Step 5—Determine the design ice load

For a circular cross-section, $D_c = D = 50$ ft (15.2 m) (see ASCE/SEI Figure 10.4-1 and Figure 4.25).

Cross-sectional area of ice, $A_i = \pi \frac{t_d}{12}\left(D_c + \frac{t_d}{12}\right) = \pi \times \frac{0.93}{12} \times \left(50.0 + \frac{0.93}{12}\right) = 12.2 \text{ ft}^2$

Design ice load $= \gamma A_i = 56 \times 12.2 = 683$ lb/ft

In S.I.:

Cross-sectional area of ice, $A_i = \pi \frac{t_d}{1{,}000}\left(D_c + \frac{t_d}{1{,}000}\right) = \pi \times \frac{23}{1{,}000} \times \left(15.2 + \frac{23}{1{,}000}\right) =$ 1.10 m^2

Design ice load $= \gamma A_i = 900 \times 9.8 \times 1.10 = 9{,}702$ N/m

Alternatively, use the equations in Table 4.6:

$$\text{Design ice load} = 14.66 t_d \left(D_c + \frac{t_d}{12}\right) = 14.66 \times 0.93 \times \left(50.0 + \frac{0.93}{12}\right) = 683 \text{ lb/ft}$$

$$\text{Design ice load} = 27.71 t_d \left(D_c + \frac{t_d}{1{,}000}\right) = 27.71 \times 23 \times \left(15.2 + \frac{23}{1{,}000}\right) = 9{,}702 \text{ N/m}$$

Step 6—Determine the wind velocity pressure, q_z

Flowchart 4.10 is used to determine q_z.

The design wind-on-ice load for tanks is determined in accordance with ASCE/SEI 29.4 based on the dimensions of the tank including ice (see Table 4.7). According to that section, q_z is evaluated at the centroid of the projected area normal to the wind, A_f. For this tank, the centroid occurs at the following height:

$$z = 25.0/2 = 12.5 \text{ ft [in S.I.: } z = 7.62/2 = 3.81 \text{ m]}$$

Step 6a—Determine the exposure category

From the design data, the Exposure Category is given as B.

Step 6b—Determine the terrain exposure constants

For Exposure B: $\alpha = 7.5$ and $z_g = 3{,}280$ ft (1,000 m) [ASCE/SEI Table 26.11-1].

Step 6c—Determine the velocity pressure exposure coefficient, K_z

For $z = 12.5$ ft (3.81 m) and exposure B: $K_z = 0.57$ (ASCE/SEI Table 26.10-1).

Alternatively, use the equations in Note 1 of ASCE/SEI Table 26.10-1:

For $z < 15 \text{ ft}: K_z = 2.41(15/z_g)^{2/\alpha} = 2.41 \times (15/3{,}280)^{2/7.5} = 0.57$

For $z < 4.57 \text{ m}: K_z = 2.41(4.57/z_g)^{2/\alpha} = 2.41 \times (4.57/1{,}000)^{2/7.5} = 0.57$

Step 6d—Determine the ground elevation factor, K_e

It is permitted to use $K_e = 1.0$ for all elevations (ASCE/SEI 26.9).

Step 6e—Determine the wind velocity pressure, q_z

Wind velocity pressure, q_z, is determined from ASCE/SEI Equations (26.10-1) and (26.10-1.SI):

$$q_z = 0.00256 K_z K_{zt} K_e V_c^2$$
$$= 0.00256 \times 0.57 \times 1.0 \times 1.0 \times 47^2 = 3.2 \text{ lb/ft}^2$$

$$q_z = 0.613 K_z K_{zt} K_e V_c^2$$
$$= 0.613 \times 0.57 \times 1.0 \times 1.0 \times 21^2 = 154.1 \text{ N/m}^2$$

Step 7—Determine the fundamental natural frequency of the structure, n_1, and the corresponding gust-effect factor, G

From the design data, $n_1 > 1$ Hz, which means the tank can be classified as rigid.

For rigid structures, G is permitted to be taken as 0.85 (ASCE/SEI 26.11.1).

Step 8—Determine the wind force coefficient, C_f

The wind force coefficient, C_f, for a round tank is determined in accordance with ASCE/SEI Figure 29.4-1 corresponding to $D\sqrt{q_z} \leq 2.5$ (in S.I.: $D\sqrt{q_z} \leq 5.3$) for all ice thicknesses, gust speeds, and structure diameters (ASCE/SEI 10.5.1; see Table 4.7).

For $h/D = 0.5$, use $C_f = 0.7$.

Step 9—Determine the wind-on-ice load, W_i

For round tanks, the wind directionality factor, $K_d = 1.0$ assuming the tank does not have a nonaxisymmetric structural system (ASCE/SEI Table 26.6-1).

$$A_f = \text{projected area normal to the wind, including ice}$$
$$= [50.0 + (2 \times 0.93/12)] \times [25.0 + (0.93/12)] = 1{,}258 \text{ ft}^2$$

From Table 4.7:

$$W_i = F = q_z K_d G C_f A_f = 3.2 \times 1.0 \times 0.85 \times 0.7 \times 1{,}258 = 2{,}395 \text{ lb}$$

In S.I.:

$$A_f = \text{projected area normal to the wind including ice}$$
$$= [15.2 + (2 \times 23/1{,}000)] \times [7.62 + (23/1{,}000)] = 116.5 \text{ m}^2$$

$$W_i = F = q_z K_d G C_f A_f = 154.1 \times 1.0 \times 0.85 \times 0.7 \times 116.5 = 10{,}682 \text{ N}$$

This force is applied at the centroid of the projected area normal to the wind, A_f, which is $[25.0 + (0.93/12)]/2 = 12.54$ ft (3.82 m) above ground level.

4.5.15 Example 4.15—Atmospheric Ice Loads on a Solid Freestanding Sign

Determine the atmospheric ice loads on a 10-ft (3.05-m) tall by 15-ft (4.57-m) wide solid freestanding sign located on the top of a commercial building in Indianapolis, IN (latitude = 39.77°, longitude = −86.16°). Assume the building is located on a relatively flat Exposure B site and the structural members of the sign consist of structural steel angles, which are L8 × 6 × 1/2. Also assume the natural fundamental frequency of the structure, n_1, is greater than 1 Hz. The top of the sign is located 70 ft (21.3 m) above the ground level and the bottom of the sign is located at the top of the roof.

The design procedure in Table 4.5 of this book is used to determine the atmospheric ice loads.

SOLUTION

Step 1—Determine the nominal ice thickness, t, and the concurrent gust speed, V_c

In accordance with IBC Table 1604.5, the sign attached to a commercial building is classified under Risk Category II.

Using Reference 4.1, $t = 1.00$ in. (25 mm) and $V_c = 41$ mi/h (18 m/s).

Step 2—Determine the height factor, f_z

$$\text{For } z = 70 \text{ ft}: \; f_z = \left(\frac{z}{33}\right)^{0.10} = \left(\frac{70}{33}\right)^{0.10} = 1.08$$

$$\text{For } z = 21.3 \text{ m}: \; f_z = \left(\frac{z}{10}\right)^{0.10} = \left(\frac{21.3}{10}\right)^{0.10} = 1.08$$

Step 3—Determine the topographic factor, K_{zt}

Because the building is located on a relatively flat site, $K_{zt} = 1.0$ (ASCE/SEI 26.8.2).

Step 4—Determine the design ice thickness, t_d

$$t_d = tf_z(K_{zt})^{0.35} = 1.00 \times 1.08 \times (1.0)^{0.35} = 1.08 \text{ in.}$$

$$t_d = tf_z(K_{zt})^{0.35} = 25 \times 1.08 \times (1.0)^{0.35} = 27 \text{ mm}$$

Step 5—Determine the design ice load

Referring to ASCE/SEI Figure 10.4-1 or Figure 4.25, the diameter of the cylinder, D_c, circumscribing the structural steel angle is the hypotenuse of the right triangle formed by the legs of the angle:

$$D_c = \sqrt{(8)^2 + (6)^2} = 10.0 \text{ in.}$$

Cross-sectional area of ice, $A_i = \pi \frac{t_d}{12}\left(D_c + \frac{t_d}{12}\right) = \pi \times \frac{1.08}{12} \times \left(\frac{10.0}{12} + \frac{1.08}{12}\right) = 0.26 \text{ ft}^2$

Design ice load $= \gamma A_i = 56 \times 0.26 = 14.6$ lb/ft

In S.I.:

$$D_c = \sqrt{(203)^2 + (152)^2} = 254 \text{ mm}$$

Cross-sectional area of ice, $A_i = \pi \frac{t_d}{1{,}000}\left(D_c + \frac{t_d}{1{,}000}\right) = \pi \times \frac{27}{1{,}000} \times \left(\frac{254}{1{,}000} + \frac{27}{1{,}000}\right) =$ 0.024 m^2

Design ice load $= \gamma A_i = 900 \times 9.8 \times 0.024 = 212$ N/m

Alternatively, use the equations in Table 4.6:

$$\text{Design ice load} = 14.66 t_d\left(D_c + \frac{t_d}{12}\right) = 14.66 \times 1.08 \times \left(\frac{10.0}{12} + \frac{1.08}{12}\right) = 14.6 \text{ lb/ft}$$

$$\text{Design ice load} = 27.71 t_d\left(D_c + \frac{t_d}{1{,}000}\right) = 27.71 \times 27 \times \left(\frac{254}{1{,}000} + \frac{27}{1{,}000}\right) = 210 \text{ N/m}$$

Step 6—Determine the wind velocity pressure, q_z

Flowchart 4.10 is used to determine q_z.

The design wind-on-ice load for solid freestanding signs is determined in accordance with ASCE/SEI 29.3 based on the dimensions of the sign including ice (see Table 4.7). According to that section, the velocity pressure is evaluated at top of the sign, that is, $q_z = q_h$.

Step 6a—Determine the exposure category

From the design data, the Exposure Category is given as B.

Step 6b—Determine the terrain exposure constants

For Exposure B: $\alpha = 7.5$ and $z_g = 3{,}280$ ft (1,000 m) [ASCE/SEI Table 26.11-1].

Step 6c—Determine the velocity pressure exposure coefficient, K_z

For $z = 70$ ft (21.3 m) and Exposure B: $K_z = 0.86$ (ASCE/SEI Table 26.10-1).

Alternatively, use the equations in Note 1 of ASCE/SEI Table 26.10-1:

For 15 ft $< z < z_g$: $K_z = 2.41(z/z_g)^{2/\alpha} = 2.41 \times (70/3{,}280)^{2/7.5} = 0.86$

For 4.57 m $< z < z_g$: $K_z = 2.41(z/z_g)^{2/\alpha} = 2.41 \times (21.3/1{,}000)^{2/7.5} = 0.86$

Step 6d—Determine the ground elevation factor, K_e

It is permitted to use $K_e = 1.0$ for all elevations (ASCE/SEI 26.9).

Step 6e—Determine the wind velocity pressure, q_h

Wind velocity pressure, q_h, is determined from ASCE/SEI Equations (26.10-1) and (26.10-1.SI):

$$q_h = 0.00256 K_h K_{zt} K_e V_c^2$$

$$= 0.00256 \times 0.86 \times 1.0 \times 1.0 \times 41^2 = 3.7 \text{ lb/ft}^2$$

$$q_h = 0.613 K_h K_{zt} K_e V_c^2$$

$$= 0.613 \times 0.86 \times 1.0 \times 1.0 \times 18^2 = 170.8 \text{ N/m}^2$$

Step 7—Determine the fundamental natural frequency of the structure, n_1, and the corresponding gust-effect factor, G

From the design data, $n_1 > 1$ Hz, which means the structure can be classified as rigid. For rigid structures, G is permitted to be taken as 0.85 (ASCE/SEI 26.11.1).

Step 8—Determine the net force coefficient, C_f

The net force coefficient, C_f, is determined in accordance with ASCE/SEI Figure 29.3-1. Load cases A, B, and C must be considered.

- Case A: Resultant wind force acts normal to the face of the sign and is applied at its geometric center

  Clearance ratio, $s/h = 10/10 = 1.0$

  Aspect ratio, $B/s = 15/10 = 1.5$

  In S.I.:

  Clearance ratio, $s/h = 3.05/3.05 = 1.0$

  Aspect ratio, $B/s = 4.57/3.05 = 1.5$

  From ASCE/SEI Figure 29.3-1, $C_f = (1.45 + 1.40)/2 = 1.43$.
- Case B: Resultant wind force acts normal to the face of the sign and is applied a distance equal to $0.2B$ from the geometric center of the sign considering the wind range at both ends of the sign

  From ASCE/SEI Figure 29.3-1, $C_f = 1.43$ for Cases A and B.
- Case C: Resultant wind forces act normal to the face of the sign and are applied at the geometric centers of the regions indicated in ASCE/SEI Figure 29.3-1.

  Because $B/s = 1.5 < 2$, Case C need not be considered (see Note 2 in ASCE/SEI Figure 29.3-1).

Step 9—Determine the wind-on-ice load, W_i

For solid freestanding signs, the wind directionality factor, $K_d = 0.85$ (ASCE/SEI Table 26.6-1).

$$A_s = \text{gross area of the solid freestanding sign including ice}$$
$$= [10.0 + (1.08/12)] \times [15.0 + (2 \times 1.08/12)] = 153.2 \text{ ft}^2$$

For load cases A and B (see Table 4.7):

$$W_i = F = q_h K_d G C_f A_s = 3.7 \times 0.85 \times 0.85 \times 1.43 \times 153.2 = 586 \text{ lb}$$

In S.I.:

$$A_s = \text{gross area of the solid freestanding sign including ice}$$
$$= [3.05 + (27/1{,}000)] \times [4.57 + (2 \times 27/1{,}000)] = 14.23 \text{ m}^2$$
$$W_i = F = q_h K_d G C_f A_s = 170.8 \times 0.85 \times 0.85 \times 1.43 \times 14.23 = 2{,}511 \text{ N}$$

This force is applied at the locations indicated in ASCE/SEI Figure 29.3-1.

4.6 References

4.1. American Society of Civil Engineers (ASCE). 2022. ASCE 7 Hazard Tool. https://asce7hazardtool.online/.

4.2. American Society of Civil Engineers (ASCE). 2021. Wind Tunnel Testing for Buildings and Other Structures, ASCE/SEI 49-21. Reston, VA.

4.3. Structural Engineers Association of Alaska. 2020. Reliability-Targeted Alaska Ground Snow Loads for the 2022 Edition of ASCE 7. Anchorage, AK.

4.4. Federal Emergency Management Agency (FEMA). 2019. Three-Dimensional Roof Snowdrifts. Washington, D.C.

4.7 Problems

4.1. Determine design snow loads on the monoslope roof in Example 4.1 for a roof slope of (a) 3 on 12 and (b) 6 on 12. Use the design data given in the example.

4.2. Determine the design snow loads on the roof of the lower building in Example 4.7 assuming the ground snow load is equal to 80 lb/ft^2 (3.83 kN/m^2) instead of 39 lb/ft^2 (1.87 kN/m^2). Use the design data given in the example.

4.3. Determine the design snow loads on the roof of the lower building in Problem 4.2 assuming the buildings are separated horizontally at a distance of 2 ft (0.61 m).

4.4. Determine the design snow loads for the buildings in Figure 4.57 given the following design data:

- Ground snow load, $p_g = 40$ lb/ft^2 (1.92 kN/m^2)
- Winter wind parameter, $W_2 = 0.45$
- Surface roughness: C
- Risk Category: II
- Thermal condition: See Figure 4.57
- Roof exposure: Fully exposed
- Roof surface: Rubber membrane
- Roof framing: All structural members in both buildings are simply supported

The width of both roofs is 75 ft (22.9 m), and both roofs have a 1.19-degree slope into the page.

Figure 4.57 Elevation of the Buildings in Problem 4.4

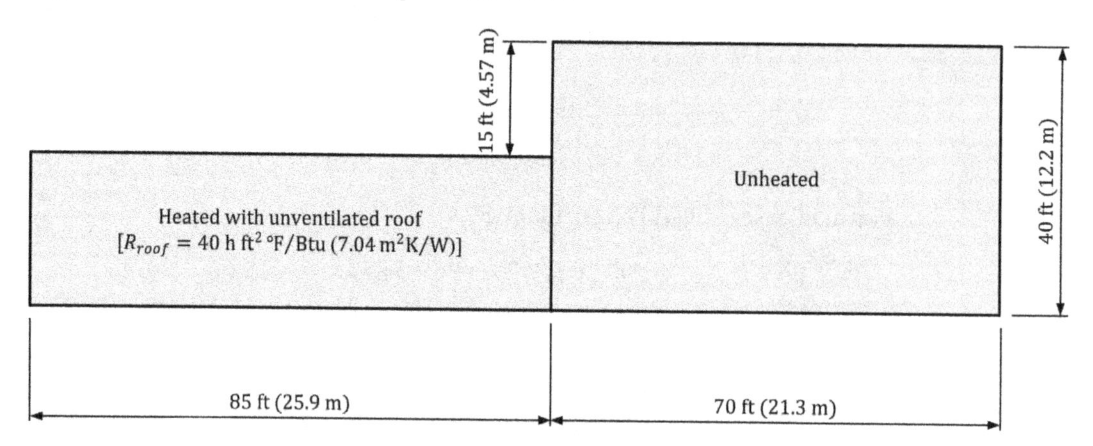

4.5. Determine the design snow loads for the building in Figure 4.58 given the following design data:

- Ground snow load, $p_g = 30$ lb/ft² (1.44 kN/m²)
- Winter wind parameter, $W_2 = 0.40$
- Surface roughness: B
- Risk Category: II
- Thermal condition: Heated structure with an unventilated roof and $R_{roof} = 25$ h ft²°F/Btu (4.40 m²K/W)
- Roof exposure: Partially exposed
- Roof surface: Smooth bituminous membrane
- Roof framing: All structural members are simply supported

Figure 4.58 Plan and Elevation of the Building in Problem 4.5

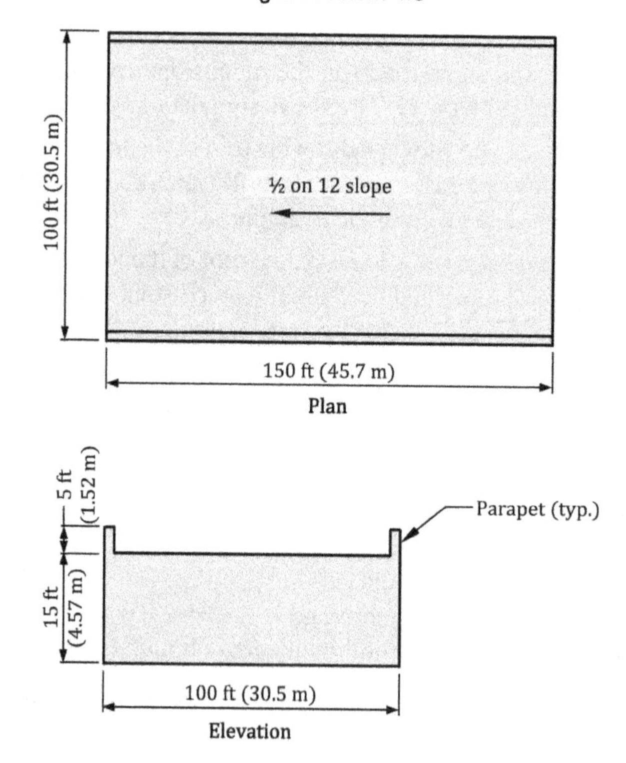

4.6. Determine the design snow loads on the upper and lower roofs of the building in Figure 4.59 given the following design data:

- Location: Fairbanks, AK
- Surface roughness: B
- Risk Category: II
- Thermal condition: Heated structure with an unventilated roof and $R_{roof} < 20$ h ft^2°F/Btu (3.52 m^{2}K/W)
- Roof exposure: Fully exposed
- Roof surface: Asphalt
- Roof framing: All structural members are simply supported

Figure 4.59 Elevation of the Building in Problem 4.6

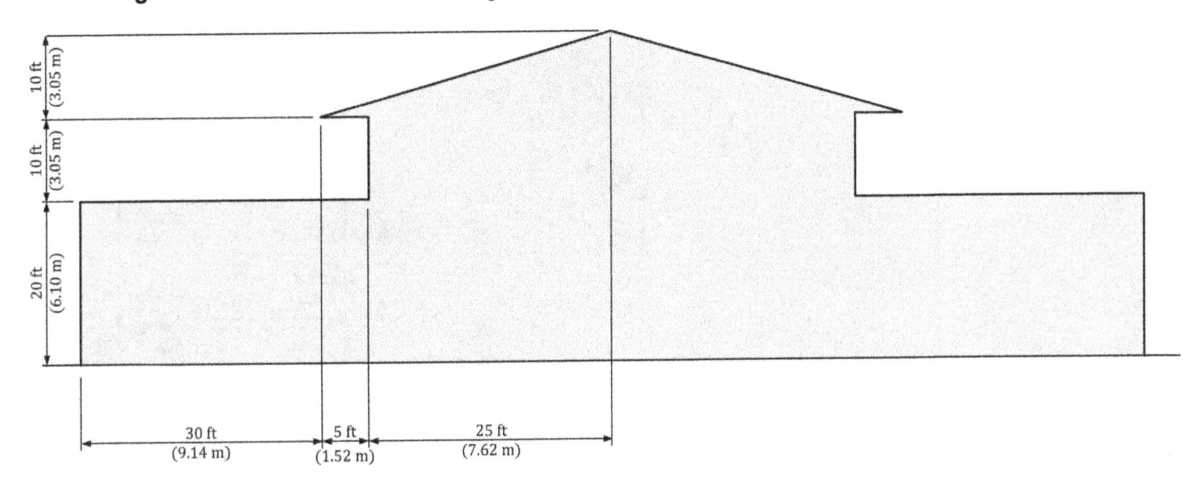

4.7. Determine the design snow loads for the building in Figure 4.60 given the following design data:

- Ground snow load, $p_g = 100$ lb/ft^2 (4.79 kN/m^2)
- Winter wind parameter, $W_2 = 0.70$
- Surface roughness: B
- Risk Category: III
- Thermal condition: Cold, ventilated roof
- Roof exposure: Partially exposed
- Roof surface: Smooth bituminous membrane
- Roof framing: All structural members are simply supported

The roof profile is an arc of a circle that has a radius of 75 ft (22.9 m).

Figure 4.60 Elevation of the Building in Problem 4.7

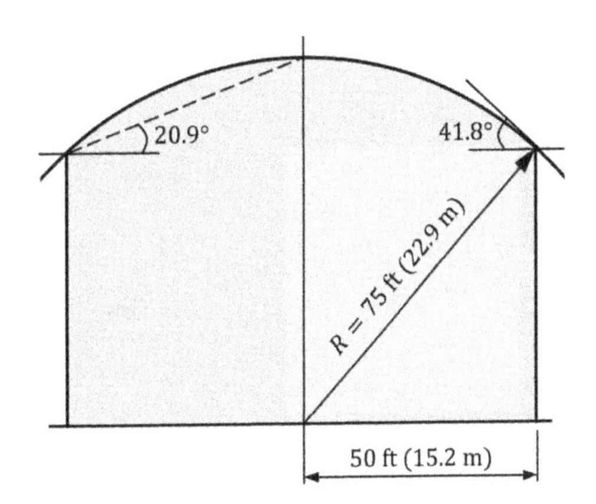

4.8. Determine the design snow loads for the building in Problem 4.7 assuming another roof abuts it within 3 ft (0.91 m) of its eave.

4.9. Determine the design snow loads for the building in Problem 4.7 assuming the following: R = 150 ft (45.7 m) instead of 75 ft (22.9 m) and the distance to the eave = 143 ft (43.6 m) instead of 50 ft (15.2 m).

4.10. Determine the design snow loads on the roof of the building and the canopy in Figure 4.61. Use the same design data as in Problem 4.6.

Figure 4.61 Elevation of the Building in Problem 4.10

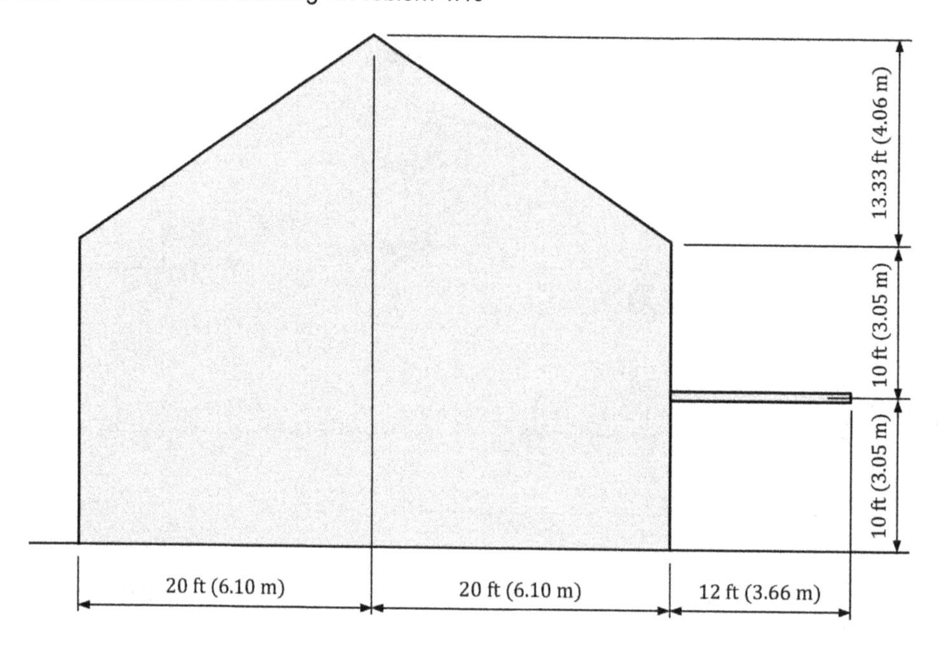

4.11. Determine the design ice loads on the support structure in Example 4.15 assuming the structural members are the following: (a) HSS6 × 0.500 and (b) HSS6 × 4 × 1/2. Use the same design data in Example 4.15.

4.12. Determine the wind-on-ice load of a round chimney that is 60 ft (18.3 m) tall and has an outside diameter of 6 ft (1.83 m). Assume the chimney is located in Milwaukee, WI, (latitude = 43.06°, longitude = −87.89°) on a relatively flat site in Exposure C.

CHAPTER 5

Wind and Tornado Loads

5.1 Introduction

5.1.1 Nature of Wind Loads

In general, wind loading is the effect of the atmosphere passing by a stationary structure attached to the earth's surface. An in-depth discussion on the mechanics of atmospheric circulations can be found in Reference 5.1.

Loads on buildings and other structures due to the effects from wind are determined by considering both atmospheric and aerodynamic effects. These effects form the basis of the methodologies given in ASCE/SEI 7 for the determination of wind loads. An elementary discussion of each effect is given below. The effects due to wind gust, which are considered a part of both atmospheric and aerodynamic effects, also play an important role in the calculation of wind loads and are covered as well.

Atmospheric Effects

Overview

The atmospheric factors that have a direct impact on the magnitude of wind loading on a building or other structure are obtained from meteorological and boundary layer effects. A brief discussion pertinent to wind loads on buildings and other structures follows.

Meteorology

Meteorology is the study of the atmosphere, and wind climatology is a branch within meteorology that focuses on the prediction of storm conditions. In particular, extreme wind speeds associated with different types of storms and the probability of occurrence of such extreme values are analyzed at specific geographical locations. This information is used in developing design wind speed maps in the IBC and ASCE/SEI 7. Wind velocity is used in calculating the maximum design wind loads expected on a building or structure during its lifespan.

The wind speed maps in the IBC and ASCE/SEI 7 contain wind speeds corresponding to 3-second gust speeds at 33 ft (10 m) above ground for exposure C (definitions for the exposure categories are given in ASCE/SEI 26.7). In nonhurricane regions in the conterminous United States, wind speeds are estimated from peak gust speed data collected at 575 meteorological stations with at least 15 years of available data. The data at each station is classified by storm type, that is, thunderstorm or nonthunderstorm. Wind speeds in hurricane regions are based on the results of a Monte Carlo simulation model (more information on the model can be found in the references in ASCE/SEI C26.5.1).

Thunderstorms, hurricanes, tornadoes, and special regional effects are the climatological events of primary interest when designing buildings and structures in the conterminous United States. The controlling climatological events producing extreme wind speeds are given in Figure 5.1. It is evident from the figure that the prevailing wind speeds are generated by thunderstorms in most of the country.

Special wind regions can have wind speeds significantly greater than those in surrounding areas. Such regional effects include wind blowing over mountain ranges or through gorges or valleys. These regions are identified as special wind regions on the wind speed maps.

Figure 5.1 Controlling Climatological Events for Extreme Wind Speeds

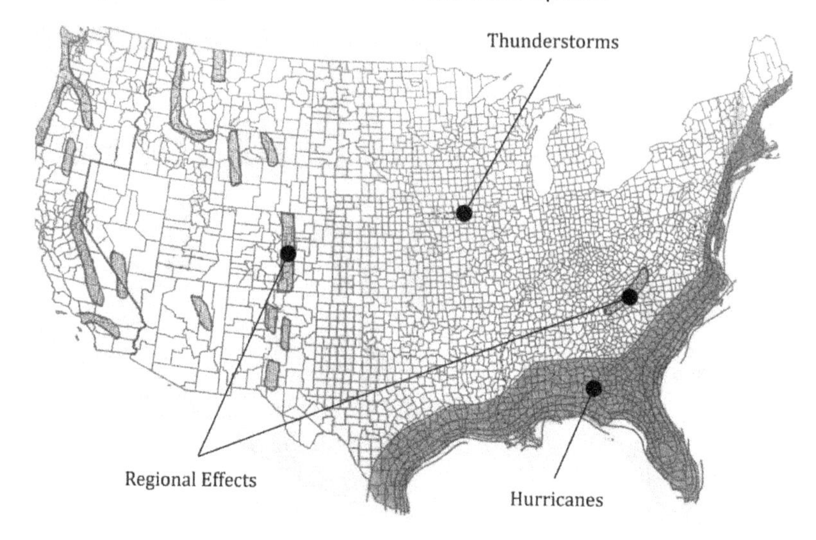

Tornado speeds are given in ASCE/SEI 32.5 for structures located in tornado-prone areas of the conterminous United States (see ASCE/SEI 32.1.1 and 32.5.2 for the conditions under which tornado loads must be considered). Information on how the tornado speed maps in ASCE/SEI 32.5 were derived is given in ASCE/SEI C32.5.

Boundary Layer Fluid Dynamics

The layer of the earth's atmosphere located from the surface of the earth to approximately 3,280 ft (1,000 m) above the surface is known as the boundary layer. The fluid dynamic effects occurring within this layer have an important impact on the magnitude of wind loads on buildings and other structures.

In general, the surface of the earth exerts a horizontal drag force on wind which impedes its flow. More frictional resistance is experienced the closer the wind flow is to the surface; thus, wind velocity is smaller at or near the ground level compared to levels above the surface. Similarly, at a given height above the surface, wind velocity is smaller over rougher surfaces compared to smoother ones because of friction. Depicted in Figure 5.2 is the variation of wind speed with respect to height and surface roughness. Both phenomena are captured in the current wind load provisions using a modified version of the power-law methodology first introduced in Reference 5.2. Wind velocities become constant above certain heights for the different roughness categories; these heights are defined as gradient heights in ASCE/SEI 7.

Aerodynamic Effects

Overview

Wind flow is disturbed due to the presence of a building or structure in its path. The resulting responses due to this disturbance are governed by the laws of aerodynamics. In the case of a building or other structure with an essentially block-like shape (which is referred to as a bluff body), bluff body aerodynamics is used to predict the effects caused by placing the bluff body in the flow of wind.

When wind comes into contact with a building or other structure, the following pressures are created:

External pressures, which act on all exterior surfaces; these pressures are caused by the effects generated when the wind strikes the building or other structure.

Internal pressures, which act on all interior surfaces; these pressures are due to leakage of air through the exterior surface to the interior space.

Figure 5.2 Variation of Wind Velocity with Respect to Height and Surface Roughness

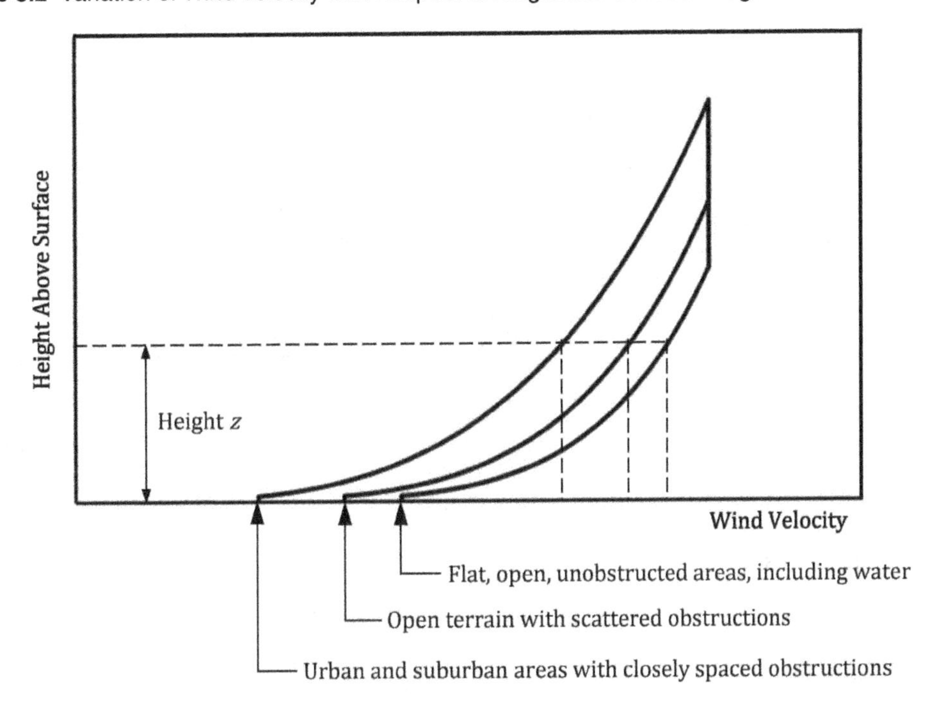

It is assumed the external and internal pressures act perpendicular to the exterior and interior surfaces of the building or other structure, respectively. A pressure is defined as positive when it acts toward the surface and negative when it acts away from the surface. Positive pressure is commonly referred to as just pressure and negative pressure is also identified as suction.

External Pressure

Idealized wind flow around a gable roof building is depicted in Figure 5.3. When the wind strikes the windward wall, a positive pressure is created on that surface, which varies with respect to height (as discussed previously, wind velocity increases at distances above the surface and, as shown in Equation (5.1) below, wind pressure is directly proportional to the square of wind velocity). Located mid-width of a wall at a distance approximately two-thirds of the windward wall height above the surface is the *stagnation point*. The pressure, p_s, at this point can be theoretically determined using Bernoulli's equation:

$$p_s = \frac{1}{2}\rho V^2 \tag{5.1}$$

In this equation, ρ is the atmospheric air density and V is the wind velocity at the elevation of the stagnation point.

A location where the wind flow is deflected by one surface of the building and separates and loses contact with the building is referred to as a *separation point* (see Figure 5.3). The void between the separated wind flow and the surface of the building or structure—commonly referred to as the *separation zone*—contains turbulent wind flow, which causes negative pressure on that surface.

The leeward wall and the side walls all have negative pressure (the negative pressure on the leeward wall is referred to as the wake). In the case of a sloped roof like the one depicted in Figure 5.3, the size of the void depends on the angle of the roof: for relatively shallow slopes, the void is relatively large and negative pressure acts over the windward portion of the roof. As the roof slope increases, the void area on the windward roof decreases. The pressure on the windward roof becomes positive when the roof angle matches or exceeds the angle of the separated wind flow; in other words, positive pressure is realized when the void of turbulent flow no longer exists on

Figure 5.3 Wind Flow around a Gable Roof Building

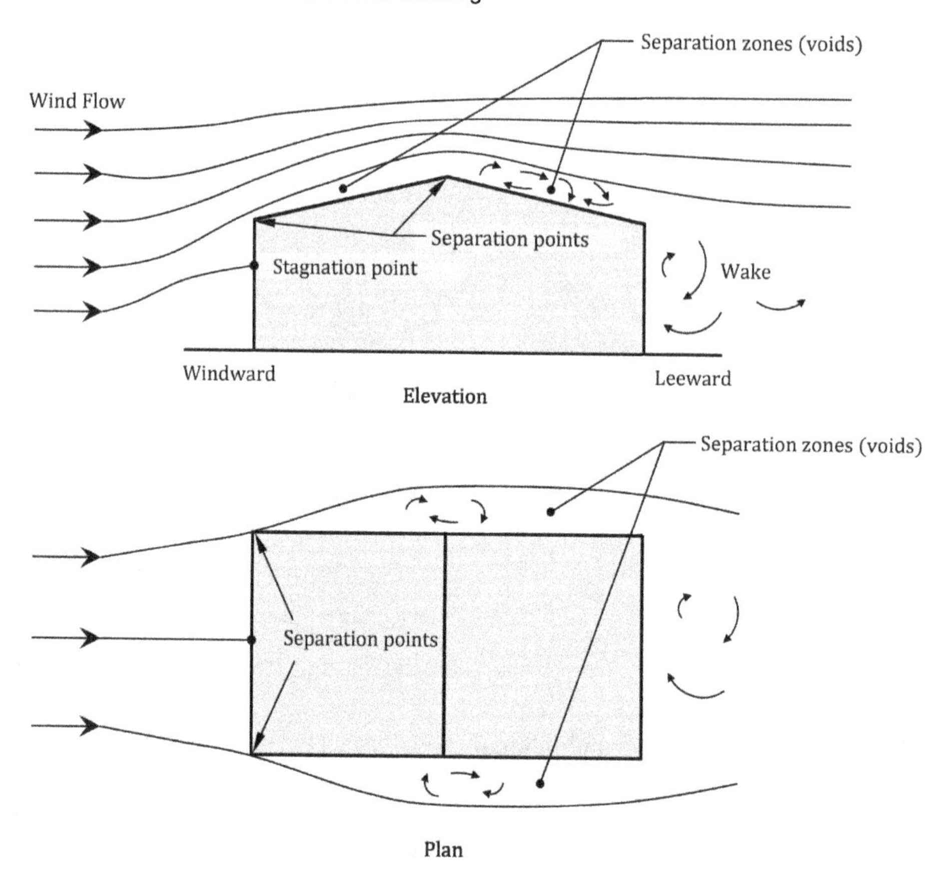

the windward area. Negative pressure occurs on the leeward area of the roof regardless of the roof slope. In the case of a flat roof, the entire roof has negative pressure (the void of turbulent flow acts over the entire area of the roof).

When the dimension of a building or structure is relatively long in the direction of wind flow, it is possible for the flow to reattach itself to the side walls or to the roof, or both. This has the effect of reducing the negative pressure on these surfaces and on the leeward wall. The reduction of pressure for such buildings is reflected in the wind design provisions in ASCE/SEI 7.

A generic representation of the external wind pressures on the walls and roof of the building in Figure 5.3 is given in Figure 5.4.

The positive pressure on the windward wall varies with respect to height and the negative pressures on all other surfaces are essentially constant. Also, the negative pressure at the windward corner of the roof can be significantly greater than the pressures on any of the other surfaces.

The dimensionless pressure, C_p, which is commonly referred to as a *pressure coefficient*, is defined as the pressure, p, at any point on the building or structure divided by the stagnation pressure:

$$C_p = \frac{p}{\rho V^2/2} \tag{5.2}$$

The external pressures vary on each surface of a building or structure based on the structural configuration and the direction of wind. For simplicity, pressure coefficients for the most part are taken as a constant over an entire surface area. One notable exception is for relatively long roofs in the direction of wind flow; in such cases, the pressure coefficients away from the windward edge of the roof are permitted to be smaller than the pressure coefficient at the windward edge (see the discussion on flow reattachment above).

Figure 5.4 Wind Pressures on a Building with a Gable Roof

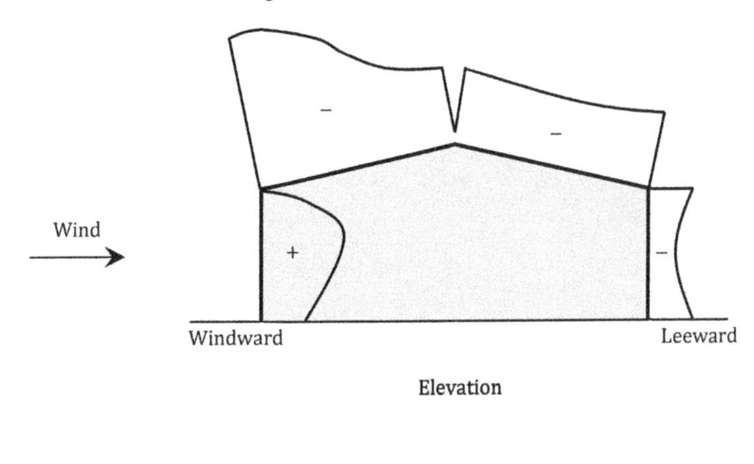

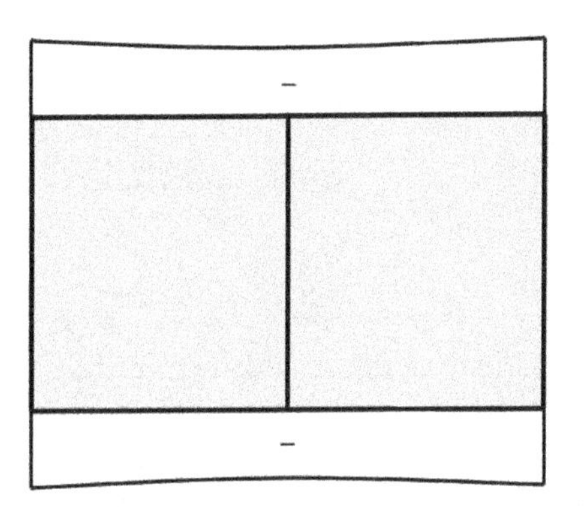

Internal Pressure and Enclosure Classification

Internal pressures act on all interior surfaces of a building or structure and are due to leakage of air through the exterior surface to the interior space. The number and size of openings in the envelope of the structure play key roles when determining the effects caused by internal pressures. ASCE/SEI 7 defines an *opening* as a hole that allows air to flow through a building envelope during a design wind event.

Air intake and exhaust vents for ventilation should always be considered openings. Doors, windows, skylights, and flexible and operable louvers likely to be closed during a design wind event need not be considered as openings. If a door, for example, must remain open during a design wind event because of the intended function of the building (e.g., a door for a hospital ambulance bay), the door should be considered an opening.

A number of different cases need to be examined to properly understand the effects internal pressure can have on a building or structure. In the case of a building with a relatively large opening on the windward wall, the wind flow will try to inflate the building resulting in positive internal pressure (see Figure 5.5). Conversely, an opening on the leeward wall, side walls or roof will try to deflate the building resulting in negative internal pressure.

Typically, there is more than just one opening in a building envelope. The size and location of these openings, which is commonly referred to as the *porosity* of the building envelope, dictates whether the internal pressure is positive or negative. The enclosure classifications defined in ASCE/SEI 7 are based on porosity.

Consider the case of a building with openings on all surfaces large enough to affect the internal pressure. Air can flow through the building without a buildup of any internal pressure. Such

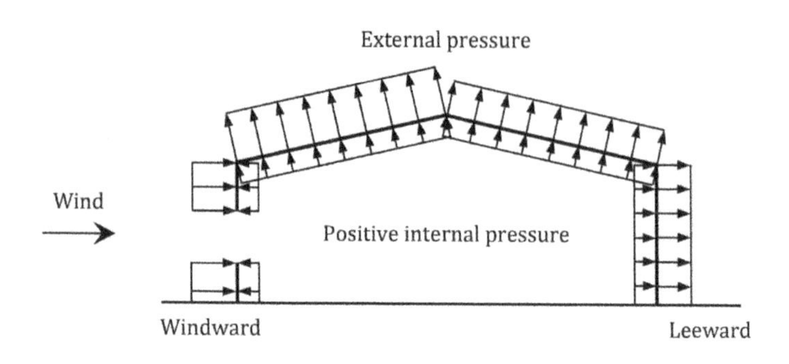

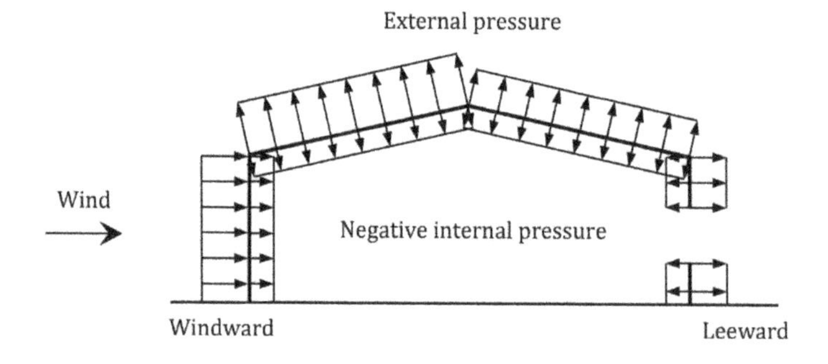

Figure 5.5 Effects of Openings on Internal Pressure Distribution

buildings are classified as *open*, and ASCE/SEI 7 defines such structures as having each wall at least 80 percent open.

A *partially enclosed* building has openings large enough to affect internal pressure and have porosity low enough to allow internal pressure to build up. A partially enclosed condition typically exists when there is a relatively large opening on the windward wall compared to the openings on all of the other surfaces of the envelope. The internal pressures in such buildings can be as large as or greater than the external pressures.

The porosity of an *enclosed* building is such that there are not sufficient openings in the envelope to allow significant airflow into the building. This results in a relatively small level of positive or negative internal pressure.

A *partially open* building is any building with openings and significant porosity that does not meet the requirements of the open, partially enclosed or enclosed classifications. Such buildings may contain large openings in two or more walls, such as in a parking garage, where air can easily pass through. Sufficient openings exist in the remaining portions of the building envelope to allow some airflow out of the building but with some buildup of internal pressure.

The effects of internal pressure are accounted for in ASCE/SEI 7 by internal pressure coefficients, the magnitudes of which are based on the enclosure classification of the building or other structure. It is evident from Figure 5.5 that internal pressure does not contribute to the overall horizontal wind pressures acting on a structure because such pressures cancel out. However, internal pressures must be considered in the design of individual components, such as walls and roof framing, and cladding; the type of internal pressure (positive or negative) resulting in the critical load combination must be used.

Both the IBC and ASCE/SEI 7 have specific requirements for the protection of openings in *wind-borne debris regions* (see Section 5.2.7 of this book for the definition of wind-borne debris regions). Openings in such regions must have impact-resistant glazing or must be protected with an impact-resistant covering that meets the requirements of impact-resistant standards.

Gust Effects

Wind velocity typically changes dramatically with time; numerous peaks and valleys normally occur over relatively short time spans. In general, the average wind speed obtained from a wind event is larger when a shorter averaging time is used. The peaks in wind velocity are called *gusts*, and these effects must be considered in design.

Gust-effect factors are used in ASCE/SEI 7 to account for this phenomenon. A gust-effect factor relates the peak to mean response in terms of an equivalent static design load or load effect. Additional information on the pioneering work in gust-effect factors can be found in Reference 5.3.

Gust-effect factors are typically less than 1.0 (except for certain types of flexible buildings) because the 3-second averaging time corresponds to a peak gust whose effects are greater than those from the gust level that has been deemed reasonable in design.

The gust-effect factors in ASCE/SEI 7 for both rigid and flexible buildings (i.e., buildings with a fundamental frequency greater than or equal to 1 Hz and less than 1 Hz, respectively) account for the effects in the along-wind direction only. In the case of flexible buildings or structures, along-wind effects due to dynamic amplification are also accounted for in the gust-effect factor.

The mass distribution, flexibility, and damping of a building or structure can have a significant impact on its response to large gusts. It is possible for the fundamental frequency of lighter, more flexible buildings and structures with relatively small amounts of inherent damping to be in the same range as the average frequencies of large gusts. When this occurs, large resonant motions can occur, which must be considered in design.

Certain types of buildings and structures—especially those that are relatively tall and slender—are susceptible to one or more of the following: (1) across-wind load effects, (2) vortex shedding, (3) instability due to galloping or flutter, or (4) dynamic torsional effects. The gust-effect factors in ASCE/SEI 7 do not account for the loading effects caused by these phenomena; wind tunnel tests must be performed in such cases to properly capture these effects.

5.1.2 Overview of Code Requirements for Wind Loads

According to IBC 1609.1.1, wind loads on buildings and structures are to be determined by the provisions of Chapters 26 through 30 of ASCE/SEI 7. Five exceptions are given in IBC 1609.1.1 that permit wind loads to be determined on certain types of structures using industry standards other than ASCE/SEI 7, and one exception is given that permits the use of wind tunnel tests conforming to the provisions of Chapter 31 of ASCE/SEI 7. A final exception deals with temporary structures which are to be designed in accordance with IBC 3103.

Wind is assumed to come from any horizontal direction and its effects are applied in the form of pressures acting normal to the surfaces of a building or other structure (IBC 1609.1.1). Positive wind pressure acts toward the surface and is commonly referred to as just pressure (see Section 5.1.1 of this book). Negative wind pressure, which is also called suction, acts away from the surface.

Positive pressure acts on the windward wall of a building and negative pressure acts on the leeward wall, the side walls, and the leeward portion of the roof (see Figure 5.6). Either positive pressure or negative pressure acts on the windward portion of the roof, depending on the slope of the roof (flatter roofs are subjected to negative pressure while more sloped roofs are subjected to positive pressure). The wind pressure on the windward face varies with respect to height and the pressures on all other surfaces are assumed to be constant.

Pressures must be considered on the *main wind force resisting system* (MWFRS) and *components and cladding* (C&C) of a building or other structure. The MWFRS consists of structural elements that have been assigned to resist the effects from the wind loads for the overall structure. Shear walls, moment frames and braced frames are a few examples of different types of MWFRSs.

The elements of the building envelope that do not qualify as part of the MWFRS are the C&C. The C&C receive the wind loads either directly or indirectly (e.g., from the cladding) and transfer the loads to the MWFRS. Examples of C&C are individual exterior walls, roof decking, roof members (joists or purlins), curtain walls, windows or doors. A discussion on how wind load propagates in a building is given in Chapter 8 of this book. Certain members must be designed for more than one type of loading; for example, a roof beam that is part of the MWFRS must be designed for the load effects associated with the MWFRS and those associated with the C&C.

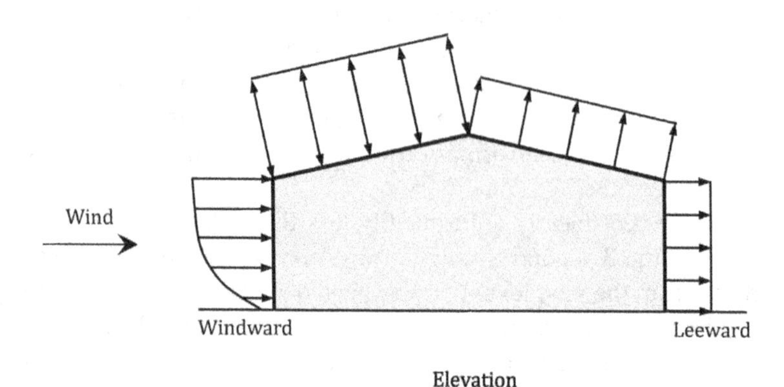

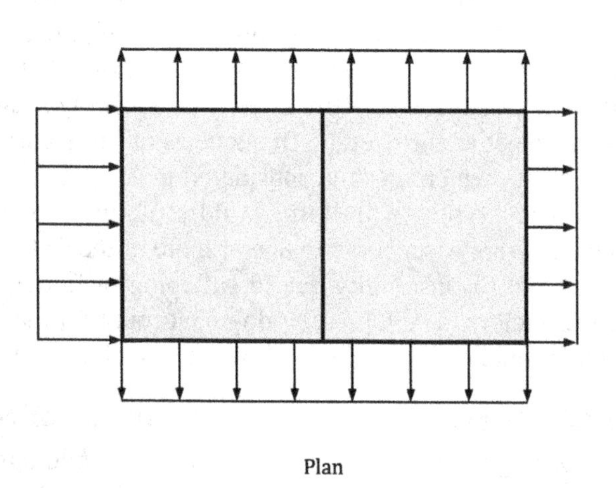

Figure 5.6 Application of Wind Pressures on a Building with a Gable or Hip Roof in Accordance with the IBC and ASCE/SEI 7

The procedures in ASCE/SEI 7 to determine wind loads on MWFRSs and C&C are given in Table 5.1. In the *directional procedure*, wind loads for specific wind directions are determined using external pressure coefficients based on wind tunnel tests of prototypical building models for the corresponding direction of wind. The provisions of this procedure cover a wide range of buildings and other structures.

Table 5.1 Summary of Wind Load Procedures in ASCE/SEI 7

System	ASCE/SEI 7 Chapter No.	Description
MWFRS	27	Directional procedure for buildings of all heights
	28	Envelope procedure for low-rise buildings
	29	Directional procedure for building appurtenances (roof overhangs and parapets) and other structures
	31	Wind tunnel procedure for any building or other structure
C&C	30	• Envelope procedure in Part 1; or • Directional procedure in Parts 2 and 3 • Building appurtenances (roof overhangs and parapets) in Part 4; and • Nonbuilding structures in Part 5
	31	Wind tunnel procedure for any building or other structure

In the *envelope procedure*, pseudo-external pressure coefficients derived from wind tunnel tests on prototypical building models rotated 360 degrees in a wind tunnel are used to determine wind pressures; the pressure coefficients envelope the maximum values obtained from all possible wind directions.

With the exception of the wind tunnel procedure, all the procedures and methods in Table 5.1 are static methods for estimating wind pressures: the magnitude of wind pressure on a building or structure depends on its size, openness, occupancy, location, and the height above ground level. Also accounted for are wind gust and local extreme pressures at various locations on a building or structure. Static methods generally yield very accurate results for low-rise buildings.

An outline of the process required for determining wind loads is given in ASCE/SEI Figure 26.1-1. These procedures, including the general requirements in ASCE/SEI Chapter 26, are covered in the following sections of this book.

5.2 General Requirements for Wind Loads

5.2.1 Overview

The following parameters for determining wind loads on the MWFRS and C&C are given in ASCE/SEI Chapter 26:

- Basic wind speed, V (ASCE/SEI Figures 26.5-1A through 26.5-1D)
- Wind directionality factor, K_d (ASCE/SEI 26.6)
- Exposure category (ASCE/SEI 26.7)
- Topographic factor, K_{zt} (ASCE/SEI 26.8)
- Ground elevation above sea level (ASCE/SEI 26.9)
- Velocity pressure (ASCE/SEI 26.10)
- Gust effect factor, G and G_f (ASCE/SEI 26.11)
- Enclosure classification (ASCE/SEI 26.12)
- Internal pressure coefficient, GC_{pi} (ASCE/SEI 26.13)

The following sections discuss these requirements and provide additional background information on fundamental concepts.

5.2.2 Wind Hazard Map

Regardless of the wind load procedure used to determine wind pressures, the basic wind speed, V, must be determined at the location of the building or other structure.

Basic wind speeds based on 3-second gusts at 33 ft (10 m) above ground for exposure C for different risk categories are given in IBC Figures 1609.3(1) through 1609.3(4) and ASCE/SEI Figures 26.5-1A through 26.5-1D. A summary of the information associated with these maps is given in Table 5.2.

Table 5.2 Summary of Basic Wind Speed Maps in the IBC and ASCE/SEI 7

Location	Figure No.		Risk Category*	Mean Recurrence Interval (years)
	IBC	ASCE/SEI 7		
• Contiguous United States • Alaska • American Samoa • Guam and Northern Mariana Islands	1609.3(1)	26.5-1A	I	300
	1609.3(2)	26.5-1B	II	700
	1609.3(3)	26.5-1C	III	1,700
	1609.3(4)	26.5-1D	IV	3,000

*See IBC Table 1604.5 for definitions of risk categories.

The crosshatched areas on the wind speed maps are designated as special wind regions which are areas where unusual wind conditions exist (see Section 5.1.1 of this book). The local building official should be consulted to obtain the local wind speed (ASCE/SEI 26.5.2). Information on how to estimate basic wind speeds from regional climatic data can be found in ASCE/SEI 26.5.3.

Additional information on the basic wind speed maps and the selection of the corresponding mean recurrence intervals can be found in ASCE/SEI C26.5.1.

Basic wind speeds for the locations in Table 5.2 as well as for Hawaii, Puerto Rico, and the U.S. Virgin Islands are permitted to be determined using the ASCE 7 Wind Design Geodatabase (Reference 5.4; ASCE/SEI 26.5.1). Wind speeds are provided for the following mean recurrence intervals based on a given risk category and site: 10, 25, 50, 100, 10,000, 100,000, and 1,000,000 years. The latitude and longitude or the address of the site can be entered, or the site can be located by clicking on the map.

The wind speeds associated with 10-, 25-, 50-, and 100-year mean recurrence intervals are typically used for serviceability applications, such as drift and habitability (see ASCE/SEI Appendix C). Wind speeds greater than those for a 3,000-year mean recurrence interval are needed for tornado design (see Section 5.6 of this book) and may be needed in performance-based wind designs and other applications (see the wind hazard maps for long return periods in ASCE/SEI Appendix F).

It is permitted to use the wind speed values from Reference 5.4 for select special wind regions in the contiguous United States (ASCE/SEI 26.5.2).

Where required, the basic design wind speeds in IBC Figures 1609.3(1) through 1609.3(4) can be converted to allowable stress design wind speeds, V_{asd}, using Equation (16-18) in IBC 1609.3.1:

$$V_{asd} = V\sqrt{0.6} \tag{5.3}$$

Values of V_{asd} are tabulated in IBC Table 1609.3.1 for various V.

Some referenced standards and some product evaluation reports have been developed and/or evaluated based on fastest-mile wind speed, which was the wind speed utilized in the 1993 and earlier editions of ASCE/SEI 7 and in the legacy codes. To facilitate coordination between the various wind speeds, ASCE/SEI Table C26.5-7 provides wind speeds corresponding to the 1993 to 2022 editions of ASCE/SEI 7.

5.2.3 Wind Directionality

The wind directionality factor, K_d, accounts for the statistical nature of wind flow and the probability of the maximum effects occurring at any particular time for any given wind direction (ASCE/SEI 26.6). In particular, this factor accounts for the following:

1. The reduced probability of maximum winds coming from any given direction.
2. The reduced probability of the maximum pressure coefficient occurring for any given wind direction.

Values of K_d are given in ASCE/SEI Table 26.6-1 as function of structure type. This factor is equal to 0.85 for the MWFRS and C&C of building structures.

5.2.4 Exposure

Wind Direction and Sectors

An exposure category must be determined upwind of a building or other structure for each wind direction considered in design (IBC 1609.4 and ASCE/SEI 26.7). Wind must be assumed to come from any horizontal direction when determining wind loads (IBC 1609.1.1 and ASCE/SEI 26.5.1). One rational way of satisfying this requirement is to assume there are eight wind directions, four perpendicular to the main axes of the building or other structure and four at 45-degree angles to the main axes. The sectors to be used to determine the exposure for a selected wind direction are given in ASCE/SEI Figure C26.7-8 (see IBC 1609.4.1, ASCE/SEI 26.7.1, and Figure 5.7).

Figure 5.7
Sectors for Determining Exposure

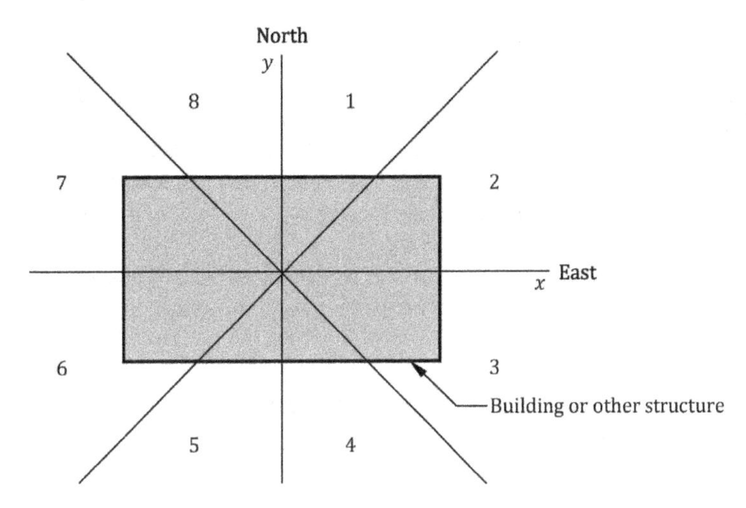

Upwind exposure in a particular direction is determined based on the terrain categories in the 45-degree sectors on each side of the wind direction axis. For wind blowing north to south, the surface roughness is determined for sectors 1 and 8 and the corresponding exposure resulting in the higher wind load effects is used; these wind loads are applied to the building in that direction. The exposures for wind coming from the east, south, and west are determined in a similar fashion. For wind blowing from the northeast (or, similarly from any of the other diagonal directions), sectors 1 and 2 are used to determine the critical exposure; full individual wind loading in the x and y directions are determined based on that exposure, and a percentage of these loads are applied simultaneously to the building (see Section 5.3.2 of this book on load cases that must be considered in wind design).

Surface Roughness Categories

Surface roughness categories are defined in IBC 1609.4.2 and ASCE/SEI 26.7.2 (see Table 5.3). These definitions are descriptive and have been purposely expressed this way so they can be applied easily—while still being sufficiently precise—in most practical applications.

Table 5.3 Surface Roughness Categories

Surface Roughness Category	Description
B	Urban and suburban areas, wooded areas or other terrain with numerous, closely spaced obstructions having the size of single-family dwellings or larger
C	Open terrain with scattered obstructions having heights generally less than 30 ft (9.14 m); this category includes flat open country and grasslands
D	Flat, unobstructed areas and water surfaces; this category includes smooth mud flats, salt flats and unbroken ice

Exposure Categories

Exposure categories are determined based on the surface roughness categories defined above and account for the boundary layer concept of surface roughness discussed in Section 5.1.1 of this book.

Definitions of the three exposure categories given in IBC 1609.4.3 and ASCE/SEI 26.7.3 are given in Table 5.4 (see ASCE/SEI Figures C26.7-1 and C26.7-2 for definitions of exposure B and D, respectively).

Table 5.4 Exposure Categories

Exposure Category	Definition
B	• For buildings and other structures with a mean roof height $h \leq 30$ ft (9.14 m): Surface roughness category B prevails in the upwind direction for a distance > 1,500 ft (457 m) • For buildings and other structures with a mean roof height $h > 30$ ft (9.14 m): Surface roughness category B prevails in the upwind direction for a distance > 2,600 ft (792 m) or 20 times the height of the building or structure, whichever is greater
C	Applies for all cases where exposures B and D do not apply
D	• Surface roughness category D prevails in the upwind direction for a distance > 5,000 ft (1,524 m) or 20 times the height of the building or structure, whichever is greater • Surface roughness immediately upwind of the site is B or C, and the building or structure is within a distance of 600 ft (183 m) or 20 times the building or structure height, whichever is greater, from an exposure D condition as defined above

Aerial photographs illustrating exposure B are given in ASCE/SEI Figures C26.7-5(a) through C26.7-5(c); similar photographs are given in ASCE/SEI Figures C26.7-6(a) and C26.7-6(b) for exposure C and ASCE/SEI Figure C26.7-7 for exposure D. Sites located along hurricane coastlines are to be assigned to exposure D (see ASCE/SEI C26.7 for a summary of the research findings that led to this requirement).

According to ASCE/SEI 26.7.3, the exposure category resulting in the largest wind forces must be used for sites located in transition zones between exposure categories. Consider the building in Figure 5.8, which is located between exposure D and exposure B or C. Exposure D

Figure 5.8 Transition Zone between Exposure Categories

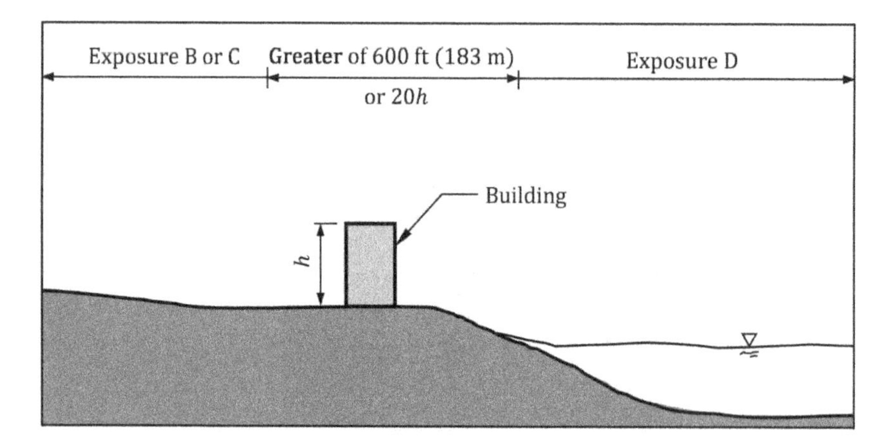

must extend the greater of 600 ft (183 m) or 20 times the building height from the point where exposure D ends (see the definition in Table 5.4). In other words, exposure D transitions to exposure B or C over the greater of those two lengths. Thus, exposure D is required to be used in this transition zone in accordance with ASCE/SEI 26.7.3; however, the exception in ASCE/SEI 26.7.3 permits an intermediate exposure category to be used provided it is determined by a rational analysis method defined in recognized literature. An example of such an analysis is given in ASCE/SEI C26.7.

A more detailed assessment of surface roughness and exposure category can be made using the mathematical procedure in ASCE/SEI C26.7. Ground surface roughness is defined in terms of a surface roughness length, z_o. Exposure categories B, C, and D correspond to a range of values of z_o which are tabulated in ASCE/SEI Table C26.7-1 (this table also includes values for the defunct exposure A as a reference for wind tunnel studies). Additional information on z_o can be found in ASCE/SEI Table C26.7-2, and ASCE/SEI Equation (C26.7-1) is permitted to be used to estimate z_o for a particular terrain.

It is possible for patches of exposure C or D to occur within exposure B. Large parking lots or lakes, freeways, and tree clearings are a few examples of such patches. Relatively large open patches can have a significant impact on the determination of the exposure category. According to the commentary, an open area can be considered an open patch when both of the following conditions are met:

- Open areas have an area greater than the minimum areas in ASCE/SEI Figure C26.7-4.
- Open areas have a length to width ratio between 0.5 and 2.0 and have one or more of the following minimum dimensions:
  1. An open area greater than or equal to approximately 164 ft (50 m) in length or width within 500 ft (152 m) of the building or structure.
  2. An open area greater than or equal to approximately 328 ft (100 m) in length or width at 1,500 ft (457 m) upwind from the building or structure.
  3. An open area greater than or equal to approximately 500 ft (152 m) in length or width at 2,600 ft (792 m) upwind from the building or structure.

Sector analysis for exposure B with upwind open patches is given in ASCE/SEI Figure C26.7-3.

Exposure Requirements

Exposure requirements pertaining to the wind load procedures in ASCE/SEI 7 are given in ASCE/SEI 26.7.4 (see Table 5.5). Wind loads must be determined based on these requirements.

It is evident from Table 5.5 that for C&C, open buildings, and low-rise buildings, wind loads are determined using the upwind exposure for the single surface roughness in one of the eight sectors that gives the greatest wind forces (see Figure 5.7). In all other cases, an exposure category is determined in each wind direction based on the upwind surface roughness, and wind loads in those directions are determined using the corresponding exposure categories.

It is common practice to use only the critical exposure category obtained from all wind directions when determining the wind loads on the MWFRS of a building or other structure; in general, this reduces the required calculations and typically yields results that are not unduly conservative.

5.2.5 Topographic Effects

Buildings or other structures sited on the upper half of an isolated hill, ridge, or escarpment can experience significantly higher wind velocities than those sited on relatively level ground. The topographic factor, K_{zt}, in ASCE/SEI 26.8 accounts for this increase in wind speed, which is commonly referred to as wind speed-up.

Table 5.5 Exposure Requirements

Wind Load Procedure	ASCE/SEI 7 Chapter	Requirements
Directional	27	• MWFRS of enclosed and partially enclosed buildings Use an exposure category determined in accordance with ASCE/SEI 26.7.3 in each wind direction considered • Open buildings with monoslope, pitched, or troughed free roofs Use the exposure category determined in accordance with ASCE/SEI 26.7.3 from the eight sectors resulting in the greatest wind loads for any wind direction at the site
Envelope	28	• MWFRS of all low-rise buildings designed using this procedure Use the exposure category determined in accordance with ASCE/SEI 26.7.3 from the eight sectors resulting in the greatest wind loads for any wind direction at the site
Directional	29	• Building appurtenances and other structures Use an exposure category determined in accordance with ASCE/SEI 26.7.3 for each wind direction considered
C&C	30	• C&C Use the exposure category determined in accordance with ASCE/SEI 26.7.3 from the eight sectors resulting in the greatest wind loads for any wind direction at the site

A two-dimensional ridge or escarpment or a three-dimensional axisymmetrical hill is described by the parameters H and L_h:

- H is the height of the hill, ridge, or escarpment relative to the upwind terrain.
- L_h is the distance upwind of the crest to where the difference in ground elevation is equal to one-half the height of the hill, ridge, or escarpment (see ASCE/SEI Figure 26.8-1).

Not every hill, ridge, or escarpment requires an increase in wind velocity. Wind speed-up must be considered only when all three conditions in ASCE/SEI 26.8.1 are satisfied:

1. The building or other structure is located as shown in ASCE/SEI Figure 26.8-1 in the upper one-half of a hill or ridge or near the crest of an escarpment.
2. $H/L_h \geq 0.2$
3. $H \geq 15$ ft (4.57 m) for exposure C and D and 60 ft (18.3 m) for exposure B.

Where all three conditions are met, K_{zt} is determined by ASCE/SEI Equation (26.8-1):

$$K_{zt} = (1 + K_1 K_2 K_3)^2 \tag{5.4}$$

Values of the multipliers K_1, K_2, and K_3 are given in ASCE/SEI Figure 26.8-1. It is assumed wind approaches the topographic feature along the direction of maximum slope, which causes the greatest increase in velocity at the crest. The multiplier K_1 accounts for the shape of the topographic feature and the maximum speed-up effect. The reduction in wind speed-up with respect to horizontal distance is accounted for in the multiplier K_2: values of K_2 decrease as the distance from the topographic feature increases (i.e., as x increases). Finally, the multiplier K_3 accounts for the reduction of wind speed-up with respect to height above the local terrain: K_3 decreases as the height above the windward terrain, z, increases.

When one or more of these conditions are not met, $K_{zt} = 1.0$ (ASCE/SEI 26.8.2).

The provisions in ASCE/SEI 26.8 are not meant to address the general case of wind flowing over hilly or complex terrain. In such cases, a wind tunnel study should be performed. Also, these provisions do not include vertical wind speed-up, which is known to exist in such cases; vertical effects determined by rational methods should be included where appropriate.

5.2.6 Ground Elevation Effects

The ground elevation factor, K_e, adjusts the velocity pressure, q_z, determined in accordance with ASCE/SEI 26.10 based on the reduced mass density of air at elevations above sea level. Values of K_e are given in ASCE/SEI Table 26.9-1; these values can be determined by the equations in note 2 of the table where z_e is the ground elevation above sea level:

$$K_e = e^{-0.0000362 z_e} \ (z_e = \text{ground elevation above sea level in ft}) \tag{5.5}$$

$$K_e = e^{-0.000119 z_e} \ (z_e = \text{ground elevation above sea level in m}) \tag{5.6}$$

Values of K_e including the air density, ρ, are given in ASCE/SEI Table C26.9-1.

The constant 0.00256 in ASCE/SEI Equation (26.10-1) or the constant 0.613 in ASCE/SEI Equation (26.10-1.SI) for q_z is used to convert a wind speed pressure based on the mass density of air for the standard atmosphere which is defined as a temperature of 59°F (15°C) and a sea level pressure of 29.92 inches (101.325 kPa) of mercury (see Section 5.2.7 of this book). Values of air density other than the standard atmosphere values are adjusted using K_e. It is permitted to take $K_e = 1.0$ for all elevations, which is conservative except for elevations below sea level; however, using $K_e = 1.0$ to calculate q_z for all areas below sea level in the United States is not unconservative.

5.2.7 Velocity Pressure

Velocity pressure is related to the atmospheric effects due to wind which are described in Section 5.1.1 of this book. The velocity pressure q_z at height z above the ground surface is determined by ASCE/SEI Equations (26.10-1) and (26.10-1.SI); these equations are essentially Bernoulli's equation, and it converts the basic wind speed, V, to a velocity pressure:

$$q_z = 0.00256 K_z K_{zt} K_e V^2 \quad (\text{lb/ft}^2) \tag{5.7}$$

$$q_z = 0.613 K_z K_{zt} K_e V^2 \quad (\text{N/m}^2) \tag{5.8}$$

At the mean roof height of a building or other structure, the velocity pressure is denoted q_h and the velocity pressure coefficient is denoted K_h, that is, the subscript changes from z to h.

The constant 0.00256 in Equation (5.7) and the constant 0.613 in Equation (5.8) are related to the mass density of air for the standard atmosphere (59°F [15°C] and a sea level pressure of 29.92 inches [101.325 kPa] of mercury), which is equal to 0.00238 slug/ft³ or 1.225 kg/m³ from ASCE/SEI Table C26.9-1 at ground elevation $z_e = 0$ ft (0 m). Thus, using Bernoulli's law for dynamic pressure, the constant is equal to one-half times the density of air times the velocity squared where the velocity is in miles per hour (meters per second) and the pressure is in pounds per square foot (Newtons per square meter):

$$\text{Constant} = \frac{1}{2} \times 0.00238 \frac{\text{slug}}{\text{ft}^3} \times \left(1 \frac{\text{mi}}{\text{hr}} \times 5{,}280 \frac{\text{ft}}{\text{mi}} \times \frac{1 \text{ hr}}{3{,}600 \text{ sec}}\right)^2 = 0.00256 \tag{5.9}$$

$$\text{Constant} = \frac{1}{2} \times 1.225 \frac{\text{kg}}{\text{m}^3} \times \frac{1 \text{ N}}{1 \frac{\text{kg m}}{\text{s}^2}} \times \left(\frac{1 \text{ m}}{\text{s}}\right)^2 = 0.613 \tag{5.10}$$

In Equations (5.7) and (5.8), K_z is the velocity pressure exposure coefficient given in ASCE/SEI Table 26.10-1; this coefficient modifies wind velocity (or wind pressure) with respect to height above ground and exposure. Values of K_z can be determined by the equations in note 1 at the bottom of ASCE/SEI Table 26.10-1:

$$K_z = \begin{cases} 2.41\left(\dfrac{15}{z_g}\right)^{2/\alpha} & \text{for } z < 15 \text{ ft} \\ 2.41\left(\dfrac{z}{z_g}\right)^{2/\alpha} & \text{for } 15 \text{ ft} \le z \le z_g \\ 2.41 & \text{for } z_g < z \le 3{,}280 \text{ ft} \end{cases} \tag{5.11}$$

In S.I.:

$$K_z = \begin{cases} 2.41\left(\dfrac{4.6}{z_g}\right)^{2/\alpha} & \text{for } z < 4.6 \text{ m} \\ 2.41\left(\dfrac{z}{z_g}\right)^{2/\alpha} & \text{for } 4.6 \text{ m} \le z \le z_g \\ 2.41 & \text{for } z_g < z \le 1{,}000 \text{ m} \end{cases} \tag{5.12}$$

Values of α and z_g are given in ASCE/SEI Table 26.11-1 based on exposure. The constant α is the 3-second gust speed power law exponent which defines the approximately parabolic shape of the wind speed profile for each exposure (see Figure 5.2). The nominal height of the atmospheric boundary layer, which is also referred to as the gradient height, is denoted as z_g. At height z_g, the value of K_z is limited to a maximum of 2.41 for a given exposure. Above this height, K_z is assumed to remain constant (see the discussion in Section 5.1.1 of this book). The pressure for each exposure is identical when it is evaluated at the top of the boundary layer, and with everything else being constant, it is equal to 2.41 times the stagnation pressure of $0.00256V^2$ (in S.I.: $0.613V^2$).

The above discussion on the determination of K_z is valid for the case of a single roughness category (i.e., uniform terrain). Procedures on how to determine K_z for a single roughness change or multiple roughness changes are given in ASCE/SEI C26.10.1.

Flowchart 5.1 in Section 5.7 of this book can be used to determine q_z and q_h.

5.2.8 Gust Effects

Gust-Effect Factors

The effects of wind gusts must be included in the design of any building or other structure (see Section 5.1.2 of this book). The gust-effect factors G and G_f for rigid and flexible structures, respectively, defined in ASCE/SEI 26.11 account for both atmospheric and aerodynamic effects in the along-wind direction.

The gust-effect factor depends on the natural frequency, n_1, of the building or other structure. The method in which the gust-effect is determined is contingent on whether the structure is rigid or flexible. By definition, a rigid building or other structure is one where $n_1 \ge 1$ Hz, and a flexible building or other structure is one where $n_1 < 1$ Hz (ASCE/SEI 26.2). Buildings meeting the conditions for a low-rise building in ASCE/SEI 26.2 (i.e., the building is enclosed, partially enclosed, or partially open with a mean roof height, h, less than or equal to 60 ft (18.3 m) and with h less than or equal to the least horizontal dimension) are permitted to be considered rigid (ASCE/SEI 26.11.2).

For rigid structures, the gust-effect factor, G, is permitted to be taken as 0.85 (ASCE/SEI 26.11.4) or may be determined using the provisions of ASCE/SEI 26.11.4. The gust-effect factor for flexible structures, G_f, accounts for along-wind loading effects due to dynamic amplification; provisions to determine G_f are given in ASCE/SEI 26.11.5. The fundamental dynamic equations for maximum along-wind displacement, root-mean-square along-wind acceleration, and maximum along-wind acceleration, which form the basis of the methodology for determining the gust-effect factor, can be found in ASCE/SEI C26.11. Also provided in this commentary section is a discussion on structural damping; recommendations on damping ratios, which are needed to determine G_f, are also included.

Any rational analysis recognized in the recognized literature to determine gust-effect factors and natural frequencies is permitted to be used (ASCE/SEI 26.11.6).

Flowchart 5.2 in Section 5.7 of this book can be used to determine the gust-effect factors for both rigid and flexible structures.

Approximate Natural Frequencies

During preliminary design, a sufficient amount of information about a building may not be available to determine n_1 using computer software. An approximate natural frequency, n_a, can be determined for structural steel, concrete, or masonry buildings using the provisions in ASCE/SEI 26.11.3 provided the height and slenderness conditions in ASCE/SEI 26.11.2.1 are satisfied:

1. Building height must be less than or equal to 300 ft (91.4 m).
2. The building height must be less than four times its effective length, L_{eff}, which is determined by ASCE/SEI Equation (26.11-1):

$$L_{eff} = \frac{\sum_{i=1}^{n} h_i L_i}{\sum_{i=1}^{n} h_i} \tag{5.13}$$

In this equation, h_i is the height above grade of level i, L_i is the building length at level i parallel to the wind direction, and n is the number of levels. Where the plan dimension in the direction of analysis does not change over the height, L_{eff} is equal to the plan dimension of the building in that direction.

For buildings satisfying the conditions in ASCE/SEI 26.11.2.1, the equations in ASCE/SEI 26.11.3 are permitted to be used to determine the approximate natural frequency, n_a (see Table 5.6).

In the equations in Table 5.6, h is the mean roof height of the building. The terms in ASCE/SEI Equations (26.11-5) and (26.11-5.SI) for concrete or masonry shear walls are defined in ASCE/SEI 26.11.3: n is the number of shear walls in the building in the direction of analysis, A_B is the base area of the building, A_i is the horizontal cross-sectional area of shear wall i, D_i is the length of shear wall i, and h_i is the height of shear wall i.

The lower-bound expressions for n_a in Table 5.6 are more appropriate than the empirical equations for natural frequency given in the seismic provisions of ASCE/SEI 7 because the latter provide higher estimates of natural frequency, which are conservative for the determination of seismic loads but are generally unconservative for the determination of wind loads.

Additional equations for natural frequency appropriate for wind design are given in ASCE/SEI C26.11.

5.2.9 Enclosure Classifications

The different types of enclosure classifications and their relationship to internal pressures are given under aerodynamic effects in Section 5.1.1 of this book. The following discussion covers definitions for each type of classification and the requirements for protecting glazed openings in wind-borne debris regions.

Table 5.6 Approximate Natural Frequency, n_a

MWFRS	ASCE/SEI Eq. No.	Equations for n_a: h in ft	Equations for n_a: h in m
Structural steel moment-resisting frame buildings	26.11-2 26.11-2.SI	$n_a = \frac{22.2}{h^{0.8}}$	$n_a = \frac{8.58}{h^{0.8}}$
Concrete moment-resisting frame buildings	26.11-3 26.11-3.SI	$n_a = \frac{43.5}{h^{0.9}}$	$n_a = \frac{14.93}{h^{0.9}}$
Structural steel and concrete buildings with lateral force-resisting systems other than moment-resisting frames	26.11-4 26.11-4.SI	$n_a = \frac{75}{h}$	$n_a = \frac{22.86}{h}$
Concrete or masonry shear wall buildings	26.11-5 26.11-5.SI	$n_a = \frac{385(C_w)^{0.5}}{h}$	$n_a = \frac{117.3(C_w)^{0.5}}{h}$
		where $C_w = \frac{100}{A_B}\sum_{i=1}^{n}\left(\frac{h}{h_i}\right)^2 \frac{A_i}{\left[1+0.83\left(\frac{h_i}{D_i}\right)^2\right]}$	

Buildings must be classified as enclosed, partially enclosed, partially open, or open in accordance with the definitions in ASCE/SEI 26.2, which are given in Table 5.7 (ASCE/SEI 26.12.1). The enclosure classification is needed in determining internal pressure coefficients.

Table 5.7 Enclosure Classifications

Classification	Definition
Enclosed building	A building complying with the following for each wall receiving positive external pressure: $A_o \leq \text{lesser of} \begin{cases} 0.01A_g \\ 4 \text{ ft}^2 \ (0.37 \text{ m}^2) \end{cases}$
Partially enclosed building	A building complying with both of the following conditions: • $A_o > \begin{cases} 1.10A_{oi} \\ \text{lesser of} \begin{cases} 4 \text{ ft}^2 \ (0.37 \text{ m}^2) \\ 0.01A_g \end{cases} \end{cases}$ • $A_{oi}/A_{gi} \leq 0.20$
Partially open building	A building not complying with the requirements for open, partially enclosed, or enclosed buildings
Open building	For each wall in the building, the following is satisfied: $A_o \geq 0.8A_g$

The parameters in Table 5.7 are defined as follows (see Figure 5.9):

- A_o = total area of openings in a wall receiving positive external pressure, ft^2 (m^2)
- A_g = gross area of wall in which A_o is identified, ft^2 (m^2)

Figure 5.9
Definition of Wall Openings for Determination of Enclosure Classification

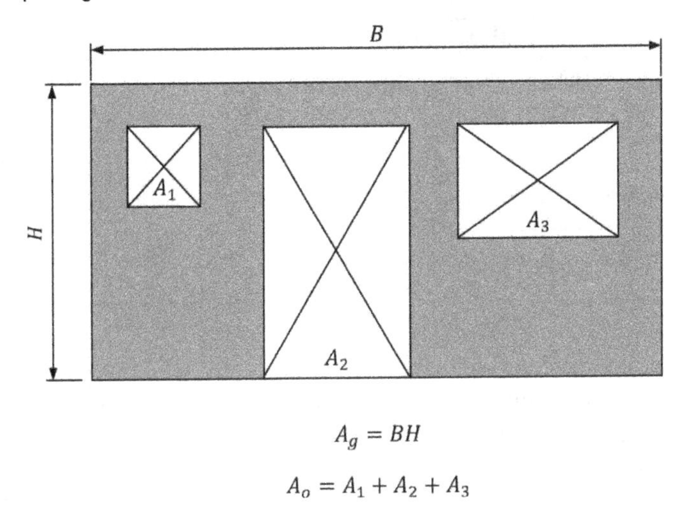

$$A_g = BH$$

$$A_o = A_1 + A_2 + A_3$$

- A_{oi} = sum of the areas of the openings in the building envelope (walls and roof) not including A_o, ft^2 (m^2)
- A_{gi} = sum of the gross surface areas of the building envelope (walls and roof) not including A_g, ft^2 (m^2)

Building satisfying both open and partially enclosed definitions must be classified as partially open (ASCE/SEI 26.12.1). Also, as noted in Table 5.7, a building must be classified as partially open where the definitions of open, partially enclosed, or enclosed are not met.

An opening is defined as a hole that allows air to flow through a building envelope during a design wind event. Air intake and exhaust vents for ventilation should always be considered openings. Doors, windows, skylights, and flexible and operable louvers likely to be closed during a design wind event need not be considered as openings. If a door, for example, must remain open during a design wind event because of the intended function of the building (e.g., a door for a hospital ambulance bay), the door should be considered an opening.

It is common practice in regions not prone to hurricanes to assume windows and other glazed units are not openings provided they have been designed for the appropriate C&C wind pressures at a particular site.

Hurricane-prone regions are located along the U.S. Atlantic Ocean and Gulf of Mexico coasts where $V > 115$ mi/h (51.4 m/s) for risk category II buildings. Hawaii, Puerto Rico, Guam, U.S. Virgin Islands, Northern Mariana Islands, and American Samoa are also designated hurricane-prone regions.

Wind-borne debris regions are in hurricane-prone regions at the following locations (IBC 202 and ASCE/SEI 26.12.3.1):

- Within 1 mile (1.6 km) of the mean high-water line where an exposure D condition exists upwind of the waterline and the basic wind speed, V, is greater than or equal to 130 mi/h (58 m/s).
- In areas where the basic wind speed, V, is greater than or equal to 140 mi/h (63 m/s).

The figures in the IBC and ASCE/SEI 7 to be used in determining wind-borne debris regions based on building classification are given in Table 5.8.

Special requirements are given in IBC 1609.2 and ASCE/SEI 26.12.3.2 for the protection of glazed openings in wind-borne debris regions. With a few exceptions (see ASCE/SEI 26.12.3.1),

Table 5.8 Figures in the IBC and ASCE/SEI 7 to be Used in Determining Wind-borne Debris Regions

Classification	Figure Nos.
• Risk category II buildings and other structures • Risk category III buildings and other structures, except health-care facilities	IBC Figure 1609.3(2) ASCE/SEI Figure 26.5-1B
Risk category III health-care facilities	IBC Figure 1609.3(3) ASCE/SEI Figure 26.5-1C
Risk category IV buildings and other structures	IBC Figure 1609.3(4) ASCE/SEI Figure 26.5-1D

all glazing must be protected with an impact-protective system (such as shutters or screens) or the glazing itself must be impact-resistant. The ASTM standards in ASCE/SEI 26.12.3.2 specify the tests that must be performed to determine the suitability of protective systems and glazing. Glazing and impact-protective systems in buildings and other structures classified as risk category IV must comply with enhanced protection requirements (see ASCE/SEI 26.12.3.2). Impact-protective systems and impact-resistant glazing that have passed the specified tests are intended to ensure the glazing will not be breached during a hurricane event.

Exception 3 in IBC 1609.2 and the exception in ASCE/SEI 26.12.3.1 permit glazing to be unprotected if it is located more than 60 ft (18.3 m) above the ground and more than 30 ft (9.14 m) above aggregate-surfaced roofs (including roofs with gravel or stone ballast) located within 1,500 ft (457 m) of the building.

5.2.10 Internal Pressure Coefficients

Internal pressure coefficients, (GC_{pi}) are given in ASCE/SEI Table 26.13-1 and are based on the enclosure classifications defined in ASCE/SEI 26.12 (see Table 5.9). These coefficients have been obtained from wind tunnel tests and full-scale data and are assumed to be valid for buildings of any height even though the wind tunnel tests were conducted primarily for low-rise buildings. Gust and aerodynamic effects are combined into one factor (GC_{pi}); therefore, the gust-effect factor must not be determined separately in the analysis (ASCE/SEI 26.11.7).

Table 5.9 Internal Pressure Coefficients, (GC_{pi}), in Accordance with ASCE/SEI 26.13

Enclosure Classification in Accordance with ASCE/SEI 26.12	(GC_{pi})
Enclosed buildings	+0.18 −0.18
Partially enclosed buildings	+0.55 −0.55
Partially open buildings	+0.18 −0.18
Open buildings	0.00

The value of (GC_{pi}) for partially enclosed buildings is approximately 3 times that for enclosed and partially open buildings. Determining the correct enclosure classification is very important to capture the appropriate effect of internal pressure on the design of applicable members in a building.

For partially enclosed buildings that contain a single, relatively large volume without any partitions, the reduction factor, R_i, determined by ASCE/SEI Equation (26.13-1) is permitted to be used to reduce the applicable internal pressure coefficient. This reduction factor is based on research that has shown the response time of internal pressure increases as the volume of a building without partitions increases; as such, the gust factor associated with the internal pressure is reduced, resulting in lower internal pressure.

5.3 Wind Loads on Main Wind Force Resisting Systems

5.3.1 Overview

Design requirements for determining wind pressures and loads on MWFRSs of buildings, building appurtenances and other structures are given in ASCE/SEI Chapters 27, 28, 29, and 31 (see Table 5.1). The provisions in ASCE/SEI Chapters 27 through 29 are discussed in the following sections. ASCE/SEI Chapter 31, which contains the requirements for wind tunnel procedures, is covered in Section 5.5 of this book.

5.3.2 Directional Procedure for Buildings (ASCE/SEI Chapter 27)

Scope

The Directional Procedure of ASCE/SEI Chapter 27 applies to the determination of wind loads on the MWFRS of enclosed, partially enclosed, partially open, and open buildings of all heights that meet the conditions and limitations given in ASCE/SEI 27.1.2 and ASCE/SEI 27.1.3, respectively.

A wide range of buildings is covered by the provisions in ASCE/SEI Chapter 27. In general, wind pressures are determined as a function of wind direction using equations appropriate for each surface of a building.

The provisions in ASCE/SEI Chapter 27 are applicable to buildings that comply with the following conditions (ASCE/SEI 27.1.2):

- The building is regular-shaped, that is, the building has no unusual geometrical irregularities in spatial form.
- The building does not have response characteristics that make it subject to across-wind loading, vortex shedding, or instability caused by galloping or flutter. In addition, the building is not located at a site where channeling effects or buffeting, in the wake of upwind obstructions, warrant special consideration.

Buildings not meeting these conditions must be designed by either recognized literature that documents such wind load effects or by the wind tunnel procedure in ASCE/SEI Chapter 31 (ASCE/SEI 27.1.3).

Reduction in wind pressure due to apparent shielding by surrounding buildings, other structures, or terrain features is not permitted (ASCE/SEI 27.1.4). Such shielding may be modified or completely removed during the lifespan of the building, which could result in significantly higher wind loads.

Minimum design wind pressures and loads are given in ASCE/SEI 27.1.5. For enclosed and partially enclosed buildings, the minimum wind loads to be used in the design of the MWFRS are equal to 16 lb/ft^2 (0.77 kN/m^2) multiplied by the wall area of the building and 8 lb/ft^2 (0.38 kN/m^2) multiplied by the roof area of the building projected onto a vertical plane normal to the wind direction. The minimum wall and roof wind loads are to be applied simultaneously. Application of minimum wind pressures are illustrated in Figure 5.10 for wind along the two primary axes of a building. For open buildings, the minimum wind load is equal to 16 lb/ft^2 (0.77 kN/m^2) multiplied by the area of the building either normal to the wind direction or projected on a plane normal to the wind direction. The application of minimum design wind pressures and loads on a building is a separate load case that must be considered in addition to the other load cases specified in ASCE/SEI Chapter 27.

Figure 5.10 Application of Minimum Design Wind Pressures in Accordance with ASCE/SEI 27.1.5

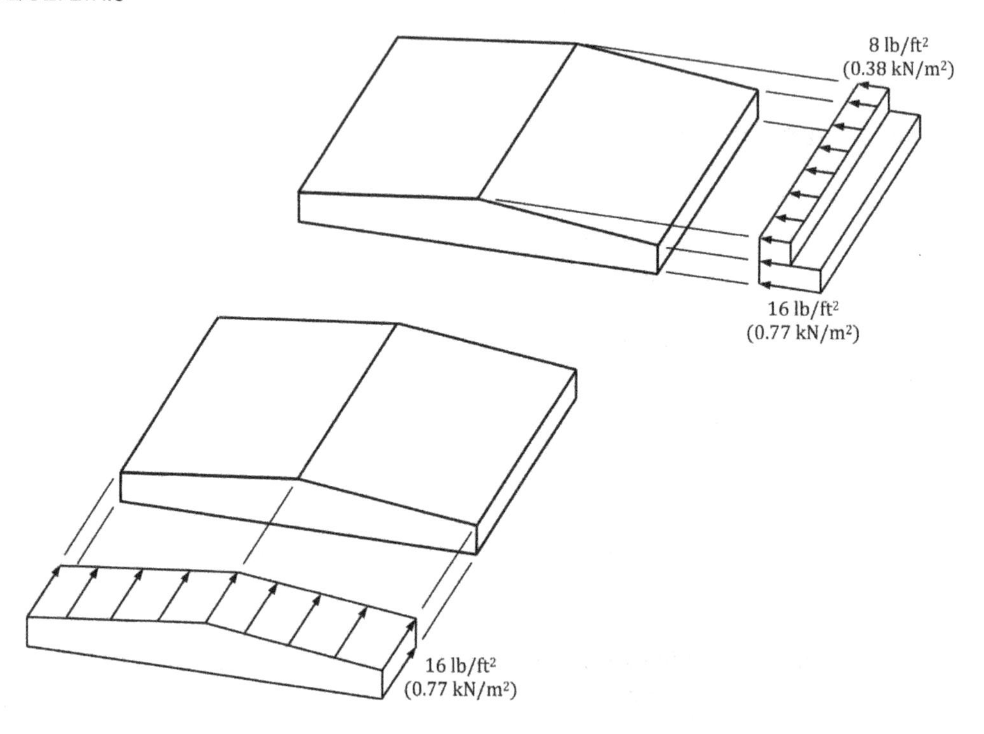

Enclosed, Partially Enclosed, and Partially Open Rigid and Flexible Buildings

Overview

The requirements in ASCE/SEI Chapter 27 are applicable to buildings with any general plan shape, height, or roof geometry that matches the figures provided in this chapter. This procedure entails the determination of velocity pressures (which are determined as a function of exposure, height, topographic effects, wind directionality, wind velocity, ground elevation, and building occupancy), gust-effect factors, external pressure coefficients, and internal pressure coefficients for each surface of a building. The steps to determine wind loads on such buildings are given in ASCE/SEI Table 27.2-1.

Design Wind Pressures

Design wind pressures, p, are determined by ASCE/SEI Equation (27.3-1) for the MWFRS of enclosed, partially enclosed, and partially open rigid and flexible buildings of all heights:

$$p = qK_dGC_p - q_iK_d(GC_{pi}) \tag{5.14}$$

This equation is used to calculate wind pressures on each surface of the building: windward wall, leeward wall, side walls, and roof. The pressures are applied simultaneously on the walls and roof, as depicted in ASCE/SEI Figure 27.3-1 (see Figure 5.11 for the application of external wind pressures on a building with a gable or hip roof).

The first part of Equation (5.14) is the external pressure contribution and the second part is the internal pressure contribution. External pressure varies with height above ground on the windward wall and is a constant on all of the other surfaces based on the mean roof height. The terms in this equation are discussed below.

Velocity pressure, q. Velocity pressure, q, which is determined in accordance with ASCE/SEI 26.10 (see Section 5.2.7 of this book), varies with respect to height on windward walls ($q = q_z$). For leeward walls, sidewalls, and roofs, the velocity pressure is a constant and is evaluated at the mean roof height, h ($q = q_h$).

Figure 5.11 Application of External Wind Pressures on a Building with a Gable or Hip Roof in Accordance with ASCE/SEI 27.3

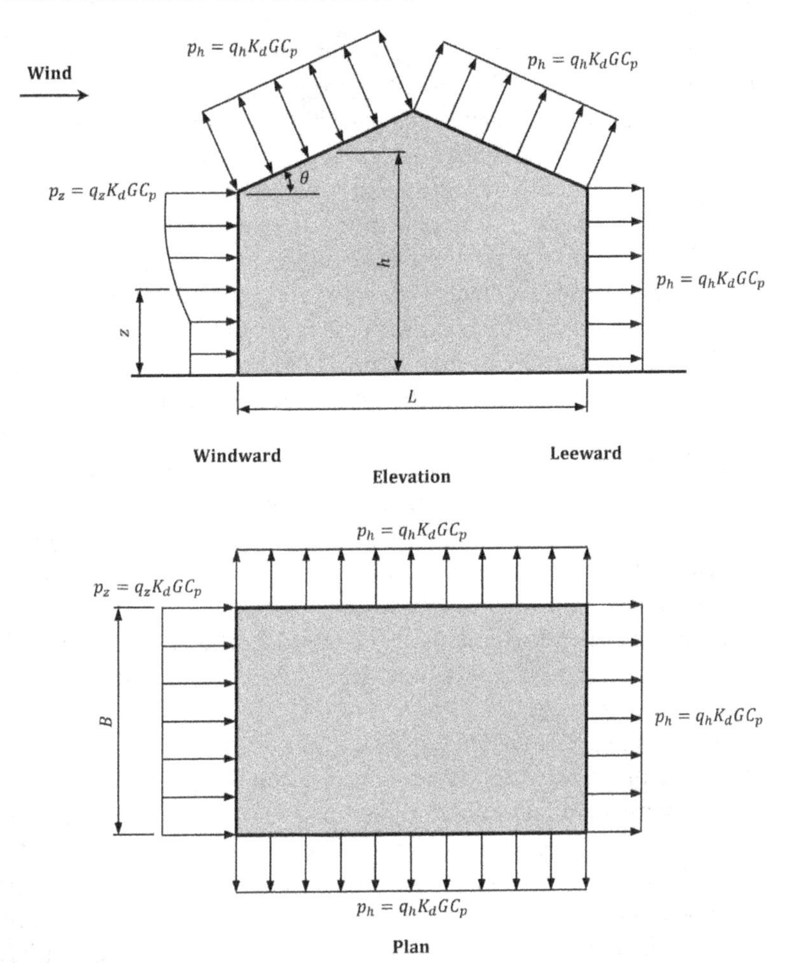

Surface	Pressure coefficient, C_p	Velocity pressure, q	Gust-effect factor	Wind pressure, p
Windward wall	0.8	$q_z = 0.00256 K_z K_{zt} K_e V^2$ (lb/ft^2) $q_z = 0.613 K_z K_{zt} K_e V^2$ (N/m^2)	• For rigid buildings, use G (ASCE/SEI 26.11.4)	$p_z = q_z K_d G C_p$
Leeward wall	L/B — C_p $0-1$ — -0.5 2 — -0.3 ≥ 4 — -0.2	$q_h = 0.00256 K_h K_{zt} K_e V^2$ (lb/ft^2) $q_h = 0.613 K_h K_{zt} K_e V^2$ (N/m^2)	• For flexible buildings, use G_f (ASCE/SEI 26.11.5)	$p_h = q_h K_d G C_p$
Side walls	-0.7			
Windward roof	See ASCE/SEI Fig. 27.3-1			
Leeward roof				

Directionality factor, K_d. The directionality factor, K_d, accounts for the statistical nature of wind flow and the probability of the maximum effects occurring at any particular time for any given wind direction (ASCE/SEI 26.6; see Section 5.2.3 of this book). This factor is equal to 0.85 for the MWFRS of buildings.

Gust-effect factor, G. The gust-effect factor, G, for rigid buildings is permitted to be taken equal to 0.85 or may be determined by ASCE/SEI Equation (26.11-6) [see Section 5.2.8 of this book]. For flexible buildings, G_f is determined in accordance with ASCE/SEI 26.11.5 and must be used in Equation (5.14) instead of G.

External pressure coefficients, C_p. External pressure coefficients, C_p, capture the aerodynamic effects discussed in Section 5.1.1 of this book, and have been determined experimentally through wind tunnel tests on buildings of various shapes and sizes. These coefficients reflect the actual wind loading on each surface of a building as a function of wind direction.

- Buildings with walls and flat, gable, hip, monoslope, or mansard roofs.
  External pressure coefficients, C_p, are given in ASCE/SEI Figure 27.3-1 for the wall and roof surfaces of enclosed, partially enclosed, and partially open buildings with flat, gable, hip, monoslope, or mansard roofs. Wall pressure coefficients are constant on windward and side walls and vary with the aspect ratio of the building, L/B, on the leeward wall where L is the plan dimension of the building parallel to the wind direction and B is the plan dimension of the building perpendicular to the wind direction.
  Roof pressure coefficients vary with the ratio of the mean roof height to the plan dimension of the building (h/L) and with the roof angle (θ) for a given wind direction (normal to the ridge or parallel to the ridge). These pressure coefficients are intended to be used with q_h, and the parallel to ridge wind direction is applicable for flat roofs. Negative roof pressures increase as h/L increases. Also, as θ increases, negative pressure decreases until a roof angle is reached where the pressure becomes positive. Where two values of C_p are listed in the figure, the windward roof is subjected to either positive or negative pressure and the building must be designed for both. Other essential information on the use of ASCE/SEI Figure 27.3-1 is given in the notes below the tabulated pressure coefficients.
- Domed roofs with a circular base and arched roofs.
  External pressure coefficients for enclosed, partially enclosed, and partially open domed roofs with a circular base and arched roofs are given in ASCE/SEI Figures 27.3-2 and 27.3-3, respectively.
  Two load cases must be considered for domed roofs. In case A, pressure coefficients are determined between various locations on the dome by linear interpolation along arcs of the dome parallel to the direction of wind; this defines maximum uplift on the dome in many cases. In case B, the pressure coefficient is assumed to be a constant value at a specific point on the dome for angles less than or equal to 25 degrees and is determined by linear interpolation from 25 degrees to other points on the dome.
  Wind tunnel tests are recommended for domes larger than 200 ft (61.0 m) in diameter and in cases where resonant response can be an issue (see ASCE/SEI C27.3.1).

Velocity pressure for internal pressure determination, q_i. The velocity pressure for internal pressure determination, q_i, is determined as follows:

- For windward walls, sidewalls, leeward walls, and roofs of enclosed buildings and partially open buildings, and for negative internal pressure evaluation in partially enclosed buildings: $q_i = q_h$.
- For positive internal pressure evaluation in partially enclosed buildings: $q_i = q_z$ where height z is defined as the level of the highest opening in the building that could affect the positive internal pressure. For buildings in wind-borne debris regions, the enclosure classification with respect to glazed openings must be in accordance with ASCE/SEI 26.12.3. For positive internal pressure evaluation, q_i may conservatively be evaluated at height h ($q_i = q_h$).

Internal pressure coefficients, (GC_{pi}). Internal pressure coefficients, (GC_{pi}), are determined in accordance with ASCE/SEI 26.13 (see Section 5.2.10 of this book). Both positive and negative values of (GC_{pi}) must be considered to establish the critical load effects.

The effects from internal pressure cancel out when evaluating the total horizontal wind pressure on the MWFRS of a building. Thus, the total horizontal pressure at any height z above ground in the direction of wind is equal to the external pressure p_z on the windward face at height z plus the external pressure p_h on the leeward face (see Figure 5.11).

Design Wind Pressures on Elevated Buildings

Requirements for buildings elevated on structural members that allow wind to pass beneath the building are given in ASCE/SEI 27.3.1.1 and are applicable to typical elevated beach houses as well as taller elevated buildings. These requirements apply in any principal wind direction where both of the following two geometric limitations are satisfied:

- The ratio of the sum of the cross-sectional areas of the supporting structural members (such as columns, piers, or walls) and any enclosed or partially enclosed space beneath the building (such as the space enclosing the elevators or stairways in the building core) and the plan cross-sectional area of the building above is less than or equal to the percentage of the plan area of the building blocked by elements or enclosures in ASCE 27.3.1.1(1), which is a function of the ratio of the plan dimensions of the building, L/B. The cross-sectional areas of the supporting structural members and the enclosed space are determined based on a horizontal cross-section through the space under the building.
- The ratio of the sum of the areas of the supporting structural elements and any enclosed, partially enclosed, or partially open spaces projected onto a vertical plane normal to the wind direction and the projected area outlined by the height of these elements and the projected width of the building above is less than or equal to 75 percent.

For buildings that satisfy these two conditions in a principal wind direction, the MWFRS must be designed for a combination of the following wind load cases:

- Wind loads on the elevated building (ASCE/SEI 27.3.1.1.1)
  Wind loads on the roof and walls of the elevated building must be determined using the methods and pressure coefficients in ASCE/SEI 27.3.1.
- Wind loads on elements below the elevated building
  Lateral wind loads on elements below the elevated building must be determined using all of the following:
  1. Projected areas of all structural elements below the elevated building perpendicular to the considered wind direction, neglecting potential shielding.
  2. A net force coefficient of 1.3 applied to the projected area of all the elements.
  3. Velocity pressure, q_z, where z is equal to the sum of the height above grade of the bottom horizontal surface of the elevated portion of the building and 25 percent of the distance from the bottom of the elevated portion of the building to the mean roof height, h.

Examples of dimensions and heights used in ASCE/SEI 27.3.1.1 for high-rise and low-rise buildings are given in ASCE/SEI Figure C27.3-1.

Use of a net force coefficient of 1.3 includes both windward and leeward pressures on the elements and is conservative for L/B ratios of about 2.5. Additional information on the applicability of this force coefficient is given in ASCE/SEI C27.3.1.1.

Vertical wind loads on the bottom surface of the elevated building must be determined using the following:

1. Roof pressure coefficients for the wind direction "normal to the ridge for $\theta < 10°$ and parallel to the ridge for all θ" given in ASCE/SEI Figure 27.3-1. Negative roof pressure coefficients in the figure denote downward loading on the bottom floor assembly and positive pressure coefficients denote upward loading.
2. The value of h to be used in determining the pressure coefficients, $C_{p,}$ in the regions from the windward edge indicated in ASCE/SEI Figure 27.3-1 for the wind direction in item 1 above must be taken as the height of the top of the elements below the elevated building.

3. Velocity pressure, q_z, where z is equal to the sum of the height above grade of the bottom horizontal surface of the elevated portion of the building and 25 percent of the distance from the bottom of the elevated portion of the building to the mean roof height, h.

For buildings that do not satisfy one or both of the conditions in ASCE/SEI 27.3.1.1 in a principal wind direction, the MWFRS must be designed for the wind loads in ASCE/SEI 27.3.1 assuming the wind is unable to pass beneath the building and that portion of the building has no openings.

The requirements in ASCE/SEI 27.3.1.1 are given in Figure 5.12.

Figure 5.12 Requirements for Wind Loads on Elevated Buildings in Accordance with ASCE/SEI 27.3.1.1

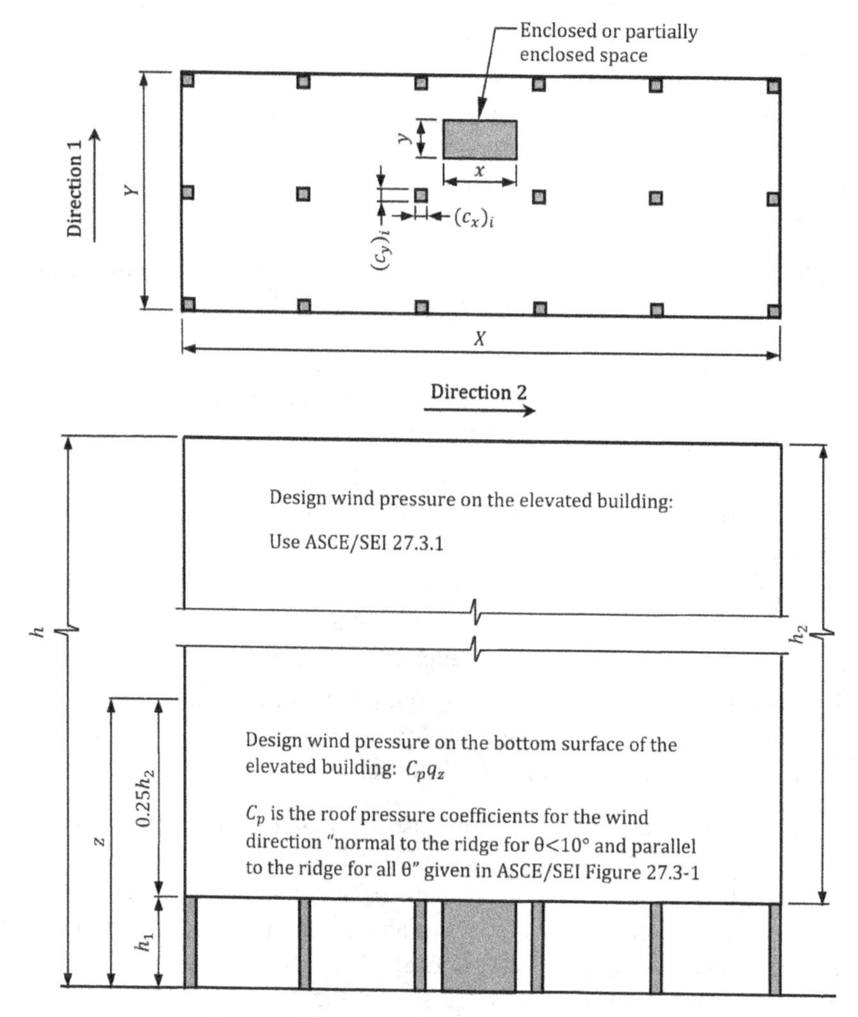

Open Buildings with Monoslope, Pitched, or Troughed Free Roofs

Design Wind Pressures, p

Net design pressure, p, for the MWFRS of open buildings with monoslope, pitched, or troughed free roofs is determined by ASCE/SEI Equation (27.3-2):

$$p = q_h K_d G C_N \tag{5.15}$$

In this equation, q_h is the velocity pressure at the mean roof height, K_d is the wind directionality factor determined in accordance with ASCE/SEI Table 26.6-1, and G is the gust-effect factor determined in accordance with ASCE/SEI 26.11.

Net pressure coefficients, C_N, are given in ASCE/SEI Figures 27.3-4 through 27.3-7 for various roof configurations. Load cases A and B are identified in the figures. Both load cases must be considered to obtain the maximum load effects for a particular roof slope and blockage configuration.

The magnitude of roof pressure in open buildings is highly dependent on the blockage configuration beneath the roof. Goods or materials stored under the roof can restrict air flow which can introduce significant upward pressures on the bottom surface of the roof. The net pressure coefficients for clear wind flow are to be used in cases where blockage is less than or equal to 50 percent. Obstructed wind flow is applicable where the blockage is greater than 50 percent. In cases where the usage below the roof is not evident, both unobstructed and obstructed load cases should be investigated.

Additional Wind Pressures

For structures with free roofs containing fascia panels where the angle of the plane of the roof is less than or equal to 5 degrees, the facia panels must be considered an inverted parapet (ASCE/SEI 27.3.2). The contribution of the wind loads on the fascia to the wind loads on the MWFRS is to be determined using the requirements in ASCE/SEI 27.3.4 with q_p in ASCE/SEI Equation (27.3-3) taken as q_h.

An additional horizontal wind load must be applied to open or partially enclosed buildings with transverse frames and pitched roofs where the angle of the plane of the roof from the horizontal is less than or equal to 45 degrees. This force is to be determined using the provisions in ASCE/SEI 28.3.7 and must act in combination with the roof loads determined in accordance with ASCE/SEI 27.3.2. This horizontal force is applied to the building parallel to the ridge of the roof (see ASCE/Figure 28.3-3).

Roof Overhangs

The positive external pressure on the bottom surface of a windward roof overhang is equal to the following (ASCE/SEI 27.3.3):

$$p_z = q_z K_d G C_p \tag{5.16}$$

where p_z and q_z are evaluated at the top of the windward wall located a distance z above ground level and the pressure coefficient C_p is equal to 0.8, which is the pressure coefficient for the windward wall (see ASCE/SEI 27.3.3).

The positive pressure on the bottom surface of the overhang is combined with the wind pressure on the top surface determined in accordance with ASCE/SEI Figure 27.3-1 (see Figure 5.13).

Provisions are not provided for wind pressures on the bottom surface of a leeward overhang.

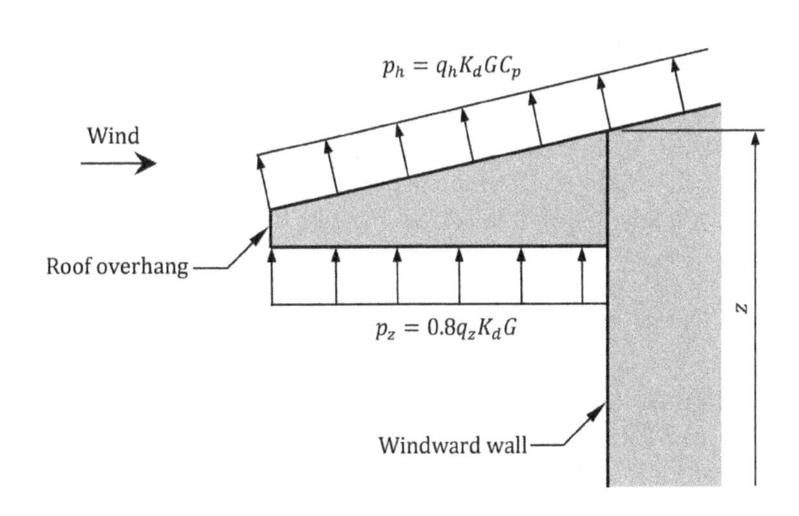

Figure 5.13 Wind Pressures on a Roof Overhang in Accordance with ASCE/SEI 27.3.3

Parapets

Design wind pressures for the effects of parapets, p_p, on the MWFRS of rigid or flexible buildings with flat, gable, or hip roofs are determined by ASCE/SEI Equation (27.3-3):

$$p_p = q_p K_d (GC_{pn}) \tag{5.17}$$

In this equation, q_p is the velocity pressure evaluated at the top of the parapet and (GC_{pn}) is the combined net pressure coefficient which is equal to +1.5 for a windward parapet and −1.0 for a leeward parapet.

As shown in Figure 5.14, a windward parapet experiences positive wall pressure on the exterior side (front surface) and negative roof pressure on the roof side (back surface). The behavior on the back surface is based on the assumption that the zone of negative pressure caused by the wind flow separation at the eave of the roof moves up to the top of the parapet, resulting in the back side of the parapet having the same negative pressure as the roof.

Figure 5.14 Wind Pressures on Parapets

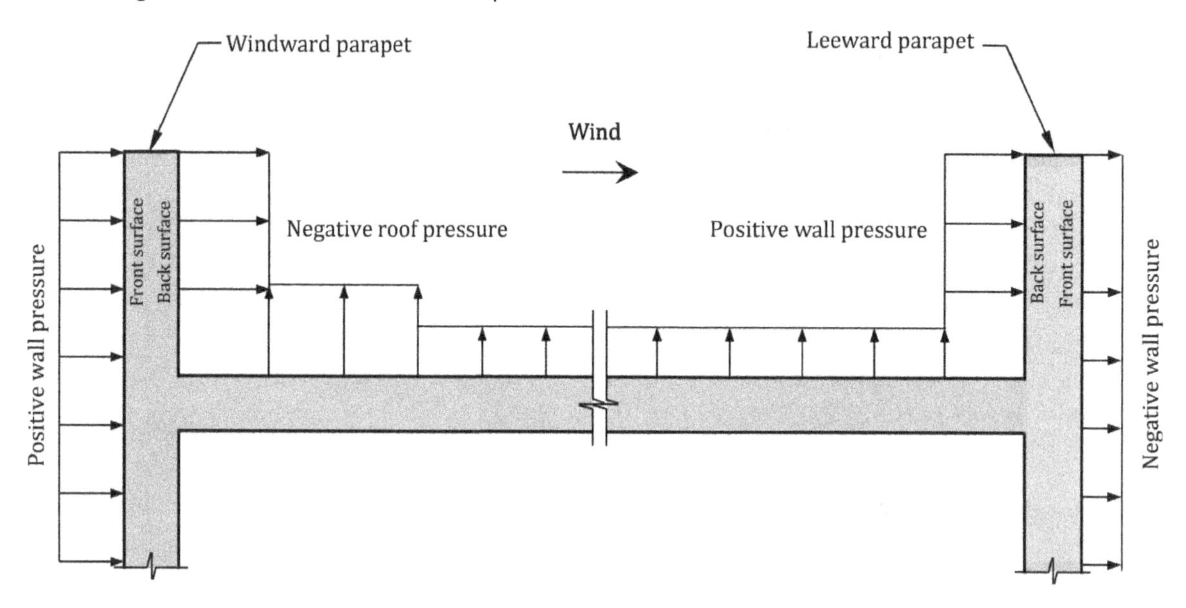

A leeward parapet experiences a positive wall pressure on the roof side (back surface) and a negative wall pressure on the exterior side (front surface). It is assumed that the windward and leeward parapets are separated a sufficient distance so that shielding by the windward parapet does not decrease the positive wall pressure on the leeward parapet.

For simplicity, the pressure p_p is the combined net pressure due to the combination of the net pressures on the front and back surfaces of the parapet, which is captured by the combined net pressure coefficients (GC_{pn}) for windward and leeward parapets (see Figure 5.15). Because wind can occur in any direction, a parapet must be designed for both sets of pressures. The internal pressures inside the parapet cancel out in the determination of the combined pressure coefficient.

The pressures determined on the parapets are combined with the external pressures on the building to obtain the total wind pressures on the MWFRS.

Design Wind Load Cases

The MWFRS of buildings of all heights subjected to the wind pressures determined by the Directional Procedure must be designed for the load cases in ASCE/SEI Figure 27.3-8

Figure 5.15 Design Wind Pressures on Parapets in Accordance with ASCE/SEI 27.3.4

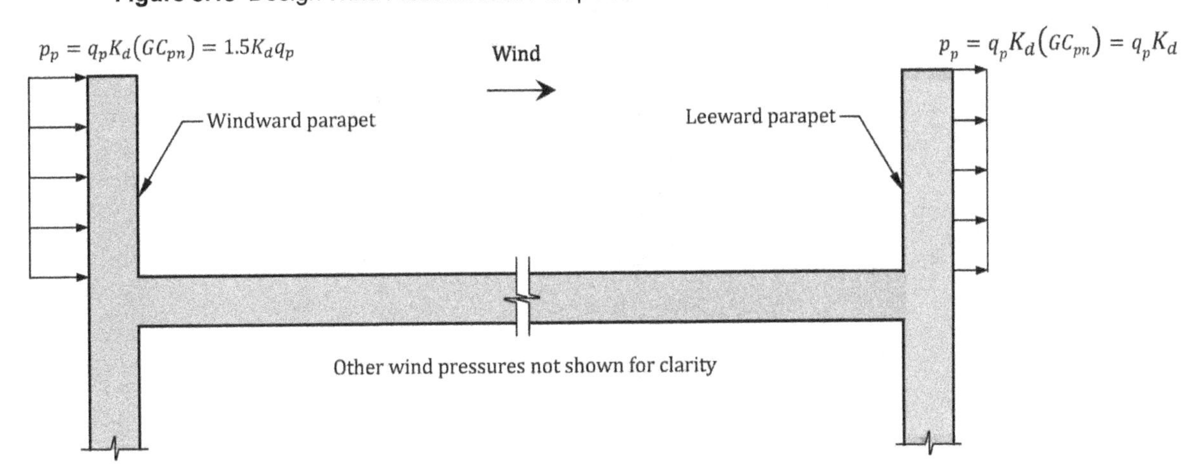

(see Figure 5.16; in this figure, plan views of the building are shown, the subscripts x and y refer to the principal axes of the building, and subscripts W and L refer to the windward and leeward faces, respectively).

In case 1, full design wind pressures are applied on the projected wall area perpendicular to each principal axis of the structure. The design wind pressures are considered separately along each principal axis. Full design wind pressures are applied on the side walls and roof areas for wind along each principal axis as specified in ASCE/SEI Figures 27.3-1 through 27.3-7. All pressures act simultaneously for each principal wind direction.

Case 2 accounts for the effects of nonuniform pressure on different faces of the building due to wind flow. Nonuniform pressures introduce torsion on the building, and this is accounted for in design by subjecting the building to 75 percent of the design wind pressures applied along the principal axis of the building plus a torsional moment, M_z, determined using an eccentricity equal to 15 percent of the appropriate plan dimension of the building. Torsional effects, which are applicable for buildings with rigid diaphragms, are determined in each principal direction separately (see note 4 in ASCE/SEI Figure 27.3-8 for application of M_z for buildings with other than rigid diaphragms). Roof pressures are equal to 75 percent of the values determined in accordance with case 1. All pressures and torsion act simultaneously for each principal wind direction.

A critical load case can occur when the design wind load acts diagonally to a building. This is accounted for in case 3, where 75 percent of the maximum design wind pressures are applied along the principal axes of a building simultaneously. The resulting pressures on any roof area defined by cases 1 and 2 must be 100 percent of the larger value of the roof pressures defined for cases 1 and 2, respectively. All pressures act simultaneously.

Case 4 considers the effects due to diagonal wind loads and torsion. Seventy five percent of the wind pressures in case 2 are applied along the principal axes of a building simultaneously, and a torsional moment is applied which is determined using 15 percent of the plan dimensions of the building. The resulting pressures on any roof area defined by cases 1 and 2 must be 100 percent of the larger value of the roof pressures defined for cases 1 and 2, respectively. All pressures act simultaneously.

The exception in ASCE/SEI 27.3.5 permits buildings that meet the requirements of Section D.1 of ASCE/SEI Appendix D to be designed for Cases 1 and 3 only. The following buildings do not need to be designed for the effects from torsion:

- One-story buildings with a mean roof height less than or equal to 30 ft (9.14 m).
- Buildings two stories or fewer framed with light-frame construction.
- Buildings two stories or fewer with flexible diaphragms.

Figure 5.16 Design Wind Load Cases on the MWFRS in Accordance with ASCE/SEI 27.3.5

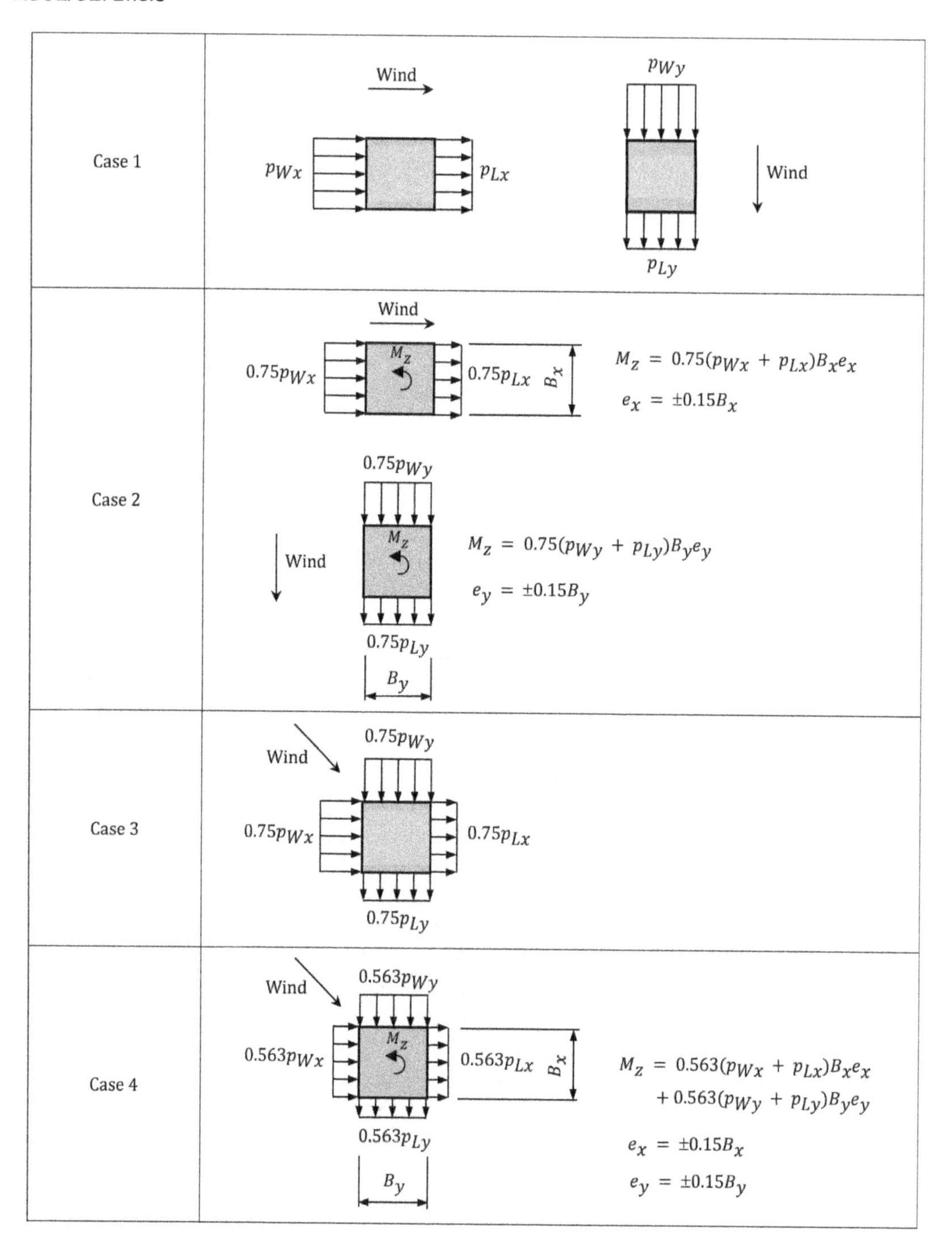

- Buildings controlled by seismic loading (see ASCE/SEI D.3.1 and D.3.2 for criteria for buildings with diaphragms that are not flexible and those that are flexible, respectively).
- Buildings classified as torsionally regular under wind loads (ASCE/SEI D.4).
- Buildings with flexible diaphragms designed for at least 1.5 times the design wind pressures in cases 1 and 3.

In the case of flexible buildings, dynamic effects can increase the effects from torsion. ASCE/SEI Equation (27.3-4) accounts for these effects. The eccentricity, e, determined by this equation is to be used in the appropriate load cases in ASCE/SEI Figure 27.3-8 in lieu of the eccentricities e_x and e_y given in that figure for rigid structures. An eccentricity must be considered for each principal axis of the building, and the sign of the eccentricity must be plus or minus, whichever causes the more severe load case.

Flowchart 5.3 in Section 5.7 of this book can be used to determine design wind pressures on the MWFRS of buildings in accordance with ASCE/SEI Chapter 27.

5.3.3 Envelope Procedure for Low-Rise Buildings (ASCE/SEI Chapter 28)

Scope

The Envelope Procedure in ASCE/SEI Chapter 28 applies to the determination of wind loads on the MWFRS of low-rise buildings that meet the conditions and limitations given in ASCE/SEI 28.1.2 and 28.1.3, respectively. According to ASCE/SEI 26.2, a low-rise building is an enclosed, partially enclosed, or partially open building that complies with the following conditions:

1. Mean roof height, h, is less than or equal to 60 ft (18.3 m).
2. Mean roof height, h, does not exceed the least horizontal dimension of the building.

The provisions in ASCE/SEI Chapter 28 are applicable to buildings that comply with the following conditions (ASCE/SEI 28.1.2):

- The building is regular-shaped, that is, the building has no unusual geometrical irregularities in spatial form.
- The building does not have response characteristics that make it subject to across-wind loading, vortex shedding, or instability caused by galloping or flutter. In addition, the building is not located at a site where channeling effects or buffeting in the wake of upwind obstructions warrant special consideration.
- The building does not have an arched, barrel, or unusually shaped roof.

The requirements in ASCE/SEI Chapter 28 account for mean and fluctuating loads caused by winds acting on rigid buildings.

Buildings not meeting these conditions must be designed by the applicable provisions in ASCE/SEI Chapter 27, recognized literature that documents such wind load effects, or by the wind tunnel procedure in ASCE/SEI Chapter 31 (ASCE/SEI 28.1.3).

Reduction in wind pressure due to apparent shielding by surrounding buildings, other structures, or terrain features is not permitted (ASCE/SEI 28.1.4). Such shielding may be modified or completely removed during the lifespan of the building, which could result in significantly higher wind loads.

Design Wind Pressures for Low-Rise Buildings

Design wind pressures, p, for the MWFRS of low-rise buildings satisfying the conditions of ASCE/SEI Chapter 28 are determined by ASCE/SEI Equations (28.3-1) and (28.3-1.SI):

$$p = q_h K_d [(GC_{pf}) - (GC_{pi})] \tag{5.18}$$

This equation is used to calculate wind pressures on the building surfaces identified in ASCE/SEI Figures 28.3-1 and 28.3-2. The velocity pressure at the mean roof height of the building, q_h, is determined in accordance with ASCE/SEI 26.10 (see Section 5.2.7 of this book).

The external pressure coefficients, (GC_{pf}), combine both a gust-effect factor and external pressure coefficients for low-rise buildings and are not allowed to be separated (ASCE/SEI 28.3.1.1). Unlike the external pressure coefficients given in ASCE/SEI Figure 27.3-1 that reflect the actual pressure on each surface of a building as a function of wind direction, the coefficients in ASCE/SEI Figures 28.3-1 and 28.3-2 for enclosed, partially enclosed, and partially open buildings are essentially "pseudo" pressure conditions that, when applied to a building, envelop the desired structural actions independent of wind direction.

The "pseudo" values of (GC_{pf}) were determined from the output of wind tunnel tests, which measured bending moments, total horizontal forces, and total uplift as a building was rotated 360 degrees in the wind tunnel (Reference 5.5). Thus, values of (GC_{pf}) produce maximum measured structural actions and are not the actual surface pressures.

Load Cases

To capture all appropriate structural actions, a building must be designed for all wind directions by considering in turn each corner of the building as the windward (or, reference) corner. These conditions are illustrated in ASCE/SEI Figure 28.3-1: at each corner, two load cases must be considered (load case 1 and load case 2), one for each range of wind direction. In general, there are two load cases with four basic scenarios for each case (see ASCE/SEI Figure C28.3-1). For symmetric buildings, some of these load cases are repetitive. Load cases 1 and 2 for the same windward corner of a low-rise building are illustrated in Figure 5.17.

A building must be evaluated for the loading patterns applied as shown in ASCE/Figure 28.3-1 and with all zones loaded simultaneously.

Figure 5.17 Basic Load Cases for Low-Rise Buildings in Accordance with ASCE/SEI Chapter 28

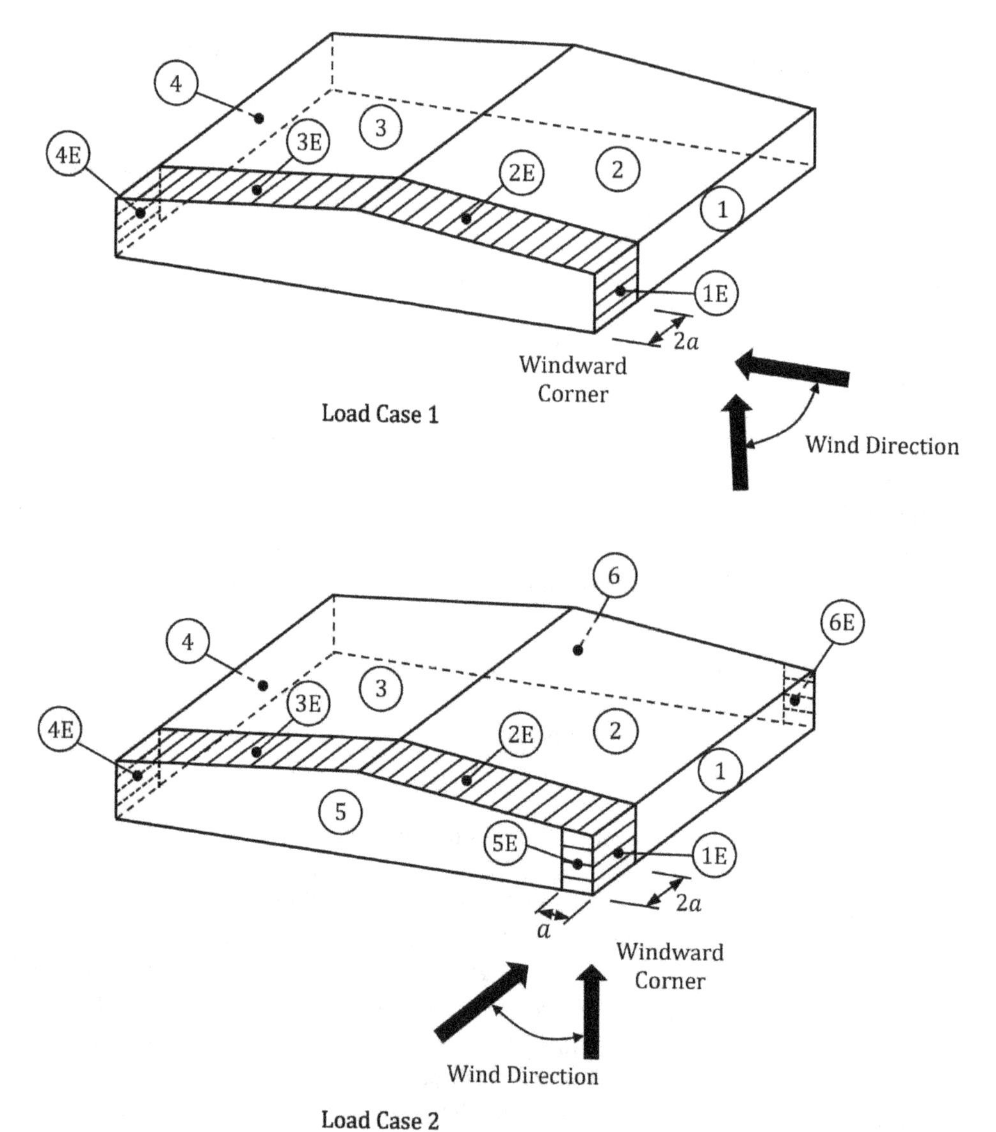

In each load case, zones 2E and 3E are located along the roof edge perpendicular to the ridge that is nearest to the corner being evaluated. The critical loading is obtained by evaluating combinations of external and internal pressures. Internal pressure loads cancel out when evaluating the total horizontal load and base shear. However, such loads can be critical in one-story buildings with moment-resisting frames and in the top story of buildings where the MWFRS consists of moment-resisting frames.

For buildings with flat roofs, roof angle, θ, is taken as zero and the boundary for zone 2/3 and zone 2E/3E must be located at the mid-width of the building.

In cases where the roof pressure coefficient, (GC_{pf}), is negative in zones 2 and 2E, that pressure coefficient must be applied in zone 2/2E for a distance from the roof edge equal to the lesser of the following: (1) 0.5 times the horizontal dimension of the building parallel to the direction of the MWFRS being designed or (2) 2.5 times the eave height at the windward wall. The remainder of zone 2/2E extending to the ridge must use the pressure coefficient for zone 3/3E.

The original and subsequent research that was performed to develop and refine this methodology was done on low-rise buildings with gable roofs. A suggested method for low-rise buildings with hip roofs is given in ASCE/SEI Figure C28.3-3.

The torsional load cases (load cases 3 and 4) in ASCE/SEI Figure 28.3-2 must be considered in the design of all low-rise buildings except for the following:

- One-story buildings with a mean roof height of less than or equal to 30 ft (9.14 m).
- One-story and two-story light-frame buildings.
- One-story and two-story buildings with flexible diaphragms.

The external pressure coefficients, (GC_{pf}), for torsional load cases must be determined in accordance with ASCE/SEI Figure 28.3-2 as follows:

- In load case 3, the pressure coefficients from load case 1 in ASCE/SEI Figure 28.3-1 must be used for zones 1 to 4 and zones 1E to 4E.
- In load case 4, the pressure coefficients from load case 2 in ASCE/SEI Figure 28.3-1 must be used for zones 1 to 6 and zones 1E to 6E.
- In zones 1T to 6T, the pressure coefficients in ASCE/SEI Figure 28.3-2 must be used.

Similar to load cases 1 and 2, the following must be satisfied for load cases 3 and 4:

- A building must be designed for all wind directions by considering in turn each corner of the building as the windward corner with loading patterns applied as shown and with all zones loaded simultaneously.
- In each load case, zones 2E and 3E are located along the roof edge perpendicular to the ridge that is nearest to the corner being evaluated. The critical loading is obtained by evaluating combinations of external and internal pressures.
- For buildings with flat roofs, roof angle, θ, is taken as zero and the boundary for zone 2/3 and zone 2E/3E must be located at the mid-width of the building.
- In cases where the roof pressure coefficient, (GC_{pf}), is negative in zones 2 and 2E, that pressure coefficient must be applied in zone 2/2E for a distance from the roof edge equal to the lesser of the following: (1) 0.5 times the horizontal dimension of the building parallel to the direction of the MWFRS being designed or (2) 2.5 times the eave height at the windward wall. The remainder of zone 2/2E extending to the ridge must use the pressure coefficient for zone 3/3E.

Total Horizontal Load

Considering the load cases discussed in the previous section, it is possible for some building configurations to have net wind pressures and forces on the roof in the direction opposite to the direction of analysis. Except for buildings utilizing moment frames as the MWFRS, the aforementioned roof loads must be neglected when calculating the total horizontal shear on the building (ASCE/SEI 28.3.3).

Parapets

Design wind pressures for the effects of parapets, p_p, on the MWFRS of low-rise buildings with flat, gable, or hip roofs are determined by ASCE/SEI Equations (28.3-2) and (28.3-2.SI):

$$p_p = q_p K_d (GC_{pn}) \tag{5.198}$$

In this equation, q_p is the velocity pressure evaluated at the top of the parapet determined in accordance with ASCE/SEI 26.10, K_d is the wind directionality factor determined by ASCE/SEI Table 26.6-1, and (GC_{pn}) is the combined net pressure coefficient, which is equal to +1.5 for a windward parapet and −1.0 for a leeward parapet (see Figure 5.15 of this book).

Roof Overhangs

Positive external pressures on the bottom surface of windward roof overhangs must be determined using the pressure coefficient $GC_p = 0.7$ in combination with the top surface pressures determined by ASCE/SEI Figure 28.3-1 (ASCE/SEI 28.3.5). Application of this pressure is similar to that shown in Figure 5.13.

Provisions are not provided for wind pressures on the bottom surface of a leeward overhang.

Minimum Design Wind Loads

Minimum design wind pressures in the design of the MWFRS for enclosed or partially enclosed low-rise buildings are given in ASCE/SEI 28.3.6. The pressures of 16 lb/ft^2 (0.77 kN/m^2) on the projected area of the walls and 8 lb/ft^2 (0.38 kN/m^2) on the projected area of the roof are considered a separate load case from any of the other required load cases (see Figure 5.10 of this book).

Horizontal Wind Loads on Open or Partially Enclosed Buildings with Transverse Frames and Pitched Roofs

The horizontal pressure, p, in the direction parallel to the roof ridge of an open or partially enclosed building with transverse frames and a pitched roof with an angle less than or equal to 45 degrees is determined by ASCE/SEI Equation (28.3-3):

$$p = q_h K_d [(GC_{pf})_{\text{windward}} - (GC_{pf})_{\text{leeward}}] K_B K_S \tag{5.20}$$

This equation is applicable to buildings with open end walls and with end walls fully or partially enclosed with cladding. This pressure, which is in the longitudinal direction (parallel to the ridge), acts in combination with the pressures on the roof determined in accordance with ASCE/SEI 27.3.2 for an open or partially enclosed building with transverse frames and a pitched roof less than 45 degrees.

On the windward end wall, the external pressure coefficients $(GC_{pf})_{\text{windward}}$ are obtained from load case 2 in ASCE/SEI Figure 28.3-1 for surfaces 5 and 5E. Similarly, for the leeward wall, external pressure coefficients $(GC_{pf})_{\text{leeward}}$ are obtained from load case 2 in ASCE/SEI Figure 28.3-1 for surfaces 6 and 6E. The terms K_B and K_S are the frame width factor and shielding factor, respectively, and are determined by the equations in ASCE/SEI 28.3.7.

The frame width factor, K_B, is determined as follows:

$$K_B = \begin{cases} 1.8 - 0.01B \text{ where } B < 100 \text{ ft (30.5 m)} \\ 0.8 \text{ where } B \geq 100 \text{ ft (30.5 m)} \end{cases} \tag{5.21}$$

where B is the width of the building perpendicular to the ridge in feet (meters).

The frame shielding factor, K_S, is determined based on the number of frames, n, perpendicular to the direction of analysis, and the solidity ratio, ϕ, which is the ratio of the effective solid area of an end wall, A_S (i.e., the projected area of any portion of the end wall exposed to the wind), to the total end wall area for an equivalent closed building, A_E (see ASCE/SEI Figure 28.3-3):

$$K_S = 0.60 + 0.073(n-3) + 1.25\phi^{1.8} \tag{5.22}$$

In buildings with 2 frames, n must be taken as 3 (ASCE/SEI 28.3.7).

The total longitudinal force, F, that must be resisted by the MWFRS due to p is determined by ASCE/SEI Equation (28.3-4):

$$F = pA_E \tag{5.23}$$

This force is applied at the centroid of A_E.

Fascia loads need not be considered as a separate load case where the fascia areas are included in the calculation of A_S.

In the transverse direction (perpendicular to the ridge), the wind pressures are determined in accordance with ASCE/SEI 27.3.2; these pressures are a separate load case from the pressures determined parallel to the ridge (see Section 5.3.2 of this book).

Flowchart 5.4 in Section 5.7 of this book can be used to determine design wind pressures on low-rise buildings in accordance with ASCE/SEI Chapter 28.

5.3.4 Directional Procedure for Building Appurtenances and Other Structures (ASCE/SEI Chapter 29)

Scope

The requirements in ASCE/SEI Chapter 29 are applicable to the determination of wind loads on the MWFRS of building appurtenances (including rooftop structures and rooftop equipment) and other structures of all heights (including solid freestanding walls, freestanding solid signs, chimneys, tanks, circular bins, silos, open signs, single-plane open frames, ground-mounted fixed-tilt solar panel systems, and trussed towers) that meet the conditions and limitations given in ASCE/SEI 29.1.2 and 29.1.3, respectively.

The provisions in ASCE/SEI Chapter 29 are applicable to appurtenances or structures that comply with the following conditions (ASCE/SEI 29.1.2):

- The structure is regular-shaped, that is, the structure has no unusual geometrical irregularities in spatial form.
- The structure does not have response characteristics that make it subject to across-wind loading, vortex shedding, or instability caused by galloping or flutter. In addition, the structure is not located at a site where channeling effects or buffeting in the wake of upwind obstructions warrant special consideration.

The requirements in ASCE/SEI Chapter 29 account for the load magnification effect caused by gusts in resonance with along-wind vibrations of flexible structures.

Structures not meeting these conditions must be designed using recognized literature that documents such wind load effects or by the wind tunnel procedure in ASCE/SEI Chapter 31 (ASCE/SEI 29.1.3).

Reduction in wind pressure due to apparent shielding by surrounding buildings, other structures, or terrain features is not permitted (ASCE/SEI 29.1.4). Such shielding may be modified or completely removed during the lifespan of the building, which could result in significantly higher wind loads.

Steps to determine wind loads on the MWFRS for rooftop equipment and other structures are given in ASCE/SEI Table 29.1-1. Similarly, steps to determine wind loads on the MWFRS of circular bins, silos, and tanks are given in ASCE/SEI Table 29.1-2.

Solid Freestanding Walls and Solid Signs

Solid Freestanding Walls and Solid Freestanding Signs

The design wind force, F, for the MWFRS of solid freestanding walls and solid freestanding signs is determined by ASCE/SEI Equations (29.3-1) and (29.3-1.SI) (walls and signs with openings comprising less than 30 percent of the gross area are classified as solid; for walls or signs with openings greater than or equal to 30 percent, the wind force must be determined in accordance with ASCE/SEI 29.4, which is covered in the next section of this book):

$$F = q_h K_d G C_f A_s \tag{5.24}$$

The velocity pressure at the mean roof height of the wall or sign, q_h, is determined in accordance with ASCE/SEI 26.10 (see Section 5.2.7 of this book), the wind directionality factor, K_d, is determined from ASCE/SEI Table 26.6-1 and the gust-effect factor, G, is determined in accordance with ASCE/SEI 26.11 (see Section 5.2.8 of the book). Net force coefficients, C_f, are given in ASCE/SEI Figure 29.3-1 as a function of the geometrical properties of the wall or sign. The term A_s is the gross area of the solid freestanding wall or sign.

In general, three cases (A, B, and C) must be investigated where F is applied at different locations on the sign or wall (see ASCE/SEI Figure 29.3-1 and Table 5.10). Case C is applicable where the width of the sign or wall, B, is greater than or equal to two times the vertical dimension of the sign or wall, s; in such cases, Case B need not be considered.

Table 5.10 Load Cases for Solid Freestanding Walls and Solid Signs

Case	Location of Resultant Force, F	
	$s/h < 1$	$s/h = 1$
A	F acts normal to the face of the wall or sign through the geometric center	F acts normal to the face of the wall or sign at a distance above the geometric center equal to 0.05 times the height of the wall or sign
B	F acts normal to the face of the wall or sign at a distance from the geometric center toward the windward edge equal to 0.2 times the horizontal dimension of the wall or sign	F acts normal to the face of the wall or sign at a distance from the geometric center toward the windward edge equal to 0.2 times the horizontal dimension of the wall or sign and a distance above the geometric center equal to 0.05 times the height of the wall or sign
C	F acts normal to the face of the wall or sign through the geometric centers of each region identified in ASCE/SEI Figure 29.3-1	F acts normal to the face of the wall or sign through the geometric centers of each region identified in ASCE/SEI Figure 29.3-1 and at a distance above the geometric center equal to 0.05 times the height of the wall or sign

Load cases A, B, and C are illustrated in Figure 5.18.

Flowchart 5.5 in Section 5.7 of this book can be used to determine the design wind forces on solid freestanding walls and solid freestanding signs.

Solid Attached Signs

The components and cladding (C&C) requirements for walls in ASCE/SEI Chapter 30 must be used to determine wind pressures on solid signs attached to the wall of a building provided the plane of the sign is parallel to and in contact with the plane of the wall and the sign does not extend beyond the side or top edges of the wall (see Section 5.4 of this book). In such cases, the internal pressure coefficient, (GC_{pi}), is set equal to zero.

This procedure is also applicable to signs attached to, but not in direct contact with, the wall provided the gap between the sign and wall is no more than 3 ft (0.91 m) and the edge of the sign is at least 3 ft (0.91 m) from the free edges of the wall (i.e., side and top edges and bottom edges of elevated walls).

Figure 5.18 Load Cases A, B, and C for Solid Freestanding Walls and Solid Freestanding Signs

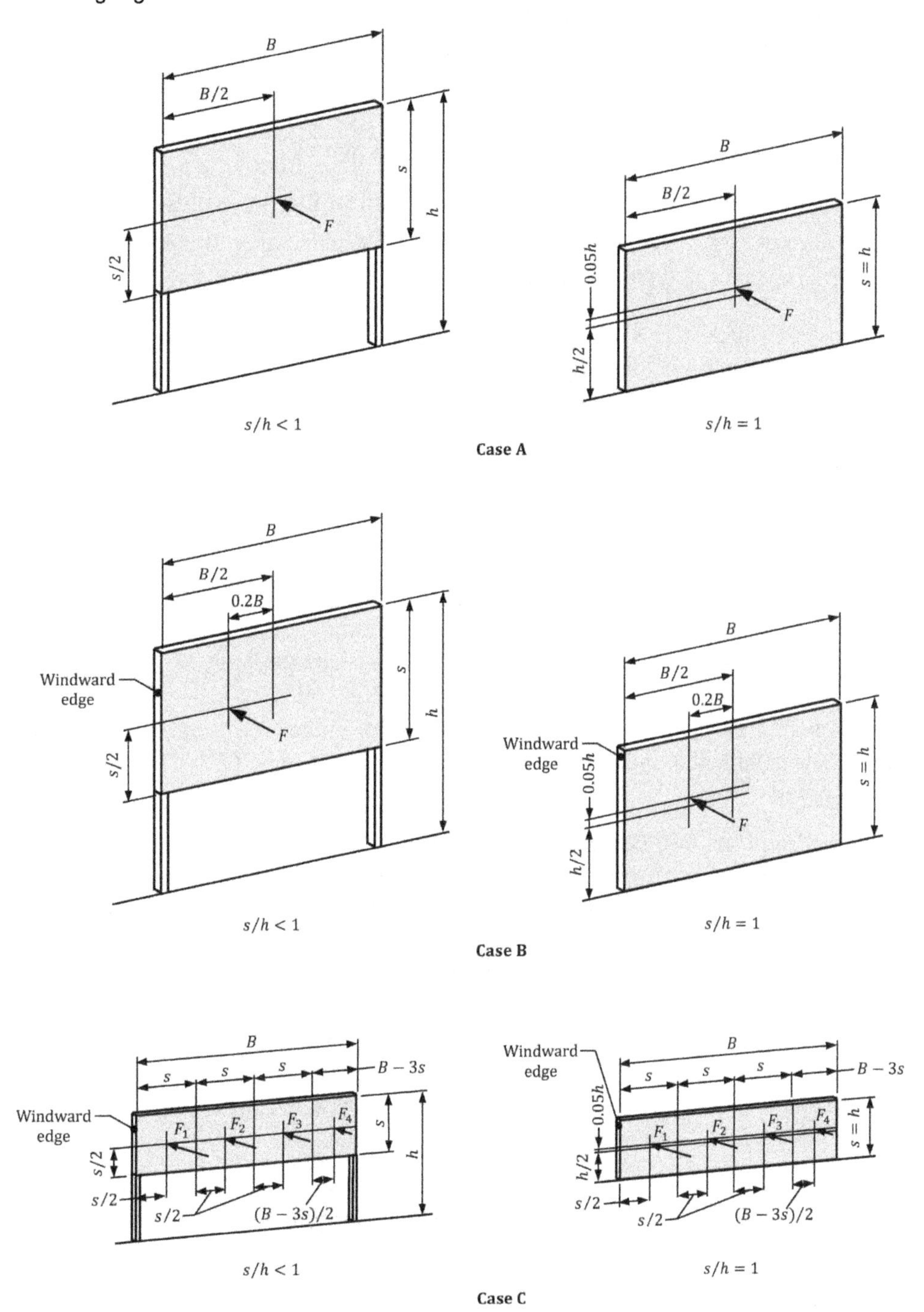

Other Structures

Chimneys, Tanks, Open Signs, Single-plane Open Frames, and Trussed Towers

The design wind force, F, for ground or roof-mounted chimneys, tanks, open signs, single-plane open frames, and trussed towers is determined by ASCE/SEI Equations (29.4-1) and (29.4-1.SI):

$$F = q_z K_d G C_f A_f \tag{5.25}$$

The velocity pressure, q_z, evaluated at height z of the centroid of the area, A_f, is determined in accordance with ASCE/SEI 26.10 (see Section 5.2.7 of this book), the wind directionality factor, K_d, is determined from ASCE/SEI Table 26.6-1 and the gust-effect factor, G, is determined in accordance with ASCE/SEI 26.11 (see Section 5.2.8 of the book). The area A_f is the projected area of the structure normal to the wind except where the force coefficient, C_f, is specified for the actual surface area.

Force coefficients, C_f, are given in the following figures:

- ASCE/SEI Figure 29.4-1: Chimneys, tanks, and similar structures
- ASCE/SEI Figure 29.4-2: Open signs and single-plane open frames
- ASCE/SEI Figure 29.4-3: Trussed towers

Values of C_f in ASCE/SEI Figure 29.4-1 are given for square, hexagonal, and round cross-sections as a function of the height of the structure, h, to the cross-sectional dimension of the section, D (D is the diameter of circular cross-sections and is the least horizontal dimension of square, hexagonal, and octagonal sections at the elevation under consideration).

The force coefficients in ASCE/SEI Figure 29.4-2 are applicable to open signs, that is signs with openings comprising 30 percent or more of the gross area. Signs not meeting this criterion are classified as solid signs and the force coefficients in ASCE/SEI Figure 29.3-1 must be used (see the previous section of this book).

The force coefficients in ASCE/SEI Figure 29.4-3 are for trussed towers with square and triangular cross-sections.

Flowchart 5.6 in Section 5.7 of this book can be used to determine the design wind force for other structures.

References are provided in ASCE/SEI 29.4 to determine G, C_f, and A_f for structures found in petrochemical and other industrial facilities not addressed in ASCE/SEI 7 and for lighting system support poles.

Rooftop Structures and Equipment for Buildings

Lateral and vertical wind forces must be determined in accordance with ASCE/SEI 29.4.1 for structures and equipment on the rooftop of buildings (excluding roof-mounted solar panels, which are covered in ASCE/SEI 29.4.3 and 29.4.4).

Lateral force, F_h, is determined by ASCE/SEI Equations (29.4-2) and (29.4-2.SI) and uplift force, F_v, is determined by ASCE/SEI Equations (29.4-3) and (29.4-3.SI) (see Table 5.11 and Figure 5.19). Forces F_h and F_v are applied at the centroids of the areas A_f and A_r, respectively.

The basic wind speed, V, used in determining q_h at the mean roof height of the building must be based on the greater of the following risk categories (ASCE/SEI 26.10.2): (1) the risk category

Table 5.11 Lateral and Uplift Forces on Rooftop Structures and Equipment for Buildings

Lateral Force, F_h	Uplift Force, F_v
$F_h = q_h K_d (GC_r) A_f$	$F_v = q_h K_d (GC_r) A_r$
A_f = vertical projected area of the rooftop structure or equipment on a plane normal to the direction of wind	A_r = horizontal projected area of the rooftop structure or equipment
$(GC_r) = \begin{cases} 1.9 & \text{for } A_f < 0.1Bh \\ 2.0 - (A_f/Bh) & \text{for } 0.1Bh \le A_f \le Bh \end{cases}$	$(GC_r) = \begin{cases} 1.5 & \text{for } A_r < 0.1BL \\ 1.0 + 0.56[1 - (A_r/BL)] & \text{for } 0.1BL \le A_r \le BL \end{cases}$

q_h = velocity pressure evaluated at the mean roof height of the building.
K_d = wind directionality factor determined from ASCE/SEI Table 26.6-1.

Figure 5.19 Lateral and Uplift Wind Forces on Rooftop Structures and Equipment

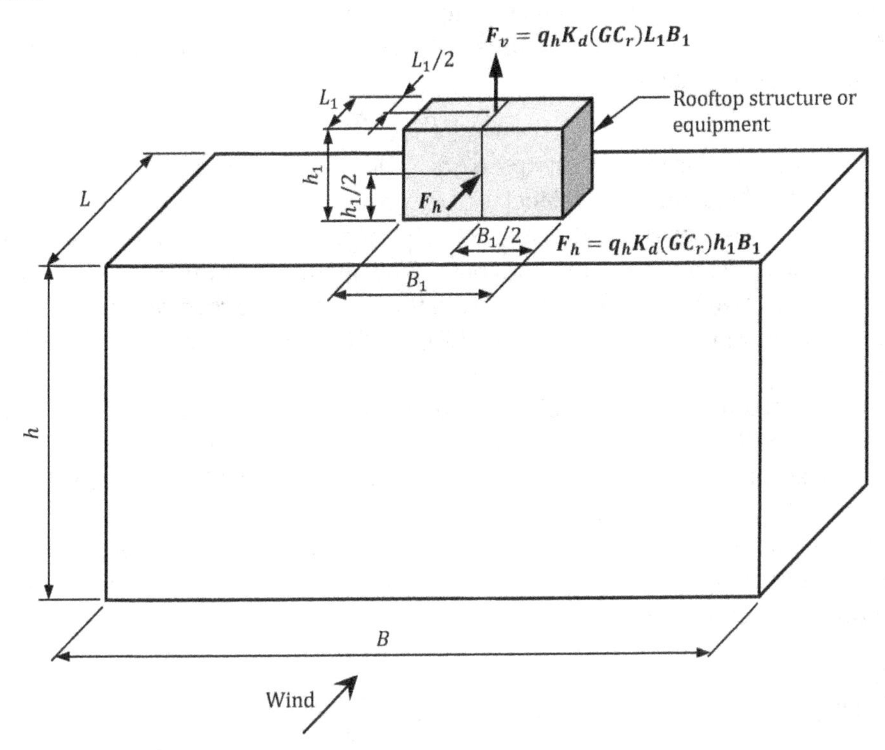

of the building on which the structure or equipment is located or (2) the risk category for any facility to which the structure or equipment provides a necessary service.

The combined gust-effect factor and pressure coefficient, (GC_r), accounts for higher wind pressures due to higher correlation of pressures across the structure surface, higher turbulence on the roof of the building, and accelerated wind speed on the roof. In the case of relatively small rooftop equipment where A_f is less than 10 percent of the windward building area, Bh, $(GC_r) = 1.9$. As the area, A_f, approaches Bh (i.e., as the projected area of the rooftop building or equipment approaches that of the windward area of the building), (GC_r) is permitted to be reduced linearly to 1.0.

The combined gust-effect factor and pressure coefficient, (GC_r), is equal to 1.5 where A_r is less than 10 percent of the building area, BL. A linear reduction to 1.0 is permitted as A_r increases to BL.

Circular Bins, Silos, and Tanks with h ≤ 120 ft (36.6 m), D ≤ 120 ft (36.6 m), and 0.25 ≤ h_c/D ≤ 4

Design wind loads for circular bins, silos, and tanks with a mean roof height, h, less than or equal to 120 ft (36.6 m), a diameter, D, less than or equal to 120 ft (36.6 m), and a ratio of solid cylinder height, h_c, to diameter, D, in the range of 0.25 and 4 are determined in accordance with ASCE/SEI 29.4.2.

A single circular bin, silo, or tank or grouped circular bins, silos, and tanks of similar size with a center-to-center spacing greater than $2D$ are treated as isolated structures. The design wind force on the external walls of an isolated structure is determined by ASCE/SEI Equations (29.4-1) and (29.4-1.SI) where the force coefficient, C_f, is equal to 0.63 based on a projected wall area equal to Dh_c:

$$F = 0.63 q_z K_d G D h_c \tag{5.26}$$

In this equation, q_z is the velocity pressure evaluated at height z, which corresponds to the centroid of the projected wall area, Dh_c (ASCE/SEI 26.10), K_d is the wind directionality factor determined from ASCE/SEI Table 26.6-1, and G is the gust-effect factor determined in accordance with ASCE/SEI 26.11.

The force coefficient, C_f, is adapted from Reference 5.6. When determining the gust-effect factor, G, for elevated structures, the natural frequency should be based on the entire structure, that is, the supported structure and the support structure.

For the roofs of isolated structures, the design wind pressure is determined by ASCE/SEI Equations (29.4-4) and (29.4-4.SI):

$$p = q_h K_d [GC_p - (GC_{pi})] \tag{5.27}$$

In this equation, q_h is the velocity pressure evaluated at mean roof height, h (ASCE/SEI 26.10).

The external pressure coefficients, C_p, are given in ASCE/SEI Figure 29.4-5 for windward and leeward portions (zones 1 and 2, respectively) of flat, conical, or dome roofs based on the angle of the plane of the roof from the horizontal. The internal pressure coefficients, (GC_{pi}), for roofed structures are given in ASCE/SEI 26.13. In the case of dome roofs where the roof angle is greater than 10 degrees, C_p is to be determined from ASCE/SEI Figure 27.3-2 for domes with circular bases.

For elevated structures, the design wind pressure, p, on the underside of the structure is determined by ASCE/SEI Equations (29.4-4) and (29.4-4.SI) where the external pressure coefficient, C_p, is to be taken as 0.8 and –0.6 where $C \leq h_c$. Where $C \leq h_c/3$, C_p is equal to $0.8(C/h) \leq 0.8$ and $-0.6(C/h) \geq -0.6$ (ASCE/SEI 29.4.2.3).

A summary of the wind loads and pressures for the external walls, roof, and, where applicable, the undersides of isolated circular bins, silos, and tanks is given in Table 5.12 based on the following conditions (see Figure 5.20):

- $0.25 \leq h_c/D \leq 4$.
- Clearance height, C, is less than or equal to h_c.
- The tank surface is moderately smooth as defined in ASCE/SEI Figure 29.4-1, that is, $D'/D < 0.02$ where D' is the depth of protruding elements (such as ribs and spoilers).

Table 5.12 Design Wind Loads and Pressures for Isolated Circular Bins, Silos, and Tanks

Element	Design Wind Load (F) or Pressure (p)
External walls	$F = 0.63 q_z K_d G D h_c$ where q_z = velocity pressure evaluated at height z, which corresponds to the centroid of the projected wall area, Dh_c (ASCE/SEI 26.10) K_d = wind directionality factor determined from ASCE/SEI Table 26.6-1 G = gust-effect factor based on the natural frequency of the entire structure (i.e., the supported structure and the support structure) [ASCE/SEI 26.11]
Roofs	$p = q_h K_d [GC_p - (GC_{pi})]$ where q_h = velocity pressure evaluated at mean roof height, h (ASCE/SEI 26.10) C_p = external pressure coefficient for zones 1 and 2 in ASCE/SEI Figure 29.4-5 for flat, dome, or conical roofs with an angle less than 10 degrees and for conical roofs with an angle between 10 and 30 degrees, inclusive* (GC_{pi}) = internal pressure coefficient for roofed structures (ASCE/SEI 26.13)
Undersides	$p = q_h K_d [GC_p - (GC_{pi})]$ where $C_p = \begin{cases} 0.8, -0.6 & \text{for } h_c/3 < C \leq h_c \\ 0.8(C/h), -0.6(C/h) & \text{for } C \leq h_c/3 \end{cases}$

*For domed roofs with a roof angle greater than 10 degrees, the external pressures must be determined from ASCE/SEI Figures 27.3-2.

Figure 5.20 Plan and Elevations of Isolated Circular Bins, Silos, and Tanks on the Ground or Supported by Columns

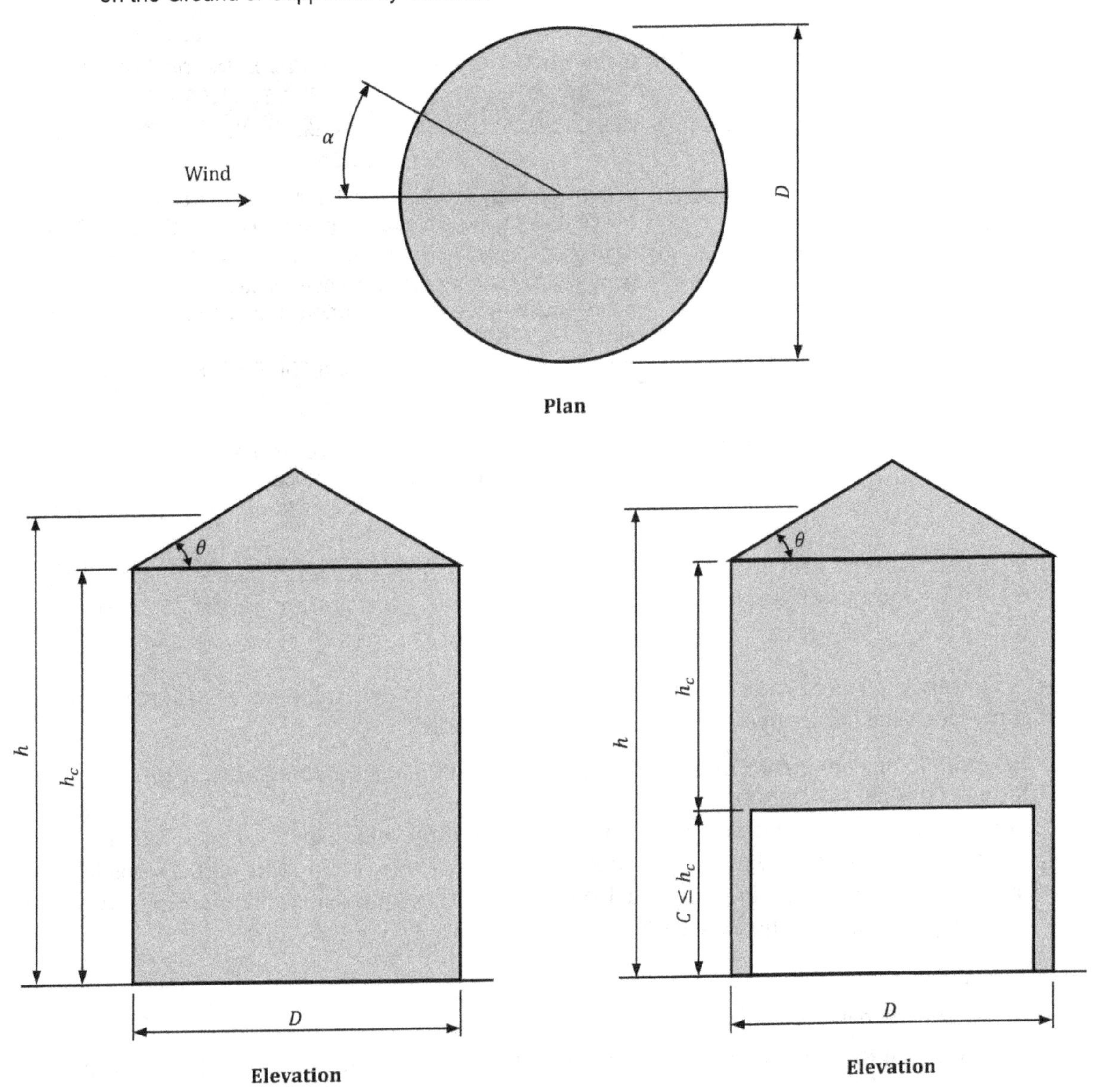

Where the center-to-center spacing of three or more structures is less than $1.25D$, the structures must be treated as grouped and the wind pressures are determined in accordance with ASCE/SEI 29.4.2.4.

ASCE/SEI Equations (29.4-4) and (29.4-4.SI) must be used to determine the net design roof pressures on grouped bins, silos, or tanks where C_p for zones 1 and 2 are to be determined from ASCE/SEI Figure 29.4-6 (ASCE/SEI 29.4.2.4). Values of C_p in this figure are greater than those in ASCE/SEI Figure 29.4-5 for roofs of isolated structures (significant increases in pressures were obtained from wind tunnel tests for grouped structures). The overall wind force on the projected walls is determined by ASCE/SEI Equations (29.4-1) and (29.4-1.SI) where the force coefficient, C_f, is determined from ASCE/SEI Figure 29.4-6. For structures that have a center-to-center spacing less than $2D$ and greater than $1.25D$, it is permitted to use linear interpolation of the pressure and force coefficients in ASCE/SEI Figures 29.4-5 and 29.4-6 to determine C_p and C_f (ASCE/SEI 29.4.2).

A summary of the wind loads and pressures for grouped circular bins, silos, and tanks is given in Table 5.13.

Table 5.13 Design Wind Loads and Pressures for Grouped Circular Bins, Silos, and Tanks

Element	Design Wind Load (F) or Pressure (p)
External walls	$F = q_z K_d G C_f D h_c$ where q_z = velocity pressure evaluated at height z, which corresponds to the centroid of the projected wall area, Dh_c (ASCE/SEI 26.10) K_d = wind directionality factor determined from ASCE/SEI Table 26.6-1 G = gust-effect factor based on the natural frequency of the entire structure (i.e., the supported structure and the support structure) [ASCE/SEI 26.11] C_f = force coefficient determined from ASCE/SEI Figure 29.4-6
Roofs	$p = q_h[GC_p - (GC_{pi})]$ where C_p = external pressure coefficient for zones 1 and 2 in ASCE/SEI Figure 29.4-6 for flat, dome, or conical roofs with an angle less than 10 degrees and for conical roofs with an angle between 10 and 30 degrees, inclusive* (GC_{pi}) = internal pressure coefficient for roofed structures (ASCE/SEI 26.11)

*For domed roofs with a roof angle greater than 10 degrees, the external pressures must be determined from ASCE/SEI Figure 27.3-2.

Flowchart 5.7 in Section 5.7 of this book can be used to determine the wind forces and pressures on isolated or grouped circular bins, silos, and tanks.

Rooftop Solar Panels for Buildings of All Heights with Flat Roofs or Gable or Hip Roofs with Slopes Less Than 7 Degrees

Provisions to determine design wind pressures on rooftop solar panels on enclosed or partially enclosed buildings of all heights with flat roofs or with gable or hip roofs with a slope less than or equal to 7 degrees are given in ASCE/SEI 29.4.3. These provisions are applicable where the following criteria are satisfied (see ASCE/SEI Figure 29.4-7):

- Panel chord length, $L_p \le 6.7$ ft (2.04 m)
- Angle the panel makes with the roof surface, $\omega \le 35$ degrees
- Height of the gap between the lower edge of the panel and the roof, $h_1 \le 2$ ft (0.61 m)
- Height of the upper edge of the panel above the roof, $h_2 \le 4$ ft (1.22 m)
- Minimum gap between all panels = 0.25 in. (6.4 mm)
- Spacing of gaps between panels ≤ 6.7 ft (2.04 m)
- Minimum horizontal clear distance between the panels and the edge of the roof $\ge$ greater of $\begin{cases} 2(h_2 - h_{pt}) \\ 4 \text{ ft (1.22 m)} \end{cases}$

The term h_{pt} is the mean parapet height above the adjacent roof surface.

The design wind pressure, p, for rooftop solar panels is determined by ASCE/SEI Equations (29.4-5) and (29.4-5.SI):

$$p = q_h K_d (GC_{rn}) \tag{5.28}$$

In this equation, q_h is the velocity pressure determined in accordance with ASCE/SEI 26.10 at the mean roof height of the building, K_d is the wind directionality factor from ASCE/SEI

Table 26.6-1, and (GC_{rn}) is the net pressure coefficient for rooftop solar panels determined by ASCE/SEI Equation (29.4-6):

$$(GC_{rn}) = \gamma_p \gamma_c \gamma_E (GC_{rn})_{nom} \tag{5.29}$$

The terms in Eqation (5.29) are defined in Table 5.14.

Table 5.14 Net Pressure Coefficients for Rooftop Solar Panels, (GC_{rn})

Parapet height factor, γ_p	$\gamma_p = \text{lesser of} \begin{cases} 1.2 \\ 0.9 + (h_{pt}/h) \end{cases}$
Panel chord factor, γ_c	$\gamma_c = \text{greater of} \begin{cases} 0.6 + 0.06\ L_p\ (L_p \text{ in ft}) \\ 0.8 \end{cases}$
Array edge factor, γ_E	• For uplift loads on exposed panels and within a distance of 1.5 L_p from the end of a row at an exposed edge of the array:* $\gamma_E = 1.5$ • For uplift loads elsewhere and for all downward loads: $\gamma_E = 1.0$
Net nominal pressure coefficient, $(GC_{rn})_{nom}$	$(GC_{rn})_{nom}$ is determined from ASCE/SEI Figure 29.4-7 based on ω and the normalized wind area, A_n, corresponding to interior, edge, and corner conditions (identified as zones 1, 2, and 3 in ASCE/SEI Figure 29.4-7, respectively)**

*A panel is defined as exposed if the distance to the roof edge, d_1, is greater than 0.5h and one of the following applies: (1) distance to the adjacent array $d_1 > \max[4h_2, 4\text{ ft (1.22 m)}]$ or (2) distance to the next adjacent panel $d_2 > \max[4h_2, 4\text{ ft (1.22 m)}]$.
**$A_n = 1{,}000A/\max[L_b, 15\text{ ft (4.57 m)}]^2$ where A is the effective wind area of the structural element of the solar panel being considered and L_b is the normalized building length, which is equal to the least of the following: (1) $0.4(hW_L)^{0.5}$, (2) h, or (3) W_S where W_L and W_S are the widths of the long and short sides of the building, respectively (see ASCE/Figure 29.4-7).

It has been shown from wind tunnel tests that parapets typically increase the wind loads on solar panels, especially for wider buildings. The parapet height factor, γ_p, accounts for this effect. In buildings without parapets, $h_{pt} = 0$ and $\gamma_p = 0.9$. In cases where $h_{pt} \geq 0.3h$, $\gamma_p = 1.2$.

The panel chord factor, γ_c, reduces the wind pressures for shorter chord lengths. This reduction factor scales down linearly from a factor of 1.0 for chord lengths of 6 ft-8 in. (2.03 m) long to 0.8 for chord lengths 3 ft-4 in. (1.02 m) long.

Solar panels are usually installed in arrays with closely spaced rows. The end rows and panels are subjected to larger wind pressures than interior panels, which are sheltered by adjacent panels. To account for these larger pressures, the array edge factor, γ_E, is applied. The value of γ_E depends on whether a panel is exposed and within a certain distance from the end of a row at an exposed edge of an array.

Panels are defined as being exposed where the distance d_1 from the panel edge to the edge of the adjacent solar array or the roof edge is greater than 0.5h and one of the following applies:

- d_1 to the adjacent array > greater of $\begin{cases} 4h_2 \\ 4\text{ ft (1.22 m)} \end{cases}$

- d_2 to the next adjacent panel > greater of $\begin{cases} 4h_2 \\ 4\text{ ft (1.22 m)} \end{cases}$

Illustrated in ASCE/SEI Figure 29.4-7 is an example of a solar panel array and the edge panels that must be designed for increased wind pressures based on the above requirements (indicated by cross-hatched areas in the figure). Possible sheltering from any rooftop structures or equipment was not considered in the analysis; γ_E must be determined based on the distance to the roof edge or adjacent solar array edge only.

The following values of γ_E are to be used for uplift and downward loads on a panel:

Uplift loads:

- For exposed panels within a distance of 1.5 L_p from the end of a row at an exposed edge of the array: $\gamma_E = 1.5$.
- For all other panels, $\gamma_E = 1.0$.

Downward loads:

- For all panels, $\gamma_E = 1.0$.

Single rows of solar panels are to be designed assuming all the panels are exposed (i.e., $\gamma_E = 1.5$).

The nominal net pressure coefficients, $(GC_{rn})_{nom}$, in ASCE/SEI Figure 29.4-7 were generated from wind tunnel test data within the range of parameters noted above for the three zones noted in the figure (zones 1, 2, and 3 correspond to interior, edge, and corner conditions, respectively), and are the same for both positive and negative wind pressures (ASCE/SEI Figure C29.4-1 can be used as a guide in identifying these three zones for buildings with irregular plan dimensions). These curves were created based on a methodology consistent with that used for C&C loads. Linear interpolation is permitted to determine $(GC_{rn})_{nom}$ in cases where ω is between 5 and 15 degrees (see note 2 in ASCE/Figure 29.4-7). The normalized wind area, A_n, in the figure is determined by the following equations:

$$A_n = \frac{1{,}000A}{\text{greater of} \begin{cases} (L_b)^2 \\ (15 \text{ ft})^2 \end{cases}} \quad (L_b \text{ in ft}) \tag{5.30}$$

$$A_n = \frac{1{,}000A}{\text{greater of} \begin{cases} (L_b)^2 \\ (4.57 \text{ m})^2 \end{cases}} \quad (L_b \text{ in m}) \tag{5.31}$$

In these equations, A is the effective wind area for the structural element of the solar panel being considered, which is equal to the greater of the following (see ASCE/SEI 26.2): (1) the tributary area of the structural element or (2) an area equal to the length of the structural element times one-third its length. The term L_b is the normalized building length and is equal to the least of the following: (1) $0.4(hW_L)^{0.5}$, (2) h, or (3) W_S where W_L and W_S are the widths of a building on its longest and shortest sides, respectively (see ASCE/SEI Figure 29.4-7).

Equations to determine $(GC_{rn})_{nom}$ are given in Table 5.15, which are from Reference 5.7. These equations can be used to facilitate determination of $(GC_{rn})_{nom}$ and aid in interpolation where ω is between 5 and 15 degrees.

Table 5.15 Equations to Determine $(GC_{rn})_{nom}$

ω (degrees)	A_n	Zone 1	Zone 2	Zone 3
0-5	1-500	$-0.4261\log_{10}(A_n)+1.500$	$-0.5743\log_{10}(A_n)+2.000$	$-0.6669\log_{10}(A_n)+2.300$
	500-5,000	$-0.2500\log_{10}(A_n)+1.025$	$-0.3000\log_{10}(A_n)+1.260$	$-0.3500\log_{10}(A_n)+1.445$
15-35	1-500	$-0.5372\log_{10}(A_n)+2.000$	$-0.8337\log_{10}(A_n)+2.900$	$-1.0004\log_{10}(A_n)+3.500$
	500-5,000	$-0.2500\log_{10}(A_n)+1.225$	$-0.2500\log_{10}(A_n)+1.325$	$-0.3000\log_{10}(A_n)+1.610$

Flowchart 5.8 in Section 5.7 of this book can be used to determine the design wind pressures on rooftop solar panels for enclosed and partially enclosed buildings with a roof slope less than or equal to 7 degrees.

In lieu of the procedure outlined above, the provisions in ASCE/SEI 29.4.4 are permitted to be used to determine wind pressures on rooftop solar panels provided all the following conditions are met (ASCE/SEI 29.4.3):

- $\omega \leq 2$ degrees
- $h_2 \leq 0.83$ ft (0.25 m)
- Minimum gap between all panels ≥ 0.25 in. (6.4 mm)
- Spacing of gaps between panels ≤ 6.7 ft (2.04 m)

Roof structures supporting solar panels must be designed for the maximum effects due to the two load cases in Table 5.16.

Table 5.16 Load Cases for Roof Structures Supporting Solar Panels

Solar panels are present	• Wind pressures on solar panels determined in accordance with ASCE/SEI 29.4.3 must be applied simultaneously with applicable roof wind pressures on areas of the roof not covered by the plan projection of the solar panels. • Wind pressures on solar panels need not be applied simultaneously to the roof C&C wind pressures determined in accordance with ASCE/SEI Chapter 30 for the roof area covered by the solar panels (i.e., roof structural members beneath the solar panels must be designed for the reactions from wind pressure on the solar panels only). • Roof structural members partially covered by solar panels must be designed for both the solar panel and the roof C&C wind pressures applied over the applicable lengths of the member.
Solar panels are not present	Roof structural members must be designed for the roof C&C wind pressures assuming the solar panels may be removed in the future.

Rooftop Solar Panels Parallel to the Roof Surface on Buildings of All Heights and Roof Slopes

Provisions to determine design wind pressures for rooftop solar panels parallel to the roof surface on enclosed and partially enclosed buildings of all heights and roof slopes are given in ASCE/SEI 29.4.4. These provisions are applicable where the following criteria are satisfied:

- Panels are parallel to the roof surface within a tolerance of 2 degrees
- Maximum height of the panel above the roof surface $h_2 \leq 0.83$ ft (0.25 m)
- Minimum gap between all panels ≥ 0.25 in. (6.4 mm)
- Spacing of gaps between panels ≤ 6.7 ft (2.0 m)
- Distance from the edge of the panel array to the roof edge, gable ridge, or hip ridge $\geq 2h_2$

The design wind pressure, p, is determined by ASCE/SEI Equations (29.4-7) and (29.4-7.SI):

$$p = q_h K_d (GC_p)\gamma_E \gamma_a \tag{5.32}$$

In this equation, q_h is the velocity pressure determined by ASCE/SEI Equation (26.10-1) at the mean roof height, h, K_d is the wind directionality factor from ASCE/SEI Table 26.6-1 and (GC_p) are the net pressure coefficients for C&C of roofs determined from ASCE/SEI Figures 30.3-2A-I through 30.3-7 or ASCE/SEI Figure 30.5-1.

Values of the array edge factor, γ_E, are defined in ASCE/SEI 29.4.4 as follows:

Uplift loads:

- For exposed panels within a distance of $2h_2$ from an edge of an array: $\gamma_E = 1.5$.
- For all other panels, $\gamma_E = 1.0$.

Downward loads:

- For all panels, $\gamma_E = 1.0$.

Panels are defined as being exposed where the distance d_1 from the panel edge to the roof edge is greater than $0.5h$ and one of the following applies:

- d_1 = distance to the adjacent solar array $> 2h_2$
- d_2 = distance to the next adjacent panel $> 2h_2$

The solar panel pressure equalization factor, γ_a, is defined in ASCE/SEI Figure 29.4-8 based on the effective wind area, A. Because of pressure equalization, wind pressures on roof-mounted, planar solar panels close to and parallel to a roof surface are typically lower than the loads on a roof without such panels, except for panels on the perimeter of the array. The solar panel pressure equalization factor accounts for this reduction.

The two load cases in Table 5.16 must be considered for roof structures supporting solar panels where the design wind pressures are determined by ASCE/SEI 29.4.4.

Flowchart 5.9 in Section 5.7 of this book can be used to determine the design wind pressure on rooftop solar panels parallel to the roof surface on buildings of all heights and roof slopes.

Ground-Mounted Fixed-Tilt Solar Panel Systems

Provisions to determine design wind pressures on ground-mounted fixed-tilt solar photovoltaic (PV) panel systems installed in rows are given in ASCE/SEI 29.4.5. These provisions are applicable where the following criteria are satisfied (see ASCE/SEI Figure 29.4-9):

- 6 ft (1.83 m) $\leq L_c \leq$ 14 ft (4.27 m) where L_c is the panel chord length
- $W_g / L_c \geq 7$ where W_g is the shortest row length in an array
- 0 degrees $\leq \omega \leq$ 60 degrees where ω is the angle between the solar panels and the ground surface
- $0.5 \leq h/L_c \leq 0.8$ where h is the mean height of a panel
- $0.20 \leq L_c/S \leq 0.60$ where S is the center-to-center row spacing
- $s_p \leq 0.014L_c$ where s_p is the gap between adjacent panels in both directions
- $S_L \leq 0.25L_c$ where S_L is the horizontal distance in the longitudinal direction of open area within a single row
- $S_T \leq 2S$ where S_T is the horizontal distance in the transverse direction of open area between adjacent rows

The requirements in this section are based on wind tunnel tests assuming the wind-induced deflections are small, that is, the system is not flexible. Flexible systems are prone to aeroelastic effects and torsional instabilities, which must be considered in design. Wind tunnel tests in accordance with ASCE/SEI Chapter 31 should be performed in such cases.

The design wind force, F_n, for ground-mounted solar panels is determined by ASCE/SEI Equations (29.4-8) and (29.4-8.SI):

$$F_n = q_h K_d [\pm(GC_{gn})]A \tag{5.33}$$

In this equation, q_h is the velocity pressure determined in accordance with ASCE/SEI 26.10 at the mean height of the panel, K_d is the wind directionality factor from ASCE/SEI Table 26.6-1,

A is the effective wind area of an element, and (GC_{gn}) is the combined static and dynamic net pressure coefficient for rooftop solar panels determined by ASCE/SEI Equation (29.4-10):

$$GC_{gn} = [\pm(GC_{gn_{\text{static}}}) \pm (GC_{gn_{\text{dynamic}}})] \tag{5.34}$$

The term $(GC_{gn_{\text{static}}})$ is the static net pressure coefficient from ASCE/SEI Figures 29.4-10 based on the effective wind area, A. Similarly, $(GC_{gn_{\text{dynamic}}})$ is the dynamic net pressure coefficient from ASCE/SEI Figure 29.4-11 based on the reduced frequency of the solar panel system, N_s.

The design moment, M_c, about the center-of-plane of panels for ground-mounted solar panels is determined by ASCE/SEI Equations (29.4-9) and (29.4-9.SI):

$$M_c = q_h K_d [\pm(GC_{gm})] A L_c \tag{5.35}$$

In this equation, (GC_{gm}) is the combined static and dynamic net pressure moment coefficient for rooftop solar panels determined by ASCE/SEI Equation (29.4-11):

$$GC_{gm} = [\pm(GC_{gm_{\text{static}}}) \pm (GC_{gm_{\text{dynamic}}})] \tag{5.36}$$

The term $(GC_{gm_{\text{static}}})$ is the static net pressure coefficient from ASCE/SEI Figures 29.4-10 based on the effective wind area, A. Similarly, $(GC_{gm_{\text{dynamic}}})$ is the dynamic net pressure coefficient from ASCE/SEI Figure 29.4-11 based on the reduced frequency of the solar panel system, N_s.

The static and dynamic net pressure coefficients are determined considering the location of the solar panel within the system. Panels at the perimeter of the system (zone 2) are subjected to higher wind pressures than panels in the interior (zone 1), which are sheltered by adjacent panels; as such, the net pressure coefficients in zone 2 are greater than those in zone 1. The boundaries for these two zones are defined in ASCE/SEI Figure 29.4-9.

The reduced frequency, N_s, is determined by ASCE/SEI Equations (29.4-12) and (29.4-12.SI):

$$N_s = 0.682 n L_c / V \tag{5.37}$$

$$N_s = n L_c / V \tag{5.38}$$

In these equations, n is the lowest natural frequency of the mode of interest in Hz and V is the basic wind speed in miles per hour (m/s).

Knowledge of the lowest natural frequency and structural damping in the primary vibration modes of the system are required to determine the dynamic net pressure coefficients in ASCE/SEI Figure 29.4-11 and the reduced frequency. According to ASCE/SEI C29.4.5, a practical approach to obtain these dynamic properties is to conduct field measurements on representative full-scale systems. In lieu of field measurements, estimates of the natural frequencies can be obtained from computer models of the system. The structural damping must be used in ASCE/SEI Figure 29.4-11 instead of the total damping including aerodynamic damping.

The wind tunnel tests for the ground-mounted systems were conducted on a flat surface in the wind tunnel. The topographic factor, K_{zt}, determined by ASCE/SEI Equation (26.8-1) should be applied at sites where the conditions in ASCE/SEI 26.8.1 are satisfied. In cases where K_{zt} is found to be greater than 1.0, zone 1 should be designed as zone 2 (ASCE/SEI C29.4.5.1).

The support posts and foundations for ground-mounted systems must be designed for the simultaneous application of F_n and M_c on the system (ASCE/SEI 29.4.5.3). The design value for the horizontal component of F_n must not be taken less than 0.1 times the vertical component of F_n.

Flowchart 5.10 in Section 5.7 of this book can be used to determine the design forces and design moments for ground-mounted fixed-tilt solar panel systems.

Parapets

ASCE/SEI 29.5 refers to ASCE/SEI 27.3.4 and 28.3.4 for wind pressures on parapets for buildings of all heights and low-rise buildings, respectively. See Sections 5.3.2 and 5.3.3 of this book for more information on these requirements.

Roof Overhangs

ASCE/SEI 29.6 refers to ASCE/SEI 27.3.3 and 28.3.5 for wind pressures on roof overhangs for buildings of all heights and for low-rise buildings, respectively. See Sections 5.3.2 and 5.3.3 of this book for more information on these requirements.

Minimum Design Wind Loading

The minimum design wind load for other structures is equal to 16 lb/ft^2 (0.77 kN/m^2) multiplied by the projected area normal to the wind, A_f (ASCE/SEI 29.7). This load case is to be applied to the structure as a separate load case in addition to the other load cases specified in ASCE/SEI Chapter 29.

5.4 Wind Loads on Components and Cladding

5.4.1 Overview

The requirements for determining wind pressures on components and cladding (C&C) elements of buildings and structures are given in ASCE/SEI Chapter 30. These requirements are permitted to be used in the design of such elements for the building types in ASCE/SEI 30.1.1 provided the conditions and limitations of ASCE/SEI 30.1.2 and 30.1.3 are satisfied.

Wind load procedures for C&C in ASCE/SEI Chapter 30 are given in Table 5.17.

The provisions in ASCE/SEI Chapter 30 are applicable to buildings that comply with the following (ASCE/SEI 30.1.2):

- The building is regular-shaped, that is, the structure has no unusual geometrical irregularities in spatial form.
- The building does not have response characteristics that make it subject to across-wind loading, vortex shedding, or instability caused by galloping or flutter. In addition, the building is not located at a site where channeling effects or buffeting in the wake of upwind obstructions warrant special consideration.

Buildings not meeting these conditions must be designed by either recognized literature that documents such wind load effects or by the wind tunnel procedure in ASCE/SEI Chapter 31 (ASCE/SEI 30.1.3).

Reduction in wind pressure due to apparent shielding by surrounding buildings, other structures, or terrain features is not permitted (ASCE/SEI 30.1.4). Such shielding may be modified or completely removed during the lifespan of the building or structure, which could result in significantly higher wind loads.

It is permitted to use the requirements in ASCE/SEI Chapter 30 to determine design wind loads on air-permeable cladding, including modular vegetative roof assemblies (ASCE/SEI 30.1.5). Loads lower than those determined by ASCE/SEI Chapter 30 are permitted where demonstrated by approved test data or recognized literature.

5.4.2 General Requirements

Minimum Design Wind Pressures

The design wind pressure for C&C elements of buildings must be greater than or equal to a net pressure of 16 lb/ft^2 (0.77 kN/m^2) acting in either direction normal to the surface (ASCE/SEI 30.2.2). This load case must be considered in addition to the other required load cases in ASCE/SEI Chapter 30.

Table 5.17 Wind Load Procedures in ASCE/SEI Chapter 30

Part	Applicability		Conditions*
	Building/Element Type	Height Limit	
1	• Enclosed • Enclosed, low-rise • Partially enclosed • Partially enclosed, low-rise • Partially open • Partially open, low-rise	$h \leq 60$ ft (18.3 m)	• Building has a flat, gable, multispan gable, hip, monoslope, stepped, or sawtooth roof • Wind pressures are determined from a wind pressure equation
2	Enclosed Partially enclosed Partially open	$h > 60$ ft (18.3 m)	• Building has a flat, pitched, gable, hip, mansard, arched or domed roof • Wind pressures are determined from a wind pressure equation
3	Open	None	Building has a pitched, monoslope, or troughed free roof
4	Building appurtenances and rooftop structures and equipment	None	See ASCE/SEI 30.8 through 30.11 for additional conditions for the various element types
5	Nonbuilding structures: Circular bins, silos, and tanks	$h \leq 120$ ft (36.6 m)	—
5	Nonbuilding structures: Rooftop solar panels	None	Buildings with flat roofs or with gable or hip roofs with roof slopes less than or equal to 7 degrees
5	Nonbuilding structures: Rooftop pavers	None	Buildings with roof slopes less than or equal to 7 degrees

*The following conditions must be satisfied for buildings designed in accordance with ASCE/SEI Chapter 30: (1) the building must be regular-shaped as defined in ASCE/SEI 26.2, (2) the building does not have response characteristics making it subject to across-wind loading, vortex shedding, instability due to galloping or flutter, and (3) the building is not located at a site where channeling effects or buffeting in the wake of upwind obstructions warrant special consideration.

Tributary Areas Greater than 700 ft^2 (65 m^2)

It is permitted to design C&C elements of buildings with tributary areas greater than 700 ft^2 (65 m^2) for wind pressures determined using the provisions for main wind force resisting systems (MWFRSs) [ASCE/SEI 30.2.3]. It is assumed localized wind effects on elements with tributary areas greater than this limit are not as pronounced as those on elements with smaller tributary areas.

External Pressure Coefficients

Combined gust-effect factor and external pressure coefficients for C&C elements, (GC_p), are given for the various elements covered in ASCE/SEI Chapter 30. According to ASCE/SEI 30.2.4, these factors and coefficients must not be separated.

5.4.3 Effective Wind Area

The effective wind area, A, is used throughout ASCE/SEI Chapter 30 to determine external pressure coefficients, which, in turn, are used to determine design wind pressures on C&C elements.

In accordance with ASCE/SEI 26.2, the effective wind area, A is determined as follows for the C&C elements in ASCE/SEI Figures 30.3-1 through 30.3-7, 30.4-1, 30.5-1 through 30.5-3, 30.7-1, 30.9-1, and 30.9-2:

$$A = \text{greater of} \begin{cases} \text{Area tributary to the element} \\ \text{Span length} \times (\text{Span length} / 3) \end{cases} \tag{5.39}$$

Two cases can arise when determining the effective wind area for C&C elements. In the first case, the effective wind area is equal to the total area tributary to the element. The second case occurs where components are spaced relatively close together (such as wall studs or roof trusses). Due to the close spacing, the load is distributed and shared among adjoining components. To account for this load distribution, an effective width equal to one-third the span length is used in determining the effective wind area, which in this case is equal to the span length multiplied by the effective width. This area is usually greater than the tributary area of the component, which is generally long and narrow. The greater of the effective areas from the two of these cases is used in determining the pressure coefficients.

In general, the effective wind area for single-unit windows, doors, and other fenestration assemblies is equal to the overall area of the assembly. For multiunit assemblies mulled together or for complex fenestration systems, it is recommended to contact the product manufacturer for guidance on the appropriate effective wind area to use when calculating design wind pressures.

The effective wind areas for a cladding system (consisting of glazing panels and mullions) and a joist system are illustrated in Figure 5.21. The glazing panels span between mullions, which are spaced a distance, s, on center. The mullions span the full story height, h_1, between floor slabs.

For rooftop solar panels, the effective wind area for a structural element in a solar panel array is equal to the greater of the following: (1) the tributary area of the element (i.e., span length times perpendicular distance to the adjacent parallel elements) and (2) the span length of the element times one-third the span length. The wind pressure is determined in accordance with ASCE/SEI Figure 29.4-7 using the calculated effective wind area, and that pressure is applied over the actual tributary area of the element. Information on the appropriate effective wind area to use for various solar array support systems can be found in ASCE/SEI C26.2.

The effective wind area for cladding fasteners must not be greater than the area tributary to an individual fastener.

5.4.4 Low-Rise Buildings (Part 1)

Overview

The requirements in Part 1 of ASCE/SEI Chapter 30 are applicable to enclosed, partially enclosed, and partially open low-rise buildings (see ASCE/SEI 26.2 for the definition of a low-rise building) and buildings with a mean roof height, h, less than or equal to 60 ft (18.3 m) that have flat, gable, multispan gable, hip, monoslope, stepped, or sawtooth roofs (see Table 5.17).

In addition to the conditions and limitations in ASCE/SEI 30.1.2 and 30.1.3, respectively, the conditions indicated on the selected figure(s) must also be satisfied (ASCE/SEI 30.3.1).

The procedure in ASCE/SEI 30.3 entails the determination of the velocity pressure at the mean roof height of the building (which is determined as a function of exposure, topographic effects, wind velocity, and building occupancy), wind directionality, combined gust-effect factors, external pressure coefficients, and internal pressure coefficients. Design wind pressures are obtained for designated zones on the walls and roof of buildings.

The overall steps that can be used to determine wind pressures on C&C of such buildings are given in ASCE/SEI Table 30.3-1.

Figure 5.21
Effective Wind Areas, A, for a Cladding System and a Joist System

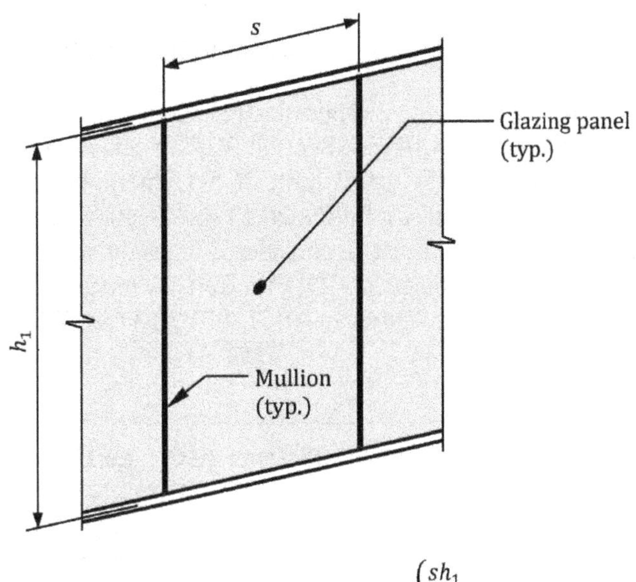

For a glazing panel: $A = \text{greater of} \begin{cases} sh_1 \\ s(s/3) \end{cases}$

For a mullion: $A = \text{greater of} \begin{cases} sh_1 \\ h_1(h_1/3) \end{cases}$

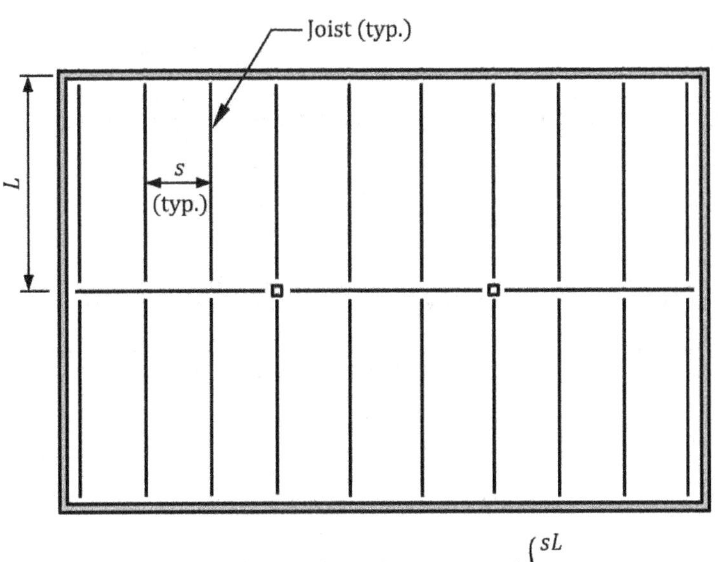

For an interior joist: $A = \text{greater of} \begin{cases} sL \\ L(L/3) \end{cases}$

Design Wind Pressures

Design wind pressures on C&C elements of low-rise buildings are determined by ASCE/SEI Equations (30.3-1) and (30.3-1.SI):

$$p = q_h K_d[(GC_p) - (GC_{pi})] \tag{5.40}$$

The velocity pressure, q_h, is determined at the mean roof height of the building, h, in accordance with ASCE/SEI 26.10 (see Section 5.2.7 of this book) and the wind directionality factor, K_d, is determined from ASCE/SEI Table 26.6-1. Internal pressure coefficients, (GC_{pi}), are determined in accordance with ASCE/SEI 26.13 (see Section 5.2.10 of this book). Both positive and negative values of (GC_{pi}) must be considered to establish the critical load effects.

External pressure coefficients, (GC_p), for C&C elements located on the walls and roof of a building are determined by the figures in ASCE/SEI 30.3 (see Table 5.18).

Table 5.18 External Pressure Coefficients, (GC_p), for Use in Part 1 of ASCE/SEI Chapter 30

Element	ASCE/SEI Figure Number
Walls	30.3-1
Bottom horizontal surface of elevated buildings	30.3-1A and 30.3-2A
Gable roofs with roof angles less than or equal to 7 degrees	30.3-2A
Gable roofs with roof angles greater than 7 degrees and less than or equal to 20 degrees	30.3-2B
Gable roofs with roof angles greater than 20 degrees and less than or equal to 27 degrees	30.3-2C
Gable roofs with roof angles greater than 27 degrees and less than or equal to 45 degrees	30.3-2D
Hip roofs with roof angles greater than 7 degrees and less than or equal to 20 degrees	30.3-2E
Hip roofs with roof angles greater than 20 degrees and less than or equal to 27 degrees	30.3-2F
Hip roofs with roof angles equal to 45 degrees	30.3-2G
Stepped roofs with roof angles less than or equal to 7 degrees	30.3-3
Multispan gable roofs	30.3-4
Monoslope roofs with roof angles greater than 3 degrees and less than or equal to 10 degrees	30.3-5A
Monoslope roofs with roof angles greater than 10 degrees and less than or equal to 30 degrees	30.3-5B
Sawtooth roofs	30.3-6
Domed roofs	30.3-7
Arched roofs	30.3-8

The equations in ASCE/SEI Tables C30.3-1 through C30.3-9 in ASCE/SEI C30.3.2 can be used to determine the external pressure coefficients, (GC_p), in ASCE/SEI Figures 30.3-1 and 30.3-2A through 30.3-2G.

Bottom Horizontal Surface of Elevated Buildings

The provisions of ASCE/SEI 30.3.2.1 are applicable to elevated buildings where the height above grade of the bottom surface of the elevated building, h_B, is greater than or equal to 2 ft (0.61 m).

The design wind pressures on the C&C elements on the bottom flat horizontal surface of an elevated building are determined by Equation (5.40) using the roof pressure coefficients, (GC_p), from ASCE/SEI Figure 30.3-2A with the following modifications:

- For determining zone dimensions in ASCE/SEI Figure 30.3-2A, set $h = h_B$. For elevated buildings with a flat bottom horizontal building surface and situated on a slope, h_B must be taken as the maximum height between the slope and the bottom of the elevated building.
- The shaded regions in ASCE/SEI Figure 30.3-1A, which correspond to areas of the horizontal surface above partially enclosed spaces and areas extending a distance a_B perpendicular to the walls beneath the elevated building with plan dimensions greater than 4 ft (1.22 m), must be designed to resist positive pressures determined by Equation (5.40) using the zone 4 wall pressure coefficients in ASCE/SEI Figure 30.3-1. The distance a_B is equal to the lesser of $0.4h_B$ and the width of the wall.

Negative pressure coefficients are to be used for downward loading on the bottom surface and positive pressure coefficients are to be used for upward loading.

Information on the derivation of the pressures to be used on the C&C elements of elevated buildings can be found in ASCE/SEI C30.3.2.1.

Flowchart 5.11 in Section 5.7 of this book can be used to determine the design wind pressures on C&C elements in accordance with Part 1 in ASCE/SEI Chapter 30 (ASCE/SEI 30.3.2).

5.4.5 Buildings with $h > 60$ ft (18.3 m) (Part 2)

Overview

The requirements in Part 2 of ASCE/SEI Chapter 30 are applicable to enclosed, partially enclosed, and partially open buildings with a mean roof height, h, greater than 60 ft (18.3 m) that have flat, pitched, gable, hip, mansard, arched, or domed roofs (see Table 5.17).

In addition to the conditions and limitations in ASCE/SEI 30.1.2 and 30.1.3, respectively, the conditions indicated on the selected figure(s) must also be satisfied (ASCE/SEI 30.4.1).

The procedure in ASCE/SEI 30.4 entails the determination of the velocity pressure (which is determined as a function of exposure, topographic effects, wind velocity, and building occupancy), wind directionality, combined gust-effect factors, external pressure coefficients, and internal pressure coefficients. Design wind pressures are obtained for designated zones on the walls and roof of buildings.

The overall steps that can be used to determine wind pressures on C&C of such buildings are given in ASCE/SEI Table 30.4-1.

Design Wind Pressures

Design wind pressures on C&C elements of buildings with $h > 60$ ft (18.3 m) are determined by ASCE/SEI Equations (30.4-1) and (30.4-1.SI):

$$p = qK_d(GC_p) - q_iK_d(GC_{pi}) \tag{5.41}$$

For windward walls, the velocity pressure, q, is determined at height z above the ground in accordance with ASCE/SEI 26.10 (see Section 5.2.7 of this book), and for leeward walls, sidewalls and roofs, q is evaluated at the mean roof height, h (i.e., $q = q_h$). The velocity pressure for internal pressure determination, q_i, is determined as follows:

- For windward walls, sidewalls, leeward walls, and roofs of enclosed buildings and partially open buildings, and for negative internal pressure evaluation in partially enclosed buildings: $q_i = q_h$.
- For positive internal pressure evaluation in partially enclosed buildings: $q_i = q_z$ where height z is defined as the level of the highest opening in the building that could affect the positive internal pressure. For positive internal pressure evaluation, q_i may conservatively be evaluated at height h $(q_i = q_h)$.

The wind directionality factor, K_d, is determined from ASCE/SEI Table 26.6-1 and the internal pressure coefficients, (GC_{pi}), are determined in accordance with ASCE/SEI 26.13 (see Section 5.2.10 of this book). Both positive and negative values of (GC_{pi}) must be considered to establish the critical load effects.

External pressure coefficients, (GC_p), for C&C elements located on the walls and roof of a building are determined by the figures in ASCE/SEI 30.4 (see Table 5.19).

Table 5.19 External Pressure Coefficients, (GC_p), for Use in Part 2 of ASCE/SEI Chapter 30

Element	ASCE/SEI Figure Number
Walls and roofs with angles less than or equal to 7 degrees	30.4-1
Buildings with roof angles greater than 7 degrees and with other roof geometries	30.3-2A through 2I, 30.3-5A and 30.3-5B
Bottom horizontal surface of elevated buildings	30.4-1A and 30.4-1
Domed roofs	30.3-7
Arched roofs	30.3-8

Bottom Horizontal Surface of Elevated Buildings

The design wind pressures on the C&C elements on the bottom flat horizontal surface of an elevated building are determined by Equation (5.41) using the roof pressure coefficients, (GC_p), from ASCE/SEI Figure 30.4-1 with the following modifications (ASCE/SEI 30.4.2.1):

- The velocity pressure, q, must be determined at a height equal to the height above grade of the bottom surface, h_B, plus 25 percent of the height of the elevated building above the bottom surface, that is, $h_B + 0.25(h - h_B)$. For elevated buildings with a flat bottom horizontal building surface and situated on a slope, h_B must be taken as the maximum height between the slope and the bottom of the elevated building.
- The shaded regions in ASCE/SEI Figure 30.4-1A, which correspond to areas of the horizontal surface above partially enclosed spaces and areas extending a distance a_B perpendicular to the walls beneath the elevated building with plan dimensions greater than 4 ft (1.22 m), must be designed to resist positive pressures determined by Equation (5.41) using the zone 4 wall pressure coefficients in ASCE/SEI Figure 30.4-1. The distance a_B is equal to the lesser of $0.4h_B$ and the width of the wall.

Negative pressure coefficients are to be used for downward loading on the bottom surface and positive pressure coefficients are to be used for upward loading.

Information on the derivation of the pressures to be used on the C&C elements of elevated buildings can be found in ASCE/SEI C30.4.2.1.

For buildings with a mean roof height greater than 60 ft (18.3 m) and less than 90 ft (27.4 m) where the mean roof height, h, is less than the least horizontal dimension of the building, the external pressure coefficients in ASCE/SEI Figures 30.3-1 through 30.3-6 for low-rise buildings are permitted to be used to calculate the design wind pressures instead of the external pressure coefficients in the figures in Table 5.19.

Flowchart 5.12 in Section 5.7 of this book can be used to determine the design wind pressures on C&C elements in accordance with Part 2 in ASCE/SEI Chapter 30 (ASCE/SEI 30.4.2).

5.4.6 Open Buildings (Part 3)

Overview

The requirements in Part 3 of ASCE/SEI Chapter 30 are applicable to open buildings that have pitched, monoslope, or troughed free roofs (see Table 5.17).

In addition to the conditions and limitations in ASCE/SEI 30.1.2 and 30.1.3, respectively, the conditions indicated on the selected figure(s) must also be satisfied (ASCE/SEI 30.5.1).

The procedure in ASCE/SEI 30.5 entails the determination of the velocity pressure at the mean roof height, h, of the building (which is determined as a function of exposure, topographic effects, wind velocity, and building occupancy), wind directionality, gust-effect factor, and net pressure coefficients.

The overall steps that can be used to determine wind pressures on C&C of such buildings are given in ASCE/SEI Table 30.5-1.

Design Wind Pressures

Design wind pressures on C&C elements of open buildings are determined by ASCE/SEI Equation (30.5-1):

$$p = q_h K_d G C_N \tag{5.42}$$

The velocity pressure, q_h, is determined at the mean roof height of the building, h, in accordance with ASCE/SEI 26.10 (see Section 5.2.7 of this book) and the wind directionality factor, K_d, is determined from ASCE/SEI Table 26.6-1. The gust-effect factor, G, is determined in accordance with ASCE/SEI 26.11 (see Section 5.2.8 of this book).

Net pressure coefficients, C_N, for C&C elements located on the roof of an open building are determined by the figures in ASCE/SEI 30.5 (see Table 5.20). These coefficients include contributions from both the top and bottom surfaces of the roof which implies the element receives pressure from both surfaces.

Table 5.20 Net Pressure Coefficients, C_N, for Use in Part 3 of ASCE/SEI Chapter 30

Element	ASCE/SEI Figure Number
Monosloped free roofs with roof angles less than or equal to 45 degrees	30.5-1
Pitched free roofs with roof angles less than or equal to 45 degrees	30.5-2
Troughed free roofs with roof angles less than or equal to 45 degrees	30.5-3

Flowchart 5.13 in Section 5.7 of this book can be used to determine the design wind pressures on C&C elements in accordance with Part 3 in ASCE/SEI Chapter 30 (ASCE/SEI 30.5.2).

5.4.7 Building Appurtenances, Rooftop Structures, and Equipment (Part 4)

Overview

Methods to determine wind pressures on C&C elements of parapets (ASCE/SEI 30.6), roof overhangs (ASCE/SEI 30.7), rooftop structures and equipment (ASCE/SEI 30.8), and attached canopies on buildings (ASCE/SEI 30.9) are given in Part 4 of ASCE/SEI Chapter 30.

The procedures in Part 4 entail the determination of velocity pressure, wind directionality, combined gust-effect factors, external pressure coefficients, and internal pressure coefficients. Design wind pressures are obtained for designated zones on the surfaces of buildings.

Design Wind Pressures

Parapets

Steps to determine design wind pressures on C&C elements of parapets are given in ASCE/SEI Table 30.6-1.

Design wind pressures are determined by ASCE/SEI Equation (30.6-1):

$$p = q_p K_d[(GC_p) - (GC_{pi})] \quad (5.43)$$

In this equation, q_p is the velocity pressure evaluated at the top of the parapet in accordance with ASCE/SEI 26.10 (see Section 5.2.7 of this books), K_d is the wind directionality factor determined from ASCE/SEI Table 26.6-1 and (GC_{pi}) are the internal pressure coefficients determined in accordance with ASCE/SEI 26.13 (see Section 5.2.10 of this book). Both positive and negative values of (GC_{pi}) must be considered to establish the critical load effects.

External pressure coefficients, (GC_p), for C&C elements of parapets are determined by the figures referenced in ASCE/SEI 30.6 (see Table 5.21).

Table 5.21 External Pressure Coefficients, (GC_p), for C&C Elements of Parapets

C&C Elements	ASCE/SEI Figure Number
Walls with $h \leq 60$ ft (18.3 m)	30.3-1
Flat, gable, and hip roofs	30.3-2A through 30.3-2C
Stepped roofs	30.3-3
Multispan gable roofs	30.3-4
Monoslope roofs	30.3-5A and 30.3-5B
Sawtooth roofs	30.3-6
Domed roofs of all heights	30.3-7
Walls and flat roofs with $h > 60$ ft (18.3 m)	30.4-1
Arched roofs	30.3-8

Load cases A and B must be considered for the windward and leeward parapets, respectively (see ASCE/SEI Figure 30.6-1 and Figure 5.22):

- Load case A (windward parapets): Windward (front surface) parapet pressure, p_1, is determined using the positive wall pressure, p_5, from zones 4 or 5 in the applicable figure in Table 5.21 (ASCE/SEI Figures 30.3-1 or 30.4-1). Leeward (back surface) parapet pressure, p_2, is determined using the negative edge or corner roof pressure, p_7, from zones 2 or 3 in the applicable figure in Table 5.21 (ASCE/SEI Figures 30.3-2A, B, or C, 30.3-3, 30.3-4, 30.3-5A or B, 30.3-6, 30.3-7, 30.4-1 or 30.3-8).
- Load case B (leeward parapets): Windward (back surface) parapet pressure, p_3, is determined using the positive wall pressure, p_5, from zones 4 or 5 in the applicable figure in Table 5.21 (ASCE/SEI Figures 30.3-1 or 30.4-1). Leeward (front surface) parapet pressure, p_4, is determined using the negative wall pressure, p_6, from zones 4 or 5 in the applicable figure in Table 5.21 (ASCE/SEI Figures 30.3-1 or 30.4-1).

Flowchart 5.14 in Section 5.7 of this book can be used to determined design wind pressures on C&C elements of parapets in accordance with ASCE/SEI 30.6.

Roof Overhangs

Steps to determine design wind pressures on C&C elements of roof overhangs are given in ASCE/SEI Table 30.7-1.

Design wind pressures are determined by ASCE/SEI Equations (30.7-1) and (30.7-1.SI):

$$p = q_h K_d[(GC_p) - (GC_{pi})] \quad (5.44)$$

In this equation, q_h is the velocity pressure evaluated at the mean roof height of the building, h, in accordance with ASCE/SEI 26.10 (see Section 5.2.7 of this books), K_d is the wind directionality

Figure 5.22 Application of Wind Pressures to C&C Elements of Parapets in Accordance with ASCE/SEI 30.6

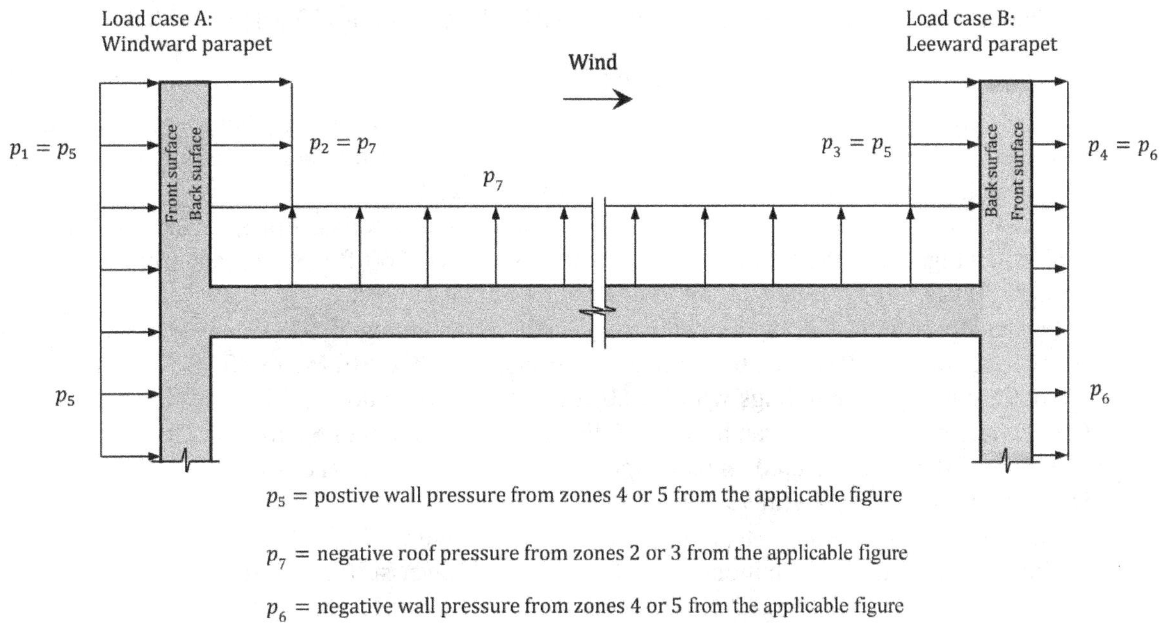

Load case A: Windward parapet

Load case B: Leeward parapet

Wind

$p_1 = p_5$ $p_2 = p_7$ p_7 $p_3 = p_5$ $p_4 = p_6$ p_5 p_6

Front surface Back surface

p_5 = postive wall pressure from zones 4 or 5 from the applicable figure

p_7 = negative roof pressure from zones 2 or 3 from the applicable figure

p_6 = negative wall pressure from zones 4 or 5 from the applicable figure

factor determined from ASCE/SEI Table 26.6-1, and (GC_{pi}) are the internal pressure coefficients determined in accordance with ASCE/SEI 26.13 (see Section 5.2.10 of this book). Both positive and negative values of (GC_{pi}) must be considered to establish the critical load effects.

External pressure coefficients, (GC_p), for C&C elements of roof overhangs are determined as follows (ASCE/SEI 30.7):

- For gable roofs with a roof angle less than or equal to 7 degrees, determine (GC_p) from ASCE/SEI Figure 30.3-2A based on the effective wind area, A.
- For all other cases, (GC_p) for the top surface of the roof overhang is the same as the (GC_p) for the roof surface, and (GC_p) for the bottom surface of the roof overhang is the same as the (GC_p) for the adjacent wall (see ASCE/SEI Figure 30.7-1).

Flowchart 5.15 in Section 5.7 of this book can be used to determine design wind pressures on C&C elements of roof overhangs in accordance with ASCE/SEI 30.7.

Rooftop Structures and Equipment for Buildings

Design wind pressures on the C&C elements of rooftop structures and equipment for buildings are determined using the requirements in ASCE/SEI 29.4.1, which are applicable in the design of the MWFRS (ASCE/SEI 30.8):

- On the walls: Design wind pressure = horizontal wind load determined by ASCE/SEI Equations (29.4-2) and (29.4-2.SI) divided by the wall surface area of the rooftop structure
- On the roof: Design wind pressure = vertical wind load determined by ASCE/SEI Equations (29.4-3) and (29.4-3.SI) divided by the horizontal projected area of the roof of the rooftop structure

The wall pressures can act inward or outward and the roof pressure acts upward.

Attached Canopies on Buildings

Steps to determine design wind pressures on C&C elements of attached canopies on buildings are given in ASCE/SEI Table 30.9-1.

Design wind pressures are determined by ASCE/SEI Equations (30.9-1) and (30.9-1.SI):

$$p = q_h K_d (GC_p) \tag{5.45}$$

In this equation, q_h is the velocity pressure evaluated at the mean roof height of the building, h, in accordance with ASCE/SEI 26.10 (see Section 5.2.7 of this books), and K_d is the wind directionality factor determined from ASCE/SEI Table 26.6-1.

Net pressure coefficients, (GC_p), for C&C elements of attached canopies are determined from ASCE/SEI Figures 30.9-1A and 30.9-1B for buildings with $h \leq 60$ ft (18.3 m) and from ASCE/SEI Figures 30.9-2A and 30.9-2B for buildings with $h > 60$ ft (18.3 m).

For canopies with two exposed surfaces, ASCE/SEI Figures 30.9-1A and 30.9-1B are both required for buildings with $h \leq 60$ ft (18.3 m). Similarly, ASCE/SEI Figures 30.9-2A and 30.9-2B are both required for buildings with $h > 60$ ft (18.3 m). For canopies with one exposed surface (e.g., where wind pressures can be applied directly to the bottom of the top surface only), ASCE/SEI Figure 30.9-1B is required for buildings with $h \leq 60$ ft (18.3 m) and ASCE/SEI Figure 30.9-2B is required for buildings with $h > 60$ ft (18.3 m).

The net pressure coefficients, (GC_p), in ASCE/SEI Figures 30.9-1A and 30.9-2A, which are applicable where the contributions from the upper and lower surfaces of the canopy are considered individually, must be used to determine wind pressures on elements and fasteners on the upper and lower surfaces. The net pressure coefficients, (GC_p), in ASCE/SEI Figures 30.9-1B and 30.9-2B, which are applicable where the contributions from the upper and lower surfaces of the canopy are considered simultaneously, must be used to determine wind pressures on the structural members of the canopy, including the connections to the building.

The equations in ASCE/SEI Tables C30.9-1 and C30.9-2 can be used to determine (GC_p) based on the effective wind area, A, for buildings with $h \leq 60$ ft (18.3 m). Similarly, the equations in ASCE/SEI Tables C30.9-3 and C30.9-4 can be used to determine (GC_p) for buildings with $h > 60$ ft (18.3 m).

For buildings with a mean roof height between 60 ft (18.3 m) and 90 ft (27.4 m), it is permitted to determine (GC_p) by linear interpolation between the value obtained from ASCE/SEI Figure 30.9-1A for a mean roof height of 60 ft (18.3 m) and the value obtained from ASCE/SEI Figure 30.9-2A for a mean roof height of 90 ft (27.4 m) for a specific h_c/h_e value where h_c is the mean canopy height and h_e is the mean eave height (see exception 1 in ASCE/SEI 30.9). The linear interpolation is permitted instead of using the value of (GC_p) from only ASCE/SEI Figure 30.9-2A. Similarly, it is permitted to determine (GC_p) by linear interpolation between the value obtained from ASCE/SEI Figure 30.9-1B for a mean roof height of 60 ft (18.3 m) and the value obtained from ASCE/SEI Figure 30.9-2B for a mean roof height of 90 ft (27.4 m) for a specific h_c/h_e value (see exception 2 in ASCE/SEI 30.9). The linear interpolation is permitted instead of using the value of (GC_p) from only ASCE/SEI Figure 30.9-2B.

Flowchart 5.16 in Section 5.7 of this book can be used to determine design wind pressures on attached canopies on buildings in accordance with ASCE/SEI 30.9.

5.4.8 Nonbuilding Structures (Part 5)

Overview

Methods to determine wind pressures on C&C elements of circular bins, silos, and tanks with $h \leq 120$ ft (36.6 m) [ASCE/SEI 30.10], rooftop solar panels for buildings of all heights with flat roofs or with gable or hip roofs with slopes less than 7 degrees (ASCE/SEI 30.11), and roof pavers for buildings of all heights with roof slopes less than 7 degrees (ASCE/SEI 30.12) are given in Part 5 of ASCE/SEI Chapter 30.

The procedures in Part 5 entail the determination of velocity pressure, wind directionality, external pressure coefficients, and internal pressure coefficients.

Design Wind Pressures

Circular Bins, Silos, and Tanks with $h \leq 120$ ft (36.6 m)

Steps to determine design wind pressures on C&C elements of circular bins, silos, and tanks with $h \leq 120$ ft (36.6 m) are given in ASCE/SEI Table 30.10-1.

Design wind pressures are determined by ASCE/SEI Equation (30.10-1):

$$p = q_h K_d[(GC_p) - (GC_{pi})] \tag{5.46}$$

In this equation, q_h is the velocity pressure evaluated at the mean roof height of the structure, h, in accordance with ASCE/SEI 26.10 (see Section 5.2.7 of this books), K_d is the wind directionality factor determined from ASCE/SEI Table 26.6-1, and (GC_{pi}) are the internal pressure coefficients determined in accordance with ASCE/SEI 26.13 (see Section 5.2.10 of this book and ASCE/SEI 30.10.3). Both positive and negative values of (GC_{pi}) must be considered to establish the critical load effects.

External pressure coefficients, (GC_p), for C&C elements are determined in accordance with ASCE/SEI 30.10.2 for external walls, ASCE/SEI 30.10.4 for roofs, and ASCE/SEI 30.10.5 for undersides.

A summary of the required (GC_p) and (GC_{pi}) for isolated circular bins, silos, and tanks is given in Table 5.22.

Table 5.22 Summary of External and Internal Pressure Coefficients for Isolated Circular Bins, Silos, and Tanks

Element	Pressure Coefficients
Walls	External pressure coefficients: $GC_{p(\alpha)} = k_b C_{(\alpha)}$* where: $k_b = \begin{cases} 1.0 \text{ for } C_{(\alpha)} \geq -0.15 \\ 1.0 - 0.55[C_{(\alpha)} + 0.15]\log_{10}(h_c/D) \text{ for } C_{(\alpha)} < -0.15 \end{cases}$ $C_{(\alpha)} = -0.5 + 0.4\cos\alpha + 0.8\cos 2\alpha + 0.3\cos 3\alpha - 0.1\cos 4\alpha - 0.05\cos 5\alpha$ h_c = height of the cylinder (see ASCE/SEI Figure 30.10-1) D = diameter of the cylinder α = angle from the wind direction to the point on a wall of a circular bin, silo, or tank in degrees
Open-topped circular bins, silos, and tanks	Internal surface of external walls: $(GC_{pi}) = -0.9 - 0.35\log_{10}(h_c/D)$
Circular bins, silos, and tanks that are not open-topped	Internal pressure coefficients determined in accordance with ASCE/SEI Table 26.13-1
Roofs or lids	External pressure coefficients, (GC_p), are determined from ASCE/SEI Figure 30.10-2 for the following (see ASCE/SEI Figure C30.10-2): • Class 1: Flat roofs and conical roofs where the roof angle is less than 10 degrees, and for domed roofs where the average roof angle is less than 10 degrees • Class 2a: Conical roofs where the roof angle is between 10 and 15 degrees inclusive • Class 2b: Conical roofs where the roof angle is greater than 15 degrees and less than or equal to 30 degrees
Undersides of circular bins, silos, and tanks	External pressure coefficients, (GC_p) [see ASCE/SEI Figure 30.10-2]: $(GC_p) = \begin{cases} 1.2 \text{ and } -0.9 \text{ for zone 3} \\ 0.8 \text{ and } -0.6 \text{ for zones 1 and 2} \end{cases}$

*Applicable to circular bins, silos, and tanks standing on the ground or supported by columns where $C \leq h_c$ (see ASCE/Figure 30.10-1) and $0.25 \leq h_c/D \leq 4.0$.

Structures are assumed to be grouped where the center-to-center spacing is less than 1.25 times the diameter of the structure, D. In such cases, the external pressure coefficients for the roof and walls of the grouped structures are given in ASCE/SEI 30.10.6 (see Table 5.23).

Table 5.23 Summary of External Pressure Coefficients for Grouped Circular Bins, Silos, and Tanks

Element	External Pressure Coefficients, (GC_p)
Roofs	ASCE/SEI Figure 30.10-3 for zones 1, 2, 3a, 3b, and 4
Walls	ASCE/SEI Figure 30.10-4 for zones 5a, 5b, 8, and 9

Flowchart 5.17 in Section 5.7 of this book can be used to determine design wind pressures on C&C elements of circular bins, silos, and tanks in accordance with ASCE/SEI 30.10.

Rooftop Solar Panels for Buildings of All Heights with Flat Roofs or Gable and Hip Roofs with Slopes Less Than 7 Degrees

Design wind pressures on the C&C elements of rooftop solar modules and panels conforming to the geometric requirements of ASCE/SEI 29.4.3 must be determined by the requirements in that section, which are applicable to the MWFRS (see ASCE/SEI 30.11 and Section 5.3.4 of this book).

Rooftop Pavers for Buildings of All Heights with Roof Slopes Less Than or Equal to 7 Degrees

Design net uplift pressures, p, on roof pavers for buildings of all heights with roof slopes less than or equal to 7 degrees are determined by ASCE/SEI Equation (30.12-1):

$$p = q_h K_d C_{L_{net}} \quad (\text{lb/ft}^2) \tag{5.47}$$

In this equation, q_h is the velocity pressure evaluated at the mean roof height of the structure, h, in accordance with ASCE/SEI 26.10 (see Section 5.2.7 of this books) and K_d is the wind directionality factor determined from ASCE/SEI Table 26.6-1.

The design net uplift pressure coefficient, $C_{L_{net}}$, must be determined by one of the following methods:

- $C_{L_{net}} = GC_p$ from ASCE/SEI Figures 30.3-2A and 30.5-1
- An approved wind tunnel method performed in accordance with ASCE/SEI Chapter 31
- Methods described in recognized literature

Using the external pressure coefficients from ASCE/SEI Figures 30.3-2A and 30.5-1 typically results in a conservative design. For buildings with roof heights less than 150 ft (45.7 m), the reference in ASCE/SEI C30.12 can be used for pavers laid directly on roof membranes as part of a ballast system.

5.5 Wind Tunnel Procedure

The Wind Tunnel Procedure in ASCE/SEI Chapter 31 can be utilized to determine design wind pressures on MWFRSs and C&C elements of any building or other structure in lieu of any of the procedures in ASCE/SEI Chapters 27 through 30, and it must be used where the conditions of these procedures are not satisfied (in particular, where a structure contains any of the characteristics defined in ASCE/SEI 27.1.3, 28.1.3, 29.1.3 or 30.1.3).

Wind tunnel tests should be used where buildings or other structures are not regularly-shaped, are flexible and/or slender, have the potential to be buffeted by upwind buildings or other structures or have the potential to be subjected to accelerated wind flow from channeling by buildings or topographic features. In the case of tall, slender buildings, only a wind tunnel test can properly capture any possible effects due to vortex shedding, galloping, or flutter. For buildings in the heart of a city, a wind tunnel test is mandatory because exposures B through D cannot properly capture the conditions in such cases. Every project has unique characteristics, and engineering judgment also plays a role in the decision-making process. When determining whether a wind tunnel test is required or not, it is always very important to keep in mind the limitations in ASCE/SEI Chapters 27 through 30, especially the general one related to along-wind response.

Of all the methods in ASCE/SEI 7 for the determination of wind pressures, the wind tunnel procedure is generally considered to produce the most accurate results. For certain types of buildings, the results from a wind tunnel test will be smaller than those from any of the other methods. On the other hand, wind tunnel tests can yield results greater than those obtained from the other methods under certain conditions; as such, it is important to understand when such tests are required to adequately design the building or other structure for the effects of wind.

Information on the basic types of wind tunnel test models commonly used is given in ASCE/SEI C31. Wind tunnel tests can also provide valuable information on snow loads, the effects of wind on pedestrians, and concentrations of air-pollutant emissions, to name a few. References providing detailed information and guidance for the determination of wind loads and other types of design data by wind tunnel tests are given in ASCE/SEI C31, including Reference 5.8.

5.6 Tornado Loads

5.6.1 Scope

The effects of tornado loads must be investigated on the MWFRS and C&C elements of risk category III and IV buildings and other structures located in the tornado-prone region, which is identified in IBC Figure 1609.5 and ASCE/SEI Figure 32.1-1 (IBC 1609.5 and ASCE/SEI 32.1.1). Such structures must be designed to resist the greater of the effects due to tornado loads determined in accordance with ASCE/SEI Chapter 32 and the applicable wind load effects determined in accordance with ASCE/SEI Chapters 26 through 31 (see Sections 5.2 through 5.5 of this book) using the load combinations in IBC 1605 and ASCE/SEI Chapter 2.

The tornado-prone region is defined as the area of the conterminous United States most vulnerable to tornadoes. This region is approximately the same as the portion of the conterminous United States east of the Continental Divide. Although tornadoes can occur to the west of the tornado-prone region, the tornado speeds associated with risk category III and IV mean recurrence intervals would produce tornado loads less than those from wind loads based on the associated basic wind speeds. Additional information on the tornado-prone region can be found in ASCE/SEI C32.1.1.

Tornado loads need not be considered in the following cases (see ASCE/SEI Figure 32.1-2):

- The building or other structure is assigned to risk category I or II in accordance with IBC Table 1604.5.
- The building or other structure is assigned to risk category III or IV and is located in a region outside the tornado-prone region identified in ASCE/SEI Figure 32.1-1.
- The building or other structure is assigned to risk category III or IV and is located in a tornado-prone region identified in ASCE/SEI Figure 32.1-1 where the tornado speed, V_T, obtained from ASCE/SEI Figures 32.5-1 and 32.5-2 is less than 60 mi/h (26.8 m/s).
- The building or other structure is assigned to risk category III or IV and is located in the tornado-prone region identified in ASCE/SEI Figure 32.1-1 where the tornado speed, V_T, obtained from ASCE/SEI Figures 32.5-1 and 32.5-2 is greater than or equal to 60 mi/h (26.8 m/s) and is less than the following threshold speeds: (1) $0.5V$ for exposure B, (2) $0.6V$ for exposure C, or (3) $0.67V$ for exposure D. The exposure category is determined in accordance with ASCE/SEI 26.7.3 (see Section 5.2.4 of this book) and V is the basic wind speed determined in accordance with ASCE/SEI 26.5 (see Section 5.2.2 of this book).

Flowchart 5.18 in Section 5.7 of this book can be used to determine cases where tornado loads must be considered or not.

The tornado load provisions in ASCE/SEI Chapter 32 are not intended to provide protection from the most violent tornadoes. In such cases, a tornado storm shelter meeting the requirements of Reference 5.9 or a tornado safe room meeting the requirements of Reference 5.10 must be provided. The requirements in these references provide a level of life safety greater than conventional buildings and structures, including those designed in accordance with ASCE/SEI Chapter 32.

5.6.2 Permitted Procedures

Procedures to determine tornado loads are adapted from the wind load procedures in ASCE/SEI Chapters 26 through 31, except for those in Chapter 28, which are not compatible with tornado load methodology (ASCE/SEI 32.1.2). Permitted procedures to determine tornado loads on the MWFRS and C&C elements of buildings and other structures are given in ASCE/SEI Figure 32.1-3, subject to the modifications in ASCE/SEI Chapter 32 (see Table 5.24).

Table 5.24 Permitted Procedures for Tornado Loads

Element	Permitted Procedure
MWFRS	Directional Procedure for buildings of all heights in ASCE/SEI Chapter 27 for buildings meeting the requirements specified in that chapter as modified by ASCE/SEI 32.15
	Directional Procedure for building appurtenances and other structures in ASCE/SEI Chapter 29 for buildings meeting the requirements specified in that chapter as modified by ASCE/SEI 32.16
	Wind Tunnel Procedure for all buildings and other structures in ASCE/SEI Chapter 31 for buildings meeting the requirements specified in that chapter as modified by ASCE/SEI 32.18
C&C	Analytical Procedures in Parts 1 through 5 of ASCE/SEI Chapter 30 for building or other structures meeting the requirements specified in those parts with all parts modified by ASCE/SEI 32.17
	Wind Tunnel Procedure for all buildings and other structures in ASCE/SEI Chapter 31 for buildings meeting the requirements specified in that chapter as modified by ASCE/SEI 32.18

A tornado design using a performance-based procedure is permitted subject to approval by the building official (ASCE/SEI 32.1.3). This procedure must conform to the requirements in ASCE/SEI 1.3.1.3. The discussion in ASCE/SEI C26.1.3 for performance-based wind design can be adapted for use in a performance-based tornado design (ASCE/SEI C32.1.3).

An outline of the overall process for the determination of tornado loads is given in ASCE/SEI Figure 32.1-3.

5.6.3 Tornado Hazard Maps

Tornado Speed

Regardless of the procedure used to determine tornado loads, the tornado speed, V_T, must be determined at the location of the building or other structure.

Tornado speeds based on 3-second gust horizontal wind speeds at 33 ft (10 m) anywhere within the effective plan area, A_e, of a building or other structure are given in ASCE/SEI Figures 32.5-1A through 32.5-1H for risk category III buildings and other structures and in ASCE/SEI Figures 32.5-2A through 32.5-2H for risk category IV buildings and other structures. A summary of the tornado speed maps is given in Table 5.25. Unlike basic wind speed, tornado speed is defined independently of terrain. Linear interpolation of tornado speed between maps using the logarithm of the effective plan area size is permitted (ASCE/SEI 32.5.1). For purposes of design, tornadic winds are assumed to come from any horizontal direction (ASCE/SEI 32.5.3).

Table 5.25 Summary of Tornado Speed Maps in ASCE/SEI 32.5

Figure No.	Risk Category	Mean Recurrence Interval (Years)	Effective Plan Area, A_e
32.5-1A	III	1,700	1 ft^2 (0.1 m^2)
32.5-1B			2,000 ft^2 (186 m^2)
32.5-1C			10,000 ft^2 (929 m^2)
32.5-1D			40,000 ft^2 (3,716 m^2)
32.5-1E			100,000 ft^2 (9,290 m^2)
32.5-1F			250,000 ft^2 (23,226 m^2)
32.5-1G			1,000,000 ft^2 (92,903 m^2)
32.5-1H			4,000,000 ft^2 (371,612 m^2)
32.5-2A	IV	3,000	1 ft^2 (0.1 m^2)
32.5-2B			2,000 ft^2 (186 m^2)
32.5-2C			10,000 ft^2 (929 m^2)
32.5-2D			40,000 ft^2 (3,716 m^2)
32.5-2E			100,000 ft^2 (9,290 m^2)
32.5-2F			250,000 ft^2 (23,226 m^2)
32.5-2G			1,000,000 ft^2 (92,903 m^2)
32.5-2H			4,000,000 ft^2 (371,612 m^2)

The mean recurrence intervals of 1,700 and 3,000 years used for risk category III and IV tornado speeds match those used for risk category III and IV basic wind speeds in ASCE/SEI Chapter 26. Tornado loads based on these mean recurrence intervals provide reasonable consistency with the reliability associated with the wind loads determined by ASCE/SEI Chapters 26 and 27 for the same mean recurrence intervals. Tornado speeds for buildings or other structures assigned to risk category I (300-year mean recurrence interval) or risk category II (700-year mean recurrence interval) are usually so low that the effects from tornado loads will be less than those from wind loads. It is for this reason that tornado loads need not be determined for risk category I and II buildings and other structures.

Tornado speed maps for mean recurrence intervals of 10,000, 100,000, 1,000,000, and 10,000,000 years are given in ASCE/SEI Appendix G. These maps may be needed for performance-based tornado designs or other applications.

Information on the determination of the effective plan area of the building or other structure, A_e, is given below.

Additional information on the development of the tornado speed maps can be found in ASCE/SEI C32.5.1.

In lieu of the maps noted above, tornado speeds are permitted to be determined using the ASCE 7 Tornado Design Geodatabase (Reference 5.4; ASCE/SEI 32.5.1). The latitude and longitude or the address of the site can be entered, or the site can be located by clicking on the map.

Effective Plan Area

In addition to the mean recurrence interval (risk category) and geographic location, tornado speed is also a function of the size and shape of the footprint of a building or other structure. Tornado width can be relatively small compared to the plan dimensions of larger buildings and structures, so for a given mean recurrence interval, the probability of a tornado strike and associated maximum tornado speeds increases with increasing plan area.

Tornado strike probability depends on both the size and shape of the footprint of the building or other structure. An effective plan area, A_e, accounts for both of these characteristics. For buildings or other structures not designated essential facilities (see IBC Table 1604.5), A_e is equal to the area of the smallest convex polygon enclosing the plan or footprint; this means A_e is always as large or larger than the actual plan area of the building or other structure (a polygon is convex where all interior angles are less than 180 degrees). For buildings with rectangular plans, a rectangle is the smallest convex polygon. For buildings with circular or oval plans, the area of the smallest convex polygon is approximately equal to the area of the circle or oval. An example of the smallest convex polygon enclosing a multisided plan area is illustrated in Figure 5.23(a) where A_e is the area of the polygon enclosed by the dashed lines (see also ASCE/SEI Figure C32.5-1 for a building with a more complex footprint). It is permitted to conservatively calculate A_e as the area of the smallest rectangle enclosing the maximum plan area, which is illustrated in Figure 5.23(b). In cases where a building or other structure has multiple, independent structural systems (e.g., buildings with expansion joints), it is permitted to calculate A_e for the entire building based on the effective plan area of the largest structurally independent portion of the building (ASCE/SEI 32.5.4.2).

For essential facilities, all the buildings and other structures required to maintain the functionality of the essential facility must be included when calculating A_e (ASCE/SEI 32.5.4.1). The smallest

Figure 5.23 Effective Plan Areas for a Building or Other Structure with a Multisided Plan Area

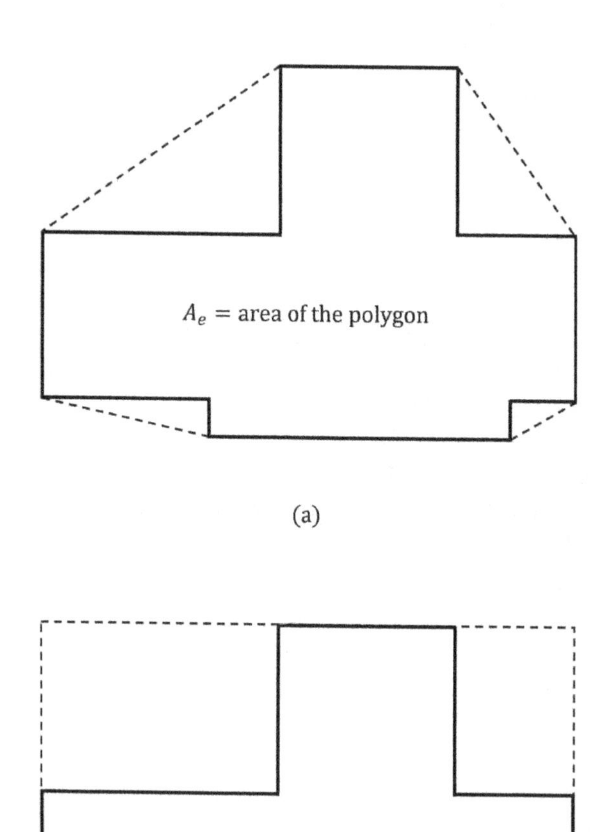

convex polygon enclosing an essential facility and its support building is illustrated in Figure 5.24(a). The area of the smallest rectangle enclosing both buildings is illustrated in Figure 5.24(b).

Figure 5.24 Effective Plan Areas of an Essential Facility and Its Support Building

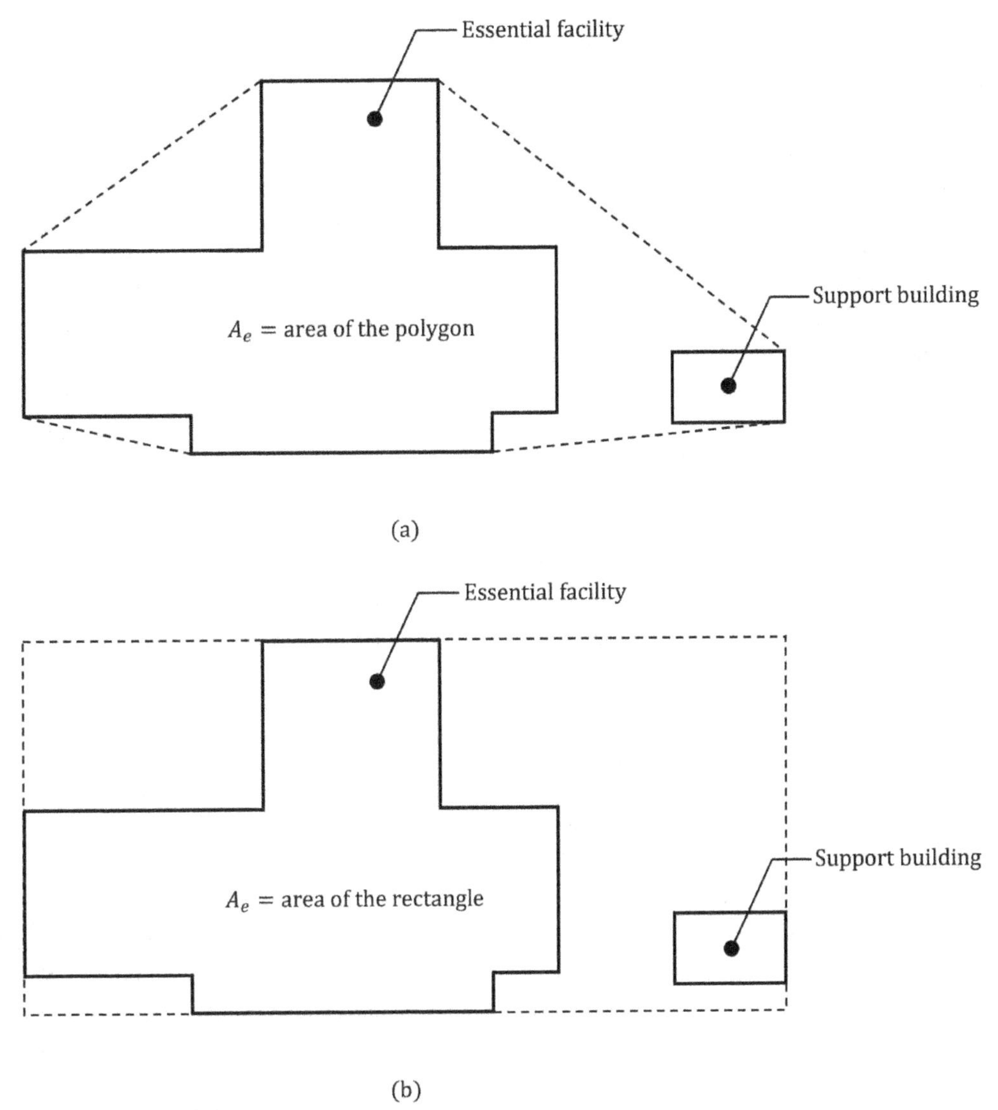

In the case of ground-mounted photovoltaic panel systems where the supporting structure is designed as a risk category III or IV structure, A_e is equal to the effective plan area of the largest structurally independent photovoltaic support structure that does not share structural components with other adjacent structures (ASCE/SEI 32.5.4.3).

5.6.4 Tornado Directionality Factor

The tornado directionality factor, K_{dT}, is determined from ASCE/SEI Table 32.6-1. For arched roofs, circular domes, and other structures, K_{dT} must be taken as the wind directionality factor, K_d, from ASCE/SEI Table 26.6-1.

Variations in tornado speed as a function of building size are captured in the modeling process used to determine the values of K_{dT}. More information on the methodology can be found in ASCE/SEI C32.6.

5.6.5 Tornado Velocity Pressure

The tornado velocity pressure q_{zT} at height z above the ground surface is determined by ASCE/SEI Equations (32.10-1) and (32.10-1.SI):

$$q_{zT} = 0.00256 K_{zTor} K_e V_T^2 \quad (\text{lb / ft}^2) \tag{5.48}$$

$$q_{zT} = 0.613 K_{zTor} K_e V_T^2 \quad (\text{N/m}^2) \tag{5.49}$$

At the mean roof height of a building or other structure, the tornado velocity pressure is denoted q_{hT} and the velocity pressure coefficient is denoted K_{hTor}, that is, the subscript changes from z to h.

The constant 0.00256 in Equation (5.48) and the constant 0.613 in Equation (5.49) are related to the mass density of air for the standard atmosphere (59°F [15°C] and a sea level pressure of 29.92 inches [101.325 kPa] of mercury), which is equal to 0.00238 slug/ft³ or 1.227 kg/m³ from ASCE/SEI Table C26.9-1 at ground elevation $z_e = 0$ ft (0 m). These constants are derived in Equations (5.9) and (5.10) in Section 5.2.7 of this book.

The tornado velocity pressure exposure coefficient, K_{zTor}, is given in ASCE/SEI Table 32.10-1; this coefficient modifies tornado speed (or tornado pressure) with respect to height above ground. Values of K_{zTor} can be determined by the equations in note 1 at the bottom of ASCE/SEI Table 32.10-1:

$$K_{zTor} = \begin{cases} 1.0 & \text{for } 0 < z \le 200 \text{ ft} \\ [(2{,}820 - z)/2{,}620]^2 & \text{for } 200 \text{ ft} < z \le 328 \text{ ft} \\ 0.90 & \text{for } z > 328 \text{ ft} \end{cases} \tag{5.50}$$

In S.I.:

$$K_{zTor} = \begin{cases} 1.0 & \text{for } 0 < z \le 61 \text{ m} \\ [(860 - z)/800]^2 & \text{for } 61 \text{ m} < z \le 100 \text{ m} \\ 0.90 & \text{for } z > 100 \text{ m} \end{cases} \tag{5.51}$$

The ground elevation factor, K_e, adjusts the tornado velocity pressure, q_{ZT}, based on the reduced mass density of air at elevations above sea level (ASCE/SEI 32.9). Values of K_e are given in ASCE/SEI Table 26.9-1; these values can be determined by the equations in note 2 of the table where z_e is the ground elevation above sea level (see Equations [5.5] and [5.6] in Section 5.2.6 of this book).

Values of K_e including the air density, ρ, are given in ASCE/SEI Table C26.9-1.

The effect of ground elevation on air density is independent of the type of windstorm, so the provisions in ASCE/SEI 26.9 for the determination of K_e for wind pressures is also appropriate for the determination of tornado pressures.

Flowchart 5.19 in Section 5.7 of this book can be used to determine the tornado velocity pressure, q_{ZT}.

5.6.6 Tornado Gust Effects

In accordance with ASCE/32.11.1, the tornado gust-effect factor, G_T, for buildings or other structures is equal to 0.85. Alternatively, G_T is permitted to be determined by ASCE/SEI Equation (26.11-6) using terrain exposure constants for exposure C (see Section 5.2.8 of this book).

Because the duration of a tornado is relatively short, the gust factor provisions of ASCE/SEI 26.11.5 for flexible or dynamically sensitive buildings and other structures do not apply. That is why the rigid building or other structure gust factor is used with tornado loads.

5.6.7 Tornado Enclosure Classification

Buildings must be classified as enclosed, partially enclosed, partially open, or open in accordance with the definitions in ASCE/SEI 26.2, which are given in Table 5.7 of this book (ASCE/SEI 32.12.2). The enclosure classification is needed in determining tornado internal pressure coefficients. For a building meeting both open and partially enclosed definitions, the building is to be classified as partially open (ASCE/SEI 32.12.1). Also, as noted in Table 5.7, a building must be classified as partially open where the definitions of open, partially enclosed, or enclosed are not met.

Other structures must be classified as sealed, as defined in ASCE/SEI 32.2, or as enclosed, partially enclosed, partially open, or open in accordance with the definitions in ASCE/SEI 26.2. A sealed other structure is defined as a structure that is completely sealed or has controlled ventilation such that tornado-induced atmospheric pressure changes will not be transmitted to the inside of the structure. Examples of sealed other structures are tanks and vessels.

Wind-borne debris hazards are much greater for tornadoes than for hurricanes. For other than essential facilities where glazed openings are not required to be protected, enclosed buildings and other structures must either be reevaluated for classification as partially enclosed assuming all unprotected glazing is breached or the glazed openings must be protected in accordance with ASCE/SEI 32.12.3.1. For essential buildings and other structures, glazed openings must be protected in accordance with ASCE/SEI 32.12.3.1 to maintain functionality of the facility.

5.6.8 Tornado Internal Pressure Coefficients

Tornado internal pressure coefficients, (GC_{piT}) are given in ASCE/SEI Table 32.13-1 and are based on the enclosure classifications defined in ASCE/SEI 32.12 (see Table 5.26). These coefficients account for the effective internal pressure induced by a tornado (i.e., the combined external and internal pressures due to atmospheric pressure change) and internal pressures due to the porosity of a building and the direct action of the wind-induced exterior pressures created by the tornado. Similar to wind gust effects, the gust-effect factor must not be determined separately in the analysis for (GC_{piT}) [ASCE/SEI 32.11.2].

Table 5.26 Tornado Internal Pressure Coefficients, (GC_{piT}) in Accordance with ASCE/SEI 32.13

Enclosure Classification in Accordance with ASCE/SEI 32.12	(GC_{piT})
Sealed other structures	+1.0
Enclosed buildings and other structures	+0.55 −0.18
Partially enclosed buildings and other structures	+0.55 −0.55
Partially open buildings and other structures	+0.18 −0.18
Open buildings and other structures	0.00

It is evident from Table 5.26 that for enclosed buildings $(GC_{piT}) = +0.55$ and -0.18 compared to $(GC_{pi}) = +0.18$ and -0.18 for enclosed buildings for wind loads. The difference in positive internal pressure is due to the contribution of atmospheric pressure change that occurs during a tornado.

For partially enclosed buildings that contain a single, relatively large volume without any partitions, it is permitted to multiply (GC_{piT}) by the reduction factor, R_i, determined by ASCE/SEI Equation (26.13-1) [ASCE/SEI 32.13.1].

5.6.9 Tornado Loads on Buildings—MWFRS

Enclosed, Partially Enclosed, and Partially Open Buildings

Overview

According to ASCE/SEI 32.15.1, the provisions in ASCE/SEI 27.3.1 must be used for the determination of tornado loads on the MWFRS of buildings of all heights, subject to the modifications in ASCE/SEI 32.15.1.

Design Tornado Pressures

Design tornado pressures, p_T, are determined by ASCE/SEI Equations (32.15-1) and (32.15-1.SI) for the MWFRS of enclosed, partially enclosed, and partially open buildings of all heights:

$$p_T = qG_T K_{dT} K_{vT} C_p - q_i(GC_{piT}) \tag{5.52}$$

This equation is used to calculate tornado pressures on the walls and roof of a building, which are applied simultaneously on all surfaces, similar to that for wind pressures (see ASCE/SEI Figure 27.3-1 or Figure 5.11 of this book).

The first part of Equation (5.52) is the external pressure contribution and the second part is the internal pressure contribution. External pressure varies with height above ground on the windward wall and is a constant on all of the other surfaces based on the mean roof height. The terms in this equation are discussed below.

Tornado velocity pressure, q_{zT}. Tornado velocity pressure, q_{zT}, which is determined in accordance with ASCE/SEI 32.10 (see Section 5.6.5 of this book), varies with respect to height on all walls ($q = q_{zT}$). For roofs, the velocity pressure is constant and is evaluated at the mean roof height, h ($q = q_{hT}$).

Tornado gust-effect factor, G_T. The tornado gust-effect factor, G_T, is equal to 0.85 or may be determined by ASCE/SEI Equation (26.11-6) [see Section 5.6.6 of this book].

Tornado directionality factor, K_{dT}. The tornado directionality factor, K_{dT}, is determined from ASCE/SEI Table 32.6-1 (see Section 5.6.4 of this book).

Tornado pressure coefficient adjustment factor, K_{vT}. The tornado pressure coefficient adjustment factor, K_{vT}, is determined from ASCE/SEI Table 32.14-1. This parameter accounts for the effects of the vertical component of the wind speed within the core of a tornado. In general, these adjustment factors modify roof uplift coefficients previously developed for boundary layer winds to account for these effects (i.e., K_{vT} is used to modify the external pressure coefficients, C_p, in ASCE/SEI 27.3.1). Additional information on the development of K_{vT} is given in ASCE/SEI C32.14.

External pressure coefficients, C_p. The external pressure coefficients, C_p, that must be used in the determination of tornado loads are given in ASCE/SEI 27.3.1 for the determination of wind loads (see ASCE/SEI 32.15.1 and Section 5.3.2 of this book).

Tornado velocity pressure for internal pressure determination, q_i. The tornado velocity pressure for internal pressure determination, q_i, is determined as follows (ASCE/SEI 32.15.1):

- For internal pressure evaluation of roofs of enclosed and partially open buildings: $q_i = q_{hT}$.
- For internal pressure evaluation of walls of enclosed and partially open buildings: $q_i = q_{zT}$.
- For internal pressure evaluation of the roof and all walls in partially enclosed buildings: $q_i = q_{z_{op}}$ where height z_{op} is defined as the level of the lowest opening in the building that could affect the positive internal pressure. Glazed openings not meeting the protection requirements in ASCE/SEI 32.12.3.1 must be considered as openings.

Tornado internal pressure coefficients, (GC_{piT}). Tornado internal pressure coefficients, (GC_{piT}), are determined in accordance with ASCE/SEI 32.13 (see Section 5.6.8 of this book). Both positive and negative values of (GC_{piT}) must be considered to establish the critical load effects.

The provisions in ASCE/SEI 27.3.1.1 apply for the determination of tornado loads on the MWFRS of elevated buildings (ASCE/SEI 32.15.1.1; see Section 5.3.2 of this book). Design tornado pressure, p_T, is determined by ASCE/SEI Equation (32.15-1) where $K_{vT} = 1.0$ for lateral loads on elements below the elevated building and vertical loads on the horizontal bottom surface of the elevated building.

Open Buildings with Monoslope, Pitched, or Troughed Free Roofs

The provisions in ASCE/SEI 27.3.2 apply for the determination of tornado loads on the MWFRS on open buildings, subject to the modifications of ASCE/SEI 32.15.2. The net design tornado pressure, p_T, is determined by ASCE/SEI Equations (32.15-2) and (32.15-2.SI):

$$p_T = q_{hT} G_T K_{dT} C_N \tag{5.53}$$

In this equation, q_{hT} is the tornado velocity pressure determined in accordance with ASCE/SEI 32.10.2 at the mean roof height, h, of the building, G_T is the tornado gust-effect factor determined in accordance with ASCE/SEI 32.11, K_{dT} is the tornado directionality factor determined from ASCE/SEI Table 32.6-1, and C_N is the net pressure coefficient determined in accordance with ASCE/SEI 27.3.2 (see Section 5.3.2 of this book).

Roof Overhangs

The provisions in ASCE/SEI 27.3.3 apply for the determination of tornado loads on roof overhangs (ASCE/SEI 32.15.3; see Section 5.3.2 of this book).

Parapets

The provisions in ASCE/SEI 27.3.4 apply for the determination of tornado loads on the MWFRS on parapets, subject to the modifications of ASCE/SEI 32.15.4. The design tornado pressure, p_{pT}, for the effects of parapets on the MWFRS is determined by ASCE/SEI Equations (32.15-3) and (32.15-3.SI):

$$p_{pT} = q_{pT} K_{dT} (GC_{pn}) \tag{5.54}$$

In this equation, q_{pT} is the tornado velocity pressure determined in accordance with ASCE/SEI 32.10.2 at the top of the parapet, K_{dT} is the tornado directionality factor determined from ASCE/SEI Table 32.6-1, and (GC_{pn}) is the combined net pressure coefficient determined in accordance with ASCE/SEI 27.3.4 (see Section 5.3.2 of this book).

Design Load Cases

The design load cases in ASCE/SEI 27.3.5 apply for the case of tornado loads. However, the exception for buildings meeting the requirements of ASCE/SEI D.1 does not apply (ASCE/SEI 32.15.5; see Figure 5.16 of this book). These load cases are likely to be conservative in cases where the tornado is smaller in plan area than the building or other structure because the maximum tornado speed does not extend across the full plan area of the building or other structure. The removal of the exception in ASCE/SEI 27.3.5 that eliminates the torsional load cases in ASCE/SEI Figure 27.3-8 is due to the asymmetry of the tornado wind field, which tends to increase torsional loads compared to other types of windstorms.

Flowchart 5.20 in Section 5.7 of this book can be used to determine design tornado pressures on the MWFRS of buildings and other structures in accordance with ASCE/SEI Chapter 32.

5.6.10 Tornado Loads on Building Appurtenances and Other Structures—MWFRS

Solid Freestanding Walls and Signs

The provisions in ASCE/SEI 29.3 apply to the determination of tornado loads on the MWFRS of solid freestanding walls and signs, subject to the modifications in ASCE/SEI 32.16.2.

The design tornado force, F_T, for the MWFRS of solid freestanding walls and solid freestanding signs is determined by ASCE/SEI Equations (32.16-1) and (32.16-1.SI) [walls and signs with openings comprising less than 30 percent of the gross area are classified as solid; for walls or signs with openings greater than or equal to 30 percent, the wind force must be determined in accordance with ASCE/SEI 32.16.3, which is covered in the next section of this book]:

$$F_T = q_{zT} G_T K_{dT} C_f A_s \tag{5.55}$$

The tornado velocity pressure, q_{zT}, at the centroid of the gross area of the sign or wall, A_s, is determined in accordance with ASCE/SEI 32.10 where the effective plan area, A_e, of the wall or sign is equal to the greater of the wall or sign length squared divided by 20 and the length multiplied by the width (see ASCE/SEI 32.16.1 and Section 5.6.5 of this book). The tornado gust-effect factor, G_T, is determined in accordance with ASCE/SEI 32.11 (see Section 5.6.6 of this book) and the tornado directionality factor, K_{dT}, is determined from ASCE/SEI Table 32.6-1 (see Section 5.6.4 of the book). Net force coefficients, C_f, are determined from ASCE/SEI Figure 29.3-1 as a function of the geometrical properties of the wall or sign.

In general, three cases (A, B, and C) must be investigated where F_T is applied at different locations on the sign or wall (see ASCE/SEI Figure 29.3-1 and Table 5.10 in Section 5.3.4 of this book). Case C is applicable where the width of the sign or wall, B, is greater than or equal to two times the vertical dimension of the sign or wall, s; in such cases, Case B need not be considered.

Flowchart 5.21 in Section 5.7 of this book can be used to determine the design tornado forces on solid freestanding walls and solid freestanding signs.

Other Structures

Overview

The provisions in ASCE/SEI 29.4 apply to the determination of tornado loads on the MWFRS of other structures, subject to the modifications in ASCE/SEI 32.16.3.

Chimneys, Tanks, Open Signs, Single-plane Open Frames, and Trussed Towers

The design wind force, F_T, for ground or roof-mounted chimneys, tanks, open signs, single-plane open frames, and trussed towers is determined by ASCE/SEI Equations (32.16-2) and (32.16-2.SI):

$$F_T = q_{zT} G_T K_{dT} C_f A_f \tag{5.56}$$

The tornado velocity pressure, q_{zT}, at the centroid of the projected area normal to the tornado load, A_f, is determined in accordance with ASCE/SEI 32.10 (see Section 5.6.5 of this book), the tornado gust-effect factor, G_T, is determined in accordance with ASCE/SEI 32.11 (see Section 5.6.6 of this book), and the tornado directionality factor, K_{dT}, is determined from ASCE/SEI Table 32.6-1 (see Section 5.6.4 of this book).

Force coefficients, C_f, are given in the following figures:

- ASCE/SEI Figure 29.4-1: Chimneys, tanks, and similar structures
- ASCE/SEI Figure 29.4-2: Open signs and single-plane open frames
- ASCE/SEI Figure 29.4-3: Trussed towers

Values of C_f in ASCE/SEI Figure 29.4-1 are given for square, hexagonal, and round cross-sections as a function of the height of the structure, h, to cross-sectional dimension of the section, D (D is the diameter of circular cross-sections and is the least horizontal dimension of square, hexagonal, and octagonal sections at the elevation under consideration).

The force coefficients in ASCE/SEI Figure 29.4-2 are applicable to open signs, that is signs with openings comprising 30 percent or more of the gross area. Signs not meeting this criterion are classified as solid signs and the force coefficients in ASCE/SEI Figure 29.3-1 must be used (see the previous section of this book).

The force coefficients in ASCE/SEI Figure 29.4-3 are for trussed towers with square and triangular cross-sections.

Damage investigations have shown that wind-borne debris can cling to trussed communication towers. An additional 40 ft^2 (3.72 m^2) of projected surface area of clinging debris at the mid-height of trussed towers and other open lattice-type structures or at 50 ft (15.2 m), whichever is less, must be included in A_f (ASCE/SEI 32.16.3.1).

Flowchart 5.22 in Section 5.7 of this book can be used to determine the design tornado force for other structures.

Rooftop Structures and Equipment for Buildings

The provisions in ASCE/SEI 29.4.1 apply to the determination of tornado loads on the MWFRS of rooftop structures and equipment (not including roof-mounted solar panels), subject to the modifications in ASCE/SEI 32.16.3.2.

The lateral design tornado force, F_{hT}, is determined by ASCE/SEI Equations (32.16-3) and (32.16-3.SI):

$$F_{hT} = q_{hT} K_{dT} (GC_r) A_f \tag{5.57}$$

In this equation, q_{hT} is the tornado velocity pressure determined in accordance with ASCE/SEI 32.10.2 at the mean roof height, h, of the building supporting the rooftop structure or equipment, K_{dT} is the tornado directionality factor determined from ASCE/SEI Table 32.6-1, and A_f is the vertical projected area of the rooftop structure or equipment (see Section 5.3.4 of this book).

The product of the external pressure coefficient and gust-effect factor, (GC_r), is determined by the following equation (ASCE/SEI 29.4.1; see Figure 5.19 of this book):

$$(GC_r) = \begin{cases} 1.9 \text{ for } A_f < 0.1Bh \\ 2.0 - (A_f/Bh) \text{ for } 0.1Bh \le A_f \le Bh \end{cases} \tag{5.58}$$

The vertical design tornado force, F_{vT}, is determined by ASCE/SEI Equations (32.16-4) and (32.16-4.SI):

$$F_{vT} = q_{hT} K_{dT} K_{vT} (GC_r) A_r \tag{5.59}$$

In this equation, K_{vT} is the tornado pressure coefficient adjustment factor determined from ASCE/SEI Table 32.14-1, A_r is the horizontal projected area of the rooftop structure or equipment (see Figure 5.19 of this book), and (GC_r) is the product of the external pressure coefficient and gust-effect factor determined by the following equation (see ASCE/SEI 29.4.1):

$$(GC_r) = \begin{cases} 1.5 \text{ for } A_r < 0.1BL \\ 1.0 + 0.56[1 - (A_r/BL)] \text{ for } 0.1BL \le A_f \le BL \end{cases} \tag{5.60}$$

Roofs of Isolated Circular Bins, Silos, and Tanks

The provisions in ASCE/SEI 29.4.2.2 apply to the determination of tornado loads on the MWFRS of roofs of isolated circular bins, silos, and tanks, subject to the modifications in ASCE/SEI 32.16.3.3.

The net design tornado pressure, p_T, is determined by ASCE/SEI Equations (32.16-5) and (32.16-5.SI):

$$p_T = q_{hT} [G_T K_{dT} K_{vT} C_p - (GC_{piT})] \tag{5.61}$$

In this equation, q_{hT} is the tornado velocity pressure determined in accordance with ASCE/SEI 32.10.2 at the mean roof height, h, of the circular bin, silo, or tank, G_T is the tornado gust-effect factor determined in accordance with ASCE/SEI 32.11 (see Section 5.6.6 of this book), K_{dT} is the tornado directionality factor determined from ASCE/SEI Table 32.6-1, K_{vT} is the tornado pressure coefficient adjustment factor determined from ASCE/SEI Table 32.14-1, and (GC_{piT}) are the tornado internal pressure coefficients determined in accordance with ASCE/SEI 32.13 (see Section 5.6.8 of this book).

The external pressure coefficients, C_p, are given in ASCE/SEI Figure 29.4-5 for windward and leeward portions (zones 1 and 2, respectively) of flat, conical or dome roofs based on the angle of the plane of the roof from the horizontal. In the case of dome roofs where the roof angle is greater than 10 degrees, C_p is to be determined from ASCE/SEI Figure 27.3-2 for domes with circular bases.

Unlike buildings, the envelope of a circular bin, silo, or tank is usually not breached during a tornado event, so the pressure differential due to atmospheric pressure change is not lessened. This, along with the vertical component of tornadic winds, can result in relatively large roof uplift pressures.

Rooftop Solar Panels for Buildings of All Heights with Flat Roofs or Gable or Hip Roofs with Slopes Less Than 7 Degrees

The provisions in ASCE/SEI 29.4.3 apply to the determination of tornado loads on the MWFRS of rooftop photovoltaic panels for buildings of all heights with flat roofs or gable or hip roofs with slopes less than 7 degrees, subject to the modifications in ASCE/SEI 32.16.3.4.

The design tornado pressure, p_T, is determined by ASCE/SEI Equations (32.16-6) and (32.16-6.SI):

$$p_T = q_{hT} K_{dT} (GC_m) \tag{5.62}$$

In this equation, q_{hT} is the tornado velocity pressure determined in accordance with ASCE/SEI 32.10.2 at the mean roof height, h, of the building supporting the solar panels, K_{dT} is the tornado directionality factor determined from ASCE/SEI Table 32.6-1, and (GC_m) are the net pressure coefficients determined in accordance with ASCE/SEI 29.4.3 [see Equation (5.29) and Table 5.14 of this book].

Rooftop Solar Panels Parallel to the Roof Surface on Buildings of All Heights and Roof Slopes

The provisions in ASCE/SEI 29.4.4 apply to the determination of tornado loads on the MWFRS of rooftop photovoltaic panels parallel to the roof surface on buildings of all heights and roof slopes, subject to the modifications in ASCE/SEI 32.16.3.5.

The design tornado pressure, p_T, is determined by ASCE/SEI Equations (32.16-7) and (32.16-7.SI):

$$p_T = q_{hT} K_{dT} K_{vT} (GC_p) \gamma_E \gamma_a \tag{5.63}$$

In this equation, q_{hT} is the tornado velocity pressure determined in accordance with ASCE/SEI 32.10.2 at the mean roof height, h, of the building supporting the solar panels, K_{dT} is the tornado directionality factor determined from ASCE/SEI Table 32.6-1, K_{vT} is the tornado pressure coefficient adjustment factor determined from ASCE/SEI Table 32.14-1, (GC_p) are the net pressure coefficients for C&C of roofs determined from ASCE/SEI Figures 30.3-2A-I through 30.3-7 or ASCE/SEI Figure 30.5-1 (see ASCE/SEI 29.4.4), γ_E is the array edge factor determined in accordance with ASCE/SEI 29.4.4, and γ_a is the solar panel pressure equalization factor determined in accordance with ASCE/SEI 29.4.4 (see Section 5.3.4 of this book).

The additional vertical component of the tornadic winds is accounted for by amplifying the net uplift pressure by K_{vT}.

5.6.11 Tornado Loads on C&C

Low-Rise Buildings

The provisions in ASCE/SEI 30.3 apply to the determination of tornado loads on C&C elements of low-rise buildings (which are defined in ASCE/SEI 26.2) and buildings with a mean roof height less than or equal to 60 ft (18.3 m), subject to the modifications in ASCE/SEI 32.17.1.

The design tornado pressure, p_T, is determined by ASCE/SEI Equations (32.17-1) and (32.17-1.SI):

$$p_T = q_{hT}[K_{dT}K_{vT}(GC_p) - (GC_{piT})] \tag{5.64}$$

In this equation, q_{hT} is the tornado velocity pressure determined in accordance with ASCE/SEI 32.10.2 at the mean roof height, h, of the building, K_{dT} is the tornado directionality factor determined from ASCE/SEI Table 32.6-1, K_{vT} is the tornado pressure coefficient adjustment factor determined from ASCE/SEI Table 32.14-1, (GC_p) are the external pressure coefficients determined in accordance with ASCE/SEI 30.3 (see Table 5.18 in Section 5.4.4 of this book), and (GC_{piT}) are the tornado internal pressure coefficients determined in accordance with ASCE/SEI 32.13 (see Section 5.6.8 of this book).

The design tornado pressure on the C&C of bottom horizontal surfaces of elevated buildings is determined by ASCE/SEI Equations (32.17-1) and (32.17-2.SI) with $K_{vT} = 1.0$ (ASCE/SEI 32.17.1.1).

Flowchart 5.23 in Section 5.7 of this book can be used to determine design tornado pressures on C&C elements of low-rise buildings in accordance with ASCE/SEI 30.3.

Buildings with $h > 60$ ft (18.3 m)

The provisions in ASCE/SEI 30.4 apply to the determination of tornado loads on C&C elements of buildings with a mean roof height greater than 60 ft (18.3 m), subject to the modifications in ASCE/SEI 32.17.2.

The design tornado pressure, p_T, is determined by ASCE/SEI Equations (32.17-2) and (32.17-2.SI):

$$p_T = qK_{dT}K_{vT}(GC_p) - q_i(GC_{piT}) \tag{5.65}$$

For external pressure on walls, $q = q_{zT}$, which is the tornado velocity pressure determined in accordance with ASCE/SEI 32.10.2 at height z above the ground. For external pressure on roofs, $q = q_{hT}$, which is the tornado velocity pressure determined in accordance with ASCE/SEI 32.10.2 at the mean roof height of the building, h.

The tornado directionality factor, K_{dT}, is determined from ASCE/SEI Table 32.6-1 and the tornado pressure coefficient adjustment factor K_{vT} is determined from ASCE/SEI Table 32.14-1.

The external pressure coefficients, (GC_p), are determined in accordance with ASCE/SEI 30.4 (see Table 5.19 in Section 5.4.5 of this book) and (GC_{piT}) are the tornado internal pressure coefficients determined in accordance with ASCE/SEI 32.13 (see Section 5.6.8 of this book).

The tornado velocity pressure for internal pressure determination, q_i, is determined as follows (ASCE/SEI 32.17.2):

- For internal pressure evaluation of roofs of enclosed and partially open buildings: $q_i = q_{hT}$.
- For internal pressure evaluation of walls of enclosed and partially open buildings: $q_i = q_{zT}$.
- For internal pressure evaluation of the roof and all walls in partially enclosed buildings: $q_i = q_{z_{op}}$ where height z_{op} is defined as the level of the lowest opening in the building that could affect the positive internal pressure. Glazed openings not meeting the protection requirements in ASCE/SEI 32.12.3.1 must be considered as openings.

The design tornado pressure on the C&C of bottom horizontal surfaces of elevated buildings is determined by ASCE/SEI Equations (32.17-1) and (32.17-2.SI) with $K_{vT} = 1.0$ where the tornado velocity pressure is determined at the height specified in ASCE/SEI 30.5.2.1, subparagraph 1 (ASCE/SEI 32.17.2.1).

Flowchart 5.24 in Section 5.7 of this book can be used to determine design tornado pressures on C&C elements of buildings with mean roof heights greater than 60 ft (18.3 m) in accordance with ASCE/SEI 30.4.

Open Buildings

The provisions in ASCE/SEI 30.5 apply to the determination of tornado loads on C&C elements of open buildings, subject to the modifications in ASCE/SEI 32.17.3.

The net design tornado pressure, p_T, is determined by ASCE/SEI Equations (32.17-3) and (32.17-3.SI):

$$p_T = q_{hT} G_T K_{dT} C_N \tag{5.66}$$

In this equation, q_{hT} is the tornado velocity pressure determined in accordance with ASCE/SEI 32.10.2 at the mean roof height, h, of the building, G_T is the tornado gust-effect factor determined in accordance with ASCE/SEI 32.11 (see Section 5.6.6 of this book) and K_{dT} is the tornado directionality factor determined from ASCE/SEI Table 32.6-1 (see Section 5.6.4 of the book).

Net pressure coefficients, C_N, for C&C elements located on the roof of an open building are determined by the figures in ASCE/SEI 30.5 (see Table 5.20 in Section 5.4.6 of this book). These coefficients include contributions from both the top and bottom surfaces of the roof, which implies the element receives pressure from both surfaces.

Building Appurtenances and Rooftop Structures and Equipment

Parapets

The provisions in ASCE/SEI 30.6 apply to the determination of tornado loads on C&C elements of parapets, subject to the modifications in ASCE/SEI 32.17.4.1.

The design tornado pressure, p_T, is determined by ASCE/SEI Equations (32.17-4) and (32.17-4.SI):

$$p_T = q_{pT}[K_{dT}(GC_p) - (GC_{piT})] \tag{5.67}$$

In this equation, q_{pT} is the tornado velocity pressure determined in accordance with ASCE/SEI 32.10.2 at the top of the parapet, K_{dT} is the tornado directionality factor determined from ASCE/SEI Table 32.6-1, (GC_p) are the external pressure coefficients determined from the figures referenced in ASCE/SEI 30.6 (see Table 5.21 in Section 5.4.7 of this book), and (GC_{piT}) are the tornado internal pressure coefficients determined in accordance with ASCE/SEI 32.13 (see Section 5.6.8 of this book).

Roof Overhangs

The provisions in ASCE/SEI 30.7 apply to the determination of tornado loads on C&C elements of roof overhangs, subject to the modifications in ASCE/SEI 32.17.4.2.

The design tornado pressure, p_T, is determined by ASCE/SEI Equations (32.17-5) and (32.17-5.SI):

$$p_T = q_{hT}[K_{dT} K_{vT}(GC_p) - (GC_{piT})] \tag{5.68}$$

In this equation, q_{hT} is the tornado velocity pressure determined in accordance with ASCE/SEI 32.10.2 at the mean roof height, h, of the building, K_{dT} is the tornado directionality factor determined from ASCE/SEI Table 32.6-1, K_{vT} is the tornado pressure coefficient adjustment factor determined from ASCE/SEI Table 32.14-1, and (GC_{piT}) are the tornado internal pressure coefficients determined in accordance with ASCE/SEI 32.13 (see Section 5.6.8 of this book).

External pressure coefficients, (GC_p), for C&C elements of roof overhangs are determined as follows (ASCE/SEI 30.7):

- For gable roofs with a roof angle less than or equal to 7 degrees, determine (GC_p) from ASCE/SEI Figure 30.3-2A based on the effective wind area, A.
- For all other cases, (GC_p) for the top surface of the roof overhang is the same as the (GC_p) for the roof surface, and (GC_p) for the bottom surface of the roof overhang is the same as the (GC_p) for the adjacent wall (see ASCE/SEI Figure 30.7-1).

Attached Canopies on Buildings with h ≤ 60 ft (18.3 m)

The provisions in ASCE/SEI 30.9 apply to the determination of tornado loads on C&C elements of attached canopies on buildings with a mean roof height, h, less than or equal to 60 ft (18.3 m), subject to the modifications in ASCE/SEI 32.17.4.3.

The design tornado pressure, p_T, is determined by ASCE/SEI Equations (32.17-6) and (32.17-6.SI):

$$p_T = q_{hT} K_{dT} (GC_p) \tag{5.69}$$

In this equation, q_{hT} is the tornado velocity pressure determined in accordance with ASCE/SEI 32.10.2 at the mean roof height, h, of the building and K_{dT} is the tornado directionality factor determined from ASCE/SEI Table 32.6-1.

Net pressure coefficients, (GC_p), for C&C elements of attached canopies are determined from ASCE/SEI Figures 30.9-1A and 30.9-1B for buildings with $h \leq 60$ ft (18.3 m).

For canopies with two exposed surfaces, ASCE/SEI Figures 30.9-1A and 30.9-1B are both required for buildings with $h \leq 60$ ft (18.3 m).

For canopies with one exposed surface (e.g., where wind pressures can be applied directly to the bottom of the top surface only), ASCE/SEI Figure 30.9-1B is required for buildings with $h \leq 60$ ft (18.3 m) [see Section 5.4.7 of this book].

Nonbuilding Structures

The provisions in ASCE/SEI 30.10 apply to the determination of tornado loads on C&C elements of isolated circular bins, silos, and tanks with a mean roof height, h, less than or equal to 120 ft (36.6 m), subject to the modifications in ASCE/SEI 32.17.5.

The design tornado pressure, p_T, is determined by ASCE/SEI Equations (32.17-7) and (32.17-7.SI):

$$p_T = q_{hT} [K_{dT} K_{vT} (GC_p) - (GC_{piT})] \tag{5.70}$$

In this equation, q_{hT} is the tornado velocity pressure determined in accordance with ASCE/SEI 32.10.2 at the mean roof height, h, of the structure and K_{dT} is the tornado directionality factor determined from ASCE/SEI Table 32.6-1.

The tornado pressure coefficient adjustment factor, K_{vT}, is determined from ASCE/SEI Table 32.14-1 as follows:

- For roof zones 1 and 2 in ASCE/SEI Figure 30.10-2, K_{vT} is equal to the zone 1 value for building roofs from ASCE/SEI Table 32.14-1.
- For roof zones 3 and 4 in ASCE/SEI Figure 30.10-2, K_{vT} is equal to the zone 2 value for building roofs from ASCE/SEI Table 32.14-1.

External pressure coefficients, (GC_p), are determined in accordance with ASCE/SEI 30.10.2 for walls, ASCE/SEI 30.10.4 for roofs, and ASCE/SEI 30.10.5 for undersides (see Table 5.22 in Section 5.4.8 of this book).

The tornado internal pressure coefficients, (GC_{piT}), are determined as follows:

- For internal surfaces of exterior walls of isolated open-topped circular bins, silos, and tanks, (GC_{piT}) is determined by ASCE/SEI Equation (30.10-5).
- In all other cases, (GC_{piT}) is determined in accordance with ASCE/SEI 32.13.

5.6.12 Wind Tunnel Procedure

The wind tunnel procedure in ASCE/SEI Chapter 31 is permitted to be used to determine external pressure coefficients and force coefficients for use with the tornado loading provisions in ASCE/SEI 32.15 through 32.17 (ASCE/SEI 32.18). Wind tunnel tests must be performed on an isolated building model (without a proximity model) in a boundary layer wind tunnel for exposure category C terrain.

Additional information on the wind tunnel procedure can be found in ASCE/SEI C32.18.

5.7 Flowcharts

A summary of the flowcharts provided in this chapter is given in Table 5.27.

Table 5.27 Summary of Flowcharts Provided in Chapter 5

Flowchart	Title
5.1	Wind Velocity Pressure, q_z (ASCE/SEI 26.10)
5.2	Gust-Effect Factors, G and G_f (ASCE/SEI 26.11)
5.3	Design Wind Pressures on the MWFRS of Buildings, p —ASCE/SEI Chapter 27
5.4	Design Wind Pressures on the MWFRS of Buildings, p —ASCE/SEI Chapter 28
5.5	Design Wind Loads on Solid Freestanding Walls and Solid Freestanding Signs, F (ASCE/SEI 29.3)
5.6	Design Wind Loads on Chimneys, Tanks, Open Signs, Single-Plane Open Frames, and Trussed Towers, F (ASCE/SEI 29.4)
5.7	Design Wind Pressures and Forces on Circular Bins, Silos, and Tanks (ASCE/SEI 29.4.2)
5.8	Design Wind Pressure on Rooftop Solar Panels in Accordance with ASCE/SEI 29.4.3
5.9	Design Wind Pressure on Rooftop Solar Panels in Accordance with ASCE/SEI 29.4.4
5.10	Design Wind Loads on Ground-Mounted Fixed-Tilt Solar Panel Systems in Accordance with ASCE/SEI 29.4.5
5.11	Design Wind Pressures on C&C Elements in Accordance with Part 1 in ASCE/SEI Chapter 30 (ASCE/SEI 30.3)
5.12	Design Wind Pressures on C&C Elements in Accordance with Part 2 in ASCE/SEI Chapter 30 (ASCE/SEI 30.4)
5.13	Design Wind Pressures on C&C Elements in Accordance with Part 3 in ASCE/SEI Chapter 30 (ASCE/SEI 30.5)
5.14	Design Wind Pressures on C&C Elements of Parapets (ASCE/SEI 30.6)
5.15	Design Wind Pressures on C&C Elements of Roof Overhangs (ASCE/SEI 30.7)
5.16	Design Wind Pressures on C&C Elements of Attached Canopies on Buildings (ASCE/SEI 30.9)
5.17	Design Wind Pressures on C&C Elements of Circular Bins, Silos, and Tanks (ASCE/SEI 30.10)
5.18	Consideration of Design Tornado Loads (ASCE/SEI 32.1.1 and 32.5.2)
5.19	Tornado Velocity Pressure, q_{zT} (ASCE/SEI 32.10)
5.20	Design Tornado Pressures on the MWFRS of Buildings, p_T—ASCE/SEI Chapter 32
5.21	Design Tornado Loads on Solid Freestanding Walls and Solid Freestanding Signs, F_T (ASCE/SEI 32.16.2)
5.22	Design Tornado Loads on Chimneys, Tanks Open Signs, Single-Plane Open Frames, and Trussed Towers, F_T (ASCE/SEI 32.16.3)
5.23	Design Tornado Pressures on C&C Elements of Low-rise Buildings (ASCE/SEI 32.17.1)
5.24	Design Tornado Pressures on C&C Elements of Buildings with $h > 60$ ft (18.3 m) (ASCE/SEI 32.17.2)

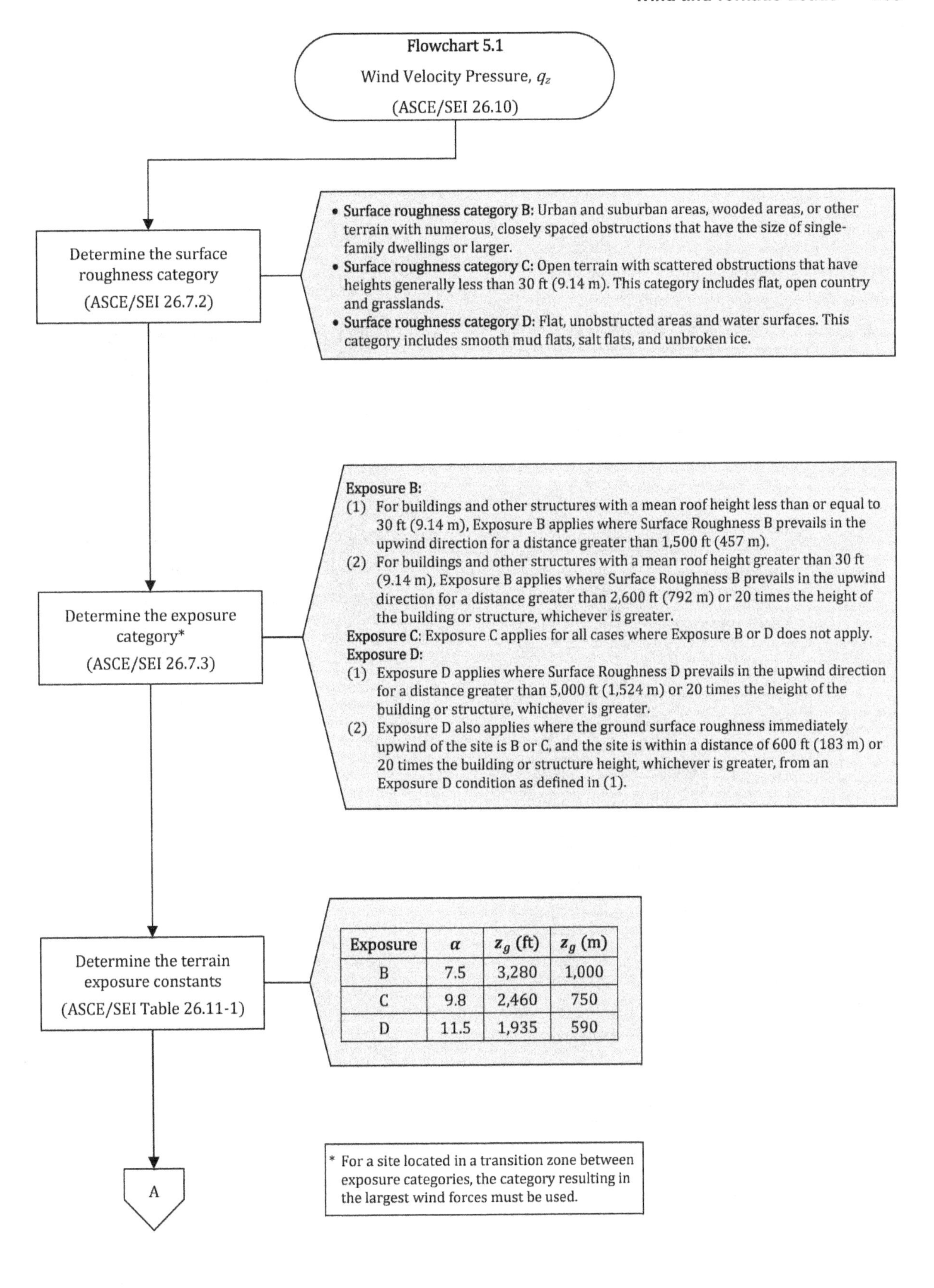

Flowchart 5.1
Wind Velocity Pressure, q_z
(ASCE/SEI 26.10)
Determine the surface roughness category
(ASCE/SEI 26.7.2)
• Surface roughness category B: Urban and suburban areas, wooded areas, or other terrain with numerous, closely spaced obstructions that have the size of single-family dwellings or larger.
• Surface roughness category C: Open terrain with scattered obstructions that have heights generally less than 30 ft (9.14 m). This category includes flat, open country and grasslands.
• Surface roughness category D: Flat, unobstructed areas and water surfaces. This category includes smooth mud flats, salt flats, and unbroken ice.
Determine the exposure category*
(ASCE/SEI 26.7.3)
Exposure B:
(1) For buildings and other structures with a mean roof height less than or equal to 30 ft (9.14 m), Exposure B applies where Surface Roughness B prevails in the upwind direction for a distance greater than 1,500 ft (457 m).
(2) For buildings and other structures with a mean roof height greater than 30 ft (9.14 m), Exposure B applies where Surface Roughness B prevails in the upwind direction for a distance greater than 2,600 ft (792 m) or 20 times the height of the building or structure, whichever is greater.
Exposure C: Exposure C applies for all cases where Exposure B or D does not apply.
Exposure D:
(1) Exposure D applies where Surface Roughness D prevails in the upwind direction for a distance greater than 5,000 ft (1,524 m) or 20 times the height of the building or structure, whichever is greater.
(2) Exposure D also applies where the ground surface roughness immediately upwind of the site is B or C, and the site is within a distance of 600 ft (183 m) or 20 times the building or structure height, whichever is greater, from an Exposure D condition as defined in (1).
Determine the terrain exposure constants
(ASCE/SEI Table 26.11-1)
A
* For a site located in a transition zone between exposure categories, the category resulting in the largest wind forces must be used.

Exposure	α	z_g (ft)	z_g (m)
B	7.5	3,280	1,000
C	9.8	2,460	750
D	11.5	1,935	590

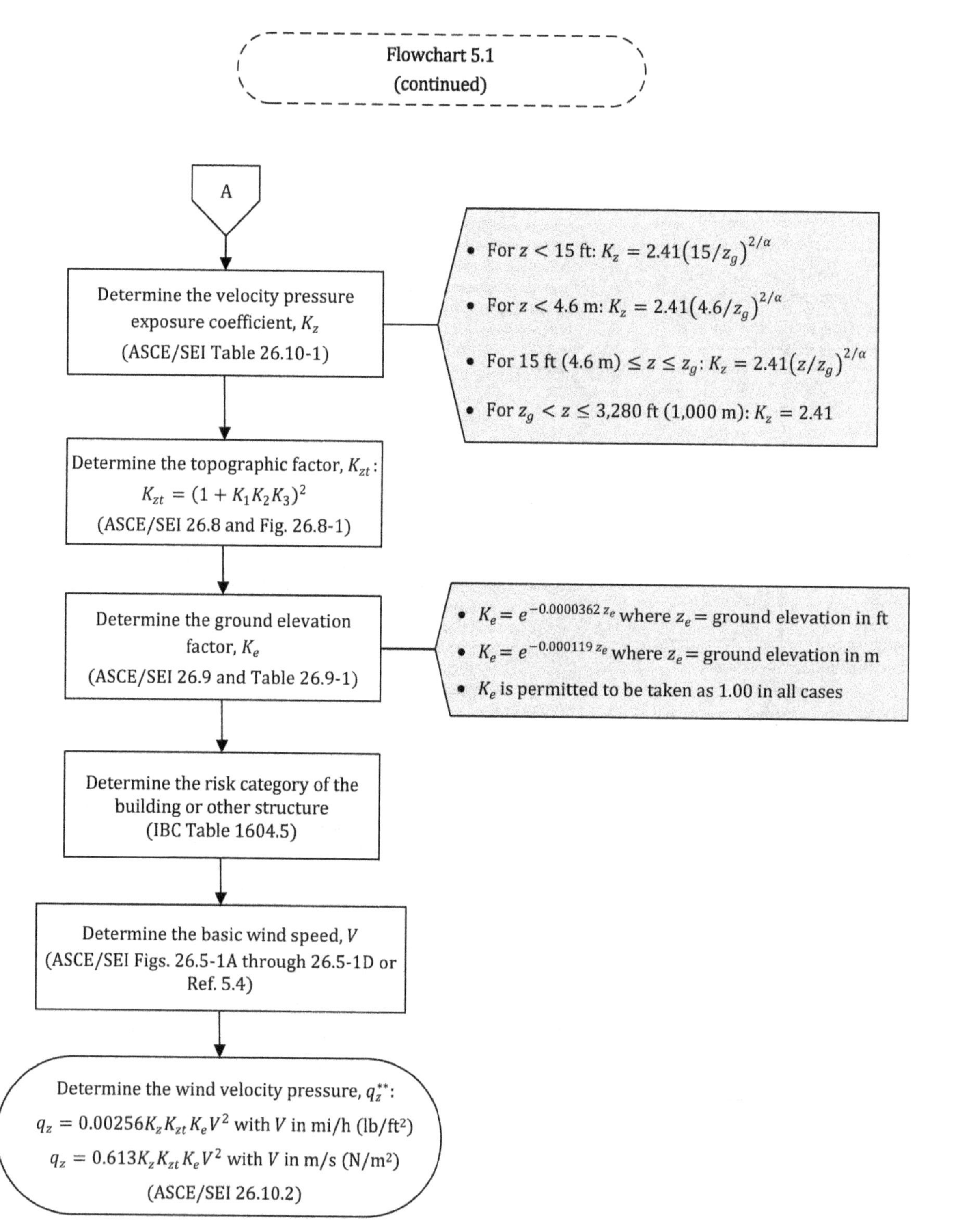

Flowchart 5.1
(continued)
A
Determine the velocity pressure exposure coefficient, K_z (ASCE/SEI Table 26.10-1)
• For z < 15 ft: $K_z = 2.41(15/z_g)^{2/\alpha}$
• For z < 4.6 m: $K_z = 2.41(4.6/z_g)^{2/\alpha}$
• For 15 ft (4.6 m) $\leq z \leq z_g$: $K_z = 2.41(z/z_g)^{2/\alpha}$
• For $z_g < z \leq$ 3,280 ft (1,000 m): $K_z = 2.41$
Determine the topographic factor, K_{zt}: $K_{zt} = (1 + K_1K_2K_3)^2$ (ASCE/SEI 26.8 and Fig. 26.8-1)
Determine the ground elevation factor, K_e (ASCE/SEI 26.9 and Table 26.9-1)
• $K_e = e^{-0.0000362\, z_e}$ where z_e = ground elevation in ft
• $K_e = e^{-0.000119\, z_e}$ where z_e = ground elevation in m
• K_e is permitted to be taken as 1.00 in all cases
Determine the risk category of the building or other structure (IBC Table 1604.5)
Determine the basic wind speed, V (ASCE/SEI Figs. 26.5-1A through 26.5-1D or Ref. 5.4)
Determine the wind velocity pressure, q_z^{**}:
$q_z = 0.00256K_zK_{zt}K_eV^2$ with V in mi/h (lb/ft²)
$q_z = 0.613K_zK_{zt}K_eV^2$ with V in m/s (N/m²)
(ASCE/SEI 26.10.2)

** Wind velocity pressure at the mean roof height, q_h, is computed as $q_h = q_z$ using $K_z = K_h$ at the mean roof height, h.

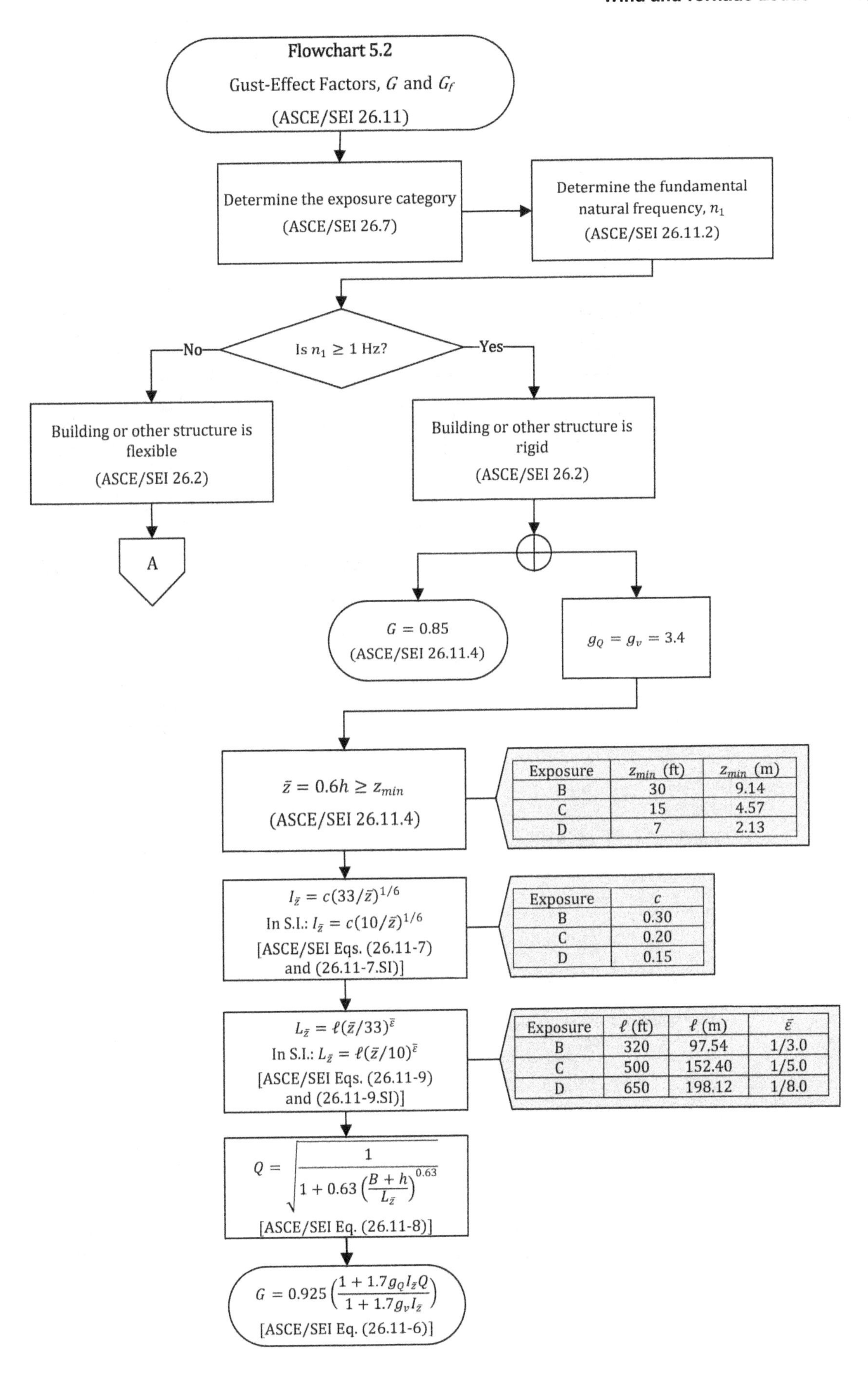

Flowchart 5.2
Gust-Effect Factors, G and G_f
(ASCE/SEI 26.11)
Determine the exposure category
(ASCE/SEI 26.7)
Determine the fundamental natural frequency, n_1
(ASCE/SEI 26.11.2)
Is $n_1 \geq 1$ Hz?
No
Yes
Building or other structure is flexible
(ASCE/SEI 26.2)
A
Building or other structure is rigid
(ASCE/SEI 26.2)
$G = 0.85$
(ASCE/SEI 26.11.4)
$g_Q = g_v = 3.4$
$\bar{z} = 0.6h \geq z_{min}$
(ASCE/SEI 26.11.4)
$I_{\bar{z}} = c(33/\bar{z})^{1/6}$
In S.I.: $I_{\bar{z}} = c(10/\bar{z})^{1/6}$
[ASCE/SEI Eqs. (26.11-7) and (26.11-7.SI)]
$L_{\bar{z}} = \ell(\bar{z}/33)^{\bar{\epsilon}}$
In S.I.: $L_{\bar{z}} = \ell(\bar{z}/10)^{\bar{\epsilon}}$
[ASCE/SEI Eqs. (26.11-9) and (26.11-9.SI)]
$Q = \sqrt{\dfrac{1}{1 + 0.63\left(\dfrac{B+h}{L_{\bar{z}}}\right)^{0.63}}}$
[ASCE/SEI Eq. (26.11-8)]
$G = 0.925\left(\dfrac{1 + 1.7 g_Q I_{\bar{z}} Q}{1 + 1.7 g_v I_{\bar{z}}}\right)$
[ASCE/SEI Eq. (26.11-6)]

Exposure	z_{min} (ft)	z_{min} (m)
B	30	9.14
C	15	4.57
D	7	2.13

Exposure	c
B	0.30
C	0.20
D	0.15

Exposure	ℓ (ft)	ℓ (m)	$\bar{\epsilon}$
B	320	97.54	1/3.0
C	500	152.40	1/5.0
D	650	198.12	1/8.0

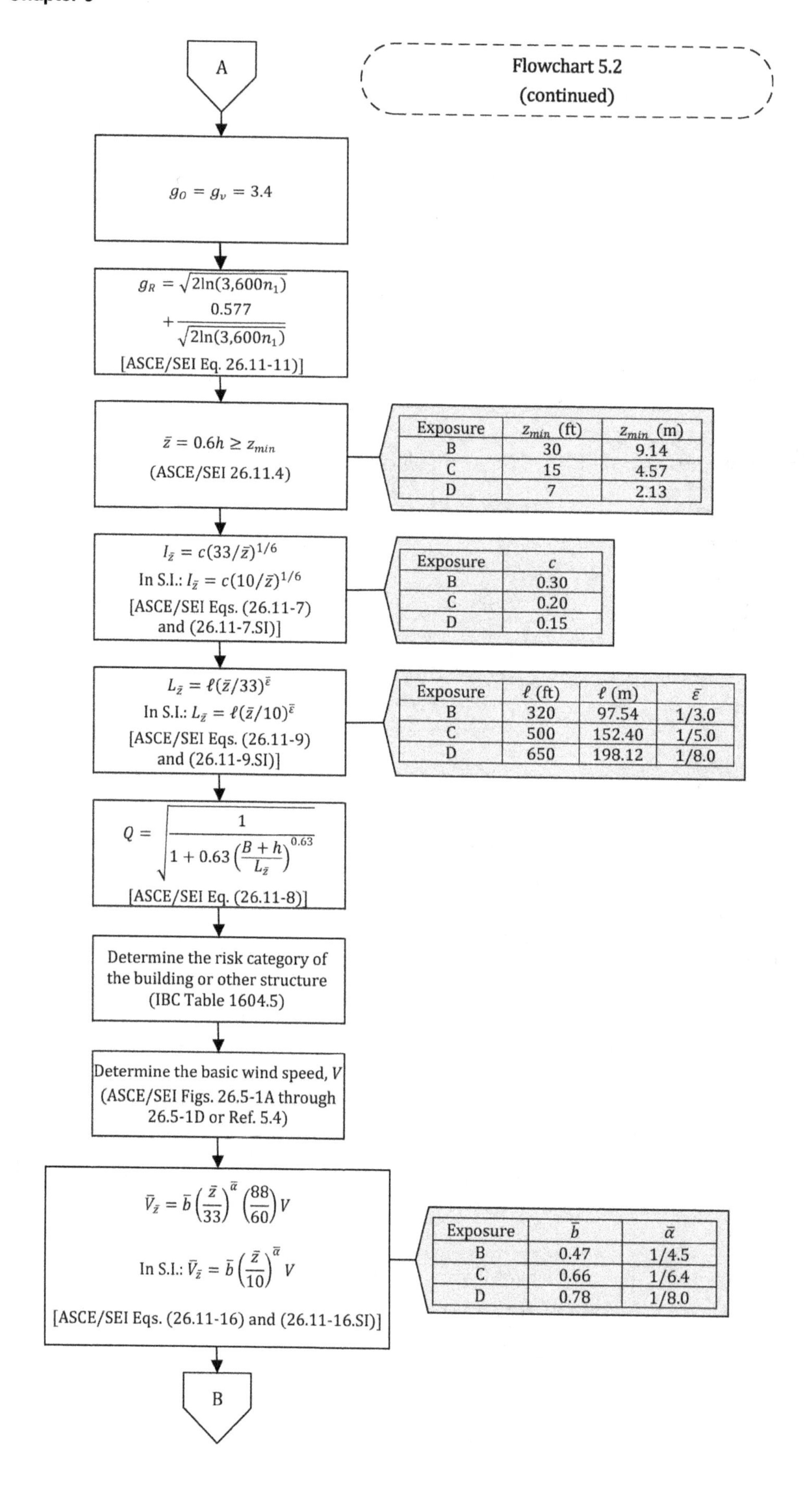

A
Flowchart 5.2
(continued)
$g_Q = g_v = 3.4$
$g_R = \sqrt{2\ln(3{,}600n_1)} + \frac{0.577}{\sqrt{2\ln(3{,}600n_1)}}$
[ASCE/SEI Eq. 26.11-11)]
$\bar{z} = 0.6h \geq z_{min}$
(ASCE/SEI 26.11.4)
Exposure | z_{min} (ft) | z_{min} (m)
B | 30 | 9.14
C | 15 | 4.57
D | 7 | 2.13
$I_{\bar{z}} = c(33/\bar{z})^{1/6}$
In S.I.: $I_{\bar{z}} = c(10/\bar{z})^{1/6}$
[ASCE/SEI Eqs. (26.11-7) and (26.11-7.SI)]
Exposure | c
B | 0.30
C | 0.20
D | 0.15
$L_{\bar{z}} = \ell(\bar{z}/33)^{\bar{\varepsilon}}$
In S.I.: $L_{\bar{z}} = \ell(\bar{z}/10)^{\bar{\varepsilon}}$
[ASCE/SEI Eqs. (26.11-9) and (26.11-9.SI)]
Exposure | ℓ (ft) | ℓ (m) | $\bar{\varepsilon}$
B | 320 | 97.54 | 1/3.0
C | 500 | 152.40 | 1/5.0
D | 650 | 198.12 | 1/8.0
$Q = \sqrt{\frac{1}{1 + 0.63\left(\frac{B+h}{L_{\bar{z}}}\right)^{0.63}}}$
[ASCE/SEI Eq. (26.11-8)]
Determine the risk category of the building or other structure (IBC Table 1604.5)
Determine the basic wind speed, V (ASCE/SEI Figs. 26.5-1A through 26.5-1D or Ref. 5.4)
$\bar{V}_{\bar{z}} = \bar{b}\left(\frac{\bar{z}}{33}\right)^{\bar{\alpha}}\left(\frac{88}{60}\right)V$
In S.I.: $\bar{V}_{\bar{z}} = \bar{b}\left(\frac{\bar{z}}{10}\right)^{\bar{\alpha}}V$
[ASCE/SEI Eqs. (26.11-16) and (26.11-16.SI)]
Exposure | $\bar{b}$ | $\bar{\alpha}$
B | 0.47 | 1/4.5
C | 0.66 | 1/6.4
D | 0.78 | 1/8.0
B

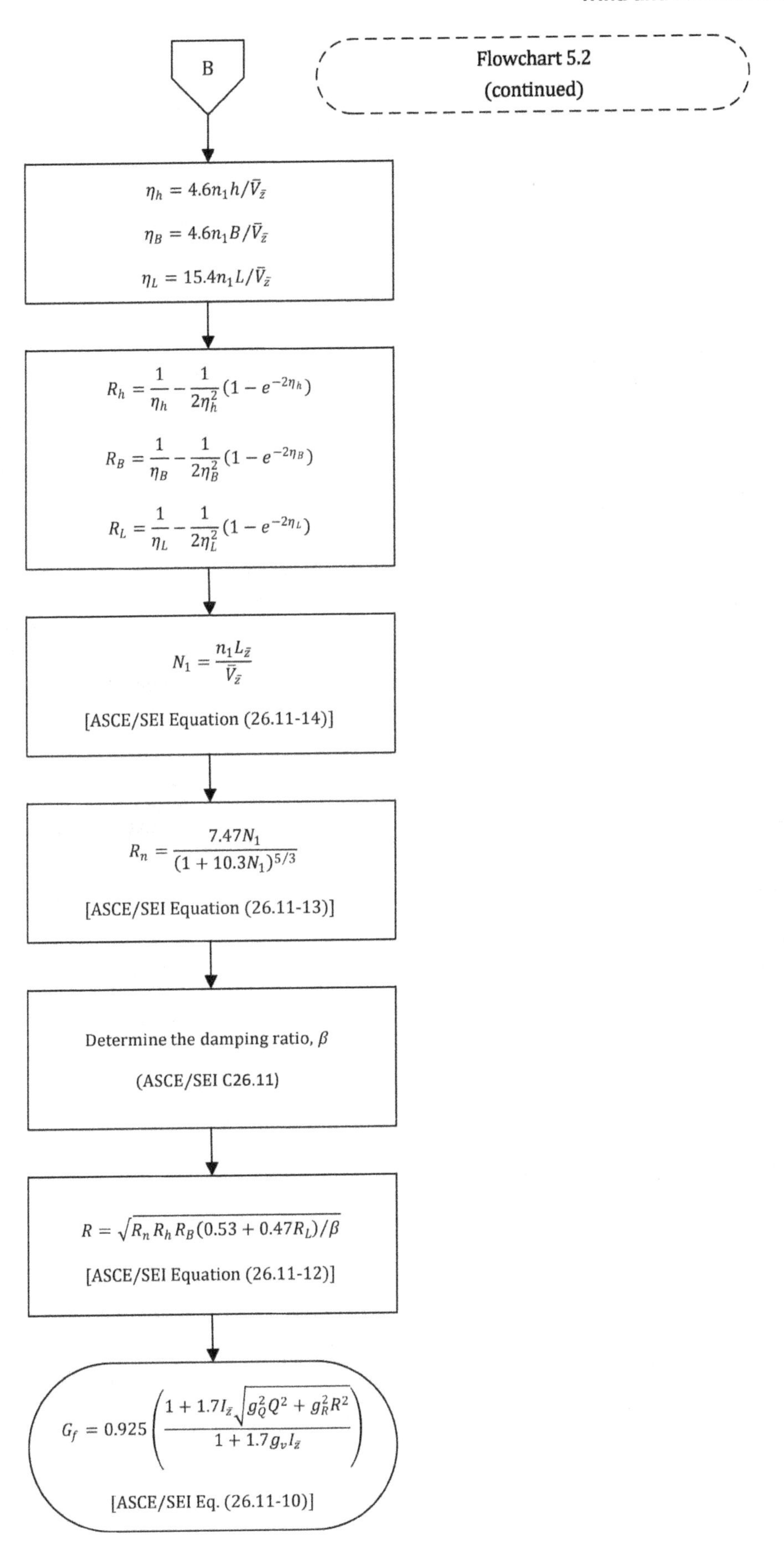
Flowchart 5.2
(continued)
B
$\eta_h = 4.6n_1h/\bar{V}_{\bar{z}}$
$\eta_B = 4.6n_1B/\bar{V}_{\bar{z}}$
$\eta_L = 15.4n_1L/\bar{V}_{\bar{z}}$
$R_h = \frac{1}{\eta_h} - \frac{1}{2\eta_h^2}(1 - e^{-2\eta_h})$
$R_B = \frac{1}{\eta_B} - \frac{1}{2\eta_B^2}(1 - e^{-2\eta_B})$
$R_L = \frac{1}{\eta_L} - \frac{1}{2\eta_L^2}(1 - e^{-2\eta_L})$
$N_1 = \frac{n_1L_{\bar{z}}}{\bar{V}_{\bar{z}}}$
[ASCE/SEI Equation (26.11-14)]
$R_n = \frac{7.47N_1}{(1 + 10.3N_1)^{5/3}}$
[ASCE/SEI Equation (26.11-13)]
Determine the damping ratio, β
(ASCE/SEI C26.11)
$R = \sqrt{R_nR_hR_B(0.53 + 0.47R_L)/\beta}$
[ASCE/SEI Equation (26.11-12)]
$G_f = 0.925\left(\frac{1 + 1.7I_{\bar{z}}\sqrt{g_Q^2Q^2 + g_R^2R^2}}{1 + 1.7g_vI_{\bar{z}}}\right)$
[ASCE/SEI Eq. (26.11-10)]

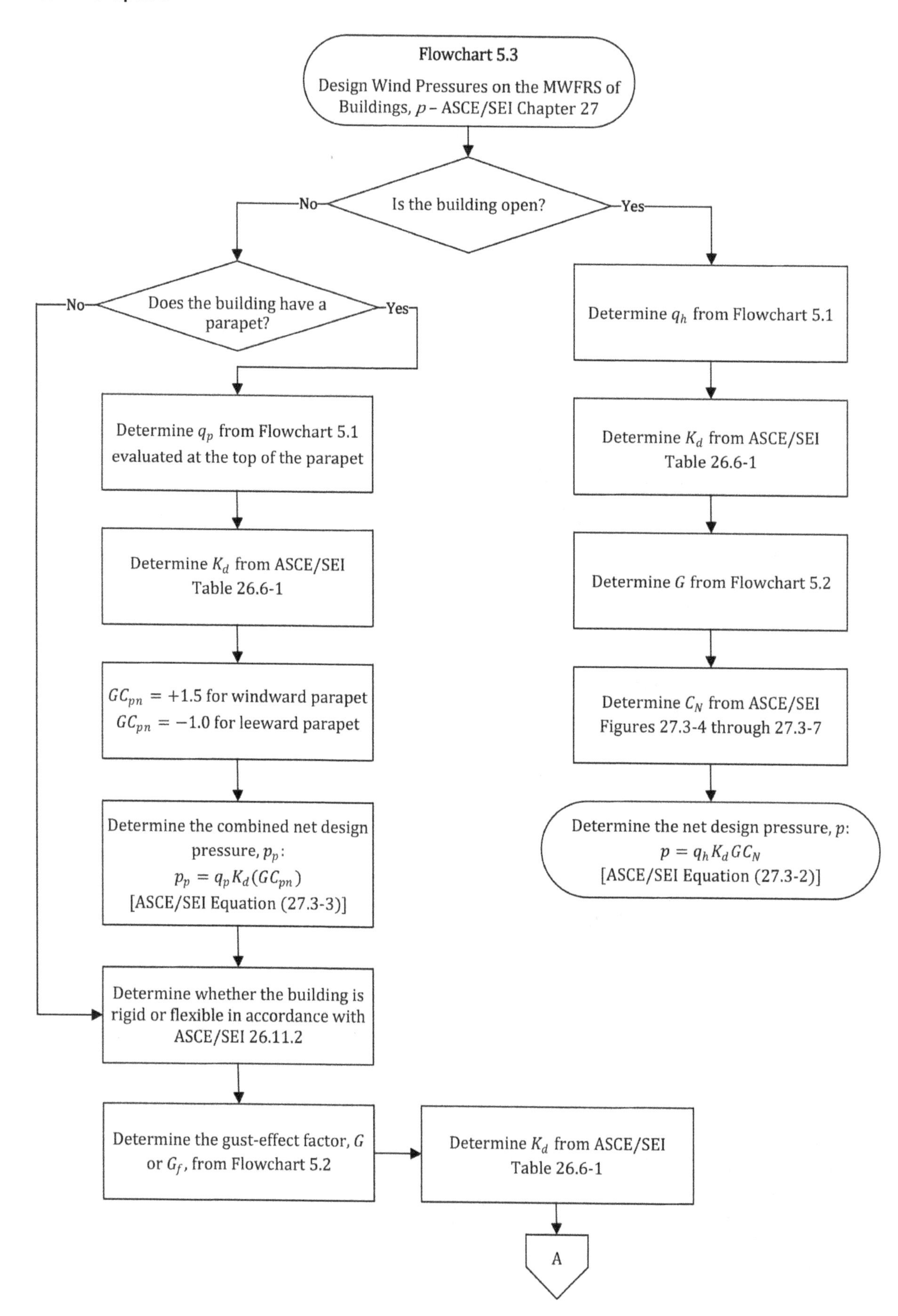
Flowchart 5.3
Design Wind Pressures on the MWFRS of Buildings, p – ASCE/SEI Chapter 27
Is the building open?
No
Yes
Does the building have a parapet?
No
Yes
Determine q_p from Flowchart 5.1 evaluated at the top of the parapet
Determine K_d from ASCE/SEI Table 26.6-1
$GC_{pn} = +1.5$ for windward parapet
$GC_{pn} = -1.0$ for leeward parapet
Determine the combined net design pressure, p_p:
$p_p = q_p K_d (GC_{pn})$
[ASCE/SEI Equation (27.3-3)]
Determine whether the building is rigid or flexible in accordance with ASCE/SEI 26.11.2
Determine the gust-effect factor, G or G_f, from Flowchart 5.2
Determine K_d from ASCE/SEI Table 26.6-1
A
Determine q_h from Flowchart 5.1
Determine K_d from ASCE/SEI Table 26.6-1
Determine G from Flowchart 5.2
Determine C_N from ASCE/SEI Figures 27.3-4 through 27.3-7
Determine the net design pressure, p:
$p = q_h K_d G C_N$
[ASCE/SEI Equation (27.3-2)]

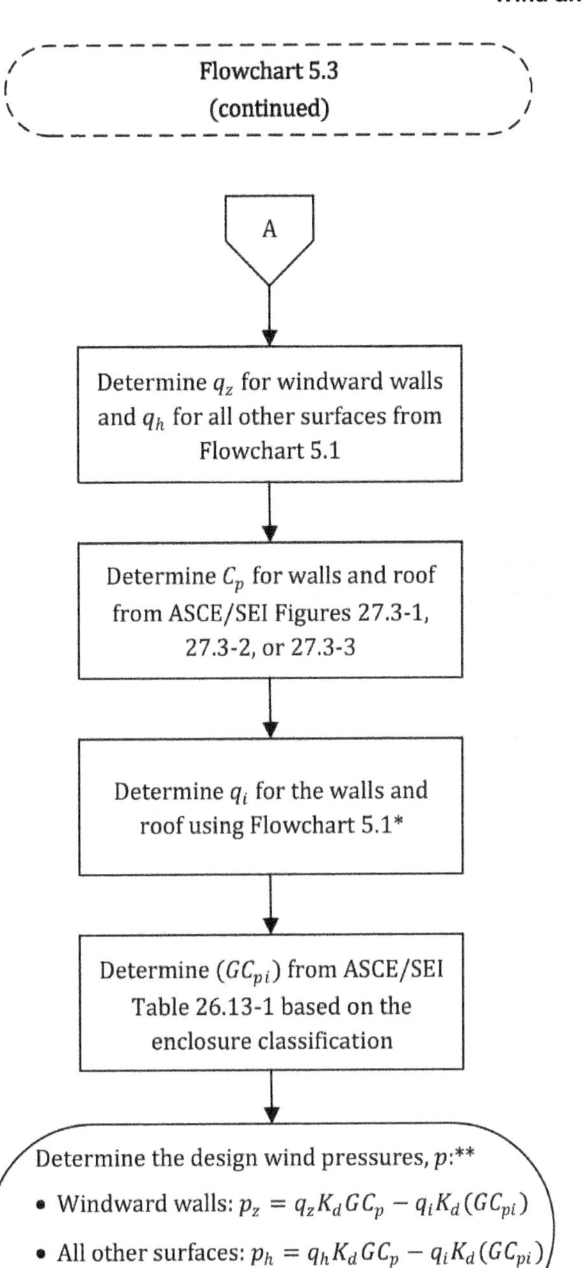

*$q_i = q_h$ or $q_i = q_z$ based on the enclosure classification (see ASCE/SEI 27.3.1). q_i may conservatively be evaluated at height h.

**Substitute G_f for G for slender buildings.

See ASCE/SEI 27.3.5 and ASCE/SEI Figure 27.3-8 for the load cases that must be considered and see ASCE/SEI 27.3.3 for roof overhangs.

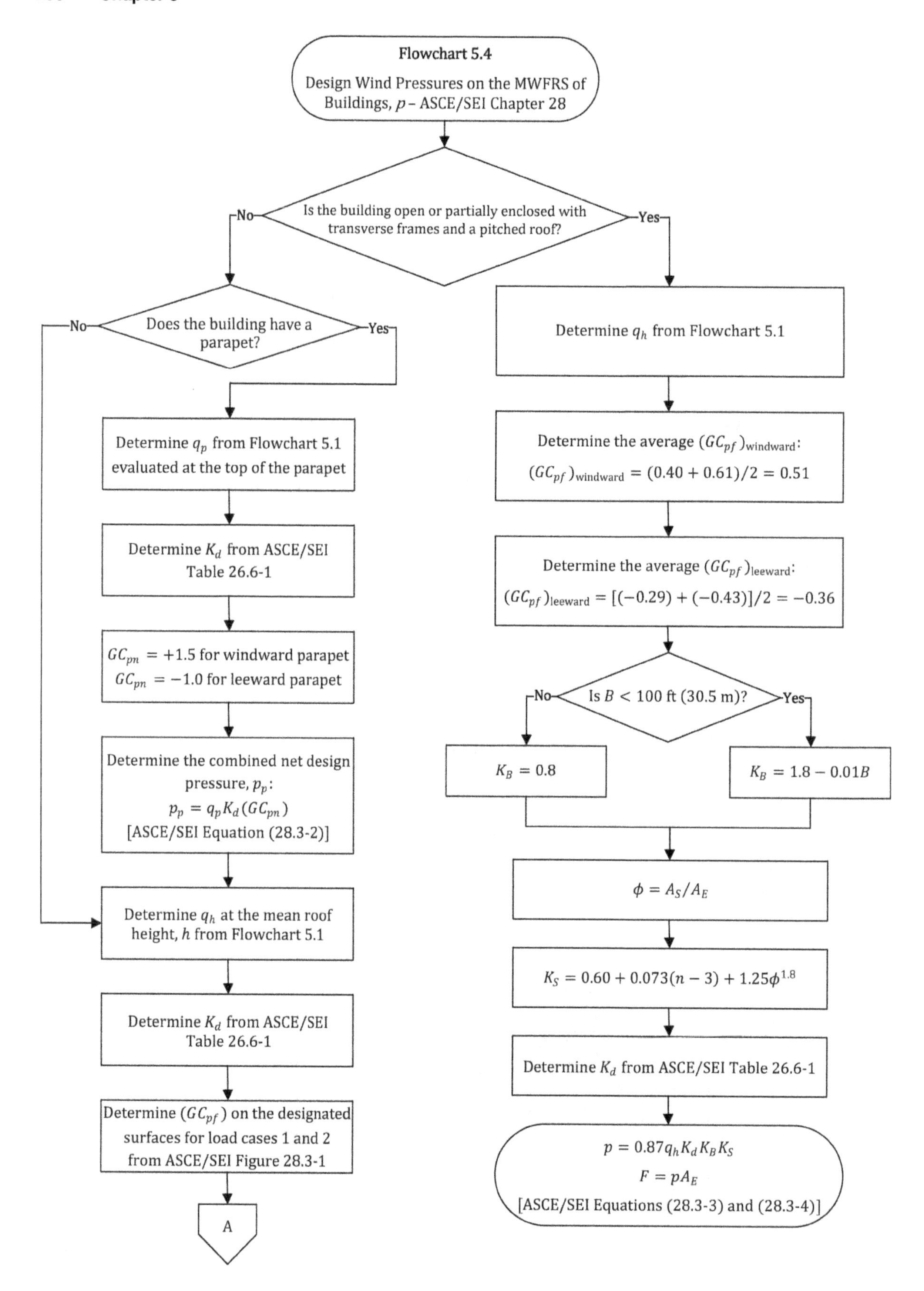

Flowchart 5.4
Design Wind Pressures on the MWFRS of Buildings, p – ASCE/SEI Chapter 28
Is the building open or partially enclosed with transverse frames and a pitched roof?
No
Yes
Does the building have a parapet?
No
Yes
Determine q_p from Flowchart 5.1 evaluated at the top of the parapet
Determine K_d from ASCE/SEI Table 26.6-1
$GC_{pn} = +1.5$ for windward parapet
$GC_{pn} = -1.0$ for leeward parapet
Determine the combined net design pressure, p_p:
$p_p = q_p K_d (GC_{pn})$
[ASCE/SEI Equation (28.3-2)]
Determine q_h at the mean roof height, h from Flowchart 5.1
Determine K_d from ASCE/SEI Table 26.6-1
Determine (GC_{pf}) on the designated surfaces for load cases 1 and 2 from ASCE/SEI Figure 28.3-1
A
Determine q_h from Flowchart 5.1
Determine the average $(GC_{pf})_{\text{windward}}$:
$(GC_{pf})_{\text{windward}} = (0.40 + 0.61)/2 = 0.51$
Determine the average $(GC_{pf})_{\text{leeward}}$:
$(GC_{pf})_{\text{leeward}} = [(-0.29) + (-0.43)]/2 = -0.36$
Is $B < 100$ ft (30.5 m)?
No
Yes
$K_B = 0.8$
$K_B = 1.8 - 0.01B$
$\phi = A_S/A_E$
$K_S = 0.60 + 0.073(n - 3) + 1.25\phi^{1.8}$
Determine K_d from ASCE/SEI Table 26.6-1
$p = 0.87 q_h K_d K_B K_S$
$F = pA_E$
[ASCE/SEI Equations (28.3-3) and (28.3-4)]

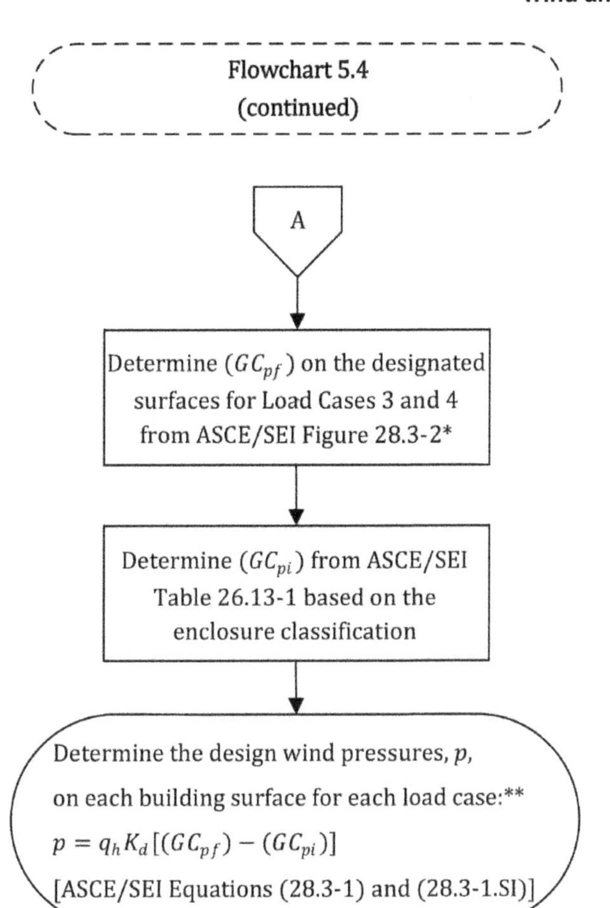
Flowchart 5.4
(continued)
A
Determine (GC_{pf}) on the designated surfaces for Load Cases 3 and 4 from ASCE/SEI Figure 28.3-2*
Determine (GC_{pi}) from ASCE/SEI Table 26.13-1 based on the enclosure classification
Determine the design wind pressures, p, on each building surface for each load case:**
$p = q_h K_d [(GC_{pf}) - (GC_{pi})]$
[ASCE/SEI Equations (28.3-1) and (28.3-1.SI)]

*See the exceptions in ASCE/SEI 28.3.2 where Load Case 3 and Load Case 4 need not be considered.

**See ASCE/SEI Figures 28.3-1 and 28.3-2 for the load cases that must be considered.

Minimum wind pressures in ASCE/SEI 28.3.6 must also be considered for enclosed and partially enclosed buildings.

See ASCE/SEI 28.3.5 for wind pressures on roof overhangs.

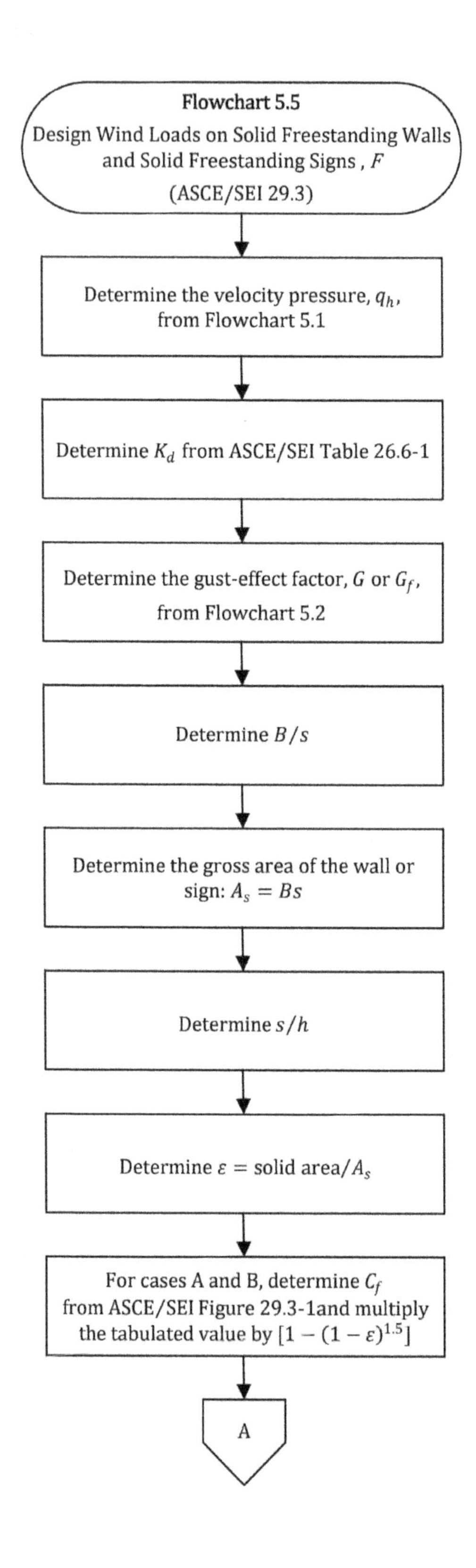
Flowchart 5.5
Design Wind Loads on Solid Freestanding Walls and Solid Freestanding Signs , F
(ASCE/SEI 29.3)
Determine the velocity pressure, q_h, from Flowchart 5.1
Determine K_d from ASCE/SEI Table 26.6-1
Determine the gust-effect factor, G or G_f, from Flowchart 5.2
Determine B/s
Determine the gross area of the wall or sign: $A_s = Bs$
Determine s/h
Determine ε = solid area/A_s
For cases A and B, determine C_f from ASCE/SEI Figure 29.3-1and multiply the tabulated value by $[1 - (1 - \varepsilon)^{1.5}]$
A

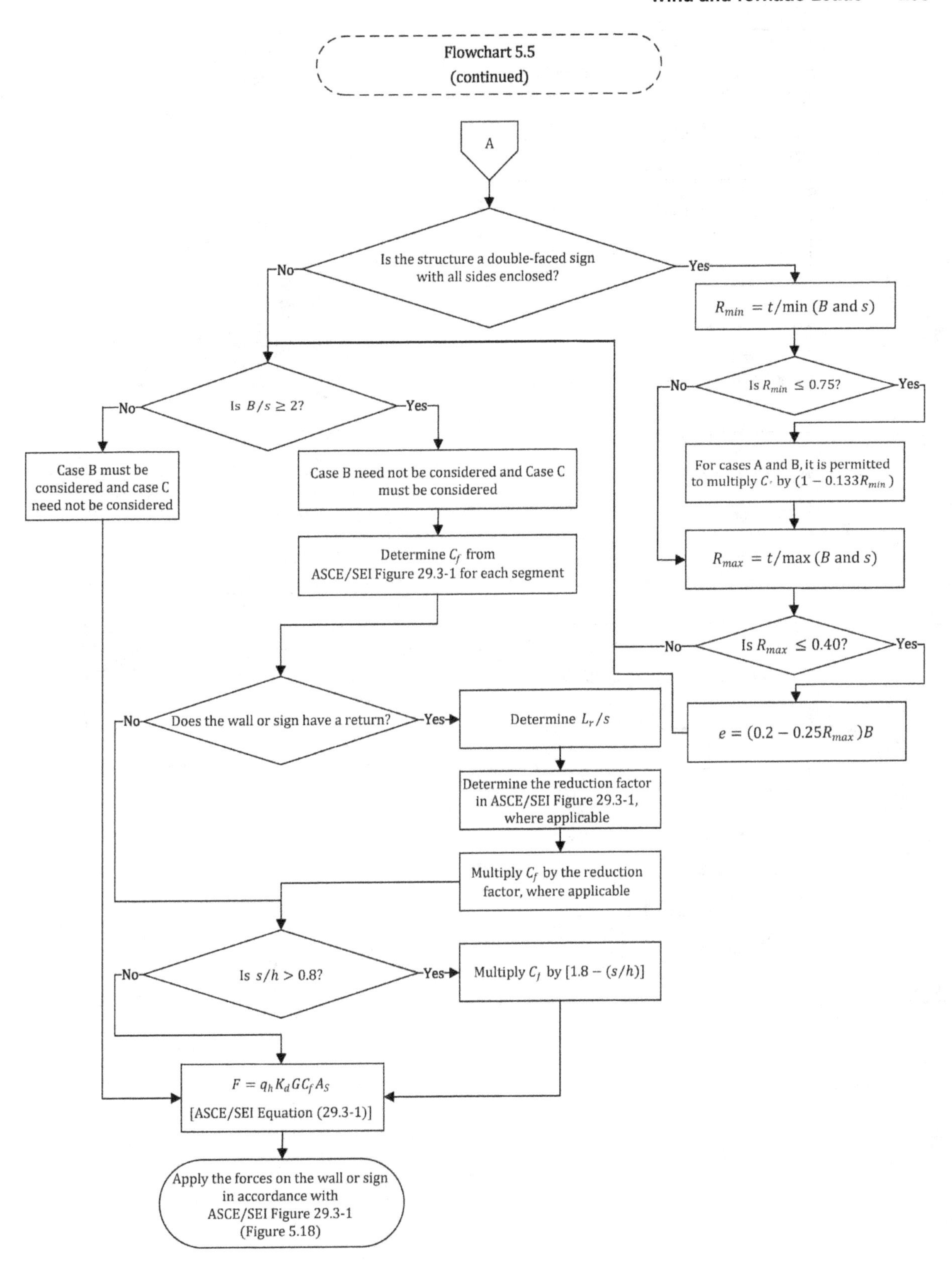

Flowchart 5.5
(continued)
A
Is the structure a double-faced sign with all sides enclosed?
No
Yes
$R_{min} = t/\min(B \text{ and } s)$
Is $R_{min} \leq 0.75$?
No
Yes
For cases A and B, it is permitted to multiply C_f by $(1 - 0.133R_{min})$
$R_{max} = t/\max(B \text{ and } s)$
Is $R_{max} \leq 0.40$?
No
Yes
$e = (0.2 - 0.25R_{max})B$
Is $B/s \geq 2$?
No
Yes
Case B must be considered and case C need not be considered
Case B need not be considered and Case C must be considered
Determine C_f from ASCE/SEI Figure 29.3-1 for each segment
Does the wall or sign have a return?
No
Yes
Determine L_r/s
Determine the reduction factor in ASCE/SEI Figure 29.3-1, where applicable
Multiply C_f by the reduction factor, where applicable
Is $s/h > 0.8$?
No
Yes
Multiply C_f by $[1.8 - (s/h)]$
$F = q_h K_d G C_f A_s$
[ASCE/SEI Equation (29.3-1)]
Apply the forces on the wall or sign in accordance with ASCE/SEI Figure 29.3-1 (Figure 5.18)

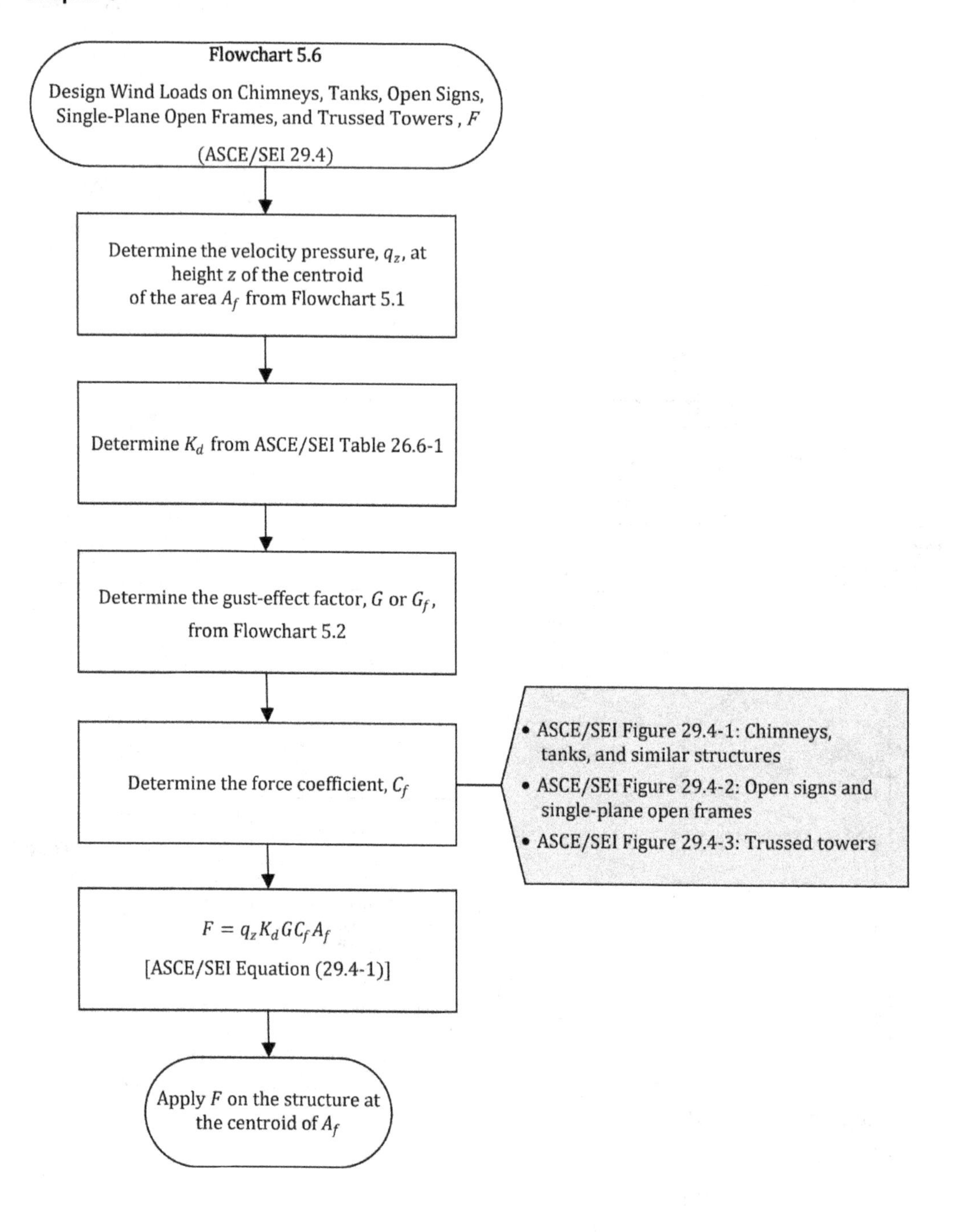
Flowchart 5.6
Design Wind Loads on Chimneys, Tanks, Open Signs, Single-Plane Open Frames, and Trussed Towers , F
(ASCE/SEI 29.4)
Determine the velocity pressure, q_z, at height z of the centroid of the area A_f from Flowchart 5.1
Determine K_d from ASCE/SEI Table 26.6-1
Determine the gust-effect factor, G or G_f, from Flowchart 5.2
Determine the force coefficient, C_f
• ASCE/SEI Figure 29.4-1: Chimneys, tanks, and similar structures
• ASCE/SEI Figure 29.4-2: Open signs and single-plane open frames
• ASCE/SEI Figure 29.4-3: Trussed towers
$F = q_z K_d G C_f A_f$
[ASCE/SEI Equation (29.4-1)]
Apply F on the structure at the centroid of A_f

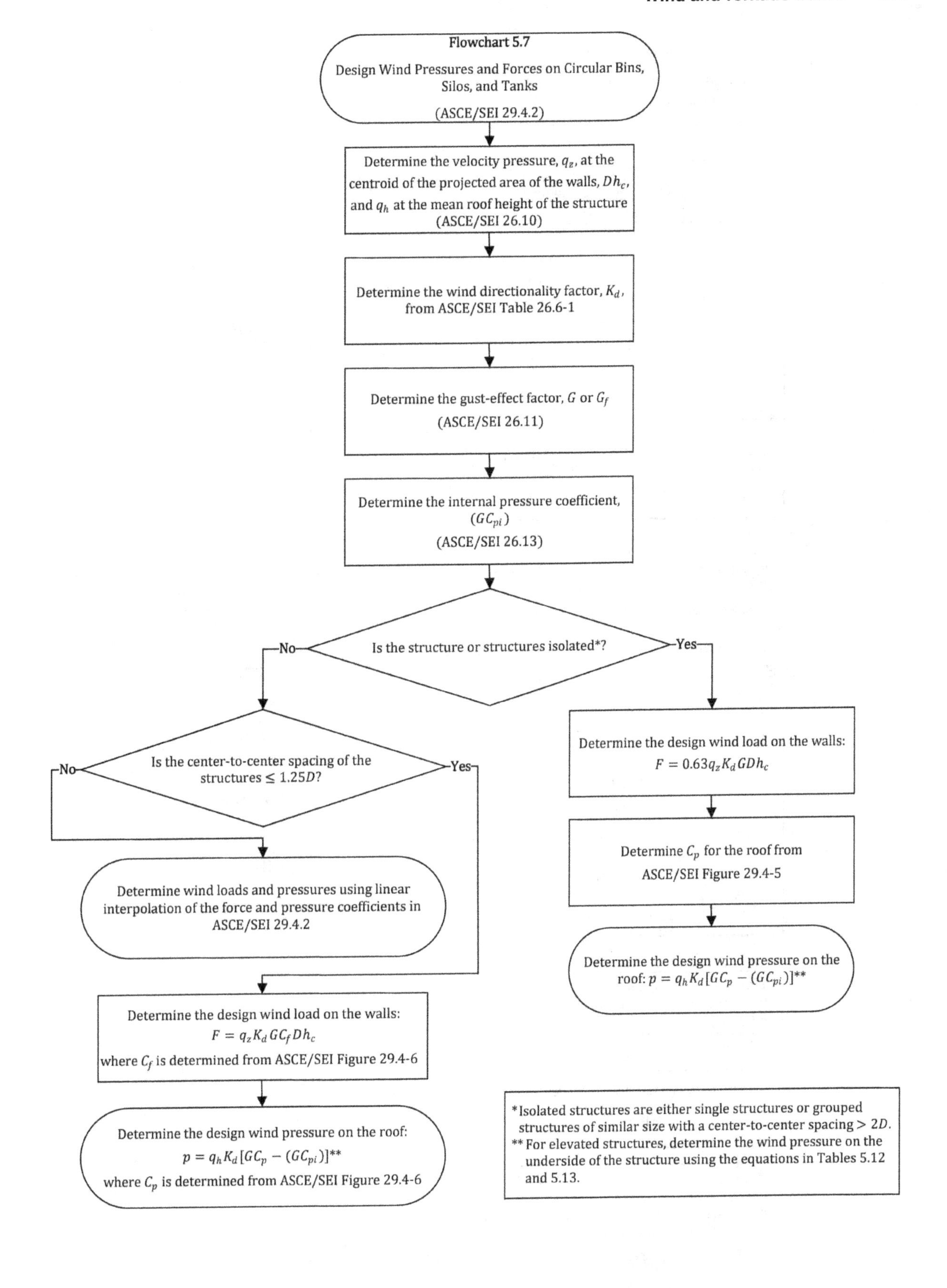

Flowchart 5.7
Design Wind Pressures and Forces on Circular Bins, Silos, and Tanks
(ASCE/SEI 29.4.2)
Determine the velocity pressure, q_z, at the centroid of the projected area of the walls, Dh_c, and q_h at the mean roof height of the structure (ASCE/SEI 26.10)
Determine the wind directionality factor, K_d, from ASCE/SEI Table 26.6-1
Determine the gust-effect factor, G or G_f (ASCE/SEI 26.11)
Determine the internal pressure coefficient, (GC_{pi}) (ASCE/SEI 26.13)
Is the structure or structures isolated*?
No
Yes
Is the center-to-center spacing of the structures $\leq 1.25D$?
No
Yes
Determine wind loads and pressures using linear interpolation of the force and pressure coefficients in ASCE/SEI 29.4.2
Determine the design wind load on the walls: $F = q_z K_d G C_f D h_c$ where C_f is determined from ASCE/SEI Figure 29.4-6
Determine the design wind pressure on the roof: $p = q_h K_d [GC_p - (GC_{pi})]$** where C_p is determined from ASCE/SEI Figure 29.4-6
Determine the design wind load on the walls: $F = 0.63 q_z K_d G D h_c$
Determine C_p for the roof from ASCE/SEI Figure 29.4-5
Determine the design wind pressure on the roof: $p = q_h K_d [GC_p - (GC_{pi})]$**
*Isolated structures are either single structures or grouped structures of similar size with a center-to-center spacing > $2D$.
** For elevated structures, determine the wind pressure on the underside of the structure using the equations in Tables 5.12 and 5.13.

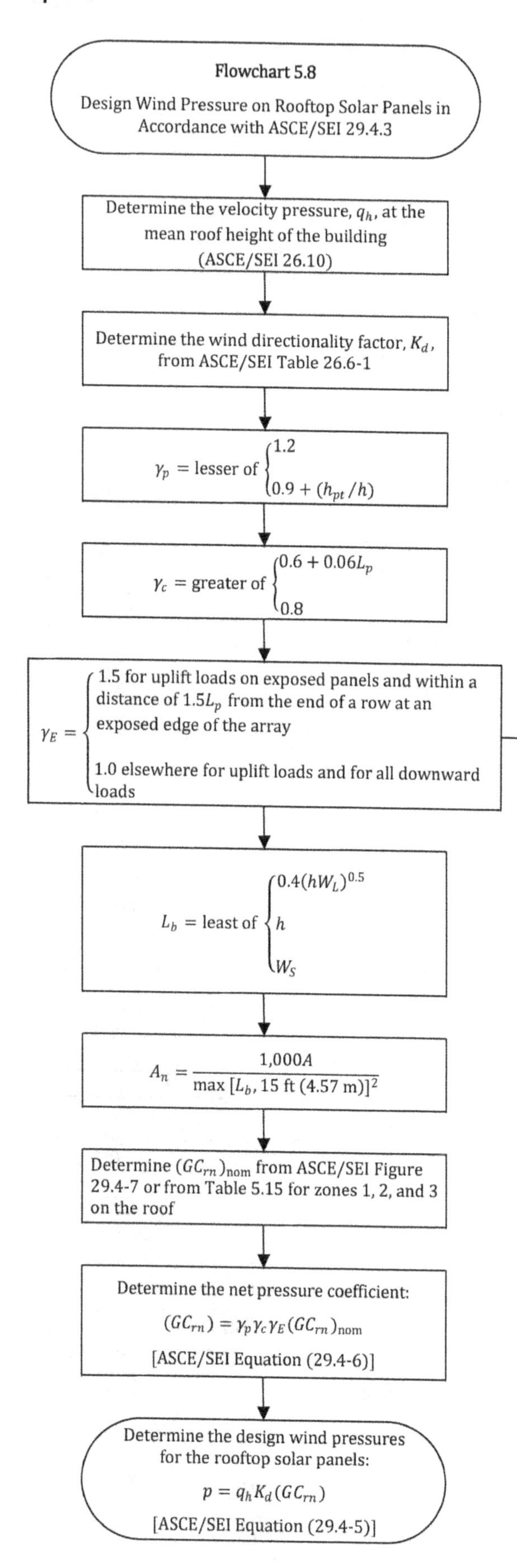
Flowchart 5.8
Design Wind Pressure on Rooftop Solar Panels in Accordance with ASCE/SEI 29.4.3
Determine the velocity pressure, q_h, at the mean roof height of the building (ASCE/SEI 26.10)
Determine the wind directionality factor, K_d, from ASCE/SEI Table 26.6-1
γ_p = lesser of $\begin{cases} 1.2 \\ 0.9 + (h_{pt}/h) \end{cases}$
γ_c = greater of $\begin{cases} 0.6 + 0.06L_p \\ 0.8 \end{cases}$
$\gamma_E = \begin{cases} \text{1.5 for uplift loads on exposed panels and within a distance of } 1.5L_p \text{ from the end of a row at an exposed edge of the array} \\ \text{1.0 elsewhere for uplift loads and for all downward loads} \end{cases}$
L_b = least of $\begin{cases} 0.4(hW_L)^{0.5} \\ h \\ W_S \end{cases}$
$A_n = \dfrac{1{,}000A}{\max\,[L_b, 15\text{ ft }(4.57\text{ m})]^2}$
Determine $(GC_{rn})_{\text{nom}}$ from ASCE/SEI Figure 29.4-7 or from Table 5.15 for zones 1, 2, and 3 on the roof
Determine the net pressure coefficient:
$(GC_{rn}) = \gamma_p \gamma_c \gamma_E (GC_{rn})_{\text{nom}}$
[ASCE/SEI Equation (29.4-6)]
Determine the design wind pressures for the rooftop solar panels:
$p = q_h K_d (GC_{rn})$
[ASCE/SEI Equation (29.4-5)]

A panel is defined as exposed if the distance to the roof edge $d_1 > 0.5h$ and one of the following applies: (1) distance to the adjacent array, $d_1 > \max\,[4h_2, 4\text{ ft }(1.22\text{ m})]$ or (2) distance to the next adjacent panel, $d_2 > \max\,[4h_2, 4\text{ ft }(1.22\text{ m})]$

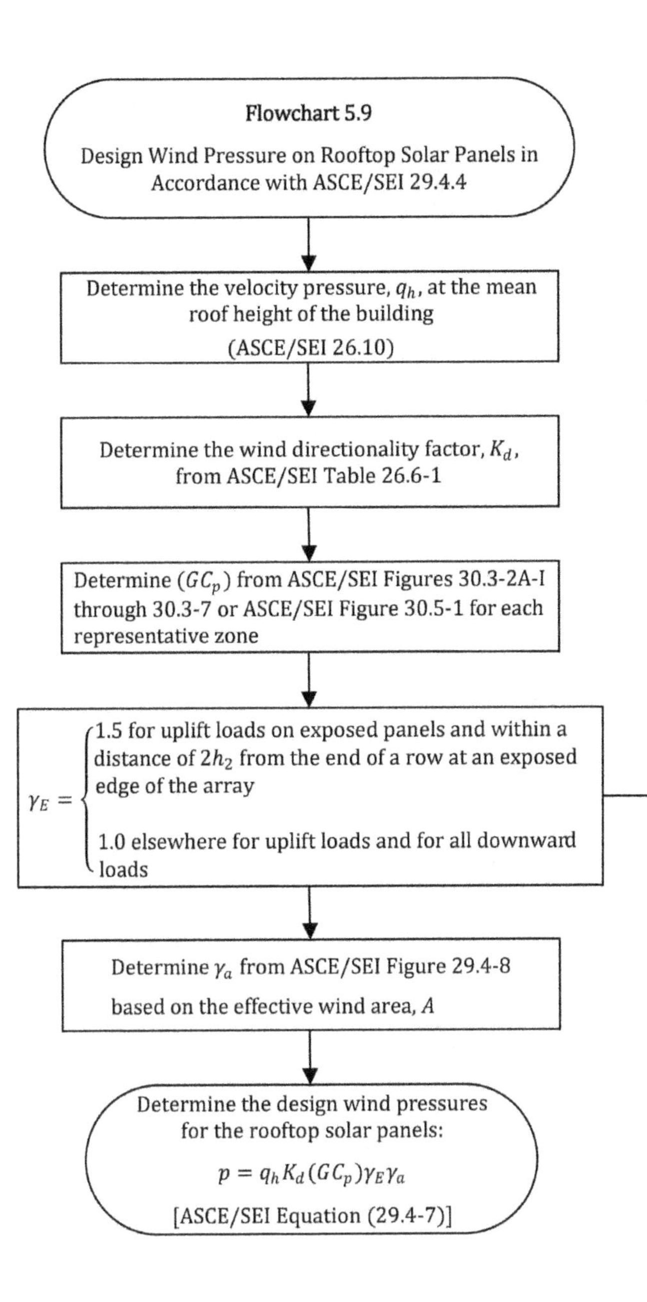
Flowchart 5.9
Design Wind Pressure on Rooftop Solar Panels in Accordance with ASCE/SEI 29.4.4
Determine the velocity pressure, q_h, at the mean roof height of the building
(ASCE/SEI 26.10)
Determine the wind directionality factor, K_d, from ASCE/SEI Table 26.6-1
Determine (GC_p) from ASCE/SEI Figures 30.3-2A-I through 30.3-7 or ASCE/SEI Figure 30.5-1 for each representative zone
$\gamma_E =$
1.5 for uplift loads on exposed panels and within a distance of $2h_2$ from the end of a row at an exposed edge of the array
1.0 elsewhere for uplift loads and for all downward loads
A panel is defined as exposed if the distance to the roof edge $d_1 > 0.5h$ and one of the following applies: (1) distance to the adjacent array, $d_1 > 2h_2$ or (2) distance to the next adjacent panel, $d_2 > 2h_2$
Determine γ_a from ASCE/SEI Figure 29.4-8
based on the effective wind area, A
Determine the design wind pressures for the rooftop solar panels:
$p = q_h K_d (GC_p) \gamma_E \gamma_a$
[ASCE/SEI Equation (29.4-7)]

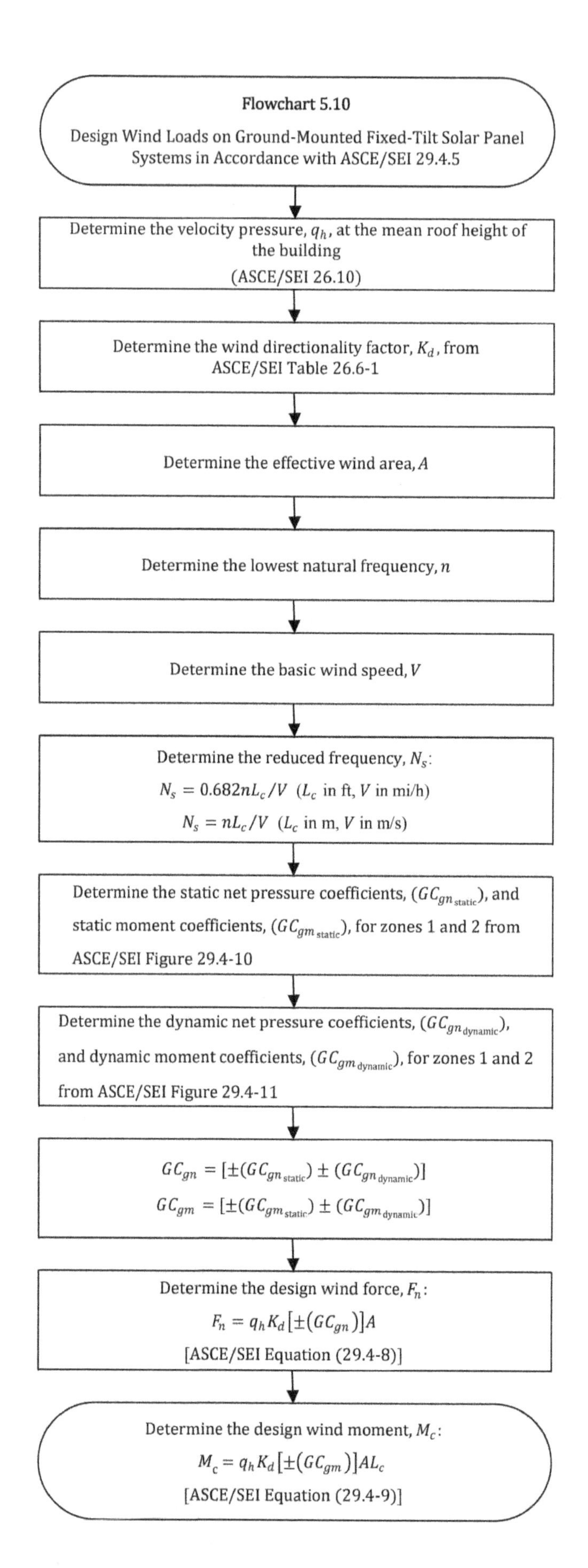

Flowchart 5.10
Design Wind Loads on Ground-Mounted Fixed-Tilt Solar Panel Systems in Accordance with ASCE/SEI 29.4.5
Determine the velocity pressure, q_h, at the mean roof height of the building
(ASCE/SEI 26.10)
Determine the wind directionality factor, K_d, from ASCE/SEI Table 26.6-1
Determine the effective wind area, A
Determine the lowest natural frequency, n
Determine the basic wind speed, V
Determine the reduced frequency, N_s:
$N_s = 0.682nL_c/V$ (L_c in ft, V in mi/h)
$N_s = nL_c/V$ (L_c in m, V in m/s)
Determine the static net pressure coefficients, $(GC_{gn_{\text{static}}})$, and static moment coefficients, $(GC_{gm_{\text{static}}})$, for zones 1 and 2 from ASCE/SEI Figure 29.4-10
Determine the dynamic net pressure coefficients, $(GC_{gn_{\text{dynamic}}})$, and dynamic moment coefficients, $(GC_{gm_{\text{dynamic}}})$, for zones 1 and 2 from ASCE/SEI Figure 29.4-11
$GC_{gn} = [\pm(GC_{gn_{\text{static}}}) \pm (GC_{gn_{\text{dynamic}}})]$
$GC_{gm} = [\pm(GC_{gm_{\text{static}}}) \pm (GC_{gm_{\text{dynamic}}})]$
Determine the design wind force, F_n:
$F_n = q_h K_d[\pm(GC_{gn})]A$
[ASCE/SEI Equation (29.4-8)]
Determine the design wind moment, M_c:
$M_c = q_h K_d[\pm(GC_{gm})]AL_c$
[ASCE/SEI Equation (29.4-9)]

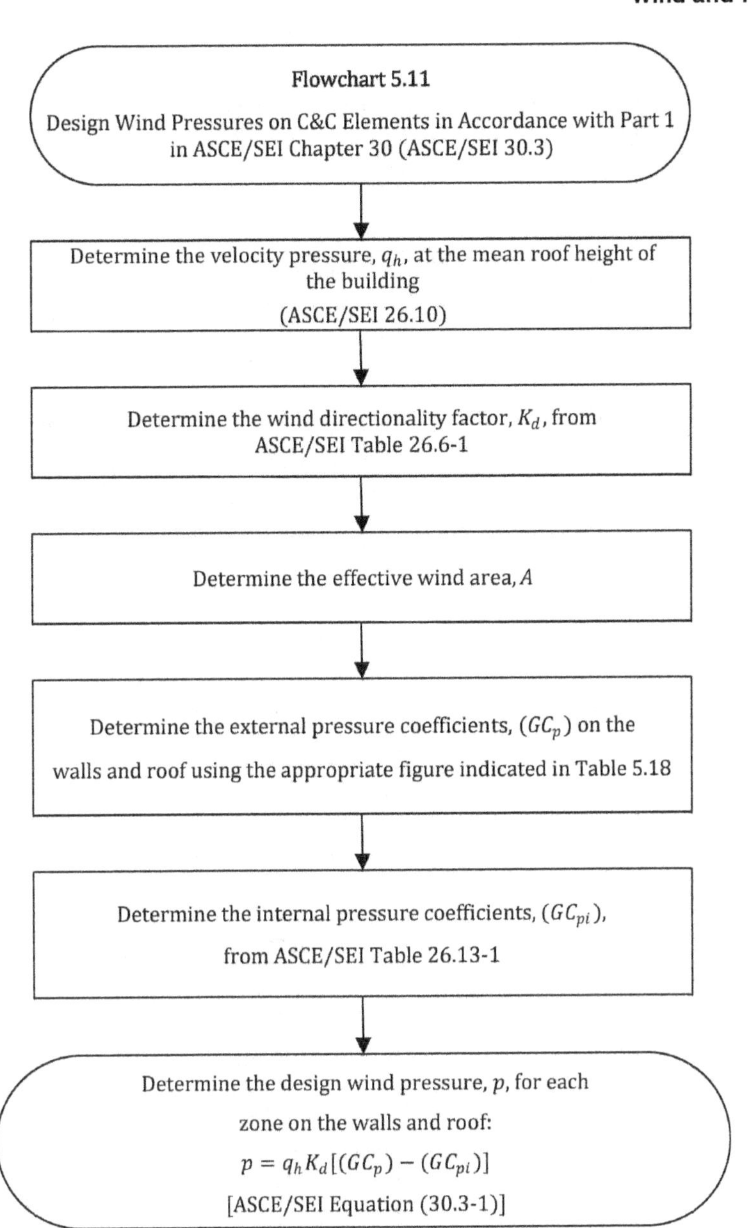
Flowchart 5.11
Design Wind Pressures on C&C Elements in Accordance with Part 1 in ASCE/SEI Chapter 30 (ASCE/SEI 30.3)
Determine the velocity pressure, q_h, at the mean roof height of the building (ASCE/SEI 26.10)
Determine the wind directionality factor, K_d, from ASCE/SEI Table 26.6-1
Determine the effective wind area, A
Determine the external pressure coefficients, (GC_p) on the walls and roof using the appropriate figure indicated in Table 5.18
Determine the internal pressure coefficients, (GC_{pi}), from ASCE/SEI Table 26.13-1
Determine the design wind pressure, p, for each zone on the walls and roof:
$p = q_h K_d[(GC_p) - (GC_{pi})]$
[ASCE/SEI Equation (30.3-1)]

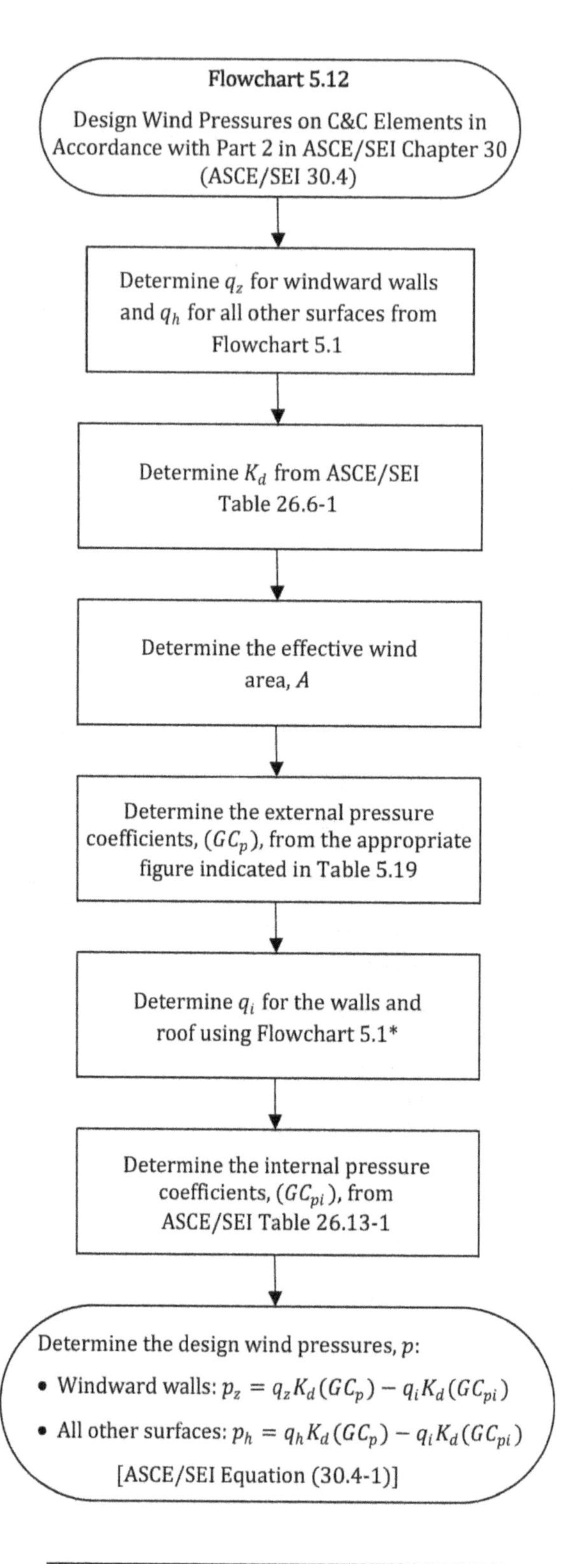
Flowchart 5.12
Design Wind Pressures on C&C Elements in Accordance with Part 2 in ASCE/SEI Chapter 30 (ASCE/SEI 30.4)
Determine q_z for windward walls and q_h for all other surfaces from Flowchart 5.1
Determine K_d from ASCE/SEI Table 26.6-1
Determine the effective wind area, A
Determine the external pressure coefficients, (GC_p), from the appropriate figure indicated in Table 5.19
Determine q_i for the walls and roof using Flowchart 5.1*
Determine the internal pressure coefficients, (GC_{pi}), from ASCE/SEI Table 26.13-1
Determine the design wind pressures, p:
• Windward walls: $p_z = q_z K_d (GC_p) - q_i K_d (GC_{pi})$
• All other surfaces: $p_h = q_h K_d (GC_p) - q_i K_d (GC_{pi})$
[ASCE/SEI Equation (30.4-1)]
*$q_i = q_h$ or $q_i = q_z$ based on the enclosure classification (see ASCE/SEI 27.3.1). q_i may conservatively be evaluated at height h.

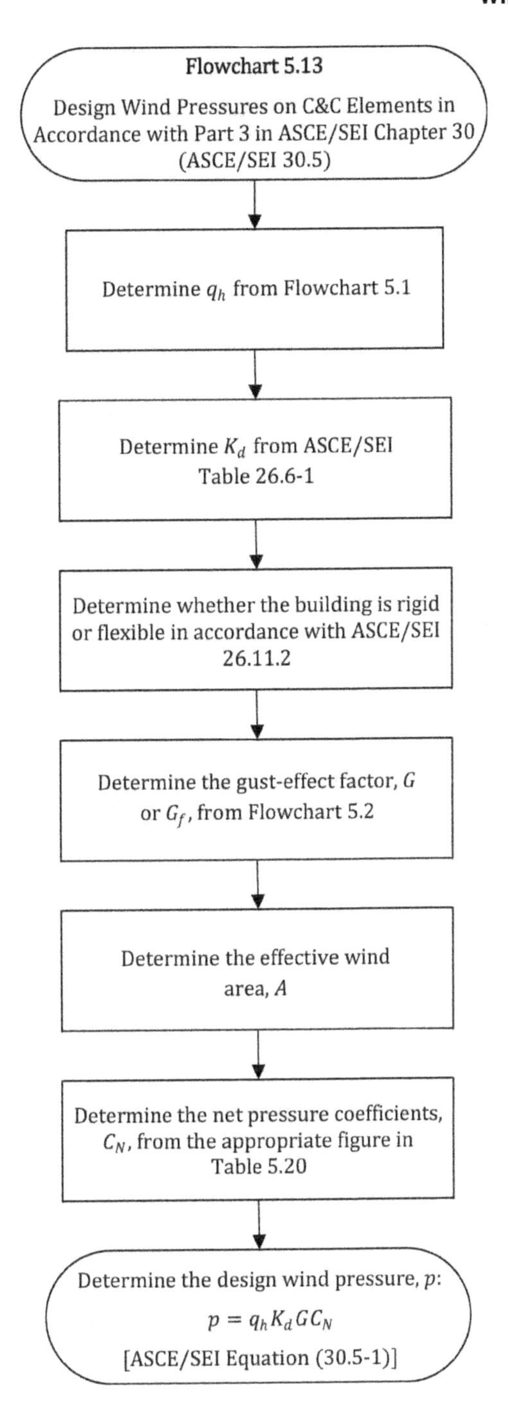
Flowchart 5.13
Design Wind Pressures on C&C Elements in Accordance with Part 3 in ASCE/SEI Chapter 30 (ASCE/SEI 30.5)
Determine q_h from Flowchart 5.1
Determine K_d from ASCE/SEI Table 26.6-1
Determine whether the building is rigid or flexible in accordance with ASCE/SEI 26.11.2
Determine the gust-effect factor, G or G_f, from Flowchart 5.2
Determine the effective wind area, A
Determine the net pressure coefficients, C_N, from the appropriate figure in Table 5.20
Determine the design wind pressure, p:
$p = q_h K_d G C_N$
[ASCE/SEI Equation (30.5-1)]

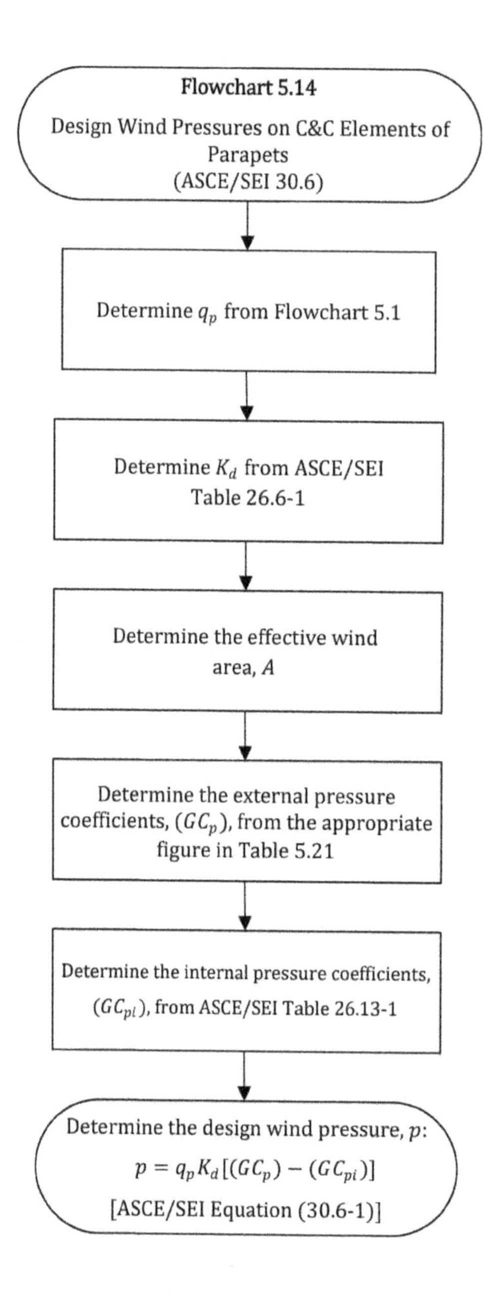

Flowchart 5.14
Design Wind Pressures on C&C Elements of Parapets
(ASCE/SEI 30.6)
Determine q_p from Flowchart 5.1
Determine K_d from ASCE/SEI Table 26.6-1
Determine the effective wind area, A
Determine the external pressure coefficients, (GC_p), from the appropriate figure in Table 5.21
Determine the internal pressure coefficients, (GC_{pi}), from ASCE/SEI Table 26.13-1
Determine the design wind pressure, p:
$p = q_p K_d [(GC_p) - (GC_{pi})]$
[ASCE/SEI Equation (30.6-1)]

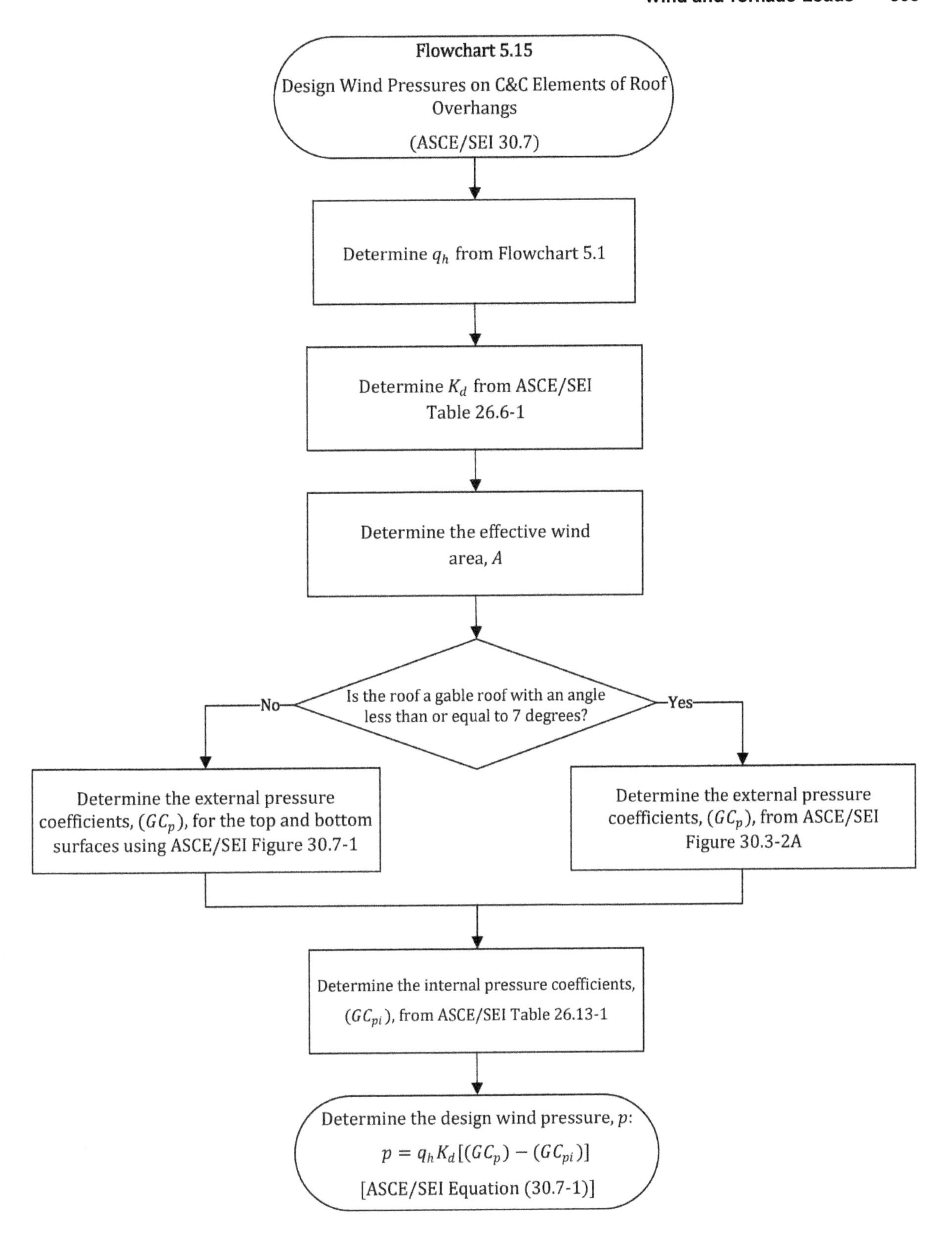
Flowchart 5.15
Design Wind Pressures on C&C Elements of Roof Overhangs
(ASCE/SEI 30.7)
Determine q_h from Flowchart 5.1
Determine K_d from ASCE/SEI Table 26.6-1
Determine the effective wind area, A
Is the roof a gable roof with an angle less than or equal to 7 degrees?
No
Yes
Determine the external pressure coefficients, (GC_p), for the top and bottom surfaces using ASCE/SEI Figure 30.7-1
Determine the external pressure coefficients, (GC_p), from ASCE/SEI Figure 30.3-2A
Determine the internal pressure coefficients, (GC_{pi}), from ASCE/SEI Table 26.13-1
Determine the design wind pressure, p:
$p = q_h K_d[(GC_p) - (GC_{pi})]$
[ASCE/SEI Equation (30.7-1)]

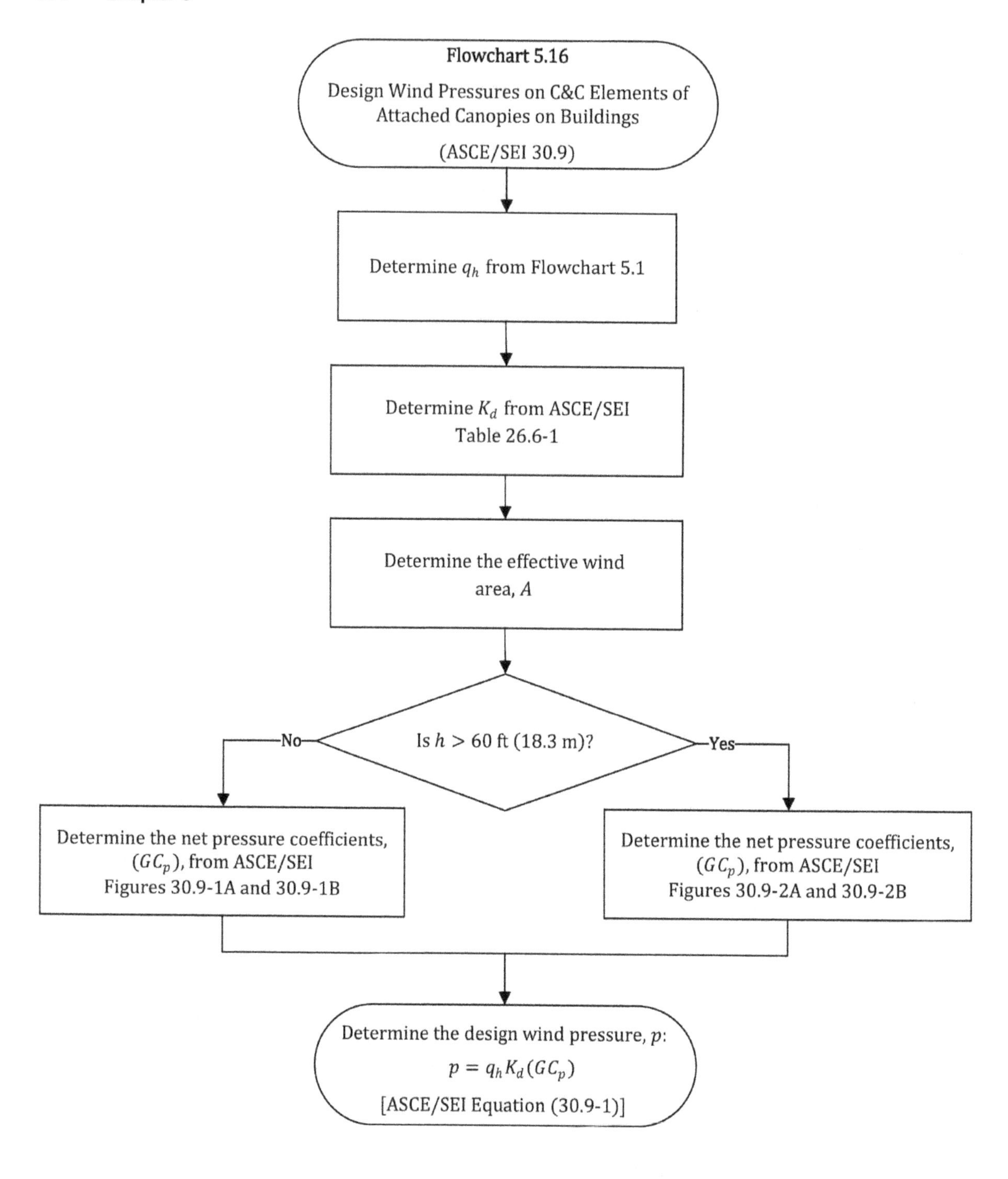

Flowchart 5.16
Design Wind Pressures on C&C Elements of Attached Canopies on Buildings
(ASCE/SEI 30.9)
Determine q_h from Flowchart 5.1
Determine K_d from ASCE/SEI Table 26.6-1
Determine the effective wind area, A
Is $h > 60$ ft (18.3 m)?
No
Yes
Determine the net pressure coefficients, (GC_p), from ASCE/SEI Figures 30.9-1A and 30.9-1B
Determine the net pressure coefficients, (GC_p), from ASCE/SEI Figures 30.9-2A and 30.9-2B
Determine the design wind pressure, p:
$p = q_h K_d (GC_p)$
[ASCE/SEI Equation (30.9-1)]

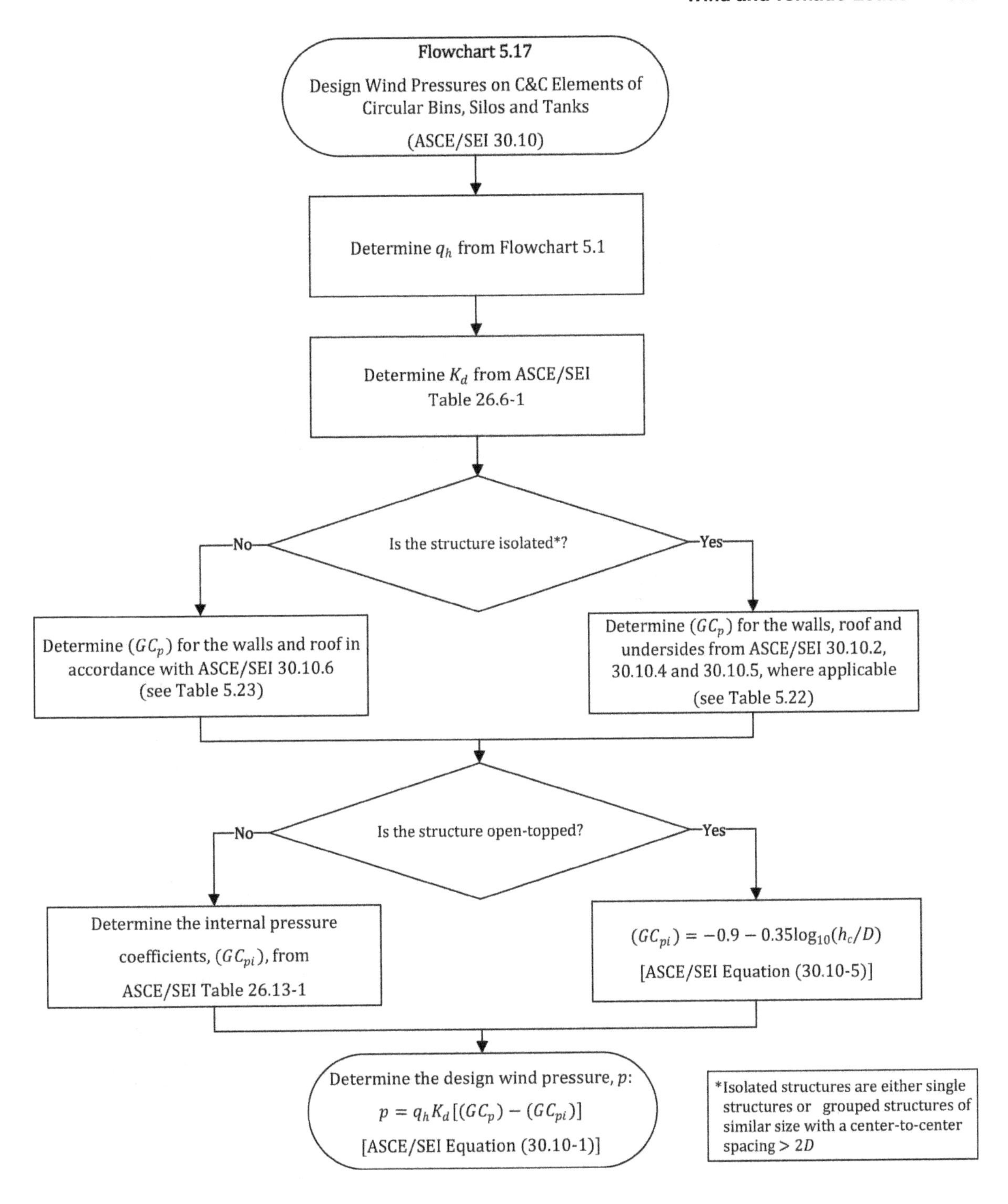
Flowchart 5.17
Design Wind Pressures on C&C Elements of Circular Bins, Silos and Tanks
(ASCE/SEI 30.10)
Determine q_h from Flowchart 5.1
Determine K_d from ASCE/SEI Table 26.6-1
Is the structure isolated*?
No
Yes
Determine (GC_p) for the walls and roof in accordance with ASCE/SEI 30.10.6 (see Table 5.23)
Determine (GC_p) for the walls, roof and undersides from ASCE/SEI 30.10.2, 30.10.4 and 30.10.5, where applicable (see Table 5.22)
Is the structure open-topped?
No
Yes
Determine the internal pressure coefficients, (GC_{pi}), from ASCE/SEI Table 26.13-1
$(GC_{pi}) = -0.9 - 0.35\log_{10}(h_c/D)$
[ASCE/SEI Equation (30.10-5)]
Determine the design wind pressure, p:
$p = q_h K_d[(GC_p) - (GC_{pi})]$
[ASCE/SEI Equation (30.10-1)]
*Isolated structures are either single structures or grouped structures of similar size with a center-to-center spacing > $2D$

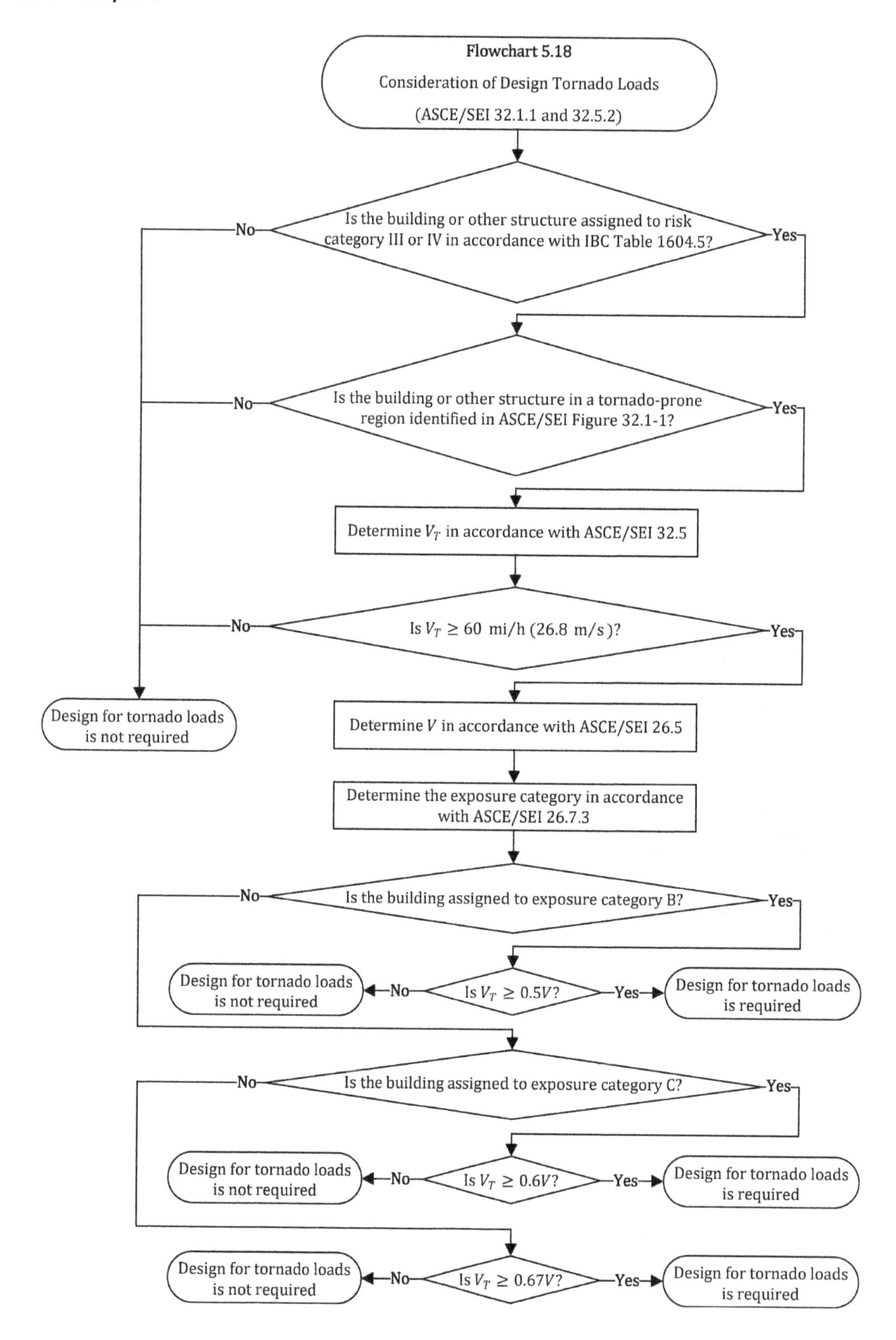
Flowchart 5.18
Consideration of Design Tornado Loads
(ASCE/SEI 32.1.1 and 32.5.2)
Is the building or other structure assigned to risk category III or IV in accordance with IBC Table 1604.5?
No
Yes
Is the building or other structure in a tornado-prone region identified in ASCE/SEI Figure 32.1-1?
No
Yes
Determine V_T in accordance with ASCE/SEI 32.5
Is $V_T \geq 60$ mi/h (26.8 m/s)?
No
Yes
Design for tornado loads is not required
Determine V in accordance with ASCE/SEI 26.5
Determine the exposure category in accordance with ASCE/SEI 26.7.3
Is the building assigned to exposure category B?
No
Yes
Design for tornado loads is not required
No
Is $V_T \geq 0.5V$?
Yes
Design for tornado loads is required
Is the building assigned to exposure category C?
No
Yes
Design for tornado loads is not required
No
Is $V_T \geq 0.6V$?
Yes
Design for tornado loads is required
Design for tornado loads is not required
No
Is $V_T \geq 0.67V$?
Yes
Design for tornado loads is required

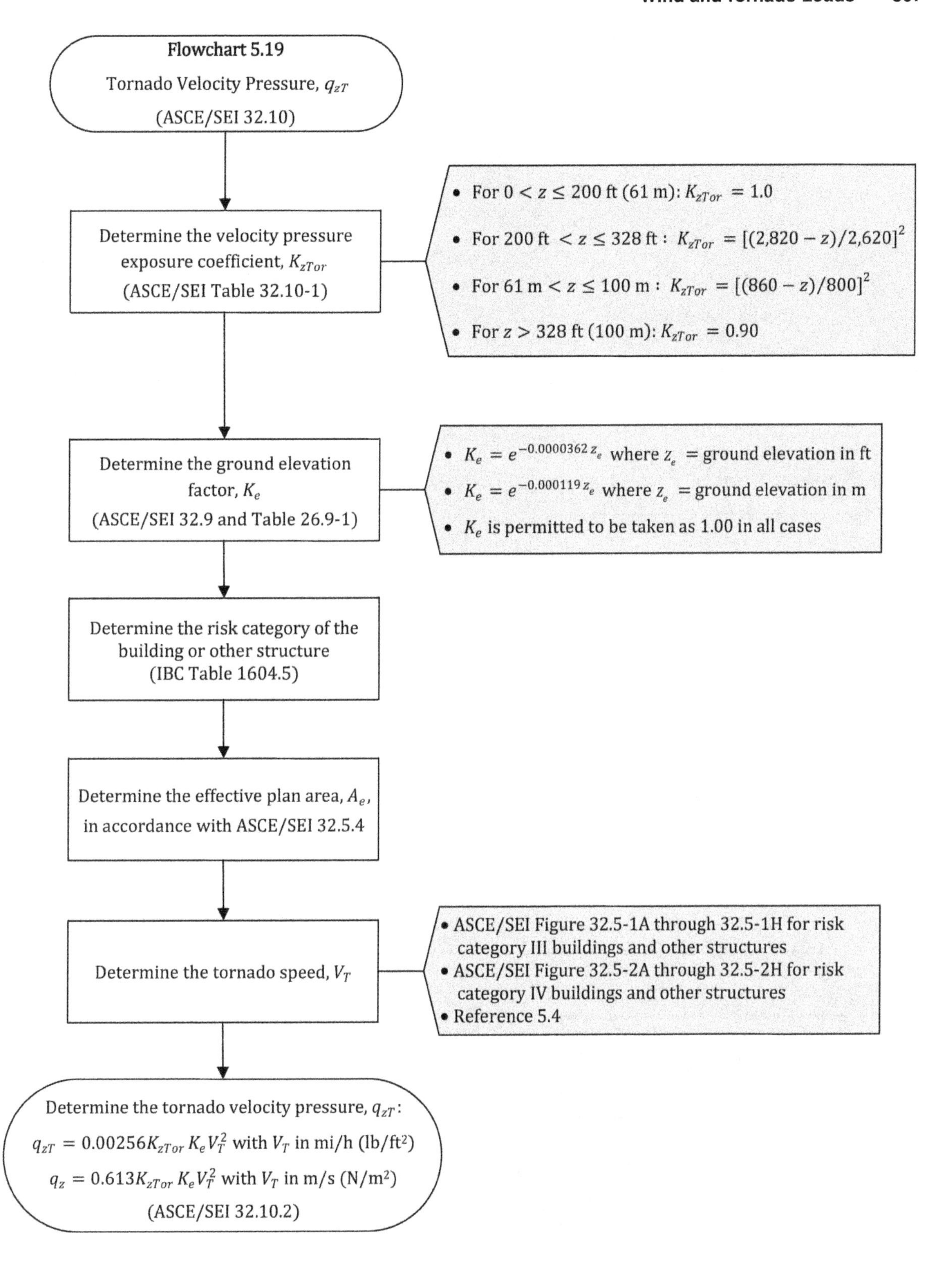

Flowchart 5.19
Tornado Velocity Pressure, q_{zT}
(ASCE/SEI 32.10)
Determine the velocity pressure exposure coefficient, K_{zTor}
(ASCE/SEI Table 32.10-1)
• For $0 < z \leq 200$ ft (61 m): $K_{zTor} = 1.0$
• For 200 ft $< z \leq 328$ ft : $K_{zTor} = [(2{,}820 - z)/2{,}620]^2$
• For 61 m $< z \leq 100$ m : $K_{zTor} = [(860 - z)/800]^2$
• For $z > 328$ ft (100 m): $K_{zTor} = 0.90$
Determine the ground elevation factor, K_e
(ASCE/SEI 32.9 and Table 26.9-1)
• $K_e = e^{-0.0000362 z_e}$ where z_e = ground elevation in ft
• $K_e = e^{-0.000119 z_e}$ where z_e = ground elevation in m
• K_e is permitted to be taken as 1.00 in all cases
Determine the risk category of the building or other structure
(IBC Table 1604.5)
Determine the effective plan area, A_e, in accordance with ASCE/SEI 32.5.4
Determine the tornado speed, V_T
• ASCE/SEI Figure 32.5-1A through 32.5-1H for risk category III buildings and other structures
• ASCE/SEI Figure 32.5-2A through 32.5-2H for risk category IV buildings and other structures
• Reference 5.4
Determine the tornado velocity pressure, q_{zT}:
$q_{zT} = 0.00256 K_{zTor} K_e V_T^2$ with V_T in mi/h (lb/ft^2)
$q_z = 0.613 K_{zTor} K_e V_T^2$ with V_T in m/s (N/m^2)
(ASCE/SEI 32.10.2)

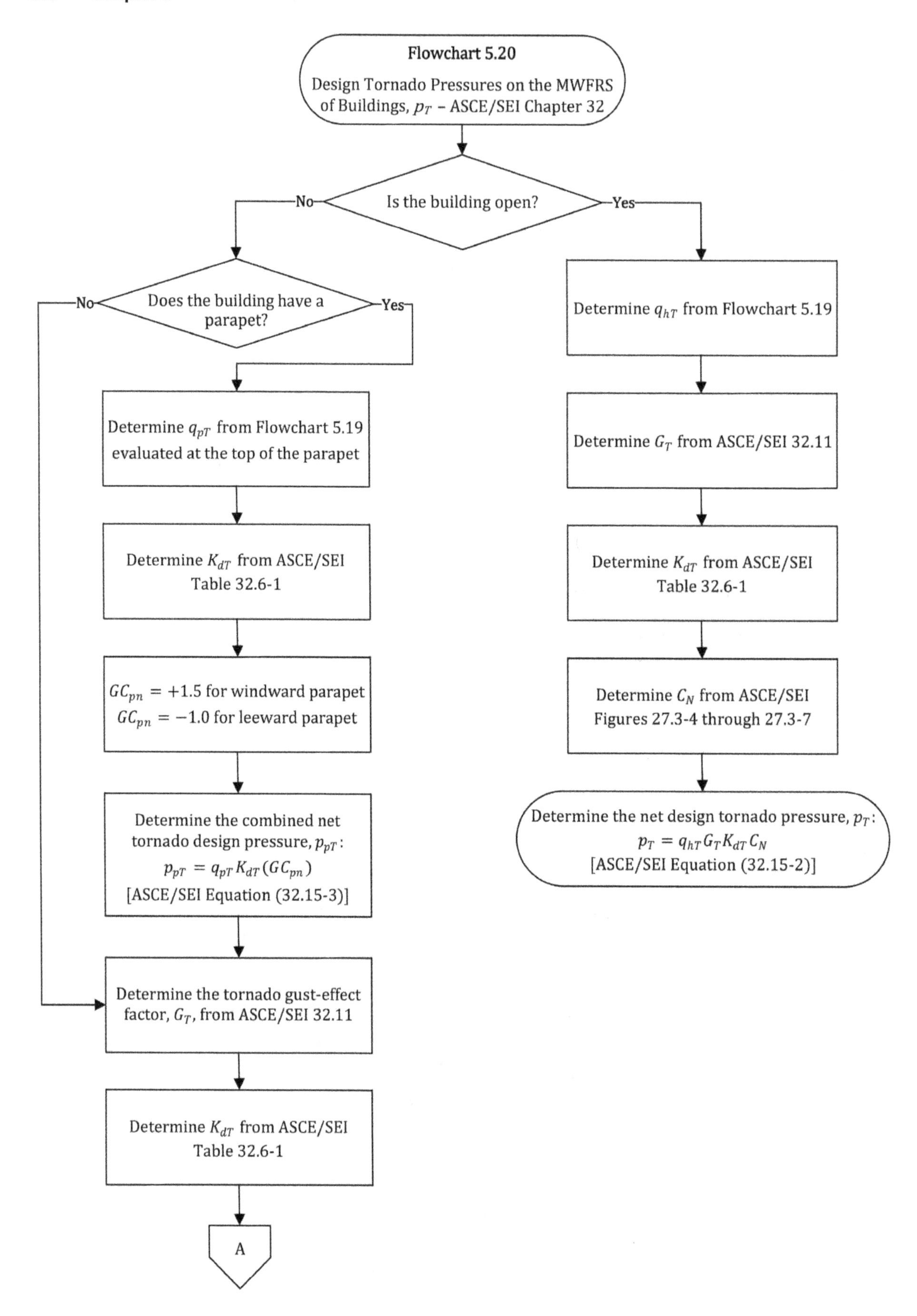

Flowchart 5.20
Design Tornado Pressures on the MWFRS of Buildings, p_T - ASCE/SEI Chapter 32
Is the building open?
No
Yes
Does the building have a parapet?
No
Yes
Determine q_{pT} from Flowchart 5.19 evaluated at the top of the parapet
Determine K_{dT} from ASCE/SEI Table 32.6-1
$GC_{pn} = +1.5$ for windward parapet
$GC_{pn} = -1.0$ for leeward parapet
Determine the combined net tornado design pressure, p_{pT}:
$p_{pT} = q_{pT} K_{dT} (GC_{pn})$
[ASCE/SEI Equation (32.15-3)]
Determine the tornado gust-effect factor, G_T, from ASCE/SEI 32.11
Determine K_{dT} from ASCE/SEI Table 32.6-1
A
Determine q_{hT} from Flowchart 5.19
Determine G_T from ASCE/SEI 32.11
Determine K_{dT} from ASCE/SEI Table 32.6-1
Determine C_N from ASCE/SEI Figures 27.3-4 through 27.3-7
Determine the net design tornado pressure, p_T:
$p_T = q_{hT} G_T K_{dT} C_N$
[ASCE/SEI Equation (32.15-2)]

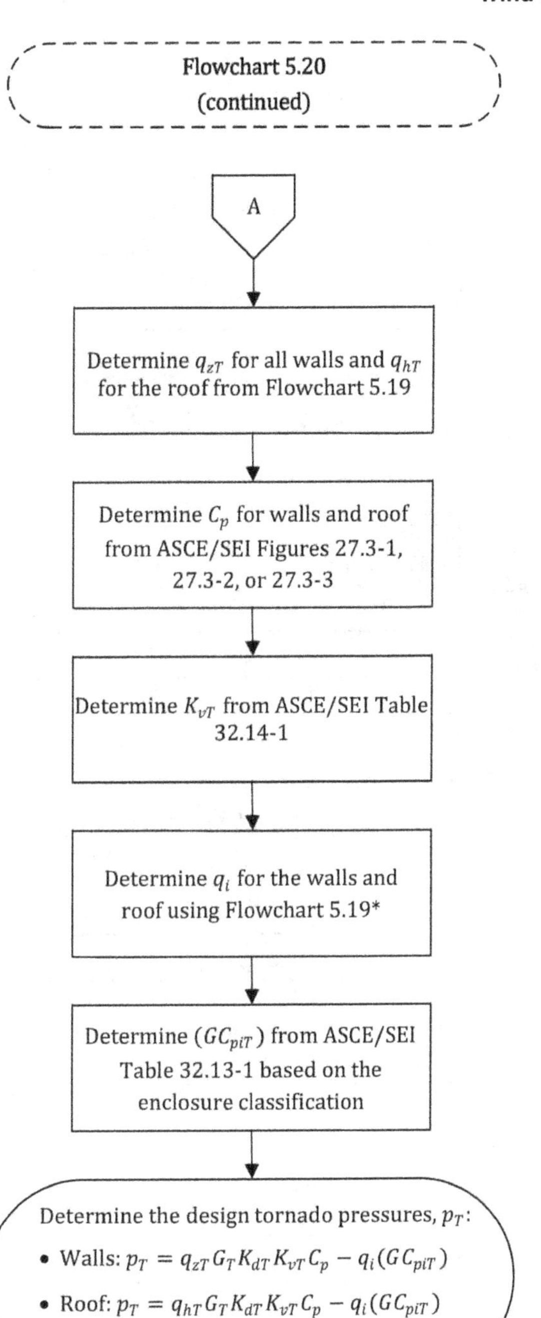

Flowchart 5.20
(continued)
A
Determine q_{zT} for all walls and q_{hT} for the roof from Flowchart 5.19
Determine C_p for walls and roof from ASCE/SEI Figures 27.3-1, 27.3-2, or 27.3-3
Determine K_{vT} from ASCE/SEI Table 32.14-1
Determine q_i for the walls and roof using Flowchart 5.19*
Determine (GC_{piT}) from ASCE/SEI Table 32.13-1 based on the enclosure classification
Determine the design tornado pressures, p_T:
• Walls: $p_T = q_{zT} G_T K_{dT} K_{vT} C_p - q_i(GC_{piT})$
• Roof: $p_T = q_{hT} G_T K_{dT} K_{vT} C_p - q_i(GC_{piT})$
[ASCE/SEI Equation (32.15-1)]

*$q_i = q_{hT}$, $q_i = q_{zT}$ or $q_i = q_{z_{op}}$ based on the enclosure classification (see ASCE/SEI 32.15.1)

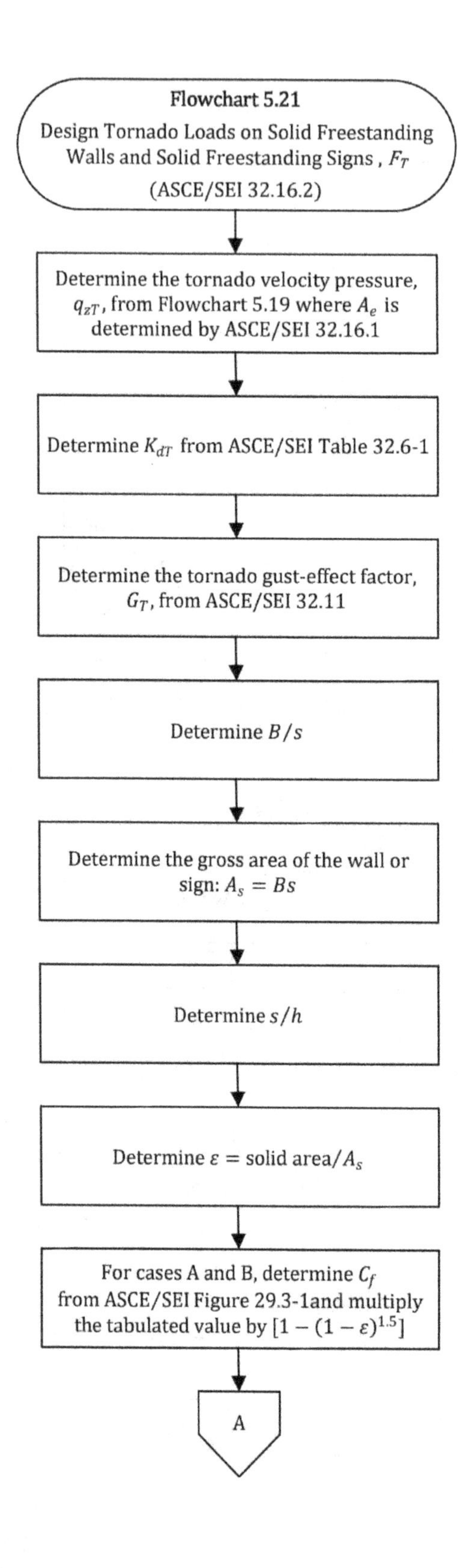

Flowchart 5.21
Design Tornado Loads on Solid Freestanding Walls and Solid Freestanding Signs , F_T
(ASCE/SEI 32.16.2)
Determine the tornado velocity pressure, q_{zT}, from Flowchart 5.19 where A_e is determined by ASCE/SEI 32.16.1
Determine K_{dT} from ASCE/SEI Table 32.6-1
Determine the tornado gust-effect factor, G_T, from ASCE/SEI 32.11
Determine B/s
Determine the gross area of the wall or sign: $A_s = Bs$
Determine s/h
Determine ε = solid area/A_s
For cases A and B, determine C_f from ASCE/SEI Figure 29.3-1and multiply the tabulated value by $[1 - (1 - \varepsilon)^{1.5}]$
A

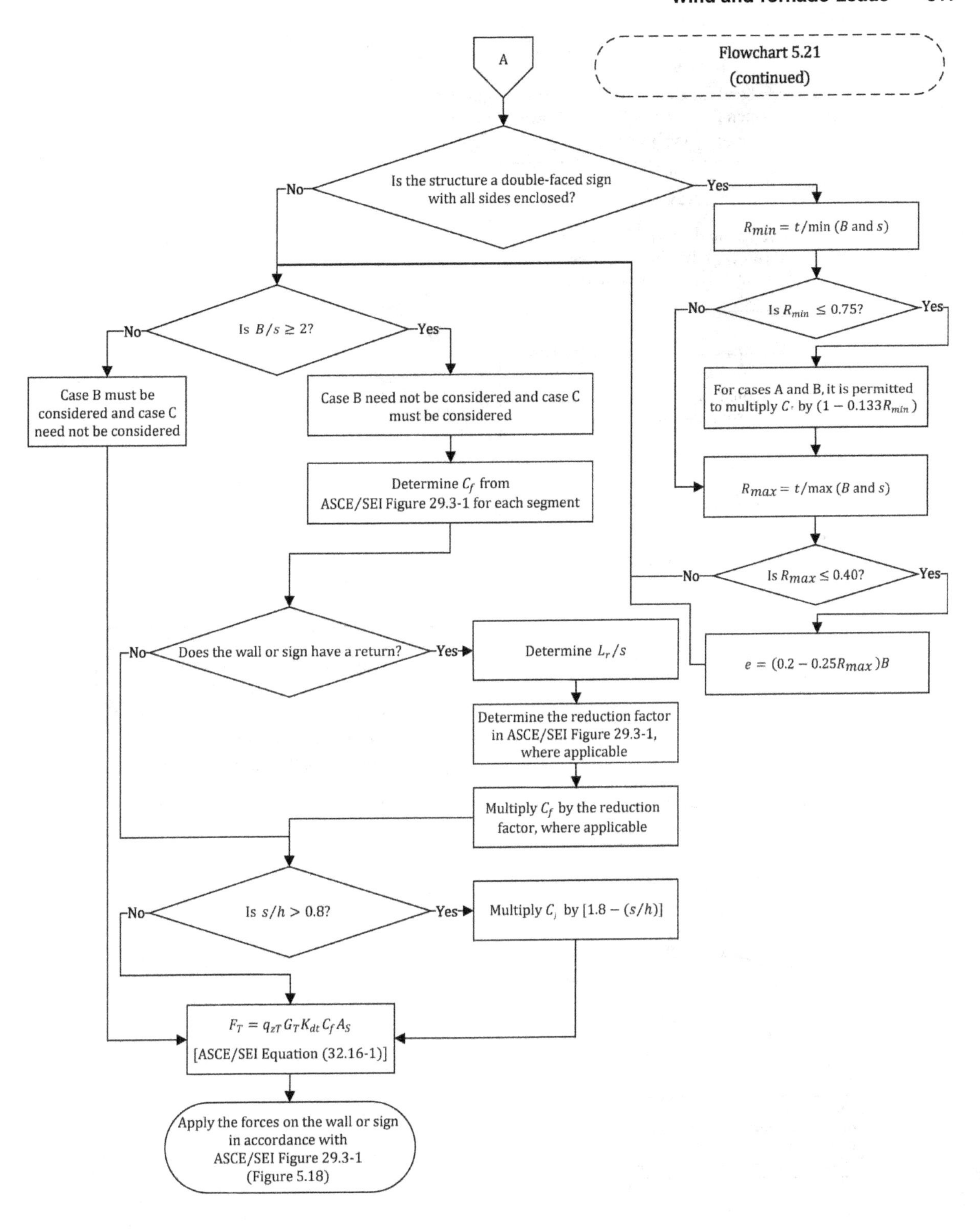

Flowchart 5.21
(continued)
A
Is the structure a double-faced sign with all sides enclosed?
No
Yes
$R_{min} = t/\min(B \text{ and } s)$
Is $R_{min} \leq 0.75$?
For cases A and B, it is permitted to multiply C_f by $(1 - 0.133R_{min})$
$R_{max} = t/\max(B \text{ and } s)$
Is $R_{max} \leq 0.40$?
$e = (0.2 - 0.25R_{max})B$
Is $B/s \geq 2$?
Case B must be considered and case C need not be considered
Case B need not be considered and case C must be considered
Determine C_f from ASCE/SEI Figure 29.3-1 for each segment
Does the wall or sign have a return?
Determine L_r/s
Determine the reduction factor in ASCE/SEI Figure 29.3-1, where applicable
Multiply C_f by the reduction factor, where applicable
Is $s/h > 0.8$?
Multiply C_f by $[1.8 - (s/h)]$
$F_T = q_{zT} G_T K_{dt} C_f A_S$
[ASCE/SEI Equation (32.16-1)]
Apply the forces on the wall or sign in accordance with ASCE/SEI Figure 29.3-1 (Figure 5.18)

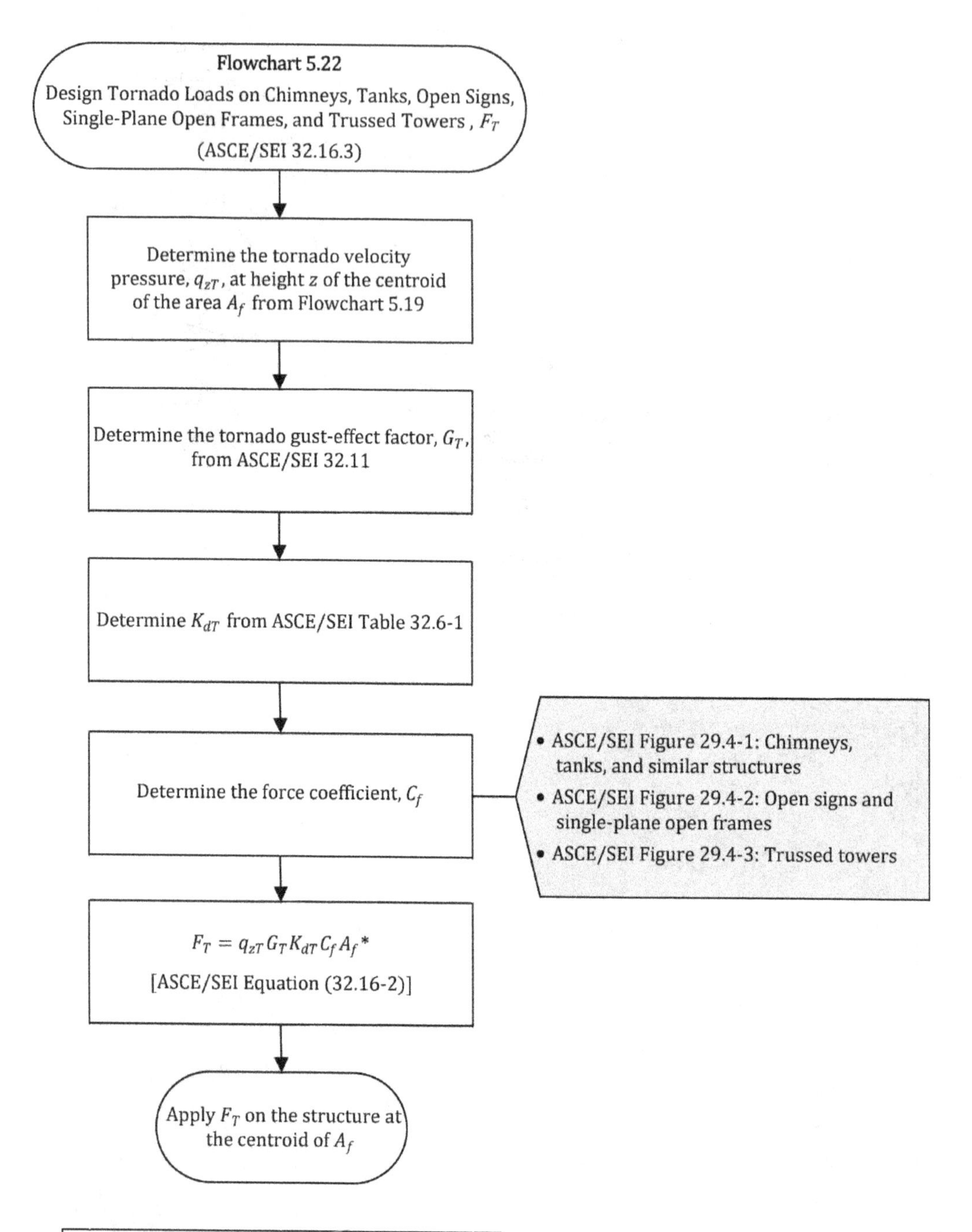
Flowchart 5.22
Design Tornado Loads on Chimneys, Tanks, Open Signs, Single-Plane Open Frames, and Trussed Towers , F_T
(ASCE/SEI 32.16.3)
Determine the tornado velocity pressure, q_{zT}, at height z of the centroid of the area A_f from Flowchart 5.19
Determine the tornado gust-effect factor, G_T, from ASCE/SEI 32.11
Determine K_{dT} from ASCE/SEI Table 32.6-1
Determine the force coefficient, C_f
• ASCE/SEI Figure 29.4-1: Chimneys, tanks, and similar structures
• ASCE/SEI Figure 29.4-2: Open signs and single-plane open frames
• ASCE/SEI Figure 29.4-3: Trussed towers
$F_T = q_{zT} G_T K_{dT} C_f A_f$*
[ASCE/SEI Equation (32.16-2)]
Apply F_T on the structure at the centroid of A_f

*An additional 40 ft^2 (3.72 m^2) of projected surface area of clinging debris at the midheight of trussed towers and other open lattice-type structures or at 50 ft (15.2 m), whichever is less, must be included in A_f

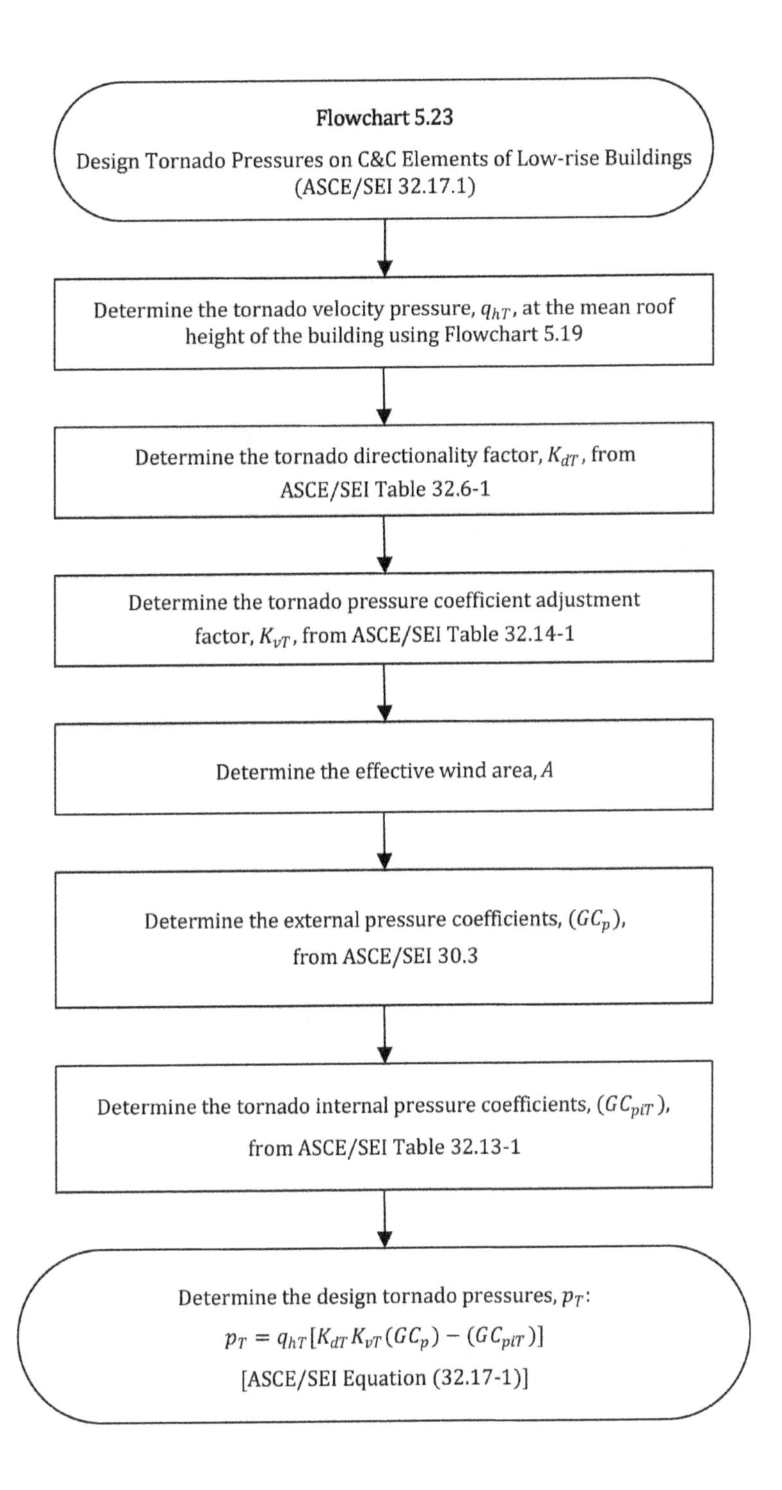

Flowchart 5.23
Design Tornado Pressures on C&C Elements of Low-rise Buildings (ASCE/SEI 32.17.1)
Determine the tornado velocity pressure, q_{hT}, at the mean roof height of the building using Flowchart 5.19
Determine the tornado directionality factor, K_{dT}, from ASCE/SEI Table 32.6-1
Determine the tornado pressure coefficient adjustment factor, K_{vT}, from ASCE/SEI Table 32.14-1
Determine the effective wind area, A
Determine the external pressure coefficients, (GC_p), from ASCE/SEI 30.3
Determine the tornado internal pressure coefficients, (GC_{piT}), from ASCE/SEI Table 32.13-1
Determine the design tornado pressures, p_T:
$p_T = q_{hT}[K_{dT}K_{vT}(GC_p) - (GC_{piT})]$
[ASCE/SEI Equation (32.17-1)]

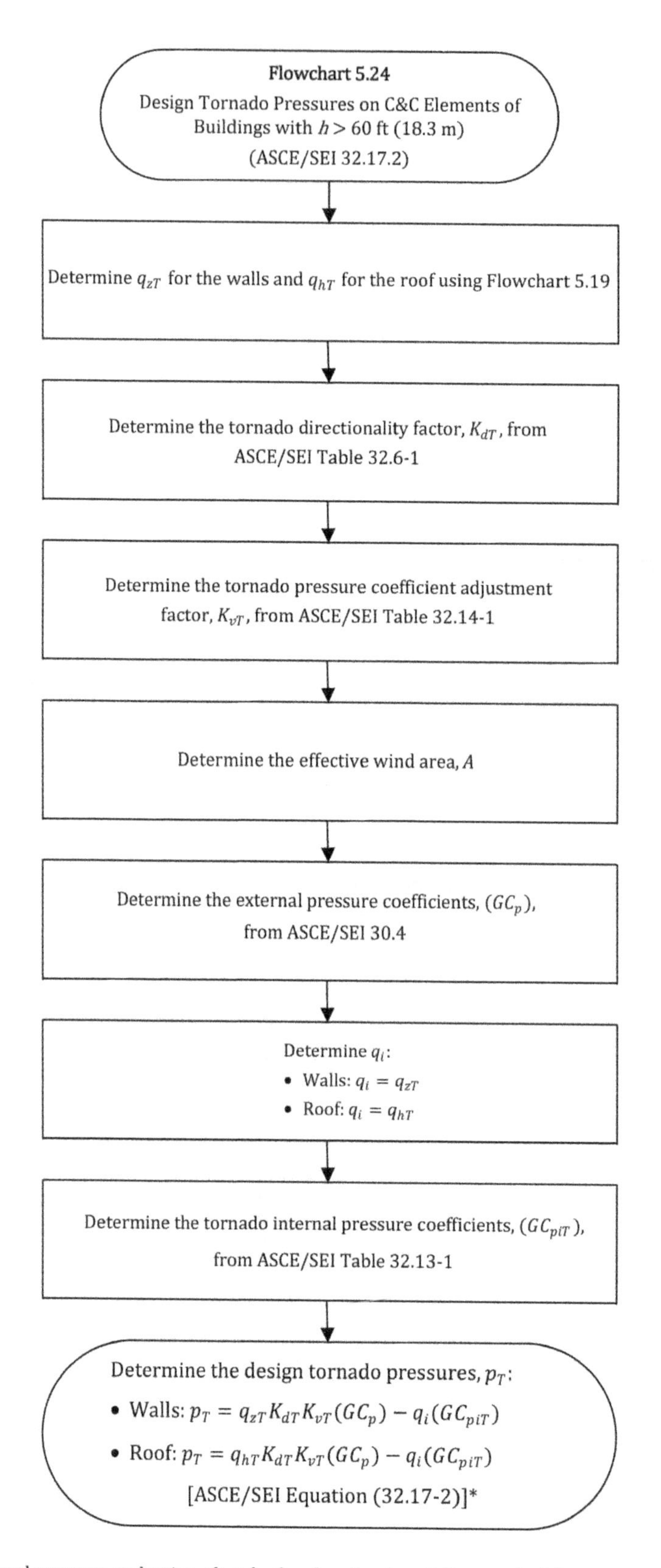
Flowchart 5.24
Design Tornado Pressures on C&C Elements of Buildings with $h > 60$ ft (18.3 m)
(ASCE/SEI 32.17.2)
Determine q_{zT} for the walls and q_{hT} for the roof using Flowchart 5.19
Determine the tornado directionality factor, K_{dT}, from ASCE/SEI Table 32.6-1
Determine the tornado pressure coefficient adjustment factor, K_{vT}, from ASCE/SEI Table 32.14-1
Determine the effective wind area, A
Determine the external pressure coefficients, (GC_p), from ASCE/SEI 30.4
Determine q_i:
• Walls: $q_i = q_{zT}$
• Roof: $q_i = q_{hT}$
Determine the tornado internal pressure coefficients, (GC_{piT}), from ASCE/SEI Table 32.13-1
Determine the design tornado pressures, p_T:
• Walls: $p_T = q_{zT}K_{dT}K_{vT}(GC_p) - q_i(GC_{piT})$
• Roof: $p_T = q_{hT}K_{dT}K_{vT}(GC_p) - q_i(GC_{piT})$
[ASCE/SEI Equation (32.17-2)]*

* $q_i = q_{hT}$ for internal pressure evaluation of roofs of enclosed and partially open buildings

$= q_{zT}$ for internal pressure evaluation of walls of enclosed and partially open buildings

$= q_{zop}$ for internal pressure evaluation of the roof and all walls in partially enclosed buildings where q_{zop} is defined as the level of the lowest opening in the building that could affect the positive internal pressure

5.8 Examples

5.8.1 Example 5.1—Determination of Enclosure Classification

Determine the enclosure classification for the building in Figure 5.25. The building has a monoslope roof in the east-west direction with a slope of 2.86 degrees. The north, south, and east walls are open. The west wall has two 25 ft by 15 ft (7.62 m by 4.57 m) openings.

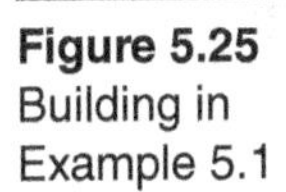

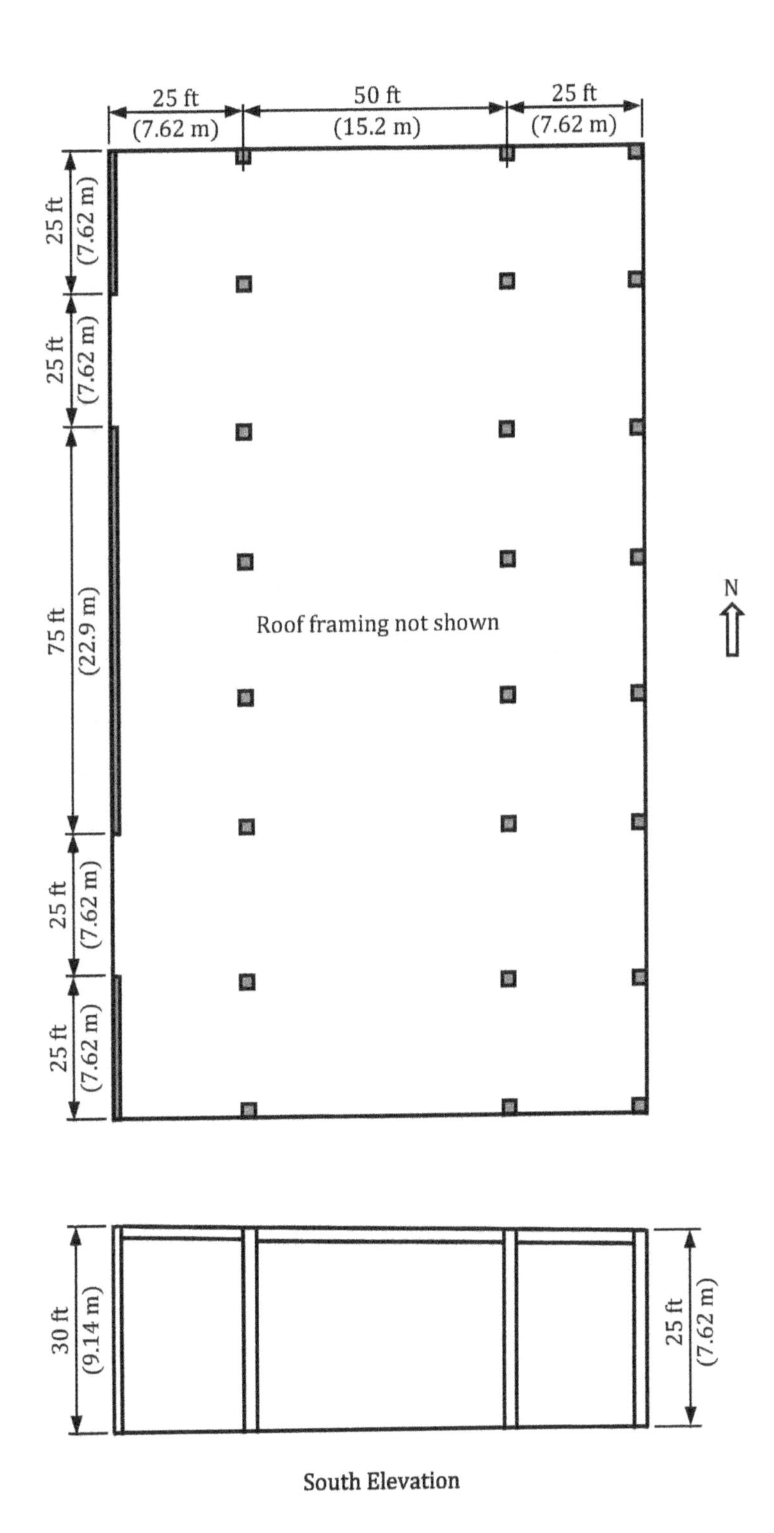

Figure 5.25 Building in Example 5.1

Two 1-ft² (0.093-m²) openings are in the roof. The opening areas in each wall and the roof are given in Table 5.28.

Table 5.28 Opening Areas in the Building in Example 5.1

Building Surface	Opening Area, A_o, ft² (m²)
North wall	$(100\times25)+(0.5\times100\times5)=2,750$ (255.5)
South wall	$(100\times25)+(0.5\times100\times5)=2,750$ (255.5)
East wall	$25\times175=4,375$ (406.5)
West wall	$2\times25\times15=750$ (69.7)
Roof	$2\times1=2$ (0.19)

Solution

Step 1—Determine if the building can be classified as open

A building is defined as open where each wall is at least 80 percent open. Wall opening percentages are given in Table 5.29. Because the west wall is less than 80 percent open, the building cannot be classified as open.

Table 5.29 Wall Opening Percentages for the Building in Example 5.1

Building Surface	Opening Area, A_o, ft² (m²)	Gross Area, A_g, ft² (m²)	$(A_o/A_g)\times100$ (%)
North wall	2,750 (255.5)	2,750 (255.5)	100
South wall	2,750 (255.5)	2,750 (255.5)	100
East wall	4,375 (406.4)	4,375 (406.4)	100
West wall	750 (69.7)	5,250 (487.7)	14.3

Step 2—Determine if the building can be classified as partially enclosed

Check if the following condition is satisfied assuming the east wall is the windward wall:

$$A_o > 1.10A_{oi}$$

For the east wall, $A_o = 4,375$ ft² (406.4 m²)

$$A_{oi} = \text{sum of openings in the building envelope not including } A_o$$
$$= 2,750+2,750+750+2 = 6,252 \text{ ft}^2$$

$$A_o = 4,375 \text{ ft}^2 < 1.10A_{oi} = 1.10\times6,252 = 6,877 \text{ ft}^2$$

In S.I.:

$$A_{oi} = 255.5+255.5+69.7+0.19 = 580.9 \text{ m}^2$$

$$A_o = 406.5 \text{ m}^2 < 1.10A_{oi} = 1.10\times580.9 = 639.0 \text{ m}^2$$

Therefore, the building cannot be classified as partially enclosed.

Step 3—Determine if the building can be classified as enclosed

The building cannot be classified as enclosed because the condition that all the walls have openings with areas less than the smaller of 4 ft^2 (0.37 m^2) and 1 percent the area of the wall is not satisfied.

Because this building does not comply with the conditions for open, partially enclosed, and enclosed buildings, the building is classified as partially open.

5.8.1 Example 5.2—Calculation of Topographic Factor, K_t

Determine the topographic factors, K_{zt}, for the building in Figure 5.26 given the following design data:

- Location: Rockford, IL (Latitude = 42.26°, Longitude = −89.05°)
- Exposure Category: C
- Risk Category: II (IBC Table 1604.5)
- Building is located on the downwind side of a two-dimensional escarpment (see Figure 5.26)

Figure 5.26 Elevation of the Building Located on the Two-Dimensional Escarpment in Example 5.2

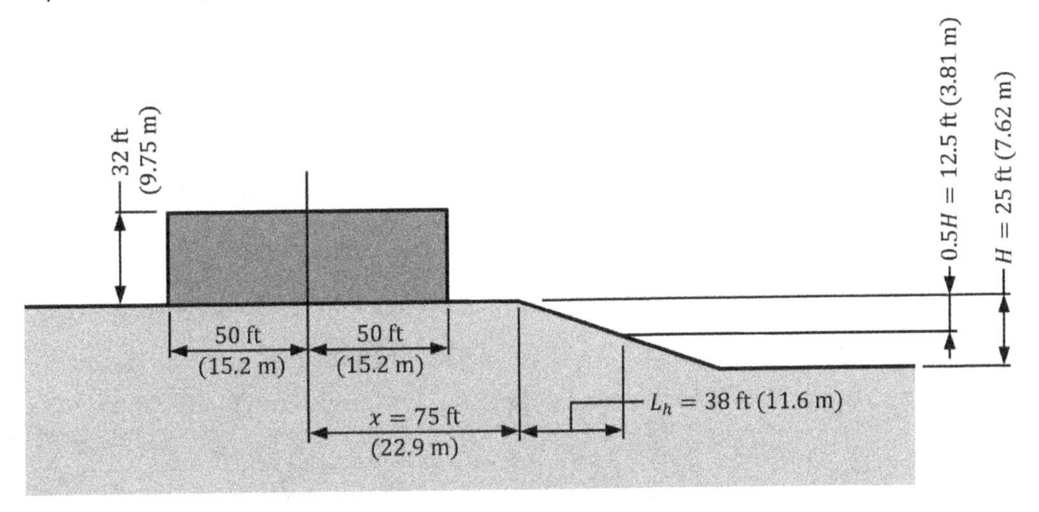

SOLUTION

Step 1—Determine if the conditions in ASCE/SEI 26.8.1 are satisfied

1. The building is located as shown in Figure 5.26 near the crest of an escarpment.
2. $H/L_h = 25/38 = 0.66 > 0.2$ (in S.I.: $7.62/11.6 = 0.66 > 0.2$).
3. $H = 25$ ft (7.62 m) > 15 ft (4.57 m) for exposure C.

Because all three conditions are satisfied, wind speed-up effects at the escarpment must be considered in design.

Step 2—Determine K_{zt} over the height of the building

The topographic factor is determined by ASCE/SEI Equation (26.8-1):

$$K_{zt} = (1 + K_1 K_2 K_3)^2$$

The multipliers K_1, K_2, and K_3 are determined from ASCE/SEI Figure 26.8-1.

It was determined in step 1 that $H/L_h = 0.66$. According to footnote b in ASCE/SEI Figure 26.8-1, where $H/L_h > 0.50$, use $H/L_h = 0.50$ when evaluating K_1 and substitute $2H$ for L_h when evaluating K_2 and K_3.

From ASCE/SEI Figure 26.8-1, $K_1 = 0.43$ for $H/L_h = 0.50$ and a two-dimensional escarpment.

$$x/L_h = x/2H = 75/(2 \times 25) = 1.5$$

In S.I.:

$$x/L_h = x/2H = 22.9/(2 \times 7.62) = 1.5$$

From ASCE/SEI Figure 26.8-1, $K_2 = 0.63$ for a two-dimensional escarpment with $x/L_h = 1.5$.

When determining K_3, z is taken midway between the height range (it is unconservative to use the top height of the range when determining K_3). For example, assume the range is from the ground level to 15 ft (4.57 m). The value of z to use in the determination of K_3 is equal to (15 + 0)/2 = 7.5 ft (2.29 m).

At this height, $z/L_h = z/2H = 7.5/(2 \times 25) = 0.15$.

In S.I.: $z/L_h = z/2H = 2.29/(2 \times 7.62) = 0.15$.

From ASCE/SEI Figure 26.8-1, $K_3 = e^{-\gamma(z/2H)} = e^{-2.5 \times 0.15} = 0.69$ for a two-dimensional escarpment where the height attenuation factor, γ, is equal to 2.5 for two-dimensional escarpments.

Thus, the topographic factor, K_{zt}, at $z = 15$ ft (4.57 m) is equal to the following:

$$K_{zt} = [1 + (0.43 \times 0.63 \times 0.69)]^2 = 1.41$$

The topographic factors over the given height ranges are given in Table 5.30.

Table 5.30 Topographic Factors, K_{zt}, for the Building in Example 5.2

Height above Ground Level z, ft (m)	Height z to Be Used in the Determination of K_3, ft (m)	z/2H	K_3	K_{zt}
32.0 (9.75)	28.5 (8.69)	0.57	0.24	1.13
25.0 (7.62)	22.5 (6.86)	0.45	0.33	1.19
20.0 (6.10)	17.5 (5.34)	0.35	0.42	1.24
15.0 (4.57)	7.5 (2.29)	0.15	0.69	1.41

5.8.3 Example 5.3—Calculation of Velocity Pressure, p_z

Determine the velocity pressures, q_z, for the building in Figure 5.26 given the design data in Example 5.2.

SOLUTION

Flowchart 5.1 is used to determine the velocity pressures.

Step 1—Determine the surface roughness category

The exposure is given as C in Example 5.2, so assume surface roughness C is present in all directions.

Step 2—Determine the exposure category

The exposure category is given as C in Example 5.2.

Step 3—Determine the terrain exposure constants

For exposure C, $\alpha = 9.8$ and $z_g = 2{,}460$ ft (750 m) from ASCE/SEI Table 26.11-1.

Step 4—Determine the velocity pressure exposure coefficient, K_z

Values of K_z are determined from ASCE/SEI 26.10-1 and are given in Table 5.31.

At height $z = h = 32.0$ ft:

$$K_z = 2.41(z/z_g)^{2/\alpha} = 2.41 \times (32.0/2{,}460)^{2/9.8} = 0.99$$

In S.I.:

$$K_z = 2.41 \times (9.75/750)^{2/9.8} = 0.99$$

Table 5.31 Velocity Pressure Exposure Coefficients, K_z and K_h for the Building in Examples 5.2 and 5.3

Height above Ground Level, z, ft (m)	K_z
32.0 (9.75)	0.99
25.0 (7.62)	0.94
20.0 (6.10)	0.90
15.0 (4.57)	0.85

Step 5—Determine the topographic factor, K_{zt}

The topographic factors over the height of the building are determined in Example 5.2 and are given in Table 5.30.

Step 6—Determine the ground elevation factor, K_e

It is permitted to take $K_e = 1.0$ for all elevations (ASCE/SEI 26.9).

Step 7—Determine the risk category

The risk category is given as II in Example 5.2.

Step 8—Determine the basic wind speed, V

For risk category II, use IBC Figure 1609.3(2) or ASCE/SEI Figure 26.5-1B (see Table 5.2 of this book).

Equivalently, use Reference 5.4 to obtain V = 108 mi/h (48 m/s) given the latitude and longitude of the site in Example 5.2.

Step 9—Determine the velocity pressures, q_z

The velocity pressures, q_z, are determined by ASCE/SEI Equations (26.10-1) and (26.10-1.SI). For example, the velocity pressure at the mean roof height of the building is equal to the following:

$$q_h = 0.00256K_hK_{zt}K_eV^2 = 0.00256 \times 0.99 \times 1.13 \times 1.0 \times 108^2 = 33.4 \text{ lb/ft}^2$$

In S.I.:

$$q_h = 0.613K_hK_{zt}K_eV^2 = 0.613 \times 0.99 \times 1.13 \times 1.0 \times 48^2/1{,}000 = 1.58 \text{ kN/m}^2$$

The velocity pressures over the height of the building are given in Table 5.32.

Note: Even though the building is located in a tornado-prone region (see ASCE/SEI Figure 32.1-1), tornado loads need not be considered because the building is assigned to risk category II (ASCE/SEI 32.1.1).

Table 5.32 Velocity Pressures, q_z and q_h, for the Building in Examples 5.2 and 5.3

Height above Ground Level, z, ft (m)	K_z	K_{zt}	q_z, lb/ft² (kN/m²)
32.0 (9.75)	0.99	1.13	33.4 (1.58)
25.0 (7.62)	0.94	1.19	33.4 (1.58)
20.0 (6.10)	0.90	1.24	33.3 (1.58)
15.0 (4.57)	0.85	1.41	35.8 (1.69)

5.8.3 Example 5.4—Calculation of Gust-Effect Factor, G

Determine the gust-effect factor, G, for the building in Figure 5.26 assuming the fundamental natural frequency of the building in both principal directions, n_1, is greater than 1 Hz. The plan dimensions of the building in both principal directions are equal to 100 ft (30.5 m).

SOLUTION

Flowchart 5.2 is used to determine the gust-effect factor.

Step 1—Determine the exposure category
The exposure category is given as C in Example 5.2.

Step 2—Determine the fundamental natural frequency of the building, n_1
It has been determined that $n_1 > 1$ Hz in both principal directions.

Step 3—Determine if the building is rigid or flexible
Because $n_1 > 1.0$ Hz, the building is classified as rigid in both principal directions (ASCE/SEI 26.2).

Step 4—Determine the gust-effect factor, G, for rigid buildings
For rigid buildings, G is permitted to be taken as 0.85 (ASCE/SEI 26.11.4).
For comparison purposes, G is determined by ASCE/SEI Equation (26.11-6).

Step 4a—Determine g_Q and g_v

$$g_Q = g_v = 3.4 \text{ (ASCE/SEI 26.11.4)}$$

Step 4b—Determine $\bar{z}$

$$\bar{z} = 0.6h = 0.6 \times 32.0 = 19.2 \text{ ft} > z_{\min} = 15.0 \text{ ft (ASCE/SEI Table 26.11-1)}$$

In S.I.:

$$\bar{z} = 0.6h = 0.6 \times 9.75 = 5.85 \text{ m} > z_{\min} = 4.57 \text{ m}$$

Step 4c—Determine $I_{\bar{z}}$ by ASCE/SEI Equations (26.11-7) and (26.11-7.SI)

$$I_{\bar{z}} = c\left(\frac{33}{\bar{z}}\right)^{1/6} = 0.20 \times \left(\frac{33}{19.2}\right)^{1/6} = 0.2189$$

In S.I.:

$$I_{\bar{z}} = c\left(\frac{10}{\bar{z}}\right)^{1/6} = 0.20 \times \left(\frac{10}{5.85}\right)^{1/6} = 0.2187$$

Step 4d—Determine $L_{\bar{z}}$ by ASCE/SEI Equations (26.11-9) and (26.11-9.SI)

$$L_{\bar{z}} = \ell\left(\frac{\bar{z}}{33}\right)^{\bar{\varepsilon}} = 500 \times \left(\frac{19.2}{33}\right)^{1/5.0} = 448.7 \text{ ft}$$

In S.I.:

$$L_{\bar{z}} = \ell\left(\frac{\bar{z}}{10}\right)^{\bar{\varepsilon}} = 152.40 \times \left(\frac{5.85}{10}\right)^{1/5.0} = 136.90 \text{ m}$$

Step 4e—Determine Q by ASCE/SEI Equation (26.11-8)

$$Q = \sqrt{\frac{1}{1 + 0.63\left(\frac{B+h}{L_{\bar{z}}}\right)^{0.63}}} = \sqrt{\frac{1}{1 + \left[0.63 \times \left(\frac{100.0+32.0}{448.7}\right)^{0.63}\right]}} = 0.88$$

In S.I.:

$$Q = \sqrt{\frac{1}{1 + 0.63\left(\frac{B+h}{L_{\bar{z}}}\right)^{0.63}}} = \sqrt{\frac{1}{1 + \left[0.63 \times \left(\frac{30.5+9.75}{136.90}\right)^{0.63}\right]}} = 0.88$$

Step 4f—Determine G by ASCE/SEI Equation (26.11-6)

$$G = 0.925\left(\frac{1 + 1.7 g_Q I_{\bar{z}} Q}{1 + 1.7 g_v I_{\bar{z}}}\right) = 0.925 \times \left[\frac{1 + (1.7 \times 3.4 \times 0.2189 \times 0.88)}{1 + (1.7 \times 3.4 \times 0.2189)}\right] = 0.86$$

In S.I.:

$$G = 0.925\left(\frac{1 + 1.7 g_Q I_{\bar{z}} Q}{1 + 1.7 g_v I_{\bar{z}}}\right) = 0.925 \times \left[\frac{1 + (1.7 \times 3.4 \times 0.2187 \times 0.88)}{1 + (1.7 \times 3.4 \times 0.2187)}\right] = 0.86$$

5.8.5 Example 5.5—Calculation of Design Wind Pressures, *p*, Using ASCE/SEI Chapter 27

Determine the design wind pressures, p, on the walls and roof of the building in Figure 5.26 using the provisions in ASCE/SEI Chapter 27 assuming the building is enclosed.

SOLUTION

Check if the building meets all the conditions and limitations of ASCE/SEI 27.1.2 and 27.1.3 so that the provisions in ASCE/SEI Chapter 27 can be used to determine the design wind pressures on the MWFRS.

The building is regularly-shaped, that is, it does not have any unusual geometric irregularities in spatial form. Also, the building does not have response characteristics that make it subject to across-wind loading or other similar effects, and it is not sited at a location where channeling effects or buffeting in the wake of upwind obstructions need to be considered.

Therefore, the provisions in ASCE/SEI Chapter 27 are permitted to be used to determine the design wind pressures on the MWFRS.

Flowchart 5.3 is used to determine the design wind pressures.

Step 1—Determine the enclosure classification of the building
The enclosure classification of the building is given as enclosed.

Step 2—Determine if the building has a parapet
The building does not have a parapet.

Step 3—Determine if the building is rigid or flexible
Because $n_1 > 1.0$ Hz (see Example 5.4), the building is classified as rigid in both principal directions (ASCE/SEI 26.2).

Step 4—Determine the gust-effect factor, G
For rigid buildings, G is permitted to be taken as 0.85 (see ASCE/SEI 26.11.4 and Example 5.4).

Step 5—Determine the wind directionality factor, K_d
For the MWFRS of a building, the wind directionality factor, K_d, is equal to 0.85 (ASCE/SEI Table 26.6-1).

Step 6—Determine the wind velocity pressures, q_z
The velocity pressures over the height of the building are determined in Example 5.3 and are given in Table 5.32.

Step 7—Determine the external pressure coefficients, C_p, for the walls and roof
The external pressure coefficients are determined from ASCE/SEI Figure 27.3-1, which are applicable in both principal directions because the building has a square plan:

- Windward wall: $C_p = 0.8$
- Leeward wall: For $L/B = 1$, $C_p = -0.5$
- Side walls: $C_p = -0.7$
- Roof: Normal to the ridge with roof angle $\theta = 0$ degrees and parallel to the ridge for all θ with $h/L = 32/100 = 0.32 < 0.5$ (In S.I.: $h/L = 9.75/30.5 = 0.32 < 0.5$)
  From the windward edge of the roof to $h = 32$ ft (9.75 m), $C_p = -0.9, -0.18$
  From $h = 32$ ft (9.75 m) to $2h = 64$ ft (19.5 m), $C_p = -0.5, -0.18$
  From $2h = 64$ ft (19.5 m) to 100 ft (30.5 m), $C_p = -0.3, -0.18$

Step 8—Determine the velocity pressure for internal pressure determination, q_i
According to ASCE/SEI 27.3.1, $q_i = q_h = 33.4$ lb/ft^2 (1.58 kN/m^2) for all surfaces of an enclosed building.

Step 9—Determine the internal pressure coefficients, (GC_{pi})
From ASCE/SEI Table 26.13-1, $(GC_{pi}) = +0.18, -0.18$ for an enclosed building.

Step 10—Determine the design wind pressures, p, on all the surfaces of the building
Design wind pressures, p, are determined by ASCE/SEI Equation (27.3-1) for an enclosed building:

$$p = qK_dGC_p - q_iK_d(GC_{pi})$$

- Windward wall:

$$p_z = (q_z \times 0.85 \times 0.85 \times 0.8) - [33.4 \times 0.85 \times (\pm 0.18)] = 0.58q_z \mp 5.1 \text{ lb/ft}^2$$

In S.I.:

$$p_z = (q_z \times 0.85 \times 0.85 \times 0.8) - [1.58 \times 0.85 \times (\pm 0.18)] = 0.58q_z \mp 0.24 \text{ kN/m}^2$$

- Leeward wall

$$p_h = [q_h \times 0.85 \times 0.85 \times (-0.5)] - [33.4 \times 0.85 \times (\pm 0.18)]$$
$$= -12.1 \mp 5.1 = -17.2 \text{ lb/ft}^2, -7.0 \text{ lb/ft}^2$$

In S.I.:

$$p_h = [q_h \times 0.85 \times 0.85 \times (-0.5)] - [1.58 \times 0.85 \times (\pm 0.18)]$$
$$= -0.57 \mp 0.24 = -0.81 \text{ kN/m}^2, -0.33 \text{ kN/m}^2$$

- Side walls

$$p_h = [q_h \times 0.85 \times 0.85 \times (-0.7)] - [33.4 \times 0.85 \times (\pm 0.18)]$$
$$= -16.9 \mp 5.1 = -22.0 \text{ lb/ft}^2, -11.8 \text{ lb/ft}^2$$

In S.I.:

$$p_h = [q_h \times 0.85 \times 0.85 \times (-0.7)] - [1.58 \times 0.85 \times (\pm 0.18)]$$
$$= -0.80 \mp 0.24 = -1.04 \text{ kN/m}^2, -0.56 \text{ kN/m}^2$$

- Roof

$$p_h = (q_h \times 0.85 \times 0.85 \times C_p) - [33.4 \times 0.85 \times (\pm 0.18)] = 24.1 C_p \mp 5.1 \text{ lb/ft}^2$$

In S.I.:

$$p_h = (q_h \times 0.85 \times 0.85 \times C_p) - [1.58 \times 0.85 \times (\pm 0.18)] = 1.14 C_p \mp 0.24 \text{ kN/m}^2$$

The roof pressure coefficient $C_p = -0.18$ may become critical where wind loads are combined with roof live loads or snow loads. Determination of wind pressures based on this pressure coefficient should be performed, but such calculations are not shown in this example.

Design wind pressures in both of the principal directions are given in Table 5.33.

Table 5.33 Design Wind Pressures, *p*, for Wind on the Building in Examples 5.2 through 5.5

Building Surface	Height above Ground Level, z, ft (m)	q, lb/ft² (kN/m²)	External Pressure qK_dGC_p, lb/ft² (kN/m²)	Internal Pressure $q_hK_d(GC_{pi})$, lb/ft² (kN/m²)	Net Pressure, p, lb/ft² (kN/m²)	
					$+(GC_{pi})$	$-(GC_{pi})$
Windward wall	32.0 (9.75)	33.4 (1.58)	19.4 (0.92)	±5.1 (±0.24)	14.3 (0.68)	24.5 (1.16)
	25.0 (7.62)	33.4 (1.58)	19.4 (0.92)	±5.1 (±0.24)	14.3 (0.68)	24.5 (1.16)
	20.0 (6.10)	33.3 (1.58)	19.3 (0.92)	±5.1 (±0.24)	14.2 (0.68)	24.4 (1.16)
	15.0 (4.57)	35.8 (1.69)	20.8 (0.98)	±5.1 (±0.24)	15.7 (0.74)	25.9 (1.22)
Leeward wall	All	33.4 (1.58)	−12.1 (−0.57)	±5.1 (±0.24)	−17.2 (−0.81)	−7.0 (−0.33)
Side walls	All	33.4 (1.58)	−16.9 (−0.80)	±5.1 (±0.24)	−22.0 (−1.04)	−11.8 (−0.56)
Roof	—*	33.4 (1.58)	−21.7 (−1.03)	±5.1 (±0.24)	−26.8 (−1.27)	−16.6 (−0.79)
	—**	33.4 (1.58)	−12.1 (−0.57)	±5.1 (±0.24)	−17.2 (−0.81)	−7.0 (−0.33)
	—†	33.4 (1.58)	−7.2 (−0.34)	±5.1 (±0.24)	−12.3 (−0.58)	−2.1 (−0.10)

*From windward edge of roof to 32 ft (9.75 m).
**From 32 ft (9.75 m) to 64 ft (19.5 m).
†From 64 ft (19.5 m) to 100 ft (30.5 m).

Note: Even though the building is located in a tornado-prone region (see ASCE/SEI Figure 32.1-1), tornado loads need not be considered because the building is assigned to risk category II (ASCE/SEI 32.1.1).

5.8.6 Example 5.6—Determination of Design Wind Pressures on a Warehouse Building Using ASCE/SEI Chapter 27—MWFRS

Determine the design wind pressures, p, on the MWFRS of the one-story warehouse building in Figure 5.27 using the provisions in ASCE/SEI Chapter 27 given the following design data:

- Location: St. Louis, MO (Latitude = 38.61°, Longitude = −90.19°)
- Surface roughness: C
- Topography: Not situated on a hill, ridge, or escarpment
- Risk category: II (IBC Table 1604.5)
- Enclosure classification: Enclosed

Figure 5.27 Plan and Elevation of the One-story Warehouse Building in Example 5.6

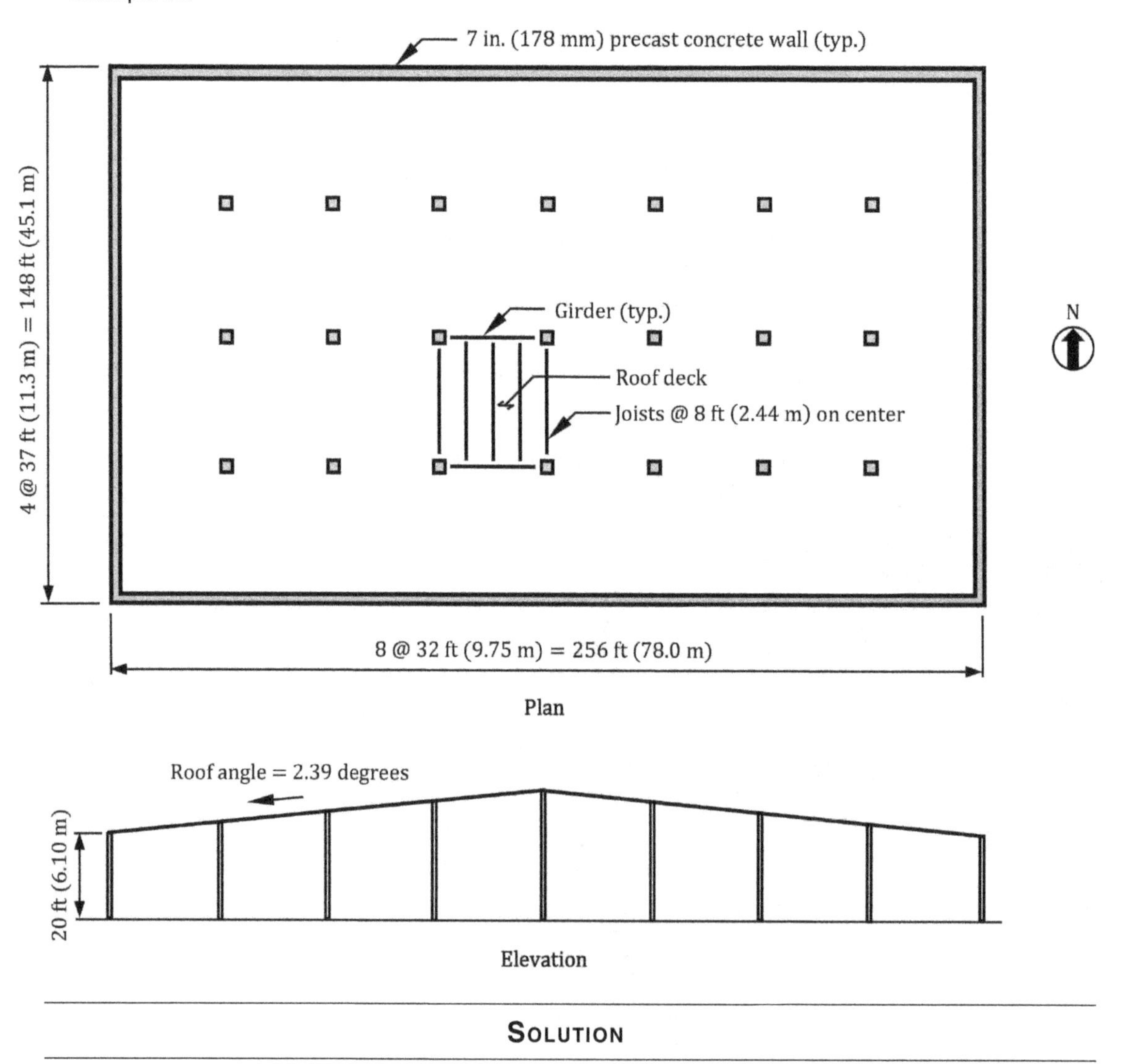

Solution

Check if the building meets all the conditions and limitations of ASCE/SEI 27.1.2 and 27.1.3 so that the provisions in ASCE/SEI Chapter 27 can be used to determine the design wind pressures on the MWFRS.

The building is regularly-shaped, that is, it does not have any unusual geometric irregularities in spatial form. Also, the building does not have response characteristics that make it subject to across-wind loading or other similar effects, and it is not sited at a location where channeling effects or buffeting in the wake of upwind obstructions need to be considered.

Therefore, the provisions in ASCE/SEI Chapter 27 are permitted to be used to determine the design wind pressures on the MWFRS.

Flowchart 5.3 is used to determine the design wind pressures.

Step 1—Determine the enclosure classification of the building

The enclosure classification of the building is given as enclosed.

Step 2—Determine if the building has a parapet

The building does not have a parapet.

Step 3—Determine whether the building is rigid or flexible

In lieu of determining the fundamental natural frequency, n_1, from a dynamic analysis of the building, check if the approximate lower bound natural frequency, n_a, given in ASCE/SEI 26.11.3 is permitted to be used:

Building height = 20 ft (6.10 m) < 300 ft (91.4 m)

Building height $= 20 \text{ ft } (6.10 \text{ m}) < 4L_{eff} = 4 \times 148 = 592 \text{ ft } (180.4 \text{ m})$ in the N-S direction

$= 20 \text{ ft } (6.10 \text{ m}) < 4L_{eff} = 4 \times 256 = 1{,}024 \text{ ft } (312.1 \text{ m})$ in the E-W direction

Because both of these limitations are satisfied, ASCE/SEI Equations (26.11-5) and (26.11-5.SI) are permitted to be used to determine n_a in both directions for the MWFRS consisting of precast concrete walls.

In the N-S direction:

A_B = base area of the building $= 148 \times 256 = 37{,}888 \text{ ft}^2 \ (3{,}520 \text{ m}^2)$

$h = h_i$ = height of walls = 20.0 ft (6.10 m)

A_i = horizontal cross-sectional area of wall $= (7/12) \times 148 = 86.3 \text{ ft}^2 \ (8.0 \text{ m}^2)$

D_i = length of wall = 148 ft (45.1 m)

$$C_w = \frac{100}{A_B} \sum_{i=1}^{n} \left(\frac{h}{h_i}\right)^2 \frac{A_i}{\left[1 + 0.83\left(\frac{h_i}{D_i}\right)^2\right]} = \frac{2 \times 100}{37{,}888} \times \left(\frac{20.0}{20.0}\right)^2 \times \frac{86.3}{\left[1 + 0.83\left(\frac{20.0}{148}\right)^2\right]} = 0.45$$

$$n_a = 385(C_w)^{0.5}/h = 385 \times (0.45)^{0.5} / 20.0 = 12.9 \text{ Hz} > 1 \text{ Hz}$$

In S.I.:

$$C_w = \frac{100}{A_B} \sum_{i=1}^{n} \left(\frac{h}{h_i}\right)^2 \frac{A_i}{\left[1 + 0.83\left(\frac{h_i}{D_i}\right)^2\right]} = \frac{2 \times 100}{3{,}520} \times \left(\frac{6.10}{6.10}\right)^2 \times \frac{8.0}{\left[1 + 0.83\left(\frac{6.10}{45.1}\right)^2\right]} = 0.45$$

$$n_a = 117.3(C_w)^{0.5}/h = 117.3 \times (0.45)^{0.5}/6.10 = 12.9 \text{ Hz} > 1 \text{ Hz}$$

Therefore, the building is rigid in the N-S direction.

Similar calculations in the E-W direction result in $n_a > 1$ Hz, so the building is rigid in that direction as well.

Note: This warehouse building is a low-rise building in accordance with the definition in ASCE/SEI 26.4 (the mean roof height is less than 60 ft (18.3 m) and the mean roof height is less than the least horizontal dimension of the building). Low-rise buildings are permitted to be considered rigid (ASCE/SEI 26.11.2).

Step 4—Determine the gust-effect factor, G

For rigid buildings, G is permitted to be taken as 0.85 (ASCE/SEI 26.11.4).

Step 5—Determine the wind directionality factor, K_d

For the MWFRS of a building, the wind directionality factor, K_d, is equal to 0.85 (ASCE/SEI Table 26.6-1).

Step 6—Determine the wind velocity pressures, q_z

Flowchart 5.1 is used to determine the wind velocity pressures.

Step 6a—Determine the surface roughness category

Surface roughness C is given in the design data and it is assumed it is present in all directions.

Step 6b—Determine the exposure category

It is assumed exposure category C is applicable in all directions.

Step 6c—Determine the terrain exposure constants

For exposure C, $\alpha = 9.8$ and $z_g = 2{,}460$ ft (750 m) from ASCE/SEI Table 26.11-1.

Step 6d—Determine the velocity pressure exposure coefficient, K_z

Values of K_z are determined from ASCE/SEI 26.10-1 and are given in Table 5.34. The mean roof height, h, is permitted to be taken at the roof eave height because the roof angle is less than 10 degrees (see the definition of mean roof height in ASCE/SEI 26.2).

Table 5.34 Velocity Pressure Exposure Coefficients, K_z and K_h, for the Warehouse Building in Example 5.6

Height above Ground Level, z, ft (m)	K_z
20.0 (6.10)	0.90
15.0 (4.57)	0.85

Step 6e—Determine the topographic factor, K_{zt}

The building is not located on a hill, ridge, or escarpment, so $K_{zt} = 1.0$ (ASCE/SEI 26.8.2).

Step 6f—Determine the ground elevation factor, K_e

It is permitted to take $K_e = 1.0$ for all elevations (ASCE/SEI 26.9).

Step 6g—Determine the risk category

The risk category is given as II in the design data.

Step 6h—Determine the basic wind speed, V

For risk category II buildings, use IBC Figure 1609.3(2) or ASCE/SEI Figure 26.5-1B (see Table 5.2 of this book).

Equivalently, use Reference 5.4 to obtain $V = 108$ mi/h (48 m/s) given the latitude and longitude of the site.

Step 6i—Determine the velocity pressures, q_z

The velocity pressures, q_z, over the height of the building are given in Table 5.35. These pressures are determined by ASCE/SEI Equations (26.10-1) and

Table 5.35 Velocity Pressures, q_z and q_h, for the Warehouse Building in Example 5.6

Height above Ground Level, z, ft (m)	K_z	q_z, lb/ft² (kN/m²)
20.0 (6.10)	0.90	26.9 (1.27)
15.0 (4.57)	0.85	25.4 (1.20)

(26.10-1.SI). For example, the velocity pressure at the mean roof height of the building is equal to the following:

$$q_h = 0.00256K_h K_{zt} K_e V^2 = 0.00256 \times 0.90 \times 1.0 \times 1.0 \times 108^2 = 26.9 \text{ lb/ft}^2$$

In S.I.:

$$q_h = 0.613K_h K_{zt} K_e V^2 = 0.613 \times 0.90 \times 1.0 \times 1.0 \times 48^2 / 1{,}000 = 1.27 \text{ kN/m}^2$$

Step 7—Determine the external pressure coefficients, C_p, for the walls and roof

The external pressure coefficients are determined from ASCE/SEI Figure 27.3-1.

For wind in the E-W direction:

- Windward wall: $C_p = 0.8$
- Leeward wall: For $L/B = 256/148 = 1.7$, $C_p = -0.36$ (by linear interpolation)
- Side walls: $C_p = -0.7$
- Roof: Normal to the ridge with roof angle $\theta < 10$ degrees and $h/L = 20/256 = 0.08 < 0.5$ (In S.I.: $h/L = 6.10/78.0 = 0.08 < 0.5$)

  From the windward edge of the roof to $h = 20$ ft (6.10 m), $C_p = -0.9, -0.18$
  From $h = 20$ ft (6.10 m) to $2h = 40$ ft (12.2 m), $C_p = -0.5, -0.18$
  From $2h = 40$ ft (12.2 m) to 256 ft (78.0 m), $C_p = -0.3, -0.18$

For wind in the N-S direction:

- Windward wall: $C_p = 0.8$
- Leeward wall: For $L/B = 148/256 = 0.6$, $C_p = -0.5$
- Side walls: $C_p = -0.7$
- Roof: Parallel to the ridge with $h/L = 20/148 = 0.14 < 0.5$ (In S.I.: $h/L = 6.10/45.1 = 0.14 < 0.5$)

  From the windward edge of the roof to $h = 20$ ft (6.10 m), $C_p = -0.9, -0.18$
  From $h = 20$ ft (6.10 m) to $2h = 40$ ft (12.2 m), $C_p = -0.5, -0.18$
  From $2h = 40$ ft (12.2 m) to 148 ft (45.1 m), $C_p = -0.3, -0.18$

Step 8—Determine the velocity pressure for internal pressure determination, q_i

According to ASCE/SEI 27.3.1, $q_i = q_h = 26.9$ lb/ft^2 (1.27 kN/m^2) for all surfaces of an enclosed building.

Step 9—Determine the internal pressure coefficients, (GC_{pi})

From ASCE/SEI Table 26.13-1, $(GC_{pi}) = +0.18, -0.18$ for an enclosed building.

Step 10—Determine the design wind pressures, p, on all the surfaces of the building

Design wind pressures, p, are determined by ASCE/SEI Equation (27.3-1) for an enclosed building:

$$p = qK_d GC_p - q_i K_d (GC_{pi})$$

For wind in the E-W direction:

- Windward wall:

$$p_z = (q_z \times 0.85 \times 0.85 \times 0.8) - [26.9 \times 0.85 \times (\pm 0.18)] = 0.58q_z \mp 4.1 \text{ lb/ft}^2$$

In S.I.:

$$p_z = (q_z \times 0.85 \times 0.85 \times 0.8) - [1.27 \times 0.85 \times (\pm 0.18)] = 0.58q_z \mp 0.19 \text{ kN/m}^2$$

- Leeward wall

$$p_h = [q_h \times 0.85 \times 0.85 \times (-0.36)] - [26.9 \times 0.85 \times (\pm 0.18)]$$

$$= -7.0 \mp 4.1 = -11.1 \text{ lb/ft}^2, -2.9 \text{ lb/ft}^2$$

In S.I.:

$$p_h = [q_h \times 0.85 \times 0.85 \times (-0.36)] - [1.27 \times 0.85 \times (\pm 0.18)]$$
$$= -0.33 \mp 0.19 = -0.52 \text{ kN/m}^2, -0.14 \text{ kN/m}^2$$

- Side walls

$$p_h = [q_h \times 0.85 \times 0.85 \times (-0.7)] - [26.9 \times 0.85 \times (\pm 0.18)]$$
$$= -13.6 \mp 4.1 = -17.7 \text{ lb/ft}^2, -9.5 \text{ lb/ft}^2$$

In S.I.:

$$p_h = [q_h \times 0.85 \times 0.85 \times (-0.7)] - [1.27 \times 0.85 \times (\pm 0.18)]$$
$$= -0.64 \mp 0.19 = -0.83 \text{ kN/m}^2, -0.45 \text{ kN/m}^2$$

- Roof

$$p_h = (q_h \times 0.85 \times 0.85 \times C_p) - [26.9 \times 0.85 \times (\pm 0.18)] = 19.4C_p \mp 4.1 \text{ lb/ft}^2$$

In S.I.:

$$p_h = \left(q_h \times 0.85 \times 0.85 \times C_p\right) - [1.27 \times 0.85 \times (\pm 0.18)] = 0.92C_p \mp 0.19 \text{ kN/m}^2$$

The roof pressure coefficient $C_p = -0.18$ may become critical where wind loads are combined with roof live loads or snow loads. Determination of wind pressures based on this pressure coefficient should be performed, but such calculations are not shown in this example.

Design wind pressures in the E-W direction are given in Table 5.36 and are illustrated in Figure 5.28 for wind in the east to west direction.

Table 5.36 Design Wind Pressures, *p*, in the E-W Direction on the MWFRS of the Warehouse Building in Example 5.6

Building Surface	Height above Ground Level, z, ft (m)	q, lb/ft² (kN/m²)	External Pressure qK_dGC_p, lb/ft² (kN/m²)	Internal Pressure $q_hK_d(GC_{pi})$, lb/ft² (kN/m²)	Net Pressure, p, lb/ft² (kN/m²)	
					$+(GC_{pi})$	$-(GC_{pi})$
Windward wall	20.0 (6.10)	26.9 (1.27)	15.6 (0.74)	±4.1 (±0.19)	11.5 (0.55)	19.7 (0.93)
	15.0 (4.57)	25.4 (1.20)	14.7 (0.70)	±4.1 (±0.19)	10.6 (0.51)	18.8 (0.89)
Leeward wall	All	26.9 (1.27)	−7.0 (−0.33)	±4.1 (±0.19)	−11.1 (−0.52)	−2.9 (−0.14)
Side walls	All	26.9 (1.27)	−13.6 (−0.64)	±4.1 (±0.19)	−17.7 (−0.83)	−9.5 (−0.45)
Roof	—*	26.9 (1.27)	−17.5 (−0.83)	±4.1 (±0.19)	−21.6 (−1.02)	−13.4 (−0.64)
	—**	26.9 (1.27)	−9.7 (−0.46)	±4.1 (±0.19)	−13.8 (−0.65)	−5.6 (−0.27)
	—†	26.9 (1.27)	−5.8 (−0.28)	±4.1 (±0.19)	−9.9 (−0.47)	−1.7 (−0.09)

*From windward edge of roof to 20 ft (6.10 m).
**From 20 ft (6.10 m) to 40 ft (12.2 m).
†From 40 ft (12.2 m) to 256 ft (78.0 m).

Figure 5.28 Design Wind Pressures for Wind in the East to West Direction on the MWFRS of the Warehouse Building in Example 5.6

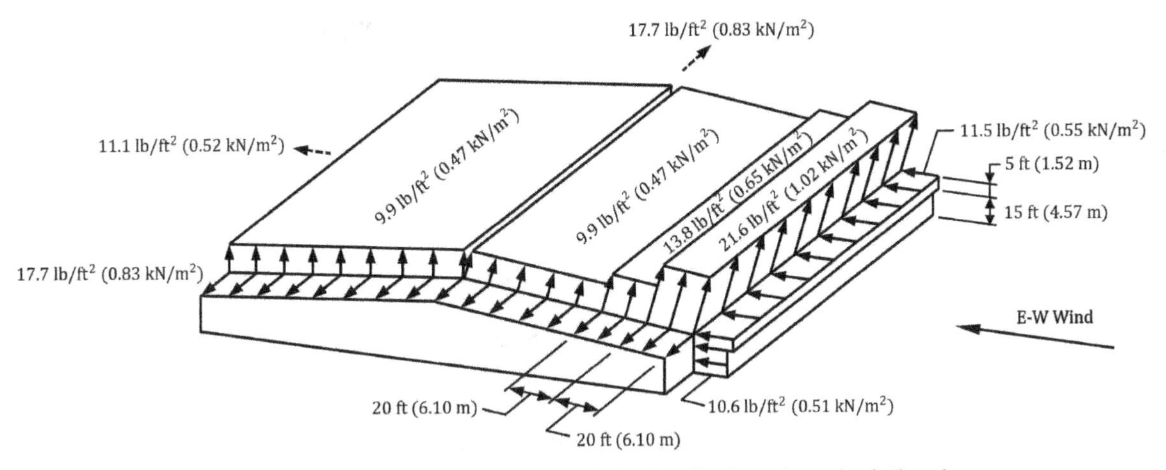

Note: dashed arrows represent uniformly distributed loads over leeward and side surfaces

Positive internal pressure

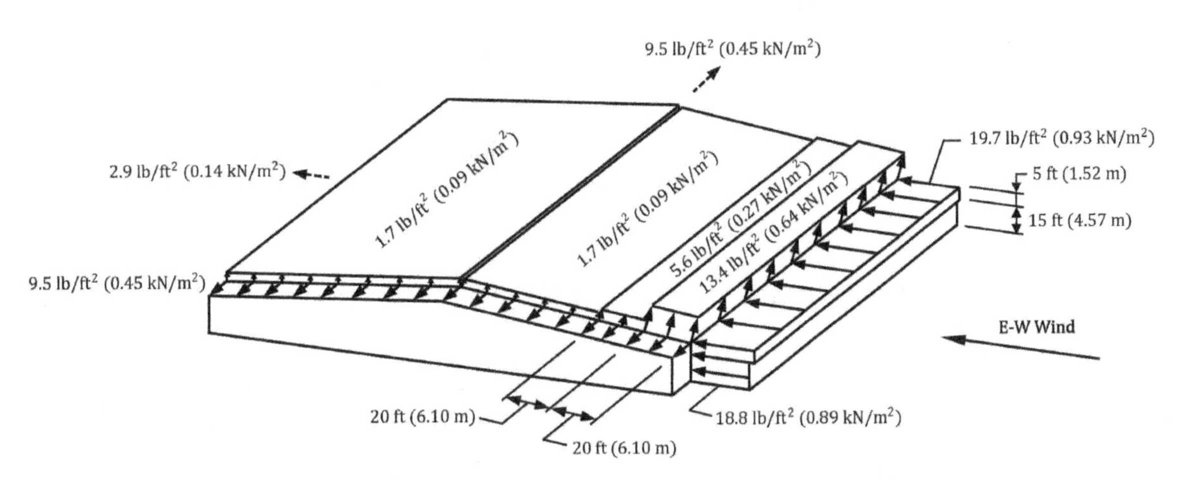

Note: dashed arrows represent uniformly distributed loads over leeward and side surfaces

Negative internal pressure

For wind in the N-S direction:

- Windward wall:

$$p_z = (q_z \times 0.85 \times 0.85 \times 0.8) - [26.9 \times 0.85 \times (\pm 0.18)] = 0.58 q_z \mp 4.1 \text{ lb/ft}^2$$

  In S.I.:

$$p_z = (q_z \times 0.85 \times 0.85 \times 0.8) - [1.27 \times 0.85 \times (\pm 0.18)] = 0.58 q_z \mp 0.19 \text{ kN/m}^2$$

- Leeward wall

$$p_h = [q_h \times 0.85 \times 0.85 \times (-0.5)] - [26.9 \times 0.85 \times (\pm 0.18)]$$
$$= -9.7 \mp 4.1 = -13.8 \text{ lb/ft}^2, -5.6 \text{ lb/ft}^2$$

In S.I.:

$$p_h = [q_h \times 0.85 \times 0.85 \times (-0.5)] - [1.27 \times 0.85 \times (\pm 0.18)]$$
$$= -0.46 \mp 0.19 = -0.65 \text{ kN/m}^2, -0.27 \text{kN/m}^2$$

- Side walls

$$p_h = [q_h \times 0.85 \times 0.85 \times (-0.7)] - [26.9 \times 0.85 \times (\pm 0.18)]$$
$$= -13.6 \mp 4.1 = -17.7 \text{ lb/ft}^2, -9.5 \text{ lb/ft}^2$$

In S.I.:

$$p_h = [q_h \times 0.85 \times 0.85 \times (-0.7)] - [1.27 \times 0.85 \times (\pm 0.18)]$$
$$= -0.64 \mp 0.19 = -0.83 \text{ kN/m}^2, -0.45 \text{ kN/m}^2$$

- Roof

$$p_h = (q_h \times 0.85 \times 0.85 \times C_p) - [26.9 \times 0.85 \times (\pm 0.18)] = 19.4C_p \mp 4.1 \text{ lb/ft}^2$$

In S.I.:

$$p_h = (q_h \times 0.85 \times 0.85 \times C_p) - [1.27 \times 0.85 \times (\pm 0.18)] = 0.92\ C_p \mp 0.19 \text{ kN/m}^2$$

The roof pressure coefficient $C_p = -0.18$ may become critical where wind loads are combined with roof live loads or snow loads. Determination of wind pressures based on this pressure coefficient should be performed, but such calculations are not shown in this example.

Design wind pressures in the N-S direction are given in Table 5.37 and are illustrated in Figure 5.29 for wind in the south to north direction.

Table 5.37 Design Wind Pressures, *p*, in the N-S Direction on the MWFRS of the Warehouse Building in Example 5.6

Building Surface	Height above Ground Level, z, ft (m)	q, lb/ft² (kN/m²)	External Pressure qK_dGC_p, lb/ft² (kN/m²)	Internal Pressure $q_hK_d(GC_{pi})$, lb/ft² (kN/m²)	Net Pressure, p, lb/ft² (kN/m²)	
					$+(GC_{pi})$	$-(GC_{pi})$
Windward wall	20.0 (6.10)	26.9 (1.27)	15.6 (0.74)	±4.1 (±0.19)	11.5 (0.55)	19.7 (0.93)
	15.0 (4.57)	25.4 (1.20)	14.7 (0.70)	±4.1 (±0.19)	10.6 (0.51)	18.8 (0.89)
Leeward wall	All	26.9 (1.27)	−9.7 (−0.46)	±4.1 (±0.19)	−13.8 (−0.65)	−5.6 (−0.27)
Side walls	All	26.9 (1.27)	−13.6 (−0.64)	±4.1 (±0.19)	−17.7 (−0.83)	−9.5 (−0.45)
Roof	—*	26.9 (1.27)	−17.5 (−0.83)	±4.1 (±0.19)	−21.6 (−1.02)	−13.4 (−0.64)
	—**	26.9 (1.27)	−9.7 (−0.46)	±4.1 (±0.19)	−13.8 (−0.65)	−5.6 (−0.27)
	—†	26.9 (1.27)	−5.8 (−0.28)	±4.1 (±0.19)	−9.9 (−0.47)	−1.7 (−0.09)

*From windward edge of roof to 20 ft (6.10 m).
**From 20 ft (6.10 m) to 40 ft (12.2 m).
†From 40 ft (12.2 m) to 148 ft (45.1 m).

The external horizontal wind pressures on the walls in both directions are shown in Figure 5.30. It is evident that the internal pressures cancel out when determining the horizontal wind pressures on the walls.

The building must be designed for the design wind load cases in ASCE/SEI Figure 27.3-8 (see Figure 5.16).

Figure 5.29 Design Wind Pressures for Wind in the South to North Direction on the MWFRS of the Warehouse Building in Example 5.6

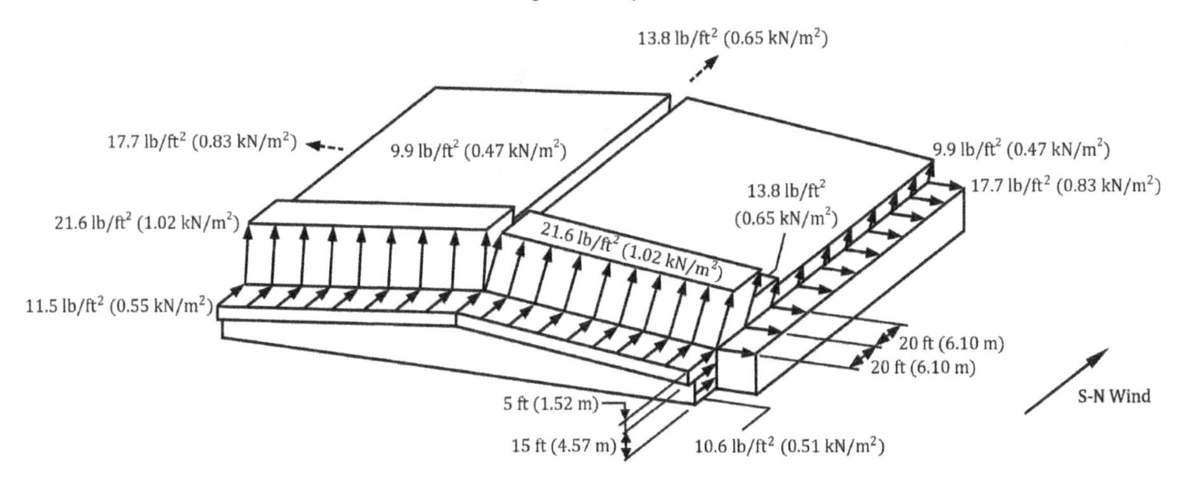

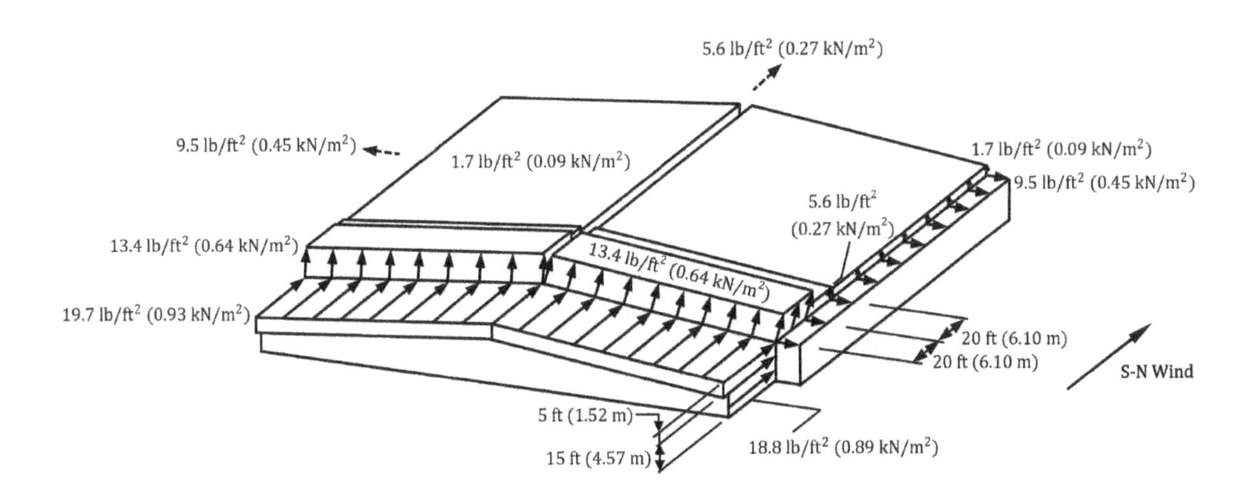

In case 1, the full design wind pressures act on the projected area perpendicular to each principal axis of the building at each level above ground and on the roof. These pressures act separately along each principal axis.

According to the exception in ASCE/SEI 27.3.5, buildings that meet the requirements of ASCE/SEI Appendix D.1 need only be designed for cases 1 and 3. Because the building is one story with a mean roof height less than 30 ft (9.14 m), only cases 1 and 3 must be considered.

In case 3, 75 percent of the windward and leeward wall pressures act simultaneously on the building at each level above ground.

Figure 5.30 External Horizontal Design Wind Pressures on the Walls of the Warehouse Building in Example 5.6

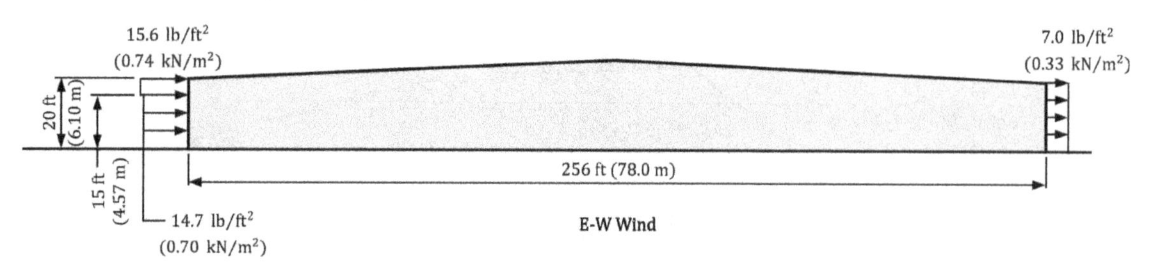

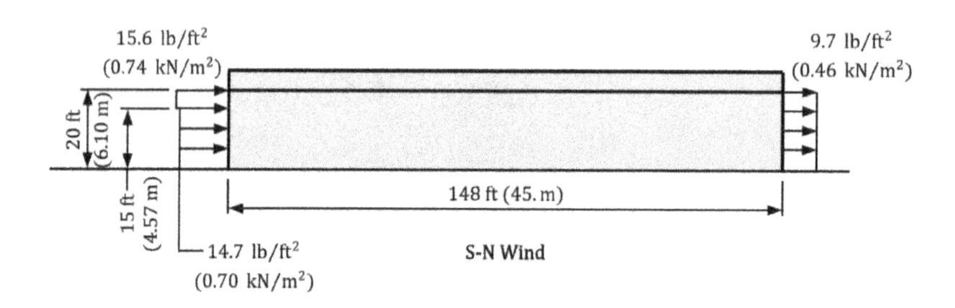

Load cases 1 and 3 are illustrated in Figure 5.31 at the mean roof height (pressures on the roof are not shown for clarity) for wind in the north to south and west to east directions. It is evident that the internal pressures cancel out when determining the horizontal wind pressures on the walls.

The minimum design wind loading in ASCE/SEI 27.1.5 must be considered as a load case in addition to the load cases above.

Note: Even though the building is located in a tornado-prone region (see ASCE/SEI Figure 32.1-1), tornado loads need not be considered because the building is assigned to risk category II (ASCE/SEI 32.1.1).

5.8.7 Example 5.7—Determination of Design Wind Pressures on a Warehouse Building Using ASCE/SEI Chapter 28—MWFRS

Determine the design wind pressures, p, on the MWFRS of the one-story warehouse building in Figure 5.27 using the provisions in ASCE/SEI Chapter 28 given the design data in Example 5.6.

SOLUTION

Check if the building meets all the conditions and limitations of ASCE/SEI 28.1.2 and 28.1.3 so that the provisions in ASCE/SEI Chapter 28 can be used to determine the design wind pressures on the MWFRS.

The building is regularly-shaped, that is, it does not have any unusual geometric irregularities in spatial form. Also, the building does not have response characteristics that make it subject to across-wind loading or other similar effects, and it is not sited at a location where channeling effects or buffeting in the wake of upwind obstructions need to be considered. Finally, the building does not have an arched, barrel, or unusually shaped roof.

The one-story warehouse building is a low-rise building based on the definition in ASCE/SEI 26.2 because the following two conditions are satisfied:

- Mean roof height = 20 ft (6.10 m) < 60 ft (18.3 m)
- Mean roof height = 20 ft (6.10 m) < least horizontal dimension = 148 ft (45.1 m)

Figure 5.31 Design Wind Load Cases at the Mean Roof Height of the Warehouse Building in Example 5.6

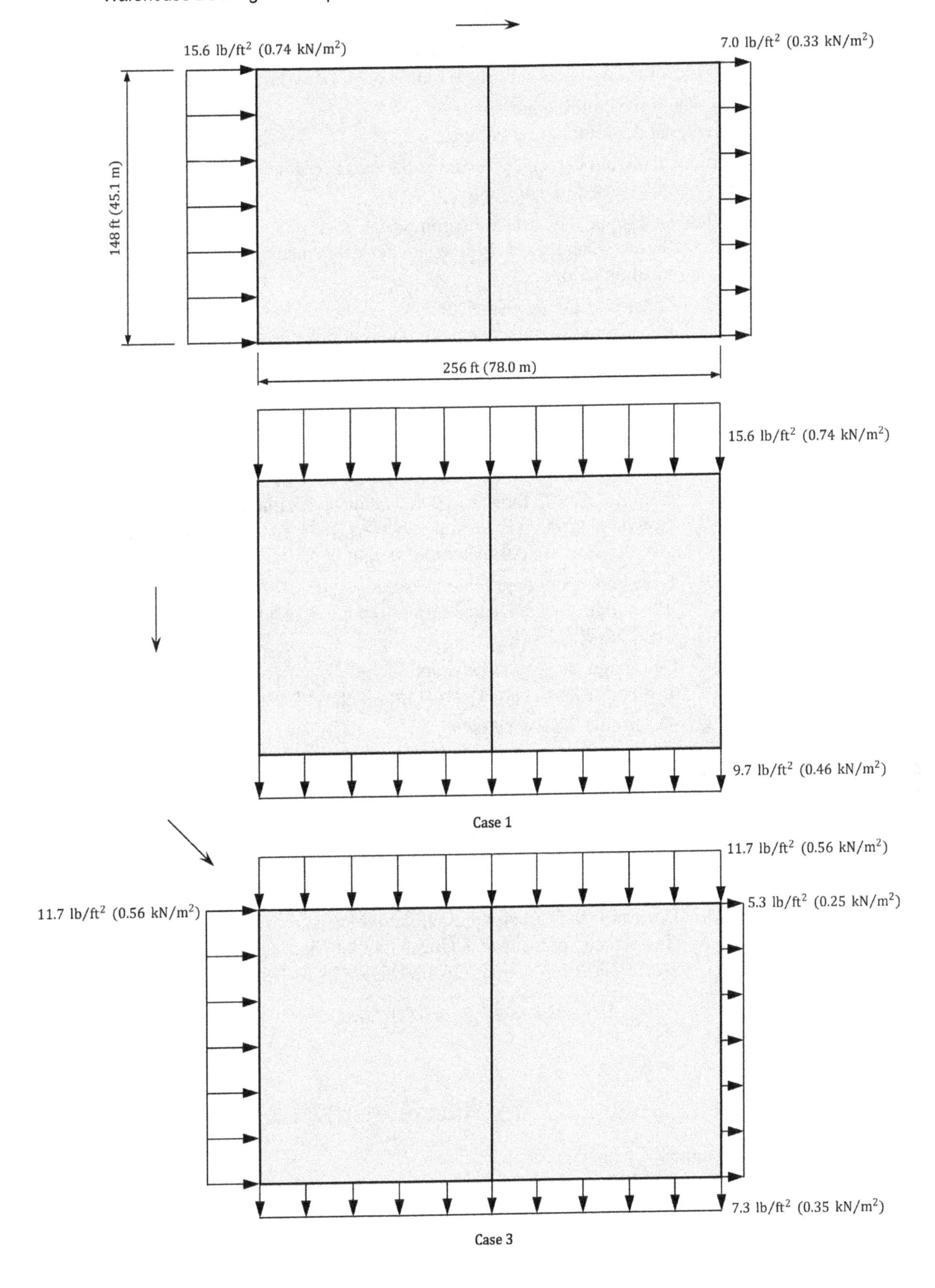

Therefore, the provisions in ASCE/SEI Chapter 28 are permitted to be used to determine the design wind pressures on the MWFRS.

Flowchart 5.4 is used to determine the design wind pressures.

Step 1—Determine the enclosure classification of the building

The enclosure classification of the building is given as enclosed.

Step 2—Determine if the building has a parapet

The building does not have a parapet.

Step 3—Determine the wind velocity pressure at the mean roof height, q_h

Flowchart 5.1 is used to determine q_h.

Step 3a—Determine the surface roughness category

Surface roughness C is given in the design data and it is assumed it is present in all directions.

Step 3b—Determine the exposure category

It is assumed exposure category C is applicable in all directions.

Step 3c—Determine the terrain exposure constants

For exposure C, $\alpha = 9.8$ and $z_g = 2{,}460$ ft (750 m) from ASCE/SEI Table 26.11-1.

Step 3d—Determine the velocity pressure exposure coefficient, K_h

The value of K_h at the mean roof height of 20 ft (6.10 m) is equal to 0.90 from ASCE/SEI Table 26.10-1. The mean roof height, h, is taken at the roof eave height, which is permitted because the roof angle is less than 10 degrees (see the definition of mean roof height in ASCE/SEI 26.2).

Step 3e—Determine the topographic factor, K_{zt}

The building is not located on a hill, ridge or escarpment, so $K_{zt} = 1.0$ (ASCE/SEI 26.8.2).

Step 3f—Determine the ground elevation factor, K_e

It is permitted to take $K_e = 1.0$ for all elevations (ASCE/SEI 26.9).

Step 3g—Determine the risk category

The risk category is given as II in the design data.

Step 3h—Determine the basic wind speed, V

For risk category II buildings, use IBC Figure 1609.3(2) or ASCE/SEI Figure 26.5-1B (see Table 5.2 of this book).

Equivalently, use Reference 5.4 to obtain $V = 108$ mi/h (48 m/s) given the latitude and longitude of the site.

Step 3i—Determine the velocity pressure at the mean roof height, q_h

The velocity pressure at the mean roof height of the building, q_h, is determined by ASCE/SEI Equations (26.10-1) and (26.10-1.SI):

$$q_h = 0.00256 K_h K_{zt} K_e V^2 = 0.00256 \times 0.90 \times 1.0 \times 1.0 \times 108^2 = 26.9 \text{ lb/ft}^2$$

In S.I.:

$$q_h = 0.613 K_h K_{zt} K_e V^2 = 0.613 \times 0.90 \times 1.0 \times 1.0 \times 48^2 / 1{,}000 = 1.27 \text{ kN/m}^2$$

Step 4—Determine the wind directionality factor, K_d

For the MWFRS of a building, the wind directionality factor, K_d, is equal to 0.85 (ASCE/SEI Table 26.6-1).

Step 5—Determine the external pressure coefficients, (GC_{pf}), for basic load cases 1 and 2
External pressure coefficients, (GC_{pf}), for basic load cases 1 and 2 are determined from ASCE/SEI Figure 28.3-1 using a roof angle between 0 and 5 degrees for wind in the E-W direction (load case 1) and a roof angle between 0 and 90 degrees for wind in the S-N direction (load case 2). The external pressure coefficients for both load cases are given in Table 5.38.

Table 5.38 External Pressure Coefficients, (GC_{pf}), for Basic Load Cases 1 and 2 for the Warehouse Building in Example 5.7

Surface	(GC_{pf})	
	Load Case 1	Load Case 2
1	0.40	−0.45
2	−0.69	−0.69
3	−0.37	−0.37
4	−0.29	−0.45
5	—	0.40
6	—	−0.29
1E	0.61	−0.48
2E	−1.07	−1.07
3E	−0.53	−0.53
4E	−0.43	−0.48
5E	—	0.61
6E	—	−0.43

Step 6—Determine the external pressure coefficients, (GC_{pf}), for load cases 3 and 4
Torsional load cases 3 and 4 need not be considered for this one-story building because the mean roof height is less than 30 ft (9.14 m) [ASCE/SEI 28.3.2].

Step 7—Determine the internal pressure coefficients, (GC_{pi})
For an enclosed building, $(GC_{pi}) = +0.18, -0.18$ (ASCE/SEI Table 26.13-1).

Step 8—Determine the design wind pressures, p
The design wind pressures, p, are determined by ASCE/SEI Equations (28.3-1) and (28.3-1.SI):

$$p = q_h K_d[(GC_{pf}) - (GC_{pi})] = (26.9 \times 0.85) \times [(GC_{pf}) - (\pm 0.18)]$$

In S.I.:

$$p = q_h K_d[(GC_{pf}) - (GC_{pi})] = (1.27 \times 0.85) \times [(GC_{pf}) - (\pm 0.18)]$$

Calculations for design wind pressure are illustrated for surface 1 for wind in the east to west and south to north directions:

- E-W direction (load case 1)
  For positive internal pressure: $p = 22.9 \times (0.40 - 0.18) = 5.0$ lb/ft^2
  For negative internal pressure: $p = 22.9 \times [0.40 - (-0.18)] = 13.3$ lb/ft^2

- S-N direction (load case 2)
  For positive internal pressure: $p = 22.9\times[(-0.45)-0.18)] = -14.4$ lb/ft^2
  For negative internal pressure: $p = 22.9\times[(-0.45)-(-0.18)] = -6.2$ lb/ft^2

In S.I.:

- E-W direction (load case 1)
  For positive internal pressure: $p = 1.08\times(0.40-0.18) = 0.24$ kN/m^2
  For negative internal pressure: $p = 1.08\times[0.40-(-0.18)] = 0.63$ kN/m^2
- S-N direction (load case 2)
  For positive internal pressure: $p = 1.08\times[(-0.45)-0.18)] = -0.68$ kN/m^2
  For negative internal pressure: $p = 1.08\times[(-0.45)-(-0.18)] = -0.29$ kN/m^2

Design wind pressures in the E-W and S-N directions are given in Table 5.39.

Table 5.39 Design Wind Pressures on the MWFRS of the Warehouse Building in Example 5.7

Surface	Design Wind Pressure, p, lb/ft^2 (kN/m^2)					
	Load Case 1 (E-W Wind)			Load Case 2 (S-N Wind)		
	(GC_{pf})	(GC_{pi})		(GC_{pf})	(GC_{pi})	
		$+(GC_{pi})$	$-(GC_{pi})$		$+(GC_{pi})$	$-(GC_{pi})$
1	0.40	5.0 (0.24)	13.3 (0.63)	−0.45	−14.4 (−0.68)	−6.2 (−0.29)
2	−0.69	−19.9 (−0.94)	−11.7 (−0.55)	−0.69	−19.9 (−0.94)	−11.7 (−0.55)
3	−0.37	−12.6 (−0.60)	−4.3 (−0.21)	−0.37	−12.6 (−0.60)	−4.3 (−0.21)
4	−0.29	−10.8 (−0.51)	−2.5 (−0.12)	−0.45	−14.4 (−0.68)	−6.2 (−0.29)
5	—	—	—	0.40	5.0 (0.24)	13.3 (0.63)
6	—	—	—	−0.29	−10.8 (−0.51)	−2.5 (−0.12)
1E	0.61	9.8 (0.47)	18.1 (0.86)	−0.48	−15.1 (−0.71)	−6.9 (−0.32)
2E	−1.07	−28.6 (−1.35)	−20.4 (−0.96)	−1.07	−28.6 (−1.35)	−20.4 (−0.96)
3E	−0.53	−16.3 (−0.77)	−8.0 (−0.38)	−0.53	−16.3 (−0.77)	−8.0 (−0.38)
4E	−0.43	−14.0 (−0.66)	−5.7 (−0.27)	−0.48	−15.1 (−0.71)	−6.9 (−0.32)
5E	—	—	—	0.61	9.8 (0.47)	18.1 (0.85)
6E	—	—	—	−0.43	−14.0 (−0.66)	−5.7 (−0.27)

The distance a is equal to the following for a building with a roof angle between 0 and 7 degrees and a least horizontal dimension less than or equal to 300 ft (91.4 m):

$$a = \text{lesser of} \begin{cases} 0.1\times\text{least horizontal dimension} = 0.1\times148 = 14.8 \text{ ft } (4.51 \text{ m}) \\ 0.4h = 0.4\times20.0 = 8.0 \text{ ft } (2.44 \text{ m}) \end{cases}$$

$$\text{Minimum } a = \text{greater of} \begin{cases} 0.04\times\text{least horizontal dimension} = 0.04\times148 = 5.9 \text{ ft } (1.80 \text{ m}) \\ 3.0 \text{ ft } (0.91 \text{ m}) \end{cases}$$

Therefore, $a = 8.0$ ft (2.44 m).

The roof pressure coefficients, (GC_{pf}), are negative in zones 2 and 2E (see Tables 5.38 or 5.39). According to ASCE/SEI 28.3.2.1, these pressure coefficients must be applied in zone 2/2E for a distance from the edge of the roof equal to 50 percent of the horizontal dimension of the building parallel to the direction of the MWFRS being designed or 2.5 times the eave height at the windward wall, whichever is less. The remainder of the zone 2/2E extending to the ridge line must use the pressure coefficients for zone 3/3E.

- E-W direction: $0.5 \times 256.0 = 128.0$ ft (39.0 m)
- S-N direction: $0.5 \times 148.0 = 74.0$ ft (22.6 m)
- $2.5 \times$ eave height $= 2.5 \times 20.0 = 50.0$ ft (15.2 m)

Therefore, in the E-W and S-N directions, zone 2/2E applies over a distance of 50.0 ft (15.2 m) from the edge of the windward roof. In the E-W direction, zone 3/3E applies over a distance $256.0 - 50.0 = 206.0$ ft (62.8 m), and in the S-N direction, zone 3/3E applies over a distance of $148.0 - 50.0 = 98.0$ ft (29.9 m).

The design pressures must be applied on the building in accordance with load cases 1 and 2 in ASCE/SEI Figure 28.3-1. The building must be evaluated with each corner taken as the windward corner with all zones loaded simultaneously. In general, eight basic load scenarios (4 for load case 1 and 4 for load case 2) must be considered individually to determine the critical loading on the structural members (see ASCE/SEI Figure C28.3-1). In addition, combinations of external and internal pressures must be evaluated to obtain the most severe loading. Considering both positive and negative internal pressure for each basic load scenario results in a total of 16 separate load scenarios. Because the building in this example is symmetric in the E-W direction and in the N-S direction, some of these load scenarios are repetitive.

Design wind pressures for one load scenario in each of the east to west and south to north directions, including positive and negative internal pressures, are given in Figures 5.32 and 5.33, respectively (in both cases, the windward corner is taken at the southeast corner of the building).

As noted in step 5, torsional load cases 3 and 4 need not be considered for this one-story building because the mean roof height is less than 30 ft (9.14 m) [ASCE/SEI 28.3.2].

The minimum design wind loads in ASCE/SEI 28.3.6 must be considered as a load case in addition to the load cases above.

Note: Even though the building is located in a tornado-prone region (see ASCE/SEI Figure 32.1-1), tornado loads need not be considered because the building is assigned to risk category II (ASCE/SEI 32.1.1).

5.8.8 Example 5.8—Determination of Design Wind Pressures on a Commercial Building Using ASCE/SEI Chapter 27—MWFRS

Determine the design wind pressures, p, on the MWFRS of the one-story commercial building in Figure 5.34 using the provisions in ASCE/SEI Chapter 27 given the following design data:

- Location: Philadelphia, PA (Latitude = 39.94°, Longitude = −75.19°)
- Surface roughness: B
- Topography: Not situated on a hill, ridge or escarpment
- Risk category: II (IBC Table 1604.5)
- Enclosure classification: Enclosed

SOLUTION

Check if the building meets all the conditions and limitations of ASCE/SEI 27.1.2 and 27.1.3 so that the provisions in ASCE/SEI Chapter 27 can be used to determine the design wind pressures on the MWFRS.

Figure 5.32 Design Wind Pressures for Wind in the East to West Direction on the MWFRS of the Warehouse Building in Example 5.7 (Load Case 1)

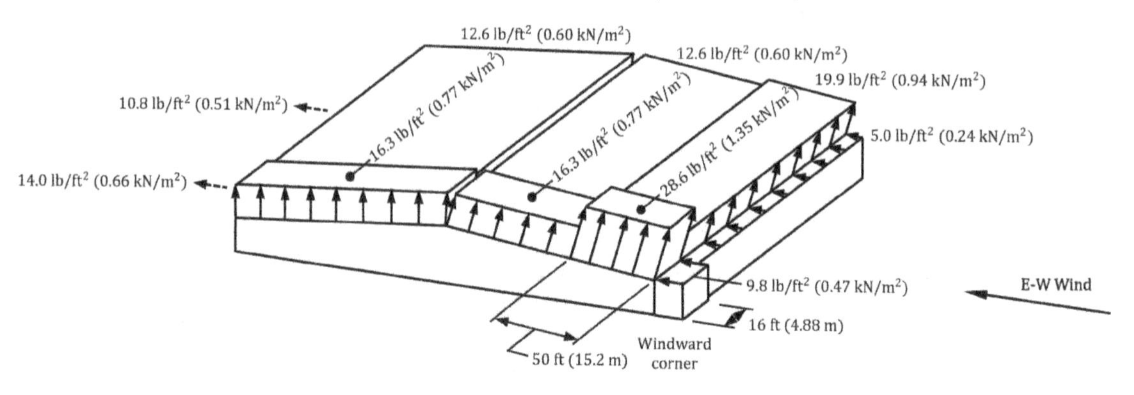

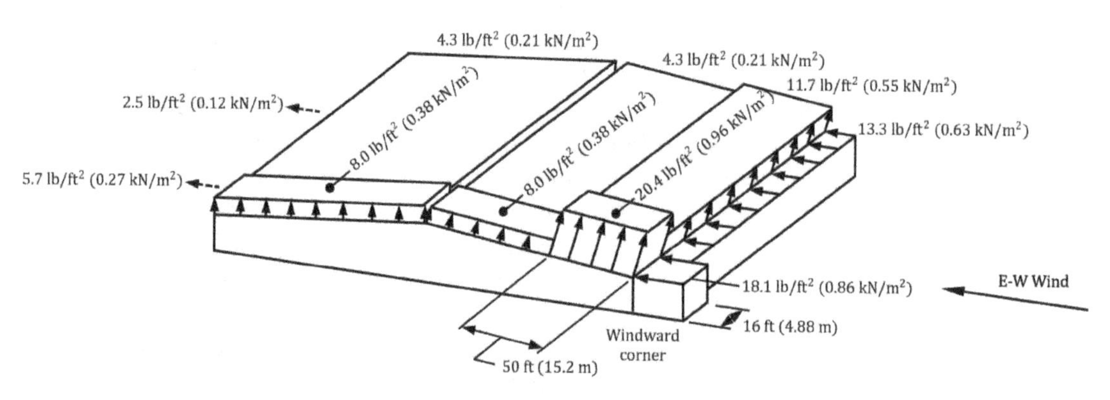

The building is regularly-shaped, that is, it does not have any unusual geometric irregularities in spatial form. Also, the building does not have response characteristics that make it subject to across-wind loading or other similar effects, and it is not sited at a location where channeling effects or buffeting in the wake of upwind obstructions need to be considered.

Therefore, the provisions in ASCE/SEI Chapter 27 are permitted to be used to determine the design wind pressures on the MWFRS.

Flowchart 5.3 is used to determine the design wind pressures.

Step 1—Determine the enclosure classification of the building

The enclosure classification of the building is given as enclosed.

Step 2—Determine if the building has a parapet

The building does not have a parapet.

Step 3—Determine whether the building is rigid or flexible

In lieu of determining the fundamental natural frequency, n_1, from a dynamic analysis of the building, check if the approximate lower bound natural frequency, n_a, given in ASCE/SEI 26.11.3 is permitted to be used:

Building height = 50 ft (15.2 m) < 300 ft (91.4 m)

Building height = 50 ft (15.2 m) $< 4L_{eff} = 4 \times 125 = 500$ ft (152.4 m) in the N-S direction

= 50 ft (15.2 m) $< 4L_{eff} = 4 \times 200 = 800$ ft (243.8 m) in the E-W direction

Figure 5.33 Design Wind Pressures for Wind in the South to North Direction on the MWFRS of the Warehouse Building in Example 5.7 (load case 2)

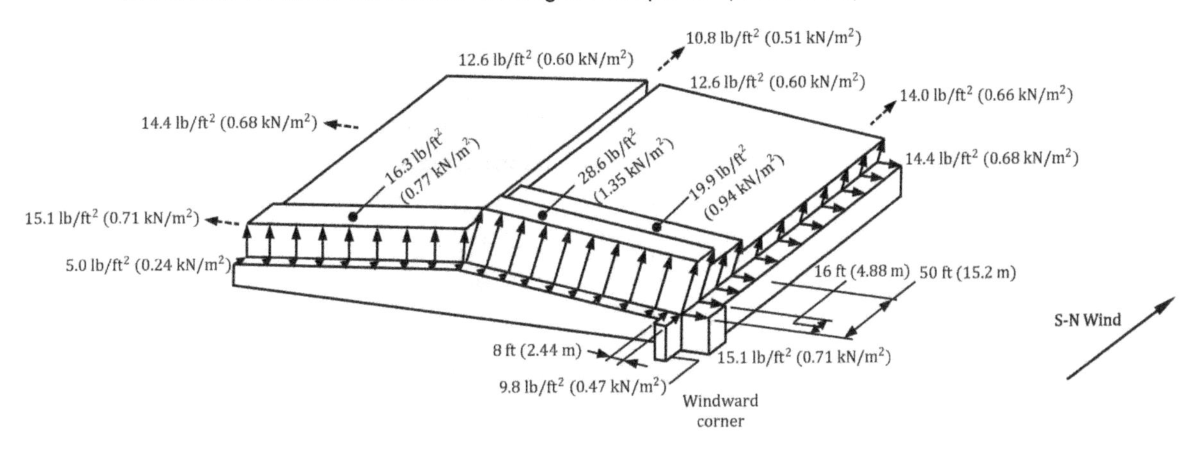

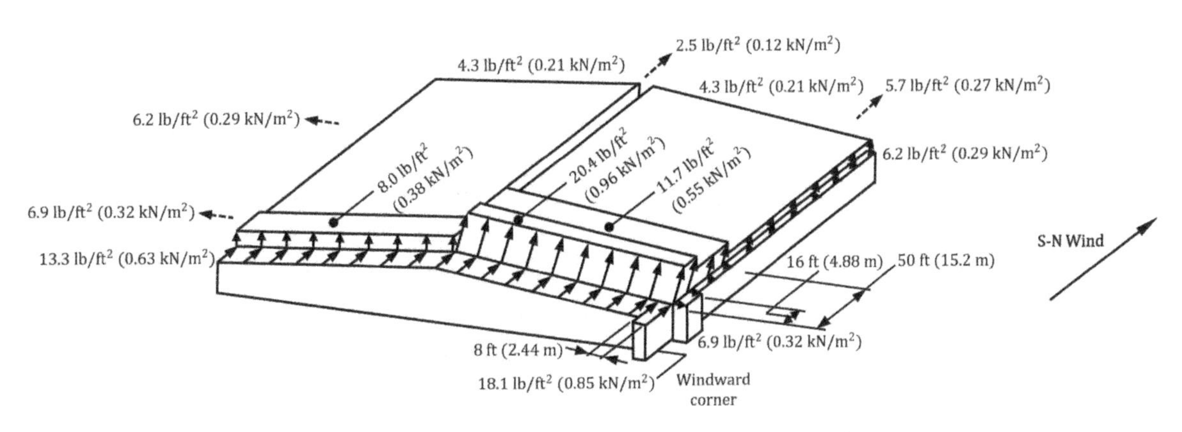

Because both of these limitations are satisfied, ASCE/SEI Equations (26.11-5) and (26.11-5.SI) are permitted to be used to determine n_a in both directions for the MWFRS consisting of precast concrete walls.

In the N-S direction:

A_B = base area of the building $= 200 \times 125 = 25{,}000$ ft^2 (2,323 m^2)

h = mean roof height $= (50+25)/2 = 37.5$ ft (11.4 m)

h_i = height of walls $= 25.0$ ft (7.62 m)

A_i = horizontal cross-sectional area of wall $= (7.625/12) \times 125.0 = 79.4$ ft^2 (7.4 m^2)

D_i = length of wall $= 125.0$ ft (38.1 m)

$$C_w = \frac{100}{A_B}\sum_{i=1}^{n}\left(\frac{h}{h_i}\right)^2 \frac{A_i}{\left[1+0.83\left(\frac{h_i}{D_i}\right)^2\right]} = \frac{2\times 100}{25{,}000}\times\left(\frac{37.5}{25.0}\right)^2\times\frac{79.4}{\left[1+0.83\left(\frac{25.0}{125.0}\right)^2\right]} = 1.38$$

$$n_a = 385(C_w)^{0.5}/h = 385\times(1.38)^{0.5}/37.5 = 12.1 \text{ Hz} > 1 \text{ Hz}$$

Figure 5.34 Plan and Elevation of the One-story Commercial Building in Example 5.8

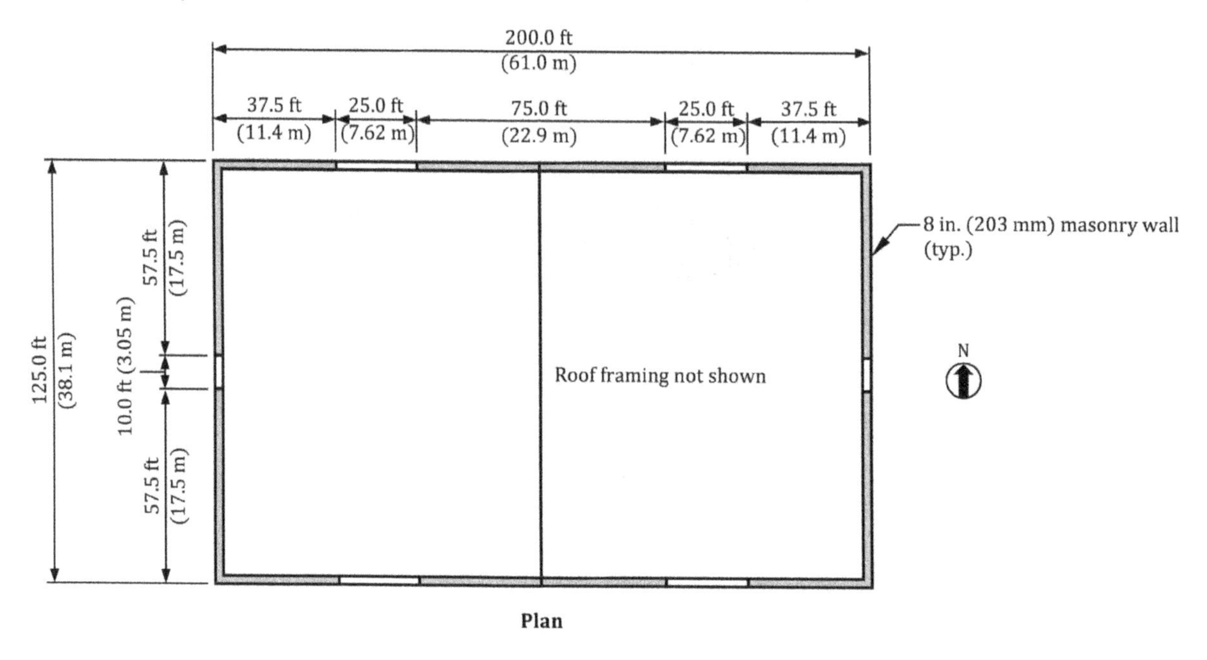

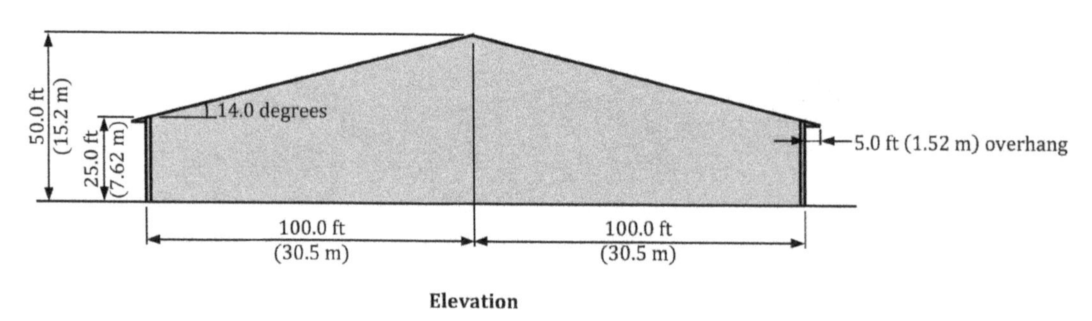

In S.I.:

$$C_w = \frac{100}{A_B}\sum_{i=1}^{n}\left(\frac{h}{h_i}\right)^2 \frac{A_i}{\left[1+0.83\left(\frac{h_i}{D_i}\right)^2\right]} = \frac{2\times 100}{2,323}\times\left(\frac{11.4}{7.62}\right)^2\times\frac{7.4}{\left[1+0.83\left(\frac{7.62}{38.1}\right)^2\right]} = 1.38$$

$$n_a = 117.3(C_w)^{0.5}/h = 117.3\times(1.38)^{0.5}/11.4 = 12.1 \text{ Hz} > 1 \text{ Hz}$$

Therefore, the building is rigid in the N-S direction.

Similar calculations in the E-W direction result in $n_a > 1$ Hz, so the building is rigid in that direction as well.

Note: This commercial building is a low-rise building in accordance with the definition in ASCE/SEI 26.4 (the mean roof height is less than 60 ft [18.3 m] and the mean roof height is less than the least horizontal dimension of the building). Low-rise buildings are permitted to be considered rigid (ASCE/SEI 26.11.2).

Step 4—Determine the gust-effect factor, G

For rigid buildings, G is permitted to be taken as 0.85 (ASCE/SEI 26.11.4).

Step 5—Determine the wind directionality factor, K_d

For the MWFRS of a building, the wind directionality factor, K_d, is equal to 0.85 (ASCE/SEI Table 26.6-1).

Step 6—Determine the wind velocity pressures, q_z

Flowchart 5.1 is used to determine the wind velocity pressures.

Step 6a—Determine the surface roughness category

Surface roughness B is given in the design data and it is assumed it is present in all directions.

Step 6b—Determine the exposure category

It is assumed exposure category B is applicable in all directions.

Step 6c—Determine the terrain exposure constants

For exposure B, $\alpha = 7.5$ and $z_g = 3{,}280$ ft (1,000 m) from ASCE/SEI Table 26.11-1.

Step 6d—Determine the velocity pressure exposure coefficient, K_z

Values of K_z are determined from ASCE/SEI 26.10-1 and are given in Table 5.40.

Table 5.40 Velocity Pressure Exposure Coefficients, K_z and K_h, for the Commercial Building in Example 5.8

Height above the Ground Level, z, ft (m)	K_z
50.0 (15.2)	0.79
37.5 (11.4)	0.73
25.0 (7.62)	0.66
15.0 (4.57)	0.57

For example, at $z = 37.5$ ft (11.4 m):

$$K_z = 2.41(z/z_g)^{2/\alpha} = 2.41 \times (37.5/3{,}280)^{2/7.5} = 0.73$$

In S.I.:

$$K_z = 2.41 \times (11.4/1{,}000)^{2/7.5} = 0.73$$

Step 6e—Determine the topographic factor, K_{zt}

The building is not located on a hill, ridge or escarpment, so $K_{zt} = 1.0$ (ASCE/SEI 26.8.2).

Step 6f—Determine the ground elevation factor, K_e

It is permitted to take $K_e = 1.0$ for all elevations (ASCE/SEI 26.9).

Step 6g—Determine the risk category

The risk category is given as II in the design data.

Step 6h—Determine the basic wind speed, V

For risk category II buildings, use IBC Figure 1609.3(2) or ASCE/SEI Figure 26.5-1B (see Table 5.2 of this book).

Equivalently, use Reference 5.4 to obtain $V = 114$ mi/h (51 m/s) given the latitude and longitude of the site.

Step 6i—Determine the velocity pressures, q_z

The velocity pressures, q_z, over the height of the building are given in Table 5.41. These pressures are determined by ASCE/SEI Equations (26.10-1) and (26.10-1.SI). For example, the velocity pressure at the mean roof height of the building is equal to the following:

$$q_h = 0.00256K_hK_{zt}K_eV^2 = 0.00256 \times 0.73 \times 1.0 \times 1.0 \times 114^2 = 24.3 \text{ lb/ft}^2$$

In S.I.:

$$q_h = 0.613K_hK_{zt}K_eV^2 = 0.613 \times 0.73 \times 1.0 \times 1.0 \times 51^2/1{,}000 = 1.16 \text{ kN/m}^2$$

Table 5.41 Velocity Pressures, q_z and q_h, for the Commercial Building in Example 5.8

Height above the Ground Level, z, ft (m)	K_z	q_z, lb/ft² (kN/m²)
50.0 (15.2)	0.79	26.3 (1.26)
37.5 (11.4)	0.73	24.3 (1.16)
25.0 (7.62)	0.66	22.0 (1.05)
15.0 (4.57)	0.57	19.0 (0.91)

Step 7—Determine the external pressure coefficients, C_p, for the walls and roof

The external pressure coefficients are determined from ASCE/SEI Figure 27.3-1.

For wind in the N-S direction:

- Windward wall: $C_p = 0.8$
- Leeward wall: For $L/B = 125/200 = 0.63$, $C_p = -0.5$
- Side walls: $C_p = -0.7$
- Roof: Parallel to the ridge with $h/L = 37.5/125 = 0.30 < 0.5$ (In S.I.: $h/L = 11.4/38.1 = 0.30 < 0.5$)
  From the windward edge of the roof to $h = 37.5$ ft (11.4 m), $C_p = -0.9, -0.18$
  From $h = 37.5$ ft (11.4 m) to $2h = 75$ ft (22.9 m), $C_p = -0.5, -0.18$
  From $2h = 75$ ft (22.9 m) to 125 ft (38.1 m), $C_p = -0.3, -0.18$

For wind in the E-W direction:

- Windward wall: $C_p = 0.8$
- Leeward wall: For $L/B = 200/125 = 1.6$, $C_p = -0.38$ (by linear interpolation)
- Side walls: $C_p = -0.7$
- Roof: Normal to the ridge with $\theta = 14$ degrees > 10 degrees and $h/L = 37.5/200 = 0.19 < 0.25$ (In S.I.: $h/L = 11.4/61.0 = 0.19 < 0.25$)

Windward: $C_p = -0.54, -0.04$ (by linear interpolation)
Leeward: $C_p = -0.46$ (by linear interpolation)

Step 8—Determine the velocity pressure for internal pressure determination, q_i

According to ASCE/SEI 27.3.1, $q_i = q_h = 24.3$ lb/ft² (1.16 kN/m²) for all surfaces of an enclosed building.

Step 9—Determine the internal pressure coefficients, (GC_{pi})

From ASCE/SEI Table 26.13-1, $(GC_{pi}) = +0.18, -0.18$ for an enclosed building.

Step 10—Determine the design wind pressures, p, on all the surfaces of the building

Design wind pressures, p, are determined by ASCE/SEI Equation (27.3-1) for an enclosed building:

$$p = qK_dGC_p - q_iK_d(GC_{pi})$$

For wind in the N-S direction:

- Windward wall:

$$p_z = (q_z \times 0.85 \times 0.85 \times 0.8) - [24.3 \times 0.85 \times (\pm 0.18)] = 0.58 q_z \mp 3.7 \text{ lb/ft}^2$$

In S.I.:

$$p_z = (q_z \times 0.85 \times 0.85 \times 0.8) - [1.16 \times 0.85 \times (\pm 0.18)] = 0.58 q_z \mp 0.18 \text{ kN/m}^2$$

- Leeward wall

$$p_h = [q_h \times 0.85 \times 0.85 \times (-0.5)] - [24.3 \times 0.85 \times (\pm 0.18)]$$

$$= -8.8 \mp 3.7 = -12.5 \text{ lb/ft}^2, -5.1 \text{ lb/ft}^2$$

In S.I.:

$$p_h = [q_h \times 0.85 \times 0.85 \times (-0.5)] - [1.16 \times 0.85 \times (\pm 0.18)]$$

$$= -0.42 \mp 0.18 = -0.60 \text{ kN/m}^2, -0.24 \text{ kN/m}^2$$

- Side walls

$$p_h = [q_h \times 0.85 \times 0.85 \times (-0.7)] - [24.3 \times 0.85 \times (\pm 0.18)]$$

$$= -12.3 \mp 3.7 = -16.0 \text{ lb/ft}^2, -8.6 \text{ lb/ft}^2$$

In S.I.:

$$p_h = [q_h \times 0.85 \times 0.85 \times (-0.7)] - [1.16 \times 0.85 \times (\pm 0.18)]$$

$$= -0.59 \mp 0.18 = -0.77 \text{ kN/m}^2, -0.41 \text{ kN/m}$$

- Roof

$$p_h = (q_h \times 0.85 \times 0.85 \times C_p) - [24.3 \times 0.85 \times (\pm 0.18)] = 17.6 C_p \mp 3.7 \text{ lb/ft}^2$$

In S.I.:

$$p_h = (q_h \times 0.85 \times 0.85 \times C_p) - [1.16 \times 0.85 \times (\pm 0.18)] = 0.84 C_p \mp 0.18 \text{ kN/m}^2$$

The roof pressure coefficient $C_p = -0.18$ may become critical where wind loads are combined with roof live loads or snow loads. Determination of wind pressures based on this pressure coefficient should be performed, but such calculations are not shown in this example.

Design wind pressures in the N-S direction are given in Table 5.42 and Figure 5.35.

For wind in the E-W direction:

- Windward wall:

$$p_z = (q_z \times 0.85 \times 0.85 \times 0.8) - [24.3 \times 0.85 \times (\pm 0.18)] = 0.58 q_z \mp 3.7 \text{ lb/ft}^2$$

In S.I.:

$$p_z = (q_z \times 0.85 \times 0.85 \times 0.8) - [1.16 \times 0.85 \times (\pm 0.18)] = 0.58 q_z \mp 0.18 \text{ kN/m}^2$$

- Leeward wall

$$p_h = [q_h \times 0.85 \times 0.85 \times (-0.38)] - [24.3 \times 0.85 \times (\pm 0.18)]$$

$$= -6.7 \mp 3.7 = -10.4 \text{ lb/ft}^2, -3.0 \text{ lb/ft}^2$$

Table 5.42 Design Wind Pressures, *p*, in the N-S Direction on the MWFRS of the Commercial Building in Example 5.8

Building Surface	Height above Ground Level, z, ft (m)	q, lb/ft² (kN/m²)	External Pressure qK_dGC_p, lb/ft² (kN/m²)	Internal Pressure $q_hK_d(GC_{pi})$, lb/ft² (kN/m²)	Net Pressure, *p*, lb/ft² (kN/m²)	
					$+(GC_{pi})$	$-(GC_{pi})$
Windward wall	50.0 (15.2)	26.3 (1.26)	15.2 (0.73)	±3.7 (±0.18)	11.5 (0.55)	18.9 (0.91)
	37.5 (11.4)	24.3 (1.16)	14.1 (0.67)	±3.7 (±0.18)	10.4 (0.49)	17.8 (0.85)
	25.0 (7.62)	22.0 (1.05)	12.7 (0.61)	±3.7 (±0.18)	9.0 (0.43)	16.4 (0.79)
	15.0 (4.57)	19.0 (0.91)	11.0 (0.53)	±3.7 (±0.18)	7.3 (0.35)	14.7 (0.71)
Leeward wall	All	24.3 (1.16)	−8.8 (−0.42)	±3.7 (±0.18)	−12.5 (−0.60)	−5.1 (−0.24)
Side walls	All	24.3 (1.16)	−12.3 (−0.59)	±3.7 (±0.18)	−16.0 (−0.77)	−8.6 (−0.41)
Roof	—*	24.3 (1.16)	−15.8 (−0.75)	±3.7 (±0.18)	−19.5 (−0.93)	−12.1 (−0.57)
	—**	24.3 (1.16)	−8.8 (−0.42)	±3.7 (±0.18)	−12.5 (−0.60)	−5.1 (−0.24)
	—†	24.3 (1.16)	−5.3 (−0.25)	±3.7 (±0.18)	−9.0 (−0.43)	−1.6 (−0.07)

*From windward edge of roof to 37.5 ft (11.4 m).
**From 37.5 ft (11.4 m) to 75 ft (22.9 m).
†From 75 ft (22.9 m) to 125 ft (38.1 m).

In S.I.:

$$p_h = [q_h \times 0.85 \times 0.85 \times (-0.38)] - [1.16 \times 0.85 \times (\pm 0.18)]$$

$$= -0.32 \mp 0.18 = -0.50 \text{ kN/m}^2, -0.14 \text{ kN/m}^2$$

- Side walls

$$p_h = [q_h \times 0.85 \times 0.85 \times (-0.7)] - [24.3 \times 0.85 \times (\pm 0.18)]$$

$$= -12.3 \mp 3.7 = -16.0 \text{ lb/ft}^2, -8.6 \text{ lb/ft}^2$$

In S.I.:

$$p_h = [q_h \times 0.85 \times 0.85 \times (-0.7)] - [1.16 \times 0.85 \times (\pm 0.18)]$$

$$= -0.59 \mp 0.18 = -0.77 \text{ kN/m}^2, -0.41 \text{ kN/m}^2$$

- Roof

$$p_h = (q_h \times 0.85 \times 0.85 \times C_p) - [24.3 \times 0.85 \times (\pm 0.18)] = 17.6C_p \mp 3.7 \text{ lb/ft}^2$$

In S.I.:

$$p_h = (q_h \times 0.85 \times 0.85 \times C_p) - [1.16 \times 0.85 \times (\pm 0.18)] = 0.84C_p \mp 0.18 \text{ kN/m}^2$$

The roof pressure coefficient $C_p = -0.18$ may become critical where wind loads are combined with roof live loads or snow loads. Determination of wind pressures based on this pressure coefficient should be performed, but such calculations are not shown in this example.

Figure 5.35 Design Wind Pressures in the N-S Direction on the MWFRS of the Commercial Building in Example 5.8

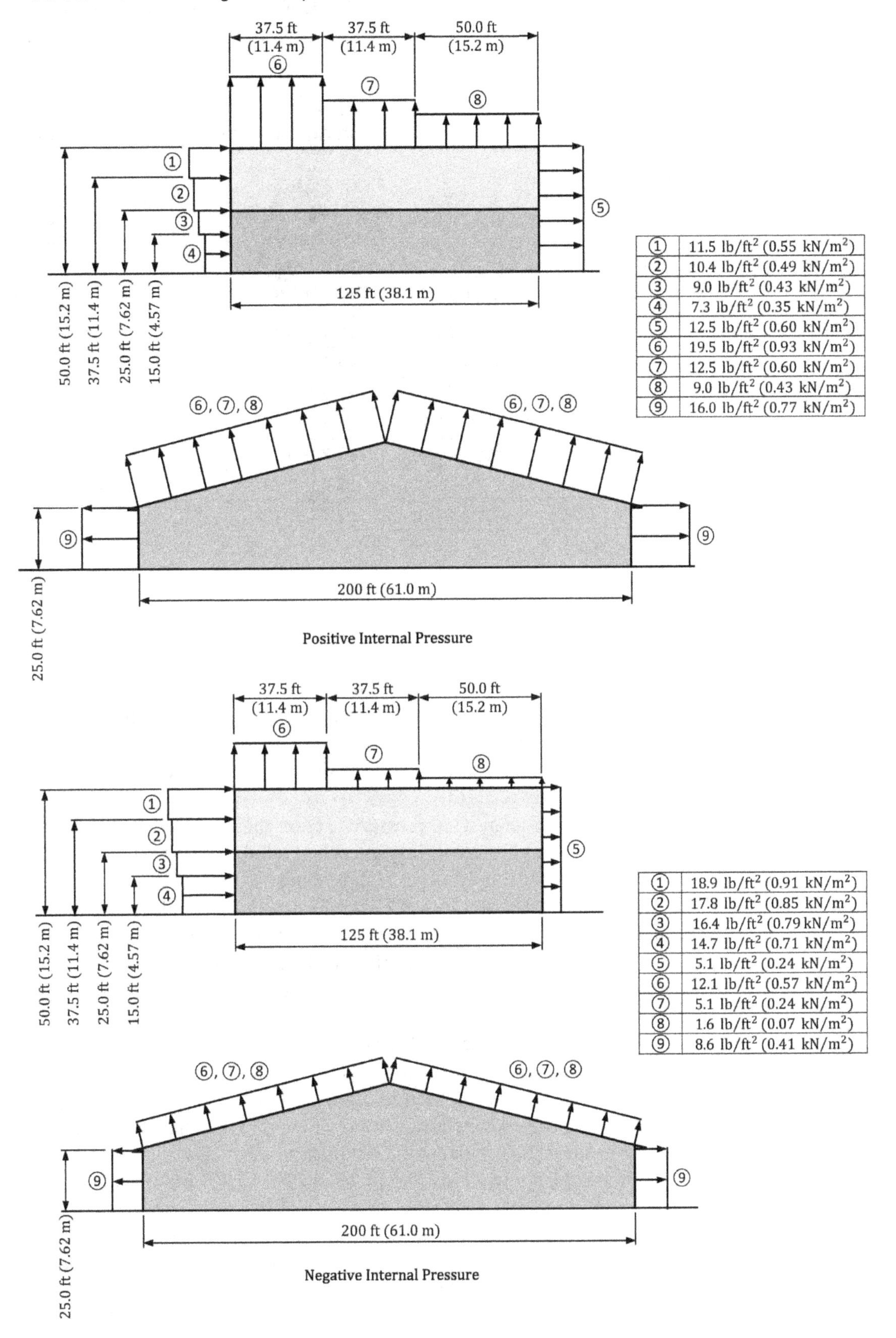

Design wind pressures in the E-W direction are given in Table 5.43 and Figure 5.36. The wind pressure in Figure 5.36 on the windward roof is for the case where the external roof pressure coefficient $C_p = -0.54$; a similar pressure diagram can be obtained using $C_p = -0.04$.

Table 5.43 Design Wind Pressures, *p*, in the E-W Direction on the MWFRS of the Commercial Building in Example 5.8

Building Surface	Height above Ground Level, z, ft (m)	q, lb/ft² (kN/m²)	External Pressure qK_dGC_p, lb/ft² (kN/m²)	Internal Pressure $q_hK_d(GC_{pi})$, lb/ft² (kN/m²)	Net Pressure, p, lb/ft² (kN/m²)	
					$+(GC_{pi})$	$-(GC_{pi})$
Windward wall	25.0 (7.62)	22.0 (1.05)	12.7 (0.61)	±3.7 (±0.18)	9.0 (0.43)	16.4 (0.79)
	15.0 (4.57)	19.0 (0.91)	11.0 (0.53)	±3.7 (±0.18)	7.3 (0.35)	14.7 (0.71)
Leeward wall	All	24.3 (1.16)	−6.7 (−0.32)	±3.7 (±0.18)	−10.4 (−0.50)	−3.0 (−0.14)
Side walls	All	24.3 (1.16)	−12.3 (−0.59)	±3.7 (±0.18)	−16.0 (−0.77)	−8.6 (−0.41)
Roof	—*	24.3 (1.16)	−9.5 (−0.45)	±3.7 (±0.18)	−13.2 (−0.63)	−5.8 (−0.27)
	—**	24.3 (1.16)	−0.7 (−0.03)	±3.7 (±0.18)	−4.4 (−0.21)	3.0 (0.15)
	—†	24.3 (1.16)	−8.1 (−0.39)	±3.7 (±0.18)	−11.8 (−0.57)	−4.4 (−0.21)

*Windward roof with external pressure coefficient $C_p = -0.54$.
**Windward roof with external pressure coefficient $C_p = -0.04$.
†Leeward roof with external pressure coefficient $C_p = -0.46$.

The external horizontal wind pressures on the walls for both directions are shown in Figure 5.37. It is evident that the internal pressures cancel out when determining the horizontal wind pressures on the walls.

The building must be designed for the design wind load cases in ASCE/SEI Figure 27.3-8 (see Figure 5.16).

In case 1, the full design wind pressures act on the projected area perpendicular to each principal axis of the building at each level above ground and on the roof. These pressures act separately along each principal axis.

According to the exception in ASCE/SEI 27.3.5, buildings that meet the requirements of ASCE/SEI Appendix D.1 need only be designed for cases 1 and 3. It can be shown that the building can be classified as torsionally regular under wind load in accordance with the definition in ASCE/SEI 26.2, so only cases 1 and 3 must be considered.

In case 3, 75 percent of the windward and leeward wall pressures act simultaneously on the building at each level above ground.

Load cases 1 and 3 are illustrated in Figure 5.38 at $z = 25$ ft (7.62 m) [pressures on the roof are not shown for clarity]. It is evident that the internal pressures cancel out when determining the horizontal wind pressures on the walls.

The minimum design wind loading in ASCE/SEI 27.1.5 must be considered as a load case in addition to the load cases above.

The positive external pressure on the bottom surface of the windward roof overhang is determined in accordance with ASCE/SEI 27.3.3 (see Figure 5.13). The velocity pressure, q_z, at the top of the wall [$z = 25$ ft (7.62 m)] is equal to 22.0 lb/ft² (1.05 kN/m²) [see Table 5.41].

Figure 5.36 Design Wind Pressures, p, in the E-W Direction on the MWFRS of the Commercial Building in Example 5.8

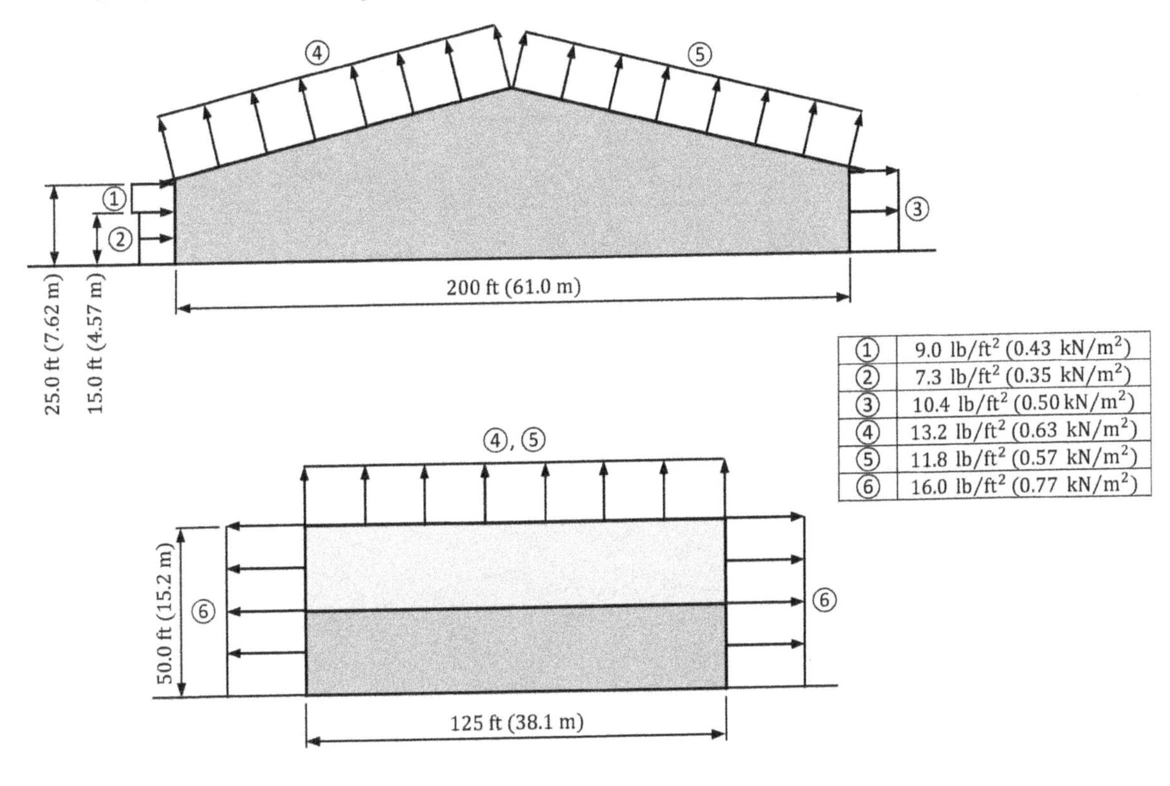

①	9.0 lb/ft^2 (0.43 kN/m^2)
②	7.3 lb/ft^2 (0.35 kN/m^2)
③	10.4 lb/ft^2 (0.50 kN/m^2)
④	13.2 lb/ft^2 (0.63 kN/m^2)
⑤	11.8 lb/ft^2 (0.57 kN/m^2)
⑥	16.0 lb/ft^2 (0.77 kN/m^2)

Positive Internal Pressure

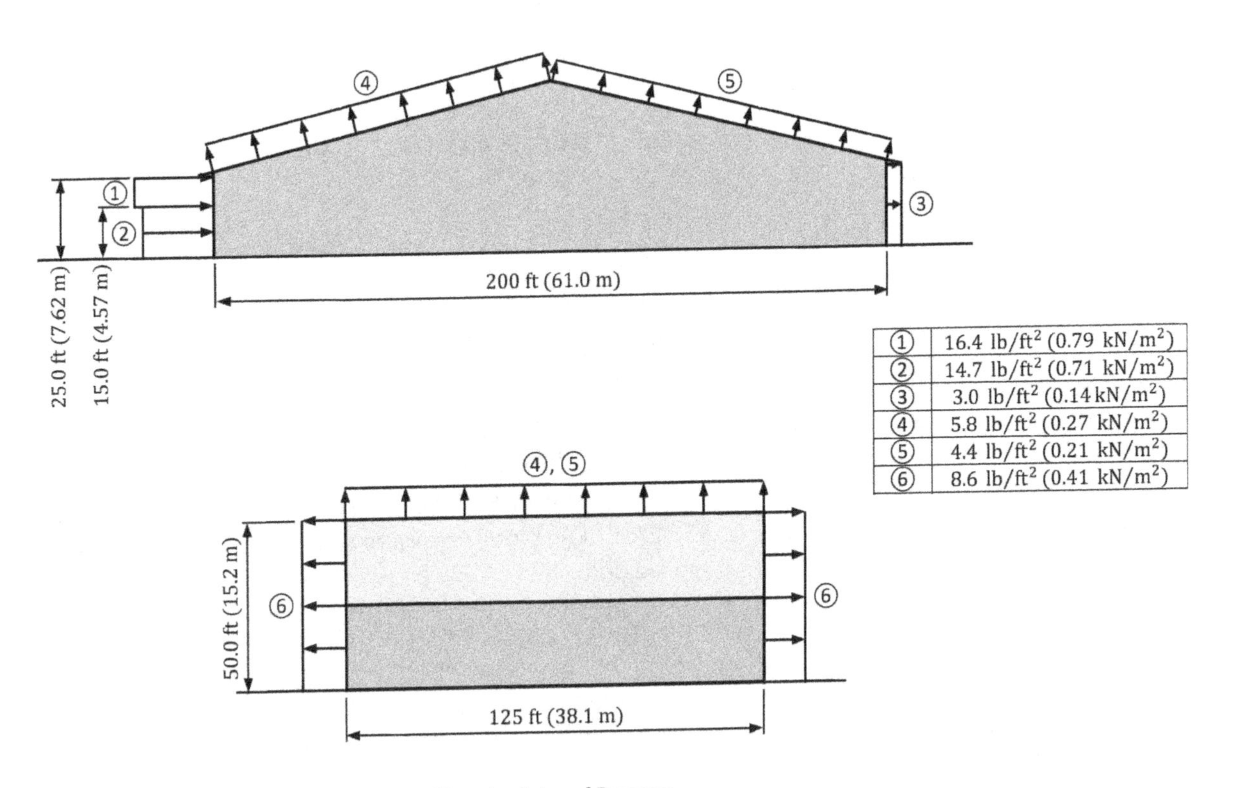

①	16.4 lb/ft^2 (0.79 kN/m^2)
②	14.7 lb/ft^2 (0.71 kN/m^2)
③	3.0 lb/ft^2 (0.14 kN/m^2)
④	5.8 lb/ft^2 (0.27 kN/m^2)
⑤	4.4 lb/ft^2 (0.21 kN/m^2)
⑥	8.6 lb/ft^2 (0.41 kN/m^2)

Negative Internal Pressure

Figure 5.37 External Horizontal Design Wind Pressures on the Walls of the Commercial Building in Example 5.8

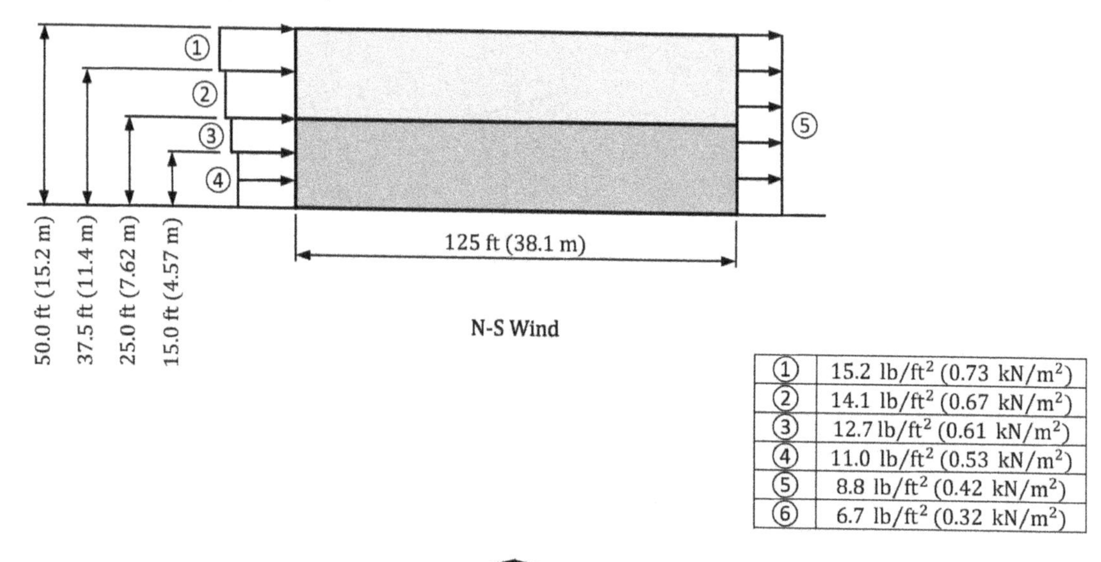

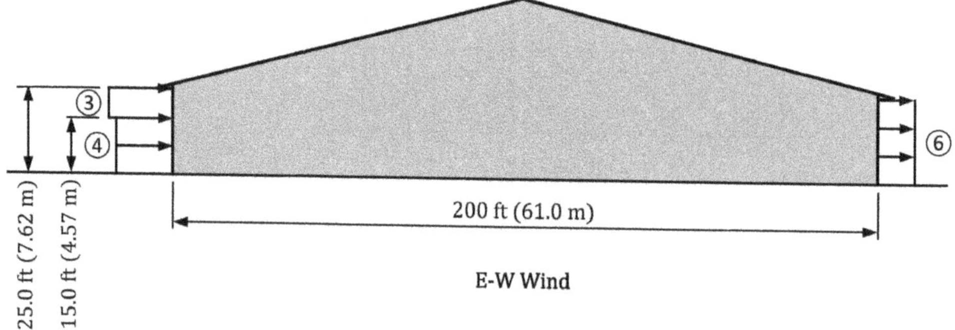

Therefore,

$$p = 0.8q_z K_d G = 0.8 \times 22.0 \times 0.85 \times 0.85 = 12.7 \text{ lb/ft}^2$$

In S.I.:

$$p = 0.8q_z K_d G = 0.8 \times 1.05 \times 0.85 \times 0.85 = 0.61 \text{ kN/m}^2$$

Note: Even though the building is located in a tornado-prone region (see ASCE/SEI Figure 32.1-1), tornado loads need not be considered because the building is assigned to risk category II (ASCE/SEI 32.1.1).

5.8.9 Example 5.9—Determination of Design Wind Pressures on a Commercial Building Using ASCE/SEI Chapter 28—MWFRS

Determine the design wind pressures, p, on the MWFRS of the one-story commercial building in Figure 5.34 using the provisions in ASCE/SEI Chapter 28 given the design data in Example 5.8 and assuming the roof diaphragm is not flexible.

Solution

Check if the building meets all the conditions and limitations of ASCE/SEI 28.1.2 and 28.1.3 so that the provisions in ASCE/SEI Chapter 28 can be used to determine the design wind pressures on the MWFRS.

Figure 5.38 Design Wind Load Cases at $z = 25$ ft (7.62 m) for the Commercial Building in Example 5.8

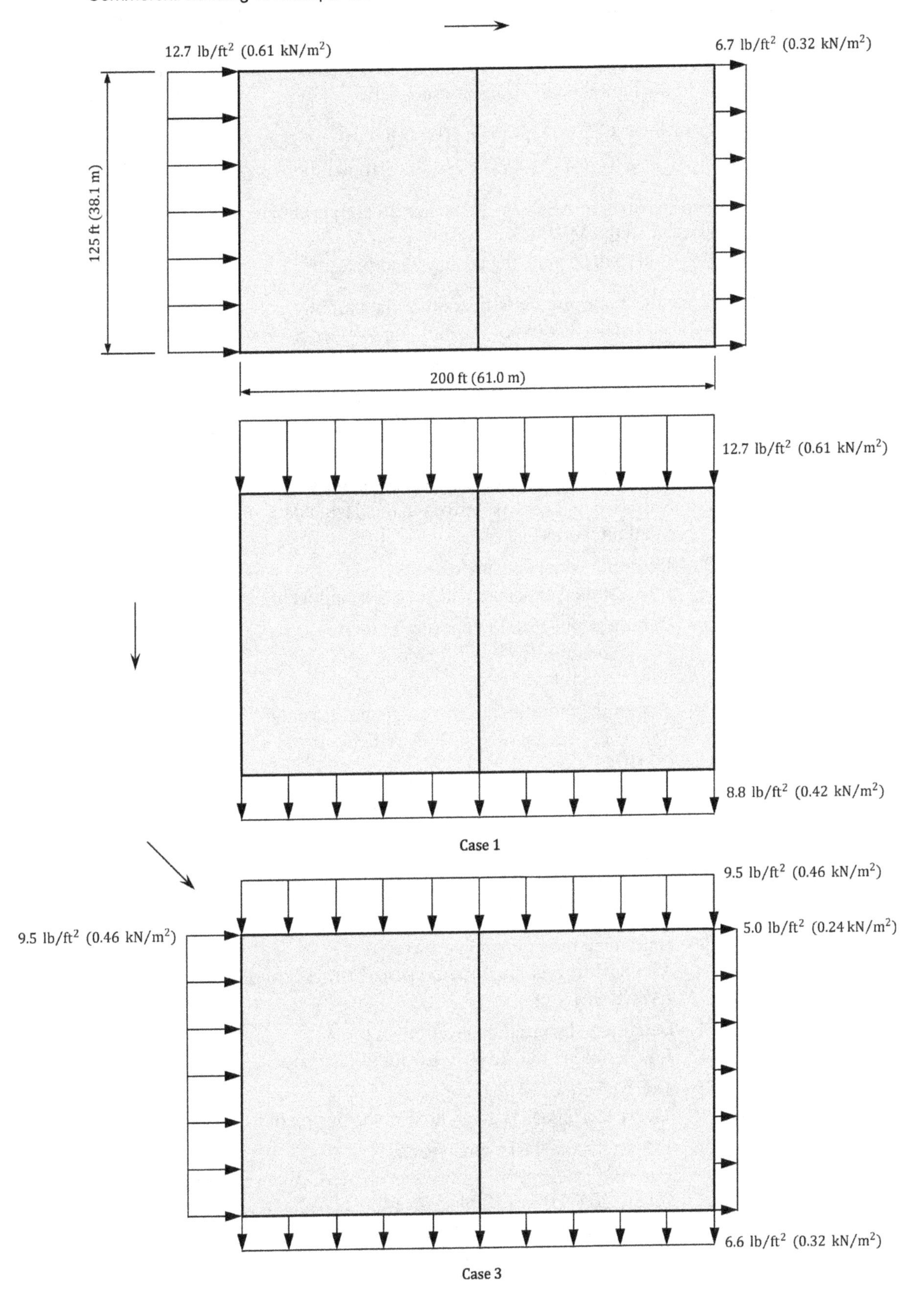

The building is regularly-shaped, that is, it does not have any unusual geometric irregularities in spatial form. Also, the building does not have response characteristics that make it subject to across-wind loading or other similar effects, and it is not sited at a location where channeling effects or buffeting in the wake of upwind obstructions need to be considered.

The one-story commercial building is a low-rise building based on the definition in ASCE/SEI 26.2 because the following two conditions are satisfied:

1. Mean roof height = 37.5 ft (11.4 m) < 60 ft (18.3 m)
2. Mean roof height = 37.5 ft (11.4 m) < least horizontal dimension = 125 ft (38.1 m)

Therefore, the provisions in ASCE/SEI Chapter 28 are permitted to be used to determine the design wind pressures on the MWFRS.

Flowchart 5.4 is used to determine the design wind pressures.

Step 1—Determine the enclosure classification of the building

The enclosure classification of the building is given as enclosed.

Step 2—Determine if the building has a parapet

The building does not have a parapet.

Step 3—Determine the wind velocity pressure at the mean roof height, q_h

Flowchart 5.1 is used to determine q_h.

Step 3a—Determine the surface roughness category

Surface roughness B is given in the design data and it is assumed it is present in all directions.

Step 3b—Determine the exposure category

It is assumed exposure category B is applicable in all directions.

Step 3c—Determine the terrain exposure constants

For exposure B, $\alpha = 7.5$ and $z_g = 3,280$ ft (1,000 m) from ASCE/SEI Table 26.11-1.

Step 3d—Determine the velocity pressure exposure coefficient, K_h

The value of K_h at $h =$ 37.5 ft (11.4 m) is determined from ASCE/SEI 26.10-1:

$$K_h = 2.41(h/z_g)^{2/\alpha} = 2.41 \times (37.5/3,280)^{2/7.5} = 0.73$$

In S.I.:

$$K_h = 2.41 \times (11.4/1,000)^{2/7.5} = 0.73$$

Step 3e—Determine the topographic factor, K_{zt}

The building is not located on a hill, ridge, or escarpment, so $K_{zt} = 1.0$ (ASCE/SEI 26.8.2).

Step 3f—Determine the ground elevation factor, K_e

It is permitted to take $K_e = 1.0$ for all elevations (ASCE/SEI 26.9).

Step 3g—Determine the risk category

The risk category is given as II in the design data.

Step 3h—Determine the basic wind speed, V

For risk category II buildings, use IBC Figure 1609.3(2) or ASCE/SEI Figure 26.5-1B (see Table 5.2 of this book).

Equivalently, use Reference 5.4 to obtain V = 114 mi/h (51 m/s) given the latitude and longitude of the site.

Step 3i—Determine the velocity pressure at the mean roof height, q_h

The velocity pressure at the mean roof height of the building, q_h, is determined by ASCE/SEI Equations (26.10-1) and (26.10-1.SI):

$$q_h = 0.00256K_hK_{zt}K_eV^2 = 0.00256\times0.73\times1.0\times1.0\times114^2 = 24.3\ \text{lb/ft}^2$$

In S.I.:

$$q_h = 0.613K_hK_{zt}K_eV^2 = 0.613\times0.73\times1.0\times1.0\times51^2/1{,}000 = 1.16\ \text{kN/m}^2$$

Step 4—Determine the wind directionality factor, K_d

For the MWFRS of a building, the wind directionality factor, K_d, is equal to 0.85 (ASCE/SEI Table 26.6-1).

Step 5—Determine the external pressure coefficients, (GC_{pf}), for basic load cases 1 and 2

External pressure coefficients, (GC_{pf}), for basic load cases 1 and 2 are determined by linear interpolation from ASCE/SEI Figure 28.3-1 with the roof angle equal to 14 degrees for wind in the E-W direction (load case 1) and are determined directly from ASCE/SEI Figure 28.3-1 with a roof angle between 0 and 90 degrees for wind in the N-S direction (load case 2). The external pressure coefficients for both load cases are given in Table 5.44.

Table 5.44 External Pressure Coefficients, (GC_{pf}), for Basic Load Cases 1 and 2 for the Commercial Building in Example 5.9

Surface	(GC_{pf})	
	Load Case 1	Load Case 2
1	0.49	−0.45
2	−0.69	−0.69
3	−0.44	−0.37
4	−0.37	−0.45
5	—	0.40
6	—	−0.29
1E	0.72	−0.48
2E	−1.07	−1.07
3E	−0.63	−0.53
4E	−0.56	−0.48
5E	—	0.61
6E	—	−0.43

Step 6—Determine the external pressure coefficients, (GC_{pf}), for torsional load cases 3 and 4

Torsional load cases must be evaluated for this building because none of the three exceptions in ASCE/SEI 28.3.2 is satisfied.

For load case 3, the external pressure coefficients for zones 1 to 4 and zones 1E to 4E are equal to those from case 1 in ASCE/SEI Figure 28.3-1 (ASCE/SEI 28.3.2.2). The external pressure coefficients for zones 1T to 4T are determined from ASCE/SEI Figure 28.3-2 by linear interpolation for a roof angle of 14 degrees.

For load case 4, the external pressure coefficients for zones 1 to 6 and zones 1E to 6E are equal to those from case 2 in ASCE/SEI Figure 28.3-1. The external pressure coefficients for zones 5T and 6T are given in ASCE/SEI Figure 28.3-2.

The external pressure coefficients for load cases 3 and 4 are given in Table 5.45.

Table 5.45 External Pressure Coefficients, (GC_{pf}), for Torsional Load Cases 3 and 4 for the Commercial Building in Example 5.9

Surface	(GC_{pf})	
	Load Case 3	Load Case 4
1	0.49	−0.45
2	−0.69	−0.69
3	−0.44	−0.37
4	−0.37	−0.45
5	—	0.40
6	—	−0.29
1E	0.72	−0.48
2E	−1.07	−1.07
3E	−0.63	−0.53
4E	−0.56	−0.48
5E	—	0.61
6E	—	−0.43
1T	0.12	—
2T	−0.17	—
3T	−0.11	—
4T	−0.09	—
5T	—	0.10
6T	—	−0.07

Step 7—Determine the internal pressure coefficients, (GC_{pi})

For an enclosed building, $(GC_{pi}) = +0.18, -0.18$ (ASCE/SEI Table 26.13-1).

Step 8—Determine the design wind pressures, p

The design wind pressures, p, are determined by ASCE/SEI Equations (28.3-1) and (28.3-1.SI):

$$p = q_h K_d[(GC_{pf}) - (GC_{pi})] = (24.3 \times 0.85) \times [(GC_{pf}) - (\pm 0.18)]$$

In S.I.:

$$p = q_h K_d[(GC_{pf}) - (GC_{pi})] = (1.16 \times 0.85) \times [(GC_{pf}) - (\pm 0.18)]$$

Calculations for design wind pressure are illustrated for surface 1 for wind in the E-W and N-S directions:

- E-W direction (load cases 1 and 3)
  For positive internal pressure: $p = 20.7 \times (0.49 - 0.18) = 6.4$ lb/ft²
  For negative internal pressure: $p = 20.7 \times [0.49 - (-0.18)] = 13.9$ lb/ft²

- N-S direction (load cases 2 and 4)
  For positive internal pressure: $p = 20.7 \times [(-0.45) - 0.18)] = -13.0$ lb/ft^2
  For negative internal pressure: $p = 20.7 \times [(-0.45) - (-0.18)] = -5.6$ lb/ft^2
  In S.I.:
- E-W direction (load cases 1 and 3)
  For positive internal pressure: $p = 0.99 \times (0.49 - 0.18) = 0.31$ kN/m^2
  For negative internal pressure: $p = 0.99 \times [0.49 - (-0.18)] = 0.66$ kN/m^2
- N-S direction (load cases 2 and 4)
  For positive internal pressure: $p = 0.99 \times [(-0.45) - 0.18)] = -0.62$ kN/m^2
  For negative internal pressure: $p = 0.99 \times [(-0.45) - (-0.18)] = -0.27$ kN/m^2
  Design wind pressures for load cases 1 and 2 and for load cases 3 and 4 are given in Tables 5.46 and 5.47, respectively.

Table 5.46 Design Wind Pressures for Load Cases 1 and 2 on the MWFRS of the Commercial Building in Example 5.9

Surface	Design Wind Pressure, p, lb/ft^2 (kN/m^2)					
	Load Case 1 (E-W Wind)			Load Case 2 (N-S Wind)		
	(GC_{pf})	(GC_{pi})		(GC_{pf})	(GC_{pi})	
		$+(GC_{pi})$	$-(GC_{pi})$		$+(GC_{pi})$	$-(GC_{pi})$
1	0.49	6.4 (0.31)	13.9 (0.66)	−0.45	−13.0 (−0.62)	−5.6 (−0.27)
2	−0.69	−18.0 (−0.86)	−10.6 (−0.51)	−0.69	−18.0 (−0.86)	−10.6 (−0.51)
3	−0.44	−12.8 (−0.61)	−5.4 (−0.26)	−0.37	−11.4 (−0.55)	−3.9 (−0.19)
4	−0.37	−11.4 (−0.55)	−3.9 (−0.19)	−0.45	−13.0 (−0.62)	−5.6 (−0.27)
5	—	—	—	0.40	4.6 (0.22)	12.0 (0.57)
6	—	—	—	−0.29	−9.7 (−0.47)	−2.3 (−0.11)
1E	0.72	11.2 (0.54)	18.6 (0.89)	−0.48	−13.7 (−0.65)	−6.2 (−0.30)
2E	−1.07	−25.9 (−1.24)	−18.4 (−0.88)	−1.07	−25.9 (−1.24)	−18.4 (−0.88)
3E	−0.63	−16.8 (−0.80)	−9.3 (−0.45)	−0.53	−14.7 (−0.70)	−7.3 (−0.35)
4E	−0.56	−15.3 (−0.73)	−7.9 (−0.38)	−0.48	−13.7 (−0.65)	−6.2 (−0.30)
5E	—	—	—	0.61	8.9 (0.43)	16.4 (0.78)
6E	—	—	—	−0.43	−12.6 (−0.60)	−5.2 (−0.25)

The distance a is determined as follows:

$$a = \text{lesser of} \begin{cases} 0.1 \times \text{least horizontal dimension} = 0.1 \times 125 = 12.5 \text{ ft (3.81 m)} \\ 0.4h = 0.4 \times 37.5 = 15.0 \text{ ft (4.57 m)} \end{cases}$$

$$\text{Minimum } a = \text{greater of} \begin{cases} 0.04 \times \text{least horizontal dimension} = 0.04 \times 125 = 5.0 \text{ ft (1.52 m)} \\ 3.0 \text{ ft (0.91 m)} \end{cases}$$

Therefore, $a = 12.5$ ft (3.81 m).

The roof pressure coefficients, (GC_{pf}), are negative in zones 2 and 2E (see Table 5.44 or 5.45). According to ASCE/SEI 28.3.2.1 and 28.3.2.2, these pressure coefficients must be applied in zone 2/2E for a distance from the edge of the roof equal to 50 percent

Table 5.47 Design Wind Pressures for Load Cases 3 and 4 on the MWFRS of the Commercial Building in Example 5.9

Surface	Design Wind Pressure, p, lb/ft^2 (kN/m^2)					
	Load Case 3 (E-W Wind)			Load Case 4 (N-S Wind)		
	(GC_{pf})	(GC_{pi})		(GC_{pf})	(GC_{pi})	
		$+(GC_{pi})$	$-(GC_{pi})$		$+(GC_{pi})$	$-(GC_{pi})$
1	0.49	6.4 (0.31)	13.9 (0.66)	−0.45	−13.0 (−0.62)	−5.6 (−0.27)
2	−0.69	−18.0 (−0.86)	−10.6 (−0.51)	−0.69	−18.0 (−0.86)	−10.6 (−0.51)
3	−0.44	−12.8 (−0.61)	−5.4 (−0.26)	−0.37	−11.4 (−0.55)	−3.9 (−0.19)
4	−0.37	−11.4 (−0.55)	−3.9 (−0.19)	−0.45	−13.0 (−0.62)	−5.6 (−0.27)
5	—	—	—	0.40	4.6 (0.22)	12.0 (0.57)
6	—	—	—	−0.29	−9.7 (−0.47)	−2.3 (−0.11)
1E	0.72	11.2 (0.54)	18.6 (0.89)	−0.48	−13.7 (−0.65)	−6.2 (−0.30)
2E	−1.07	−25.9 (−1.24)	−18.4 (−0.88)	−1.07	−25.9 (−1.24)	−18.4 (−0.88)
3E	−0.63	−16.8 (−0.80)	−9.3 (−0.45)	−0.53	−14.7 (−0.70)	−7.3 (−0.35)
4E	−0.56	−15.3 (−0.73)	−7.9 (−0.38)	−0.48	−13.7 (−0.65)	−6.2 (−0.30)
5E	—	—	—	0.61	8.9 (0.43)	16.4 (0.78)
6E	—	—	—	−0.43	−12.6 (−0.60)	−5.2 (−0.25)
1T	0.12	−1.2 (−0.06)	6.2 (0.30)	—	—	—
2T	−0.17	−7.3 (−0.35)	0.2 (0.01)	—	—	—
3T	−0.11	−6.0 (−0.29)	1.5 (0.07)	—	—	—
4T	−0.10	−5.6 (−0.27)	1.9 (0.09)	—	—	—
5T	—	—	—	0.10	−1.7 (−0.08)	5.8 (0.28)
6T	—	—	—	−0.07	−5.2 (−0.25)	2.3 (0.11)

of the horizontal dimension of the building parallel to the direction of the MWFRS being designed or 2.5 times the eave height at the windward wall, whichever is less. The remainder of the zone 2/2E extending to the ridge line must use the pressure coefficients for zone 3/3E.

For this building:

- E-W direction: $0.5 \times 200.0 = 100.0$ ft (30.5 m)
- N-S direction: $0.5 \times 125.0 = 62.5$ ft (19.1 m)
- $2.5 \times \text{eave height} = 2.5 \times 25.0 = 62.5$ ft (19.1 m)

Therefore, in the E-W and N-S directions, zone 2/2E applies over a distance of 62.5 ft (19.1 m) from the edge of the windward roof. In the E-W direction, zone 3/3E applies over a distance of $100.0 - 62.5 = 37.5$ ft (11.4 m) on the windward roof and over the entire leeward roof.

The design pressures are to be applied on the building in accordance with load cases 1 and 2 in ASCE/SEI Figure 28.3-1 and load cases 3 and 4 in ASCE/SEI 28.3-2. The building must be evaluated with each corner taken as the windward corner with all zones loaded simultaneously. Because the building is symmetric about each principal axis, some of the scenarios are repetitive.

Design wind pressures for one basic load scenario for wind in the east to west direction (load case 1) and in the south to north direction (load case 2), including positive and negative internal pressures, are given in Figures 5.39 and 5.40, respectively (in both cases, the windward corner is taken at the southeast corner of the building).

Similarly, design wind pressures for one torsional load scenario for wind in the east to west direction (load case 3) and in the south to north direction (load case 4),

Figure 5.39 Design Wind Pressures for Wind in the East to West Direction on the MWFRS of the Commercial Building in Example 5.9—Load Case 1

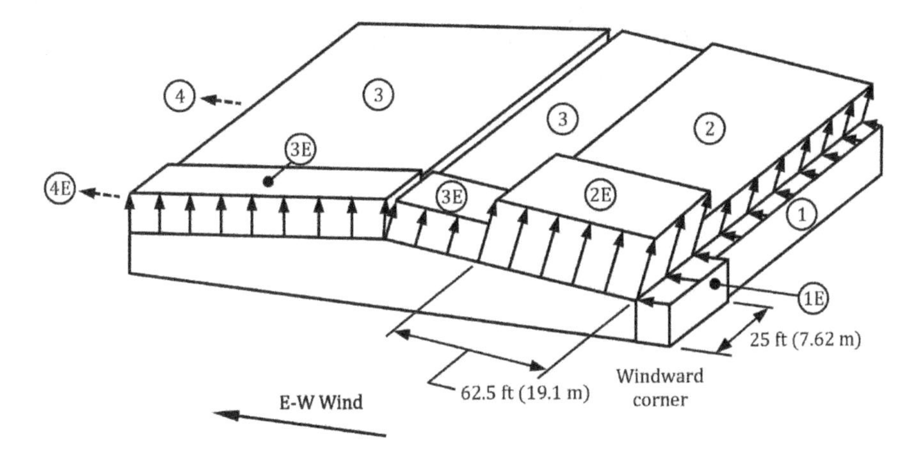

Surface	p, lb/ft² (kN/m²)
1	6.4 (0.31)
2	18.0 (0.86)
3	12.8 (0.61)
4	11.4 (0.55)
1E	11.2 (0.54)
2E	25.9 (1.24)
3E	16.8 (0.80)
4E	15.3 (0.73)

Note: dashed arrows represent uniformly distributed loads over surfaces 4 and 4E

Positive internal pressure

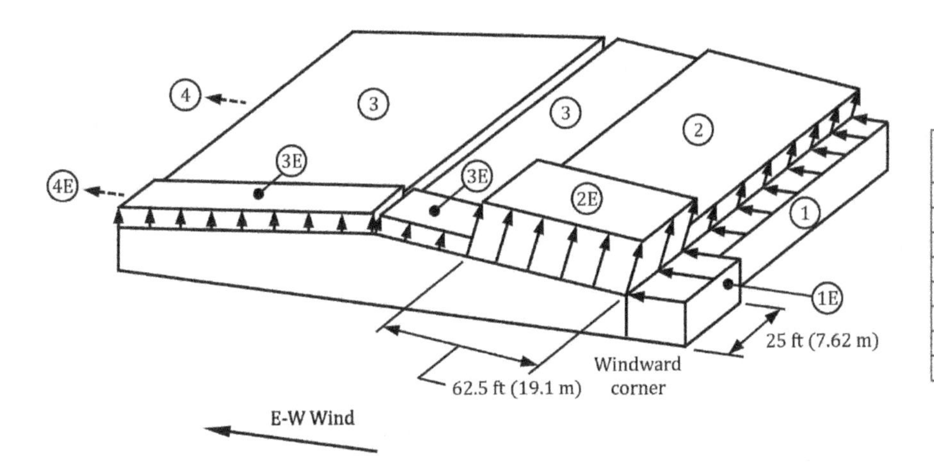

Surface	p, lb/ft² (kN/m²)
1	13.9 (0.66)
2	10.6 (0.51)
3	5.4 (0.26)
4	3.9 (0.19)
1E	18.6 (0.89)
2E	18.4 (1.88)
3E	9.3 (0.45)
4E	7.9 (0.38)

Note: dashed arrows represent uniformly distributed loads over surfaces 4 and 4E

Negative internal pressure

Figure 5.40 Design Wind Pressures for Wind in the South to North Direction on the MWFRS of the Commercial Building in Example 5.9—Load Case 2

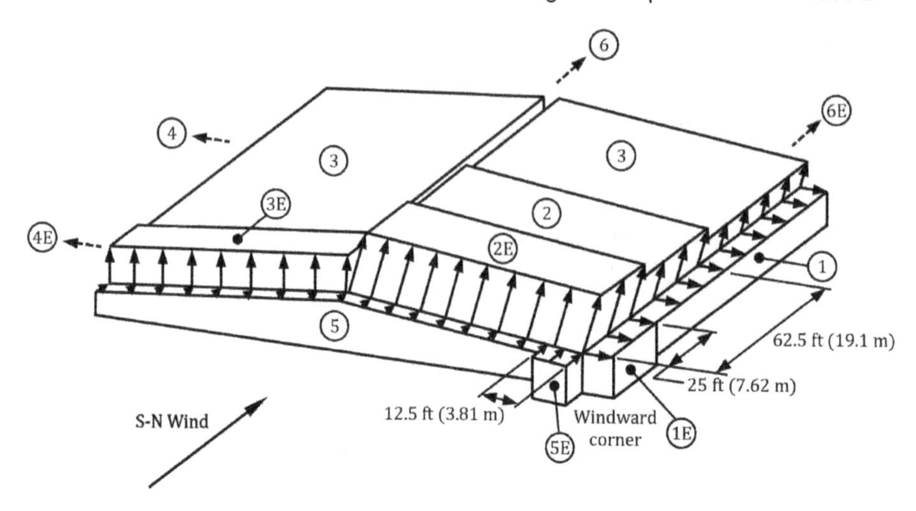

Surface	p, lb/ft^2 (kN/m^2)
1	13.0 (0.62)
2	18.0 (0.86)
3	11.4 (0.55)
4	13.0 (0.62)
5	4.6 (0.22)
6	9.7 (0.47)
1E	13.7 (0.65)
2E	25.9 (1.24)
3E	14.7 (0.70)
4E	13.7 (0.65)
5E	8.9 (0.43)
6E	12.6 (0.60)

Note: dashed arrows represent uniformly distributed loads over surfaces 4, 4E, 6, and 6E

Positive internal pressure

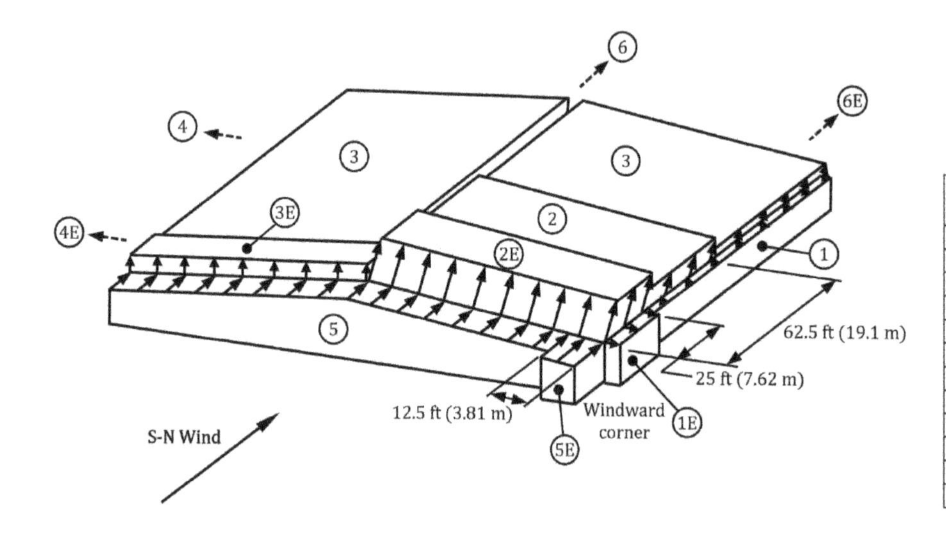

Surface	p, lb/ft^2 (kN/m^2)
1	5.6 (0.27)
2	10.6 (0.51)
3	3.9 (0.19)
4	5.6 (0.27)
5	12.0 (0.57)
6	2.3 (0.11)
1E	6.2 (0.30)
2E	18.4 (0.88)
3E	7.3 (0.35)
4E	6.2 (0.30)
5E	16.4 (0.78)
6E	5.2 (0.25)

Note: dashed arrows represent uniformly distributed loads over surfaces 4, 4E, 6, and 6E

Negative internal pressure

including positive and negative internal pressures, are given in Figures 5.41 and 5.42, respectively (in both cases, the windward corner is taken at the southwest corner of the building).

The minimum design wind loads in ASCE/SEI 28.3.6 must be considered as a load case in addition to the load cases above.

The positive external pressure on the bottom surface of the windward roof overhang is determined in accordance with ASCE/SEI 28.3.5. The velocity pressure, q_h, at the mean roof height is equal to 24.3 lb/ft^2 (1.16 kN/m^2).

Figure 5.41 Design Wind Pressures for Wind in the East to West Direction on the MWFRS of the Commercial Building in Example 5.9—Load Case 3

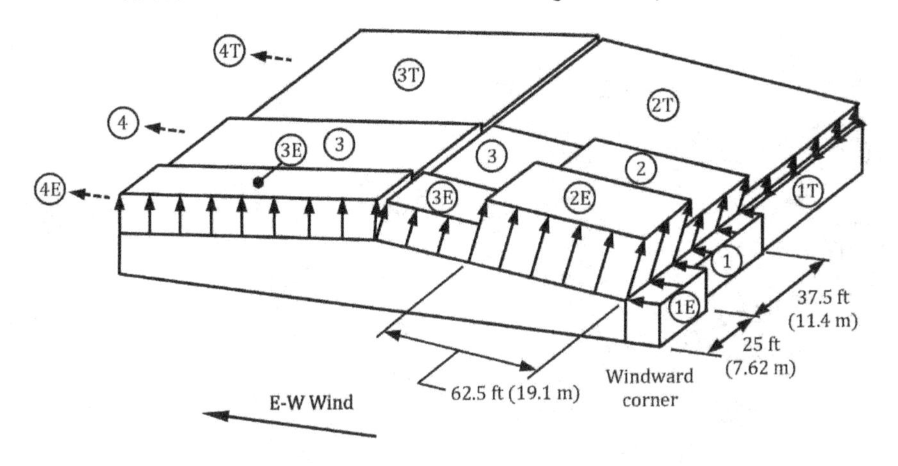

Surface	p, lb/ft^2 (kN/m^2)
1	6.4 (0.31)
2	18.0 (0.86)
3	12.8 (0.61)
4	11.4 (0.55)
1E	11.2 (0.54)
2E	25.9 (1.24)
3E	16.8 (0.80)
4E	15.3 (0.73)
1T	1.2 (0.06)
2T	7.3 (0.35)
3T	6.0 (0.29)
4T	5.8 (0.28)

Note: dashed arrows represent uniformly distributed loads over surfaces 4, 4E, and 4T

Positive internal pressure

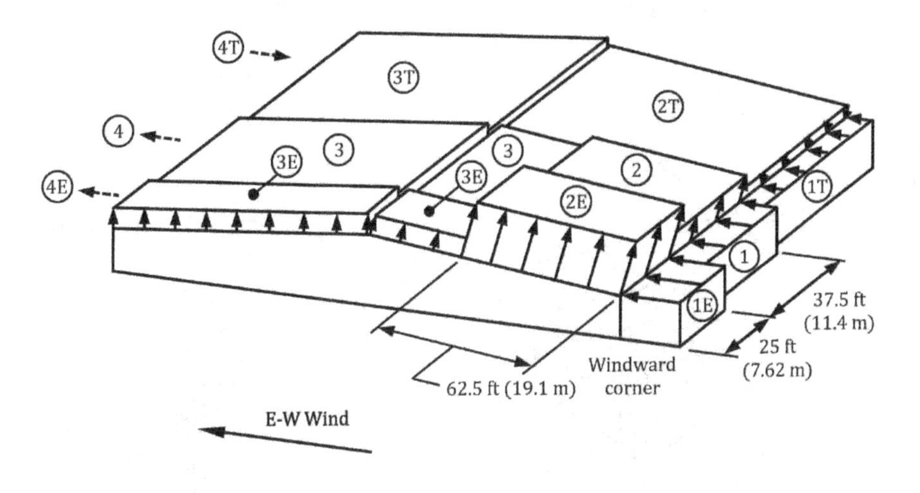

Surface	p, lb/ft^2 (kN/m^2)
1	13.9 (0.66)
2	10.6 (0.51)
3	5.4 (0.26)
4	3.9 (0.19)
1E	18.6 (0.89)
2E	18.4 (0.88)
3E	9.3 (0.45)
4E	7.9 (0.38)
1T	6.2 (0.30)
2T	0.2 (0.01)
3T	1.5 (0.07)
4T	1.7 (0.08)

Note: dashed arrows represent uniformly distributed loads over surfaces 4, 4E, and 4T

Negative internal pressure

Therefore,

$$p = 0.7 q_h K_d = 0.7 \times 24.3 \times 0.85 = 14.5 \text{ lb/ft}^2$$

In S.I.:

$$p = 0.7 q_h K_d = 0.7 \times 1.16 \times 0.85 = 0.69 \text{ kN/m}^2$$

Note: Even though the building is located in a tornado-prone region (see ASCE/SEI Figure 32.1-1), tornado loads need not be considered because the building is assigned to risk category II (ASCE/SEI 32.1.1).

Figure 5.42 Design Wind Pressures for Wind in the South to North Direction on the MWFRS of the Commercial Building in Example 5.9—Load Case 4

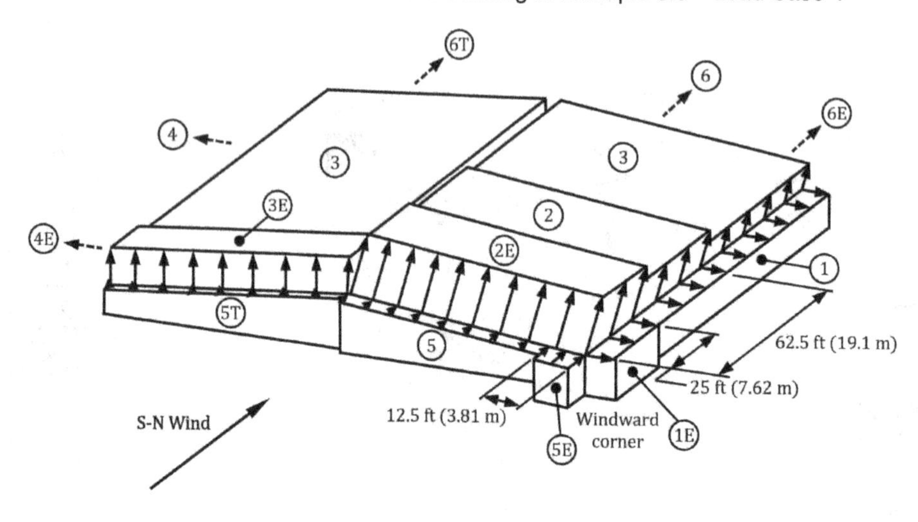

Surface	p, lb/ft^2 (kN/m^2)
1	13.0 (0.62)
2	18.0 (0.86)
3	11.4 (0.55)
4	13.0 (0.62)
5	4.6 (0.22)
6	9.7 (0.47)
1E	13.7 (0.65)
2E	25.9 (1.24)
3E	14.7 (0.70)
4E	13.7 (0.65)
5E	8.9 (0.43)
6E	12.6 (0.60)
5T	1.7 (0.08)
6T	5.2 (0.25)

Note: dashed arrows represent uniformly distributed loads over surfaces 4, 4E, 6, 6E, and 6T

Positive internal pressure

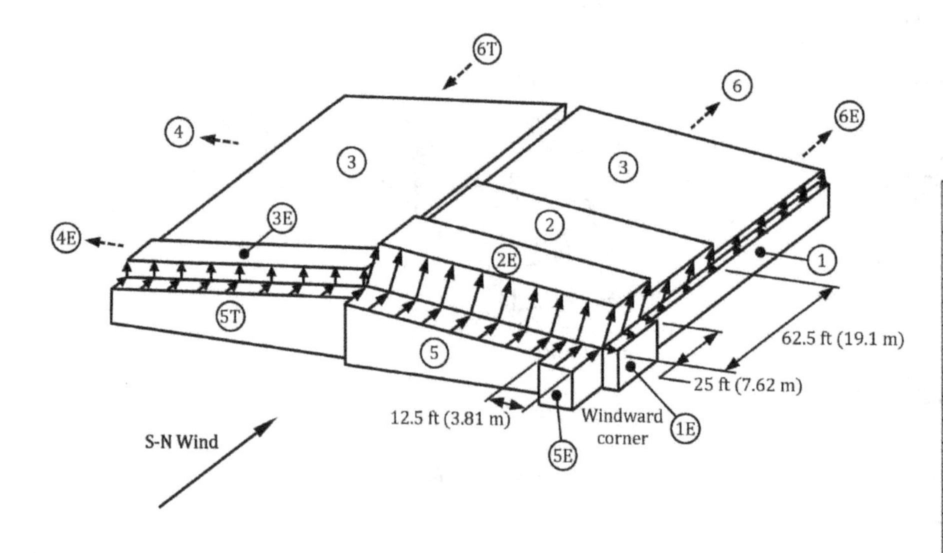

Surface	p, lb/ft^2 (kN/m^2)
1	5.6 (0.27)
2	10.6 (0.51)
3	3.9 (0.19)
4	5.6 (0.27)
5	12.0 (0.57)
6	2.3 (0.11)
1E	6.2 (0.30)
2E	18.4 (0.88)
3E	7.3 (0.35)
4E	6.2 (0.30)
5E	16.4 (0.78)
6E	5.2 (0.25)
5T	5.8 (0.28)
6T	2.3 (0.11)

Note: dashed arrows represent uniformly distributed loads over surfaces 4, 4E, 6, 6E, and 6T

Negative internal pressure

5.8.10 Example 5.10—Determination of Design Wind Pressures on a Residential Building Using ASCE/SEI Chapter 27—MWFRS

Determine the design wind pressures, p, on the MWFRS of the three-story residential building in Figure 5.43 using the provisions in ASCE/SEI Chapter 27 given the following design data:

- Location: Houston, TX (Latitude = 29.74°, Longitude = −95.38°)
- Surface roughness: B
- Topography: Not situated on a hill, ridge or escarpment
- Risk category: II (IBC Table 1604.5)

Figure 5.43 Plan and Elevations of the Three-Story Residential Building in Example 5.10

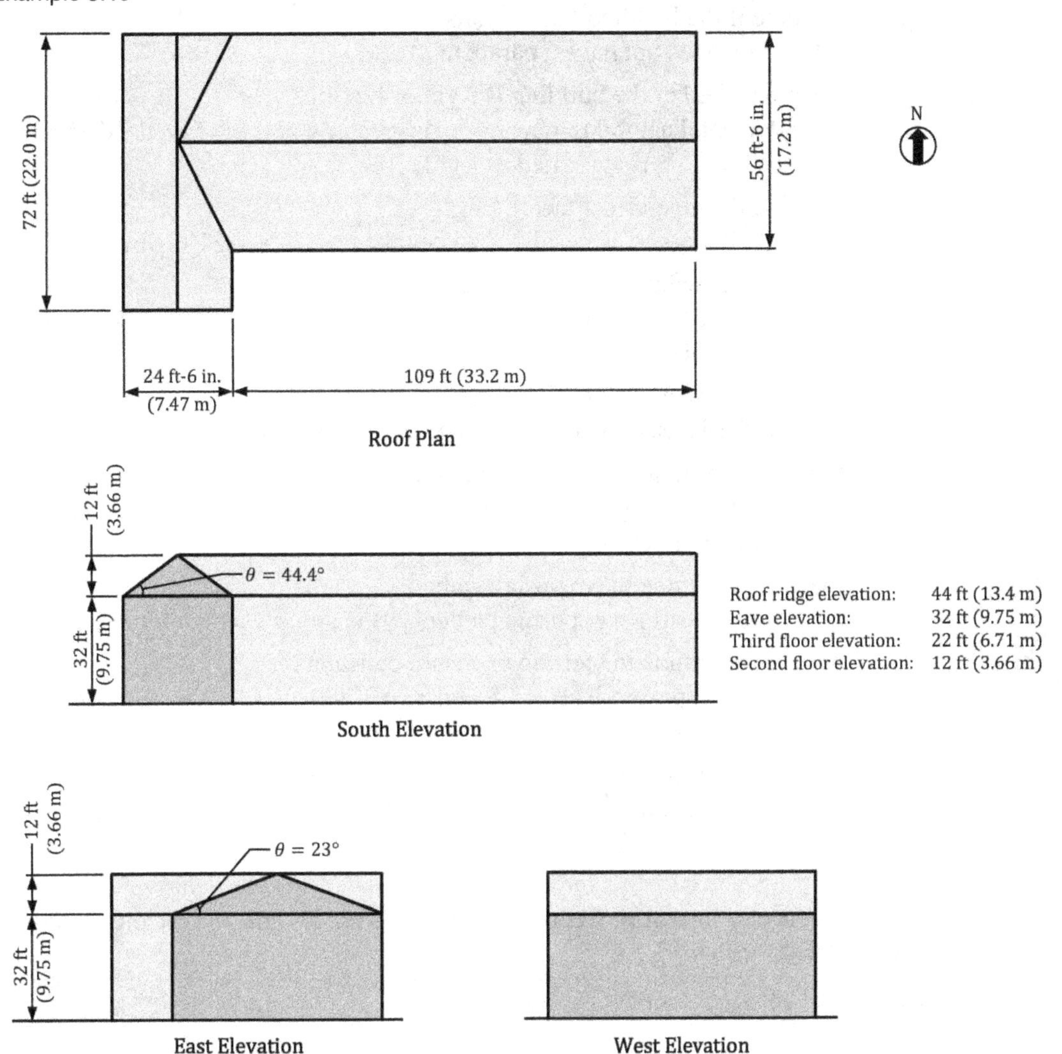

- Enclosure classification: Enclosed
- Fundamental natural frequency, $n_1 > 1$ Hz

Solution

Check if the building meets all the conditions and limitations of ASCE/SEI 27.1.2 and 27.1.3 so that the provisions in ASCE/SEI Chapter 27 can be used to determine the design wind pressures on the MWFRS.

The building is regularly-shaped, that is, it does not have any unusual geometric irregularities in spatial form. Also, the building does not have response characteristics that make it subject to across-wind loading or other similar effects, and it is not sited at a location where channeling effects or buffeting in the wake of upwind obstructions need to be considered.

Therefore, the provisions in ASCE/SEI Chapter 27 are permitted to be used to determine the design wind pressures on the MWFRS.

Flowchart 5.3 is used to determine the design wind pressures.

Note: Even though the building is less than 60 ft (18.3 m) in height, it is not recommended to use the provisions of ASCE/SEI Chapter 28 because L-, T-, and U-shaped buildings are considered to be outside the scope of that method.

Step 1—Determine the enclosure classification of the building
The enclosure classification of the building is given as enclosed.

Step 2—Determine if the building has a parapet
The building does not have a parapet.

Step 3—Determine whether the building is rigid or flexible
The fundamental natural frequency, n_1, is greater than 1 Hz (see the design data), which means the building is classified as rigid.

Step 4—Determine the gust-effect factor, G
For rigid buildings, G is permitted to be taken as 0.85 (ASCE/SEI 26.11.4).

Step 5—Determine the wind directionality factor, K_d
For the MWFRS of a building, the wind directionality factor, K_d, is equal to 0.85 (ASCE/SEI Table 26.6-1).

Step 6—Determine the wind velocity pressures, q_z
Flowchart 5.1 is used to determine the wind velocity pressures.

Step 6a—Determine the surface roughness category
Surface roughness B is given in the design data and it is assumed it is present in all directions.

Step 6b—Determine the exposure category
It is assumed exposure category B is applicable in all directions.

Step 6c—Determine the terrain exposure constants
For exposure B, $\alpha = 7.5$ and $z_g = 3{,}280$ ft (1,000 m) from ASCE/SEI Table 26.11-1.

Step 6d—Determine the velocity pressure exposure coefficient, K_z
Values of K_z are determined from ASCE/SEI 26.10-1 and are given in Table 5.48.

Table 5.48 Velocity Pressure Exposure Coefficients, K_z and K_h, for the Residential Building in Example 5.10

Height above the Ground Level, z, ft (m)	K_z
44 (13.4)	0.76
38 (11.6)	0.73
32 (9.75)	0.70
22 (6.71)	0.64
12 (3.66)	0.57

For example, at the mean roof height $h = \dfrac{44+32}{2} = 38.0$ ft (11.6 m):

$$K_z = 2.41(z/z_g)^{2/\alpha} = 2.41 \times (38.0/3{,}280)^{2/7.5} = 0.73$$

In S.I.:

$$K_z = 2.41 \times (11.6/1{,}000)^{2/7.5} = 0.73$$

Step 6e—Determine the topographic factor, K_{zt}
The building is not located on a hill, ridge, or escarpment, so $K_{zt} = 1.0$ (ASCE/SEI 26.8.2).

Step 6f—Determine the ground elevation factor, K_e
It is permitted to take $K_e = 1.0$ for all elevations (ASCE/SEI 26.9).

Step 6g—Determine the risk category

The risk category is given as II in the design data.

Step 6h—Determine the basic wind speed, V

For risk category II buildings, use IBC Figure 1609.3(2) or ASCE/SEI Figure 26.5-1B (see Table 5.2 of this book).

Equivalently, use Reference 5.4 to obtain V = 134 mi/h (60 m/s) given the latitude and longitude of the site.

Step 6i—Determine the velocity pressures, q_z

The velocity pressures, q_z, over the height of the building are given in Table 5.49. These pressures are determined by ASCE/SEI Equations (26.10-1) and (26.10-1.SI). For example, the velocity pressure at the mean roof height of the building is equal to the following:

$$q_h = 0.00256 K_h K_{zt} K_e V^2 = 0.00256 \times 0.73 \times 1.0 \times 1.0 \times 134^2 = 33.6 \text{ lb/ft}^2$$

In S.I.:

$$q_h = 0.613 K_h K_{zt} K_e V^2 = 0.613 \times 0.73 \times 1.0 \times 1.0 \times 60^2 / 1{,}000 = 1.61 \text{ kN/m}^2$$

Table 5.49 Velocity Pressures, q_z and q_h, for the Residential Building in Example 5.10

Height above the Ground Level, z, ft (m)	K_z	q_z, lb/ft² (kN/m²)
44 (13.4)	0.76	34.9 (1.68)
38 (11.6)	0.73	33.6 (1.61)
32 (9.75)	0.70	32.2 (1.55)
22 (6.71)	0.64	29.4 (1.41)
12 (3.66)	0.57	26.2 (1.26)

Step 7—Determine the external pressure coefficients, C_p, for the walls and roof

Because the building is not symmetric, all four wind directions normal to the walls must be considered.

Identification marks for each surface of the building are given in Figure 5.44.

Figure 5.44 Identification Marks for the Surfaces of the Residential Building in Example 5.10

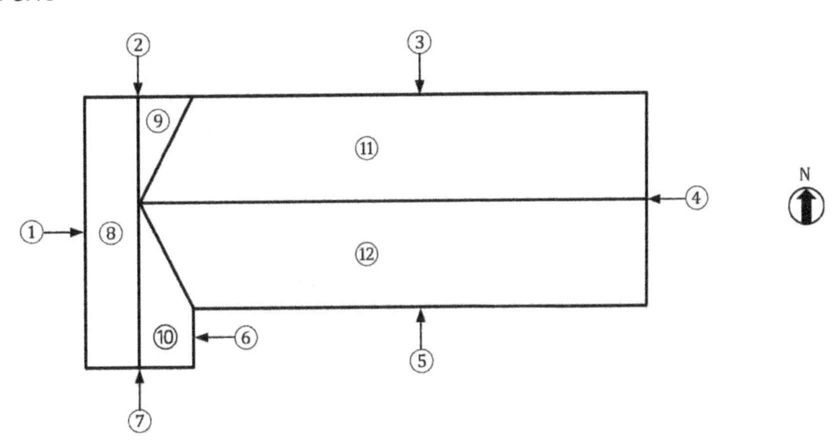

External pressure coefficients for wind in all four directions, which are obtained from ASCE/SEI Figure 27.3-1, are given in Tables 5.50 through 5.53.

Table 5.50 External Pressure Coefficients, C_p, for Wind from West to East

Surface(s)	Type		C_p
1	Windward wall		0.80
2, 3, 5, 7	Side wall		−0.70
4, 6	Leeward wall*		−0.32
8	Windward roof**		0.40
9, 10	Leeward roof**		−0.60
11, 12	Roof parallel to ridge†	Windward edge [located 12.25 ft (3.73 m) from the west wall] to $h = 38$ ft (11.6 m)	−0.90, −0.18
		$h = 38$ ft (11.6 m) to $2h = 76$ ft (23.2 m)	−0.50, −0.18
		$2h = 76$ ft (23.2 m) to the east wall	−0.30, −0.18

*Obtained by linear interpolation using $L/B = 133.5/72 = 1.9$ (In. S.I.: $L/B = 40.7/22.0 = 1.9$).
**Normal to ridge with $\theta = 44$ degrees and $h/L = 38/133.5 = 0.29$ (In S.I.: $h/L = 11.6/40.7 = 0.29$).
†Parallel to ridge for all θ and $h/L < 0.5$. The smaller uplift pressures on the roof due to $C_p = -0.18$ may govern the design when combined with roof live load or snow loads. This pressure is not shown in this example, but in general must be considered.

Table 5.51 External Pressure Coefficients, C_p, for Wind from East to West

Surface(s)	Type		C_p
1	Leeward wall*		−0.32
2, 3, 5, 7	Side wall		−0.70
4, 6	Windward wall		0.80
8	Leeward roof**		−0.60
9, 10	Windward roof**		0.40
11, 12	Roof parallel to ridge†	Windward edge to $h = 38$ ft (11.6 m)	−0.90, −0.18
		$h = 38$ ft (11.6 m) to $2h = 76$ ft (23.2 m)	−0.50, −0.18
		$2h = 76$ ft (23.2 m) to 121.25 ft (37.0 m)	−0.30, −0.18

*Obtained by linear interpolation using $L/B = 133.5/72 = 1.9$ (In. S.I.: $L/B = 40.7/22.0 = 1.9$).
**Normal to ridge with $\theta = 44$ degrees and $h/L = 38/133.5 = 0.29$ (In S.I.: $h/L = 11.6/40.7 = 0.29$).
†Parallel to ridge for all θ and $h/L < 0.5$. The smaller uplift pressures on the roof due to $C_p = -0.18$ may govern the design when combined with roof live load or snow loads. This pressure is not shown in this example, but in general must be considered.

Step 8—Determine the velocity pressure for internal pressure determination, q_i

According to ASCE/SEI 27.3.1, $q_i = q_h = 33.6$ lb/ft² (1.61 kN/m²) for all surfaces of an enclosed building.

Step 9—Determine the internal pressure coefficients, (GC_{pi})

From ASCE/SEI Table 26.13-1, $(GC_{pi}) = +0.18, -0.18$ for an enclosed building.

Step 10—Determine the design wind pressures, p, on all the surfaces of the building

Design wind pressures, p, are determined by ASCE/SEI Equation (27.3-1) for an enclosed building:

$$p = qK_d GC_p - q_i K_d (GC_{pi})$$

- Windward walls:

$$p_z = (q_z \times 0.85 \times 0.85 \times 0.8) - [33.6 \times 0.85 \times (\pm 0.18)] = 0.58 q_z \mp 5.1 \text{ lb/ft}^2$$

Table 5.52 External Pressure Coefficients, C_p, for Wind from North to South

Surface(s)	Type		C_p
1, 4, 6	Side wall		−0.70
2, 3	Windward wall		0.80
5, 7	Leeward wall*		−0.50
8, 9	Roof parallel to ridge**	Windward edge to $h/2 = 19$ ft (5.79 m)	−0.92, −0.18
		$h/2 = 19$ ft (5.79 m) to $h = 38$ ft (11.6 m)	−0.89, −0.18
		$h = 38$ ft (11.6 m) to end	−0.51, −0.18
10, 12	Leeward roof†		−0.60
11	Windward roof†		0.11, −0.35

*Obtained using $L/B = 72/133.5 = 0.54$ (In. S.I.: $L/B = 22.0/40.7 = 0.54$).
**Parallel to ridge for all θ and $h/L = 38/72 = 0.53$ (In S.I.: $h/L = 11.6/22.0 = 0.53$). Values of C_p are obtained by linear interpolation. The smaller uplift pressures on the roof due to $C_p = -0.18$ may govern the design when combined with roof live load or snow loads. This pressure is not shown in this example, but in general must be considered.
†Normal to ridge with $\theta = 23$ degrees and $h/L = 38/72 = 0.53$ (In S.I.: $h/L = 11.6/22.0 = 0.53$). For the windward roof, values of C_p are obtained by double linear interpolation.

Table 5.53 External Pressure Coefficients, C_p, for Wind from South to North

Surface(s)	Type		C_p
1, 4, 6	Side wall		−0.70
2, 3	Leeward wall*		−0.50
5, 7	Windward wall		0.80
8, 10	Roof parallel to ridge**	Windward edge to $h/2 = 19$ ft (5.79 m)	−0.92, −0.18
		$h/2 = 19$ ft (5.79 m) to $h = 38$ ft (11.6 m)	−0.89, −0.18
		$h = 38$ ft (11.6 m) to end	−0.51, −0.18
9, 11	Leeward roof†		−0.60
12	Windward roof†		0.11, −0.35

*Obtained using $L/B = 72/133.5 = 0.54$ (In. S.I.: $L/B = 22.0/40.7 = 0.54$).
**Parallel to ridge for all θ and $h/L = 38/72 = 0.53$ (In S.I.: $h/L = 11.6/22.0 = 0.53$). Values of C_p are obtained by linear interpolation. The smaller uplift pressures on the roof due to $C_p = -0.18$ may govern the design when combined with roof live load or snow loads. This pressure is not shown in this example, but in general must be considered.
†Normal to ridge with $\theta = 23$ degrees and $h/L = 38/72 = 0.53$ (In S.I.: $h/L = 11.6/22.0 = 0.53$). For the windward roof, values of C_p are obtained by double linear interpolation.

In S.I.:

$$p_z = (q_z \times 0.85 \times 0.85 \times 0.8) - [1.61 \times 0.85 \times (\pm 0.18)] = 0.58q_z \mp 0.25 \text{ kN/m}^2$$

- Leeward walls, side walls, windward roof, and leeward roof

$$p_h = (q_h \times 0.85 \times 0.85 \times C_p) - [33.6 \times 0.85 \times (\pm 0.18)] = 24.3C_p \mp 5.1 \text{ lb/ft}^2$$

In S.I.:

$$p_h = (q_h \times 0.85 \times 0.85 \times C_p) - [1.61 \times 0.85 \times (\pm 0.18)] = 1.16C_p \mp 0.25 \text{ kN/m}^2$$

Design wind pressures in all four wind directions are given in Tables 5.54 through 5.57.

The building must be designed for the design wind load cases in ASCE/SEI Figure 27.3-8 (see Figure 5.16). Because this building is not symmetric, all four wind directions must be considered when combining wind loads according to ASCE/SEI Figure 27.3-8.

In case 1, the full design wind pressures act on the projected area perpendicular to each principal axis of the structure. These pressures are assumed to act separately along each principal axis. The wind pressures on the windward and leeward walls given in Tables 5.54 through 5.57 fall under case 1. Full design wind pressures are also applied on the roof.

Assuming none of the requirements in ASCE/SEI D.1 are satisfied, this building must be designed for cases 2 and 4 in addition to cases 1 and 3 (ASCE/SEI 27.3.5). In case 2, 75 percent of the design wind pressures on the windward and leeward walls are applied on the projected area perpendicular to each principal axis of the building along with a torsional moment. Roof pressures are equal to 75 percent of those in case 1. The wind pressures and torsional moment are applied separately for each principal axis.

The wind pressures and torsional moment at 32.0 ft (9.75 m) above ground level for case 2 are determined as follows using net pressures based on negative internal pressure:

For west-to-east wind (see Table 5.54):

$$0.75p_{Wx} = 0.75 \times 23.8 = 17.9 \text{ lb/ft}^2 \text{ (surface 1)}$$

$$0.75p_{Lx} = 0.75 \times 2.7 = 2.0 \text{ lb/ft}^2 \text{ (surfaces 4 and 6)}$$

$$e_x = \pm 0.15B_x = \pm 0.15 \times 72.0 = \pm 10.8 \text{ ft}$$

$$M_z = 0.75(p_{Wx} + p_{Lx})B_x e_x = (17.9 + 2.0) \times 72.0 \times (\pm 10.8) \times 5.0/1{,}000 = \pm 77.4 \text{ ft-kips}$$

Table 5.54 Design Wind Pressures, *p*, for Wind from West to East on the MWFRS of the Residential Building in Example 5.10

Surface(s)	Height above Ground Level, z, ft (m)	q, lb/ft² (kN/m²)	External Pressure $qK_d\,GC_p$, lb/ft² (kN/m²)	Internal Pressure $q_hK_d\,(GC_{pi})$, lb/ft² (kN/m²)	Net Pressure, p, lb/ft² (kN/m²)	
					$+(GC_{pi})$	$-(GC_{pi})$
1	32 (9.75)	32.2 (1.55)	18.7 (0.90)	±5.1 (±0.25)	13.6 (0.65)	23.8 (1.15)
	22 (6.71)	29.4 (1.41)	17.1 (0.82)	±5.1 (±0.25)	12.0 (0.57)	22.2 (1.07)
	12 (3.66)	26.2 (1.26)	15.2 (0.73)	±5.1 (±0.25)	10.1 (0.48)	20.3 (0.98)
2, 3, 5, 7	All	33.6 (1.61)	−17.0 (−0.81)	±5.1 (±0.25)	−22.1 (−1.06)	−11.9 (−0.56)
4, 6	All	33.6 (1.61)	−7.8 (−0.37)	±5.1 (±0.25)	−12.9 (−0.62)	−2.7 (−0.12)
8	38 (11.6)	33.6 (1.61)	9.7 (0.46)	±5.1 (±0.25)	4.6 (0.21)	14.8 (0.71)
9, 10	38 (11.6)	33.6 (1.61)	−14.6 (−0.70)	±5.1 (±0.25)	−19.7 (−0.95)	−9.5 (−0.45)
11, 12	38 (11.6)	33.6 (1.61)	−21.9 (−1.04)*	±5.1 (±0.25)	−27.0 (−1.29)	−16.8 (−0.79)
	38 (11.6)	33.6 (1.61)	−12.2 (−0.58)**	±5.1 (±0.25)	−17.3 (−0.83)	−7.1 (−0.33)
	38 (11.6)	33.6 (1.61)	−7.3 (−0.35)†	±5.1 (±0.25)	−12.4 (−0.60)	−2.2 (−0.10)

*From windward edge of roof [located 12.25 ft (3.73 m) from the west wall] to 38 ft (11.6 m).
**From 38 ft (11.6 m) to 76 ft (23.2 m).
†From 76 ft (23.2 m) to the east wall.

Table 5.55 Design Wind Pressures, *p*, for Wind from East to West on the MWFRS of the Residential Building in Example 5.10

Surface(s)	Height above Ground Level, z, ft (m)	q, lb/ft² (kN/m²)	External Pressure $qK_d\,GC_p$, lb/ft² (kN/m²)	Internal Pressure $q_hK_d\,(GC_{pi})$, lb/ft² (kN/m²)	Net Pressure, p, lb/ft² (kN/m²)	
					$+(GC_{pi})$	$-(GC_{pi})$
1	All	33.6 (1.61)	−7.8 (−0.37)	±5.1 (±0.25)	−12.9 (−0.62)	−2.7 (−0.12)
2, 3, 5, 7	All	33.6 (1.61)	−17.0 (−0.81)	±5.1 (±0.25)	−22.1 (−1.06)	−11.9 (−0.56)
4	44 (13.4)	34.9 (1.68)	20.2 (0.97)	±5.1 (±0.25)	15.1 (0.72)	25.3 (1.22)
	38 (11.6)	33.6 (1.61)	19.5 (0.93)	±5.1 (±0.25)	14.4 (0.68)	24.6 (1.18)
	32 (9.75)	32.2 (1.55)	18.7 (0.90)	±5.1 (±0.25)	13.6 (0.65)	23.8 (1.15)
	22 (6.71)	29.4 (1.41)	17.1 (0.82)	±5.1 (±0.25)	12.0 (0.57)	22.2 (1.07)
	12 (3.66)	26.2 (1.26)	15.2 (0.73)	±5.1 (±0.25)	10.1 (0.48)	20.3 (0.98)
6	32 (9.75)	32.2 (1.55)	18.7 (0.90)	±5.1 (±0.25)	13.6 (0.65)	23.8 (1.15)
	22 (6.71)	29.4 (1.41)	17.1 (0.82)	±5.1 (±0.25)	12.0 (0.57)	22.2 (1.07)
	12 (3.66)	26.2 (1.26)	15.2 (0.73)	±5.1 (±0.25)	10.1 (0.48)	20.3 (0.98)
8	38 (11.6)	33.6 (1.61)	−14.6 (−0.70)	±5.1 (±0.25)	−19.7 (−0.95)	−9.5 (−0.45)
9, 10	38 (11.6)	33.6 (1.61)	9.7 (0.46)	±5.1 (±0.25)	4.6 (0.21)	14.8 (0.71)
11, 12	38 (11.6)	33.6 (1.61)	−21.9 (−1.04)*	±5.1 (±0.25)	−27.0 (−1.29)	−16.8 (−0.79)
	38 (11.6)	33.6 (1.61)	−12.2 (−0.58)**	±5.1 (±0.25)	−17.3 (−0.83)	−7.1 (−0.33)
	38 (11.6)	33.6 (1.61)	−7.3 (−0.35)†	±5.1 (±0.25)	−12.4 (−0.60)	−2.2 (−0.10)

*From windward edge of roof to 38 ft (11.6 m).
**From 38 ft (11.6 m) to 76 ft (23.2 m).
†From 76 ft (23.2 m) to 121.5 ft (37.0 m).

In S.I.:

$$0.75p_{Wx} = 0.75\times 1.15 = 0.86 \text{ kN/m}^2 \text{ (surface 1)}$$

$$0.75p_{Lx} = 0.75\times 0.12 = 0.09 \text{ kN/m}^2 \text{ (surfaces 4 and 6)}$$

$$e_x = \pm 0.15B_x = \pm 0.15\times 22.0 = \pm 3.30 \text{ m}$$

$$M_z = 0.75(p_{Wx}+p_{Lx})B_xe_x = (0.86+0.09)\times 22.0\times(\pm 3.30)\times 1.52 = \pm 104.8 \text{ kN-m}$$

For north-to-south wind (see Table 5.56):

$$0.75p_{Wy} = 0.75\times 23.8 = 17.9 \text{ lb/ft}^2 \text{ (surfaces 2 and 3)}$$

$$0.75p_{Ly} = 0.75\times 7.1 = 5.3 \text{ lb/ft}^2 \text{ (surfaces 5 and 7)}$$

$$e_y = \pm 0.15B_y = \pm 0.15\times 133.5 = \pm 20.0 \text{ ft}$$

$$M_z = 0.75(p_{Wy}+p_{Ly})B_ye_y = (17.9+5.3)\times 133.5\times(\pm 20.0)\times 5.0/1{,}000 = \pm 309.7 \text{ ft-kips}$$

Table 5.56 Design Wind Pressures, *p*, for Wind from North to South on the MWFRS of the Residential Building in Example 5.10

Surface(s)	Height above Ground Level, z, ft (m)	q, lb/ft² (kN/m²)	External Pressure $qK_d GC_p$, lb/ft² (kN/m²)	Internal Pressure $q_h K_d (GC_{pi})$, lb/ft² (kN/m²)	Net Pressure, p, lb/ft² (kN/m²)	
					$+(GC_{pi})$	$-(GC_{pi})$
2	44 (13.4)	34.9 (1.68)	20.2 (0.97)	±5.1 (±0.25)	15.1 (0.72)	25.3 (1.22)
	38 (11.6)	33.6 (1.61)	19.5 (0.93)	±5.1 (±0.25)	14.4 (0.68)	24.6 (1.18)
	32 (9.75)	32.2 (1.55)	18.7 (0.90)	±5.1 (±0.25)	13.6 (0.65)	23.8 (1.15)
	22 (6.71)	29.4 (1.41)	17.1 (0.82)	±5.1 (±0.25)	12.0 (0.57)	22.2 (1.07)
	12 (3.66)	26.2 (1.26)	15.2 (0.73)	±5.1 (±0.25)	10.1 (0.48)	20.3 (0.98)
3	32 (9.75)	32.2 (1.55)	18.7 (0.90)	±5.1 (±0.25)	13.6 (0.65)	23.8 (1.15)
	22 (6.71)	29.4 (1.41)	17.1 (0.82)	±5.1 (±0.25)	12.0 (0.57)	22.2 (1.07)
	12 (3.66)	26.2 (1.26)	15.2 (0.73)	±5.1 (±0.25)	10.1 (0.48)	20.3 (0.98)
5, 7	All	33.6 (1.61)	−12.2 (−0.58)	±5.1 (±0.25)	−17.3 (−0.83)	−7.1 (−0.33)
8, 9	38 (11.6)	33.6 (1.61)	−22.3 (−1.07)*	±5.1 (±0.25)	−27.4 (−1.32)	−17.2 (−0.82)
	38 (11.6)	33.6 (1.61)	−21.6 (−1.04)**	±5.1 (±0.25)	−26.7 (−1.29)	−16.5 (−0.79)
	38 (11.6)	33.6 (1.61)	−12.4 (−0.59)†	±5.1 (±0.25)	−17.5 (−0.84)	−7.3 (−0.34)
10, 12	38 (11.6)	33.6 (1.61)	−14.6 (−0.70)	±5.1 (±0.25)	−19.7 (−0.95)	−9.5 (−0.45)
11	38 (11.6)	33.6 (1.61)	2.7 (0.13)	±5.1 (±0.25)	−2.4 (−0.12)	7.8 (0.38)
			−8.5 (−0.41)	±5.1 (±0.25)	−13.6 (−0.66)	−3.4 (−0.16)

*From windward edge of roof to 19 ft (5.79 m).
**From 19 ft (5.79 m) to 38 ft (11.6 m).
†From 38 ft (11.6 m) to end.

In S.I.:

$$0.75 p_{Wy} = 0.75 \times 1.15 = 0.86 \text{ kN/m}^2 \text{ (surfaces 2 and 3)}$$

$$0.75 p_{Ly} = 0.75 \times 0.33 = 0.25 \text{ kN/m}^2 \text{ (surfaces 5 and 7)}$$

$$e_y = \pm 0.15 B_y = \pm 0.15 \times 40.7 = \pm 6.11 \text{ m}$$

$$M_z = 0.75(p_{Wy} + p_{Ly}) B_y e_y = (0.86 + 0.25) \times 40.7 \times (\pm 6.11) \times 1.52 = \pm 419.6 \text{ kN-m}$$

Similar calculations can be made for east-to-west wind and for south-to-north wind. All four load combinations must be considered for case 2 because the building is not symmetric. Similar calculations can also be made at other elevations.

In case 3, 75 percent of the wind pressures of case 1 are applied to the building simultaneously. This accounts for wind along the diagonal of the building. Like in case 2, four load combinations must be considered for case 3.

In case 4, 75 percent of the wind pressures and torsional moments in case 2 act simultaneously on the building. Like all other cases, four load combinations must be considered for case 4.

The minimum design wind loading in ASCE/SEI 27.1.5 must be considered as a load case in addition to the load cases above.

Table 5.57 Design Wind Pressures, p, for Wind from South to North on the MWFRS of the Residential Building in Example 5.10

Surface(s)	Height above Ground Level, z, ft (m)	q, lb/ft² (kN/m²)	External Pressure $qK_d GC_p$, lb/ft² (kN/m²)	Internal Pressure $q_h K_d (GC_{pi})$, lb/ft² (kN/m²)	Net Pressure, p, lb/ft² (kN/m²)	
					$+(GC_{pi})$	$-(GC_{pi})$
1, 4, 6	All	33.6 (1.61)	−17.0 (−0.81)	±5.1 (±0.25)	−22.1 (−1.06)	−11.9 (−0.56)
2, 3	All	33.6 (1.61)	−12.2 (−0.58)	±5.1 (±0.25)	−17.3 (−0.83)	−7.1 (−0.33)
5	32 (9.75)	32.2 (1.55)	18.7 (0.90)	±5.1 (±0.25)	13.6 (0.65)	23.8 (1.15)
	22 (6.71)	29.4 (1.41)	17.1 (0.82)	±5.1 (±0.25)	12.0 (0.57)	22.2 (1.07)
	12 (3.66)	26.2 (1.26)	15.2 (0.73)	±5.1 (±0.25)	10.1 (0.48)	20.3 (0.98)
7	44 (13.4)	34.9 (1.68)	20.2 (0.97)	±5.1 (±0.25)	15.1 (0.72)	25.3 (1.22)
	38 (11.6)	33.6 (1.61)	19.5 (0.93)	±5.1 (±0.25)	14.4 (0.68)	24.6 (1.18)
	32 (9.75)	32.2 (1.55)	18.7 (0.90)	±5.1 (±0.25)	13.6 (0.65)	23.8 (1.15)
	22 (6.71)	29.4 (1.41)	17.1 (0.82)	±5.1 (±0.25)	12.0 (0.57)	22.2 (1.07)
	12 (3.66)	26.2 (1.26)	15.2 (0.73)	±5.1 (±0.25)	10.1 (0.48)	20.3 (0.98)
8, 10	38 (11.6)	33.6 (1.61)	−22.3 (−1.07)*	±5.1 (±0.25)	−27.4 (−1.32)	−17.2 (−0.82)
	38 (11.6)	33.6 (1.61)	−21.6 (−1.04)**	±5.1 (±0.25)	−26.7 (−1.29)	−16.5 (−0.79)
	38 (11.6)	33.6 (1.61)	−12.4 (−0.59)†	±5.1 (±0.25)	−17.5 (−0.84)	−7.3 (−0.34)
9, 11	38 (11.6)	33.6 (1.61)	−14.6 (−0.70)	±5.1 (±0.25)	−19.7 (−0.95)	−9.5 (−0.45)
12	38 (11.6)	33.6 (1.61)	2.7 (0.13)	±5.1 (±0.25)	−2.4 (−0.12)	7.8 (0.38)
			−8.5 (−0.41)	±5.1 (±0.25)	−13.6 (−0.66)	−3.4 (−0.16)

*From windward edge of roof to 19 ft (5.79 m).
**From 19 ft (5.79 m) to 38 ft (11.6 m).
†From 38 ft (11.6 m) to end.

The four combinations of wind loads that must be considered for case 3 are given in Figure 5.45 at 32.0 ft (9.75 m) above ground level. External pressures on the wall surfaces are given in the figure.

Note: Even though the building is located in a tornado-prone region (see ASCE/SEI Figure 32.1-1), tornado loads need not be considered because the building is assigned to risk category II (ASCE/SEI 32.1.1).

5.8.11 Example 5.11—Determination of Design Wind Pressures on a Hotel Using ASCE/SEI Chapter 27—MWFRS

Determine the design wind pressures, p, on the MWFRS of the six-story hotel in Figure 5.46 using the provisions in ASCE/SEI Chapter 27 given the following design data:

- Location: Miami, FL (Latitude = 25.77°, Longitude = −80.19°)
- Surface roughness: D
- Topography: Not situated on a hill, ridge or escarpment
- Risk category: II (IBC Table 1604.5)
- Enclosure classification: Enclosed
- Structural system: Steel reinforced concrete moment frames in both directions

Figure 5.45 Design Wind Load Cases at 32.0 ft (9.75 m) above Ground Level for Case 3 for the Residential Building in Example 5.10

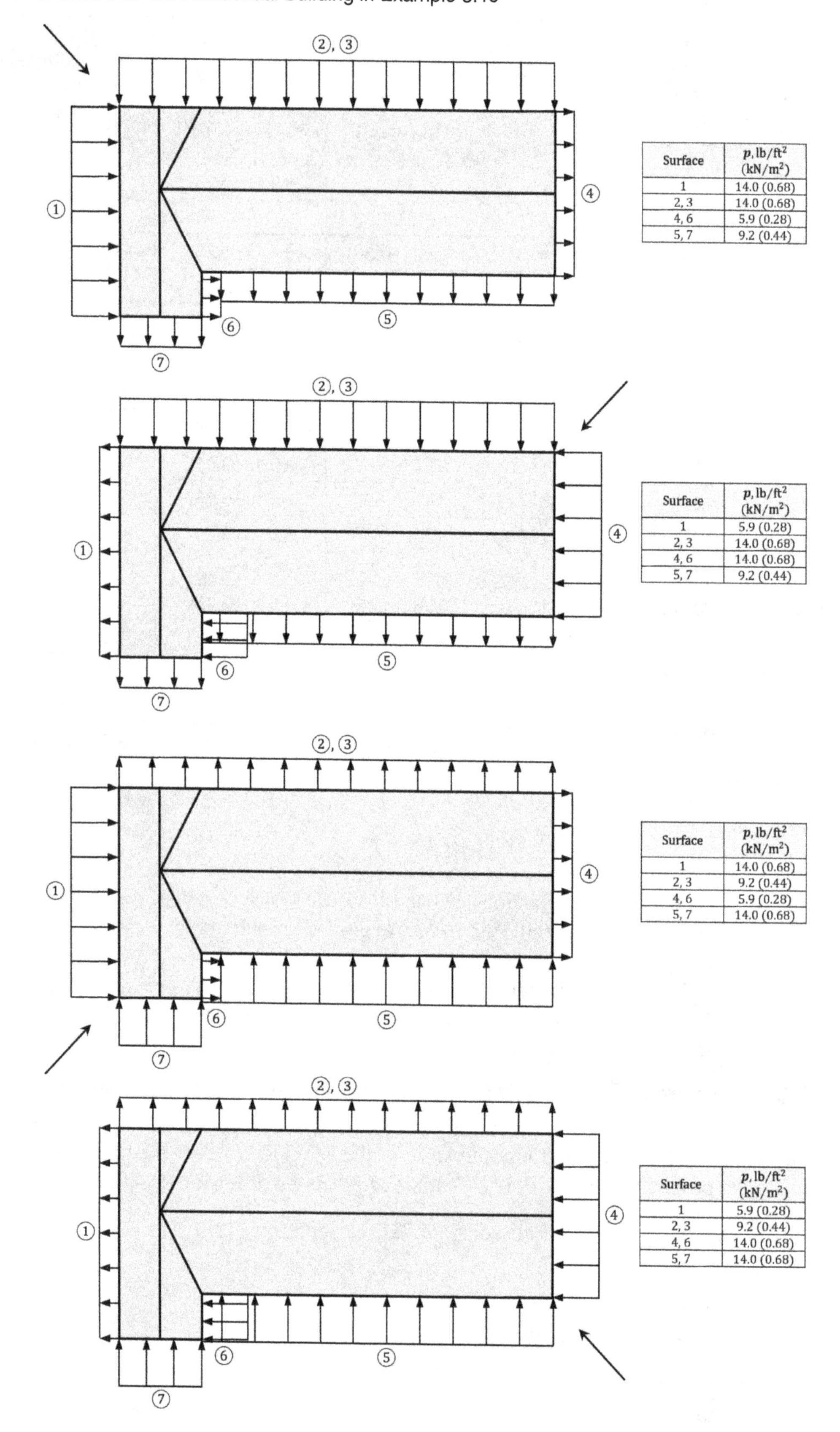

Surface	p, lb/ft^2 (kN/m^2)
1	14.0 (0.68)
2, 3	14.0 (0.68)
4, 6	5.9 (0.28)
5, 7	9.2 (0.44)

Surface	p, lb/ft^2 (kN/m^2)
1	5.9 (0.28)
2, 3	14.0 (0.68)
4, 6	14.0 (0.68)
5, 7	9.2 (0.44)

Surface	p, lb/ft^2 (kN/m^2)
1	14.0 (0.68)
2, 3	9.2 (0.44)
4, 6	5.9 (0.28)
5, 7	14.0 (0.68)

Surface	p, lb/ft^2 (kN/m^2)
1	5.9 (0.28)
2, 3	9.2 (0.44)
4, 6	14.0 (0.68)
5, 7	14.0 (0.68)

Figure 5.46 Plan and Elevation of the Six-Story Hotel in Example 5.11

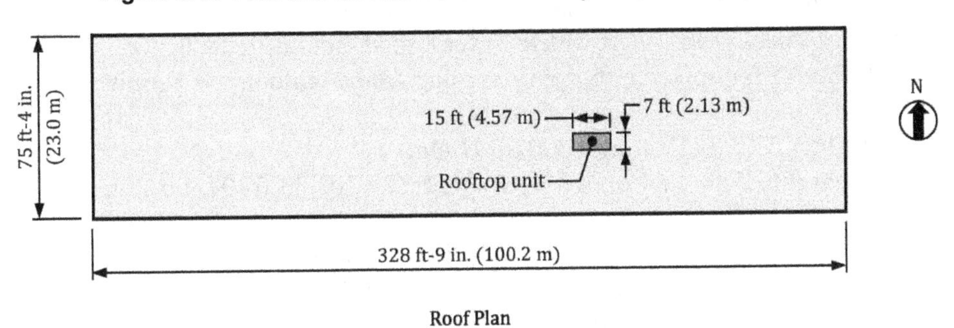

Roof Plan

15 ft (4.57 m)

6 ft (1.83 m)

63 ft-6 in. (19.4 m)

Typical floor height: 10 ft (3.05 m)

Sixth floor height: 13.5 ft (4.12 m)

North/South Elevation

Solution

Check if the building meets all the conditions and limitations of ASCE/SEI 27.1.2 and 27.1.3 so that the provisions in ASCE/SEI Chapter 27 can be used to determine the design wind pressures on the MWFRS.

The building is regularly-shaped, that is, it does not have any unusual geometric irregularities in spatial form. Also, the building does not have response characteristics that make it subject to across-wind loading or other similar effects, and it is not sited at a location where channeling effects or buffeting in the wake of upwind obstructions need to be considered.

Therefore, the provisions in ASCE/SEI Chapter 27 are permitted to be used to determine the design wind pressures on the MWFRS.

Flowchart 5.3 is used to determine the design wind pressures.

Step 1—Determine the enclosure classification of the building

From IBC Figure 1609.3(2), ASCE/SEI Figure 26.5-1B or Reference 5.4, the basic wind speed, V, for this risk category II building is equal to 169 mi/h (76 m/s) given the latitude and longitude of the site.

This building is located in a wind-borne debris region because the basic wind speed is greater than 140 mi/h (63 m/s) [see item 2 in ASCE/SEI 26.12.3.1]. Glazing in buildings located in wind-borne debris regions must be protected with an impact-protective system or must be impact-resistant glazing (ASCE/SEI 26.12.3.2).

It is assumed that impact-resistant glazing is provided over the entire height of the building, which means the building is classified as enclosed.

Step 2—Determine if the building has a parapet

The building does not have a parapet.

Step 3—Determine whether the building is rigid or flexible

In lieu of determining the fundamental natural frequency, n_1, from a dynamic analysis of the building, check if the approximate lower bound natural frequency, n_a, given in ASCE/SEI 26.11.3 for concrete moment-resisting frame buildings is permitted to be used:

Building height = 63.5 ft (19.4 m) < 300 ft (91.4 m)

Building height = 63.5 ft (19.4 m) $< 4L_{eff} = 4 \times 75.33 = 301.33$ ft (91.9 m) in the N-S direction

= 63.5 ft (19.4 m) $< 4L_{eff} = 4 \times 328.75 = 1{,}315.0$ ft (400.8 m) in the E-W direction

Because both of these limitations are satisfied, ASCE/SEI Equations (26.11-3) and (26.11-3.SI) are permitted to be used to determine n_a in both directions:

$$n_a = 43.5/h^{0.9} = 43.5/63.5^{0.9} = 1.04 \text{ Hz} > 1 \text{ Hz}$$

In S.I.:

$$n_a = 14.93/h^{0.9} = 14.93/19.4^{0.9} = 1.04 \text{ Hz} > 1 \text{ Hz}$$

Because $n_a > 1$ Hz, the building is defined as rigid.

Step 4—Determine the gust-effect factor, G

For rigid buildings, G is permitted to be taken as 0.85 (ASCE/SEI 26.11.4).

Step 5—Determine the wind directionality factor, K_d

For the MWFRS of a building, the wind directionality factor, K_d, is equal to 0.85 (ASCE/SEI Table 26.6-1).

Step 6—Determine the wind velocity pressures, q_z

Flowchart 5.1 is used to determine the wind velocity pressures.

Step 6a—Determine the surface roughness category

Surface roughness D is given in the design data and it is assumed it is present in all directions.

Step 6b—Determine the exposure category

It is assumed exposure category D is applicable in all directions.

Step 6c—Determine the terrain exposure constants

For exposure D, $\alpha = 11.5$ and $z_g = 1{,}935$ ft (590 m) from ASCE/SEI Table 26.11-1.

Step 6d—Determine the velocity pressure exposure coefficient, K_z

Values of K_z are determined from ASCE/SEI Table 26.10-1 and are given in Table 5.58 (the equations in footnote 1 of ASCE/SEI Table 26.10-1 are used to determine K_z).

For example, at z = 63.5 ft (19.4 m):

$$K_z = 2.41(z/z_g)^{2/\alpha} = 2.41 \times (63.5/1{,}935)^{2/11.5} = 1.33$$

In S.I.:

$$K_z = 2.41 \times (19.4/590)^{2/11.5} = 1.33$$

Step 6e—Determine the topographic factor, K_{zt}

The building is not located on a hill, ridge, or escarpment, so $K_{zt} = 1.0$ (ASCE/SEI 26.8.2).

Table 5.58 Velocity Pressure Exposure Coefficients, K_z and K_h, for the Hotel in Example 5.11

Height above the Ground Level, z, ft (m)	K_z
63.5 (19.4)	1.33
60.0 (18.3)	1.32
50.0 (15.2)	1.28
40.0 (12.2)	1.23
30.0 (9.14)	1.17
25.0 (7.62)	1.13
20.0 (6.10)	1.09
15.0 (4.57)	1.04

Step 6f—Determine the ground elevation factor, K_e

It is permitted to take $K_e = 1.0$ for all elevations (ASCE/SEI 26.9).

Step 6g—Determine the risk category

The risk category is given as II in the design data.

Step 6h—Determine the basic wind speed, V

For risk category II buildings, use IBC Figure 1609.3(2) or ASCE/SEI Figure 26.5-1B (see Table 5.2 of this book).

Equivalently, use Reference 5.4 to obtain $V = 169$ mi/h (76 m/s) given the latitude and longitude of the site.

Step 6i—Determine the velocity pressures, q_z

The velocity pressures, q_z, over the height of the building are given in Table 5.59. These pressures are determined by ASCE/SEI Equations (26.10-1) and (26.10-1.SI). For example, the velocity pressure at the mean roof height of the building is equal to the following:

$$q_h = 0.00256 K_h K_{zt} K_e V^2 = 0.00256 \times 1.33 \times 1.0 \times 1.0 \times 169^2 = 97.3 \text{ lb/ft}^2$$

In S.I.:

$$q_h = 0.613 K_h K_{zt} K_e V^2 = 0.613 \times 1.33 \times 1.0 \times 1.0 \times 76^2 / 1{,}000 = 4.71 \text{ kN/m}^2$$

Table 5.59 Velocity Pressures, q_z and q_h, for the Hotel in Example 5.11

Height above the Ground Level, z, ft (m)	K_z	q_z, lb/ft² (kN/m²)
63.5 (19.4)	1.33	97.3 (4.71)
60.0 (18.3)	1.32	96.5 (4.67)
50.0 (15.2)	1.28	93.6 (4.53)
40.0 (12.2)	1.23	89.9 (4.36)
30.0 (9.14)	1.17	85.6 (4.14)
25.0 (7.62)	1.13	82.6 (4.00)
20.0 (6.10)	1.09	79.7 (3.86)
15.0 (4.57)	1.04	76.0 (3.68)

Step 7—Determine the external pressure coefficients, C_p, for the walls and roof

The external pressure coefficients are determined from ASCE/SEI Figure 27.3-1.

For wind in the N-S direction:

- Windward wall: $C_p = 0.8$
- Leeward wall: For $L/B = 75.33/328.75 = 0.23$, $C_p = -0.5$ (In S.I.: $L/B = 23.0/100.2 = 0.23$)
- Side walls: $C_p = -0.7$
- Roof: Normal to the ridge with $\theta < 10$ degrees and parallel to ridge for all θ with $h/L = 63.5/75.33 = 0.84$ (In S.I.: $h/L = 19.4/23.0 = 0.84$)

1. From the windward edge of the roof to $h/2 = 31.75$ ft (9.68 m):
   It is permitted to obtain C_p for $h/L = 0.84$ by linear interpolation using the values of C_p for $h/L \le 0.5$ and $h/L \ge 1.0$ (see note 2 in ASCE/SEI Figure 27.3-1).

   For $h/L \le 0.5$, $C_p = -0.9, -0.18$
   For $h/L \ge 1.0$, $C_p = -1.3, -0.18$

   In accordance with footnote b in ASCE/SEI Figure 27.3-1, it is permitted to linearly reduce $C_p = -1.3$ based on the area over which it is applicable, which in this case is equal to $31.75 \times 328.75 = 10{,}438$ ft² (971.9 m²). Because the area is greater than 1,000 ft² (92.9 m²), $C_p = 0.8 \times (-1.3) = -1.04$.

   $$\text{For } h/L = 0.84,\ C_p = -0.9 + \frac{[(-1.04) - (-0.9)] \times (0.84 - 0.5)}{1.0 - 0.5} = -1.0 \text{ and } -0.18$$

2. From $h/2 = 31.75$ ft (9.68 m) to $h = 63.5$ ft (19.4 m):

   For $h/L \le 0.5$, $C_p = -0.9, -0.18$
   For $h/L \ge 1.0$, $C_p = -0.7, -0.18$

   Footnote b in ASCE/SEI Figure 27.3-1 is not applicable in this horizontal distance range.

   $$\text{For } h/L = 0.84,\ C_p = -0.7 + \frac{[(-0.9) - (-0.7)] \times (1.0 - 0.84)}{1.0 - 0.5} = -0.76 \text{ and } -0.18$$

3. From $h = 63.5$ ft (19.4 m) to 75.33 ft (23.0 m):

   For $h/L \le 0.5$, $C_p = -0.5, -0.18$
   For $h/L \ge 1.0$, $C_p = -0.7, -0.18$

   Footnote b in ASCE/SEI Figure 27.3-1 is not applicable in this horizontal distance range.

   $$\text{For } h/L = 0.84,\ C_p = -0.5 + \frac{[(-0.7) - (-0.5)] \times (0.84 - 0.5)}{1.0 - 0.5} = -0.64 \text{ and } -0.18$$

For wind in the E-W direction:

- Windward wall: $C_p = 0.8$
- Leeward wall: For $L/B = 328.75/75.33 = 4.4$, $C_p = -0.2$ (In S.I.: $L/B = 100.2/23.0 = 4.4$)
- Side walls: $C_p = -0.7$
- Roof: Normal to the ridge with $\theta < 10$ degrees and parallel to ridge for all θ with $h/L = 63.5/328.75 = 0.19$ (In S.I.: $h/L = 19.4/100.2 = 0.19$)

  From the windward edge of the roof to $h = 63.5$ ft (19.4 m): $C_p = -0.9, -0.18$
  From $h = 63.5$ ft (19.4 m) to $2h = 127.0$ ft (38.8 m): $C_p = -0.5, -0.18$
  From $2h = 127.0$ ft (38.8 m) to 328.75 ft (100.2 m): $C_p = -0.3, -0.18$

Step 8—Determine the velocity pressure for internal pressure determination, q_i

According to ASCE/SEI 27.3.1, $q_i = q_h = 97.3$ lb/ft^2 (4.71 kN/m^2) for all surfaces of an enclosed building.

Step 9—Determine the internal pressure coefficient, (GC_{pi})

From ASCE/SEI Table 26.13-1, $(GC_{pi}) = +0.18, -0.18$ for an enclosed building.

Step 10—Determine the design wind pressures, p, on all the surfaces of the building

Design wind pressures, p, are determined by ASCE/SEI Equation (27.3-1) for an enclosed building:

$$p = qK_dGC_p - q_iK_d(GC_{pi})$$

For wind in the N-S direction:

- Windward wall:

$$p_z = (q_z \times 0.85 \times 0.85 \times 0.8) - [97.3 \times 0.85 \times (\pm 0.18)] = 0.58q_z \mp 14.9 \text{ lb/ft}^2$$

In S.I.:

$$p_z = (q_z \times 0.85 \times 0.85 \times 0.8) - [4.71 \times 0.85 \times (\pm 0.18)] = 0.58q_z \mp 0.72 \text{ kN/m}^2$$

- Leeward wall

$$p_h = [q_h \times 0.85 \times 0.85 \times (-0.5)] - [97.3 \times 0.85 \times (\pm 0.18)]$$

$$= -35.2 \mp 14.9 = -50.1 \text{ lb/ft}^2, -20.3 \text{ lb/ft}^2$$

In S.I.:

$$p_h = [q_h \times 0.85 \times 0.85 \times (-0.5)] - [4.71 \times 0.85 \times (\pm 0.18)]$$

$$= -1.70 \mp 0.72 = -2.42 \text{ kN/m}^2, -0.98 \text{ kN/m}^2$$

- Side walls

$$p_h = [q_h \times 0.85 \times 0.85 \times (-0.7)] - [97.3 \times 0.85 \times (\pm 0.18)]$$

$$= -49.2 \mp 14.9 = -64.1 \text{ lb/ft}^2, -34.3 \text{ lb/ft}^2$$

In S.I.:

$$p_h = [q_h \times 0.85 \times 0.85 \times (-0.7)] - [4.71 \times 0.85 \times (\pm 0.18)]$$

$$= -2.38 \mp 0.72 = -3.10 \text{ kN/m}^2, -1.66 \text{ kN/m}^2$$

- Roof

$$p_h = (q_h \times 0.85 \times 0.85 \times C_p) - [97.3 \times 0.85 \times (\pm 0.18)] = 70.3C_p \mp 14.9 \text{ lb/ft}^2$$

In S.I.:

$$p_h = (q_h \times 0.85 \times 0.85 \times C_p) - [4.71 \times 0.85 \times (\pm 0.18)] = 3.40C_p \mp 0.72 \text{ kN/m}^2$$

The roof pressure coefficient $C_p = -0.18$ may become critical where wind loads are combined with roof live loads or snow loads. Determination of wind pressures based on this pressure coefficient should be performed, but such calculations are not shown in this example.

Design wind pressures in the N-S direction are given in Table 5.60 and the external design wind pressures are given in Figure 5.47.

Table 5.60 Design Wind Pressures, *p*, in the N-S Direction on the MWFRS of the Hotel in Example 5.11

Building Surface	Height above Ground Level, z, ft (m)	q, lb/ft² (kN/m²)	External Pressure $qK_d\,GC_p$, lb/ft² (kN/m²)	Internal Pressure $q_hK_d\,(GC_{pi})$, lb/ft² (kN/m²)	Net Pressure, p, lb/ft² (kN/m²)	
					$+(GC_{pi})$	$-(GC_{pi})$
Windward wall	63.5 (19.4)	97.3 (4.71)	56.2 (2.72)	±14.9 (±0.72)	41.3 (2.00)	71.1 (3.44)
	60.0 (18.3)	96.5 (4.67)	55.8 (2.70)	±14.9 (±0.72)	40.9 (1.98)	70.7 (3.42)
	50.0 (15.2)	93.6 (4.53)	54.1 (2.62)	±14.9 (±0.72)	39.2 (1.90)	69.0 (3.34)
	40.0 (12.2)	89.9 (4.36)	52.0 (2.52)	±14.9 (±0.72)	37.1 (1.80)	66.9 (3.24)
	30.0 (9.14)	85.6 (4.14)	49.5 (2.39)	±14.9 (±0.72)	34.6 (1.67)	64.4 (3.11)
	25.0 (7.62)	82.6 (4.00)	47.7 (2.31)	±14.9 (±0.72)	32.8 (1.59)	62.6 (3.03)
	20.0 (6.10)	79.7 (3.86)	46.1 (2.23)	±14.9 (±0.72)	31.2 (1.51)	61.0 (2.95)
	15.0 (4.57)	76.0 (3.68)	43.9 (2.13)	±14.9 (±0.72)	29.0 (1.41)	58.8 (2.85)
Leeward wall	All	97.3 (4.71)	−35.2 (−1.70)	±14.9 (±0.72)	−50.1 (−2.42)	−20.3 (−0.98)
Side walls	All	97.3 (4.71)	−49.2 (−2.38)	±14.9 (±0.72)	−64.1 (−3.10)	−34.3 (−1.66)
Roof	—(1)	97.3 (4.71)	−70.3 (−3.40)	±14.9 (±0.72)	−85.2 (−4.12)	−55.4 (−2.68)
	—(2)	97.3 (4.71)	−53.4 (−2.59)	±14.9 (±0.72)	−68.3 (−3.31)	−38.5 (−1.87)
	—(3)	97.3 (4.71)	−45.0 (−2.18)	±14.9 (±0.72)	−59.9 (−2.90)	−30.1 (−1.46)

*From windward edge of roof to 31.75 ft (9.68 m).
**From 31.75 ft (9.68 m) to 63.5 ft (19.4 m).
†From 63.5 ft (19.4 m) to 75.33 ft (23.0 m).

Figure 5.47 External Design Wind Pressures in the N-S Direction on the MWFRS of the Hotel in Example 5.11

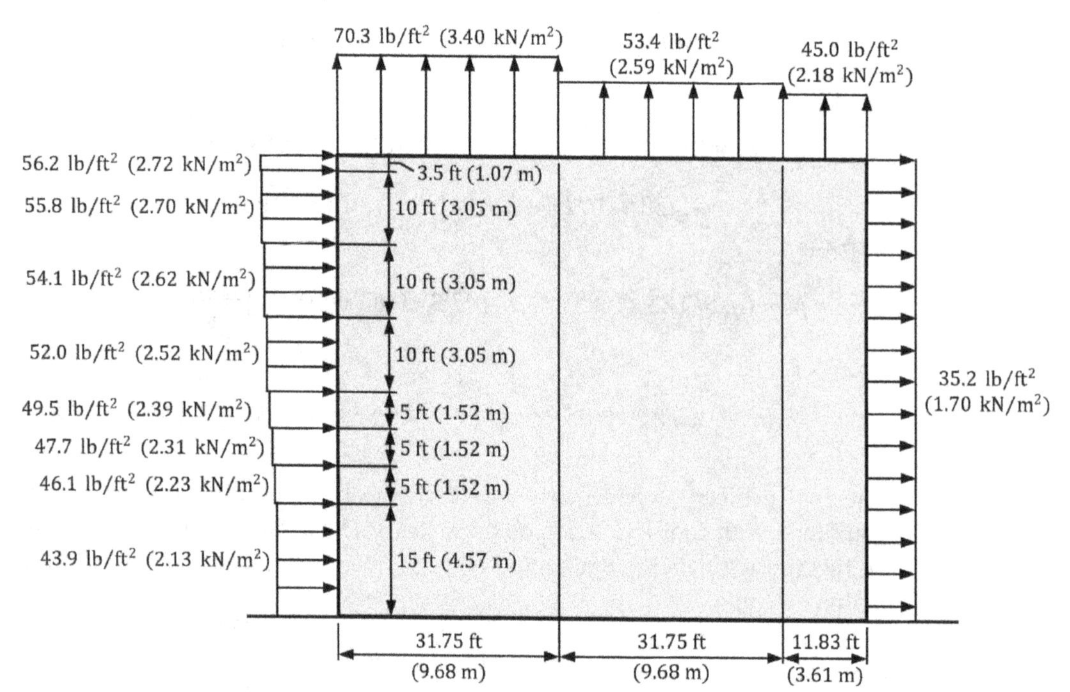

For wind in the E-W direction:

- Windward wall:

$$p_z = (q_z \times 0.85 \times 0.85 \times 0.8) - [97.3 \times 0.85 \times (\pm 0.18)] = 0.58q_z \mp 14.9 \text{ lb/ft}^2$$

In S.I.:

$$p_z = (q_z \times 0.85 \times 0.85 \times 0.8) - [4.71 \times 0.85 \times (\pm 0.18)] = 0.58q_z \mp 0.72 \text{ kN/m}^2$$

- Leeward wall

$$p_h = [q_h \times 0.85 \times 0.85 \times (-0.2)] - [97.3 \times 0.85 \times (\pm 0.18)]$$

$$= -14.1 \mp 14.9 = -29.0 \text{ lb/ft}^2, \ 0.8 \text{ lb/ft}^2$$

In S.I.:

$$p_h = [q_h \times 0.85 \times 0.85 \times (-0.2)] - [4.71 \times 0.85 \times (\pm 0.18)]$$

$$= -0.68 \mp 0.72 = -1.40 \text{ kN/m}^2, \ 0.04 \text{ kN/m}^2$$

- Side walls

$$p_h = [q_h \times 0.85 \times 0.85 \times (-0.7)] - [97.3 \times 0.85 \times (\pm 0.18)]$$

$$= -49.2 \mp 14.9 = -64.1 \text{ lb/ft}^2, \ -34.3 \text{ lb/ft}^2$$

In S.I.:

$$p_h = [q_h \times 0.85 \times 0.85 \times (-0.7)] - [4.71 \times 0.85 \times (\pm 0.18)]$$

$$= -2.38 \mp 0.72 = -3.10 \text{ kN/m}^2, \ -1.66 \text{ kN/m}^2$$

- Roof

$$p_h = (q_h \times 0.85 \times 0.85 \times C_p) - [97.3 \times 0.85 \times (\pm 0.18)] = 70.3C_p \mp 14.9 \text{ lb/ft}^2$$

In S.I.:

$$p_h = (q_h \times 0.85 \times 0.85 \times C_p) - [1.27 \times 0.85 \times (\pm 0.18)] = 3.40C_p \mp 0.72 \text{ kN/m}^2$$

The roof pressure coefficient $C_p = -0.18$ may become critical where wind loads are combined with roof live loads or snow loads. Determination of wind pressures based on this pressure coefficient should be performed, but such calculations are not shown in this example.

Design wind pressures in the E-W direction are given in Table 5.61 and the external design wind pressures are given in Figure 5.48.

When considering horizontal wind forces on the MWFRS, the effects from the internal pressure cancel out. On the roof, the effects from internal pressure add directly to those from the external pressure.

The building must be designed for the design wind load cases in ASCE/SEI Figure 27.3-8 (see Figure 5.16).

In case 1, the full design wind pressures act on the projected area perpendicular to each principal axis of the building at each level above ground and on the roof. These pressures act separately along each principal axis. The external wind pressures on the windward and leeward walls given in Tables 5.60 and 5.61 fall under Case 1. Full design wind pressures are also applied on the roof.

Assuming none of the requirements in ASCE/SEI D.1 are satisfied, this building must be designed for cases 2 and 4 in addition to cases 1 and 3 (ASCE/SEI 27.3.5). In case 2, 75 percent of the design wind pressures on the windward and leeward walls are

Table 5.61 Design Wind Pressures, p, in the E-W Direction on the MWFRS of the Hotel in Example 5.11

Building Surface	Height above Ground Level, z, ft (m)	q, lb/ft² (kN/m²)	External Pressure $qK_d\,GC_p$, lb/ft² (kN/m²)	Internal Pressure $q_hK_d\,(GC_{pi})$, lb/ft² (kN/m²)	Net Pressure, p, lb/ft² (kN/m²)	
					$+(GC_{pi})$	$-(GC_{pi})$
Windward wall	63.5 (19.4)	97.3 (4.71)	56.2 (2.72)	±14.9 (±0.72)	41.3 (2.00)	71.1 (3.44)
	60.0 (18.3)	96.5 (4.67)	55.8 (2.70)	±14.9 (±0.72)	40.9 (1.98)	70.7 (3.42)
	50.0 (15.2)	93.6 (4.53)	54.1 (2.62)	±14.9 (±0.72)	39.2 (1.90)	69.0 (3.34)
	40.0 (12.2)	89.9 (4.36)	52.0 (2.52)	±14.9 (±0.72)	37.1 (1.80)	66.9 (3.24)
	30.0 (9.14)	85.6 (4.14)	49.5 (2.39)	±14.9 (±0.72)	34.6 (1.67)	64.4 (3.11)
	25.0 (7.62)	82.6 (4.00)	47.7 (2.31)	±14.9 (±0.72)	32.8 (1.59)	62.6 (3.03)
	20.0 (6.10)	79.7 (3.86)	46.1 (2.23)	±14.9 (±0.72)	31.2 (1.51)	61.0 (2.95)
	15.0 (4.57)	76.0 (3.68)	43.9 (2.13)	±14.9 (±0.72)	29.0 (1.41)	58.8 (2.85)
Leeward wall	All	97.3 (4.71)	−14.1 (−0.68)	±14.9 (±0.72)	−29.0 (−1.40)	0.8 (0.04)
Side walls	All	97.3 (4.71)	−49.2 (−2.38)	±14.9 (±0.72)	−64.1 (−3.10)	−34.3 (−1.66)
Roof	—*	97.3 (4.71)	−63.3 (−3.06)	±14.9 (±0.72)	−78.2 (−3.78)	−48.4 (−2.34)
	—**	97.3 (4.71)	−35.2 (−1.70)	±14.9 (±0.72)	−50.1 (−2.42)	−20.3 (−0.98)
	—†	97.3 (4.71)	−21.1 (−1.02)	±14.9 (±0.72)	−36.0 (−1.74)	−6.2 (−0.30)

*From windward edge of roof to 63.5 ft (19.4 m).
**From 63.5 ft (19.4 m) to 127.0 ft (38.7 m).
†From 127.0 ft (38.7 m) to 328.75 ft (100.2 m).

Figure 5.48 External Design Wind Pressures, p, in the E-W Direction on the MWFRS of the Hotel in Example 5.11

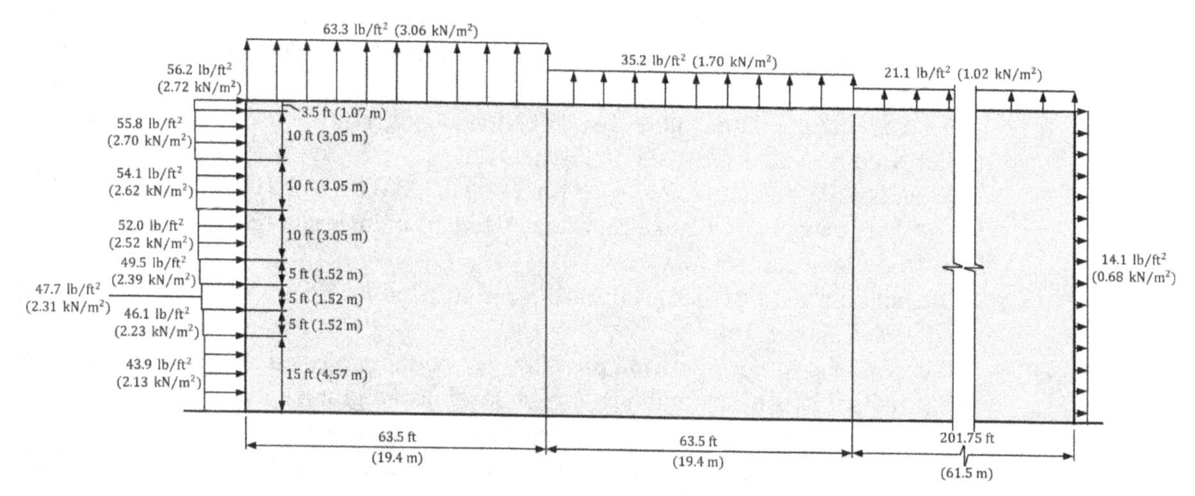

applied on the projected area perpendicular to each principal axis of the building along with a torsional moment. Roof pressures are equal to 75 percent of those in case 1. The wind pressures and torsional moment are applied separately for each principal axis.

The wind pressures and torsional moment at the mean roof height are determined as follows using external pressures:

For N-S wind (see Table 5.60):

$$0.75p_{Wy} = 0.75 \times 56.2 = 42.2 \text{ lb/ft}^2 \text{ (windward wall)}$$

$$0.75p_{Ly} = 0.75 \times 35.2 = 26.4 \text{ lb/ft}^2 \text{ (leeward wall)}$$

$$e_y = \pm 0.15B_y = \pm 0.15 \times 328.75 = \pm 49.3 \text{ ft}$$

$$M_z = 0.75(p_{Wy} + p_{Ly})B_y e_y = (42.2 + 26.4) \times 328.75 \times (\pm 49.3) \times 3.5/1{,}000$$

$$= \pm 3{,}891.4 \text{ ft-kips}$$

In S.I.:

$$0.75p_{Wy} = 0.75 \times 2.72 = 2.04 \text{ kN/m}^2 \text{ (windward wall)}$$

$$0.75p_{Ly} = 0.75 \times 1.70 = 1.28 \text{ kN/m}^2 \text{ (leeward wall)}$$

$$e_y = \pm 0.15B_y = \pm 0.15 \times 100.2 = \pm 15.0 \text{ m}$$

$$M_z = 0.75(p_{Wy} + p_{Ly})B_y e_y = (2.04 + 1.28) \times 100.2 \times (\pm 15.0) \times 1.07 = \pm 5{,}339.3 \text{ kN-m}$$

For E-W wind (see Table 5.61):

$$0.75p_{Wx} = 0.75 \times 56.2 = 42.2 \text{ lb/ft}^2 \text{ (windward wall)}$$

$$0.75p_{Lx} = 0.75 \times 14.1 = 10.6 \text{ lb/ft}^2 \text{ (leeward wall)}$$

$$e_x = \pm 0.15B_x = \pm 0.15 \times 75.33 = \pm 11.3 \text{ ft}$$

$$M_z = 0.75(p_{Wx} + p_{Lx})B_x e_x = (42.2 + 10.6) \times 73.33 \times (\pm 11.3) \times 3.5/1{,}000 = \pm 157.3 \text{ ft-kips}$$

In S.I.:

$$0.75p_{Wx} = 0.75 \times 2.72 = 2.04 \text{ kN/m}^2 \text{ (windward wall)}$$

$$0.75p_{Lx} = 0.75 \times 0.68 = 0.51 \text{ kN/m}^2 \text{ (leeward wall)}$$

$$e_x = \pm 0.15B_x = \pm 0.15 \times 23.0 = \pm 3.45 \text{ m}$$

$$M_z = 0.75(p_{Wx} + p_{Lx})B_x e_x = (2.04 + 0.51) \times 23.0 \times (\pm 3.45) \times 1.07 = \pm 216.5 \text{ kN-m}$$

In case 3, 75 percent of the wind pressures of case 1 are applied to the building simultaneously. This accounts for wind along the diagonal of the building.

In case 4, 75 percent of the wind pressures and torsional moments in case 2 act simultaneously on the building.

The minimum design wind loading in ASCE/SEI 27.1.5 must be considered as a load case in addition to the load cases above.

Load cases 1 through 4 for wind pressures acting on the projected area at the mean roof height are given in Figure 5.49. Similar loading diagrams can be obtained below the mean roof height.

Note: Even though the building is located in a tornado-prone region (see ASCE/SEI Figure 32.1-1), tornado loads need not be considered because the building is assigned to risk category II (ASCE/SEI 32.1.1).

Figure 5.49 Design Wind Load Cases at the Mean Roof Height of the Hotel in Example 5.11

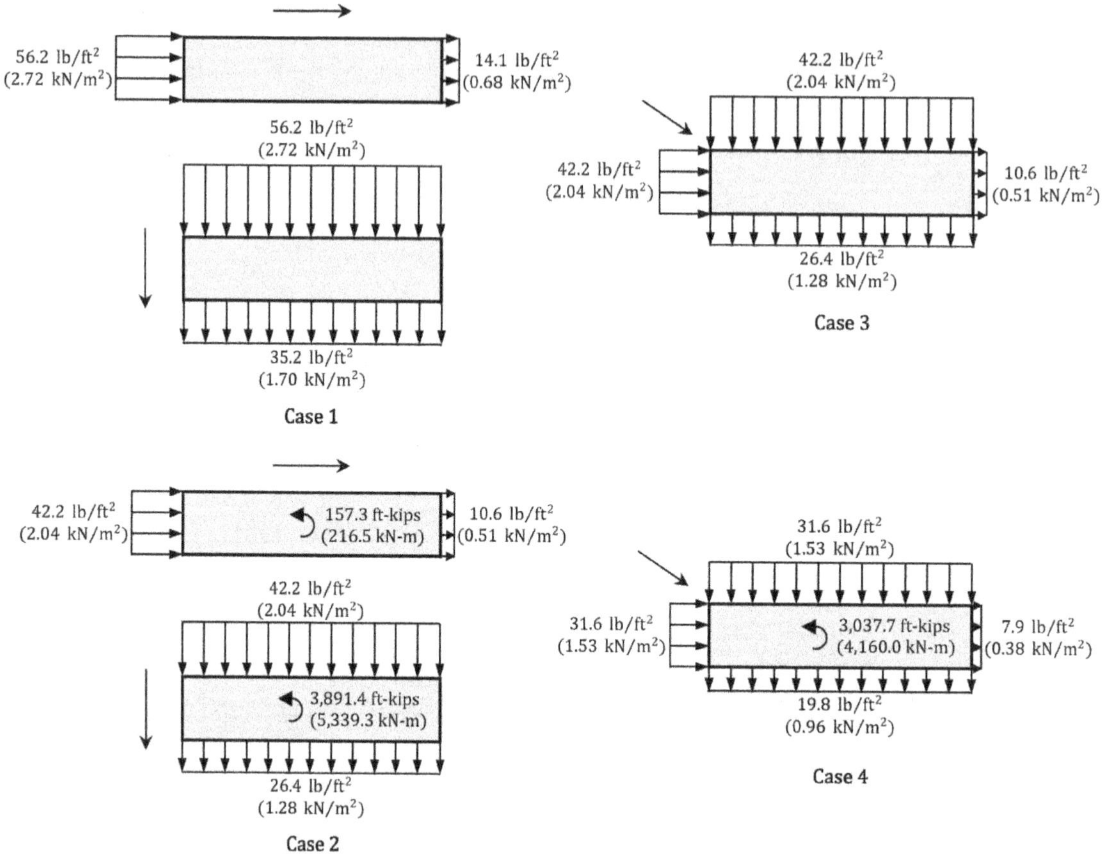

5.8.12 Example 5.12—Determination of Design Wind Pressures on an Office Building Using ASCE/SEI Chapter 27—MWFRS

Determine the design wind pressures, p, on the MWFRS of the fifteen-story office building in Figure 5.50 using the provisions in ASCE/SEI Chapter 27 given the following design data:

- Basic wind speed: $V = 105$ mi/h (47 m/s)
- Surface roughness: B
- Topography: Not situated on a hill, ridge, or escarpment
- Risk category: II (IBC Table 1604.5)
- Enclosure classification: Enclosed
- Structural system: Combination of steel reinforced concrete shear walls and moment frames

SOLUTION

Check if the building meets all the conditions and limitations of ASCE/SEI 27.1.2 and 27.1.3 so that the provisions in ASCE/SEI Chapter 27 are permitted to be used to determine the design wind pressures on the MWFRS.

The building is regularly-shaped, that is, it does not have any unusual geometric irregularities in spatial form. Also, the building does not have response characteristics that make it subject to across-wind loading or other similar effects, and it is not sited at a location where channeling effects or buffeting in the wake of upwind obstructions need to be considered.

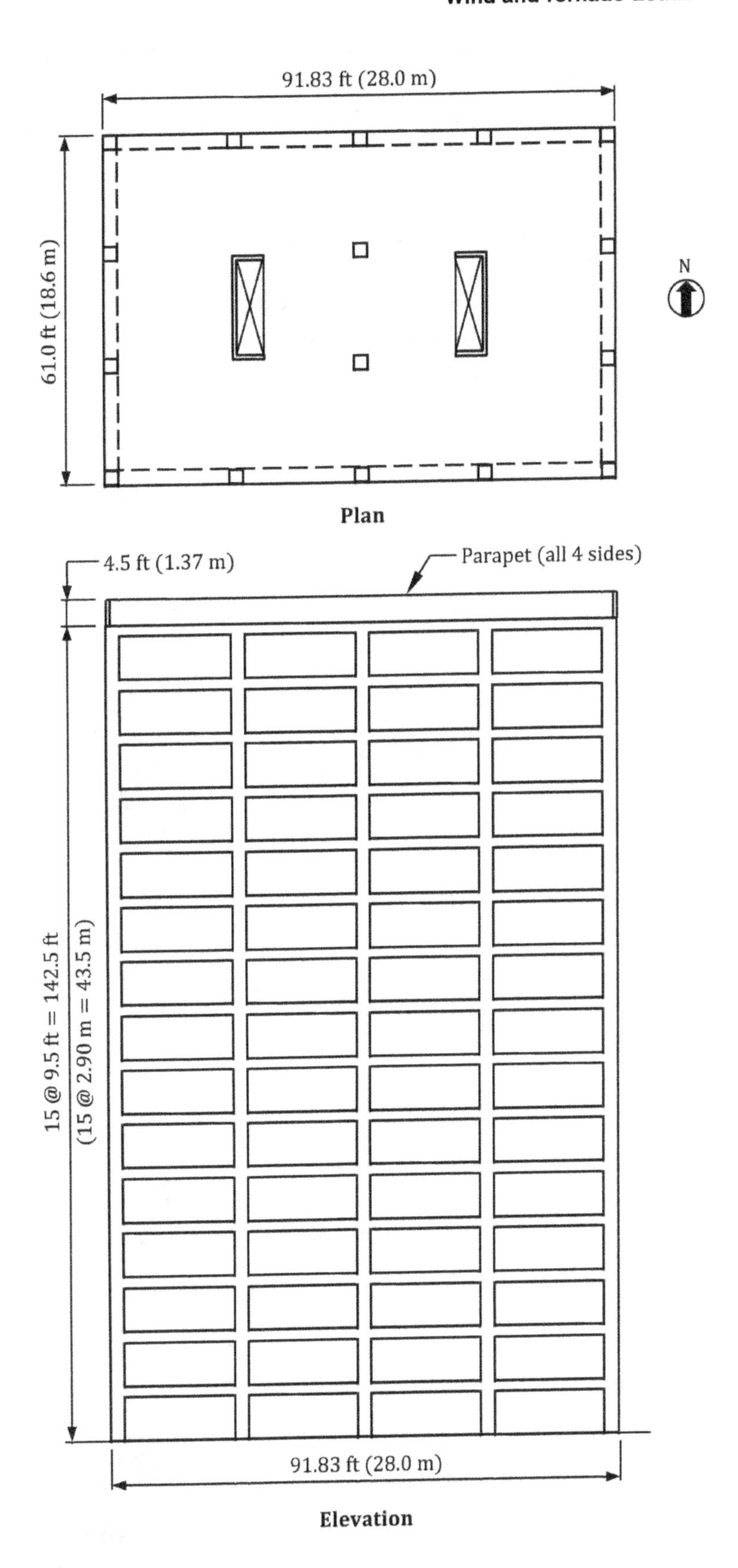

Figure 5.50 Plan and Elevation of the Fifteen-story Office Building in Example 5.12

Therefore, the provisions in ASCE/SEI Chapter 27 are permitted to be used to determine the design wind pressures on the MWFRS.

Flowchart 5.3 is used to determine the design wind pressures.

Step 1—Determine the enclosure classification of the building

The enclosure classification of the building is given as enclosed.

Step 2—Determine if the building has a parapet

The building does have a parapet and the combined net design pressure, p_p, is determined in accordance with ASCE/SEI 27.3.4.

Step 2a—Determine the velocity pressure, q_p, at the top of the parapet by ASCE/SEI Equations (26.10-1) and (26.10-1.SI) [see Flowchart 5.1]

$$q_p = 0.00256 K_z K_{zt} K_e V^2 \text{ (lb/ft}^2\text{)}$$

$$q_p = 0.613 K_z K_{zt} K_e V^2 \text{ (kN/m}^2\text{)}$$

From ASCE/SEI Table 26.10-1 and ASCE/SEI Table 26.11-1 for exposure B:

$$K_z = 2.41(z/z_g)^{2/\alpha} = 2.41 \times (147.0/3{,}280)^{2/7.5} = 1.05$$

In S.I.: $$K_z = 2.41 \times (44.8/1{,}000)^{2/7.5} = 1.05$$

The building is not located on a hill, ridge, or escarpment, so $K_{zt} = 1.0$ (ASCE/SEI 26.8.2).

It is permitted to take $K_e = 1.0$ for all elevations (ASCE/SEI 26.9).

From the design data, $V = 105$ mi/h (47 m/s).

Therefore,

$$q_p = 0.00256 \times 1.05 \times 1.0 \times 1.0 \times 105^2 = 29.6 \text{ lb/ft}^2$$

In S.I.: $$q_p = 0.613 \times 1.05 \times 1.0 \times 1.0 \times 47^2/1{,}000 = 1.42 \text{ kN/m}^2$$

Step 2b—Determine the wind directionality factor, K_d, from ASCE/SEI Table 26.6-1

For building structures, $K_d = 0.85$.

Step 2c—Determine the combined net pressure coefficients, (GC_{pn})

Windward parapet: $(GC_{pn}) = +1.5$

Leeward parapet: $(GC_{pn}) = -1.0$

Step 2d—Determine the combined net pressure, p_p, from ASCE/SEI Equation (27.3-3)

$$p_p = q_p K_d (GC_{pn})$$

Windward parapet: $p_p = 29.6 \times 0.85 \times 1.5 = 37.7$ lb/ft^2

Leeward parapet: $p_p = 29.6 \times 0.85 \times (-1.0) = -25.2$ lb/ft^2

In S.I.:

Windward parapet: $p_p = 1.42 \times 0.85 \times 1.5 = 1.81$ kN/m^2

Leeward parapet: $p_p = 1.42 \times 0.85 \times (-1.0) = -1.21$ kN/m^2

Step 3—Determine whether the building is rigid or flexible

In lieu of determining the fundamental natural frequency, n_1, from a dynamic analysis of the building, check if the approximate lower bound natural frequency, n_a, given in ASCE/SEI 26.11.3 is permitted to be used:

Building height = 142.5 ft (43.5 m) < 300 ft (91.4 m)
Building height = 142.5 ft (43.5 m) < $4L_{eff} = 4 \times 61.0 = 244.0$ ft (74.4 m) in the N-S direction
= 142.5 ft (43.5 m) < $4L_{eff} = 4 \times 91.83 = 367.3$ ft (112.0 m) in the E-W direction

Because both of these limitations are satisfied, ASCE/SEI Equations (26.11-4) and (26.11-4.SI) are permitted to be used to determine n_a in both directions for the MWFRS consisting of shear walls and moment frames.

$$n_a = 75/h = 75/142.5 = 0.53 \text{ Hz} < 1.0 \text{ Hz}$$

In S.I.:

$$n_a = 22.86/h = 22.86/43.5 = 0.53 \text{ Hz} < 1.0 \text{ Hz}$$

Therefore, the building is flexible in both directions.

Step 4—Determine the gust-effect factor, G_f
For flexible buildings, G_f is determined by ASCE/SEI Equation (26.11-10).
Calculations for G_f in both principal directions are given in Table 5.62.

Step 5—Determine the wind directionality factor, K_d
For the MWFRS of a building, the wind directionality factor, K_d, is equal to 0.85 (ASCE/SEI Table 26.6-1).

Step 6—Determine the wind velocity pressures, q_z
Flowchart 5.1 is used to determine the wind velocity pressures.

Step 6a—Determine the surface roughness category
Surface roughness B is given in the design data and it is assumed it is present in all directions.

Step 6b—Determine the exposure category
It is assumed exposure category B is applicable in all directions.

Step 6c—Determine the terrain exposure constants
For exposure B, $\alpha = 7.5$ and $z_g = 3{,}280$ ft (1,000 m) from ASCE/SEI Table 26.11-1.

Step 6d—Determine the velocity pressure exposure coefficient, K_z
Values of K_z are determined from ASCE/SEI 26.10-1 and are given in Table 5.63 based on the height of the floor levels above ground.

For example, at z = 142.5 ft (43.5 m):

$$K_z = 2.41(z/z_g)^{2/\alpha} = 2.41 \times (142.5/3{,}280)^{2/7.5} = 1.04$$

In S.I.:

$$K_z = 2.41 \times (43.5/1{,}000)^{2/7.5} = 1.04$$

Step 6e—Determine the topographic factor, K_{zt}
The building is not located on a hill, ridge or escarpment, so $K_{zt} = 1.0$ (ASCE/SEI 26.8.2).

Step 6f—Determine the ground elevation factor, K_e
It is permitted to take $K_e = 1.0$ for all elevations (ASCE/SEI 26.9).

Step 6g—Determine the risk category
The risk category is given as II in the design data.

Step 6h—Determine the basic wind speed, V
From the design data, V = 105 mi/h (47 m/s).

Table 5.62 Determination of Gust-Effect Factor, G_f, for the Building in Example 3.6

Calculations	ASCE/SEI Reference
$g_Q = g_v = 3.4$	26.11.5
$g_R = \sqrt{2\ \ln(3{,}600n_1)} + [0.577/\sqrt{2\ \ln(3{,}600n_1)}] = 4.04$	Eq. (26.11-11)
$z_{min} = 30.0$ ft (9.14 m)	Table 26.11-1
$\bar{z} = \text{greater of} \begin{cases} 0.6h = 85.5 \text{ ft (26.1 m)} \\ z_{min} = 30.0 \text{ ft (9.14 m)} \end{cases}$	26.11.4
$\ell = 320.0$ ft (97.5 m)	Table 26.11-1
$c = 0.30$	Table 26.11-1
$\bar{\varepsilon} = 1/3.0$	Table 26.11-1
$I_{\bar{z}} = c(33/\bar{z})^{1/6} = 0.26$ [In S.I.: $I_{\bar{z}} = c(10/\bar{z})^{1/6} = 0.26$]	Eq. (26.11-7)
$L_{\bar{z}} = \ell(\bar{z}/33)^{\bar{\varepsilon}} = 439.5$ ft [In S.I.: $L_{\bar{z}} = \ell(\bar{z}/10)^{\bar{\varepsilon}} = 134.2$ m]	Eq. (26.11-9)
$Q = \sqrt{\dfrac{1}{1+0.63\left(\dfrac{B+h}{L_{\bar{z}}}\right)^{0.63}}} = \begin{cases} 0.84 \text{ in the N-S direction with } B = 91.83 \text{ ft (28.0 m)} \\ 0.85 \text{ in the E-W direction with } B = 61.0 \text{ ft (18.6 m)} \end{cases}$	Eq. (26.11-8)
Damping ratio $\beta = 0.015$ for concrete buildings	C26.11
$\bar{b} = 0.47$	Table 26.11-1
$\bar{\alpha} = 1/4.5$	Table 26.11-1
$\bar{V}_{\bar{z}} = \bar{b}\left(\dfrac{\bar{z}}{33}\right)^{\bar{\alpha}}\left(\dfrac{88}{60}\right)V = 89.4$ ft/s [In S.I.: $\bar{V}_{\bar{z}} = \bar{b}\left(\dfrac{\bar{z}}{10}\right)^{\bar{\alpha}} V = 27.3$ m/s]	Eq. (26.11-16)
$N_1 = n_1 L_{\bar{z}}/\bar{V}_{\bar{z}} = 2.61$	Eq. (26.11-14)
$R_n = \dfrac{7.47N_1}{(1+10.3N_1)^{5/3}} = 0.08$	Eq. (26.11-13)
$\eta_h = 4.6n_1 h/\bar{V}_{\bar{z}} = 3.89$	Eq. (26.11-15b)
$R_h = \dfrac{1}{\eta_h} - \dfrac{1}{2\eta_h^2}(1-e^{-2\eta_h}) = 0.22$	Eq. (26.11-15a)
$\eta_B = 4.6n_1 B/\bar{V}_{\bar{z}} = \begin{cases} 2.50 \text{ in the N-S direction with } B = 91.83 \text{ ft (28.0 m)} \\ 1.67 \text{ in the E-W direction with } B = 61.0 \text{ ft (18.6 m)} \end{cases}$	Eq. (26.11-15b)

(continued)

Table 5.62 Determination of Gust-Effect Factor, G_f, for the Building in Example 3.6 (*Continued*)

Calculations	ASCE/SEI Reference
$R_B = \frac{1}{\eta_B} - \frac{1}{2\eta_B^2}(1-e^{-2\eta_B}) = \begin{cases} 0.32 \text{ in the N-S direction} \\ 0.43 \text{ in the E-W direction} \end{cases}$	Eq. (26.11-15a)
$\eta_L = 15.4 n_1 L/\bar{V}_{\bar{z}} = \begin{cases} 5.57 \text{ in the N-S direction with } L = 61.0 \text{ ft (18.6 m)} \\ 8.38 \text{ in the E-W direction with } L = 91.83 \text{ ft (28.0 m)} \end{cases}$	Eq. (26.11-15b)
$R_L = \frac{1}{\eta_L} - \frac{1}{2\eta_L^2}(1-e^{-2\eta_L}) = \begin{cases} 0.16 \text{ in the N-S direction} \\ 0.11 \text{ in the E-W direction} \end{cases}$	Eq. (26.11-15a)
$R = \sqrt{R_n R_h R_B (0.53 + 0.47 R_L)/\beta} = \begin{cases} 0.48 \text{ in the N-S direction} \\ 0.54 \text{ in the E-W direction} \end{cases}$	Eq. (26.11-12)
$G_f = 0.925\left(\frac{1+1.7 I_{\bar{z}}\sqrt{g_Q^2 Q^2 + g_R^2 R^2}}{1+1.7 g_v I_{\bar{z}}}\right) = \begin{cases} 0.93 \text{ in the N-S direction} \\ 0.96 \text{ in the E-W direction} \end{cases}$	Eq. (26.11-10)

Table 5.63 Velocity Pressure Exposure Coefficients, K_z and K_h, for the Office Building in Example 5.12

Height above the Ground Level, z, ft (m)	K_z
142.5 (43.5)	1.04
133.0 (40.6)	1.03
123.5 (37.7)	1.01
114.0 (34.8)	0.98
104.5 (31.9)	0.96
95.0 (29.0)	0.94
85.5 (26.1)	0.91
76.0 (23.2)	0.88
66.5 (20.3)	0.85
57.0 (17.4)	0.82
47.5 (14.5)	0.78
38.0 (11.6)	0.73
28.5 (8.70)	0.68
19.0 (5.80)	0.61
9.5 (2.90)	0.57

Step 6i—Determine the velocity pressures, q_z

The velocity pressures, q_z, over the height of the building are given in Table 5.64. These pressures are determined by ASCE/SEI Equations (26.10-1) and (26.10-1.SI). For example, the velocity pressure at the mean roof height of the building is equal to the following:

$$q_h = 0.00256 K_h K_{zt} K_e V^2 = 0.00256 \times 1.04 \times 1.0 \times 1.0 \times 105^2 = 29.4 \text{ lb/ft}^2$$

In S.I.:

$$q_h = 0.613 K_h K_{zt} K_e V^2 = 0.613 \times 1.04 \times 1.0 \times 1.0 \times 47^2 / 1{,}000 = 1.41 \text{ kN/m}^2$$

Table 5.64 Velocity Pressures, q_z and q_h, for the Office Building in Example 5.12

Height above the Ground Level, z, ft (m)	K_z	q_z, lb/ft² (kN/m²)
142.5 (43.5)	1.04	29.4 (1.41)
133.0 (40.6)	1.03	29.1 (1.40)
123.5 (37.7)	1.01	28.5 (1.37)
114.0 (34.8)	0.98	27.7 (1.33)
104.5 (31.9)	0.96	27.1 (1.30)
95.0 (29.0)	0.94	26.5 (1.27)
85.5 (26.1)	0.91	25.7 (1.23)
76.0 (23.2)	0.88	24.8 (1.19)
66.5 (20.3)	0.85	24.0 (1.15)
57.0 (17.4)	0.82	23.1 (1.11)
47.5 (14.5)	0.78	22.0 (1.06)
38.0 (11.6)	0.73	20.6 (0.99)
28.5 (8.70)	0.68	19.2 (0.92)
19.0 (5.80)	0.61	17.2 (0.83)
9.5 (2.90)	0.57	16.1 (0.77)

Step 7—Determine the external pressure coefficients, C_p, for the walls and roof

The external pressure coefficients are determined from ASCE/SEI Figure 27.3-1.

For wind in the N-S direction:

- Windward wall: $C_p = 0.8$
- Leeward wall: For $L/B = 61.0/91.83 = 0.66$, $C_p = -0.5$
- Side walls: $C_p = -0.7$
- Roof: Normal to the ridge with $\theta < 10$ degrees and $h/L = 142.5/61.0 = 2.3$ (In S.I.: $h/L = 43.5/18.6 = 2.3$)

From the windward edge of the roof to 61.0 ft (18.6 m), $C_p = -1.3, -0.18$

The value of C_p equal to (−1.3) is permitted to be reduced linearly with the area over which it is applicable in accordance with footnote b in ASCE/SEI Figure 27.3-1.

Area over which (−1.3) is applicable $= 61.0 \times 91.83 = 5{,}602$ ft² (520.8 m²) > 1,000 ft² (92.9 m²)

Therefore, $C_p = 0.8 \times (-1.3) = -1.04$

For wind in the E-W direction:

- Windward wall: $C_p = 0.8$
- Leeward wall: For $L/B = 91.83/61.0 = 1.5$, $C_p = -0.4$ (by linear interpolation)
- Side walls: $C_p = -0.7$
- Roof: Parallel to the ridge with $h/L = 142.5/91.83 = 1.6$ (In S.I.: $h/L = 43.5/28.0 = 1.6$)

From the windward edge of the roof to $h/2 = 71.25$ ft (21.7 m): $C_p = -1.3, -0.18$
Area over which (−1.3) is applicable = $71.25 \times 61.0 = 4{,}346$ ft^2 (403.9 m^2) > 1,000 ft^2 (92.9 m^2)
Therefore, $C_p = 0.8 \times (-1.3) = -1.04$
From $h/2 = 71.25$ ft (21.7 m) to 91.83 ft (28.0 m): $C_p = -0.7, -0.18$

Step 8—Determine the velocity pressure for internal pressure determination, q_i
According to ASCE/SEI 27.3.1, $q_i = q_h = 29.4$ lb/ft^2 (1.41 kN/m^2) for all surfaces of an enclosed building.

Step 9—Determine the internal pressure coefficients, (GC_{pi})
From ASCE/SEI Table 26.13-1, $(GC_{pi}) = +0.18, -0.18$ for an enclosed building.

Step 10—Determine the design wind pressures, p, on all the surfaces of the building
Design wind pressures, p, are determined by ASCE/SEI Equation (27.3-1) for an enclosed building:

$$p = qK_dGC_p - q_iK_d(GC_{pi}) \text{ where } G = G_f$$

For wind in the N-S direction:

- Windward wall:

$$p_z = (q_z \times 0.85 \times 0.93 \times 0.8) - [29.4 \times 0.85 \times (\pm 0.18)] = 0.63q_z \mp 4.5 \text{ lb/ft}^2$$

In S.I.:

$$p_z = (q_z \times 0.85 \times 0.93 \times 0.8) - [1.41 \times 0.85 \times (\pm 0.18)] = 0.63q_z \mp 0.22 \text{ kN/m}^2$$

- Leeward wall

$$p_h = [q_h \times 0.85 \times 0.93 \times (-0.5)] - [29.4 \times 0.85 \times (\pm 0.18)]$$
$$= -11.6 \mp 4.5 = -16.1 \text{ lb/ft}^2, -7.1 \text{ lb/ft}^2$$

In S.I.:

$$p_h = [q_h \times 0.85 \times 0.93 \times (-0.5)] - [1.41 \times 0.85 \times (\pm 0.18)]$$
$$= -0.56 \mp 0.22 = -0.78 \text{ kN/m}^2, -0.34 \text{ kN/m}^2$$

- Side walls

$$p_h = [q_h \times 0.85 \times 0.93 \times (-0.7)] - [29.4 \times 0.85 \times (\pm 0.18)]$$
$$= -16.3 \mp 4.5 = -20.8 \text{ lb/ft}^2, -11.8 \text{ lb/ft}^2$$

In S.I.:

$$p_h = [q_h \times 0.85 \times 0.93 \times (-0.7)] - [1.41 \times 0.85 \times (\pm 0.18)]$$
$$= -0.78 \mp 0.22 = -1.00 \text{ kN/m}^2, -0.56 \text{ kN/m}^2$$

- Roof

$$p_h = (q_h \times 0.85 \times 0.93 \times C_p) - [29.4 \times 0.85 \times (\pm 0.18)] = 23.2C_p \mp 4.5 \text{ lb/ft}^2$$

In S.I.:

$$p_h = (q_h \times 0.85 \times 0.93 \times C_p) - [1.41 \times 0.85 \times (\pm 0.18)] = 1.12C_p \mp 0.22 \text{ kN/m}^2$$

The roof pressure coefficient $C_p = -0.18$ may become critical where wind loads are combined with roof live loads or snow loads. Determination of wind pressures based on this pressure coefficient should be performed, but such calculations are not shown in this example.

Design wind pressures in the N-S direction are given in Table 5.65 and Figure 5.51.

Table 5.65 Design Wind Pressures, *p*, in the N-S direction on the MWFRS of the Office Building in Example 5.12

Building Surface	Height above Ground Level, z, ft (m)	q, lb/ft² (kN/m²)	External Pressure $qK_d GC_p$, lb/ft² (kN/m²)	Internal Pressure $q_h K_d (GC_{pi})$, lb/ft² (kN/m²)	Net Pressure, p, lb/ft² (kN/m²)	
					$+(GC_{pi})$	$-(GC_{pi})$
Windward wall	142.5 (43.5)	29.4 (1.41)	18.6 (0.89)	±4.5 (±0.22)	14.1 (0.67)	23.1 (1.11)
	133.0 (40.6)	29.1 (1.40)	18.4 (0.88)	±4.5 (±0.22)	13.9 (0.66)	22.9 (1.10)
	123.5 (37.7)	28.5 (1.37)	18.0 (0.87)	±4.5 (±0.22)	13.5 (0.65)	22.5 (1.09)
	114.0 (34.8)	27.7 (1.33)	17.5 (0.84)	±4.5 (±0.22)	13.0 (0.62)	22.0 (1.06)
	104.5 (31.9)	27.1 (1.30)	17.1 (0.82)	±4.5 (±0.22)	12.6 (0.60)	21.6 (1.04)
	95.0 (29.0)	26.5 (1.27)	16.8 (0.80)	±4.5 (±0.22)	12.3 (0.58)	21.3 (1.02)
	85.5 (26.1)	25.7 (1.23)	16.3 (0.78)	±4.5 (±0.22)	11.8 (0.56)	20.8 (1.00)
	76.0 (23.2)	24.8 (1.19)	15.7 (0.75)	±4.5 (±0.22)	11.2 (0.53)	20.2 (0.97)
	66.5 (20.3)	24.0 (1.15)	15.2 (0.73)	±4.5 (±0.22)	10.7 (0.51)	19.7 (0.95)
	57.0 (17.4)	23.1 (1.11)	14.6 (0.70)	±4.5 (±0.22)	10.1 (0.48)	19.1 (0.92)
	47.5 (14.5)	22.0 (1.06)	13.9 (0.67)	±4.5 (±0.22)	9.4 (0.45)	18.4 (0.89)
	38.0 (11.6)	20.6 (0.99)	13.0 (0.63)	±4.5 (±0.22)	8.5 (0.41)	17.5 (0.85)
	28.5 (8.70)	19.2 (0.92)	12.1 (0.58)	±4.5 (±0.22)	7.6 (0.36)	16.6 (0.80)
	19.0 (5.80)	17.2 (0.83)	10.9 (0.52)	±4.5 (±0.22)	6.4 (0.30)	15.4 (0.74)
	9.5 (2.90)	16.1 (0.77)	10.2 (0.49)	±4.5 (±0.22)	5.7 (0.27)	14.7 (0.71)
Leeward wall	All	29.4 (1.41)	−11.6 (−0.56)	±4.5 (±0.22)	−16.1 (−0.78)	−7.1 (−0.34)
Side walls	All	29.4 (1.41)	−16.3 (−0.78)	±4.5 (±0.22)	−20.8 (−1.00)	−11.8 (−0.56)
Roof	—*	29.4 (1.41)	−24.2 (−1.16)	±4.5 (±0.22)	−28.7 (−1.38)	−19.7 (−0.94)

*From windward edge of roof to 61.0 ft (18.6 m).

Figure 5.51 Design Wind Pressures in the N-S Direction on the MWFRS of the Office Building in Example 5.12

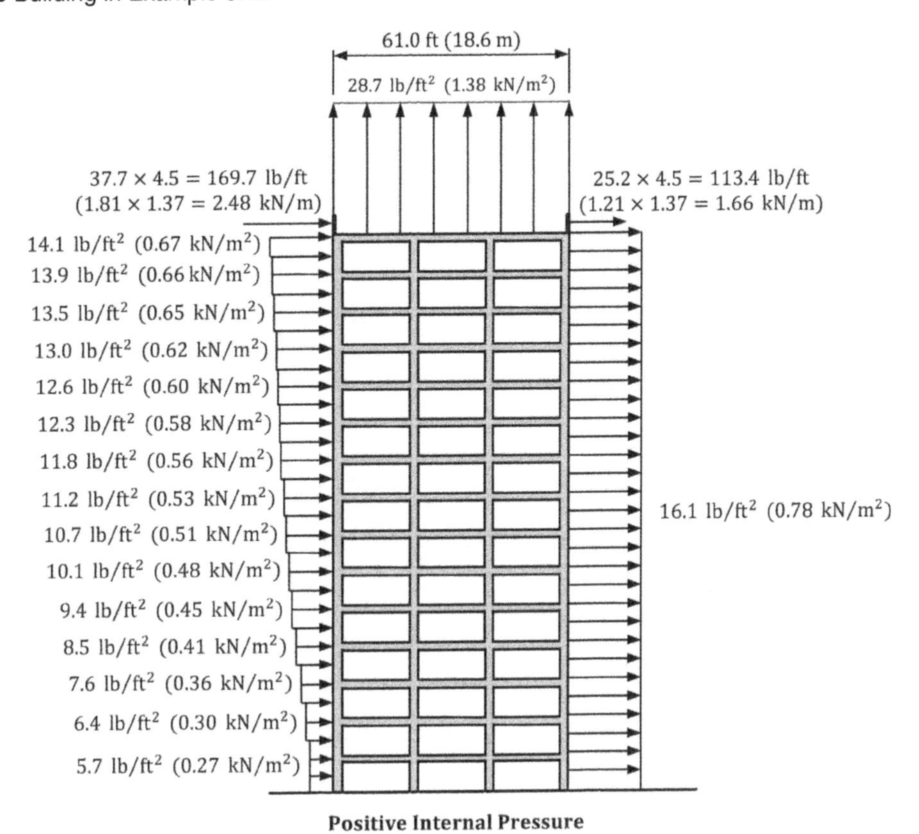

Positive Internal Pressure

19.7 lb/ft² (0.94 kN/m²)

37.7 × 4.5 = 169.7 lb/ft
(1.81 × 1.37 = 2.48 kN/m)

25.2 × 4.5 = 113.4 lb/ft
(1.21 × 1.37 = 1.66 kN/m)

23.1 lb/ft² (1.11 kN/m²)
22.9 lb/ft² (1.10 kN/m²)
22.5 lb/ft² (1.09 kN/m²)
22.0 lb/ft² (1.06 kN/m²)
21.6 lb/ft² (1.04 kN/m²)
21.3 lb/ft² (1.02 kN/m²)
20.8 lb/ft² (1.00 kN/m²)
20.2 lb/ft² (0.97 kN/m²)
19.7 lb/ft² (0.95 kN/m²)
19.1 lb/ft² (0.92 kN/m²)
18.4 lb/ft² (0.89 kN/m²)
17.5 lb/ft² (0.85 kN/m²)
16.6 lb/ft² (0.80 kN/m²)
15.4 lb/ft² (0.74 kN/m²)
14.7 lb/ft² (0.71 kN/m²)

7.1 lb/ft² (0.34 kN/m²)

Negative Internal Pressure

For wind in the E-W direction:

- Windward wall:

$$p_z = (q_z \times 0.85 \times 0.96 \times 0.8) - [29.4 \times 0.85 \times (\pm 0.18)] = 0.65q_z \mp 4.5 \text{ lb/ft}^2$$

In S.I.:

$$p_z = (q_z \times 0.85 \times 0.96 \times 0.8) - [1.41 \times 0.85 \times (\pm 0.18)] = 0.65q_z \mp 0.22 \text{ kN/m}^2$$

- Leeward wall

$$p_h = [q_h \times 0.85 \times 0.96 \times (-0.4)] - [29.4 \times 0.85 \times (\pm 0.18)]$$

$$= -9.6 \mp 4.5 = -14.1 \text{ lb/ft}^2, -5.1 \text{ lb/ft}^2$$

In S.I.:

$$p_h = [q_h \times 0.85 \times 0.96 \times (-0.4)] - [1.41 \times 0.85 \times (\pm 0.18)]$$

$$= -0.46 \mp 0.22 = -0.68 \text{ kN/m}^2, -0.24 \text{ kN/m}^2$$

- Side walls

$$p_h = [q_h \times 0.85 \times 0.96 \times (-0.7)] - [29.4 \times 0.85 \times (\pm 0.18)]$$

$$= -16.8 \mp 4.5 = -21.3 \text{ lb/ft}^2, -12.3 \text{ lb/ft}^2$$

In S.I.:

$$p_h = [q_h \times 0.85 \times 0.96 \times (-0.7)] - [1.41 \times 0.85 \times (\pm 0.18)]$$

$$= -0.81 \mp 0.22 = -1.03 \text{ kN/m}^2, -0.59 \text{ kN/m}^2$$

- Roof

$$p_h = (q_h \times 0.85 \times 0.96 \times C_p) - [29.4 \times 0.85 \times (\pm 0.18)] = 24.0C_p \mp 4.5 \text{ lb/ft}^2$$

In S.I.:

$$p_h = (q_h \times 0.85 \times 0.96 \times C_p) - [1.41 \times 0.85 \times (\pm 0.18)] = 1.15C_p \mp 0.22 \text{ kN/m}^2$$

The roof pressure coefficient $C_p = -0.18$ may become critical where wind loads are combined with roof live loads or snow loads. Determination of wind pressures based on this pressure coefficient should be performed, but such calculations are not shown in this example.

Design wind pressures in the E-W direction are given in Table 5.66 and Figure 5.52.

The external horizontal wind pressures on the walls for both directions are shown in Figure 5.53. It is evident that the internal pressures cancel out when determining the horizontal wind pressures on the walls.

The building must be designed for the wind load cases defined in ASCE/SEI Figure 27.3-8 (see Figure 5.16).

In case 1, the full design wind pressures act on the projected area perpendicular to each principal axis of the building at each level above ground. These pressures act separately along each principal axis. The windward and leeward pressures in Figure 5.53 fall under case 1.

Table 5.66 Design Wind Pressures, p, in the E-W Direction on the MWFRS of the Office Building in Example 5.12

Building Surface	Height above Ground Level, z, ft (m)	q, lb/ft² (kN/m²)	External Pressure $qK_d\,GC_p$, lb/ft² (kN/m²)	Internal Pressure $q_hK_d\,(GC_{pi})$, lb/ft² (kN/m²)	Net Pressure, p, lb/ft² (kN/m²)	
					$+(GC_{pi})$	$-(GC_{pi})$
Windward wall	142.5 (43.5)	29.4 (1.41)	19.2 (0.92)	±4.5 (±0.22)	14.7 (0.70)	23.7 (1.14)
	133.0 (40.6)	29.1 (1.40)	19.0 (0.91)	±4.5 (±0.22)	14.5 (0.69)	23.5 (1.13)
	123.5 (37.7)	28.5 (1.37)	18.6 (0.89)	±4.5 (±0.22)	14.1 (0.67)	23.1 (1.11)
	114.0 (34.8)	27.7 (1.33)	18.1 (0.87)	±4.5 (±0.22)	13.6 (0.65)	22.6 (1.09)
	104.5 (31.9)	27.1 (1.30)	17.7 (0.85)	±4.5 (±0.22)	13.2 (0.63)	22.2 (1.07)
	95.0 (29.0)	26.5 (1.27)	17.3 (0.83)	±4.5 (±0.22)	12.8 (0.61)	21.8 (1.05)
	85.5 (26.1)	25.7 (1.23)	16.8 (0.80)	±4.5 (±0.22)	12.3 (0.58)	21.3 (1.02)
	76.0 (23.2)	24.8 (1.19)	16.2 (0.78)	±4.5 (±0.22)	11.7 (0.56)	20.7 (1.00)
	66.5 (20.3)	24.0 (1.15)	15.7 (0.75)	±4.5 (±0.22)	11.2 (0.53)	20.2 (0.97)
	57.0 (17.4)	23.1 (1.11)	15.1 (0.73)	±4.5 (±0.22)	10.6 (0.51)	19.6 (0.95)
	47.5 (14.5)	22.0 (1.06)	14.4 (0.69)	±4.5 (±0.22)	9.9 (0.47)	18.9 (0.91)
	38.0 (11.6)	20.6 (0.99)	13.5 (0.65)	±4.5 (±0.22)	9.0 (0.43)	18.0 (0.87)
	28.5 (8.70)	19.2 (0.92)	12.5 (0.60)	±4.5 (±0.22)	8.0 (0.38)	17.0 (0.82)
	19.0 (5.80)	17.2 (0.83)	11.2 (0.54)	±4.5 (±0.22)	6.7 (0.32)	15.7 (0.76)
	9.5 (2.90)	16.1 (0.77)	10.5 (0.50)	±4.5 (±0.22)	6.0 (0.28)	15.0 (0.72)
Leeward wall	All	29.4 (1.41)	−9.6 (−0.46)	±4.5 (±0.22)	−14.1 (−0.68)	−5.1 (−0.24)
Side walls	All	29.4 (1.41)	−16.8 (−0.81)	±4.5 (±0.22)	−21.3 (−1.03)	−12.3 (−0.59)
Roof	—*	29.4 (1.41)	−25.0 (−1.20)	±4.5 (±0.22)	−29.5 (−1.42)	−20.5 (−0.98)
	—**	29.4 (1.41)	−16.8 (−0.81)	±4.5 (±0.22)	−21.3 (−1.03)	−12.3 (−0.59)

*From windward edge of roof to $h/2 = 71.25$ ft (21.7 m).
**From $h/2 = 71.25$ ft (21.7 m) to 91.83 ft (28.0 m).

In case 2, 75 percent of the design wind pressures on the windward and leeward walls are applied on the projected area perpendicular to each principal axis of the building along with a torsional moment. The wind pressures and torsional moments, which vary over the height of the building, are applied separately for each principal axis.

The torsional moment, M_z, is calculated using the eccentricity, e, for flexible buildings, which is determined by ASCE/SEI Equation (27.3-4):

$$e = \frac{e_Q + 1.7I_{\bar{z}}\sqrt{(g_Q Q e_Q)^2 + (g_R R e_R)^2}}{1 + 1.7I_{\bar{z}}\sqrt{(g_Q Q)^2 + (g_R R)^2}}$$

where e_Q = eccentricity e determined for rigid buildings in ASCE/SEI Figure 27.3-8
e_R = distance between the elastic shear center and the center of mass of each floor

The other terms in this equation are given in Table 5.62.

Figure 5.52 Design Wind Pressures, p, in the E-W Direction on the MWFRS of the Office Building in Example 5.12

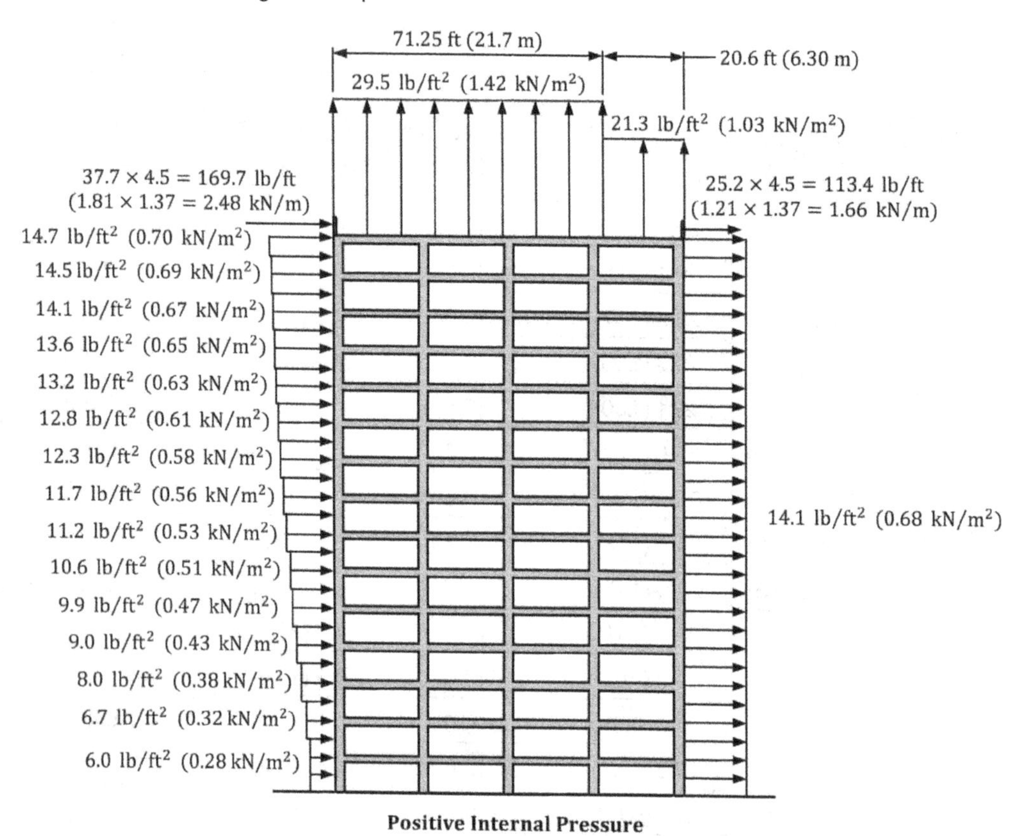

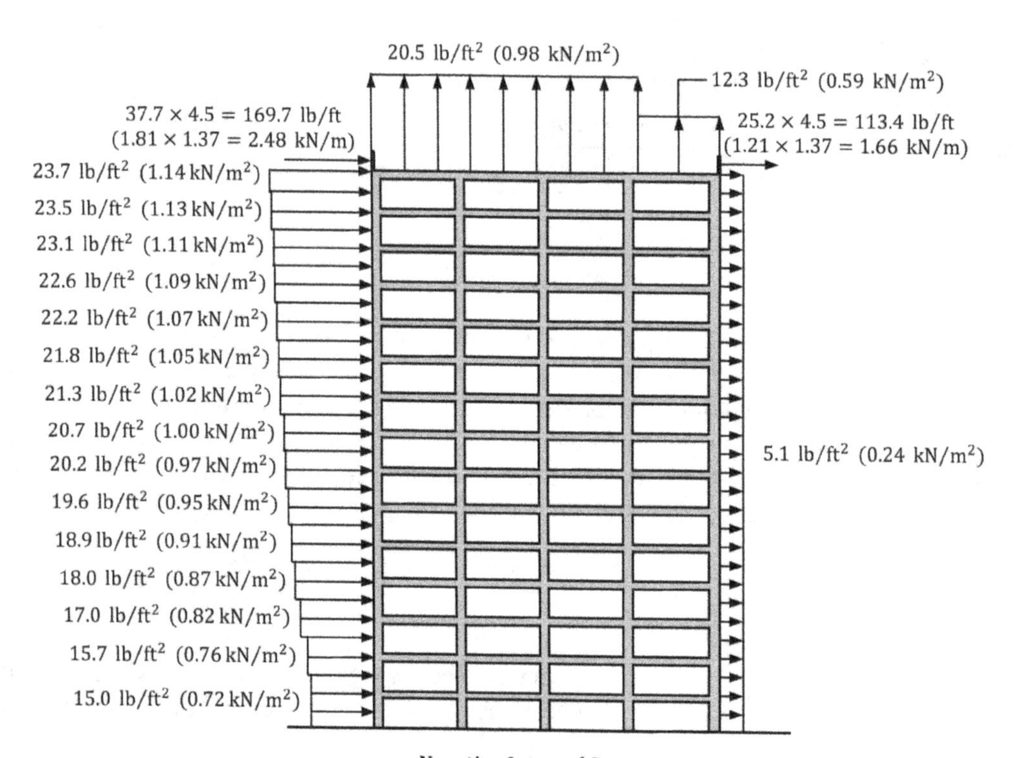

Figure 5.53 External Horizontal Design Wind Pressures on the Walls of the Office Building in Example 5.12

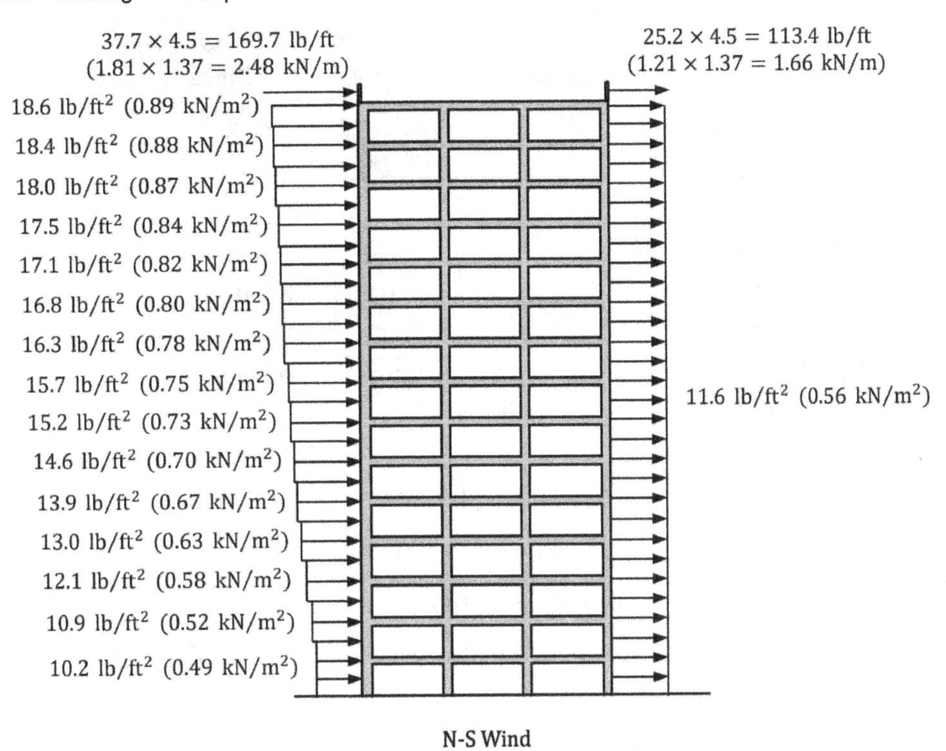

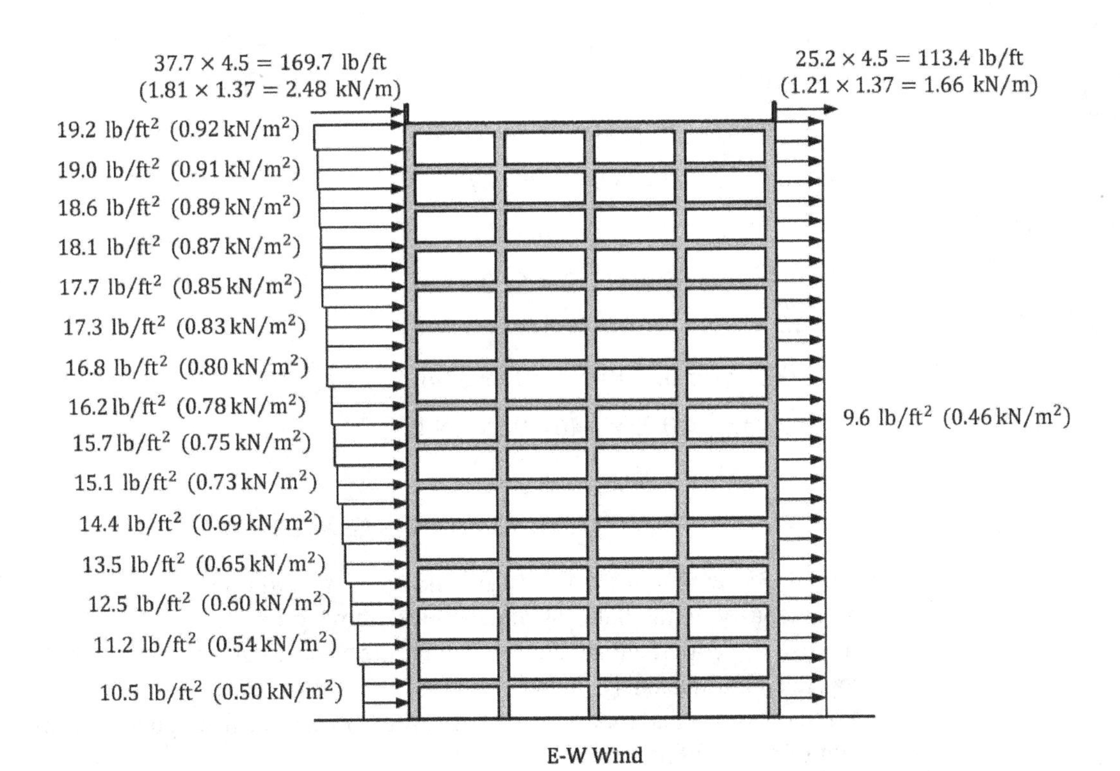

Calculations for e are given in Table 5.67 for both principal wind directions. Because the locations of the elastic shear center and center of mass coincide in both principal directions, $e_R = 0$.

Table 5.67 Calculations for Eccentricity, e, for the Office Building in Example 5.12

Wind Direction	e_Q	$I_{\bar{z}}$	g_Q	Q	g_R	R	e
N-S	0.15 × 91.83 = 13.8 ft (4.20 m)	0.26	3.4	0.84	4.04	0.48	12.4 ft (3.77 m)
E-W	0.15 × 61.0 = 9.2 ft (2.79 m)	0.26	3.4	0.85	4.04	0.54	8.1 ft (2.46 m)

The external wind pressures and torsional moment at the mean roof height of the building for case 2 are as follows:

- N-S wind

$$0.75p_{Wy} = 0.75 \times 18.6 = 14.0 \text{ lb/ft}^2$$

$$0.75p_{Ly} = 0.75 \times 11.6 = 8.7 \text{ lb/ft}^2$$

$$M_z = 0.75(p_{Wy} + p_{Ly})B_y e$$

$$= (14.0 + 8.7) \times 91.83 \times 12.4 \times (9.5/2)/1{,}000 = 122.8 \text{ ft-kips}$$

In S.I.:

$$0.75p_{Wy} = 0.75 \times 0.89 = 0.67 \text{ kN/m}^2$$

$$0.75p_{Ly} = 0.75 \times 0.56 = 0.42 \text{ kN/m}^2$$

$$M_z = 0.75(p_{Wy} + p_{Ly})B_y e$$

$$= (0.67 + 0.42) \times 28.0 \times 3.77 \times (2.90/2) = 166.8 \text{ kN-m}$$

- E-W wind

$$0.75p_{Wx} = 0.75 \times 19.2 = 14.4 \text{ lb/ft}^2$$

$$0.75p_{Lx} = 0.75 \times 9.6 = 7.2 \text{ lb/ft}^2$$

$$M_z = 0.75(p_{Wx} + p_{Lx})B_x e$$

$$= (14.4 + 7.2) \times 61.0 \times 8.1 \times (9.5/2)/1{,}000 = 50.7 \text{ ft-kips}$$

In S.I.:

$$0.75p_{Wx} = 0.75 \times 0.92 = 0.69 \text{ kN/m}^2$$

$$0.75p_{Lx} = 0.75 \times 0.46 = 0.35 \text{ kN/m}^2$$

$$M_z = 0.75(p_{Wx} + p_{Lx})B_x e$$

$$= (0.69 + 0.35) \times 18.6 \times 2.46 \times (2.90/2) = 69.0 \text{ kN-m}$$

In case 3, 75 percent of the windward and leeward wall pressures in Figure 5.53 act simultaneously on the building at each level above ground.

In case 4, 75 percent of the wind pressures and torsional moments defined in case 2 act simultaneously on the building.

Wind pressures and torsional moments for load cases 1 through 4 at the mean roof height of the building are given in Figure 5.54.

The minimum design wind loading in ASCE/SEI 27.1.5 must be considered as a load case in addition to the load cases above.

Figure 5.54 Load Cases 1 through 4 at the Mean Roof Height of the Office Building in Example 5.12

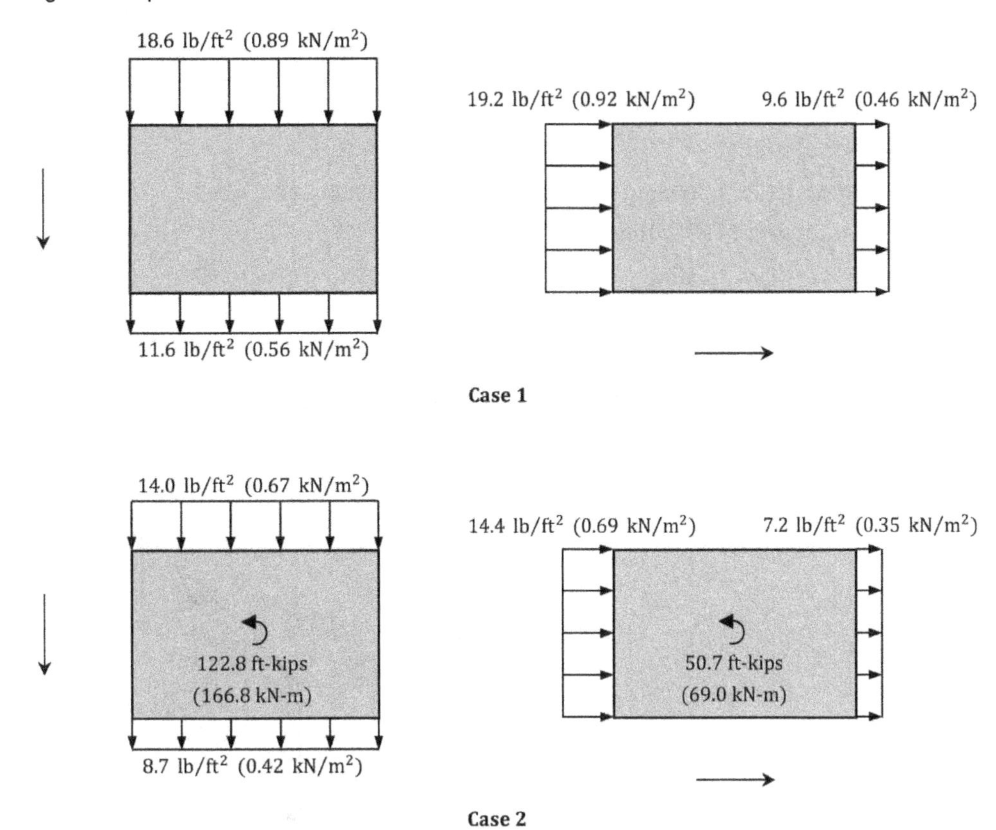

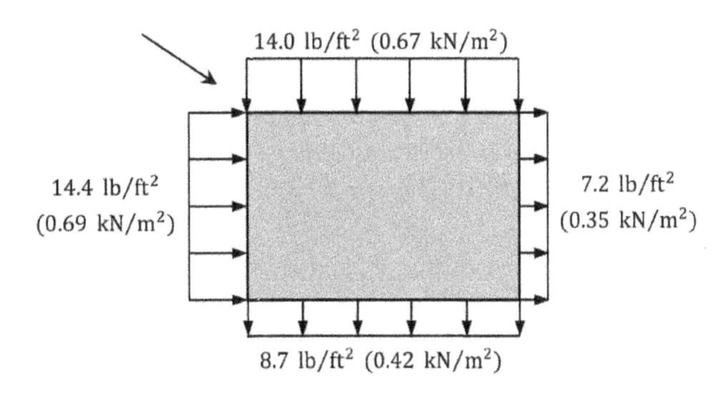

Case 3

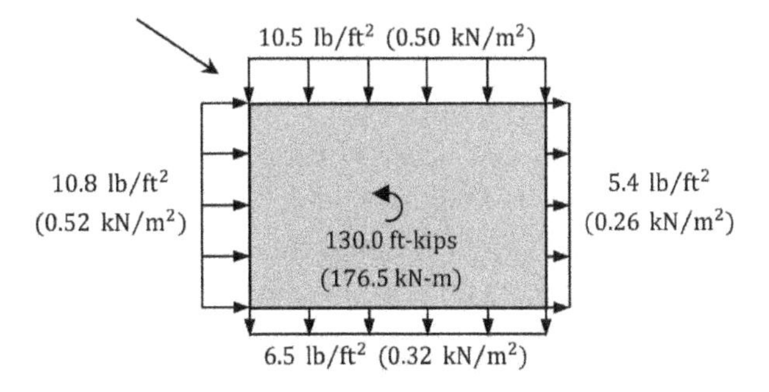

Case 4

5.8.13 Example 5.13—Determination of Design Wind Pressures on an Open Agricultural Building Using ASCE/SEI Chapter 27—MWFRS

Determine the design wind pressures, p, on the MWFRS of the open agricultural building in Figure 5.55 using the provisions in ASCE/SEI Chapter 27 given the following design data:

- Location: Ames, IA (Latitude = 42.04°, Longitude = −93.71°)
- Surface roughness: C
- Topography: Not situated on a hill, ridge, or escarpment
- Risk category: I (IBC Table 1604.5)
- Enclosure classification: Open

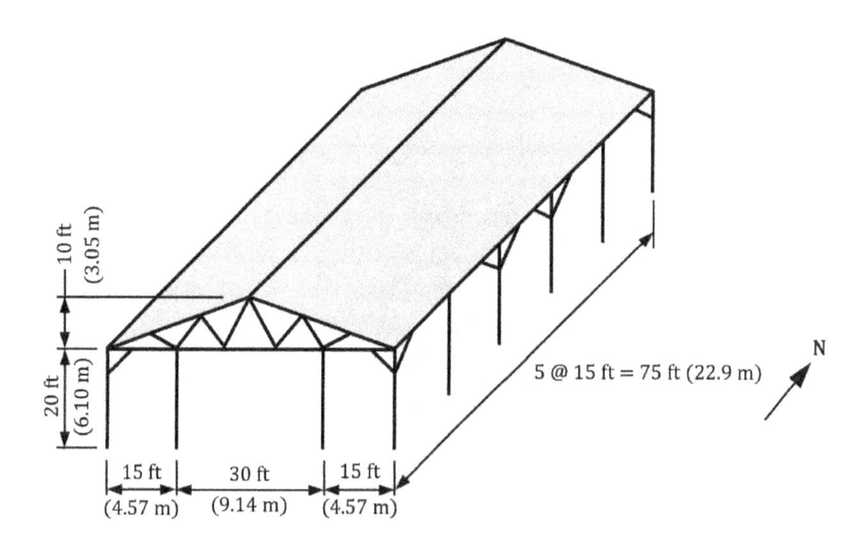

Figure 5.55 Open Agricultural Building in Example 5.13

Solution

Check if the building meets all the conditions and limitations of ASCE/SEI 27.1.2 and 27.1.3 so that the provisions in ASCE/SEI Chapter 27 are permitted to be used to determine the design wind pressures on the MWFRS.

The building is regularly-shaped, that is, it does not have any unusual geometric irregularities in spatial form. Also, the building does not have response characteristics that make it subject to across-wind loading or other similar effects, and it is not sited at a location where channeling effects or buffeting in the wake of upwind obstructions need to be considered.

Therefore, the provisions in ASCE/SEI Chapter 27 are permitted to be used to determine the design wind pressures on the MWFRS.

Flowchart 5.3 is used to determine the design wind pressures.

Step 1—Determine the enclosure classification of the building

The enclosure classification of the building is given as open.

Step 2—Determine the wind velocity pressure, q_h

Flowchart 5.1 is used to determine q_h.

Step 2a—Determine the surface roughness category

Surface roughness C is given in the design data and it is assumed it is present in all directions.

Step 2b—Determine the exposure category

It is assumed exposure category C is applicable in all directions.

Step 2c—Determine the terrain exposure constants

For exposure C, $\alpha = 9.8$ and $z_g = 2,460$ ft (750 m) from ASCE/SEI Table 26.11-1.

Step 2d—Determine the velocity pressure exposure coefficient, K_h

The velocity exposure coefficient, K_h, is determined from ASCE/SEI Table 26.10-1.

Mean roof height $= \dfrac{20+30}{2} = 25$ ft (7.62 m)

$$K_h = 2.41(z/z_g)^{2/\alpha} = 2.41(25/2,460)^{2/9.8} = 0.94$$

In S.I.:

$$K_h = 2.41(7.62/750)^{2/9.8} = 0.94$$

Step 2e—Determine the topographic factor, K_{zt}

The building is not located on a hill, ridge, or escarpment, so $K_{zt} = 1.0$ (ASCE/SEI 26.8.2).

Step 2f—Determine the ground elevation factor, K_e

It is permitted to take $K_e = 1.0$ for all elevations (ASCE/SEI 26.9).

Step 2g—Determine the risk category

The risk category is given as I in the design data.

Step 2h—Determine the basic wind speed, V

For risk category I buildings, use IBC Figure 1609.3(1) or ASCE/SEI Figure 26.5-1A (see Table 5.2 of this book).

Equivalently, use Reference 5.4 to obtain V = 103 mi/h (46 m/s) given the latitude and longitude of the site.

Step 2i—Determine the velocity pressures, q_z

The velocity pressure at the mean roof height, q_h, is determined by ASCE/SEI Equations (26.10-1) and (26.10-1.SI):

$$q_h = 0.00256 K_h K_{zt} K_e V^2 = 0.00256 \times 0.94 \times 1.0 \times 1.0 \times 103^2 = 25.5 \text{ lb/ft}^2$$

In S.I.:

$$q_h = 0.613 K_h K_{zt} K_e V^2 = 0.613 \times 0.94 \times 1.0 \times 1.0 \times 46^2/1,000 = 1.22 \text{ kN/m}^2$$

Step 3—Determine the wind directionality factor, K_d

For the MWFRS of a building structure, $K_d = 0.85$ from ASCE/SEI Table 26.6-1.

Step 4—Determine the gust-effect factor, G

It has been determined that $n_a > 1$ Hz, which means the building is defined as rigid. For rigid buildings, G is permitted to be taken as 0.85 (ASCE/SEI 26.11.4).

Step 5—Determine the net pressure coefficients, C_N, for the roof

The net pressure coefficients, C_{NW} and C_{NL}, are determined from ASCE/SEI Figure 27.3-5 for open buildings with pitched roofs. These net pressure coefficients include contributions from both the top and bottom surfaces of the roof (see note 1 in ASCE/SEI Figure 27.3-5).

The wind pressures on the roof depend on the level of wind flow restriction below the roof. Clear wind flow implies that little (less than or equal to 50 percent) or no portion of the cross-section below the roof is blocked by goods or materials (see note 2 in ASCE/SEI Figure 27.3-5). Obstructed wind flow means that a significant portion (more than 50 percent) of the cross-section is blocked. Because the usage of the space below the roof is not known, wind pressures will be determined for both situations.

$$\text{Roof angle} = \tan^{-1}\left(\frac{10}{30}\right) = 18.4 \text{ degrees}$$

In S.I.:

$$\text{Roof angle} = \tan^{-1}\left(\frac{3.05}{9.14}\right) = 18.4 \text{ degrees}$$

- Wind in the E-W direction ($\gamma = 0°, 180°$)
  Net pressure coefficients are given in Table 5.68 for load cases A and B. Values of C_{NW} and C_{NL} are obtained by linear interpolation (see note 3 in ASCE/SEI Figure 27.3-5).

Table 5.68 Windward and Leeward Net Pressure Coefficients, C_{NW} and C_{NL}, for Wind in the E-W Direction on the Open Agricultural Building in Example 5.13

Load Case	Clear Wind Flow		Obstructed Wind Flow	
	C_{NW}	C_{NL}	C_{NW}	C_{NL}
A	1.10	−0.17	−1.20	−1.09
B	0.01	−0.96	−0.69	−1.65

- Wind in the N-S direction ($\gamma = 90°, 270°$)
  Net pressure coefficients, C_N, at various distances from the windward edge of the roof, which are obtained from ASCE/SEI Figure 27.3-7, are given in Table 5.69.

Table 5.69 Net Pressure Coefficients C_N for Wind in the N-S Direction on the Open Agricultural Building in Example 5.13

Horizontal Distance from the Windward Edge of the Roof	Load Case	Clear Wind Flow	Obstructed Wind Flow
$\leq h = 25$ ft (7.62 m)	A	−0.8	−1.2
	B	0.8	0.5
$> h = 25$ ft (7.62 m) $\leq 2h = 50$ ft (15.2 m)	A	−0.6	−0.9
	B	0.5	0.5
$> 2h = 50$ ft (15.2 m)	A	−0.3	−0.6
	B	0.3	0.3

Step 6—Determine the net design pressures, p

The net design pressures, p, are determined by ASCE/SEI Equation (27.3-2):

$$p = q_h K_d G C_N = 25.5 \times 0.85 \times 0.85 \times C_N = 18.42 C_N \text{ lb/ft}^2$$

In S.I.:

$$p = 1.22 \times 0.85 \times 0.85 \times C_N = 0.882 C_N \text{ kN/m}^2$$

Net design wind pressures are given in Tables 5.70 and 5.71 for wind in the E-W and N-S directions, respectively. These pressures act perpendicular to the roof surface.

Table 5.70 Net Design Wind Pressures on the Open Agricultural Building in Example 5.13 for Wind in the E-W Direction, lb/ft² (kN/m²)

Load Case	Clear Wind Flow		Obstructed Wind Flow	
	Windward	Leeward	Windward	Leeward
A	20.3 (0.97)	−3.1 (−0.15)	−22.1 (−1.06)	−20.1 (−0.96)
B	0.2 (0.01)	−17.7 (−0.85)	−12.7 (−0.61)	−30.4 (−1.46)

Table 5.71 Net Design Wind Pressures on the Open Agricultural Building in Example 5.13 for Wind in the N-S Direction, lb/ft² (kN/m²)

Horizontal Distance from the Windward Edge of the Roof	Load Case	Clear Wind Flow	Obstructed Wind Flow
$\leq h = 25$ ft (7.62 m)	A	−14.7 (−0.71)	−22.1 (−1.06)
	B	14.7 (0.71)	9.2 (0.44)
$> h = 25$ ft (7.62 m) $\leq 2h = 50$ ft (15.2 m)	A	−11.1 (−0.53)	−16.6 (−0.79)
	B	9.2 (0.44)	9.2 (0.44)
$> 2h = 50$ ft (15.2 m)	A	−5.5 (−0.27)	−11.1 (−0.53)
	B	5.5 (0.27)	5.5 (0.27)

For this open building with transverse frames and a pitched roof with an angle less than or equal to 45 degrees, a wind pressure in the N-S direction (parallel to the ridge) must be applied to the structure and must be resisted by the MWFRS in that direction (ASCE/SEI 27.3.2). The design wind pressure is determined in accordance with ASCE/SEI 28.3.7.

Flowchart 5.4 is used to determine the pressure, p.

Step 1—Determine q_h

It was determined previously that $q_h = 25.5$ lb/ft² (1.22 kN/m²).

Step 2—Determine the external pressure coefficients, (GC_{pf})

The external pressure coefficients, (GC_{pf}), for the windward and leeward sides are determined from load case 2 in ASCE/SEI Figure 28.3-1. Average values of the coefficients are used on the windward end (surfaces 5 and 5E) and the leeward end (surfaces 6 and 6E).

$$(GC_{pf})_{\text{windward}} = (0.40 + 0.61)/2 = 0.51$$

$$(GC_{pf})_{\text{leeward}} = [(-0.29) + (-0.43)]/2 = -0.36$$

Step 3—Determine the frame width factor, K_B

The width of the frame, B, is equal to 60 ft (18.3 m), which is less than 100 ft (30.5 m). Therefore, K_B is equal to the following:

$$K_B = 1.8 - 0.01B = 1.8 - (0.01 \times 60) = 1.2$$

Step 4—Determine the solidity ratio, ϕ

The solidity ratio, ϕ, is equal to the projected area of any portion of the end wall exposed to wind, A_S, divided by the total end wall area, A_E, assuming the building is enclosed.

Assume $\phi = A_S / A_E = 0.1$.

Step 5—Determine the shielding factor, K_S

With the number of frames, n, equal to 6, K_S is equal to the following:

$$K_S = 0.60 + 0.073(n-3) + 1.25\phi^{1.8} = 0.60 + [0.073 \times (6-3)] + (1.25 \times 0.1^{1.8}) = 0.84$$

Step 6—Determine the wind directionality factor, K_d

From ASCE/SEI Table 26.6-1, $K_d = 0.85$ for building structures.

Step 7—Determine the design wind pressure, p, and total longitudinal force, F

The design wind pressure, p, is determined by ASCE/SEI Equation (28.3-3):

$$p = q_h K_d \left[(GC_{pf})_{\text{windward}} - (GC_{pf})_{\text{leeward}} \right] K_B K_S$$

$$= 25.5 \times 0.85 \times [0.51 - (-0.36)] \times 1.2 \times 0.84 = 19.0 \text{ lb/ft}^2$$

In S.I.:

$$p = 1.22 \times 0.85 \times [0.51 - (-0.36)] \times 1.2 \times 0.84 = 0.91 \text{ kN/m}^2$$

The total longitudinal force, F, in the N-S direction that must be resisted by the MWFRS in that direction is determined by ASCE/SEI Equation (28.3-4):

$$F = pA_E = 19.0 \times \{(20 \times 60) + [0.5 \times (60 \times 10)]\} / 1{,}000 = 28.5 \text{ kips}$$

In S.I.:

$$F = 0.91 \times \{(6.10 \times 18.3) + [0.5 \times (18.3 \times 3.05)]\} = 127.0 \text{ kN}$$

This force is applied at the centroid of the area A_E.

Note: Even though the building is located in a tornado-prone region (see ASCE/SEI Figure 32.1-1), tornado loads need not be considered because the building is assigned to risk category I (ASCE/SEI 32.1.1).

5.8.14 Example 5.14—Determination of Design Wind Forces on a Solid Freestanding Wall Using ASCE/SEI Chapter 29

Determine the design wind forces on the solid freestanding masonry privacy wall depicted in Figure 5.56 using the provisions in ASCE/SEI Chapter 29 given the following design data:

- Basic wind speed: $V = 88$ mi/h (39 m/s)
- Surface roughness: B
- Topography: Not situated on a hill, ridge, or escarpment
- Risk category: I (IBC Table 1604.5)

Figure 5.56 Solid Freestanding Wall in Example 5.14

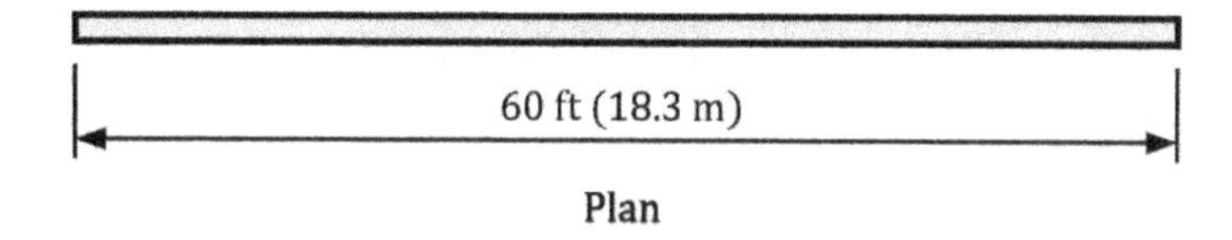

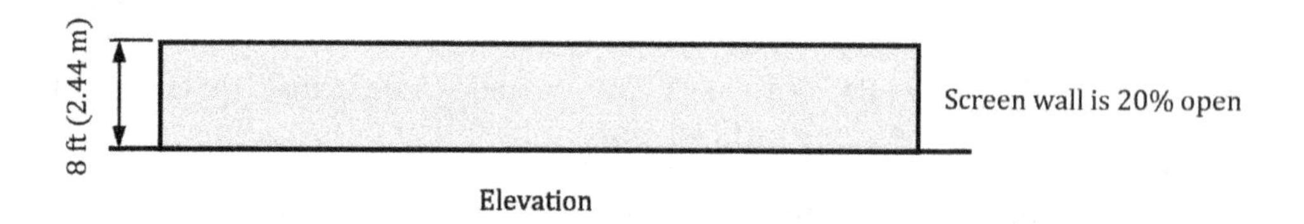

Solution

Check if the solid freestanding wall meets all the conditions and limitations of ASCE/SEI 29.1.2 so that the provisions in ASCE/SEI Chapter 29 are permitted to be used to determine the design wind forces.

The wall is regularly-shaped, that is, it does not have any unusual geometric irregularities in spatial form. Also, the wall does not have response characteristics that make it subject to across-wind loading or other similar effects, and it is not sited at a location where channeling effects or buffeting in the wake of upwind obstructions need to be considered.

Therefore, the provisions in ASCE/SEI Chapter 29 are permitted to be used to determine the design wind forces on the wall.

Flowchart 5.5 is used to determine the design wind forces.

Step 1—Determine the wind velocity pressure, q_h

Flowchart 5.1 is used to determine q_h.

Step 1a—Determine the surface roughness category

Surface roughness B is given in the design data and it is assumed it is present in all directions.

Step 1b—Determine the exposure category

It is assumed exposure category B is applicable in all directions.

Step 1c—Determine the terrain exposure constants

For exposure B, $\alpha = 7.5$ and $z_g = 3{,}280$ ft (1,000 m) from ASCE/SEI Table 26.11-1.

Step 1d—Determine the velocity pressure exposure coefficient, K_h

The velocity pressure exposure coefficient, K_h, is determined from ASCE/SEI Table 26.10-1.

For exposure B and $h = 8$ ft (2.44 m), $K_h = 0.57$ from ASCE/SEI Table 26.10-1.

Step 1e—Determine the topographic factor, K_{zt}

The wall is not located on a hill, ridge, or escarpment, so $K_{zt} = 1.0$ (ASCE/SEI 26.8.2).

Step 1f—Determine the ground elevation factor, K_e

It is permitted to take $K_e = 1.0$ for all elevations (ASCE/SEI 26.9).

Step 1g—Determine the risk category

The risk category is given as I in the design data.

Step 1h—Determine the basic wind speed, V

From the design data, $V = 88$ mi/h (39 m/s).

Step 1i—Determine the velocity pressure, q_h

The velocity pressure at the top of the wall, q_h, is determined by ASCE/SEI Equations (26.10-1) and (26.10-1.SI):

$$q_h = 0.00256 K_h K_{zt} K_e V^2 = 0.00256 \times 0.57 \times 1.0 \times 1.0 \times 88^2 = 11.3 \text{ lb/ft}^2$$

In S.I.:

$$q_h = 0.613 K_h K_{zt} K_e V^2 = 0.613 \times 0.57 \times 1.0 \times 1.0 \times 39^2 / 1{,}000 = 0.53 \text{ kN/m}^2$$

Step 2—Determine the wind directionality factor, K_d

For a solid freestanding wall, $K_d = 0.85$ from ASCE/SEI Table 26.6-1.

Step 3—Determine the gust-effect factor, G

It has been determined that $n_a > 1$ Hz, which means the wall is defined as rigid.
For rigid structures, G is permitted to be taken as 0.85 (ASCE/SEI 26.11.4).

Step 4—Determine the length to height ratio B/s

$$B/s = 60/8 = 7.5$$

In S.I.: $B/s = 18.3/2.44 = 7.5$

Step 5—Determine the gross area of the wall, A_s

$$A_s = Bs = 60 \times 8 = 480 \text{ ft}^2$$

In S.I.: $A_s = 18.3 \times 2.44 = 44.7 \text{ m}^2$

Step 6—Determine the clearance ratio s/h

For a ground-supported wall, $s/h = 1.0$.

Step 7—Determine the ratio of solid area to gross area, ε

The wall is 20 percent open, so $\varepsilon = 1 - 0.2 = 0.8$.

Step 8—Determine the net force coefficient, C_f, for cases A and B

The net force coefficient, C_f, for cases A and B is determined from ASCE/SEI Figure 29.3-1.

$$\text{For } s/h = 1.0 \text{ and } B/s = 7.5,\ C_f = 1.30 + \frac{(1.35 - 1.30) \times (10 - 7.5)}{(10 - 5)} = 1.33$$

Because the wall has openings equal to 20 percent of the gross area, which is less than 30 percent, it is permitted to multiply C_f by the following reduction factor (see note 1 in ASCE/SEI Figure 29.3-1:

Reduction factor $= 1 - (1 - \varepsilon)^{1.5} = 1 - (1 - 0.8)^{1.5} = 0.91$

Therefore, $C_f = 0.91 \times 1.33 = 1.21$.

Because $B/s = 7.5 > 2$, case B need not be considered (see note 2 in ASCE/SEI Figure 29.3-1).

Step 9—Determine if case C must be considered

Case C must be considered because $B/s = 7.5 > 2$ (see note 2 in ASCE/SEI Figure 29.3-1).

Step 10—Determine the net force coefficients, C_f, for case C

The net force coefficients, C_f, for case C are determined from ASCE/SEI Figure 29.3-1 based on the horizontal distance from the windward edge of the wall.

The net force coefficients are given in Table 5.72, which are obtained by linear interpolation.

Step 11—Determine the reduction factors on the net force coefficients, C_f, for case C

Because $s/h = 1.0 > 0.8$, the net force coefficients, C_f, for case C are permitted to be multiplied by the following reduction factor (see note 3 in ASCE/SEI Figure 29.3-1):

Table 5.72 Net Force Coefficients, C_f, for Case C for the Solid Freestanding Wall in Example 5.14

Horizontal Distance from Windward Edge	C_f
0 to $s = 8$ ft (2.44 m)	3.48
$s = 8$ ft (2.44 m) to $2s = 16$ ft (4.88 m)	2.28
$2s = 16$ ft (4.88 m) to $3s = 24$ ft (7.32 m)	1.68
$3s = 24$ ft (7.32 m) to 60 ft (18.3 m)	1.05

Reduction factor $= 1.8 - (s/h) = 1.8 - 1.0 = 0.8$
The porosity reduction factor determined in step 8 is also applicable.
Reduced net force coefficients are given in Table 5.73.

Table 5.73 Reduced Net Force Coefficients, C_f, for Case C for the Solid Freestanding Wall in Example 5.14

Horizontal Distance from Windward Edge	C_f
0 to $s = 8$ ft (2.44 m)	2.53
$s = 8$ ft (2.44 m) to $2s = 16$ ft (4.88 m)	1.66
$2s = 16$ ft (4.88 m) to $3s = 24$ ft (7.32 m)	1.22
$3s = 24$ ft (7.32 m) to 60 ft (18.3 m)	0.76

Step 12—Determine the design wind forces, F

The design wind forces, F, are determined by ASCE/SEI Equations (29.3-1) and (29.3-1.SI):

$$F = q_h K_d G C_f A_s$$

For case A:

$$F = 11.3 \times 0.85 \times 0.85 \times 1.21 \times (8 \times 60) / 1{,}000 = 4.7 \text{ kips}$$

In S.I.:

$$F = 0.53 \times 0.85 \times 0.85 \times 1.21 \times (2.44 \times 18.3) = 20.7 \text{ kN}$$

For case C:

- From 0 to $s = 8$ ft (2.44 m):

$$F = 11.3 \times 0.85 \times 0.85 \times 2.53 \times (8 \times 8) / 1{,}000 = 1.3 \text{ kips}$$

In S.I.:

$$F = 0.53 \times 0.85 \times 0.85 \times 2.53 \times (2.44 \times 2.44) = 5.8 \text{ kN}$$

- From $s = 8$ ft (2.44 m) to $2s = 16$ ft (4.88 m):

$$F = 11.3 \times 0.85 \times 0.85 \times 1.66 \times (8 \times 8) / 1{,}000 = 0.9 \text{ kips}$$

In S.I.:

$$F = 0.53 \times 0.85 \times 0.85 \times 1.66 \times (2.44 \times 2.44) = 3.8 \text{ kN}$$

- From $2s = 16$ ft (4.88 m) to $3s = 24$ ft (7.32 m):

$$F = 11.3 \times 0.85 \times 0.85 \times 1.22 \times (8 \times 8) / 1{,}000 = 0.6 \text{ kips}$$

In S.I.:

$$F = 0.53 \times 0.85 \times 0.85 \times 1.22 \times (2.44 \times 2.44) = 2.8 \text{ kN}$$

- From $3s = 24$ ft (7.32 m) to 60 ft (18.3 m):

$$F = 11.3 \times 0.85 \times 0.85 \times 0.76 \times (8 \times 36)/1{,}000 = 1.8 \text{ kips}$$

In S.I.:

$$F = 0.53 \times 0.85 \times 0.85 \times 0.76 \times (2.44 \times 11.0) = 7.8 \text{ kN}$$

The design wind forces for cases A, B, and C are given in Figure 5.57 where case B is shown for illustration purposes only.

The minimum design wind loading in ASCE/SEI 29.7 must be considered as a load case in addition to the load cases above.

Figure 5.57 Design Wind Forces, *F*, for the Solid Freestanding Wall in Example 5.14

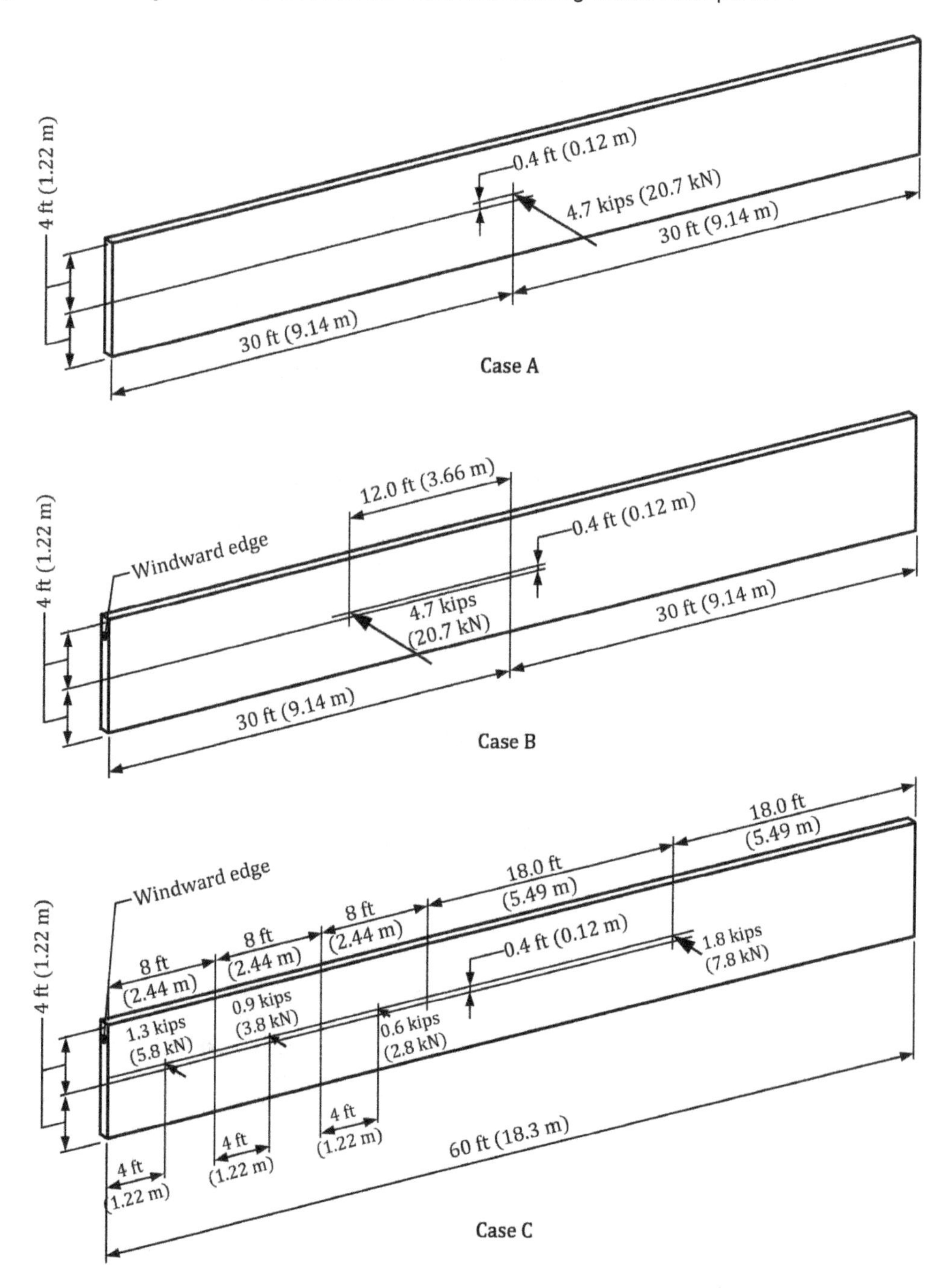

5.8.15 Example 5.15—Determination of Design Wind Forces on a Solid Sign and Support Structure Using ASCE/SEI Chapter 29

Determine the design wind forces on the solid sign and support structure depicted in Figure 5.58 using the provisions in ASCE/SEI Chapter 29 given the following design data:

- Location: Fargo, ND (Latitude = 46.876°, Longitude = −96.743°)
- Surface roughness: C
- Topography: Not situated on a hill, ridge or escarpment
- Risk category: I (IBC Table 1604.5)

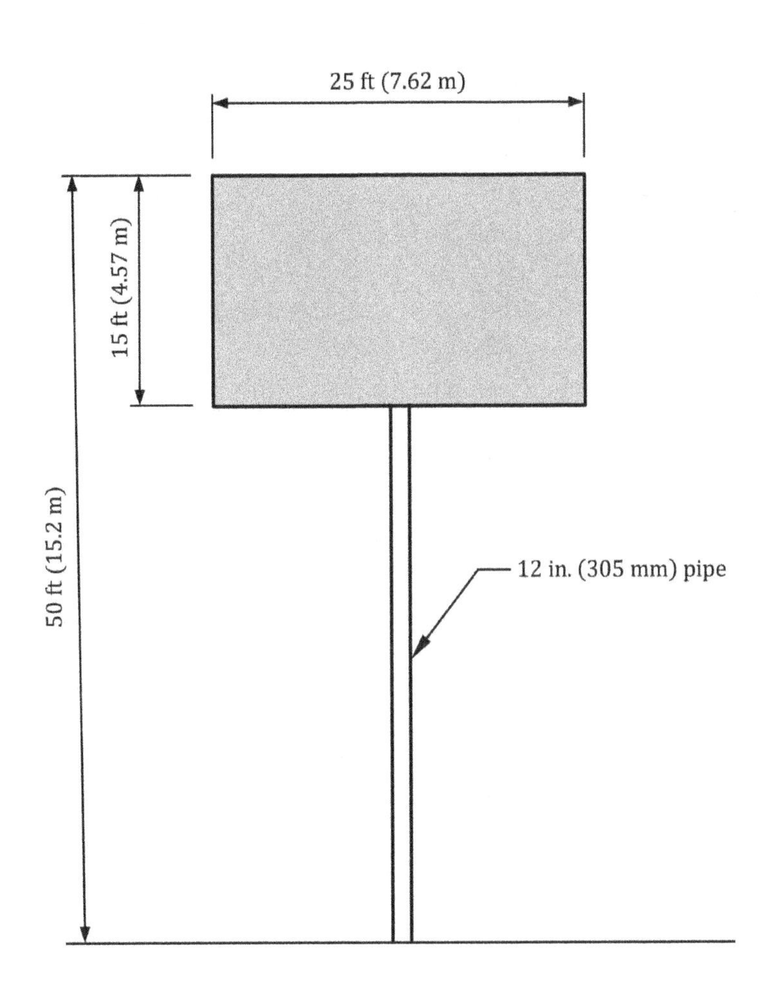

Figure 5.58 Solid Sign and Support Structure in Example 5.15

Solution

Check if the combined sign and support structure meets all the conditions and limitations of ASCE/SEI 29.1.2 so that the provisions in ASCE/SEI Chapter 29 are permitted to be used to determine the design wind forces.

The combined structure is regularly-shaped, that is, it does not have any unusual geometric irregularities in spatial form. Also, it does not have response characteristics that make it subject to across-wind loading or other similar effects, and it is not sited at a location where channeling effects or buffeting in the wake of upwind obstructions need to be considered.

Therefore, the provisions in ASCE/SEI Chapter 29 are permitted to be used to determine the design wind forces on the sign and on the support structure.

Flowchart 5.5 is used to determine the design wind forces on the sign.

Step 1—Determine the wind velocity pressure, q_h

Flowchart 5.1 is used to determine q_h.

Step 1a—Determine the surface roughness category

Surface roughness C is given in the design data and it is assumed it is present in all directions.

Step 1b—Determine the exposure category

It is assumed exposure category C is applicable in all directions.

Step 1c—Determine the terrain exposure constants

For exposure C, $\alpha = 9.8$ and $z_g = 2{,}460$ ft (750 m) from ASCE/SEI Table 26.11-1.

Step 1d—Determine the velocity pressure exposure coefficient, K_h

The velocity pressure exposure coefficient, K_h, is determined from ASCE/SEI Table 26.10-1.

For exposure C and $h = 50$ ft (15.2 m), $K_h = 1.09$ from ASCE/SEI Table 26.10.1.

Step 1e—Determine the topographic factor, K_{zt}

The sign is not located on a hill, ridge, or escarpment, so $K_{zt} = 1.0$ (ASCE/SEI 26.8.2).

Step 1f—Determine the ground elevation factor, K_e

It is permitted to take $K_e = 1.0$ for all elevations (ASCE/SEI 26.9).

Step 1g—Determine the risk category

The risk category is given as I in the design data.

Step 1h—Determine the basic wind speed, V

For risk category I structures, use IBC Figure 1609.3(1) or ASCE/SEI Figure 26.5-1A (see Table 5.2 of this book).

Equivalently, use Reference 5.4 to obtain V = 103 mi/h (46 m/s) given the latitude and longitude of the site.

Step 1i—Determine the velocity pressure, q_h

The velocity pressure at the top of the sign, q_h, is determined by ASCE/SEI Equations (26.10-1) and (26.10-1.SI):

$$q_h = 0.00256 K_h K_{zt} K_e V^2 = 0.00256 \times 1.09 \times 1.0 \times 1.0 \times 103^2 = 29.6 \text{ lb/ft}^2$$

In S.I.:

$$q_h = 0.613 K_h K_{zt} K_e V^2 = 0.613 \times 1.09 \times 1.0 \times 1.0 \times 46^2 / 1{,}000 = 1.41 \text{ kN/m}^2$$

Step 2—Determine the wind directionality factor, K_d

For a solid sign, $K_d = 0.85$ from ASCE/SEI Table 26.6-1.

Step 3—Determine the gust-effect factor, G

It has been determined that $n_a > 1$ Hz for the combined structure, which means the combined structure is defined as rigid.

For rigid structures, G is permitted to be taken as 0.85 (ASCE/SEI 26.11.4).

Step 4—Determine the length to height ratio B/s

$$B/s = 25/15 = 1.67$$

In S.I.:

$$B/s = 7.62/4.57 = 1.67$$

Step 5—Determine the gross area of the sign, A_s

$$A_s = Bs = 25 \times 15 = 375 \text{ ft}^2$$

In S.I.:

$$A_s = Bs = 7.62 \times 4.57 = 34.8 \text{ m}^2$$

Step 6—Determine the clearance ratio s/h

$$s/h = 15/50 = 0.3$$

In S.I.:

$$s/h = 4.57/15.2 = 0.3$$

Step 7—Determine the ratio of solid area to gross area, ε

Because the sign is solid, $\varepsilon = 1.0$.

Step 8—Determine the net force coefficient, C_f, for cases A and B

The net force coefficient, C_f, for cases A and B is determined from ASCE/SEI Figure 29.3-1.

For $s/h = 0.3$ and $B/s = 1.67$, $C_f = 1.80$

Case B must be considered because $B/s = 1.67 < 2$ (see note 2 in ASCE/SEI Figure 29.3-1).

Step 9—Determine if case C must be considered

Case C need not be considered because $B/s = 1.67 < 2$ (see note 2 in ASCE/SEI Figure 29.3-1).

Step 10—Determine the design wind forces, F, on the sign

The design wind forces, F, on the sign are determined by ASCE/SEI Equations (29.3-1) and (29.3-1.SI):

$$F = q_h K_d G C_f A_s$$

For cases A and B:

$$F = 29.6 \times 0.85 \times 0.85 \times 1.80 \times 375/1{,}000 = 14.4 \text{ kips}$$

In S.I.:

$$F = 1.41 \times 0.85 \times 0.85 \times 1.80 \times 34.8 = 63.9 \text{ kN}$$

The design wind forces for cases A and B are given in Figure 5.59.

Flowchart 5.6 is used to determine the design wind forces on the pipe.

Step 1—Determine the wind velocity pressure, q_z, at the centroid of the projected area, A_f

Flowchart 5.1 is used to determine q_z.

Step 1a—Determine the surface roughness category

Surface roughness C is given in the design data and it is assumed it is present in all directions.

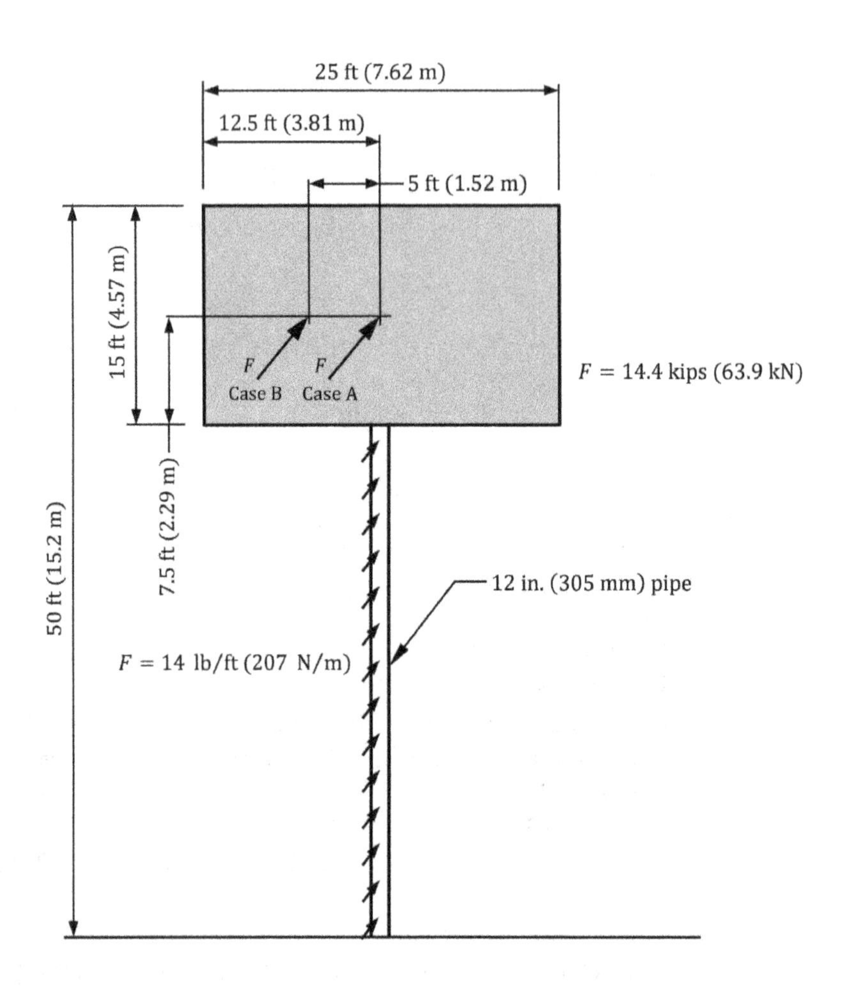

Figure 5.59 Design Wind Forces, *F*, for the Solid Sign and Pipe in Example 5.15

Step 1b—Determine the exposure category

It is assumed exposure category C is applicable in all directions.

Step 1c—Determine the terrain exposure constants

For exposure C, $\alpha = 9.8$ and $z_g = 2{,}460$ ft (750 m) from ASCE/SEI Table 26.11-1.

Step 1d—Determine the velocity pressure exposure coefficient, K_z

The velocity pressure exposure coefficient, K_z, is determined from ASCE/SEI Table 26.10-1. The top of the pipe is connected to the underside of the sign structure; therefore, $z = (50 - 15)/2 = 17.5$ ft (5.33 m).

$$K_z = 2.41(z/z_g)^{2/\alpha} = 2.41(17.5/2{,}460)^{2/9.8} = 0.88$$

In S.I.:

$$K_z = 2.41(5.33/750)^{2/9.8} = 0.88$$

Step 1e—Determine the topographic factor, K_{zt}

The structure is not located on a hill, ridge, or escarpment, so $K_{zt} = 1.0$ (ASCE/SEI 26.8.2).

Step 1f—Determine the ground elevation factor, K_e

It is permitted to take $K_e = 1.0$ for all elevations (ASCE/SEI 26.9).

Step 1g—Determine the risk category

The risk category is given as I in the design data.

Step 1h—Determine the basic wind speed, V

For risk category I structures, use IBC Figure 1609.3(1) or ASCE/SEI Figure 26.5-1A (see Table 5.2 of this book).

Equivalently, use Reference 5.4 to obtain V = 103 mi/h (46 m/s) given the latitude and longitude of the site.

Step 1i—Determine the velocity pressure, q_z

The velocity pressure, q_z, is determined by ASCE/SEI Equations (26.10-1) and (26.10-1.SI):

$$q_z = 0.00256 K_z K_{zt} K_e V^2 = 0.00256 \times 0.88 \times 1.0 \times 1.0 \times 103^2 = 23.9 \text{ lb/ft}^2$$

In S.I.:

$$q_z = 0.613 K_z K_{zt} K_e V^2 = 0.613 \times 0.88 \times 1.0 \times 1.0 \times 46^2 / 1{,}000 = 1.14 \text{ kN/m}^2$$

Step 2—Determine the wind directionality factor, K_d

For a round cross-section, $K_d = 1.0$ from ASCE/SEI Table 26.6-1.

Step 3—Determine the gust-effect factor, G

It has been determined that $n_a > 1$ Hz for the combined structure, which means the combined structure is defined as rigid.

For rigid structures, G is permitted to be taken as 0.85 (ASCE/SEI 26.11.4).

Step 4—Determine the force coefficient, C_f

The force coefficient, C_f, is determined from ASCE/SEI Figure 29.4-1.

For a round cross-section:

$$D\sqrt{q_z} = (12/12) \times \sqrt{23.9} = 4.9 > 2.5$$

$$h/D = (50 - 15)/(12/12) = 35.0$$

Assuming a moderately smooth surface, $C_f = 0.7$.

In S.I.:

$$D\sqrt{q_z} = (305/1{,}000) \times \sqrt{1.14 \times 1{,}000} = 10.3 > 5.3$$

$$h/D = (15.2 - 4.57)/(305/1{,}000) = 34.8$$

Assuming a moderately smooth surface, $C_f = 0.7$.

Step 5—Determine the design wind force, F, on the pipe

The design wind force, F, on the pipe is determined by ASCE/SEI Equations (29.4-1) and (29.4-1.SI):

$$F = q_z K_d G C_f A_f = 23.9 \times 1.0 \times 0.85 \times 0.7 \times (12/12) = 14 \text{ lb/ft}$$

In S.I.:

$$F = (1.14 \times 1{,}000) \times 1.0 \times 0.85 \times 0.7 \times (305/1{,}000) = 207 \text{ N/m}$$

The design wind force on the pipe is given in Figure 5.59.

The minimum design wind loading in ASCE/SEI 29.7 must be considered as a load case in addition to the load cases above for the solid sign and pipe.

Note: Even though the structure is located in a tornado-prone region (see ASCE/SEI Figure 32.1-1), tornado loads need not be considered because the structure is assigned to risk category I (ASCE/SEI 32.1.1).

5.8.16 Example 5.16—Determination of Design Wind Forces on Rooftop Equipment Using ASCE/SEI Chapter 29

Determine the design wind forces on the equipment on the roof of the hotel depicted in Figure 5.46 using the provisions in ASCE/SEI Chapter 29. Use the design data in Example 5.11.

SOLUTION

Step 1—Determine the wind velocity pressure, q_h

From Example 5.11, $q_h = 97.3$ lb/ft^2 (4.71 kN/m^2).

Step 2—Determine the gust-effect factor, G

From Example 5.11, $G = 0.85$.

Step 3—Determine the wind directionality factor, K_d

For rooftop equipment, $K_d = 0.85$ from ASCE/SEI Table 26.6-1.

Step 4—Determine the resultant lateral force, F_h

The resultant lateral force, F_h, is determined by ASCE/SEI Equations (29.4-2) and (29.4-2.SI):

$$F_h = q_h K_d (GC_r) A_f$$

$$A_f = \begin{cases} 6 \times 7 = 42 \text{ ft}^2 \text{ (smaller face)} \\ 6 \times 15 = 90 \text{ ft}^2 \text{ (larger face)} \end{cases}$$

Because $0.1Bh = 0.1 \times 75.33 \times 63.5 = 478.4 \text{ ft}^2 > A_f$, $(GC_r) = 1.9$.

$$F_h = \begin{cases} 97.3 \times 0.85 \times 1.9 \times 42/1{,}000 = 6.6 \text{ kips (smaller face)} \\ 97.3 \times 0.85 \times 1.9 \times 90/1{,}000 = 14.1 \text{ kips (larger face)} \end{cases}$$

In S.I.:

$$A_f = \begin{cases} 1.83 \times 2.13 = 3.9 \text{ m}^2 \text{ (smaller face)} \\ 1.83 \times 4.57 = 8.4 \text{ ft}^2 \text{ (larger face)} \end{cases}$$

Because $0.1Bh = 0.1 \times 23.0 \times 19.4 = 44.6 \text{ m}^2 > A_f$, $(GC_r) = 1.9$.

$$F_h = \begin{cases} 4.71 \times 0.85 \times 1.9 \times 3.9 = 29.7 \text{ kN (smaller face)} \\ 4.71 \times 0.85 \times 1.9 \times 8.4 = 63.9 \text{ kN (smaller face)} \end{cases}$$

These forces act perpendicular to the respective faces of the equipment.

Step 5—Determine the vertical uplift force, F_v

The vertical uplift force, F_v, is determined by ASCE/SEI Equations (29.4-3) and (29.4-3.SI):

$$F_v = q_h K_d (GC_r) A_r$$

$$A_r = 15 \times 7 = 105 \text{ ft}^2$$

Because $0.1BL = 0.1 \times 75.33 \times 328.75 = 2{,}476.5 \text{ ft}^2 > A_r$, $(GC_r) = 1.5$.

$$F_v = 97.3 \times 0.85 \times 1.5 \times 105/1{,}000 = 13.0 \text{ kips}$$

In S.I.:

$$A_r = 4.57 \times 2.13 = 9.7 \text{ m}^2$$

Because $0.1BL = 0.1 \times 23.0 \times 100.2 = 230.5 \text{ m}^2 > A_r$, $(GC_r) = 1.5$.

$$F_v = 4.71 \times 0.85 \times 1.5 \times 9.7 = 58.3 \text{ kN}$$

This force acts perpendicular to the top surface of the equipment.

Note: Even though the building is located in a tornado-prone region (see ASCE/SEI Figure 32.1-1), tornado loads need not be considered on the rooftop equipment because the building is assigned to risk category II (ASCE/SEI 32.1.1).

5.8.17 Example 5.17—Determination of Design Wind Pressures on an Isolated Circular Tank Using ASCE/SEI Chapter 29

Determine the design wind pressures on the isolated circular tank depicted in Figure 5.60 using the provisions in ASCE/SEI Chapter 29 given the following design data:

- Location: Reno, NV (Latitude = 39.52°, Longitude = −119.71°)
- Surface roughness: C
- Topography: Not situated on a hill, ridge or escarpment
- Risk category: IV (IBC Table 1604.5; the tank is part of a public utility facility required as an emergency backup facility for a risk category IV building)

Figure 5.60 Isolated Circular Tank in Example 5.17

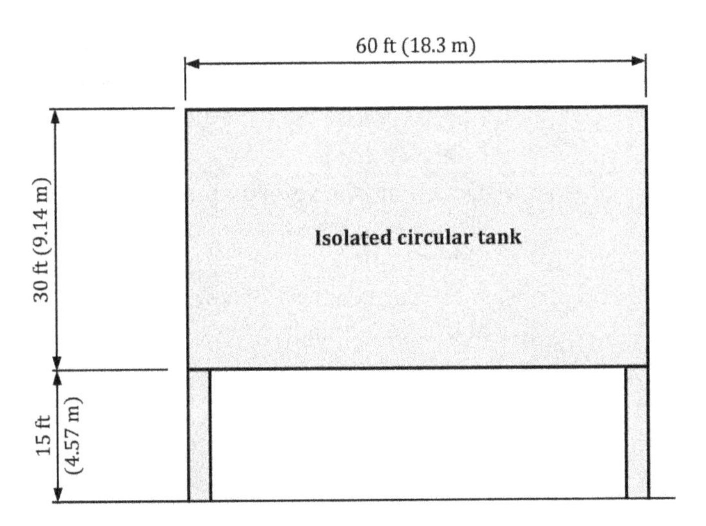

SOLUTION

Check if the tank meets all the conditions and limitations of ASCE/SEI 29.1.2 so that the provisions in ASCE/SEI Chapter 29 are permitted to be used to determine the design wind pressures.

The tank is regularly-shaped, that is, it does not have any unusual geometric irregularities in spatial form. Also, it does not have response characteristics that make it subject to across-wind loading or other similar effects, and it is not sited at a location where channeling effects or buffeting in the wake of upwind obstructions need to be considered.

Also, the geometric requirements of ASCE/SEI 29.4.2 are satisfied (see Figure 5.20):

- $h = C + h_c = 15 + 30 = 45$ ft < 120 ft
- $D = 60$ ft < 120 ft
- $0.25 < h_c / D = 30/60 = 0.5 < 4$
- $C = 15$ ft $< h_c = 30$ ft

In S.I.:

- $h = C + h_c = 4.57 + 9.14 = 13.7$ m < 36.6 m
- $D = 18.3$ m < 36.6 ft
- $0.25 < h_c / D = 9.14/18.3 = 0.5 < 4$
- $C = 4.57$ m $< h_c = 9.14$ m

Therefore, the provisions in ASCE/SEI Chapter 29 are permitted to be used to determine the design wind pressures on the MWFRS of the isolated tank.

Flowchart 5.7 is used to determine the design wind pressures.

Step 1—Determine the wind velocity pressures, q_z and q_h

Flowchart 5.1 is used to determine q_z at the centroid of the projected areas of the walls, Dh_c, and q_h at the mean roof height of the tank.

Step 1a—Determine the surface roughness category

Surface roughness C is given in the design data and it is assumed it is present in all directions.

Step 1b—Determine the exposure category

It is assumed exposure category C is applicable in all directions.

Step 1c—Determine the terrain exposure constants

For exposure C, $\alpha = 9.8$ and $z_g = 2{,}460$ ft (750 m) from ASCE/SEI Table 26.11-1.

Step 1d—Determine the velocity pressure exposure coefficient, K_z

The velocity pressure exposure coefficient, K_z, is determined from ASCE/SEI Table 26.10-1.

At the centroid of Dh_c, which is located [15+(30/2)] = 30 ft (9.14 m) above the ground:

$$K_z = 2.41(z/z_g)^{2/\alpha} = 2.41(30/2{,}460)^{2/9.8} = 0.98$$

In S.I.:

$$K_z = 2.41(9.14 / 750)^{2/9.8} = 0.98$$

At the mean roof height of 45 ft (13.7 m) above the ground:

$$K_h = 2.41(z/z_g)^{2/\alpha} = 2.41(45/2{,}460)^{2/9.8} = 1.07$$

In S.I.:

$$K_h = 2.41(13.7/750)^{2/9.8} = 1.07$$

Step 1e—Determine the topographic factor, K_{zt}

The tank is not located on a hill, ridge or escarpment, so $K_{zt} = 1.0$ (ASCE/SEI 26.8.2).

Step 1f—Determine the ground elevation factor, K_e

It is permitted to take $K_e = 1.0$ for all elevations (ASCE/SEI 26.9).

Step 1g—Determine the risk category

The risk category is given as IV in the design data.

Step 1h—Determine the basic wind speed, V

For risk category IV structures, use IBC Figure 1609.3(4) or ASCE/SEI Figure 26.5-1D (see Table 5.2 of this book).

Equivalently, use Reference 5.4 to obtain $V = 107$ mi/h (48 m/s) given the latitude and longitude of the site.

Step 1i—Determine the velocity pressures, q_z and q_h

The velocity pressures are determined by ASCE/SEI Equations (26.10-1) and (26.10-1.SI):

At the centroid of Dh_c:

$$q_z = 0.00256 K_z K_{zt} K_e V^2 = 0.00256 \times 0.98 \times 1.0 \times 1.0 \times 107^2 = 28.7 \text{ lb/ft}^2$$

In S.I.:

$$q_z = 0.613 K_z K_{zt} K_e V^2 = 0.613 \times 0.98 \times 1.0 \times 1.0 \times 48^2/1{,}000 = 1.38 \text{ kN/m}^2$$

At the mean roof height:

$$q_h = 0.00256 K_h K_{zt} K_e V^2 = 0.00256 \times 1.07 \times 1.0 \times 1.0 \times 107^2 = 31.4 \text{ lb/ft}^2$$

In S.I.:

$$q_h = 0.613 K_h K_{zt} K_e V^2 = 0.613 \times 1.07 \times 1.0 \times 1.0 \times 48^2/1{,}000 = 1.51 \text{ kN/m}^2$$

Step 2—Determine the wind directionality factor, K_d

For the MWFRS of a round section, $K_d = 1.0$ from ASCE/SEI Table 26.6-1.

Step 3—Determine the gust-effect factor, G

It has been determined that $n_a > 1$ Hz for the tank, which means the tank is rigid. For rigid structures, G is permitted to be taken as 0.85 (ASCE/SEI 26.11.4).

Step 4—Determine the internal pressure coefficients, (GC_{pi})

Assuming the tank is enclosed, $(GC_{pi}) = +0.18, -0.18$ (ASCE/SEI Table 26.13-1).

Step 5—Determine the design wind load on the walls, F

The design wind load on the walls, F, is determined by ASCE/SEI Equations (29.4-1) and (29.4-1.SI):

$$F = q_z K_d G C_f A_f$$

It is permitted to use a force coefficient, C_f, equal to 0.63 instead of determining it in accordance with ASCE/SEI Figure 29.4-1 provided all the following conditions are met:

(a) $0.25 < h_c/D = 30/60 = 0.5 < 4$ (In S.I.: $0.25 < h_c/D = 9.14/18.3 = 0.5 < 4$).

(b) $C = 15$ ft $< h_c = 30$ ft (In S.I.: $C = 4.57$ m $< h_c = 9.14$ m).

(c) Assume the tank is moderately smooth as defined in ASCE/SEI Figure 29.4-1, that is, the ratio of the depth of protruding elements (such as ribs and spoiler), D', to the diameter of the tank, D, is less than 0.02.

Because conditions (a) through (c) in ASCE/SEI 29.4.2.1 are satisfied, $C_f = 0.63$.

$$F = q_z K_d G C_f A_f = 28.7 \times 1.0 \times 0.85 \times 0.63 \times (30 \times 60)/1{,}000 = 27.7 \text{ kips}$$

In S.I.:

$$F = 1.38 \times 1.0 \times 0.85 \times 0.63 \times (9.14 \times 18.3) = 123.6 \text{ kN}$$

Step 6—Determine the external pressure coefficients for the roof, C_p

The external pressure coefficients for the roof, C_p, are determined from ASCE/SEI Figure 29.4-5.

In zone 1, $C_p = -0.8$ (zone 1 is the area of the roof adjacent to the windward edge of the roof)

In zone 2, $C_p = -0.5$.

Because $h_c/D = 0.5$, the horizontal extent of zone 1 from the windward edge of the roof, b, is equal to $0.5D = 30$ ft (9.14 m). The horizontal extent of zone 2 is also 30 ft (9.14 m).

Step 7—Determine the design wind pressures on the roof, p

The design wind pressures on the roof, p, are determined by ASCE/SEI Equations (29.4-4) and (29.4-4.SI):

$$p = q_h K_d [GC_p - (GC_{pi})]$$

The design wind pressures for zones 1 and 2 are given in Table 5.74.

Table 5.74 Design Wind Pressures on the Roof of the Isolated Tank in Example 5.17

Zone	Design Wind Pressure, p, lb/ft² (kN/m²)	
	$+(GC_{pi})$	$-(GC_{pi})$
1	−27.0 (−1.30)	−15.7 (−0.76)
2	−19.0 (−0.91)	−7.7 (−0.37)

The pressure, p, on the underside of the tank is determined by ASCE/SEI Equation (29.4-4) using external pressure coefficients, C_p, equal to 0.8 and −0.6 because $C = 15\text{ ft} > h_c/3 = 10$ ft (In S.I.: $C = 4.57\text{ m} > h_c/3 = 3.05$ m) [ASCE/SEI 29.4.2.3]:

- $C_p = 0.8$

  For positive internal pressure:

$$p = 31.4 \times 1.0 \times [(0.85 \times 0.8) - 0.18] = 15.7 \text{ lb/ft}^2$$

In S.I.:

$$p = 1.51 \times 1.0 \times [(0.85 \times 0.8) - 0.18] = 0.76 \text{ kN/m}^2$$

For negative internal pressure:

$$p = 31.4 \times 1.0 \times [(0.85 \times 0.8) - (-0.18)] = 27.0 \text{ lb/ft}^2$$

In S.I.:

$$p = 1.51 \times 1.0 \times [(0.85 \times 0.8) - (-0.18)] = 1.30 \text{ kN/m}^2$$

- $C_p = -0.6$

  For positive internal pressure:

$$p = 31.4 \times 1.0 \times \{[0.85 \times (-0.6)] - 0.18\} = -21.7 \text{ lb/ft}^2$$

In S.I.:

$$p = 1.51 \times 1.0 \times \{[0.85 \times (-0.6)] - 0.18\} = -1.04 \text{ kN/m}^2$$

For negative internal pressure:

$$p = 31.4 \times 1.0 \times \{[0.85 \times (-0.6)] - (-0.18)\} = -10.4 \text{ lb/ft}^2$$

In S.I.:

$$p = 1.51 \times 1.0 \times \{[0.85 \times (-0.6)] - (-0.18)\} = -0.50 \text{ kN/m}^2$$

The design wind force on the walls, the maximum design wind pressures on the roof, and the design wind pressure on the underside of the tank for $C_p = 0.8$ with negative internal pressure are given in Figure 5.61.

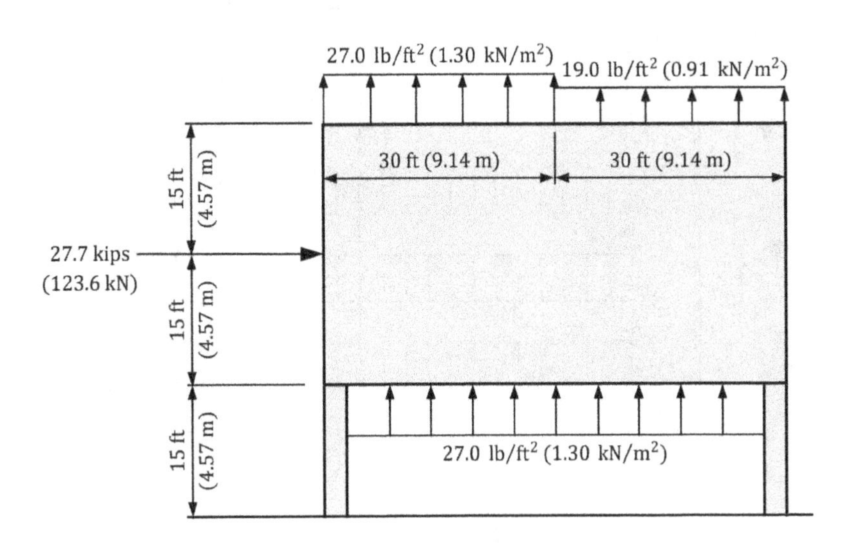

Figure 5.61 Design Wind Force and Pressures on the Isolated Tank in Example 5.17

5.8.18 Example 5.18—Determination of Design Wind Pressures on Rooftop Solar Panels Using ASCE/SEI 29.4.3

Solar panels are supported by the roof of a one-story commercial building, which has a 2.5-ft (0.76-m) parapet on all sides (see Figure 5.62). Determine the design wind pressures on the solar panels using the requirements in ASCE/SEI 29.4.3 and the following design data:

- Indianapolis, IN (Latitude = 39.775°, Longitude = −86.144°)
- Surface roughness: B
- Topography: Not situated on a hill, ridge, or escarpment
- Risk category: II (IBC Table 1604.5)
- Enclosure classification: Enclosed
- Solar panel dimensions: 3.25 ft (0.99 m) by 6.5 ft (1.98 m)
- Orientation of solar panels: Portrait
- Gap between panels in the E-W direction: 1.0 in. (25.4 mm)
- Spacing of gaps between panels in the N-S direction: 6.0 ft (1.83 m)

Figure 5.62 Rooftop Solar Panels in Example 5.18

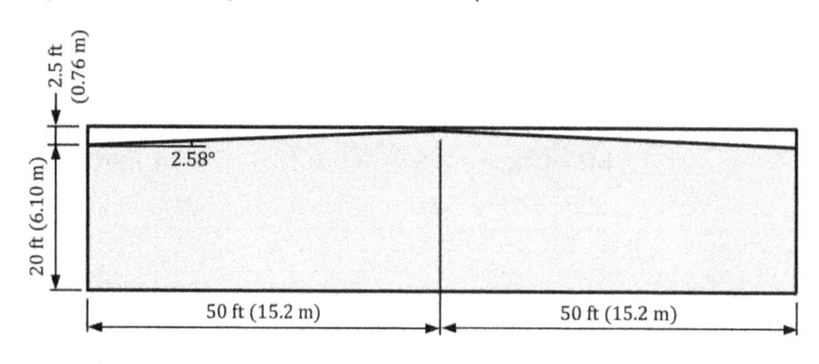

Elevation

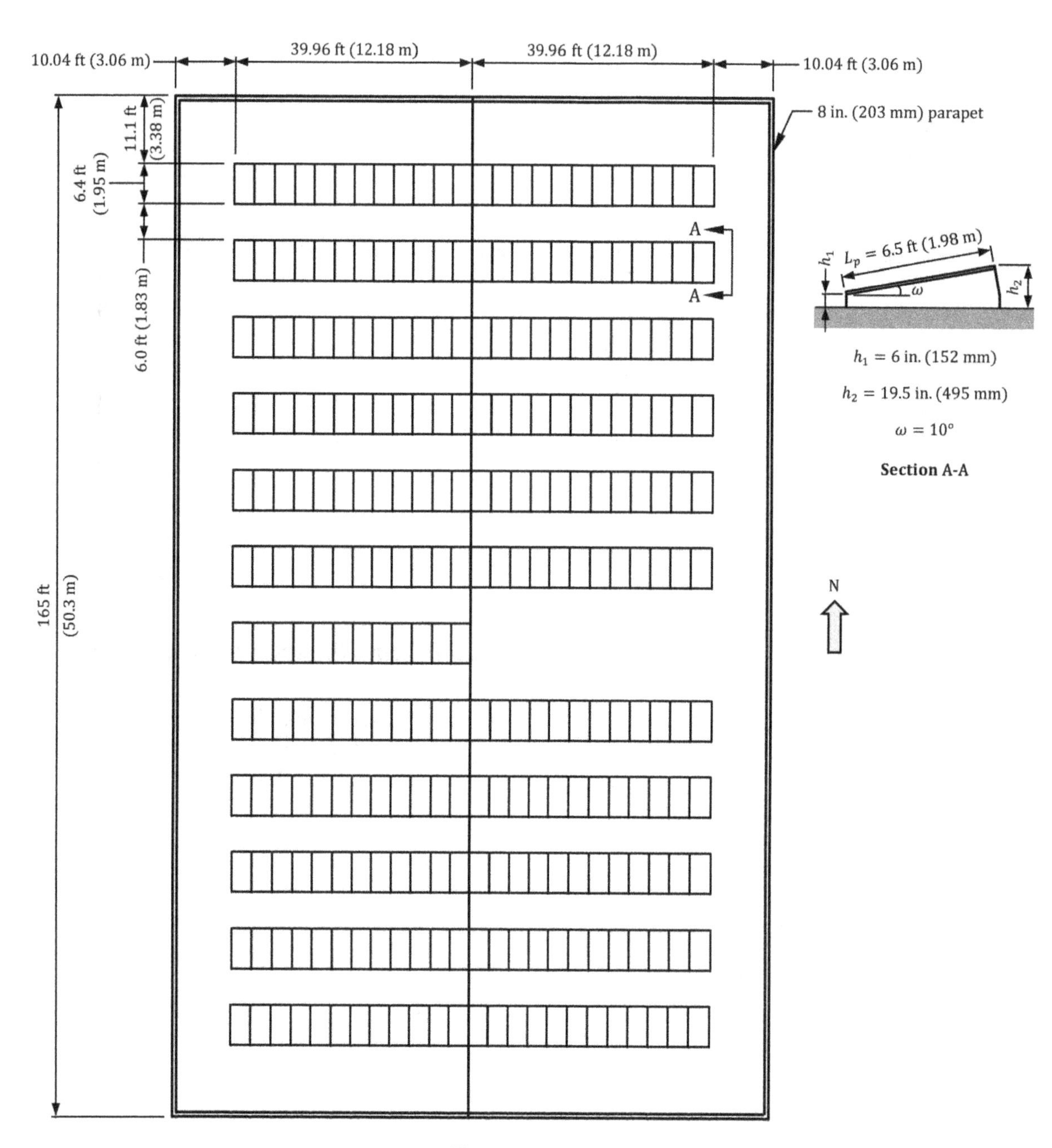

Plan

Solution

Check if the limitations in ASCE/SEI 29.4.3 are satisfied:

- Building is enclosed and has a roof slope $= 2.58$ degrees < 7 degrees
- Panel chord length $L_p = 6.5$ ft (1.98 m) < 6.7 ft (2.04 m)
- Angle the panel makes with the roof surface $\omega = 10$ degrees < 35 degrees
- Height of the panel above the roof at the lower edge of the panel $h_1 = 0.5$ ft (0.15 m) < 2 ft (0.61 m)
- Height of the panel above the roof at the upper edge of the panel $h_2 = 1.63$ ft (0.50 m) < 4.0 ft (1.22 m)
- Gap between panels (E-W direction) $= 1.0$ in. (25.4 mm) > 0.25 in. (6.4 mm)
- Spacing of gaps between panels (N-S direction) $= 6.0$ ft (1.83 m) < 6.7 ft (2.04 m)
- Minimum horizontal clear distance between the panels and the edge of the roof =

$$10.04 \text{ ft } (3.06 \text{ m}) > \text{greater of} \begin{cases} 2(h_2 - h_{pt}) = 2 \times (1.63 - 2.5) < 0 \\ 4 \text{ ft } (1.22 \text{ m}) \end{cases}$$

Therefore, the provisions in ASCE/SEI 29.4.3 are permitted to be used to determine the design wind p ressures on the rooftop solar panels.

Flowchart 5.8 is used to determine the design wind pressures.

Step 1—Determine the wind velocity pressure, q_h

Flowchart 5.1 is used to determine q_h at the mean roof height of the building.

Step 1a—Determine the surface roughness category

Surface roughness B is given in the design data and it is assumed it is present in all directions.

Step 1b—Determine the exposure category

It is assumed exposure category B is applicable in all directions.

Step 1c—Determine the terrain exposure constants

For exposure B, $\alpha = 7.5$ and $z_g = 3{,}280$ ft (1,000 m) from ASCE/SEI Table 26.11-1.

Step 1d—Determine the velocity pressure exposure coefficient, K_h

The velocity pressure exposure coefficient, K_h, is determined from ASCE/SEI Table 26.10-1.

At the mean roof height of 20 ft (6.10 m) above the ground:

$$K_h = 2.41(z/z_g)^{2/\alpha} = 2.41(20/3{,}280)^{2/7.5} = 0.62$$

In S.I.:

$$K_h = 2.41(6.10/1{,}000)^{2/7.5} = 0.62$$

Step 1e—Determine the topographic factor, K_{zt}

The structure is not located on a hill, ridge, or escarpment, so $K_{zt} = 1.0$ (ASCE/SEI 26.8.2).

Step 1f—Determine the ground elevation factor, K_e

It is permitted to take $K_e = 1.0$ for all elevations (ASCE/SEI 26.9).

Step 1g—Determine the risk category

The risk category is given as II in the design data.

Step 1h—Determine the basic wind speed, V

For risk category II structures, use IBC Figure 1609.3(2) or ASCE/SEI Figure 26.5-1B (see Table 5.2 of this book).

Equivalently, use Reference 5.4 to obtain $V = 106$ mi/h (47 m/s) given the latitude and longitude of the site.

Step 1i—Determine the velocity pressure, q_h

The velocity pressure, q_h, is determined by ASCE/SEI Equations (26.10-1) and (26.10-1.SI):

$$q_h = 0.00256 K_h K_{zt} K_e V^2 = 0.00256 \times 0.62 \times 1.0 \times 1.0 \times 106^2 = 17.8 \text{ lb/ft}^2$$

In S.I.:

$$q_h = 0.613 K_h K_{zt} K_e V^2 = 0.613 \times 0.62 \times 1.0 \times 1.0 \times 47^2/1{,}000 = 0.84 \text{ kN/m}^2$$

Step 2—Determine the wind directionality factor, K_d

For building structures, $K_d = 0.85$ from ASCE/SEI Table 26.6-1.

Step 3—Determine the parapet height factor, γ_p

$$\gamma_p = \text{lesser of } \begin{cases} 1.2 \\ 0.9 + (h_{pt}/h) = 0.9 + (2.5/20) = 1.03 \end{cases}$$

In S.I.:

$$\gamma_p = \text{lesser of } \begin{cases} 1.2 \\ 0.9 + (h_{pt}/h) = 0.9 + (0.76/6.10) = 1.03 \end{cases}$$

Step 4—Determine the panel chord factor, γ_c

$$\gamma_c = \text{greater of } \begin{cases} 0.6 + 0.06 L_p = 0.6 + (0.06 \times 6.5) = 0.99 \\ 0.8 \end{cases}$$

Step 5—Determine the array edge factor, γ_E

A panel is defined as exposed where the distance to the roof edge $d_1 > 0.5h = 10$ ft (3.05 m) and one of the following applies:

1. Distance to the adjacent array $d_1 >$ greater of $\begin{cases} 4h_2 = 4 \times 1.63 = 6.5 \text{ ft (1.98 m)} \\ 4 \text{ ft (1.22 m)} \end{cases}$

2. Distance to the next adjacent panel $d_2 >$ greater of $\begin{cases} 4h_2 = 4 \times 1.63 = 6.5 \text{ ft (1.98 m)} \\ 4 \text{ ft (1.22 m)} \end{cases}$

Exposed and nonexposed panels are identified in Figure 5.63.

The array edge factor, γ_E, is equal to the following:

$$\gamma_E = \begin{cases} 1.5 & \text{for uplift loads on exposed panels and within a distance of } 1.5L_p \text{ from the end of a row at an exposed edge of the array} \\ 1.0 & \text{elsewhere for uplift loads and for all downward loads} \end{cases}$$

Step 6—Determine the normalized building length, L_b

$$L_b = \text{least of } \begin{cases} 0.4(hW_L)^{0.5} = 0.4 \times (20.0 \times 165.0)^{0.5} = 23.0 \text{ ft} \\ h = 20.0 \text{ ft} \\ W_S = 100 \text{ ft} \end{cases}$$

Figure 5.63 Exposed and Nonexposed Rooftop Solar Panels in Example 5.18

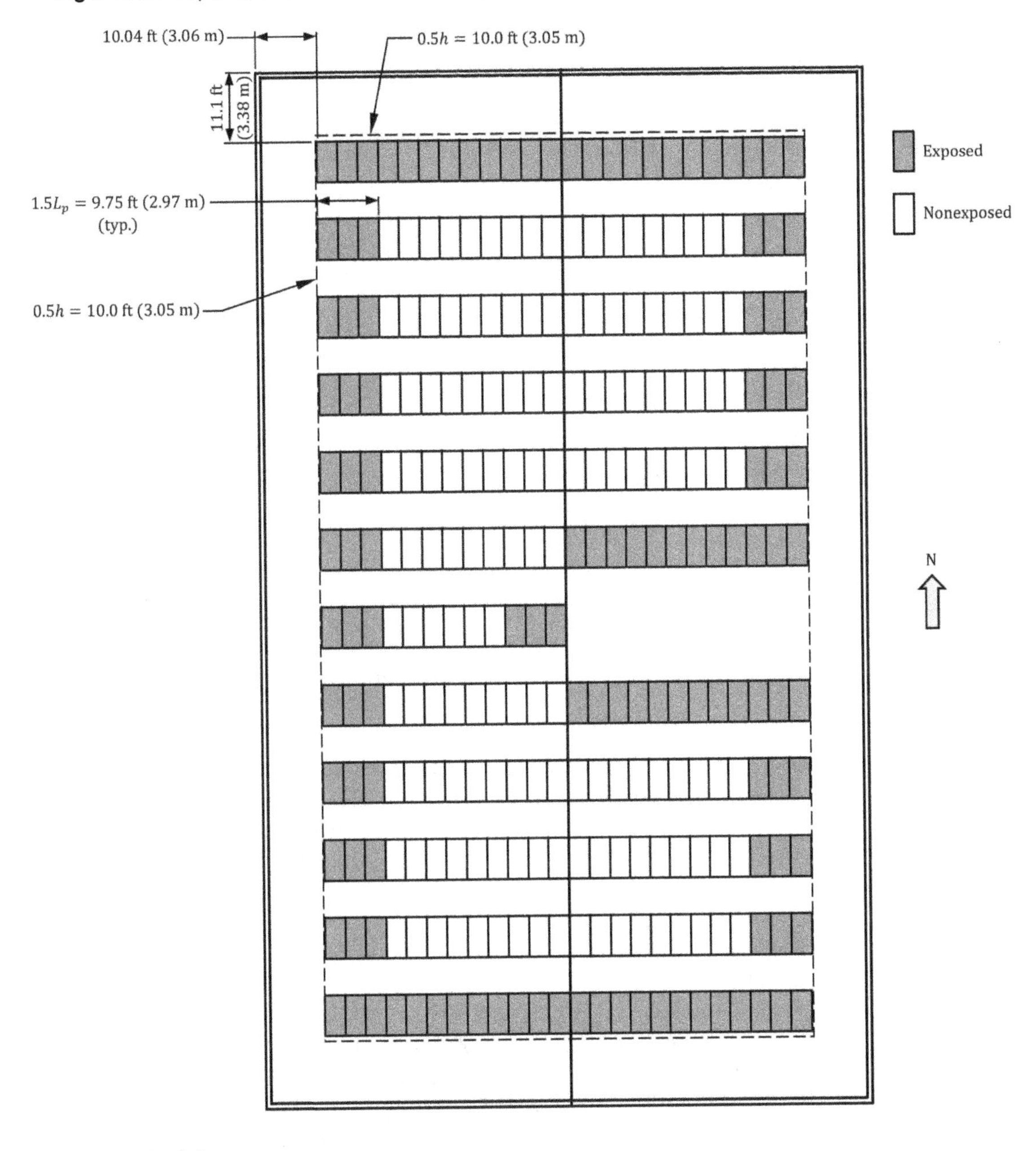

In S.I.:

$$L_b = \text{least of} \begin{cases} 0.4(hW_L)^{0.5} = 0.4 \times (6.10 \times 50.3)^{0.5} = 7.00 \text{ m} \\ h = 6.10 \text{ m} \\ W_S = 30.5 \text{ m} \end{cases}$$

Step 7—Determine the normalized wind area, A_n

The normalized wind area, A_n, is determined by the equation in note 3 of ASCE/SEI Figure 29.4-7:

$$A_n = \frac{1{,}000A}{\max[L_b,\ 15 \text{ ft } (4.57 \text{ m})]^2}$$

where $L_b = 20.0 \text{ ft } (6.10 \text{ m}) > 15.0 \text{ ft } (4.57 \text{ m})$

The effective wind area, A, for a solar panel, which is defined in ASCE/SEI 26.2, is equal to the area of a panel:

$$A = 3.25 \times 6.5 = 21.1 \text{ ft}^2 \ (1.96 \text{ m}^2)$$

Assuming a panel is supported at its four corners, A for a structural support is equal to $21.1/4 = 5.3$ ft^2 (0.49 m^2).

For a solar panel:

$$A_n = \frac{1{,}000 \times 21.1}{20.0^2} = 52.8$$

In S.I.:

$$A_n = \frac{1{,}000 \times 1.96}{6.10^2} = 52.7$$

For a structural support:

$$A_n = \frac{1{,}000 \times 5.3}{20.0^2} = 13.3$$

In S.I.:

$$A_n = \frac{1{,}000 \times 0.49}{6.10^2} = 13.2$$

Step 8—Determine the nominal net pressure coefficients, $(GC_{rn})_{nom}$

Values of $(GC_{rn})_{nom}$ are determined from ASCE/SEI Figure 29.4-7 for the solar panels and structural supports for zones 1, 2, and 3 identified in the figure (see Figure 5.64). The widths of zones 2 and 3 in this example are equal to $2h = 40.0$ ft (12.2 m).

Nominal net pressure coefficients, $(GC_{rn})_{nom}$, are given in Table 5.75. The coefficients are determined by the equations in Table 5.15. Linear interpolation was used to obtain the values in the table for $\omega = 10$ degrees (see note 2 in ASCE/SEI Figure 29.4-7). For example, for a structural support in zone 3:

- For $\omega = 5$ degrees: $(GC_{rn})_{nom} = -0.6669 \log_{10}(13.3) + 2.300 = 1.55$
- For $\omega = 15$ degrees: $(GC_{rn})_{nom} = -1.0004 \log_{10}(13.3) + 3.500 = 2.38$
- For $\omega = 10$ degrees: $(GC_{rn})_{nom} = (1.55 + 2.38)/2 = 1.97$

Step 9—Determine the net pressure coefficients, (GC_{rn})

The net pressure coefficients, (GC_{rn}), are determined by ASCE/SEI Equation (29.4-6):

$$(GC_{rn}) = \gamma_p \gamma_c \gamma_E (GC_{rn})_{nom} = 1.03 \times 0.99 \times \gamma_E (GC_{rn})_{nom} = 1.02 \gamma_E (GC_{rn})_{nom}$$

The net pressure coefficients are given in Table 5.76 for the solar panels and structural supports in each zone.

Step 10—Determine the design wind pressures, p

The design wind pressures, p, are determined by ASCE/SEI Equations (29.4-5) and (29.4-5.SI):

$$p = q_h K_d (GC_{rn}) = 15.1(GC_{rn}) \text{ lb/ft}^2$$

$$p = q_h K_d (GC_{rn}) = 0.71(GC_{rn}) \text{ kN/m}^2$$

Figure 5.64 Roof Zones for Determining Nominal Net Pressure Coefficients, $(GC_{rn})_{nom}$

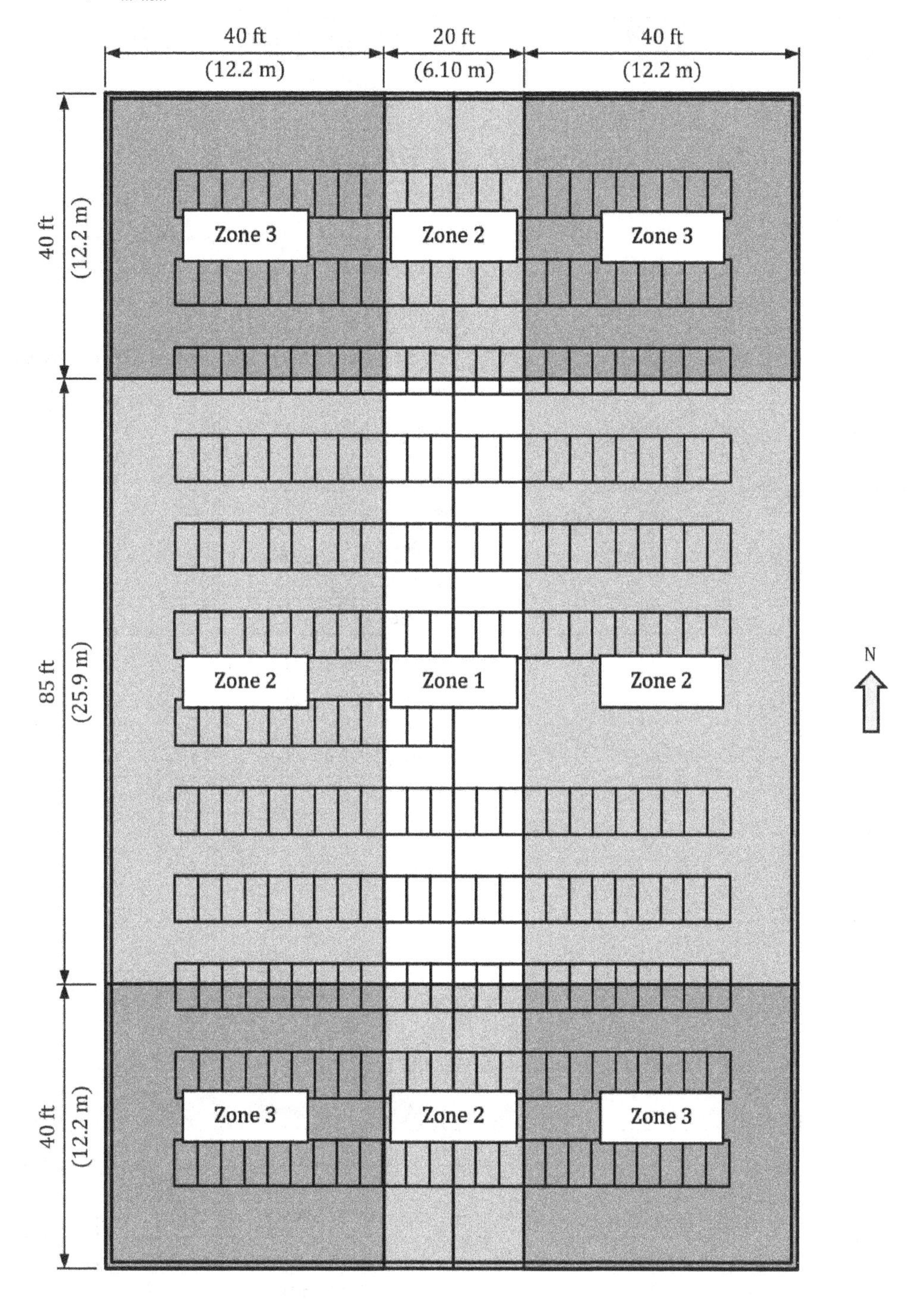

The design wind pressures are given in Table 5.77. These pressures are normal to the surface of the solar panels.

The minimum design wind pressure for other structures is equal to 16 lb/ft^2 (0.77 kN/m^2) in accordance with ASCE/SEI 29.7. The downward design wind pressures on the exposed solar panels in zone 1 and the uplift and downward design wind pressures on the nonexposed solar panels in zone 1 must be increased to the minimum prescribed design wind pressure.

Table 5.75 Nominal Net Pressure Coefficients, $(GC_{rn})_{nom}$, for the Rooftop Solar Panels in Example 5.18

Element	A_n	Zone	$(GC_{rn})_{nom}$
Solar panel	52.8	1	0.93
		2	1.24
		3	1.47
Structural support	13.3	1	1.21
		2	1.66
		3	1.97

Table 5.76 Net Pressure Coefficients, (GC_{rn}), for the Rooftop Solar Panels in Example 5.18

Element	Zone	$(GC_{rn})_{nom}$	Panel Type	(GC_{rn})	
				Uplift Loads	Downward Loads
Solar panel	1	0.93	Exposed	1.42	0.95
			Nonexposed	0.95	0.95
	2	1.24	Exposed	1.90	1.27
			Nonexposed	1.27	1.27
	3	1.47	Exposed	2.25	1.50
			Nonexposed	1.50	1.50
Structural support	1	1.21	Exposed	1.85	1.23
			Nonexposed	1.23	1.23
	2	1.66	Exposed	2.54	1.69
			Nonexposed	1.69	1.69
	3	1.97	Exposed	3.01	2.01
			Nonexposed	2.01	2.01

Table 5.77 Design Wind Pressures for the Rooftop Solar Panels and Structural Supports in Example 5.18

Element	Zone	Panel Type	Uplift Loads		Downward Loads	
			(GC_{rn})	p, lb/ft² (kN/m²)	(GC_{rn})	p, lb/ft² (kN/m²)
Solar panel	1	Exposed	1.42	21.4 (1.01)	0.95	14.4 (0.68)
		Nonexposed	0.95	14.4 (0.68)	0.95	14.4 (0.68)
	2	Exposed	1.90	28.7 (1.35)	1.27	19.2 (0.90)
		Nonexposed	1.27	19.2 (0.90)	1.27	19.2 (0.90)
	3	Exposed	2.25	34.0 (1.60)	1.50	22.7 (1.07)
		Nonexposed	1.50	22.7 (1.07)	1.50	22.7 (1.07)
Structural support	1	Exposed	1.85	27.9 (1.31)	1.23	18.6 (0.87)
		Nonexposed	1.23	18.6 (0.87)	1.23	18.6 (0.87)
	2	Exposed	2.54	38.4 (1.80)	1.69	25.5 (1.20)
		Nonexposed	1.69	25.5 (1.20)	1.69	25.5 (1.20)
	3	Exposed	3.01	45.5 (2.14)	2.01	30.4 (1.43)
		Nonexposed	2.01	30.4 (1.43)	2.01	30.4 (1.43)

The design wind loads on the solar panels and the structural supports are determined by multiplying the uplift and downward design wind pressures in Table 5.77 by the respective effective wind areas, A, determined in step 7. For example, the design wind loads, F, on a structural support for an exposed solar panel in zone 3 are equal to the following:

$$\text{Uplift load: } F = 45.5 \times 5.3 = 241 \text{ lb}$$

$$\text{Downward load: } F = 30.4 \times 5.3 = 161 \text{ lb}$$

In S.I.:

$$\text{Uplift load: } F = 2.14 \times 0.49 \times 1{,}000 = 1{,}049 \text{ N}$$

$$\text{Downward load: } F = 1.43 \times 0.49 \times 1{,}000 = 701 \text{ N}$$

These loads are transferred to the roof structure.
The roof structure must be designed for the load cases given in ASCE/SEI 29.4.3.

5.8.19 Example 5.19—Determination of Design Wind Pressures on the C&C of a Warehouse Building Using ASCE/SEI Chapter 30, Part 1

Determine design wind pressures on (1) the 7-in. (178 mm) precast concrete wall and (2) a roof joist for the warehouse building in Figure 5.27 using the provisions in Part 1 of ASCE/SEI Chapter 30.

SOLUTION

Check if the building meets all the conditions and limitations of ASCE/SEI 30.1.2 and 30.1.3 and the additional conditions of ASCE/SEI 30.3 so that the provisions in Part 1 of ASCE/SEI Chapter 30 for low-rise buildings (ASCE/SEI 30.3) can be used to determine the wind pressures on the C&C.

The building is regularly-shaped, that is, it does not have any unusual geometric irregularities in spatial form. Also, the building does not have response characteristics that make it subject to across-wind loading or other similar effects, and it is not sited at a location where channeling effects or buffeting in the wake of upwind obstructions need to be considered.

The building is enclosed with a gable roof. The mean roof height, h, is equal to 20 ft (6.10 m), which is less than 60 ft (18.3 m), and the building is also a low-rise building as defined in ASCE/SEI 26.2.

Therefore, the provisions in Part 1 of ASCE/SEI Chapter 30 are permitted to be used to determine the design wind pressures on the C&C.

Flowchart 5.11 is used to determine the design wind pressures.

Part 1—Design wind pressures on the precast concrete walls

Step 1—Determine the wind velocity pressure, q_h

The velocity pressure, q_h, at the mean roof height of the building is determined in Example 5.6 and is equal to 26.9 lb/ft^2 (1.27 kN/m^2).

Step 2—Determine the wind directionality factor, K_d

For building structures, $K_d = 0.85$ from ASCE/SEI Table 26.6-1.

Step 3—Determine the effective wind area, A

The effective wind area, A, for a wall is equal to the span length (wall height) multiplied by an effective width that need not be less than one-third the span length (ASCE/SEI 26.2):

$$A = 20 \times (20/3) = 133.3 \text{ ft}^2 \ (12.4 \text{ m}^2)$$

The span length corresponding to the east and west walls is used because this results in a smaller A, and, thus, larger pressures.

Step 4—Determine the external pressure coefficients, (GC_p)

The external pressure coefficients, (GC_p), are determined using ASCE/SEI Figure 30.3-1 for the walls based on the effective wind area, A. In lieu of ASCE/SEI Figure 30.3-1, the equations in ASCE/SEI Table C30.3-1 can be used to determine (GC_p) for zones 4 and 5:

- Zone 4 with $10\ \text{ft}^2\ (0.93\ \text{m}^2) < A = 133.3\ \text{ft}^2\ (12.4\ \text{m}^2) < 500\ \text{ft}^2\ (46.5\ \text{m}^2)$
  Positive $(GC_p) = 1.1766 - 0.1766 \log A = 0.80$
  Negative $(GC_p) = -1.2766 + 0.1766 \log A = -0.90$
- Zone 5 with $10\ \text{ft}^2\ (0.93\ \text{m}^2) < A = 133.3\ \text{ft}^2\ (12.4\ \text{m}^2) < 500\ \text{ft}^2\ (46.5\ \text{m}^2)$
  Positive $(GC_p) = 1.1766 - 0.1766 \log A = 0.80$
  Negative $(GC_p) = -1.7532 + 0.3532 \log A = -1.00$

According to note 5 in ASCE/SEI Figure 30.3-1, values of (GC_p) for walls are permitted to be reduced by 10 percent where the roof slope, θ, is less than 10 degrees. The roof slope is 2.39 degrees for this building, so this note is applicable. The modified external pressure coefficients for the walls are given in Table 5.78.

Table 5.78 Modified External Pressure Coefficients, (GC_p), for the Walls in Example 5.19

Zone	(GC_p)	
	Positive	Negative
4	0.72	–0.81
5	0.72	–0.90

The width of zone 5 is equal to a (ASCE/SEI Figure 30.3-1):

$$a = \text{lesser of} \begin{cases} 0.1 \times \text{least horizontal dimension} = 0.1 \times 148 = 14.8\ \text{ft}\ (4.51\ \text{m}) \\ 0.4h = 0.4 \times 20 = 8.0\ \text{ft}\ (2.44\ \text{m}) \end{cases}$$

$$\text{Minimum } a = \text{greater of} \begin{cases} 0.04 \times \text{least horizontal dimension} = 0.04 \times 148 = 5.9\ \text{ft}\ (1.80\ \text{m}) \\ 3.0\ \text{ft}\ (0.91\ \text{m}) \end{cases}$$

Therefore, $a = 8.0$ ft (2.44 m).

Step 5—Determine the internal pressure coefficients, (GC_{pi})

For an enclosed building, $(GC_{pi}) = +0.18, -0.18$ (ASCE/SEI Table 26.13-1).

Step 6—Determine the design wind pressures, p

The design wind pressures, p, are determined by ASCE/SEI Equations (30.3-1) and (30.3-1.SI):

$$p = q_h K_d[(GC_p) - (GC_{pi})]$$

$$= (26.9 \times 0.85) \times [(GC_p) - (\pm 0.18)] = 22.9 \times [(GC_p) - (\pm 0.18)]\ \text{lb/ft}^2$$

In S.I.:

$$p = (1.27 \times 0.85) \times [(GC_p) - (\pm 0.18)] = 1.08 \times [(GC_p) - (\pm 0.18)]\ \text{kN/m}^2$$

Maximum design wind pressures for positive and negative internal pressures are given in Table 5.79. These pressures act perpendicular to the face of the walls.

Table 5.79 Maximum Design Wind Pressures, p, for the Walls in Example 5.19

Zone	(GC_p)	p, lb/ft² (kN/m²)
4	0.72	20.6 (0.97)
	−0.81	−22.7 (−1.07)
5	0.72	20.6 (0.97)
	−0.90	−24.7 (−1.17)

For example, the maximum negative design wind pressure in zone 5 is equal to the following:

$$p = 22.9 \times [(-0.90) - (0.18)] = -24.7 \text{ lb/ft}^2$$

In S.I.:

$$p = 1.08 \times [(-0.90) - (0.18)] = -1.17 \text{ kN/m}^2$$

The maximum design wind pressures in Table 5.79 are greater than the minimum design wind pressure of 16 lb/ft² (0.77 kN/m²) in ASCE/SEI 30.2.2.

Part 2—Design wind pressures on the roof joists

Step 1—Determine the wind velocity pressure, q_h

The velocity pressure, q_h, at the mean roof height of the building is determined in Example 5.6 and is equal to 26.9 lb/ft² (1.27 kN/m²).

Step 2—Determine the wind directionality factor, K_d

For building structures, $K_d = 0.85$ from ASCE/SEI Table 26.6-1.

Step 3—Determine the effective wind area, A

The effective wind area, A, for the roof joists is equal to the greater of the tributary area of a joist and the span length multiplied by an effective width that need not be less than one-third the span length (ASCE/SEI 26.2):

For the interior joists:

$$A = \text{greater of} \begin{cases} 8.0 \times 37.0 = 296 \text{ ft}^2 \ (27.5 \text{ m}^2) \\ 37.0 \times (37.0/3) = 456.3 \text{ ft}^2 \ (42.4 \text{ m}^2) \end{cases}$$

For the edge joists:

$$A = \text{greater of} \begin{cases} [(8.0/2) + 5.0] \times 37.0 = 333 \text{ ft}^2 \ (30.9 \text{ m}^2) \\ 37.0 \times (37.0/3) = 456.3 \text{ ft}^2 \ (42.4 \text{ m}^2) \end{cases}$$

Step 4—Determine the external pressure coefficients, (GC_p)

The external pressure coefficients, (GC_p), are determined from ASCE/SEI Figure 30.3-2A for the joists based on the effective wind area, A. In lieu of ASCE/SEI Figure 30.3-2A, the equations in ASCE/SEI Table C30.3-2 can be used to determine (GC_p) for zones 1′, 1, 2, and 3:

- Zone 1′
  For $A = 456.3 \text{ ft}^2 \ (42.4 \text{ m}^2) > 100 \text{ ft}^2 \ (9.3 \text{ m}^2)$: positive $(GC_p) = 0.20$
  For $100 \text{ ft}^2 \ (9.3 \text{ m}^2) < A = 456.3 \text{ ft}^2 \ (42.4 \text{ m}^2) < 1{,}000 \text{ ft}^2 \ (92.9 \text{ m}^2)$: negative (GC_p) without overhang $= -1.9000 + 0.5000 \log A = -0.57$
- Zone 1
  For $A = 456.3 \text{ ft}^2 \ (42.4 \text{ m}^2) > 100 \text{ ft}^2 \ (9.3 \text{ m}^2)$: positive $(GC_p) = 0.20$
  For $10 \text{ ft}^2 \ (0.93 \text{ m}^2) < A = 456.3 \text{ ft}^2 \ (42.4 \text{ m}^2) < 500 \text{ ft}^2 \ (46.5 \text{ m}^2)$: negative (GC_p) without overhang $= -2.1120 + 0.4120 \log A = -1.02$

- Zone 2
  For $A = 456.3$ ft^2 (42.4 m^2) > 100 ft^2 (9.3 m^2): positive $(GC_p) = 0.20$
  For 10 ft^2 (0.93 m^2) < $A = 456.3$ ft^2 (42.4 m^2) < 500 ft^2 (46.5 m^2): negative (GC_p) without overhang $= -2.8297 + 0.5297 \log A = -1.42$
- Zone 3
  For $A = 456.3$ ft^2 (42.4 m^2) > 100 ft^2 (9.3 m^2): positive $(GC_p) = 0.20$
  For 10 ft^2 (0.93 m^2) < $A = 456.3$ ft^2 (42.4 m^2) < 500 ft^2 (46.5 m^2): negative (GC_p) without overhang $= -4.2595 + 1.0595 \log A = -1.44$

Step 5—Determine the internal pressure coefficients, (GC_{pi})

For an enclosed building, $(GC_{pi}) = +0.18, -0.18$ (ASCE/SEI Table 26.13-1).

Step 6—Determine the design wind pressures, p

The design wind pressures, p, are determined by ASCE/SEI Equations (30.3-1) and (30.3-1.SI):

$$p = q_h K_d[(GC_p) - (GC_{pi})]$$

$$= (26.9 \times 0.85) \times [(GC_p) - (\pm 0.18)] = 22.9 \times [(GC_p) - (\pm 0.18)] \text{ lb/ft}^2$$

In S.I.:

$$p = (1.27 \times 0.85) \times [(GC_p) - (\pm 0.18)] = 1.08 \times [(GC_p) - (\pm 0.18)] \text{ kN/m}^2$$

Maximum design wind pressures for positive and negative internal pressures are given in Table 5.80. These pressures are applied normal to the joists and act over the tributary area of each joist, which is equal to $8.0 \times 37.0 = 296$ ft^2 (27.5 m^2). If the tributary area were greater than 700 ft^2 (65.0 m^2), the provisions for MWFRSs are permitted to be used to determine the design wind pressures (ASCE/SEI 30.2.3).

Table 5.80 Maximum Design Wind Pressures, *p*, for the Roof Joists in Example 5.19

Zone	(GC_p)	p, lb/ft^2 (kN/m^2)
1′	0.20	8.7 (0.41)
	−0.57	−17.2 (−0.81)
1	0.20	8.7 (0.41)
	−1.02	−27.5 (−1.30)
2	0.20	8.7 (0.41)
	−1.42	−36.6 (−1.73)
3	0.20	8.7 (0.41)
	−1.44	−37.1 (−1.75)

For example, the maximum negative design wind pressure in zone 3 is equal to the following:

$$p = 22.9 \times [(-1.44) - (0.18)] = -37.1 \text{ lb/ft}^2$$

In S.I.:

$$p = 1.08 \times [(-1.44) - (0.18)] = -1.75 \text{ kN/m}^2$$

The positive design wind pressures in Table 5.80 must be increased to the minimum design wind pressure of 16 lb/ft^2 (0.77 kN/m^2) in ASCE/SEI 30.2.2.

For simplicity, zones 2 and 3 on the roof are conservatively combined into one zone using the maximum design wind pressures from zone 3. Loading diagrams for type A roof joists located in combined zones 2 and 3 are given in Figure 5.65. Similar loading diagrams can be obtained for the other joist types indicated in the figure.

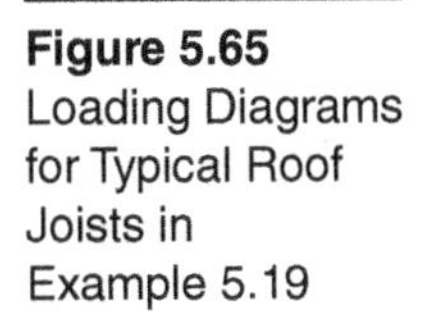

Figure 5.65 Loading Diagrams for Typical Roof Joists in Example 5.19

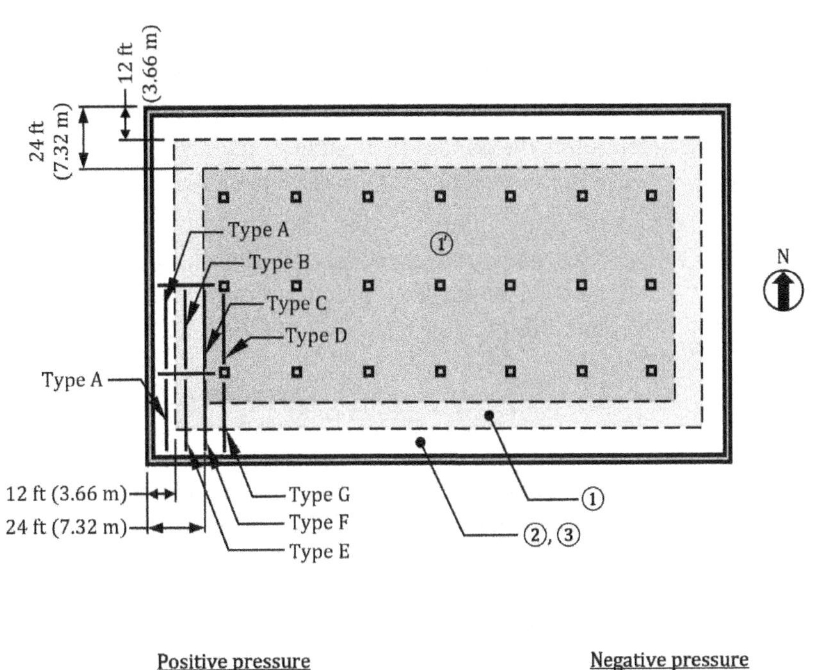

5.8.20 Example 5.20—Determination of Design Wind Pressures on the C&C of a Residential Building Using ASCE/SEI Chapter 30, Part 1

Determine design wind pressures on (1) a typical wall stud in the third story assuming the studs are spaced 16 in. (406 mm) on center, (2) a roof truss assuming the trusses are spaced 2 ft (0.61 m) on center, and (3) a typical sheathing panel on the roof assuming the panels are 4 ft by 8 ft (1.22 m by 2.44 m) sheets for the residential building in Figure 5.43 using the provisions in Part 1 of ASCE/SEI Chapter 30.

SOLUTION

Check if the building meets all the conditions and limitations of ASCE/SEI 30.1.2 and 30.1.3 and the additional conditions of ASCE/SEI 30.3 so that the provisions in Part 1 of ASCE/SEI Chapter 30 for low-rise buildings (ASCE/SEI 30.3) can be used to determine the wind pressures on the C&C.

The building is regularly-shaped, that is, it does not have any unusual geometric irregularities in spatial form. Also, the building does not have response characteristics that make it subject to across-wind loading or other similar effects, and it is not sited at a location where channeling effects or buffeting in the wake of upwind obstructions need to be considered.

The building is enclosed with a gable roof with a mean roof height, h, equal to 38 ft (11.6 m), which is less than 60 ft (18.3 m).

Therefore, the provisions in Part 1 of ASCE/SEI Chapter 30 are permitted to be used to determine the design wind pressures on the C&C.

Flowchart 5.11 is used to determine the design wind pressures.

Part 1—Design wind pressures on a typical wall stud

Step 1—Determine the wind velocity pressure, q_h

The velocity pressure, q_h, at the mean roof height of the building is determined in Example 5.10 and is equal to 33.6 lb/ft^2 (1.61 kN/m^2).

Step 2—Determine the wind directionality factor, K_d

For building structures, $K_d = 0.85$ from ASCE/SEI Table 26.6-1.

Step 3—Determine the effective wind area, A

The effective wind area, A, is defined as the greater of the tributary area of the wall stud and the span length (wall height) multiplied by an effective width that need not be less than one-third the span length (ASCE/SEI 26.2):

$$A = \text{greater of } \begin{cases} (16/12) \times 10.0 = 13.3 \text{ ft}^2 \ (1.2 \text{ m}^2) \\ 10.0 \times (10.0/3) = 33.3 \text{ ft}^2 \ (3.1 \text{ m}^2) \end{cases}$$

Step 4—Determine the external pressure coefficients, (GC_p)

The external pressure coefficients, (GC_p), are determined using ASCE/SEI Figure 30.3-1 for the walls based on the effective wind area, A. In lieu of ASCE/SEI Figure 30.3-1, the equations in ASCE/SEI Table C30.3-1 are permitted to be used to determine (GC_p) for zones 4 and 5:

- Zone 4 with $10 \text{ ft}^2 \ (0.93 \text{ m}^2) < A = 33.3 \text{ ft}^2 \ (3.10 \text{ m}^2) < 500 \text{ ft}^2 \ (46.5 \text{ m}^2)$
  Positive $(GC_p) = 1.1766 - 0.1766 \log A = 0.91$
  Negative $(GC_p) = -1.2766 + 0.1766 \log A = -1.01$
- Zone 5 with $10 \text{ ft}^2 \ (0.93 \text{ m}^2) < A = 33.3 \text{ ft}^2 \ (3.10 \text{ m}^2) < 500 \text{ ft}^2 \ (46.5 \text{ m}^2)$
  Positive $(GC_p) = 1.1766 - 0.1766 \log A = 0.91$
  Negative $(GC_p) = -1.7532 + 0.3532 \log A = -1.22$

The width of zone 5 is equal to a (ASCE/SEI Figure 30.3-1):

$$a = \text{lesser of } \begin{cases} 0.1 \times \text{least horizontal dimension} = 0.1 \times 72 = 7.2 \text{ ft (2.20 m)} \\ 0.4h = 0.4 \times 38 = 15.2 \text{ ft (4.63 m)} \end{cases}$$

$$\text{Minimum } a = \text{greater of } \begin{cases} 0.04 \times \text{least horizontal dimension} = 0.04 \times 72 = 2.9 \text{ ft (0.88 m)} \\ 3.0 \text{ ft (0.91 m)} \end{cases}$$

Therefore, $a = 7.2$ ft (2.20 m).

Step 5—Determine the internal pressure coefficients, (GC_{pi})

For an enclosed building, $(GC_{pi}) = +0.18, -0.18$ (ASCE/SEI Table 26.13-1).

Step 6—Determine the design wind pressures, p

The design wind pressures, p, are determined by ASCE/SEI Equations (30.3-1) and (30.3-1.SI):

$$\begin{aligned} p &= q_h K_d [(GC_p) - (GC_{pi})] \\ &= (33.6 \times 0.85) \times [(GC_p) - (\pm 0.18)] = 28.6 \times [(GC_p) - (\pm 0.18)] \text{ lb/ft}^2 \end{aligned}$$

In S.I.:

$$p = (1.61 \times 0.85) \times [(GC_p) - (\pm 0.18)] = 1.37 \times [(GC_p) - (\pm 0.18)] \text{ kN/m}^2$$

Maximum design wind pressures for positive and negative internal pressures are given in Table 5.81. These pressures act perpendicular to the wall studs.

Table 5.81 Maximum Design Wind Pressures, *p*, for the Wall Studs in Example 5.20

Zone	(GC_p)	p, lb/ft² (kN/m²)
4	0.91	31.1 (1.49)
	−1.01	−34.0 (−1.63)
5	0.91	31.1 (1.49)
	−1.22	−40.0 (−1.92)

For example, the maximum negative design wind pressure in zone 5 is equal to the following:

$$p = 28.6 \times [(-1.22) - (0.18)] = -40.0 \text{ lb/ft}^2$$

In S.I.:

$$p = 1.37 \times [(-1.22) - (0.18)] = -1.92 \text{ kN/m}^2$$

The maximum design wind pressures in Table 5.81 are greater than the minimum design wind pressure of 16 lb/ft² (0.77 kN/m²) in ASCE/SEI 30.2.2.

Part 2—Design wind pressures on the roof trusses

Step 1—Determine the wind velocity pressure, q_h

The velocity pressure, q_h, at the mean roof height of the building is determined in Example 5.10 and is equal to 33.6 lb/ft² (1.61 kN/m²).

Step 2—Determine the wind directionality factor, K_d

For building structures, $K_d = 0.85$ from ASCE/SEI Table 26.6-1.

Step 3—Determine the effective wind area, A

The effective wind area, A, is defined as the greater of the tributary area of a truss and the span length multiplied by an effective width that need not be less than one-third the span length (ASCE/SEI 26.2):

For trusses spanning in the N-S direction:

$$A = \text{greater of} \begin{cases} 2.0 \times 56.5 = 113 \text{ ft}^2 \ (10.5 \text{ m}^2) \\ 56.5 \times (56.5/3) = 1{,}064 \text{ ft}^2 \ (98.9 \text{ m}^2) \end{cases}$$

For trusses spanning in the E-W direction:

$$A = \text{greater of} \begin{cases} 2.0 \times 24.5 = 49 \text{ ft}^2 \ (4.6 \text{ m}^2) \\ 24.5 \times (24.5/3) = 200 \text{ ft}^2 \ (18.6 \text{ m}^2) \end{cases}$$

Step 4—Determine the external pressure coefficients, (GC_p)

- Trusses spanning in the N-S direction
  The external pressure coefficients, (GC_p), for trusses spanning in the N-S direction are determined using ASCE/SEI Figure 30.3-2C for an enclosed building with a gable roof with 20 degrees < $\theta = 23$ degrees < 27 degrees and $A = 1{,}064$ ft^2 (42.4 m^2).
  The pressure coefficients from ASCE/SEI Figure 30.3-2C are given in Table 5.82.

Table 5.82 External Pressure Coefficients, (GC_p), for the Roof Trusses Spanning in the N-S in Example 5.20

Zone	(GC_p)	
	Positive	Negative
1	0.30	−0.80
2	0.30	−1.20
3	0.30	−1.40

- Trusses spanning in the E-W direction
  The external pressure coefficients, (GC_p), for trusses spanning in the E-W direction are determined using ASCE/SEI Figure 30.3-2D for an enclosed building with a gable roof with 27 degrees < $\theta = 44.4$ degrees < 45 degrees and $A = 200$ ft^2 (18.6 m^2).
  The pressure coefficients from ASCE/SEI Figure 30.3-2D are given in Table 5.83.

Table 5.83 External Pressure Coefficients, (GC_p), for the Roof Trusses Spanning in the E-W in Example 5.20

Zone	(GC_p)	
	Positive	Negative
1	0.50	−0.80
2	0.50	−1.00
3	0.50	−1.00

Step 5—Determine the internal pressure coefficients, (GC_{pi})

For an enclosed building, $(GC_{pi}) = +0.18, -0.18$ (ASCE/SEI Table 26.13-1).

Step 6—Determine the design wind pressures, p

The design wind pressures, p, are determined by ASCE/SEI Equations (30.3-1) and (30.3-1.SI):

$$p = q_h K_d[(GC_p) - (GC_{pi})]$$

$$= (33.6 \times 0.85) \times [(GC_p) - (\pm 0.18)] = 28.6 \times [(GC_p) - (\pm 0.18)] \text{ lb/ft}^2$$

In S.I.:

$$p = (1.61 \times 0.85) \times [(GC_p) - (\pm 0.18)] = 1.37 \times [(GC_p) - (\pm 0.18)] \text{ kN/m}^2$$

- Trusses spanning in the N-S direction
  Maximum design wind pressures for positive and negative internal pressures are given in Table 5.84. These pressures are applied normal to the trusses and act over the tributary area of each truss, which is equal to $2.0 \times 56.5 = 113$ ft^2 (10.5 m^2). If the tributary area were greater than 700 ft^2 (65.0 m^2), the provisions for MWFRSs are permitted to be used to determine the design wind pressures (ASCE/SEI 30.2.3).

Table 5.84 Maximum Design Wind Pressures, *p*, for the Roof Trusses Spanning in the N-S Direction in Example 5.20

Zone	(GC_p)	p, lb/ft² (kN/m²)
1	0.30	13.7 (0.66)
	−0.80	−28.0 (−1.34)
2	0.30	13.7 (0.66)
	−1.20	−39.5 (−1.89)
3	0.30	13.7 (0.66)
	−1.40	−45.2 (−2.16)

For example, the maximum negative design wind pressure in zone 3 is equal to the following:

$$p = 28.6 \times [(-1.40) - (0.18)] = -45.2 \text{ lb/ft}^2$$

In S.I.:

$$p = 1.37 \times [(-1.40) - (0.18)] = -2.16 \text{ kN/m}^2$$

The positive design wind pressures in Table 5.84 must be increased to the minimum design wind pressure of 16 lb/ft² (0.77 kN/m²) in ASCE/SEI 30.2.2.

- Trusses spanning in the E-W direction
  Maximum design wind pressures for positive and negative internal pressures are given in Table 5.85. These pressures are applied normal to the trusses and act over the tributary area of each truss, which is equal to $2.0 \times 24.5 = 49$ ft² (4.6 m²).

Table 5.85 Maximum Design Wind Pressures, *p*, for the Roof Trusses Spanning in the E-W Direction in Example 5.20

Zone	(GC_p)	p, lb/ft² (kN/m²)
1	0.50	19.5 (0.93)
	−0.80	−28.0 (−1.34)
2 and 3	0.50	19.5 (0.93)
	−1.00	−33.8 (−1.62)

For example, the maximum negative design wind pressure in zone 3 is equal to the following:

$$p = 28.6 \times [(-1.00) - (0.18)] = -33.8 \text{ lb/ft}^2$$

In S.I.:

$$p = 1.37 \times [(-1.00) - (0.18)] = -1.62 \text{ kN/m}^2$$

Loading diagrams for typical trusses located in various zones are given in Figure 5.66.

Part 3—Design wind pressures on a typical roof sheathing panel

Step 1—Determine the wind velocity pressure, q_h

The velocity pressure, q_h, at the mean roof height of the building is determined in Example 5.10 and is equal to 33.6 lb/ft² (1.61 kN/m²).

Figure 5.66 Loading Diagrams for Typical Roof Trusses in Example 5.20

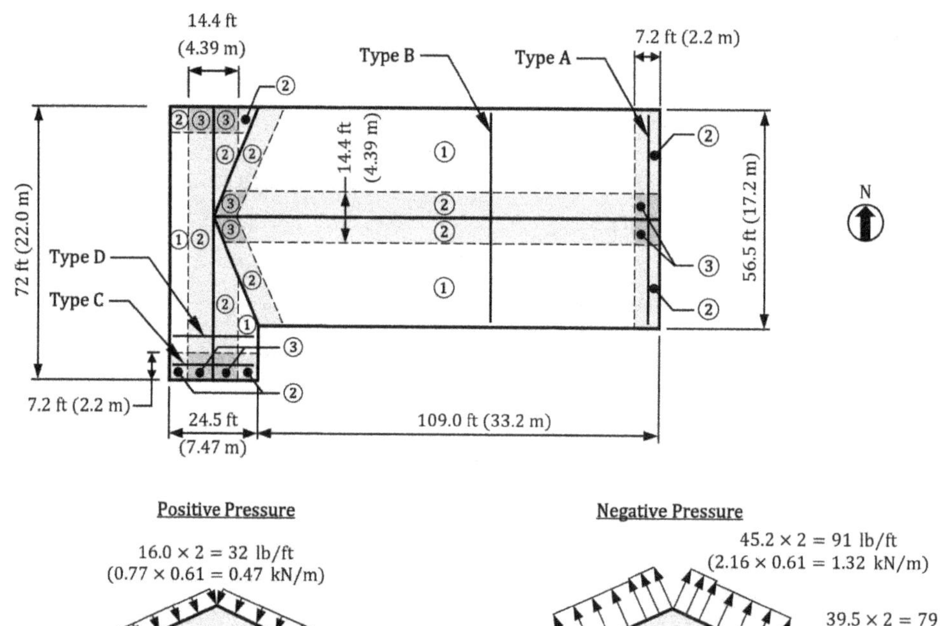

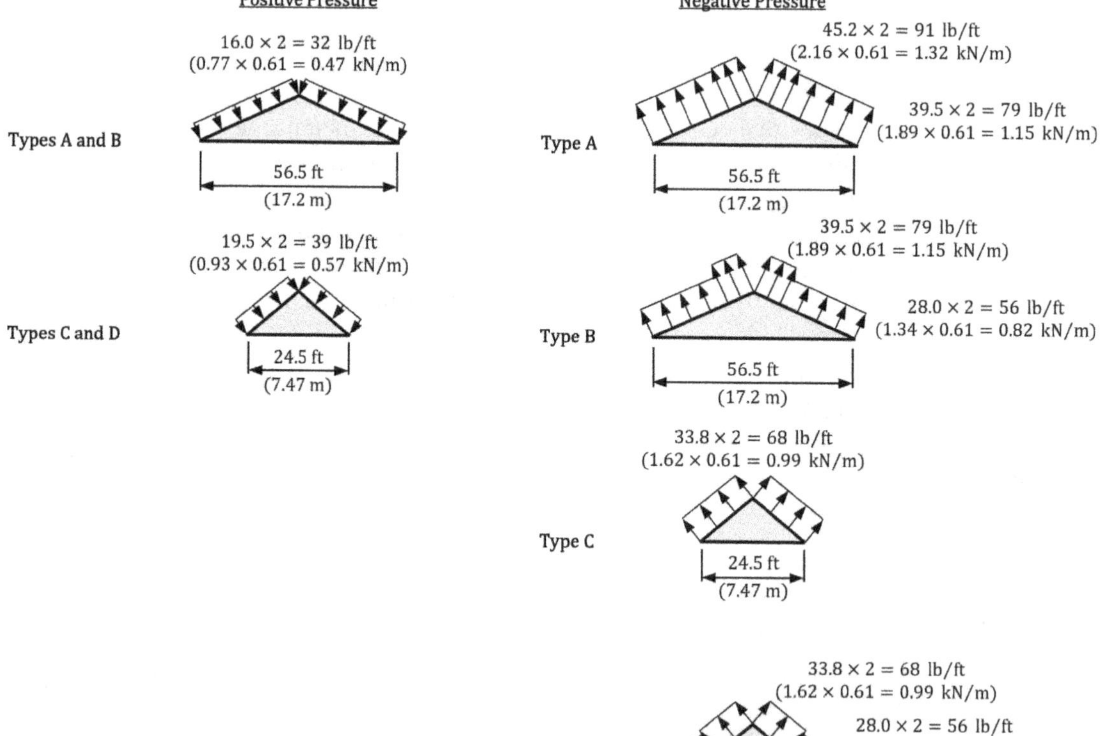

Step 2—Determine the wind directionality factor, K_d

For building structures, $K_d = 0.85$ from ASCE/SEI Table 26.6-1.

Step 3—Determine the effective wind area, A

The effective wind area, A, is defined as the greater of the tributary area of a sheathing panel and the span length multiplied by an effective width that need not be less than one-third the span length (ASCE/SEI 26.2):

$$A = \text{greater of} \begin{cases} 2.0 \times 4.0 = 8 \text{ ft}^2 \ (0.74 \text{ m}^2) \\ 2.0 \times (2.0/3) = 1.3 \text{ ft}^2 \ (0.12 \text{ m}^2) \end{cases}$$

Step 4—Determine the external pressure coefficients, (GC_p)

- Sheathing panels on the east wing

  The external pressure coefficients, (GC_p), for sheathing panels on the east wing are determined using ASCE/SEI Figure 30.3-2C for an enclosed building with a gable roof with 20 degrees $< \theta = 23$ degrees < 27 degrees based on the effective wind area, $A = 8$ ft^2 (0.74 m^2).

  The pressure coefficients from ASCE/SEI Figure 30.3-2C are given in Table 5.86.

Table 5.86 External Pressure Coefficients, (GC_p), for Typical Roof Sheathing Panels on the East Wing of the Building in Example 5.20

Zone	(GC_p)	
	Positive	Negative
1	0.60	−1.50
2	0.60	−2.50
3	0.60	−3.00

- Sheathing panels on the west wing

  The external pressure coefficients, (GC_p), sheathing panels on the west wing are determined using ASCE/SEI Figure 30.3-2D for an enclosed building with a gable roof with 27 degrees $< \theta = 44.4$ degrees < 45 degrees based on the effective wind area, $A = 8$ ft^2 (0.74 m^2).

  The pressure coefficients from ASCE/SEI Figure 30.3-2D are given in Table 5.87.

Table 5.87 External Pressure Coefficients, (GC_p), for Typical Roof Sheathing Panels on the West Wing of the Building in Example 5.20

Zone	(GC_p)	
	Positive	Negative
1	0.90	−1.80
2	0.90	−2.00
3	0.90	−2.50

Step 5—Determine the internal pressure coefficients, (GC_{pi})

For an enclosed building, $(GC_{pi}) = +0.18, -0.18$ (ASCE/SEI Table 26.13-1).

Step 6—Determine the design wind pressures, p

The design wind pressures, p, are determined by ASCE/SEI Equations (30.3-1) and (30.3-1.SI):

$$p = q_h K_d[(GC_p) - (GC_{pi})]$$

$$= (33.6 \times 0.85) \times [(GC_p) - (\pm 0.18)] = 28.6 \times [(GC_p) - (\pm 0.18)] \text{ lb/ft}^2$$

In S.I.:

$$p = (1.61 \times 0.85) \times [(GC_p) - (\pm 0.18)] = 1.37 \times [(GC_p) - (\pm 0.18)] \text{ kN/m}^2$$

- Sheathing panels on the east wing

  Maximum design wind pressures for positive and negative internal pressures are given in Table 5.88. These pressures are applied normal to the sheathing panels and act over the tributary area of the panel, which is equal to $2.0 \times 4.0 = 8$ ft^2 (0.74 m^2). If the tributary area were greater than 700 ft^2 (65.0 m^2), the provisions for MWFRSs are permitted to be used to determine the design wind pressures (ASCE/SEI 30.2.3).

Table 5.88 Maximum Design Wind Pressures, *p*, for the Roof Sheathing Panels on the East Wing of the Building in Example 5.20

Zone	(GC_p)	p, lb/ft² (kN/m²)
1	0.60	22.3 (1.07)
	–1.50	–48.0 (–2.30)
2	0.60	22.3 (1.07)
	–2.50	–76.7 (–3.67)
3	0.60	22.3 (1.07)
	–3.00	–91.0 (–4.36)

For example, the maximum negative design wind pressure in zone 3 is equal to the following:

$$p = 28.6 \times [(-3.00) - (0.18)] = -91.0 \text{ lb/ft}^2$$

In S.I.:

$$p = 1.37 \times [(-3.00) - (0.18)] = -4.36 \text{ kN/m}^2$$

- Sheathing panels on the west wing
  Maximum design wind pressures for positive and negative internal pressures are given in Table 5.89. These pressures are applied normal to the sheathing panels and act over the tributary area of the panel, which is equal to $2.0 \times 4.0 = 8 \text{ ft}^2 (0.74 \text{ m}^2)$.

Table 5.89 Maximum Design Wind Pressures, *p*, for the Roof Sheathing Panels on the West Wing of the Building in Example 5.20

Zone	(GC_p)	p, lb/ft² (kN/m²)
1	0.90	30.9 (1.48)
	–1.80	–56.6 (–2.71)
2	0.90	30.9 (1.48)
	–2.00	–62.4 (–2.99)
	0.90	30.9 (1.48)
	–2.50	–76.7 (–3.67)

For example, the maximum negative design wind pressure in zone 3 is equal to the following:

$$p = 28.6 \times [(-2.50) - (0.18)] = -76.7 \text{ lb/ft}^2$$

In S.I.:

$$p = 1.37 \times [(-2.50) - (0.18)] = -3.67 \text{ kN/m}^2$$

5.8.21 Example 5.21—Determination of Design Wind Pressures on the C&C of an Office Building Using ASCE/SEI Chapter 30, Part 2

Determine design wind pressures on the cladding in the fifteenth story of the office building in Figure 5.50 assuming the cladding consists of mullions spanning 9.5 ft (2.90 m) between the floor slabs and glazing panels supported by the mullions, which are spaced 5 ft (1.52 m) on center, using the provisions in Part 2 of ASCE/SEI Chapter 30.

SOLUTION

Check if the building meets all the conditions and limitations of ASCE/SEI 30.1.2 and 30.1.3 and the additional conditions of ASCE/SEI 30.4 so that the provisions in Part 2 of ASCE/SEI Chapter 30 for buildings with a mean roof height greater than 60 ft (18.3 m) [ASCE/SEI 30.4] can be used to determine the wind pressures on the C&C.

The building is regularly shaped, that is, it does not have any unusual geometric irregularities in spatial form. Also, the building does not have response characteristics that make it subject to across-wind loading or other similar effects, and it is not sited at a location where channeling effects or buffeting in the wake of upwind obstructions need to be considered.

The building is enclosed with a flat roof with a mean roof height, h, equal to 142.5 ft (43.5 m), which is greater than 60 ft (18.3 m).

Therefore, the provisions in Part 2 of ASCE/SEI Chapter 30 are permitted to be used to determine the design wind pressures on the C&C.

Flowchart 5.12 is used to determine the design wind pressures.

Step 1—Determine the wind velocity pressure, q_h

The velocity pressure, q_h, at the mean roof height of the building is determined in Example 5.12 and is equal to 29.4 lb/ft^2 (1.41 kN/m^2) [the velocity pressure at the mean roof height is used because the design wind pressures on the cladding are to be determined for the fifteenth story].

Step 2—Determine the wind directionality factor, K_d

For building structures, $K_d = 0.85$ from ASCE/SEI Table 26.6-1.

Step 3—Determine the effective wind area, A

The effective wind area, A, is defined as the greater of the tributary area of the element and the span length multiplied by an effective width that need not be less than one-third the span length (ASCE/SEI 26.2).

For the mullions (see Figure 5.21):

$$A = \text{greater of} \begin{cases} 9.5 \times 5.0 = 47.5 \text{ ft}^2 \ (4.4 \text{ m}^2) \\ 9.5 \times (9.5/3) = 30.1 \text{ ft}^2 \ (2.8 \text{ m}^2) \end{cases}$$

For the glazing panels:

$$A = \text{greater of} \begin{cases} 9.5 \times 5.0 = 47.5 \text{ ft}^2 \ (4.4 \text{ m}^2) \\ 5.0 \times (5.0/3) = 8.3 \text{ ft}^2 \ (0.8 \text{ m}^2) \end{cases}$$

Therefore, the effective wind areas are the same for the mullions and glazing panels.

Step 4—Determine the external pressure coefficients, (GC_p)

The external pressure coefficients, (GC_p), are determined using ASCE/SEI Figure 30.4-1 for the cladding elements based on the effective wind area, A (see Table 5.90).

Table 5.90 External Pressure Coefficients, (GC_p), for the Cladding of the Building in Example 5.21

Zone	(GC_p)	
	Positive	Negative
4	0.82	−0.85
5	0.82	−1.59

The width of zone 5 is equal to a (ASCE/SEI Figure 30.4-1):

$$a = \text{greater of} \begin{cases} 0.1 \times \text{least horizontal dimension} = 0.1 \times 61.0 = 6.1 \text{ ft (1.86 m)} \\ 3.0 \text{ ft (0.91 m)} \end{cases}$$

Step 5—Determine the velocity pressure for internal pressure determination, q_i

According to ASCE/SEI 30.4.2, $q_i = q_h = 29.4$ lb/ft^2 (1.41 kN/m^2) for all surfaces of an enclosed building.

Step 6—Determine the internal pressure coefficients, (GC_{pi})

For an enclosed building, $(GC_{pi}) = +0.18, -0.18$ (ASCE/SEI Table 26.13-1).

Step 7—Determine the design wind pressures, p_h

The design wind pressures, p, are determined by ASCE/SEI Equations (30.4-1) and (30.4-1.SI):

$$p_h = q_h K_d (GC_p) - q_i K_d (GC_{pi})$$

$$= [(29.4 \times 0.85) \times (GC_p)] - [(29.4 \times 0.85) \times (\pm 0.18)] = [25.0 \times (GC_p)] \mp 4.5 \text{ lb/ft}^2$$

In S.I.:

$$p_h = [(1.41 \times 0.85) \times (GC_p)] - [(1.41 \times 0.85) \times (\pm 0.18)] = [1.20 \times (GC_p)] \mp 0.22 \text{ kN/m}^2$$

Maximum design wind pressures for positive and negative internal pressures are given in Table 5.91. The maximum positive pressure is obtained with negative internal pressure. Similarly, the maximum negative pressure is obtained with positive internal pressure. These pressures act perpendicular to the elements of the cladding system.

Table 5.91 Maximum Design Wind Pressures, *p*, for the Elements of the Cladding System in Example 5.21

Zone	p_h, lb/ft^2 (kN/m^2)	
	Positive	Negative
4	25.0 (1.20)	–25.8 (–1.24)
5	25.0 (1.20)	–44.3 (–2.13)

For example, the maximum negative design wind pressure in zone 5 is equal to the following:

$$p_h = [25.0 \times (-1.59)] - 4.5 = -44.3 \text{ lb/ft}^2$$

In S.I.:

$$p_h = [1.20 \times (-1.59)] - 0.22 = -2.13 \text{ kN/m}^2$$

The maximum design wind pressures in Table 5.91 are greater than the minimum design wind pressure of 16 lb/ft^2 (0.77 kN/m^2) in ASCE/SEI 30.2.2.

5.8.22 Example 5.22—Determination of Design Wind Pressures on the C&C of an Open Agricultural Building Using ASCE/SEI Chapter 30, Part 3

Determine design wind pressures on the roof trusses of the open agricultural building in Figure 5.55 using the provisions in Part 3 of ASCE/SEI Chapter 30. Assume the trusses are spaced 3 ft (0.91 m) on center.

Solution

Check if the building meets all the conditions and limitations of ASCE/SEI 30.1.2 and 30.1.3 and the additional conditions of ASCE/SEI 30.5 so that the provisions in Part 3 of ASCE/SEI Chapter 30 for open buildings can be used to determine the wind pressures on the C&C.

The building is regularly-shaped, that is, it does not have any unusual geometric irregularities in spatial form. Also, the building does not have response characteristics that make it subject to across-wind loading or other similar effects, and it is not sited at a location where channeling effects or buffeting in the wake of upwind obstructions need to be considered.

The building is open with a pitched free roof.

Therefore, the provisions in Part 3 of ASCE/SEI Chapter 30 are permitted to be used to determine the design wind pressures on the C&C.

Flowchart 5.13 is used to determine the design wind pressures.

Step 1—Determine the wind velocity pressure, q_h

The velocity pressure, q_h, at the mean roof height of the building is determined in Example 5.13 and is equal to 25.5 lb/ft^2 (1.22 kN/m^2).

Step 2—Determine the wind directionality factor, K_d

For building structures, $K_d = 0.85$ from ASCE/SEI Table 26.6-1.

Step 3—Determine whether the building is rigid or flexible

It is determined in Example 5.13 that the building is rigid.

Step 4—Determine the gust-effect factor, G

For rigid buildings, $G = 0.85$ in accordance with ASCE/SEI 26.11.4.

Step 5—Determine the effective wind area, A

The effective wind area, A, is defined as the greater of the tributary area of the element and the span length multiplied by an effective width that need not be less than one-third the span length (ASCE/SEI 26.2).

$$A = \text{greater of} \begin{cases} 60.0 \times 3.0 = 180 \text{ ft}^2 \text{ (16.7 m}^2) \\ 60.0 \times (60.0/3) = 1{,}200 \text{ ft}^2 \text{ (111.5 m}^2) \end{cases}$$

Step 6—Determine the net pressure coefficients, C_N

The net pressure coefficients, C_N, are determined using ASCE/SEI Figure 30.5-2 for pitched roofs with a slope, θ, equal to 18.4 degrees, which is less than 45 degrees, and $h/L = 25/60 = 0.42$ which is between the limits of 0.25 and 1.0.

Like in the case of the MWFRS, C_N depends on whether the wind flow is clear or obstructed. Both situations are examined in this example.

The width of zones 2 and 3 is equal to a (ASCE/SEI Figure 30.5-2):

$$a = \text{lesser of} \begin{cases} 0.1 \times \text{least horizontal dimension} = 0.1 \times 60 = 6.0 \text{ ft (1.83 m)} \\ 0.4h = 0.4 \times 25 = 10.0 \text{ ft (3.05 m)} \end{cases}$$

$$\text{Minimum } a = \text{greater of} \begin{cases} 0.04 \times \text{least horizontal dimension} = 0.04 \times 60 = 2.4 \text{ ft (0.73 m)} \\ 3.0 \text{ ft (0.91 m)} \end{cases}$$

Therefore, $a = 6.0$ ft (1.83 m).

The effective wind area, A, which is equal to 1,200 ft^2 (111.5 m^2), is greater than $4a^2 = 4 \times 6.0^2 = 144$ ft^2 (13.4 m^2); this information is also needed to determine the net pressure coefficients.

Net pressure coefficients, C_N, are given in Table 5.92. Linear interpolation was used to determine these values for a roof angle of 18.4 degrees and an effective wind area greater than $4a^2$ (see note 3 in ASCE/SEI Figure 30.5-2).

Table 5.92 Net Pressure Coefficients, C_N, for the Open Agricultural Building in Example 5.22

Zone	Clear Wind Flow		Obstructed Wind Flow	
1	1.15	−1.06	0.50	−1.51
2	1.15	−1.06	0.50	−1.51
3	1.15	−1.06	0.50	−1.51

Step 7—Determine the net design wind pressures, p

The net design wind pressures, p, are determined by ASCE/SEI Equation (30.5-1):

$$p = q_h K_d G C_N = 25.5 \times 0.85 \times 0.85 \times C_N = 18.4 C_N \text{ lb/ft}^2$$

In S.I.:

$$p = 1.22 \times 0.85 \times 0.85 \times C_N = 0.88 C_N \text{ kN/m}^2$$

For clear wind flow in zones 1, 2, and 3:

$$p = 25.5 \times 0.85 \times 0.85 \times 1.15 = 21.2 \text{ lb/ft}^2$$

$$p = 25.5 \times 0.85 \times 0.85 \times (-1.06) = -19.5 \text{ lb/ft}^2$$

In S.I.:

$$p = 1.22 \times 0.85 \times 0.85 \times 1.15 = 1.01 \text{ kN/m}^2$$

$$p = 1.22 \times 0.85 \times 0.85 \times (-1.06) = -0.93 \text{ kN/m}^2$$

For obstructed wind flow in zones 1, 2, and 3:

$$p = 25.5 \times 0.85 \times 0.85 \times 0.5 = 9.2 \text{ lb/ft}^2$$

$$p = 25.5 \times 0.85 \times 0.85 \times (-1.51) = -27.8 \text{ lb/ft}^2$$

In S.I.:

$$p = 1.22 \times 0.85 \times 0.85 \times 0.5 = 0.44 \text{ kN/m}^2$$

$$p = 1.22 \times 0.85 \times 0.85 \times (-1.51) = -1.33 \text{ kN/m}^2$$

In the case of obstructed wind flow, the net positive pressure must be increased to 16 lb/ft^2 (0.77 kN/m^2) to satisfy the minimum requirements of ASCE/SEI 30.2.2.

The loading diagrams on a typical interior roof truss for clear wind flow are given in Figure 5.67. Similar loading diagrams can be obtained for obstructed wind flow.

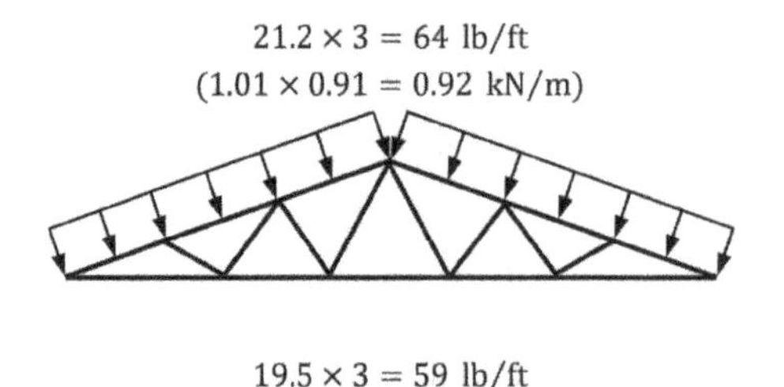

Figure 5.67 Loading Diagrams for Clear Wind Flow on a Typical Interior Roof Truss for the Open Agricultural Building in Example 5.22

5.8.23 Example 5.23—Determination of Design Wind Pressures on the C&C of a Parapet on an Office Building Using ASCE/SEI Chapter 30, Part 4

Determine design wind pressures on the C&C of a parapet on the office building in Figure 5.50 using the provisions in Part 4 of ASCE/SEI Chapter 30.

Solution

Check if the building meets all the conditions and limitations of ASCE/SEI 30.1.2 and 30.1.3 so that the provisions in Part 4 of ASCE/SEI Chapter 30 for parapets in ASCE/SEI 30.6 can be used to determine the wind pressures on the C&C.

The building is regularly-shaped, that is, it does not have any unusual geometric irregularities in spatial form. Also, the building does not have response characteristics that make it subject to across-wind loading or other similar effects, and it is not sited at a location where channeling effects or buffeting in the wake of upwind obstructions need to be considered.

Therefore, the provisions in Part 4 of ASCE/SEI Chapter 30 are permitted to be used to determine the design wind pressures on the C&C of the parapet.

Flowchart 5.14 is used to determine the design wind pressures.

Step 1—Determine the wind velocity pressure, q_p

The velocity pressure, q_p, at the top of the parapet is determined in Example 5.12 and is equal to 29.6 lb/ft^2 (1.42 kN/m^2).

Step 2—Determine the wind directionality factor, K_d

For building structures, $K_d = 0.85$ from ASCE/SEI Table 26.6-1.

Step 3—Determine the effective wind area, A

The effective wind area, A, is defined in this case as the span length multiplied by an effective width that need not be less than one-third the span length (ASCE/SEI 26.2):

$$A = 4.5 \times (4.5/3) = 6.8 \text{ ft}^2 \ (0.63 \text{ m}^2)$$

Step 4—Determine the external pressure coefficients, (GC_p)

The external pressure coefficients,(GC_p), are determined using ASCE/SEI Figure 30.4-1 for walls where the mean roof height is greater than 60 ft (18.3 m).

Load cases A and B must be considered (see Figure 5.22).

- Load case A (windward parapet):
  Windward (front) surface: $(GC_p) = 0.9$ for zone 5
  Leeward (back) surface: $(GC_p) = -3.2$ for zone 3
- Load case B (leeward parapet):
  Windward (back) surface: $(GC_p) = 0.9$ for zone 5
  Leeward (front) surface: $(GC_p) = -1.8$ for zone 5
  The width of zone 5 is equal to a (ASCE/SEI Figure 30.4-1):

$$a = \text{greater of} \begin{cases} 0.1 \times \text{least horizontal dimension} = 0.1 \times 61 = 6.1 \text{ ft } (1.86 \text{ m}) \\ 3.0 \text{ ft } (0.91 \text{ m}) \end{cases}$$

Step 5—Determine the internal pressure coefficients, (GC_{pi})

From ASCE/SEI Table 26.13-1, $(GC_{pi}) = +0.18, -0.18$ for an enclosed building.

Step 6—Determine the design wind pressures, p

The design wind pressures, p, are determined by ASCE/SEI Equation (30.6-1):

$$p = q_p K_d[(GC_p) - (GC_{pi})] = 29.6 \times 0.85 \times [(GC_p) - (\pm 0.18)] = 25.2(GC_p) \mp 4.5 \text{ lb/ft}^2$$

In S.I.:

$$p = 1.42 \times 0.85 \times [(GC_p) - (\pm 0.18)] = 1.21(GC_p) \mp 0.22 \text{ kN/m}^2$$

- Load case A (windward parapet):
  Windward (front) surface: $(GC_p) = (25.2 \times 0.9) \mp 4.5 = 18.2 \text{ lb/ft}^2, 27.2 \text{ lb/ft}^2$
  Leeward (back) surface: $(GC_p) = [25.2 \times (-3.2)] \mp 4.5 = -85.1 \text{ lb/ft}^2, -76.1 \text{ lb/ft}^2$

  In S.I.:
  Windward (front) surface: $(GC_p) = (1.21 \times 0.9) \mp 0.22 = 0.87 \text{ kN/m}^2, 1.31 \text{ kN/m}^2$
  Leeward (back) surface: $(GC_p) = [1.21 \times (-3.2)] \mp 0.22 = -4.09 \text{ kN/m}^2, -3.65 \text{ kN/m}^2$
- Load case B (leeward parapet):
  Windward (back) surface: $(GC_p) = (25.2 \times 0.9) \mp 4.5 = 18.2 \text{ lb/ft}^2, 27.2 \text{ lb/ft}^2$
  Leeward (front) surface: $(GC_p) = [25.2 \times (-1.8)] \mp 4.5 = -49.9 \text{ lb/ft}^2, -40.9 \text{ lb/ft}^2$

  In S.I.:
  Windward (back) surface: $(GC_p) = (1.21 \times 0.9) \mp 0.22 = 0.87 \text{ kN/m}^2, 1.31 \text{ kN/m}^2$
  Leeward(front)surface: $(GC_p) = [1.21 \times (-1.8)] \mp 0.22 = -2.40 \text{ kN/m}^2, -1.96 \text{ kN/m}^2$

The design wind pressures are applied perpendicular to the faces of the parapets.

5.8.24 Example 5.24—Determination of Design Wind Pressures on the C&C of an Attached Canopy on a Building Using ASCE/SEI Chapter 30, Part 4

Determine design wind pressures on the C&C of the attached canopy on the building in Figure 5.68 using the provisions in Part 4 of ASCE/SEI Chapter 30. The cantilevered structural members of the canopy are spaced 3 ft (0.91 m) on center. It has been determined that the velocity pressure at the mean roof height of the building, q_h, is equal to 30 lb/ft² (1.44 kN/m²).

Figure 5.68 Attached Canopy on the Building in Example 5.24

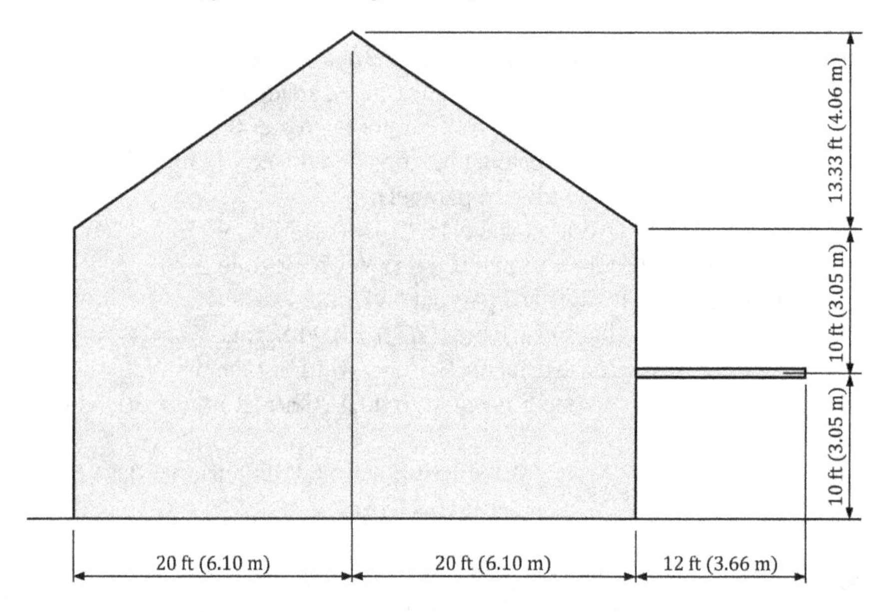

Solution

Check if the building meets all the conditions and limitations of ASCE/SEI 30.1.2 and 30.1.3 so that the provisions in Part 4 of ASCE/SEI Chapter 30 for attached canopies in ASCE/SEI 30.9 can be used to determine the wind pressures on the C&C.

The building is regularly-shaped, that is, it does not have any unusual geometric irregularities in spatial form. Also, the building does not have response characteristics that make it subject to across-wind loading or other similar effects, and it is not sited at a location where channeling effects or buffeting in the wake of upwind obstructions need to be considered.

Therefore, the provisions in Part 4 of ASCE/SEI Chapter 30 are permitted to be used to determine the design wind pressures on the C&C of the attached canopy.

Flowchart 5.16 is used to determine the design wind pressures.

Step 1—Determine the wind velocity pressure, q_h

It has been determined that the velocity pressure at the mean roof height of the building, q_h, is equal to 30 lb/ft² (1.44 kN/m²).

Step 2—Determine the wind directionality factor, K_d

For building structures, $K_d = 0.85$ from ASCE/SEI Table 26.6-1.

Step 3—Determine the effective wind area, A

The effective wind area, A, is defined as the greater of the tributary area of the element and the span length multiplied by an effective width that need not be less than one-third the span length (ASCE/SEI 26.2).

$$A = \text{greater of } \begin{cases} 12.0 \times 3.0 = 36 \text{ ft}^2 \ (3.4 \text{ m}^2) \\ 12.0 \times (12.0/3) = 48 \text{ ft}^2 \ (4.5 \text{ m}^2) \end{cases}$$

Step 4—Determine the net pressure coefficients, (GC_p)

Because the mean roof height of the building, which is equal to 26.67 ft (8.13 m), is less than 60 ft (18.3 m), the external pressure coefficients, (GC_p), are determined using ASCE/SEI Figure 30.9-1A considering separate upper and lower surfaces of the canopy and ASCE/SEI Figure 30.9-1B considering simultaneous contributions from the upper and lower surfaces of the canopy.

In lieu of the aforementioned figures, the external pressure coefficients are permitted to be determined using the equations in ASCE/SEI Table C30.9-1 on separate surfaces and ASCE/SEI Table C30.9-2 considering simultaneous contributions from the upper and lower surfaces.

- Separate upper and lower surfaces

Upper surface: $(GC_p) = -1.55 + 0.4000 \log A = -1.55 + 0.4000 \log 48 = -0.88$

$$(GC_p) = 1.0 - 0.2000 \log A = 1.0 - 0.2000 \log 48 = 0.66$$

Lower surface: $(GC_p) = -0.95 + 0.1500 \log A = -0.95 + 0.1500 \log 48 = -0.70$

$$(GC_p) = 1.0 - 0.2000 \log A = 1.0 - 0.2000 \log 48 = 0.66$$

- Simultaneous contributions from the upper and lower surfaces

$$h_c / h_e = 10.0/20.0 = 0.5$$

In S.I.: $h_c / h_e = 3.05/6.10 = 0.5$

$$(GC_p) = -0.70 + 0.1000 \log A = -0.70 + 0.1000 \log 48 = -0.53$$

$$(GC_p) = 1.15 - 0.2500 \log A = 1.15 - 0.2500 \log 48 = 0.73$$

Step 5—Determine the design wind pressures, p

The design wind pressures are determined by ASCE/SEI Equations (30.9-1) and (30.9-1.SI):

$$p = q_h K_d (GC_p) = 30.0 \times 0.85 \times (GC_p) = 25.5(GC_p) \text{ lb/ft}^2$$

In S.I.:

$$p = 1.44 \times 0.85 \times (GC_p) = 1.22(GC_p) \text{ kN/m}^2$$

- Separate upper and lower surfaces

Upper surface: $p = 25.5 \times (-0.88) = -22.4 \text{ lb/ft}^2$

$p = 25.5 \times 0.66 = 16.8 \text{ lb/ft}^2$

Lower surface: $p = 25.5 \times (-0.70) = -17.9 \text{ lb/ft}^2$

$p = 25.5 \times 0.66 = 16.8 \text{ lb/ft}^2$

In S.I.:

Upper surface: $p = 1.22 \times (-0.88) = -1.07 \text{ kN/m}^2$

$p = 1.22 \times 0.66 = 0.81 \text{ kN/m}^2$

Lower surface: $p = 1.22 \times (-0.70) = -0.85 \text{ kN/m}^2$

$p = 1.22 \times 0.66 = 0.81 \text{ kN/m}^2$

- Simultaneous contributions from the upper and lower surfaces

$$p = 25.5 \times (-0.53) = -13.5 \text{ lb/ft}^2$$

$$p = 25.5 \times 0.73 = 18.6 \text{ lb/ft}^2$$

In S.I.:

$$p = 1.22 \times (-0.53) = -0.65 \text{ kN/m}^2$$

$$p = 1.22 \times 0.73 = 0.89 \text{ kN/m}^2$$

The negative design wind pressure for simultaneous contributions from the upper and lower surfaces must be increased to -16.0 lb/ft^2 (-0.77 kN/m^2) to satisfy the minimum design wind pressure requirement in accordance with ASCE/SEI 30.2.2.

5.8.25 Example 5.25—Determination of Design Wind Pressures on the C&C of an Isolated Circular Tank Using ASCE/SEI Chapter 30, Part 5

Determine design wind pressures on the C&C of the isolated circular tank in Figure 5.60 using the provisions in Part 5 of ASCE/SEI Chapter 30.

SOLUTION

Check if the tank meets all the conditions and limitations of ASCE/SEI 30.1.2 and 30.1.3 so that the provisions in Part 5 of ASCE/SEI Chapter 30 for circular bins, silos, and tanks in ASCE/SEI 30.10 can be used to determine the wind pressures on the C&C.

The tank is regularly-shaped, that is, it does not have any unusual geometric irregularities in spatial form. Also, it does not have response characteristics that make it subject to across-wind loading or other similar effects, and it is not sited at a location where channeling effects or buffeting in the wake of upwind obstructions need to be considered.

Also, the overall height of the tank, which is equal to 45 ft (13.7 m), is less than 120 ft (36.6 m).

Therefore, the provisions in Part 5 of ASCE/SEI Chapter 30 are permitted to be used to determine the design wind pressures on the C&C of the tank.

Flowchart 5.17 is used to determine the design wind pressures.

Step 1—Determine the wind velocity pressure, q_h

The velocity pressure, q_h, at the top of tank is determined in Example 5.17 and is equal to 31.4 lb/ft^2 (1.51 kN/m^2).

Step 2—Determine the wind directionality factor, K_d

For round structures, $K_d = 1.0$ from ASCE/SEI Table 26.6-1.

Step 3—Determine the external pressure coefficients, (GC_p)

- External walls

External pressure coefficients, (GC_p), for the external walls of an isolated tank are determined in accordance with ASCE/SEI Figure 30.10-1.

The aspect ratio of the tank, h_c/D, is equal to 0.5 (see Figures 5.20 and 5.60).

The maximum positive pressure coefficient, (GC_p), is equal to 1.00 for a wind angle, α, of 0 degrees. Similarly, the maximum negative pressure coefficient is equal to −1.10 for a wind angle, α, of 75 degrees or 90 degrees.

- Roof

External pressure coefficients, (GC_p), for the roof of an isolated tank are determined in accordance with ASCE/SEI Figure 30.10-2.

For tanks with flat roofs, zones 1, 2, and 3 are applicable. Because the framing for the roof structure is not given, the external pressure coefficients are determined based on an effective wind area, A, less than or equal to 100 ft^2 (9.3 m^2):

Zone 1: $(GC_p) = 0.30, -0.80$

Zone 2: $(GC_p) = 0.30, -0.50$

Zone 3: $(GC_p) = 0.30, -1.20$

Line A in ASCE/Figure 30.10-2 is used to determine the positive pressure coefficients because the slope of the tank roof is less than 10 degrees (see note 6 in ASCE/SEI Figure 30.10-2).

Width of zone 1: $b = 0.5D = 30$ ft (9.14 m) for $h_c/D = 0.5$.

Width of zone 3: $a = 0.1 \times$ least horizontal dimension $= 0.1D = 6$ ft (1.83 m).

- Underside

External pressure coefficients, (GC_p), for the underside of an isolated tank are determined in accordance with ASCE/SEI 30.10.5.

Zones 1 and 2 in ASCE/SEI Figure 30.10-2: $(GC_p) = 0.80, -0.60$

Zone 3 in ASCE/SEI Figure 30.10-2: $(GC_p) = 1.20, -0.90$

Step 4—Determine the internal pressure coefficients, (GC_{pi})

From ASCE/SEI Table 26.13-1, $(GC_{pi}) = +0.18, -0.18$ for an enclosed circular tank that is not open-topped.

Step 5—Determine the design wind pressure, p

The design wind pressure, p, is determined by ASCE/SEI Equation (30.10-1):

$$p = q_h K_d[(GC_p) - (GC_{pi})] = 31.4 \times 1.0 \times [(GC_p) - (\pm 0.18)] = 31.4(GC_p) \mp 5.7 \text{ lb/ft}^2$$

In S.I.:

$$p = 1.51 \times 1.0 \times [(GC_p) - (\pm 0.18)] = 1.51(GC_p) \mp 0.27 \text{ kN/m}^2$$

Maximum design wind pressures on the C&C of the isolated circular tank are given in Table 5.93. These pressures are applied perpendicular to the surfaces of the tank.

Table 5.93 Maximum Design Wind Pressures on the C&C of the Isolated Circular Tank in Example 5.25

Element		p, lb/ft^2 (kN/m^2)	
		Positive	Negative
External walls		37.1 (1.78)	−40.2 (−1.93)
Roof	Zone 1	15.1 (0.72)	−30.8 (−1.48)
	Zone 2	15.1 (0.72)	−21.4 (−1.03)
	Zone 3	15.1 (0.72)	−43.4 (−2.08)
Underside	Zones 1 and 2	30.8 (1.48)	−24.5 (−1.18)
	Zone 3	43.4 (2.08)	−34.0 (−1.63)

The positive design wind pressure for all zones on the roof must be increased to 16.0 lb/ft^2 (0.77 kN/m^2) to satisfy the minimum design wind pressure requirement in accordance with ASCE/SEI 30.2.2.

5.8.26 Example 5.26—Determination of Design Tornado Pressures on an Essential Facility Using ASCE/SEI Chapter 32—MWFRS

Determine the design tornado pressures, p_T, on the MWFRS of the five-story hospital (structure 1) in Figure 5.69 using the provisions in ASCE/SEI Chapter 32 given the following design data:

- Location: Springfield, MO (Latitude = 37.20°, Longitude = −93.28°)
- Surface roughness: B
- Topography: Not situated on a hill, ridge, or escarpment

Figure 5.69 Plan of the 5-story Hospital and Support Structures in Example 5.26

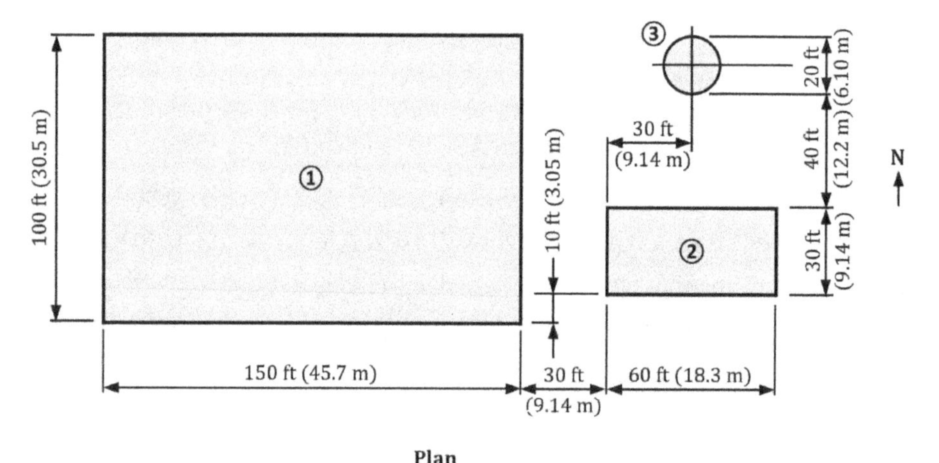

Structure	Description
1	• 5-story hospital with a flat roof • Typical story height = 10 ft (3.05 m)
2	One-story power-generating station that provides emergency backup for the hospital
3	Ground-supported tank that maintains the functionality of the hospital [h_c = 10 ft (3.05 m)]

- Risk category: IV (IBC Table 1604.5)
- Enclosure classification: Enclosed

Structure 2 is a one-story power-generating station that provides emergency backup for the hospital and structure 3 is a ground-supported tank that maintains the functionality of the hospital.

SOLUTION

Check if the essential facility must be designed for the effects of tornado loads (see Flowchart 5.18):

1. The building is assigned to risk category IV in accordance with IBC Table 1604.5.
2. The building is located in a tornado-prone region identified in ASCE/SEI Figure 32.1-1.
3. Determine the tornado speed, V_T.
   The tornado speed, V_T, is determined based on the effective plan area of the building, A_e, which is defined in ASCE/SEI 32.5.4.1 for essential facilities.
   Because structures 2 and 3 are required to maintain the functionality of structure 1, A_e must be equal to the area of the smallest convex polygon enclosing both the essential facility and all the buildings and other structures that maintain the functionality of the essential facility [ASCE/SEI 32.5.4.1; see Figure 5.70(a)]. Alternatively, A_e can be conservatively calculated as the area of the smallest rectangle that encloses the maximum plan area of the facility [ASCE/SEI C32.5.4; see Figure 5.70(b)]:

$$A_e = (150 + 30 + 60) \times 100 = 24{,}000 \text{ ft}^2 \, (2{,}230 \text{ m}^2)$$

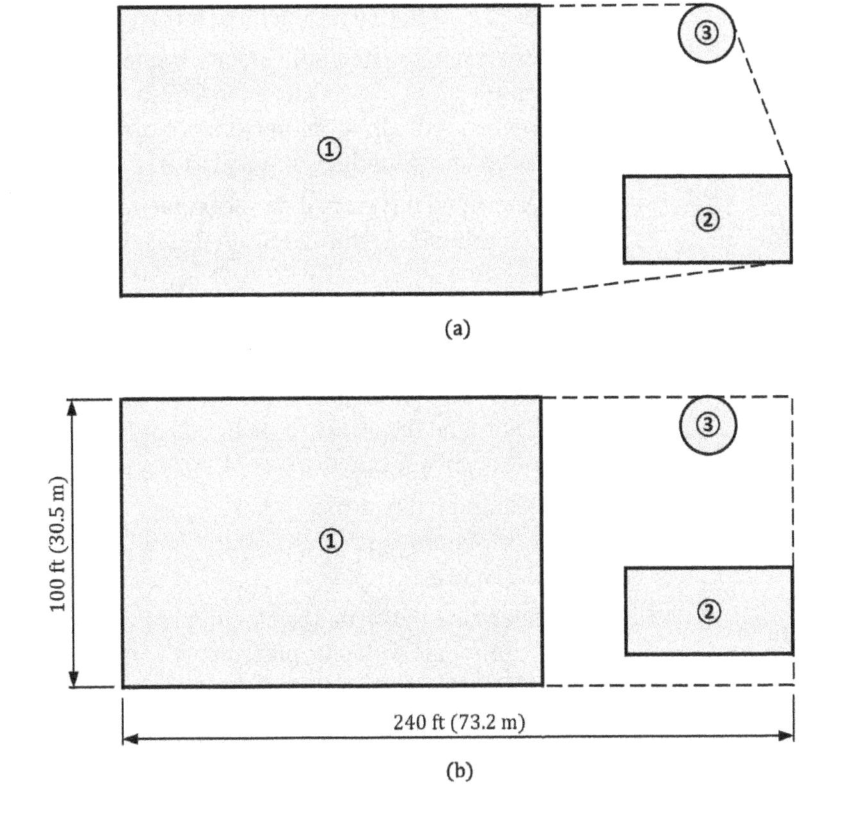

Figure 5.70 Effective Plan Area, A_e, for the Essential Facility in Example 5.26 Determined Using (a) the Smallest Convex Polygon Enclosing the Facility and (b) a Rectangle Enclosing the Facility

The tornado hazard map in ASCE/SEI Figure 32.5-2D with A_e equal to 40,000 ft^2 (3,716 m^2) is used to determine $V_T = 103$ mi/h (46 m/s) for this risk category IV building because this is the next available mapped A_e greater than the calculated A_e (ASCE/SEI 32.5.1). Alternatively, V_T can be obtained using Reference 5.4.

4. $V_T = 103$ mi/h (46 m/s) $>$ 60 mi/h (26.8 m/s) [see ASCE/SEI 32.5.2]
5. Determine the basic wind speed, V, in accordance with ASCE/SEI 26.5.
   The basic wind speed, V, is determined from ASCE/SEI Figure 26.5-1D or Reference 5.4: $V = 120$ mi/h (53 m/s).
6. Determine the exposure category in accordance with ASCE/SEI 26.7.3.
   Surface roughness B is given in the design data and it is assumed it is present in all directions. It is assumed that exposure category B is applicable in all directions.
7. $V_T = 103$ mi/h (46 m/s) $> 0.5V = 0.5 \times 120 = 60$ mi/h (26.8 m/s) [see ASCE/SEI 32.5.2]

Based on the requirements in ASCE/SEI 32.1.1 and 32.5.2, design for tornado loads is required. Flowchart 5.20 is used to determine the design tornado pressures.

Step 1—Determine if the building has a parapet.
The hospital does not have a parapet.

Step 2—Determine the tornado gust-effect factor, G_T
The tornado gust-effect factor, G_T, is determined in accordance with ASCE/SEI 32.11. In lieu of determining G_T by ASCE/SEI Equation (26.11-6) for rigid structures using exposure C terrain exposure constants, it is permitted to take G_T equal to 0.85 (ASCE/SEI 32.11.1).

Step 3—Determine the tornado directionality factor, K_{dT}
The tornado directionality factor, K_{dT}, is determined from ASCE/SEI Table 32.6-1. For the MWFRS of building structures, $K_{dT} = 0.80$.

Step 4—Determine the tornado velocity pressure, q_{zT}
Flowchart 5.19 is used to determine the tornado velocity pressure, q_{zT}.

Step 4a—Determine the tornado velocity pressure exposure coefficient, K_{zTor}
Values of K_{zTor} are determined from ASCE/SEI Table 32.10-1. Because the elevations of all floors and the roof of the hospital are less than 200 ft (61.0 m) above the ground level, $K_{zTor} = 1.0$.

Step 4b—Determine the ground elevation factor, K_e
The ground elevation factor, K_e, is determined in accordance with ASCE/SEI 26.9 (ASCE/SEI 32.9). It is permitted to take $K_e = 1.0$ for all elevations (ASCE/SEI 26.9).

Step 4c—Determine the risk category
The risk category is IV for the hospital.

Step 4d—Determine the effective plan area of the building, A_e
It is determined above that $A_e = 24{,}000$ ft^2 (2,230 m^2).

Step 4e—Determine the tornado speed, V_T
It is determined above that $V_T = 103$ mi/h (46 m/s) for $A_e = 40{,}000$ ft^2 (3,716 m^2).

Step 4f—Determine the tornado velocity pressure, q_{zT}
The tornado velocity pressure, q_{zT}, is determined by ASCE/SEI Equations (32.10-1) and (32.10-1.SI):

$$q_{zT} = 0.00256 K_{zTor} K_e V_T^2 = 0.00256 \times 1.0 \times 1.0 \times 103^2 = 27.2 \text{ lb/ft}^2$$

In S.I.:

$$q_{zT} = 0.613K_{zTor}K_eV_T^2 = 0.613 \times 1.0 \times 1.0 \times 46^2/1{,}000 = 1.30 \text{ kN/m}^2$$

The tornado velocity pressure is constant over the height of the hospital because $K_{zTor} = 1.0$ over the height.

Step 5—Determine the external pressure coefficients, C_p

The external pressure coefficients, C_p, are determined in accordance with ASCE/SEI Figure 27.3-1 (ASCE/SEI 32.15.1).

N-S direction:

- Windward wall: $C_p = 0.8$
- Leeward wall: For $L/B = 100.0/150.0 = 0.67$, $C_p = -0.5$
- Side walls: $C_p = -0.7$
- Roof: Normal to the ridge with $\theta < 10$ degrees and $h/L = 50.0/100.0 = 0.5$ (In S.I.: $h/L = 15.2/30.5 = 0.5$)
  From the windward edge of the roof to $h = 50.0$ ft (15.2 m), $C_p = -0.9, -0.18$
  From $h = 50.0$ ft (15.2 m) to $2h = 100.0$ ft (30.5 m), $C_p = -0.5, -0.18$

E-W direction:

- Windward wall: $C_p = 0.8$
- Leeward wall: For $L/B = 150.0/100.0 = 1.5$, $C_p = -0.4$ (by linear interpolation)
- Side walls: $C_p = -0.7$
- Roof: Parallel to the ridge with $h/L = 50.0/150.0 = 0.33$ (In S.I.: $h/L = 15.2/45.7 = 0.33$)
  From the windward edge of the roof to $h = 50.0$ ft (15.2 m), $C_p = -0.9, -0.18$
  From $h = 50.0$ ft (15.2 m) to $2h = 100.0$ ft (30.5 m), $C_p = -0.5, -0.18$
  From $2h = 100.0$ ft (30.5 m) to 150.0 ft (45.7 m), $C_p = -0.3, -0.18$

Step 6—Determine the tornado pressure coefficient adjustment factors, K_{vT}

The tornado pressure coefficient adjustment factors, K_{vT}, are determined from ASCE/SEI Table 32.14-1 for buildings (see Table 5.94).

Table 5.94 Tornado Pressure Coefficient Adjustment Factors, K_{vT}, for the Hospital in Example 5.26

Pressure Type	K_{vT}
Negative pressures on roofs (MWFRS)	1.1
Positive pressures on roofs	1.0
Wall pressures	1.0

Step 7—Determine the tornado velocity pressure for internal pressure determination, q_i

The tornado velocity pressure for internal pressure determination, q_i, is determined in accordance with ASCE/SEI 32.15.1:

- For roofs of enclosed buildings: $q_i = q_{hT} = 27.2 \text{ lb/ft}^2$ (1.30 kN/m^2)
- For walls of enclosed buildings: $q_i = q_{zT}$
  Because $K_{zTor} = 1.0$ over the height of the building, $q_i = q_{hT} = 27.2 \text{ lb/ft}^2$ (1.30 kN/m^2).

Step 8—Determine the tornado internal pressure coefficients, (GC_{piT})

The tornado internal pressure coefficients, (GC_{piT}), are determined from ASCE/SEI Table 32.13-1:

For an enclosed building, $(GC_{piT}) = +0.55, -0.18$.

Step 9—Determine the tornado design pressures, p_T

The tornado design pressures, p_T, are determined by ASCE/SEI Equations (32.15-1) and (32.15-1.SI):

$$p_T = qG_T K_{dT} K_{vT} C_p - q_i(GC_{piT})$$

N-S direction:

- Windward wall:

$$p_T = (27.2 \times 0.85 \times 0.80 \times 1.0 \times 0.80) - [27.2 \times (+0.55, -0.18)]$$
$$= 14.8 - 15.0 = -0.2 \text{ lb/ft}^2,\ 14.8 + 4.9 = 19.7 \text{ lb/ft}^2$$

In S.I.:

$$p_T = (1.30 \times 0.85 \times 0.80 \times 1.0 \times 0.80) - [1.30 \times (+0.55, -0.18)]$$
$$= 0.71 - 0.72 = -0.01 \text{ kN/m}^2,\ 0.71 + 0.23 = 0.94 \text{ kN/m}^2$$

- Leeward wall

$$p_T = [27.2 \times 0.85 \times 0.80 \times 1.0 \times (-0.50)] - [27.2 \times (+0.55, -0.18)]$$
$$= -9.3 - 15.0 = -24.3 \text{ lb/ft}^2,\ -9.3 + 4.9 = -4.4 \text{ lb/ft}^2$$

In S.I.:

$$p_T = [1.30 \times 0.85 \times 0.80 \times 1.0 \times (-0.50)] - [1.30 \times (+0.55, -0.18)]$$
$$= -0.44 - 0.72 = -1.16 \text{ kN/m}^2,\ -0.44 + 0.23 = -0.21 \text{ kN/m}^2$$

- Side walls

$$p_T = [27.2 \times 0.85 \times 0.80 \times 1.0 \times (-0.70)] - [27.2 \times (+0.55, -0.18)]$$
$$= -13.0 - 15.0 = -28.0 \text{ lb/ft}^2,\ -13.0 + 4.9 = -8.1 \text{ lb/ft}^2$$

In S.I.:

$$p_T = [1.30 \times 0.85 \times 0.80 \times 1.0 \times (-0.70)] - [1.30 \times (+0.55, -0.18)]$$
$$= -0.62 - 0.72 = -1.34 \text{ kN/m}^2,\ -0.62 + 0.23 = -0.39 \text{ kN/m}^2$$

- Roof

From the windward edge of the roof to 50 ft (15.2 m):

$$p_T = [27.2 \times 0.85 \times 0.80 \times 1.1 \times (-0.90)] - [27.2 \times (+0.55, -0.18)]$$
$$= -18.3 - 15.0 = -33.3 \text{ lb/ft}^2,\ -18.3 + 4.9 = -13.4 \text{ lb/ft}^2$$

In S.I.:

$$p_T = [1.30 \times 0.85 \times 0.80 \times 1.1 \times (-0.90)] - [1.30 \times (+0.55, -0.18)]$$
$$= -0.88 - 0.72 = -1.60 \text{ kN/m}^2,\ -0.88 + 0.23 = -0.65 \text{ kN/m}^2$$

From 50 ft (15.2 m) to 100 ft (30.5 m):

$$p_T = [27.2 \times 0.85 \times 0.80 \times 1.1 \times (-0.50)] - [27.2 \times (+0.55, -0.18)]$$
$$= -10.2 - 15.0 = -25.2 \text{ lb/ft}^2,\ -10.2 + 4.9 = -5.3 \text{ lb/ft}^2$$

In S.I.:

$$p_T = [1.30 \times 0.85 \times 0.80 \times 1.1 \times (-0.50)] - [1.30 \times (+0.55, -0.18)]$$
$$= -0.49 - 0.72 = -1.21 \text{ kN/m}^2,\ -0.49 + 0.23 = -0.26 \text{ kN/m}^2$$

Design tornado pressures in the N-S direction are given in Figure 5.71.

Figure 5.71 Design Tornado Pressures in the N-S Direction on the MWFRS of the Hospital in Example 5.26

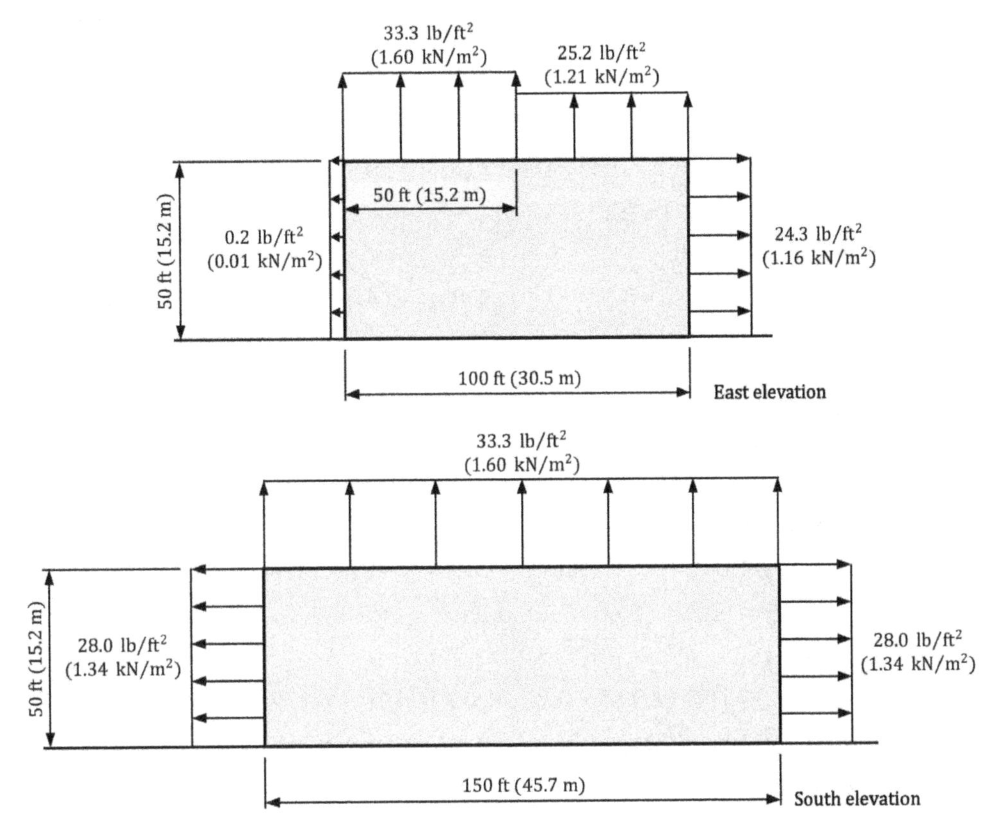

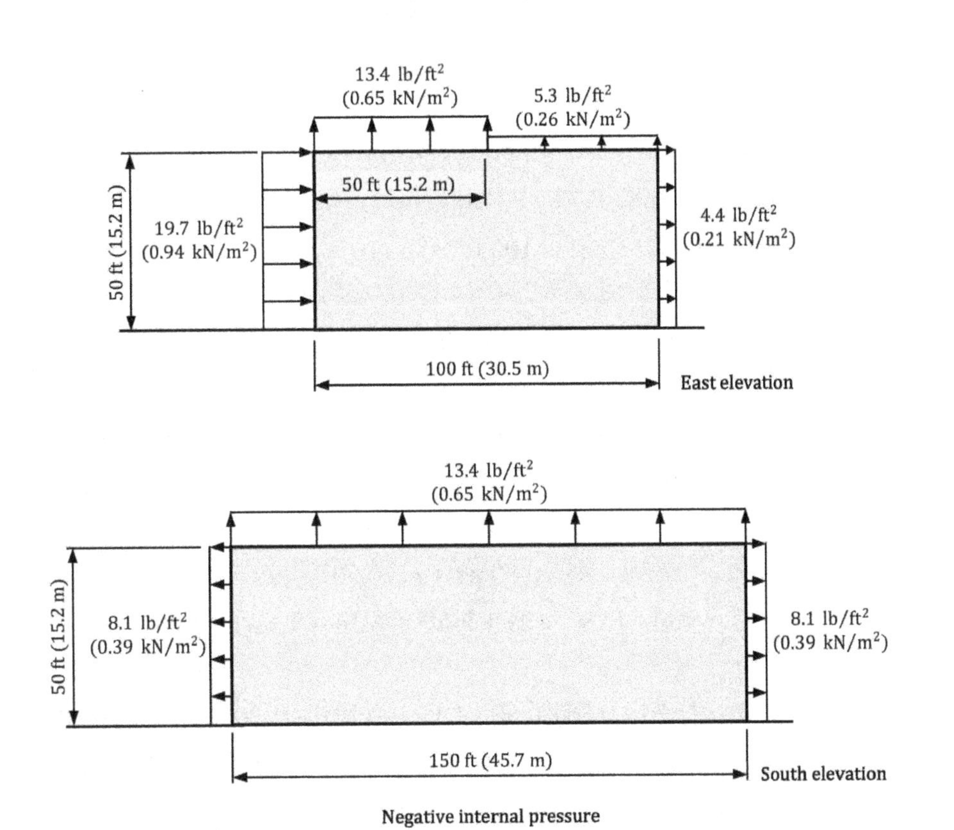

E-W direction:

- Windward wall:

$$p_T = (27.2\times 0.85\times 0.80\times 1.0\times 0.80) - [27.2\times(+0.55,-0.18)]$$
$$= 14.8 - 15.0 = -0.2 \text{ lb/ft}^2,\ 14.8 + 4.9 = 19.7 \text{ lb/ft}^2$$

In S.I.:

$$p_T = (1.30\times 0.85\times 0.80\times 1.0\times 0.80) - [1.30\times(+0.55,-0.18)]$$
$$= 0.71 - 0.72 = -0.01 \text{ kN/m}^2,\ 0.71 + 0.23 = 0.94 \text{ kN/m}^2$$

- Leeward wall

$$p_T = [27.2\times 0.85\times 0.80\times 1.0\times(-0.4)] - [27.2\times(+0.55,-0.18)]$$
$$= -7.4 - 15.0 = -22.4 \text{ lb/ft}^2,\ -7.4 + 4.9 = -2.5 \text{ lb/ft}^2$$

In S.I.:

$$p_T = [1.30\times 0.85\times 0.80\times 1.0\times(-0.4)] - [1.30\times(+0.55,-0.18)]$$
$$= -0.35 - 0.72 = -1.07 \text{ kN/m}^2,\ -0.35 + 0.23 = -0.12 \text{ kN/m}^2$$

- Side walls

$$p_T = [27.2\times 0.85\times 0.80\times 1.0\times(-0.70)] - [27.2\times(+0.55,-0.18)]$$
$$= -13.0 - 15.0 = -28.0 \text{ lb/ft}^2,\ -13.0 + 4.9 = -8.1 \text{ lb/ft}^2$$

In S.I.:

$$p_T = [1.30\times 0.85\times 0.80\times 1.0\times(-0.70)] - [1.30\times(+0.55,-0.18)]$$
$$= -0.62 - 0.72 = -1.34 \text{ kN/m}^2,\ -0.62 + 0.23 = -0.39 \text{ kN/m}^2$$

- Roof

From the windward edge of the roof to 50 ft (15.2 m):

$$p_T = [27.2\times 0.85\times 0.80\times 1.1\times(-0.90)] - [27.2\times(+0.55,-0.18)]$$
$$= -18.3 - 15.0 = -33.3 \text{ lb/ft}^2,\ -18.3 + 4.9 = -13.4 \text{ lb/ft}^2$$

In S.I.:

$$p_T = [1.30\times 0.85\times 0.80\times 1.1\times(-0.90)] - [1.30\times(+0.55,-0.18)]$$
$$= -0.88 - 0.72 = -1.60 \text{ kN/m}^2,\ -0.88 + 0.23 = -0.65 \text{ kN/m}^2$$

From 50 ft (15.2 m) to 100 ft (30.5 m):

$$p_T = [27.2\times 0.85\times 0.80\times 1.1\times(-0.50)] - [27.2\times(+0.55,-0.18)]$$
$$= -10.2 - 15.0 = -25.2 \text{ lb/ft}^2,\ -10.2 + 4.9 = -5.3 \text{ lb/ft}^2$$

In S.I.:

$$p_T = [1.30\times 0.85\times 0.80\times 1.1\times(-0.50)] - [1.30\times(+0.55,-0.18)]$$
$$= -0.49 - 0.72 = -1.21 \text{ kN/m}^2,\ -0.49 + 0.23 = -0.26 \text{ kN/m}^2$$

From 100 ft (30.5 m) to 150.0 ft (45.7 m):

$$p_T = [27.2\times 0.85\times 0.80\times 1.1\times(-0.30)] - [27.2\times(+0.55,-0.18)]$$
$$= -6.1 - 15.0 = -21.1 \text{ lb/ft}^2,\ -6.1 + 4.9 = -1.2 \text{ lb/ft}^2$$

In S.I.:

$$p_T = [1.30\times 0.85\times 0.80\times 1.1\times(-0.30)] - [1.30\times(+0.55,-0.18)]$$
$$= -0.29 - 0.72 = -1.01 \text{ kN/m}^2,\ -0.29 + 0.23 = -0.06 \text{ kN/m}^2$$

Design tornado pressures in the E-W direction are given in Figure 5.72.

Figure 5.72 Design Tornado Pressures in the E-W Direction on the MWFRS of the Hospital in Example 5.26

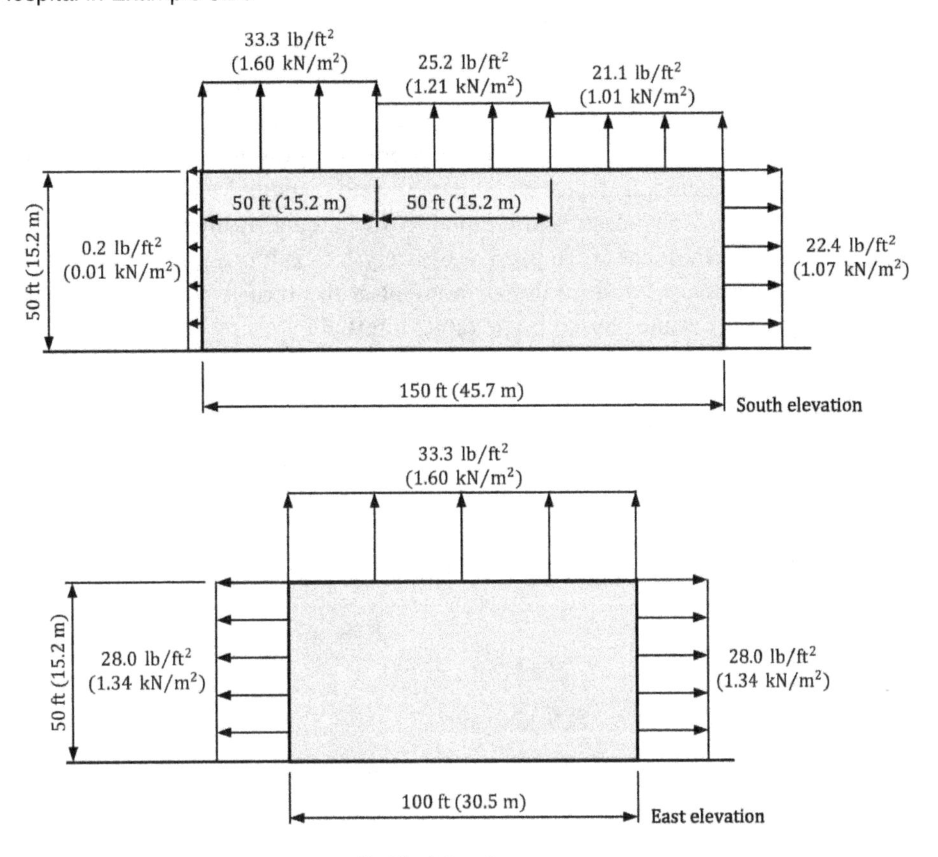

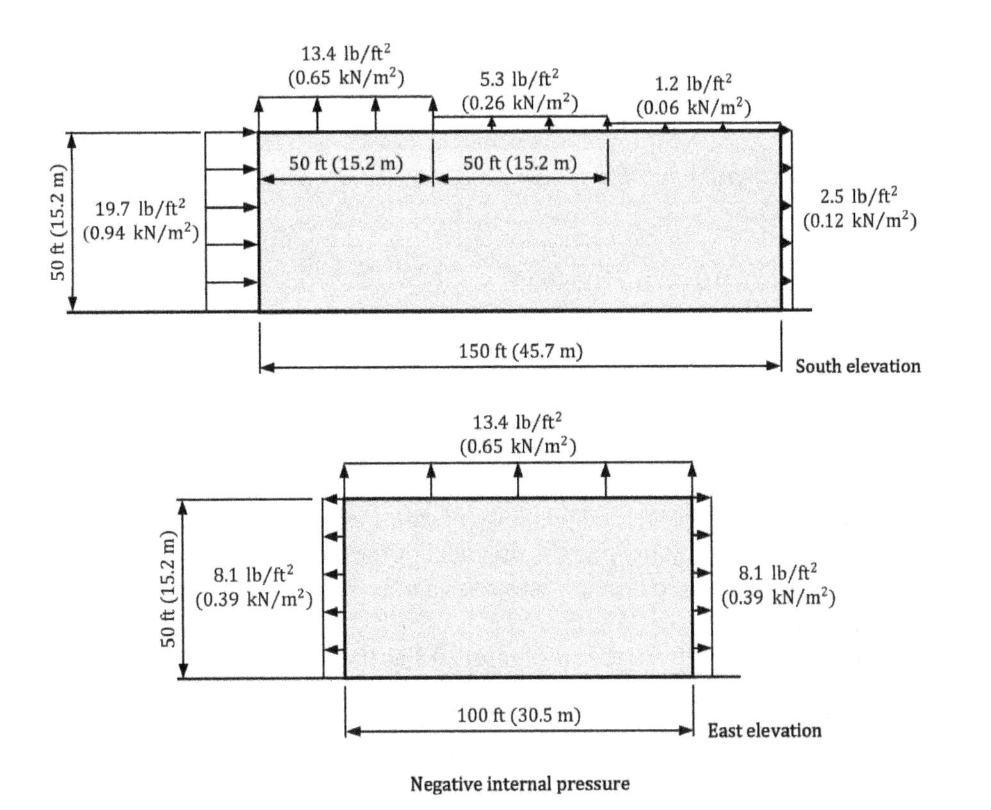

The building must be designed for the design load cases in ASCE/SEI Figure 27.3-8 (ASCE/SEI 32.15.5; see Figure 5.16). All four load cases must be considered (the exception for buildings meeting the requirements of ASCE/SEI Appendix D do not apply).

In case 1, the full design tornado pressures act on the projected area perpendicular to each principal axis of the structure. These pressures are assumed to act separately along each principal axis. Full design wind pressures are also applied on the roof.

In case 2, 75 percent of the design tornado pressures on the windward and leeward walls are applied on the projected area perpendicular to each principal axis of the building along with a torsional moment. Roof pressures are equal to 75 percent of those in case 1. The tornado pressures and torsional moment are applied separately for each principal axis.

The tornado pressures and torsional moment at the roof level for case 2 are determined as follows using net pressures based on negative internal pressure:

E-W direction:

$$0.75p_{Wx} = 0.75 \times 19.7 = 14.8 \text{ lb/ft}^2$$

$$0.75p_{Lx} = 0.75 \times 2.5 = 1.9 \text{ lb/ft}^2$$

$$e_x = \pm 0.15B_x = \pm 0.15 \times 100.0 = \pm 15.0 \text{ ft}$$

$$M_z = 0.75(p_{Wx} + p_{Lx})B_x e_x = (14.8 + 1.9) \times 100.0 \times (\pm 15.0) \times 5.0/1{,}000 = \pm 125.3 \text{ ft-kips}$$

In S.I.:

$$0.75p_{Wx} = 0.75 \times 0.94 = 0.71 \text{ kN/m}^2$$

$$0.75p_{Lx} = 0.75 \times 0.12 = 0.09 \text{ kN/m}^2$$

$$e_x = \pm 0.15B_x = \pm 0.15 \times 30.5 = \pm 4.58 \text{ m}$$

$$M_z = 0.75(p_{Wx} + p_{Lx})B_x e_x = (0.71 + 0.09) \times 30.5 \times (\pm 4.58) \times 1.52 = \pm 169.9 \text{ kN-m}$$

N-S direction:

$$0.75p_{Wy} = 0.75 \times 19.7 = 14.8 \text{ lb/ft}^2$$

$$0.75p_{Ly} = 0.75 \times 4.4 = 3.3 \text{ lb/ft}^2$$

$$e_y = \pm 0.15B_y = \pm 0.15 \times 150.0 = \pm 22.5 \text{ ft}$$

$$M_z = 0.75(p_{Wy} + p_{Ly})B_y e_y = (14.8 + 3.3) \times 150.0 \times (\pm 22.5) \times 5.0/1{,}000 = \pm 305.4 \text{ ft-kips}$$

In S.I.:

$$0.75p_{Wy} = 0.75 \times 0.94 = 0.71 \text{ kN/m}^2$$

$$0.75p_{Ly} = 0.75 \times 0.21 = 0.16 \text{ kN/m}^2$$

$$e_y = \pm 0.15B_y = \pm 0.15 \times 45.7 = \pm 6.86 \text{ m}$$

$$M_z = 0.75(p_{Wy} + p_{Ly})B_y e_y = (0.71 + 0.16) \times 45.7 \times (\pm 6.86) \times 1.52 = \pm 414.6 \text{ kN-m}$$

Similar calculations can be made at other elevations.

In case 3, 75 percent of the tornado pressures of case 1 are applied to the building simultaneously. This accounts for the direction along the diagonal of the building.

In case 4, 75 percent of the tornado pressures and torsional moments in case 2 act simultaneously on the building.

Cases 1 through 4 are illustrated in Figure 5.73 at the roof level of the hospital for negative internal pressure.

The MWFRS of the hospital must be designed and constructed to resist the greater of the tornado loads determined above and the wind loads determined in accordance with ASCE/SEI Chapter 27 using the load combinations in ASCE/SEI Chapter 2 (ASCE/SEI 32.1.1).

Figure 5.73 Design Tornado Load Cases at the Roof Level for the Hospital in Example 5.26

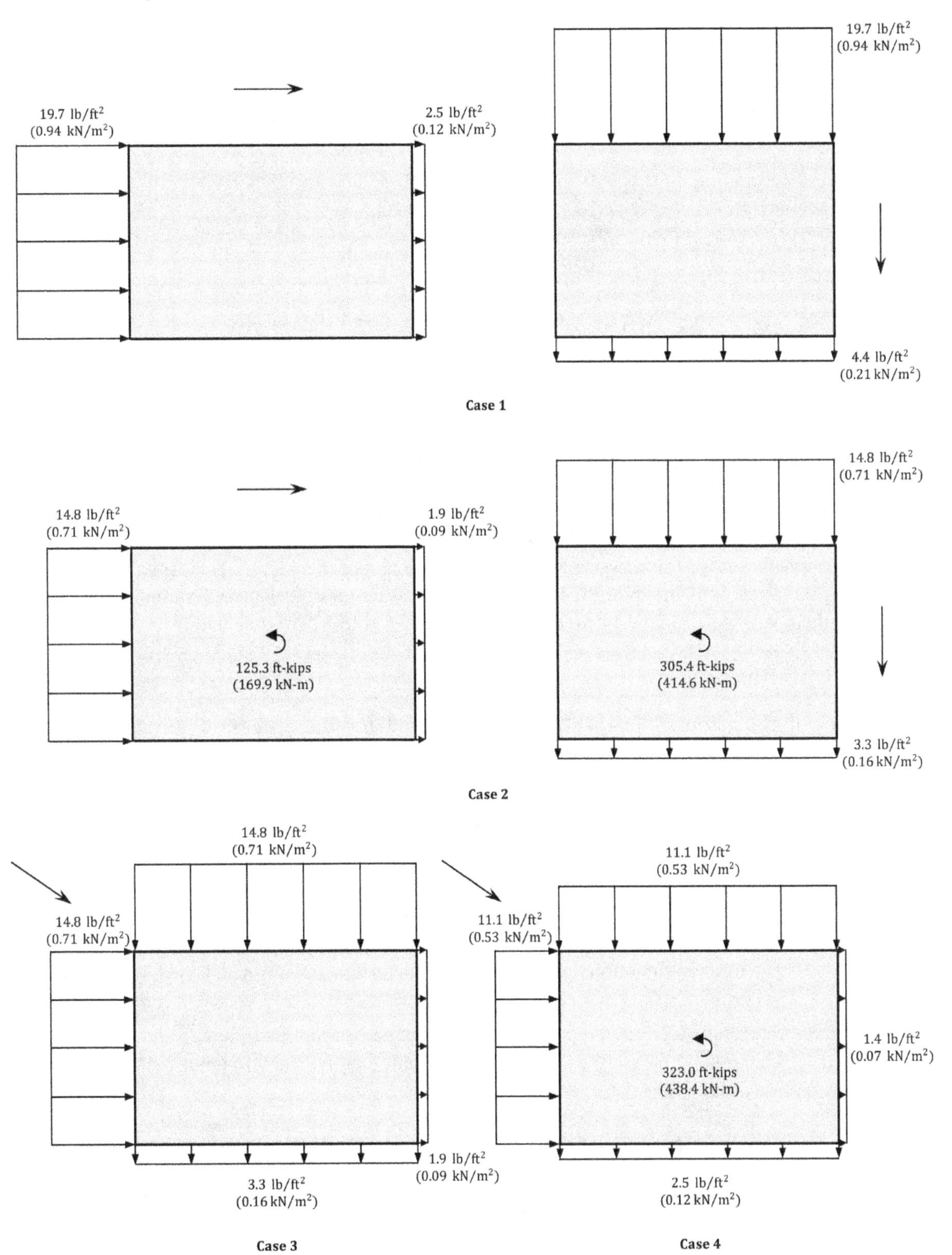

A comparison of the design tornado and wind pressures for the windward and leeward walls over the height of the building is given in Table 5.95 for negative internal pressure in the N-S direction. The design wind loads are determined using the provisions in ASCE/SEI 27.3.1. The minimum design wind pressure of 16 lb/ft^2 (0.77 kN/m^2) in accordance with ASCE/SEI must also be considered.

Table 5.95 Comparison of Design Tornado and Wind Wall Pressures for Negative Internal Pressure in the N-S Direction for the Hospital in Example 5.26

Height above the Ground Level, z, ft (m)	Windward Wall Pressure		Leeward Wall Pressure	
	Wind, p, lb/ft^2 (kN/m^2)	Tornado, p_T, lb/ft^2 (kN/m^2)	Wind, p, lb/ft^2 (kN/m^2)	Tornado, p_T, lb/ft^2 (kN/m^2)
50 (15.2)	21.3 (1.02)	19.7 (0.94)	−6.1 (0.29)	−4.4 (0.21)
40 (12.2)	20.3 (0.97)	19.7 (0.94)	−6.1 (0.29)	−4.4 (0.21)
30 (9.14)	19.1 (0.92)	19.7 (0.94)	−6.1 (0.29)	−4.4 (0.21)
20 (6.10)	17.6 (0.84)	19.7 (0.94)	−6.1 (0.29)	−4.4 (0.21)
10 (3.05)	16.7 (0.80)	19.7 (0.94)	−6.1 (0.29)	−4.4 (0.21)

Similar comparisons can be performed for positive internal pressure in the N-S direction and for negative and positive internal pressures in the E-W direction.

5.8.27 Example 5.27—Determination of Design Tornado Pressures on an Essential Facility Using ASCE/SEI Chapter 32—C&C

Determine the design tornado pressures, p_T, on a typical roof member of the hospital in Figure 5.69 using the provisions in ASCE/SEI Chapter 32 given the design data in Example 5.26. Typical roof members span 30 ft (9.14 m) and are spaced 12.5 ft (3.81 m) on center.

SOLUTION

Check if the essential facility must be designed for the effects of tornado loads (see Flowchart 5.18). It is shown in Example 5.26 that based on the requirements in ASCE/SEI 32.1.1 and 32.5.2, design for tornado loads is required.

Step 1—Determine the tornado velocity pressure, q_{hT}

The tornado velocity pressure, q_{hT}, at the mean roof height of the building is determined in accordance with ASCE/SEI 32.10.2.

From Example 5.26, $q_{hT} = 27.2$ lb/ft^2 (1.30 kN/m^2).

Step 2—Determine the tornado directionality factor, K_{dT}

The tornado directionality factor, K_{dT}, is determined from ASCE/SEI Table 32.6-1.

For the C&C of essential facilities, $K_{dT} = 1.0$.

Step 3—Determine the tornado pressure coefficient adjustment factors, K_{vT}

The tornado pressure coefficient adjustment factors, K_{vT}, are determined from ASCE/SEI Table 32.14-1 for buildings (see Table 5.96).

Table 5.96 Tornado Pressure Coefficient Adjustment Factors, K_{vT}, for the Hospital in Example 5.27

Pressure Type		K_{vT}
Negative pressures on roofs (C&C with a roof slope less than 7 degrees)	Zones 1′ and 1	1.2
	Zone 2	1.05
	Zone 3	1.05
Positive pressures on roofs		1.0

Step 4—Determine the effective wind area, A

The effective wind area, A, is defined as the greater of the tributary area of the element and the span length multiplied by an effective width that need not be less than one-third the span length (ASCE/SEI 26.2).

$$A = \text{greater of} \begin{cases} 12.5 \times 30.0 = 375 \text{ ft}^2 \ (34.8 \text{ m}^2) \\ 30.0 \times (30.0/3) = 300 \text{ ft}^2 \ (27.9 \text{ m}^2) \end{cases}$$

Step 5—Determine the external pressure coefficients, C_p

The external pressure coefficients, C_p, are determined from ASCE/SEI 30.3 (ASCE/SEI 32.17.1).

The external pressure coefficients, (GC_p), are determined from ASCE/SEI Figure 30.3-2A for the roof members based on the effective wind area, A. In lieu of ASCE/SEI Figure 30.3-2A, the equations in ASCE/SEI Table C30.3-2 can be used to determine (GC_p) for zones 1′, 1, 2 and 3:

- Zone 1′
  For $A = 375 \text{ ft}^2 \ (34.8 \text{ m}^2) > 100 \text{ ft}^2 \ (9.3 \text{ m}^2)$: positive $(GC_p) = 0.20$
  For $100 \text{ ft}^2 \ (9.3 \text{ m}^2) < A = 375 \text{ ft}^2 \ (34.8 \text{ m}^2) < 1{,}000 \text{ ft}^2 \ (92.9 \text{ m}^2)$: negative (GC_p) without overhang $= -1.9000 + 0.5000 \log A = -0.61$
- Zone 1
  For $A = 375 \text{ ft}^2 \ (34.8 \text{ m}^2) > 100 \text{ ft}^2 \ (9.3 \text{ m}^2)$: positive $(GC_p) = 0.20$
  For $10 \text{ ft}^2 \ (0.93 \text{ m}^2) < A = 375 \text{ ft}^2 \ (34.8 \text{ m}^2) < 500 \text{ ft}^2 \ (46.5 \text{ m}^2)$: negative (GC_p) without overhang $= -2.1120 + 0.4120 \log A = -1.05$
- Zone 2
  For $A = 375 \text{ ft}^2 \ (34.8 \text{ m}^2) > 100 \text{ ft}^2 \ (9.3 \text{ m}^2)$: positive $(GC_p) = 0.20$
  For $10 \text{ ft}^2 \ (0.93 \text{ m}^2) < A = 375 \text{ ft}^2 \ (34.8 \text{ m}^2) < 500 \text{ ft}^2 \ (46.5 \text{ m}^2)$: negative (GC_p) without overhang $= -2.8297 + 0.5297 \log A = -1.47$
- Zone 3
  For $A = 375 \text{ ft}^2 \ (34.8 \text{ m}^2) > 100 \text{ ft}^2 \ (9.3 \text{ m}^2)$: positive $(GC_p) = 0.20$
  For $10 \text{ ft}^2 \ (0.93 \text{ m}^2) < A = 375 \text{ ft}^2 \ (34.8 \text{ m}^2) < 500 \text{ ft}^2 \ (46.5 \text{ m}^2)$: negative (GC_p) without overhang $= -4.2595 + 1.0595 \log A = -1.53$

Step 6—Determine the internal pressure coefficients, (GC_{piT})

For an enclosed building, $(GC_{piT}) = +0.55, -0.18$ (ASCE/SEI Table 32.13-1).

Step 7—Determine the design tornado pressures, p_T

The design tornado pressures, p_T, are determined by ASCE/SEI Equations (32.17-1) and (32.17-1.SI):

$$p_T = q_{hT}[K_{dT}K_{vT}(GC_p) - (GC_{piT})]$$
$$= 27.2 \times [1.0 \times K_{vT}(GC_p) - (+0.55, -0.18)] \text{ lb/ft}^2$$

In S.I.:

$$p_T = 1.30 \times [1.0 \times K_{vT}(GC_p) - (+0.55, -0.18)] \text{ kN/m}^2$$

- Zone 1′

Positive (GC_p): $p_T = 27.2 \times [(1.0 \times 1.0 \times 0.20) - (+0.55, -0.18)]$
$= -9.5 \text{ lb/ft}^2, 10.3 \text{ lb/ft}^2$

Negative (GC_p): $p_T = 27.2 \times \{[1.0 \times 1.2 \times (-0.61)] - (+0.55, -0.18)\}$
$= -34.9 \text{ lb/ft}^2, -15.0 \text{ lb/ft}^2$

In S.I.:

Positive (GC_p): $p_T = 1.30 \times [(1.0 \times 1.0 \times 0.2) - (+0.55, -0.18)]$
$= -0.46 \text{ kN/m}^2, \ 0.49 \text{ kN/m}^2$

Negative (GC_p): $p_T = 1.30 \times \{[1.0 \times 1.2 \times (-0.61)] - (+0.55, -0.18)\}$
$= -1.67 \text{ kN/m}^2, \ -0.72 \text{ kN/m}^2$

- Zone 1

Positive (GC_p): $p_T = 27.2 \times [(1.0 \times 1.0 \times 0.2) - (+0.55, -0.18)]$
$= -9.5 \text{ lb/ft}^2, \ 10.3 \text{ lb/ft}^2$

Negative (GC_p): $p_T = 27.2 \times \{[1.0 \times 1.2 \times (-1.05)] - (+0.55, -0.18)\}$
$= -49.2 \text{ lb/ft}^2, \ -29.4 \text{ lb/ft}^2$

In S.I.:

Positive (GC_p): $p_T = 1.30 \times [(1.0 \times 1.0 \times 0.2) - (+0.55, -0.18)]$
$= -0.46 \text{ kN/m}^2, \ 0.49 \text{ kN/m}^2$

Negative (GC_p): $p_T = 1.30 \times \{[1.0 \times 1.2 \times (-1.05)] - (+0.55, -0.18)\}$
$= -2.35 \text{ kN/m}^2, \ -1.40 \text{ kN/m}^2$

- Zone 2

Positive (GC_p): $p_T = 27.2 \times [(1.0 \times 1.0 \times 0.2) - (+0.55, -0.18)]$
$= -9.5 \text{ lb/ft}^2, \ 10.3 \text{ lb/ft}^2$

Negative (GC_p): $p_T = 27.2 \times \{[1.0 \times 1.05 \times (-1.47)] - (+0.55, -0.18)\}$
$= -56.9 \text{ lb/ft}^2, \ -37.1 \text{ lb/ft}^2$

In S.I.:

Positive (GC_p): $p_T = 1.30 \times [(1.0 \times 1.0 \times 0.2) - (+0.55, -0.18)]$
$= -0.46 \text{ kN/m}^2, \ 0.49 \text{ kN/m}^2$

Negative (GC_p): $p_T = 1.30 \times \{[1.0 \times 1.05 \times (-1.47)] - (+0.55, -0.18)\}$
$= -2.72 \text{ kN/m}^2, \ -1.77 \text{ kN/m}^2$

- Zone 3

Positive (GC_p): $p_T = 27.2 \times [(1.0 \times 1.0 \times 0.2) - (+0.55, -0.18)]$
$= -9.5 \text{ lb/ft}^2, \ 10.3 \text{ lb/ft}^2$

Negative (GC_p): $p_T = 27.2 \times \{[1.0 \times 1.05 \times (-1.53)] - (+0.55, -0.18)\}$
$= -58.7 \text{ lb/ft}^2, \ -38.8 \text{ lb/ft}^2$

In S.I.:

Positive (GC_p): $p_T = 1.30 \times [(1.0 \times 1.0 \times 0.2) - (+0.55, -0.18)]$
$= -0.46 \text{ kN/m}^2, \ 0.49 \text{ kN/m}^2$

Negative (GC_p): $p_T = 1.30 \times \{[1.0 \times 1.05 \times (-1.53)] - (+0.55, -0.18)\}$
$= -2.80 \text{ kN/m}^2, \ -1.86 \text{ kN/m}^2$

The design tornado pressures act perpendicular to the roof members and act over the tributary area, which is equal to $12.5 \times 30.0 = 375 \text{ ft}^2$ (34.8 m^2).

The roof members of the hospital must be designed and constructed to resist the greater of the tornado loads determined above and the wind loads determined in accordance with ASCE/SEI 30.3 using the load combinations in ASCE/SEI Chapter 2 (ASCE/SEI 32.1.1).

A comparison of the maximum design tornado and wind pressures for zones 1′, 1, 2, and 3 is given in Table 5.97. The design wind loads are determined using the provisions in ASCE/SEI 30.3. The minimum design wind pressure of 16 lb/ft² (0.77 kN/m²) in accordance with ASCE/SEI must also be considered.

Table 5.97 Comparison of Design Tornado and Wind C&C Pressures for the Hospital in Example 5.27

Zone	Wind, p, lb/ft² (kN/m²)		Tornado, p_T, lb/ft² (kN/m²)	
	Positive	Negative	Positive	Negative
1′	9.4 (0.45)	−19.6 (−0.94)	10.3 (0.49)	−34.9 (−1.67)
1	9.4 (0.45)	−30.5 (−1.46)	10.3 (0.49)	−49.2 (−2.35)
2	9.4 (0.45)	−40.7 (−1.95)	10.3 (0.49)	−56.9 (−2.72)
3	9.4 (0.45)	−42.4 (−2.03)	10.3 (0.49)	−58.7 (−2.80)

5.8.28 Example 5.28—Determination of Design Tornado Pressures on an Essential Facility Using ASCE/SEI Chapter 32—Other Structure

Determine the design tornado pressures, p_T, on the MWFRS of the tank in Figure 5.69 using the provisions in ASCE/SEI Chapter 32 given the design data in Example 5.26. Assume the tank is sealed.

SOLUTION

- Design tornado force, F_T, on the external walls of the isolated circular tank.
  The design tornado force, F_T, on the external walls is determined in accordance with ASCE/SEI 32.16.3.

Step 1—Determine the tornado velocity pressure, q_{zT}

The tornado velocity pressure, q_{zT}, at the centroid of the projected area, A_f, normal to the tornadic wind [$z = 10.0/2 = 5$ ft (1.52 m)] is determined by ASCE/SEI Equations (32.10-1) and (32.10-1.SI) [see Flowchart 5.19]:

$$q_{zT} = 0.00256K_{zTor}K_eV_T^2 = 0.00256 \times 1.0 \times 1.0 \times 103^2 = 27.2 \text{ lb/ft}^2 \text{ } (1.30 \text{ kN/m}^2).$$

In S.I.:

$$q_{zT} = 0.613K_{zTor}K_eV_T^2 = 0.613 \times 1.0 \times 1.0 \times 46^2/1{,}000 = 1.30 \text{ kN/m}^2$$

Step 2—Determine the tornado gust-effect factor, G_T

The tornado gust-effect factor, G_T, is determined in accordance with ASCE/SEI 32.11.

In lieu of determining G_T by ASCE/SEI Equation (26.11-6) for rigid structures using exposure C terrain exposure constants, it is permitted to take G_T equal to 0.85 (ASCE/SEI 32.11.1).

Step 3—Determine the tornado directionality factor, K_{dT}

The tornado directionality factor, K_{dT}, is determined from ASCE/SEI Table 32.6-1.

For other structures, K_{dT} must be determined from ASCE/SEI Table 26.6-1. Therefore, $K_{dT} = 1.0$ for round tanks.

Step 4—Determine the force coefficient, C_f

The force coefficient, C_f, is determined from ASCE/SEI 29.4 (ASCE/SEI 32.16.3).

For external walls of isolated circular tanks that satisfy the three conditions in ASCE/SEI 29.4.2.1, it is permitted to take C_f equal to 0.63.

Step 5—Determine the projected area, A_f

The projected area, A_f, normal to the tornadic wind is equal to $h_c D$:

$$A_f = 10.0 \times 20.0 = 200 \text{ ft}^2 \ (18.6 \text{ m}^2)$$

Step 6—Determine the design tornado force, F_T

The design tornado force, F_T, is determined by ASCE/SEI Equations (32.16-2) and (32.16-2.SI):

$$F_T = q_{zT} G_T K_{dT} C_f A_f = 27.2 \times 0.85 \times 1.0 \times 0.63 \times 200/1{,}000 = 2.9 \text{ kips}$$

In S.I.:

$$F_T = 1.30 \times 0.85 \times 1.0 \times 0.63 \times 18.6 = 13.0 \text{ kN}$$

- Design tornado pressures, p_T, on the roof

The design tornado pressure, p_T, on the roof of an isolated circular tank is determined in accordance with ASCE/SEI 32.16.3.3.

Step 1—Determine the tornado velocity pressure, q_{hT}

It is determined above that $q_{zT} = 27.2$ lb/ft^2 (1.30 kN/m^2). Because $K_{zTor} = 1.0$ over the height of the tank, $q_{hT} = 27.2$ lb/ft^2 (1.30 kN/m^2).

Step 2—Determine the tornado gust-effect factor, G_T

The tornado gust-effect factor, G_T, is determined in accordance with ASCE/SEI 32.11.

In lieu of determining G_T by ASCE/SEI Equation (26.11-6) for rigid structures using exposure C terrain exposure constants, it is permitted to take G_T equal to 0.85 (ASCE/SEI 32.11.1).

Step 3—Determine the tornado directionality factor, K_{dT}

The tornado directionality factor, K_{dT}, is determined from ASCE/SEI Table 32.6-1.

For other structures, K_{dT} must be determined from ASCE/SEI Table 26.6-1. Therefore, $K_{dT} = 1.0$ for round tanks.

Step 4—Determine the tornado pressure coefficient adjustment factor, K_{vT}

The tornado pressure coefficient adjustment factor, K_{vT}, is determined from ASCE/SEI Table 32.14-1.

For negative (uplift) pressures on the MWFRS of roofs of tanks, $K_{vT} = 1.1$.

Step 5—Determine the external pressure coefficients, C_p

The external pressure coefficients, C_p, are determined from ASCE/SEI Figure 29.4-5 for roofs:

For Zone 1, $C_p = -0.8$.
For Zone 2, $C_p = -0.5$.
For $h_c/D = 0.5$, the horizontal extent of zone 1 is equal to $0.5D = 10$ ft (3.05 m).

Step 6—Determine the tornado internal pressure coefficient, (GC_{piT})

The tornado internal pressure coefficient, (GC_{piT}), is determined from ASCE/SEI Table 32.13.1.

For sealed other structures, $(GC_{piT}) = +1.0$.

Step 7—Determine the design tornado pressures, p_T, on the roof

The design tornado pressures, p_T, on the roof are determined by ASCE/SEI Equations (32.16-5) and (32.16-5.SI):

$$p_T = q_{hT}[G_T K_{dT} K_{vT} C_p - (GC_{piT})]$$

For Zone 1:

$$p_T = 27.2 \times \{[0.85 \times 1.0 \times 1.1 \times (-0.8)] - 1.0\} = -47.6 \text{ lb/ft}^2$$

In S.I.:

$$p_T = 1.30 \times \{[0.85 \times 1.0 \times 1.1 \times (-0.8)] - 1.0\} = -2.27 \text{ kN/m}^2$$

For Zone 2:

$$p_T = 27.2 \times \{[0.85 \times 1.0 \times 1.1 \times (-0.5)] - 1.0\} = -39.9 \text{ lb/ft}^2$$

In S.I.:

$$p_T = 1.30 \times \{[0.85 \times 1.0 \times 1.1 \times (-0.5)] - 1.0\} = -1.91 \text{ kN/m}^2$$

The external walls and roof members of the tank must be designed and constructed to resist the greater of the tornado loads determined above and the wind loads determined in accordance with ASCE/SEI 29.4.2 using the load combinations in ASCE/SEI Chapter 2 (ASCE/SEI 32.1.1).

From ASCE/SEI Equations (29.4-1) and (29.4-1.SI):

$$F = 2.3 \text{ kips } (10.2 \text{ kN})$$

From ASCE/SEI Equations (29.4-4) and (29.4-4.SI):

$$\text{For zone 1, } p = -18.2 \text{ lb/ft}^2 \ (-0.87 \text{ kN/m}^2)$$

$$\text{For zone 2, } p = -12.8 \text{ lb/ft}^2 \ (-0.61 \text{ kN/m}^2)$$

5.9 References

5.1. Simiu, E. and Scanlan, R.H. 1996. *Wind Effects on Structures: Fundamentals and Applications to Design*, 3rd Ed. John Wiley & Sons, Inc., New York, NY.

5.2. Davenport, A.G. 1960. "Rationale for Determining Design Wind Velocities." *Journal of the Structural Division*, American Society of Civil Engineers, 88:39–68.

5.3. Smith, B.S. and Coull, A. 1991. *Tall Building Structures: Analysis and Design*, John Wiley & Sons, Inc., New York, NY.

5.4. American Society of Civil Engineers (ASCE). 2022. ASCE 7 Hazard Tool. https://asce7hazardtool.online/.

5.5. Davenport, A.G., Surry, D., and Stathopoulos, T. 1978. *Wind Loads on Low-Rise Buildings*, Final Report on Phase III, BLWT-SS4. University of Western Ontario, London, Ontario, Canada.

5.6. Standards Australia. 2011. *Structural Design Actions—Wind Actions*, AS/NZS 1170.2:2011. Sydney, New South Wales, AU.

5.7. Structural Engineers Association of California (SEAOC). 2017. *Wind Design for Solar Arrays*, Report SEAOC PV2-2017, Sacramento, CA.

5.8. American Society of Civil Engineers (ASCE). 2021. *Wind Tunnel Testing for Buildings and Other Structures*, ASCE/SEI 49-21, Reston, VA.

5.9. International Code Council (ICC). 2020. *ICC/NSSA Standard for the Design and Construction of Storm Shelters*, ICC 500, Washington, D.C.

5.10. Federal Emergency Management Agency (FEMA). 2021. *Safe Rooms for Tornadoes and Hurricanes—Guidance for Community and Residential Safe Rooms*, FEMA P-361, Washington, D.C.

5.10 Problems

5.1. Determine design wind pressures on the MWFRS of the one-story warehouse building in Example 5.6 in both directions using ASCE/SEI Chapter 27 assuming the roof slope is equal to (a) 4.76 degrees and (b) 14.0 degrees.

5.2. Determine design wind pressures on (1) a wall and (2) an interior joist of the one-story warehouse building in Example 5.6 using Part 1 of ASCE/SEI Chapter 30 assuming the roof slope is equal to (a) 4.76 degrees and (b) 14.0 degrees.

5.3. Determine the design wind pressures on the MWFRS in the E-W direction of the two-story building in Figure 5.74 using ASCE/SEI Chapter 27 given the design data in Table 5.98. The first floor is located 20 ft (6.10 m) above the ground level and the roof, which has an angle less than 5 degrees from the horizontal, is located 32 ft (9.75 m) above the ground level. Use the floor heights as the reference heights and assume the building is rigid.

Figure 5.74 Plan of the Two-story Building in Problem 5.3

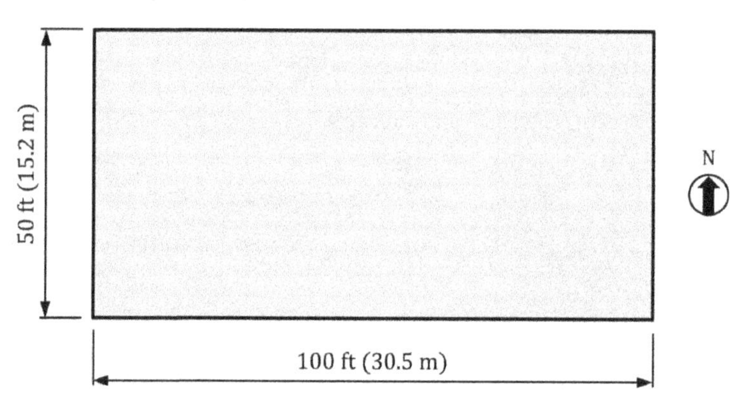

Table 5.98 Design Data for the Two-story Building in Problem 5.3

Basic wind speed, V	115 mi/h (52 m/s)
Exposure category	B
Topography	Not situated on a hill, ridge, or escarpment
Risk category	II
Enclosure classification	Enclosed

5.4. Determine the design wind pressures on the MWFRS in the E-W direction of the building in Figure 5.74 using ASCE/SEI Chapter 27 given the design data in Table 5.98 assuming the building has five stories, each with a story height of 15 ft (4.57 m). The roof angle is less than 5 degrees from the horizontal. Use the floor heights as the reference heights and assume the building is rigid.

5.5. Determine the design wind pressures on the MWFRS of the building in Problem 5.3 using ASCE/SEI Chapter 28.

5.6. Determine the topographic factor, K_{zt}, for a building with a mean roof height of 20 ft (6.10 m) located in the upper one-half of a two-dimensional ridge with the following properties: $H = 100$ ft (30.5 m), $L_h = 40$ ft (12.2 m), and $x = 0$. Assume the exposure category is C.

Figure 5.75 The Three-Story Police Station in Problem 5.7

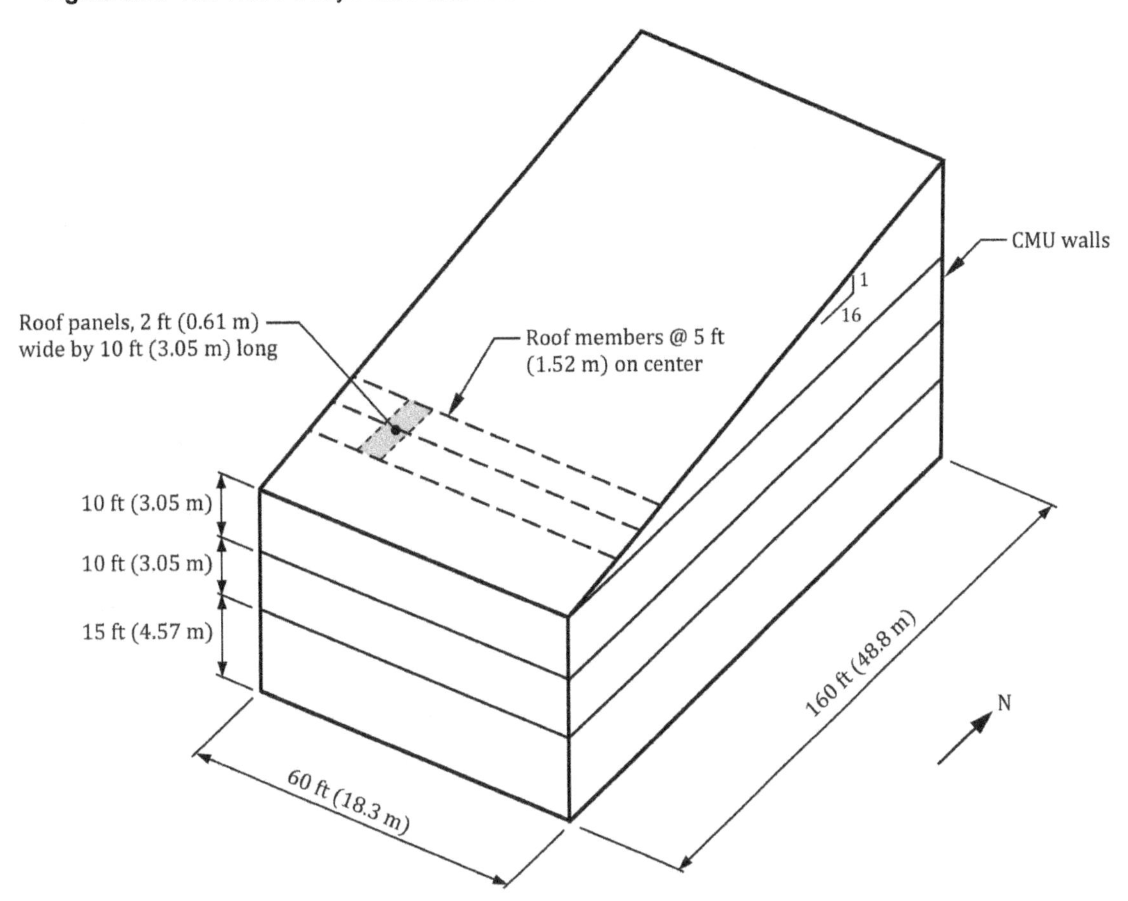

5.7. Determine the design wind pressures on the MWFRS in both the N-S and E-W directions of the three-story police station in Figure 5.75 using ASCE/SEI Chapter 27 given the design data in Table 5.99. The CMU walls around the perimeter of the building have a nominal thickness of 8 in. (203 mm).

Table 5.99 Design Data for the Three-Story Police Station in Problem 5.7

Basic wind speed, *V*	130 mi/h (58 m/s)
Exposure category	C
Topography	Not situated on a hill, ridge or escarpment
Risk category	IV
Enclosure classification	Enclosed

5.8. Determine the design wind pressures on (a) a CMU wall located in the first story, (b) a typical roof member, and (c) a typical roof panel of the police station in Problem 5.7 using Part 1 in ASCE/SEI Chapter 30.

5.9. Assume there is a roof overhang on the south elevation of the police station in Problem 5.7 that is 5 ft (1.52 m) in plan. Determine the design wind pressure on the roof overhang using ASCE/SEI Chapter 27.

5.10. Determine the enclosure classification of the police station in Problem 5.7 assuming the following openings: (1) one 3 ft-6 in. by 7 ft-0 in. (0.31 m by 2.13 m) opening and one 20 ft-0 in. by 8 ft-0 in. (6.10 m by 2.44 m) opening in each of the east and west walls in the first story and (2) two 3 ft-6 in. by 7 ft-0 in. (0.31 m by 2.13 m) openings in the south wall. The north wall and the roof do not have openings.

5.11. Assume the police station in Problem 5.7 has a mechanical unit on the roof that has a height of 8 ft (2.44 m) and base dimensions of 15 ft by 15 ft (4.57 m by 4.57 m). Determine the wind forces on the MWFRS of the mechanical unit using ASCE/SEI Chapter 29.

5.12. Determine the design wind forces on a steel reinforced concrete chimney using ASCE/SEI Chapter 29 assuming exposure C and a basic wind speed of 115 mi/h (52 m/s). The chimney, which is ground supported, has a 6 ft (1.83 m) diameter and has a height of 60 ft (18.3 m) above the ground level. A preliminary analysis indicates that the approximate

Figure 5.76 Plan of the 20-Story Ambulatory Care Facility in Problem 5.13

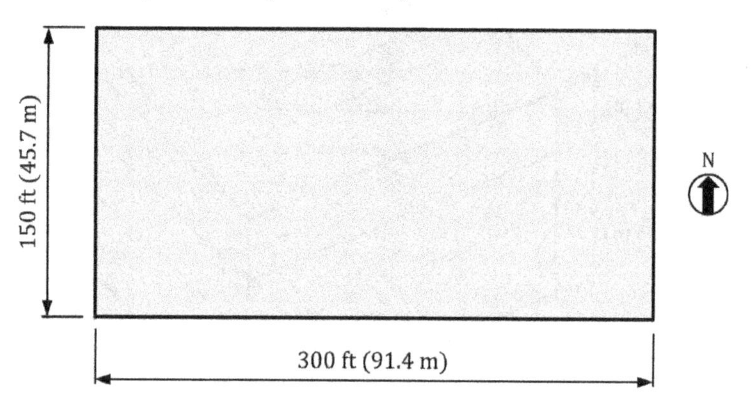

fundamental frequency of the chimney is 0.3 Hz and the damping ratio is 2 percent. Assume the response characteristics are predominantly along-wind.

5.13. Determine the design tornado pressures in the N-S direction on the MWFRS of the 20-story ambulatory care facility that has emergency treatment facilities in Figure 5.76 given the design data in Table 5.100 using ASCE/SEI Chapter 32. A typical story height is 12 ft (3.66 m) and the roof is primarily flat.

Table 5.100 Design Data for the 20-Story Ambulatory Care Facility in Problem 5.13

Location	Latitude = 42.055624° Longitude = −87.889355°
Exposure category	B
Topography	Not situated on a hill, ridge, or escarpment
Risk category	IV
Enclosure classification	Enclosed

5.14. Determine the design tornado pressures on a typical roof member of the 20-story ambulatory care facility in Problem 5.13 using ASCE/SEI Chapter 32. Assume the roof members have 30 ft (9.14 m) spans and are spaced 10 ft (3.05 m) on center.

CHAPTER 6

Earthquake Loads

6.1 Introduction

6.1.1 Nature of Earthquake Ground Motion

The earth's crust consists of a number of large and fairly stable rock slabs, which are referred to as *tectonic plates* (see Figure 6.1). For the most part, the plates are prevented from moving relative to each other due to friction at their boundaries. Over time, stress builds up along these boundaries, and an earthquake is generated when energy is released by the following mechanisms:

1. The built-up stress along a plate boundary exceeds the frictional force that locks the plates together resulting in slip along the fault lines.
2. The stress at an internal location in the plate exceeds the strength of the rock causing the rock to fracture.

Faults are locations where earthquakes have occurred previously and are generally found at the plate boundaries although they can occur within the interior of a plate. A fault is considered

Figure 6.1 Major Tectonic Plates (Courtesy of the U.S. Geological Survey)

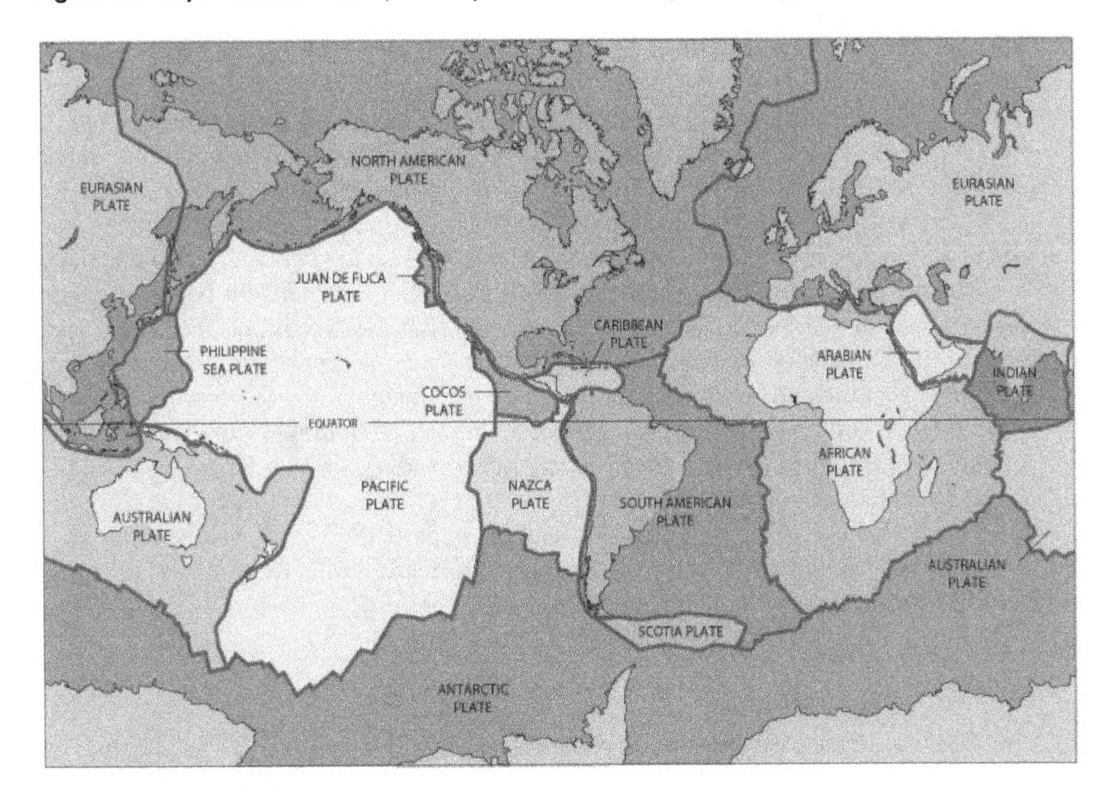

to be active if evidence exists that it has moved within the past 10,000 years. New faults are created as the stress patterns in the earth's crust change over time.

Slippage can occur at the surface of the earth or many feet beneath it. *Ground fault ruptures* are created when the slippage extends to the surface. In such cases, abrupt vertical and lateral offsets occur; surface offsets can range from a few inches (about 50 mm) to a dozen feet (about 3.4 m) depending on the magnitude of the earthquake. Because it is very difficult to design structures to accommodate such movement, building should not be located above or in close proximity to known faults.

The energy released during an earthquake radiates out in the form of random vibrations in all directions from the area where slippage occurs. Ground shaking is felt on the surface due to these vibrations. As the ground displaces, buildings move and undergo a series of oscillations. Depending on the size of the earthquake, ground shaking can last from a few seconds to several minutes. Deformations are produced in the structure due to ground shaking and both structural and nonstructural members must be designed for the ensuing effects.

Significant damage or complete failure of a building can occur due to landslides and soil liquefaction caused by ground shaking. Tsunamis, which are seismically induced water waves, can have a devastating impact on structures. An earthquake-induced fire is another example of a damaging event created by earthquakes.

6.1.2 Seismic Hazard Analysis

Magnitude and *intensity* measure different characteristics of earthquakes. Magnitude is a measure of the energy released at the source of an earthquake and is determined using seismographs. Intensity measures the strength of ground shaking produced by an earthquake at a specific location.

The Modified Mercalli Intensity (MMI) scale is a qualitative measure of earthquake intensity. Descriptions of effects on people, structures, and the natural environment are provided for 12 intensity levels, which range from low-intensity earthquakes that are not felt except by a very few under especially favorable conditions (MMI I) to high-intensity earthquakes that cause total damage (MMI XII).

The MMI scale is not directly useful for the design of structures because it is not based on quantifiable effects due to ground shaking. A mathematical relationship that quantifies earthquake shaking is an acceleration response spectrum, which gives peak accelerations a structure would experience if subjected to specific earthquake motion. A response spectrum provides the maximum response value for elastic, single-degree-of-freedom oscillators as a function of structure period, thereby eliminating the need to account for the total response history for every period of interest.

A generic representation of an acceleration response spectrum obtained from recordings of an actual earthquake is depicted in Figure 6.2. The horizontal axis of the spectrum is the period, T, of a structure, which is related to its dynamic properties (methods on how to determine T are given in Section 6.3.8 of this book). The vertical axis of the spectrum is the acceleration a structure would experience based on its period. The spectral acceleration, S_a, is derived from a response spectrum and is usually expressed as a percentage of the acceleration due to gravity, g.

Structures with longer periods (e.g., flexible, high-rise buildings) would experience relatively smaller accelerations than structures with shorter periods (such as stiff, low-rise buildings) when subjected to the earthquake in Figure 6.2.

Ground shaking at a particular site and for a particular earthquake is unique, as is the corresponding response spectrum. The following factors affect the shape and amplitude of response spectra:

- Magnitude of the earthquake
- Depth of the earthquake
- Distance of the earthquake from the site
- Soil type

Figure 6.2 Acceleration Response Spectrum

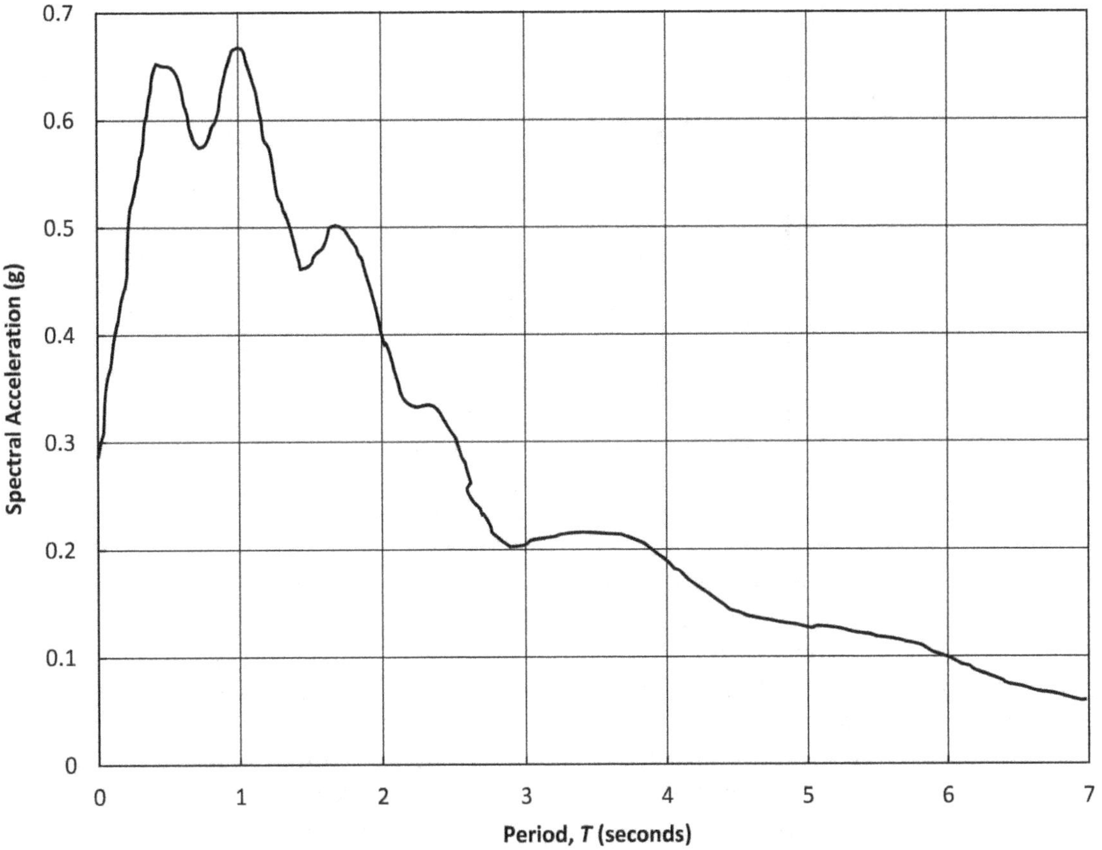

The U.S. Geological Survey (USGS) periodically performs national seismic hazard analyses to determine spectral response accelerations for different recurrence intervals. Mapped values of acceleration parameters are archived in the USGS Seismic Design Geodatabase (Reference 6.1) at gridded locations across the United States for user-specified site locations (latitude and longitude) and site class (soil type). Additional information on these acceleration parameters is given in the next section.

The IBC and ASCE/SEI 7 provide a design response spectrum to facilitate representation of these complex ground motions, which, in turn, facilitates the design of structures for earthquake motions (see Section 6.2.1 of this book for more information). Design response spectra are basically smoothed and normalized approximations of many different recorded ground motions.

Acceleration parameters can be plotted as a function of *annual frequency of exceedance*, which is the number of times in any one year that, on average, a specific site can experience ground shaking greater than or equal to a specific acceleration. Such plots are commonly referred to as *hazard curves* where the annual frequency of exceedance is on the vertical axis and the acceleration is on the horizontal axis. The average return period is the inverse of the annual frequency of exceedance and is the number of years likely to elapse between two events, each producing ground shaking that is at least the value indicated for a specific acceleration on the horizontal axis.

A generic representation of a hazard curve for a particular site is given in Figure 6.3. Ground shaking producing a spectral acceleration equal to $0.32g$ has an annual frequency of exceedance of approximately 0.01, which is equivalent to an average return period of $1/0.01 = 100$ years. At an annual frequency of exceedance of 0.001 (1,000-year return period), the spectral acceleration

Figure 6.3 Hazard Curve for Spectral Acceleration

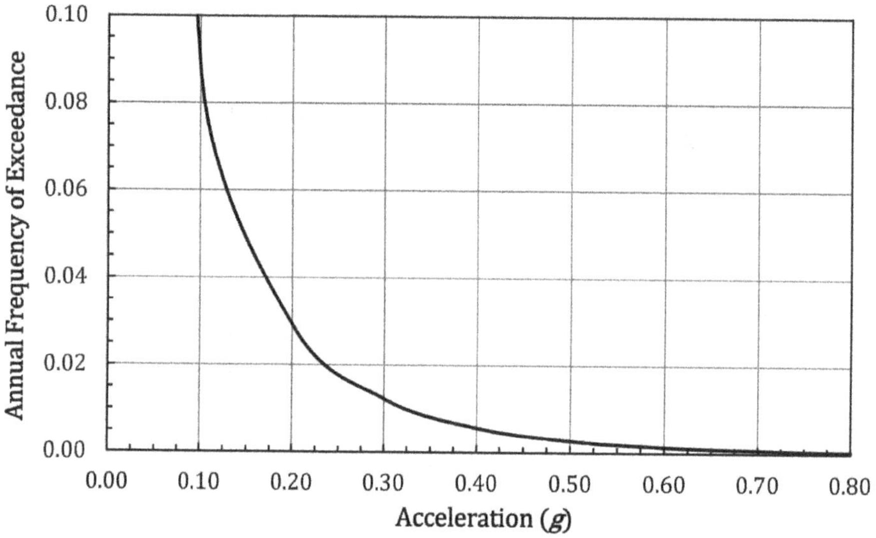

is equal to approximately $0.60g$. For a spectral acceleration of $0.80g$, the annual frequency of exceedance is about 0.0004 (2,500-year return period).

All hazard curves have a form similar to the one in Figure 6.3. In general, low-intensity earthquakes (low accelerations) occur relatively frequently while high-intensity earthquakes (large accelerations) occur rarely.

Methods on how to calculate the effects of earthquakes on structural and nonstructural members and components of a structure are covered in subsequent sections of this chapter. In these methods, such effects are determined using static earthquake loads. The purpose of these loads is to produce the same deformations that would occur in the structure (when multiplied by a deflection amplification factor) if it were subjected to the design-basis ground motion.

The following section provides information on the reference earthquake shaking stipulated in the IBC and ASCE/SEI 7; this reference shaking is used to determine earthquake effects on structures.

6.1.3 Reference Earthquake Shaking

Any structure will collapse if subjected to sufficiently strong ground shaking. However, it is virtually impossible to predict for a particular earthquake the magnitude and direction of the ground motion that can cause collapse. Variability in material strengths and structural member sizes and quality of workmanship are a few of the many other variables that make collapse prediction essentially unattainable.

A *structural fragility function* can be used in determining an estimate of the acceleration that will cause collapse of a structure. A fragility curve provides probabilities of structural collapse as a function of the spectral response acceleration where the probability of collapse is on the vertical axis and the spectral acceleration at period T is on the horizontal axis. For example, a building may have a 10 percent chance of collapsing if subjected to an acceleration of $0.5g$ and a 50 percent chance of collapsing if subjected to an acceleration of $0.9g$. More information on fragility functions for structural collapse can be found in Reference 6.2, including an example of a fragility curve (Figure C16.4-1).

The risk associated with collapse is a function of the fragility of a structure and the seismic hazard at the site, the latter of which is represented by a hazard curve (see the discussion in Section 6.1.2 of this book on hazard). The mathematical combination of these two functions gives the probability that a structure will collapse in any given year or number of years. Reference 6.2, which forms the basis of the earthquake provisions in the IBC and ASCE/SEI 7, establishes the probability at 1 percent or less in 50 years that a structure with ordinary occupancy will collapse due to an earthquake. Additional information is provided in Section 6.1.4 of this book.

Ground motion based on uniform risk, as described above, is referred to as *probabilistic ground motion*. At most locations within the United States, this shaking corresponds to a mean recurrence interval of approximately 2,500 years. A recurrence interval that is somewhat shorter is expected at sites where earthquakes occur relatively frequently while a recurrence level that is longer is anticipated at sites where earthquakes rarely occur.

The probabilistic ground motion can be severe in regions of the United States with major active faults capable of producing large earthquakes; at certain locations, values of the parameters determined by the probabilistic approach far exceed those corresponding to the strongest ground motion ever recorded in those areas. Thus, a *deterministic* estimate of the ground shaking is used in those regions rather than the risk-based shaking. Estimates of the ground motion parameters are determined using the maximum earthquake that can be generated by the nearby faults. In particular, acceleration values are established using the near-source 84th-percentile 5-percent damped ground motions in the direction of maximum horizontal response. Deterministic ground shaking implies a somewhat higher level of collapse risk than the 1 percent probability of collapse in 50 years associated with the probabilistic ground shaking.

The *risk-targeted maximum considered earthquake* (MCE_R) *ground motion* is the shaking from the probabilistic ground motion except at locations where the deterministic ground motion governs, as described above. For nearly all of the United States where probabilistic ground motion is applicable, there is a small probability (approximately 10 percent or less) that structures with ordinary occupancies (risk category II structures; see IBC Table 1604.5) will collapse when subjected to such shaking. This ground motion is computed such that for structures with a typical collapse fragility when subjected to various ground motions, there is an overall 1 percent chance of collapse in 50 years. The combination of these two probabilities defines the MCE_R for the probabilistic approach. In areas near faults that produce frequent, large earthquakes, MCE_R ground shaking is determined by assuming a characteristic earthquake for that fault occurs and then computing the maximum ground motion reduction from the fault to the site using the 84th percentile level. Methods on how to determine MCE_R ground shaking are given in Section 6.2.1 of this book.

6.1.4 Overview of Seismic Requirements

According to IBC 1613.1, the effects of earthquake motion on structures and their components must be determined in accordance with ASCE/SEI 7 Chapters 11, 12, 13, 15, 17, and 18, as applicable.

The seismic provisions in the IBC and ASCE/SEI 7 are based on the requirements in Reference 6.2. This resource document contains the fundamental seismic requirements, which are presented in a format that can be readily adopted into a code. The history on the creation and subsequent development of this document can be found in Reference 6.3.

The chapters in ASCE/SEI 7 referenced by the 2024 IBC for seismic load provisions are given in Table 6.1. The primary focus in this chapter is on the provisions in ASCE/SEI Chapters 11, 12, 13, 15, 20, 21, and 22.

Table 6.1 Chapters in ASCE/SEI 7 Referenced by the IBC for Seismic Load Provisions

Chapter	Title
11	Seismic Design Criteria
12	Seismic Design Requirements for Building Structures
13	Seismic Design Requirements for Nonstructural Components
15	Seismic Design Requirements for Nonbuilding Structures
20	Site Classification Procedure for Seismic Design
21	Site-Specific Ground Motion Procedures for Seismic Design
22	Seismic Ground Motion and Long-Period Transition Maps

Six exceptions to seismic design requirements are given in IBC 1613.1:

1. Detached one- and two-family dwellings assigned to Seismic Design Category (SDC) A, B, or C.
2. The seismic force-resisting system (SFRS) of wood-frame buildings that conform to IBC 2308.
3. Agricultural storage structures intended only for incidental human occupancy.
4. Structures covered under other regulations, such as vehicular bridges, electrical transmission towers, hydraulic structures, buried utility lines, and nuclear reactors.
5. References within ASCE/SEI 7 to Chapter 14 do not apply, except as specifically required within the IBC.
6. Temporary structures complying with IBC 3103.6.1.4.

The first and third exceptions reflect the low seismic risks and the low risk to loss of human life, respectively, associated with these occupancies. The second exception recognizes that wood-frame seismic design requirements of IBC 2308 substantially meet the intent of conventional construction (wood-frame) provisions included in Reference 6.2. The types of structures listed under the fourth exception have unique design and performance issues and are not covered in the IBC and ASCE/SEI 7 because the requirements in these documents were developed for buildings and building-like structures. The fifth exception is given because the IBC contains structural material provisions in Chapters 19 through 23.

The seismic requirements of the IBC need not be applied to the structures identified in exceptions 1, 2, 3, 4, and 6.

ASCE/SEI 11.1.2 contains exemptions, which are essentially the same as the IBC exceptions with the following differences:

1. The first exemption in ASCE/SEI 11.1.2 also permits detached one- and two-family dwellings to be exempt where the mapped, short period, spectral response acceleration parameter, S_S, is less than 0.4.
2. In the second exemption, the International Residential Code® (IRC®) is referenced in ASCE/SEI 11.1.2 instead of IBC 2308. This exemption recognizes that the wood-frame seismic design requirements of the IRC for structures up to two stories substantially meet the intent of conventional construction (wood-frame) provisions included in Reference 6.2.
3. ASCE/SEI 11.1.2 includes an exemption for piers and wharves not accessible to the general public, which is not included in IBC 1613.1.
4. IBC 1613.1 includes an exception for temporary structures complying with IBC 3103.6.1.4, which is not included in ASCE/SEI 11.1.2.

Flowchart 6.1 in Section 6.6 of this book can be used to determine where the seismic provisions of the IBC and ASCE/SEI 7 must be considered.

Seismic requirements for existing buildings are given in Reference 6.4. Included are provisions related to additions, alterations, repairs, and change of occupancy of structures.

6.2 Seismic Design Criteria

6.2.1 Seismic Ground Motion Values

Near-Fault Sites

Structures located close to a zone of fault rupture can be subjected to very damaging ground motion. Restrictive design criteria must be satisfied where such ground motion can occur.

Because the distance from the zone of fault rupture at which these effects can be experienced is dependent on a number of factors (including the rupture type, depth of fault, magnitude, and direction of fault rupture), a precise definition of what constitutes a near-fault site is difficult to create in general terms. According to ASCE/SEI 11.4.1, a site satisfying either of the following two conditions is classified as a near-fault site:

- Sites located 9.5 miles (15 km) or less from the surface projection of a known active fault capable of producing M_W7 or larger events.
- Sites located 6.25 miles (10 km) or less from the surface projection of a known active fault capable of producing M_W6 or larger events but smaller than M_W7 events.

The moment magnitude scale denoted by M_W is a measure of an earthquake's magnitude (i.e., size or strength) based on its seismic moment, which is a measure of the work done by the faulting of an earthquake.

A method of determining the distance of a site from a fault where the fault plane dips at an angle relative to the ground surface is illustrated in Figure 6.4 (see ASCE/SEI Figure C11.4-1).

Two exceptions are given in ASCE/SEI 11.4.1. In the first exception, faults with estimated slip rates along the fault less than 0.04 inch (1 mm) per year need not be used to determine whether a site is a near-fault site. In the second exception, surface projections used in the determination of near-fault site classification need not include portions of the fault at depths greater than or equal to 6.25 miles (10 km) [see Figure 6.4].

Figure 6.4 Near-Fault Criteria of ASCE/SEI 11.4.1

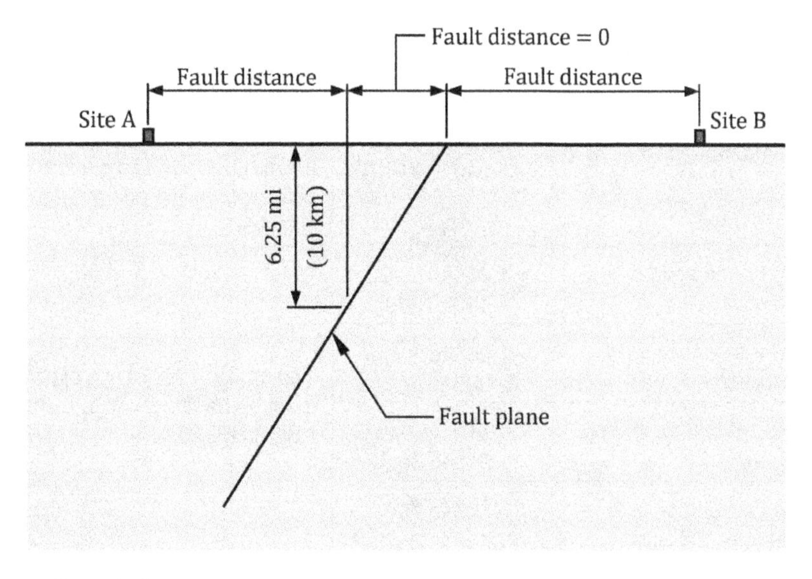

Near-fault site:

Fault distance $\leq$ { 9.5 mi (15 km) from an M_W7 or larger active fault; 6.25 mi (10 km) from an M_W6 or larger active fault but smaller than M_W7 events }

Site Class Definitions

According to ASCE/SEI 11.4.2, the soil at a site must be classified in accordance with ASCE/SEI Chapter 20 (the site class is needed to determine the MCE_R spectral response acceleration parameters at a site). Nine site classes are defined in ASCE/SEI Table 20.2-1 (see Table 6.2) based on the average shear wave velocity parameter, $\overline{v}_s$, which is derived from the measured shear wave

Table 6.2 Site Classification

Site Class	$\bar{v}_s$, ft/s (m/s)[1]	Notes
A. Hard rock	> 5,000 (1,524)	• Shear wave velocity measurement must be on site or on profiles of the same rock type in the same formation with an equal or greater degree of weathering and fracturing. • Measurements of the shear wave velocity to depths less than 100 ft (30.5 m) are permitted to determine $\bar{v}_s$ where hard rock conditions are known to be continuous to a depth of 100 ft (30.5 m). • Site class A is not permitted to be assigned to a site where more than 10 ft (3.05 m) of soil is present between the rock surface and the bottom of the foundation (spread footing or mat).
B. Medium hard rock	> 3,000 to 5,000 (> 914 to 1,524)	• Site Class B can only be assigned to a site where the shear wave velocity is measured on site. • Site Class B is not permitted to be assigned to a site where more than 10 ft (3.05 m) of soil is present between the rock surface and the bottom of the foundation (spread footing or mat).
BC. Soft rock	> 2,100 to 3,000 (> 640 to 914)	Where shear wave velocity data are not available and the site condition is estimated by a geotechnical engineer, engineering geologist or seismologist as site class B or BC on the basis of site geology, the site must be classified as site class BC.
C. Very dense sand or hard clay	> 1,450 to 2,100 (> 442 to 640)	The assignment of site class C, CD, D, and DE must be based on $\bar{v}_s$ determined in accordance with ASCE/SEI 20.4.
CD. Dense sand or very stiff clay	> 1,000 to 1,450 (> 305 to 442)	
D. Medium dense sand or stiff clay	> 700 to 1,000 (> 213 to 305)	
DE. Loose sand or medium stiff clay	> 500 to 700 (> 152 to 213)	
E. Very loose sand or soft clay	≤ 500 (≤ 152)	A site must be classified as site class E where it does not qualify under the criteria for site class F and there is a total thickness of soft clay greater than 10 ft (3.05 m), regardless of $\bar{v}_s$ determined in accordance with ASCE/SEI 20.4.[2]
F. Soils requiring a site response analysis in accordance with ASCE/SEI 21.1	—	Site Class F must be assigned where any of the conditions in ASCE/SEI 20.2.1 are satisfied.

[1]The average shear wave velocity parameter, $\bar{v}_s$, is derived from the measured shear wave velocity profile from the ground surface to a depth of 100 ft (30.5 m). Where shear wave velocity is not measured, an estimation of $\bar{v}_s$ is permitted to be determined in accordance ASCE/SEI 20.3.

[2]A soft clay layer is defined as follows: (a) undrained shear strength, $s_u < 500$ lb/ft^2 (23.9 kN/m^2), (b) soil moisture content, $w \geq 40$ percent, and (c) plasticity index, $PI > 20$.

velocity profile from the ground surface to a depth of 100 ft (30.5 m) [see ASCE/SEI Equation (20.4-1)]:

$$\bar{v}_s = \sum_{i=1}^{n} d_i \Big/ \sum_{i=1}^{n} (d_i / v_{si}) \tag{6.1}$$

where d_i = thickness of a soil layer between 0 and 100 ft (30.5 m)
v_{si} = measured shear wave velocity associated with a soil layer
n = number of distinct soil layers in the upper 100 ft (30.5 m) of the site

It may not be possible to acquire measured shear wave velocities for the entire 100-ft (30.5-m) depth (e.g., a layer of hard rock may be encountered 25 ft [7.62 m] below the surface). At such locations or in cases where it is not feasible to measure the shear wave velocity to a depth of 100 ft (30.5 m), it is permitted to estimate shear wave velocities using correlations with suitable geotechnical parameters, including standard penetration test (SPT) blow counts, shear strength, overburden pressure, void ratio or cone penetration test (CPT) tip resistance measured at the site (see ASCE/SEI 20.3 for additional information).

Where soil data are available to depths more than 50 ft (15.2 m) and less than 100 ft (30.5 m) below the surface at a site and it is unlikely that soft layers of soil are present below the maximum profile depth, it is permitted to extend the shear wave velocity of the last layer in the profile to 100 ft (30.5 m) for the calculation of $\bar{v}_s$ by ASCE/SEI Equation (20.4-1) [ASCE/SEI 20.3]. Where data are not available below 50 ft (15.2 m), a default site class in accordance with ASCE/SEI 20.1 must be used unless another site class can be justified based on site geology.

At sites where soil properties are not known in sufficient detail to determine the site class in accordance with ASCE/SEI Chapter 20, MCE_R spectral response accelerations must be based on the most critical spectral response acceleration at each period corresponding to site class C, site class CD, and site class D, unless the building official or geotechnical data, determine that site class DE, E, or F soils are present at the site. The site class generating the most critical accelerations at the site is referred to as the *default site class*.

Risk-Targeted Maximum Considered Earthquake (MCE_R) Spectral Response Acceleration Parameters

For a site class determined in accordance with ASCE/SEI 11.4.2 (see previous section), the following risk-targeted maximum considered earthquake (MCE_R) spectral response acceleration parameters are obtained from Reference 6.1 (ASCE/SEI 11.4.3):

- S_S = MCE_R, 5 percent-damped, spectral response acceleration parameter at a period of 0.2 s for Site Class BC site conditions as determined in accordance with ASCE/SEI 11.4.3
- S_1 = MCE_R, 5 percent-damped, spectral response acceleration parameter at a period of 1 s for Site Class BC site conditions as determined in accordance with ASCE/SEI 11.4.3
- S_{MS} = MCE_R, 5 percent-damped, spectral response acceleration parameter at short periods adjusted for site class effects as determined in accordance with ASCE/SEI 11.4.3
- S_{M1} = MCE_R, 5 percent-damped, spectral response acceleration parameter at a period of 1 s adjusted for site class effects as determined in accordance with ASCE/SEI 11.4.3

As noted in Section 6.1.2 of this book, mapped values of the spectral response acceleration parameters are obtained from the USGS Seismic Design Geodatabase at gridded locations in the United States based on site location (latitude and longitude) and site class. These mapped spectral accelerations correspond to probabilistic ground motion except at locations where deterministic ground motion governs (see Section 6.1.3 of this book). Maps of spectral response acceleration parameters S_{MS} and S_{M1} for default site conditions and based on the most critical spectral response acceleration of Site Classes C, CD, and D are given in ASCE/SEI Chapter 22 for the conterminous United States, Alaska, Hawaii, Puerto Rico, the U.S. Virgin Islands, Guam, Northern Mariana and American Samoa.

In cases where a site-specific study in accordance with ASCE/SEI 11.4.7 is required (structures on Site Class F sites) or such a study is chosen for a particular site, S_{MS} and S_{M1} must be determined in accordance with ASCE/SEI 21.4 instead of being obtained directly from Reference 6.1 (see the exception in ASCE/SEI 11.4.3). Similarly, S_S and S_1 must be determined from the site-specific MCE_R response spectrum defined in ASCE/SEI 21.2.3.

Design Spectral Acceleration Parameters

Design, 5 percent-damped, spectral response acceleration parameters at short periods, S_{DS}, and at 1-s periods, S_{D1}, are determined by ASCE/SEI Equations (11.4-1) and (11.4-2), respectively:

$$S_{DS} = 2S_{MS}/3 \tag{6.2}$$

$$S_{D1} = 2S_{M1}/3 \tag{6.3}$$

Design of buildings and other structures is based on earthquake demands that are two-thirds of the MCE_R ground motions. Values of these acceleration parameters are obtained from Reference 6.1 for a user-specified site location and site class and are based on the multi-period design response spectrum in ASCE/SEI 11.4.5 (see the discussion below).

The value of S_{DS} must be determined in accordance with ASCE/SEI 12.14.8.1 for structures designed using the simplified alternative design procedure in ASCE/SEI 12.14 (see Section 6.3.14 of this book). Values of S_{D1} need not be determined in such cases.

Design Response Spectrum

Where a site-specific ground motion analysis is not required or is not performed in accordance with ASCE/SEI 11.4.7, a design response spectrum for a site must be determined using the multi-period design response spectrum provisions in ASCE/SEI 11.4.5.1 (ASCE/SEI 11.4.5). In this method, the 5 percent-damped design spectral response acceleration parameters, S_a, are obtained directly from Reference 6.1 for the following 22 structure periods, T, based on site location and site class: 0.0 s, 0.01 s, 0.02 s, 0.03 s, 0.05 s, 0.075 s, 0.1 s, 0.15 s, 0.2 s, 0.25 s, 0.3 s, 0.4 s, 0.5 s, 0.75 s, 1.0 s, 1.5 s, 2.0 s, 3.0 s, 4.0 s, 5.0 s, 7.5 s, and 10 s (ASCE/SEI 11.4.5.1). The S_a corresponding to these periods are equal to two-thirds of the multi-period 5 percent-damped MCE_R accelerations, which are also obtained from Reference 6.1.

It is permitted to determine values of S_a for any period less than or equal to 10 s and not equal to one of the discrete periods listed above by linear interpolation between the values of S_a listed above. Values of S_a for any period greater than 10 s are determined by the following equations:

$$S_a = \begin{cases} 10(S_a)_{10}/T \text{ where } T \le T_L \\ 10(S_a)_{10}T_L/T^2 \text{ where } T > T_L \end{cases} \tag{6.4}$$

In this equation, $(S_a)_{10}$ is the value of S_a at the period of 10 s obtained from Reference 6.1 and T_L is the long-period transition period determined from the maps in ASCE/SEI Chapter 22 or from Reference 6.1.

An example of a multi-period design response spectrum is given in Figure 6.5 for a risk category II structure on a site in Charleston, SC, that is classified as site class C.

In cases where values of the multi-period 5 percent-damped MCE_R response spectrum are not available from Reference 6.1, it is permitted to determine the design response spectrum using the two-period design response spectrum provisions in ASCE/SEI 11.4.5.2 (see exception 2 in ASCE/SEI 11.4.5). In this method, the 5 percent-damped design spectral response accelerations, S_a, are determined by ASCE/SEI Equations (11.4-3), (11.4-4) and (11.4-5) [see Figure 6.6].

For periods up to the period T_0, which is equal to $0.2(S_{D1}/S_{DS})$, the design spectral response acceleration, S_a, is determined by ASCE/SEI Equation (11.4-3):

$$S_a = S_{DS}[0.4 + 0.6(T/T_0)] \tag{6.5}$$

Figure 6.5 Example of a Multi-Period Design Response Spectrum

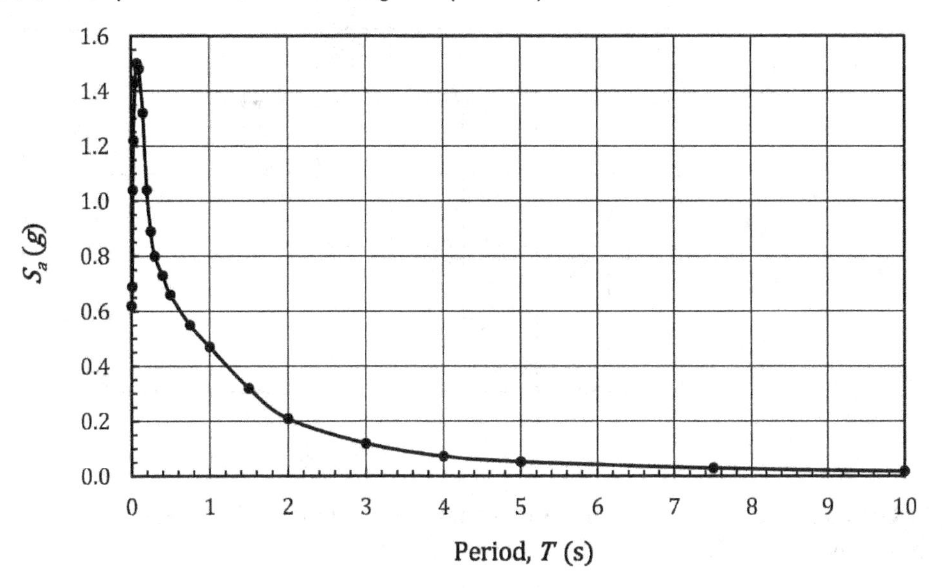

Figure 6.6 Two-Period Design Response Spectrum

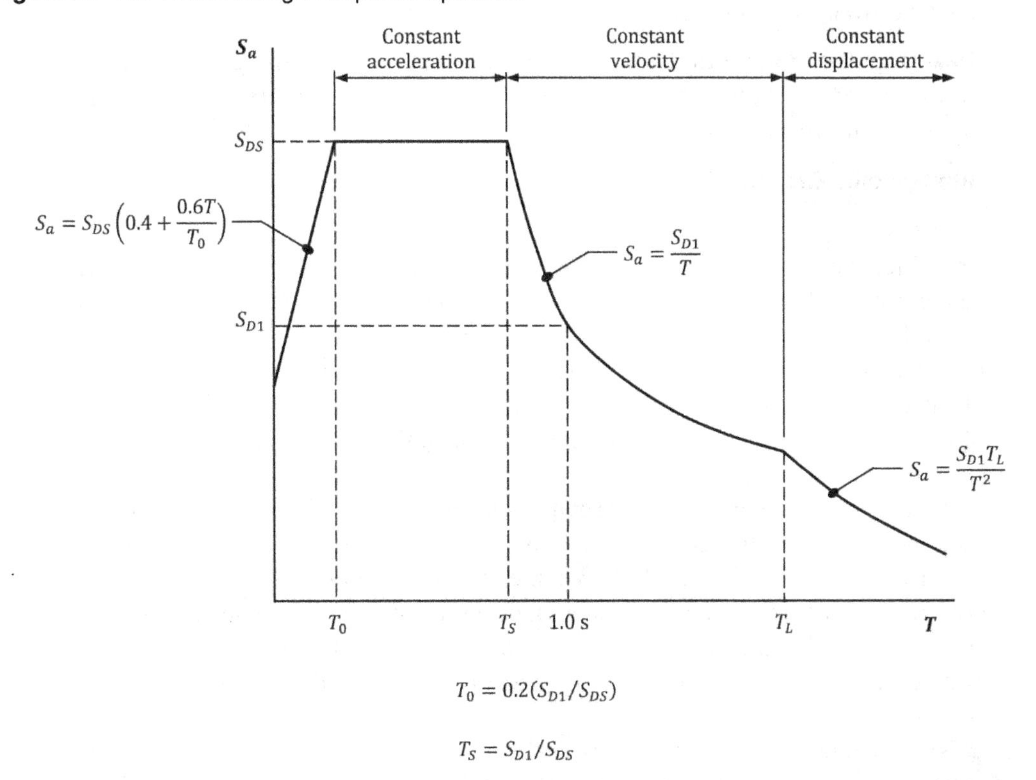

The constant acceleration portion of the spectrum is between the periods T_0 and T_S, where $T_S = S_{D1}/S_{DS}$. In this region, S_a is a constant and is equal to S_{DS}.

The constant velocity region is between T_S and T_L where the response acceleration is proportional to the inverse of the period, T. In this region, S_a is determined by ASCE/SEI Equation (11.4-4):

$$S_a = S_{D1}/T \tag{6.6}$$

Figure 6.7 Example of a Two-Period Design Response Spectrum

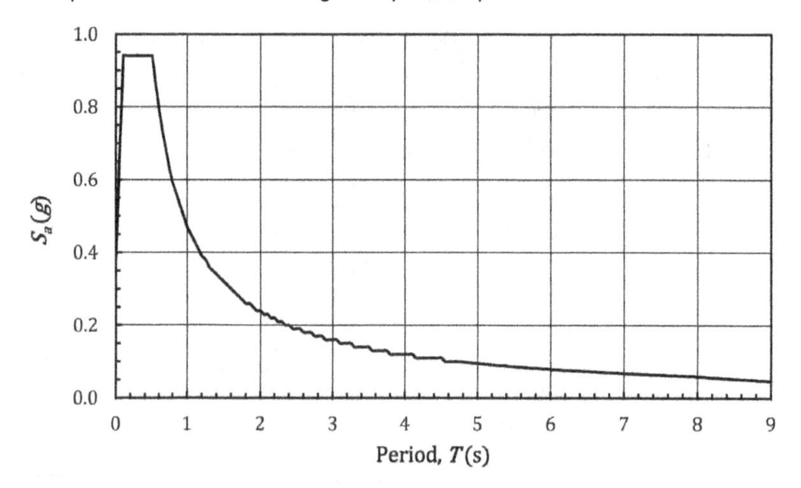

The constant displacement region occurs where the period is greater than T_L, and S_a is determined by ASCE/SEI Equation (11.4-5):

$$S_a = S_{D1}T_L/T^2 \tag{6.7}$$

The two-period design response spectrum for the risk category II structure in Charleston, SC, noted above is given in Figure 6.7.

Risk-Targeted Maximum Considered Earthquake (MCE_R) Response Spectrum

In cases where a risk-targeted MCE_R response spectrum is required, it is determined by multiplying the values from the applicable design response spectrum method by 1.5 (ASCE/SEI 11.4.6).

Site-Specific Ground Motion Procedures

Overview

The objective of a site-specific ground motion analysis is to determine seismic ground motion values for local seismic and site conditions with higher confidence than is possible using the general procedure in ASCE/SEI 11.4 (ASCE/SEI C11.4.7). Because the site-specific procedures in ASCE/SEI Chapter 21 are the same as those used by the USGS to develop the mapped values of MCE_R ground motion, site-specific ground motions should not differ significantly from those of the mapped values of MCE_R ground motions unless there are significant differences in local seismic and site conditions.

A site response analysis in accordance with ASCE/SEI 21.1 or a ground motion hazard analysis in accordance with ASCE/SEI 21.2 is permitted to be used to determine ground motions for any structure (ASCE/SEI 11.4.7). When either of these procedures is used, the design response spectrum must be determined in accordance with ASCE/SEI 21.3, the design acceleration parameters must be determined in accordance with ASCE/SEI 21.4 and where required, the MCE_G peak ground acceleration parameter, PGA_M, must be determined in accordance with ASCE/SEI 21.5 (ASCE/SEI 11.4.7).

For structures on site class F sites, a site response analysis in accordance with ASCE/SEI 21.1 must be performed unless one or more of the following exceptions in ASCE/SEI 20.2.1 is satisfied:

1. Structures with $T \leq 0.5$ s and located on sites with liquefiable soils. In such cases, a site class is permitted to be determined in accordance with ASCE/SEI 20.2 and the corresponding response spectrum is determined from ASCE/SEI 11.4 for the purpose of determining spectral accelerations only.
2. Structures assigned to Seismic Design Category (SDC) A and B where the SDC is based on the values S_{DS} and S_{D1} at a site where the soil profile includes very high plasticity clays

with a thickness greater than 25 ft (7.62 m) and a plasticity index (*PI*) greater than 75 in a soil profile that would otherwise be classified as site class CD, D, DE, or E.

3. Structures assigned to SDC A and B where the SDC is based on the values S_{DS} and S_{D1} at a site where the soil profile includes very thick soft/medium stiff clays with a thickness greater than 120 ft (36.6 m) and an undrained shear strength, s_u, less than 1,000 lb/ft^2 (47.9 kN/m^2).

Site Response Analysis

In a site response analysis, a soil profile model is developed using an MCE_R response spectrum developed for a base condition consisting of bedrock or, where bedrock is very deep, firm soil conditions below softer surficial layers using the procedure in ASCE/SEI 11.4.6 or 21.2 (ASCE/SEI 21.1.1). Where ASCE/SEI 11.4.6 is used, a site condition representative of the geological conditions at the base (i.e., represented by a base-condition average shear wave velocity, $\overline{v}_s$) must be used to develop the MCE_R base response spectrum. In the model, at least five recorded or simulated horizontal ground motion acceleration time histories must be selected that have magnitudes and fault distances consistent with those that control the MCE_R ground motion. These base ground motion time histories are input to the soil profile model at the site as outcropping motions, and surface ground motion time histories are calculated. The calculated MCE_R ground motion response spectrum must be greater than or equal to the MCE_R response spectrum of the base motion multiplied by the average surface-to-base response spectral ratios (calculated period by period) obtained from the site response analyses (ASCE/SEI 21.1.3).

Additional information on conducting a site response analysis is given in ASCE/SEI C21.1 including websites with ground motion acceleration histories. A list of frequently used computer programs that can perform such an analysis is given in ASCE/SEI C21.1.3.

Risk-Targeted Maximum Considered Earthquake (MCE_R) *Ground Motion Hazard Analysis*

Site-specific MCE_R ground motions are determined in accordance with ASCE/SEI 21.2 based on site-specific probabilistic and site-specific deterministic ground motions, both of which are defined in terms of the 5 percent damped spectral response in the maximum direction of the horizontal response. The results of a site-specific MCE_R ground motion hazard analysis are used in determining the MCE_R response spectrum (ASCE/SEI 21.2), the design response spectrum (ASCE/SEI 21.3), and the site-specific design accelerations S_{MS}, S_{M1}, S_{DS}, and S_{D1} (ASCE/SEI 21.4).

The analysis must account for the following:

- Regional tectonic setting, geology and seismicity
- Expected recurrence rates and maximum magnitudes of earthquakes on known faults and source zones
- Characteristics of ground motion near source effects, if any, on ground motions
- Effects of subsurface site conditions on ground motions

Most ground motion relations are defined in terms of average (geometric mean) horizontal response. Because maximum response in the horizontal plane is greater than average response by an amount that varies with period, it may be reasonably estimated by factoring the average response by period-dependent factors by 1.2 for periods less than or equal to 0.2 s, 1.25 for a period of 1.0 s, and 1.3 for periods greater than or equal to 10 s unless it can be shown other scale factors more closely represent the maximum response.

Probabilistic (MCE_R) Ground Motions. Except at locations near highly active faults, a probabilistic seismic hazard analysis method and subsequent computations of risk-targeted probabilistic ground motions based on the output of the analysis method are permitted to define (MCE_R) ground motion. The primary output of a probabilistic seismic hazard analysis is a hazard curve, which provides mean annual frequencies of exceeding various user-specified ground motion amplitudes. Risk-targeted probabilistic ground motions are derived from hazard curves using one of the methods described in ASCE/SEI C21.2.1.

Deterministic (MCE_R) Ground Motions. Deterministic ground motions must be based on characteristic earthquakes on all known active faults in a region. The magnitude of a characteristic earthquake on a given fault should be a best estimate of the maximum magnitude capable for that fault but not less than the largest magnitude that has occurred historically on the fault. According to the exception in ASCE/SEI 21.2.2, the deterministic ground motion response spectrum need not be calculated where the probabilistic ground motion response spectrum in ASCE/SEI 21.2.1 is less than the deterministic lower limit response spectrum in ASCE/SEI Table 21.2-1 at all response periods for a given site class, which is determined in accordance with ASCE/SEI 11.4.2. Additional information on this method, including examples of scenario earthquakes from hazard deaggregations at a site, can be found in ASCE/SEI C21.2.2.

Site-Specific MCE_R Response Spectrum. Once the values of the S_a are determined from the probabilistic and deterministic methods, the site-specific MCE_R spectral response acceleration, S_{aM}, is set equal to the lesser of the values obtained from these two methods for a given period (ASCE/SEI 21.2.3). According to the exception in ASCE/SEI 21.2.3, the site-specific MCE_R response spectrum is permitted to be taken equal to the MCE_R response spectrum obtained from the USGS Seismic Design Geodatabase for a given site class.

The site-specific MCE_R spectral acceleration at any period must not be taken less than 80 percent of the MCE_R response spectrum determined from ASCE/SEI 11.4.6 for a given site class. This lower limit was established, among other things, to prevent the possibility of site-specific studies generating unreasonably low ground motions from potential misapplication of site-specific procedures or misinterpretation or mistakes in the quantification of the basic inputs to these procedures.

Where a site has been classified as site class F that requires a site-specific analysis in accordance with ASCE/SEI 11.4.7, the site-specific MCE_R spectral acceleration at any period must not be taken less than 80 percent of the MCE_R response spectrum determined from ASCE/SEI 11.4.6 for site class E (ASCE/SEI 21.2.3.1). According to the exception in this section, where a different site class can be justified using the classification procedures in ASCE/SEI 20.3.3, a lower limit equal to 80 percent of S_{aM} for the justified site class is permitted to be used. This exception permits a site class other than site class E to be used in establishing this limit when a site is classified as site class F, which eliminates the possibility of an overly conservative design spectrum on sites that would normally be classified as site class C or D.

Design Response Spectrum

The design spectral response acceleration, S_a, is determined by ASCE/SEI Equation (21.3-1):

$$S_a = \frac{2}{3} S_{aM} \tag{6.8}$$

where S_{aM} is the MCE_R spectral response acceleration obtained from ASCE/SEI 21.1 or 21.2.

Design Acceleration Parameters

Requirements on how to determine the parameters S_{DS} and S_{D1} where the site-specific procedure is used to determine the design ground motion in accordance with ASCE/SEI 21.3 are as follows (ASCE/SEI 21.4):

- $S_{DS} = 0.9S_a$ at any T within the range of 0.2 to 5 s, inclusive.
- S_{D1} = maximum value of $0.9TS_a$ for periods from 1 to 2 s for sites with $\overline{v}_s > 1{,}450$ ft/s (442 m/s) and for periods from 1 to 5 s for sites with $\overline{v}_s \le 1{,}450$ ft/s (442 m/s), but not less than S_a at 1 s.

The parameters S_{MS} and S_{M1} must be taken as 1.5 times S_{DS} and S_{D1}, respectively.

Where a site-specific ground motion procedure is used in conjunction with the equivalent lateral force (ELF) procedure of ASCE/SEI 12.8, S_a at T is permitted to replace S_{D1}/T in

ASCE/SEI Equation (12.8-4) and $S_{D1}T_L/T^2$ in ASCE/SEI Equation (12.8-5). The following are also relevant:

- S_{DS} determined in accordance with ASCE/SEI 21.4 is permitted to be used in ASCE/SEI Equations (12.8-3), (12.8-6), (15.4-1), and (15.4-3).
- The mapped value of S_1 must be used in ASCE/SEI Equations (12.8-6), (15.4-2), and (15.4-4).

Maximum Considered Earthquake Geometric Mean (MCE$_G$) Peak Ground Acceleration
According to ASCE/SEI 21.5.3, the site specific MCE$_G$ peak ground acceleration, PGA$_M$, must be taken as the lesser of the following:

- The probabilistic geometric mean peak ground acceleration of ASCE/SEI 21.5.1 (which is the geometric mean peak ground acceleration with a 2-percent probability of exceedance within a 50-year period)
- The deterministic geometric mean peak ground acceleration of ASCE/SEI 21.5.2 (which is the largest 84th-percentile geometric mean peak ground acceleration for characteristic earthquakes on all known active faults within the site region), which must be greater than or equal to the value of PGA$_G$ in ASCE/SEI Table 21.2-1 for a given site class (ASCE/SEI 21.5.2).

The site-specific MCE$_G$ peak ground acceleration adjusted for site class effects, PGA$_M$, which is used for evaluation of liquefaction, lateral soil movement, seismic settlements, and other soil-related issues (see ASCE/SEI 11.8.3), must be greater than or equal to 80 percent of PGA$_M$ obtained from the USGS Seismic Design Geodatabase for a given site class.

6.2.2 Importance Factor and Risk Category

Risk categories are defined in IBC Table 1604.5 and ASCE/SEI Table 1.5-1. These categories are based on the occupancy of a building or other structure and are used to relate the criteria for maximum environmental loads or distortions specified in the code or referenced standards to the consequence that would occur to the structure and its occupants if such loads were exceeded.

A seismic importance factor, I_e, is assigned to a building or structure in accordance with ASCE/SEI Table 1.5-2 based on its risk category (see Table 6.3). Larger values of I_e are assigned to more important risk categories, such as assembly and essential facilities, to increase the likelihood that such structures would suffer less damage and continue to function during and following a design earthquake. The risk category of a structure is also used in determining the SDC (see Section 6.2.3 of this book).

Table 6.3 Importance Factor, I_e

Risk Category	I_e
I, II	1.00
III	1.25
IV	1.50

Requirements pertaining to operational access to a risk category IV structure through an adjacent structure are given in ASCE/SEI 11.5.2. In such cases, the adjacent structure must conform to the requirements for risk category IV structures. In addition, where operational access is less than 10 ft (3.05 m) from an interior lot line or another structure on the same lot, protection from potential falling debris from adjacent structures must be provided by the owner of the risk category IV structure.

6.2.3 Seismic Design Category (SDC)

All buildings and structures must be assigned to a SDC in accordance with IBC 1613.2 or ASCE/SEI 11.6. In general, SDC is a function of the risk category and the design spectral response acceleration parameters at the site.

Six SDCs are defined ranging from A (minimal seismic risk) to F (highest seismic risk). As the SDC of a structure increases, so do the strength and detailing requirements.

The SDC of a building or structure is assigned as follows:

- Where $S_1 \geq 0.75$: $\begin{cases} \text{Risk category I, II or III structures are assigned to SDC E} \\ \text{Risk category IV structures are assigned to SDC F} \end{cases}$

- Where $S_1 < 0.75$: $\begin{cases} \text{Determine the SDC as a function of } S_{DS} \text{ by ASCE/SEI Table 11.6-1} \\ \text{Determine the SDC as a function of } S_{D1} \text{ by ASCE/SEI Table 11.6-2} \end{cases}$

The more severe SDC of the two determined by ASCE/SEI Tables 11.6-1 and 11.6-2 governs.

The SDC may be determined by ASCE/SEI Table 11.6-1 based solely on S_{DS} in cases where $S_1 < 0.75$ provided all the following conditions in ASCE/SEI 11.6 are satisfied:

1. In each of the two orthogonal directions, the approximate fundamental period of the structure, T_a, determined in accordance with ASCE/SEI 12.8.2.1 is less than $0.8T_S = 0.8S_{D1}/S_{DS}$.
2. In each of the two orthogonal directions, the fundamental period of the structure, T, used to calculate the story drift is less than $T_S = S_{D1}/S_{DS}$.
3. ASCE/SEI Equation (12.8-2) is used to determine the seismic response coefficient, C_s.
4. The diaphragms are rigid in accordance with ASCE/SEI 12.3.1; or, for diaphragms that are not rigid, the horizontal distance between vertical elements of the seismic force-resisting system (SFRS) does not exceed 40 ft (12.2 m).

This exception should always be considered—especially when determining the SDC of relatively stiff low-rise buildings—because it is possible for a building to be assigned to a lower SDC. As noted above, a lower SDC generally means less stringent design and detailing requirements.

For structures that are permitted to be designed in accordance with the alternate simplified design procedure in ASCE/SEI 12.14, the SDC is permitted to be determined from ASCE/SEI Table 11.6-1 alone, using the value of S_{DS} determined in ASCE/SEI 12.14.8.1. However, in cases where $S_1 \geq 0.75$, the building or structure must be assigned to SDC E. The provisions for this method are given in Section 6.3.14 of this book.

In lieu of ASCE/SEI 11.6, the SDC is permitted to be determined using IBC Figures 1613.2(1) through 1613.2(7) based on the risk category of the building or other structure (IBC 1613.2). The maps in these figures are based on the multi-period response spectra in ASCE/SEI 11.4.5.1 and the default site conditions defined in ASCE/SEI 11.4.2.1 (i.e., site class C, CD, and D). Where the building official or a geotechnical investigation has determined site class DE, E, or F are present at a site, the SDC maps are not permitted to be used; the SDC must be determined in accordance ASCE/SEI 11.6 instead. For site classes A, B, and BC, the maps will provide a conservative upper bound assignment of the SDC. In such cases, it may be beneficial to determine the SDC by ASCE/SEI 11.6 for comparison purposes.

The SDC is a trigger mechanism for many seismic requirements, including the following:

- Permissible seismic force-resisting systems.
- Limitations on building height.
- Consideration of structural irregularities.
- The need for additional special inspections, structural testing, and structural observation for seismic resistance.

Flowchart 6.2 in Section 6.6 of this book can be used to determine the SDC of a building or structure.

6.2.4 Design Requirements for SDC A

Structures assigned to SDC A need only comply with the general structural integrity requirements of ASCE/SEI 1.4 (ASCE/SEI 11.7). The lateral force-resisting system must be proportioned to resist a lateral force, F_x, at each floor level equal to 1 percent of the total dead load, W_x, located or assigned at that floor level (see Figure 6.8).

Figure 6.8 General Structural Integrity Requirements of ASCE/SEI 1.4 for Structures Assigned to SDC A

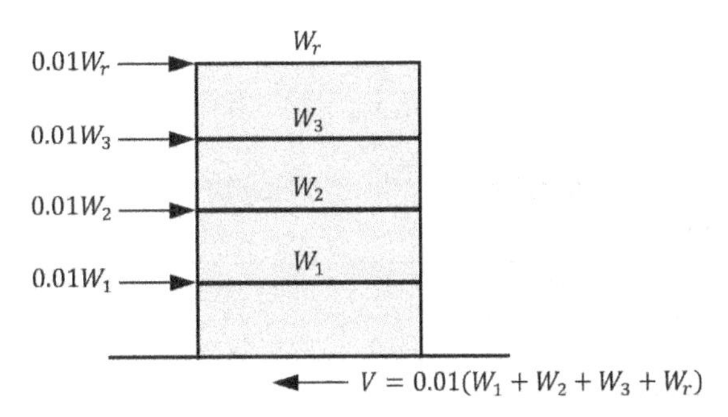

According to ASCE/SEI 1.4.2, the lateral forces are to be applied independently in each of two orthogonal directions.

Requirements for load path connections, connection to supports, and anchorage of structural walls are given in ASCE/SEI 1.4.1, 1.4.3, and 1.4.4, respectively.

Nonstructural components of a building or structure assigned to SDC A are automatically exempt from any seismic design requirements (ASCE/SEI 11.7). Also, tanks assigned to risk category IV must satisfy the freeboard requirement in ASCE/SEI 15.6.5.1.

6.2.5 Geologic Hazards and Geotechnical Investigation

Site Limitation for SDC D and F

Any structure assigned to SDC E or F must not be located where there is a potential for an active fault to cause rupture of the ground surface at that location (ASCE/SEI 11.8.1). It is very difficult to design structures to accommodate large movement associated with ground fault ruptures; thus, buildings are not permitted at these locations.

Geotechnical Investigation Report Requirements for Seismic Design Categories C through F

For structures assigned to SDC C, D, E or F, the geotechnical report for a particular site must include an evaluation of the following potential geologic and seismic hazards:

- Slope instability
- Liquefaction
- Total and differential settlement
- Surface displacement due to faulting or seismically induced lateral spreading or lateral flow

Recommendations on the type of foundation and other methods to mitigate any of the aforementioned hazards must also be included in the geotechnical report.

The exception in ASCE/SEI 11.8.2 states that a site-specific geotechnical report is not required at locations where evaluations of nearby sites with similar soil conditions have been performed previously; in such cases, the nearby information can be utilized for the proposed site provided the building official approves it.

Additional Geotechnical Investigation Report Requirements for Seismic Design Categories D through F

For structures assigned to SDC D, E or F, the geotechnical report must include the information in ASCE/SEI 11.8.3 in addition to the information discussed above (see Table 6.4).

It is essential to have a clear plan on how to control the potential effects from liquefaction and soil strength loss. Appropriate foundation and lateral force-resisting systems must be established in the early design stages to counteract any potential damaging effects.

Table 6.4 Additional Geotechnical Investigation Report Requirements for Seismic Design Categories D through F

Requirement	Remarks
The determination of dynamic seismic lateral earth pressures on basement and retaining walls due to design earthquake ground motions.	Dynamic lateral pressures are considered to be an earthquake load, *E*, which is superimposed on the pre-existing lateral earth load, *H*, during ground shaking.
The potential for liquefaction, seismically induced permanent ground displacement and soil strength loss evaluated for site peak ground acceleration, earthquake magnitude, and source characteristics consistent with the maximum considered earthquake geometric mean (MCE_G) peak ground acceleration.	Peak accelerations are to be determined using one of the following methods: (1) a site-specific study that accounts for soil amplification effects as specified in the site-specific ground motion procedures of ASCE/SEI 11.4.7 (see Section 6.2.1 of this book) or (2) the peak ground acceleration PGA_M determined from the USGS Seismic Design Geodatabase for the applicable site class.
Assessment of potential consequences of liquefaction, seismically induced permanent ground displacement and soil strength loss, including, but not limited to the following: 1. Estimation of total and differential settlement 2. Lateral soil movement 3. Lateral soil loads on foundations 4. Reduction in soil bearing capacity and lateral soil reactions 5. Soil downdrag and reduction in axial and lateral soil reaction for pile foundations 6. Increases in soil lateral pressures on retaining walls 7. Flotation of buried structures	Because the effects from liquefaction and soil strength loss can be disastrous, it is very important to have a clear understanding of how the effects will impact the structure and its foundation.
Discussion of mitigation measures, such as, but not limited to the following: 1. Selection of appropriate foundation type and depth 2. Selection of appropriate structural systems to accommodate anticipated displacements and forces 3. Ground stabilization 4. Any combinations of these measures and how they must be considered in the design of the structure	—

6.2.6 Vertical Ground Motions for Seismic Design

The requirements in ASCE/SEI 11.9 may be used in lieu of the requirements in ASCE/SEI 12.4.2.2 for the effects of vertical design earthquake motions on a structure. The requirements in ASCE/SEI 11.9.2 only apply to structures assigned to SDC C, D, E, or F and located in the conterminous United States at or west of the −105 degree longitude, Alaska, Hawaii, Puerto Rico, US Virgin Islands, Guam and Northern Mariana Islands, and American Samoa. The requirements are applicable to only the aforementioned regions because models for vertical-component response spectra are available only in these regions.

These requirements are meant for the design of structures where a significant response to vertical ground motion is possible (such as many types of nonbuilding structures, like those in ASCE/SEI Chapter 15). Additional information can be found in ASCE/SEI C11.9.

6.3 Seismic Design Requirements for Building Structures

6.3.1 Structural Design Basis

Basic requirements for seismic analysis and design of building structures are given in ASCE/SEI 12.1. A summary of these requirements is given in Table 6.5.

Table 6.5 Basic Requirements for Seismic Analysis and Design of Building Structures

	Requirements	ASCE/SEI Section No.
Basic requirements	• The building structure must have a complete lateral and vertical force-resisting system capable of providing adequate strength, stiffness, and energy dissipation. • Design ground motions are assumed to occur along any horizontal direction. • The adequacy of the structural system must be demonstrated through the construction of a mathematical model and evaluation of the model for the effects of the design ground motions. • The design seismic forces must be determined using one of the applicable procedures indicated in ASCE/SEI 12.6.	12.1.1
Member design, connection design, and deformation limit	• Individual structural members must be designed to resist the effects due to the design ground motion. • Connections must develop the strength of the connected members or the forces indicated in ASCE/SEI 12.1.1. • The deformation of the structure subject to the design seismic forces must not exceed the prescribed limits.	12.1.2
Continuous load path and interconnection	• A continuous load path, or paths, with adequate strength and stiffness must be provided to transfer all forces from the point of application to the final point of resistance. • All parts of the structure between separation joints must be interconnected to form a continuous path to the seismic force-resisting system (SFRS). The connection must be capable of transmitting the seismic force F_p induced by the parts being connected. • Any smaller portion of a structure must be tied to the remainder of the structure with elements having a design strength capable of transmitting a seismic force equal to the greater of (1) 0.133 S_{DS} times the weight of the smaller portion or (2) 0.05 times the portion's weight. This connection force does not apply to the overall design of the SFRS. • Connection design forces need not exceed the maximum forces that the structural system can deliver to the connection.	12.1.3

(*Continued*)

Table 6.5 Basic Requirements for Seismic Analysis and Design of Building Structures (*Continued*)

	Requirements	ASCE/SEI Section No.
Connection to supports	• For each beam, girder, or truss in a structure, a positive connection must be provided for resisting a horizontal force acting parallel to the member, either directly to its supporting elements or to slabs designed as diaphragms. • The member's supporting element must be connected to the diaphragm in cases where the connection is through the diaphragm. • The connection must have a minimum design strength equal to 0.05 times the dead plus live load reaction.	12.1.4
Foundation design	• Foundations must be designed to resist the forces developed and to accommodate the movements imparted to the structure and foundation by the design ground motions. • The following must be included in the determination of the foundation design criteria: (1) dynamic natures of the forces, (2) expected ground motion, (3) design basis for the strength and energy dissipation capacity of the structure, and (4) the dynamic properties of the soil. • The design and construction of foundations must comply with ASCE/SEI 12.13. • The weights of foundations must be considered as dead loads in accordance with ASCE/SEI 3.1.2 when calculating load combinations using ASCE/SEI 2.3 or 2.4. The dead loads are permitted to include overlying fill and paving materials.	12.1.5

6.3.2 Structural System Selection

Selection and Limitations

The basic lateral and vertical SFRSs in a building structure must conform to one of the types indicated in ASCE/SEI Table 12.2-1 or a combination of the systems indicated in ASCE/SEI Table 12.2-1 as permitted in ASCE/SEI 12.2.2, 12.2.3, and 12.2.4. The general categories of the SFRSs are given in Table 6.6.

The primary materials of construction are included under the general categories of the SFRSs in ASCE/SEI Table 12.2-1. The SFRSs are categorized according to the quality and extent of the seismic-resistant detailing that must be using in the design of the structure:

- "Special" systems provide superior seismic resistance, which is accomplished through extensive design and detailing of the structural members.
- "Ordinary" systems have basic design and detailing requirements.
- "Intermediate" systems provide seismic resistance that is better than ordinary systems, but not as good as special systems.

Design and detailing requirements for these three categories are given in the material standards for each type of material.

A summary of the seismic parameters in ASCE/SEI Table 12.2-1 is given Table 6.7.

The basic seismic analysis procedure uses response spectra representative of, but significantly less than, the anticipated ground motions (see Section 6.2.1 of this book). Therefore, at the MCE_R level of ground shaking, structural members are expected to yield or behave inelastically.

Relationships between the seismic parameters are illustrated in the inelastic force-deformation curve in ASCE/SEI Figure C12.1-1. The structure responds elastically until the first hinge forms. As the horizontal forces increase, a properly designed and detailed structure will respond inelastically

Table 6.6 General Categories of the SFRSs in ASCE/SEI Table 12.2-1

SFRS	Description
Bearing wall system	A bearing wall system is a structural system where bearing walls support all or a major portion of the vertical loads. Some or all of the walls also provide resistance to the seismic forces.
Building frame system	In a building frame system, an essentially complete space frame provides support for the vertical loads. Shear walls or braced frames provide resistance to the seismic forces.
Moment-resisting frame system	A moment-resisting frame system is a structural system with an essentially complete space frame that supports the vertical loads. Seismic forces are resisted primarily by flexural action of the frame members through the joints. The entire space frame or selected portions of the frame may be designated as the SFRS. Requirements for deformation compatibility must be satisfied for structures assigned to SDC D, E, or F (ASCE/SEI 12.12.5).
Dual system	In a dual system, an essentially complete space frame provides support for the vertical loads. Moment-resisting frames and shear walls or braced frames provide resistance to seismic forces in accordance with their rigidities (see ASCE/SEI 12.2.5.1). Additionally, the moment-resisting frames must act as a backup for the walls or braces; this is accomplished by requiring that the moment frames be capable of resisting at least 25 percent of the design seismic forces. Dual systems are also referred to as shear wall-frame interactive systems.
Cantilever column system	In this system, the vertical forces and the seismic forces are resisted entirely by columns acting as cantilevers from their base. This system is usually used in one-story buildings or in the top story of a multistory building. Severe restrictions are placed on the use of this system because it has performed poorly in past earthquakes.
Steel systems not specifically detailed for seismic resistance, excluding cantilever column systems	—

Table 6.7 Seismic Parameters in ASCE/SEI Table 12.2-1

Seismic Parameter	Description
Response modification coefficient, R	This coefficient accounts for the ability of a SFRS to respond to ground shaking in a ductile manner without loss of load-carrying capacity. In other words, R is an approximate way of accounting for the effective damping and energy dissipation that can be mobilized during inelastic response to ground shaking. It represents the ratio of the forces that would develop under the ground motion specified in ASCE/SEI 7 if the structure had responded to the prescribed design forces in an entirely linear-elastic manner. A system that has no ability to respond in a ductile manner has an R-value equal to 1; the only such system is a cantilevered column system consisting of ordinary reinforced concrete moment frames (see ASCE/SEI Table 12.2-1). Systems capable of highly ductile response have an R-value equal to 8.

(*Continued*)

Table 6.7 Seismic Parameters in ASCE/SEI Table 12.2-1 (*Continued*)

Seismic Parameter	Description
Overstrength factor, Ω_0	The purpose of this factor is to amplify the prescribed seismic forces for use in design of the following types of structural elements: (1) elements whose action cannot provide reliable inelastic response or energy dissipation and (2) elements required to remain essentially elastic to maintain the structural integrity of the structure. In essence, it is an approximate way of providing a nominal degree of protection against undesirable behavior. For most of the structural systems given in the table, Ω_0 ranges from 2 to 3.
Deflection amplification factor, C_d	This factor is used to increase the elastic lateral displacements, δ_e, determined for a structure using the design earthquake forces to the lateral displacements expected during the design earthquake ground motion. It is evident from ASCE/SEI Table 12.2-1 that C_d is equal to or slightly less than the corresponding R-value for a given SFRS. The more ductile a system is (i.e., the greater the R-value), the greater the difference is between the values of R and C_d.

with a series of plastic hinges forming due to the redundancy built into the SFRS. A yielding mechanism occurs at the strength level V_y, which is larger than the design seismic force, V_s.

Limitations on SDC and building height are also given in ASCE/SEI Table 12.2-1. Some SFRSs can be utilized in structures assigned to any SDC with no height limitations, while others are permitted in structures up to certain heights. The least ductile systems are not permitted in the higher SDCs regardless of height.

The use of SFRSs not in ASCE/SEI Table 12.2-1 is permitted provided the requirements in ASCE/SEI 12.2.1 are satisfied, including approval by the building official (ASCE/SEI 12.2.1.1).

Combinations of Framing Systems in Different Directions

It is permitted to use different SFRSs along each of the two orthogonal axes of a structure provided the respective R, Ω_0, and C_d values and structural system limitations in ASCE/SEI Table 12.2-1 are used in each direction (ASCE/SEI 12.2.2).

It is possible for one of the systems to have more restrictive limitations on the structural system or the structural height. If this occurs, the more restrictive limitations govern for the entire building. Consider the building in Figure 6.9, which is assigned to SDC D. In the north-south direction, the SFRS consists of special reinforced concrete moment frames at the two ends of the building (moment-resisting frame system C5 in ASCE/SEI Table 12.2-1) where $R = 8$, $\Omega_0 = 3$,

Figure 6.9 Combinations of Framing Systems in Different Directions

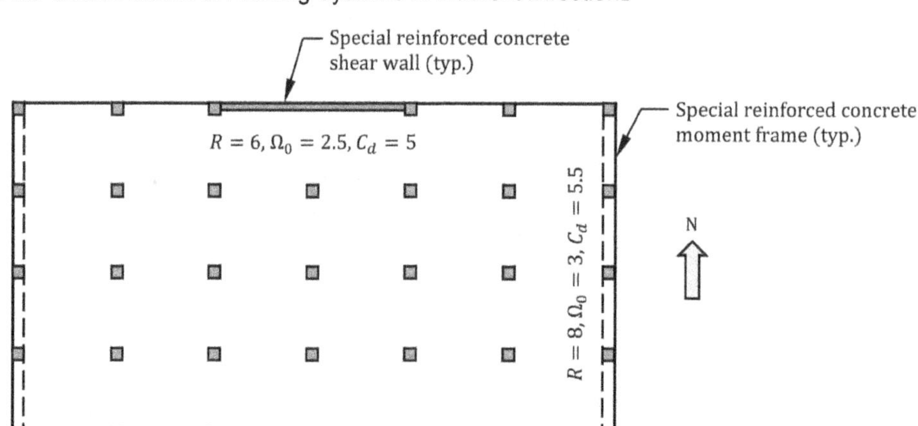

and $C_d = 5.5$. In the east-west direction, the SFRS consists of special reinforced concrete shear walls (building frame system B4 in ASCE/SEI Table 12.2-1) where $R = 6$, $\Omega_0 = 2.5$, and $C_d = 5$. The respective R, Ω_0, and C_d values are used in each orthogonal direction. However, the building height is limited to 160 ft (48.8 m) based on the structural height limit of the special reinforced concrete shear wall system (there is no height limit for the special reinforced concrete moment frame system; see ASCE/SEI Table 12.2-1).

Combinations of Framing Systems in the Same Direction

Overview

Where different SFRSs are used in the same direction to resist seismic forces (other than combinations considered as dual systems), the most stringent structural system limitations in ASCE/SEI Table 12.2-1 apply, except as permitted by ASCE/SEI 12.2.3. The intent of these requirements is to prevent concentration of inelastic behavior in the lower stories of a structure.

Vertical Combinations of Systems

Requirements for structures with a vertical combination of SFRSs in the same direction are given in ASCE/SEI 12.2.3.1 and are summarized in Figure 6.10.

Figure 6.10 Vertical Combinations of SFRSs in Accordance with ASCE/SEI 12.2.3.1

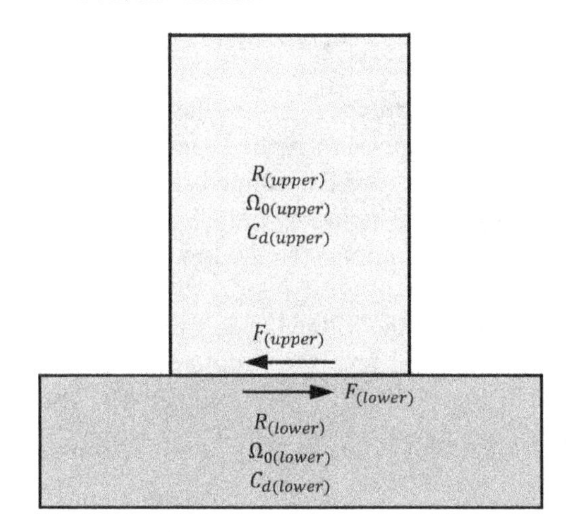

1. $R_{(lower)} < R_{(upper)}$
   (a) Calculate forces and drifts on the upper system using R, Ω_0, and C_d of the upper system
   (b) Calculate forces and drifts on the lower system using R, Ω_0, and C_d of the lower system
   (c) $F_{(lower)} = [R_{(upper)}/R_{(lower)}]F_{(upper)}$
2. $R_{(upper)} < R_{(lower)}$
   Calculate forces and drifts on both system using R, Ω_0, and C_d of the upper system

Three exceptions to these requirements are given in ASCE/SEI 12.2.3.1:

- Rooftop structures less than or equal to two stories in height and weighing less than 10 percent of the total structure weight.
- Other supported structural systems with a weight less than or equal to 10 percent of the weight of the structure.
- Detached one- and two-family dwellings of light-frame construction.

Two-Stage Analysis Procedure for Vertical Combinations of Systems

A two-stage equivalent lateral force procedure is permitted to be used for structures with a flexible upper portion supported by a rigid lower portion provided the conditions in ASCE/SEI 12.2.3.2 are satisfied (see Figure 6.11). An example of where this analysis procedure may be used is a reinforced concrete podium supporting a structure of light-frame construction.

In a two-stage analysis, the upper portion of the structure is analyzed as a separate structure for seismic forces and drifts using $R_{(upper)}$, $\rho_{(upper)}$, $C_{d(upper)}$, and $\Omega_{0(upper)}$. Either the equivalent lateral

Figure 6.11 Requirements for a Two-Stage Analysis Procedure for Vertical Combinations of Systems in Accordance with ASCE/SEI 12.2.3.2

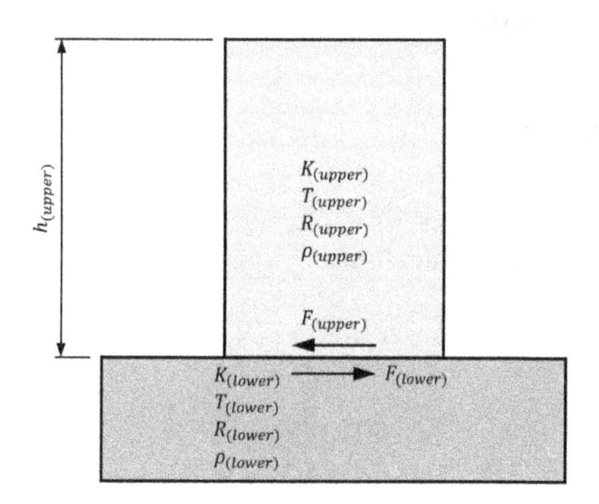

(a) Stiffness: $K_{(lower)} \geq 10K_{(upper)}$ where K is calculated in accordance with ASCE/SEI 12.2.3.2(a)

(b) $T_{(structure)} \leq 1.1T_{(upper)}$

(c) Upper portion must be designed as a separate structure using $R_{(upper)}$ and $\rho_{(upper)}$

(d) Lower portion must be designed as a separate structure using $R_{(lower)}$ and $\rho_{(lower)}$ while meeting the requirements of ASCE/SEI 12.2.3.1 where $F_{(lower)} = [(R_{(upper)}/\rho_{(upper)})/(R_{(lower)}/\rho_{(lower)})]F_{(upper)} \geq F_{(upper)}$

(e) Upper portion must be analyzed using the equivalent lateral force procedure (ASCE/SEI 12.8) or the modal response spectrum procedure (ASCE/SEI 12.9.1) and the lower portion must be analyzed using the equivalent lateral force procedure (ASCE/SEI 12.8)

(f) $h_{(upper)} \leq$ height limit in ASCE/SEI Table 12.2-1 for the SFRS used

(g) Where horizontal irregularity type 4 or vertical irregularity type 3 exist at the interface, the reactions from the upper portion must be amplified in accordance with ASCE/SEI 12.3.3.4, 12.10.1.1, and 12.10.3.3, in addition to the amplification required in item (d) above.

force procedure in ASCE/SEI 12.8 or the modal response spectrum procedure in ASCE/SEI 12.9.1 is permitted to be used.

The seismic forces resulting from horizontal seismic loading, $F_{(upper)}$, at the base of the upper portion are transferred to the top of the lower portion and scaled up by the ratio of $(R_{(upper)}/\rho_{(upper)})/(R_{(lower)}/\rho_{(lower)}) \geq 1.0$ where $R_{(lower)}$ and $\rho_{(lower)}$ are determined in accordance with ASCE/SEI 12.2.3.1. All other forces from the upper portion, including gravity forces and seismic forces resulting from vertical seismic loading, are directly applied to the lower portion without scaling.

The lower portion is designed as a separate structure for the following seismic forces: (1) seismic forces of the lower portion using the equivalent lateral force procedure in ASCE/SEI 12.8 with the appropriate values of $R_{(lower)}$, $\rho_{(lower)}$, $C_{d(lower)}$, and $\Omega_{0(lower)}$ determined by ASCE/SEI 12.2.3.1 and (2) the scaled seismic transfer force from the upper portion, $F_{(lower)} = (R_{(upper)}/\rho_{(upper)})/(R_{(lower)}/\rho_{(lower)})F_{(upper)} \geq F_{(upper)}$.

Where horizontal structural irregularity type 4 (out-of-plane offset; see ASCE/SEI Table 12.3-1) or vertical structural irregularity type 3 (in-plane discontinuity; see ASCE/SEI Table 12.3-2) is present at the transition from the upper portion to the lower portion of a building, the requirement for the overstrength effects is associated with the expected peak capacity of the upper portion. Therefore, the reactions transferred from the upper portion to the lower portion must include the overstrength effects in ASCE/SEI 12.3.3.4 for elements supporting discontinued walls or frames and in ASCE/SEI 12.10.1.1 and 12.10.3.3 for diaphragm transfer forces.

Horizontal Combinations of SFRSs

For structures utilizing different SFRSs in the same horizontal direction, the least value of R and the corresponding C_d and Ω_0 are to be used in design in that direction (ASCE/SEI 12.2.3.3). The SFRSs for the building in Figure 6.12 in the north-south direction consist of ordinary reinforced concrete shear walls ($R = 5$, $\Omega_0 = 2.5$, $C_d = 4.5$) and ordinary reinforced concrete moment frames ($R = 3$, $\Omega_0 = 3$, $C_d = 2.5$). Assuming a rigid diaphragm, the structure must be designed using $R = 3$, $\Omega_0 = 3$, and $C_d = 2.5$ in this direction (note: in lieu of utilizing separate SFRSs in this direction, the shear wall-frame interactive system with ordinary reinforced concrete moment frames and ordinary reinforced concrete shear walls in ASCE/SEI Table 12.2-1 may be used with $R = 4.5$, $\Omega_0 = 2.5$, and $C_d = 4$ provided all the appropriate requirements are satisfied).

Figure 6.12 Horizontal Combinations of SFRSs

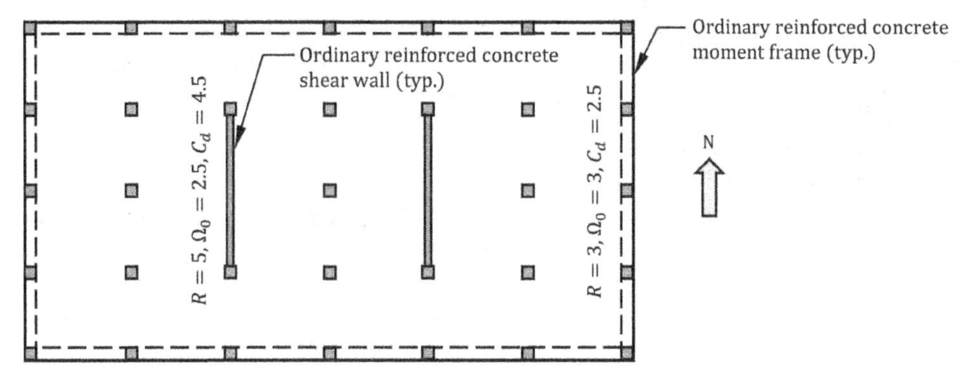

For design in the N-S direction, use $R = 3, \Omega_0 = 3, C_d = 2.5$

According to the exception in ASCE/SEI 12.2.3.3, risk category I or II buildings two stories or less in height with light-frame construction or flexible diaphragms are permitted to be designed using the least value of R for the different SFRSs in each independent line of resistance. In other words, each SFRS in that direction is permitted to be designed for R of that system. The value of R that must be used in the design of the diaphragms is the least R for any of the SFRSs in that direction.

Two-Stage Analysis Procedure for One-Story Structures with Flexible Diaphragms and Rigid Vertical Elements

A two-stage equivalent lateral force procedure for the determination of seismic design forces in vertical elements of one-story structures with flexible diaphragms supported by rigid vertical elements is given in ASCE/SEI 12.2.3.4 (see ASCE/SEI 12.3.1 for the definition of a flexible diaphragm condition). This procedure is permitted to be used for structures complying with the requirements in ASCE/SEI 12.10.4.

This method, which is based on the numerical studies in Reference 6.5, includes the diaphragm response in the analysis; as such, seismic design forces to the vertical elements of the SFRS are reduced compared to methods where seismic forces to vertical elements are based on the approximate period of the vertical elements, irrespective of diaphragm response.

Seismic design forces to the vertical elements in each horizontal direction are equal to the sum of the following forces:

1. **Force contributed by the effective seismic weight tributary to the diaphragm.** This force is equal to the reaction from the diaphragm seismic design force, F_{px}, determined by ASCE/SEI Equation (12.10-15) amplified by $R_{\text{diaph}}/(R/\rho) \geq 1.0$ where R_{diaph} is the response modification coefficient given in ASCE/SEI 12.10.4.2.1 based on the type of flexible diaphragm, R is the response modification coefficient for the vertical SFRS and ρ is the redundancy factor for the vertical SFRS.
2. **Force contributed by the effective seismic weight tributary to the in-plane rigid vertical elements of the SFRS.** This force is determined in accordance with the equivalent lateral force procedure in ASCE/SEI 12.8. Seismic forces are applied to the vertical elements not entering through the horizontal diaphragm. Typically, these forces are from the vertical element's self-weight and other effective seismic weights in the same plane as the vertical element. This total effective seismic weight induces seismic design forces associated with the in-plane vertical element's period and response modification coefficient.

Additional information on these alternative diaphragm design provisions can be found in Section 6.3.10 of this book.

Combination Framing Detailing Requirements

Structural members common to different SFRSs used to resist seismic forces in any direction must be designed using the detailing requirements of ASCE/SEI Chapter 12 for the largest R of the connected SFRSs (ASCE/SEI 12.2.4). The intent is to preserve the integrity of all the SFRSs when subjected to an earthquake event.

System-Specific Requirements

Requirements for dual systems, cantilever column systems, inverted pendulum-type structures, and other systems are given in ASCE/SEI 12.2.5. These requirements must be satisfied in addition to the requirements in ASCE/SEI Table 12.2-1 and other requirements in ASCE/SEI Chapter 12.

Increased height limits for particular systems meeting the requirements in ASCE/SEI 12.2.5.4 are also given in ASCE/SEI 12.2.5.

For shear wall-frame interactive systems, the shear strength of the shear walls must be at least 75 percent of the design story shear at each story (ASCE/SEI 12.2.5.8). The frames, which are backup for the walls, must be capable of resisting at least 25 percent of the design story shear in each story.

6.3.3 Diaphragm Flexibility, Configuration Irregularities, and Redundancy

Diaphragm Flexibility

The manner in which horizontal seismic forces get transferred to the vertical elements of the SFRS depends on the flexibility of the diaphragm. Descriptions of flexible, rigid, and semirigid diaphragms are given in Table 6.8.

Diaphragms are permitted to be idealized as flexible or rigid based on construction type (see ASCE/SEI 12.3.1.1, ASCE/SEI 12.3.1.2 and Table 6.9).

For diaphragms not satisfying the conditions in ASCE/SEI 12.3.1.1 for flexible diaphragms and ASCE/SEI 12.3.1.2 for rigid diaphragms, ASCE/SEI Equation (12.3-1) can be used to

Table 6.8 Types of Diaphragms

Diaphragm Type	Description
Flexible	In-plane deflection of the diaphragm due to horizontal forces is relatively large compared to the deflection of the vertical elements of the SFRS. The diaphragm does not undergo rigid body rotation when subjected to horizontal forces. Horizontal forces are distributed to the vertical elements of the SFRS in proportion to the mass of the diaphragm tributary to the vertical elements of the SFRS (for diaphragms of uniform material and weight, horizontal forces can be distributed in proportion to the area tributary to the vertical elements of the SFRS).
Rigid	In-plane deflection of the diaphragm due to horizontal forces is relatively small compared to the deflection of the vertical elements of the SFRS. The diaphragm displaces and rotates as a rigid body when subjected to horizontal forces. Horizontal forces are distributed to the vertical elements of the SFRS in proportion to the relative rigidities (stiffnesses) and their location with respect to the center of rigidity (CR).
Semirigid	In-plane deflection of the diaphragm due to horizontal forces is of the same order of magnitude as the deflection of the vertical elements of the SFRS. Horizontal forces are distributed to the vertical elements of the SFRS based on a structural analysis that explicitly includes the in-plane stiffness of the diaphragm.

Table 6.9 Idealized Diaphragm Flexibility

Flexible	Untopped steel decking or wood structural panels provided any of the following conditions are satisfied: 1. The SFRS consists of the following: (a) steel braced frames, (b) steel and concrete composite braced frames, or (c) concrete, masonry, steel, or steel and concrete composite shear walls. 2. The building is a one- or two-family dwelling. 3. Structures of light-frame construction where both of the following conditions are satisfied: (a) topping of concrete or similar materials is not placed over wood structural panel diaphragms except for nonstructural topping less than or equal to 1.5 in. (38 mm) thick and (b) each line of vertical elements of the SFRS complies with the allowable story drift requirements in ASCE/SEI Table 12.12-1.
Rigid	Concrete slabs or concrete-filled metal deck with span-to-depth ratios of 3 or less (see Figure 6.13) in structures that do not have a type 2, 3, 4, or 5 horizontal structural irregularity.

Figure 6.13 Definition of Span-to-Depth Ratio for a Diaphragm

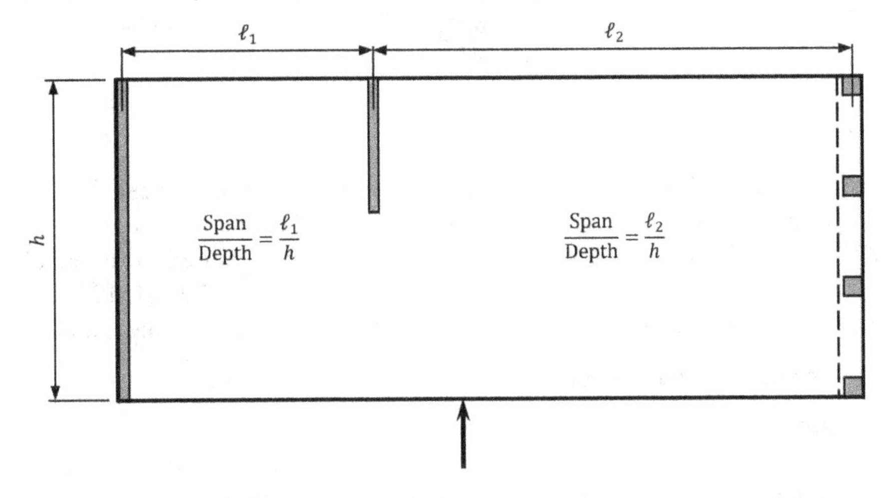

determine whether or not a diaphragm can be idealized as flexible. A diaphragm is permitted to be idealized as flexible where the following equation is satisfied:

$$\frac{\delta_{MDD}}{\Delta_{ADVE}} > 2 \tag{6.9}$$

In this equation, δ_{MDD} is the computed maximum in-plane deflection of the diaphragm subjected to lateral loads and Δ_{ADVE} is the average drift of adjoining vertical elements of the SFRS over the story below the diaphragm under consideration subjected to a tributary lateral load equivalent to that used in the computation of δ_{MDD} (see Figure 6.14). Diaphragms satisfying this equation can be idealized as flexible.

Diaphragms in structures of light-frame construction located on a hillside must be modeled as rigid or semirigid where either of the following conditions are satisfied (see ASCE/SEI 12.3.1.4 and ASCE/SEI Figure C12.3-1a):

- Diaphragms are seismically braced on one or more sides directly by a foundation wall or by a foundation stem wall.
- Diaphragms are seismically braced by a framed wall with a clear wall height of 2 ft (0.61 m) or less and are seismically braced on other sides by framed walls with any wall having a clear height of more than 7 ft (2.13 m).

Figure 6.14 Definition of a Flexible Diaphragm in Accordance with ASCE/SEI 12.3.1.3

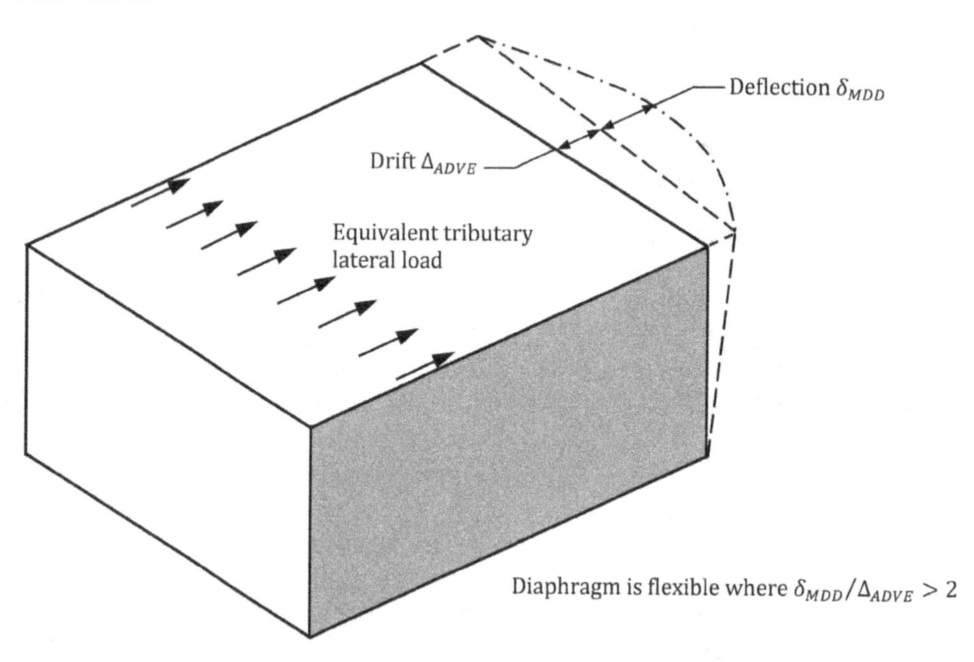

Structures modeled with flexible diaphragms in such cases have been found in numerical studies to provide seismic performance markedly less than minimum performance levels in the ASCE/SEI 7 seismic provisions. Other diaphragms are permitted to be modeled as semirigid or idealized as flexible or rigid in accordance with ASCE/SEI 12.3.1.1, 12.3.1.2, or 12.3.1.3.

Flowchart 6.3 in Section 6.6 of this book can be used to determine diaphragm flexibility.

Irregular and Regular Classification

Overview

A regular structure can be defined as one that possesses a distribution of mass, stiffness, and strength that results in a uniform sway when subjected to ground shaking. In other words, the lateral displacement in each story on each side of the structure will be about the same.

The dissipation of earthquake energy is uniform throughout a regular structure; this results in light and well-distributed damage in the structure. In contrast, structures that are irregular can suffer extreme damage at only one or a few locations; this can result in localized failure of structural members and can lead to a loss of the structure's ability to survive the ground shaking. Past earthquakes have revealed that irregular structures suffer greater damage than those that are regular.

Structures are classified as regular or irregular based on the criteria in ASCE/SEI 12.3.2. Structural configurations can be divided into horizontal and vertical types for purposes of defining irregularities. Both types of irregularities are covered below.

Horizontal Irregularity

Overview Structures with one or more of the irregularities defined in ASCE/SEI Table 12.3-1 are designated as having a horizontal structural irregularity. Buildings with reentrant corners may be classified as horizontally irregular depending on the geometry of the reentrant corner. Buildings without reentrant corners and with symmetric geometric shapes may still be classified as horizontally irregular based on the distribution and location of the SFRS or on an abrupt discontinuity or variation of diaphragm stiffness.

The horizontal irregularities in ASCE/SEI Table 12.3-1 are covered below. Accidental torsion, which is defined in ASCE/SEI 12.8.4.2.2, must be applied to structures with diaphragms that are not flexible to determine if a horizontal irregularity exists (ASCE/SEI 12.8.4.2.1).

Torsional Irregularity (Type 1) The conditions under which a torsional irregularity exists are given in Table 6.10 (see ASCE/SEI Table 12.3-1). Included in the table are ASCE/SEI reference sections that correspond to the requirements that must be satisfied based on the SDC assigned to the building.

Table 6.10 Horizontal Torsional Irregularity

Description	ASCE/SEI Reference Section	SDC Application
• More than 75% of any story's lateral strength below the diaphragm is provided at or on one side of the center of mass (CM)[1] • The torsional irregularity ratio (TIR) exceeds 1.2[2]	Table 12.3-1a, 12.3.3.5	B, C, D, E, F
	12.5.1.2	C, D, E, F
	12.7.3	B, C, D, E, F
	12.8.4.3	C, D, E, F
	12.8.6	C, D, E, F
	16.3.4	B, C, D, E, F

[1]Story lateral strength = total strength of all seismic-resisting elements sharing the story shear for the direction under consideration.
[2]$TIR = \Delta_{max}/\Delta_{avg}$.

Studies have concluded that the performance of a building is adversely affected where a significant percentage of the lateral strength is located on one side of the center of mass (CM) in the direction of analysis (see Reference 6.6). A horizontal torsional irregularity is defined to exist where more than 75 percent of any story's lateral strength below the diaphragm is provided at or on one side of the CM. The story lateral strength is equal to the total strength of the seismic-resisting elements sharing the story shear in the direction of analysis. A potential negative outcome of uneven strength distribution is that yielding on one side of the CM may cause the center of rigidity (CR) to move farther away from the CM, thus increasing the torsional response.

A horizontal torsional irregularity is also defined to exist where the torsional irregularity ratio, TIR, is greater than 1.2. This ratio is determined as follows for each story and for each accidental torsion case [see ASCE/SEI Equation (12.3-2) and Figure 6.15)]:

$$TIR = \Delta_{max} / \Delta_{avg} \tag{6.10}$$

where Δ_{max} = maximum story drift determined in accordance with ASCE/SEI 12.8.6 at the building's edge subjected to lateral forces determined from the equivalent lateral force procedure in ASCE/SEI 12.8 including the application of accidental torsion in accordance with ASCE/SEI 12.8.4 with the torsional amplification factor, A_x, equal to 1.0

Δ_{avg} = average of the story drifts at the two opposing edges of the building determined using the same loading and diaphragm rigidity that was applied in the determination of Δ_{max}

For structures with rigid or semirigid diaphragms, it is permitted to assume the diaphragm is rigid when determining Δ_{max} and Δ_{avg}. In such cases, TIR is the maximum value from the values calculated for each story and each direction. The TIR does not apply to structures with flexible diaphragms.

Figure 6.15 Torsional Irregularity

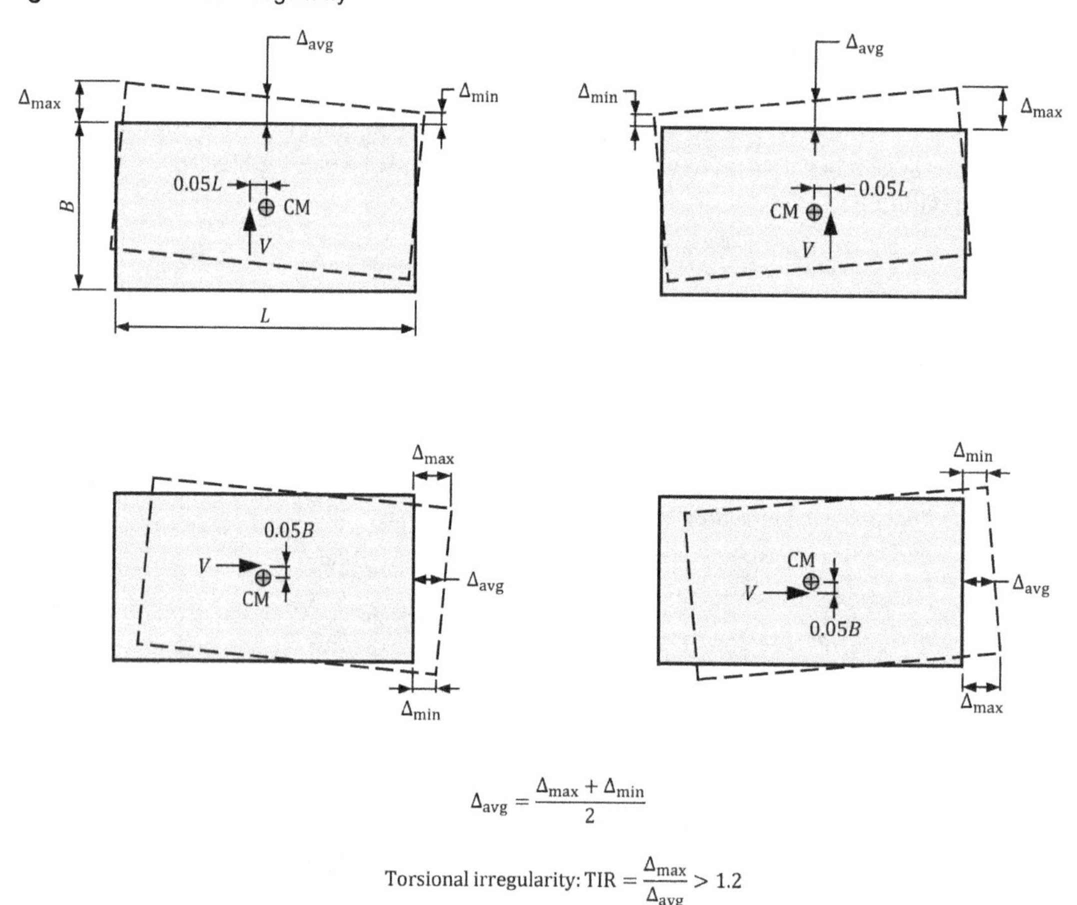

The requirements that must be satisfied for horizontal torsional irregularity where TIR > 1.2, 1.4, or 1.6 are given in ASCE/SEI Table 12.3-1a. These requirements are cumulative; for example, in cases where TIR > 1.4, the requirements for TIR > 1.2 must also be satisfied. According to note b in ASCE/SEI Table 12.3-1a, the requirements for TIR > 1.2 must also be satisfied in cases where TIR ≤ 1.2 and more than 75 percent of any story's lateral strength below the diaphragm is provided at or on one side of the CM.

The building in Figure 6.15 has a symmetrical geometric shape in plan without reentrant corners or wings. Such configurations may still be classified as having a torsional irregularity due to the distribution of the SFRSs and their relationship to the CM in plan. A torsional irregularity is defined to exist in the entire SFRS of multistory structures where it has been determined in at least one story that TIF > 2 under at least one of the four possible accidental eccentricities (see Figure 6.15).

Plan configurations with wings, such as L-, H-, or Y-shapes, are very susceptible to this type of irregularity even where symmetrical in plan.

Accidental torsion need not be included in the determination of the design seismic forces and the design story drifts for structures where it has been determined a torsional irregularity does not exist.

Reentrant Corner Irregularity (Type 2) A reentrant corner irregularity is defined to exist where both plan projections of a structure beyond a reentrant corner are greater than 20 percent of the plan dimension of the structure in the given direction (see Figure 6.16). The requirements in ASCE/SEI 12.3.3.5 must be satisfied for structures assigned to SDC D, E, or F.

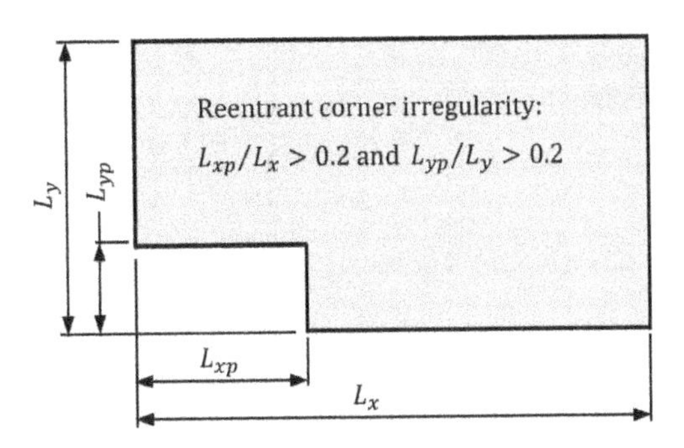

Figure 6.16 Reentrant Corner Irregularity

It is evident that a concentration of forces can occur at a relatively large reentrant corner which can lead to significant local damage.

Diaphragm Discontinuity Irregularity (Type 3) A diaphragm discontinuity irregularity is defined to exist where a diaphragm has an abrupt discontinuity or variation in stiffness. This includes diaphragms with the following: (a) a cutout or open area that is greater than 25 percent of the gross enclosed diaphragm area (see Figure 6.17) or (b) a change in effective diaphragm stiffness of more than 50 percent from one story to the next (see ASCE/SEI Figure C12.3-1b).

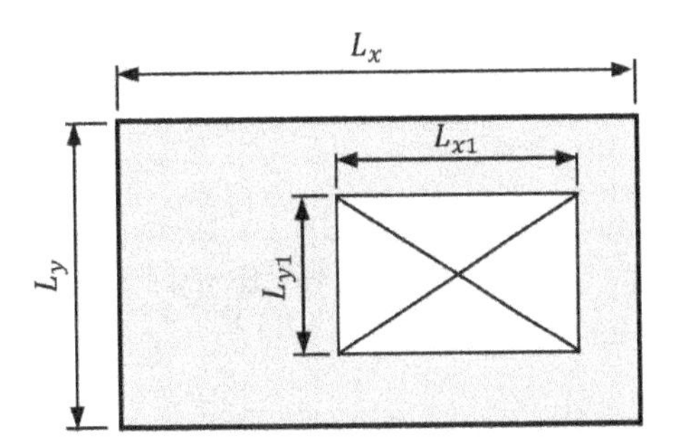

Figure 6.17 Diaphragm Discontinuity Irregularity

Significant variations in stiffness due to relatively large openings or abrupt changes in the diaphragm rigidity affect the distribution of seismic forces to the vertical elements of the SFRS. Torsional forces can be created that must be considered in design; these types of torsional forces are not accounted for in the force distribution that is normally used for regular buildings. Typical opening sizes for elevators and stairways in buildings are generally not large enough to cause significant differences in diaphragm stiffness.

The requirements in ASCE/SEI 12.3.3.5 must be satisfied for structures assigned to SDC D, E, or F.

Out-of-Plane Offset Irregularity (Type 4) An out-of-plane offset irregularity is defined to exist where there is a discontinuity in the path of lateral force resistance. An out-of-plane offset of at least one of the vertical elements in the SFRS is an example of such a discontinuity. Illustrated in Figure 6.18 is an out-of-plane offset of a shear wall that is part of the SFRS (a similar such offset would occur if instead of a wall, a column, or a brace was present at that location). This offset is the most critical discontinuity of its type because forces must be adequately transferred through horizontal members at the location of the offset.

The requirements in ASCE/SEI 12.3.3.5, 12.7.3, and 16.3.4 must be satisfied for structures assigned to SDC B, C, D, E, and F.

Nonparallel System Irregularity (Type 5) A nonparallel system irregularity is defined to exist where vertical lateral force-resisting elements are not parallel to the major orthogonal axes of the SFRS (see Figure 6.19).

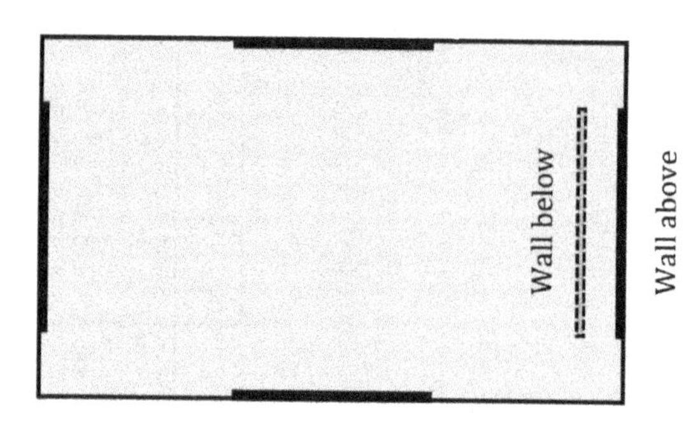

Figure 6.18 Out-of-Plane Offset Irregularity

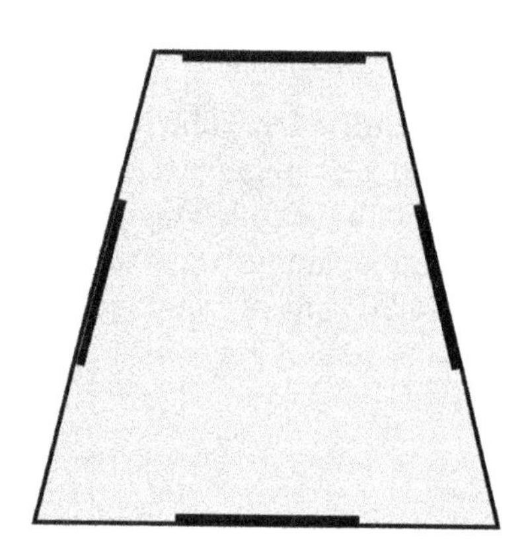

Figure 6.19 Nonparallel System Irregularity

The requirements in ASCE/SEI 12.5.3 must be satisfied for structures assigned to SDC C, D, E, and F, and the requirements in ASCE/SEI 12.5.4, 12.7.3, and 16.3.4 must be satisfied for structures assigned to SDC B, C, D, E, and F.

Vertical Irregularities

Overview Structures with one or more of the irregularities defined in ASCE/SEI Table 12.3-2 are designated as having a vertical structural irregularity. The sections with limitations and requirements that must be satisfied for the various SDCs are also given in this table. The types of vertical irregularities are covered below.

Stiffness–Soft Story Irregularity (Type 1a) A stiffness–soft story irregularity is defined to exist where there is a story in which the lateral stiffness is less than 70 percent of that in the story above or, where there are at least three stories above, less than 80 percent of the average stiffness of the three stories above.

An example of a stiffness–soft story irregularity is illustrated in Figure 6.20. This type of irregularity commonly occurs in the lower level of a building that has a lobby with a taller story height than the typical floors above.

Stiffness–Extreme Soft Story Irregularity (Type 1b) A stiffness–extreme soft story irregularity is defined to exist where there is a story in which the lateral stiffness is less than 60 percent of that in the story above or, where there are at least three stories above, less than 70 percent of the average stiffness of the three stories above (see Figure 6.20).

The requirements in ASCE/SEI 12.3.3.1 must be satisfied for structures assigned to SDC E and F.

Vertical Geometric Irregularity (Type 2) A vertical geometric irregularity is defined to exist where the horizontal dimension of the SFRS in any story is more than 130 percent of that in an adjacent story (see Figure 6.21). This irregularity applies regardless of whether the larger dimension is above or below the smaller one.

Figure 6.20 Stiffness–Soft and Stiffness–Extreme Soft Story Irregularity

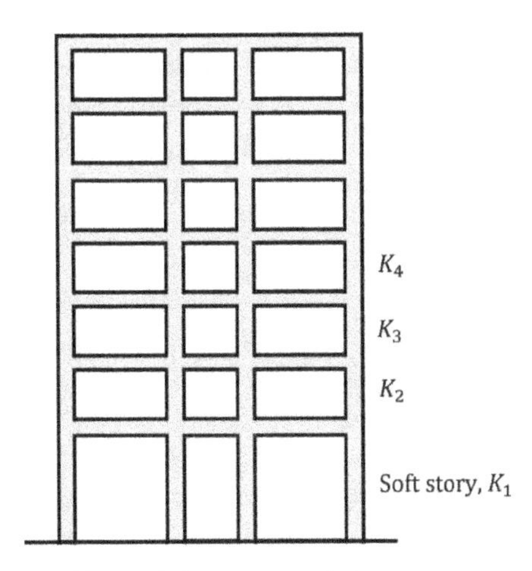

- Stiffness—Soft story irregularity:
  $K_1 < 0.7K_2$ or $K_1 < 0.8(K_2 + K_3 + K_4)/3$
- Stiffness—Extreme soft story irregularity:
  $K_1 < 0.6K_2$ or $K_1 < 0.7(K_2 + K_3 + K_4)/3$

Figure 6.21 Vertical Geometric Irregularity

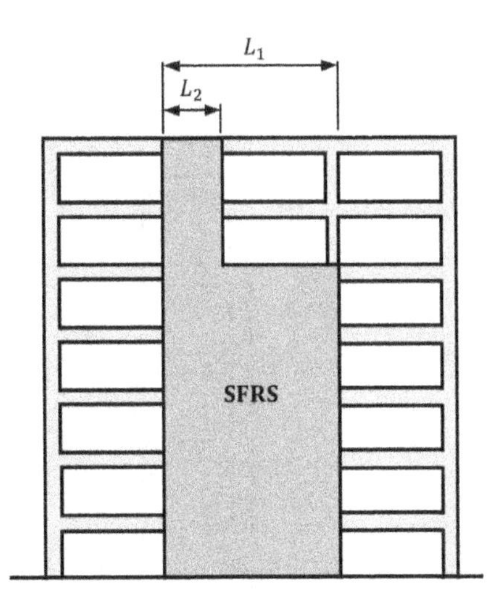

Vertical geometric irregularity: $L_1 > 1.3L_2$

In-Plane Discontinuity in Vertical Lateral Force-Resisting Element Irregularity (Type 3) This type of irregularity is defined to exist where there is an in-plane offset of a vertical seismic force-resisting element resulting in overturning demands on a supporting beam, column, truss, or slab (see Figure 6.22).

The requirements in ASCE/SEI 12.3.3.4 must be satisfied for structures assigned to SDC B, C, D, E, and F, and in ASCE/SEI 12.3.3.5 for structures assigned to SDC D, E, and F.

Discontinuity in Lateral Strength–Weak Story Irregularity (Type 4a) This type of irregularity is defined to exist where a story lateral strength is less than that in the story above (see Figure 6.23). The story lateral strength is the total lateral strength of all SFRS elements resisting the story shear in the direction of analysis.

Buildings with a weak story irregularity usually develop all of their inelastic behavior at the weak story along with the resulting damage, which could lead to a possible collapse.

Figure 6.22 In-plane Discontinuity in Vertical Lateral Force-Resisting Element Irregularity

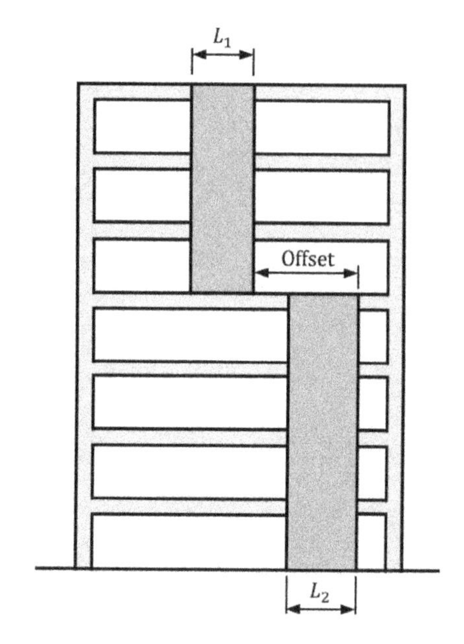

Figure 6.23 Discontinuity in Lateral Strength–Weak Story Irregularity

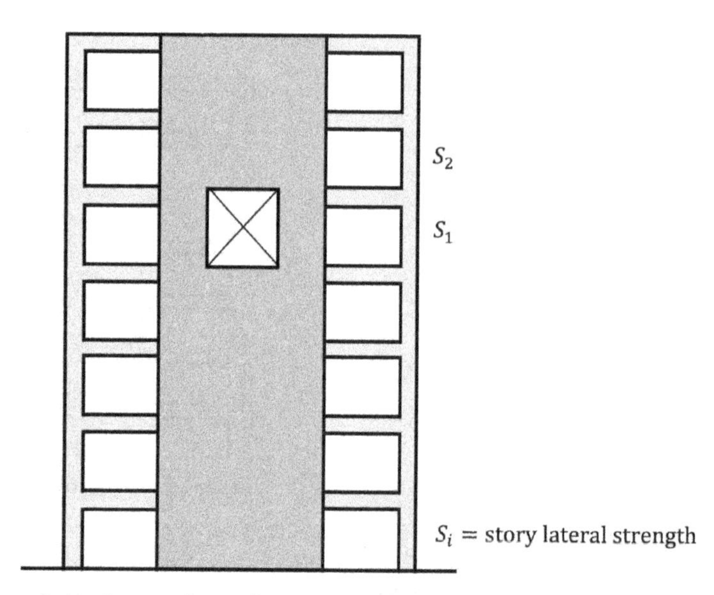

- Lateral strength–weak story irregularity:

  $S_1 < S_2$

- Lateral strength–extreme weak story irregularity:

  $S_1 < 0.65S_2$

The requirements in ASCE/SEI 12.3.3.1 must be satisfied for structures assigned to SDC E and F.

Discontinuity in Lateral Strength–Extreme Weak Story Irregularity (Type 4b) This irregularity is defined the same as that of Type 4a except the limit is 65 percent instead of 100 percent (see Figure 6.23).

Two exceptions are given in ASCE/SEI 12.3.2.2 related to vertical irregularities:

1. Types 1a and 1b vertical irregularities do not apply where no design story drift ratio determined using the design seismic forces is greater than 130 percent of the story drift ratio of the next story above. Note that torsional effects need not be considered in the calculation of the story drifts when applying this exception. Also, the design story drift ratio relationship for the top two stories is not required to be evaluated.
2. Types 1a and 1b vertical irregularities need not be considered for one-story buildings assigned to any SDC or for two-story buildings assigned to SDC B, C, or D.

Limitations and Additional Requirements for Systems with Structural Irregularities

As noted previously, most of the earthquake provisions in ASCE/SEI 7 are applicable to regular structures. The provisions in ASCE/SEI 12.3.3 are meant to account for detrimental effects that can be caused by irregularities and to prohibit or limit certain types of irregularities in areas of high seismic risk.

Limitations and additional requirements for systems with different types of structural irregularities are given in Table 6.11.

The types of horizontal and vertical irregularities prohibited in SDCs D through F in ASCE/SEI 12.3.3.1 and 12.3.3.2 are based on poor performance in past earthquakes. These irregularities have the tendency to concentrate large inelastic demands in certain portions of a structure, which could lead to collapse. As such, they should be avoided wherever possible, even in cases where they are permitted.

The main purpose of the requirement in ASCE/SEI 12.3.3.4 for designing elements supporting discontinuous walls or frames using the load combinations including overstrength in ASCE/SEI 12.4.3 is to protect the supporting elements from overload caused by overstrength of a discontinued seismic force-resisting element (columns, beams, slabs, or trusses). The connection between the discontinuous element and the supporting member must be adequate to transfer the forces for which the discontinuous elements are required to be designed. Examples of framing systems where this requirement must be satisfied are given in Figure 6.24. Indicated in the figure are the elements that must be designed using the load combinations including overstrength.

The increase in forces due to the irregularities listed for SDCs D, E, and F in ASCE/SEI 12.3.3.5 are intended to account for the difference between the loads that are distributed using the equivalent lateral force procedure in ASCE/SEI 12.8 and the load distribution including the irregularities. The difference in load distribution is especially important for the connection of the diaphragm to the vertical elements of the SFRS. No additional increase is warranted where the load combinations with overstrength factor apply.

Redundancy

Overview

Redundancy has long been recognized as a desirable attribute to have in any structure. The redundancy factor, ρ, is a measure of the redundancy inherent in a structure. A higher degree of redundancy exists where there are multiple paths to resist the lateral forces. This factor reduces the response modification coefficient, R, for less redundant structures, which, in turn, increases the seismic demand on the system. A redundancy factor equal to 1.0 means a structure is sufficiently redundant, and no increase in the seismic load effects is warranted.

The value of ρ is either 1.0 or 1.3 and is to be determined in each of the two orthogonal directions for all structures.

The following sections contain information on how to determine ρ in accordance with ASCE/SEI 12.3.4.

Table 6.11 Limitations and Additional Requirements for Systems with Structural Irregularities

Item	SDC	Irregularity Type	Limitation / Requirement	Exceptions
Prohibited vertical irregularities (ASCE/SEI 12.3.3.1 and 12.3.3.2)	D	Vertical irregularity Type 4b	Not permitted	—
	E or F	Vertical irregularity Type 1b, 4a or 4b		Structures assigned to SDC E or F that have vertical irregularity Type 4a are permitted where the story lateral strength is not less than 80% of that in the story above
Extreme weak stories (ASCE/SEI 12.3.3.3)	B or C	Vertical irregularity Type 4b	Structures are limited to two stories or 30 ft (9.14 m) in structural height, h_n	The limit does not apply where the weak story is capable of resisting a total seismic force equal to Ω_0 times the design seismic force determined in accordance with ASCE/SEI 12.8
Elements supporting discontinuous walls or frames (ASCE/SEI 12.3.3.4)	B – F	• Horizontal irregularity Type 4 • Vertical irregularity Type 3	Elements supporting discontinuous walls or frames must be designed to resist the seismic load effects including overstrength in accordance with ASCE/SEI 12.4.3	—
Increase in forces caused by irregularities (ASCE/SEI 12.3.3.5)	D – F	• Horizontal irregularity Type 1, 2, 3, or 4 • Vertical irregularity Type 3	Design forces determined in accordance with ASCE/SEI 12.10.1.1 must be increased by 25% at each diaphragm level where the irregularity occurs for the following elements of the SFRS: • Connections of diaphragms to vertical elements and collectors • Collectors and their connections, including connections to the vertical elements of the SFRS	Forces calculated using the seismic load effects including overstrength of ASCE/SEI 12.4.3 need not be increased

Conditions where $\rho = 1.0$

Conditions where ρ is equal to 1.0 are as follows (ASCE/SEI 12.3.4.1):

1. Structures assigned to SDC B or C
2. Drift calculations and P-delta effects.
3. Design of nonstructural components.
4. Design of nonbuilding structures that are not similar to buildings.

Figure 6.24 Elements Supporting Discontinuous Walls or Frames

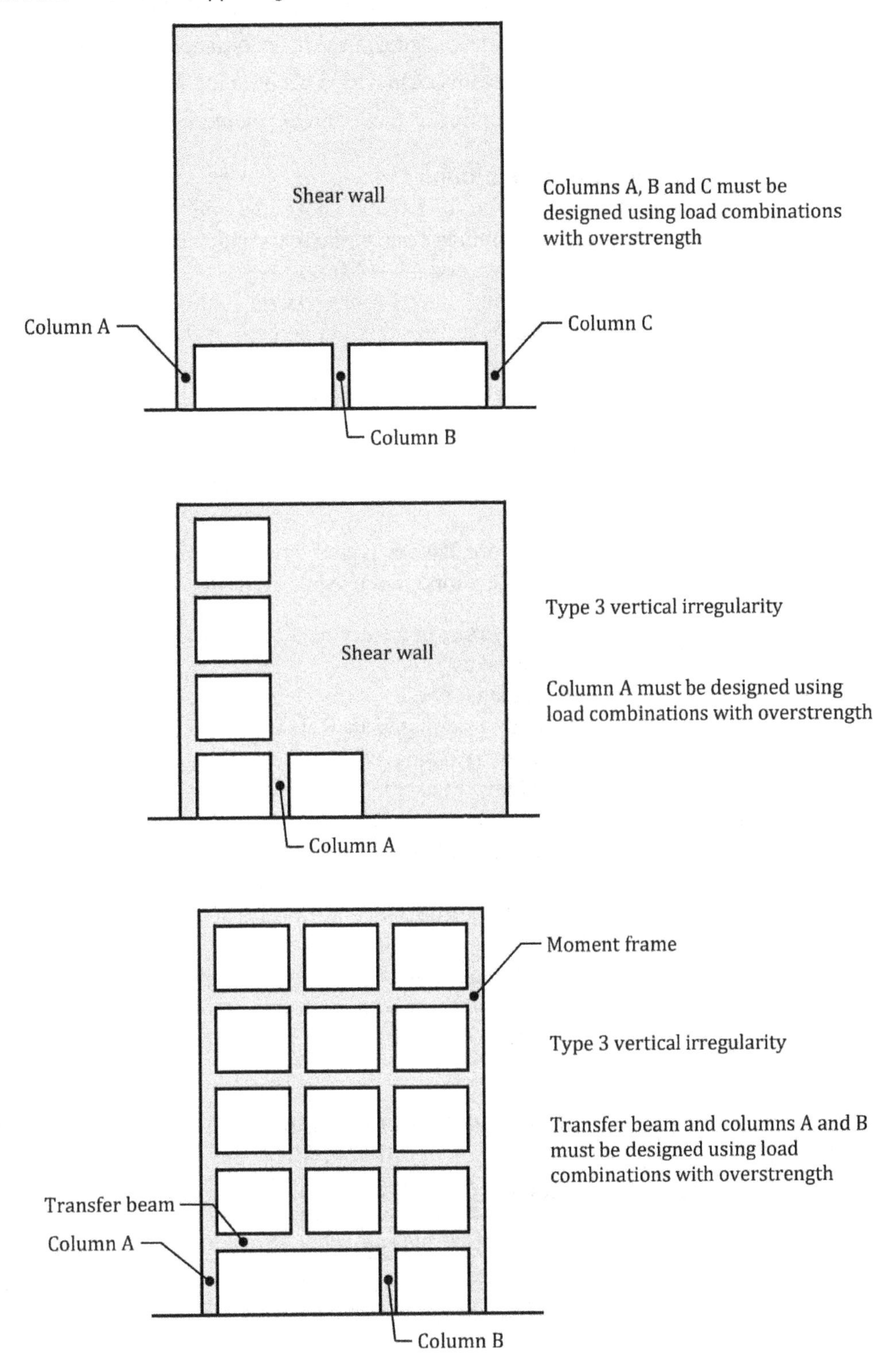

5. Design of collector elements, splices and their connections for which the seismic load effects including overstrength of ASCE/SEI 12.4.3 are used.
6. Design of members or connections where the seismic load effects, including overstrength of ASCE/SEI 12.4.3 are required for design.

7. Diaphragm seismic design forces determined by ASCE/SEI Equation (12.10-1), including the limits imposed by ASCE/SEI Equations (12.10-2) and (12.10-3).
8. Diaphragm seismic design forces determined in accordance with ASCE/SEI 12.10.4.
9. Structures with damping systems designed in accordance with ASCE/SEI Chapter 18.
10. Design of structural walls for out-of-plane forces, including their anchorage.

Redundancy Factor for SDC D *through* F

For structures assigned to SDC D, E, or F that do not satisfy any of the conditions in ASCE/SEI 12.3.4.1, ρ must be set equal to 1.3 unless one of the two conditions in ASCE/SEI 12.3.4.2 is met, in which case it is permitted to be set equal to 1.0.

In the first condition, an individual lateral force-resisting element or connection is removed in each story where the story shear is greater than 35 percent of the base shear in the direction of analysis to determine its effect on the remaining structure. Based on the element or connection removal, the following conditions must be met in order for ρ to be taken as 1.0:

1. The structure has at least two bays of seismic force-resisting framing on each side of the CM.
2. The reduction in lateral strength of the story in the direction of analysis is less than or equal to 35 percent.
3. The resisting system does not have a type 1 horizontal irregularity in accordance with ASCE/SEI Table 12.3-1 with a torsional irregularity ratio, TIR, greater than 1.4.

If the removal of elements, one by one, in a story results in one or more of the conditions above not being satisfied, ρ must be taken as 1.3.

The requirement of having at least two bays of seismic force-resisting framing on each side of the CM is based on the study conducted in Reference 6.6, which showed that structural configurations where all vertical resisting elements occurred on the same side of the CM should include a ρ of 1.3 to achieve the same probability of collapse as structures with redundant configurations.

The 35-percent value in the second requirement is based on unpublished parametric studies where it was indicated that stories with story shears equal to at least 35 percent of the base shear include all stories of low-rise buildings up to about 6 stories and approximately 87 percent of the stories of tall buildings. The intent of this limit is to exclude the penthouses of most buildings and the uppermost stories of tall buildings from the redundancy requirements.

The elements and connections that must be removed are given in ASCE/SEI Table 12.3-3 for the lateral force-resisting elements indicated in the table (see Table 6.12). Structures with

Table 6.12 Requirements for Each Story Resisting More Than 35 Percent of the Base Shear in Accordance with the First Condition in ASCE/SEI 12.3.4.2

Lateral Force-Resisting Element	Condition
Braced frames or light-frame walls with flat-strap bracing	Removal of an individual brace or connection
Moment frames	Loss of moment resistance at the beam-column connections at both ends of a single beam
Shear walls or wall piers with a height-to-length ratio h_{wall}/L_{wall} or h_{wp}/L_{wp}, respectively, greater than 1.0 (see Figure 6.25)	Removal of a shear wall bay, wall pier or connections*
Cantilever columns	Loss of moment resistance at the base of any single cantilever column

*A shear wall bay is defined as the length of the wall divided by the story height (rounded down).

Figure 6.25 Shear Wall and Wall Pier Dimensions to be Used in the Determination of Height-to-Length Ratios

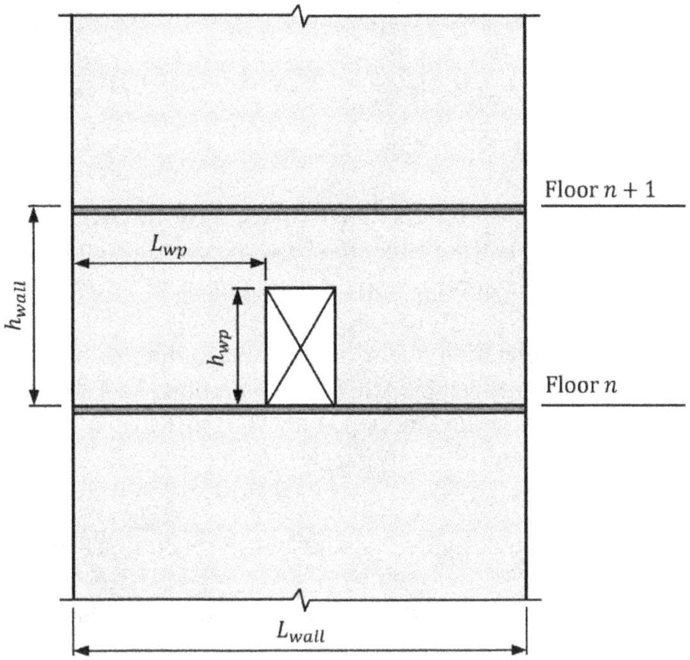

elements in SFRSs not listed in this table need not consider the requirements in this first condition.

In the second condition, ρ is permitted to be taken as 1.0 when the following are satisfied:

1. The structure does not have any horizontal irregularities in accordance with ASCE/SEI Table 12.3-1 in the direction of analysis at all levels of the building.
2. The SFRSs must consist of at least two bays of seismic force-resisting perimeter framing on each side of the structure at each story resisting more than 35 percent of the base shear.

As part of the parametric study noted above, simplified braced frame and moment frame systems were investigated to determine their sensitivity to the analytical redundancy criteria. The requirements in this condition are consistent with the results of the study.

The redundancy factor, ρ, must be taken as 1.3 for the entire structure in both orthogonal directions for structures assigned to SDC D, E or F that have a type 1 horizontal irregularity where the torsional irregularity ratio, TIR, is greater than 1.4 in both orthogonal directions at any given level (ASCE/SEI 12.3.4.2.1). The requirements of ASCE/SEI 12.3.4.2 apply for all other structures, including structures with a type 1 horizontal irregularity.

Flowchart 6.4 in Section 6.6 of this book can be used to determine the redundancy factor, ρ.

6.3.4 Seismic Load Effects and Combinations

Overview

Unless otherwise exempted, all structural members, including those that are not part of the SFRS, must be designed using the seismic load effects in ASCE/SEI 12.4. In general, these effects are produced in the structural members due to the application of horizontal and vertical forces on the structure (ASCE/SEI 12.4.2).

Certain structural members must be designed to account for the seismic load effects including overstrength (ASCE/SEI 12.4.3). The conditions under which these effects are applicable are covered below.

Seismic Load Effect, *E*

Seismic load effect, E, in the load combinations defined in ASCE/SEI 2.3 and 2.4 consists of effects from horizontal (E_h) and vertical (E_v) seismic forces determined by ASCE/SEI Equations (12.4-3) and (12.4-4a), respectively:

$$E_h = \rho Q_E \tag{6.11}$$

$$E_v = 0.2 S_{DS} D \tag{6.12}$$

In Equation (6.11), Q_E are the effects (axial forces, shear forces, and bending moments) obtained from the structural analysis from either the base shear, V, which is distributed over the height of the structure, or the seismic force acting on a component of a structure, F_p (ASCE/SEI 12.4). The effect from the vertical seismic forces, E_v, is based on an assumed effective vertical acceleration of $0.2S_{DS}$ times the acceleration due to gravity. According to the first exception in ASCE/SEI 12.4.2.2, E_v must be determined by ASCE/SEI Equation (12.4-4b) where the option to incorporate the effects of vertical ground motions by ASCE/SEI 11.9 is required:

$$E_v = 0.3 S_{av} D \tag{6.13}$$

In this equation, S_{av} is the design vertical response spectral acceleration determined by ASCE/SEI 11.9.3.

The second exception in ASCE/SEI 12.4.2.2 permits E_v to be taken as zero for either of the following cases:

1. In ASCE/SEI Equations (12.4-1), (12.4-2), (12.4-5), and (12.4-6) for structures assigned to SDC B.
2. In ASCE/SEI Equation (12.4-2) when determining demands on the soil-structure interface of foundations.

The proper sign convention must be used in the load combinations with E. In strength design load combination 6 in ASCE/SEI 2.3.6 or the allowable stress design load combinations 8 and 9 in ASCE/SEI 2.4.5, E is determined by ASCE/SEI Equation (12.4-1):

$$E = E_h + E_v = \rho Q_E + 0.2 S_{DS} D \tag{6.14}$$

In these load combinations, effects from the dead load, D, are additive to those due to E.

In strength design load combination 7 in ASCE/SEI 2.3.6 or the allowable stress design load combination 10 in ASCE/SEI 2.4.5, E is determined by ASCE/SEI (12.4-2):

$$E = E_h - E_v = \rho Q_E - 0.2 S_{DS} D \tag{6.15}$$

In these load combinations, the effects of E counteract those from D.

Basic load combinations for strength design and allowable stress design based on SDC are given in Table 6.13.

Seismic Load Effects Including Overstrength

The seismic load effect including overstrength, E_m, used in the load combinations in ASCE/SEI 2.3 and 2.4 consists of effects of horizontal seismic forces including overstrength (E_{mh}) and vertical seismic load effect (E_v), the latter of which is determined by Equation (6.12). The horizontal seismic load effect with overstrength is determined by ASCE/SEI Equation (12.4-7):

$$E_{mh} = \Omega_0 Q_E \tag{6.16}$$

where Ω_0 is the overstrength factor given in ASCE/SEI Table 12.2-1 as a function of the SFRS (ASCE/SEI 12.4.3).

Table 6.13 Seismic Load Combinations

ASCE/SEI 7 Load Combination		Load Combination
Strength Design*		
6	SDC B	$1.2D + Q_E + L + 0.15S$
7		$0.9D + Q_E$
6	SDC C	$(1.2 + 0.2S_{DS})D + Q_E + L + 0.15S$
7		$(0.9 - 0.2S_{DS})D + Q_E$
6	SDC D, E, and F	$(1.2 + 0.2S_{DS})D + \rho Q_E + L + 0.15S$
7		$(0.9 - 0.2S_{DS})D + \rho Q_E$
Allowable Stress Design**		
8	SDC B	$D + 0.7Q_E$
9		$D + 0.525Q_E + 0.75L + 0.1S$
10		$0.6D + 0.7Q_E$
8	SDC C	$(1.0 + 0.14S_{DS})D + 0.7Q_E$
9		$(1.0 + 0.105S_{DS})D + 0.525Q_E + 0.75L + 0.1S$
10		$(0.6 - 0.14S_{DS})D + 0.7Q_E$
8	SDC D, E, and F	$(1.0 + 0.14S_{DS})D + 0.7\rho Q_E$
9		$(1.0 + 0.105S_{DS})D + 0.525\rho Q_E + 0.75L + 0.1S$
10		$(0.6 - 0.14S_{DS})D + 0.7\rho Q_E$

*Notes for strength design load combinations:

(1) Load factor on L in combination 6 is permitted to equal 0.5 for all occupancies in which L_o in ASCE/SEI Table 4.3-1 is less than or equal to 100 lb/ft^2 (4.79 kN/m^2), with the exception of garages or areas occupied as places of public assembly.

(2) In combination 6, S must be taken as the flat roof snow load, p_f, or the sloped roof snow load, p_s.

(3) Where fluid loads, F, are present, they must be included with the same load factor as dead load, D, in combinations 6 and 7.

(4) Where load H is present, it must be included as follows:
  (a) Where the effect of H adds to the primary variable load effect, include H with a load factor of 1.6.
  (b) Where the effect of H resists the primary variable load effect, include H with a load factor of 0.9 where the load H is permanent or a load factor of 0 for all other conditions.

**Notes for allowable stress design load combinations:

(1) In combination 9, S must be taken as the flat roof snow load, p_f, or the sloped roof snow load, p_s.

(2) In combination 10, it is permitted to replace $0.6D$ with $0.9D$ for the design of special reinforced masonry shear walls where the walls satisfy the requirement in ASCE/SEI 14.4.2.

(3) Where fluid loads, F, are present, they must be included in combinations 8, 9, and 10 with the same factor as that used for dead load, D.

(4) Where load H is present, it must be included as follows:
  (a) Where the effect of H adds to the primary variable load effect, include H with a load factor of 1.0.
  (b) Where the effect of H resists the primary variable load effect, include H with a load factor of 0.6 where the load H is permanent or a load factor of 0 for all other conditions.

In the strength design load combination 6 in ASCE/SEI 2.3.6 or the allowable stress load combinations 8 and 9 in ASCE/SEI 2.4.5, E must be taken equal to E_m determined by ASCE/SEI Equation (12.4-5):

$$E_m = E_{mh} + E_v = \Omega_0 Q_E + 0.2S_{DS}D \tag{6.17}$$

In the strength design load combination 7 in ASCE/SEI 2.3.6 or the allowable stress load combination 10 in ASCE/SEI 2.4.5, E must be taken equal to E_m determined by ASCE/SEI Equation (12.4-6):

$$E_m = E_{mh} - E_v = \Omega_0 Q_E - 0.2S_{DS}D \tag{6.18}$$

Where capacity-limited design is required, the capacity-limited horizontal seismic load, E_{cl}, which is the maximum force that can develop in the element determined by a rational, plastic mechanism analysis, must be substituted for E_{mh} in the above equations (ASCE/SEI 12.4.3.2).

Basic combinations for strength design with overstrength and allowable stress design with overstrength based on SDC are given in Table 6.14. These combinations pertain only to the following structural elements:

- Elements supporting discontinuous walls or frames in structures assigned to SDC B through F.
- Collector elements and their connections, including connections to vertical elements in structures assigned to SDC C through F.

Table 6.14 Seismic Load Combinations with Overstrength

ASCE/SEI 7 Load Combination		Load Combination
Strength Design*		
6	SDC B	$1.2D + \Omega_0 Q_E + L + 0.15S$
7	SDC B	$0.9D + \Omega_0 Q_E$
6	SDC C, D, E, and F	$(1.2 + 0.2S_{DS})D + \Omega_0 Q_E + L + 0.15S$
7	SDC C, D, E, and F	$(0.9 - 0.2S_{DS})D + \Omega_0 Q_E$
Allowable Stress Design**		
8	SDC B	$D + 0.7\Omega_0 Q_E$
9	SDC B	$D + 0.525\Omega_0 Q_E + 0.75L + 0.1S$
10	SDC B	$0.6D + 0.7\Omega_0 Q_E$
8	SDC C, D, E, and F	$(1.0 + 0.14S_{DS})D + 0.7\Omega_0 Q_E$
9	SDC C, D, E, and F	$(1.0 + 0.105S_{DS})D + 0.525\Omega_0 Q_E + 0.75L + 0.1S$
10	SDC C, D, E, and F	$(0.6 - 0.14S_{DS})D + 0.7\Omega_0 Q_E$

*Notes for strength design load combinations:

(1) Load factor on L in combination 6 is permitted to equal 0.5 for all occupancies in which L_o in ASCE/SEI Table 4.3-1 is less than or equal to 100 lb/ft^2 (4.79 kN/m^2), with the exception of garages or areas occupied as places of public assembly.
(2) In combination 6, S must be taken as the flat roof snow load, p_f, or the sloped roof snow load, p_s.
(3) Where fluid loads, F, are present, they must be included with the same load factor as dead load, D, in combinations 6 and 7.
(4) Where load H is present, it must be included as follows:
  (a) Where the effect of H adds to the primary variable load effect, include H with a load factor of 1.6.
  (b) Where the effect of H resists the primary variable load effect, include H with a load factor of 0.9 where the load is permanent or a load factor of 0 for all other conditions.

**Notes for allowable stress design load combinations:

(1) In combination 9, S must be taken as the flat roof snow load, p_f, or the sloped roof snow load, p_s.
(2) In combination 10, it is permitted to replace $0.6D$ with $0.9D$ for the design of special reinforced masonry shear walls where the walls satisfy the requirement in ASCE/SEI 14.4.2.
(3) Where fluid loads, F, are present, they must be included in combinations 8, 9, and 10 with the same factor as that used for dead load, D.
(4) Where load H is present, it must be included as follows:
  (a) Where the effect of H adds to the primary variable load effect, include H with a load factor of 1.0.
  (b) Where the effect of H resists the primary variable load effect, include H with a load factor of 0.6 where the load is permanent or a load factor of 0 for all other conditions.

Allowable stresses are permitted to be determined using an allowable stress increase factor of 1.2 where seismic loads including overstrength in ASCE/SEI 12.4.3 are used in load combinations 8, 9, and 10; this increase is not to be combined with any other increases in allowable stresses or load combination reductions permitted in any of the provisions (ASCE/SEI 2.4.5).

Minimum Upward Force for Horizontal Cantilevers

Horizontal cantilever members in buildings assigned to SDC D, E, or F must be designed for a supplemental basic load combination consisting of a minimum net upward force of $0.2D$ for strength design or $0.14D$ for allowable stress design. This requirement is meant to provide minimum strength in the upward direction and to account for any dynamic amplification of vertical ground motion due to the vertical flexibility of the cantilever.

6.3.5 Direction of Loading

Loading Criteria

Earthquakes can produce ground motion in any direction and a structure must be designed to resist the maximum possible effects. Seismic forces must be applied to the structure in directions producing the most critical load effects on the structural members (ASCE/SEI 12.5).

In lieu of an analysis that determines the critical direction of loading for each element of the SFRS, it is permitted to use one of the following procedures subject to the conditions in ASCE/SEI 12.5.2, 12.5.3, and 12.5.4 for SDC B, C, and D through F, respectively (ASCE/SEI 12.5.1):

- Independent Directional Procedure
  Design seismic forces are permitted to be applied independently in each of two orthogonal directions on a structure.
- Orthogonal Directional Combination Procedure
  Design seismic forces must be applied in two orthogonal directions simultaneously using one of the methods in Table 6.15. The loading requirements are based on the analysis method used to determine the design seismic forces.

Table 6.15 Orthogonal Directional Combination Procedure

Analysis Method	Requirements
• Equivalent lateral force procedure in ASCE/SEI 12.8 • Modal response spectrum analysis procedure in ASCE/SEI 12.9.1	• The effects of 100% of the design seismic forces in one direction must be combined with the effects of 30% of the design seismic forces in the perpendicular direction. • Each axis must be subjected to 100% of the design seismic forces acting in the positive and negative directions (i.e., sidesway to the right and to the left must be considered for each axis).
• Linear response history procedure in ASCE/SEI 12.9.2 • Nonlinear response history procedure in ASCE/SEI Chapter 16	Orthogonal pairs of ground motion acceleration histories must be applied simultaneously to the structure.

Accidental torsion must be included when applying the orthogonal loads in this procedure. According to item 3 in ASCE/SEI 12.8.4.2.2, a 5 percent displacement of the center of mass, which is equal to 0.05 times the dimension of the structure perpendicular to the direction of analysis, need not be applied in both of the orthogonal directions simultaneously. However, it must be applied in the direction that produces the greatest effect for each element considered.

Based on these requirements, eight loading scenarios must be considered in each orthogonal direction. The following equations can be used to determine the combined horizontal seismic effects for simultaneous application of 100 percent of the design seismic forces in one direction and 30 percent of the design seismic forces in the perpendicular direction:

$$E_{h,x} = \rho_x(\pm Q_{Ex} \pm Q_{Ex,t}) \pm 0.3\rho_y Q_{Ey} \tag{6.19}$$

$$E_{h,y} = \rho_y(\pm Q_{Ey} \pm Q_{Ey,t}) \pm 0.3\rho_x Q_{Ex} \tag{6.20}$$

where $E_{h,x}$ = combined horizontal seismic load effect caused by 100 percent of the design seismic forces applied in the x-direction including accidental torsion and 30 percent of the design seismic forces applied in the y-direction without accidental torsion

$E_{h,y}$ = combined horizontal seismic load effect caused by 100 percent of the design seismic forces applied in the y-direction including accidental torsion and 30 percent of the design seismic forces applied in the x-direction without accidental torsion

ρ_x = redundancy factor of the structure in the x-direction

ρ_y = redundancy factor of the structure in the y-direction

Q_{Ex} = horizontal seismic load effect caused by 100 percent of the design seismic forces applied in the x-direction without accidental torsion

$Q_{Ex,t}$ = horizontal seismic load effect caused by 100 percent of the design seismic forces applied in the x-direction assuming the CM is displaced a distance equal to 5 percent of the dimension of the structure in the y-direction (i.e., including accidental torsion)

Q_{Ey} = horizontal seismic load effect caused by 100 percent of the design seismic forces applied in the y-direction without accidental torsion

$Q_{Ey,t}$ = horizontal seismic load effect caused by 100 percent of the design seismic forces applied in the y-direction assuming the CM is displaced a distance equal to 5 percent of the dimension of the structure in the x-direction (i.e., including accidental torsion)

The "plus-minus" symbol in the above equation signifies the following: (1) seismic load effects must be determined for the design seismic forces applied to the structure in both the positive and negative directions (i.e., sidesway to the right and to the left) and (2) the CM must be displaced in the positive and negative directions.

The eight required loading scenarios for each orthogonal direction are given in Figures 6.26 and 6.27. In the figures, V_x and V_y are the design seismic forces in the x- and y-directions, respectively.

Seismic load effects $E_{h,x}$ and $E_{h,y}$ are combined with the effects due to dead loads, D, live loads, L, snow loads, S (where applicable) and vertical seismic load effects, E_v, using load combinations 6 and 7 in ASCE/SEI 2.3.6.

The following equations are applicable where seismic load effects with overstrength are required:

$$E_{mh,x} = \Omega_0[(\pm Q_{Ex} \pm Q_{Ex,t}) \pm 0.3Q_{Ey}] \tag{6.21}$$

$$E_{mh,y} = \Omega_0[(\pm Q_{Ey} \pm Q_{Ey,t}) \pm 0.3Q_{Ex}] \tag{6.22}$$

where $E_{mh,x}$ = combined seismic load effect caused by 100 percent of the design seismic forces applied in the x-direction including accidental torsion and 30 percent of the design seismic forces applied in the y-direction without accidental torsion

$E_{mh,y}$ = combined seismic load effect caused by 100 percent of the design seismic forces applied in the y-direction including accidental torsion and 30 percent of the design seismic forces applied in the x-direction without accidental torsion

Ω_0 = overstrength factor determined from ASCE/SEI Table 12.2-1 based on the SFRS

Figure 6.26 Required Load Scenarios in the x-Direction for the Orthogonal Directional Combination Procedure

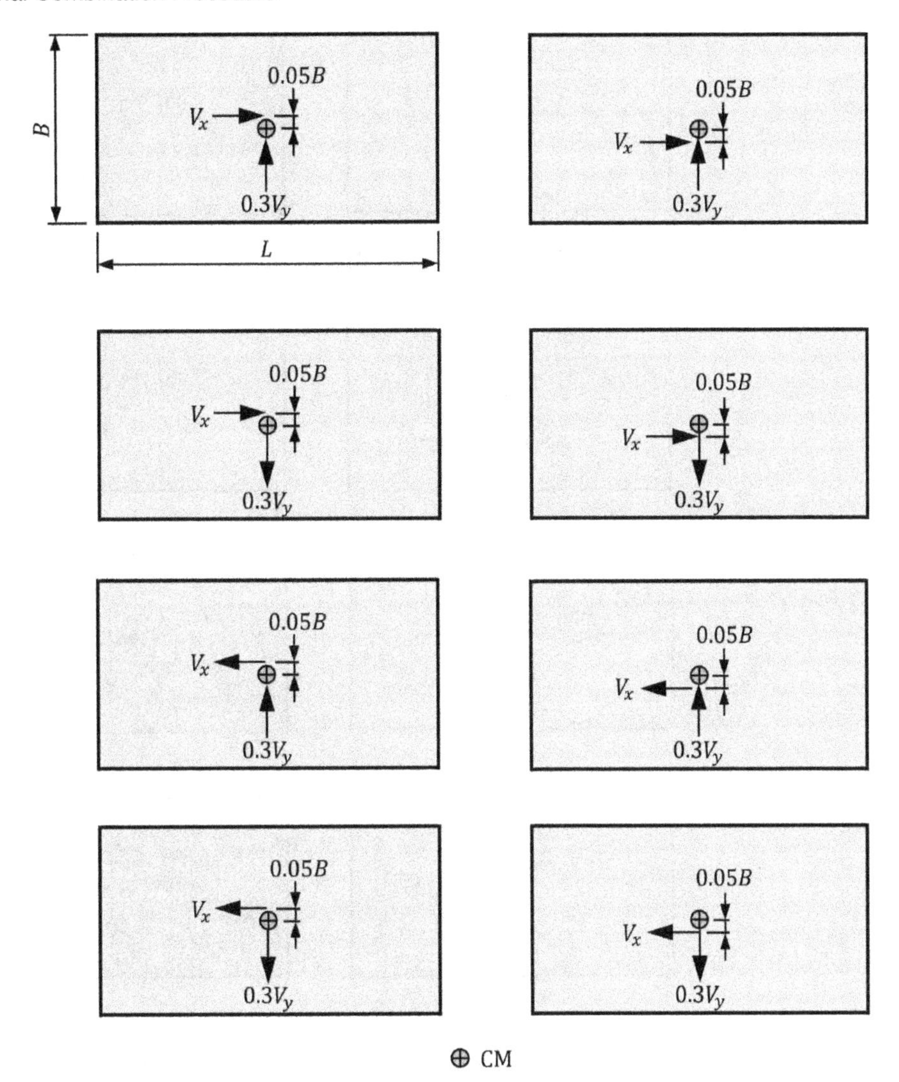

Unlike the redundancy factor, which is directional, the overstrength factor, Ω_0, is independent of the direction of analysis and applies only to a specific structural element (e.g., a collector element) that is part of the SFRS.

Seismic load effects $E_{mh,x}$ and $E_{mh,y}$ are combined with the effects due to dead loads, D, live loads, L, snow loads, S (where applicable) and vertical seismic load effects, E_v, using load combinations 6 and 7 in ASCE/SEI 2.3.6.

Procedure Limitations

The independent directional and orthogonal directional combination procedures are permitted to be used for structures assigned to any SDC with the following limitations:

- For structures assigned to SDC C with a type 5 horizontal irregularity, the independent directional procedure is not permitted (ASCE/SEI 12.5.3).
- For structures assigned to SDC D through F, the independent directional procedure is not permitted where one or more of the following conditions exist (ASCE/SEI 12.5.4):
  (a) The structure has a type 1 horizontal irregularity.
  (b) The structure has a type 5 horizontal irregularity.

Figure 6.27 Required Load Scenarios in the *y*-Direction for the Orthogonal Directional Combination Procedure

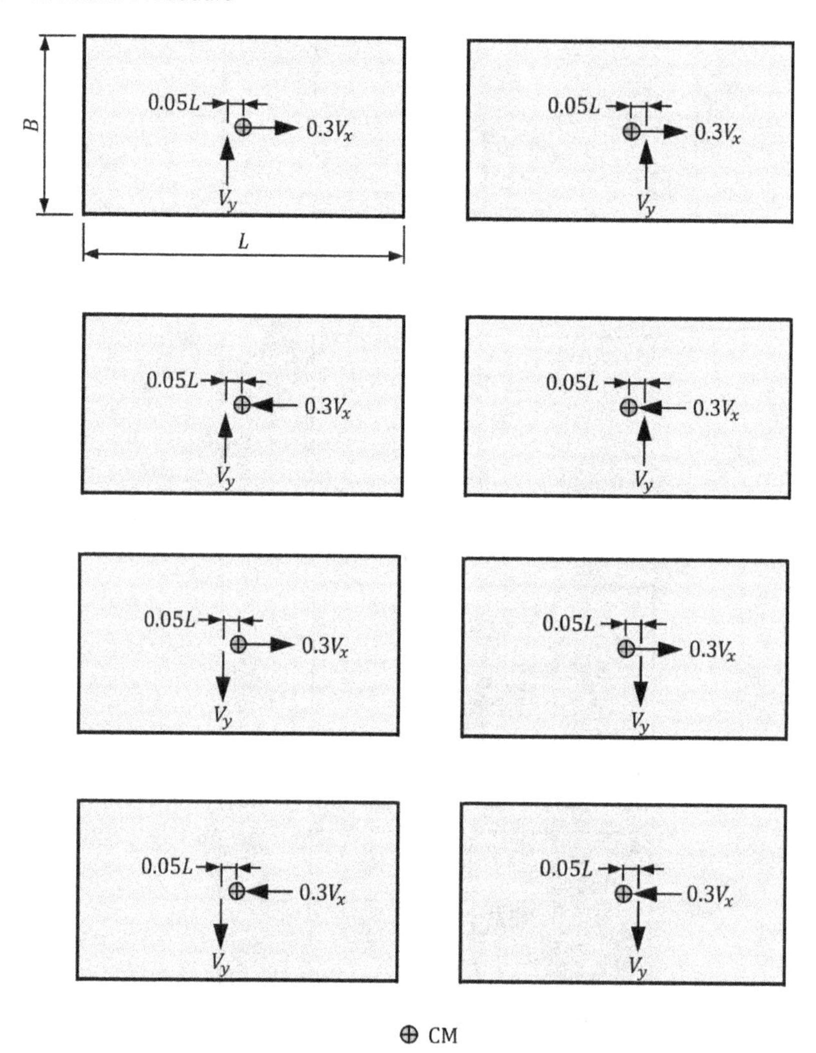

(c) A column that is part of two or more intersecting SFRSs and is subjected to an axial load due to seismic forces acting along either principal plan axis exceeding 20 percent of the axial design strength of the column.

(d) A concrete shear wall with connected intersecting wall planes and an axial demand exceeding 20 percent of the axial design strength at any location in the wall due to seismic forces acting along either principal plan axis.

(e) A masonry shear wall with connected intersecting wall planes and an axial demand exceeding 20 percent of the axial design strength at any location in the wall due to seismic forces acting along either principal plan axis.

(f) A wood or cold-formed steel light-frame shear wall system with connected intersecting wall planes and an axial load exceeding 20 percent of the axial design strength of the wall element at any location within the wall due to seismic forces acting along either principal plan axis.

(g) Cantilever column systems.

The following exception is also applicable: Where items (a) and (b) do not apply, the independent directional procedure is permitted to be used for components of the SFRS not listed in items (c) through (g).

The orthogonal directional combination procedure must be used in the following two cases:

1. Foundation elements and connections supporting elements of structures with any of conditions (c) through (g)
2. Foundations and connections supporting SFRS elements that are not coplanar.

6.3.6 Analysis Procedure Selection

The following procedures are permitted to be used to satisfy the structural analysis requirements in ASCE/SEI Chapter 12 (ASCE/SEI 12.6):

- Equivalent lateral force (ELF) procedure (ASCE/SEI 12.8)
- Modal response spectrum analysis (MRSA) [ASCE/SEI 12.9.1]
- Linear response history analysis (LRHA) [ASCE/SEI 12.9.2]
- An analysis procedure approved by the building official

In addition to the above procedures, the nonlinear response history analysis in ASCE/SEI Chapter 16 is permitted to be used for any structure. The simplified procedure in ASCE/SEI 12.14 may also be used for bearing wall and building frame systems that conform to the limitations in ASCE/SEI 12.14.1.1.

The appropriate procedure to use is related to the complexity of a structure. For example, it is unlikely that a MRSA or LRHA procedure will be utilized for regular one- or two-story buildings in usual situations. On the other hand, a tall, slender building with a number of transfers may warrant a more advanced analysis procedure.

Additional information on these analysis procedures is given in the following sections.

6.3.7 Modeling Criteria

Overview

Requirements pertaining to the construction of an adequate model for the purposes of seismic load analysis are given in ASCE/SEI 12.7. Included are provisions for foundation modeling, effective seismic weight, structural modeling, and interaction effects.

Foundation Modeling

It is permitted to assume that the base of a structure is fixed for purposes of determining the seismic loads. However, where the flexibility of the foundation is considered in the analysis, the requirements of ASCE/SEI 12.13.3 (foundation load-deformation characteristics) or ASCE/SEI Chapter 19 (soil-structure interaction for seismic design) must be satisfied.

The base of a structure is defined as the level at which the horizontal seismic ground motions are considered to be imparted to the structure (ASCE/SEI 11.2). Additional information on where the base occurs for a number of common situations can be found in ASCE/SEI C11.2.

Additional recommendations for foundation modeling are given in Reference 6.7.

Effective Seismic Weight

In general, a structure accelerates laterally during an earthquake, which leads to its mass producing inertial forces. These inertial forces accumulate over the height of the structure and produce the design base shear, V.

The definition of the effective seismic weight, W, which is used in determining the base shear, V, is given in ASCE/SEI 12.7.2. In addition to the dead load of the structure above the base, the following loads above the base must be included in W:

1. A minimum of 25 percent of the floor live load in areas used for storage.
2. The actual partition weight or a minimum weight of 10 lb/ft^2 (0.48 kN/m^2), whichever is greater, where provisions for partitions are made or where required in accordance with ASCE/SEI 4.3.2.

3. Total operating weight of permanent equipment.
4. Fifteen percent of the uniform design snow load, regardless of roof slope, where the flat roof snow load, p_f, determined in accordance with ASCE/SEI 7.3 exceeds 45 lb/ft^2 (2.16 kN/m^2) [see Chapter 4 of this book].
5. Weight of landscaping and other materials at roof gardens and similar areas, as defined in ASCE/SEI 3.1.4.
6. Weight of fluids and bulk material expected to be present during normal use.

It is clear from the list that only the weight that is strongly coupled or physically tied to the structure is considered effective. Live loads due to human occupancy and loose furniture, to name a few, need not be considered effective in such cases.

The probability that a portion of live load in storage areas is present during an earthquake is not small, especially in densely packed storage areas; that is why 25 percent of the floor live load is considered effective.

The full snow load is not considered to be part of the effective seismic weight because it is unlikely that maximum snow load and maximum earthquake load will occur simultaneously.

The following exceptions are given pertaining to effective seismic weight:

(a) The weight in items 1, 3, 5, and 6 need not be included in W where the total weight of these items adds no more than 5 percent to the effective seismic weight at that level.

(b) Floor live load in public garages and open parking structures need not be included in W.

(c) The weights of fluids included as effective seismic weight are permitted to be reduced in accordance with a rational fluid-structure interaction analysis.

Structural Modeling

A mathematical model of the structure that includes member stiffness and strength must be constructed to determine forces in the structural members and story drifts due to ground shaking.

Cracked section properties must be used when analyzing concrete and masonry structures. In steel moment frames, the contribution of panel zone deformations to displacement and drift must be included. These factors must be incorporated in the model to obtain more realistic values of lateral deformations.

A three-dimensional analysis is required for structures that have horizontal irregularity types 1, 4, or 5. In such cases, a minimum of three degrees of freedom (translation in two orthogonal plan directions and rotation about the vertical axis) must be included at each level of the structure. Structures with flexible diaphragms and type 4 horizontal structural irregularities need not be analyzed using a three-dimensional model. For diaphragms that are not classified as rigid or flexible in accordance with ASCE/SEI 12.3.1, the model must include representation of the diaphragm stiffness characteristics and sufficient degrees of freedom to account for the participation of the diaphragm in the structure's dynamic response.

Additional information on structural modeling can be found in ASCE/SEI C12.7.3.

Interaction Effects

Interaction effects between moment-resisting frames that are part of the SFRS and rigid elements that are not part of the SFRS that enclose or adjoin the moment-resisting frame must be considered in the analysis to prevent any unexpected detrimental effects from occurring during an earthquake.

Consider the situation in Figure 6.28 where masonry is used as infill between the reinforced concrete columns in a moment-resisting frame. The masonry fits tightly against the columns. When subjected to ground shaking, hinges will most likely form at the tops of the columns and at the top of the masonry rather than at the top and bottom of the columns, as would be expected, because the masonry is much stiffer than the columns. The shear forces in the columns increase by a factor of H/h. If this increase is not accounted for in design, an abrupt shear failure of the columns can occur.

Figure 6.28 Interaction Effects between Members of the SFRS and Rigid Elements

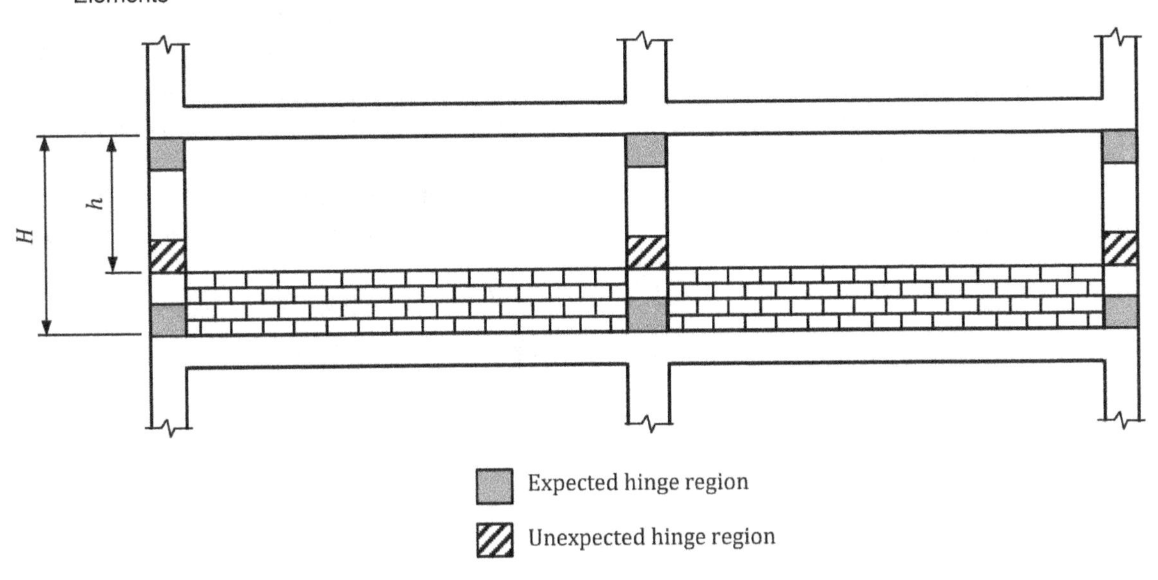

6.3.8 Equivalent Lateral Force Procedure

Overview

The equivalent lateral force (ELF) procedure is a simplified method of determining the effects of ground motion on a structure. Structures are modeled as a single-degree-of-freedom system with 100 percent mass participation in the fundamental mode. The effects of inelastic dynamic response are determined by using a linear static analysis: a design base shear, V, is determined, which is distributed over the height of the structure at each floor level. The structure is analyzed for these static forces, which are distributed to the members of the SFRS considering the flexibility of the diaphragms.

The provisions of the ELF procedure are in ASCE/SEI 12.8 and are discussed below.

Seismic Base Shear

Overview

The seismic base shear, V, in a given direction is determined by multiplying the seismic response coefficient, C_s, by the effective seismic weight, W [see ASCE/SEI Equation (12.8-1)]:

$$V = C_s W \tag{6.23}$$

Two methods are given in ASCE/SEI 12.8.1.1 to determine the seismic response coefficient, C_s, which are discussed in the next section.

Calculation of Seismic Response Coefficient

Method 1 in ASCE/SEI 12.8.1.1 is permitted to be used to determine C_s where the design spectral response acceleration parameter, S_a, has been obtained from the multi-period design response spectrum provisions in ASCE/SEI 11.4.5.1 or from the site-specific ground motion procedures for seismic design in ASCE/SEI Chapter 21 (see Section 6.2 of this book). In this method, C_s is determined by ASCE/SEI Equation (12.8-2):

$$C_s = \frac{S_a}{\left(\dfrac{R}{I_e}\right)} \tag{6.24}$$

In this equation, S_a is the design spectral response acceleration parameter defined in ASCE/SEI 11.4.5.1 corresponding to T, which is the period of the structure determined in accordance with ASCE/SEI 12.8.2, R is the response modification coefficient in ASCE/SEI Table 12.2-1 for the given SFRS and I_e is the importance factor determined in accordance with ASCE/SEI 11.5 based on the risk category of the structure. Where T of the structure is less than the period at which S_a is maximum, the maximum value of S_a must be used in the determination of C_s. This means there is no transition to peak ground acceleration, which is illustrated in ASCE/SEI Figure 11.4-1 for the two-period design response spectrum in ASCE/SEI 11.4.5.2; in other words, the design response spectrum using method 1 of the ELF procedure is horizontal from T equal to zero to T corresponding to the maximum S_a.

The equations for C_s in ASCE/SEI 12.8.1.1 form the design response spectrum in method 2:

- Constant acceleration region where $0 \le T \le T_S = S_{D1}/S_{DS}$ [ASCE/SEI Equation (12.8-3)]:

$$C_s = \frac{S_{DS}}{\left(\frac{R}{I_e}\right)} \tag{6.25}$$

- Constant velocity region where $T_S < T \le T_L$ [ASCE/SEI Equation (12.8-4)]:

$$C_s = \frac{S_{D1}}{T\left(\frac{R}{I_e}\right)} \tag{6.26}$$

- Constant displacement region where $T > T_L$ [ASCE/SEI Equation (12.8-5)]:

$$C_s = \frac{S_{D1}T_L}{T^2\left(\frac{R}{I_e}\right)} \tag{6.27}$$

In the constant acceleration region of the spectrum, V is independent of T, while in the constant velocity region, V is inversely proportional to T. The constant displacement region is generally valid for tall and flexible structures.

It is also permitted to determine C_s by method 2 where S_a is available. Method 2 must be used in cases where values of the multi-period, 5 percent-damped MCE_R response spectrum are not available from Reference 6.1 (exception 2 in ASCE/SEI 11.4.5).

Regardless of which method is used, C_s must not be taken less than the value determined by ASCE/SEI Equation (12.8-6):

$$C_s = 0.044 S_{DS} I_e \ge 0.01 \tag{6.28}$$

This equation provides a minimum base shear as a function of S_{DS} and I_e. In no case is V permitted to be less than 1 percent of the effective seismic weight, W.

For structures located where $S_1 \ge 0.6$, which applies to sites near active faults, an additional lower bound for C_s is determined by ASCE/SEI Equation (12.8-7):

$$C_s = \frac{0.5S_1}{\left(\frac{R}{I_e}\right)} \tag{6.29}$$

The design response spectrum in accordance with method 2 of the ELF procedure is depicted in Figure 6.29.

Figure 6.29 Design Response Spectrum in Accordance with Method 2 in ASCE/SEI 12.8.1.1

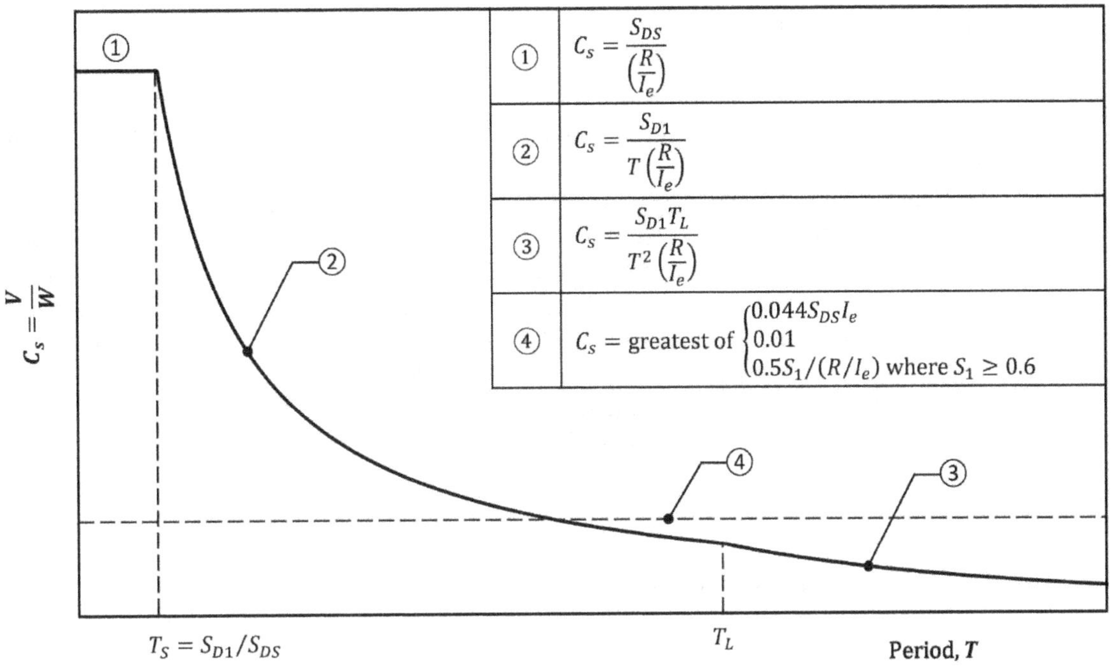

①	$C_s = \dfrac{S_{DS}}{\left(\dfrac{R}{I_e}\right)}$
②	$C_s = \dfrac{S_{D1}}{T\left(\dfrac{R}{I_e}\right)}$
③	$C_s = \dfrac{S_{D1}T_L}{T^2\left(\dfrac{R}{I_e}\right)}$
④	$C_s = \text{greatest of} \begin{cases} 0.044S_{DS}I_e \\ 0.01 \\ 0.5S_1/(R/I_e) \text{ where } S_1 \geq 0.6 \end{cases}$

$C_s = \dfrac{V}{W}$

$T_S = S_{D1}/S_{DS}$ T_L **Period, *T***

Soil-Structure Interaction Reduction

The dynamic response of a structure can be significantly affected by soil-structure interaction. A reduction due to the consideration of soil-structure interaction is permitted provided the analysis conforms to the requirements in ASCE/SEI Chapter 19 or to other generally accepted procedures approved by the building official (ASCE/SEI 12.8.1.2).

Maximum S_{DS} Value in the Determination of C_s

Values of C_s and the vertical seismic load effect, E_v, which is defined in ASCE/SEI 12.4.2.2, are permitted to be calculated using a value of S_{DS} equal to 1.0, but not less than 70 percent of the value of S_{DS} determined by the design response spectrum provisions in ASCE/SEI 11.4.5, provided all of the criteria in ASCE/SEI 12.8.1.3 are satisfied:

1. The structure does not have any horizontal or vertical irregularities, as defined in ASCE/SEI 12.3.2.
2. The structure has less than or equal to five stories above the lower of the base or grade plane defined in ASCE/SEI 11.2. Where present, mezzanine levels must be considered a story for purposes of this limit.
3. The fundamental period of the structure, T, determined in accordance with ASCE/SEI 12.8.2, is less than or equal to 0.5 s.
4. The structure meets the requirements necessary for the redundancy factor, ρ, to be taken as 1.0.
5. The site soil properties are not classified as site class E or F.
6. The structure is classified as risk category I or II.
7. The response modification coefficient. R, is greater than or equal to 3.

This cap on the maximum value of S_{DS} reflects engineering judgment about the performance of code-complying, regular, low-rise buildings in past earthquakes.

Period Determination

The fundamental period of a structure, T, in the direction of analysis must be determined based on the structural properties and deformational characteristics of the SFRS using fundamental principles of a dynamic analysis (ASCE/SEI 12.8.2).

To prevent the use of an unusually low base shear in the design of a structure that is overly flexible because of mass and stiffness inaccuracies in the analytical model used to determine T, the fundamental period used to determine V must not exceed the approximate fundamental period, T_a, times the upper limit coefficient, C_u, given in ASCE/SEI Table 12.8-1 as a function of S_{D1}.

The upper limit coefficients, C_u, in ASCE/SEI Table 12.8-1 remove the conservatism of the lower-bound equations used to determine T_a. As is evident from the values of C_u in the table, the limiting periods are larger in regions of lower seismicity because structures in these areas usually have longer periods (i.e., are more flexible) than those in regions of higher seismicity.

In the preliminary design stage, some of the information needed to determine T from a dynamic analysis may not be known. The equations in ASCE/SEI 12.8.2.1 can be used to determine an approximate fundamental period, T_a, for a variety of structure types (see Table 6.16). These equations provide a lower-bound value for the approximate period, which, in turn, provides an upper-bound value of V.

Table 6.16 Approximate Period, T_a

Structure Type		T_a
Moment-resisting frame systems in which the frames resist 100 percent of the required seismic forces and are not enclosed or adjoined by components that are more rigid and will prevent the frames from deflecting when subjected to seismic forces	Steel	$0.028h_n^{0.8}$ (h_n in ft) $0.0724h_n^{0.8}$ (h_n in m)
	Concrete	$0.016h_n^{0.9}$ (h_n in ft) $0.0466h_n^{0.9}$ (h_n in m)
Steel eccentrically braced frame in accordance with ASCE/SEI Table 12.2-1 lines B1 or D1		$0.030h_n^{0.75}$ (h_n in ft) $0.0731h_n^{0.75}$ (h_n in m)
Steel buckling-restrained braced frames		$0.030h_n^{0.75}$ (h_n in ft) $0.0731h_n^{0.75}$ (h_n in m)
Structures less than or equal to 12 stories above the base defined in ASCE/SEI 11.2 where the SFRS consists entirely of concrete or steel moment-resisting frames where the average story height is greater than or equal to 10 ft (3.05 m)		$0.1\,N$
Masonry or concrete shear wall structures not exceeding 120 ft (36.6 m) in height		$\dfrac{C_q h_n}{\sqrt{C_w}}$ where $C_q = 0.0019$ ft (0.00058 m) $C_w = \dfrac{100}{A_B}\sum_{i=1}^{x}\dfrac{A_i}{\left[1+0.83\left(\dfrac{h_n}{D_i}\right)^2\right]}$
All other structural systems		$0.020h_n^{0.75}$ (h_n in ft) $0.0488h_n^{0.75}$ (h_n in m)

The terms in Table 6.16 are as follows:

h_n = vertical distance from the base to the highest level of the SFRS of a structure (for structures with pitched or sloped roofs, h_n is from the base to the average height of the roof)

N = number of stories above the base of a structure

A_B = area of base of structure, ft^2 (m^2)

A_i = web area of shear wall i, ft^2 (m^2)

D_i = length of shear wall i, ft (m)

x = number of shear walls in a building effective in resisting lateral forces in the direction of analysis

A summary of the periods to be used when calculating the seismic forces on a structure is as follows:

- Where the calculated $T \le T_a$: use $T = T_a$
- Where $T_a <$ calculated $T < C_u T_a$: use $T =$ calculated T
- Where the calculated $T \ge C_u T_a$: use $T = C_u T_a$

Vertical Distribution of Seismic Forces

Once the seismic base shear, V, has been determined, it is distributed over the height of a structure in accordance with ASCE/SEI Equations (12.8-12) and (12.8-13):

$$F_x = C_{vx} V \tag{6.30}$$

$$C_{vx} = w_x h_x^k / \sum_{i=1}^{n} w_i h_i^k \tag{6.31}$$

The terms in these equations are as follows:

- F_x is the lateral seismic force induced at level x of the building (see Figure 6.30)
- w_i and w_x are the portions of W located or assigned to level i or x

Figure 6.30 Vertical Distribution of Seismic Forces in Accordance with ASCE/SEI 12.8.3

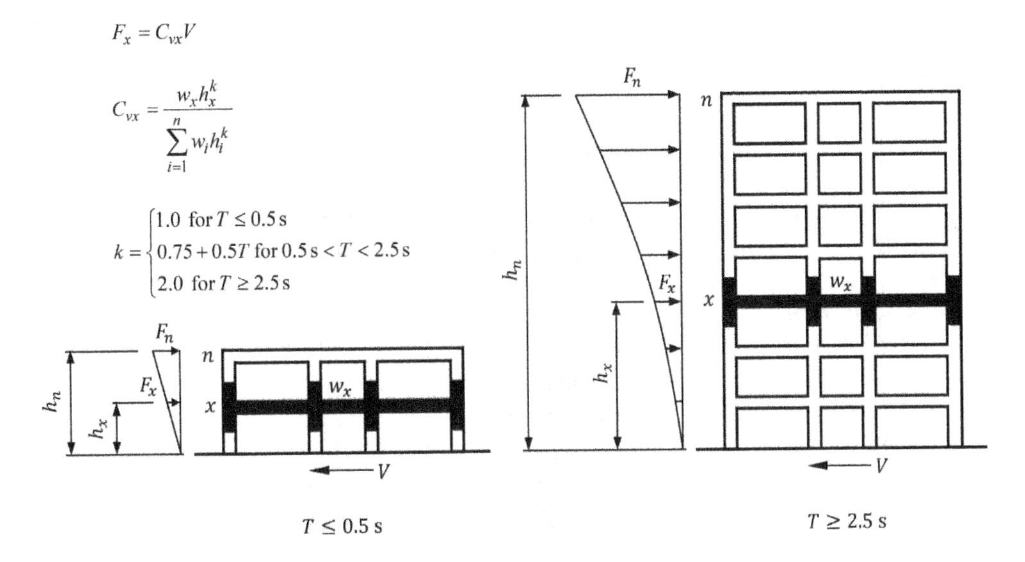

- h_i and h_x are the heights from the base of the structure to level i or x
- k is the exponent related to the structure period: $k = \begin{cases} 1.0 \text{ for } T \leq 0.5 \text{ s} \\ 0.75 + 0.5T \text{ for } 0.5 \text{ s} < T < 2.5 \text{ s} \\ 2.0 \text{ for } T \geq 2.5 \text{ s} \end{cases}$

The exponent related to the structure period, k, is intended to approximate the effects of higher modes, which are usually more dominant in structures with a longer fundamental period of vibration (i.e., in structures that are more flexible). It is permitted to take $k = 2.0$ for structures with a period between 0.5 and 2.5 s instead of using the equation given above for that period range.

For structures with a fundamental period less than or equal to 0.5 s, V is distributed linearly over the height, varying from zero at the base to a maximum value at the top (see Figure 6.30). Where T is greater than 2.5 s, the distribution is parabolic. As noted above, for a period between these two values, a linear interpolation between a linear and parabolic distribution is permitted or a parabolic distribution may be utilized.

The lateral forces, F_x, determined by Equations (6.30) and (6.31) are not the inertial forces that occur in the structure at any particular time during an actual earthquake. Rather, they provide design story shears consistent with enveloped results obtained from more refined analyses.

Horizontal Distribution of Forces

Overview

The design seismic story shear, V_x, in story x is the sum of the lateral force, F_i, induced at the floor level supported by that story plus the sum of the lateral forces of all the floor levels above, $\sum F_i$, including the roof [see ASCE/SEI Equation (12.8-14)]. The design story shear is distributed to the vertical elements of the SFRS based on the in-plane stiffness of the diaphragm.

Flexible Diaphragms

For flexible diaphragms, it is assumed the horizontal seismic forces are distributed to the vertical elements of the SFRS in proportion to the mass tributary to the vertical elements. For diaphragms of uniform thickness and material, horizontal force allocation to the vertical elements of the SFRS is in proportion to the areas tributary to the vertical elements. Flexible diaphragms are not capable of distributing horizontal seismic forces in proportion to the rigidities (or, stiffnesses) of the vertical elements of the SFRS. Also, torsion generated by the application of the horizontal seismic forces need not be considered in flexible diaphragms.

Consider the flexible diaphragm in Figure 6.31. The horizontal seismic force, V_y, is transferred through the web of the diaphragm to the shear walls, which act as supports for the diaphragm. Walls 1 and 2 each resist 50 percent of the seismic force. A uniform shear distribution in the diaphragm occurs along the depth of the diaphragm adjacent to the walls and is equal to $(V_y/2)/L$. Collector elements are required at wall 1 because the wall does not extend the full depth of the diaphragm.

The chords at the top and bottom edges of the diaphragm are perpendicular to the direction of the seismic force and resist the tension and compression forces induced in the diaphragm due to bending, that is, $T_u = C_u = M_{max}/L$ where M_{max} is the maximum bending moment in the diaphragm due to the horizontal design seismic force applied to the diaphragm.

Rigid Diaphragms

For rigid diaphragms, horizontal forces are distributed to the vertical elements of the SFRS in proportion to their relative rigidities and their locations with respect to the center of rigidity (CR). Thus, the location of the CR must be determined on a floor/roof level prior to horizontal force allocation. By definition, the CR is the point on a floor/roof level where the equivalent story stiffness is assumed to be located. It is often referred to as the stiffness centroid. For buildings with rigid diaphragms, application of a lateral force through the CR produces only rigid body

Figure 6.31 Horizontal Distribution of Forces in a Flexible Diaphragm

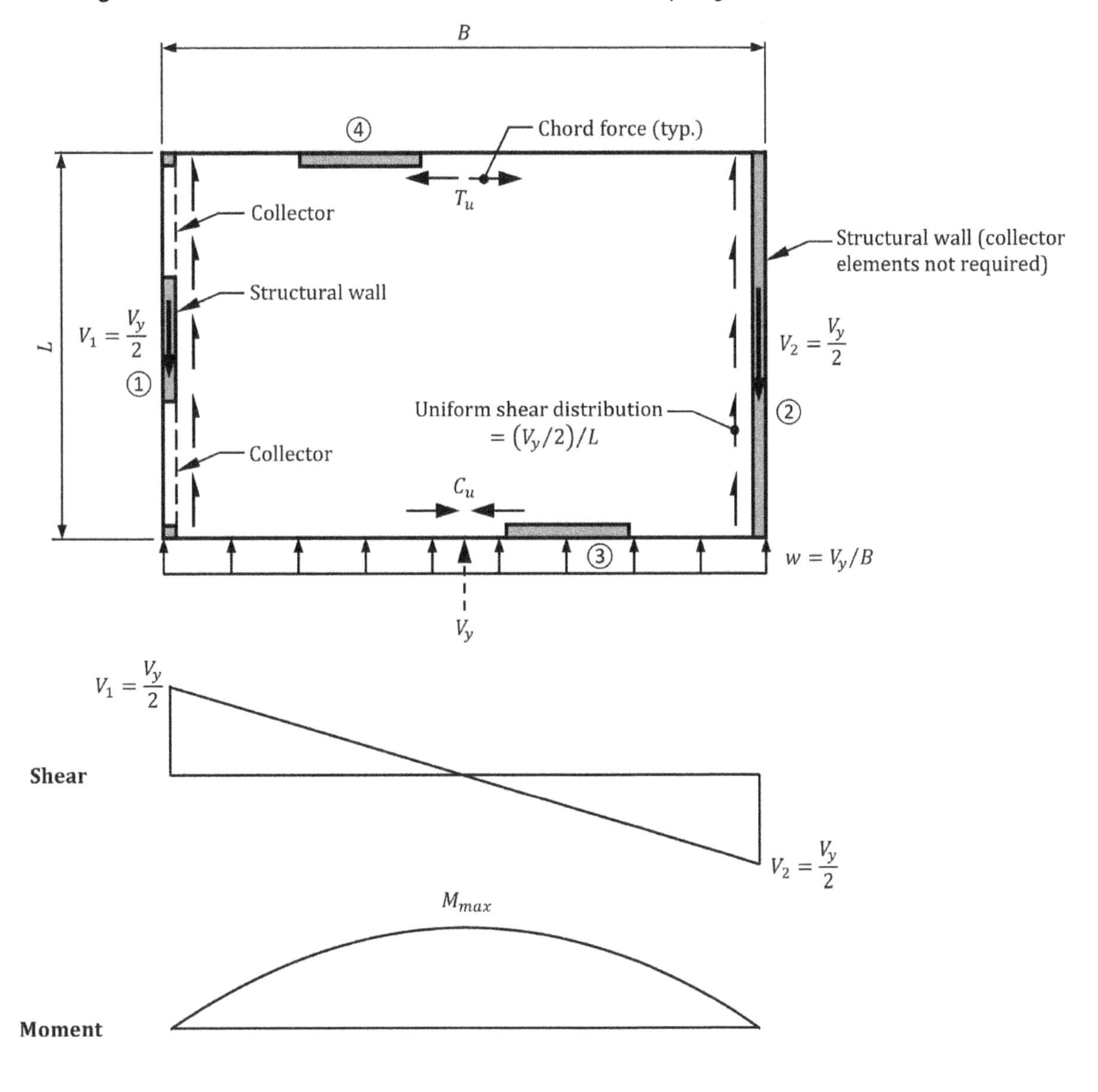

displacement of the story. Displacement and rotation occur where the lateral force is applied at any point other than the CR. Depending on the structural layout, the CR can be at different locations on different levels in a building.

The location of the CR in the x-direction (x_{cr}) and in the y-direction (y_{cr}) can be determined using the equations in Figure 6.32. The terms $(k_i)_y$ and $(k_i)_x$ are the in-plane stiffnesses of lateral force-resisting elements i in the y- and x-directions, respectively, and x_i and y_i are the distances in the x- and y-directions from an arbitrarily selected origin to the centroids of the lateral force-resisting elements i. All elements of the SFRS parallel to the direction of analysis are included in the equations to determine the location of the CR; it is commonly assumed that out-of-plane resistance of a lateral force-resisting element is negligible compared to its in-plane resistance. Thus, when determining the location of the CR in the x-direction, only the stiffnesses of walls 1 and 2 are considered. A portion of wall 3 (i.e., an effective flange width) may be included when determining the stiffness of wall 1. Similarly, only the stiffnesses of wall 3 (including an effective flange width of wall 1, if desired) and wall 4 are considered when determining the location of the CR in the y-direction.

Figure 6.32 Determination of the Location of the CR

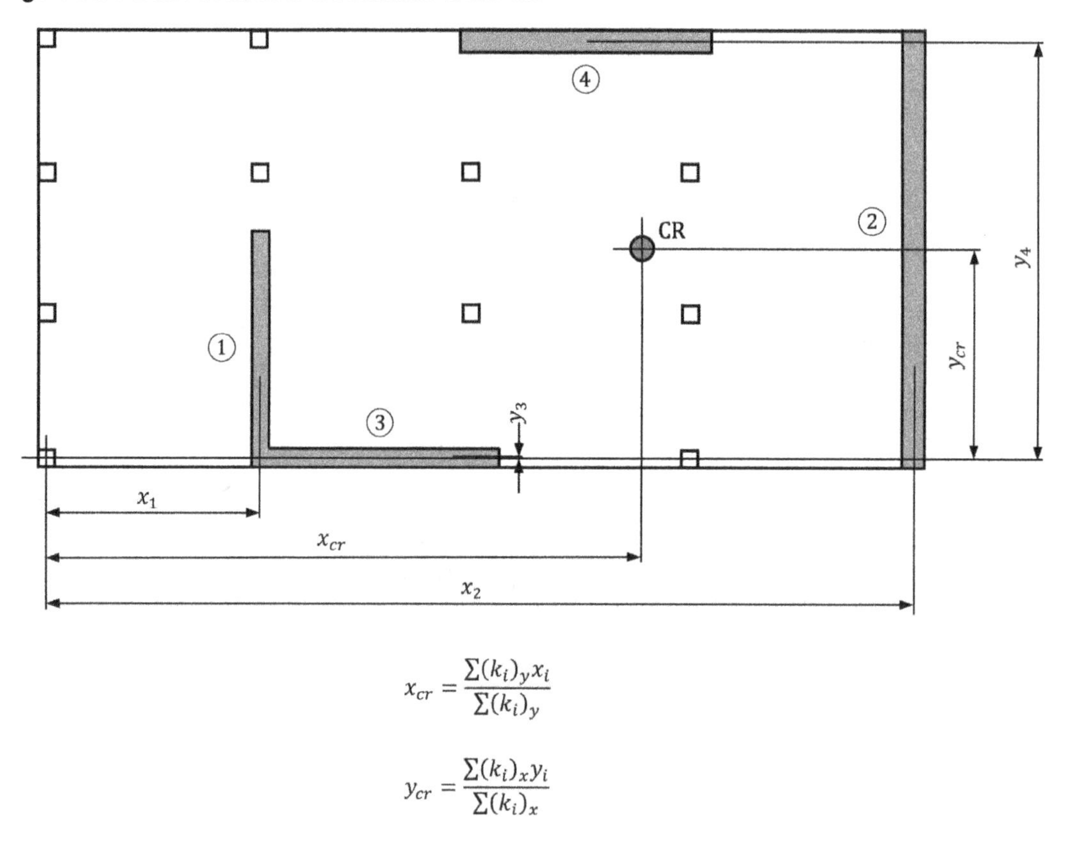

$$x_{cr} = \frac{\Sigma(k_i)_y x_i}{\Sigma(k_i)_y}$$

$$y_{cr} = \frac{\Sigma(k_i)_x y_i}{\Sigma(k_i)_x}$$

In-plane stiffness, k_i, can be obtained by any rational method. Approximate in-plane stiffnesses of solid walls, rigid frames, and braced frames are given in Figure 6.33.

The center of mass (CM) is the location on a level of a building where the mass of the entire story is assumed to be concentrated. The seismic force for the story is assumed to act at this location.

The equations in Figure 6.34 can be used to determine the portion of the total story shear, V_y, in the y-direction that is resisted by element i of the SFRS considering both direct and torsional shear forces, which is designated $(V_i)_y$. In these equations, $\bar{x}_i$ = perpendicular distance from the centroid of element i to the CR parallel to the x-axis and $\bar{y}_i$ = perpendicular distance from the centroid of element i to the CR parallel to the y-axis.

Term 1 is the portion of the total story shear resisted by element i based on relative stiffness. Only the elements of the MWFRS parallel to the y-direction share this force (see Figure 6.34, which indicates term 1 forces for walls 1 and 2 in the direction resisting the applied seismic force, V_y). Term 2 is the shear force resisted by element i that is generated by the inherent torsional moment, $M_t = V_y e_x$ (ASCE/SEI 12.8.4.1). The elements of the MWFRS in both directions share the shear forces generated by this moment (see Figure 6.34, which indicates term 2 forces for all four walls in the directions resisting the applied clockwise torsional moment). The forces in walls 3 and 4 are calculated using term 2 with the numerator equal to $\bar{y}_i(k_i)_x$ instead of $\bar{x}_i(k_i)_y$. The denominator of this term is the torsional stiffness of all the walls and is analogous to the polar moment of inertia of a section. Term 2 is equal to zero in this equation when V_y acts through the CR; in such cases, the floor translates as a rigid body and the elements of the MWFRS all displace an equal amount horizontally.

Term 2 forces may be either positive or negative depending on the location of the element with respect to the CR. For example, for seismic forces in the y-direction, the term 2 force for wall 2 acts

Figure 6.33 Approximate In-Plane Stiffnesses of Solid Walls, Rigid Frames, and Braced Frames

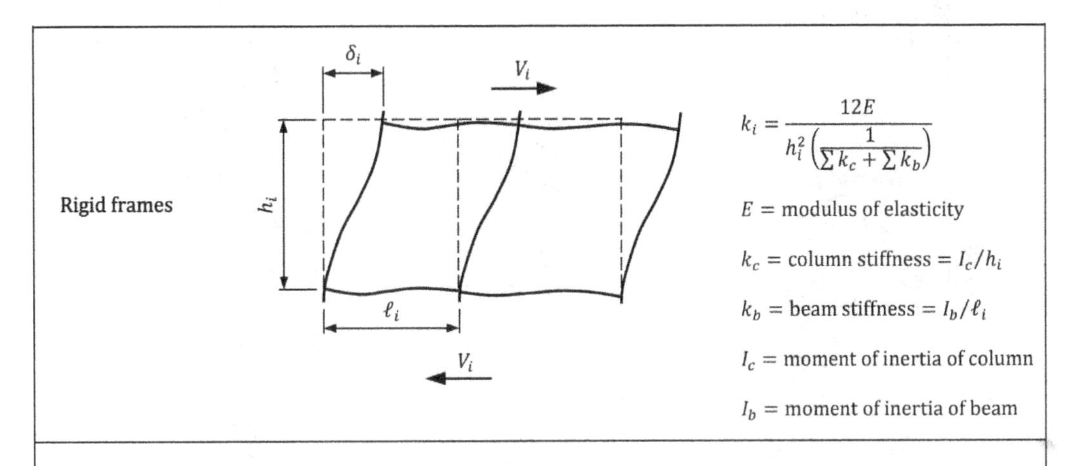

δ_i
V_i
Rigid frames
h_i
ℓ_i
V_i
$k_i = \dfrac{12E}{h_i^2\left(\dfrac{1}{\sum k_c + \sum k_b}\right)}$
E = modulus of elasticity
k_c = column stiffness = I_c/h_i
k_b = beam stiffness = I_b/ℓ_i
I_c = moment of inertia of column
I_b = moment of inertia of beam

Solid walls or piers

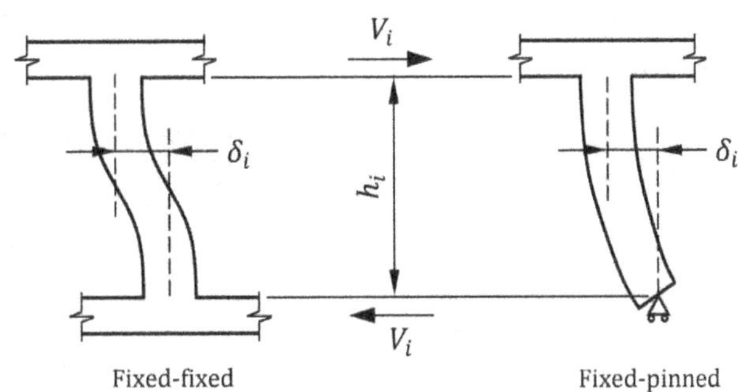

$$k_i = \frac{k_{Fi}k_{Vi}}{k_{Fi} + k_{Vi}}$$

E = modulus of elasticity

G = shear modulus

k_{Fi} = flexural stiffness = V_i/δ_{Fi}

k_{Vi} = shear stiffness = V_i/δ_{Vi}

I_i = moment of inertia of wall or pier

A_i = area of wall or pier

Support condition	Flexural deflection, δ_{Fi}	Shear deflection, δ_{Vi}
Fixed-fixed	$\dfrac{V_i h_i^3}{12EI_i}$	$\dfrac{1.2V_i h_i}{GA_i}$
Fixed-pinned	$\dfrac{V_i h_i^3}{3EI_i}$	$\dfrac{1.2V_i h_i}{GA_i}$

Figure 6.33 *(Continued)*

Braced Bents		
Single diagonal	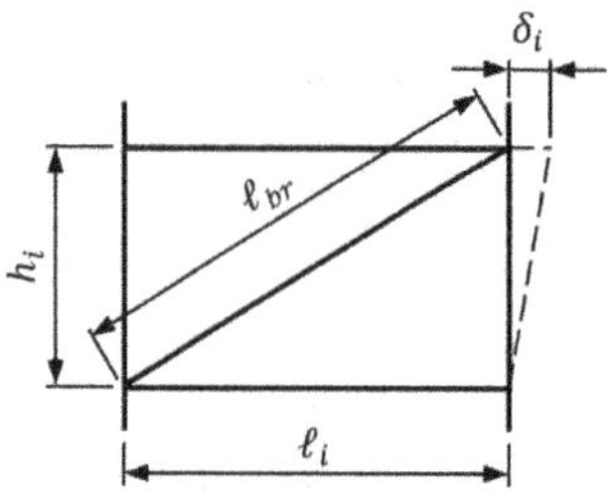	$k_i = \dfrac{E}{\left(\dfrac{\ell_{br}^3}{A_{br}\ell_i^2} + \dfrac{\ell_i}{A_b}\right)}$ E = modulus of elasticity A_{br} = area of brace A_b = area of upper beam
Double diagonal	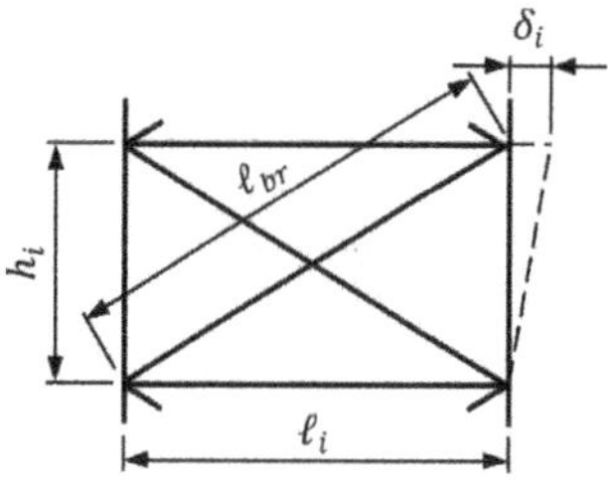	$k_i = \dfrac{2E}{\left(\dfrac{\ell_{br}^3}{A_{br}\ell_i^2}\right)}$ E = modulus of elasticity A_{br} = area of brace
K-brace	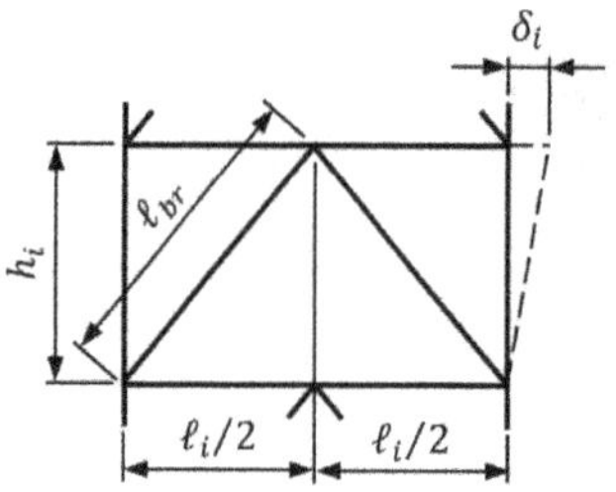	$k_i = \dfrac{E}{\left(\dfrac{2\ell_{br}^3}{A_{br}\ell_i^2} + \dfrac{\ell_i}{4A_b}\right)}$ E = modulus of elasticity A_{br} = area of brace A_b = area of upper beam
Knee brace	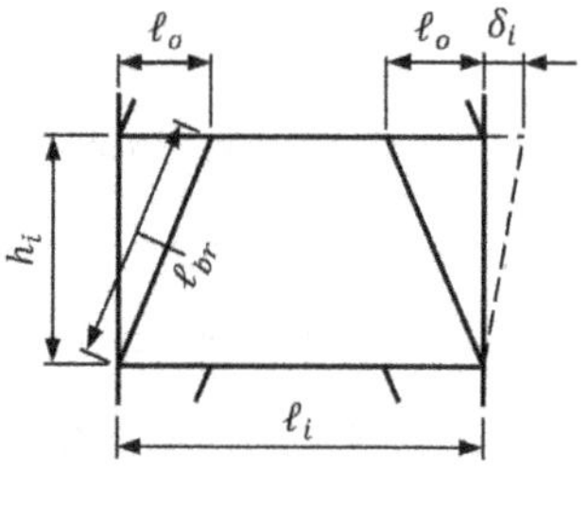	$k_i = \dfrac{E}{\left[\dfrac{\ell_{br}^3}{2A_{br}\ell_o^2} + \dfrac{\ell_o}{2A_b} + \dfrac{h_i^2(\ell_i - 2\ell_o)^2}{12I_b\ell_i}\right]}$ E = modulus of elasticity A_{br} = area of brace A_b = area of upper beam I_b = moment of inertia of upper beam
Offset diagonal	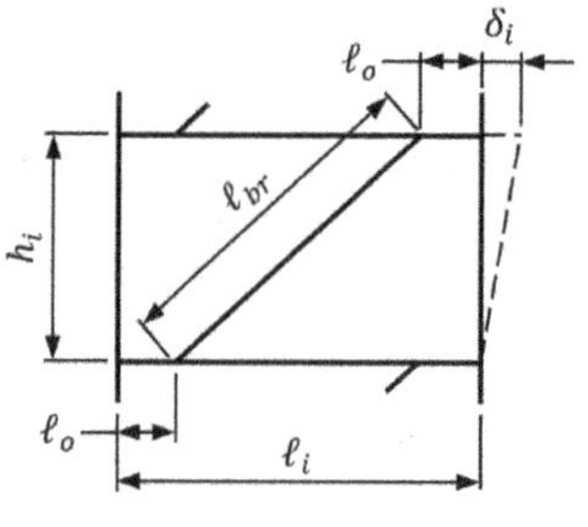	$k_i = \dfrac{E}{\left[\dfrac{\ell_{br}^3}{A_{br}(\ell_i - 2\ell_o)^2} + \dfrac{\ell_i - 2\ell_o}{A_b} + \dfrac{h_i^2\ell_o^2}{3I_b\ell_i}\right]}$ E = modulus of elasticity A_{br} = area of brace A_b = area of upper beam I_b = moment of inertia of upper beam

Figure 6.34 Horizontal Distribution of Forces in a Rigid Diaphragm

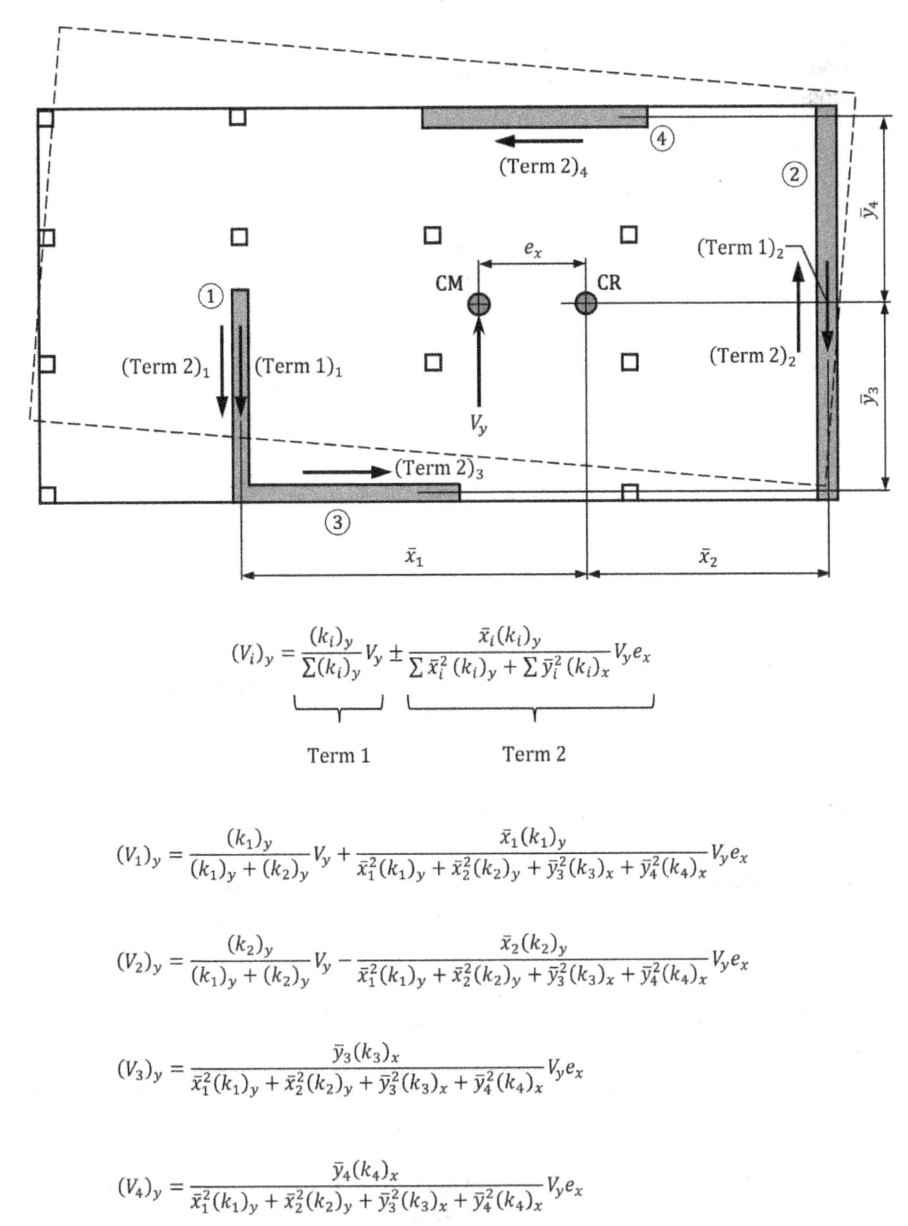

in the opposite direction of the term 1 force (see Figure 6.34). To properly capture the maximum force effects on an element, seismic forces acting in the opposite direction must also be considered.

Equations for horizontal force distribution in the x-direction for seismic story shear V_x are similar to those in Figure 6.34.

The accidental torsional moment, M_{ta}, must be included for diaphragms that are not flexible when considering whether a horizontal irregularity in ASCE/SEI Table 12.3-1 exists or not. Accidental torsional moments are determined based on the assumption that the CM is displaced each way from its actual location by a distance equal to 5 percent of the dimension of the structure perpendicular to the direction of analysis.

A floor or roof diaphragm that is not flexible is illustrated in Figure 6.35. The seismic story shear, V_y, acts through the CM and the inherent torsional moment, M_t, is equal to $V_y e_x$ where e_x is

Figure 6.35 Inherent and Accidental Torsional Moments

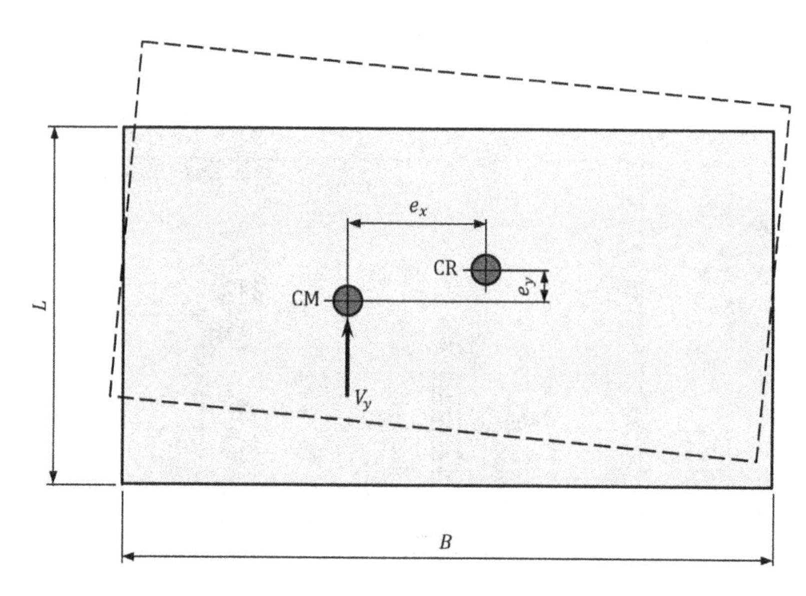

Inherent torsional moment: $M_t = V_y e_x$

Accidental torsional moment: $M_{ta} = V_y(0.05B)^*$

Total torsional moment: $M_t + M_{ta}$

$^*M_{ta}$ to be included only when required by ASCE/SEI 12.8.4.2

the eccentricity between the CM and the CR perpendicular to the direction of analysis. The accidental torsional moment, M_{ta}, is equal to $V_y(0.05B)$, and the total torsional moment for purposes of analysis is equal to $M_t + M_{ta}$. Similar calculations can be performed for the seismic story shear, V_x, in the perpendicular direction using the eccentricity e_y.

In cases where forces are applied concurrently in orthogonal directions, the required 5 percent displacement of the CM need not be applied in both of the orthogonal directions simultaneously; rather, it must be applied in the direction producing the greatest effects for each element considered (see Section 6.3.7 of this book).

Accidental torsional moments need not be included in the analysis and design of structures, except in the following cases (ASCE/SEI 12.8.4.2.1):

- Structures assigned to SDC B with a type 1 horizontal structural irregularity and a torsional irregularity ratio, TIR, greater than 1.4.
- Structures assigned to SDC C, D, E, or F with a type 1 horizontal structural irregularity.

Accidental torsional moments must be amplified in accordance with ASCE/SEI 12.8.4.3 in structures assigned to SDC C, D, E, or F where a type 1 torsional irregularity exists. At each level, M_{ta} must be multiplied by the torsional amplification factor, A_x, which is determined by ASCE/SEI Equation (12.8-15):

$$1.0 \le A_x = \left(\frac{\delta_{\text{max}}}{1.2\delta_{\text{avg}}}\right)^2 \le 3.0 \tag{6.32}$$

In this equation, δ_{max} is the maximum displacement at level x determined assuming $A_x = 1.0$ and δ_{avg} is the average of the displacements at the extreme points of the structure at level x assuming $A_x = 1.0$ (see Figure 6.36). Setting $A_x = 1.0$ when determining the displacements at this stage eliminates the need for iterations in the calculation of A_x.

Figure 6.36 Torsional Amplification Factor, A_x

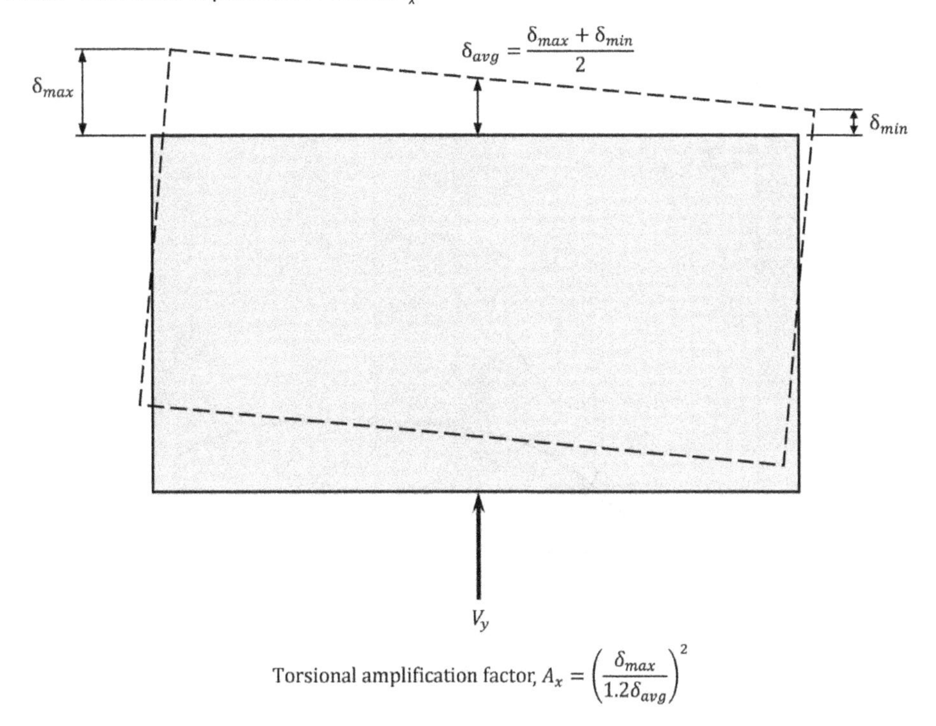

Overturning

The structure must be designed to resist the overturning effects caused by the seismic forces (ASCE/SEI 12.8.5).

The overturning effect on a vertical lateral force-resisting element is determined based on the overturning moment at the level of interest, which is equal to the lateral seismic force, F_x, times the height from the base of the structure to the level of the horizontal lateral force-resisting element that transfers forces to the vertical element, summed over each story. Each vertical lateral force-resisting element resists its portion of overturning based on its relative stiffness with respect to all vertical lateral force-resisting elements in the structure.

Overturning forces are resisted by dead loads and can be combined with dead and live loads or other loads, in accordance with the load combinations in ASCE/SEI 2.3.7. Critical load combinations may occur for some structural members where the effects from gravity and seismic loads counteract.

Displacement and Drift Determination

Overview

The following displacements and drifts are defined in ASCE/SEI 12.8.6, which are used in various applications:

- Design earthquake displacement, δ_{DE}
- Maximum considered earthquake displacement, δ_{MCE}
- Design story drift, Δ

The force-displacement relationships between elastic response, response to design-level forces, and expected inelastic response of a structure are given in Figure 6.37. These relationships are used to compute drift (see below). Structures designed to remain elastic during an earthquake would be subjected to a force equal to V_E and a corresponding deflection equal to δ_E. It has been shown that the maximum displacement of an inelastic system is approximately equal to that of

Figure 6.37 Displacements Used to Calculate Drift

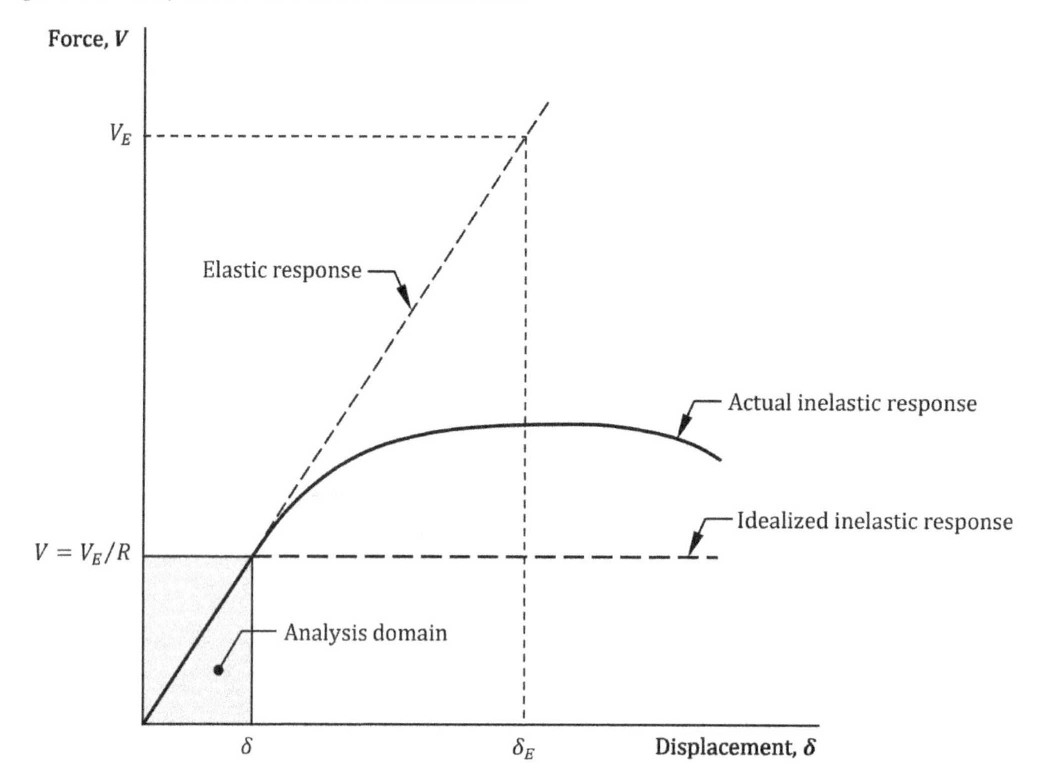

an elastic system with the same initial stiffness (this is commonly referred to as the equal displacement approximation rule of seismic response).

Because seismic forces are reduced by the response modification coefficient, R, which is given in ASCE/SEI Table 12.2-1 based on the SFRS, the corresponding displacements are representative of an elastic system and must be amplified to account for inelastic response when calculating drifts. The deflection amplification factor, C_d, which is given in ASCE/SEI Table 12.2-1 based on the SFRS, amplifies the displacements determined from an elastic analysis using code-prescribed forces; the resulting displacements represent the expected inelastic displacements for the design-level earthquake. It is evident from ASCE/SEI Table 12.2-1 that C_d is typically less than R. Note that C_d is a story-level amplification factor, which means it does not represent displacement amplification of the elastic response of a structure.

More information on these displacements and drift is given in the following sections, along with corresponding provisions for minimum base shear, load combinations, and structure period.

Design Earthquake Displacement

Design earthquake displacement, δ_{DE}, is determined by ASCE/SEI Equation (12.8-16):

$$\delta_{DE} = \frac{C_d \delta_e}{I_e} + \delta_{di} \tag{6.33}$$

where C_d = deflection amplification factor in ASCE/SEI 12.2-1 based on the SFRS

δ_e = elastic displacement due to the design seismic forces, including the effects of accidental torsion and torsional amplification, where required

I_e = importance factor determined in accordance with ASCE/SEI 11.5.1 based on the risk category of the structure

δ_{di} = displacement due to diaphragm deformation corresponding to the design earthquake, including the diaphragm forces in ASCE/SEI 12.10

This displacement, which is calculated at the location of the element or component being evaluated, is utilized in the following provisions:

- Structural separation (ASCE/SEI 12.12.2)
- Deformation compatibility (ASCE/SEI 12.12.4)
- Nonstructural components (ASCE/SEI 13.3.2)

Elastic displacement, δ_e, includes the overall displacement and rotation of the diaphragm (as part of the SFRS; see ASCE/SEI Figure C12.8-1b), and δ_{di} is the displacement due to deformation of the diaphragm, which may be elastic or inelastic under the expected demands. Force-displacement characteristics of different types of diaphragms are given in ASCE/SEI C12.10.3.5, which can be used as a guide in determining δ_{di}. Diaphragm inelastic response models are given in ASCE Figure C12.10-8 for diaphragm systems not expected to exhibit a distinct yield point (such as a wood or steel deck diaphragm) and diaphragm systems exhibiting a distinct yield point (such as a cast-in-place reinforced concrete diaphragm).

The importance factor, I_e, is included because the allowable story drifts in ASCE/SEI 12.12.1 are more stringent for structures assigned to higher risk categories.

Maximum Considered Earthquake Displacement

Maximum considered earthquake displacement, δ_{MCE}, at the location of an element or component is determined by ASCE/SEI Equation (12.8-17):

$$\delta_{\text{MCE}} = 1.5\left(\frac{R\delta_e}{I_e} + \delta_{di}\right) \tag{6.34}$$

This displacement is similar to δ_{DE} with two differences:

- A factor of 1.5 is used, which corrects for the two-third factor used in the calculation of the seismic base shear, V (the two-third factor reduces the base shear from the value based on the MCE_R ground motion; see ASCE/SEI 11.4.4 and Section 6.2.1 of this book).
- Elastic displacements, δ_e, are multiplied by the response modification coefficient, R, instead of the deflection amplification factor, C_d ; this accounts for cases where values of C_d less than R may substantially underestimate displacements for many SFRSs.

Instead of using ASCE/SEI Equation (12.8-17), δ_{MCE} is permitted to be determined in accordance with applicable provisions in ASCE/SEI Chapter 16, 17, or 18.

Maximum considered earthquake displacement is used for members spanning between structures (ASCE/SEI 12.12.3).

Design Story Drift

Design story drift, Δ, is the interstory drift at each story corresponding to the design earthquake, and is equal to the difference in design earthquake displacements, δ_{DE}, at the centers of mass at the top and bottom of the story under consideration for structures without torsional irregularity (see Figure 6.38 for a two-dimensional representation of story drifts in a frame structure with a rigid diaphragm where the displacement at the CM is essentially the same as at the diaphragm edges). Where the centers of mass do not align vertically by more than 5 percent of the width of the diaphragm extents, it is permitted to calculate the deflection for the bottom of the story at the point on the floor that is vertically aligned with the location of the CM at the top of the story (see the three-dimensional view in ASCE/SEI Figure C12.8-1b). This situation can arise, for example, where a building has story offsets and the diaphragm extents at the top of the story are smaller than those at the bottom of the story. The diaphragm deflection, δ_{di}, in ASCE/SEI Equation (12.8-16) is permitted to be neglected when determining Δ (ASCE/SEI 12.8.6.5).

Figure 6.38 Design Story Drift Determination

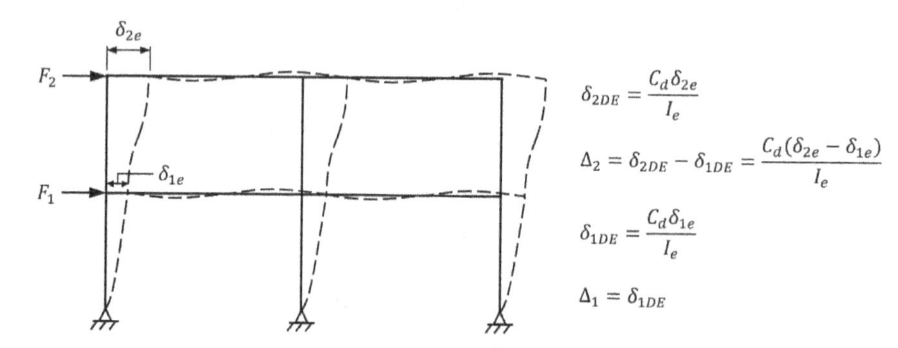

For structures assigned to SDC C, D, E, or F with type 1 (torsional) horizontal irregularities, diaphragm rotation must also be considered in the determination of Δ. In such cases, Δ is determined using the largest difference of the δ_{DE} of vertically aligned points at the top and bottom of the story along any edges of the diaphragm extents.

A summary of the aforementioned displacements and drifts is given in ASCE/SEI Table C12.7-1.

Minimum Base Shear and Load Combinations for Computing Displacement and Drift

When conducting a drift analysis, the applicable load combinations in ASCE/SEI 2.3.1 and 2.3.6 must be used based on an elastic analysis of the structure with the prescribed seismic design forces determined in accordance with ASCE/SEI 12.8 considering the following modifications (ASCE/SEI 12.8.6.1):

- A load factor of 1.0 must be used on the horizontal seismic load effect, E_h, in the presence of expected gravity loads.
- Minimum expected gravity loads must be taken as greater than or equal to $1.0D + 0.5L$, with $L = 0.8L_o$ for live loads greater than 100 lb/ft^2 (4.79 kN/m^2) and $L = 0.4L_o$ for all other live loads where L_o is the unreduced design live load in ASCE/SEI Table 4.3-1.

According to the exception in ASCE/SEI 12.8.6, the minimum seismic base shear determined by ASCE/SEI Equation (12.8-6) need not be considered when determining drifts because this equation represents a minimum strength that needs to be provided to the system and is not related to drift.

Period for Computing Displacement and Drift

When conducting a drift analysis, it is permitted to determine the elastic displacements, δ_e, using design seismic forces based on the computed fundamental period, T, of the structure without the upper limit C_uT_a, which is specified in ASCE/SEI 12.8.2 (ASCE/SEI 12.8.6.2).

In cases where the design response spectrum in ASCE/SEI 11.4.6 or the corresponding equations in ASCE/SEI 12.8.1 are used and the fundamental period of the structure, T, is less than the long-period transition period, T_L, displacements increase with increasing period, even though forces may decrease. The upper limit of C_uT_a on T is prescribed so that design forces are not underestimated. However, if the lateral forces used to determine drifts are inconsistent with the forces corresponding to T, then displacements can be overestimated. To account for this, it is permitted to determine displacements using forces consistent with the computed fundamental period of the structure, without the upper limit prescribed in ASCE/SEI 12.8.2.

P-Delta Effects

As a structure deflects horizontally due to seismic forces, the vertical loads are displaced from their original position and additional effects are introduced into the structural members (i.e.,

P-delta effects), which cause the structure to displace further (see ASCE/SEI Figure C12.8-7). If sufficient strength and stiffness are not provided, the deflections can continue to increase, leading to overall instability of the structure.

Member forces and story drifts induced by P-delta effects must be considered in member design where such effects are significant (ASCE/SEI 12.8.7). In lieu of a more refined analysis, P-delta effects need not be considered in a story where the stability coefficient, θ, determined by ASCE/SEI Equation (12.8-18) is less than or equal to 0.10:

$$\theta = \frac{P_x / h_{sx}}{V_x / \Delta_{xe}} \tag{6.35}$$

where P_x = the total vertical design load at and above level x
h_{sx} = story height below level x
V_x = seismic force acting between levels x and $x-1$ determined in accordance with ASCE/SEI 12.8 or 12.9
Δ_{xe} = elastic story drift due to the seismic design shear, V_x, calculated at the locations specified in ASCE/SEI 12.8.6 using δ_e (not δ_{DE})

Load combination 6 results in the greatest vertical design load including earthquake effects (and, thus, has the greatest potential for inclusion of P-delta effects in the analysis), and that load must be used in ASCE/SEI Equation (12.8-18):

$$P_x = (1.2 + 0.2S_{DS})D + \rho Q_E + L + 0.15S$$

where ρ is permitted to be taken equal to 1.0 when evaluating P-delta effects (ASCE/SEI 12.3.4.1, item 2).

According to ASCE/SEI 12.8.7, it is permitted to take the load factor on D equal to 1.0; this results in the dead load effect being fairly close to best estimates of the arbitrary point-in-time value for dead load. Also, the load factor on L (which is the reduced live load), is permitted to be taken as 0.5 for certain occupancies in accordance with exception 1 in ASCE/SEI 2.3.6. Finally, the vertical seismic component, $E_v = 0.2S_{DS}D$, need not be considered. Therefore, the following equation must be used to determine P_x:

$$P_x = D + \rho Q_E + (1.0 \text{ or } 0.5)L + 0.15S \tag{6.36}$$

Where P-delta effects must be considered (i.e., $0.10 < \theta \leq \theta_{max}$), the displacement and member forces must be increased accordingly using a rational analysis. Alternatively, it is permitted to multiply displacements and member forces by $1.0/(1-\theta)$. Rational analysis methods are presented in ASCE/SEI C12.8.7.

A structure is considered to be potentially unstable when the following equation is satisfied at any story in the structure [see ASCE/SEI Equation (12.8-19)]:

$$\theta > \theta_{max} = \frac{0.5}{\beta C_d} \leq 0.25 \tag{6.37}$$

The term β is the ratio of the shear demand to the shear capacity of the story under consideration, which is permitted to be conservatively taken as 1.0 and must not be taken less than $1.25/\Omega_0$. The value of θ_{max} determined by ASCE/SEI Equation (12.8-19) need not be taken less than 0.1.

ASCE/SEI Equation (12.8-19) must be satisfied even where an automated analysis is used to determine the P-delta effects. However, the value of θ determined by ASCE/SEI Equation (12.8-18) using the results of the P-delta analysis is permitted to be divided by $(1+\theta)$ prior to checking ASCE/SEI Equation (12.8-19).

Flowchart 6.5 in Section 6.6 of this book can be used to determine seismic base shear and its vertical distribution on a SFRS of a structure in accordance with the ELF procedure in ASCE/SEI 12.8.

6.3.9 Linear Dynamic Analysis

Modal Response Spectrum Analysis

In the modal response spectrum analysis, a structure is broken down into a number of single-degree-of-freedom systems. Each system has its own mode shape and natural period of vibration.

For seismic design, the number of modes corresponds to the number of mass degrees of freedom of the structure. Even for a simple low-rise building, the number of mass degrees of freedom is immense. Fortunately, some simple idealizations can be made to significantly reduce the number of degrees of freedom while still preserving a meaningful solution. For example, in the case of planar structures with rigid diaphragms, the number of mass degrees of freedom can be taken as one per story (horizontal translation in the direction of analysis) with the mass of the story concentrated in the floor and roof systems.

The three modes of vibration for a three-story building are illustrated in Figure 6.39. For three-dimensional structures with rigid diaphragms, three mass degrees of freedom can be utilized per story (two horizontal translations and rotation about the vertical axis). The number of modes that is required for design is discussed below.

Figure 6.39 Modes of Vibration for an Idealized Planar Structure

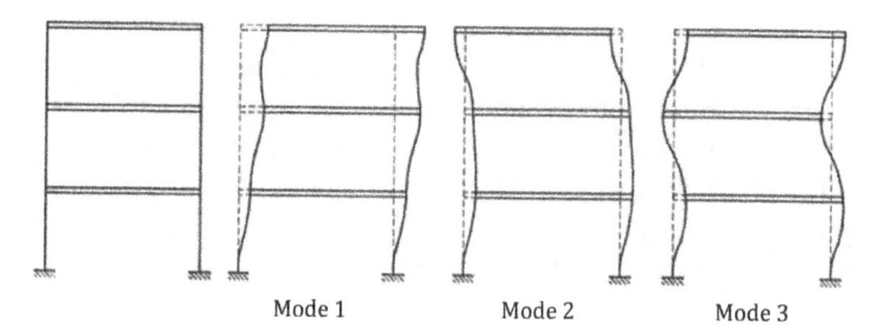

Details on how to determine modal periods, mode shapes, and modal participation are given in numerous standard references on dynamic analysis and are not covered here (see e.g., Reference 6.8).

A modal response spectrum analysis is permitted for any structure (see ASCE/SEI 12.6). The requirements in ASCE/SEI 12.9.1 must be satisfied for this method of analysis (see Table 6.17).

Linear Response History Analysis

Requirements for the linear response history analysis method are given in ASCE/SEI 12.9.2. In this method, a three-dimensional model of a structure is constructed in accordance with ASCE/SEI 12.9.2.2, and the model is subjected to suites of spectrally matched acceleration histories compatible with the design response spectrum for the site. The system response is obtained by simultaneous solution of the full set of equations of motion. Additional information on the proper application of this method is given in ASCE/SEI C12.9.2.

6.3.10 Diaphragms, Chords, and Collectors

Overview

A diaphragm is a horizontal or sloped system that transmits lateral forces to the vertical elements of the SFRS. Horizontal seismic forces are transferred to the elements of the SFRS based on the flexibility of the diaphragm (see Section 6.3.8 of this book).

Table 6.17 Requirements for the Modal Response Spectrum Analysis

Item	Requirements
Number of modes	• Analysis must include a sufficient number of modes to obtain a combined modal mass participation of 100% of the structure's mass. It is permitted to represent all modes with periods less than 0.05 s in a single rigid body mode that has a period of 0.05 s. • Alternatively, at least 90% of the actual mass in each orthogonal horizontal direction must participate in the response to obtain a distribution of forces and displacements sufficient for design.
Modal response parameters	• The forces and deflections are determined for each mode using the properties of each mode and the response spectra defined in ASCE/SEI 11.4.5 or the site-specific response spectrum defined ASCE/SEI 21.2 (see Section 6.2.1 of this book). Regardless of the spectrum used, the spectral ordinates must be divided by (R/I_e). • Displacement and drift quantities must be multiplied by (C_d/I_e).
Combined response parameters	The results from the analysis of each mode must be combined using one of the following methods: • Square root of the sum of the squares (SRSS) • Complete quadratic combination (CQC) • Complete quadratic combination as modified by ASCE 4 (CQC-4) [Reference 6.9] • Approved equivalent approach The CQC or CQC-4 methods must be used where closely spaced modes have significant cross-correlation of translational and torsional response.
Scaling design values of combined response	• Scaling of forces Where the combined response for the modal base shear, V_t, determined in accordance with ASCE/SEI 12.9.1.4 is less than 100% of the calculated base shear, V, by the ELF procedure, the forces must be scaled by the factor V/V_t. Where the calculated fundamental period, T, exceeds C_uT_a in a given direction, C_uT_a must be used instead of T in the analysis in that direction. • Scaling of drifts Drifts must be scaled by the factor (C_sW/V_t) when $V_t < C_sW$ where C_s is determined by ASCE/SEI Equation (12.8-7).
Horizontal shear distribution	• Distribution of the horizontal forces to the SFRSs must be in accordance with ASCE/SEI 12.8.4 (see Section 6.3.8 of this book). • Accidental torsion must be accounted for by applying a static accidental torsional moment, M_{ta}, determined in accordance with ASCE/SEI 12.8.4.2 and 12.8.4.3, to the mathematical model and combining the results with the scaled design values determined in accordance with ASCE/SEI 12.9.1.4. • The effects of accidental torsion may be included in the dynamic analysis model in lieu of applying M_{ta} for structures without a type 1 (torsional) horizontal irregularity. • Amplification of torsion in accordance with ASCE/SEI 12.8.4.3 is not required where accidental torsion effects are included in the dynamic analysis model and TIR $\leq$1.6. • Where TIR >1.6, accidental torsion must be added as a static load case in accordance with ASCE/SEI 12.8.4.3.
P-delta effects	P-delta effects in accordance with ASCE/SEI 12.8.7 must be determined. The base shear used to determine displacements and drifts must be determined in accordance with ASCE/SEI 12.8.6.
Soil-structure interaction reduction	Reduction in forces associated with soil structure interaction is permitted provided the requirements of ASCE/SEI Chapter 19 are used. Any generally accepted procedures approved by the building official are also permitted.
Structural modeling	• A three-dimensional mathematical model of the structure is required, which must be constructed in accordance with ASCE/SEI 12.7.3. • In structures without rigid diaphragms, the model must include stiffness characteristics of the diaphragm and any additional dynamic degrees of freedom needed to accurately account for the participation of the diaphragm in the dynamic response of the structure.

Diaphragms also transfer gravity loads to floor or roof members (beams, joists, and columns). When properly attached to the top surface of horizontal framing members, diaphragms increase the flexural lateral stability of such members.

Diaphragms, chords, and collectors must be designed in accordance with ASCE/SEI 12.10.1 and 12.10.2. Precast concrete diaphragms, including chords and collectors, in structures assigned to SDC C, D, E, or F must be designed in accordance with ASCE/SEI 12.10.3 [see exception (a) in ASCE/SEI 12.10].

The requirements in ASCE/SEI 12.10.3 are permitted to be used for the design of precast concrete diaphragms in structures assigned to SDC B, cast-in-place concrete diaphragms, cast-in-place concrete equivalent precast diaphragms, wood-sheathed diaphragms supported by wood diaphragm framing, bare steel deck diaphragms, and concrete-filled steel deck diaphragms.

Diaphragms, chords, and collectors in one-story structures conforming to the limitations in ASCE/SEI 12.10.4.1 are permitted to be designed in accordance with ASCE/SEI 12.10.4.

Diaphragm Design

Diaphragms must be designed for the shear and bending stresses caused by the seismic design forces (ASCE/SEI 12.10.1). The resulting shear and tension forces must be less than or equal to the corresponding design strengths of the diaphragm, including at discontinuities, such as openings and reentrant corners (see ASCE/SEI Figures C12.10-1 and C12.10-2, respectively).

For structures assigned to SDC B and higher, floor and roof diaphragms must be designed to resist the effects from the larger of the following forces (ASCE/SEI 12.10.1.1):

- The lateral seismic force, F_x, determined from the vertical distribution of the seismic base shear, V (see Figure 6.30).
- The diaphragm design force, F_{px}, determined by ASCE/SEI Equation (12.10-1), including the minimum and maximum limits in ASCE/SEI Equations (12.10-2) and (12.10-3), respectively:

$$F_{px} = \frac{\sum_{i=x}^{n} F_i}{\sum_{i=x}^{n} w_i} w_{px}$$

$$\geq 0.2 S_{DS} I_e w_{px}$$

$$\leq 0.4 S_{DS} I_e w_{px}$$

In the equation for F_{px}, w_i is the portion of the effective seismic weight, W, located at or assigned to level i and w_{px} is the weight tributary to the diaphragm at level x. For structures with walls, the weights of the walls parallel to the direction of analysis are often not included in w_{px} because these weights do not contribute to the diaphragm shear forces. However, including the weights of these walls in w_{px} is conservative, which means $w_{px} = w_x$ at level x.

The diaphragm forces are inertial forces and are applied at the CM at each level. The CM can occur at different locations on different levels, and this needs to be accounted for in the analysis.

The redundancy factor, ρ, is permitted to be set equal to 1.0 in the determination of F_{px} by ASCE/SEI Equations (12.10-1) through (12.10-3) [item 7 in ASCE/SEI 12.3.4.1].

In addition to the above inertial forces, diaphragms must also be designed for transfer forces from the vertical elements of the SFRS. For structures with a type 4 horizontal structural irregularity (out-of-plane offset irregularity), the transfer forces from the vertical seismic force-resisting elements above the diaphragm to the horizontally offset vertical seismic force-resisting elements below the diaphragm must be increased by the overstrength factor, Ω_0 (see Figure 6.40). The modified transfer force is added to the diaphragm inertial force, F_{px}, originating on the diaphragm below.

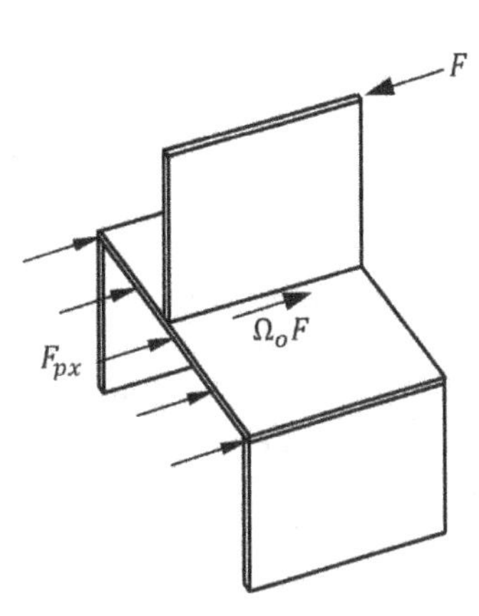

Figure 6.40 Seismic Transfer Force due to Out-of-Plane Offset in the SFRS

A similar discontinuity is illustrated in Figure 6.41 where reinforced concrete shear walls that are part of the SFRS above transfer forces through the reinforced concrete podium slab (diaphragm) to the reinforced concrete basement walls below. In addition to the inertial force of the podium slab, the diaphragm force includes the static soil pressure and dynamic seismic pressure, all of which get transferred to the basement walls parallel to the direction of analysis.

Figure 6.41 Transfer Forces to a Podium Slab

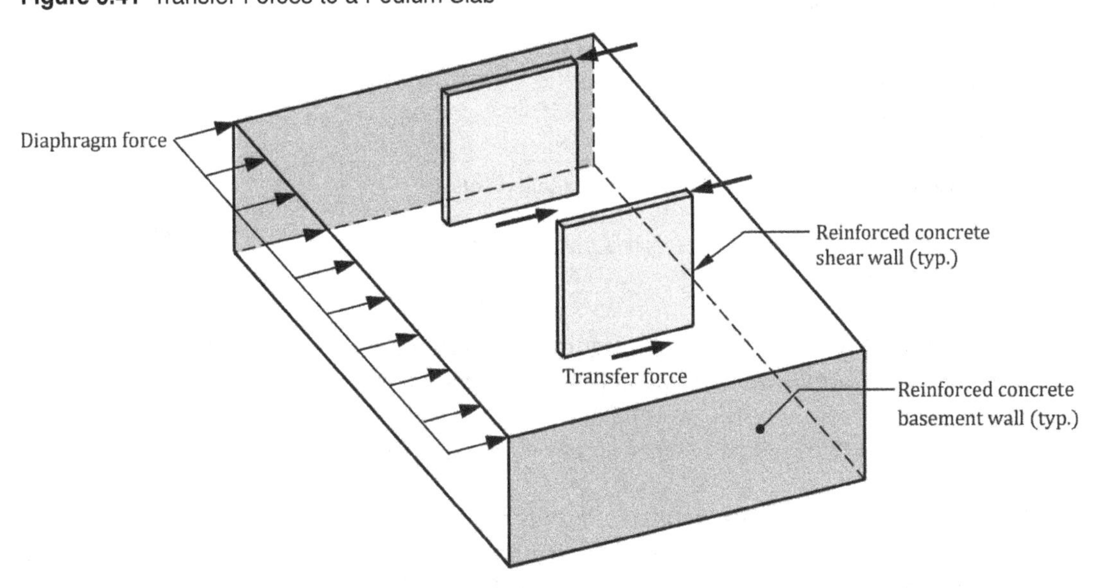

According to the exception in ASCE/SEI 12.10.1.1, it is permitted to use $\Omega_0 = 1.0$ in the determination of transfer forces for one- and two-family dwellings of light-frame construction.

Where transfer forces must be considered, the redundancy factor, ρ, for the structure above must be applied to the transfer force resisted by the diaphragm below, except where the transfer force must be increased by Ω_0.

For structures assigned to SDC D, E, or F that possess a type 1, 2, 3, or 4 horizontal structural irregularity or a type 3 vertical structural irregularity, the diaphragm forces determined by ASCE/SEI 12.10.1.1 must be increased by 25 percent at each diaphragm level where the irregularity occurs for the following elements of the SFRS (ASCE/SEI 12.3.3.5):

- Connections of diaphragms to the vertical elements of the SFRS and to collectors.
- Collectors and their connections, including connections to vertical elements of the SFRS.

According to the exception in this section, these forces need not be increased by 25 percent where the forces must be increased by Ω_0.

Collector Elements

Collectors must be designed for the combined effects due to gravity loads (flexure and shear) and lateral loads (axial tension and compression forces caused by in-plane shear transfer in the diaphragm due to seismic forces, which are determined in accordance with ASCE/SEI 12.10.1). Once the forces are determined, the diaphragm is modeled and analyzed using the methods in ASCE/SEI 12.3 (see Section 6.3.3 of this book), and the axial forces in a collector can be determined based on the selected method of analysis.

Collectors and their connections to the vertical elements of the SFRS in buildings assigned to SDC C, D, E, or F must be designed to resist the effects from the maximum of the three forces given in ASCE/SEI 12.10.2.1 (see Table 6.18). Loads are combined using load combinations 6 and 7 in ASCE/SEI 2.3.6.

Table 6.18 Collector Design Forces for Structures Assigned to SDC C, D, E, or F

Force Requirement	Load Combinations
Forces are calculated using the seismic load effects including overstrength of ASCE/SEI 12.4.3 with seismic forces determined by the ELF procedure of ASCE/SEI 12.8 or the modal response spectrum analysis procedure of ASCE/SEI 12.9.1.	In load combinations 6 and 7, use $E_{mh} = \Omega_o Q_E$ where Q_E is determined using F_x.
Forces are calculated using the seismic load effects including overstrength of ASCE/SEI 12.4.3 with seismic forces determined by ASCE/SEI Equation (12.10-1).	In load combinations 6 and 7, use $E_{mh} = \Omega_o Q_E$ where Q_E is determined using F_{px}.
Forces are calculated using the load combinations of ASCE/SEI 2.3.6 with seismic forces determined by ASCE/SEI Equation (12.10-2), which is the lower-limit diaphragm force.	In load combinations 6 and 7, use $E_h = \rho Q_E$ where Q_E is determined using $F_{px} = 0.2S_{DS}I_e w_{px}$.

The purpose of the overstrength requirements is to help ensure inelastic behavior occurs in the ductile elements of the SFRS and not in the collectors or their connections, which must perform as intended during a seismic event.

Collectors are designed using whichever of these three methods produces the critical combined effects.

Where applicable, transfer forces must also be considered when determining forces in a collector (ASCE/SEI 12.10.2.1). As discussed above, seismic transfer forces must be increased by the overstrength factor, Ω_o, in structures with a type 4 horizontal structural irregularity.

The exception in ASCE/SEI 12.10.2.1 permits collector elements and their connections in structures or portions thereof braced entirely by wood light-frame shear walls to be designed using load combinations with seismic forces determined in accordance with ASCE/SEI 12.10.1.1.

Alternative Design Provisions for Diaphragms, Including Chords and Collectors

Overview

The requirements in ASCE/SEI 12.10.3 must be used in the design of precast concrete diaphragms, including chords and collectors, in structures assigned to SDC C, D, E, or F and are permitted to be used for the design of precast concrete diaphragms in structures assigned to SDC B, cast-in-place concrete diaphragms, cast-in-place concrete equivalent precast diaphragms, wood-sheathed diaphragms supported by wood diaphragm framing, bare steel deck diaphragms, and concrete-filled steel deck diaphragms. The following requirements are also applicable:

1. Footnote b in ASCE/SEI Table 12.2-1 pertaining to the permitted reduction of Ω_0 for structures with flexible diaphragms is not applicable.
2. ASCE/SEI 12.3.3.5 pertaining to the increase of forces caused by irregularities for SDC D through F is not applicable.
3. Item 5 in ASCE/SEI 12.3.4.1 pertaining to conditions where ρ is permitted to be taken as 1.0 is replaced with the following: "Design of diaphragms, including chords, collectors, and their connections to the vertical elements" are used.
4. Item 7 in ASCE/SEI 12.3.4.1 is not applicable.

Design

Diaphragms, including chords, collectors, and their connections to the vertical elements must be designed in two orthogonal directions to resist the in-plane design seismic force, F_{px}, determined in accordance with ASCE/SEI 12.10.3.2. Transfer of forces at diaphragm discontinuities, such as openings and reentrant corners, must also be considered in design.

Design Seismic Forces for Diaphragms, Including Chords and Collectors

The in-plane design seismic force, F_{px}, is determined by ASCE/SEI Equation (12.10-4) and must be taken greater than or equal to the minimum force in ASCE/SEI Equation (12.10-5):

$$F_{px} = \frac{C_{px}}{R_s} w_{px} \geq 0.2 S_{DS} I_e w_{px} \tag{6.38}$$

The design acceleration coefficient, C_{px}, is determined from ASCE/SEI Figure 12.10-2 and depends on the number of stories in the building, N, the height of the floor level of interest from the base, h_x, and the structural height of the building, h_n. Design acceleration coefficients C_{p0}, C_{pi}, and C_{pn} determined by ASCE/SEI Equations (12.10-6), (12.10-8), (12.10-9), and (12.10-7), respectively, are needed in the determination of C_{px}. The following parameters are also required:

1. The modal contribution factors Γ_{m1} and Γ_{m2} determined by ASCE/SEI Equations (12.10-13) and (12.10-14), respectively.
2. The higher mode seismic response coefficient, C_{s2}, determined by ASCE/SEI Equations (12.10-10), (12.10-11), (12.10-12a), and (12.10-12b).
3. The mode shape factor, z_s, which is based on the SFRS (ASCE/SEI 12.10.3.2).

The diaphragm design force reduction factor, R_s, is determined in accordance ASCE/SEI Table 12.10-1 and depends on the type of diaphragm system.

The provisions in ASCE/SEI 12.10.3 are based on experimental and analytical data and observations of building performance in past earthquakes, and consider both of the following:

- The determination and vertical distribution of diaphragm seismic forces based on the assumption of near-elastic diaphragm response (which is defined by the design acceleration coefficients C_{p0}, C_{pi}, and C_{pn} in ASCE/SEI Figure 12.10-2).
- The reduction of the near-elastic diaphragm forces based on the ductility and inelastic deformation capacity of the diaphragm system (which is captured in the diaphragm design force reduction factor, R_s).

This method improves the distribution of diaphragm strength over the height of a structure and among structures with different types of SFRSs.

In-depth information on the derivation of this method can be found in ASCE/SEI C12.10.3.

Transfer Forces

In addition to the inertia forces, F_{px}, determined by ASCE/SEI Equations (12.10-4) and (12.10-5), diaphragms must be designed for all applicable transfer forces (ASCE/SEI 12.10.3.3). Transfer forces from the vertical seismic force-resisting elements above the diaphragm to the vertical seismic force-resisting elements below the diaphragm must be increased by the overstrength factor, Ω_0, in structures with a type 4 horizontal structural irregularity (see Figure 6.40). In the case of one- and two-family dwellings of light-frame construction, it is permitted to use $\Omega_0 = 1.0$ (see the exception in ASCE/SEI 12.10.3.3).

Collectors—**SDCs C** *through* **F**

Collectors and their connections, including connections to vertical elements of the SFRS, in structures assigned to SDC C, D, E, or F must be designed for the effects due to 1.5 times F_{px} determined in accordance with ASCE/SEI 12.10.3.2 (ASCE/SEI 12.10.3.4). Where transfer forces are applicable, the effects from the transfer forces must also be multiplied by 1.5. A multiplier of 1.5 is used instead of Ω_0 to increase collector forces because the diaphragm forces are more accurately determined by ASCE/SEI Equation (12.10-4) than by the applicable diaphragm force determined in accordance with ASCE/SEI 12.10.2.1.

Three exceptions to these requirements are given in ASCE/SEI 12.10.3.4:

- Transfer forces required to be increased by Ω_0 need not be further increased by 1.5.
- In moment frame and braced frame SFRSs, collector forces need not exceed the lateral strength of the corresponding frame line below the collector, considering only the moment frames or braced frames. Also, diaphragm design forces need not exceed the forces corresponding to the collector forces so determined.
- In structures or portions thereof braced entirely by light-frame shear walls, collector elements and their connections need only be designed to resist the diaphragm seismic design forces without the 1.5 multiplier.

Diaphragm Design Force Reduction Factor

The diaphragm design force reduction factor, R_s, reduces the near-elastic diaphragm forces, $C_{px}w_x$, and accounts for diaphragm overstrength and/or the inelastic deformation capacity of the diaphragm.

This factor is typically greater than 1.0 for diaphragms with inelastic deformation capacity sufficient to permit inelastic response when subjected to the design earthquake. This means F_{px} will be less than the force demand for a diaphragm that remains linear-elastic under the design earthquake. For diaphragm systems that do not have sufficient inelastic deformation capacity, R_s is less than 1.0, which means the diaphragm will have a near-elastic response under the design earthquake.

Values of R_s for various diaphragm systems are given in ASCE/SEI Table 12.10-1. Some systems have values for both shear-controlled and flexure-controlled diaphragms, which are defined in ASCE/SEI 11.2 as follows:

- Flexure-controlled diaphragm: Diaphragm with a flexural yielding mechanism, which limits the maximum forces that develop in the diaphragm, and a design shear strength or factored nominal shear capacity greater than the shear corresponding to the nominal flexural strength.
- Shear-controlled diaphragm: Diaphragm that does not meet the requirements of a flexure-controlled diaphragm.

An example of a flexure-controlled diaphragm is a cast-in-place concrete diaphragm with steel reinforcement where the flexural yielding mechanism is yielding of the chord tension reinforcement along the edges of the diaphragm perpendicular to the direction of analysis.

Shear-controlled diaphragms fall into the following two main categories: (1) diaphragms that cannot develop a flexural mechanism because of aspect ratio, chord member strength, or other constraints and (2) diaphragms intended to yield in shear instead of in flexure. A typical wood-sheathed diaphragm is an example of diaphragm that falls into the second category.

The value of R_s to be used for a precast diaphragm designed and detailed in accordance with Reference 6.10 depends on the design option selected: elastic, basic, or reduced. These design options are defined in Reference 6.11, and the performance targets are given in ASCE/SEI Table C12.10-1.

In-depth information on the derivation of R_s for the various diaphragm systems is given in ASCE/SEI C12.10.3.5.

Flowchart 6.6 in Section 6.6 of this book can be used to determine the design seismic diaphragm force, F_{px}, in accordance with ASCE/SEI 12.10.3.

Alternative Diaphragm Design Provisions for One-Story Structures with Flexible Diaphragms and Rigid Vertical Elements

Overview

The provisions in ASCE/SEI 12.10.4 are permitted to be used to determine diaphragm design forces, including design forces for chords, collectors, and their in-plane connections to the vertical elements of the SFRS, for one-story rigid-wall, flexible diaphragm (RWFD) structures, subject to the limitations in ASCE/SEI 12.10.4.1 (see the next section).

These provisions are applicable to wood structural panel and bare steel deck diaphragms with concrete shear walls, precast concrete shear walls, masonry shear walls, steel concentrically braced frames, steel and concrete composite braced frames, or steel and concrete composite shear walls. It is assumed any combination of the aforementioned diaphragm and vertical systems is an RWFD structure.

The requirements in ASCE/SEI 12.10.4 are based on the results of numerical studies in Reference 6.5 and recent research on steel deck diaphragms. The dynamic response of RWFD structures is dominated by the dynamic response of and inelastic behavior in the diaphragm. Unlike the methods in ASCE/SEI 12.10.1 and 12.10.3, design seismic forces for the diaphragm in this method are based on the period of the diaphragm instead of on the period of the SFRS. The studies indicated improved seismic performance can be obtained for these types of structures, especially where the diaphragms span more than 100 ft (30.5 m).

Limitations

Diaphragms in one-story structures are permitted to be designed in accordance with ASCE/SEI 12.10.4 provided the following limitations are satisfied, which are based on the studies in Reference 6.5 (ASCE/SEI 12.10.4.1):

1. All portions of the diaphragm must be designed using the provisions in ASCE/SEI 12.10.4 in both orthogonal directions.

   Performance of a diaphragm using different design provisions in different directions has not been studied, so the resulting seismic performance is not known.

2. The diaphragm must consist of either of the following: (a) a wood structural panel diaphragm designed in accordance with AWC SDPWS and fastened to wood framing members or wood nailers with sheathing nailing in accordance with AWC SDPWS Section 4.2 nominal shear capacity tables or (b) a bare (untopped) steel deck diaphragm meeting the requirements of AISI S400 and AISI S310.

   The diaphragms listed here are consistent with those in the study in Reference 6.5.

3. Toppings of concrete or similar materials that can affect diaphragm strength or stiffness are not permitted.

4. The diaphragm must not contain any of the horizontal structural irregularities in ASCE/SEI Table 12.3-1, except for type 2, which is permitted.

   Diaphragms with any of the horizontal structural irregularities other than type 2 (reentrant corner irregularity) are prohibited because they fall outside the scope of the study. Type 2 irregularities are permitted because it is judged this type of irregularity will not influence building performance in a manner that would make use of the provisions inappropriate.

5. The diaphragm must be rectangular in shape and must be divisible into rectangular segments for purpose of seismic design, with vertical elements of the SFRS or collectors provided at each end of each rectangular segment span.

   Diaphragm segments for transverse seismic forces with each segment supported on all sides by concrete or masonry shear walls are illustrated in ASCE/SEI Figure C12.10.4-1(a). ASCE/SEI Figure C12.10.4-1(b) illustrates the same concept with two building plans in which, for transverse design, the diaphragm segments span to a combination of walls and collectors. This limitation also prohibits application of this design method to nonrectangular diaphragm segments such as triangular, trapezoidal or curved configurations.

6. The vertical elements of the SFRS must be one or more of the following: (a) concrete shear walls, (b) precast concrete shear walls, (c) masonry shear walls, (d) steel concentrically braced frames, (e) steel and concrete composite braced frames, or (f) steel and concrete composite shear walls.

   This list functions as the definition of rigid vertical elements for purposes of ASCE/SEI 12.10.4. The provisions do not distinguish between ordinary, intermediate, or special vertical elements; all are considered sufficiently rigid and are permitted.

7. The vertical elements of the SFRS must be designed in accordance with the ELF procedure in ASCE/SEI 12.8 and are permitted to be designed using the two-stage analysis procedure in ASCE/SEI 12.2.3.4, where applicable.

   This limitation serves as a reminder that the provisions of ASCE/SEI 12.10.4 addresses only the design of the diaphragm; they are not intended or permitted for the design of the vertical elements of the SFRS.

Design

Diaphragms, including chords, collectors, and their connections to vertical elements of the SFRS must be designed in two orthogonal directions to resist the seismic design force, F_{px}, determined by ASCE/SEI Equation (12.10-15):

$$F_{px} = C_{s\text{-diaph}}(w_{px}) \tag{6.39}$$

Seismic weight tributary to the diaphragm, w_{px}, is the total seismic weight at level x minus the weight of the walls resisting the seismic forces in the direction of analysis.

The seismic response coefficient, $C_{s\text{-diaph}}$, is determined by ASCE/SEI Equations (12.10-16a) and (12.10-16b):

$$C_{s\text{-diaph}} = \frac{S_{DS}}{\dfrac{R_{\text{diaph}}}{I_e}} \le \frac{S_{D1}}{T_{\text{diaph}}\left(\dfrac{R_{\text{diaph}}}{I_e}\right)} \tag{6.40}$$

The response modification coefficient, R_{diaph}, is equal to the following based on the type of diaphragm:

$$R_{\text{diaph}} = \begin{cases} 4.5 \text{ for wood structural panel diaphragms} \\ 4.5 \text{ for bare steel deck diaphragms meeting the special requirements in AISI S400} \\ 1.5 \text{ for all other bare steel deck diaphragms} \end{cases} \tag{6.41}$$

The period of the diaphragm, T_{diaph}, is equal to the following based on the type of diaphragm:

$$T_{\text{diaph}} = \begin{cases} 0.002L_{\text{diaph}} & \text{for wood structural panel diaphragms} \\ 0.001L_{\text{diaph}} & \text{for profiled steel deck panel diaphragms} \end{cases} \tag{6.42}$$

where L_{diaph} is the span of the horizontal diaphragm or diaphragm segment being considered, measured between vertical elements of the SFRS or collectors that provide support to the diaphragm or diaphragm segment in each orthogonal direction.

Diaphragm periods derived from analytical models or other sources are not permitted; the applicable T_{diaph} in Equation (6.42) must be used in determining F_{px}.

Equation (6.40) forms the basis of the response acceleration spectrum for these systems. Where T_{diaph} is greater than $T_S = S_{D1}/S_{DS}$, $C_{s\text{-diaph}}$ is determined from the velocity-controlled potion of the spectrum. This method is unlike the methods that depend on the approximate period and response modification factor of the vertical elements of the SFRS. No lower bound on $C_{s\text{-diaph}}$ is set because the numerical studies investigated values of $C_{s\text{-diaph}}$ below ASCE/SEI Equation (12.10-2) and found acceptable performance.

Diaphragm design shear forces are computed for each diaphragm segment using F_{px} determined by ASCE/SEI Equation (12.10-15). Where the diaphragm segment span, L_{diaph}, is less than 100 ft (30.5 m), the design shear force in the diaphragm must be determined using $1.5F_{px}$. The shear is amplified by 1.5 over the entire diaphragm span. Where L_{diaph} is greater than or equal to 100 ft (30.5 m), the design shear force in the diaphragm must be determined using $1.5F_{px}$ across an amplified shear boundary with a width equal to $0.1L_{\text{diaph}}$ at each supporting end of the diaphragm segment span under consideration. The diaphragm shear requirements of ASCE/SEI 12.10.4.2.2 are illustrated in Figure 6.42 for a one-story building where the roof diaphragm is divided into two diaphragm segments.

Figure 6.42 Diaphragm Shear Requirements in Accordance with ASCE/SEI 12.10.4.2.2

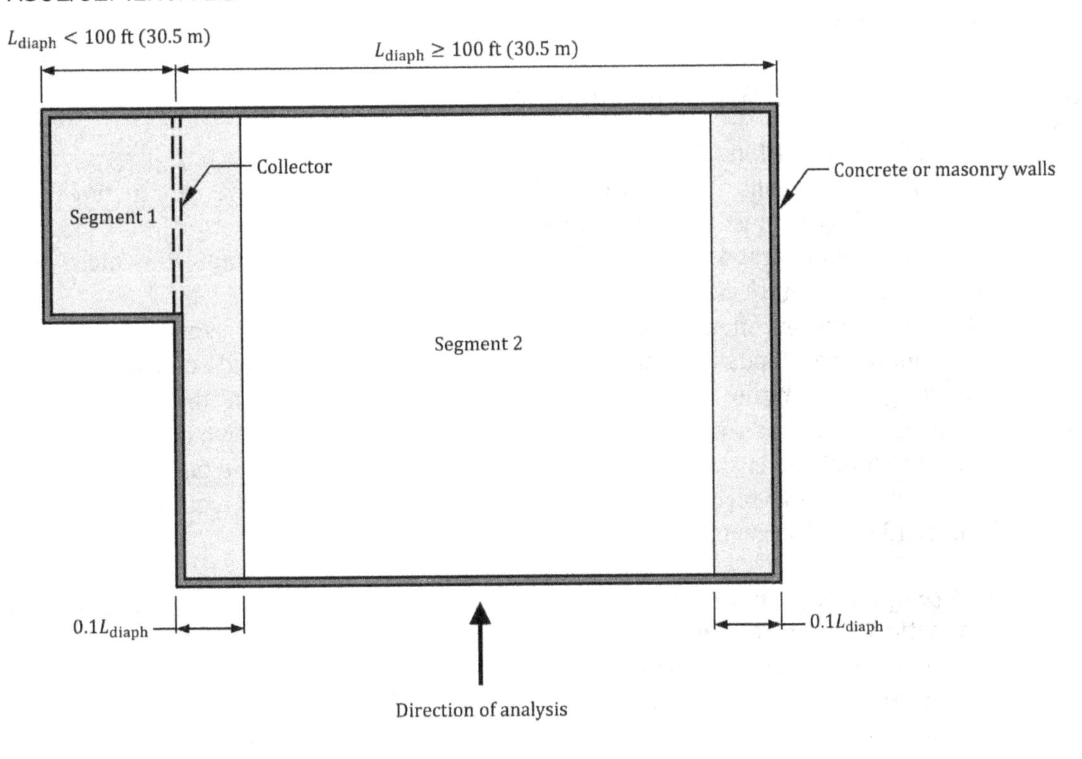

Diaphragm chords must be provided at each end of each diaphragm segment to resist the maximum tension and compression chord forces caused by the maximum bending moment in the diaphragm. For the roof diaphragm in Figure 6.31, chord tension and compression forces, T and C can be determined by the following equation:

$$T = C = \frac{M_{\max}}{L} = \frac{V_y B}{8L} \tag{6.43}$$

In this equation, $V_y = F_{px}$, B is the span of the diaphragm and L is the depth of the diaphragm in the direction of analysis.

Collectors and their connections to the vertical elements of the SFRS must be designed based on F_{px}. In structures assigned to SDC C through F, the load combinations including overstrength in ASCE/SEI 12.4.3 must be used where the diaphragm overstrength factor, $\Omega_{0-\text{diaph}} = 2 \leq R_{\text{diaph}}$. In such cases, the shear amplification of 1.5 in ASCE/SEI 12.10.4.2.2 need not be applied.

The deflection amplification factor, $C_{d\text{-diaph}}$, is determined as follows:

$$C_{d\text{-diaph}} = \begin{cases} 3.0 \text{ for wood structural panel diaphragms} \\ 3.0 \text{ for bare steel deck diaphragms meeting the special requirements in AISI S400} \\ 1.5 \text{ for all other bare steel deck diaphragms} \end{cases} \tag{6.44}$$

Diaphragm deflections must be calculated using F_{px}. Information on the derivation of the deflection amplification factor is given in ASCE/SEI C12.10.4.2.

Where the diaphragm design forces are determined in accordance with ASCE/SEI 12.10.4, the following requirements do not apply:

- Footnote b in ASCE/SEI Table 12.2-1
- ASCE/SEI 12.3.3.5

Flowchart 6.7 in Section 6.6 of this book can be used to determine the design seismic diaphragm force, F_{px}, in accordance with ASCE/SEI 12.10.4.

6.3.11 Structural Walls and Their Anchorage

Design for Out-of-Plane Forces

In addition to in-plane forces, structural walls and their anchorage in structures assigned to SDC B and higher must be designed for an out-of-plane force, F_p, equal to $0.4S_{DS}I_e$ times the weight of the structural wall or 0.10 times the weight of the structural wall, whichever is greater (ASCE/SEI 12.11.1). This force acts perpendicular to the face of the wall.

Because walls are often subjected to local deformations due to material shrinkage, temperature changes and foundation settlement, structural wall elements and connections to supporting framing systems require some degree of ductility to accommodate these deformations while providing the required strength for combined gravity and seismic forces.

Nonstructural walls need not be designed for this requirement but must be designed in accordance with the seismic design for nonstructural component requirements in ASCE/SEI Chapter 13 (see Section 6.4 of this book).

Anchorage of Structural Walls and Transfer of Design Forces into Diaphragms or Other Supporting Structural Elements

Structural walls in structures assigned to SDC B and higher must be adequately anchored to the roof and floor members that provide lateral support for the walls. The anchorage of structural walls must provide a direct connection that is capable of resisting the force F_p determined by ASCE/SEI Equation (12.11-1):

$$F_p = 0.4S_{DS}k_aI_eW_p \geq \text{greater of} \begin{cases} 0.2\,k_aI_eW_pA_{\text{trib}} \\ 5\text{ lb/ft}^2\ (0.24\text{ kN/m}^2) \times A_{\text{trib}} \end{cases} \tag{6.45}$$

where W_p is the weight of the wall tributary to the anchor and A_{trib} is the area of the wall tributary to the anchor.

The amplification factor for diaphragm flexibility, k_a, which accounts for amplification of out-of-plane accelerations caused by diaphragm flexibility, is determined by ASCE/SEI Equation (12.11-2):

$$k_a = \begin{cases} 1.0 + \dfrac{L_f}{100} \leq 2.0 \text{ for flexible diaphragms} \\ 1.0 \text{ for diaphragms that are not flexible} \end{cases} \tag{6.46}$$

where L_f is the span, in feet, of a flexible diaphragm that provides lateral support for the wall; the span is measured between vertical elements that provide lateral support to the diaphragm in the direction of analysis ($L_f = 0$ for rigid diaphragms).

The purpose of this requirement is to help prevent separation of structural walls from the roof and floors. The force F_p applies only to the design of the anchorage or connection of the wall to the structure and not to the overall design of the wall. This force must be applied for both tension (out-of-plane) and sliding (in-plane) where applicable (e.g., where roof or floor framing is not perpendicular to the anchored walls).

In a structure where all the diaphragms are not flexible, the value of F_p determined by Equation (6.45) is permitted to be multiplied by the factor $(1+2z/h)/3$ where anchorage is not located at the roof. The height of the anchor above the base of the structure is designated z and h is the height of the roof above the base. This reduction factor can be conservatively set equal to 1.0 to ensure that a smaller than appropriate anchorage force is not used in design. In this case, F_p must be greater than or equal to that required by ASCE/SEI 12.11.1 with a minimum anchorage force equal to the greater of $0.2W_p$ and 5 lb/ft^2 (0.24 kN/m^2) times A_{trib}.

Where the anchor spacing is greater than 4 ft (1.22 m) along the length of the wall, the section of wall spanning between the anchors must be designed to resist the local out-of-plane bending caused by F_p.

Additional Requirements for Anchorage of Concrete and Masonry Structural Walls to Diaphragms in Structures Assigned to SDC C through F

Requirements for anchorage of concrete and masonry walls to diaphragms in structures assigned to SDC C through F are given in ASCE/SEI 12.11.2.2. The main purpose of these requirements is to ensure that a continuous load path exists that will survive the ground shaking. Additional information is given in ASCE/SEI C12.11.2.2.

6.3.12 Drift and Deformation

Overview

Story drift must be controlled for both structural and nonstructural reasons. Limiting the drift helps limit inelastic strain in structural members and helps control overall structural stability by reducing P-delta effects. Drift control is also needed to limit damage to nonstructural elements such as partitions, elevator and stair enclosures, and glass, to name a few.

Story Drift Limit

Design drifts are determined in accordance with ASCE/SEI 12.8.6 (see Section 6.3.8 of this book). These drifts must be less than or equal to the allowable story drift, Δ_a, in ASCE/SEI Table 12.12-1. The drift limits depend on the risk category: drift limits are generally more restrictive for risk categories III and IV and are meant to provide a higher level of performance for more important structures.

Drift limits also depend on the type of structure. For many structures, the drift limit is 2 percent of the story height. For low-rise structures (four stories or less) where the interior walls, partitions, ceilings, and exterior wall systems have been designed to accommodate story drifts, the drift limits are not as stringent as those for other types of structures.

Satisfying strength requirements may result in a structure that also satisfies story drift limits. Moment-resisting frames or tall, slender structures are often controlled by drift, and member sizes of the SFRS generally have to be increased to satisfy prescribed drift limits.

For structures assigned to SDC D, E, or F with SFRSs consisting of only moment-resisting frames, the design story drift, Δ, must be less than or equal to Δ_a/ρ for any story (ASCE/SEI 12.12.1.1). Given the inherent flexibility of moment-resisting systems, this provision essentially penalizes structures utilizing nonredundant systems in regions of high seismic risk.

Structural Separation

Portions of a structure or adjacent structures must have sufficient distance between them so that they can respond independently to ground motion without impacting.

The design earthquake displacement, δ_{DE}, which is determined in accordance with ASCE/SEI 12.8.6 (see Section 6.3.8 of this book), must be used in determining separation distances (ASCE/SEI 12.12.2). The minimum separation distance, δ_{SS}, between adjacent structures on the same property is determined by ASCE/SEI Equation (12.12-2):

$$\delta_{SS} = \sqrt{(\delta_{DE1})^2 + (\delta_{DE2})^2} \tag{6.47}$$

where δ_{DE1} and δ_{DE2} are the design earthquake displacements of the adjacent structures at their adjacent edges (see Figure 6.43).

The above requirements are applicable to single buildings with one or more internal separation joints (commonly referred to seismic joints) or to adjacent buildings on the same property where sufficient information is known about both structures so that the required displacements can be calculated.

Where a new building will be located in proximity to an existing structure (i.e., it adjoins a property line), it is unlikely information about the existing structure is available. In such cases, separation calculations must be performed based on the maximum deflection of the new building.

Figure 6.43 Structural Separation Requirements in Accordance with ASCE/SEI 12.12.2 for Adjacent Structures on the Same Property

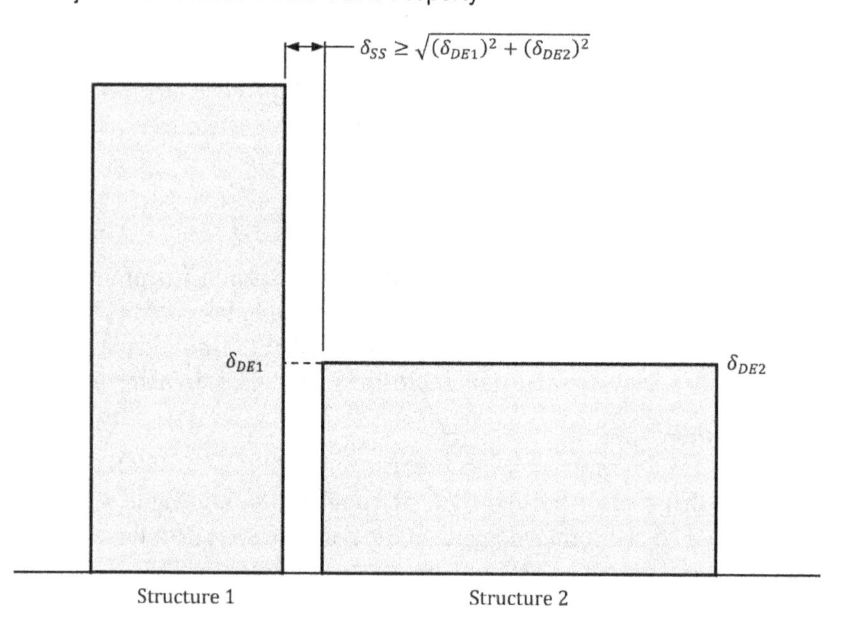

For example, if the taller building in Figure 6.43 is the new building, it must be set back a distance of δ_{DE1} from the property line.

Members Spanning between Structures

For members connected to adjacent structures or to seismically separated portions of structures, the gravity connections or supports of such members must be designed to accommodate the maximum anticipated relative displacements. These displacements must be calculated using the maximum considered earthquake displacements, δ_{MCE}, determined in accordance with ASCE/SEI 12.8.6 (see Section 6.3.8 of this book) assuming the two structures are moving in opposite directions (ASCE/SEI 12.12.3). Unlike the requirements for structural separation where a modal combination is used to determine the displacement (which represents a probable value of displacement), this provision requires the absolute sum of displacements of the two structures be used.

Deformation Compatibility for Seismic Design Categories D through F

Certain members or components in a structure are assigned to be part of the SFRS, which means these elements must be designed for the combined effects due to gravity and seismic loads. During an earthquake, all the structural members undergo displacement due to the ground motion regardless if they are part of the SFRS or not.

For structures assigned to SDC D through F, structural members not included in the SFRS must be designed for deformation compatibility: these members must be designed to adequately resist gravity load effects when subjected to the design earthquake displacement, δ_{DE}, and the associated drifts determined in accordance with ASCE/SEI 12.8.6 (ASCE/SEI 12.12.4; see Section 6.3.8 of this book). The story drifts induce seismic load effects into these members that must be accounted for in design.

Reinforced concrete frame members not designated as part of the SFRS must be designed and detailed in accordance with the provisions in Section 18.14 of Reference 6.10.

The stiffening effects of adjoining rigid structural and nonstructural elements must be considered in the design of the members not included in the SFRS.

6.3.13 Foundation Design

General requirements for the design of various types of foundation systems are given in ASCE/SEI 12.13 and IBC Chapter 18. Special seismic requirements are based on the SDC.

Overturning effects at the soil-foundation interface are permitted to be reduced by 25 percent for foundations of structures that satisfy both of the following conditions (ASCE/SEI 12.13.4):

- The structure is designed in accordance with the ELF procedure in ASCE/SEI 12.8.
- The structure is not an inverted pendulum or cantilevered column type structure.

For foundations of structures designed in accordance with ASCE/SEI 12.9, a 10 percent reduction in overturning effects is permitted. A modal response spectrum analysis more accurately reflects the actual distribution of base shear compared to the ELF procedure, so the permitted reduction in overturning is smaller.

Both strength design and allowable stress design of foundation elements are permitted (see ASCE/SEI 12.13.5 and 12.13.6, respectively). For strength design, the nominal foundation geotechnical capacity, Q_{ns}, is determined by one of the following methods (ASCE/SEI 12.13.5.1):

- Presumptive load-bearing values.
- By a registered design professional based on geotechnical site investigations that include field and laboratory testing.
- By in situ testing of prototype foundations.

Once Q_{ns} has been determined, the design strength is determined by multiplying Q_{ns} by the resistance factors, ϕ, for strength design in ASCE/SEI Table 12.13-1. Resistance factors are provided for vertical resistance (bearing strength and pile friction) and lateral resistance (lateral bearing pressure and sliding by friction or cohesion). For a linear seismic analysis in accordance with ASCE/SEI 12.8 or 12.9, the required strength is obtained by using the strength design load combinations in ASCE/SEI 2.3; these factored load effects, including reductions permitted in ASCE/SEI 12.13.4, must be less than or equal to the corresponding design strength, ϕQ_{ns}.

In the allowable stress method, the factored load effects determined by the allowable stress design load combinations in ASCE/SEI 2.4 must be less than or equal to the allowable foundation load capacities, Q_{as}, which are determined by geotechnical investigations required by the building official. The reduction of foundation overturning effects in ASCE/SEI 12.13.4 are permitted to be used.

Provisions for foundations subject to seismically induced soil displacement or strength are given in ASCE/SEI 12.13.9.

6.3.14 Simplified Alternative Structural Design Criteria for Simple Bearing Wall or Building Frame Systems

Overview

The simplified method in ASCE/SEI 12.14 is permitted to be used in lieu of other analytical procedures in ASCE/SEI Chapter 12 for the analysis and design of simple bearing wall or building frame systems provided the twelve limitations of ASCE/SEI 12.14.1.1 are satisfied (see Table 6.19). Even though twelve conditions must be met, the procedure is applicable to a wide range of relatively stiff, low-rise structures assigned to risk category I or II that possess SFRSs arranged in a torsionally resistant, regular layout.

Table 6.19 Limitations of the Simplified Alternative Seismic Procedure of ASCE/SEI 12.14

Number	Limitation
1	Risk category I or II structures
2	Site Class A through DE
3	Structure must not exceed three stories above grade
4	SFRS must be either a bearing wall or building frame system as indicated in ASCE/SEI Table 12.14-1
5	Two lines of walls or frames are required in each of two major axis directions where at least one line of resistance must be provided on each side of the CM in each direction
6	The CM in each story must be located not further from the geometric centroid of the diaphragm than 10 percent of the length of the diaphragm parallel to the eccentricity
7	ASCE/SEI Equation (12.14-1) must be satisfied for structures with cast-in-place concrete diaphragms with overhangs beyond the outside line of shear walls or braced frames, and ASCE/SEI Equation (12.14-2) must be satisfied for all other diaphragm overhangs
8	For structures with diaphragms that are not flexible, the forces must be apportioned to the vertical elements of the SFRS as if the diaphragms were flexible and the additional requirements in 8(a) and 8(b) must be satisfied
9	Lines of walls or frames must be oriented at angles of no more than 15 degrees from the major orthogonal horizontal axes of the building
10	The simplified procedure must be used for each major orthogonal horizontal axis of the building
11	System irregularities caused by in-plane or out-of-plane offsets of lateral force-resisting elements are not permitted except for the building type given in the exception
12	Story lateral strength of any story must not be less than 80% of that in the story above where the story lateral stiffness is the total lateral strength of all seismic force-resisting system elements resisting the story shear in the direction of analysis

Design Basis

The structure must have complete lateral and vertical force-resisting systems with adequate strength to resist the effects of the design seismic forces, which must be distributed to the elements of the systems using the linear elastic analysis in ASCE/SEI 12.14.8 (see below). A continuous load path or paths must be provided with adequate strength and stiffness to transfer all forces from the point of application to the final point of resistance. The foundation must be designed to resist the forces from the superstructure.

Seismic Load Effects

Seismic load effect, E, in the simplified method is defined the same as in ASCE/SEI 12.4.2 and consists of the effects of both horizontal and vertical motions E_h and E_v, respectively (see ASCE/SEI 12.14.3.1). In the simplified method, it is assumed the structure possesses a reasonable level of redundancy (see the fifth limitation in Table 6.19), so the redundancy factor, ρ, is equal to 1.0 and E_h is equal to Q_E [see ASCE/SEI Equation (12.14-5)].

The vertical seismic load effect, E_v, is permitted to be taken as zero for either of the following conditions:

1. In ASCE/SEI Equations (12.14-3), (12.14-4), (12.14-7), and (12.14-8) where $S_{DS} \leq 0.125$.
2. In ASCE/SEI Equation (12.14-4) where determining demands on the soil-structure interface of foundations.

Horizontal and vertical effects E_h and E_v must be used in the strength design load combinations 6 and 7 in ASCE/SEI 2.3.6 and the allowable stress design load combinations 8, 9, and 10 in ASCE/SEI 2.4.5.

Seismic load effect, E_m, in the simplified method is defined the same as in ASCE/SEI 12.4.3 and consists of the effects of both horizontal and vertical ground motions E_{mh} and E_v, respectively (see ASCE/SEI 12.14.3.2). In the simplified method, $\Omega_0 = 2.5$, which is consistent with the values of Ω_0 in ASCE/SEI Table 12.2-1 for bearing wall and building frame systems [see ASCE/SEI Equation (12.14-9)].

Horizontal effect E_h need not be taken larger than the capacity-limited horizontal seismic load effect, E_{cl}, which is defined in ASCE/SEI 11.3.

Horizontal and vertical effects E_{mh} and E_v must be used in the strength design load combinations 6 and 7 in ASCE/SEI 2.3.6 and the allowable stress design load combinations 8, 9, and 10 in ASCE/SEI 2.4.5.

Seismic Force-Resisting Systems

Selection and Limitations

Parts of ASCE/SEI Table 12.2-1 are reproduced in ASCE/SEI Table 12.14-1 for the systems permitted to be designed by the simplified method. The sections pertaining to detailing requirements, the response modification coefficient, R, and system limitations for SDCs B through E are given in ASCE/SEI Table 12.14-1. The system overstrength factor, Ω_0, is taken as 2.5 for these systems, so it is not included in the table. Also, because drift calculations are not required (see ASCE/SEI 12.14.8.5 and the discussion below), the deflection amplification factor, C_d, is not in ASCE/SEI Table 12.14-1.

Combinations of Framing Systems

Horizontal and vertical combinations of framing systems are permitted in the simplified method (see ASCE/SEI 12.14.4.2). The value of R used in design must be the least value of any of the SFRSs in the direction of analysis.

Diaphragm Flexibility

Untopped steel deck, wood structural panels, and similar panelized construction are permitted to be considered flexible diaphragms (ASCE/SEI 12.14.5), and lateral load is distributed to the vertical elements of the SFRS using tributary area (ASCE/SEI 12.14.8.3.1).

For diaphragms that are not flexible, a rigidity analysis is required, which includes torsional moments resulting from eccentricity between the locations of CM and CR (ASCE/SEI 12.14.8.3.2). Accidental torsion and dynamic amplification of torsion need not be included in the analysis because the simplified method is applicable to regular structures with essentially a uniform distribution of lateral stiffness.

Application of Loading

Design seismic forces, which are determined in accordance with ASCE/SEI 12.14.8 (see below), are permitted to be applied separately in each of the orthogonal directions (ASCE/SEI 12.14.6). In other words, two separate analyses are acceptable, and the combination of load effects from the two directions need not be considered.

Design and Detailing Requirements

The design and detailing requirements in ASCE/SEI 12.14.7 are given in Table 6.20.

Table 6.20 Design and Detailing Requirements in ASCE/SEI 12.14.7

ASCE/SEI Section Number		Requirement	Exceptions/Remarks
12.14.7.1	Connections	All parts of the structure must be adequately interconnected to achieve a continuous load path through the structure down to the foundation.	Design and detailing requirements are independent of the SDC. The requirements for interconnection in ASCE/SEI 12.14.7.1 are somewhat more stringent than those in ASCE/SEI 12.1.3, while the requirements for connections to supports are the same in ASCE/SEI 12.1.4 and 12.14.7.1. Foundation design requirements are the same in ASCE/SEI 12.14.7 and 12.1.5.
12.14.7.2	Openings or reentrant building corners	Reinforcement must be provided at the edges of openings in shear walls, diaphragms, or other plate-type elements and at reentrant corners.	Shear walls of wood structural panels are permitted where designed in accordance with AWC SDPWS for perforated shear walls or ANSI/AISI S400 for type II shear walls.
12.14.7.3	Collector elements	Collector elements and their connections must be designed to resist the seismic load effects including an overstrength factor equal to 2.5.	In structures braced entirely by light-frame shear walls, it is permitted to design collectors and their connections to resist the forces in ASCE/SEI 12.14.7.4.
12.14.7.4	Diaphragms	Diaphragms must be designed to resist the design seismic force, F_x, determined by ASCE/SEI Equation (12.14-13). Where applicable, transfer forces from the SFRS above must be added to the diaphragm design force.	—
12.14.7.5	Anchorage of structural walls	Structural walls must be anchored to floors, roofs, and members providing out-of-plane lateral support for the wall. The anchorage must resist the force determined by ASCE/SEI Equation (12.14-10).	The equation to determine the design force in individual anchors is the same as that in ASCE/SEI 12.11.2.1.

(Continued)

Table 6.20 Design and Detailing Requirements in ASCE/SEI 12.14.7 (*Continued*)

ASCE/SEI Section Number		Requirement	Exceptions/Remarks
12.14.7.6	Bearing walls and shear walls	Exterior and interior bearing walls and shear walls and their anchorage must be designed to resist a force normal to the surface equal to $0.4S_{DS}W_c \geq 0.1W_c$.	The requirements for out-of-plane forces on exterior and interior structural walls are the same as those in ASCE/SEI 12.11.1.
12.14.7.7	Anchorage of nonstructural components	All portions or components of the structure must be anchored to resist the seismic force F_p determined in accordance with ASCE/SEI Chapter 13, where required.	—

Simplified Lateral Force Analysis Procedure

The requirements in ASCE/SEI 12.14.8 for the simplified lateral force analysis procedure are given in Table 6.21.

Flowchart 6.8 in Section 6.6 of this book can be used to determine the design seismic forces using the simplified alternative structural design criteria in ASCE/SEI 12.14.

Table 6.21 Simplified Lateral Force Analysis Procedure

ASCE/SEI Section Number		Requirement	Exceptions/Remarks
12.14.8	Modeling	A linear mathematical model of the structure must be constructed. For purposes of analysis, the structure is permitted to be fixed at the base.	—
12.14.8.1	Seismic base shear	Seismic base shear, V, is determined by ASCE/SEI Equation (12.14-12): $V = \frac{FS_{DS}}{R}W$ where $F = \begin{cases} 1.0 \text{ for one-story buildings} \\ 1.1 \text{ for two-story buildings} \\ 1.2 \text{ for three-story buildings} \end{cases}$ $S_{DS} = 2F_aS_S/3$ In calculating S_{DS}, S_s is determined in accordance with ASCE/SEI 11.4.4 but need not be taken larger than 1.5. $F_a = \begin{cases} 1.0 \text{ for rock sites} \\ 1.4 \text{ for soil sites} \end{cases}$ Sites are permitted to be considered rock where there is no more than 10 ft (3.05 m) of soil between the rock surface and the bottom of the foundation. W = effective seismic weight defined in ASCE/SEI 12.14.8.1 R = response modification coefficient in ASCE/SEI Table 12.14-1.	Seismic base shear, V, which is independent of the building period, is increased by 10 and 20% for two-story and three-story buildings, respectively. These increases primarily account for the method used for vertical distribution of V, which is based on tributary weight, and have been shown by parametric studies to be adequate without being overly conservative. Only a basic geotechnical investigation is needed; 100-ft (30.5-m) deep borings and seismic shear velocity tests are not necessary.

(*Continued*)

Table 6.21 Simplified Lateral Force Analysis Procedure (*Continued*)

ASCE/SEI Section Number		Requirement	Exceptions/Remarks
12.14.8.2	Vertical distribution	Seismic forces at each level are determined by ASCE/SEI Equation (12.14-13): $$F_x = \frac{w_x}{W}V$$ where w_x is the portion of W at level x	The vertical distribution of the seismic forces is linear.
12.14.8.3	Horizontal shear distribution	The design seismic story shear is determined by ASCE/SEI Equation (12.14-14): $$V_x = \sum_{i=x}^{n} F_i$$ where F_i is the portion of V induced at level i. For structures with flexible diaphragms, the seismic design story shear is distributed to the vertical elements of the SFRS based on the area tributary to the vertical element. For structures with diaphragms that are not flexible, the seismic design story shear is distributed to the vertical elements of the SFRS based on the relative lateral stiffnesses of the vertical elements and the diaphragm. The inherent torsional moment, M_t, resulting from the eccentricity between the locations of the CM and CR must be included in the analysis.	These requirements are the same as those for the ELF procedure. A two-dimensional analysis is permitted for structures with flexible diaphragms.
12.14.8.4	Overturning	The structure must be designed to resist the effects from overturning caused by the seismic forces. Foundations of structure must be designed for not less than 75% of the foundation overturning moment, M_f, at the foundation-soil interface.	These requirements are the same as those for the ELF procedure. Included is the 25% reduction that can be applied when considering foundation overturning.
12.14.8.5	Drift limits and building separation	Structural drift need not be calculated. Where drift calculations are needed to determine structural separation between buildings or from property lines, for design of cladding, or for other design requirements, the drift must be taken as 1% of the structural height, h_n, unless computed to be less.	The simplified method does not require a drift check because it is assumed that bearing wall and building frame systems within the prescribed height range will not require such a check (unlike moment frame systems where drift is a major concern in design). Accordingly, calculations for P-delta effects need not be considered.

6.4 Seismic Design Requirements for Nonstructural Components

6.4.1 Overview

Minimum seismic design criteria for nonstructural components, including their supports and attachments, are given in ASCE/SEI Chapter 13. Requirements are provided for architectural components, mechanical and electrical components, and anchorage. Nonstructural components include the following:

- Components in or supported by a structure.
- Components outside of a structure (except for nonbuilding structures within the scope of ASCE/SEI Chapter 15) permanently connected to the mechanical or electrical systems.
- Components part of the egress system of a structure.

A nonstructural component weighing greater than or equal to 20 percent of the combined effective seismic weight, W, of the nonstructural component and the supporting structure as defined in ASCE/SEI 12.7.2 must be designed in accordance with ASCE/SEI 13.2.9 (ASCE/SEI 13.1.1).

Nonstructural components must be assigned to the same SDC as the following (ASCE/SEI 13.1.2):

1. The structure they occupy or are supported by.
2. The structure to which they are permanently connected by mechanical or electrical systems.
3. For parts of an egress system, the structure it serves.

Where one or more the above criteria is applicable, the nonstructural component must be assigned to the highest SDC.

The component importance factor, I_p, is equal to 1.0 except where the four conditions in ASCE/SEI 13.1.3 apply, in which it is equal to 1.5.

A list of nonstructural components exempt from the requirements of ASCE/SEI Chapter 13 is given in ASCE/SEI Table 13.1-1 (see ASCE/SEI 13.1.4). It is assumed these nonstructural components and systems can achieve the required performance goals based on their inherent strength and stability, the lower level of earthquake demand, or both.

Storage racks, tanks, and other nonbuilding structures supported by other structures must be designed in accordance with ASCE/SEI Chapter 15. Where ASCE/SEI 15.3 stipulates that seismic forces be determined in accordance with ASCE/SEI Chapter 13 and values of the component resonance ductility factor, C_{AR}, and component strength factor, R_{po}, are not given in ASCE/SEI Table 13.5-1 for architectural components or ASCE/SEI Table 13.6-1 for mechanical and electrical components, the ratio C_{AR}/R_{po} in ASCE/SEI Equation (13.3-1), which is used to determine the horizontal seismic design force, F_p, must be taken equal to $2.5/R$ where the response modification coefficient, R, for the nonbuilding structure is obtained from ASCE/SEI Table 15.4-1 for nonbuilding structures similar to buildings or ASCE/SEI Table 15.4-2 for nonbuilding structures not similar to buildings (ASCE/SEI 13.1.6).

6.4.2 General Design Requirements

Applicable Requirements for Nonstructural Components

The requirements in ASCE/SEI Table 13.2-1 applicable to the supports and attachments of architectural, mechanical, and electrical components are given in Table 6.22.

Load Combinations

Nonstructural components, including their supports and attachments, must comply with the basic requirements in ASCE/SEI 1.3. In addition, they must be designed using the applicable load combinations in ASCE/SEI 2.3 or 2.4.

For purposes of combining load effects, the horizontal seismic design force, F_p, determined by ASCE/SEI Equation (13.3-1) must be used in ASCE/SEI 12.4.2.1, that is, $E_h = \rho Q_E = 1.0 \times F_p = F_p$ where $\rho = 1.0$ for nonstructural components (item 3 in ASCE/SEI 12.3.4.1).

Similarly, F_p must be used in ASCE/SEI 12.4.3.1, that is, $E_{mh} = \Omega_0 Q_E = \Omega_{0p} F_p$ where the design load combinations including overstrength are applicable (these load combinations are only applicable to the design of anchors in concrete and masonry in accordance with ASCE/SEI 13.4.2).

Table 6.22 Requirements for Supports and Attachments of Architectural, Mechanical, and Electrical Components

Nonstructural Element	Requirements				
	General Design (ASCE/SEI 13.2)	Force and Displacement (ASCE/SEI 13.3)	Attachment (ASCE/SEI 13.4)	Architectural Component (ASCE/SEI 13.5)	Mechanical and Electrical Component (ASCE/SEI 13.6)
Architectural components and supports and attachments for architectural components	X	X	X	X	
Mechanical and electrical components	X	X	X		X
Supports and attachments for mechanical and electrical components	X	X	X		X

The anchorage overstrength factor, Ω_{0p}, is given in ASCE/SEI Table 13.5-1 for architectural components and ASCE/SEI Table 13.6-1 for mechanical and electrical components.

The vertical seismic load effect, E_v, in ASCE/SEI 12.4.2.2 is equal to $0.2S_{DS}D = 0.2S_{DS}W_p$ where W_p is the component operating weight.

Supported Nonstructural Components with Greater Than or Equal to 20 Percent Combined Weight

Requirements are given in ASCE/SEI 13.2.9 where the weight of a nonstructural component is greater than or equal to 20 percent of the combined effective seismic weight, W, of the nonstructural component and the supporting structure. In such cases, the nonstructural component and the supporting structure must be designed for forces and displacements determined in accordance with ASCE/SEI Chapter 12 or ASCE/SEI 15.5 for nonbuilding structures similar to buildings, as appropriate. The analysis must include the combined structural characteristics of the component and supporting structure with the response modification coefficient, R, of the combined system equal to the lesser of $0.40(C_{AR}/R_{po})$ of the nonstructural component or R of the supporting structure. Where the ratio of the fundamental period of the nonstructural component to the supporting structure is less than 0.5 or greater than 2.0, the following are applicable:

- The supporting structure is permitted to be designed in accordance with ASCE/SEI Chapter 12 or ASCE/SEI 15.5, as appropriate.
- The supported nonstructural component must satisfy the requirements of ASCE/SEI Chapter 13 as if the weight of the nonstructural component were less than 20 percent of the combined effective seismic weight, W, of the nonstructural component and supporting structure.

6.4.3 Seismic Demands on Nonstructural Components

Horizontal Seismic Design Forces

The horizontal seismic design force, F_p, which is applied to the center of gravity of the component, is determined by ASCE/SEI Equation (13.3-1) with the upper and lower limits determined by ASCE/SEI Equations (13.3-2) and (13.3-3), respectively:

$$F_p = 0.4S_{DS}I_pW_p\left(\frac{H_f}{R_\mu}\right)\left(\frac{C_{AR}}{R_{po}}\right) \tag{6.48}$$

$$\geq 0.3S_{DS}I_pW_p$$

$$\leq 1.6S_{DS}I_pW_p$$

The terms in Equation (6.48) are defined in Table 6.23.

Table 6.23 Terms in the Determination of the Horizontal Seismic Design Force, F_p

Term	Definition
S_{DS}	Design spectral response acceleration at short periods determined in accordance with ASCE/SEI 11.4.4
I_p	Component importance factor, which is equal to 1.0 or 1.5 (see ASCE/SEI 13.1.3)
W_p	Component operating weight
H_f	Factor for force amplification as a function of height determined in ASCE/SEI 13.3.1.1
R_μ	Structure ductility reduction factor determined in ASCE/SEI 13.3.1.2
C_{AR}	Component resonance ductility factor determined in ASCE/SEI 13.3.1.3
R_{po}	Component strength factor determined in ASCE/SEI 13.3.1.4

Because nonstructural components can be attached at different heights in a structure, F_p should be determined for each point of attachment considering the prescribed minimum and maximum values. The component weight, W_p, at each point should be the weight tributary to the point of attachment.

The force F_p must be applied in the direction that produces the critical effects on the component, the component supports and attachments. This requirement is permitted to be satisfied by using the more severe of the following load cases:

1. Case 1: A combination of 100 percent of F_p in any one horizontal direction and 30 percent of F_p in a perpendicular horizontal direction applied simultaneously.
2. Case 2: The combination from case 1 rotated 90 degrees.

The factor for force amplification, H_f, approximates the dynamic amplification of ground acceleration by the supporting structure, and is equal to 1.0 for components supported at or below grade. For components supported above grade, H_f is permitted to be determined by ASCE/SEI Equations (13.3-4) or (13.3-5):

$$H_f = 1 + a_1\left(\frac{z}{h}\right) + a_2\left(\frac{z}{h}\right)^{10} \tag{6.49}$$

$$H_f = 1 + 2.5\left(\frac{z}{h}\right) \tag{6.50}$$

where $a_1 = 1/T_a \leq 2.5$

$a_2 = 1-(0.4/T_a)^2$

T_a = the lowest approximate fundamental period of the supporting building or nonbuilding structure in either orthogonal direction

To determine H_f by ASCE/SEI Equation (13.3-4), T_a must be known. For buildings, T_a is determined by ASCE/SEI Equation (12.8-8). Where the SFRS of the building is unknown, the approximate period parameters for "all other structures" is permitted to be used in ASCE/SEI Equation (12.8-8) to determine T_a.

For nonbuilding structures, T_a is permitted to be taken as one of the following:

1. The period of the nonbuilding structure, T, determined using the structural properties and deformation characteristics of the resisting elements in a properly substantiated analysis in accordance with ASCE/SEI 12.8.2.
2. The period of the nonbuilding structure, T, determined by ASCE/SEI Equation (15.4-6).
3. The approximate period, T_a, determined by ASCE/SEI Equation (12.8-8) using the approximate period parameters for "all other structural systems."

In building structures, the gravity system, partitions, and cladding all act to reduce the fundamental period, which increases the seismic force on nonstructural components. That is why T_a determined by ASCE/SEI Equation (12.8-8), which generally underestimates the period, is used in determining H_f for building structures instead of T. Because nonbuilding structures typically do not have partitions and cladding and the gravity system is less extensive than that for buildings, the period of the nonbuilding structure, T, is permitted to be used in determining H_f because it is close to the actual period of the structure suitable for component lateral force calculations.

For structures with combinations of SFRSs, the SFRS producing the lowest value of T_a must be used in determining H_f.

The structure ductility reduction factor, R_μ, is determined by ASCE/SEI Equation (13.3-6) for components supported above grade:

$$R_\mu = \left(\frac{1.1R}{I_e\Omega_0}\right)^{1/2} \geq 1.3 \tag{6.51}$$

In this equation, I_e is the importance factor determined in accordance with ASCE/SEI 11.5.1 for the building or nonbuilding structure supporting the component. The response modification coefficient, R, and the overstrength factor, Ω_0, for the building structure or nonbuilding structure supporting the component are obtained from ASCE/SEI Table 12.2-1 for buildings, ASCE/SEI Table 15.4-1 for nonbuilding structures similar to buildings or ASCE/SEI Table 15.4-2 for nonbuilding structures not similar to buildings.

For components supported at or below grade, $R_\mu = 1.0$. In cases where the SFRS of the building or nonbuilding structure is not specified or is not listed in ASCE/SEI Tables 12.2-1, 15.4-1 or 15.4-2, R_μ must be taken as 1.3 for components above grade.

Where a building or nonbuilding structure has combinations of SFRSs in different directions or has vertical combinations of SFRSs, R_μ for the entire structure must be based on the SFRS with the lowest R_μ.

For nonbuilding structures similar to buildings, some of the structure types in ASCE/SEI Table 15.4-1 have seismic coefficients corresponding to "with permitted height increase" and "with unlimited height" (see, e.g., intermediate reinforced concrete moment frames). In such cases, the value of R_μ is permitted to be determined using R and Ω_0 for the structure type "with permitted height increase."

The component resonance ductility factor, C_{AR}, is determined in accordance with ASCE/SEI 13.3.1.3. This factor is the ratio of the peak component acceleration and the peak floor acceleration. A summary of the provisions that must be used to determine C_{AR} is given in Table 6.24.

Table 6.24 Component Resonance Ductility Factor, C_{AR}

Nonstructural Component	ASCE/SEI Reference
Architectural components	ASCE/SEI Table 13.5-1
Mechanical and electrical equipment*	ASCE/SEI Table 13.6-1
Equipment support structures or platforms**	ASCE/SEI 13.6.4.6
Distribution systems†	ASCE/SEI Table 13.6-1
Distribution system supports	ASCE/SEI 13.6.4.7

*C_{AR} for mechanical and electrical equipment mounted on equipment support structures or platforms must not be less than C_{AR} for the equipment support structure or platform itself.
**The weight of supported mechanical and electrical components must be included when calculating the component operating weight, W_p, of the equipment support structures or platforms.
†Distribution systems include piping, ducts, and raceways.

The component strength factor, R_{po}, is given in ASCE/SEI Table 13.5-1 for architectural components and ASCE/SEI Table 13.6-1 for mechanical and electrical components. This factor accounts for the effect of component reserve strength.

Nonlinear Response History Analysis

The nonlinear response history analysis procedures in ASCE/SEI Chapters 16, 17, and 18 are permitted to determine the horizontal seismic design force, F_p, for nonstructural components instead of by ASCE/SEI Equation (13.3-1). If the model does not explicitly include the dynamic properties of the nonstructural component, F_p must be determined by ASCE/SEI Equation (13.3-7):

$$F_p = I_p W_p a_i \left(\frac{C_{AR}}{R_{po}} \right) \tag{6.52}$$

In this equation, a_i is the maximum acceleration at level i obtained from the nonlinear response history analysis at the design earthquake ground motion using a suite of not less than seven ground motions. The value of a_i to be used in ASCE/SEI Equation (13.3-7) must be the mean of the maximum values of acceleration at the center of mass of the support level obtained from each analysis. The upper and lower limits of F_p in ASCE/SEI Equations (13.3-2) and (13.3-3), respectively, are applicable in this type of analysis as well.

Vertical Seismic Force

Components, including their supports and attachments, must be designed for the vertical seismic force, E_v, determined in accordance with ASCE/SEI 12.4.2.2 (see Section 6.2.6 of this book). The vertical seismic force must act concurrently with the horizontal force, F_p, determined in accordance with ASCE/SEI 13.3.1.

Lay-in access floor panels and lay-in ceiling panels are exempt from the concurrent vertical seismic force requirement.

Nonseismic-loads greater than the horizontal force, F_p, determined in accordance with ASCE/SEI 13.3.1 govern the strength design of components. However, any and all detailing requirements and limitations in ASCE/SEI Chapter 13 must be satisfied in such cases.

Seismic Relative Displacements

The requirements for seismic relative displacements in ASCE/SEI 13.3.2 are applicable in the design of cladding, stairways, windows, piping systems, sprinkler components and other components connected to one structure at multiple levels (case 1) or to two or more adjoining structures (case 2).

Seismic relative displacements, D_{pI}, are determined by ASCE/SEI Equation (13.3-8):

$$D_{pI} = D_p I_e \quad (6.53)$$

In this equation, D_p is the displacement determined in accordance with ASCE/SEI 13.3.2.1 for displacements within structures and ASCE/SEI 13.3.2.2 for displacements between structures. The seismic importance factor, I_e, for the structure is determined in accordance with ASCE/SEI 11.5.1.

Two displacement equations are given for case 1 and case 2 (see Table 6.25). ASCE/SEI Equation (13.3-9), which is applicable to displacements within structures (case 1), and ASCE/SEI Equation (13.3-11), which is applicable to displacements between structures (case 2), produce elastic structural displacements unreduced by R. Because the actual displacements may not be known during the design phase, ASCE Equations (13.3-10) and (13.3-12) provide upper-bound displacements based on structural drift limits.

Table 6.25 Seismic Relative Displacements

Case 1: Displacements within structures (ASCE/SEI 13.3.2.1)	• $D_p = \delta_{xA} - \delta_{ya} \le (h_x - h_y)\Delta_{aA}/h^*_{sx}$ • D_p is permitted to be determined using the linear dynamic procedures in ASCE/SEI 12.9
Case 2: Displacements between structures (ASCE/SEI 13.3.2.2)	$D_p = \lvert\delta_{xA}\rvert + \lvert\delta_{yB}\rvert \le \dfrac{h_x \Delta_{aA}}{h_{sx}} + \dfrac{h_y \Delta^*_{aB}}{h_{sx}}$

*Upper limits on D_p in ASCE/SEI Equations (13.3-10) and (13.3-12) are applicable where the story drift associated with the design earthquake displacement is less than or equal to the allowable story drift in ASCE/SEI 12.12-1. These upper-limit equations do not apply where single-story structures are designed in accordance with note a of ASCE/SEI Table 12.12-1.

The terms in Table 6.25 are as follows:

δ_{xA} = deflection at building level x of structure A at the design earthquake displacement, determined in accordance with ASCE/SEI Equation (12.8-16)

δ_{yA} = deflection at building level y of structure A at the design earthquake displacement, determined in accordance with ASCE/SEI Equation (12.8-16)

δ_{yB} = deflection at building level y of structure B at the design earthquake displacement, determined in accordance with ASCE/SEI Equation (12.8-16)

h_x = height of level x to which the upper connection point is attached

h_y = height of level y to which the lower connection point is attached

h_{sx} = story height used in the definition of the allowable drift, Δ_a, in ASCE/SEI Table 12.12-1

Δ_{aA} = allowable story drift for structure A defined in ASCE/SEI Table 12.12-1

Δ_{aB} = allowable story drift for structure B defined in ASCE/SEI Table 12.12-1

An example of seismic relative displacements within a structure is given in Figure 6.44 for a glazing system. The piping system connecting the two buildings in Figure 6.45 is an example of seismic relative displacements between two structures.

The effects of seismic relative displacements must be considered in combination with displacements due to other loads.

Figure 6.44 Seismic Relative Displacements within a Structure

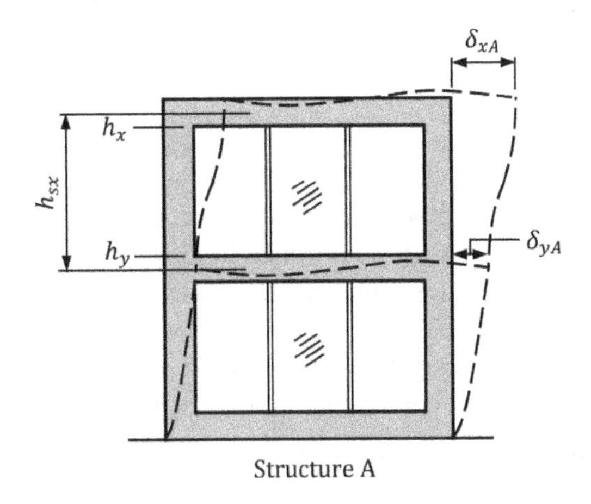

Figure 6.45 Seismic Relative Displacements between Structures

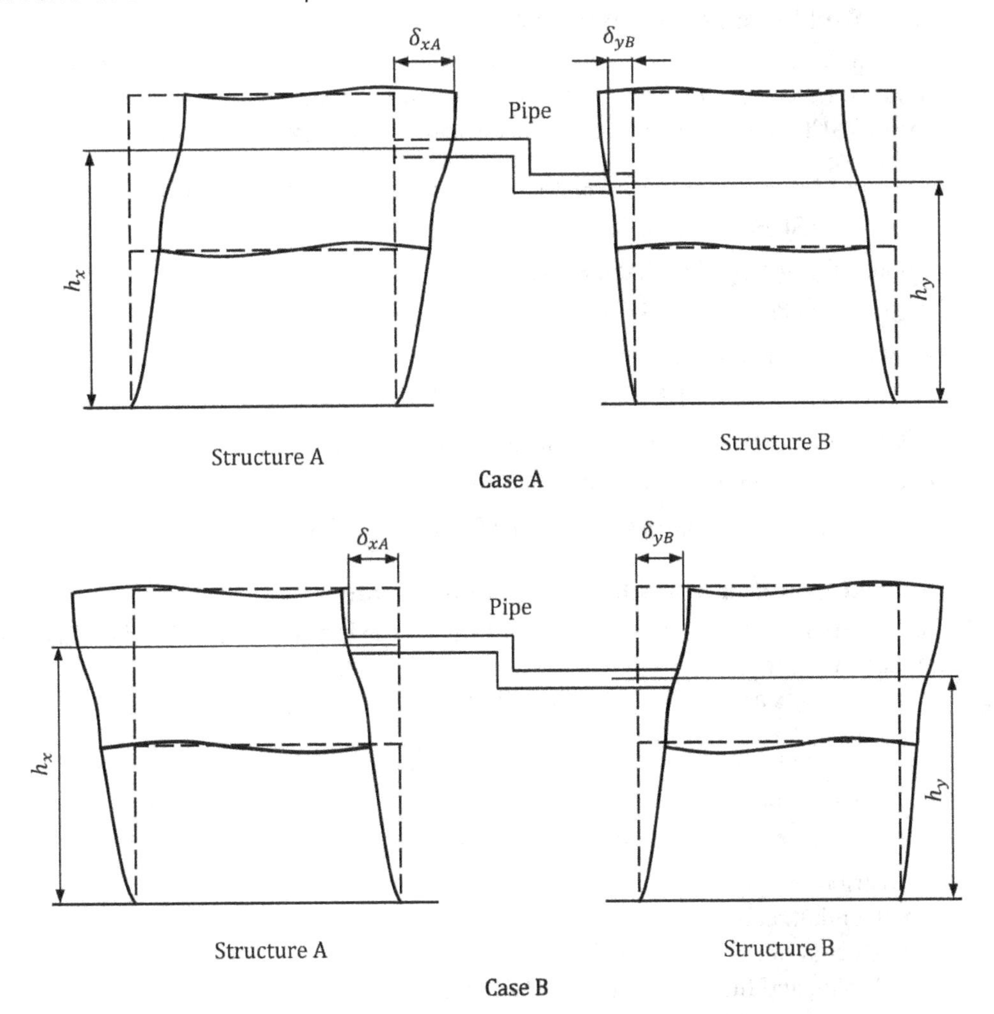

Component Period

The fundamental period, T_p, of a nonstructural component, including its supports and attachments to the structure, is permitted to be determined by ASCE/SEI Equation (13.3-13):

$$T_p = 2\pi\sqrt{\frac{W_p}{K_p g}} \tag{6.54}$$

In this equation, W_p is the component operating weight, K_p is the combined stiffness of the component, supports and attachments determined in terms of load per unit deflection at the center of gravity of the component and g is the acceleration due to gravity.

In lieu of ASCE/SEI Equation (13.3-13), T_p is permitted to be determined from experimental test data or by a properly substantiated analysis.

Additional information on component periods can be found in ASCE/SEI C13.3.3.

6.4.4 Nonstructural Component Anchorage and Attachment

Components and their supports must be attached or anchored to the structure in accordance with the requirements in ASCE/SEI 13.4. Forces in the attachments must be determined using the component forces and displacements given in ASCE/SEI 13.3.1 and 13.3.2, respectively. Anchors in concrete or masonry elements must satisfy the provisions of ASCE/SEI 13.4.2, which includes designing the anchors using the seismic load effects including overstrength.

6.4.5 Architectural Components

Design and detailing requirements for architectural components are given in ASCE/SEI 13.5. Architectural components and attachments must be designed for the seismic component forces in ASCE/SEI 13.3.1 (ASCE/SEI 13.5.2). Requirements are given for the following:

- Exterior nonstructural wall elements and connections (ASCE/SEI 13.5.3)
- Glass (ASCE/SEI 13.5.4)
- Out-of-plane bending (ASCE/SEI 13.5.5)
- Suspended ceilings (ASCE/SEI 13.5.6)
- Access floors (ASCE/SEI 13.5.7)
- Partitions (ASCE/SEI 13.5.8)
- Glass in glazed curtain walls, glazed storefronts, and glazed partitions (ASCE/SEI 13.5.9)
- Egress stairs and ramps (ASCE/SEI 13.5.10)
- Penthouses and rooftop structures (ASCE/SEI 13.5.11)

6.4.6 Mechanical and Electrical Components

The requirements of ASCE/SEI 13.6 must be satisfied for mechanical and electrical components and their supports.

Requirements are provided for the following:

- Mechanical components (ASCE/SEI 13.6.2)
- Electrical components (ASCE/SEI 13.6.3)
- Component supports (ASCE/SEI 13.6.4)
- Distribution systems
  - Conduit, cable trays, and raceways (ASCE/SEI 13.6.5)
  - Duct systems (ASCE/SEI 13.6.6)
  - Piping and tubing systems (ASCE/SEI 13.6.7)
  - Trapezes with a combination of systems (ASCE/SEI 13.6.8)
- Utility and service lines (ASCE/SEI 13.6.9)

- Boilers and pressure vessels (ASCE/SEI 13.6.10)
- Elevators and escalators (ASCE/SEI 13.6.11)
- Rooftop solar panels (ASCE/SEI 13.6.12)
- Other mechanical and electrical components (ASCE/SEI 13.6.13)

Flowchart 6.9 in Section 6.6 of this book can be used to determine the design seismic forces for nonstructural components in accordance with ASCE/SEI 13.3.1.

6.5 Seismic Design Requirements for Nonbuilding Structures

6.5.1 Overview

Unoccupied nonbuilding structures and buildings whose primary purpose is to enclose equipment or machinery and whose occupants are engaged in maintenance or monitoring of the equipment, machinery or their processes must be designed and detailed to resist the minimum seismic forces in ASCE/SEI Chapter 15.

The selection of a structural analysis procedure for a nonbuilding structure is based on its similarity to buildings. Nonbuilding structures similar to buildings exhibit behavior similar to building structures; however, their function and performance are different. Structural analysis procedures for such buildings must be selected in accordance with ASCE/SEI 12.6, which are applicable to building structures (ASCE/SEI 15.1.3). Guidelines and recommendations on the use of these methods are given in ASCE/SEI C15.1.3.

Nonbuilding structures not similar to buildings exhibit behavior that is markedly different than that of building structures. Most of these types of structures have reference documents that address their unique structural performance and behavior. Such reference documents are permitted to be used to analyze the structure (ASCE/SEI 15.1.3). In addition, the following procedures may be used: equivalent lateral force procedure (ASCE/SEI 12.8), linear dynamic analysis procedures (ASCE/SEI 12.9), and nonlinear response history analysis procedure (ASCE/SEI Chapter 16). Guidelines and recommendations on the proper analysis method to utilize for nonbuilding structures not similar to buildings are given in ASCE/SEI C15.1.3.

Requirements for nonbuilding structures sensitive to vertical ground motion, including tanks, vessels, hanging structures, and nonbuilding structures incorporating horizontal cantilevers, are given in ASCE/SEI 15.1.4.

6.5.2 Nonbuilding Structures Connected by Nonstructural Components to Other Adjacent Structures

The requirements in ASCE/SEI 15.2 must be satisfied where nonbuilding structures are connected by nonstructural components to other adjacent structures. This scenario occurs often in industrial buildings, such as pipe racks and equipment structures.

The analysis must include the combination of the structural characteristics of the nonbuilding structure, the adjacent structure, and the connecting nonstructural components. Design seismic forces must be determined in accordance with ASCE/SEI Chapter 12 or ASCE/SEI 15.5, as appropriate.

According to the exception in ASCE/SEI 15.2.1, regular nonbuilding structures connected to regular adjacent structures are permitted to be designed independently where any of the following conditions is satisfied:

1. Ratio of the fundamental period of the nonbuilding structure to that of the adjacent connected structure in the direction of analysis is greater than 0.9 and less than 1.1.
2. Ratio of the fundamental period of the nonbuilding structure to that of the adjacent connected structure in the direction of analysis is greater than 0.8 and less than 1.2, and the ratio of the effective seismic weight of the nonbuilding structure and the adjacent structure is greater than 0.8 and less than 1.2.

3. Ratio of the stiffness of the connecting nonstructural components to that of the nonbuilding structure in the direction of analysis and the ratio of the stiffness of the connecting nonstructural components to that of the adjacent structure in the direction of analysis are both less than 0.2.

These exceptions only apply to structures with vertical mass and stiffness regularity with similar fundamental periods.

Nonstructural components spanning between the nonbuilding structures must satisfy all the applicable requirements in ASCE/SEI Chapter 13, including the seismic relative displacement requirements in ASCE/SEI 13.3.2.

Additional information on the application of these provisions can be found in ASCE/SEI C15.2.

6.5.3 Nonbuilding Structures Supported by Other Structures

The requirements in ASCE/SEI 15.3 are applicable to the nonbuilding structures not similar to buildings in ASCE/SEI Table 15.4-2 that are supported by other structures and that are not part of the SFRS. The design method that must be used depends on the weight of the nonbuilding structure relative to the weight of the combined nonbuilding and supporting structure:

- Where the weight of the supported nonbuilding structure is less than 20 percent of the combined effective seismic weights of the nonbuilding structure and the supporting structure, the design seismic forces of the nonbuilding structure must be determined in accordance with the nonstructural component requirements in ASCE/SEI Chapter 13 where values of the component resonance ductility factor, C_{AR}, and the component strength factor, R_{po}, must be determined in accordance with ASCE/SEI 13.1.6 (ASCE/SEI 15.3.1; see Section 6.4.2 of this book). The appropriate requirements in ASCE/SEI Chapter 12 or ASCE/SEI 15.5 must be used in the design of the supporting structure where the weight of the nonbuilding structure must be included in the determination of the effective seismic weight, W.

  The following two exceptions are given related to these requirements:

  1. Where the ratio of the fundamental period of the nonbuilding structure to that of the supporting structure (including the lumped weight of the nonbuilding structure) is greater than 2.0, the supporting structure is permitted to be designed in accordance with ASCE/SEI Chapter 12 or ASCE/SEI 15.5 assuming the nonbuilding structure is modeled as attached to a rigid base.
  2. Where the ratio of the fundamental period of the nonbuilding structure to that of the supporting structure (including the lumped weight of the nonbuilding structure) is less than 0.5, the supporting structure is permitted to be designed in accordance with ASCE/SEI Chapter 12 or ASCE/SEI 15.5 with the weight of the nonbuilding structure included in W. The supported nonbuilding structure must be designed in accordance with the requirements in ASCE/SEI 15.3.1 for supported nonbuilding structures with less than 20 percent of the combined weight.

- Where the weight of the nonbuilding structure is greater than or equal to 20 percent of the combined effective seismic weights of the nonbuilding structure and the supporting structure, an analysis combining the structural characteristics of both the nonbuilding structure and the supporting structure must be performed to determine the seismic design forces (ASCE/SEI 15.3.2). The combined structure must be designed in accordance with ASCE/SEI 15.5 (which is applicable to nonbuilding structures similar to buildings) with the response modification coefficient, R, for the combined system taken as the lesser of the R values of the nonbuilding structure or the supporting structure. The nonbuilding structure and attachments must be designed for the forces determined for the nonbuilding structure in the combined analysis.

The two exceptions in ASCE/SEI 15.3.1 are also applicable to the requirements in ASCE/SEI 15.3.2.

Where nonstructural components are supported by nonbuilding structures, the nonstructural components must be designed in accordance with ASCE/SEI Chapter 13 (ASCE/SEI 15.3.3).

6.5.4 Structural Design Requirements

Design Basis

Basic coefficients and minimum design forces to be used in the determination of design seismic forces for nonbuilding structures are given in ASCE/SEI 15.4. Also provided are height limits and restrictions for nonbuilding structures.

Nonbuilding structures must be designed in accordance with the requirements in ASCE/SEI 15.5 for nonbuilding structures similar to buildings or ASCE/SEI 15.6 for nonbuilding structures not similar to buildings. The seismic lateral forces determined by these requirements must be greater than or equal to those determined by the equivalent lateral force (ELF) procedure in ASCE/SEI 12.8 considering the additions and exceptions in ASCE/SEI 15.4.1, which are summarized in Table 6.26.

Rigid Nonbuilding Structures

Nonbuilding structures with a fundamental period, T, less than 0.06 s are defined to be rigid, and the total design lateral seismic base shear, V, in such cases is determined by ASCE/SEI Equation (15.4-5):

$$V = 0.30 S_{DS} W I_e \tag{6.55}$$

where W is the operating weight of the nonbuilding structure.

The base shear determined by this equation is distributed over the height of the structure in accordance with ASCE/SEI 12.8.3.

Loads

The seismic effective weight, W, for nonbuilding structures must include the following where applicable (ASCE/SEI 15.4.3):

- Dead load and other loads defined in ASCE/SEI 12.7.2.
- Weight of normal operating contents for items such as tanks, vessels, bins, hoppers, and the contents of piping.
- Snow and ice loads where these loads constitute 25 percent or more of W or where required by the building official based on local environmental characteristics.

Fundamental Period

In lieu of determining the fundamental period, T, of a nonbuilding structure using a substantiated dynamic analysis, it is permitted to calculate T by ASCE/SEI Equation (15.4-6):

$$T = 2\pi \sqrt{\frac{\sum w_i \delta_i^2}{g \sum f_i \delta_i}} \tag{6.56}$$

In this equation, f_i represents any lateral force distribution in accordance with the principles of structural mechanics applied over the height of the structure and δ_i are the elastic deflections caused by those forces.

ASCE/SEI Equations (12.8-7) through (12.8-10) are not permitted to be used for determining T (ASCE/SEI 15.4.4).

Table 6.26 Structural Design Requirements for Nonbuilding Structures

Additions and Exceptions in ASCE/SEI 15.4.1	Requirement
Selection of SFRS	• Nonbuilding structures similar to buildings 1. A SFRS must be selected from those given in ASCE/SEI Table 12.2-1 or 15.4-1 based on SDC and subject to the limitations given in the table. 2. The seismic coefficients given in ASCE/SEI Table 15.4-1 must be used to determine the seismic base shear, V, element design forces, and design story drifts. 3. Design and detailing requirements must comply with the sections referenced in the selected table. • Nonbuilding structures not similar to buildings 1. A SFRS must be selected from those given in ASCE/SEI Table 15.4-2 based on SDC and subject to the limitations given in the table. 2. The seismic coefficients given in ASCE/SEI Table 15.4-2 must be used to determine the seismic base shear, V, element design forces, and design story drifts. 3. Design and detailing requirements must comply with the sections referenced in ASCE/SEI Table 15.4-2.
Minimum seismic response coefficient, C_s	• For nonbuilding systems that have an R value provided in ASCE/SEI Table 15.4-2, the minimum C_s determined by ASCE/SEI Equation (12.8-5) must be replaced by that determined by ASCE/SEI Equation (15.4-1): $C_s = 0.044 S_{DS} I_e \geq 0.03$. • For nonbuilding structures located where $S_1 \geq 0.6$, the minimum C_s determined by ASCE/SEI Equation (12.8-6) must be replaced by that determined by ASCE/SEI Equation (15.4-2): $C_s = 0.8 S_1/(R/I_e)$. • See the exception in ASCE/SEI 15.4.1(2) for tanks and vessels designed in accordance with the cited reference documents.
Seismic importance factor, I_e	The value of I_e must be the largest value determined by the following (ASCE/SEI 15.4.1.1): • Applicable reference document in ASCE/SEI Chapter 23. • The largest value from ASCE/SEI Table 1.5-2 based on the risk category. • As specified elsewhere in ASCE/SEI Chapter 15.
Vertical distribution of lateral seismic forces	Vertical distribution of lateral seismic forces must be determined by one of the following: • The requirements in ASCE/SEI 12.8.3. • The procedures in ASCE/SEI 12.9.1. • In accordance with the applicable reference document.
Accidental torsion	The accidental torsion requirements of ASCE/SEI 12.8.4.2 need not be accounted for in the analysis of the following structures provided the mass locations for the structure, any contents, and any supported structural or nonstructural elements (including but not limited to piping and stairs) that could contribute to the mass or stiffness of the structure are accounted for and quantified in the analysis: • Rigid nonbuilding structures. • Nonbuilding structures not similar to buildings and designed with R values less than or equal to 3.5.

Table 6.26 Structural Design Requirements for Nonbuilding Structures (*Continued*)

Additions and Exceptions in ASCE/SEI 15.4.1	Requirement
Accidental torsion (*continued*)	• Nonbuilding structures similar to buildings and designed with R values less than or equal to 3.5 provided one of the following conditions is met: 1. The calculated center of rigidity (CR) at each diaphragm is greater than 5 percent of the plan dimension of the diaphragm in each direction from the calculated center of mass (CM) of the diaphragm. 2. The structure does not have a type 1 horizontal torsional irregularity in ASCE/SEI Table 12.3-1 and the structure has at least two lines of lateral resistance in each of two major axis directions. At least one line of lateral resistance must be provided at a distance greater than or equal to 20 percent of the structure's plan dimension from the CM on each side of the CM. Structures designed in accordance with the above requirements must be analyzed using a three-dimensional representation of the structure in accordance with ASCE/SEI 12.7.3.
Minimum design seismic force for nonbuilding structures containing liquids, gases, and granular solids supported at the base	The minimum design seismic force must be greater than or equal to that required by the applicable reference document.
Applicability of reference documents	Where a reference document provides a basis for earthquake-resistant design of a particular type of nonbuilding structure covered by ASCE/SEI Chapter 15, the reference document is not permitted to be used unless the following limitations are satisfied: • The seismic ground accelerations and seismic coefficients must be in conformance with the requirements in ASCE/SEI 11.4. • The values for total lateral force and total base overturning moment must not be less than 80% of the base shear and overturning moment values, each adjusted for soil-structure interaction obtained from the applicable requirements in ASCE/SEI 7.
Reduction of design seismic base shear	The design seismic base shear, V, is permitted to be reduced in accordance with ASCE/SEI 19.2 to account for the effects of foundation damping from soil-structure interaction. The reduced base shear must not be taken less than 0.7 V.
Load combinations	The effects on the nonbuilding structure caused by gravity loads and seismic forces must be combined in accordance with the factored load combinations in ASCE/SEI 2.3, unless otherwise noted in ASCE/SEI Chapter 15.
Seismic load effects including overstrength	The design seismic forces on nonbuilding structures must be determined in accordance with ASCE/SEI 12.4.3 where specifically required by ASCE/SEI Chapter 15.

Drift Limit and P-Delta Effects

For nonbuilding structures, deflection, drifts, and structure separation are calculated using strength-level seismic forces to be compatible with the prescribed definition of the seismic load and the definition of the deflection amplification factors in ASCE/SEI Tables 15.4-1 and 15.4-2.

According to ASCE/SEI 15.4.5, the drift limits in ASCE/SEI 12.12.1 need not apply to nonbuilding structures where a rational analysis indicates larger drifts can be exceeded without adversely affecting structural stability or any attached or interconnected components and elements (such as walkways and piping).

For nonbuilding structures similar to buildings, P-delta effects must be evaluated using the requirements in ASCE/SEI 12.8.7. For nonbuilding structures not similar to buildings, the requirements in ASCE/SEI 12.8.7 do not apply; in such cases, P-delta effects must be based on displacements determined by an elastic analysis multiplied by (C_d/I_e) using the appropriate C_d from ASCE/SEI Table 15.4-2.

Site-Specific Response Spectra

Site-specific analyses must be utilized for specific types of nonbuilding structures where required by a reference document or the building official (ASCE/SEI 15.4.8). The analysis must be developed in accordance with ASCE/SEI 11.4.7 and must account for local seismicity and geology, expected recurrence intervals, and magnitudes of seismic events from known seismic hazards.

Anchors in Concrete or Masonry

Requirements for anchors in concrete and masonry used for nonbuilding structure anchorage are given in Table 6.27 (ASCE/SEI 15.4.9).

Table 6.27 Requirements for Anchors in Concrete and Masonry

Material	Requirement
Concrete	Design in accordance with ACI 318 (Reference 6.10)
Masonry	Design in accordance with TMS 402 (Reference 6.12)

In lieu of designing masonry anchors in accordance with Reference 6.12, it permitted to use either of the exceptions in ASCE/SEI 15.4.9.2.

Requirements for post-installed anchors and ASTM F1554 anchors are given in ASCE/SEI 15.4.9.3 and 15.4.9.4, respectively.

Requirements for Nonbuilding Structure Foundations on Liquefiable Sites

Where nonbuilding structures are located on liquefiable sites, the foundations must comply with the requirements in ASCE/SEI 12.13.9 and 15.4.10.1 (ASCE/SEI 15.4.10).

6.5.5 Nonbuilding Structures Similar to Buildings

Requirements are given in ASCE/SEI 15.5 for nonbuilding structures similar to buildings. The requirements for the following structures must be satisfied in addition to the design and detailing requirements outlined previously:

- Pipe racks (ASCE/SEI 15.5.2)
- Storage racks (ASCE/SEI 15.5.3)
- Electrical power-generating facilities (ASCE/SEI 15.5.4)
- Structural towers for tanks and vessels (ASCE/SEI 15.5.5)
- Piers and wharves (ASCE/SEI 15.5.6)

6.5.6 Nonbuilding Structures Not Similar to Buildings

Requirements are given in ASCE/SEI 15.6 for nonbuilding structures not similar to buildings. The requirements for the following structures must be satisfied in addition to the design and detailing requirements outlined previously:

- Earth-retaining structures (ASCE/SEI 15.6.1)
- Trussed towers, chimneys, and stacks (ASCE/SEI 15.6.2)
- Amusement structures (ASCE/SEI 15.6.3)
- Special hydraulic structures (ASCE/SEI 15.6.4)
- Secondary containment systems (ASCE/SEI 15.6.5)
- Telecommunication towers (ASCE/SEI 15.6.6)
- Steel tubular support structures for onshore wind turbine generator systems (ASCE/SEI 15.6.7)
- Ground-supported cantilever walls or fences (ASCE/SEI 15.6.8)
- Reinforced concrete tabletop structure for rotating equipment and process vessels or drums (ASCE/SEI 15.6.9)

6.5.7 Tanks and Vessels

Comprehensive seismic design requirements are given in ASCE/SEI 15.7 for tanks, vessels, bins, silos and similar containers storing liquids, gases, and granular solids.

Flowchart 6.10 in Section 6.6 of this book can be used to determine the design seismic forces for nonbuilding structures in accordance with ASCE/SEI Chapter 15.

6.6 Flowcharts

A summary of the flowcharts provided in this chapter is given in Table 6.28.

Table 6.28 Summary of Flowcharts Provided in Chapter 6

Flowchart	Title
6.1	Consideration of Seismic Design Requirements (IBC 1613.1 and ASCE/SEI 11.1.2)
6.2	Determination of Seismic Design Category (SDC) (IBC 1613.2 and ASCE/SEI 11.6)
6.3	Diaphragm Flexibility (ASCE/SEI 12.3)
6.4	Redundancy Factor, ρ (ASCE/SEI 12.3.4)
6.5	Equivalent Lateral Force (ELF) Procedure (ASCE/SEI 12.8)
6.6	Diaphragm Design Seismic Force (ASCE/SEI 12.10.3)
6.7	Diaphragm Design Seismic Force (ASCE/SEI 12.10.4)
6.8	Simplified Alternative Structural Design Criteria (ASCE/SEI 12.14)
6.9	Nonstructural Components (ASCE/SEI 13.3.1)
6.10	Nonbuilding Structures (ASCE/SEI Chapter 15)

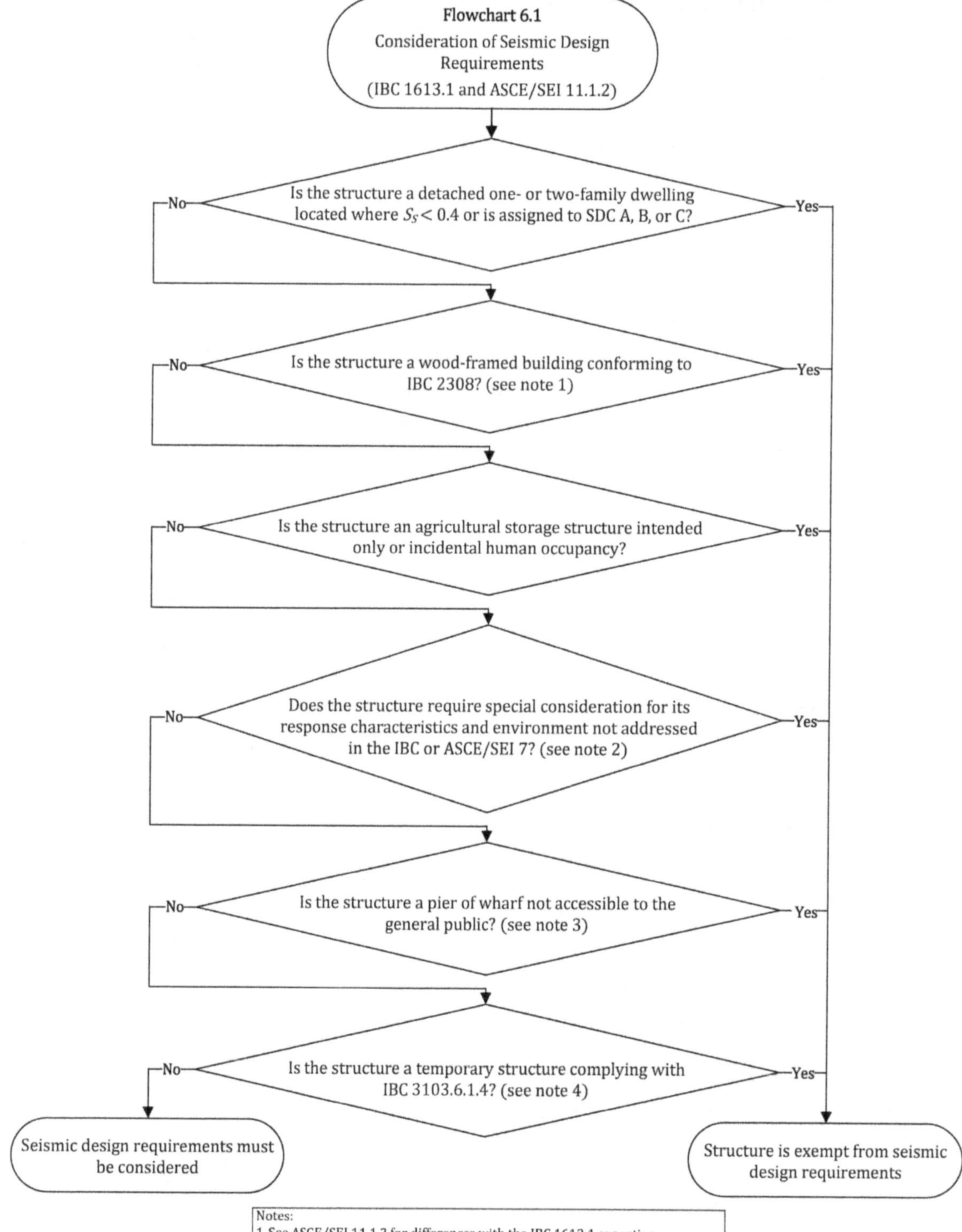
Flowchart 6.1
Consideration of Seismic Design Requirements
(IBC 1613.1 and ASCE/SEI 11.1.2)
Is the structure a detached one- or two-family dwelling located where S_S < 0.4 or is assigned to SDC A, B, or C?
No
Yes
Is the structure a wood-framed building conforming to IBC 2308? (see note 1)
No
Yes
Is the structure an agricultural storage structure intended only or incidental human occupancy?
No
Yes
Does the structure require special consideration for its response characteristics and environment not addressed in the IBC or ASCE/SEI 7? (see note 2)
No
Yes
Is the structure a pier of wharf not accessible to the general public? (see note 3)
No
Yes
Is the structure a temporary structure complying with IBC 3103.6.1.4? (see note 4)
No
Yes
Seismic design requirements must be considered
Structure is exempt from seismic design requirements

Notes:
1. See ASCE/SEI 11.1.2 for differences with the IBC 1613.1 exception.
2. Examples of such structures are included IBC 1613.1 and ASCE/SEI 11.1.2.
3. Piers and wharves are not in the exceptions in IBC 1613.1.
4. Temporary structures are not in the exemptions in ASCE/SEI 11.1.2.

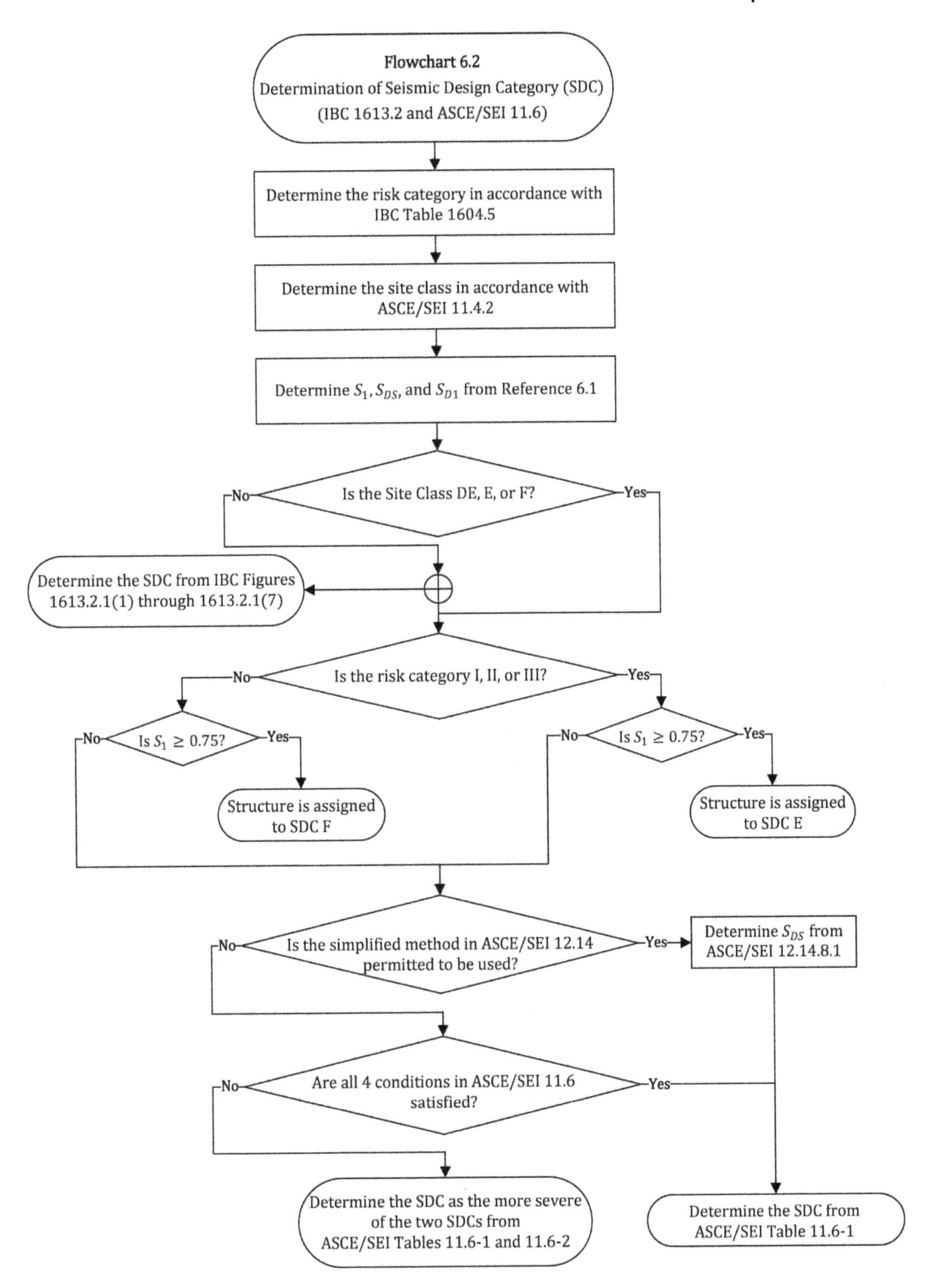
Flowchart 6.2
Determination of Seismic Design Category (SDC)
(IBC 1613.2 and ASCE/SEI 11.6)
Determine the risk category in accordance with IBC Table 1604.5
Determine the site class in accordance with ASCE/SEI 11.4.2
Determine S_1, S_{DS}, and S_{D1} from Reference 6.1
Is the Site Class DE, E, or F?
No
Yes
Determine the SDC from IBC Figures 1613.2.1(1) through 1613.2.1(7)
Is the risk category I, II, or III?
No
Yes
Is $S_1 \geq 0.75$?
No
Yes
Structure is assigned to SDC F
Is $S_1 \geq 0.75$?
No
Yes
Structure is assigned to SDC E
Is the simplified method in ASCE/SEI 12.14 permitted to be used?
No
Yes
Determine S_{DS} from ASCE/SEI 12.14.8.1
Are all 4 conditions in ASCE/SEI 11.6 satisfied?
No
Yes
Determine the SDC as the more severe of the two SDCs from ASCE/SEI Tables 11.6-1 and 11.6-2
Determine the SDC from ASCE/SEI Table 11.6-1

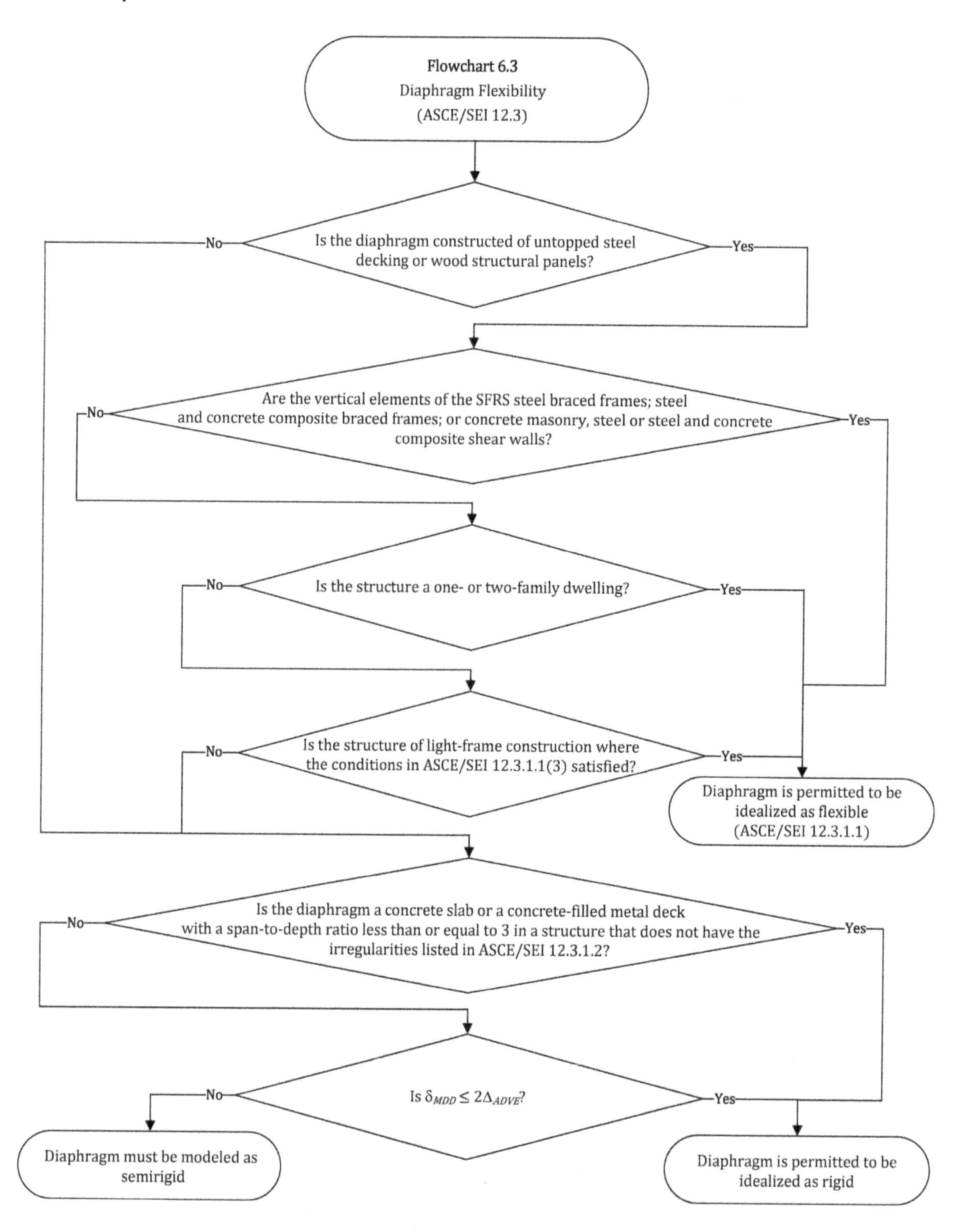
Flowchart 6.3
Diaphragm Flexibility
(ASCE/SEI 12.3)
Is the diaphragm constructed of untopped steel decking or wood structural panels?
No
Yes
Are the vertical elements of the SFRS steel braced frames; steel and concrete composite braced frames; or concrete masonry, steel or steel and concrete composite shear walls?
No
Yes
Is the structure a one- or two-family dwelling?
No
Yes
Is the structure of light-frame construction where the conditions in ASCE/SEI 12.3.1.1(3) satisfied?
No
Yes
Diaphragm is permitted to be idealized as flexible (ASCE/SEI 12.3.1.1)
Is the diaphragm a concrete slab or a concrete-filled metal deck with a span-to-depth ratio less than or equal to 3 in a structure that does not have the irregularities listed in ASCE/SEI 12.3.1.2?
No
Yes
Is $\delta_{MDD} \leq 2\Delta_{ADVE}$?
No
Yes
Diaphragm must be modeled as semirigid
Diaphragm is permitted to be idealized as rigid

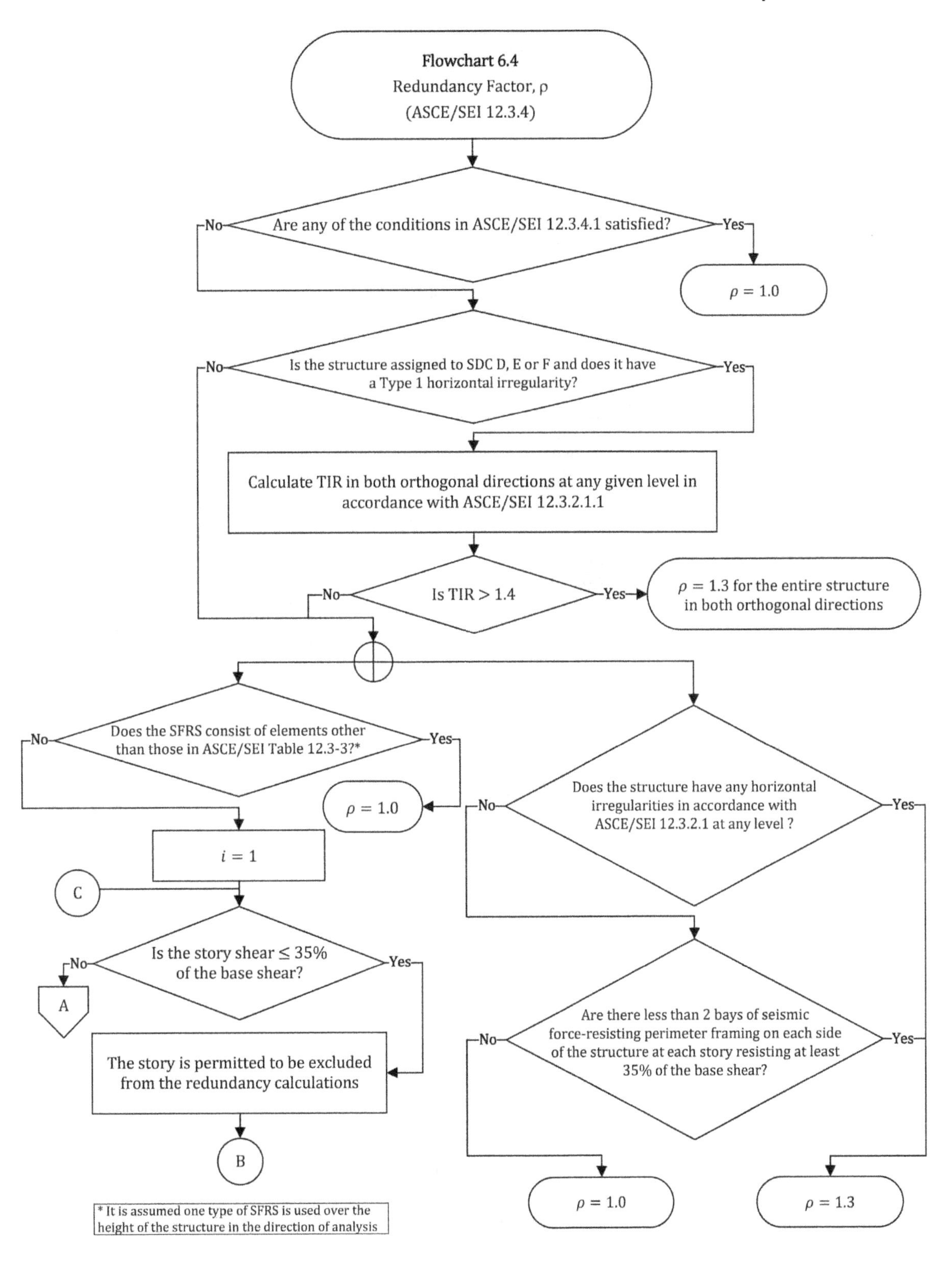

Flowchart 6.4
Redundancy Factor, ρ
(ASCE/SEI 12.3.4)
Are any of the conditions in ASCE/SEI 12.3.4.1 satisfied?
No
Yes
ρ = 1.0
Is the structure assigned to SDC D, E or F and does it have a Type 1 horizontal irregularity?
No
Yes
Calculate TIR in both orthogonal directions at any given level in accordance with ASCE/SEI 12.3.2.1.1
Is TIR > 1.4
No
Yes
ρ = 1.3 for the entire structure in both orthogonal directions
Does the SFRS consist of elements other than those in ASCE/SEI Table 12.3-3?*
No
Yes
ρ = 1.0
i = 1
C
Is the story shear ≤ 35% of the base shear?
No
Yes
A
The story is permitted to be excluded from the redundancy calculations
B
* It is assumed one type of SFRS is used over the height of the structure in the direction of analysis
Does the structure have any horizontal irregularities in accordance with ASCE/SEI 12.3.2.1 at any level ?
No
Yes
Are there less than 2 bays of seismic force-resisting perimeter framing on each side of the structure at each story resisting at least 35% of the base shear?
No
Yes
ρ = 1.0
ρ = 1.3

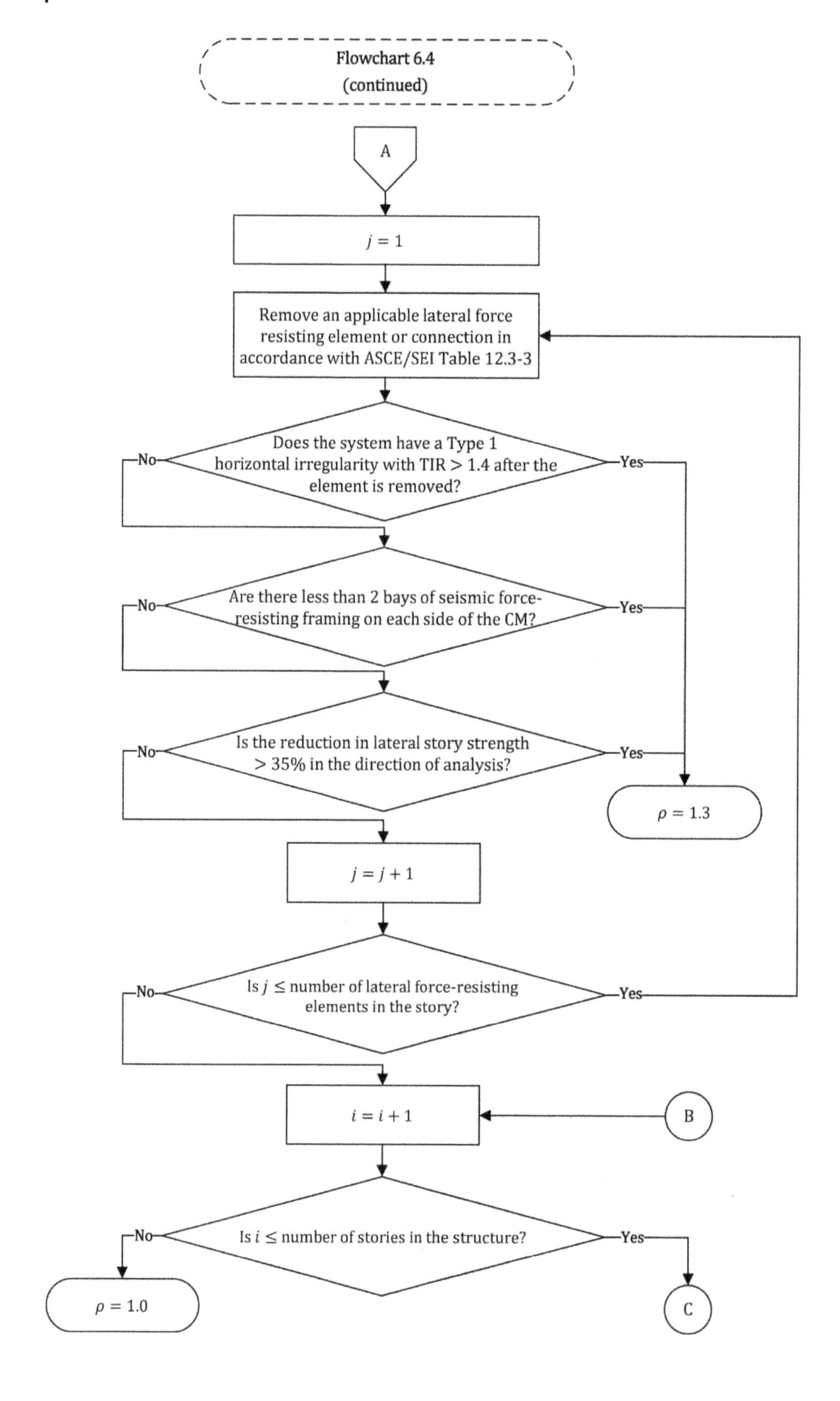
Flowchart 6.4
(continued)
A
$j = 1$
Remove an applicable lateral force resisting element or connection in accordance with ASCE/SEI Table 12.3-3
Does the system have a Type 1 horizontal irregularity with TIR > 1.4 after the element is removed?
No
Yes
Are there less than 2 bays of seismic force-resisting framing on each side of the CM?
No
Yes
Is the reduction in lateral story strength > 35% in the direction of analysis?
No
Yes
$\rho = 1.3$
$j = j + 1$
Is $j \leq$ number of lateral force-resisting elements in the story?
No
Yes
$i = i + 1$
B
Is $i \leq$ number of stories in the structure?
No
Yes
$\rho = 1.0$
C

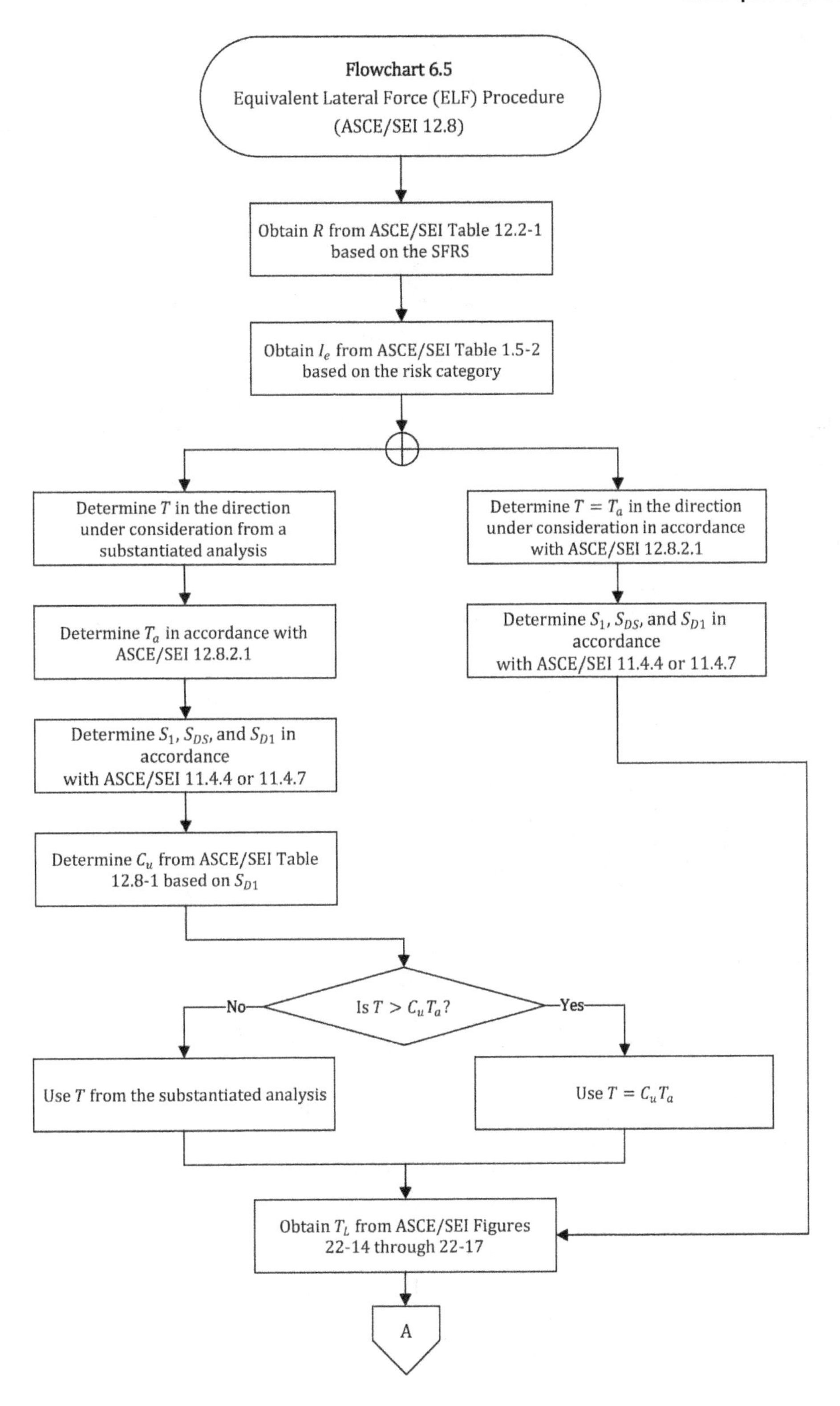
Flowchart 6.5
Equivalent Lateral Force (ELF) Procedure
(ASCE/SEI 12.8)
Obtain R from ASCE/SEI Table 12.2-1 based on the SFRS
Obtain I_e from ASCE/SEI Table 1.5-2 based on the risk category
Determine T in the direction under consideration from a substantiated analysis
Determine $T = T_a$ in the direction under consideration in accordance with ASCE/SEI 12.8.2.1
Determine T_a in accordance with ASCE/SEI 12.8.2.1
Determine S_1, S_{DS}, and S_{D1} in accordance with ASCE/SEI 11.4.4 or 11.4.7
Determine S_1, S_{DS}, and S_{D1} in accordance with ASCE/SEI 11.4.4 or 11.4.7
Determine C_u from ASCE/SEI Table 12.8-1 based on S_{D1}
Is $T > C_u T_a$?
No
Yes
Use T from the substantiated analysis
Use $T = C_u T_a$
Obtain T_L from ASCE/SEI Figures 22-14 through 22-17
A

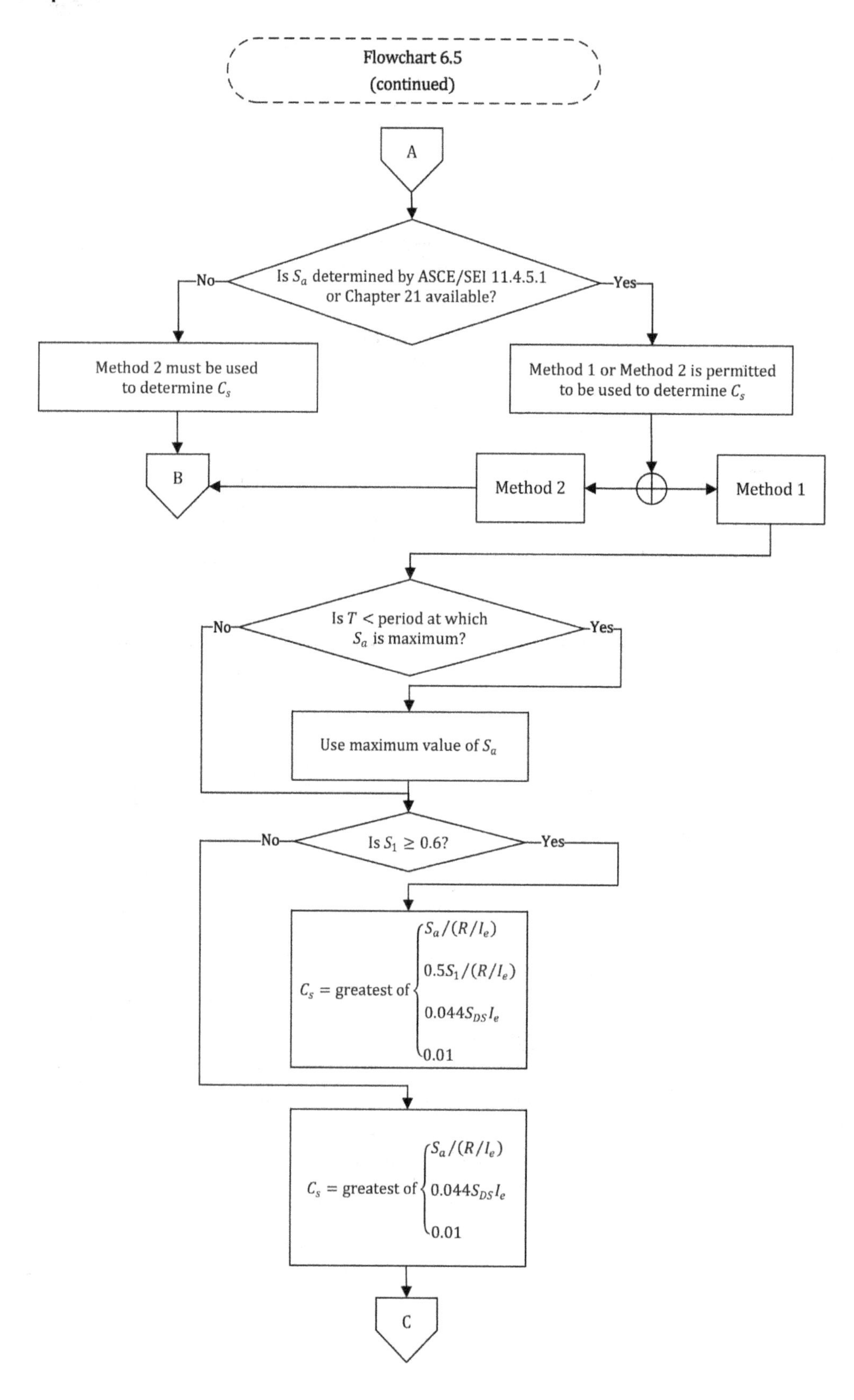

Flowchart 6.5
(continued)
A
Is S_a determined by ASCE/SEI 11.4.5.1 or Chapter 21 available?
No
Yes
Method 2 must be used to determine C_s
Method 1 or Method 2 is permitted to be used to determine C_s
B
Method 2
Method 1
Is $T <$ period at which S_a is maximum?
No
Yes
Use maximum value of S_a
Is $S_1 \geq 0.6$?
No
Yes
C_s = greatest of $\begin{cases} S_a/(R/I_e) \\ 0.5S_1/(R/I_e) \\ 0.044S_{DS}I_e \\ 0.01 \end{cases}$
C_s = greatest of $\begin{cases} S_a/(R/I_e) \\ 0.044S_{DS}I_e \\ 0.01 \end{cases}$
C

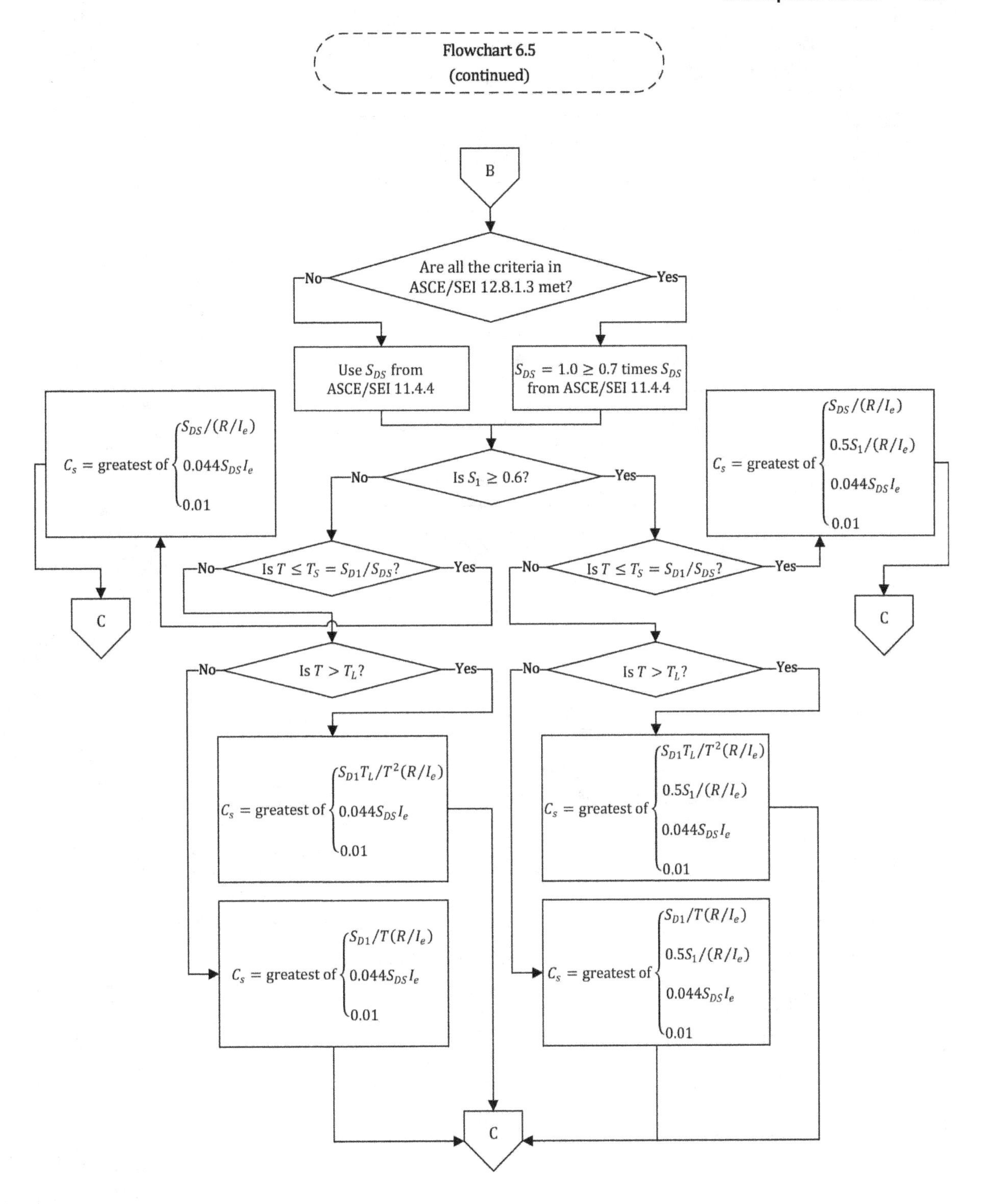

Flowchart 6.5
(continued)
B
Are all the criteria in ASCE/SEI 12.8.1.3 met?
No
Yes
Use S_{DS} from ASCE/SEI 11.4.4
$S_{DS} = 1.0 \geq 0.7$ times S_{DS} from ASCE/SEI 11.4.4
Is $S_1 \geq 0.6$?
No
Yes
C_s = greatest of $S_{DS}/(R/I_e)$, $0.044S_{DS}I_e$, 0.01
C_s = greatest of $S_{DS}/(R/I_e)$, $0.5S_1/(R/I_e)$, $0.044S_{DS}I_e$, 0.01
Is $T \leq T_S = S_{D1}/S_{DS}$?
Is $T > T_L$?
C_s = greatest of $S_{D1}T_L/T^2(R/I_e)$, $0.044S_{DS}I_e$, 0.01
C_s = greatest of $S_{D1}T_L/T^2(R/I_e)$, $0.5S_1/(R/I_e)$, $0.044S_{DS}I_e$, 0.01
C_s = greatest of $S_{D1}/T(R/I_e)$, $0.044S_{DS}I_e$, 0.01
C_s = greatest of $S_{D1}/T(R/I_e)$, $0.5S_1/(R/I_e)$, $0.044S_{DS}I_e$, 0.01
C

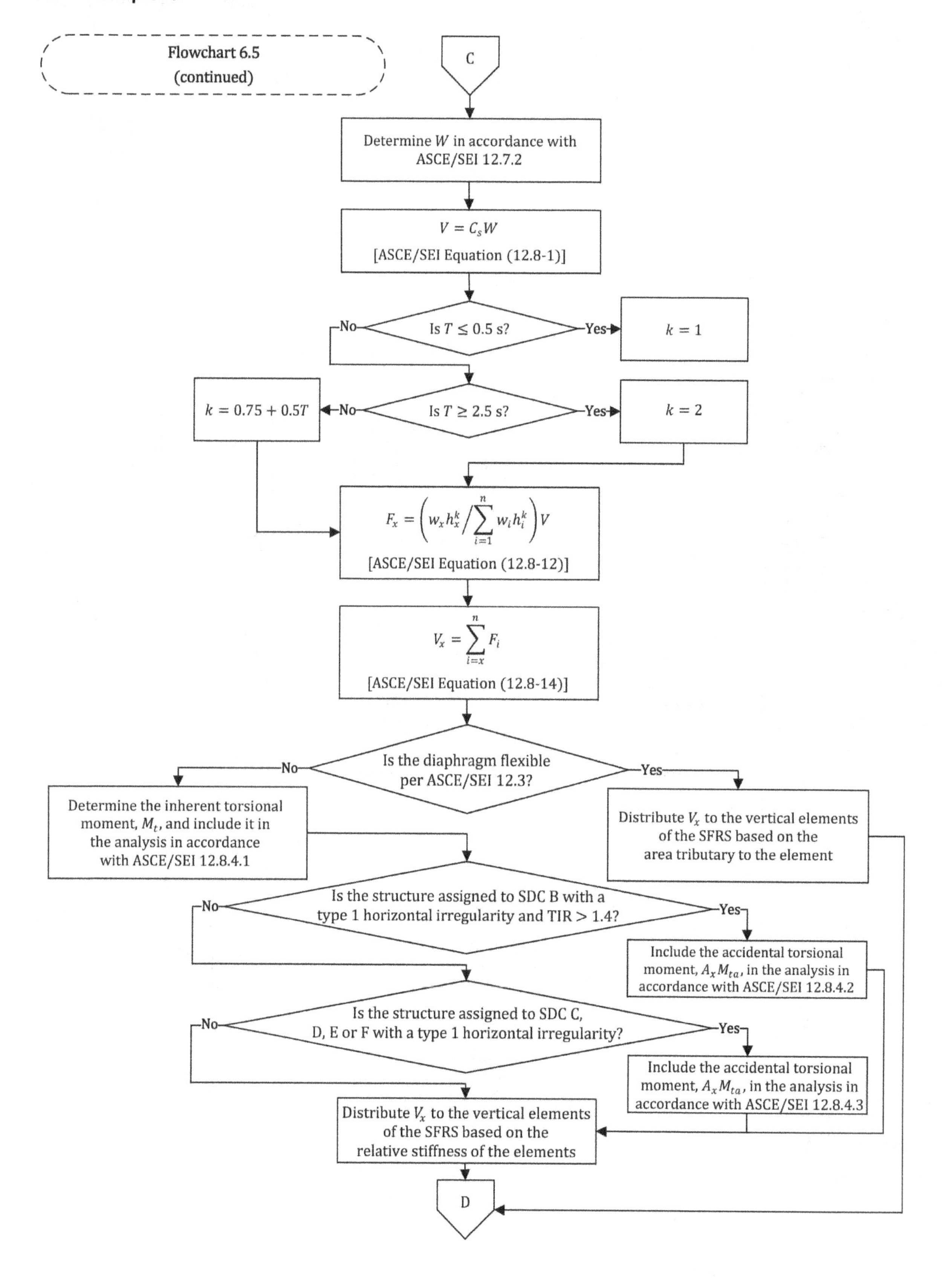

Flowchart 6.5
(continued)
C
Determine W in accordance with ASCE/SEI 12.7.2
$V = C_s W$
[ASCE/SEI Equation (12.8-1)]
Is $T \leq 0.5$ s?
No
Yes
$k = 1$
Is $T \geq 2.5$ s?
No
Yes
$k = 0.75 + 0.5T$
$k = 2$
$F_x = \left(w_x h_x^k / \sum_{i=1}^{n} w_i h_i^k \right) V$
[ASCE/SEI Equation (12.8-12)]
$V_x = \sum_{i=x}^{n} F_i$
[ASCE/SEI Equation (12.8-14)]
Is the diaphragm flexible per ASCE/SEI 12.3?
No
Yes
Determine the inherent torsional moment, M_t, and include it in the analysis in accordance with ASCE/SEI 12.8.4.1
Distribute V_x to the vertical elements of the SFRS based on the area tributary to the element
Is the structure assigned to SDC B with a type 1 horizontal irregularity and TIR > 1.4?
No
Yes
Include the accidental torsional moment, $A_x M_{ta}$, in the analysis in accordance with ASCE/SEI 12.8.4.2
Is the structure assigned to SDC C, D, E or F with a type 1 horizontal irregularity?
No
Yes
Include the accidental torsional moment, $A_x M_{ta}$, in the analysis in accordance with ASCE/SEI 12.8.4.3
Distribute V_x to the vertical elements of the SFRS based on the relative stiffness of the elements
D

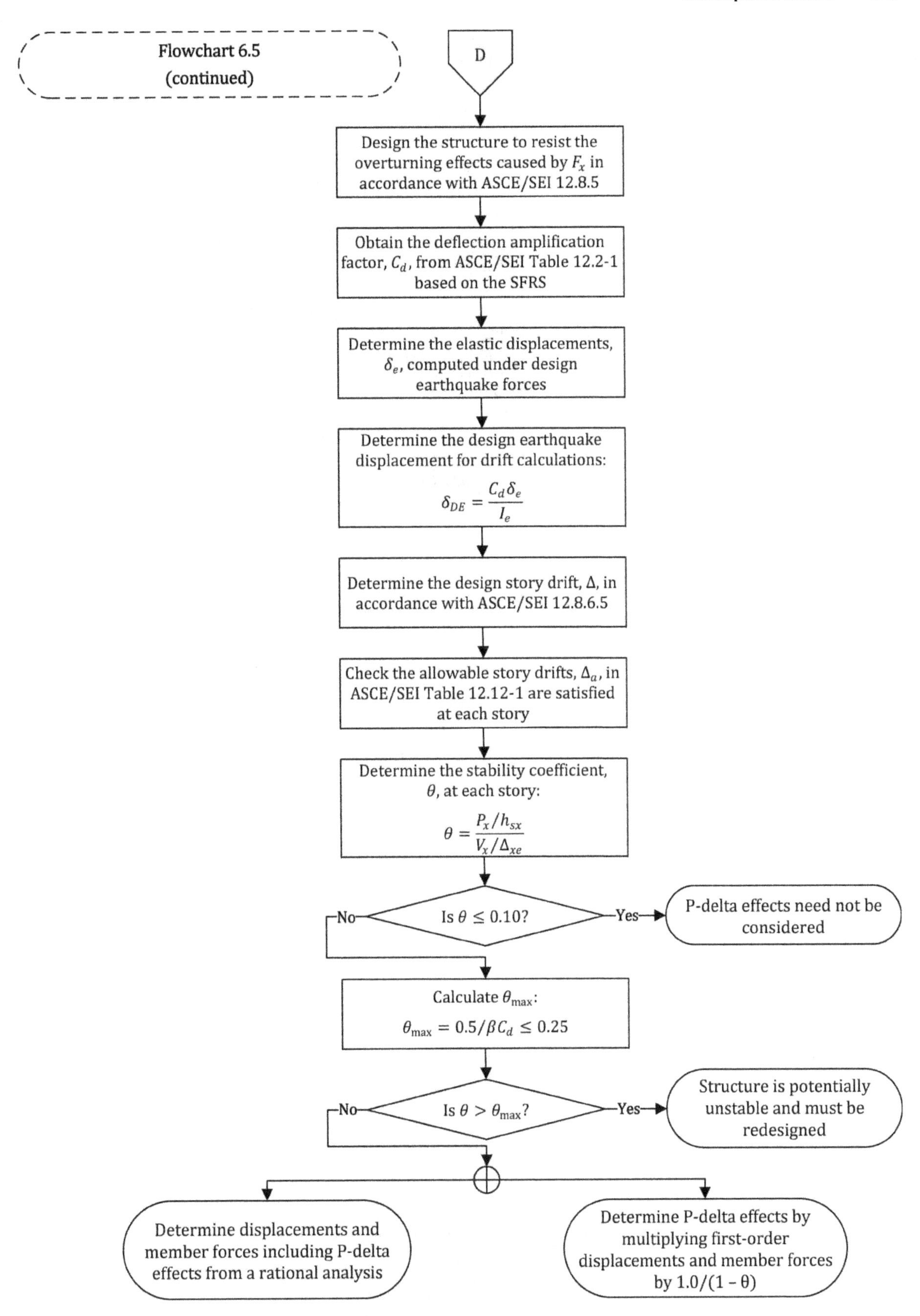
Flowchart 6.5
(continued)
D
Design the structure to resist the overturning effects caused by F_x in accordance with ASCE/SEI 12.8.5
Obtain the deflection amplification factor, C_d, from ASCE/SEI Table 12.2-1 based on the SFRS
Determine the elastic displacements, δ_e, computed under design earthquake forces
Determine the design earthquake displacement for drift calculations:
$\delta_{DE} = \frac{C_d \delta_e}{I_e}$
Determine the design story drift, Δ, in accordance with ASCE/SEI 12.8.6.5
Check the allowable story drifts, Δ_a, in ASCE/SEI Table 12.12-1 are satisfied at each story
Determine the stability coefficient, θ, at each story:
$\theta = \frac{P_x / h_{sx}}{V_x / \Delta_{xe}}$
Is $\theta \leq 0.10$?
No
Yes
P-delta effects need not be considered
Calculate θ_{max}:
$\theta_{max} = 0.5/\beta C_d \leq 0.25$
Is $\theta > \theta_{max}$?
No
Yes
Structure is potentially unstable and must be redesigned
Determine displacements and member forces including P-delta effects from a rational analysis
Determine P-delta effects by multiplying first-order displacements and member forces by 1.0/(1 - θ)

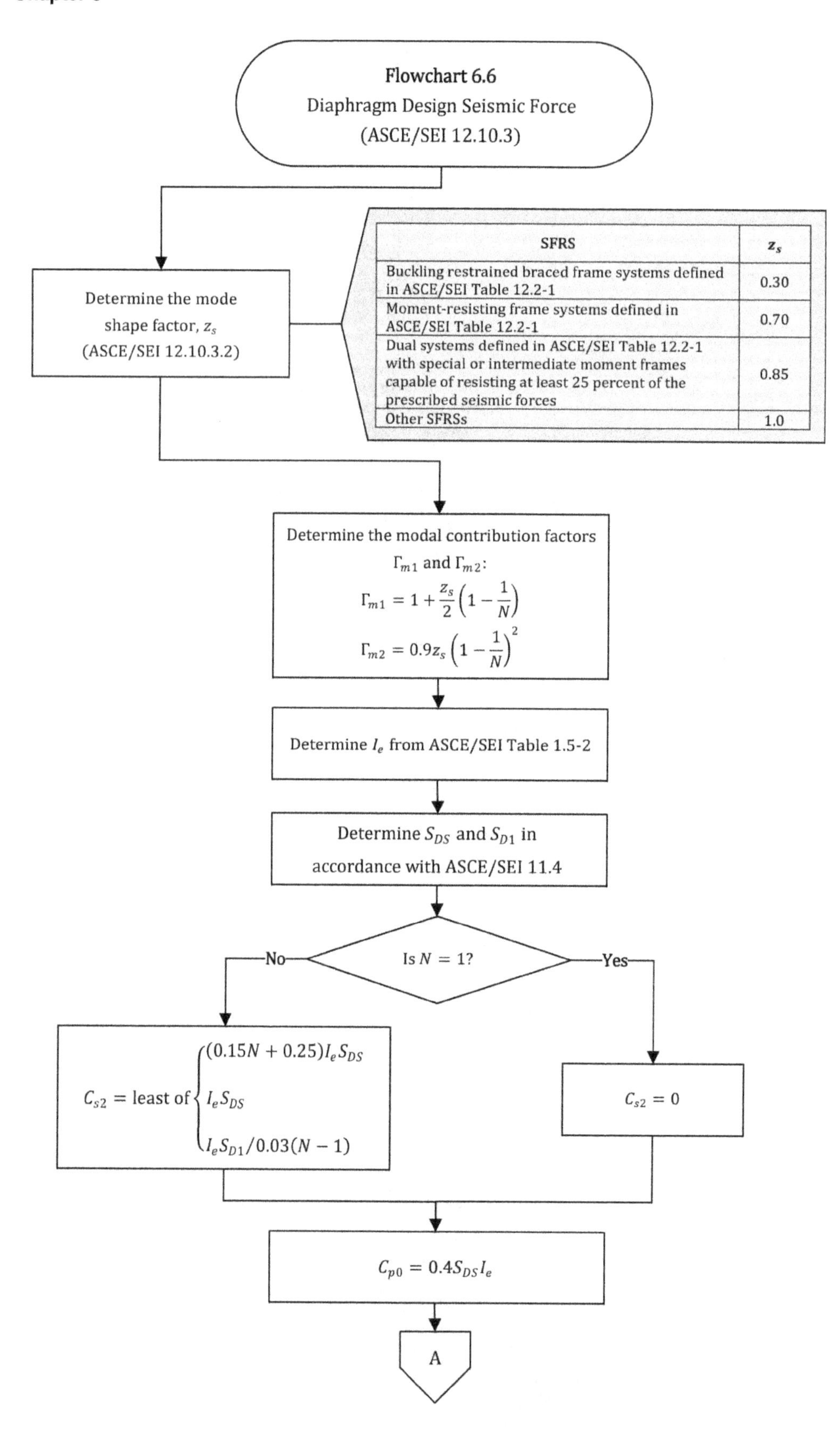

Flowchart 6.6
Diaphragm Design Seismic Force
(ASCE/SEI 12.10.3)
Determine the mode shape factor, z_s (ASCE/SEI 12.10.3.2)
SFRS
z_s
Buckling restrained braced frame systems defined in ASCE/SEI Table 12.2-1
0.30
Moment-resisting frame systems defined in ASCE/SEI Table 12.2-1
0.70
Dual systems defined in ASCE/SEI Table 12.2-1 with special or intermediate moment frames capable of resisting at least 25 percent of the prescribed seismic forces
0.85
Other SFRSs
1.0
Determine the modal contribution factors Γ_{m1} and Γ_{m2}:
$\Gamma_{m1} = 1 + \frac{z_s}{2}\left(1 - \frac{1}{N}\right)$
$\Gamma_{m2} = 0.9z_s\left(1 - \frac{1}{N}\right)^2$
Determine I_e from ASCE/SEI Table 1.5-2
Determine S_{DS} and S_{D1} in accordance with ASCE/SEI 11.4
Is $N = 1$?
No
Yes
$C_{s2} = \text{least of} \begin{cases} (0.15N + 0.25)I_eS_{DS} \\ I_eS_{DS} \\ I_eS_{D1}/0.03(N-1) \end{cases}$
$C_{s2} = 0$
$C_{p0} = 0.4S_{DS}I_e$
A

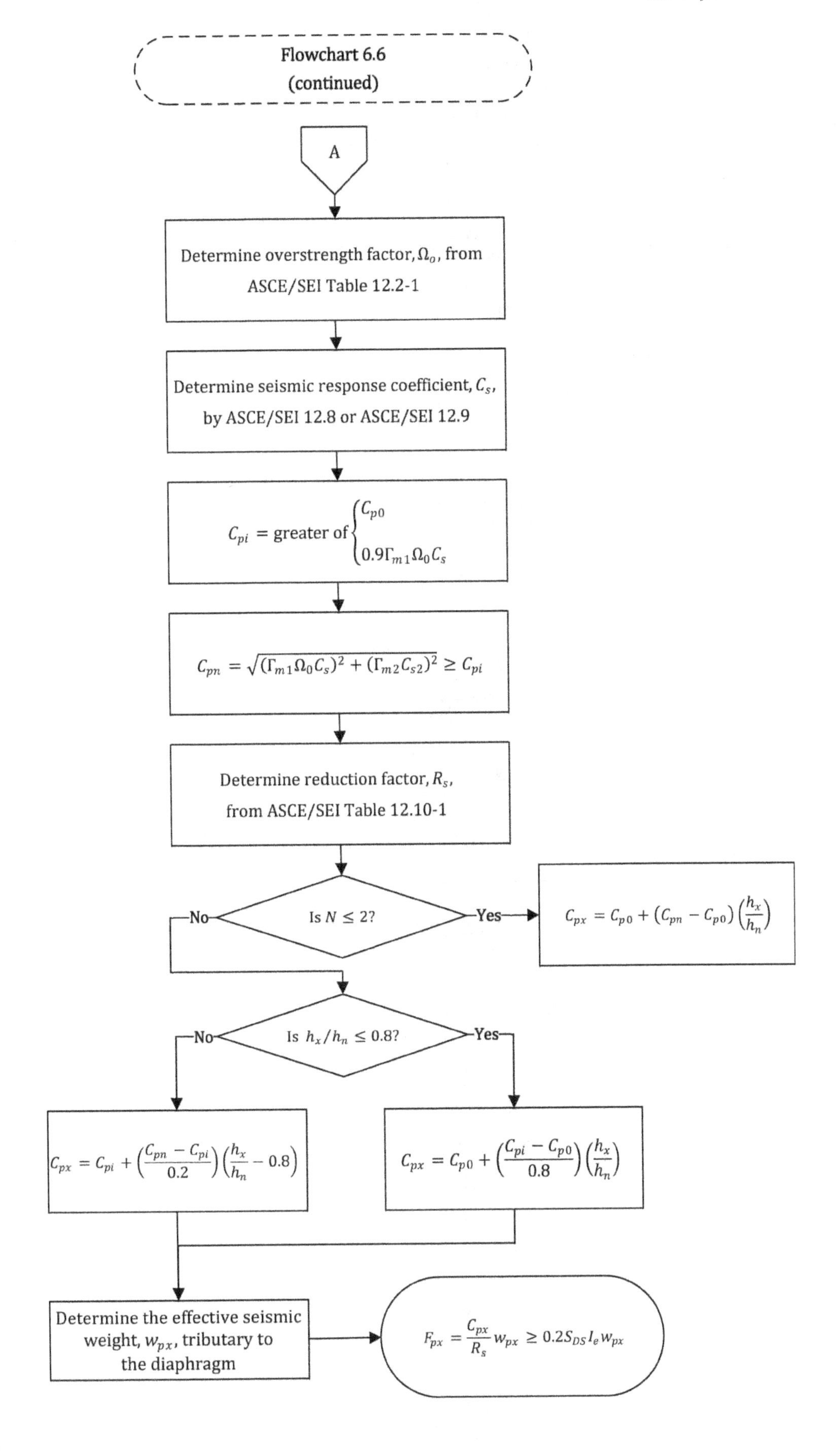

Flowchart 6.6
(continued)
A
Determine overstrength factor, Ω_o, from ASCE/SEI Table 12.2-1
Determine seismic response coefficient, C_s, by ASCE/SEI 12.8 or ASCE/SEI 12.9
C_{pi} = greater of $\begin{cases} C_{p0} \\ 0.9\Gamma_{m1}\Omega_0 C_s \end{cases}$
$C_{pn} = \sqrt{(\Gamma_{m1}\Omega_0 C_s)^2 + (\Gamma_{m2}C_{s2})^2} \geq C_{pi}$
Determine reduction factor, R_s, from ASCE/SEI Table 12.10-1
Is $N \leq 2$?
No
Yes
$C_{px} = C_{p0} + (C_{pn} - C_{p0})\left(\frac{h_x}{h_n}\right)$
Is $h_x/h_n \leq 0.8$?
No
Yes
$C_{px} = C_{pi} + \left(\frac{C_{pn} - C_{pi}}{0.2}\right)\left(\frac{h_x}{h_n} - 0.8\right)$
$C_{px} = C_{p0} + \left(\frac{C_{pi} - C_{p0}}{0.8}\right)\left(\frac{h_x}{h_n}\right)$
Determine the effective seismic weight, w_{px}, tributary to the diaphragm
$F_{px} = \frac{C_{px}}{R_s} w_{px} \geq 0.2 S_{DS} I_e w_{px}$

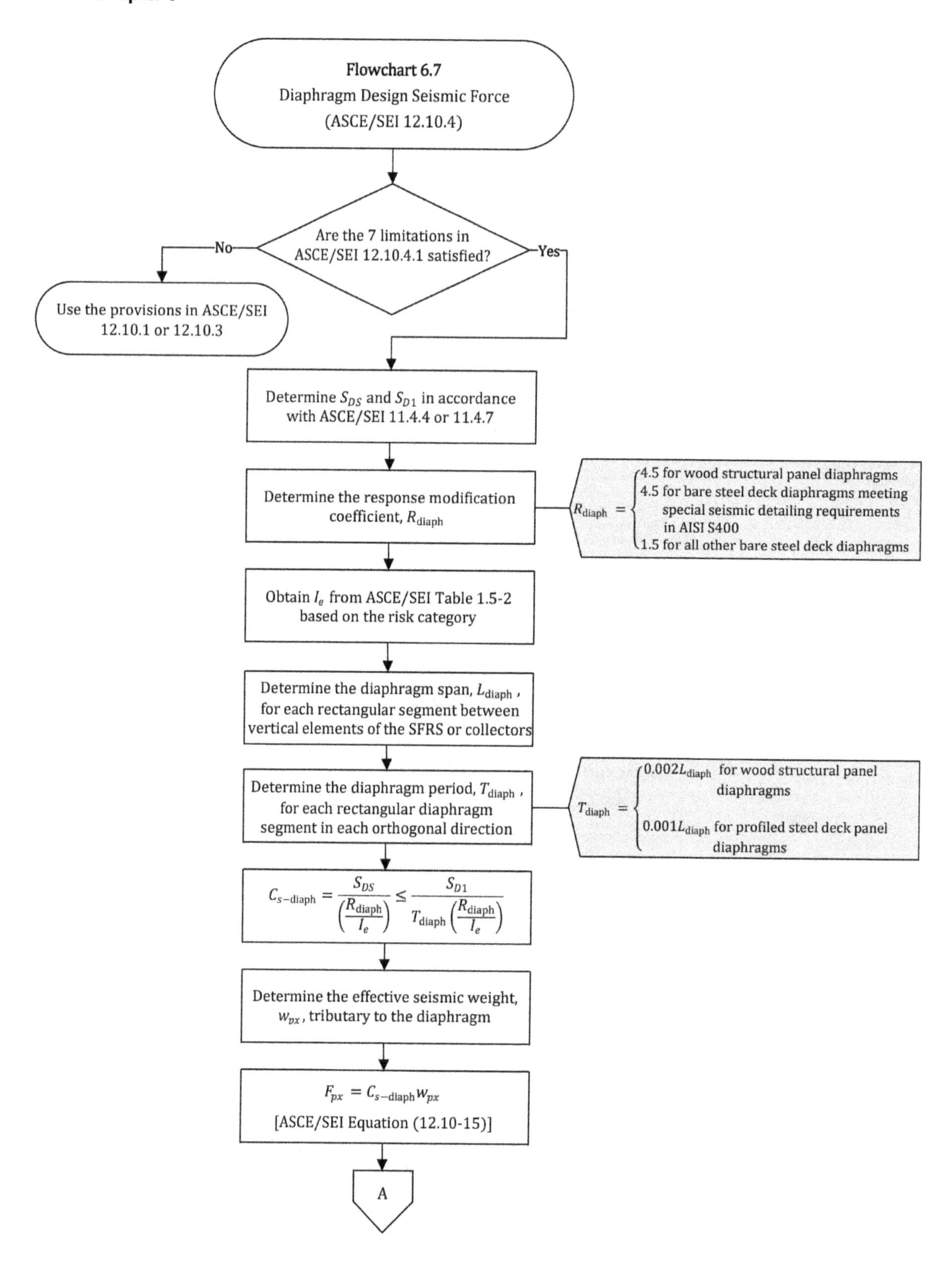

Flowchart 6.7
Diaphragm Design Seismic Force
(ASCE/SEI 12.10.4)
Are the 7 limitations in ASCE/SEI 12.10.4.1 satisfied?
No
Yes
Use the provisions in ASCE/SEI 12.10.1 or 12.10.3
Determine S_{DS} and S_{D1} in accordance with ASCE/SEI 11.4.4 or 11.4.7
Determine the response modification coefficient, R_{diaph}
$R_{\text{diaph}} = \begin{cases} 4.5 \text{ for wood structural panel diaphragms} \\ 4.5 \text{ for bare steel deck diaphragms meeting special seismic detailing requirements in AISI S400} \\ 1.5 \text{ for all other bare steel deck diaphragms} \end{cases}$
Obtain I_e from ASCE/SEI Table 1.5-2 based on the risk category
Determine the diaphragm span, L_{diaph}, for each rectangular segment between vertical elements of the SFRS or collectors
Determine the diaphragm period, T_{diaph}, for each rectangular diaphragm segment in each orthogonal direction
$T_{\text{diaph}} = \begin{cases} 0.002L_{\text{diaph}} \text{ for wood structural panel diaphragms} \\ 0.001L_{\text{diaph}} \text{ for profiled steel deck panel diaphragms} \end{cases}$
$C_{s-\text{diaph}} = \dfrac{S_{DS}}{\left(\dfrac{R_{\text{diaph}}}{I_e}\right)} \leq \dfrac{S_{D1}}{T_{\text{diaph}}\left(\dfrac{R_{\text{diaph}}}{I_e}\right)}$
Determine the effective seismic weight, w_{px}, tributary to the diaphragm
$F_{px} = C_{s-\text{diaph}} w_{px}$
[ASCE/SEI Equation (12.10-15)]
A

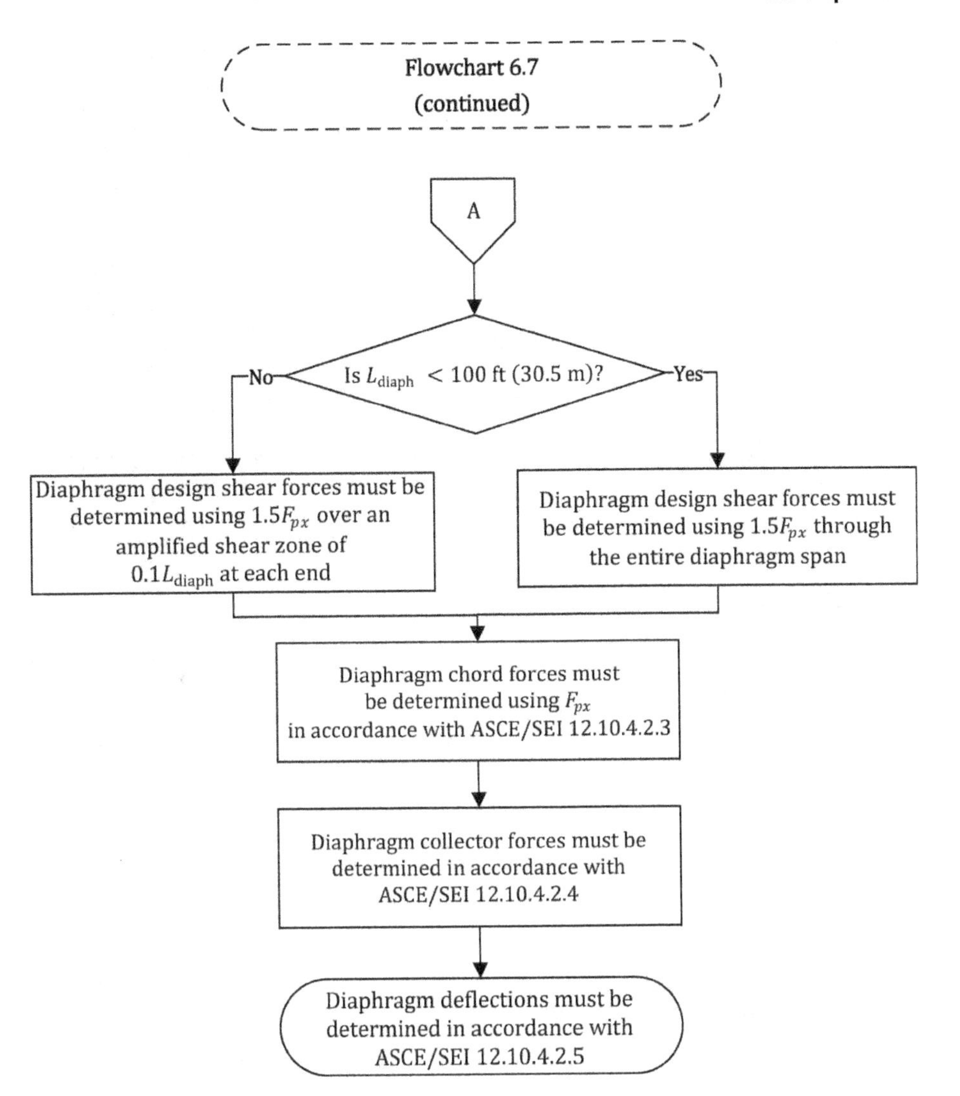

Flowchart 6.7
(continued)
A
Is $L_{\text{diaph}} < 100$ ft (30.5 m)?
No
Yes
Diaphragm design shear forces must be determined using $1.5F_{px}$ over an amplified shear zone of $0.1L_{\text{diaph}}$ at each end
Diaphragm design shear forces must be determined using $1.5F_{px}$ through the entire diaphragm span
Diaphragm chord forces must be determined using F_{px} in accordance with ASCE/SEI 12.10.4.2.3
Diaphragm collector forces must be determined in accordance with ASCE/SEI 12.10.4.2.4
Diaphragm deflections must be determined in accordance with ASCE/SEI 12.10.4.2.5

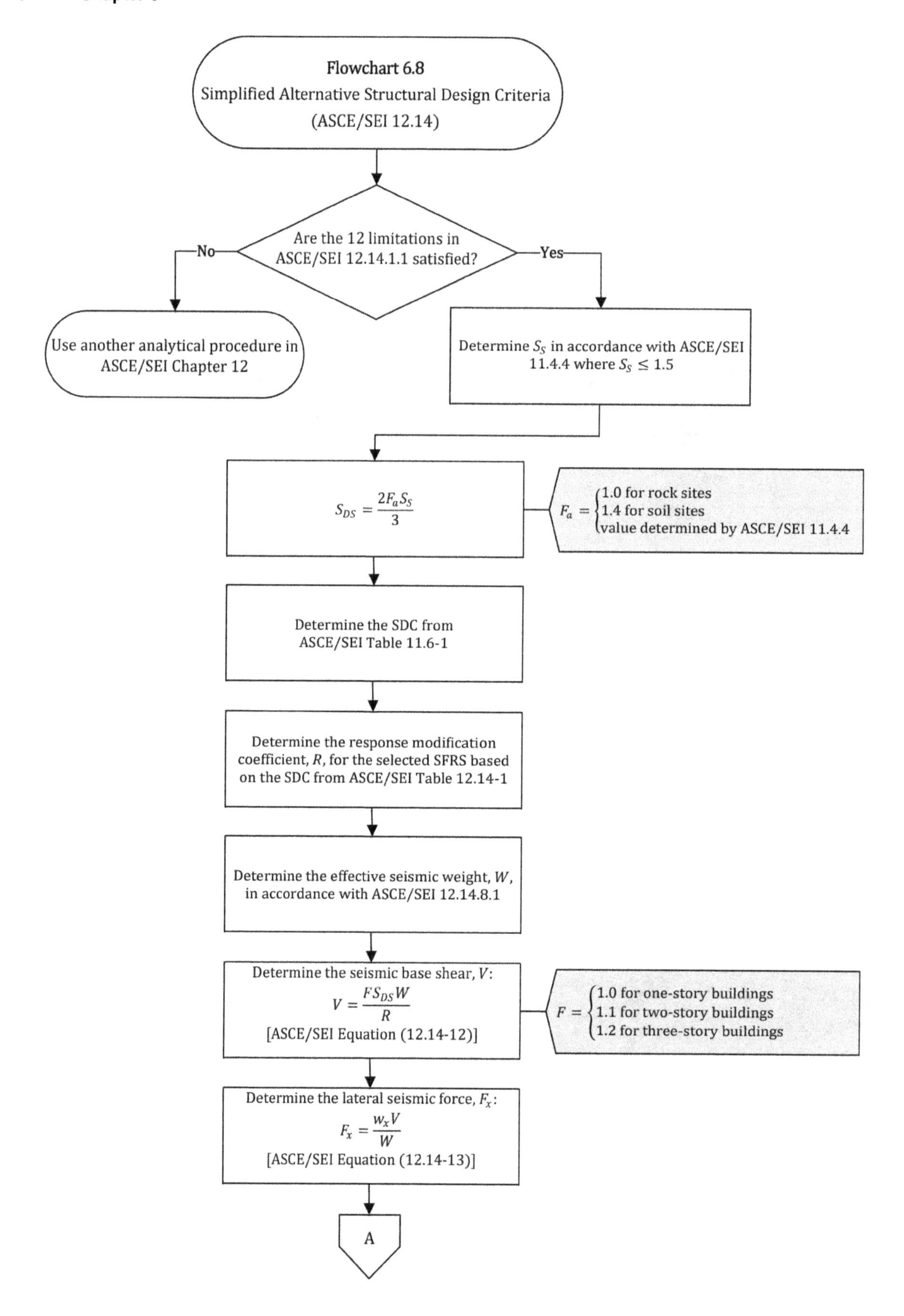
Flowchart 6.8
Simplified Alternative Structural Design Criteria
(ASCE/SEI 12.14)
Are the 12 limitations in ASCE/SEI 12.14.1.1 satisfied?
No
Yes
Use another analytical procedure in ASCE/SEI Chapter 12
Determine S_S in accordance with ASCE/SEI 11.4.4 where $S_S \leq 1.5$
$S_{DS} = \frac{2F_a S_S}{3}$
$F_a = \begin{cases} 1.0 \text{ for rock sites} \\ 1.4 \text{ for soil sites} \\ \text{value determined by ASCE/SEI 11.4.4} \end{cases}$
Determine the SDC from ASCE/SEI Table 11.6-1
Determine the response modification coefficient, R, for the selected SFRS based on the SDC from ASCE/SEI Table 12.14-1
Determine the effective seismic weight, W, in accordance with ASCE/SEI 12.14.8.1
Determine the seismic base shear, V:
$V = \frac{F S_{DS} W}{R}$
[ASCE/SEI Equation (12.14-12)]
$F = \begin{cases} 1.0 \text{ for one-story buildings} \\ 1.1 \text{ for two-story buildings} \\ 1.2 \text{ for three-story buildings} \end{cases}$
Determine the lateral seismic force, F_x:
$F_x = \frac{w_x V}{W}$
[ASCE/SEI Equation (12.14-13)]
A

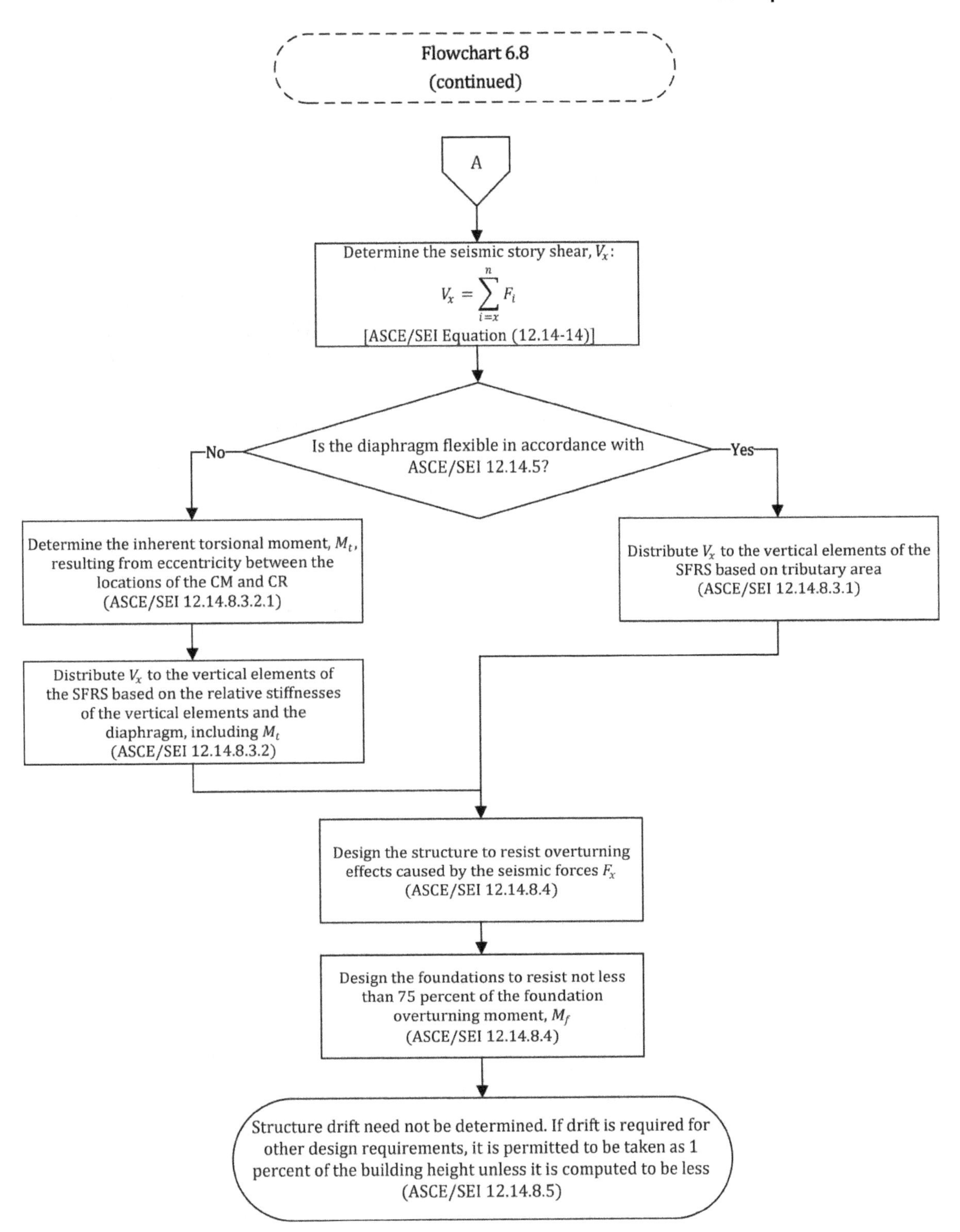
Flowchart 6.8
(continued)
A
Determine the seismic story shear, V_x:
$V_x = \sum_{i=x}^{n} F_i$
[ASCE/SEI Equation (12.14-14)]
Is the diaphragm flexible in accordance with ASCE/SEI 12.14.5?
No
Yes
Determine the inherent torsional moment, M_t, resulting from eccentricity between the locations of the CM and CR (ASCE/SEI 12.14.8.3.2.1)
Distribute V_x to the vertical elements of the SFRS based on tributary area (ASCE/SEI 12.14.8.3.1)
Distribute V_x to the vertical elements of the SFRS based on the relative stiffnesses of the vertical elements and the diaphragm, including M_t (ASCE/SEI 12.14.8.3.2)
Design the structure to resist overturning effects caused by the seismic forces F_x (ASCE/SEI 12.14.8.4)
Design the foundations to resist not less than 75 percent of the foundation overturning moment, M_f (ASCE/SEI 12.14.8.4)
Structure drift need not be determined. If drift is required for other design requirements, it is permitted to be taken as 1 percent of the building height unless it is computed to be less (ASCE/SEI 12.14.8.5)

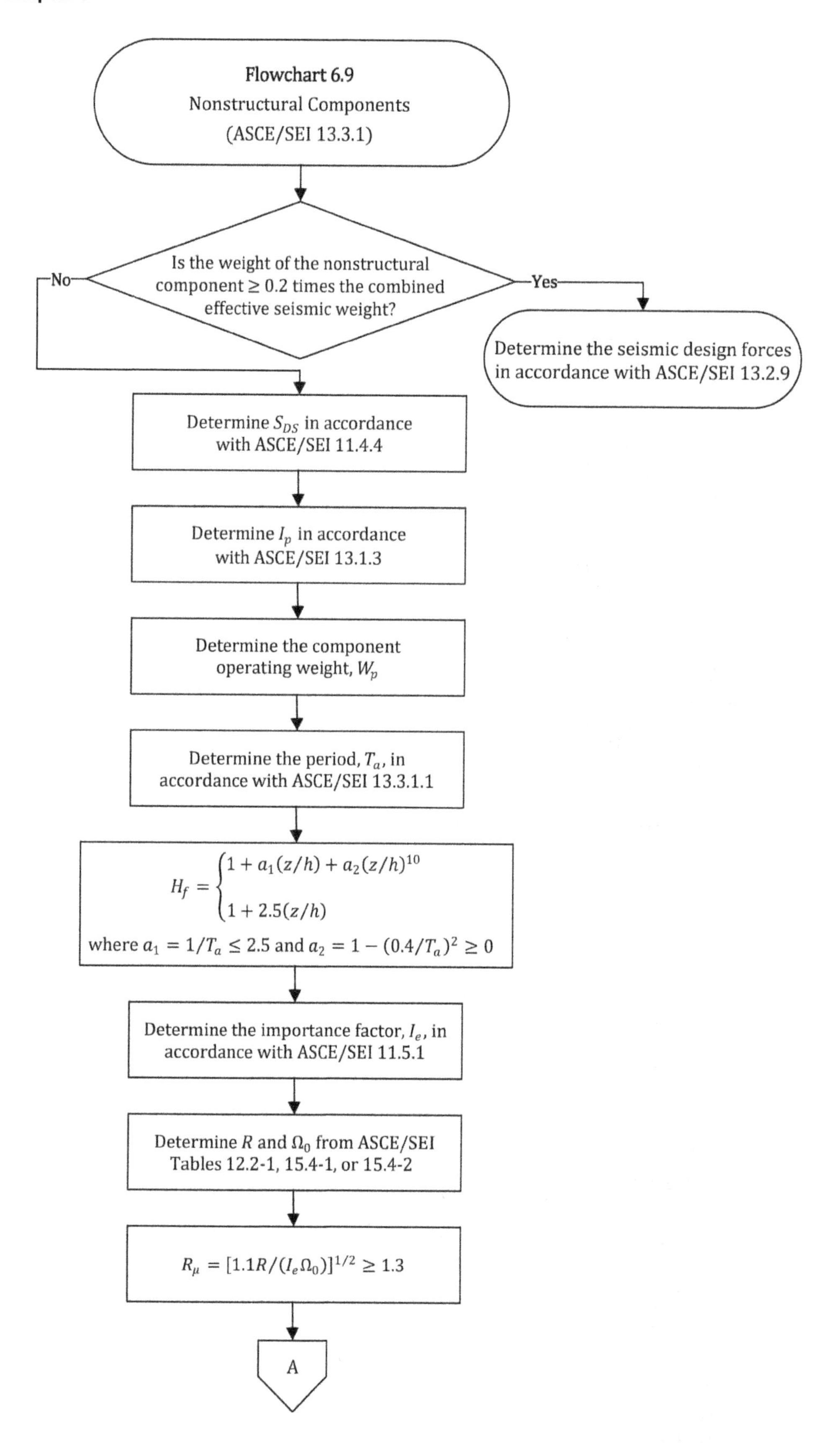
Flowchart 6.9
Nonstructural Components
(ASCE/SEI 13.3.1)
Is the weight of the nonstructural component ≥ 0.2 times the combined effective seismic weight?
No
Yes
Determine the seismic design forces in accordance with ASCE/SEI 13.2.9
Determine S_{DS} in accordance with ASCE/SEI 11.4.4
Determine I_p in accordance with ASCE/SEI 13.1.3
Determine the component operating weight, W_p
Determine the period, T_a, in accordance with ASCE/SEI 13.3.1.1
$H_f = \begin{cases} 1 + a_1(z/h) + a_2(z/h)^{10} \\ 1 + 2.5(z/h) \end{cases}$
where $a_1 = 1/T_a \leq 2.5$ and $a_2 = 1 - (0.4/T_a)^2 \geq 0$
Determine the importance factor, I_e, in accordance with ASCE/SEI 11.5.1
Determine R and Ω_0 from ASCE/SEI Tables 12.2-1, 15.4-1, or 15.4-2
$R_\mu = [1.1R/(I_e\Omega_0)]^{1/2} \geq 1.3$
A

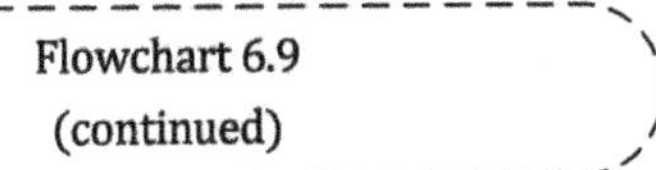
Flowchart 6.9
(continued)

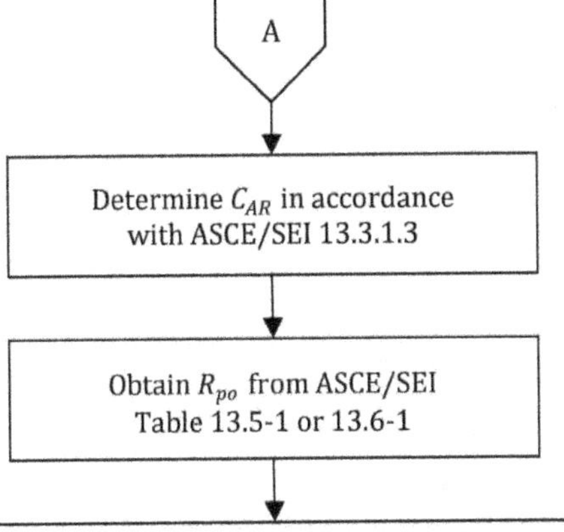
A
Determine C_{AR} in accordance with ASCE/SEI 13.3.1.3
Obtain R_{po} from ASCE/SEI Table 13.5-1 or 13.6-1

Determine the horizontal seismic design force, F_p:

$$F_p = 0.4 S_{DS} I_p W_p \left(\frac{H_f}{R_\mu}\right)\left(\frac{C_{AR}}{R_{po}}\right)$$

$$\geq 0.3 S_{DS} I_p W_p$$

$$\leq 1.6 S_{DS} I_p W_p$$

[ASCE/SEI Equations (13.3-1), (13.3-2), and (13.3-3)]

Determine the vertical seismic force, E_v:

$$E_v = 0.2 S_{DS} W_p$$

Consider seismic relative displacements in accordance with ASCE/SEI 13.3.2 where required

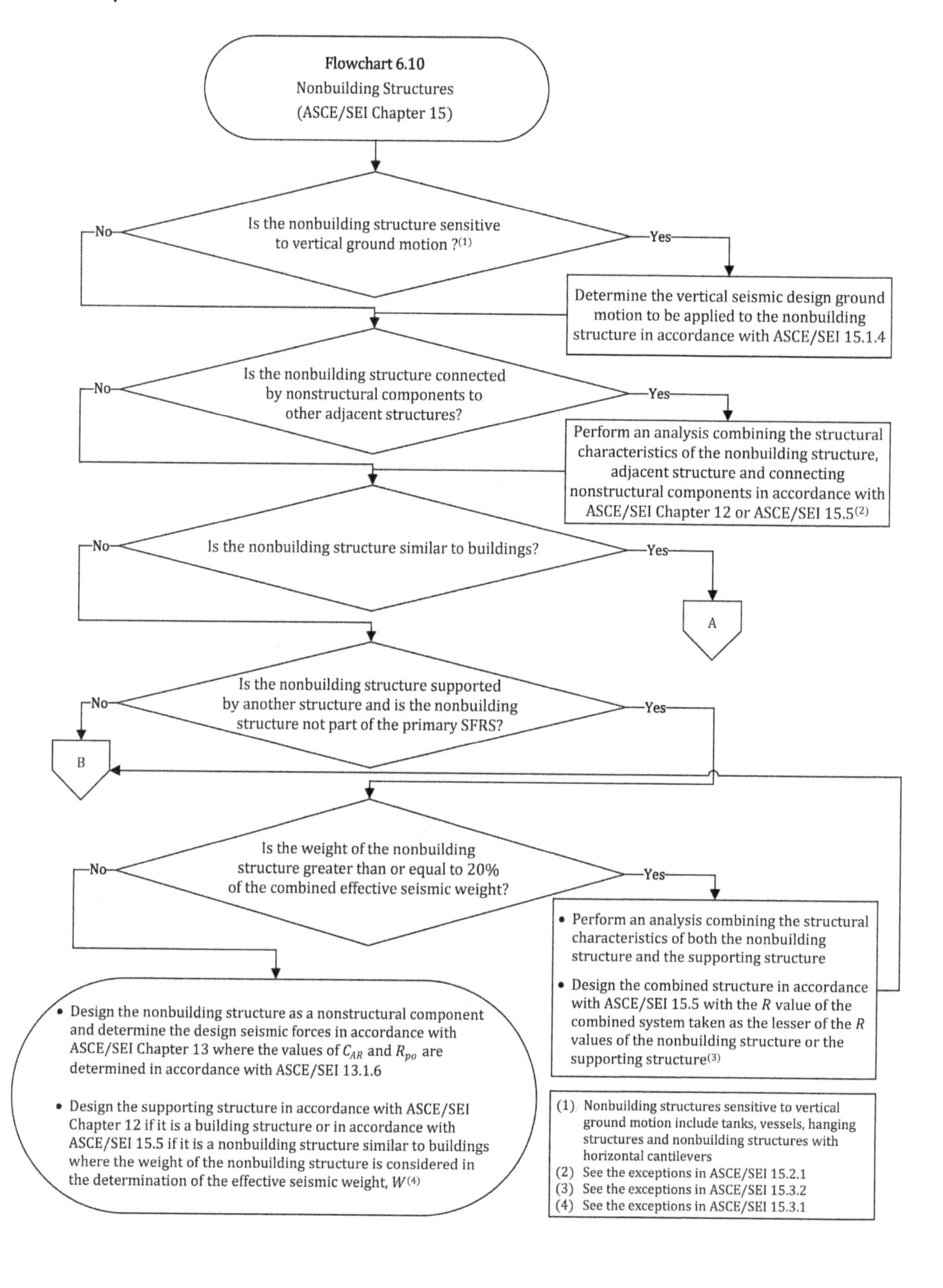
Flowchart 6.10
Nonbuilding Structures
(ASCE/SEI Chapter 15)
Is the nonbuilding structure sensitive to vertical ground motion ?(1)
No
Yes
Determine the vertical seismic design ground motion to be applied to the nonbuilding structure in accordance with ASCE/SEI 15.1.4
Is the nonbuilding structure connected by nonstructural components to other adjacent structures?
No
Yes
Perform an analysis combining the structural characteristics of the nonbuilding structure, adjacent structure and connecting nonstructural components in accordance with ASCE/SEI Chapter 12 or ASCE/SEI 15.5(2)
Is the nonbuilding structure similar to buildings?
No
Yes
A
Is the nonbuilding structure supported by another structure and is the nonbuilding structure not part of the primary SFRS?
No
Yes
B
Is the weight of the nonbuilding structure greater than or equal to 20% of the combined effective seismic weight?
No
Yes
• Perform an analysis combining the structural characteristics of both the nonbuilding structure and the supporting structure
• Design the combined structure in accordance with ASCE/SEI 15.5 with the R value of the combined system taken as the lesser of the R values of the nonbuilding structure or the supporting structure(3)
• Design the nonbuilding structure as a nonstructural component and determine the design seismic forces in accordance with ASCE/SEI Chapter 13 where the values of C_{AR} and R_{po} are determined in accordance with ASCE/SEI 13.1.6
• Design the supporting structure in accordance with ASCE/SEI Chapter 12 if it is a building structure or in accordance with ASCE/SEI 15.5 if it is a nonbuilding structure similar to buildings where the weight of the nonbuilding structure is considered in the determination of the effective seismic weight, W(4)
(1) Nonbuilding structures sensitive to vertical ground motion include tanks, vessels, hanging structures and nonbuilding structures with horizontal cantilevers
(2) See the exceptions in ASCE/SEI 15.2.1
(3) See the exceptions in ASCE/SEI 15.3.2
(4) See the exceptions in ASCE/SEI 15.3.1

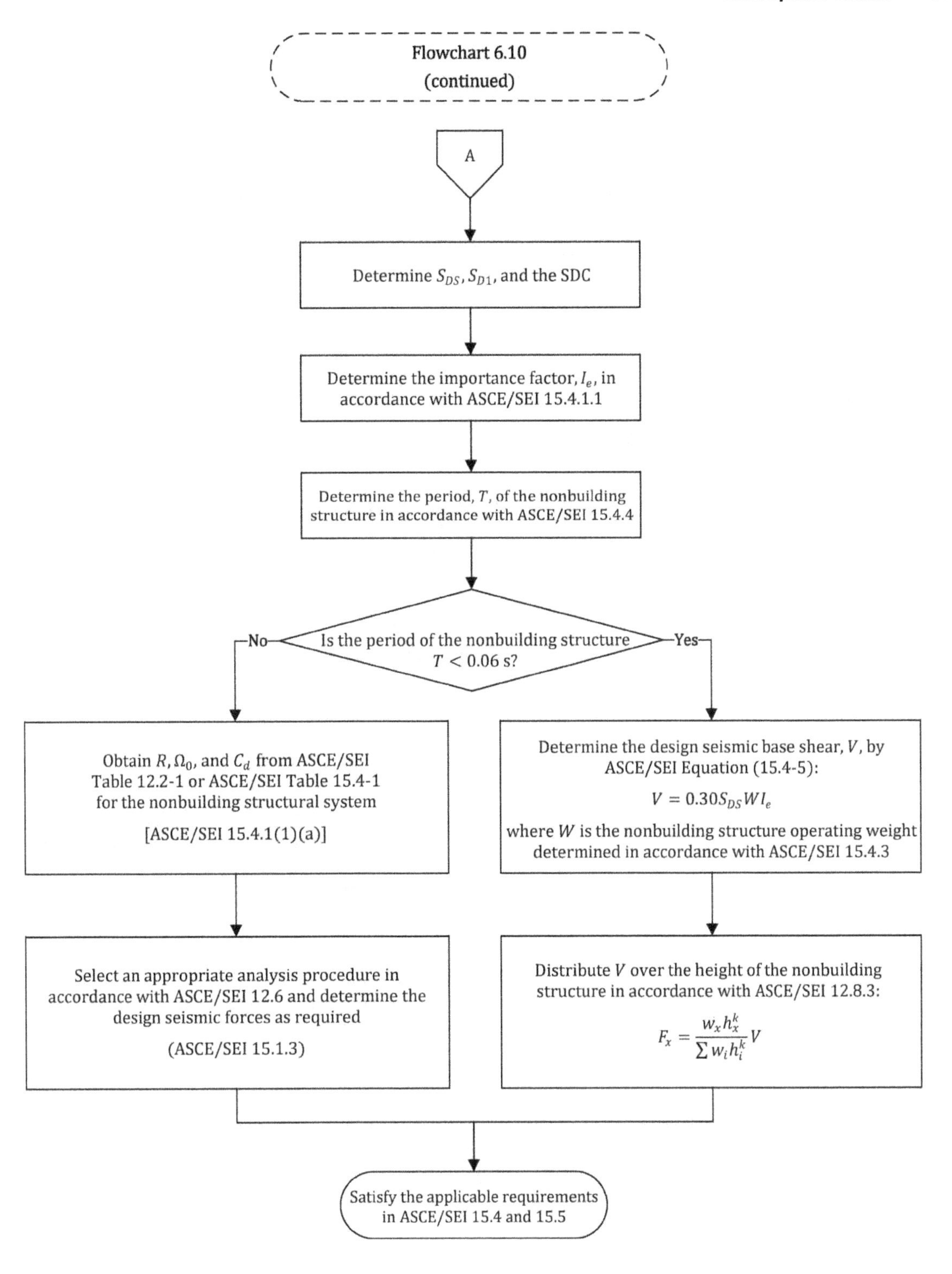

Flowchart 6.10
(continued)
A
Determine S_{DS}, S_{D1}, and the SDC
Determine the importance factor, I_e, in accordance with ASCE/SEI 15.4.1.1
Determine the period, T, of the nonbuilding structure in accordance with ASCE/SEI 15.4.4
Is the period of the nonbuilding structure $T < 0.06$ s?
No
Yes
Obtain R, Ω_0, and C_d from ASCE/SEI Table 12.2-1 or ASCE/SEI Table 15.4-1 for the nonbuilding structural system
[ASCE/SEI 15.4.1(1)(a)]
Determine the design seismic base shear, V, by ASCE/SEI Equation (15.4-5):
$V = 0.30 S_{DS} W I_e$
where W is the nonbuilding structure operating weight determined in accordance with ASCE/SEI 15.4.3
Select an appropriate analysis procedure in accordance with ASCE/SEI 12.6 and determine the design seismic forces as required
(ASCE/SEI 15.1.3)
Distribute V over the height of the nonbuilding structure in accordance with ASCE/SEI 12.8.3:
$F_x = \frac{w_x h_x^k}{\sum w_i h_i^k} V$
Satisfy the applicable requirements in ASCE/SEI 15.4 and 15.5

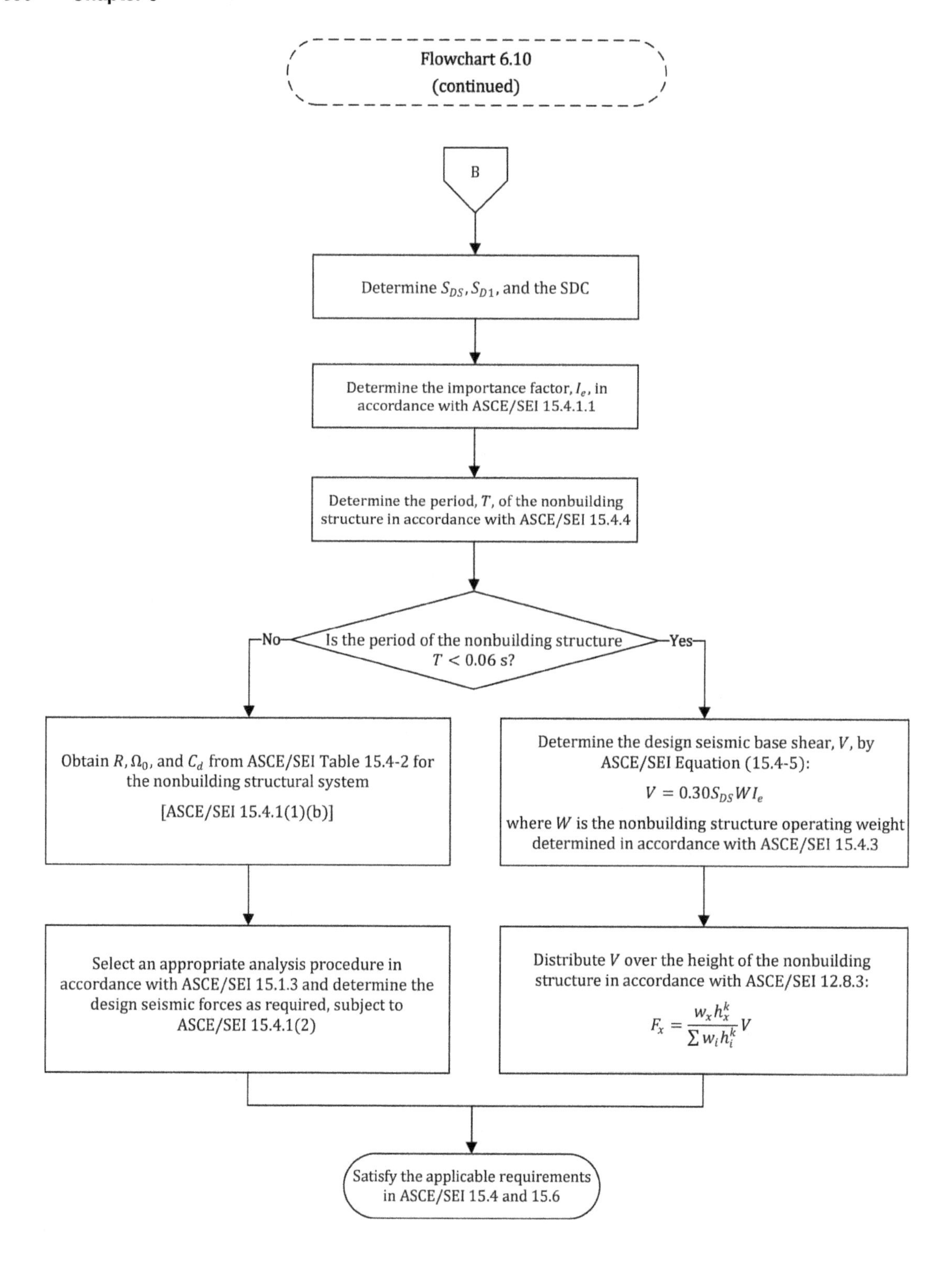

Flowchart 6.10
(continued)
B
Determine S_{DS}, S_{D1}, and the SDC
Determine the importance factor, I_e, in accordance with ASCE/SEI 15.4.1.1
Determine the period, T, of the nonbuilding structure in accordance with ASCE/SEI 15.4.4
Is the period of the nonbuilding structure $T < 0.06$ s?
No
Yes
Obtain R, Ω_0, and C_d from ASCE/SEI Table 15.4-2 for the nonbuilding structural system
[ASCE/SEI 15.4.1(1)(b)]
Determine the design seismic base shear, V, by ASCE/SEI Equation (15.4-5):
$V = 0.30 S_{DS} W I_e$
where W is the nonbuilding structure operating weight determined in accordance with ASCE/SEI 15.4.3
Select an appropriate analysis procedure in accordance with ASCE/SEI 15.1.3 and determine the design seismic forces as required, subject to ASCE/SEI 15.4.1(2)
Distribute V over the height of the nonbuilding structure in accordance with ASCE/SEI 12.8.3:
$F_x = \frac{w_x h_x^k}{\sum w_i h_i^k} V$
Satisfy the applicable requirements in ASCE/SEI 15.4 and 15.6

6.7 Examples

6.7.1 Example 6.1—Determination of Design Response Spectrum

Determine the design response spectrum for a commercial building (risk category II) located in Nashville, TN, at the following coordinates: latitude = 36.135°, longitude = −86.780. It has been determined the site class is D.

SOLUTION

The multi-period design response spectrum is determined using Reference 6.1. Values of the 5-percent damped, design spectral response acceleration parameters, S_a, at the periods, T, defined in ASCE/SEI 11.4.5.1(1) are given in Table 6.29 and are plotted in Figure 6.46.

Table 6.29 Design Spectral Response Acceleration Parameters, S_a, for the Commercial Building in Example 6.1

Period, T(s)	S_a
0.00	0.140
0.01	0.150
0.02	0.200
0.03	0.230
0.05	0.290
0.075	0.340
0.10	0.380
0.15	0.400
0.20	0.400
0.25	0.390
0.30	0.370
0.40	0.330
0.50	0.310
0.75	0.270
1.0	0.230
1.5	0.160
2.0	0.120
3.0	0.073
4.0	0.050
5.0	0.038
7.5	0.024
10.0	0.016

Other seismic design parameters for this site obtained from Reference 6.1 are given in Table 6.30.

Figure 6.46 Design Spectral Response Acceleration Parameters, S_a, for the Commercial Building in Example 6.1

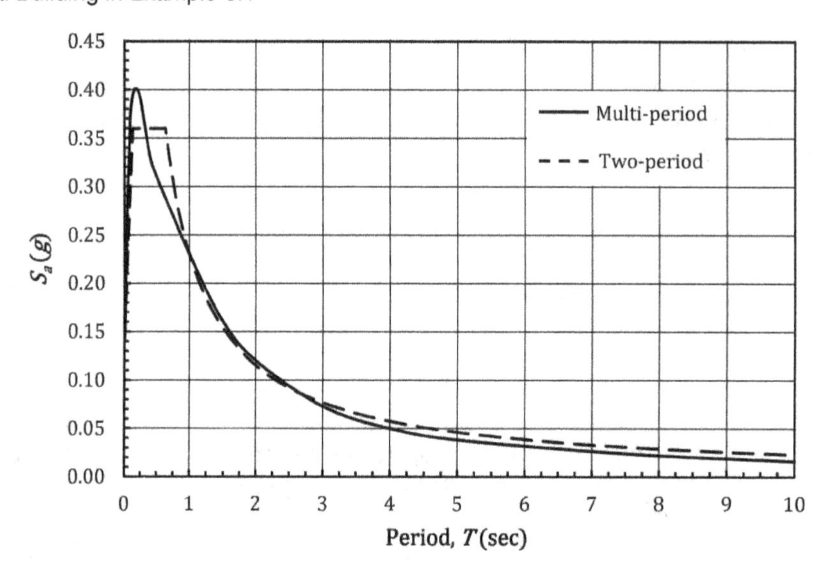

Table 6.30 Seismic Design Parameters for the Commercial Building in Example 6.1

S_{MS}	0.54
S_{M1}	0.34
S_{DS}	0.36
S_{D1}	0.23
S_S	0.46
S_1	0.16
T_L	12.0 s

For comparison purposes only, the two-period design response spectrum in ASCE/SEI 11.4.5.2 is plotted in Figure 6.46 given the parameters in Table 6.30. This design spectrum is permitted to be used where values of the multi-period response spectrum are not available. The equations in ASCE/SEI 11.4.5.2 are used to construct the two-period design response spectrum (see Figure 6.6):

- For $T < T_0 = 0.2(S_{D1}/S_{DS}) = 0.2 \times (0.23/0.36) = 0.13$ s:

$$S_a = S_{DS}\left(0.4 + \frac{0.6T}{T_0}\right) = 0.36 \times \left(0.4 + \frac{0.6T}{0.13}\right)$$

- For $T_0 = 0.13\text{ s} \le T \le T_S = S_{D1}/S_{DS} = 0.64$ s:

$$S_a = S_{DS} = 0.36$$

- For $T_S = 0.64\text{ s} < T \le T_L = 12.0$ s:

$$S_a = \frac{S_{D1}}{T} = \frac{0.23}{T}$$

- For $T > T_L = 12.0$ s:

$$S_a = \frac{S_{D1}T_L}{T^2} = \frac{0.23 \times 12.0}{T^2} = \frac{2.76}{T^2}$$

6.7.2 Example 6.2—Determination of Seismic Design Category

Determine the seismic design category (SDC) for the commercial building in Example 6.1. Preliminary analysis indicates the approximate period, T_a, is equal to 0.7 s. Assume the simplified method in ASCE/SEI 12.14 is not permitted to be used.

Flowchart 6.2 is used to determine the SDC for this building.

SOLUTION

Step 1—Determine the risk category

In accordance with IBC Table 1604.5, this commercial building is assigned to risk category II.

Step 2—Determine the site class in accordance with ASCE/SEI 11.4.2

From Example 6.1, the site class is given as D.

Step 3—Determine S_1, S_{DS}, and S_{D1} from Reference 6.1

These acceleration parameters are determined in Example 6.1 (see Table 6.30):

- $S_1 = 0.16$
- $S_{DS} = 0.36$
- $S_{D1} = 0.23$

Step 4—Determine if the building is assigned to SDC E

Because $S_1 < 0.75$, the building is not assigned to SDC E.

Step 5—Determine if the building is assigned to SDC F

Because the building is assigned to risk category II, it is not assigned to SDC F.

Step 6—Determine if the simplified method in ASCE/SEI 12.14 is permitted to be used

According to the problem statement, this method is not permitted to be used.

Step 7—Determine if all four conditions in ASCE/SEI 11.6 are satisfied

Check if the first condition is satisfied:

$$T_a < 0.8T_S$$

From the problem statement, $T_a = 0.7\ \text{s} > 0.8 \times (S_{D1}/S_{DS}) = 0.51\ \text{s}$.

Because this condition is not satisfied, it is not permitted to determine the SDC from ASCE/SEI Table 11.6-1 alone.

Step 8—Determine the SDC from ASCE/SEI Tables 11.6-1 and 11.6-2

From ASCE/SEI Table 11.6-1 with $0.33 < S_{DS} = 0.36 < 0.50$ and risk category II, the SDC is C.

From ASCE/SEI Table 11.6-2 with $S_{D1} = 0.23 > 0.20$ and risk category II, the SDC is D. Therefore, the SDC is D for this commercial building.

In lieu of determining the SDC using ASCE/SEI Tables 11.6-1 and 11.6-2, it is permitted to determine it using IBC Figures 1613.2.1(1) through 1613.2.1(7) because the site class is not DE, E, or F.

Given the latitude and longitude of the site in Example 6.1, the SDC is D from IBC Figure 1613.2.1(2), which is based on default site conditions.

6.7.3 Example 6.3—Combinations of Framing Systems in the Same Direction: Vertical Combinations

Determine the governing design coefficients R, Ω_0, and C_d in the north-south direction for the office building in Figure 6.47. The SFRS consists of ordinary reinforced concrete shear walls supported by ordinary reinforced concrete moment frames. Assume the office building is assigned to SDC B.

Figure 6.47 Vertical Combination of Framing Systems for the Office Building in Example 6.3

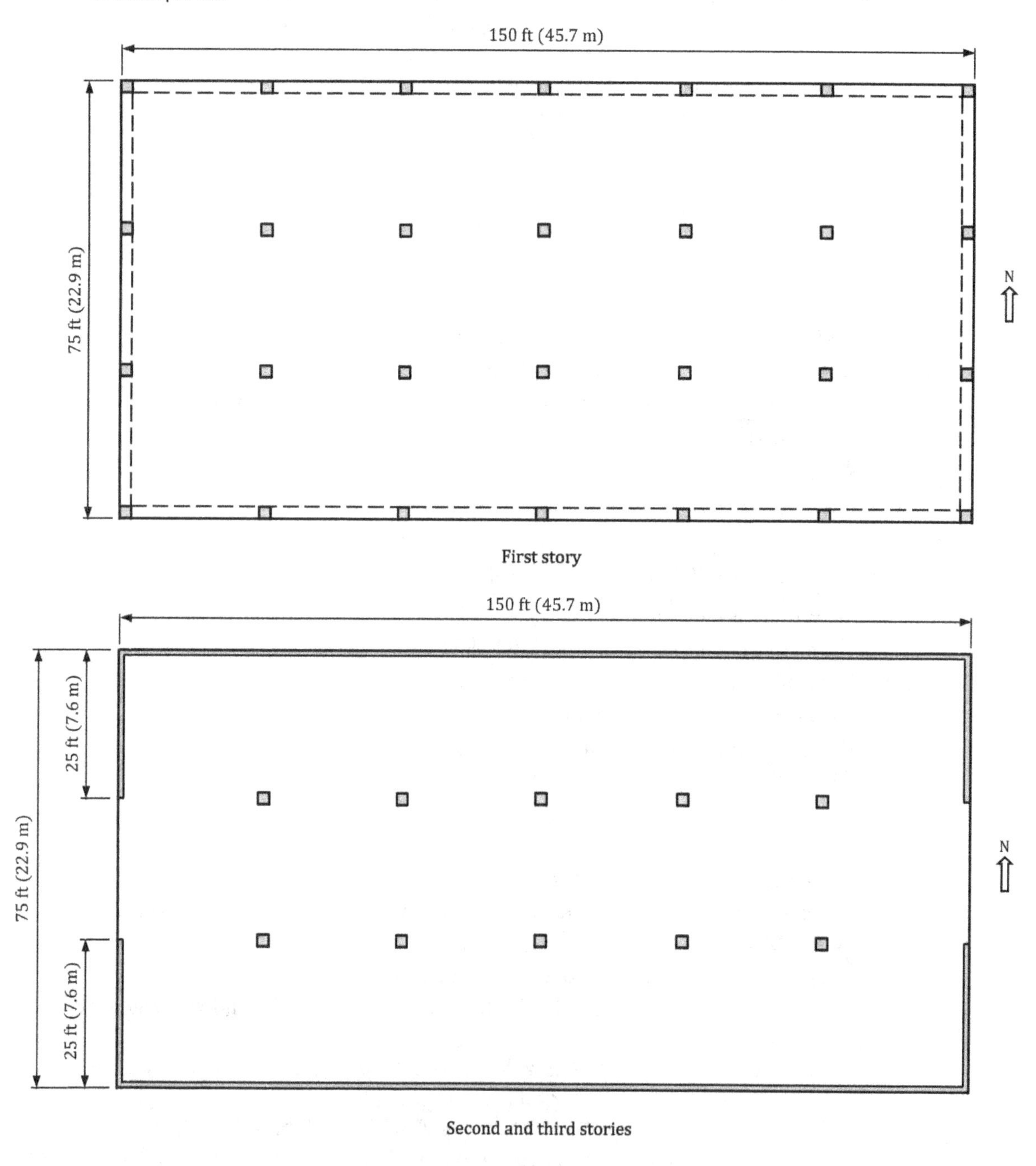

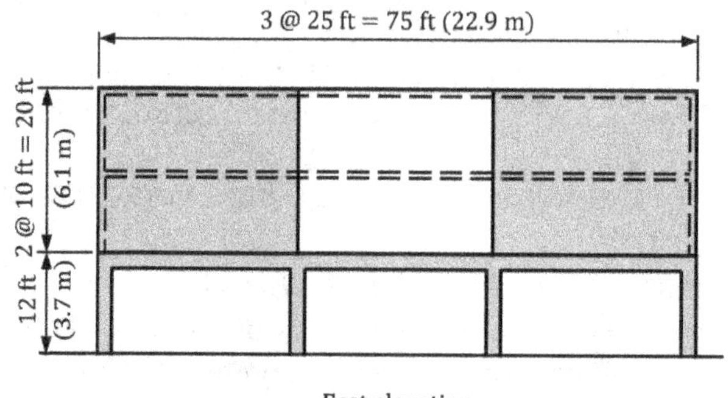

SOLUTION

The requirements in ASCE/SEI 12.2.3 are used to determine the governing design coefficients.

Because the upper rigid portion is supported by a flexible lower portion, a two-stage analysis procedure is not permitted (ASCE/SEI 12.2.3.2). Therefore, use the requirements in ASCE/SEI 12.2.3.1.

Step 1—Determine R, Ω_0, and C_d for each system (ASCE/SEI Table 12.2-1)

For the ordinary reinforced concrete shear walls (system B6): $R = 5$, $\Omega_0 = 2.5$, and $C_d = 4.5$.

For the ordinary reinforced concrete moment frames (system C7): $R = 3$, $\Omega_0 = 3$, and $C_d = 2.5$.

Step 2—Determine the governing design coefficients

The design coefficients for the upper system are permitted to be used to calculate the forces and drifts on the upper system because R of the lower system is less than R of the upper system (see Figure 6.10). For the design of the lower system, the design coefficients of the lower system must be used.

The force transferred from the shear walls to the moment frames must be increased by the ratio R of the upper system to R of the lower system, that is, by the factor $5/3 = 1.67$.

6.7.4 Example 6.4—Combinations of Framing Systems in the Same Direction: Vertical Combinations

Determine the governing design coefficients R, Ω_0, and C_d in the north-south direction for the office building in Figure 6.48. The SFRS consists of ordinary reinforced concrete moment frames supported by ordinary reinforced concrete shear walls. Design data are as follows:

- Latitude = 41.590°, longitude = –93.621°
- Soil classification D
- Seismic design category B
- Superimposed dead load on the roof level = 10 lb/ft^2 (0.48 kN/m^2)
- Superimposed dead load on a typical floor = 20 lb/ft^2 (0.96 kN/m^2)
- Cladding weight = 8 lb/ft^2 (0.38 kN/m^2)
- Normal weight concrete with a density of 150 lb/ft^3 (23.6 kN/m^3)

SOLUTION

The requirements in ASCE/SEI 12.2.3 are used to determine the governing design coefficients.

Step 1—Determine if a two-stage analysis procedure can be used

The vertical combination of SFRSs in this building consists of a flexible upper portion supported by a rigid lower portion. A two-stage equivalent lateral force procedure is permitted to be used provided the seven conditions in ASCE/SEI 12.2.3.2 are satisfied:

(a) The stiffness of the lower portion must be at least 10 times the stiffness of the upper portion.

For purposes of determining the stiffnesses of the upper and lower portions of the structure, the base shear must be calculated and distributed vertically according to the ELF procedure in ASCE/SEI 12.8. Using these forces, the stiffness for each portion is calculated as the ratio of the base shear for that portion to the elastic

Figure 6.48 Vertical Combination of Framing Systems for the Office Building in Example 6.4

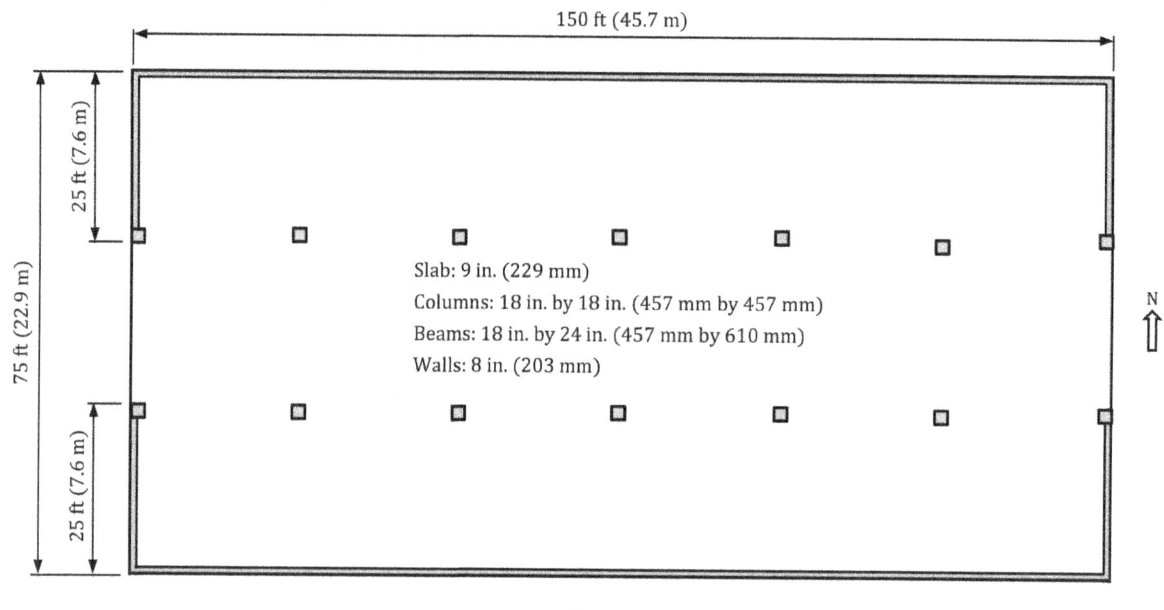

First story

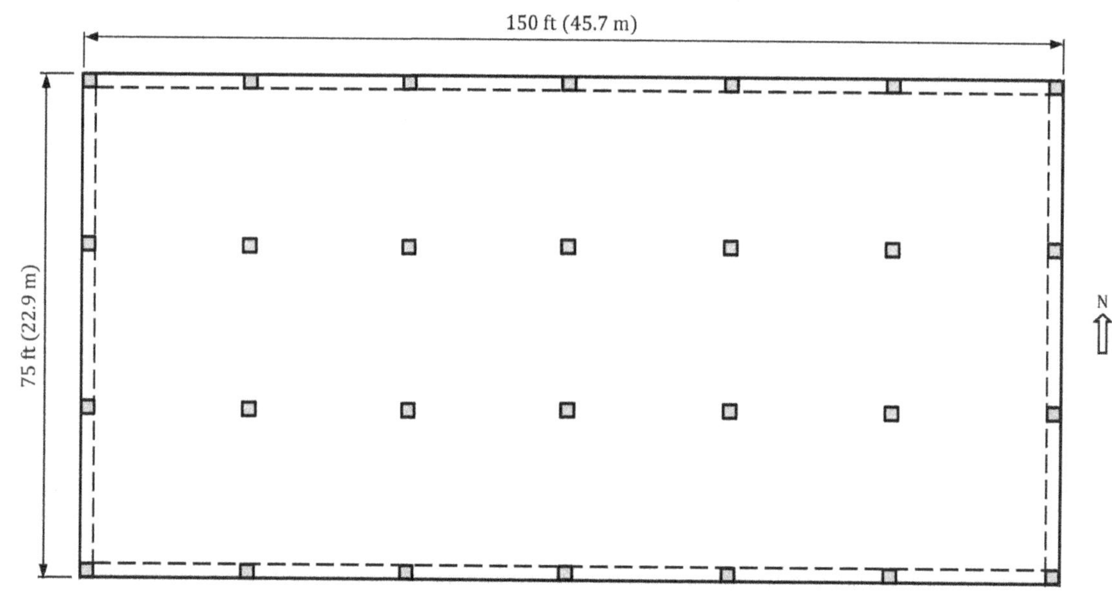

Second and third stories

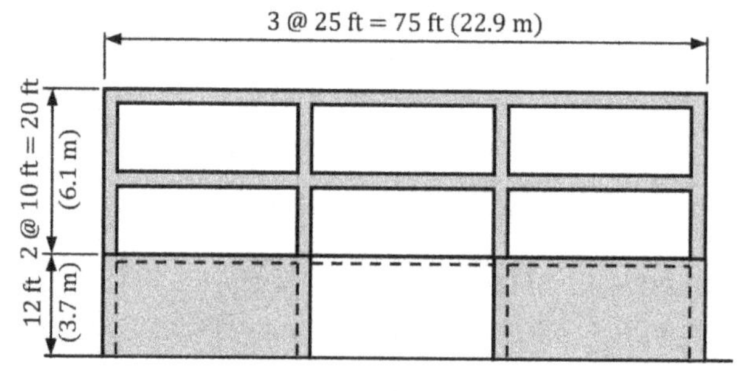

East elevation

displacement, δ_e, calculated at the top of that portion, considering the portion is fixed at its base. For the lower portion, the applied forces must include the reactions from the upper portion, modified as required in item (d) in ASCE/SEI 12.2.3.2.

Flowchart 6.5 in Section 6.6 of this book is used to determine the seismic base shear and the lateral seismic forces.

Step 1a—Determine the response modification coefficient, R

For the ordinary reinforced concrete moment frames (system C7), $R = 3$ (ASCE/SEI Table 12.2-1).

For the ordinary reinforced concrete shear walls (system B6), $R = 5$.

Because the upper portion (ordinary reinforced concrete moment frames) has a lower value of R than the lower portion (ordinary reinforced concrete shear walls), $R = 3$ must be used to determine the seismic base shear [ASCE/SEI 12.2.3.1(2)].

Step 1b—Obtain the importance factor, I_e

For risk category II, $I_e = 1.00$ (ASCE/SEI Table 1.5-2).

Step 1c—Determine the fundamental period of the combined structure, T

The period of the combined structure is equal to 0.26 s from a dynamic analysis.

Step 1d—Determine T_L from ASCE/SEI Figures 22-14 through 22-17

In lieu of using ASCE/SEI Figures 22-14 through 22-17, $T_L = 12$ s from Reference 6.1.

Step 1e—Determine S_1, S_{DS}, and S_{D1}

For the given latitude and longitude and site class, $S_1 = 0.059$, $S_{DS} = 0.093$, and $S_{D1} = 0.085$ from Reference 6.1.

Step 1f—Determine the design spectral acceleration parameter, S_a

From Reference 6.1, using the multi-period design spectrum, $S_a = 0.094$ at $T = 0.25$ s, which is adequate for a period of 0.26 s.

Maximum $S_a = 0.100$ occurs at $T = 0.50$ s, which is greater than the structure period of 0.26 s. Therefore, use $S_a = 0.100$ (ASCE/SEI 12.8.1.1).

Step 1g—Determine the seismic response coefficient, C_s

Because $S_1 = 0.059 < 0.60$, C_s is determined as follows:

$$C_s = \text{greatest of} \begin{cases} S_a/(R/I_e) = 0.10/(3/1.00) = 0.033 \\ 0.044 S_{DS} I_e = 0.044 \times 0.093 \times 1.00 = 0.004 \\ 0.01 \end{cases}$$

Step 1h—Determine the effective seismic weight, W, in accordance with ASCE/SEI 12.7.2

Slab weight $= (9/12) \times 150 \times 150 \times 75/1{,}000 = 1{,}266$ kips (5,632 kN)

Column weight tributary to each of the first elevated floor and the roof $= 28 \times (18^2/144) \times 150 \times (10/2)/1{,}000 = 47$ kips (210 kN)

Column weight tributary to the second elevated floor $= 28 \times (18^2/144) \times 150 \times 10/1{,}000 = 95$ kips (420 kN)

Wall weight tributary to the first elevated floor $= (8/12) \times 150 \times [(2 \times 150) + (4 \times 25)] \times (12/2)/1{,}000 = 240$ kips (1,068 kN)

Beam weight on each of the second elevated floor and the roof $= 18 \times [(18 \times 24)/144] \times 150 \times 25 - (18/12)/1{,}000 = 190$ kips (847 kN)

Superimposed dead load on the roof $= 10 \times 150 \times 75/1{,}000 = 113$ kips (500 kN)

Superimposed dead load on a typical floor = 20 × 150 × 75/1,000 = 225 kips (1,001 kN)

Cladding weight tributary to the first elevated floor = 8 × 11 × 450/1,000 = 40 kips (176 kN)

Cladding weight tributary to the second elevated floor = 8 × 10 × 450/1,000 = 36 kips (160 kN)

Cladding weight tributary to the roof = 8 × 5 × 450/1,000 = 18 kips (80 kN)

Effective seismic weight tributary to the first elevated floor, w_1 = 1,266 + 47 + 240 + 225 + 40 = 1,818 kips (8,087 kN)

Effective seismic weight tributary to the second elevated floor, w_2 = 1,266 + 95 + 190 + 225 + 36 = 1,812 kips (8,060 kN)

Effective seismic weight tributary to the roof level, w_R = 1,266 + 47 + 190 + 113 + 18 = 1,634 kips (7,668 kN)

$$W = 1{,}818 + 1{,}812 + 1{,}634 = 5{,}264 \text{ kips (23,815 kN)}$$

Step 1i—Determine the seismic base shear, V

The seismic base shear, V, is determined by ASCE/SEI Equation (12.8-1):

$$V = C_s W = 0.033 \times 5{,}264 = 173.7 \text{ kips (772.7 kN)}$$

Step 1j—Determine the exponent related to the structure period, k

Because $T = 0.26\ s < 0.5\ s$, $k = 1.0$.

Step 1k—Determine the lateral seismic force, F_x, at each level

Lateral seismic forces F_x are determined by ASCE/SEI Equations (12.8-12) and (12.8-13):

$$\Sigma w_i h_i^k = (1{,}634 \times 32) + (1{,}812 \times 22) + (1{,}818 \times 12)$$
$$= 113{,}968 \text{ kips-ft (154,520 kN-m)}$$

$$F_x = \frac{w_x h_x^k}{\Sigma w_i h_i^k} V$$

$$F_R = \frac{1{,}634 \times 32}{113{,}968} \times 173.7 = 79.7 \text{ kips (354.5 kN)}$$

$$F_2 = \frac{1{,}812 \times 22}{113{,}968} \times 173.7 = 60.8 \text{ kips (270.3 kN)}$$

$$F_1 = \frac{1{,}818 \times 12}{113{,}968} \times 173.7 = 33.2 \text{ kips (147.9 kN)}$$

Forces F_R and F_2 are applied to the upper portion of the structure at the roof and second elevated floors, respectively. Assuming the base of the upper portion is fixed, the total elastic lateral displacement at the roof level, δ_e, is equal to 0.22 in. (6 mm).

The stiffness of the upper portion = (79.7 + 60.8)/0.22 = 639 kips/in. (11.9 kN/mm). For analysis of the lower portion, the force F_1 is applied at the first elevated floor level. According to ASCE/SEI 12.2.3.2(a), the force transferred from the upper portion to the lower portion must be determined in accordance with ASCE/SEI 12.2.3.2(d): the reactions from the upper portion must be amplified by the ratio of the R/ρ of the upper portion over R/ρ of the lower portion where this ratio must not be less than 1.0.

In this example, $[(R_{(upper)}/\rho_{(upper)})/(R_{(lower)}/\rho_{(lower)})] = [(3/1.0)/(5/1.0)] = 0.60 < 1.0$ where $\rho = 1.0$ for structures assigned to SDC B (ASCE/SEI 12.3.4.1, item 1).

Therefore, the force transferred from the upper portion to the lower portion is equal to $79.7 + 60.8 = 140.5$ kips (625.0 kN).

The total elastic lateral displacement at the first elevated floor level, δ_e, due to a total force of 140.5 + 33.2 = 173.7 kips (772.7 kN) is equal to 0.01 in. (0.3 mm).

The stiffness of the lower portion $= 173.7/0.01 = 17{,}370$ kips/in. (3,042 kN/mm)

Ratio of lower portion to the upper portion = 17,370/639 = 27 > 10

(b) The period of the entire structure must be less than or equal to 1.1 times the period of the upper portion considered as a separate structure supported at the transition from the upper to the lower portion.

The approximate period of the moment frame in the upper portion is determined using ASCE/SEI Equation (12.8-8) and the approximate period parameters in ASCE/SEI Table 12.8-2:

For concrete moment-resisting frames, $C_t = 0.016$ (in S.I.: 0.0466) and $x = 0.9$.

$$T_a = C_t h_n^x = 0.016 \times 20^{0.9} = 0.24 \text{ s}$$

In S.I.: $T_a = 0.0466 \times 6.10^{0.9} = 0.24$ s

The period of the combined structure obtained from a dynamic analysis of the entire structure is equal to 0.26 s, which is equal to 1.1 times the period of the upper structure (i.e., $1.1 \times 0.24 = 0.26$ s).

(c) The upper portion must be designed as a separate structure using the appropriate value of R and ρ.

For the ordinary reinforced concrete moment frame (system C7): $R = 3$, $\Omega_0 = 3$, and $C_d = 2.5$ with no height limit for SDC B (ASCE/SEI Table 12.2-1).

For SDC B, $\rho = 1.0$ [(ASCE/SEI 12.3.4.1(1)].

(d) The lower portion must be designed as a separate structure using the appropriate value of R and ρ while meeting the requirements of ASCE/SEI 12.2.3.1. The reactions from the upper portion must be those determined from the analysis of the upper portion where the effects of the horizontal seismic load, E_h, are amplified by the ratio of R/ρ of the upper portion by R/ρ of the lower portion, which must be greater than or equal to 1.0 (see Figure 6.11).

For the ordinary reinforced concrete shear wall (system B6): $R = 5$, $\Omega_0 = 2.5$, and $C_d = 4.5$ with no height limit for SDC B (ASCE/SEI Table 12.2-1).
Amplification factor $= (3/1.0)/(5/1.0) = 0.6 < 1.0$, use 1.0

(e) The upper portion must be analyzed using the ELF or modal response spectrum procedure, and the lower portion must be analyzed using the ELF procedure.

In this example, both portions are analyzed using the ELF procedure.

(f) The structure height of the upper portion must not exceed the height limits of ASCE/SEI Table 12.2-1 for the SFRS used where the height is measured from the base of the upper portion.

As noted in item (c) above, the ordinary reinforced concrete moment frame does not have a height limit for SDC B.

(g) Where horizontal irregularity type 4 or vertical irregularity type 3 exists at the transition from the upper to the lower portion, the reactions from the upper portion must be amplified in accordance with ASCE/SEI 12.3.3.4, 12.10.1.1, and 12.10.3.3 in addition to the amplification required by item (d) above.

Amplification of the forces is not necessary because the building does not possess the aforementioned discontinuities.

Therefore, a two-stage equivalent lateral force procedure is permitted to be used.

Step 2—Determine the governing design coefficients

Design the upper ordinary reinforced concrete moment frames using $R = 3$, $\Omega_0 = 3$, and $C_d = 2.5$.

Design the lower ordinary reinforced concrete shear walls using $R = 5$, $\Omega_0 = 2.5$, and $C_d = 4.5$. This system is subjected to the combined effects of the amplified base shear from the upper portion and the lateral force due to the base shear from the lower portion (see Figure 6.49).

Figure 6.49 Seismic Load Effects on the Upper and Lower Portions of the Office Building in Example 6.4

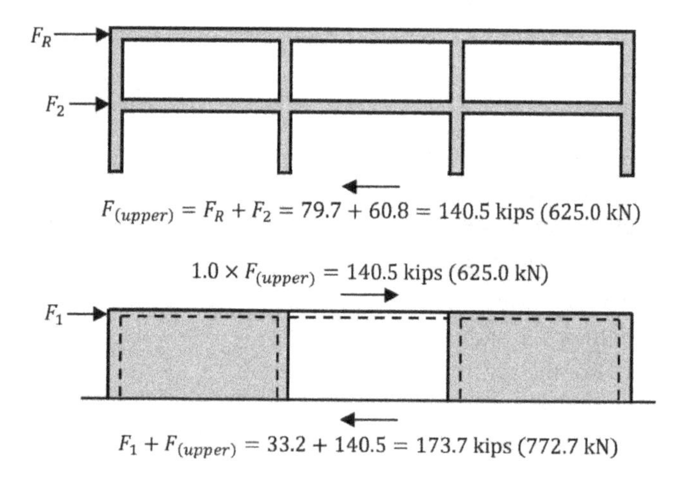

6.7.5 Example 6.5—Combinations of Framing Systems in the Same Direction: Horizontal Combinations

Determine the governing design coefficients R, Ω_0, and C_d in the north-south direction for the one-story warehouse building in Figure 6.50 assuming the roof diaphragm is flexible. Assume the building is assigned to SDC C.

Solution

The requirements in ASCE/SEI 12.2.3.3 are used to determine the governing design coefficients.

Step 1—Determine R, Ω_0, and C_d for each system (ASCE/SEI Table 12.2-1)

For the intermediate precast shear walls (system B9): $R = 5$, $\Omega_0 = 2.5$, and $C_d = 4.5$.

For the steel ordinary concentrically braced frames (system B3): $R = 3.25$, $\Omega_0 = 2$, and $C_d = 3.25$.

Step 2—Determine the governing design coefficients

Because of the horizontal combinations of SFRSs in the north-south direction, the requirements in ASCE/SEI 12.2.3.3 are applicable.

Figure 6.50 Horizontal Combination of Framing Systems for the Warehouse Building in Example 6.5

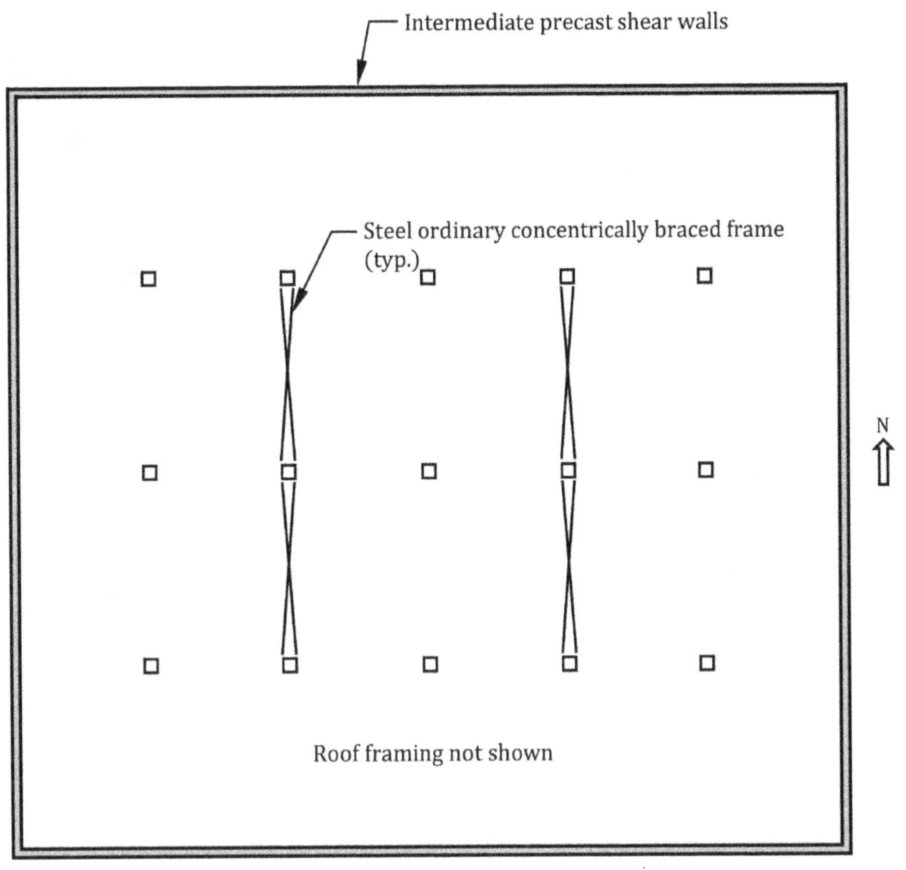

Check if the conditions in the exception in ASCE/SEI 12.2.3.3 are satisfied:

- The building must be assigned to risk category I or II.
  Because of the warehouse occupancy, the building is assigned to risk category II in accordance with ASCE/SEI Table 1.5-1.
- The building must be no more than two stories above the grade plane.
  The building in this example is one story above the grade plane.
- The building must be of light-frame construction or must have flexible diaphragms.
  The roof diaphragm is given as flexible.

Therefore, all the conditions in the exception are satisfied and the intermediate precast shear walls are permitted to be designed using $R = 5$ and the steel ordinary concentrically braced frames are permitted to be designed using $R = 3.25$. The diaphragm must be designed using the least value of R in that direction, that is, using $R = 3.25$.

If the diaphragm was rigid instead of flexible, all conditions of the exception are not satisfied and the intermediate precast shear walls and the steel ordinary concentrically braced frames must both be designed using $R = 3.25$.

6.7.6 Example 6.6—Horizontal Irregularities

Determine whether the 10-story residential building in Figure 6.51 possesses any of the horizontal structural irregularities in ASCE/SEI Table 12.3-1. Design data are given in Table 6.31.

Figure 6.51 Typical Floor Plan and Elevation for the 10-Story Residential Building in Example 6.6

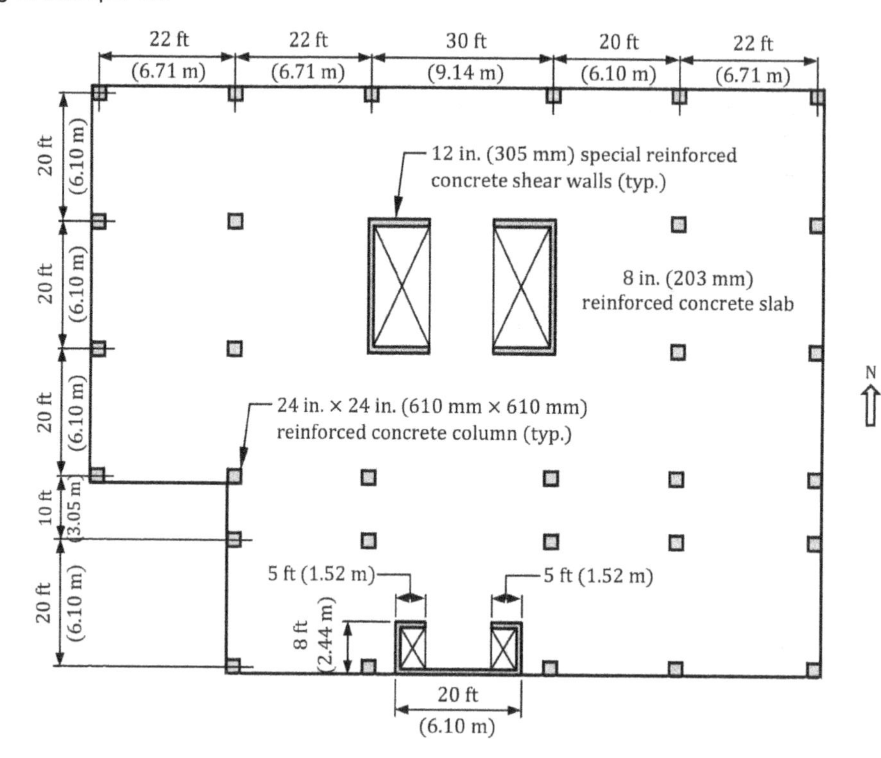

Typical floor plan

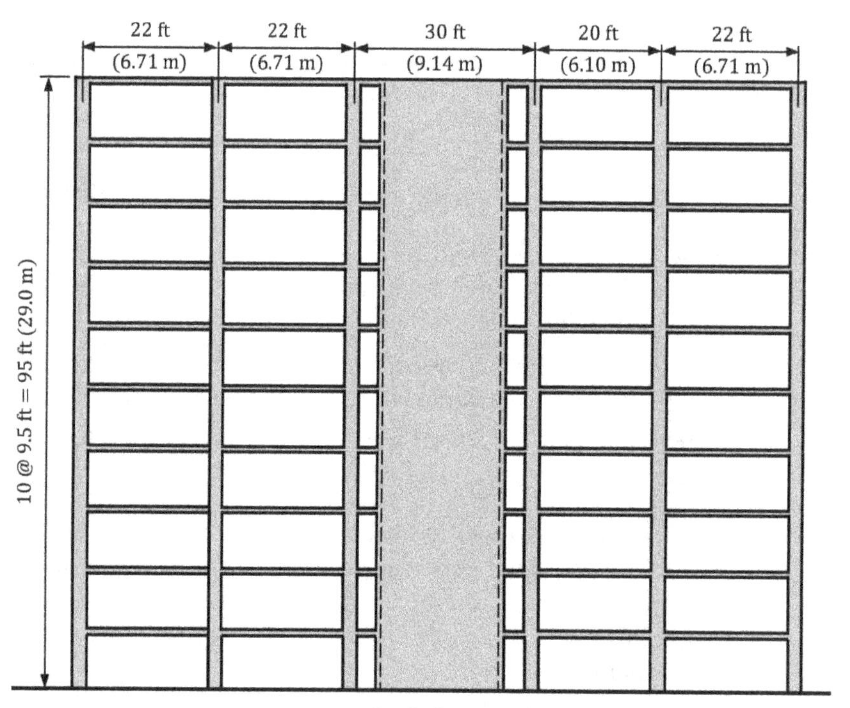

South elevation

Table 6.31 Design Data for the 10-Story Residential Building in Example 6.6

Location	Latitude: 38.602°, Longitude: –90.203°
Soil classification	Site class D
Occupancy	Residential occupancy where less than 300 people congregate in one area
Material	Cast-in-place, reinforced concrete
SFRS	Building frame system with special reinforced concrete shear walls (system B4)

Solution

- Determine the SDC.

Flowchart 6.2 in Section 6.6 of this book is used to determine the SDC.

Step 1—Determine the risk category in accordance with IBC Table 1604.5

For a residential occupancy where less than 300 people congregate in one area, the risk category is II.

Step 2—Determine the site class in accordance with ASCE/SEI 11.4.2

The site class is given as D.

Step 3—Determine the acceleration parameters S_1, S_{DS}, and S_{D1} from Reference 6.1

For the given latitude and longitude of the site, risk category II, and site class D, the multi-period design spectrum is available from Reference 6.1 where $S_1 = 0.18$, $S_{DS} = 0.45$, and $S_{D1} = 0.25$.

Step 4—Determine if the building can be assigned to SDC F

Because the risk category is II, the building is not assigned to SDC F.

Step 5—Determine if the building can be assigned to SDC E

Because $S_1 = 0.18 < 0.75$, the building is not assigned to SDC E.

Step 6—Determine if the simplified method in ASCE/SEI 12.14 is permitted to be used

Because the building has more than three stories above grade, the simplified method is not permitted to be used [ASCE/SEI 12.14.1.1(3)].

Step 7—Check if all four conditions in ASCE/SEI 11.6 are satisfied

Use ASCE/SEI Equation (12.8-8) with approximate period parameters for "All other structural systems" to determine the approximate fundamental period, T_a:

$$T_a = C_t h_n^x = 0.02 \times (95.0)^{0.75} = 0.61 \text{ s}$$

In S.I.:

$$T_a = 0.0488 \times (29.0)^{0.75} = 0.61 \text{ s}$$

$$T_a = 0.61 \text{ s} > 0.8T_S = 0.8(S_{D1}/S_{DS}) = 0.8 \times (0.25/0.45) = 0.44 \text{ s}$$

Therefore, the first of the four conditions is not satisfied and the SDC is not permitted to be determined by ASCE/SEI Table 11.6-1 alone.

Step 8—Determine the SDC using ASCE/SEI Tables 11.6-1 and 11.6-2

With $0.33 < S_{DS} = 0.45 < 0.50$ and risk category II, the SDC is C in accordance with ASCE/SEI Table 11.6-1.

With $S_{D1} = 0.25 > 0.20$ and risk category II, the SDC is D in accordance with ASCE/SEI Table 11.6-2.

Therefore, the SDC is D for this building.

- Determine if any horizontal irregularities exist.

Step 1—Determine if a torsional irregularity exists

Story drifts are needed to ascertain whether torsional irregularity or an extreme torsional irregularity exists. The seismic forces from the ELF procedure in ASCE/SEI 12.8 are used to determine the displacements and story drifts for this purpose. Flowchart 6.5 is used to determine the seismic base shear, V, and the lateral seismic forces, F_x, at the floor and roof levels.

Step 1a—Obtain R from ASCE/SEI Table 12.2-1

For a building frame system with special reinforced concrete shear walls (system B4), $R = 6$.

Step 1b—Obtain I_e from ASCE/SEI Table 1.5-2

For a residential building, the risk category is II (IBC Table 1604.5), and $I_e = 1.00$.

Step 1c—Determine T in the direction under consideration from a substantiated analysis

A three-dimensional model of the building was constructed in accordance with ASCE/SEI 12.7 using Reference 6.13 assuming rigid diaphragms at all levels. From the analysis, it was found that $T = 1.22$ s in the north-south direction and $T = 1.75$ s in the east-west direction.

Step 1d—Determine the approximate fundamental period, T_a, in accordance with ASCE/SEI 12.8.2.1

Use ASCE/SEI Equation (12.8-8) with approximate period parameters for "All other structural systems" to determine T_a:

$$T_a = C_t h_n^x = 0.02 \times (95.0)^{0.75} = 0.61 \text{ s}$$

In S.I.:

$$T_a = 0.0488 \times (29.0)^{0.75} = 0.61 \text{ s}$$

Step 1e—Determine S_{DS} and S_{D1} in accordance with ASCE/SEI 11.4.4 or 11.4.7

For the given latitude and longitude of the site, risk category II and site class D, the multi-period design spectrum is available from Reference 6.1 where $S_{DS} = 0.45$ and $S_{D1} = 0.25$.

Step 1f—Determine coefficient C_u from ASCE/SEI Table 12.8-1 based on S_{D1}

The coefficient for upper limit on calculated period, C_u, is obtained from ASCE/SEI Table 12.8-1. For $0.2 < S_{D1} = 0.25 < 0.3$, $C_u = 1.45$ by linear interpolation.

Step 1g—Determine if $T > C_u T_a$

From step 1c, $T = 1.22$ s in the north-south direction and $T = 1.75$ s in the east-west direction.

From steps 1d and 1f, $C_u T_a = 1.45 \times 0.61 = 0.89 \text{ s} < T$ in both directions.

Therefore, $T = 0.89$ s must be used in the determination of the seismic forces in both orthogonal directions.

Step 1h—Determine T_L from ASCE/SEI Figures 22-14 through 22-17

In lieu of using ASCE/SEI Figures 22-14 through 22-17, $T_L = 12$ s from Reference 6.1.

Step 1i—Determine the permitted method to determine C_s

Because the design spectral acceleration parameter, S_a, is available from Reference 6.1 for a period, $T = 0.89$ s, Method 1 or Method 2 in

ASCE/SEI 12.8.1.1 is permitted to be used to determine C_s. Method 1 is selected in this example.

Step 1j—Determine the design spectral acceleration parameter, S_a

From Reference 6.1, $S_a = 0.29$ at $T = 0.75$ s and $S_a = 0.25$ at $T = 1.0$ s. Therefore, $S_a = 0.27$ at $T = 0.89$ s by linear interpolation.

Maximum $S_a = 0.55$ occurs at $T = 0.1$ s, which is less than $T = 0.89$ s. Therefore, use $S_a = 0.27$ in the determination of C_s.

Step 1k—Determine the spectral response acceleration parameter at a period of 1 s, S_1

From Reference 6.1, $S_1 = 0.18$.

Step 1l—Determine the seismic response coefficient, C_s

Because $S_1 = 0.18 < 0.6$, C_s is determined as follows:

$$C_s = \text{greatest of} \begin{cases} S_a/(R/I_e) = 0.27/(6/1.00) = 0.045 \\ 0.044 S_{DS} I_e = 0.044 \times 0.45 \times 1.00 = 0.020 \\ 0.01 \end{cases}$$

Step 1m—Determine the effective seismic weight, W, in accordance with ASCE/SEI 12.7.2.

The effective seismic weight, W, is the summation of the effective dead loads at each level and includes the weight of the structural members, superimposed dead loads, the weight of the cladding, and the weight of the mechanical equipment on the roof (details of the weight calculations are not shown here).

The story weights are given in Table 6.32.

Table 6.32 Seismic Forces and Seismic Story Shears for the Residential Building in Example 6.6

Level	Story Weight, w_x (kips)	Height, h_x (ft)	$w_x h_x^k$	Lateral Force, F_x (kips)	Story Shear, V_x (kips)
R	1,116	95.0	263,592	100.7	100.7
10	1,286	85.5	267,670	102.3	203.0
9	1,286	76.0	232,390	88.8	291.8
8	1,286	66.5	197,983	75.7	367.5
7	1,286	57.0	164,547	62.9	430.4
6	1,286	47.5	132,213	50.5	480.9
5	1,286	38.0	101,154	38.7	519.6
4	1,286	28.5	71,623	27.4	547.0
3	1,286	19.0	44,030	16.8	563.8
2	1,286	9.5	19,165	7.3	571.1
Σ	12,690		1,494,367	571.1	

1 ft = 0.3048 m; 1 kip = 4.4482 kN.

Step 1n—Determine the seismic base shear, V

The seismic base shear, V, is determined by ASCE/SEI Equation (12.8-1):

$$V = C_s W = 0.045 \times 12{,}690 = 571.1 \text{ kips}$$

In S.I.:

$$V = C_s W = 0.045 \times 56,448 = 2,540.2 \text{ kN}$$

Step 1o—Determine the exponent related to the structure period, k

Because $0.5 \text{ s} < T = 0.89 \text{ s} < 2.5 \text{ s}$, $k = 0.75 + 0.5T = 1.20$.

Step 1p—Determine the lateral seismic force, F_x, at each level

Lateral seismic forces F_x are determined by ASCE/SEI Equations (12.8-12) and (12.8-13) and are given in Table 6.32 along with the story shears, V_x.

For example, at the roof level:

$$C_{vx} = \frac{w_x h_x^k}{\sum w_i h_i} = \frac{1,116 \times (95.0)^{1.2}}{1,494,367} = 0.1764$$

$$F_x = C_{vx} V = 0.1764 \times 571.1 = 100.7 \text{ kips}$$

In S.I.:

$$C_{vx} = \frac{4,964.2 \times (29.0)^{1.2}}{1,597,569} = 0.1767$$

$$F_x = 0.1767 \times 2,540.2 = 448.9 \text{ kN}$$

Step 1q—Check the conditions for torsional irregularities

A type 1 torsional irregularity occurs where either of the following conditions exist (ASCE/SEI Table 12.3-1):

- More than 75 percent of any story's lateral strength below the diaphragm is provided at or on one side of the CM.
- The torsional irregularity ratio (TIR) exceeds 1.2.

The second condition is checked in this example. The TIR is defined as follows [ASCE Equation (12.3-2)]:

$$\text{TIR} = \frac{\Delta_{\text{max}}}{\Delta_{\text{avg}}}$$

where Δ_{max} is the maximum story drift at the building's edge subject to F_x including accidental torsion and Δ_{avg} is the average of the story drifts at the two opposing edges of the building determined using the same loading and diaphragm rigidity used for the determination of Δ_{max} (see Figure 6.15).

The building was analyzed in the north-south direction assuming rigid diaphragms at all levels and including reduced stiffness properties due to cracking of the concrete (ASCE/SEI 12.7.3). The forces are applied at the CM at each level, which is displaced in each orthogonal direction 5 percent of the dimension of the structure perpendicular to the direction of the applied forces; this accidental torsion, M_{ta}, which is in addition to the inherent torsion, M_t, due to the CM and CR being at different locations, must be applied when determining if a horizontal irregularity exists (ASCE/SEI 12.8.4.2).

The design earthquake displacements, δ_{DE}, and drifts at all levels are given in Table 6.33 for seismic forces in the north-south direction where δ_{DE} is determined by ASCE/SEI Equation (12.8-16):

$$\delta_{DE} = \frac{C_d \delta_e}{I_e} + \delta_{di}$$

Table 6.33 Lateral Displacements and Story Drifts in the North-South Direction for the Residential Building in Example 6.6

Story	$(\delta_{DE})_1$ (in.)	Δ_1 (in.)	$(\delta_{DE})_2$ (in.)	Δ_2 (in.)	Δ_{avg} (in.)	TIR = $\Delta_{max}/\Delta_{avg}$
10	8.60	1.00	4.65	0.60	0.80	1.25
9	7.60	1.05	4.05	0.60	0.83	1.27
8	6.55	1.05	3.45	0.60	0.83	1.27
7	5.50	1.05	2.85	0.60	0.83	1.27
6	4.45	1.05	2.25	0.55	0.80	1.31
5	3.40	1.00	1.70	0.50	0.75	1.33
4	2.40	0.85	1.20	0.45	0.65	1.31
3	1.55	0.75	0.75	0.35	0.55	1.36
2	0.80	0.55	0.40	0.25	0.40	1.38
1	0.25	0.25	0.15	0.15	0.20	1.25

1 in. = 25.4 mm.

The elastic displacements, δ_e, are determined using design seismic forces based on the computed fundamental period of the structure ($T = 1.22$ s) without the upper limit, $C_u T_a$, specified in ASCE/SEI 12.8.2 (ASCE/SEI 12.8.6.2).

The deflection amplification factor, C_d, is equal to 5 for building frame systems with special reinforced concrete shear walls (ASCE/SEI Table 12.2-1) and the displacement due to diaphragm deformation corresponding to the design earthquake, δ_{di}, is permitted to be neglected in determining the design story drift (ASCE/SEI 12.8.6.5). The design story drifts, Δ, are the difference of the δ_{DE} at the top and bottom of the story under consideration.

For example, in the 10th story:

$$\Delta_1 = 8.60 - 7.60 = 1.00 \text{ in.}$$

$$\Delta_2 = 4.65 - 4.05 = 0.60 \text{ in.}$$

$$\Delta_{avg} = \frac{\Delta_1 + \Delta_2}{2} = \frac{1.00 + 0.60}{2} = 0.80 \text{ in.}$$

$$\text{TIR} = \Delta_{max}/\Delta_{avg} = 1.00/0.80 = 1.25$$

In S.I.:

$$\Delta_1 = 218.4 - 193.0 = 25.4 \text{ mm}$$

$$\Delta_2 = 118.1 - 102.9 = 15.2 \text{ mm}$$

$$\Delta_{avg} = \frac{25.4 + 15.2}{2} = 20.3 \text{ mm}$$

$$\Delta_{max}/\Delta_{avg} = 25.4/20.3 = 1.25$$

A type 1 torsional irregularity exists in all stories in the north-south direction because TIR is greater than 1.2 (see Table 6.33).

According to ASCE/SEI 12.8.4.3, the accidental torsional moments, M_{ta}, must be increased by the torsional amplification factor, A_x, determined by ASCE/SEI Equation (12.8-15) where a torsional irregularity exists:

$$A_x = \left(\frac{\delta_{max}}{1.2\delta_{avg}}\right)^2$$

For example, in the 10th story:

$$A_x = \left[\frac{8.60}{1.2 \times \left(\frac{8.60+4.65}{2}\right)}\right]^2 = 1.17 > 1.0$$

In S.I.:

$$A_x = \left[\frac{218.4}{1.2 \times \left(\frac{218.4+118.1}{2}\right)}\right]^2 = 1.17 > 1.0$$

Step 2—Determine if a reentrant corner irregularity exists

According to ASCE/SEI Table 12.3-1, a reentrant corner irregularity exists where both plan projections of the structure beyond a reentrant corner are greater than 20 percent of the plan dimension of the structure in the given direction (see ASCE/SEI Table 12.3-1 and Figure 6.16).

$$L_{xp} = 22.0 \text{ ft} < 0.2L_x = 0.2 \times 118.0 = 23.6 \text{ ft}$$

$$L_{yp} = 30.0 \text{ ft} > 0.2L_y = 0.2 \times 92.0 = 18.4 \text{ ft}$$

In S.I.:

$$L_{xp} = 6.71 \text{ m} < 0.2L_x = 0.2 \times 36.0 = 7.20 \text{ m}$$

$$L_{yp} = 9.14 \text{ m} > 0.2L_y = 0.2 \times 28.0 = 5.60 \text{ m}$$

Therefore, a reentrant corner irregularity does not exist at all levels because only one plan projection of the structure beyond the reentrant corner is greater than 20 percent of the plan dimension of the structure in the given direction.

Step 3—Determine if a diaphragm discontinuity irregularity exists

A diaphragm discontinuity does not exist because the diaphragms do not have an abrupt discontinuity or variation in stiffness, including a cutout or open area greater than 25 percent of the gross enclosed diaphragm area or a change in effective stiffness of more than 50 percent from one story to the next (see Figure 6.17).

Step 4—Determine if an out-of-plane offset irregularity exists

An out-of-plane offset irregularity does not exist because there are no out-of-plane offsets of the shear walls (see Figure 6.18).

Step 5—Determine if a nonparallel system irregularity exists

A nonparallel system irregularity does not exist because all the shear walls are parallel to a major orthogonal axis of the building (see Figure 6.19).

6.7.7 Example 6.7—Vertical Irregularities

Determine whether the four-story emergency treatment facility in Figure 6.52 possesses any of the vertical structural irregularities in ASCE/SEI Table 12.3-2. Design data are given in Table 6.34.

Figure 6.52 Typical Floor Plan and Elevation for the Four-Story Emergency Treatment Facility in Example 6.7

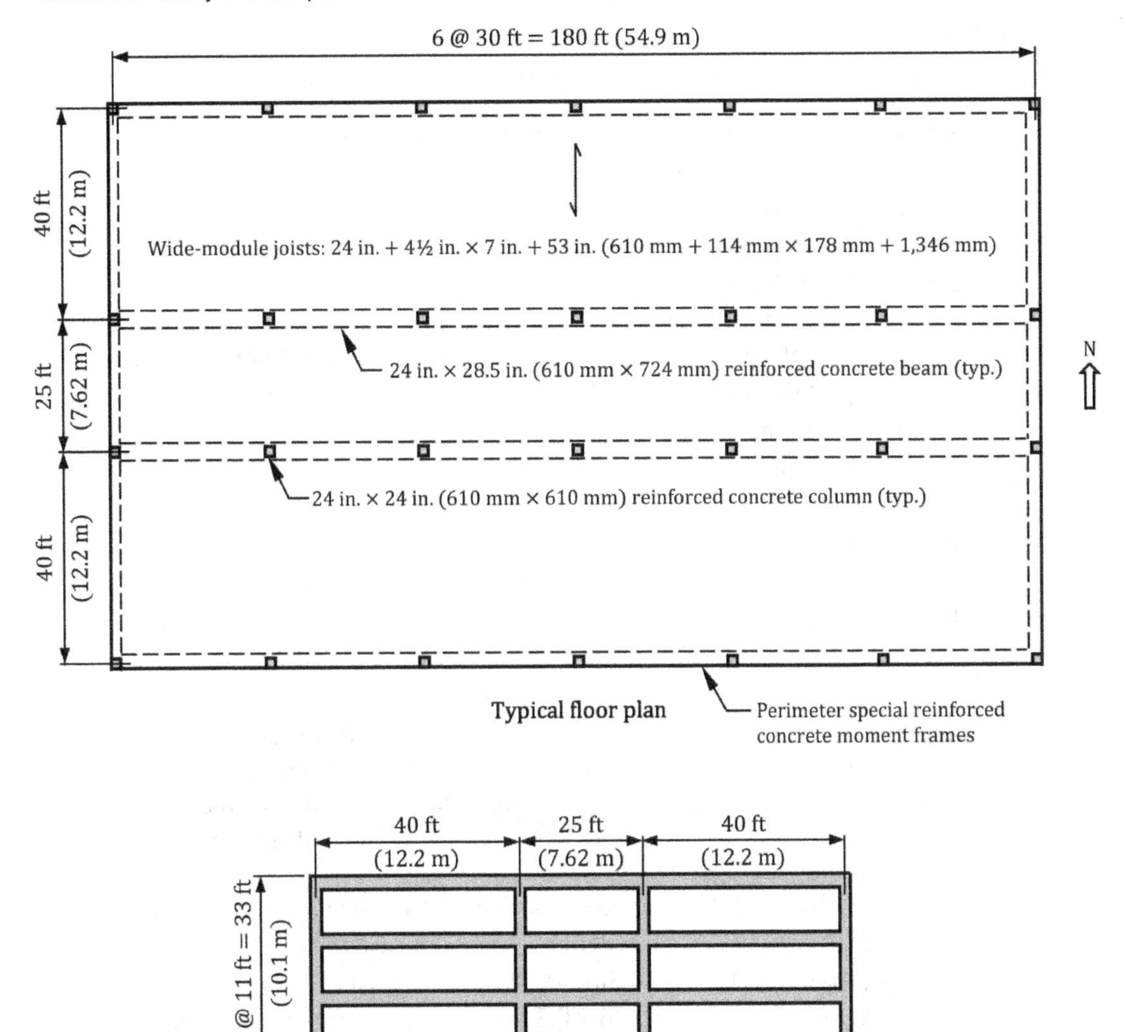

Table 6.34 Design Data for the Four-Story Emergency Treatment Facility in Example 6.7

Location	Latitude: 37.145°, Longitude: –93.278°
Soil classification	Site class D
Occupancy	Essential facility
Material	Cast-in-place, reinforced concrete
SFRS	Special reinforced concrete moment frames (system C5)

Solution

- Determine the SDC.
  Flowchart 6.2 in Section 6.6 of this book is used to determine the SDC.

Step 1—Determine the risk category in accordance with IBC Table 1604.5
For an essential facility, the risk category is IV.

Step 2—Determine the site class in accordance with ASCE/SEI 11.4.2
The site class is given as D.

Step 3—Determine the acceleration parameters S_1, S_{DS}, and S_{D1} from Reference 6.1
For the given latitude and longitude of the site, risk category IV and site class D, the multi-period design spectrum is available from Reference 6.1 where $S_1 = 0.13$, $S_{DS} = 0.26$, and $S_{D1} = 0.18$.

Step 4—Determine if the building can be assigned to SDC F
Because the risk category is IV and $S_1 = 0.13 < 0.75$, the building is not assigned to SDC F.

Step 5—Determine if the building can be assigned to SDC E
Because the risk category is IV, the building is not assigned to SDC E.

Step 6—Determine if the simplified method in ASCE/SEI 12.14 is permitted to be used
Because the building has more than three stories above grade, the simplified method is not permitted to be used.

Step 7—Check if all four conditions in ASCE/SEI 11.6 are satisfied
From a three-dimensional model of the building constructed in accordance with ASCE/SEI 12.7, it is found that $T = 1.70$ s in the north-south direction and $T = 1.40$ s in the east-west direction. Both of these periods are greater than $T_s = S_{D1}/S_{DS} = 0.18/0.26 = 0.69$ s.

Therefore, the second of the four conditions is not satisfied and the SDC is not permitted to be determined by ASCE/SEI Table 11.6-1 alone.

Step 8—Determine the SDC using ASCE/SEI Tables 11.6-1 and 11.6-2
With $0.167 < S_{DS} = 0.26 < 0.33$ and risk category IV, the SDC is C in accordance with ASCE/SEI Table 11.6-1.

With $0.133 < S_{D1} = 0.18 < 0.20$ and risk category IV, the SDC is D in accordance with ASCE/SEI Table 11.6-2.

Therefore, the SDC is D for this building.

- Determine if any vertical irregularities exist.

Step 1—Determine if a stiffness-soft story irregularity exists
Lateral story stiffnesses are needed to determine whether a soft story irregularity exists. Story stiffness can be obtained by dividing the story shear by the story drift in the direction of analysis. Therefore, the lateral displacements at the CM for each story are needed. The seismic forces from the ELF procedure in ASCE/SEI 12.8 are used to determine the displacements and story drifts for this purpose (see Flowchart 6.5).

Step 1a—Obtain R from ASCE/SEI Table 12.2-1
For special reinforced concrete moment frames (system C5), $R = 8$.

Step 1b—Obtain I_e from ASCE/SEI Table 1.5-2
For an essential facility, the risk category is IV (IBC Table 1604.5), and $I_e = 1.50$.

Step 1c—Determine T in the direction under consideration from a substantiated analysis

A three-dimensional model of the building was constructed in accordance with ASCE/SEI 12.7 using Reference 6.13 assuming rigid diaphragms at all levels. From the analysis, it was found that $T = 1.70$ s in the north-south direction and $T = 1.40$ s in the east-west direction.

Step 1d—Determine the approximate fundamental period, T_a, in accordance with ASCE/SEI 12.8.2.1

Use ASCE/SEI Equation (12.8-8) with approximate period parameters for "concrete moment-resisting frames" to determine T_a:

$$T_a = C_t h_n^x = 0.016 \times (45.0)^{0.9} = 0.49 \text{ s}$$

In S.I.:

$$T_a = 0.0466 \times (13.7)^{0.9} = 0.49 \text{ s}$$

Step 1e—Determine S_{DS} and S_{D1} in accordance with ASCE/SEI 11.4.4 or 11.4.7

For the given latitude and longitude of the site, risk category IV and site class D, the multi-period design spectrum is available from Reference 6.1 where $S_{DS} = 0.26$ and $S_{D1} = 0.18$.

Step 1f—Determine coefficient C_u from ASCE/SEI Table 12.8-1 based on S_{D1}

The coefficient for upper limit on calculated period, C_u, is obtained from ASCE/SEI Table 12.8-1. For $0.15 < S_{D1} = 0.18 < 0.2$, $C_u = 1.54$ by linear interpolation.

Step 1g—Determine if $T > C_u T_a$

From step 1c, $T = 1.70$ s in the north-south direction and $T = 1.40$ s in the east-west direction.

From steps 1d and 1f, $C_u T_a = 1.54 \times 0.49 = 0.76 \text{ s} < T$ in both directions.

Therefore, $T = 0.76$ s must be used in the determination of the seismic forces in both orthogonal directions.

Step 1h—Determine T_L from ASCE/SEI Figures 22-14 through 22-17

In lieu of using ASCE/SEI Figures 22-14 through 22-17, $T_L = 12$ s from Reference 6.1.

Step 1i—Determine the permitted method to determine C_s

Because the design spectral acceleration parameter, S_a, is available from Reference 6.1 for a period, $T = 0.76$ s, Method 1 or Method 2 in ASCE/SEI 12.8.1.1 is permitted to be used to determine C_s. Method 1 is selected in this example.

Step 1j—Determine the design spectral acceleration parameter, S_a

From Reference 6.1, $S_a = 0.20$ at $T = 0.75$ s, which is slightly conservative for a period of 0.76 s.

Maximum $S_a = 0.29$ occurs at $T = 0.25$ s, which is less than the period of the structure, $T = 0.76$ s. Therefore, use $S_a = 0.20$ in the determination of C_s.

Step 1k—Determine the spectral response acceleration parameter at a period of 1 s, S_1

From Reference 6.1, $S_1 = 0.13$.

Step 1l—Determine the seismic response coefficient, C_s

Because $S_1 = 0.13 < 0.6$, C_s is determined as follows:

$$C_s = \text{greatest of} \begin{cases} S_a/(R/I_e) = 0.20/(8/1.50) = 0.038 \\ 0.044 S_{DS} I_e = 0.044 \times 0.26 \times 1.50 = 0.017 \\ 0.01 \end{cases}$$

Step 1m—Determine the effective seismic weight, W, in accordance with ASCE/SEI 12.7.2

The effective seismic weight, W, is the summation of the effective dead loads at each level and includes the weight of the structural members, superimposed dead loads, the weight of the cladding and the weight of the mechanical equipment on the roof (details of the weight calculations are not shown here).

The story weights are given in Table 6.35.

Table 6.35 Seismic Forces and Seismic Story Shears for the Essential Facility in Example 6.7

Level	Story Weight, w_x (kips)	Height, h_x (ft)	$w_x h_x^k$	Lateral Force, F_x (kips)	Story Shear, V_x (kips)
R	2,561	45.0	189,034	161.6	161.6
4	2,623	34.0	141,049	120.5	282.1
3	2,623	23.0	90,688	77.5	359.6
2	2,640	12.0	43,760	37.4	397.0
Σ	10,447		464,531	397.0	

1 ft = 0.3048 m; 1 kip = 4.4482 kN.

Step 1n—Determine the seismic base shear, V

The seismic base shear, V, is determined by ASCE/SEI Equation (12.8-1):

$$V = C_s W = 0.038 \times 10{,}447 = 397.0 \text{ kips}$$

In S.I.:

$$V = C_s W = 0.038 \times 46{,}471 = 1{,}765.9 \text{ kN}$$

Step 1o—Determine the exponent related to the structure period, k

Because $0.5 \text{ s} < T = 0.76 \text{ s} < 2.5 \text{ s}$, $k = 0.75 + 0.5T = 1.13$.

Step 1p—Determine the lateral seismic force, F_x, at each level

Lateral seismic forces F_x are determined by ASCE/SEI Equations (12.8-12) and (12.8-13) and are given in Table 6.35 along with the story shears, V_x.

For example, at the second level:

$$C_{vx} = \frac{w_x h_x^k}{\sum w_i h_i} = \frac{2{,}640 \times (12.0)^{1.13}}{464{,}531} = 0.0942$$

$$F_x = C_{vx} V = 0.0942 \times 397.0 = 37.4 \text{ kips}$$

In S.I.:

$$C_{vx} = \frac{11{,}743.3 \times (3.66)^{1.13}}{539{,}724} = 0.0942$$

$$F_x = 0.0942 \times 1{,}765.9 = 166.4 \text{ kN}$$

Step 1q—Check the conditions for a stiffness-soft story irregularity

The building was analyzed in the north-south and east-west directions assuming rigid diaphragms at all levels and including reduced stiffness properties due to cracking of the concrete (ASCE/SEI 12.7.3). The seismic forces are applied at the CM at each level. Because the analysis is not checking for horizontal irregularities, the CM need not be displaced in each orthogonal direction 5 percent of the dimension of the structure perpendicular to the direction of the applied forces (ASCE/SEI 12.8.4.2). The design earthquake displacements, δ_{DE}, and the corresponding story drifts, Δ, are given in Table 6.36 where δ_{DE} is determined by ASCE/SEI Equation (12.8-16):

$$\delta_{DE} = \frac{C_d \delta_e}{I_e} + \delta_{di}$$

Table 6.36 Lateral Displacements and Story Drifts in the North-South and East-West Directions for the Essential Facility in Example 6.7

Story	North-South		East-West	
	$(\delta_{DE})_1$ in. (mm)	Δ, in. (mm)	δ_{DE} in. (mm)	Δ, in. (mm)
4	5.57 (141.6)	0.77 (19.5)	3.81 (96.8)	0.55 (13.9)
3	4.80 (122.1)	1.28 (32.6)	3.26 (82.9)	0.91 (23.1)
2	3.52 (89.5)	1.65 (41.8)	2.35 (59.8)	1.14 (29.0)
1	1.87 (47.7)	1.87 (47.7)	1.21 (30.8)	1.21 (30.8)

The elastic displacement, δ_e, are determined using the design seismic forces based on the computed fundamental periods of the structure without the upper limit, $C_u T_a$, specified in ASCE/SEI 12.8.2 (ASCE/SEI 12.8.6.2).

The deflection amplification factor, C_d, is equal to 5.5 for special reinforced concrete moment frames (ASCE/SEI Table 12.2-1) and the displacement due to diaphragm deformation corresponding to the design earthquake, δ_{di}, is permitted to be neglected in determining the design story drift (ASCE/SEI 12.8.6.5). The design story drifts, Δ, are the difference of the δ_{DE} at the top and bottom of the story under consideration.

Story stiffness, k_x, is determined by dividing the story shear, V_x, determined based on the computed fundamental periods by the story drift, Δ, from Table 6.36. Story stiffnesses in the north-south and east-west directions are given in Table 6.37.

Table 6.37 Story Stiffnesses in the North-South and East-West Directions for the Essential Facility in Example 6.7

Story	North-South			East-West		
	V_x, kips (kN)	Δ, in. (mm)	k_x, kips/in. (kN/mm)	V_x, kips (kN)	Δ, in. (mm)	k_x, kips/in. (kN/mm)
4	103.1 (458.6)	0.77 (19.5)	133.9 (23.5)	112.0 (502.2)	0.55 (13.9)	205.3 (36.1)
3	170.5 (758.4)	1.28 (32.6)	133.2 (23.3)	189.9 (844.7)	0.91 (23.1)	208.7 (36.6)
2	206.6 (919.0)	1.65 (41.8)	125.2 (22.0)	233.6 (1,039.1)	1.14 (29.0)	204.9 (35.8)
1	219.4 (975.9)	1.87 (47.7)	117.3 (20.5)	250.7 (1,115.2)	1.21 (30.8)	207.2 (36.2)

A type 1a soft story exits in the first story where one of the two following conditions is satisfied:

$$k_1 < 0.7k_2$$

$$k_1 < 0.8\left(\frac{k_2 + k_3 + k_4}{3}\right)$$

Check if a type 1a soft story irregularity exists in the first story:

- North-south direction

$$k_1 = 117.3 \text{ kips/in.} > 0.7k_2 = 0.7 \times 125.2 = 87.6 \text{ kips/in.}$$

$$k_1 = 117.3 \text{ kips/in.}$$

$$> 0.8\left(\frac{k_2 + k_3 + k_4}{3}\right) = 0.8 \times \left(\frac{125.2 + 133.2 + 133.9}{3}\right) = 104.6 \text{ kips/in.}$$

In S.I.:

$$k_1 = 20.5 \text{ kN/mm} > 0.7k_2 = 0.7 \times 22.0 = 15.4 \text{ kN/mm}$$

$$k_1 = 20.5 \text{ kN/mm}$$

$$> 0.8\left(\frac{k_2 + k_3 + k_4}{3}\right) = 0.8 \times \left(\frac{22.0 + 23.7 + 23.5}{3}\right) = 18.5 \text{ kN/mm}$$

Therefore, a type 1a soft story irregularity does not exist in the first story for analysis in the north-south direction.

- East-west direction

$$k_1 = 207.2 \text{ kips/in.} > 0.7k_2 = 0.7 \times 204.9 = 143.4 \text{ kips/in.}$$

$$k_1 = 207.2 \text{ kips/in.}$$

$$> 0.8\left(\frac{k_2 + k_3 + k_4}{3}\right) = 0.8 \times \left(\frac{204.9 + 208.7 + 205.3}{3}\right) = 165.0 \text{ kips/in.}$$

In S.I.:

$$k_1 = 36.2 \text{ kN/mm} > 0.7k_2 = 0.7 \times 35.8 = 25.1 \text{ kN/mm}$$

$$k_1 = 36.2 \text{ kN/mm}$$

$$> 0.8\left(\frac{k_2 + k_3 + k_4}{3}\right) = 0.8 \times \left(\frac{35.8 + 36.6 + 36.1}{3}\right) = 28.9 \text{ kN/mm}$$

Therefore, a type 1a soft story irregularity does not exist in the first story for analysis in the east-west direction.

It is evident from the above calculations that a type 1b extreme soft story irregularity does not exist in the first story in either direction.

Step 2—Determine if a vertical geometric irregularity exist

Because the horizontal dimensions of the SFRSs are the same over the height of the building, a vertical geometric irregularity does not exist.

Step 3—Determine if an in-plane discontinuity in vertical lateral force-resisting element irregularity exists

There are no in-plane offsets of the vertical SFRSs in the building, so this irregularity does not exist.

Step 4—Determine if discontinuity in lateral strength–weak story or extreme weak story irregularity exists

A weak story irregularity exists where a story lateral strength is less than that in the story above. The story lateral strength is the total lateral strength of all seismic force-resisting elements resisting the story shear in the direction under consideration. Similarly, an extreme weak story irregularity exists where a story lateral strength is less than 65 percent of that in the story above.

Because the SFRSs consist of special moment frames, the sums of the column shear forces in the first and second stories are used to determine if a weak story irregularity exists in the first story. The shear forces are calculated based on the nominal flexural strengths of the beams and columns in the special moment frames. Assuming the members have been designed such that a strong column-weak beam condition exists at the joints, the beams yield first, which means the column shear forces can be calculated using the nominal flexural strengths of the beams. Based on the member sizes, areas of negative and positive flexural reinforcement and material properties, the nominal negative and positive flexural strengths of the beams are equal to the following (calculations not shown here):

$$\left(M_n^-\right)_{\text{beam}} = 373.5 \text{ ft-kips } (506.4 \text{ kN-m})$$

$$(M_n^+)_{\text{beam}} = 297.8 \text{ ft-kips } (403.8 \text{ kN-m})$$

Check for a weak story irregularity in the north-south direction:

Free-body diagrams of edge and interior columns in the first story are given in Figure 6.53. Assuming the columns have uniform stiffness over the height of the building, the beam nominal flexural strengths are distributed above and below the joint in proportion to the inverse of the clear column heights. At the base of the columns, the bending moments are equal to the strength of the foundations supporting the columns, which are determined using the bending moments transferred to the foundation from the lateral analysis. Similar free-body diagrams for the columns in the second floor are given in Figure 6.54. Because the story heights for the second and third floor are the same, one-half of the beam flexural strengths are transferred to the columns above and below the joint.

The first- and second-story strengths are equal to the sum of the column shear forces in those stories:

First-story strength $= 4 \times (33.2 + 55.7) = 355.6$ kips

Second-story strength $= 4 \times (44.3 + 79.5) = 495.2$ kips

$$\frac{\text{First-story strength}}{\text{Second-story strength}} = \frac{355.6}{495.2} = 0.72 < 1.0$$

In S.I.:

First-story strength $= 4 \times (147.6 + 247.8) = 1{,}581.6$ kN

Second-story strength $= 4 \times (194.9 + 353.7) = 2{,}194.4$ kN

$$\frac{\text{First-story strength}}{\text{Second-story strength}} = \frac{1{,}581.6}{2{,}194.4} = 0.72 < 1.0$$

Therefore, a weak story irregularity exists in the north-south direction. An extreme weak story does not exist because the ratio of the first story to second story strengths is greater than 0.65 (see ASCE/SEI Table 12.3-2).

In the east-west direction, the ratio of the first story to second story strengths is greater than 1.0, which means a weak story irregularity does not exist in that direction.

Figure 6.53 Free-Body Diagrams of the First-Story Columns in the Essential Facility in Example 6.7

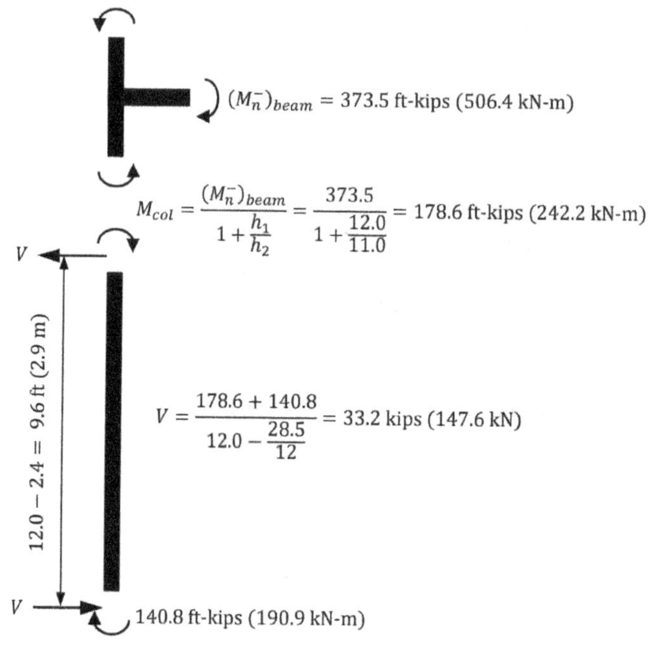

$(M_n^-)_{beam} = 373.5$ ft-kips (506.4 kN-m)
$M_{col} = \frac{(M_n^-)_{beam}}{1+\frac{h_1}{h_2}} = \frac{373.5}{1+\frac{12.0}{11.0}} = 178.6$ ft-kips (242.2 kN-m)
V
$12.0 - 2.4 = 9.6$ ft (2.9 m)
$V = \frac{178.6 + 140.8}{12.0 - \frac{28.5}{12}} = 33.2$ kips (147.6 kN)
V
140.8 ft-kips (190.9 kN-m)
Edge column

$(M_n^+)_{beam} = 297.8$ ft-kips (403.8 kN-m)

$(M_n^-)_{beam} = 373.5$ ft-kips (506.4 kN-m)

$$M_{col} = \frac{(M_n^-)_{beam} + (M_n^+)_{beam}}{1+\frac{h_1}{h_2}} = \frac{373.5 + 297.8}{1+\frac{12.0}{11.0}} = 321.1 \text{ ft-kips (435.4 kN-m)}$$

V

$12.0 - 2.4 = 9.6$ ft (2.9 m)

$$V = \frac{321.1 + 215.0}{12.0 - \frac{28.5}{12}} = 55.7 \text{ kips (247.8 kN)}$$

V

215.0 ft-kips (291.5 kN-m)

Interior column

Figure 6.54 Free-Body Diagrams of the Second-Story Columns in the Essential Facility in Example 6.7

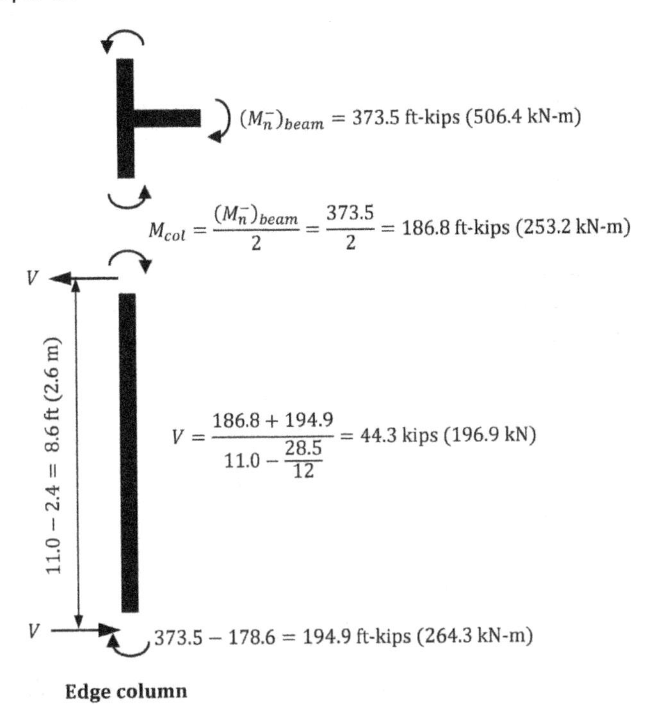

$(M_n^-)_{beam} = 373.5$ ft-kips (506.4 kN-m)
$M_{col} = \frac{(M_n^-)_{beam}}{2} = \frac{373.5}{2} = 186.8$ ft-kips (253.2 kN-m)
V
$11.0 - 2.4 = 8.6$ ft (2.6 m)
$V = \frac{186.8 + 194.9}{11.0 - \frac{28.5}{12}} = 44.3$ kips (196.9 kN)
V
$373.5 - 178.6 = 194.9$ ft-kips (264.3 kN-m)
Edge column

$(M_n^+)_{beam} = 297.8$ ft-kips (403.8 kN-m)

$(M_n^-)_{beam} = 373.5$ ft-kips (506.4 kN-m)

$$M_{col} = \frac{(M_n^-)_{beam} + (M_n^+)_{beam}}{2} = \frac{373.5 + 297.8}{2} = 335.7 \text{ ft-kips (455.2 kN-m)}$$

V

$11.0 - 2.4 = 8.6$ ft (2.6 m)

$$V = \frac{335.7 + 350.2}{11.0 - \frac{28.5}{12}} = 79.5 \text{ kips (353.7 kN)}$$

V

$297.8 + 373.5 - 321.1 = 350.2$ ft-kips (474.8 kN-m)

Interior column

Structures assigned to SDC D with vertical irregularity type 4a (weak story irregularity) are permitted; structures assigned to SDC D with vertical irregularity type 4b (extreme weak story irregularity) are not permitted (ASCE/SEI 12.3.3.2).

6.7.8 Example 6.8—Horizontal and Vertical Irregularities

Determine whether the seven-story office in Figure 6.55 possesses any of the horizontal or vertical structural irregularities in ASCE/SEI Tables 12.3-1 and 12.3-2. Design data are given in Table 6.38.

Figure 6.55 Typical Floor Plans and Elevations for the 7-Story Office Building in Example 6.8

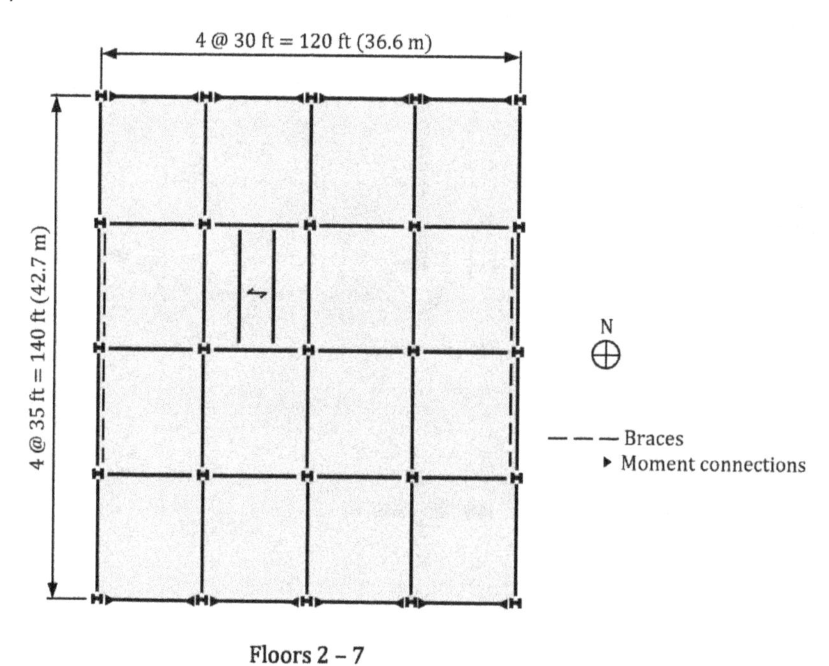

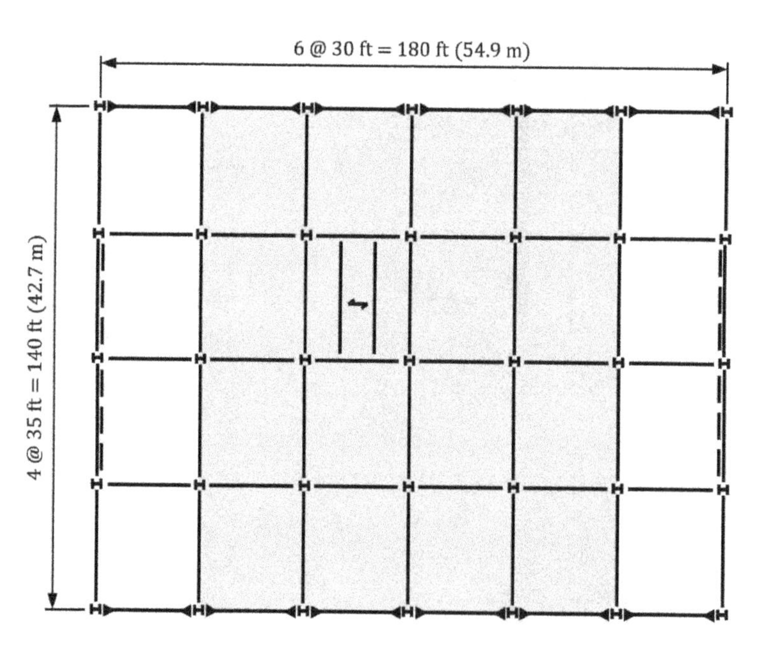

Figure 6.55 *(Continued)*

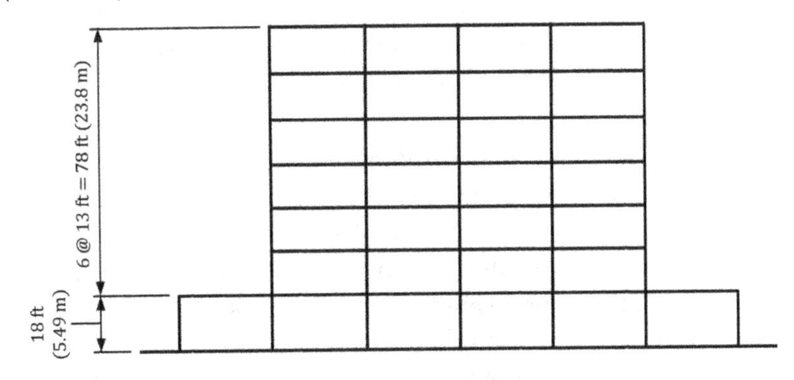

North or South Elevation

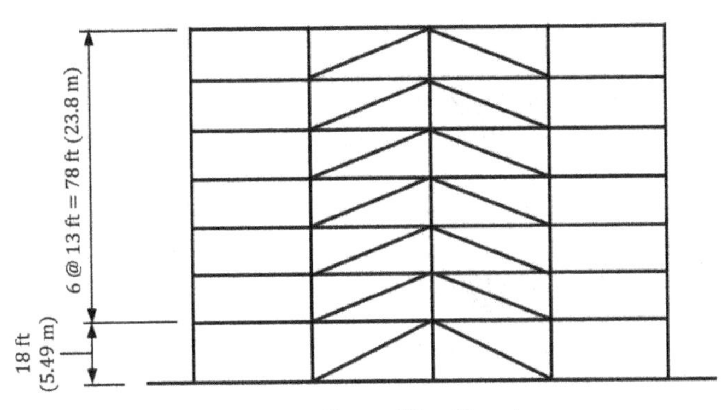

East or West Elevation

Table 6.38 Design Data for the 7-Story Office Building in Example 6.8

Location	Latitude: 35.130°, Longitude: −90.050°
Soil classification	Site class D
Occupancy	Office occupancy where less than 300 people congregate in one area
Material	Structural steel
SFRS	Steel special concentrically braced frames (system B2) and steel special moment frames (system C1)

Solution

- Determine the SDC.
  Flowchart 6.2 in Section 6.6 of this book is used to determine the SDC.

  Step 1—Determine the risk category in accordance with IBC Table 1604.5
  For an office occupancy, the risk category is II.

  Step 2—Determine the site class in accordance with ASCE/SEI 11.4.2
  The site class is given as D.

  Step 3—Determine the acceleration parameters S_1, S_{DS}, and S_{D1} from Reference 6.1
  For the given latitude and longitude of the site, risk category II and site class D, the multi-period design spectrum is available from Reference 6.1 where $S_1 = 0.35$, $S_{DS} = 0.74$, and $S_{D1} = 0.47$.

Step 4—Determine if the building can be assigned to SDC F

Because the risk category is II, the building is not assigned to SDC F.

Step 5—Determine if the building can be assigned to SDC E

Because the risk category is II and $S_1 = 0.35 < 0.75$, the building is not assigned to SDC E.

Step 6—Determine if the simplified method in ASCE/SEI 12.14 is permitted to be used

Because the building has more than three stories above grade, the simplified method is not permitted to be used.

Step 7—Check if all 4 conditions in ASCE/SEI 11.6 are satisfied

Check if the approximate period, T_a, is less than $0.8T_S$.

In the north-south direction, the concentrically braced frame falls under "All other structural systems" in ASCE/SEI Table 12.8-2. Therefore, T_a is determined by ASCE/SEI Equation (12.8-8):

$$T_a = C_t h_n^x = 0.02 \times (96.0)^{0.75} = 0.61 \text{ s}$$

In S.I.: $T_a = 0.0488 \times (29.3)^{0.75} = 0.61$ s

In the east-west direction, values of C_t and x for steel moment-resisting frames are used:

$$T_a = 0.028 \times (96.0)^{0.8} = 1.08 \text{ s}$$

In S.I.: $T_a = 0.0724 \times (29.3)^{0.8} = 1.08$ s

$$T_S = S_{D1}/S_{DS} = 0.47/0.74 = 0.64 \text{ s}$$

$0.8T_S = 0.8 \times 0.64 = 0.51$ s, which is less than the approximate periods in both the north-south and east-west directions.

Therefore, the first of the 4 conditions is not satisfied and the SDC is not permitted to be determined by ASCE/SEI Table 11.6-1 alone.

Step 8—Determine the SDC using ASCE/SEI Tables 11.6-1 and 11.6-2

With $S_{DS} = 0.74 > 0.50$ and risk category II, the SDC is D in accordance with ASCE/SEI Table 11.6-1.

With $S_{D1} = 0.47 > 0.20$ and risk category II, the SDC is D in accordance with ASCE/SEI Table 11.6-2.

Therefore, the SDC is D for this building.

- Determine if any horizontal irregularities exist.

Step 1—Determine if a torsional irregularity exists

Story drifts are needed to ascertain whether torsional irregularity or an extreme torsional irregularity exists. The seismic forces from the ELF procedure in ASCE/SEI 12.8 are used to determine the displacements and story drifts for this purpose. Flowchart 6.5 is used to determine the seismic base shear, V, and the lateral seismic forces, F_x, at the floor and roof levels.

Step 1a—Obtain R from ASCE/SEI Table 12.2-1

In the north-south direction, steel special concentrically braced frames are required because the building is assigned to SDC D (system B2 in ASCE/SEI Table 12.2-1). Therefore, $R = 6$ and the height of the building, which is equal to 96 ft (29.3 m), is less than the limiting height of 160 ft (48.8 m) for SDC D.

In the east-west direction, steel special moment frames are required (system C1 in ASCE/SEI Table 12.2-1) and $R = 8$. There is no height limit for this system in SDC D.

Step 1b—Obtain I_e from ASCE/SEI Table 1.5-2

For an office occupancy, the risk category is II (IBC Table 1604.5), and $I_e = 1.00$.

Step 1c—Determine T in the direction under consideration from a substantiated analysis

In lieu of determining the fundamental periods of the building, the approximate fundamental periods, T_a, will be used to determine the seismic forces, which is permitted in ASCE/SEI 12.8.2.

Step 1d—Determine the approximate fundamental period, T_a, in accordance with ASCE/SEI 12.8.2.1

In the north-south direction, the concentrically braced frame falls under "All other structural systems in ASCE/SEI Table 12.8-2. Therefore, T_a is determined by ASCE/SEI Equation (12.8-8):

$$T_a = C_t h_n^x = 0.02 \times (96.0)^{0.75} = 0.61 \text{ s}$$

In S.I.: $T_a = 0.0488 \times (29.3)^{0.75} = 0.61$ s

In the east-west direction, values of C_t and x for steel special moment-resisting frames are used:

$$T_a = 0.028 \times (96.0)^{0.8} = 1.08 \text{ s}$$

In S.I.: $$T_a = 0.0724 \times (29.3)^{0.8} = 1.08 \text{ s}$$

Step 1e—Determine S_{DS} and S_{D1} in accordance with ASCE/SEI 11.4.4 or 11.4.7

For the given latitude and longitude of the site, risk category II and site class D, the multi-period design spectrum is available from Reference 6.1 where $S_{DS} = 0.74$ and $S_{D1} = 0.47$.

Step 1f—Determine coefficient C_u from ASCE/SEI Table 12.8-1 based on S_{D1}

Not applicable.

Step 1g—Determine if $T > C_u T_a$

Not applicable.

Step 1h—Determine T_L from ASCE/SEI Figures 22-14 through 22-17

In lieu of using ASCE/SEI Figures 22-14 through 22-17, $T_L = 12$ s from Reference 6.1.

Step 1i—Determine the permitted method to determine C_s

Because the design spectral acceleration parameter, S_a, is available from Reference 6.1 for the periods determined in step 1d above, Method 1 or Method 2 in ASCE/SEI 12.8.1.1 is permitted to be used to determine C_s. Method 1 is selected in this example.

Step 1j—Determine the design spectral acceleration parameter, S_a

- North-south direction

  From Reference 6.1, $S_a = 0.59$ at $T = 0.50$ s and $S_a = 0.54$ at $T = 0.75$ s. Therefore, $S_a = 0.57$ at $T = 0.61$ s by linear interpolation.

- East-west direction

  From Reference 6.1, $S_a = 0.47$ at $T = 1.00$ s and $S_a = 0.33$ at $T = 1.50$ s. Therefore, $S_a = 0.45$ at $T = 1.08$ s by linear interpolation.

  Maximum $S_a = 0.85$ occurs at $T = 0.15$ s, which is less than the period of the structure in both directions. Therefore, use the values of S_a above in the determination of C_s.

Step 1k—Determine the spectral response acceleration parameter at a period of 1 s, S_1

From Reference 6.1, $S_1 = 0.35$.

Step 1l—Determine the seismic response coefficient, C_s

Because $S_1 = 0.35 < 0.6$, C_s is determined as follows:

- North-south direction

$$C_s = \text{greatest of} \begin{cases} S_a/(R/I_e) = 0.57/(6/1.00) = 0.095 \\ 0.044 S_{DS} I_e = 0.044 \times 0.74 \times 1.00 = 0.033 \\ 0.01 \end{cases}$$

- East-west direction

$$C_s = \text{greatest of} \begin{cases} S_a/(R/I_e) = 0.45/(8/1.00) = 0.056 \\ 0.044 S_{DS} I_e = 0.044 \times 0.74 \times 1.00 = 0.033 \\ 0.01 \end{cases}$$

Step 1m—Determine the effective seismic weight, W, in accordance with ASCE/SEI 12.7.2

The effective seismic weight, W, is the summation of the effective dead loads at each level and includes the weight of the structural members, superimposed dead loads, the weight of the cladding and the weight of the mechanical equipment on the roof (details of the weight calculations are not shown here).

The story weights are given in Tables 6.39 and 6.40.

Step 1n—Determine the seismic base shear, V

The seismic base shear, V, is determined by ASCE/SEI Equation (12.8-1).

- North-south direction

$$V = C_s W = 0.095 \times 9{,}960 = 946.2 \text{ kips}$$

In S.I.:

$$V = C_s W = 0.095 \times 44{,}304 = 4{,}208.9 \text{ kN}$$

- East-west direction

$$V = C_s W = 0.056 \times 9{,}960 = 557.8 \text{ kips}$$

Table 6.39 Seismic Forces and Seismic Story Shears in the North-South Direction for the Office Building in Example 6.8

Level	Story Weight, w_x (kips)	Height, h_x (ft)	$w_x h_x^k$	Lateral Force, F_x (kips)	Story Shear, V_x (kips)
R	1,018	96.0	128,515	180.0	180.0
6	1,381	83.0	149,423	209.3	389.3
5	1,381	70.0	124,738	174.7	564.0
4	1,381	57.0	100,328	140.5	704.5
3	1,381	44.0	76,252	106.8	811.3
2	1,381	31.0	52,606	73.8	885.1
1	2,037	18.0	43,609	61.1	946.2
Σ	9,960		675,471	946.2	

1 ft = 0.3048 m; 1 kip = 4.4482 kN.

Table 6.40 Seismic Forces and Seismic Story Shears in the East-West Direction for the Office Building in Example 6.8

Level	Story Weight, w_x (kips)	Height, h_x (ft)	$w_x h_x^k$	Lateral Force, F_x (kips)	Story Shear, V_x (kips)
R	1,018	96.0	367,179	117.1	117.1
6	1,381	83.0	412,862	131.7	248.8
5	1,381	70.0	331,414	105.7	354.5
4	1,381	57.0	254,527	81.2	435.7
3	1,381	44.0	182,074	58.1	493.8
2	1,381	31.0	115,891	37.0	530.8
1	2,037	18.0	84,780	27.0	557.8
Σ	9,960		1,748,727	557.8	

1 ft = 0.3048 m; 1 kip = 4.4482 kN.

In S.I.:

$$V = C_s W = 0.056 \times 44{,}304 = 2{,}481.0 \text{ kN}$$

Step 1o—Determine the exponent related to the structure period, k

- North-south direction: Because $0.5 \text{ s} < T = 0.61 \text{ s} < 2.5 \text{ s}$, $k = 0.75 + 0.5T = 1.06$.
- East-west direction: Because $0.5 \text{ s} < T = 1.08 \text{ s} < 2.5 \text{ s}$, $k = 0.75 + 0.5T = 1.29$.

Step 1p—Determine the lateral seismic force, F_x, at each level

Lateral seismic forces F_x are determined by ASCE/SEI Equations (12.8-12) and (12.8-13) and are given in Tables 6.39 and 6.40 along with the story shears, V_x, for the north-south and east-west directions, respectively.

For example, at the roof level in the north-south direction:

$$C_{vx} = \frac{w_x h_x^k}{\sum w_i h_i} = \frac{1{,}018 \times (96.0)^{1.06}}{675{,}471} = 0.1903$$

$$F_x = C_{vx} V = 0.1903 \times 946.2 = 180.0 \text{ kips}$$

In S.I.:

$$C_{vx} = \frac{4{,}528.3 \times (29.3)^{1.06}}{853{,}839} = 0.1903$$

$$F_x = 0.1903 \times 4{,}208.9 = 801.0 \text{ kN}$$

Step 1q—Check the conditions for torsional irregularities

A type 1 torsional irregularity occurs where either of the following conditions exist (ASCE/SEI Table 12.3-1):

- More than 75 percent of any story's lateral strength below the diaphragm is provided at or on one side of the CM.
- The torsional irregularity ratio (TIR) exceeds 1.2.

The second condition is checked in this example. The TIR is defined as follows [ASCE Equation (12.3-2)]:

$$TIR = \frac{\Delta_{max}}{\Delta_{avg}}$$

where Δ_{max} is the maximum story drift at the building's edge subject to F_x including accidental torsion and Δ_{avg} is the average of the story drifts at the two opposing edges of the building determined using the same loading and diaphragm rigidity used for the determination of Δ_{max} (see Figure 6.15).

Three-dimensional analyses were performed independently in the north-south and east-west directions assuming the floors and roof are constructed of concrete-filled metal deck (which is permitted to be idealized as a rigid diaphragm provided all the conditions in ASCE/SEI 12.3.1.2 are satisfied) at all levels. The forces are applied at the CM at each level, which is displaced in each orthogonal direction 5 percent of the dimension of the structure perpendicular to the direction of the applied forces; this accidental torsion, M_{ta}, which is in addition to the inherent torsion, M_t, due to the CM and CR being at different locations, must be applied when determining if a horizontal irregularity exists where diaphragms are not flexible (ASCE/SEI 12.8.4.2).

A summary of the design earthquake displacements, δ_{DE}, and drifts at all levels is given in Tables 6.41 and 6.42 for seismic forces in the north-south and east-west directions, respectively, where δ_{DE} are determined by ASCE/SEI Equation (12.8-16):

$$\delta_{DE} = \frac{C_d \delta_e}{I_e} + \delta_{di}$$

The elastic displacement, δ_e, are determined using the lateral forces, F_x, in Tables 6.39 and 6.40 for analysis in the north-south and east-west directions, respectively.

The deflection amplification factor, C_d, is equal to 5 for steel special concentrically braced frames and is equal to 5.5 for steel special moment frames (ASCE/SEI Table 12.2-1). The displacement due to

Table 6.41 Lateral Displacements and Story Drifts in the North-South Direction for the Office Building in Example 6.8

Story	$(\delta_{DE})_1$ (in.)	Δ_1 (in.)	$(\delta_{DE})_2$ (in.)	Δ_2 (in.)	Δ_{avg} (in.)	TIR = $\Delta_{max}/\Delta_{avg}$
7	8.65	0.75	6.70	0.65	0.70	1.07
6	7.90	0.90	6.05	0.80	0.85	1.06
5	7.00	1.20	5.25	1.10	1.15	1.04
4	5.80	1.15	4.15	1.05	1.10	1.05
3	4.65	1.20	3.10	0.90	1.05	1.14
2	3.45	0.95	2.20	0.65	0.80	1.19
1	2.50	2.50	1.55	1.55	2.03	1.23

1 in. = 25.4 mm.

Table 6.42 Lateral Displacements and Story Drifts in the East-West Direction for the Office Building in Example 6.8

Story	$(\delta_{DE})_1$ (in.)	Δ_1 (in.)	$(\delta_{DE})_2$ (in.)	Δ_2 (in.)	Δ_{avg} (in.)	TIR = $\Delta_{max}/\Delta_{avg}$
7	18.45	1.23	17.96	1.16	1.20	1.03
6	17.22	1.71	16.80	1.62	1.67	1.02
5	15.51	2.42	15.18	2.45	2.44	1.00
4	13.09	3.79	12.73	3.65	3.72	1.02
3	9.30	4.10	9.08	3.99	4.05	1.01
2	5.20	3.05	5.09	3.00	3.03	1.01
1	2.15	2.15	2.09	2.09	2.12	1.01

1 in. = 25.4 mm.

diaphragm deformation corresponding to the design earthquake, δ_{di}, is permitted to be neglected in determining the design story drift (ASCE/SEI 12.8.6.5). The design story drifts, Δ, are the difference of the δ_{DE} at the top and bottom of the story under consideration.

For example, in the seventh story in the north-south direction:

$$\Delta_1 = 8.65 - 7.90 = 0.75 \text{ in.}$$
$$\Delta_2 = 6.70 - 6.05 = 0.65 \text{ in.}$$
$$\Delta_{avg} = \frac{\Delta_1 + \Delta_2}{2} = \frac{0.75 + 0.65}{2} = 0.70 \text{ in.}$$
$$\text{TIR} = \Delta_{max}/\Delta_{avg} = 0.75/0.70 = 1.07$$

In S.I.:

$$\Delta_1 = 219.7 - 200.7 = 19.0 \text{ mm}$$
$$\Delta_2 = 170.2 - 153.7 = 16.5 \text{ mm}$$
$$\Delta_{avg} = \frac{19.0 + 16.5}{2} = 17.8 \text{ mm}$$
$$\Delta_{max}/\Delta_{avg} = 19.0/17.8 = 1.07$$

A type 1 torsional irregularity exists in the first story in the north-south direction because TIR is greater than 1.2 (see Table 6.41).

According to ASCE/SEI 12.8.4.3, the accidental torsional moments, M_{ta}, must be increased by the torsional amplification factor, A_x, determined by ASCE/SEI Equation (12.8-15) where a torsional irregularity exists:

$$A_x = \left(\frac{\delta_{max}}{1.2\delta_{avg}}\right)^2 = \left[\frac{2.50}{1.2 \times \left(\frac{2.50 + 1.55}{2}\right)}\right]^2 = 1.06 > 1.0$$

In S.I.:

$$A_x = \left[\frac{63.5}{1.2 \times \left(\frac{63.5 + 39.4}{2}\right)}\right]^2 = 1.06 > 1.0$$

Assuming the CM and the CR are at the same location, the accidental torsional moment, M_{ta}, at the first story is equal to the following:

$$M_{ta} = A_x F_1 e = 1.06 \times 61.1 \times (0.05 \times 180.0) = 583 \text{ ft-kips}$$

In S.I.: $M_{ta} = 1.06 \times 271.8 \times (0.05 \times 54.9) = 791$ kN-m

Step 2—Determine if a reentrant corner irregularity exists

By inspection, this irregularity does not exist.

Step 3—Determine if a diaphragm discontinuity irregularity exists

A diaphragm discontinuity does not exist because the diaphragms do not have an abrupt discontinuity or variation in stiffness, including a cutout or open area greater than 25 percent of the gross enclosed diaphragm area or a change in effective stiffness of more than 50 percent from one story to the next (see Figure 6.17).

Step 4—Determine if an out-of-plane offset irregularity exists

In the first story, the steel special concentrically braced frames occur at the east and west faces of the building. Above the first story, there is a 30-foot (9.14-m) offset of the braced frames on both sides of the building.

Therefore, a type 4 out-of-plane offset irregularity exists. As such, the diaphragms are not permitted to be idealized as rigid as originally assumed (ASCE/SEI 12.3.1.2). The diaphragms must be modeled as semirigid in subsequent analyses.

The forces from the braced frames above the first story must be transferred through the diaphragm at the first elevated level to the braced frames in the first story.

Step 4—Determine if a nonparallel system irregularity exists

A nonparallel system irregularity does not exist because all the braced frames and moment-resisting frames are parallel to a major orthogonal axis of the building (see Figure 6.19).

- Determine if any vertical irregularities exist

Step 1—Determine if a stiffness-soft story irregularity exists

Lateral story stiffnesses are needed to determine whether a soft story irregularity exists. Story stiffness can be obtained by dividing the story shear by the story drift in the direction of analysis. Therefore, the lateral displacements at the CM for the lower four stories are needed, which are given in Table 6.43 (soft story irregularities must be investigated considering the relative lateral stiffnesses of adjacent stories and the average of the lateral stiffnesses of three stories above, where applicable; see ASCE/SEI Table 12.3-2). Using the lateral displacements at the CM provides an average lateral story stiffness.

Table 6.43 Lateral Displacements and Story Drifts in the North-South and East-West Directions for the Office Building in Example 6.8

	North-South		East-West	
Story	δ_{DE}, in. (mm)	Δ, in. (mm)	δ_{DE}, in. (mm)	Δ, in. (mm)
4	4.95 (125.7)	1.20 (30.4)	12.90 (327.7)	3.69 (93.7)
3	3.75 (95.3)	0.95 (24.2)	9.21 (234.0)	4.09 (104.1)
2	2.80 (71.1)	1.10 (27.9)	5.12 (129.9)	3.01 (76.3)
1	1.70 (43.2)	1.70 (43.2)	2.11 (53.6)	2.11 (53.6)

Story stiffness, k_x, is determined by dividing the story shears, V_x, from Tables 6.39 and 6.40 by the story drifts, Δ, from Table 6.43. Story stiffnesses in the north-south and east-west directions are given in Table 6.44.

Table 6.44 Story Stiffnesses in the North-South and East-West Directions for the Office Building in Example 6.8

	North-South			East-West		
Story	V_x, kips (kN)	Δ, in. (mm)	k_x, kips/in. (kN/mm)	V_x, kips (kN)	Δ, in. (mm)	k_x, kips/in. (kN/mm)
4	704.5 (3,133.8)	1.20 (30.4)	587.1 (103.1)	435.7 (1,938.1)	3.69 (93.7)	118.1 (20.7)
3	811.3 (3,608.8)	0.95 (24.2)	854.0 (149.1)	493.8 (2,196.5)	4.09 (104.1)	120.7 (21.0)
2	885.1 (3,937.1)	1.10 (27.9)	804.6 (141.1)	530.8 (2,361.1)	3.01 (76.3)	176.4 (31.0)
1	946.2 (4,208.9)	1.70 (43.2)	556.6 (97.4)	557.8 (2,481.2)	2.11 (53.6)	264.4 (46.3)

A type 1a soft story exits in the first story where one of the two following conditions is satisfied:

$$k_1 < 0.7k_2$$

$$k_1 < 0.8\left(\frac{k_2 + k_3 + k_4}{3}\right)$$

Check if a type 1a soft story irregularity exists in the first story:

- North-south direction

$$k_1 = 556.6 \text{ kips/in.} < 0.7k_2 = 0.7 \times 804.6 = 563.2 \text{ kips/in.}$$

In S.I.:

$$k_1 = 97.4 \text{ kN/mm} < 0.7k_2 = 0.7 \times 141.1 = 98.8 \text{ kN/mm}$$

Therefore, a type 1a soft story irregularity exists in the first story for analysis in the north-south direction.

- East-west direction

$$k_1 = 264.4 \text{ kips/in.} > 0.7k_2 = 0.7 \times 176.4 = 123.5 \text{ kips/in.}$$

$$k_1 = 264.4 \text{ kips/in.}$$

$$> 0.8\left(\frac{k_2 + k_3 + k_4}{3}\right) = 0.8 \times \left(\frac{176.4 + 120.7 + 118.1}{3}\right) = 110.7 \text{ kips/in.}$$

In S.I.:

$$k_1 = 46.3 \text{ kN/mm} > 0.7k_2 = 0.7 \times 31.0 = 21.7 \text{ kN/mm}$$

$$k_1 = 46.3 \text{ kN/mm}$$

$$> 0.8\left(\frac{k_2 + k_3 + k_4}{3}\right) = 0.8 \times \left(\frac{31.0 + 21.0 + 20.7}{3}\right) = 19.4 \text{ kN/mm}$$

Therefore, a type 1a soft story irregularity does not exist in the first story for analysis in the east-west direction.

It is evident from the above calculations that a type 1b extreme soft story irregularity does not exist in the first story in either direction.

Step 2—Determine if a vertical geometric irregularity exist

A vertical geometric irregularity is considered to exist where the horizontal dimension of the SFRS in any story is 1.3 times that in an adjacent story.

The setbacks above the first story must be investigated for the moment-resisting frames in the east-west direction located at the north and south faces of the building:

$$\frac{\text{Width of SFRS in the first story}}{\text{Width of the SFRS above the first story}} = \frac{180.0}{120.0} = 1.5 > 1.3$$

In S.I.:

$$\frac{\text{Width of SFRS in the first story}}{\text{Width of the SFRS above the first story}} = \frac{54.9}{36.6} = 1.5 > 1.3$$

Therefore, a type 3 vertical geometric irregularity exists.

Step 3—Determine if an in-plane discontinuity in vertical lateral force-resisting element irregularity exists

There are no in-plane offsets of the vertical SFRSs in the building, so this irregularity does not exist.

Step 4—Determine if discontinuity in lateral strength–weak story or extreme weak story irregularity exists

A weak story irregularity exists where a story lateral strength is less than that in the story above. The story lateral strength is the total lateral strength of all seismic force-resisting elements resisting the story shear in the direction under consideration. Similarly, an extreme weak story irregularity exists where a story lateral strength is less than 65 percent of that in the story above.

Check if there is a weak story irregularity in the first story.

Assuming the same beam, column, and brace sizes are used in the first and second stories in the north-south direction, the shear strengths in these stories are almost identical; therefore, no weak story irregularity exists in the north-south direction.

In the east-west direction, the story strength is equal to the sum of the column shear forces in the moment-resisting frames in that story when the member moment capacity is developed by lateral loading. It is assumed in this example that the same column and beam sections are used in the moment-resisting frames in the first and second stories.

Assume the following nominal flexural strengths (note: the assumed nominal flexural strengths of the beams and columns are based on preliminary member sizes):

Beams: $M_{nb} = 525$ ft-kips (711.8 kN-m)
Columns: $M_{nc} = 550$ ft-kips (745.7 kN-m)

- First story shear strength

  Check the corner columns for strong column-weak beam condition:

$$\Sigma M_c = 2 \times 550 = 1{,}100 \text{ ft-kips} > \Sigma M_b = 525 \text{ ft-kips}$$

The maximum shear force that can develop in each corner column is based on the moment capacity of the beam because it is less than that of the column at the top of the column. At the base of the column, it is assumed the full moment capacity of the column can be developed. Therefore,

$$V_{\text{corner}} = \frac{\dfrac{525}{1 + \dfrac{18.0}{13.0}} + 550}{18.0} = 42.8 \text{ kips}$$

Check the interior columns for strong column-weak beam condition:

$$\Sigma M_c = 2 \times 550 = 1{,}100 \text{ ft-kips} > \Sigma M_b = 2 \times 525 = 1{,}050 \text{ ft-kips}$$

The maximum shear force that can develop in each interior column is based on the moment capacity of the beam because it is less than that of the column at the top of the column:

$$V_{\text{interior}} = \frac{\dfrac{2 \times 525}{1 + \dfrac{18.0}{13.0}} + 550}{18.0} = 55.0 \text{ kips}$$

Total first-story shear strength $= (4 \times 42.8) + (10 \times 55.0) = 721.2$ kips

In S.I.:

Check the corner columns for strong column-weak beam condition:

$$\Sigma M_c = 2 \times 745.7 = 1{,}491.4 \text{ kN-m} > \Sigma M_b = 711.8 \text{ kN-m}$$

The maximum shear force that can develop in each corner column is based on the moment capacity of the beam because it is less than that of the column at the top of the column. At the base of the column, it is assumed the full moment capacity of the column can be developed. Therefore,

$$V_{\text{corner}} = \frac{\dfrac{711.8}{1 + \dfrac{5.49}{3.96}} + 745.7}{5.49} = 190.2 \text{ kN}$$

Check the interior columns for strong column-weak beam condition:

$$\Sigma M_c = 2 \times 745.7 = 1{,}491.4 \text{ kN-m} > \Sigma M_b = 2 \times 711.8 = 1{,}423.6 \text{ kN-m}$$

The maximum shear force that can develop in each interior column is based on the moment capacity of the beam because it is less than that of the column at the top of the column:

$$V_{\text{interior}} = \frac{\dfrac{2 \times 711.8}{1 + \dfrac{5.49}{3.96}} + 745.7}{5.49} = 244.5 \text{ kN}$$

Total first-story shear strength $= (4 \times 190.2) + (10 \times 244.5) = 3{,}205.8$ kN

- Second story shear strength

$$V_{\text{corner}} = \frac{\dfrac{525}{2} + \dfrac{525}{1 + \dfrac{13.0}{18.0}}}{13.0} = 43.6 \text{ kips}$$

$$V_{\text{interior}} = \frac{525 + \dfrac{2 \times 525}{1 + \dfrac{13.0}{18.0}}}{13.0} = 87.3 \text{ kips}$$

Total second-story shear strength $= (4 \times 43.6) + (6 \times 87.3) = 698.2$ kips

In. S.I.:

$$V_{\text{corner}} = \frac{\dfrac{711.8}{2} + \dfrac{711.8}{1+\dfrac{3.96}{5.49}}}{3.96} = 194.3 \text{ kN}$$

$$V_{\text{interior}} = \frac{711.8 + \dfrac{2\times 711.8}{1+\dfrac{3.96}{5.49}}}{3.96} = 388.6 \text{ kN}$$

$$\text{Total second-story shear strength} = (4\times 194.3)+(6\times 388.6) = 3{,}108.8 \text{ kN}$$

$$\text{First-story shear strength} = 721.1 \text{ kips} > \text{Second-story strength} = 698.2 \text{ kips}$$

In S.I.:

$$\text{First-story shear strength} = 3{,}205.8 \text{ kN} > \text{Second-story strength} = 3{,}108.8 \text{ kN}$$

Therefore, a weak story irregularity does not exist in the first story in the east-west direction.

6.7.9 Example 6.9—Redundancy Factor

Determine the redundancy factor, ρ, in the north-south and east-west directions for the 10-story residential building in Figure 6.56. Assume the building is assigned to SDC D.

Figure 6.56 Typical Floor Plan for the Residential Building in Example 6.9

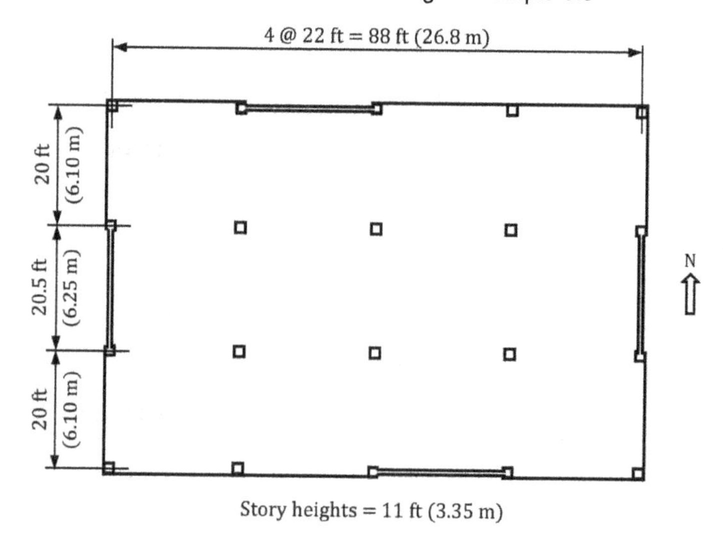

SOLUTION

The requirements in ASCE/SEI 12.3.4.2(1) and in ASCE/SEI 12.3.4.2(2) are both used to determine ρ for illustration purposes.

Step 1—Determine ρ using the requirements in ASCE/SEI 12.3.4.2(1)

For each story in a building where the story shear is greater than 35 percent of the base shear in the direction of interest, it is permitted to take $\rho = 1.0$ where all three conditions in ASCE/SEI 12.3.4.2(1) are met with the notional removal of any lateral force-resisting element or connection given in ASCE/SEI Table 12.3-3. Buildings with a lateral force-resisting system not listed in ASCE/SEI Table 12.3-3 need not consider this requirement.

According to ASCE/SEI Table 12.3-3, shear walls must have a height-to-length ratio greater than 1.0 in order for the requirements in ASCE/SEI 12.3.4.2(1) to be considered.

For the shear walls in the north-south direction: $h_w/\ell_w = 11.0/20.5 = 0.54$
For the shear walls in the east-west direction: $h_w/\ell_w = 11.0/22.0 = 0.50$

In S.I.:

For the shear walls in the north-south direction: $h_w/\ell_w = 3.35/6.25 = 0.54$
For the shear walls in the east-west direction: $h_w/\ell_w = 3.35/6.70 = 0.50$

Because the height-to-length ratios of the shear walls is less than 1.0, $\rho = 1.0$ in both directions based on the requirements in ASCE/SEI Table 12.3-3.

Step 2—Determine ρ using the requirements in ASCE/SEI 12.3.4.2(2)

It is permitted to take $\rho = 1.0$ for structures that do not possess any horizontal irregularities in ASCE/SEI Table 12.3-1 in the direction of interest at all levels and with at least two bays of seismic force-resisting perimeter framing on each side of the structure at each story resisting more than 35 percent of the base shear.

For structures with shear walls, the number of bays is equal to the length of the shear wall divided by the story height (ASCE/SEI Table 12.3-3):

In the north-south direction: Number of bays per wall = $20.5/11.0 = 1.9 < 2.0$
In the east-west direction: Number of bays per wall = $22.0/11.0 = 2.0$

In S.I.:

In the north-south direction: Number of bays per wall = $6.25/3.35 = 1.9 < 2.0$
In the east-west direction: Number of bays per wall = $6.70/3.35 = 2.0$

Therefore, it is permitted to take $\rho = 1.0$ in the east-west direction; in the north-south direction, $\rho = 1.3$ because the provided number of bays is less than 2.

6.7.10 Example 6.10—Equivalent Lateral Force (ELF) Procedure

Determine the seismic forces in the north-south direction for the 10-story residential building in Figure 6.56 using the ELF procedure. Design data are given in Table 6.45.

Table 6.45 Design Data for the 10-Story Residential Building in Example 6.10

Location	Latitude: 34.045°, Longitude: −118.258°
Soil classification	Site class CD
Occupancy	Residential occupancy where less than 300 people congregate in one area
Material	Cast-in-place, reinforced concrete
SFRS	Building frame with special reinforced concrete shear walls (system B4)

Solution

- Determine the SDC.
  Flowchart 6.2 in Section 6.6 of this book is used to determine the SDC.

  Step 1—Determine the risk category in accordance with IBC Table 1604.5
  For a residential occupancy, the risk category is II.

Step 2—Determine the site class in accordance with ASCE/SEI 11.4.2
The site class is given as CD.

Step 3—Determine the acceleration parameters S_1, S_{DS}, and S_{D1} from Reference 6.1
For the given latitude and longitude of the site, risk category II and site class CD, the multi-period design spectrum is available from Reference 6.1 where $S_1 = 0.73$, $S_{DS} = 1.57$, and $S_{D1} = 0.93$.

Step 4—Determine if the building can be assigned to SDC F
Because the risk category is II, the building is not assigned to SDC F.

Step 5—Determine if the building can be assigned to SDC E
Because the risk category is II and $S_1 = 0.73 < 0.75$, the building is not assigned to SDC E.

Step 6—Determine if the simplified method in ASCE/SEI 12.14 is permitted to be used
Because the building has more than three stories above grade, the simplified method is not permitted to be used.

Step 7—Check if all four conditions in ASCE/SEI 11.6 are satisfied
From a three-dimensional model of the building constructed in accordance with ASCE/SEI 12.7 assuming rigid diaphragms at all levels, it is found that $T = 0.77$ s in the north-south direction. This period greater than $T_S = S_{D1}/S_{DS} = 0.93/1.57 = 0.59$ s.

Therefore, the second of the four conditions is not satisfied and the SDC is not permitted to be determined by ASCE/SEI Table 11.6-1 alone.

Step 8—Determine the SDC using ASCE/SEI Tables 11.6-1 and 11.6-2
With $S_{DS} = 1.57 > 0.50$ and risk category II, the SDC is D in accordance with ASCE/SEI Table 11.6-1.

With $S_{D1} = 0.93 > 0.20$ and risk category II, the SDC is D in accordance with ASCE/SEI Table 11.6-2.

Therefore, the SDC is D for this building.

- Determine the seismic forces.

Flowchart 6.5 is used to determine the seismic forces.

Step 1—Obtain R from ASCE/SEI Table 12.2-1
For a building frame system with special reinforced concrete shear walls (system B4), $R = 6$.

Step 2—Obtain I_e from ASCE/SEI Table 1.5-2
For residential occupancy, the risk category is II (IBC Table 1604.5), and $I_e = 1.00$.

Step 3—Determine T in the direction under consideration from a substantiated analysis
A three-dimensional model of the building was constructed in accordance with ASCE/SEI 12.7 using Reference 6.13 assuming rigid diaphragms at all levels. From the analysis, it was found that $T = 0.77$ s in the north-south direction.

Step 4—Determine the approximate fundamental period, T_a, in accordance with ASCE/SEI 12.8.2.1

Use ASCE/SEI Equation (12.8-8) with approximate period parameters for "All other structural systems" to determine T_a:

$$T_a = C_t h_n^x = 0.02 \times (110.0)^{0.75} = 0.68 \text{ s}$$

In S.I.:

$$T_a = 0.0488 \times (33.5)^{0.75} = 0.68 \text{ s}$$

Step 5—Determine S_{DS} and S_{D1} in accordance with ASCE/SEI 11.4.4 or 11.4.7

For the given latitude and longitude of the site, risk category II and site class CD, the multi-period design spectrum is available from Reference 6.1 where $S_{DS} = 1.57$ and $S_{D1} = 0.93$.

Step 6—Determine coefficient C_u from ASCE/SEI Table 12.8-1 based on S_{D1}

The coefficient for upper limit on calculated period, C_u, is obtained from ASCE/SEI Table 12.8-1. For $S_{D1} = 0.93 > 0.4$, $C_u = 1.4$.

Step 7—Determine if $T > C_u T_a$

From step 3, $T = 0.77$ s in the north-south direction.

From steps 4 and 6, $C_u T_a = 1.4 \times 0.68 = 0.95 \text{ s} > T$.

Therefore, use $T = 0.77$ s in the determination of the seismic forces.

Step 8—Determine T_L from ASCE/SEI Figures 22-14 through 22-17

In lieu of using ASCE/SEI Figures 22-14 through 22-17, $T_L = 8$ s from Reference 6.1.

Step 9—Determine the permitted method to determine C_s

Because the design spectral acceleration parameter, S_a, is available from Reference 6.1 for $T = 0.77$ s, Method 1 or Method 2 in ASCE/SEI 12.8.1.1 is permitted to be used to determine C_s. Method 1 is selected in this example.

Step 10—Determine the design spectral acceleration parameter, S_a

From Reference 6.1, $S_a = 1.17$ at $T = 0.75$ s and $S_a = 0.93$ at $T = 1.0$ s. From linear interpolation, $S_a = 1.15$ at $T = 0.77$ s.

Maximum $S_a = 1.74$ occurs at $T = 0.3$ s, which is less than the period of the structure, $T = 0.77$ s. Therefore, use $S_a = 1.15$ in the determination of C_s.

Step 11—Determine the spectral response acceleration parameter at a period of 1 s, S_1

From Reference 6.1, $S_1 = 0.73$.

Step 12—Determine the seismic response coefficient, C_s

Because $S_1 = 0.73 > 0.6$, C_s is determined as follows:

$$C_s = \text{greatest of} \begin{cases} S_a/(R/I_e) = 1.15/(6/1.00) = 0.192 \\ 0.5S_1/(R/I_e) = (0.5 \times 0.73)/(6/1.00) = 0.061 \\ 0.044S_{DS}I_e = 0.044 \times 1.57 \times 1.00 = 0.069 \\ 0.01 \end{cases}$$

Step 13—Determine the effective seismic weight, W, in accordance with ASCE/SEI 12.7.2

The effective seismic weight, W, is the summation of the effective dead loads at each level and includes the weight of the structural members, superimposed dead loads, the weight of the cladding and the weight of the mechanical equipment on the roof (details of the weight calculations are not shown here).

The story weights are given in Table 6.46.

Step 14—Determine the seismic base shear, V

The seismic base shear, V, is determined by ASCE/SEI Equation (12.8-1):

$$V = C_s W = 0.192 \times 8{,}016 = 1{,}539.1 \text{ kips}$$

In S.I.:

$$V = C_s W = 0.192 \times 35{,}656.9 = 6{,}846.1 \text{ kN}$$

Step 15—Determine the exponent related to the structure period, k

Because $0.5 \text{ s} < T = 0.77 \text{ s} < 2.5 \text{ s}$, $k = 0.75 + 0.5T = 1.14$.

Table 6.46 Seismic Forces and Seismic Story Shears for the Residential Building in Example 6.10

Level	Story Weight, w_x (kips)	Height, h_x (ft)	$w_x h_x^k$	Lateral Force, F_x (kips)	Story Shear, V_x (kips)
R	708	110.0	150,391	265.7	265.7
10	812	99.0	152,961	270.3	536.0
9	812	88.0	133,741	236.3	772.3
8	812	77.0	114,856	203.0	975.3
7	812	66.0	96,347	170.2	1,145.5
6	812	55.0	78,265	138.3	1,283.8
5	812	44.0	60,687	107.2	1,391.0
4	812	33.0	43,718	77.3	1,468.3
3	812	22.0	27,537	48.7	1,517.0
2	812	11.0	12,495	22.1	1,539.1
Σ	8,016		870,998	1,539.1	

1 ft = 0.3048 m; 1 kip = 4.4482 kN.

Step 16—Determine the lateral seismic force, F_x, at each level

Lateral seismic forces F_x are determined by ASCE/SEI Equations (12.8-12) and (12.8-13) and are given in Table 6.46 along with the story shears, V_x.

For example, at the second level:

$$C_{vx} = \frac{w_x h_x^k}{\sum w_i h_i} = \frac{812 \times (11.0)^{1.14}}{870,998} = 0.0143$$

$$F_x = C_{vx} V = 0.0143 \times 1,539.1 = 22.1 \text{ kips}$$

In S.I.:

$$C_{vx} = \frac{3,612.0 \times (3.35)^{1.14}}{1,002,219} = 0.0143$$

$$F_x = 0.0143 \times 6,846.1 = 97.9 \text{ kN}$$

6.7.11 Example 6.11—Story Drift

Determine the story drifts for seismic forces in the north-south direction for the 10-story residential building in Figure 6.56 and check the allowable story drift requirements. Design data are given in Table 6.45.

Solution

Step 1—Determine the design earthquake displacement, δ_{DE}

The design earthquake displacement, δ_{DE}, is determined by ASCE/SEI Equation (12.8-16):

$$\delta_{DE} = \frac{C_d \delta_e}{I_e} + \delta_{di}$$

where C_d = deflection amplification factor from ASCE/SEI Table 12.2-1
δ_e = elastic displacements determined using the design earthquake forces, including the effects of accidental torsion and torsion amplification as applicable
I_e = importance factor determined in accordance with ASCE/SEI 11.5.1
δ_{di} = displacement due to diaphragm deformation corresponding to the design earthquake, including diaphragm forces determined in accordance with ASCE/SEI 12.10

For a building frame with special reinforced concrete shear walls (system B4), $C_d = 5$. The elastic displacements, δ_e, at the CM due to the seismic forces F_x in Table 6.46 are obtained from the three-dimensional analysis of the structure and are given in Table 6.47. The following information is applicable in the determination of δ_e (ASCE/SEI 12.8.6.1):

Table 6.47 Elastic and Design Earthquake Displacements and Design Story Drifts for the Residential Building in Example 6.11

Story	δ_e, in. (mm)	δ_{DE}, in. (mm)	Δ, in. (mm)
10	3.08 (78.2)	15.40 (391.0)	2.05 (52.0)
9	2.67 (67.8)	13.35 (339.0)	2.05 (52.0)
8	2.26 (57.4)	11.30 (287.0)	2.05 (52.0)
7	1.85 (47.0)	9.25 (235.0)	1.95 (49.5)
6	1.46 (37.1)	7.30 (185.5)	1.85 (47.0)
5	1.09 (27.7)	5.45 (138.5)	1.70 (43.0)
4	0.75 (19.1)	3.75 (95.5)	1.50 (38.5)
3	0.45 (11.4)	2.25 (57.0)	1.15 (29.0)
2	0.22 (5.6)	1.10 (28.0)	0.75 (19.0)
1	0.07 (1.8)	0.35 (9.0)	0.35 (9.0)

1. As shown in Example 6.10, the seismic forces F_x are determined using the computed fundamental period of the structure, which is less than the upper limit of $C_u T_a$, so the requirement in ASCE/SEI 12.8.6.2 pertaining to the period for computing displacement and drift is automatically satisfied.
2. The load combinations given in ASCE/SEI 12.8.6.1 are utilized when determining δ_e.
3. The seismic forces F_x used in the determination of δ_e are greater than the minimum seismic force determined by ASCE/SEI Equation (12.8-6), so the requirement in ASCE/SEI 12.8.6.1 pertaining to minimum base shear is automatically satisfied.

Because the structure does not possess any horizontal irregularities, an accidental torsional moment, M_{ta}, is not required to be applied in the analysis (ASCE/SEI 12.8.4.2.1).

Also given in Table 6.47 are the design earthquake displacements, δ_{DE}, and the design story drifts, Δ, which are equal to δ_{DE} at the top of a story minus δ_{DE} at the bottom of a story. The diaphragm deformation, δ_{di}, need not be included in δ_{DE} when determining design story drifts (ASCE/SEI 12.8.6.5). Because the structure does not possess a type 1 horizontal irregularity, the δ_{DE} can be taken at the center of mass at the top and bottom of story (the centers of mass all align vertically in the structure; ASCE/SEI 12.8.6.5). The importance factor $I_e = 1.00$ for a residential occupancy with a risk category of II.

Step 2—Determine the allowable story drifts, Δ_a

The design story drifts, Δ, must be less than or equal to the allowable story drifts, Δ_a, given in ASCE/SEI Table 12.12-1.

For risk category II and "all other structures, $\Delta_a = 0.020h_{sx} = 0.020 \times 11.0 \times 12 = 2.64$ in. (67.1 mm) for all stories.

It is evident from Table 6.47 that $\Delta < \Delta_a$ in all stories, so the drift limits are satisfied in the north-south direction.

6.7.12 Example 6.12—P-Delta Effects

Determine the P-delta effects for seismic forces in the north-south direction for the 10-story residential building in Figure 6.56. Assume the live load on the roof is equal to 20 lb/ft^2 (0.96 kN/m^2) and is equal to 40 lb/ft^2 (1.92 kN/m^2) on the typical floors. Design data are given in Table 6.45.

Solution

Step 1—Determine the P-delta effects

In lieu of including P-delta effects in the computer program analysis, the following procedure can be used to determine if P-delta effects need to be considered.

The stability coefficient, θ, is determined by ASCE/SEI Equation (12.8-18):

$$\theta = \frac{P_x / h_{sx}}{V_x / \Delta_{xe}}$$

In this equation, P_x is the total vertical design load at and above level x determined in accordance with ASCE/SEI 12.8.6.1 where no individual load factor need exceed 1.0 [see Equation (6.36)], h_{sx} is the story height below level x, V_x is the seismic force acting between levels x and $(x-1)$ and Δ_{xe} is the elastic story drift due to V_x calculated at the locations specified in ASCE/SEI 12.8.6 using δ_e.

P-delta calculations for the north-south direction are given in Table 6.48. The floor live load is reduced in accordance with ASCE/SEI 4.7.2. Story seismic forces, V_x, are given in Table 6.46.

Table 6.48 P-Delta Effects in the North-South Direction for the Residential Building in Example 6.12

Level	h_{sx}, ft (m)	P_x, kips (kN)	V_x, kips (kN)	δ_e, in. (mm)	Δ_{xe}, in. (mm)	θ
R	11.0 (3.4)	820.5 (3,649.7)	265.7 (1,181.9)	3.08 (78.2)	0.41 (10.4)	0.0096
10	11.0 (3.4)	1,677.5(7,461.9)	536.0 (2,384.2)	2.67 (67.8)	0.41 (10.4)	0.0097
9	11.0 (3.4)	2,534.5(11,274.1)	772.3 (3,435.3)	2.26 (57.4)	0.41 (10.4)	0.0102
8	11.0 (3.4)	3,391.5(15,086.3)	975.3(4,338.3)	1.85 (47.0)	0.39 (9.9)	0.0103
7	11.0 (3.4)	4,248.5(18,898.5)	1,145.5(5,095.4)	1.46 (37.1)	0.37 (9.4)	0.0104
6	11.0 (3.4)	5,105.5(22,710.7)	1,283.8(5,710.6)	1.09 (27.7)	0.34 (8.6)	0.0102
5	11.0 (3.4)	5,962.5(26,522.9)	1,391.0(6,187.4)	0.75 (19.1)	0.30 (7.7)	0.0097
4	11.0 (3.4)	6,819.5(30,335.1)	1,468.3(6,531.3)	0.45 (11.4)	0.23 (5.8)	0.0081
3	11.0 (3.4)	7,676.5(34,147.3)	1,517.0(6,747.9)	0.22 (5.6)	0.15 (3.8)	0.0058
2	11.0 (3.4)	8,533.5(37,959.5)	1,539.1(6,846.2)	0.07 (1.8)	0.07 (1.8)	0.0029

Sample calculations:

In this example, the axial forces due to earthquake effects, P_{Q_E}, and snow load effects, P_S, are equal to zero.

- Roof level

  $P_D = 708.0$ kips (3,149.3 kN) from Table 6.46

  $P_L = 20 \times 90.0 \times 62.5/1{,}000 = 112.5$ kips (500.4 kN)

  $P_{\text{roof}} = P_D + P_L = 820.5$ kips (3,649.7 kN)

- Typical floor

  $P_D = 812.0$ kips (3,612.0 kN) from Table 6.46

  $P_L = 0.4 \times 0.5 \times 40 \times 90.0 \times 62.5 / 1{,}000 = 45.0$ kips (200.2 kN) where the 0.4 factor is the maximum live load reduction determined in accordance with ASCE/SEI 4.7.2 and the factor 0.5 is permitted to be used because $L_o = 40$ lb/ft² (1.92 kN/m²)

  < 100 lb/ft² (4.79 kN/m²) [see exception 1 in ASCE/SEI 2.3.6].

  $P_D + P_L = 857.0$ kips (3,812.2 kN)

- Level 10

$$P_{10} = P_{\text{roof}} + (P_D + P_L)_{10} = 820.5 + 857.0 = 1{,}677.5 \text{ kips } (7{,}461.9 \text{ kN})$$

$$\theta = \frac{P_x / h_{sx}}{V_x / \Delta_{xe}} = \frac{1{,}677.5/(11.0 \times 12)}{536.0/0.41} = 0.0097$$

In S.I.:

$$\theta = \frac{7{,}461.9/(3.35 \times 1{,}000)}{2{,}384.2/10.4} = 0.0097$$

Step 2—Determine if P-delta effects must be considered

P-delta effects need not be considered where the stability coefficient, θ, determined by ASCE/SEI Equation (12.8-18) is less than or equal to 0.10 (ASCE/SEI 12.8.7).

It is evident from Table 6.48 that P-delta effects need not be considered at any of the levels because $\theta < 0.10$. Also, $\theta < \theta_{\max}$ where $\theta_{\max}$ is determined by ASCE/SEI Equation (12.8-19):

$$\theta_{\max} = \frac{0.5}{\beta C_d} = \frac{0.5}{1.0 \times 5} = 0.1000 < 0.25$$

where the ratio of shear demand to design shear capacity, β, is conservatively taken as 1.0 in accordance with ASCE/SEI 12.8.7.

6.7.13 Example 6.13—Diaphragm Design Forces (ASCE/SEI 12.10.1)

Determine the diaphragm design forces in the north-south direction for the 10-story residential building in Figure 6.56 using the requirements in ASCE/SEI 12.10.1. Design data are given in Table 6.45.

Solution

The diaphragm design seismic force, F_{px}, at level x is determined by ASCE/SEI Equation (12.10-1), including the minimum and maximum limits in ASCE/SEI Equations (12.10-2) and (12.10-3), respectively:

$$F_{px} = \frac{\sum_{i=x}^{n} F_i}{\sum_{i=x}^{n} w_i} w_{px}$$

$$\geq 0.2 S_{DS} I_e w_{px} = 0.3140 w_{px}$$

$$\leq 0.4 S_{DS} I_e w_{px} = 0.6280 w_{px}$$

where $I_e = 1.00$ and $S_{DS} = 1.57$ (see Example 6.10).

In these equations, w_i is the weight tributary to level i and w_{px} is the weight tributary to the diaphragm at level x. For structures with walls like the building in this example, the weights of the walls parallel to the direction of analysis are often not included in w_{px} because these weights do not contribute to the diaphragm shear forces. However, including the weights of these walls in w_{px} is conservative. In this example, $w_{px} = w_x$ at level x.

Design diaphragm forces in the north-south direction are given in Table 6.49.

Table 6.49 Design Seismic Diaphragm Forces for the Residential Building in Example 6.13

Level	w_i (kips)	w_{px} (kips)	F_i (kips)	Σw_i (kips)	ΣF_i (kips)	F_{px} (kips)	Min. F_{px} (kips)	Max. F_{px} (kips)	Design F_{px} (kips)
R	708	708	265.7	708	265.7	265.7	222.3	444.6	265.7
10	812	812	270.3	1,520	536.0	286.3	255.0	509.9	286.3
9	812	812	236.3	2,332	772.3	268.9	255.0	509.9	268.9
8	812	812	203.0	3,144	975.3	251.9	255.0	509.9	255.0
7	812	812	170.2	3,956	1,145.5	235.1	255.0	509.9	255.0
6	812	812	138.3	4,768	1,283.8	218.6	255.0	509.9	255.0
5	812	812	107.2	5,580	1,391.0	202.4	255.0	509.9	255.0
4	812	812	77.3	6,392	1,468.3	186.5	255.0	509.9	255.0
3	812	812	48.7	7,204	1,517.0	171.0	255.0	509.9	255.0
2	812	812	22.1	8,016	1,539.1	155.9	255.0	509.9	255.0

1 kip = 4.4482 kN.

For example, at level 2:

$$F_{p2} = \frac{1{,}539.1}{8{,}016} \times 812 = 155.9 \text{ kips } (693.5 \text{ kN}) > 22.1 \text{ kips } (98.3 \text{ kN})$$

Minimum $F_{p2} = 0.2 \times 1.57 \times 1.00 \times 812 = 255.0$ kips (1,134.2 kN)

Maximum $F_{p2} = 0.4 \times 1.57 \times 1.0 \times 812 = 509.9$ kips (2,268.3 kN)

Diaphragm design force = 255.0 kips (1,134.2 kN)

6.7.14 Example 6.14—Alternative Diaphragm Design Forces (ASCE/SEI 12.10.3)

Determine the diaphragm design forces in the north-south direction for the 10-story residential building in Figure 6.56 using the requirements in ASCE/SEI 12.10.3. Design data are given in Table 6.45.

SOLUTION

Flowchart 6.6 in Section 6.6 of this book is used to determine the diaphragm design seismic forces, F_{px}, at each level.

Step 1—Determine the mode shape factor, z_s

A building frame system with special reinforced concrete shear walls is utilized as the SFRS in this building, which is not specifically listed in ASCE/SEI 12.10.3.2.1. In such cases, $z_s = 1.0$.

Step 2—Determine the modal contribution factors, Γ_{m1} and Γ_{m2}

The modal contribution factors, Γ_{m1} and Γ_{m2}, are determined by ASCE/SEI Equations (12.10-13) and (12.10-14), respectively:

$$\Gamma_{m1} = 1 + \frac{z_s}{2}\left(1 - \frac{1}{N}\right) = 1 + \left[\frac{1.0}{2} \times \left(1 - \frac{1}{10}\right)\right] = 1.45$$

$$\Gamma_{m2} = 0.9 z_s \left(1 - \frac{1}{N}\right)^2 = 0.9 \times 1.0 \times \left(1 - \frac{1}{10}\right)^2 = 0.73$$

where N is the number of stories in the building above the base, which is equal to 10.

Step 3—Determine the importance factor, I_e

For risk category II buildings, $I_e = 1.00$ from ASCE/SEI Table 1.5-2.

Step 4—Determine S_{DS} and S_{D1}

From Example 6.10, $S_{DS} = 1.57$ and $S_{D1} = 0.93$.

Step 5—Determine the higher mode seismic response coefficient, C_{s2}

Because $N = 10 > 2$, C_{s2} is the least of the values determined by ASCE/SEI Equations (12.10-10), (12.10-11), and (12.10-12a):

$$C_{s2} = \text{least of} \begin{cases} (0.15N + 0.25) I_e S_{DS} = [(0.15 \times 10) + 0.25] \times 1.00 \times 1.57 = 2.748 \\ I_e S_{DS} = 1.00 \times 1.57 = 1.57 \\ \dfrac{I_e S_{D1}}{0.03(N-1)} = \dfrac{1.00 \times 0.93}{0.03 \times (10-1)} = 3.444 \end{cases}$$

Step 6—Determine the design acceleration coefficient, C_{p0}

The design acceleration coefficient at the structure base, C_{p0}, is determined by ASCE/SEI Equation (12.10-6):

$$C_{p0} = 0.4 S_{DS} I_e = 0.4 \times 1.57 \times 1.00 = 0.628$$

Step 7—Determine the overstrength factor, Ω_o

The overstrength factor, Ω_o, is obtained from ASCE/SEI Table 12.2-1.

For a building frame system with special reinforced concrete shear walls, $\Omega_0 = 2.5$.

Step 8—Determine the seismic response coefficient, C_s

The seismic response coefficient, C_s, must be determined in accordance with ASCE/SEI 12.8 or 12.9.

This coefficient is determined by the requirements in ASCE/SEI 12.8 in this example and is equal to 0.192 (see step 12 under the section titled "Determine the seismic forces" in Example 6.10).

Step 9—Determine the design acceleration coefficient, C_{pi}

The design acceleration coefficient at 80 percent of the structural height above the base, C_{pi}, is the greater of the values determined by ASCE/SEI Equations (12.10-8) and (12.10-9):

$$C_{pi} = \text{greater of} \begin{cases} C_{p0} = 0.628 \\ 0.9\Gamma_{m1}\Omega_0 C_s = 0.9 \times 1.45 \times 2.5 \times 0.192 = 0.626 \end{cases}$$

Step 10—Determine the design acceleration coefficient, C_{pn}

The design acceleration coefficient at the structural height, C_{pn}, is determined by ASCE/SEI Equation (12.10-7):

$$C_{pn} = \sqrt{(\Gamma_{m1}\Omega_0 C_s)^2 + (\Gamma_{m2} C_{s2})^2} \geq C_{pi}$$

$$= \sqrt{(1.45 \times 2.5 \times 0.192)^2 + (0.73 \times 1.57)^2} = 1.341 > C_{pi} = 0.628$$

Step 11—Determine the diaphragm design force reduction factor, R_s

The diaphragm design force reduction factor, R_s, is obtained from ASCE/SEI Table 12.10-1.

Cast-in-place reinforced concrete slabs designed in accordance with ACI 318 are typically controlled by in-plane flexure (i.e., flexure-controlled). Thus, $R_s = 2.0$.

Step 12—Determine the design acceleration coefficients, C_{px}

The design acceleration coefficient at level x, C_{px}, is determined from ASCE/SEI Figure 12.10-2.

For $h_x \leq 0.8h_n = 88$ ft (26.8 m) with $C_{pi} = C_{p0} = 0.628$ in this example:

$$C_{px} = C_{p0} + \left(\frac{C_{pi} - C_{p0}}{0.8}\right)\left(\frac{h_x}{h_n}\right) = 0.628$$

For $h_x > 0.8h_n = 88$ ft (26.8 m) [see Flowchart 6.6]:

$$C_{px} = C_{pi} + \left(\frac{C_{pn} - C_{pi}}{0.2}\right)\left(\frac{h_x}{h_n} - 0.8\right) = 0.628 + \left[\left(\frac{1.341 - 0.628}{0.2}\right) \times \left(\frac{h_x}{110.0} - 0.8\right)\right]$$

Values of the design acceleration coefficient, C_{px}, at the floor and roof levels are given in Table 6.50.

Table 6.50 Design Acceleration Coefficients, C_{px}, for the 10-Story Residential Building in Example 6.14

Level	Height above Grade, ft (m)	C_{px}
R	110 (33.5)	1.341
9	99 (30.2)	0.985
8	88 (26.8)	0.628
7	77 (23.5)	0.628
6	66 (20.1)	0.628
5	55 (16.8)	0.628
4	44 (13.4)	0.628
3	33 (10.1)	0.628
2	22 (6.71)	0.628
1	11 (3.35)	0.628

For example, at level 9 with $h_x = 99$ ft (30.2 m) > $0.8h_n = 88$ ft (26.8 m):

$$C_{px} = 0.628 + \left[\left(\frac{1.341 - 0.628}{0.2}\right) \times \left(\frac{99.0}{110.0} - 0.8\right)\right] = 0.985$$

Step 13—Determine the diaphragm design seismic force, F_{px}

The diaphragm design seismic force, F_{px}, at level x is determined by ASCE/SEI Equation (12.10-4) with the minimum F_{px} determined by ASCE/SEI Equation (12.10-5):

$$F_{px} = \frac{C_{px}}{R_s} w_{px} \geq 0.2 S_{DS} I_e w_{px} = 0.2 \times 1.57 \times 1.00 \times w_{px} = 0.314 w_{px}$$

Diaphragm seismic design forces, F_{px}, at the floor and roof levels are given in Table 6.51 where the weights tributary to the diaphragm, w_{px}, are given in Table 6.49.

Table 6.51 Diaphragm Design Seismic Forces, F_{px}, for the 10-Story Residential Building in Example 6.14

Level	Height above Grade, ft (m)	C_{px}	w_{px}, kips	F_{px}, kips	Minimum F_{px}, kips	Design F_{px}, kips
R	110 (33.5)	1.341	708	474.7	222.3	474.7
9	99 (30.2)	0.985	812	399.9	255.0	399.9
8	88 (26.8)	0.628	812	255.0	255.0	255.0
7	77 (23.5)	0.628	812	255.0	255.0	255.0
6	66 (20.1)	0.628	812	255.0	255.0	255.0
5	55 (16.8)	0.628	812	255.0	255.0	255.0
4	44 (13.4)	0.628	812	255.0	255.0	255.0
3	33 (10.1)	0.628	812	255.0	255.0	255.0
2	22 (6.71)	0.628	812	255.0	255.0	255.0
1	11 (3.35)	0.628	812	255.0	255.0	255.0

1 kip = 4.4482 kN.

For example, at level 9:

$$F_{px} = \frac{C_{px}}{R_s} w_{px} = \frac{0.985}{2} \times 812 = 399.9 \text{ kips (1,778.9 kN)}$$

$$\text{Minimum } F_{px} = 0.314 w_{px} = 0.314 \times 812 = 255.0 \text{ kips (1,134.2 kN)}$$

In the upper two levels of the building, the design seismic forces, F_{px}, are significantly greater than the forces determined by the method in ASCE/SEI 12.10.1 (see Table 6.49). Minimum design forces govern over the other levels.

6.7.15 Example 6.15—Diaphragm Design Forces (ASCE/SEI 12.10.4)

Determine the diaphragm design forces, shear forces, and chord forces for the one-story commercial building in Figure 6.57 using the requirements in ASCE/SEI 12.10.4. The walls extend the full depth of the diaphragm in both orthogonal directions, so collectors are not required. Design data are given in Table 6.52. The mean roof height of the building above grade is 25 ft (7.62 m) and there is a 3.5-ft (1.07-m) parapet around the perimeter.

Figure 6.57 One-Story Commercial Building in Example 6.15

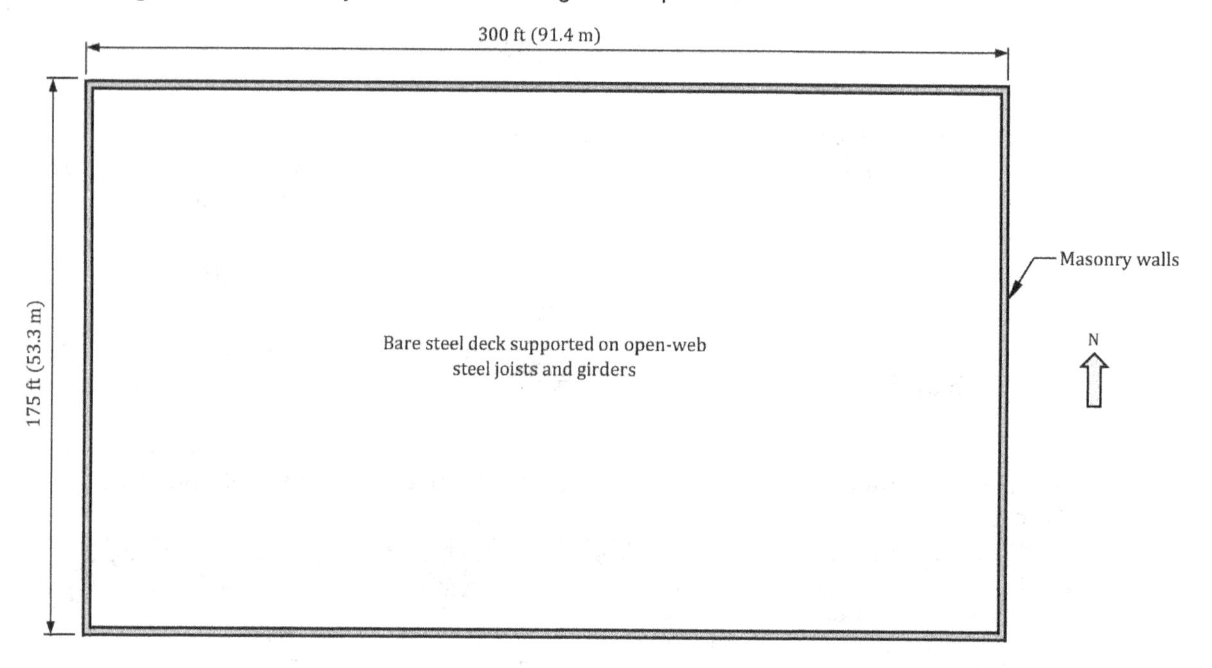

Table 6.52 Design Data for the One-Story Commercial Building in Example 6.15

Location	Latitude: 37.726°, Longitude: −89.217°
Soil classification	Site class BC
Occupancy	Commercial occupancy where less than 300 people congregate in one area
Roof	• Bare steel deck diaphragm not meeting the special seismic detailing requirements of AISI S400 • Weight of framing and superimposed dead load = 19 lb/ft^2 (0.91 kN/m^2)
SFRS	• 8-in. (203-mm) fully-grouted masonry walls • Wall weight = 86 lb/ft^2 (4.12 kN/m^2)

Solution

Flowchart 6.7 in Section 6.6 of this book is used to determine the design forces.

Step 1—Check if the 7 limitations in ASCE/SEI 12.10.4.1 are satisfied

1. All portions of the diaphragm must be designed using the provisions in ASCE/SEI 12.10.4 in both orthogonal directions.
   This limitation is satisfied, as shown below.
2. The diaphragm must consist of either a wood structural panel diaphragm satisfying the requirements in ASCE/SEI 12.10.4.1(2) or a bare (untopped) steel deck diaphragm meeting the requirements in AISI S400 and AISI S310.
   In the design data, the diaphragm is given as a bare steel deck diaphragm.
3. Toppings of concrete or similar material that affect diaphragm strength or stiffness are not permitted over the wood or bare steel deck diaphragm.
   No toppings are used on this diaphragm.

4. The diaphragm must not contain horizontal structural irregularities as specified in ASCE/SEI Table 12.3-1, except that horizontal structural irregularity type 2 is permitted.
   No horizontal irregularities are present.
5. The diaphragm must be rectangular in shape or must be divisible into rectangular segments for purpose of seismic design, with vertical elements of the SFRS or collectors provided at each end of each rectangular segment span.
   There is one rectangular diaphragm with walls around the perimeter of the building, which serve as the SFRS.
6. The vertical elements of the SFRS must be one or more of the following: concrete shear walls, precast concrete shear walls, masonry shear walls, steel concentrically braced frames, steel and concrete composite braced frames, or steel and concrete composite shear walls.
   In the design data, the SFRS consists of masonry shear walls.
7. The vertical elements of the SFRS must be designed in accordance with the ELF procedure in ASCE/SEI 12.8, except it is permitted to use the two-stage analysis procedure in ASCE/SEI 12.2.3.4 where applicable.
   The masonry walls are designed using ASCE/SEI 12.8.

Because all seven limitations in ASCE/SEI 12.10.4.1 are satisfied, the provisions in ASCE/SEI 12.10.4 are permitted to be used to determine the design seismic forces.

Step 2—Determine the acceleration parameters S_{DS} and S_{D1}

For the given latitude and longitude of the site, risk category II and site class BC, the multi-period design spectrum is available from Reference 6.1 where $S_{DS} = 0.64$ and $S_{D1} = 0.19$.

Step 3—Determine the response modification coefficient, R_{diaph}

For a bare steel deck diaphragm not meeting the special seismic detailing requirements of AISI S400, $R_{\text{diaph}} = 1.5$ (ASCE/SEI 12.10.4.2).

Step 4—Obtain the importance factor, I_e

For risk category II buildings, $I_e = 1.00$ from ASCE/SEI Table 1.5-2.

Step 5—Determine the diaphragm spans, L_{diaph}

Diaphragm spans, L_{diaph}, are determined for each rectangular segment between vertical elements of the SFRS or collectors.

For seismic forces in the north-south direction: $L_{\text{diaph}} = 300$ ft (91.4 m)

For seismic forces in the east-west direction: $L_{\text{diaph}} = 175$ ft (53.3 m)

Step 6—Determine the diaphragm periods, T_{diaph}

Diaphragm periods, T_{diaph}, are determined for each rectangular diaphragm segment in each orthogonal direction. For profiled steel deck panel diaphragms, $T_{\text{diaph}} = 0.001L_{\text{diaph}}$ (ASCE/SEI 12.10.4.2.1).

For seismic forces in the north-south direction: $T_{\text{diaph}} = 0.001 \times 300 = 0.30$ s

For seismic forces in the east-west direction: $T_{\text{diaph}} = 0.001 \times 175 = 0.18$ s

Step 7—Determine the seismic response coefficients, $C_{s\text{-diaph}}$

The seismic response coefficient, $C_{s\text{-diaph}}$, is determined by ASCE/SEI Equations (12.10-16a) and (12.10-16b):

$$C_{s\text{-diaph}} = \frac{S_{DS}}{\left(\dfrac{R_{\text{diaph}}}{I_e}\right)} \leq \frac{S_{D1}}{T_{\text{diaph}}\left(\dfrac{R_{\text{diaph}}}{I_e}\right)}$$

For seismic forces in the north-south direction:

$$C_{s\text{-diaph}} = \frac{0.64}{\left(\frac{1.5}{1.00}\right)} = 0.427 > \frac{0.19}{0.30 \times \left(\frac{1.5}{1.00}\right)} = 0.422, \text{ use } C_{s\text{-diaph}} = 0.422$$

For seismic forces in the east-west direction:

$$C_{s\text{-diaph}} = \frac{0.64}{\left(\frac{1.5}{1.00}\right)} = 0.427 < \frac{0.19}{0.18 \times \left(\frac{1.5}{1.00}\right)} = 0.704, \text{ use } C_{s\text{-diaph}} = 0.427$$

Step 8—Determine the effective seismic weight, w_{px}

The effective seismic weight, w_{px}, tributary the diaphragm is equal to the total seismic weight minus the weight of the walls resisting the seismic forces:

- Roof weight $= 19 \times 300 \times 175/1{,}000 = 998$ kips (4,437 kN)
- Wall weight tributary to the roof $= 86 \times \left(\frac{25.0}{2} + 3.5\right)/1{,}000 = 1.4$ kips/ft (20.4 kN/m)
- Weight of walls on the north-south faces $= 2 \times 1.4 \times 300 = 840$ kips (3,737 kN)
- Weight of walls on the east-west faces $= 2 \times 1.4 \times 175 = 490$ kips (2,180 kN)
- Total seismic weight $= 998 + 840 + 490 = 2{,}328$ kips (10,354 kN)

For seismic forces in the north-south direction: $w_{px} = 2{,}328 - 490 = 1{,}838$ kips (8,174 kN)

For seismic forces in the east-west direction: $w_{px} = 2{,}328 - 840 = 1{,}488$ kips (6,617 kN)

Step 9—Determine the diaphragm design seismic force, F_{px}

The diaphragm design seismic force, F_{px}, is determined by ASCE/SEI Equation (12.10-15):

$$F_{px} = C_{s\text{-diaph}} w_{px}$$

For seismic forces in the north-south direction: $F_{px} = 0.422 \times 1{,}838 = 776$ kips (3,450 kN)

For seismic forces in the east-west direction: $F_{px} = 0.427 \times 1{,}488 = 635$ kips (2,826 kN)

Step 10—Determine the diaphragm design shear forces

Because $L_{\text{diaph}} > 100$ ft (30.5 m) in both orthogonal directions, the diaphragm shear forces must be amplified to 1.5 times the shear force calculated using F_{px} over an amplified shear boundary zone equal to at least 10 percent of L_{diaph} (ASCE/SEI 12.10.4.2.2).

For seismic forces in the north-south direction, the design diaphragm force, $F_{px} = 776$ kips (3,450 kN) can be transformed into a uniformly distributed load, w (see Figure 6.58):

$$w = 776/300 = 2.59 \text{ kips/ft}$$

Maximum shear force, V, in the diaphragm without amplification:

$$V = 2.59 \times 300/2 = 389 \text{ kips}$$

Maximum shear flow, v, in the diaphragm without amplification:

$$v = 389/175 = 2.22 \text{ kips/ft}$$

Maximum shear flow, v, in the diaphragm with amplification (ASCE/SEI 12.10.4.2.2):

$$v = 1.5 \times 2.22 = 3.33 \text{ kips/ft}$$

The width of the amplified shear boundary zone at each end of the diaphragm is equal to 10 percent of L_{diaph}, that is, $0.1 \times 300 = 30$ ft.

Figure 6.58 Diaphragm Shear Forces in the North-South Direction for the One-Story Commercial Building in Example 6.15

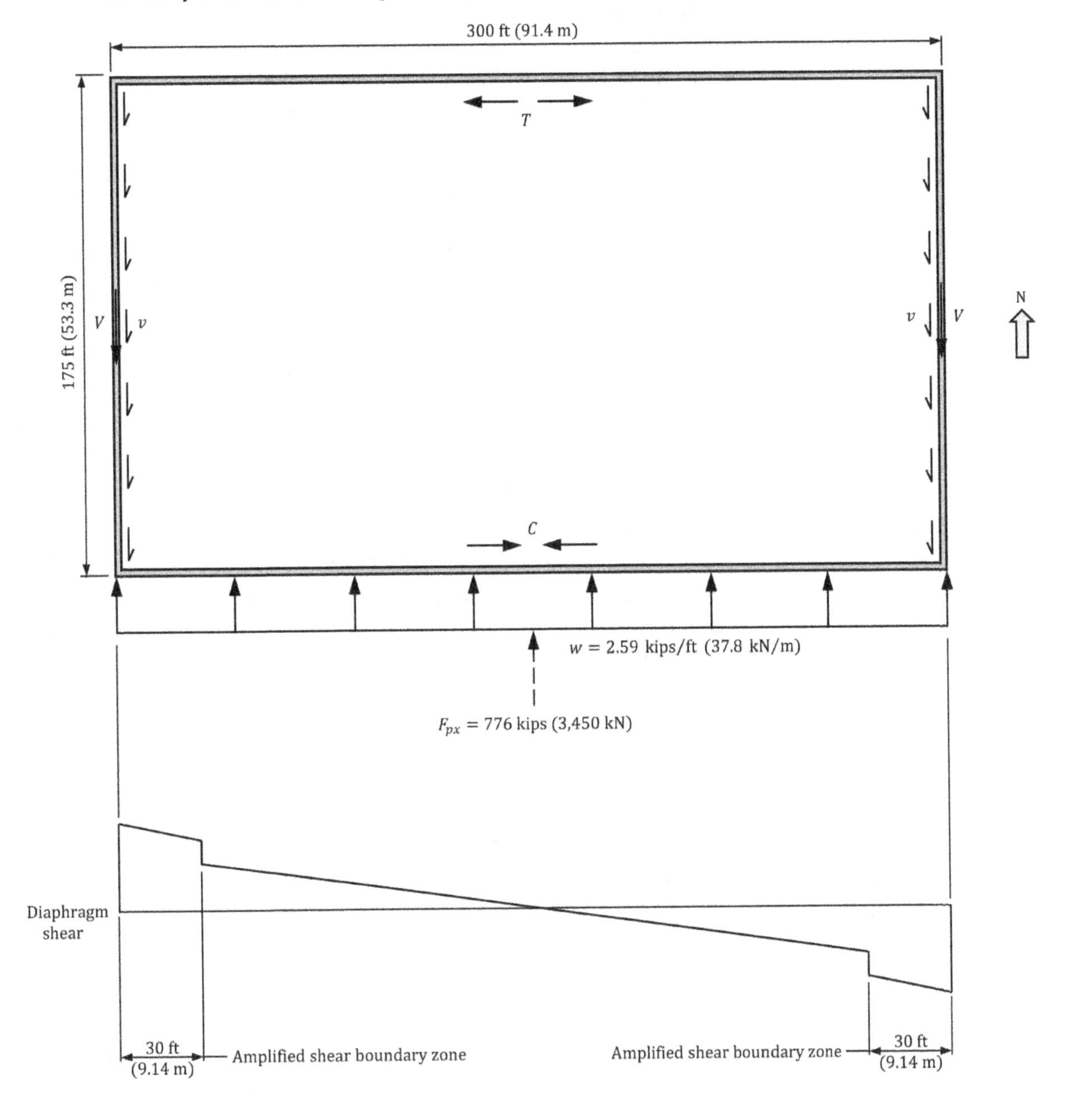

In S.I.:

$$w = 3{,}450/91.4 = 37.8 \text{ kN/m}$$

Maximum shear force, V, in the diaphragm without amplification:

$$V = 37.8 \times 91.4/2 = 1{,}728 \text{ kN}$$

Maximum shear flow, v, in the diaphragm without amplification:

$$v = 1{,}728/53.3 = 32.4 \text{ kN/m}$$

Maximum shear flow, v, in the diaphragm with amplification (ASCE/SEI 2.10.4.2.2):

$$v = 1.5 \times 32.4 = 48.6 \text{ kN/m}$$

The width of the amplified shear boundary zone at each end of the diaphragm is equal to 10 percent of L_{diaph}, that is, $0.1 \times 91.4 = 9.14$ m.

Similar calculations can be performed for seismic forces in the east-west direction.

Step 11—Determine the design diaphragm chord forces

For seismic forces in the north-south direction, tension and compression design diaphragm chord forces, T and C, are determined using the maximum bending moment in the diaphragm based on F_{px} (ASCE/SEI 12.10.4.2.3):

$$M = w(L_{\text{diaph}})^2/8 = 2.59 \times 300^2/8 = 29{,}138 \text{ ft-kips}$$

$$T = C = M/d = 29{,}138/175 = 167 \text{ kips}$$

In S.I.:

$$M = w(L_{\text{diaph}})^2/8 = 37.8 \times 91.4^2/8 = 39{,}473 \text{ kN-m}$$

$$T = C = M/d = 39{,}473/53.3 = 741 \text{ kN}$$

Similar calculations can be performed for seismic forces in the east-west direction.

6.7.16 Example 6.16—Horizontal Distribution of Seismic Forces—Rigid Diaphragm

Determine the design seismic forces in the elements of the SFRS in the north-south direction for the one-story commercial building in Figure 6.59 assuming the diaphragm is rigid. The height of the building is 15.0 ft (4.57 m). Design data are given in Table 6.53. The weight of the cladding on all sides of the building is 8 lb/ft^2 (0.38 kN/m^2).

Figure 6.59 Roof Plan of the Commercial Building in Example 6.16

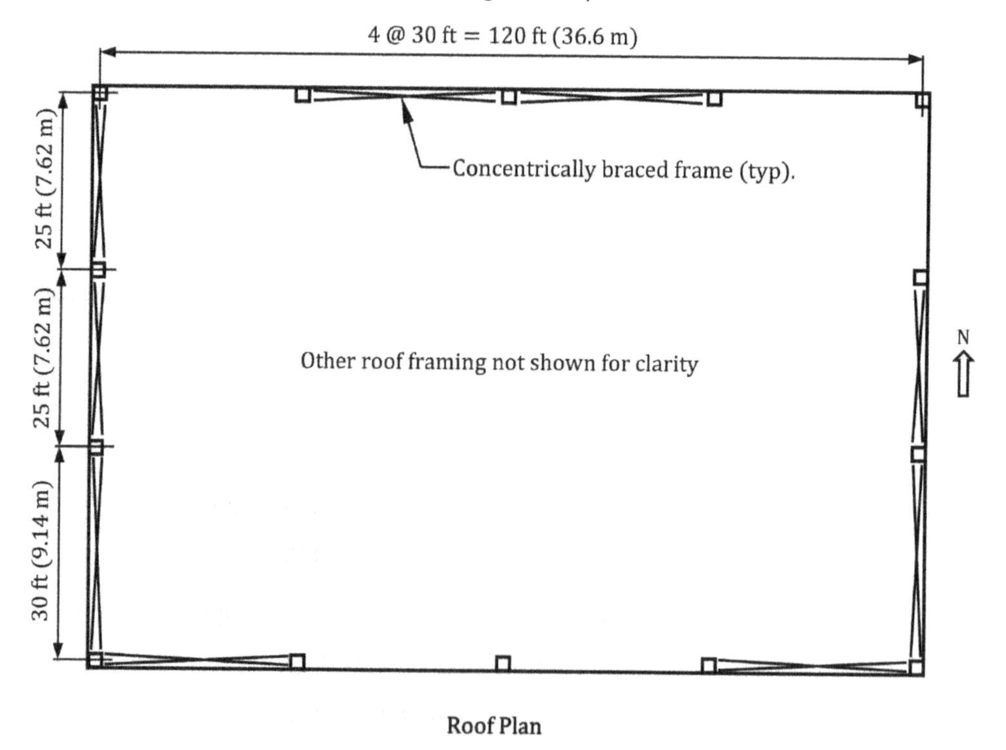

Table 6.53 Design Data for the One-Story Commercial Building in Example 6.16

Location	Latitude: 40.700°, Longitude: −111.801°
Soil classification	Site class D
Occupancy	Commercial occupancy where less than 300 people congregate in one area
Roof	Weight of framing and superimposed dead load = 54 lb/ft² (2.59 kN/m²)
SFRS	Steel special concentrically braced frames

SOLUTION

- Determine the SDC.

Flowchart 6.2 in Section 6.6 of this book can be used to determine the SDC.

Step 1—Determine the risk category in accordance with IBC Table 1604.5
For a commercial occupancy, the risk category is II.

Step 2—Determine the site class in accordance with ASCE/SEI 11.4.2
The site class is given as D.

Step 3—Determine the acceleration parameters S_1, S_{DS}, and S_{D1} from Reference 6.1
For the given latitude and longitude of the site, risk category II and site class D, the multi-period design spectrum is available from Reference 6.1 where $S_1 = 0.43$, $S_{DS} = 0.91$, and $S_{D1} = 0.67$.

Step 4—Determine if the building can be assigned to SDC F
Because the risk category is II, the building is not assigned to SDC F.

Step 5—Determine if the building can be assigned to SDC E
Because the risk category is II and $S_1 = 0.43 < 0.75$, the building is not assigned to SDC E.

Step 6—Determine if the simplified method in ASCE/SEI 12.14 is permitted to be used
The simplified method is not used in this example.

Step 7—Check if all four conditions in ASCE/SEI 11.6 are satisfied
This check is not performed in this example.

Step 8—Determine the SDC using ASCE/SEI Tables 11.6-1 and 11.6-2
With $S_{DS} = 0.91 > 0.50$ and risk category II, the SDC is D in accordance with ASCE/SEI Table 11.6-1.

With $S_{D1} = 0.67 > 0.20$ and risk category II, the SDC is D in accordance with ASCE/SEI Table 11.6-2.

Therefore, the SDC is D for this building.

- Determine the base shear, V.

Flowchart 6.5 is used to determine V.

Step 1—Obtain R from ASCE/SEI Table 12.2-1
For a building frame with steel special concentrically braced frames (system B2), $R = 6$.

Step 2—Obtain I_e from ASCE/SEI Table 1.5-2
For commercial occupancy, the risk category is II (IBC Table 1604.5), and $I_e = 1.00$.

Step 3—Determine T in the direction under consideration from a substantiated analysis
In lieu of determining T from a substantiated analysis, use the approximate fundamental period determined in step 4 below.

Step 4—Determine the approximate fundamental period, T_a, in accordance with ASCE/SEI 12.8.2.1

Use ASCE/SEI Equation (12.8-8) with approximate period parameters for "All other structural systems" to determine T_a:

$$T_a = C_t h_n^x = 0.02 \times (15.0)^{0.75} = 0.15 \text{ s}$$

In S.I.:

$$T_a = 0.0488 \times (4.57)^{0.75} = 0.15 \text{ s}$$

Step 5—Determine S_{DS} and S_{D1} in accordance with ASCE/SEI 11.4.4 or 11.4.7

For the given latitude and longitude of the site, risk category II and site class D, the multi-period design spectrum is available from Reference 6.1 where $S_{DS} = 0.91$ and $S_{D1} = 0.67$.

Step 6—Determine coefficient C_u from ASCE/SEI Table 12.8-1 based on S_{D1}

Not applicable.

Step 7—Determine if $T > C_u T_a$

Not applicable.

Step 8—Determine T_L from ASCE/SEI Figures 22-14 through 22-17

In lieu of using ASCE/SEI Figures 22-14 through 22-17, $T_L = 8$ s from Reference 6.1.

Step 9—Determine the permitted method to determine C_s

Because the design spectral acceleration parameter, S_a, is available from Reference 6.1 for a period, $T = 0.15$ s, Method 1 or Method 2 in ASCE/SEI 12.8.1.1 is permitted to be used to determine C_s. Method 1 is selected in this example.

Step 10—Determine the design spectral acceleration parameter, S_a

From Reference 6.1, $S_a = 0.78$ at $T = 0.15$ s.

Maximum $S_a = 1.01$ occurs at $T = 0.4$ s, which is greater than the period of the structure, $T = 0.15$ s. Therefore, use $S_a = 1.01$ in the determination of C_s (ASCE/SEI 12.8.1.1).

Step 11—Determine the spectral response acceleration parameter at a period of 1 s, S_1

From Reference 6.1, $S_1 = 0.43$.

Step 12—Determine the seismic response coefficient, C_s

Because $S_1 = 0.43 < 0.6$, C_s is determined as follows:

$$C_s = \text{greatest of} \begin{cases} S_a/(R/I_e) = 1.01/(6/1.00) = 0.168 \\ 0.044 S_{DS} I_e = 0.044 \times 0.91 \times 1.00 = 0.040 \\ 0.01 \end{cases}$$

Step 13—Determine the effective seismic weight, W, in accordance with ASCE/SEI 12.7.2

The effective seismic weight, W, is the summation of the effective dead loads tributary to the roof level and includes the weight of the structural members and superimposed dead loads at the roof level, the weight of the steel braces and columns and the weight of the cladding tributary to the roof:

Weight of the roof members and superimposed dead load = $54 \times 122 \times 82/1{,}000 =$ 540 kips (2,403 kN) assuming the column dimensions are 2 ft by 2 ft (0.61 m by 0.61 m)

Weight of steel braces and columns $\cong$ 20 kips (89 kN)

Weight of cladding = $8 \times (15/2) \times [2 \times (122 + 82)]/1{,}000 = 25$ kips (109 kN)

$$W = 540 + 20 + 25 = 585 \text{ kips } (2{,}601 \text{ kN})$$

Step 14—Determine the seismic base shear, V

The seismic base shear, V, is determined by ASCE/SEI Equation (12.8-1):

$$V = C_s W = 0.168 \times 585 = 98.3 \text{ kips}$$

In S.I.:

$$V = C_s W = 0.168 \times 2{,}601 = 437.0 \text{ kN}$$

Step 15—Determine the exponent related to the structure period, k

Because $T = 0.15 \text{ s} < 0.5 \text{ s}$, $k = 1$.

Step 16—Determine the lateral seismic force, F_x, at the roof level

Because the building has one story, the lateral force F_x at the roof level is equal to $V = 98.3$ kips (437.0 kN).

- Determine the diaphragm force, F_{px}, at the roof level

  The diaphragm force is determined using the requirements in ASCE/SEI 12.10.1.

  Because the building has one story, $F_{px} = V = 98.3$ kips (437.0 kN).

  Minimum $F_{px} = 0.2 S_{DS} I_e w_{px} = 0.2 \times 0.91 \times 1.00 \times 585 = 106.5$ kips (473.6 kN)

  Maximum $F_{px} = 0.4 S_{DS} I_e w_{px} = 213.0$ kips (947.2 kN)

  Therefore, $F_{px} = 106.5$ kips (473.6 kN) at the roof level.

- Determine the distribution of the design seismic diaphragm force to the elements of the SFRS

  Step 1—Determine the relative stiffnesses of the elements of the SFRS

  Because the roof diaphragm is rigid, the design seismic diaphragm force is distributed to the elements of the SFRS based on their relative rigidities.

  Approximate relative stiffnesses of the concentrically braced frames are determined using Figure 6.33. Assuming a double diagonal configuration where the sizes of the members are the same for each frame, the approximate stiffness, k_i, can be determined by the following equation (the area of the braces, A_{br}, and the modulus of elasticity, E, are the same for all frames, so both cancel out from relative stiffness calculations):

$$k_i = \frac{2E}{\left(\dfrac{\ell_{br}^3}{A_{br}\ell_i^2}\right)} = \frac{2\ell_i^2}{\ell_{br}^3}$$

For the 25-ft frames:

$$\ell_i = 25.0 \text{ ft}$$

$$\ell_{br} = \sqrt{25.0^2 + 15.0^2} = 29.2 \text{ ft}$$

$$k_i = \frac{2 \times 25.0^2}{29.2^3} = 0.0502$$

For the 30-ft frames:

$$\ell_i = 30.0 \text{ ft}$$

$$\ell_{br} = \sqrt{30.0^2 + 15.0^2} = 33.5 \text{ ft}$$

$$k_i = \frac{2 \times 30.0^2}{33.5^3} = 0.0479$$

In S.I.:
For the 7.62-m frames:

$$\ell_i = 7.62 \text{ m}$$

$$\ell_{br} = \sqrt{7.62^2 + 4.57^2} = 8.89 \text{ m}$$

$$k_i = \frac{2 \times 7.62^2}{8.89^3} = 0.1653$$

For the 9.14-m frames:

$$\ell_i = 9.14 \text{ m}$$

$$\ell_{br} = \sqrt{9.14^2 + 4.57^2} = 10.2 \text{ m}$$

$$k_i = \frac{2 \times 9.14^2}{10.2^3} = 0.1574$$

Step 2—Determine the location of the CM

The majority of the weight tributary to the roof is due to the roof framing. The CM is located from the west and south frame centerlines as follows:

$$x_{\text{cm}} = 60.0 \text{ ft } (18.3 \text{ m})$$

$$y_{\text{cm}} = 40.0 \text{ ft } (12.2 \text{ m})$$

Step 3—Determine the location of the CR

The CR is determined using the intersection of the centerlines of the west and south frames as the origin. The equations in Figure 6.32 are used to determine the location of the CR:

$$x_{cr} = \frac{\sum (k_i)_y x_i}{\sum (k_i)_y} = \frac{(0.0502 + 0.0479) \times 120.0}{(3 \times 0.0502) + (2 \times 0.0479)} = 47.8 \text{ ft}$$

$$y_{cr} = \frac{\sum (k_i)_x y_i}{\sum (k_i)_x} = \frac{2 \times 0.0479 \times 80.0}{4 \times 0.0479} = 40.0 \text{ ft}$$

In S.I.:

$$x_{cr} = \frac{\sum (k_i)_y x_i}{\sum (k_i)_y} = \frac{(0.1653 + 0.1574) \times 36.6}{(3 \times 0.1653) + (2 \times 0.1574)} = 14.6 \text{ m}$$

$$y_{cr} = \frac{\sum (k_i)_x y_i}{\sum (k_i)_x} = \frac{2 \times 0.1574 \times 24.4}{4 \times 0.1574} = 12.2 \text{ m}$$

As expected, y_{cr} is located halfway between the north and south frames due to symmetry (see Figure 6.60).

Step 4—Distribute the diaphragm seismic forces to the vertical elements of the SFRS

The diaphragm force F_{pR} at the roof level acts through the CM. Because the diaphragm is rigid, the distribution of the diaphragm force to the vertical elements of the SFRS must consider the effects of the inherent torsional moment, M_t, due to the eccentricity, e_x, between the locations of the CM and CR. Accidental torsion,

Figure 6.60 Location of the CM and CR for the Commercial Building in Example 6.16

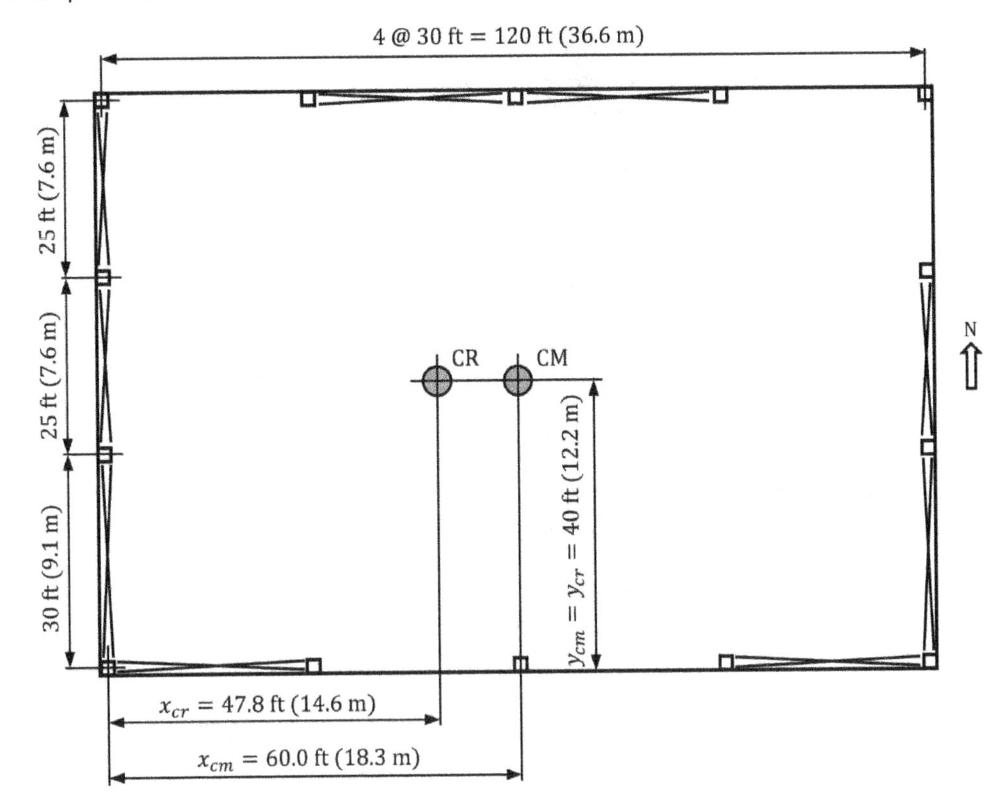

M_{ta}, need not be considered because a type 1 horizontal irregularity does not exist for this structure (ASCE/SEI 12.8.4.2.1).

Force calculations are given in Figure 6.61 for seismic forces in the north direction using the equations in Figure 6.34. The forces indicated in the figure are in the directions that resist the in-plane seismic force and inherent torsional moment. Forces indicated by a solid arrow are term 1 forces (forces due to F_{pR}) and those indicated by a dashed arrow are term 2 forces (forces due to M_t).

For example, the forces in the braced frame labeled 2 are determined as follows:

Term 1 force:

$$(V_2)_1 = \frac{(k_2)_y F_{pr}}{\sum (k_i)_y} = \frac{0.0502 \times 106.5}{0.2464} = 21.70 \text{ kips}$$

Term 2 force:

$$\bar{x}_2^2 (k_2)_y = 47.8^2 \times 0.0502 = 114.7$$

$$(V_2)_2 = \frac{-\bar{x}_2 (k_2)_y F_{pr} e_x}{\sum \bar{x}_i^2 (k_i)_y + \sum \bar{y}_i^2 (k_i)_x} = \frac{-47.8 \times 0.0502 \times 106.5 \times 12.2}{850.2 + 306.4} = -2.70 \text{ kips}$$

Total force $V_2 = 21.70 - 2.70 = 19.00$ kips

In S.I.:

Term 1 force:

$$(V_2)_1 = \frac{(k_2)_y F_{pr}}{\sum (k_i)_y} = \frac{0.1653 \times 473.6}{0.8107} = 96.57 \text{ kN}$$

Figure 6.61 Force Allocation to the Braced Frames for the Commercial Building in Example 6.16

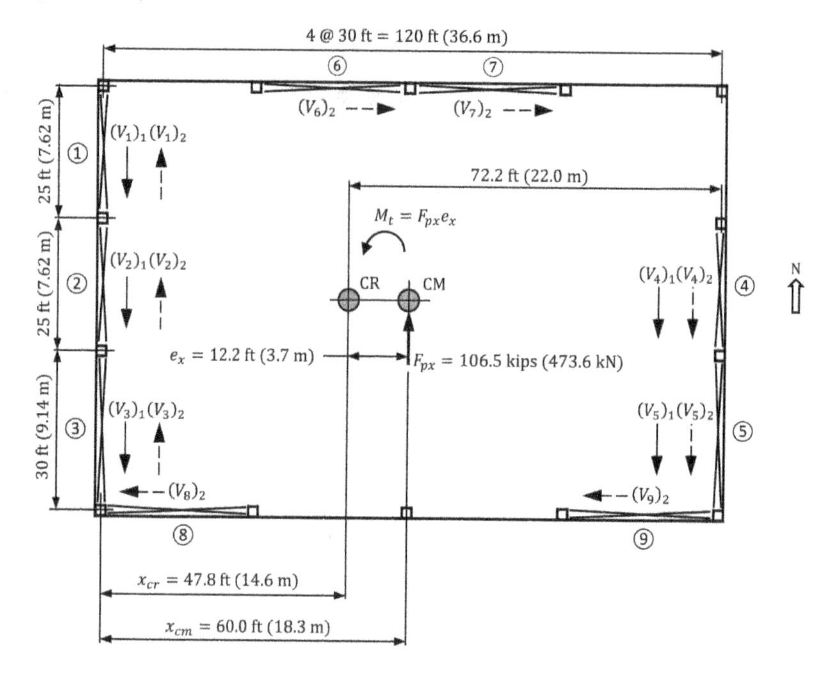

Braced Frame	x_i	y_i	$(k_i)_y$	$(k_i)_x$	$\bar{x}_i$	$\bar{y}_i$	$\bar{x}_i^2(k_i)_y$	$\bar{y}_i^2(k_i)_x$	Term 1 $(V_i)_1 = \frac{(k_i)_y F_{px}}{\Sigma(k_i)_y}$	Term 2 $(V_i)_2 = \left(\frac{\bar{x}_i(k_i)_y F_{px} e_x}{\Sigma \bar{x}_i^2(k_i)_y + \Sigma \bar{y}_i^2(k_i)_x}\right)^*$	$V_i = (V_i)_1 + (V_i)_2$
	(ft)	(ft)			(ft)	(ft)			(kips)	(kips)	(kips)
1	0.0	—	0.0502	—	47.8	—	114.7	—	21.70	−2.70	19.00
2	0.0	—	0.0502	—	47.8	—	114.7	—	21.70	−2.70	19.00
3	0.0	—	0.0479	—	47.8	—	109.4	—	20.70	−2.57	18.13
4	120.0	—	0.0502	—	72.2	—	261.7	—	21.70	4.07	25.77
5	120.0	—	0.0479	—	72.2	—	249.7	—	20.70	3.89	24.59
6	—	80.0	—	0.0479	—	40.0	—	76.6	—	2.15	2.15
7	—	80.0	—	0.0479	—	40.0	—	76.6	—	2.15	2.15
8	—	0.0	—	0.0479	—	40.0	—	76.6	—	−2.15	−2.15
9	—	0.0	—	0.0479	—	40.0	—	76.6	—	−2.15	−2.15
		Σ	0.2464	0.1916			850.2	306.4			106.49

Braced Frame	x_i	y_i	$(k_i)_y$	$(k_i)_x$	$\bar{x}_i$	$\bar{y}_i$	$\bar{x}_i^2(k_i)_y$	$\bar{y}_i^2(k_i)_x$	Term 1 $(V_i)_1 = \frac{(k_i)_y F_{px}}{\Sigma(k_i)_y}$	Term 2 $(V_i)_2 = \left(\frac{\bar{x}_i(k_i)_y F_{px} e_x}{\Sigma \bar{x}_i^2(k_i)_y + \Sigma \bar{y}_i^2(k_i)_x}\right)^*$	$V_i = (V_i)_1 + (V_i)_2$
	(m)	(m)			(m)	(m)			(kN)	(kN)	(kN)
1	0.0	—	0.1653	—	14.6	—	35.24	—	96.57	−11.95	84.62
2	0.0	—	0.1653	—	14.6	—	35.24	—	96.57	−11.95	84.62
3	0.0	—	0.1574	—	14.6	—	33.55	—	91.95	−11.38	80.57
4	36.6	—	0.1653	—	22.0	—	80.01	—	96.57	18.00	114.57
5	36.6	—	0.1574	—	22.0	—	76.18	—	91.95	17.14	109.09
6	—	24.4	—	0.1574	—	12.2	—	23.43	—	9.51	9.51
7	—	24.4	—	0.1574	—	12.2	—	23.43	—	9.51	9.51
8	—	0.0	—	0.1574	—	12.2	—	23.43	—	−9.51	−9.51
9	—	0.0	—	0.1574	—	12.2	—	23.43	—	−9.51	−9.51
		Σ	0.8107	0.6296			260.22	93.72			473.47

*For frames 7, 8, and 9, replace $\bar{x}_i(k_i)_y$ with $\bar{y}_i(k_i)_x$.

Term 2 force:

$$\bar{x}_2^2(k_2)_y = 14.6^2 \times 0.1653 = 35.24$$

$$(V_2)_2 = \frac{-\bar{x}_2(k_2)_y F_{pr} e_x}{\sum \bar{x}_i^2(k_i)_y + \sum \bar{y}_i^2(k_i)_x} = \frac{-14.6 \times 0.1653 \times 473.6 \times 3.70}{260.22 + 93.72} = -11.95 \text{ kN}$$

Total force $V_2 = 96.57 - 11.95 = 84.62$ kN

The calculations are performed with 2 decimal places to demonstrate that the summation of the V_i must equal F_{px} (considering roundoff).

The term 2 force is negative because it acts in the opposite direction of the term 1 force for F_{px} acting in the north direction. However, the diaphragm must also be designed for F_{px} acting in the south direction, so the maximum force that braced frame 2 is subjected to is $21.70+2.70=24.40$ kips ($96.57+11.95=108.5$ kN).

Similar calculations can be performed for seismic forces in the east-west direction.

6.7.17 Example 6.17—Horizontal Distribution of Seismic Forces—Flexible Diaphragm

Determine the seismic forces in the elements of the SFRS in the north-south direction for the one-story commercial building in Figure 6.62 assuming the diaphragm is flexible. The height of the building is 15.0 ft (4.57 m) and there is a 2.0-ft (0.61-m) tall parapet around the perimeter of the building. Design data are given in Table 6.54.

SOLUTION

- Determine the SDC.
  Flowchart 6.2 in Section 6.6 of this book can be used to determine the SDC.

  Step 1—Determine the risk category in accordance with IBC Table 1604.5
  For a commercial occupancy, the risk category is II.

Figure 6.62 Roof Plan of the Commercial Building in Example 6.17

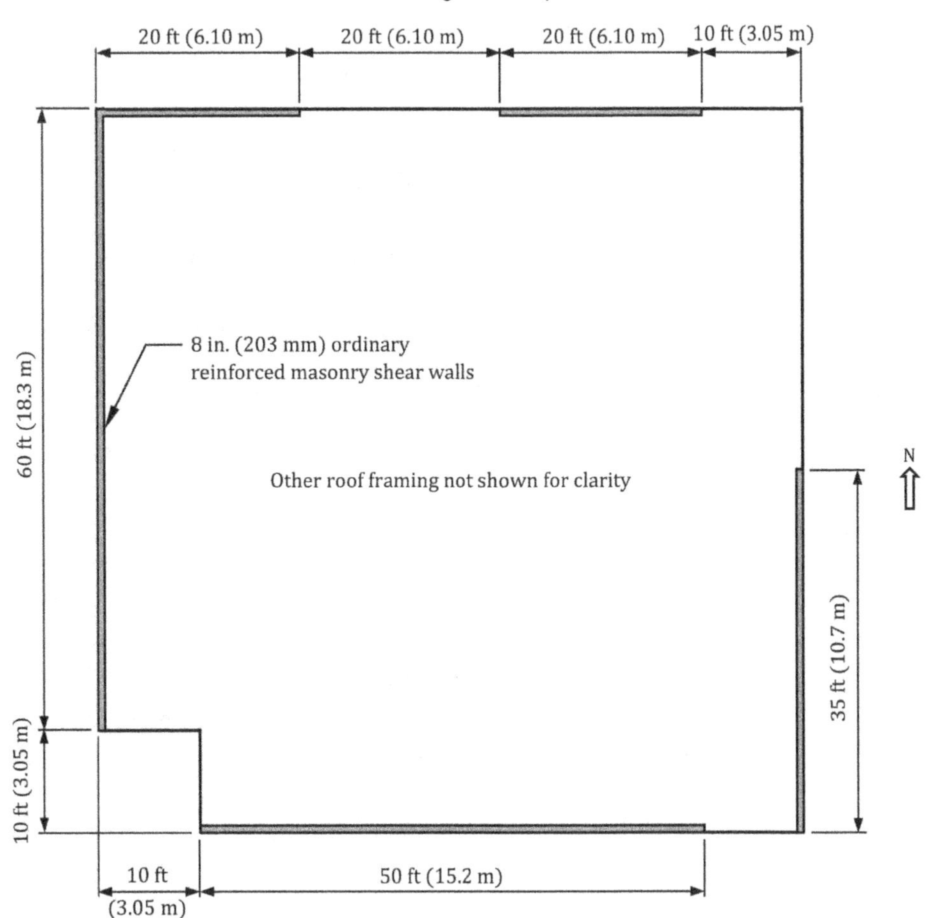

Table 6.54 Design Data for the Commercial Building in Example 6.17

Location	Latitude: 36.641°, Longitude: –93.219°
Soil classification	Site class C
Occupancy	Commercial occupancy where less than 300 people congregate in one area
SFRS	Ordinary reinforced masonry shear walls (system B19)
Weight of roof framing and superimposed dead load	19 lb/ft^2 (0.91 kN/m^2)
Weight of 8 in. (203 mm) concrete masonry walls, fully grouted	86 lb/ft^2 (4.12 kN/m^2)

Step 2—Determine the site class in accordance with ASCE/SEI 11.4.2
The site class is given as C.

Step 3—Determine the acceleration parameters S_1, S_{DS}, and S_{D1} from Reference 6.1
For the given latitude and longitude of the site, risk category II and site class C, the multi-period design spectrum is available from Reference 6.1 where $S_1 = 0.13$, $S_{DS} = 0.25$, and $S_{D1} = 0.12$.

Step 4—Determine if the building can be assigned to SDC F
Because the risk category is II, the building is not assigned to SDC F.

Step 5—Determine if the building can be assigned to SDC E
Because the risk category is II and $S_1 = 0.13 < 0.75$, the building is not assigned to SDC E.

Step 6—Determine if the simplified method in ASCE/SEI 12.14 is permitted to be used
The simplified method is not used in this example.

Step 7—Check if all 4 conditions in ASCE/SEI 11.6 are satisfied
This check is not performed in this example.

Step 8—Determine the SDC using ASCE/SEI Tables 11.6-1 and 11.6-2
With $0.167 < S_{DS} = 0.25 < 0.33$ and risk category II, the SDC is B in accordance with ASCE/SEI Table 11.6-1.

With $0.067 < S_{D1} = 0.12 < 0.133$ and risk category II, the SDC is B in accordance with ASCE/SEI Table 11.6-2.

Therefore, the SDC is B for this building.

- Determine the base shear, V.
Flowchart 6.5 is used to determine V.

Step 1—Obtain R from ASCE/SEI Table 12.2-1
For a building frame system with ordinary reinforced masonry shear walls (system B19), $R = 2$.

Step 2—Obtain I_e from ASCE/SEI Table 1.5-2
For commercial occupancy, the risk category is II (IBC Table 1604.5), and $I_e = 1.00$.

Step 3—Determine T in the direction under consideration from a substantiated analysis
In lieu of determining T from a substantiated analysis, use the approximate fundamental period determined in step 4 below.

Step 4—Determine the approximate fundamental period, T_a, in accordance with ASCE/SEI 12.8.2.1

Use ASCE/SEI Equation (12.8-8) with approximate period parameters for "All other structural systems" to determine T_a:

$$T_a = C_t h_n^x = 0.02 \times (15.0)^{0.75} = 0.15 \text{ s}$$

In S.I.:

$$T_a = 0.0488 \times (4.57)^{0.75} = 0.15 \text{ s}$$

Step 5—Determine S_{DS} and S_{D1} in accordance with ASCE/SEI 11.4.4 or 11.4.7
For the given latitude and longitude of the site, risk category II and site class C, the multi-period design spectrum is available from Reference 6.1 where $S_{DS} = 0.25$ and $S_{D1} = 0.12$.

Step 6—Determine coefficient C_u from ASCE/SEI Table 12.8-1 based on S_{D1}
Not applicable.

Step 7—Determine if $T > C_u T_a$
Not applicable.

Step 8—Determine T_L from ASCE/SEI Figures 22-14 through 22-17
In lieu of using ASCE/SEI Figures 22-14 through 22-17, $T_L = 12$ s from Reference 6.1.

Step 9—Determine the permitted method to determine C_s
Because the design spectral acceleration parameter, S_a, is available from Reference 6.1 for a period, $T = 0.15$ s, Method 1 or Method 2 in ASCE/SEI 12.8.1.1 is permitted to be used to determine C_s. Method 1 is selected in this example.

Step 10—Determine the design spectral acceleration parameter, S_a
From Reference 6.1, $S_a = 0.32$ at $T = 0.15$ s. Maximum S_a occurs at this period; therefore, use $S_a = 0.32$ in the determination of C_s (ASCE/SEI 12.8.1.1).

Step 11—Determine the spectral response acceleration parameter at a period of 1 s, S_1
From Reference 6.1, $S_1 = 0.13$.

Step 12—Determine the seismic response coefficient, C_s
Because $S_1 = 0.13 < 0.6$, C_s is determined as follows:

$$C_s = \text{greatest of} \begin{cases} S_a/(R/I_e) = 0.32/(2/1.00) = 0.160 \\ 0.044 S_{DS} I_e = 0.044 \times 0.25 \times 1.00 = 0.011 \\ 0.01 \end{cases}$$

Step 13—Determine the effective seismic weight, W, in accordance with ASCE/SEI 12.7.2
The effective seismic weight, W, is the summation of the effective dead loads tributary to the roof level and includes the weight of the structural members and superimposed dead loads at the roof level and the weight of the walls:

Weight of the roof members and superimposed dead load

$$= 19 \times 4{,}800/1{,}000 = 91 \text{ kips (406 kN)}$$

Weight of walls parallel to the direction of analysis

$$= 86 \times [(15/2) + 2] \times (60 + 35)/1{,}000 = 78 \text{ kips (345 kN)}$$

Weight of walls perpendicular to the direction of analysis

$$= 86 \times [(15/2) + 2] \times (50 + 20 + 20)/1{,}000 = 74 \text{ kips (327 kN)}$$

$$W = 91 + 78 + 74 = 243 \text{ kips (1,078 kN)}$$

Step 14—Determine the seismic base shear, V
The seismic base shear, V, is determined by ASCE/SEI Equation (12.8-1):

$$V = C_s W = 0.160 \times 243 = 38.9 \text{ kips}$$

In S.I.:

$$V = C_s W = 0.160 \times 1{,}078 = 173.0 \text{ kN}$$

Step 15—Determine the exponent related to the structure period, k
Because $T = 0.15 \text{ s} < 0.5 \text{ s}$, $k = 1$.

Step 16—Determine the lateral seismic force, F_x, at the roof level
Because the building has one story, the lateral force F_x at the roof level is equal to $V = 38.9$ kips (173.0 kN).

- Determine the diaphragm force, F_{px}, at the roof level
The diaphragm force ,F_{px}, is determined using the requirements in ASCE/SEI 12.10.1 considering the weight of the roof framing, the superimposed dead load and the weight of the masonry walls perpendicular to the direction of analysis:

$$F_{px} = \left(\frac{V}{W}\right) w_{px} = \left(\frac{38.9}{243}\right) \times (91 + 74) = 26.4 \text{ kips (117.6 kN)}$$

Minimum $F_{px} = 0.2 S_{DS} I_e w_{px} = 0.2 \times 0.25 \times 1.00 \times 165 = 8.3$ kips (36.7 kN)

Maximum $F_{px} = 0.4 S_{DS} I_e w_{px} = 16.5$ kips (73.3 kN)

Therefore, $F_{px} = 16.5$ kips (73.3 kN) at the roof level.

- Determine the distribution of the design diaphragm force to the elements of the SFRS
The diaphragm force, F_{px}, is converted to a uniformly distributed load over the width of the diaphragm perpendicular to the direction of analysis:

$$w = \frac{16.5}{70} = 0.236 \text{ kips/ft (3.44 kN/m)}$$

Because the diaphragm is idealized as flexible, it is assumed the two walls in the direction of analysis each resist 50 percent of the design diaphragm force (see Figure 6.63):

$$V_1 = V_2 = \frac{0.236 \times 70}{2} = 8.3 \text{ kips (36.7 kN)}$$

- Determine the design diaphragm chord forces
For seismic forces in the north-south direction, tension and compression design diaphragm chord forces, T and C, are determined using the maximum bending moment in the diaphragm based on F_{px} (ASCE/SEI 12.10.4.2.3):

$$M = wB^2/8 = 0.236 \times 70^2/8 = 114.6 \text{ ft-kips}$$

$$T = C = M/d = 144.6/70 = 2.1 \text{ kips}$$

In S.I.:

$$M = wB^2/8 = 3.44 \times 21.3^2/8 = 195.1 \text{ kN-m}$$

$$T = C = M/d = 195.1/21.3 = 9.2 \text{ kN}$$

Similar calculations can be performed for seismic forces in the east-west direction.

- Determine the design axial forces in the collector element at wall 2
The unit shear forces, net shear forces and axial forces at wall 2 are given in Figure 6.64.

Figure 6.63 Lateral Force Allocation to the Masonry Walls in the Commercial Building in Example 6.17

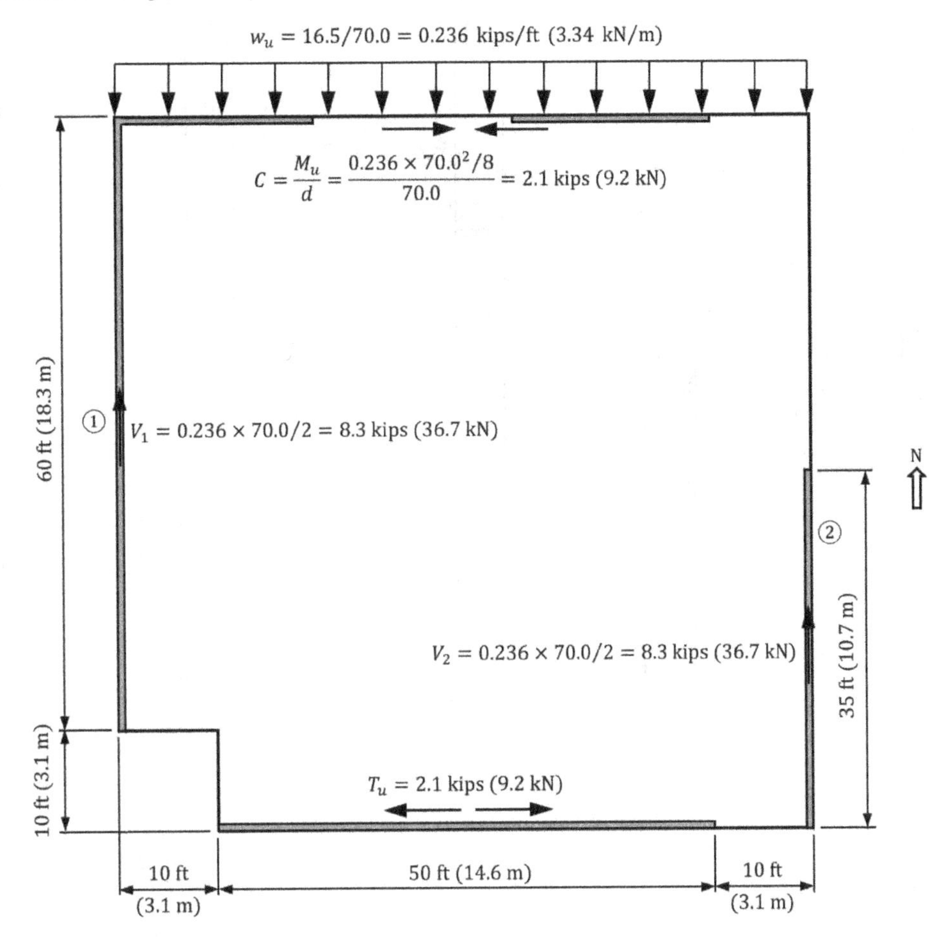

The unit shear force in the diaphragm is determined by dividing the force in the wall, V_2, by the overall depth of the diaphragm in the direction of analysis. Similarly, the unit shear force in wall 2 is determined by dividing V_2 by the length of the wall. The net shear forces are determined by algebraically combining the unit shear forces. The axial forces in the collector are determined by calculating the area under the net shear diagram:

$$\text{Axial force} = 118.6 \times 35.0 = 4{,}151 \text{ lb (18.4 kN)}$$

6.7.18 Example 6.18—Structural Wall Out-of-Plane Forces

Determine the out-of-plane seismic force on a masonry wall in Example 6.17. Design data are given in Table 6.54.

SOLUTION

Structural walls must be designed for a seismic force, F_p, normal to the surface equal to $0.4S_{DS}I_e$ times the weight of the wall, w_p (ASCE/SEI 12.11.1). The minimum F_p is equal to 10 percent times w_p.

Weight of wall $w_p = 86.0 \times 17.0 = 1{,}462$ lb per foot width of wall

$$F_p = 0.4S_{DS}I_e w_p = 0.4 \times 0.25 \times 1.00 \times 1{,}462 = 146 \text{ lb per foot width of wall}$$

$$= 0.1w_p = 0.1 \times 1{,}462 = 146 \text{ lb per foot width of wall}$$

This force acts at the centroid of the wall (see Figure 6.65).

Figure 6.64 Unit and Net Shear Forces in the Diaphragm and Axial Forces in the Collector at Wall 2

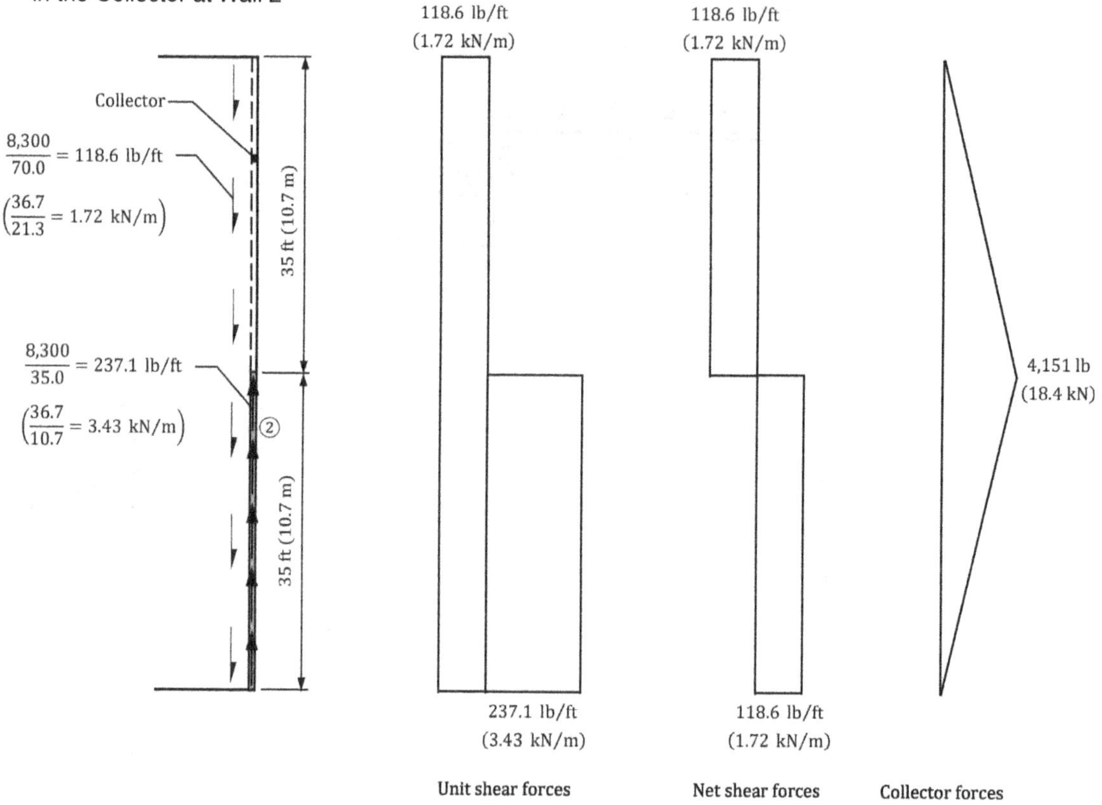

Collector
8,300/70.0 = 118.6 lb/ft
(36.7/21.3 = 1.72 kN/m)
8,300/35.0 = 237.1 lb/ft
(36.7/10.7 = 3.43 kN/m)
2
35 ft (10.7 m)
35 ft (10.7 m)
118.6 lb/ft
(1.72 kN/m)
118.6 lb/ft
(1.72 kN/m)
237.1 lb/ft
(3.43 kN/m)
118.6 lb/ft
(1.72 kN/m)
4,151 lb
(18.4 kN)
Unit shear forces
Net shear forces
Collector forces

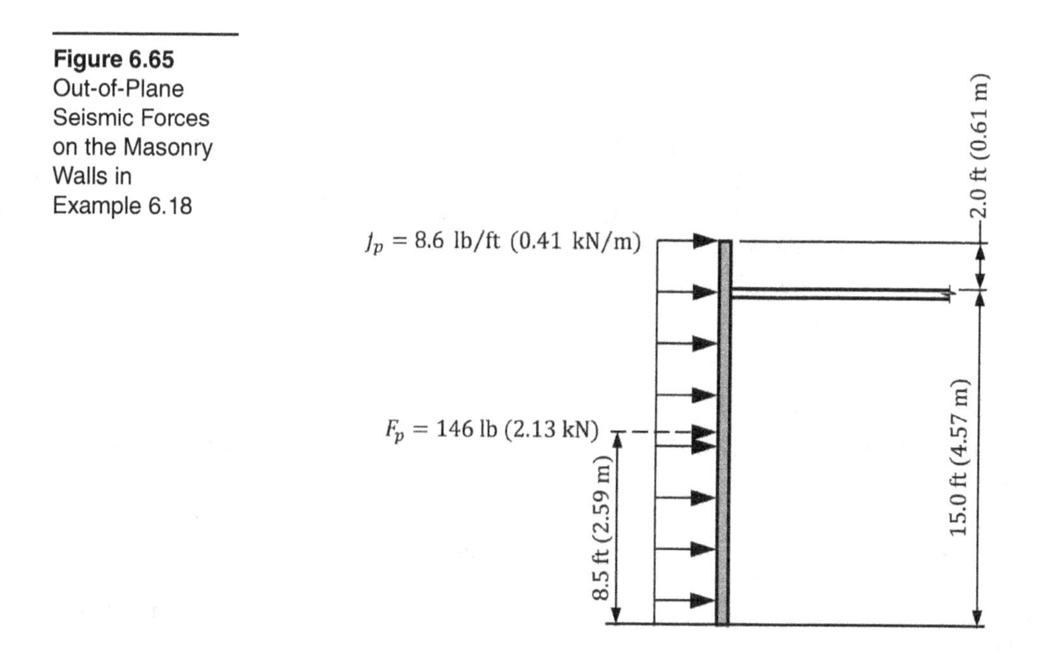

Figure 6.65 Out-of-Plane Seismic Forces on the Masonry Walls in Example 6.18

f_p = 8.6 lb/ft (0.41 kN/m)
F_p = 146 lb (2.13 kN)
8.5 ft (2.59 m)
2.0 ft (0.61 m)
15.0 ft (4.57 m)

The out-of-plane seismic force can also be expressed as a uniformly distributed load over the height of the wall:

$$f_p = 146/17.0 = 8.6 \text{ lb/ft per foot width of wall}$$

In S.I.:

Weight of wall $w_p = 4.12 \times 5.18 = 21.3$ kN per meter width of wall

$$F_p = 0.4S_{DS}I_e w_p = 0.4 \times 0.25 \times 1.00 \times 21.3 = 2.13 \text{ kN per meter width of wall}$$

$$= 0.1w_p = 0.1 \times 21.3 = 2.13 \text{ kN per meter width of wall}$$

$$f_p = 2.13/5.18 = 0.41 \text{ kN/m per meter width of wall}$$

6.7.19 Example 6.19—Wall Anchorage Forces

Determine the anchorage force between the masonry wall and the roof diaphragm in Example 6.17. Design data are given in Table 6.54.

Solution

Wall anchorage force, F_p, is determined by ASCE/SEI Equation (12.11-1):

$$F_p = 0.4S_{DS}k_a I_e W_p \geq \text{greater of} \begin{cases} 0.2k_a I_e W_p \\ 5 \text{ lb/ft}^2 \ (0.24 \text{ kN/m}^2) \end{cases}$$

The amplification factor for diaphragm flexibility, k_a, is determined by ASCE/SEI Equation (12.11-2) for connections to flexible diaphragms:

$$k_a = 1.0 + \frac{L_f}{100} = 1.0 + \frac{70.0}{100} = 1.7 < 2.0$$

where L_f = span of a flexible diaphragm providing lateral support of the wall = 70.0 ft (21.3 m) [see Figure 6.62].

W_p = weight of the wall tributary to the anchor $= 86.0 \times [(15.0/2) + 2.0] = 817$ lb/ft

Assume the anchors are spaced 3.0 ft on center.

$$F_p = 0.4 \times 0.25 \times 1.7 \times 1.00 \times 817 \times 3.0 = 417 \text{ lb}$$

$$< \text{greater of} \begin{cases} 0.2 \times 1.7 \times 1.00 \times 817 \times 3.0 = 833 \text{ lb} \\ 5.0 \times 3.0 \times 9.5 = 143 \text{ lb} \end{cases}$$

In S.I.:

W_p = weight of the wall tributary to the anchor $= 4.12 \times [(4.57/2) + 0.61] = 11.9$ kN/m

Assume the anchors are spaced 0.91 m on center.

$$F_p = 0.4 \times 0.25 \times 1.7 \times 1.00 \times 12.0 \times 0.91 = 1.86 \text{ kN/m}$$

$$< \text{greater of} \begin{cases} 0.2 \times 1.7 \times 1.00 \times 12.0 \times 0.91 = 3.71 \text{ kN} \\ 0.24 \times 0.91 \times 2.90 = 0.63 \text{ kN} \end{cases}$$

Use $F_p = 833$ lb (3.71 kN).

6.7.20 Example 6.20—Story Drift Limit

For the four-story essential facility in Example 6.7 (see Figure 6.52), check the story drift limits in the north-south direction. The design earthquake displacements, δ_{DE}, are given in Table 6.36.

Solution

For convenience, the design earthquake displacements, δ_{DE}, and the corresponding story drifts, Δ, in the north-south direction for the essential facility in Example 6.7 are given in Table 6.55. The displacement due to diaphragm deformation corresponding to the design earthquake, δ_{di}, is

Table 6.55 Design Earthquake Displacements and Story Drifts for the Essential Facility in Example 6.7

Story	δ_{DE}, in. (mm)	Δ, in. (mm)
4	5.57 (141.6)	0.77 (19.5)
3	4.80 (122.1)	1.28 (32.6)
2	3.52 (89.5)	1.65 (41.8)
1	1.87 (47.7)	1.87 (47.7)

permitted to be neglected in determining the design story drift (ASCE/SEI 12.8.6.5), and the design story drifts, Δ, are the difference of the δ_{DE} at the top and bottom of the story under consideration. Additional details on the calculation of δ_{DE} and Δ are given in Example 6.7.

Allowable story drifts, Δ_a, are determined from ASCE/SEI Table 12.12-1.

For "all other structures," $\Delta_a = 0.010h_{sx}$ for buildings assigned to risk category IV where h_{sx} is the story height below level x. Because the SFRS consists solely of moment frames in a building assigned to SDC D, $\Delta_a = 0.010h_{sx}/\rho$ for any story where ρ is the redundancy factor determined in accordance with ASCE/SEI 12.3.4.2 (ASCE/SEI 12.12.1.1).

It is evident that $\rho = 1.0$ because the condition in ASCE/SEI 12.3.4.2(2) is satisfied, including type 1 torsional irregularity (calculations including accidental torsion are not shown here).

Therefore, for the first story, $\Delta_a = 0.010h_{sx}/\rho = 0.010 \times 12.0 \times 12.0/1.0 = 1.44$ in.

For the other stories, $\Delta_a = 0.010 \times 11.0 \times 12.0/1.0 = 1.32$ in.

In S.I.:

First story: $\Delta_a = 0.010 \times 3.66 \times 1{,}000/1.0 = 36.6$ mm
Other stories: $\Delta_a = 0.010 \times 3.35 \times 1{,}000/1.0 = 33.5$ mm

Allowable drift limits are not satisfied in the first and second stories of this building in the north-south direction. Either the drifts must be decreased (which means the structure must be made laterally stiffer) or the allowable story drift can be taken as $0.015h_{sx}/\rho$, which is applicable for structures without masonry shear walls four stories or less above the base defined in ASCE/SEI 11.2 where the interior walls, partitions, and ceilings are designed to accommodate the drifts associated with δ_{DE} (see the first structure type in ASCE/SEI Table 12.12-1). It can be shown the allowable story drift requirements are satisfied if the interior elements are designed for δ_{DE}.

6.7.21 Example 6.21—Simplified Alternative Structural Design Criteria for a Simple Bearing Wall Building

Determine the seismic base shear and vertical distribution of seismic forces for the two-story commercial building in Figure 6.66 using the simplified alternative structural design criteria in ASCE/SEI 12.14 given the design data in Table 6.56. Assume the diaphragms are flexible at all levels. The story heights are equal to 12 ft (3.66 m).

SOLUTION

Flowchart 6.8 in Section 6.6 of this book is used to determine the seismic design forces.

Step 1—Check if the 12 limitations in ASCE/SEI 12.14.1.1 are satisfied

A check of the limitations is given in Table 6.57.

Because all 12 limitations of ASCE/SEI 12.14.1.1 are satisfied, the simplified design procedure is permitted to be used.

Figure 6.66 Typical Plan of the Two-Story Commercial Building in Example 6.21

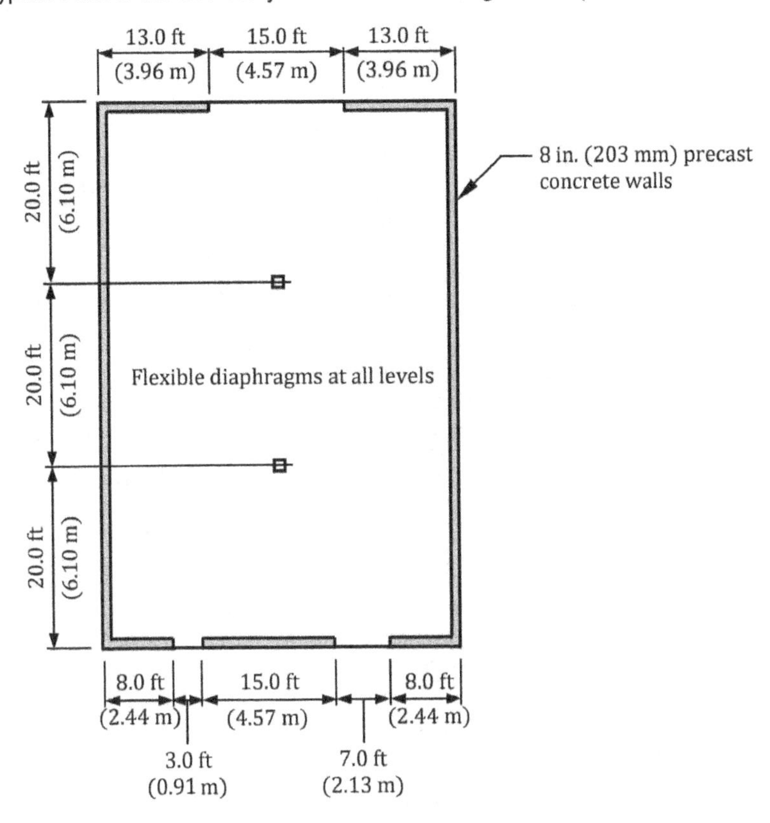

Table 6.56 Design Data for the Two-Story Commercial Building in Example 6.21

Location	Latitude: 45.492°, Longitude: −122.635°
Soil classification	Site class C
Occupancy	Commercial occupancy where less than 300 people congregate in one area
SFRS	Bearing wall system with intermediate precast shear walls (system A5 in ASCE/SEI Table 12.14-1)
Weight of roof framing and superimposed dead load	20 lb/ft² (0.96 kN/m²)
Weight of floor framing and superimposed dead load	30 lb/ft² (1.44 kN/m²)
Weight of 8 in. (203 mm) precast shear walls	100 lb/ft² (4.79 kN/m²)

Step 2—Determine the mapped spectral response acceleration at short periods, S_S

From Reference 6.1, $S_S = 0.95 < 1.5$ given the latitude and longitude of the site and site class (see ASCE/SEI 12.14.8.1).

Step 3—Determine S_{DS}

The design spectral response acceleration at short periods, S_{DS}, is determined in accordance with ASCE/SEI 12.14.8.1:

$$S_{DS} = \frac{2F_a S_S}{3} = \frac{2 \times 1.4 \times 0.95}{3} = 0.89$$

where $F_a = 1.4$ for soil sites (note: $F_a = 1.12$ and $S_{DS} = 0.71$ in accordance with ASCE/SEI 11.4.4).

Table 6.57 Check of the Limitations in ASCE/SEI 12.14.1.1 for the Commercial Building in Example 6.21

	Limitation	Discussion
1	The structure must qualify for risk category I or II in accordance with ASCE/SEI Table 1.5-1.	The building is assigned to risk category II given the commercial occupancy where less than 300 people congregate in one area.
2	The site class defined in ASCE/SEI Chapter 20 must not be site class E or F.	The site class for this building is C (see Table 6.56).
3	The structure must not exceed three stories above the grade plane.	The structure is two stories.
4	The SFRS must be either a bearing wall system or a building frame system, as indicated in ASCE/SEI Table 12.14-1.	The building has a bearing wall system consisting of intermediate precast concrete shear walls (see Table 6.56).
5	The structure must have at least two lines of lateral resistance in each of two major axis directions. At least one line of resistance must be provided on each side of the CM in each direction.	Precast shear walls are provided along two lines at the perimeter of the building in both directions with at least one wall on each side of the CM (see Figure 6.66).
6	The CM in each story must be located not further from the geometric centroid of the diaphragm than 10% of the length of the diaphragm parallel to the eccentricity.	The CM and the geometric centroid of the diaphragm are essentially at the same location.
7	For structures with cast-in-place concrete diaphragms, overhangs beyond the outside line of shear walls or braced frames must satisfy ASCE/SEI Equation (12.14-1).	This limitation is not applicable because cast-in-place concrete diaphragms are not used in this building.
8	For buildings with diaphragms that are not flexible, the forces must be apportioned to the vertical elements of the SFRS as if the diaphragms were flexible.	The diaphragms in this building are flexible.
9	Lines of resistance of the SFRS must be oriented at angles of no more than 15 degrees from alignment with the major orthogonal axes of the building.	The shear walls in both directions are parallel to the major axes.
10	The simplified design procedure must be used for each major horizontal axis direction of the building.	The simplified design procedure is used in both directions.
11	System irregularities caused by in-plane or out-of-plane offsets of the lateral force-resisting elements are not permitted.	This building does not have any irregularities.
12	The story lateral strength of any story must not be less than 80% of the story above where the story lateral strength is the total lateral strength of all seismic force-resisting system elements resisting the story shear for the direction under consideration.	In each orthogonal direction, the story lateral strengths of both stories are equal in a given direction because the thickness, length and arrangement of the shear walls are the same in both stories.

Step 4—Determine the SDC

For risk category II and $S_{DS} = 0.89 > 0.50$, the SDC is D (see ASCE/SEI 11.6, which permits the SDC to be determined from ASCE/SEI Table 11.6-1 alone where the alternate simplified design procedure is used).

Step 5—Determine the response modification coefficient, R

For a bearing wall system with intermediate precast shear walls (system A5), $R = 4$ (ASCE/SEI Table 12.14-1).

The height limit for this SFRS is 40.0 ft (12.2 m) for buildings assigned to SDC D, which is greater than the building height of 24.0 ft (7.32 m).

Step 6—Determine the effective seismic weight, W

Weight of roof framing and superimposed dead load $= 20.0 \times (41.0 \times 60.0)/1{,}000 =$ 49 kips

Weight of floor framing and superimposed dead load $= 30.0 \times (41.0 \times 60.0)/1{,}000 =$ 74 kips

Weight of precast walls tributary to the roof and floor diaphragms assuming full-story wall openings $= 100 \times (12.0 + 6.0) \times [(2 \times 60.0) + 26.0 + 31.0]/1{,}000 = 319$ kips

$$W = 49 + 74 + 319 = 442 \text{ kips}$$

In S.I.:

Weight of roof framing and superimposed dead load $= 0.96 \times (12.5 \times 18.3) = 219.6$ kN

Weight of floor framing and superimposed dead load $= 1.44 \times (12.5 \times 18.3) = 329.4$ kN

Weight of precast walls tributary to the roof and floor diaphragms $= 4.79 \times (3.66 + 1.83) \times [(2 \times 18.3) + 7.92 + 9.45] = 1{,}419.3$ kN

$$W = 219.6 + 329.4 + 1{,}419.3 = 1{,}968.3 \text{ kN}$$

Step 7—Determine the seismic base shear, V

The seismic base shear, V, is determined by ASCE/SEI Equation (12.14-12):

$$V = \frac{FS_{DS}}{R} W = \frac{1.1 \times 0.89}{4} \times 442 = 108 \text{ kips}$$

where $F = 1.1$ for a two-story building (ASCE/SEI 12.14.8.1).

In S.I.:

$$V = \frac{1.1 \times 0.89}{4} \times 1{,}968.3 = 482 \text{ kN}$$

Step 8—Determine the lateral seismic force, F_x

The lateral seismic forces at the floor and roof level are determined by ASCE/SEI Equation (12.14-13):

$$F_x = \frac{w_x}{W} V$$

where w_x is the portion of the effective seismic weight, W, at level x.

- Roof level

$$w_R = 49 + \{100 \times 6.0 \times [(2 \times 60.0) + 26.0 + 31.0]/1{,}000\} = 155 \text{ kips}$$

$$F_R = \frac{155}{442} \times 108 = 38 \text{ kips}$$

In S.I.:

$$w_R = 219.6 + \{4.79 \times 1.83 \times [(2 \times 18.3) + 7.92 + 9.45]\} = 692.7 \text{ kN}$$

$$F_R = \frac{692.7}{1{,}968.3} \times 482 = 170 \text{ kN}$$

- Second floor

$$w_2 = 74 + \{100 \times 12.0 \times [(2 \times 60.0) + 26.0 + 31.0]/1{,}000\} = 287 \text{ kips}$$

$$F_2 = \frac{287}{442} \times 108 = 70 \text{ kips}$$

In S.I.:

$$w_2 = 329.4 + \{4.79 \times 3.66 \times [(2 \times 18.3) + 7.92 + 9.45]\} = 1{,}275.6 \text{ kN}$$

$$F_2 = \frac{1{,}275.6}{1{,}968.3} \times 482 = 312 \text{ kN}$$

6.7.22 Example 6.22—Nonstructural Component—Parapet

Determine the design seismic forces on the 2-ft (0.61-m) parapet of the one-story commercial building in Example 6.17 (see Figure 6.62). Design data are given in Table 6.54.

Solution

Flowchart 6.9 in Section 6.6 of this book is used to determine the seismic design forces.

Step 1—Check if the weight of the parapet is greater than or equal to 0.2 times the combined effective seismic weight (ASCE/SEI 13.1.1)

Parapet weight $= 86 \times 2 \times (95 + 90)/1{,}000 = 32$ kips (142 kN)

Combined effective seismic weight = 243 kips (1,078 kN) [see Example 6.17]

Parapet weight/combined effective seismic weight $32/243 = 0.13 < 0.20$ (In S.I.: $142/1{,}078 = 0.13 < 0.20$)

Therefore, the requirements in ASCE/SEI 13.2.9 are not applicable.

Step 2—Determine S_{DS} in accordance with ASCE/SEI 11.4.4

From Example 6.17, $S_{DS} = 0.25$.

Step 3—Determine I_p

The component importance factor, I_p, is determined in accordance with ASCE/SEI 13.1.3.

Because none of the conditions in ASCE/SEI 13.1.3 that require $I_p = 1.5$ apply to this parapet, $I_p = 1.0$.

Step 4—Determine the component operating weight, W_p

From Example 6.17, the 8-in. (203-mm) concrete masonry weighs 86 lb/ft^2 (4.12 kN/m^2).

$$W_p = 86 \times 2.0 = 172 \text{ lb per foot width of parapet}$$

In S.I.: $W_p = 4.12 \times 0.61 = 2.5$ kN per meter width of parapet

Step 5—Determine the period, T_a

The lowest approximate fundamental period of the supporting building in either horizontal direction, T_a, is determined by ASCE/SEI Equation (12.8-8) in Example 6.17 and is equal to 0.15 s (see ASCE/SEI 13.3.1.1).

Step 6—Determine H_f

The factor for force amplification as a function of height, H_f, is permitted to be determined by ASCE/SEI Equations (13.3-4) or (13.3-5):

$$H_f = \begin{cases} 1 + a_1\left(\frac{z}{h}\right) + a_2\left(\frac{z}{h}\right)^{10} = 1 + \left[2.5 \times \left(\frac{15.0}{15.0}\right)\right] + 0 = 3.5 \\ 1 + 2.5\left(\frac{z}{h}\right) = 1 + 2.5\left(\frac{15.0}{15.0}\right) = 3.5 \end{cases}$$

where $a_1 = 1/T_a = 1/0.15 = 6.7 > 2.5$, use 2.5

$a_2 = 1 - (0.4/T_a)^2 = 1 - (0.4/0.15)^2 = -6.1 < 0$, use 0

z = height above the base of the building to the point of attachment of the parapet = 15.0 ft (4.57 m)

h = average roof height of the building with respect to the base = 15.0 ft (4.57 m)

Step 7—Determine I_e

The importance factor for the building, I_e, is determined in accordance with ASCE/SEI 11.5.1.

For commercial occupancy, the risk category is II and $I_e = 1.00$.

Step 8—Determine R and Ω_0

The response modification factor, R, and the overstrength factor, Ω_0, for the building are determined from ASCE/SEI Table 12.2-1.

For a building frame system with ordinary reinforced masonry shear walls (system B19), $R = 2$ and $\Omega_0 = 2.5$.

Step 9—Determine R_μ

The structure ductility reduction factor, R_μ, is determined by ASCE/SEI Equation (13.3-6):

$$R_\mu = \left(\frac{1.1R}{I_e\Omega_0}\right)^{1/2} = \left(\frac{1.1 \times 2}{1.00 \times 2.5}\right)^{1/2} = 0.94 < 1.3, \text{ use } 1.3$$

Step 10—Determine C_{AR}

The component resonance ductility factor, C_{AR}, is determined in accordance with ASCE/SEI 13.3.1.3.

For architectural components, C_{AR} is obtained from ASCE/SEI Table 13.5-1 and is equal to 2.2 for a cantilevered parapet that is braced to the structural frame below its center of mass and is supported above the grade plane by a structure.

Step 11—Obtain R_{po}

The component strength factor, R_{po}, is obtained from ASCE/SEI Table 13.5-1 for architectural components.

For a cantilevered parapet that is braced to the structural frame below its center of mass and is supported above the grade plane by a structure, $R_{po} = 1.5$.

Step 12—Determine F_p

The horizontal seismic force on the parapet, F_p, is determined by ASCE/SEI Equation (13.3-1):

$$F_p = 0.4S_{DS}I_pW_p\left(\frac{H_f}{R_\mu}\right)\left(\frac{C_{AR}}{R_{po}}\right) = 0.4 \times 0.25 \times 1.0 \times 172 \times \left(\frac{3.5}{1.3}\right) \times \left(\frac{2.2}{1.5}\right) = 68 \text{ lb per foot}$$

$$\text{Minimum } F_p = 0.3S_{DS}I_pW_p = 0.3 \times 0.25 \times 1.0 \times 172 = 13 \text{ lb per foot}$$

$$\text{Maximum } F_p = 1.6S_{DS}I_pW_p = 1.6 \times 0.25 \times 1.0 \times 172 = 69 \text{ lb per foot}$$

In S.I.:

$$F_p = 0.4 \times 0.25 \times 1.0 \times 2.5 \times \left(\frac{3.5}{1.3}\right) \times \left(\frac{2.2}{1.5}\right) = 1 \text{ kN per meter}$$

$$\text{Minimum } F_p = 0.3 \times 0.25 \times 1.0 \times 2.5 = 0.2 \text{ kN per meter}$$
$$\text{Maximum } F_p = 1.6 \times 0.25 \times 1.0 \times 2.5 = 1 \text{ kN per meter}$$

Use $F_p = 68$ lb per foot (1 kN per meter).

The horizontal seismic load effect, E_h, is determined by ASCE/SEI Equation (12.4-3):

$$E_h = \rho Q_E = \rho F_p = 1.0 \times 68 = 68 \text{ lb per foot (1 kN per meter)}$$

where the redundancy factor, ρ, is equal to 1.0 for the design of nonstructural components (item in ASCE/SEI 12.3.4.1).

The out-of-plane seismic force can be expressed as a uniformly distributed load over the height of the parapet:

$$f_p = 68/2.0 = 34 \text{ lb/ft per foot width of parapet}$$

In S.I.: $$f_p = 1/0.61 = 2 \text{ kN/m per meter width of parapet}$$

Step 13—Determine E_v

The vertical seismic force, E_v, is determined in accordance with ASCE/SEI 12.4.2.2:

$$E_v = 0.2 S_{DS} W_p = 0.2 \times 0.25 \times 172 = 9 \text{ lb per foot}$$

In S.I.: $$E_v = 0.2 \times 0.25 \times 2.5 = 0.13 \text{ kN per meter}$$

The horizontal and vertical seismic forces on the parapet are illustrated in Figure 6.67.

Figure 6.67 Seismic Forces on the Parapet in Example 6.22

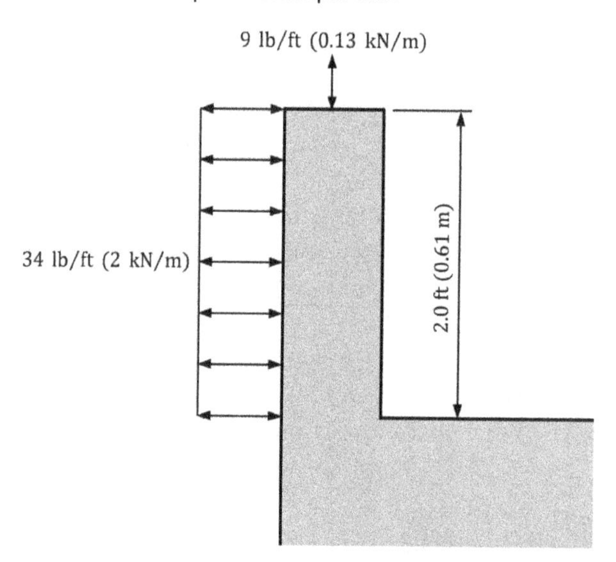

6.7.23 Example 6.23—Nonstructural Component—Air Handling Unit

A 2,000-lb (9-kN) air handling unit (AHU) is mounted directly on the roof of the one-story commercial building in Example 6.17. Determine the design seismic force on the AHU. Design data are given in Table 6.54.

Solution

Flowchart 6.9 in Section 6.6 of this book is used to determine the seismic design forces.

Step 1—Check if the weight of the AHU is greater than or equal to 0.2 times the combined effective seismic weight

By inspection, the weight of the AHU is less than 0.2 times the combined effective seismic weight.

Therefore, the requirements in ASCE/SEI 13.2.9 are not applicable.

Step 2—Determine S_{DS} in accordance with ASCE/SEI 11.4.4

From Example 6.17, $S_{DS} = 0.25$.

Step 3—Determine I_p

The component importance factor, I_p, is determined in accordance with ASCE/SEI 13.1.3.

Because none of the conditions in ASCE/SEI 13.1.3 that require $I_p = 1.5$ apply to this AHU, $I_p = 1.0$.

Step 4—Determine the component operating weight, W_p

From the design data, $W_p = 2{,}000$ lb (9 kN).

Step 5—Determine the period, T_a

The lowest approximate fundamental period of the supporting building in either horizontal direction, T_a, is determined by ASCE/SEI Equation (12.8-8) in Example 6.17 and is equal to 0.15 s.

Step 6—Determine H_f

The factor for force amplification with height, H_f, is permitted to be determined by ASCE/SEI Equations (13.3-4) or (13.3-5):

$$H_f = \begin{cases} 1 + a_1\left(\frac{z}{h}\right) + a_2\left(\frac{z}{h}\right)^{10} = 1 + \left[2.5 \times \left(\frac{15.0}{15.0}\right)\right] + 0 = 3.5 \\ 1 + 2.5\left(\frac{z}{h}\right) = 1 + 2.5\left(\frac{15.0}{15.0}\right) = 3.5 \end{cases}$$

where $a_1 = 1/T_a = 1/0.15 = 6.7 > 2.5$, use 2.5

$a_2 = 1 - (0.4/T_a)^2 = 1 - (0.4/0.15)^2 = -6.1 < 0$, use 0

z = height above the base of the building to the point of attachment of the AHU = 15.0 ft (4.57 m)

h = average roof height of the building with respect to the base = 15.0 ft (4.57 m)

Step 7—Determine I_e

The importance factor for the building, I_e, is determined in accordance with ASCE/SEI 11.5.1.

For commercial occupancy, the risk category is II and $I_e = 1.00$.

Step 8—Determine R and Ω_0

The response modification factor, R, and the overstrength factor, Ω_0, for the building are determined from ASCE/SEI Table 12.2-1.

For a building frame system with ordinary reinforced masonry shear walls (system B19), $R = 2$ and $\Omega_0 = 2.5$.

Step 9—Determine R_μ

The structure ductility reduction factor, R_μ, is determined by ASCE/SEI Equation (13.3-6):

$$R_\mu = \left(\frac{1.1R}{I_e\Omega_0}\right)^{1/2} = \left(\frac{1.1 \times 2}{1.00 \times 2.5}\right)^{1/2} = 0.94 < 1.3, \text{ use } 1.3$$

Step 10—Determine C_{AR}

The component resonance ductility factor, C_{AR}, is determined in accordance with ASCE/SEI 13.3.1.3.

For mechanical components, C_{AR} is obtained from ASCE/SEI Table 13.6-1 and is equal to 1.4 for air-side HVACR fans, air handlers, air conditioning units, cabinet heaters, air distribution boxes, and other mechanical components constructed of sheet metal framing supported above the grade plane by a structure.

Step 11—Obtain R_{po}

The component strength factor, R_{po}, is obtained from ASCE/SEI Table 13.6-1 for mechanical components.

For this AHU, $R_{po} = 2$.

Step 12—Determine F_p

The horizontal seismic force on the AHU, F_p, is determined by ASCE/SEI Equation (13.3-1):

$$F_p = 0.4S_{DS}I_pW_p\left(\frac{H_f}{R_\mu}\right)\left(\frac{C_{AR}}{R_{po}}\right) = 0.4\times0.25\times1.0\times2{,}000\times\left(\frac{3.5}{1.3}\right)\times\left(\frac{1.4}{2}\right) = 377\text{ lb}$$

$$\text{Minimum } F_p = 0.3S_{DS}I_pW_p = 0.3\times0.25\times1.0\times2{,}000 = 150\text{ lb}$$

$$\text{Maximum } F_p = 1.6S_{DS}I_pW_p = 1.6\times0.25\times1.0\times2{,}000 = 800\text{ lb}$$

In S.I.:

$$F_p = 0.4\times0.25\times1.0\times9\times\left(\frac{3.5}{1.3}\right)\times\left(\frac{1.4}{2}\right) = 2\text{ kN}$$

$$\text{Minimum } F_p = 0.3\times0.25\times1.0\times9 = 1\text{ kN}$$

$$\text{Maximum } F_p = 1.6\times0.25\times1.0\times9 = 4\text{ kN}$$

Use $F_p = 377$ lb (2 kN).

The horizontal seismic load effect, E_h, is determined by ASCE/SEI Equation (12.4-3):

$$E_h = \rho Q_E = \rho F_p = 1.0\times377 = 377\text{ lb (2 kN)}$$

where the redundancy factor, ρ, is equal to 1.0 for the design of nonstructural components (item 3 in ASCE/SEI 12.3.4.1).

Step 13—Determine E_v

The vertical seismic force, E_v, is determined in accordance with ASCE/SEI 12.4.2.2:

$$E_v = 0.2S_{DS}W_p = 0.2\times0.25\times2{,}000 = 100\text{ lb}$$

In S.I.: $E_v = 0.2\times0.25\times9 = 0.5$ kN

The horizontal and vertical seismic forces on the AHU are illustrated in Figure 6.68.

Figure 6.68 Seismic Forces on the AHU in Example 6.23

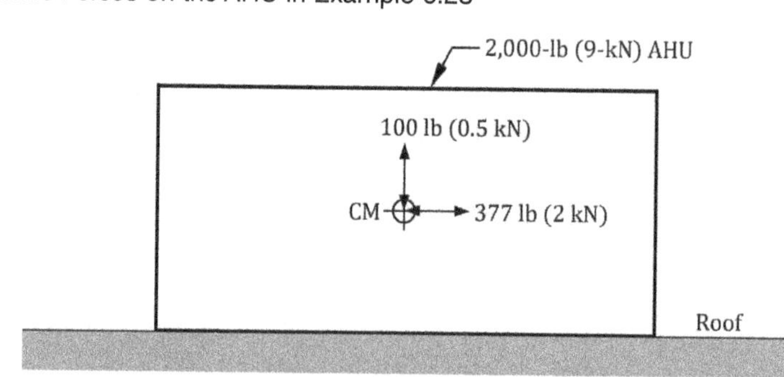

According to footnote b in ASCE/SEI Table 13.6-1, if the AHU were to be anchored to concrete or masonry, the force transferred from the AHU to the concrete or masonry must be equal to $\Omega_{0p}F_p$ where the anchorage overstrength factor, Ω_{0p}, is given in ASCE/SEI Table 13.6-1, which is equal to 2 for an AHU.

6.7.24 Example 6.24—Nonstructural Component—Exterior Nonstructural Wall

Assume the four-story emergency treatment facility in Example 6.7 has cladding on the exterior of the building that consists of 8-in. (203-mm) thick precast concrete wall panels that are 20-ft (6.1 m) wide and weigh 100 lb/ft^2 (4.79 kN/m^2). Determine the design seismic force on the precast concrete wall panels and their connections to the building in the fourth story assuming the connections to the structure are made at the roof and floor levels at four points (i.e., the wall panels span vertically between the roof and floor levels with two connections per level). The fourth story height is 11.0 ft (3.35 m) [see Figure 6.52]. Design data are given in Table 6.34.

SOLUTION

Flowchart 6.9 in Section 6.6 of this book is used to determine the seismic design forces.

Step 1—Check if the weight of a single precast concrete wall panel is greater than or equal to 0.2 times the combined effective seismic weight

By inspection, the weight of a single precast concrete wall panel is less than 0.2 times the combined effective seismic weight.

Therefore, the requirements in ASCE/SEI 13.2.9 are not applicable.

Step 2—Determine S_{DS} in accordance with ASCE/SEI 11.4.4

From Example 6.7, $S_{DS} = 0.26$.

Step 3—Determine I_p

The component importance factor, I_p, is determined in accordance with ASCE/SEI 13.1.3.

Assuming the failure of the precast concrete wall panels would impair the continued operation of the risk category IV structure, $I_p = 1.5$.

Step 4—Determine the component operating weight, W_p

From the design data, the weight of the precast concrete wall panel is $W_p = 100$ lb/ft^2 (4.8 kN/m^2).

Step 5—Determine the period, T_a

The lowest approximate fundamental period of the supporting building in either horizontal direction, T_a, is determined by ASCE/SEI Equation (12.8-8) in Example 6.7 and is equal to 0.49 s.

Step 6—Determine H_f

The factor for force amplification as a function of height, H_f, is permitted to be determined by ASCE/SEI Equations (13.3-4) or (13.3-5):

$$H_f = \begin{cases} 1 + a_1\left(\frac{z}{h}\right) + a_2\left(\frac{z}{h}\right)^{10} \\ 1 + 2.5\left(\frac{z}{h}\right) \end{cases}$$

where $a_1 = 1/T_a$ and $a_2 = 1 - (0.4/T_a)^2$.

Values of H_f are calculated at the top and bottom of the precast concrete wall panel in the fourth story, that is, at $z = 45.0$ ft (13.7 m) and $z = 34.0$ ft (10.4 m).

At $z = 45.0$ ft (13.7 m):

$$H_f = \begin{cases} 1 + \left[2.04 \times \left(\frac{45.0}{45.0}\right)\right] + \left[0.33 \times \left(\frac{45.0}{45.0}\right)^{10}\right] = 3.4 \\ 1 + 2.5\left(\frac{45.0}{45.0}\right) = 3.5 \end{cases}$$

where $a_1 = 1/0.49 = 2.04 < 2.5$ and $a_2 = 1 - (0.4/0.49)^2 = 0.33 > 0$.

At $z = 34.0$ ft (10.4 m):

$$H_f = \begin{cases} 1 + \left[2.04 \times \left(\frac{34.0}{45.0}\right)\right] + \left[0.33 \times \left(\frac{34.0}{45.0}\right)^{10}\right] = 2.6 \\ 1 + 2.5\left(\frac{34.0}{45.0}\right) = 2.9 \end{cases}$$

In this example, $H_f = 3.4$ and $H_f = 2.6$ will be used at the top and bottom of the fourth story, respectively.

Step 7—Determine I_e

The importance factor for the building, I_e, is determined in accordance with ASCE/SEI 11.5.1.

For an essential facility, the risk category is IV and $I_e = 1.50$.

Step 8—Determine R and Ω_0

The response modification factor, R, and the overstrength factor, Ω_0, for the building are determined from ASCE/SEI Table 12.2-1.

For special reinforced concrete moment frames (system C5), $R = 8$ and $\Omega_0 = 3$.

Step 9—Determine R_μ

The structure ductility reduction factor, R_μ, is determined by ASCE/SEI Equation (13.3-6):

$$R_\mu = \left(\frac{1.1R}{I_e\Omega_0}\right)^{1/2} = \left(\frac{1.1 \times 8}{1.50 \times 3}\right)^{1/2} = 1.4 > 1.3$$

Step 10—Determine C_{AR}

The component resonance ductility factor, C_{AR}, is determined in accordance with ASCE/SEI 13.3.1.3.

For architectural components, C_{AR} is obtained from ASCE/SEI Table 13.5-1. The following factors are applicable for exterior nonstructural wall elements and connections supported above the grade plane by a structure:

- For wall elements and body of wall panel connections, $C_{AR} = 1.0$
- For fasteners of the connecting system, $C_{AR} = 2.8$

Step 11—Obtain R_{po}

The component strength factor, R_{po}, is obtained from ASCE/SEI Table 13.5-1 for architectural components.

For wall elements, body of wall panel connections and fasteners of the connecting system, $R_{po} = 1.5$.

Step 12—Determine F_p

The horizontal seismic force, F_p, is determined by ASCE/SEI Equation (13.3-1).

- Wall elements and body of wall panel connections

For wall panels spanning between two floors, the seismic force on the panels can be taken as the average of the seismic forces at the attachment locations.

$$W_p = 100 \times 20.0 \times 11.0 = 22{,}000 \text{ lb}$$

$$F_{p(\text{upper})} = 0.4 S_{DS} I_p W_p \left(\frac{H_f}{R_\mu}\right)\left(\frac{C_{AR}}{R_{po}}\right)$$

$$= 0.4 \times 0.26 \times 1.5 \times 22{,}000 \times \left(\frac{3.4}{1.4}\right) \times \left(\frac{1.0}{1.5}\right) = 5{,}557 \text{ lb}$$

$$F_{p(\text{lower})} = 0.4 \times 0.26 \times 1.5 \times 22{,}000 \times \left(\frac{2.6}{1.4}\right) \times \left(\frac{1.0}{1.5}\right) = 4{,}249 \text{ lb}$$

Minimum $F_p = 0.3 S_{DS} I_p W_p = 0.3 \times 0.26 \times 1.5 \times 22{,}000 = 2{,}574$ lb

Maximum $F_p = 1.6 S_{DS} I_p W_p = 1.6 \times 0.26 \times 1.5 \times 22{,}000 = 13{,}728$ lb

Therefore, average $F_p = \dfrac{5{,}557 + 4{,}249}{2} = 4{,}903$ lb

In S.I.:

$$W_p = 4.79 \times 6.10 \times 3.35 = 98 \text{ kN}$$

$$F_{p(\text{upper})} = 0.4 \times 0.26 \times 1.5 \times 98 \times \left(\frac{3.4}{1.4}\right) \times \left(\frac{1.0}{1.5}\right) = 25 \text{ kN}$$

$$F_{p(\text{lower})} = 0.4 \times 0.26 \times 1.5 \times 98 \times \left(\frac{2.6}{1.4}\right) \times \left(\frac{1.0}{1.5}\right) = 19 \text{ kN}$$

Minimum $F_p = 0.3 S_{DS} I_p W_p = 0.3 \times 0.26 \times 1.5 \times 98 = 12$ kN

Maximum $F_p = 1.6 S_{DS} I_p W_p = 1.6 \times 0.26 \times 1.5 \times 98 = 61$ kN

Therefore, average $F_p = \dfrac{25 + 19}{2} = 22$ kN

The horizontal seismic load effect, E_h, is determined by ASCE/SEI Equation (12.4-3):

$$E_h = \rho Q_E = \rho F_p = 1.0 \times 4{,}903 = 4{,}903 \text{ lb (22 kN)}$$

where the redundancy factor, ρ, is equal to 1.0 for the design of nonstructural components (item 3 in ASCE/SEI 12.3.4.1).

- Fasteners of the connecting system
  The seismic forces on the fasteners of the connecting system can be determined based on the distance to the uppermost fasteners in the story, which is conservative for the fasteners at the bottom of the precast concrete wall panel.

$$F_p = 0.4 S_{DS} I_p W_p \left(\frac{H_f}{R_\mu}\right)\left(\frac{C_{AR}}{R_{po}}\right) = 0.4 \times 0.26 \times 1.5 \times 22{,}000 \times \left(\frac{3.4}{1.4}\right) \times \left(\frac{2.8}{1.5}\right) = 15{,}558 \text{ lb}$$

Minimum $F_p = 0.3 S_{DS} I_p W_p = 0.3 \times 0.26 \times 1.5 \times 22{,}000 = 2{,}574$ lb

Maximum $F_p = 1.6 S_{DS} I_p W_p = 1.6 \times 0.26 \times 1.5 \times 22{,}000 = 13{,}728$ lb

Therefore, $F_p = 13{,}728$ lb.

In S.I.:

$$F_p = 0.4 \times 0.26 \times 1.5 \times 98 \times \left(\frac{3.4}{1.4}\right) \times \left(\frac{2.8}{1.5}\right) = 69 \text{ kN}$$

Minimum $F_p = 0.3 S_{DS} I_p W_p = 0.3 \times 0.26 \times 1.5 \times 98 = 12$ kN

Maximum $F_p = 1.6 S_{DS} I_p W_p = 1.6 \times 0.26 \times 1.5 \times 98 = 61$ kN

Therefore, $F_p = 61$ kN.

The horizontal seismic load effect, E_h, is determined by ASCE/SEI Equation (12.4-3):

$$E_h = \rho Q_E = \rho F_p = 1.0 \times 13{,}728 = 13{,}728 \text{ lb (61 kN)}$$

where the redundancy factor, ρ, is equal to 1.0 for the design of nonstructural components (item 3 in ASCE/SEI 12.3.4.1).

Step 13—Determine E_v

The vertical seismic force, E_v, is determined in accordance with ASCE/SEI 12.4.2.2:

$$E_v = 0.2 S_{DS} W_p = 0.2 \times 0.26 \times 22{,}000 = 1{,}144 \text{ lb}$$

In S.I.: $E_v = 0.2 \times 0.26 \times 98 = 5 \text{ kN}$

The horizontal and vertical seismic forces on the precast concrete wall act through the CM of the element with the horizontal seismic force acting perpendicular to the face of the panel.

6.7.25 Example 6.25—Nonstructural Component—Egress Stair

Determine the design seismic forces on a precast concrete egress stairway and its connections in the third story of a four-story precast concrete parking structure (see Figure 6.69). The stairway is not part of the SFRS, which is a building frame system with special reinforced concrete shear walls (system B4 in ASCE/SEI Table 12.2-1).

Figure 6.69 Precast Concrete Egress Stairway in Example 6.25

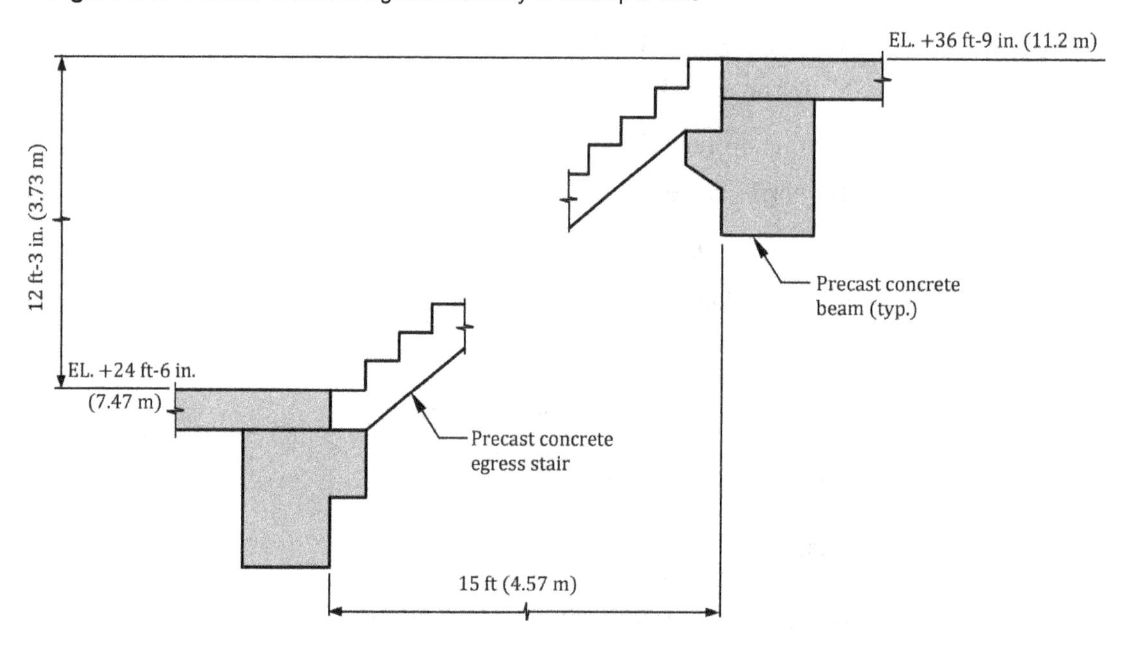

The top of the stairway is connected to a precast beam where the connection resists forces in the vertical direction and in the horizontal direction parallel and perpendicular to the plane of the stair risers. The bottom of the stairway is also connected to a precast beam where the connection resists forces in the vertical direction and in the horizontal direction parallel to the plane of the stair risers; the connection permits movement in the horizontal direction perpendicular to the plane of the stair risers, which means horizontal forces cannot be resisted in that direction.

The parking structure, which has typical story heights of 12 ft-3 in. (3.73 m), is assigned to SDC D where $S_{DS} = 0.73$ and $S_{D1} = 0.45$. The weight of the stairway is 9.6 kips (42.7 kN) and the superimposed dead load over the entire stair run is equal to 1.0 kip (4.5 kN).

SOLUTION

Flowchart 6.9 in Section 6.6 of this book is used to determine the seismic design forces.

Step 1—Check if the weight of the precast concrete stairway is greater than or equal to 0.2 times the combined effective seismic weight

By inspection, the weight the precast concrete stairway is less than 0.2 times the combined effective seismic weight.

Therefore, the requirements in ASCE/SEI 13.2.9 are not applicable.

Step 2—Determine S_{DS} in accordance with ASCE/SEI 11.4.4

From the design data, $S_{DS} = 0.73$.

Step 3—Determine I_p

The component importance factor, I_p, is determined in accordance with ASCE/SEI 13.1.3.

Because this is an egress stairway, $I_p = 1.5$.

Step 4—Determine the component operating weight, W_p

The operating weight, W_p, is equal to the weight of the stair plus the superimposed dead load over the entire stair run:

$$W_p = 9.6 + 1.0 = 10.6 \text{ kips } (47.2 \text{ kN})$$

Step 5—Determine the period, T_a

The lowest approximate fundamental period of the supporting building in either horizontal direction, T_a, is determined by ASCE/SEI Equation (12.8-8) assuming "all other structural systems":

$$T_a = C_t h_n^x = 0.02 \times (49.0)^{0.75} = 0.37 \text{ s}$$

In S.I.:

$$T_a = 0.0488 \times (14.9)^{0.75} = 0.37 \text{ s}$$

Step 6—Determine H_f

The factor for force amplification with height, H_f, is permitted to be determined by ASCE/SEI Equations (13.3-4) or (13.3-5):

$$H_f = \begin{cases} 1 + a_1\left(\frac{z}{h}\right) + a_2\left(\frac{z}{h}\right)^{10} \\ 1 + 2.5\left(\frac{z}{h}\right) \end{cases}$$

where $a_1 = 1/T_a = 1/0.37 = 2.7 > 2.5$, use 2.5

$a_2 = 1 - (0.4/T_a)^2 = 1 - (0.4/0.37)^2 = -0.16 < 0$, use 0

In the direction of analysis perpendicular to the plane of the stair risers, $z = 36.75$ ft (11.2 m) because the seismic forces are imparted to the stair at the top connection only due to the type of connection at the bottom of the stair. In the direction parallel to the plane of the stair risers, an average z is used, that is, $z = (36.75 + 24.5)/2 = 30.6$ ft (9.33 m), because seismic forces are imparted to the stair at both the top and bottom connections.

At $z = 36.75$ ft (11.2 m):

$$H_f = \begin{cases} 1 + a_1\left(\frac{z}{h}\right) + a_2\left(\frac{z}{h}\right)^{10} = 1 + \left[2.5 \times \left(\frac{36.75}{49.0}\right)\right] + 0 = 2.9 \\ 1 + 2.5\left(\frac{z}{h}\right) = 1 + \left[2.5 \times \left(\frac{36.75}{49.0}\right)\right] = 2.9 \end{cases}$$

At $z = 30.6$ ft (9.33 m):

$$H_f = \begin{cases} 1+\left[2.5\times\left(\frac{30.6}{49.0}\right)\right]+0=2.6 \\ 1+\left[2.5\times\left(\frac{30.6}{49.0}\right)\right]=2.6 \end{cases}$$

Step 7—Determine I_e

The importance factor for the building, I_e, is determined in accordance with ASCE/SEI 11.5.1.

For a parking garage occupancy, the risk category is II and $I_e = 1.00$.

Step 8—Determine R and Ω_0

The response modification factor, R, and the overstrength factor, Ω_0, for the building are determined from ASCE/SEI Table 12.2-1.

For a building frame system with special reinforced concrete shear walls (system B4), $R = 6$ and $\Omega_0 = 2.5$.

Step 9—Determine R_μ

The reduction factor for ductility of the supporting structure, R_μ, is determined by ASCE/SEI Equation (13.3-6):

$$R_\mu = \left(\frac{1.1R}{I_e\Omega_0}\right)^{1/2} = \left(\frac{1.1\times 6}{1.00\times 2.5}\right)^{1/2} = 1.6 > 1.3$$

Step 10—Determine C_{AR}

The component resonance ductility factor, C_{AR}, is determined in accordance with ASCE/SEI 13.3.1.3.

For architectural components, C_{AR} is obtained from ASCE/SEI Table 13.5-1.

The following factors are applicable for elements supported above the grade plane by a structure:

- For egress stairways not part of the building SFRS, $C_{AR} = 1.0$
- For egress stairs and ramp fasteners and attachments, $C_{AR} = 2.2$ for elements supported above the grade plane of the structure

Step 11—Obtain R_{po}

The component strength factor, R_{po}, is obtained from ASCE/SEI Table 13.5-1 for architectural components.

For egress stairways not part of the building SFRS and egress stairs and ramp fasteners and attachments, $R_{po} = 1.5$.

Step 12—Determine F_p and horizontal connection forces

The horizontal seismic force, F_p, is determined by ASCE/SEI Equation (13.3-1):

$$F_p = 0.4S_{DS}I_pW_p\left(\frac{H_f}{R_\mu}\right)\left(\frac{C_{AR}}{R_{po}}\right)$$

This force must be determined for directions of analysis perpendicular and parallel to the plane of the stair risers.

- Direction of analysis perpendicular to the plane of the stair risers

  For the egress stairway:

$$F_p = 0.4S_{DS}I_pW_p\left(\frac{H_f}{R_\mu}\right)\left(\frac{C_{AR}}{R_{po}}\right) = 0.4\times 0.73\times 1.5\times 10.6\times\left(\frac{2.9}{1.6}\right)\times\left(\frac{1.0}{1.5}\right) = 5.6 \text{ kips}$$

Minimum $F_p = 0.3S_{DS}I_pW_p = 0.3\times 0.73\times 1.5\times 10.6 = 3.5$ kips

Maximum $F_p = 1.6S_{DS}I_pW_p = 1.6\times 0.73\times 1.5\times 10.6 = 18.6$ kips

In S.I.:

$$F_p = 0.4 \times 0.73 \times 1.5 \times 47.2 \times \left(\frac{2.9}{1.6}\right) \times \left(\frac{1.0}{1.5}\right) = 25.0 \text{ kN}$$

Minimum $F_p = 0.3 \times 0.73 \times 1.5 \times 47.2 = 15.5$ kN

Maximum $F_p = 1.6 \times 0.73 \times 1.5 \times 47.2 = 82.7$ kN

Use $F_p = 5.6$ kips (25.0 kN).

The horizontal seismic load effect, E_h, is determined by ASCE/SEI Equation (12.4-3):

$$E_h = \rho Q_E = \rho F_p = 1.0 \times 5.6 = 5.6 \text{ kips (25.0 kN)}$$

where the redundancy factor, ρ, is equal to 1.0 for the design of nonstructural components (item 3 in ASCE/SEI 12.3.4.1).

For the fasteners and attachments:

$$F_p = 0.4 S_{DS} I_p W_p \left(\frac{H_f}{R_\mu}\right)\left(\frac{C_{AR}}{R_{po}}\right) = 0.4 \times 0.73 \times 1.5 \times 10.6 \times \left(\frac{2.9}{1.6}\right) \times \left(\frac{2.2}{1.5}\right) = 12.3 \text{ kips}$$

Minimum $F_p = 0.3 S_{DS} I_p W_p = 0.3 \times 0.73 \times 1.5 \times 10.6 = 3.5$ kips

Maximum $F_p = 1.6 S_{DS} I_p W_p = 1.6 \times 0.73 \times 1.5 \times 10.6 = 18.6$ kips

In S.I.:

$$F_p = 0.4 \times 0.73 \times 1.5 \times 47.2 \times \left(\frac{2.9}{1.6}\right) \times \left(\frac{2.2}{1.5}\right) = 55.0 \text{ kN}$$

Minimum $F_p = 0.3 \times 0.73 \times 1.5 \times 47.2 = 15.5$ kN

Maximum $F_p = 1.6 \times 0.73 \times 1.5 \times 47.2 = 82.7$ kN

Use $F_p = 12.3$ kips (55.0 kN)

Calculated values of F_p for fasteners and attachments are multiplied by the overstrength factor, Ω_{0p}, for nonductile anchorage to concrete (see footnote a in ASCE/SEI Table 13.5-1) where $\Omega_{0p} = 1.75$ for fasteners and attachments of egress stairs (see ASCE/SEI Table 13.5-1):

Connection force at the top of the stair $= 1.75 \times 12.3 = 21.5$ kips (96.3 kN)

- Direction of analysis parallel to the plane of the stair risers

For the egress stairway:

$$F_p = 0.4 S_{DS} I_p W_p \left(\frac{H_f}{R_\mu}\right)\left(\frac{C_{AR}}{R_{po}}\right) = 0.4 \times 0.73 \times 1.5 \times 10.6 \times \left(\frac{2.6}{1.6}\right) \times \left(\frac{1.0}{1.5}\right) = 5.0 \text{ kips}$$

Minimum $F_p = 0.3 S_{DS} I_p W_p = 0.3 \times 0.73 \times 1.5 \times 10.6 = 3.5$ kips

Maximum $F_p = 1.6 S_{DS} I_p W_p = 1.6 \times 0.73 \times 1.5 \times 10.6 = 18.6$ kips

In S.I.:

$$F_p = 0.4 \times 0.73 \times 1.5 \times 47.2 \times \left(\frac{2.6}{1.6}\right) \times \left(\frac{1.0}{1.5}\right) = 22.4 \text{ kN}$$

Minimum $F_p = 0.3 \times 0.73 \times 1.5 \times 47.2 = 15.5$ kN

Maximum $F_p = 1.6 \times 0.73 \times 1.5 \times 47.2 = 82.7$ kN

Use $F_p = 5.0$ kips (22.4 kN).

The horizontal seismic load effect, E_h, is determined by ASCE/SEI Equation (12.4-3):

$$E_h = \rho Q_E = \rho F_p = 1.0 \times 5.0 = 5.0 \text{ kips (22.4 kN)}$$

where the redundancy factor, ρ, is equal to 1.0 for the design of nonstructural components (item 3 in ASCE/SEI 12.3.4.1).

For the fasteners and attachments:

$$F_p = 0.4 S_{DS} I_p W_p \left(\frac{H_f}{R_\mu}\right)\left(\frac{C_{AR}}{R_{po}}\right) = 0.4 \times 0.73 \times 1.5 \times 10.6 \times \left(\frac{2.6}{1.6}\right) \times \left(\frac{2.2}{1.5}\right) = 11.1 \text{ kips}$$

Minimum $F_p = 0.3 S_{DS} I_p W_p = 0.3 \times 0.73 \times 1.5 \times 10.6 = 3.5$ kips

Maximum $F_p = 1.6 S_{DS} I_p W_p = 1.6 \times 0.73 \times 1.5 \times 10.6 = 18.6$ kips

In S.I.:

$$F_p = 0.4 \times 0.73 \times 1.5 \times 47.2 \times \left(\frac{2.6}{1.6}\right) \times \left(\frac{2.2}{1.5}\right) = 49.3 \text{ kN}$$

Minimum $F_p = 0.3 \times 0.73 \times 1.5 \times 47.2 = 15.5$ kN

Maximum $F_p = 1.6 \times 0.73 \times 1.5 \times 47.2 = 82.7$ kN

Use F_p = 11.1 kips (49.3 kN)

Calculated values of F_p for fasteners and attachments are multiplied by the overstrength factor, Ω_{0p}, for nonductile anchorage to concrete (see footnote a in ASCE/SEI Table 13.5-1) where $\Omega_{0p} = 1.75$ for fasteners and attachments of egress stairs (see ASCE/SEI Table 13.5-1):

$$\text{Connection force at the top and bottom of the stair} = 1.75 \times 11.1 = 19.4 \text{ kips (86.3 kN)}$$

Step 13—Determine E_v

The vertical seismic force, E_v, is determined in accordance with ASCE/SEI 12.4.2.2:

$$E_v = 0.2 S_{DS} W_p = 0.2 \times 0.73 \times 10.6 = 1.6 \text{ kips}$$

In S.I.: $E_v = 0.2 \times 0.73 \times 47.2 = 6.9$ kN

The horizontal and vertical seismic forces on the egress stairway perpendicular and parallel to the stair are illustrated in Figures 6.70 and 6.71, respectively.

Step 14—Consideration of seismic relative displacements

Egress stairs not part of the SFRS must be detailed to accommodate the seismic relative displacement, D_{pI}, defined in ASCE/SEI 13.3.2, including diaphragm deformation (ASCE/SEI 13.5.10).

Assuming the results of an elastic analysis of the parking structure is not available, seismic displacements can be determined based on the allowable story drift requirements in ASCE/SEI 12.12.1.

The allowable story drift, Δ_a, for a risk category II structure is equal to $0.020h_{sx}$ where it is assumed any interior nonstructural elements are not designed to accommodate the drifts associated with the design earthquake displacements, δ_{DE}, determined by ASCE/SEI Equation (12.8-16).

For the 12 ft-3 in. (3.73 m) story height:

$$\Delta_a = 0.020 \times 12.25 \times 12 = 2.9 \text{ in. (75 mm)}$$

This allowable drift is the allowable story drift, Δ_{aA}, for the parking structure (structure A) defined in ASCE/SEI 13.3.2.1.

Figure 6.70 Seismic Forces Perpendicular to the Plane of the Stair Risers for the Egress Stairway in Example 6.25

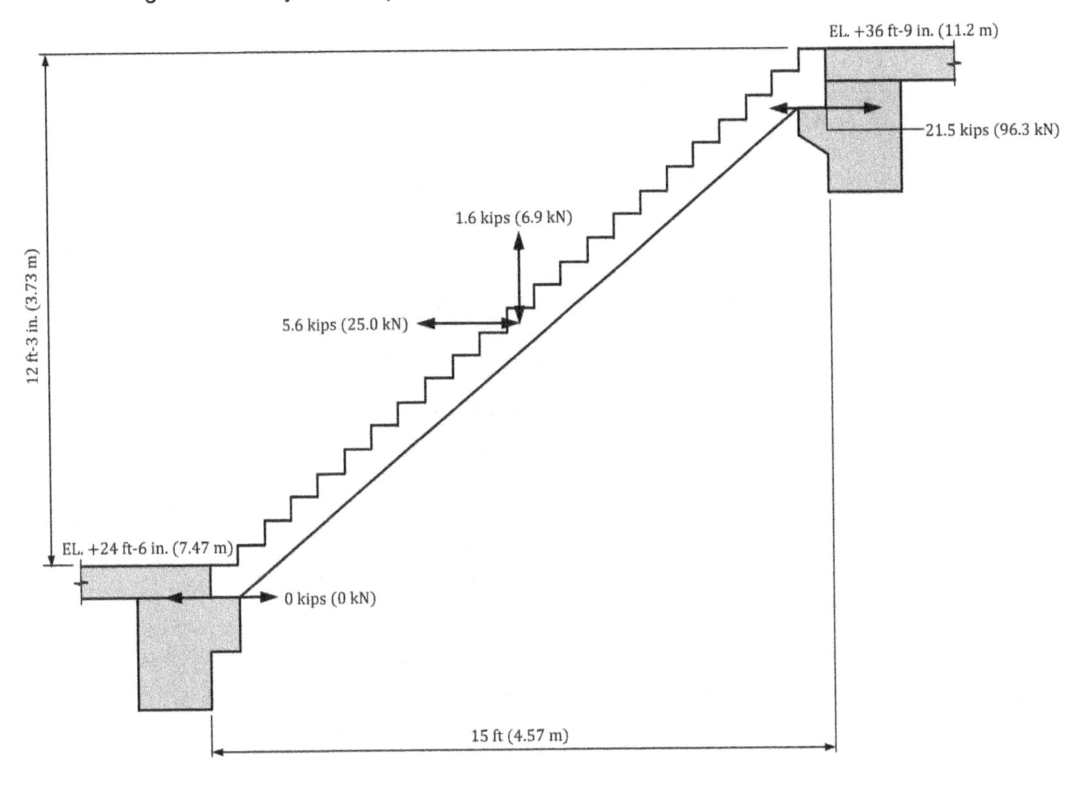

Figure 6.71 Seismic Forces Parallel to the Plane of the Stair Risers for the Egress Stairway in Example 6.25

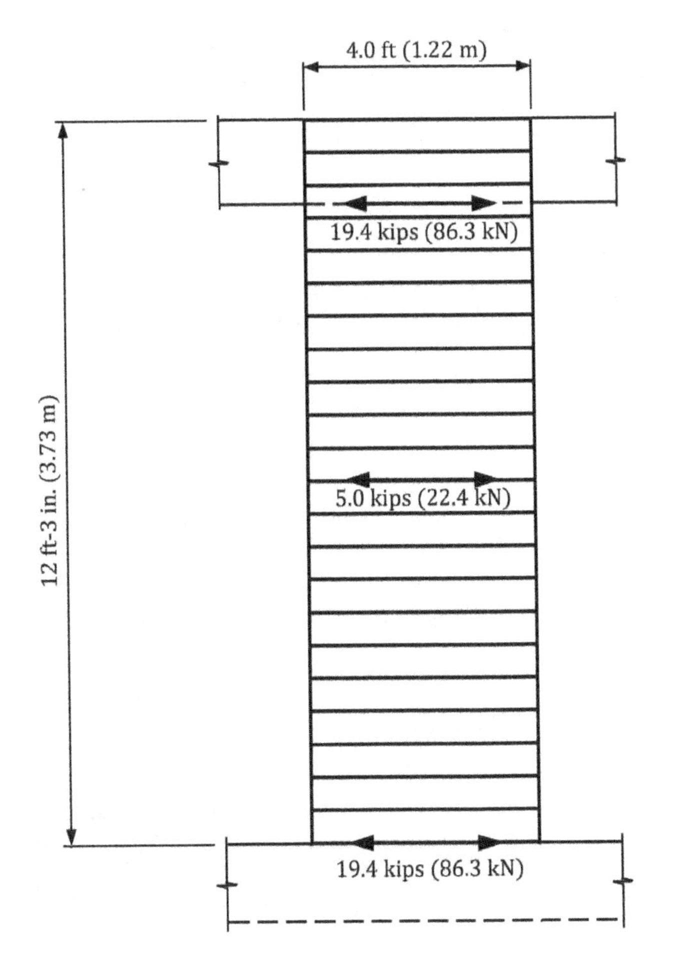

The elevations of the connection points for this stair are as follows:

$$h_x = 36.75 \text{ ft } (11.2 \text{ m})$$

$$h_y = 24.5 \text{ ft } (7.47 \text{ m})$$

The relative seismic displacement, D_p, that the stair must be designed to accommodate is determined by ASCE/SEI Equation (13.3-10):

$$D_p = \frac{(h_x - h_y)\Delta_{aA}}{h_{sx}} = \frac{(36.75 - 24.5) \times 2.9}{12.25} = 2.9 \text{ in. } (75 \text{ mm})$$

The seismic relative displacement, D_{pI}, is equal to D_p times I_e [see ASCE/SEI Equation (13.3-8)]:

$$D_{pI} = 2.9 \times 1.0 = 2.9 \text{ in. } (75 \text{ mm})$$

The sliding connection at the bottom of the stair must be provided with restraints so that the displacement perpendicular to the plane of the stair risers does not exceed D_{pI} (this displacement can act in any direction, so the connection must be able to accommodate a total range equal to $2D_{pI}$). If restraints are not provided at the bottom of the stair, the connection must be designed to accommodate a displacement equal to $1.5D_{pI} = 1.5 \times 2.9 = 4.4$ in. (112 mm) [ASCE/SEI 13.5.10(b)].

If sliding or ductile connections are not provided to accommodate seismic relative displacements, the stair must be included in the structural model of the parking structure as prescribed in ASCE/SEI 12.7.3 and it must be designed using $\Omega_0 = 2.5$ of the special reinforced concrete shear walls, which are the SFRS of the parking garage (see the exception in ASCE/SEI 13.5.10).

6.7.26 Example 6.26—Nonbuilding Structure Similar to a Building

Determine the seismic forces on the nonbuilding structure in Figure 6.72. Design data are given in Table 6.58. The operating weight of each transformer is equal to 30 kips (133.5 kN). The transformers need not remain operational after a design seismic event.

SOLUTION

Based on its framing system, this nonbuilding structure is similar to a building.

Flowchart 6.10 in Section 6.6 of this book is used to determine the seismic design forces.

Step 1—Check if the nonbuilding structure is sensitive to vertical ground motion

Because this nonbuilding structure is not any of the nonbuilding types listed in ASCE/SEI 15.1.4, it is not sensitive to vertical ground motion, so the requirements in ASCE/SEI 15.1.4 need not be satisfied.

Step 2—Check if the nonbuilding structure is connected by nonstructural components to any adjacent structures

This nonbuilding structure is a standalone structure and is not connected to any adjacent structures, so the requirements in ASCE/SEI 15.2 need not be satisfied.

Step 3—Determine S_{DS}, S_{D1}, and the SDC

Given the latitude and longitude of the site, risk category II and the site class, $S_{DS} = 1.63$ and $S_{D1} = 0.80$ from Reference 6.1.

The SDC is determined using Flowchart 6.2 in Section 6.6 of this book.

Step 3a—Determine the risk category in accordance with IBC Table 1604.5

The risk category is II for this nonbuilding structure.

Figure 6.72 Plan and Elevation of the Nonbuilding Structure in Example 6.26

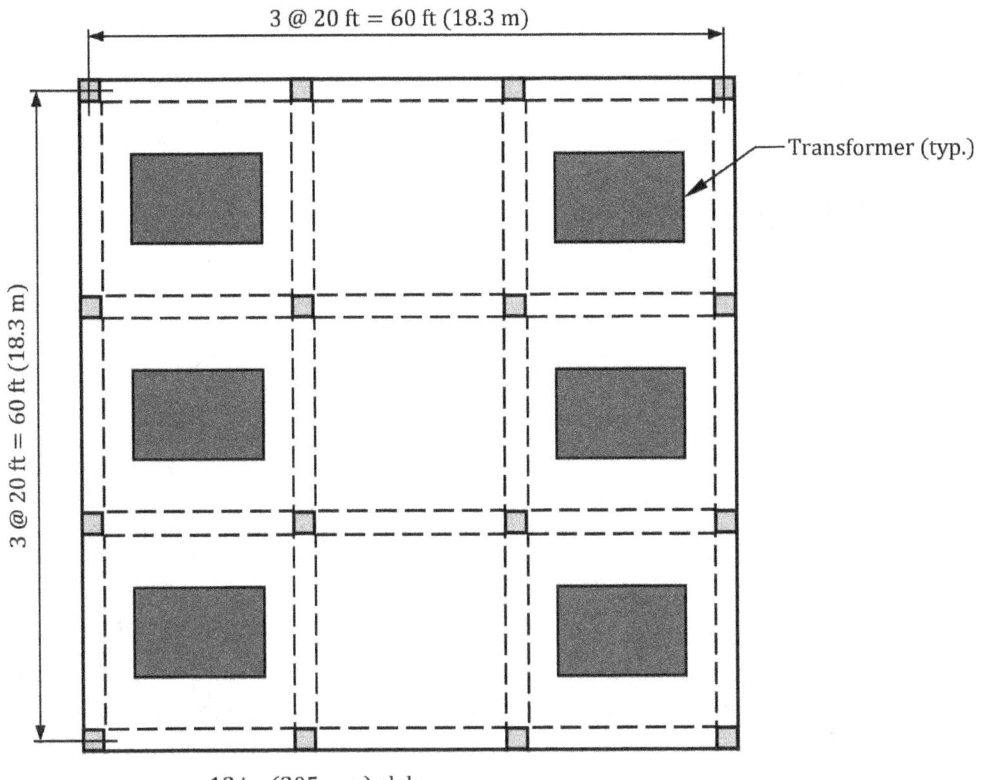

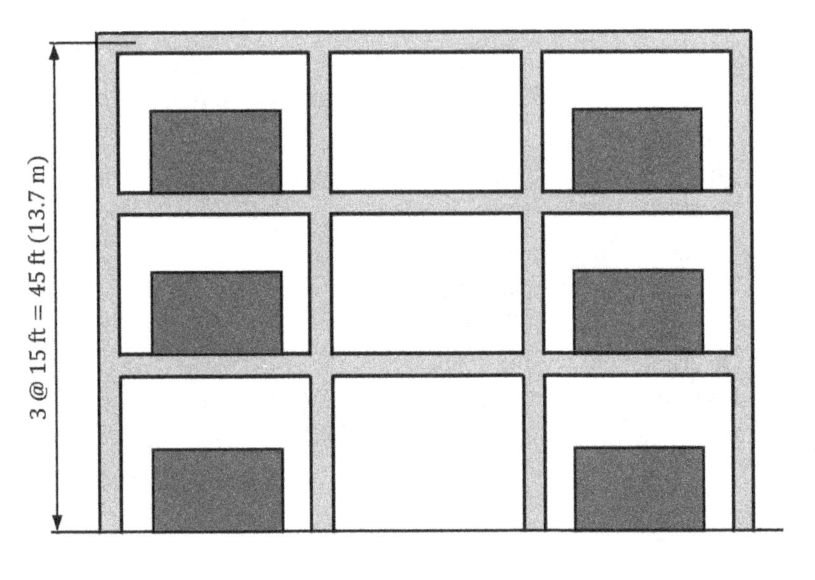

Table 6.58 Design Data for the Nonbuilding Structure in Example 6.26

Location	Latitude: 37.870°, Longitude: −122.280°
Soil classification	Site class C
Material	Cast-in-place, reinforced concrete
SFRS	Special reinforced concrete moment frames in both directions
Superimposed dead load on all levels	20 lb/ft² (0.96 kN/m²)

Step 3b—Determine the site class in accordance with ASCE/SEI 11.4.2
The site class is C. (see Table 6.58).

Step 3c—Determine the acceleration parameters S_1, S_{DS}, and S_{D1} from Reference 6.1
For the given latitude and longitude of the site, risk category II and site class C, the multi-period design spectrum is available from Reference 6.1 where $S_1 = 0.84$, $S_{DS} = 1.63$, and $S_{D1} = 0.80$.

Step 3d—Determine if the building can be assigned to SDC F
Because the risk category is II, the building is not assigned to SDC F.

Step 3e—Determine if the building can be assigned to SDC E
Because $S_1 = 0.80 > 0.75$, the building is assigned to SDC E.

Step 4—Determine the period, T
A three-dimensional analysis of the nonbuilding structure was performed using Reference 6.13. From the analysis, the fundamental period, T, is equal to 0.63 s.

Step 5—Obtain R, Ω_0, and C_d
For special reinforced concrete moment frames $R = 8$, $\Omega_0 = 3$, and $C_d = 5.5$ with no height limits (see ASCE/SEI Table 15.4-1 for nonbuilding structures similar to buildings).

Step 6—Determine the design seismic forces
According to ASCE/SEI 15.1.3, structural analysis procedures for nonbuilding structures similar to buildings must be selected in accordance with ASCE/SEI 12.6. The equivalent lateral force procedure in ASCE/SEI 12.8 is selected for this structure.
Flowchart 6.5 is used to determine the design seismic forces.

Step 6a—Obtain the response modification coefficient, R
The response modification coefficient, R, is equal to 8 for a special reinforced concrete moment frame system (ASCE/SEI Table 15.4-1).

Step 6b—Obtain the importance factor, I_e
The importance factor, I_e, is equal to 1.00 for a risk category II structure (ASCE/SEI Table 1.5-2).

Step 6c—Determine the period, T
From step 4, the fundamental period, T, is equal to 0.63 s.

Step 6d—Determine the approximate period, T_a
The approximate period, T_a, is determined by ASCE/SEI Equation (12.8-8):

$$T_a = C_t h_n^x$$

where $C_t = 0.016$ (0.0466) and $x = 0.9$ for concrete moment-resisting frames.
Therefore,

$$T_a = 0.016 \times (45.0)^{0.9} = 0.49 \text{ s}$$

In S.I.:

$$T_a = 0.0466 \times (13.7)^{0.9} = 0.49 \text{ s}$$

Step 6e—Determine C_u

The coefficient for the upper limit on the calculated period, C_u, is determined from ASCE/SEI Table 12.8-1 based on S_{D1}.

For $S_{D1} = 0.80 > 0.4$, $C_u = 1.4$.

Step 6f—Determine if $T > C_u T_a$

From step 4, $T = 0.63$ s.

From steps 6d and 6e, $C_u T_a = 1.4 \times 0.49 = 0.69 \text{ s} > T$.

Therefore, use $T = 0.63$ s in the determination of the seismic forces.

Step 6g—Obtain the long-period transition period, T_L

From Reference 6.1, $T_L = 8$ s.

Step 6h—Determine the permitted method to determine C_s

Because the design spectral acceleration parameter, S_a, is available from Reference 6.1 for a period, $T = 0.63$ s, Method 1 or Method 2 in ASCE/SEI 12.8.1.1 is permitted to be used to determine C_s. Method 1 is selected in this example.

Step 6i—Determine the design spectral acceleration parameter, S_a

From Reference 6.1, $S_a = 1.32$ at $T = 0.50$ s and $S_a = 1.01$ at $T = 0.75$ s. From linear interpolation, $S_a = 1.16$ at $T = 0.63$ s.

Maximum $S_a = 1.81$ occurs at $T = 0.25$ s, which is less than the period of the structure, $T = 0.63$ s. Therefore, use $S_a = 1.16$ in the determination of C_s.

Step 6j—Determine the spectral response acceleration parameter at a period of 1 s, S_1

From Reference 6.1, $S_1 = 0.84$.

Step 6k—Determine the seismic response coefficient, C_s

Because $S_1 = 0.84 > 0.6$, C_s is determined as follows:

$$C_s = \text{greatest of} \begin{cases} S_a/(R/I_e) = 1.16/(8/1.00) = 0.145 \\ 0.5S_1/(R/I_e) = (0.5 \times 0.84)/(8/1.00) = 0.053 \\ 0.044 S_{DS} I_e = 0.044 \times 1.63 \times 1.00 = 0.072 \\ 0.01 \end{cases}$$

Step 6l—Determine the effective seismic weight, W, in accordance with ASCE/SEI 12.7.2

The effective seismic weight, W, is the summation of the effective dead loads at each level and includes the weight of the structural members, superimposed dead loads, and the weight of the transformers (details of the weight calculations are not shown here).

The story weights are given in Table 6.59.

Table 6.59 Seismic Forces and Seismic Story Shears for the Nonbuilding Structure in Example 6.26

Level	Story Weight, w_x (kips)	Height, h_x (ft)	$w_x h_x^k$	Lateral Force, F_x (kips)	Story Shear, V_x (kips)
R	918	45.0	53,924	206.8	206.8
3	989	30.0	37,646	144.3	351.2
2	989	15.0	17,931	68.8	419.9
Σ	2,896		109,501	419.9	

1 ft = 0.3048 m; 1 kip = 4.4482 kN.

Step 6m—Determine the seismic base shear, V

The seismic base shear, V, is determined by ASCE/SEI Equation (12.8-1):

$$V = C_s W = 0.145 \times 2{,}896 = 419.9 \text{ kips}$$

In S.I.:

$$V = C_s W = 0.145 \times 12{,}882.1 = 1{,}867.9 \text{ kN}$$

Step 6n—Determine the exponent related to the structure period, k

Because $0.5 \text{ s} < T = 0.63 \text{ s} < 2.5 \text{ s}$, $k = 0.75 + 0.5T = 1.07$.

Step 6o—Determine the lateral seismic force, F_x, at each level

Lateral seismic forces F_x are determined by ASCE/SEI Equations (12.8-12) and (12.8-13) and are given in Table 6.59 along with the story shears, V_x.

For example, at the second level:

$$C_{vx} = \frac{w_x h_x^k}{\sum w_i h_i} = \frac{989 \times (15.0)^{1.07}}{109{,}501} = 0.164$$

$$F_x = C_{vx} V = 0.164 \times 419.9 = 68.8 \text{ kips}$$

In S.I.:

$$C_{vx} = \frac{4{,}399.3 \times (4.57)^{1.07}}{136{,}348} = 0.164$$

$$F_x = 0.164 \times 1{,}867.9 = 306.3 \text{ kN}$$

6.7.27 Example 6.27—Nonbuilding Structure Similar to a Building

Determine the seismic forces on the nonbuilding structure supporting the storage bin in Figure 6.73. Dashed lines represent the location of braced frames. Design data are as follows: $S_1 = 0.60$, $S_{DS} = 1.00$, $S_{D1} = 0.68$, $T_L = 12$ s, and the seismic design category is D.

SOLUTION

Based on its framing system, this nonbuilding structure is similar to a building.

Flowchart 6.10 in Section 6.6 of this book is used to determine the seismic design forces.

Step 1—Check if the nonbuilding structure is sensitive to vertical ground motion

Because this nonbuilding structure is not any of the nonbuilding types listed in ASCE/SEI 15.1.4, it is not sensitive to vertical ground motion, so the requirements in ASCE/SEI 15.1.4 need not be satisfied.

Step 2—Check if the nonbuilding structure is connected by nonstructural components to any adjacent structures

This nonbuilding structure is a standalone structure and is not connected to any adjacent structures, so the requirements in ASCE/SEI 15.2 need not be satisfied.

Step 3—Determine S_{DS}, S_{D1}, and the SDC

From the design data, $S_{DS} = 1.00$, $S_{D1} = 0.68$, and the SDC is D.

Step 4—Determine the period, T

In lieu of a more rigorous analysis, T is determined by ASCE/SEI Equation (15.4-6):

$$T = 2\pi \sqrt{\frac{\sum w_i \delta_i^2}{g \sum f_i \delta_i}}$$

Figure 6.73
The Nonbuilding Structure in Example 6.27

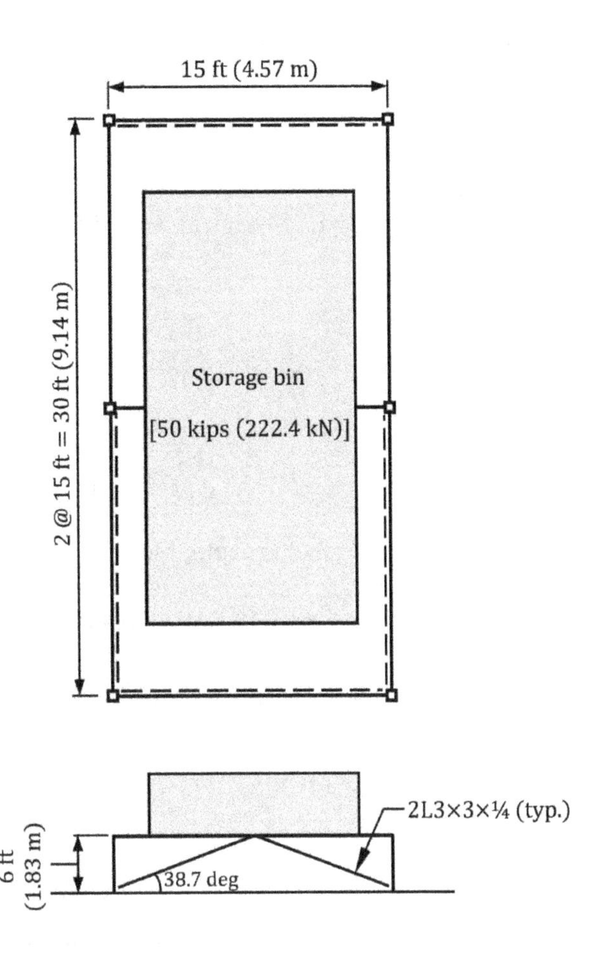

The above equation can be reduced to the following for nonbuilding structures with one level:

$$T = 2\pi\sqrt{\frac{W}{gk}}$$

where k is the lateral stiffness of the structure.

The stiffness can be determined by applying a unit horizontal load to the top of the structural frame; this load does not produce any load in the columns. Assuming the elastic shortening of the beams are negligible, only the braces contribute to that lateral stiffness of the structure in either direction of analysis.

From statics, the axial force, u, in each brace due to the unit horizontal load (which is shared between the two sets of braced frames in each direction of analysis) is equal to the following:

$$u = \frac{(1.0/2)}{2 \times \cos 38.7} = 0.32$$

The horizontal deflection, δ, due to the unit horizontal load can be obtained from the following equation (which is from the virtual work method):

$$\delta = \frac{\sum u^2 L}{AE}$$

where L = brace length = $\sqrt{7.5^2 + 6.0^2} = 9.61$ ft (2.93 m)
A = area of the 2L3 × 3 × ¼ braces = 2.88 in.^{2} (1,858 mm^2)
E = modulus of elasticity of the braces = 29,000 kips/in.^{2} (200 kN/mm^2)

Therefore, for four braces in the direction of analysis:

$$\delta = \frac{\sum u^2 L}{AE} = \frac{4 \times 0.32^2 \times (9.61 \times 12)}{2.88 \times 29{,}000} = 0.00057 \text{ in. } (0.01448 \text{ mm})$$

$$k = \frac{1}{\delta} = 1{,}754 \text{ kips/in. } (307{,}172 \text{ kN/m})$$

Thus,

$$T = 2\pi\sqrt{\frac{w}{gk}} = 2\pi\sqrt{\frac{50.0}{386.1 \times 1{,}754}} = 0.05 \text{ s}$$

In. S.I.:

$$T = 2\pi\sqrt{\frac{w}{gk}} = 2\pi\sqrt{\frac{222.4}{9.8 \times 307{,}172}} = 0.05 \text{ s}$$

The weight of the steel is negligible compared to the weight of the storage bin, so it is not included in w.

Because $T = 0.05 \text{ s} < 0.06 \text{ s}$, the nonbuilding structure is defined as rigid.

Step 4—Determine the design seismic forces

For rigid nonbuilding structures, the design seismic base shear, V, must be determined by ASCE/SEI Equation (15-4.5):

$$V = 0.30 S_{DS} W I_e = 0.30 \times 1.00 \times 50.0 \times 1.0 = 15.0 \text{ kips}$$

In S.I.:

$$V = 0.30 \times 1.00 \times 222.4.0 \times 1.0 = 66.7 \text{ kN}$$

Because this is a one-story structure, this base shear is applied as a concentrated force at the top of the braced frame.

6.7.28 Example 6.28—Nonbuilding Structure Not Similar to a Building

Determine the seismic forces on the 8-in. (203-mm) thick ground-supported reinforced concrete privacy wall in Figure 6.74. The wall is anchored into a continuous spread footing. Design data are as follows: $S_1 = 0.34$, $S_{DS} = 0.73$, $S_{D1} = 0.45$, $T_L = 12$ s, and the seismic design category is D. The wall weighs 100 lb/ft² (4.79 kN/m²).

SOLUTION

This nonbuilding structure is not similar to a building.

Flowchart 6.10 in Section 6.6 of this book is used to determine the design seismic forces.

Step 1—Check if the nonbuilding structure is sensitive to vertical ground motion

Because this nonbuilding structure is not any of the nonbuilding types listed in ASCE/SEI 15.1.4, it is not sensitive to vertical ground motion, so the requirements in ASCE/SEI 15.1.4 need not be satisfied.

Step 2—Check if the nonbuilding structure is connected by nonstructural components to any adjacent structures

This nonbuilding structure is a standalone structure and is not connected to any adjacent structures, so the requirements in ASCE/SEI 15.2 need not be satisfied.

Step 3—Determine S_{DS}, S_{D1}, and the SDC

From the design data, $S_{DS} = 0.73$, $S_{D1} = 0.45$, and the SDC is D.

Figure 6.74 The Reinforced Concrete Privacy Wall in Example 6.28

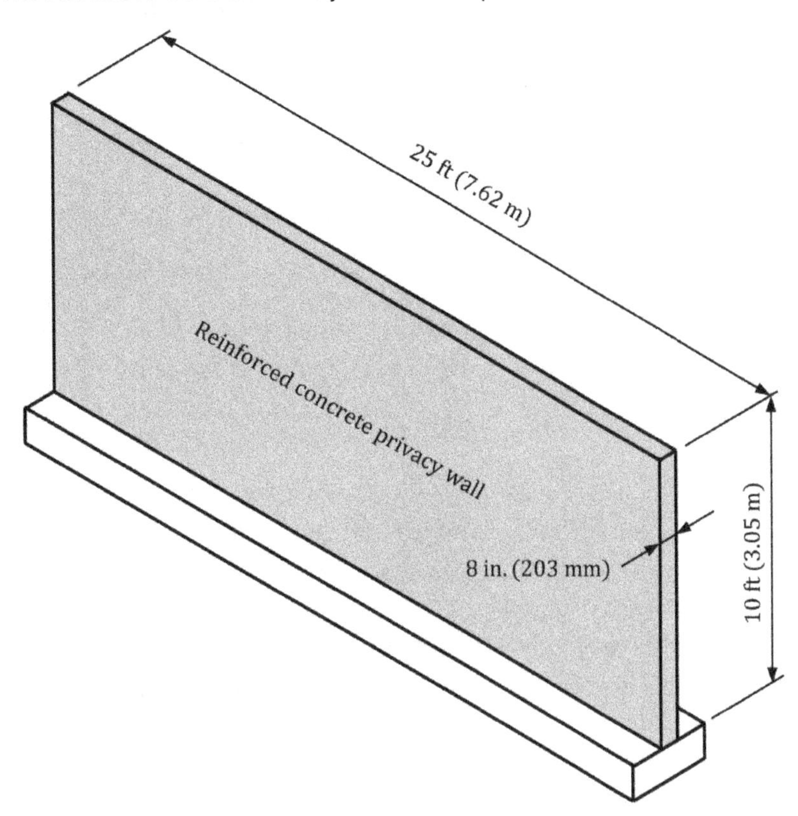

Step 4—Determine the period, T

From a dynamic analysis of the cantilever wall, $T = 0.20$ s > 0.06 s.

Step 5—Obtain R, Ω_0, and C_d

For ground-supported cantilever walls, $R = 1.25$, $\Omega_0 = 2$, and $C_d = 2.5$ (see ASCE/SEI Table 15.4-2 for nonbuilding structures not similar to buildings).

Step 6—Determine the design seismic forces

According to ASCE/SEI 15.1.3, the equivalent lateral force procedure in ASCE/SEI 12.8 is permitted to be used to determine the design seismic forces for nonbuilding structures not similar to buildings subject to the requirements in ASCE/SEI 15.4. Method 2 is used to determine the design seismic forces.

Flowchart 6.5 in Section 6.6 of this book is used to determine the design seismic forces.

Step 6a—Determine T_S

$$T_S = S_{D1}/S_{DS} = 0.45/0.73 = 0.62 \text{ s}$$

Step 6b—Determine C_s

Because $T = 0.20$ s $< T_S = 0.62$ s, C_s is the greatest of the following:

$$C_s = \text{greatest of} \begin{cases} S_{DS}/(R/I_e) = 0.73/(1.25/1.00) = 0.584 \\ 0.044 S_{DS} I_e = 0.044 \times 0.730 \times 1.0 = 0.032 \\ 0.03 \text{ [see ASCE/SEI 15.4.1(2)]} \end{cases}$$

Step 6c—Determine the effective seismic weight, W

The effective seismic weight, W, is the weight of the wall:

$$W = 100 \times 10.0 \times 25.0/1{,}000 = 25.0 \text{ kips}$$

In S.I.: $W = 4.79 \times 3.05 \times 7.62 = 111.3$ kN

Step 6d—Determine the seismic base shear, V

The seismic base shear, V, is determined by ASCE/SEI Equation (12.8-1):

$$V = C_s W = 0.584 \times 25.0 = 14.6 \text{ kips}$$

In S.I.: $V = 0.584 \times 111.3 = 65.0$ kN

This horizontal force is applied at the center of mass of the wall.

6.8 References

6.1. American Society of Civil Engineers (ASCE). 2022. ASCE 7 Hazard Tool. https://asce7hazardtool.online/. (Accessed May 24, 2023).

6.2. Building Seismic Safety Council (BSSC). 2020. *NEHRP Recommended Seismic Provisions for New Buildings and Other Structures*, FEMA P-2082-1. Federal Emergency Management Agency, Washington, D.C.

6.3. Building Seismic Safety Council (BSSC). 2010. *Earthquake-Resistant Design Concepts—An Introduction to the NEHRP Recommended Seismic Provisions for New Buildings and Other Structures*, FEMA P-749. Federal Emergency Management Agency, Washington, D.C.

6.4. International Code Council (ICC). 2024. *International Existing Building Code*, Washington, D.C.

6.5. Applied Technology Council (ATC). 2021. *Seismic Design of Rigid Wall-Flexible Diaphragm Buildings—An Alternative Procedure*, FEMA P-1026, Second edition, Federal Emergency Management Agency, Washington, D.C.

6.6. Applied Technology Council (ATC). 2018. *Assessing Seismic Performance of Buildings with Configuration Irregularities—Calibrating Current Standards and Practices*, FEMA P-2012, Federal Emergency Management Agency, Washington, D.C.

6.7. Stewart, J., Crouse, C., Hutchinson, T., Lizundia, B., Naeim, F., and Ostadan, F. (2012). *Soil-Structure Interaction for Building Structures*, Grant/Contract Reports (NISTGCR), National Institute of Standards and Technology, Gaithersburg, MD, [online], https://www.nist.gov/publications/soil-structure-interaction-building-structures (Accessed May 24, 2023).

6.8. Chopra, A.K. 2017. *Dynamics of Structures*, Prentice Hall, Upper Saddle River, NJ.

6.9. American Society of Civil Engineers (ASCE). 2017. *Seismic Analysis of Safety-Related Nuclear Structures*, Reston, VA.

6.10. American Concrete Institute (ACI). 2019. *Building Code Requirements for Structural Concrete (ACI 318-19) and Commentary (ACI 318R-19)*, Farmington Hills, MI.

6.11. American Concrete Institute (ACI). 2018. *Code Requirements for the Design of Precast Concrete Diaphragms for Earthquake Motions (ACI 550.5) and Commentary (ACI 550.5R)*, Farmington Hills, MI.

6.12. The Masonry Society (TMS). 2016. *Building Code Requirements and Specification for Masonry Structures*, TMS 402/602-16, Longmont, CO.

6.13. Computers and Structures, Inc. (CSI). 2021. *ETABS–Integrated Analysis, Design and Drafting of Building Systems*, Walnut Creek, CA.

6.9 Problems

6.1. A typical floor plan of a five-story residential building is shown in Figure 6.75. Determine the SDC of the structure for the cities identified in Table 6.60 assuming site class D.

Figure 6.75 Typical Plan of the Five-Story Residential Building in Problem 6.1

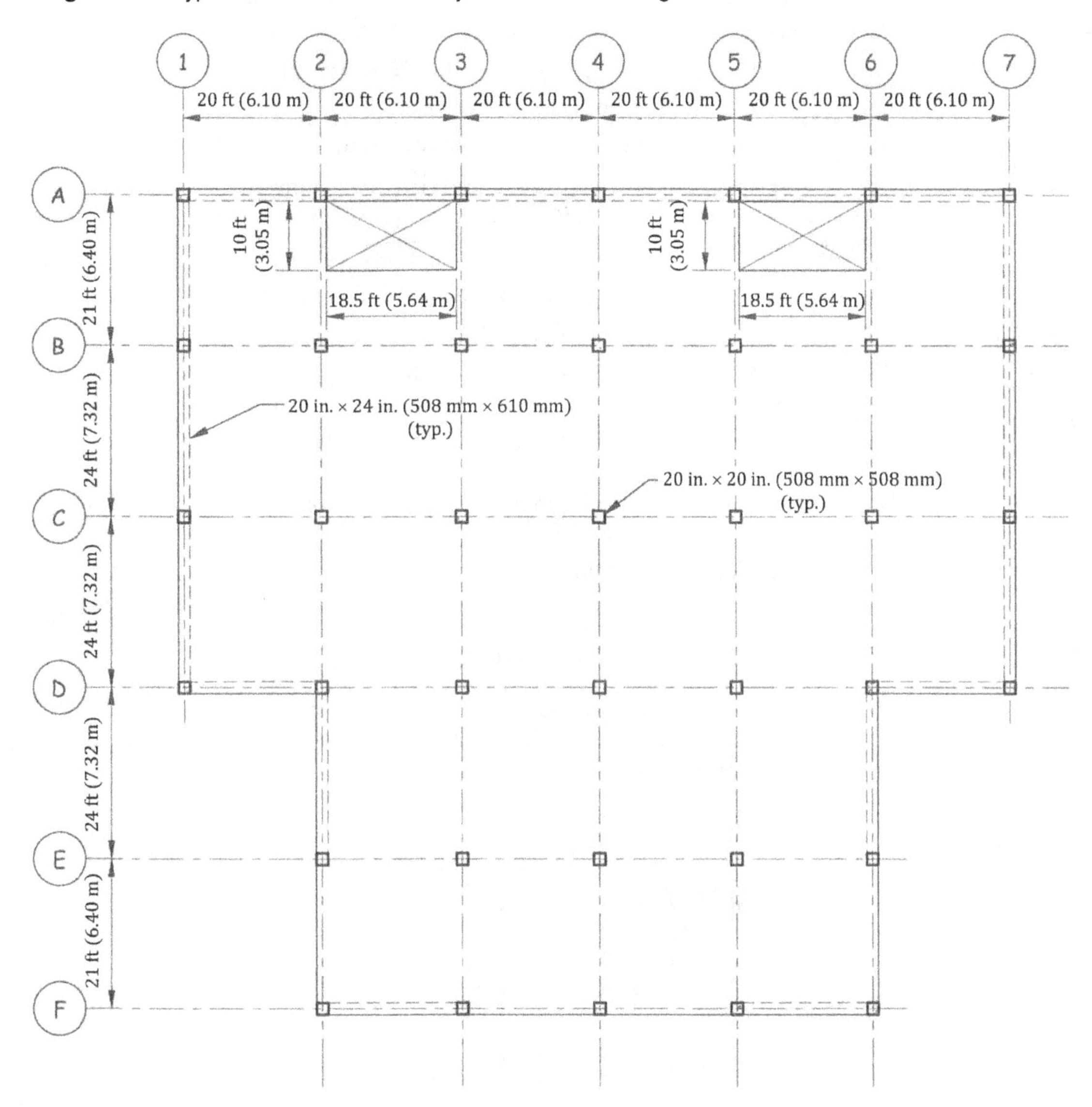

Table 6.60 Locations for Problem 6.1

City	Latitude	Longitude
Berkeley, CA	37.87	−122.28
Boston, MA	42.36	−71.06
Chicago, IL	41.88	−87.63
Denver, CO	39.74	−104.99
Houston, TX	29.76	−95.37
New York, NY	40.72	−74.00

6.2. Given the information in Problem 6.1, determine V and F_x in both directions using the equivalent lateral force procedure for the building located in Berkeley, CA. The first-story height is 12 ft (3.66 m) and typical floor heights are 10 ft (3.05 m). Assume a 9-inch-thick (229-mm thick) reinforced concrete slab at all levels and that all concrete is normal weight [density = 150 lb/ft^3 (23.6 kN/m^3)]. Also assume a superimposed dead load of 30 lb/ft^2 (1.44 kN/m^2) on all floors and 10 lb/ft^2 (0.48 kN/m^2) on the roof. A glass curtain wall system is used on all faces of the building, which weighs 8 lb/ft^2 (0.38 kN/m^2). The SFRS consists of moment-resisting frames in both directions.

6.3. Repeat Problem 6.2 for the building located in New York, NY.

6.4. Given the information in Problems 6.1 and 6.2, determine the diaphragm forces for the building located in New York, NY, using the requirements in ASCE/SEI 12.10.1.

6.5. Given the information in Problems 6.1 and 6.2, determine the diaphragm forces for the building located in New York, NY, using the requirements in ASCE/SEI 12.10.3.

6.6. A typical floor in a 20-story structural steel-framed office building is shown in Figure 6.76. Determine V and F_x in both directions using Method 2 of the equivalent lateral force procedure given the following design data:

- $S_s = 0.46$, $S_1 = 0.20$, $S_{MS} = 0.414$, $S_{M1} = 0.160$, $S_{DS} = 0.276$, $S_{D1} = 0.107$ and $T_L = 6$ s
- Site Class B
- Dead load per floor (includes superimposed dead load and cladding) = 1,000 kips (4,448 kN)
- Dead load on roof (includes mechanical equipment and cladding) = 900 kips (4,003 kN)
- All story heights are 13 ft (3.96 m)
- SFRS: steel concentrically braced frames as indicated in Figure 6.76

Figure 6.76 Typical Plan of the 20-Story Office Building in Problem 6.6

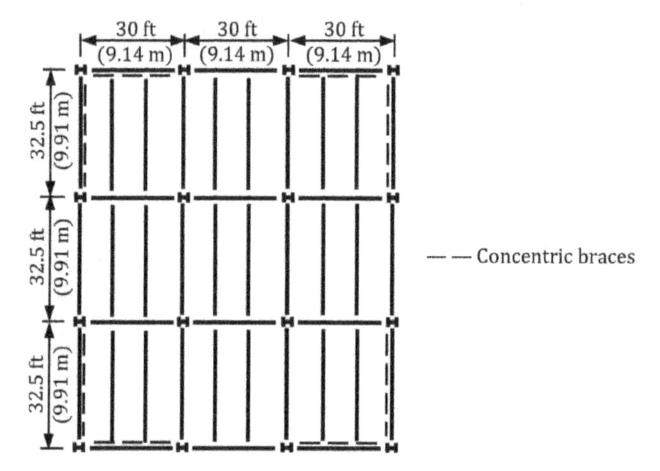

6.7. The roof of a one-story commercial building is shown in Figure 6.77. Determine if the simplified alternative method of ASCE/SEI 12.14 can be used to determine the seismic forces given the following design data:

- $S_s = 0.28$ and $S_1 = 0.10$
- Site Class C
- All walls are 10-in. thick (254-mm) and are made of the same concrete
- Roof structure consists of open-web joists and bare steel deck [weight = 20 lb/ft^2 (0.96 kN/m^2), which includes superimposed dead loads].

Figure 6.77 Roof Plan of the One-Story Commercial Building in Problem 6.7

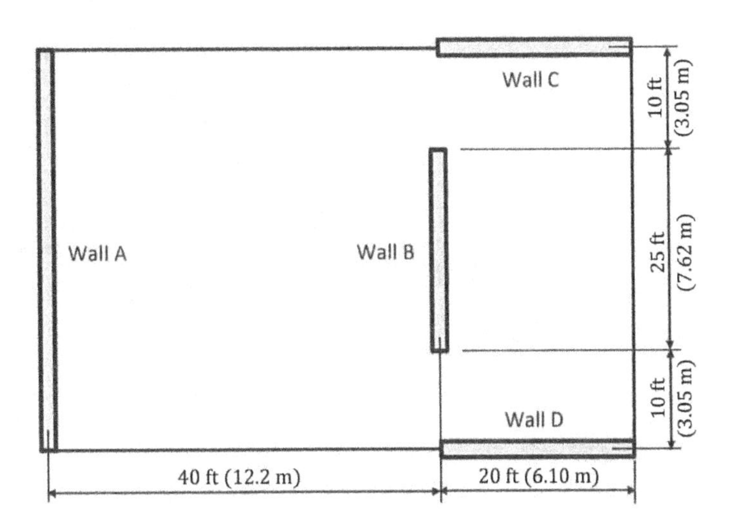

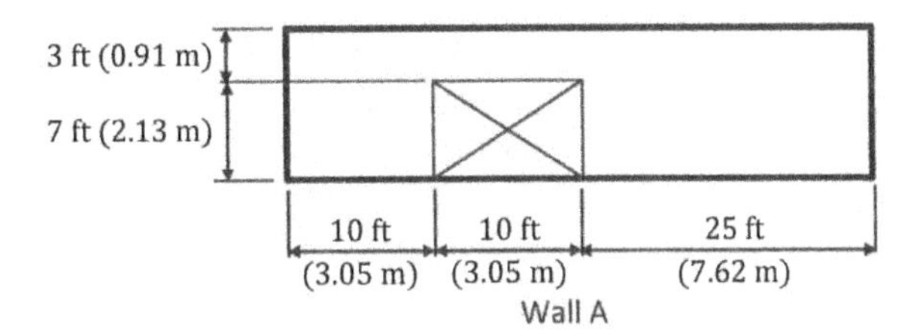

Determine the seismic base shear assuming the simplified alternative method can be used. For analysis purposes, assume the walls are fixed at the top and bottom.

6.8. Given the information in Problem 6.7, determine the diaphragm forces using the requirements in ASCE/SEI 12.10.4 assuming the following: (1) $S_{DS} = 0.26$ and $S_{D1} = 0.10$, (2) (2) Wall B is at the perimeter of the building (i.e, it is located 20 ft (6.10 m) to the right of where it is located in Figure 6.77), and (3) bare steel deck does not meet the special seismic detailing requirements of AISI S400.

6.9. A 15-kip (66.7-kN) cooling tower that is braced below its center of mass is supported on the roof of a multistory office building with a SFRS consisting of a dual system with special reinforced concrete shear walls and special reinforced concrete moment frames. The roof elevation is 200 ft (61.0 m) above the base of the building. Determine the design seismic force using $S_{DS} = 1.1$. Assume the weight of the cooling tower is less than 0.2 times the combined effective seismic weight and the approximate fundamental period of the supporting building is 2 s.

6.10. A glass curtain wall panel spans 12 ft (3.66 m) between floor levels in an office building with a SFRS consisting of special steel moment-resisting frames. It has been determined that the deflections at the top and bottom of one of the floors are $\delta_{xA} = 1.2$ in. (31 mm) and $\delta_{yA} = 0.8$ in. (20 mm), respectively. These deflections occur 48 ft (14.6 m) and 36 ft (11.0 m) above the base of the structure, respectively. Determine the seismic relative displacement.

CHAPTER 7

Flood and Tsunami Loads

7.1 Introduction

7.1.1 Overview of Flood and Tsunami Loads

In general, *flooding* is the overflow of excess water from a body of water (river, stream, lake, ocean, etc.) onto adjoining land. Depending on local topography, one or more bodies of water can contribute to flooding at a particular site. Storms usually generate the most significant flood hazards in areas other than those where tsunamis need to be considered.

Flooding is defined in IBC 202 as a general or temporary condition of partial or complete inundation of normally dry land due to the following:

1. The overflow of inland or tidal waters; or
2. The unusual or rapid accumulation or runoff of surface waters from any source.

Along coastlines and the shorelines of large lakes, flooding is caused by wind-driven surges and waves that push water onshore, while along streams and rivers, flooding results from the accumulation of rainfall runoff that drains from upland watersheds.

To correctly characterize the potential hazards caused by floodwaters, the following must be investigated at any site regardless of the source:

- Origin of flooding
- Frequency of flooding
- Depth of floodwaters
- Velocity of floodwaters
- Direction of floodwaters
- Duration of flooding
- Effects due to waves
- Effects due to erosion and scour
- Effects due to flood-borne debris

Buildings can be subjected to a variety of load types that can be exerted on buildings due to flooding. The effects of these loads can be intensified by erosion and scour, which can lower the ground surface around foundations, causing loss of bearing capacity and resistance to uplift due to lateral loads. Because it is not always possible to quantify each of the above parameters for a particular site, conservative estimates need to be made to define the corresponding flood loads.

A *tsunami* is basically a series of waves with variable long periods, which typically result from tsunamigenic earthquakes. These types of earthquakes occur along major subduction zone plate boundaries, such as those bordering the Pacific Ocean, where a tectonic plate is subducting under an adjoining plate (see ASCE/SEI Figure C6.7-2). Tsunamis can also be generated by landslides near a coast or by underwater and undersea volcanic eruptions.

The effects on structures due to this coastal flooding hazard can be catastrophic, especially for one- and two-family dwellings and low-rise buildings of light-frame construction. According to ASCE/SEI C6.1, it is not practical to include tsunami design requirements for these types of low-rise structures because, based on forensic engineering surveys of major tsunami events of the past decade, such structures do not survive significant tsunami loading.

Tsunami loads and effects on structures are based on inundation depth and flow velocities of water during inflow and outflow stages at the site. The types of loads that must be considered are similar to those required for flooding. Because the characteristics of future tsunamis cannot be adequately determined using only historical records, a probabilistic design approach is used instead of a deterministic one.

7.1.2 Overview of Code Requirements

All structures and portions of structures located in flood hazard areas must be designed and constructed to resist the effects of flood hazards and flood loads (IBC 1612.1). The design and construction of buildings and structures located in flood hazard areas, including coastal high hazard areas and Coastal A Zones, must be in accordance with ASCE/SEI Chapter 5 and Reference 7.1 (IBC 1612.2).

Where a building or structure is located in more than one flood zone or is partially located in a flood zone, the entire building or structure must be designed and constructed according to the requirements of the more restrictive flood zone.

The design and construction of risk category III and IV buildings and structures located in the tsunami design zones defined in the Tsunami Design Geodatabase (Reference 7.3) must be in accordance with ASCE/SEI Chapter 6 (IBC 1615.1). Tsunami vertical evacuation planning criteria are given in IBC Appendix M for those coastal communities where a tsunami hazard, as shown in a tsunami design zone map, is evident.

Hazards and loads that must be considered for structures located in flood hazard areas and in tsunami-prone regions are given in the following sections.

7.2 Flood Loads

7.2.1 Terminology

Terminology used throughout this section related to flood loads is given in Table 7.1.

7.2.2 Flood Hazard Areas

The first step in the design for flood loads is to determine if a building or structure is located in a flood hazard area or not. By definition, a flood hazard area is the greater of the following two areas (IBC 202):

1. The area within a floodplain that is subjected to a 1-percent or greater chance of flooding in any year.
2. The area designated as a flood hazard area on a community's flood hazard map, or otherwise legally designated.

The first of these two areas is typically acquired from flood insurance rate maps (FIRMs), which are prepared by the Federal Emergency Management Agency (FEMA) through the National Flood Insurance Program (NFIP). The NFIP is a voluntary program whose goal is to reduce the loss of life and the damage caused by floods, to help victims recover from floods, and to promote an equitable distribution of costs among those who are protected by flood insurance and the general public. Conducting flood hazard studies and providing FIRMs and flood

Table 7.1 Terminology Related to Flood Loads

500-year flood elevation	The water surface elevation that corresponds to the 0.2% annual chance flood (commonly referred to as the 500-year flood). This elevation is commonly specified in the Flood Insurance Study (FIS) or similar study.
A Zone	• A high-risk area within a special flood hazard area (SFHA) shown on flood insurance rate maps (FIRMs) that is not subjected to high-velocity wave action. • This zone is also referred to as the minimal wave action (MiWA) area where wave heights are less than 1.5 ft (0.46 m). • This zone is designated on FIRMs as A, AE, A1 through A30, A99, AR, AO, and AH. See Figure C1-3 in Reference 7.1.
Base flood	A flood that has a 1-percent chance of being equaled or exceeded in any given year. This is commonly referred to as the extent of the 100-year floodplain.
Base flood elevation (BFE)	• Height to which flood waters will rise during passage or occurrence of the base flood relative to the datum. • The shape and nature of the floodplain (ground contours and the presence of any buildings, bridges, and culverts) are used to obtain the BFE along rivers and streams. • Along coastal areas, a BFE includes wave heights and is established considering historical storm and wind patterns. See Figure C1-3 in Reference 7.1.
Breakaway wall	Any type of wall subjected to flooding not required to provide structural support to a building or structure and designed and constructed to collapse under base flood or lesser conditions in such a way that it allows free passage of floodwaters and does not damage the structure or supporting foundation system. Loads on breakaway walls are given in ASCE/SEI 5.3.3.
Breaking wave height	Wave heights of depth-limited breaking waves (i.e., wave heights limited by the depth of the water) in Coastal A Zones and V Zones.
Coastal A Zone	• An area within a SFHA landward of a V Zone or landward of an open coast without mapped V Zones (such as the shorelines of the Great Lakes). A schematic of a Coastal A Zone is given in Figure C1-1 of Reference 7.1. Also see Figure C1-3 in Reference 7.1. • This area is referred to as the moderate wave action (MoWA) area by Federal Emergency Management Agency (FEMA) flood mappers and is between the limit of moderate wave action (LiMWA) and the V Zone. • The principal source of flooding is from astronomical tides, storm surges, seiches, or tsunamis and not from riverine flooding. • Wave forces and erosion potential should be taken into consideration when designing a structure for flood loads in such zones. • Stillwater depth must be greater than or equal to 2.0 ft (0.61 m) and breaking wave heights are between 1.5 ft (0.46 m) and 3.0 ft (0.91 m) during the base flood. • This zone is not delineated by FEMA as a separate zone on FIRMs.
Coastal high hazard area (V Zone)	• An area within a SFHA with the following characteristics: 1. An area extending offshore to the inland limit of a primary frontal dune along an open coast. 2. An area subject to high-velocity wave action from storms or seismic sources. • This zone is designated on FIRMs as V, VE, VO, or V1 through V30. See Figure C1-3 in Reference 7.1.

(*continued*)

Table 7.1 Terminology Related to Flood Loads (*Continued*)

Design flood	• The base flood that is identified on the community's FIRM where a community has chosen to adopt minimum National Flood Insurance Program (NFIP) building elevation requirements. • The flood corresponding to the area designated as a flood hazard area (FHA) on a community's flood hazard map or otherwise legally designated where a community has chosen to exceed minimum NFIP building elevation requirements.
Design flood elevation (DFE)	• The elevation of the design flood, including wave height (where applicable), relative to the datum specified on a community's flood hazard map. • For communities that have adopted the minimum NFIP requirements, the DFE is identical to the BFE. • The DFE is greater than the BFE in communities that have adopted requirements that exceed minimum NFIP requirements.
Digital flood insurance rate map (DFIRM)	A FIRM produced in a digital format.
Flood or flooding	A general and temporary condition of partial or complete inundation of normally dry land from the following (IBC 202): 1. The overflow of inland or tidal waters. 2. The unusual and rapid accumulation or runoff of surface waters from any source.
Flood exposure	Flood exposures are sources of flooding and are grouped into the following categories: 1. River, riverine, fluvial flooding: Rivers, lakes, manmade drainage channels, smaller watercourses overflowing due to upstream heavy rains, melting snow, and dam releases. 2. Alluvial fan flooding: Flooding that occurs in areas at the base of steep-sloped areas. As the water exits the steep area, it fans out into the flat areas in a random manner. 3. Coastal flooding: Oceans, bays, estuaries, and rivers affected by coastal waters overflowing due to abnormal high tides, coastal storms, high winds, or tsunamis. 4. Storm water flooding: Storm water flooding is caused by an accumulation of runoff on land and paved areas from rainfall before it enters a stream, river, body of water, or a manmade drainage system.
Flood hazard boundary map (FHBM)	The first flood risk map prepared by FEMA for a community, which identifies FHAs based on the approximation of land areas in the community having a 1% or greater chance of flooding in any given year.
Flood hazard area (FHA)	An area defined as the greater of the following two areas (IBC 202): 1. The area within a floodplain subjected to a 1% or greater chance of flooding in any year (base flood); or, 2. The area designated as an FHA on a community's flood hazard map, or otherwise legally designated.
Flood insurance rate map (FIRM)	• The official map of a community prepared by FEMA through the NFIP, which delineates both the special flood hazard areas and the risk premium zones applicable to the community. • A FIRM shows FHAs along bodies of water where there is a risk of flooding by a base flood. Examples of coastal and riverine FIRMs are given in Figure C1-3 of Reference 7.1.
Flood insurance study (FIS)	• A report prepared by FEMA to document the examination, evaluation, and determination of flood hazards. • Included in a FIS are the FIRM, the flood boundary and floodway map (FBFM), the BFE and supporting technical data.
Floodway	A channel of a river, creek, or other watercourse and adjacent land areas that must be reserved to discharge the base flood waters without cumulatively increasing the water surface elevation by more than a designated height (IBC 202). A floodway schematic for a SFHA is given in Figure C1-2 of Reference 7.1.

(*continued*)

Table 7.1 Terminology Related to Flood Loads (*Continued*)

Freeboard	Additional depth of water above the BFE approved by a local jurisdiction.
High-risk flood hazard area	An FHA where one or more of the following hazards are known to occur (Reference 7.1): • Alluvial fan flooding • Flash floods • Mudslides • Ice jams • High-velocity flows • High-velocity wave action • Breaking wave heights greater than or equal to 1.5 ft (0.46 m) • Erosion
Limit of moderate wave action (LiMWA)	• The inland limit of the area expected to receive 1.5-ft (0.46-m) or greater breaking waves during the base flood. • The boundary between the MoWA and MiWA areas.
Minimal wave action (MiWA) area	• The area between the LiMWA and the landward limit of an A Zone. • Wave heights are less than 1.5 ft (0.46 m) during the base flood.
Moderate wave action (MoWA) area	The area between the LiMWA and the V Zone (see the definition for coastal A Zone).
National Flood Insurance Program (NFIP)	A program administered by FEMA that enables property owners in participating communities to purchase federally backed flood insurance as financial protection against flood losses in exchange for state and community floodplain management regulations that reduce future flood damage (Reference 7.1).
Special flood hazard area (SFHA)	The land area subject to flood hazards, which are shown on a FIRM or other flood hazard map as A Zones (A, AE, A1 through A30, A99, AR, AO, AH) or V Zones (V, VO, VE, or V1 through V30). See Figure C1-3 in Reference 7.1.
Stillwater elevation (SWEL)	• In riverine and lake areas, the BFE published in the FIS and shown on the FIRM, unless the local jurisdiction has adopted a more severe design flood. • In coastal areas, the average water level including waves, which is published in the FIS. • The SWEL must be referenced to the same datum used in establishing the BFE and the DFE.
Stillwater flood depth	The vertical distance between the eroded ground elevation and the SWEL (Reference 7.1).
V Zone	See the definition for coastal high-hazard area.
X Zone	• Formerly identified as B and C Zones on older FIRMs, an area outside of the flood hazard zone (Figure C1-3 of Reference 7.1). • X Zones are identified as follows: 1. Shaded X Zone (formerly B zones): Identifies areas subject to flooding having a 0.2% probability of being equaled or exceeded during any given year (500-year flood). 2. Unshaded X Zone (formerly C Zones): Identifies areas of minimal flood risk where the annual exceedance probability of flooding is less than 0.2%.

insurance studies (FISs) for participating communities are major activities undertaken by the NFIP. Included in FISs are the FIRM, the flood boundary and floodway map (FBFM), the base flood elevation (BFE), and supporting technical data.

A FIRM is the official map of a community on which FEMA has delineated both the special flood hazard areas and the risk premium zones that are applicable to the community. In general, FIRMs show flood hazard areas along bodies of water where there is a risk of flooding by a base flood, that is, a flood having a 1-percent chance of being equaled or exceeded in any given year. Sample FIRMs are shown in Figure C1-3 of Reference 7.1 and Figure 7.1.

Figure 7.1 Sample Flood Insurance Rate Map (Source: FEMA)

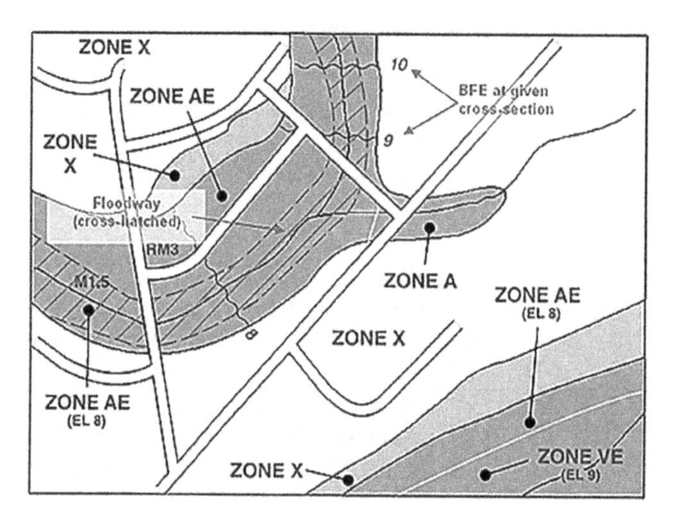

A FIRM that is produced in a digital format is designated a digital flood insurance rate map (DFIRM). A sample DFIRM is given in Figure 7.2. FIRMs and DFIRMs for specific areas can be obtained from the FEMA Flood Map Service Center website (Reference 7.2).

Figure 7.2 Sample Digital Flood Insurance Rate Map (Source: FEMA)

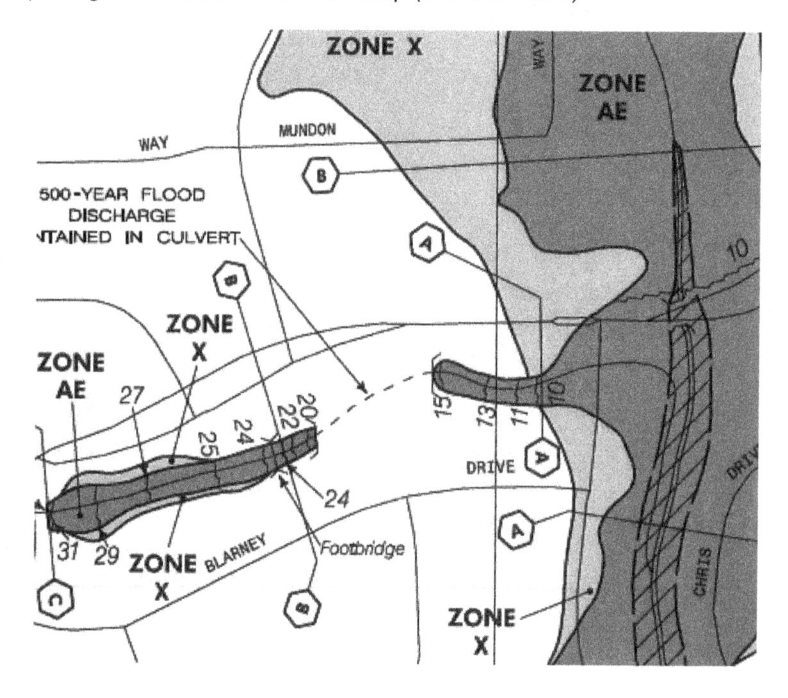

In regard to the return period for floods, the term "100-year flood" is a misnomer. Contrary to popular belief, it is not the flood that will occur once every 100 years, but the flood elevation that has a 1-percent chance of being equaled or exceeded in any given year. The "100-year flood" could occur more than once in a relatively short period of time. The flood elevation that has a 1-percent chance of being equaled or exceeded in any given year is the standard used by most government agencies and the NFIP for floodplain management and to determine the need for flood insurance.

In addition to showing the extent of flood hazards, the FIRMs and DFIRMs also show base flood elevations (BFEs) and floodways. The BFE is the height to which flood waters will rise during passage or occurrence of the base flood relative to the datum that is used on the flood hazard map. Statistical methods and computer methods that consider the shape and nature of the floodplain (ground contours and the presence of any buildings, bridges, and culverts) were used by FEMA to obtain the BFEs along rivers and streams. In such cases, the BFEs are provided next to river cross-sections on the flood hazard maps (see Figure 7.2). Along coastal areas, BFEs include wave heights and are established considering historical storm and wind patterns (see Figure 7.3). The elevations in parentheses below the zone designation are the BFEs (in feet) with respect to either the North American Vertical Datum of 1988 (NAVD 88 or NAVD) or the National Geodetic Vertical Datum of 1929 (NGVD 29 or NGVD). The FIRM identifies which datum was

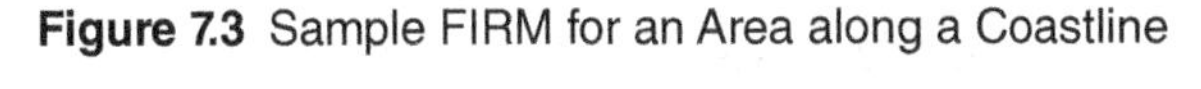

Figure 7.3 Sample FIRM for an Area along a Coastline

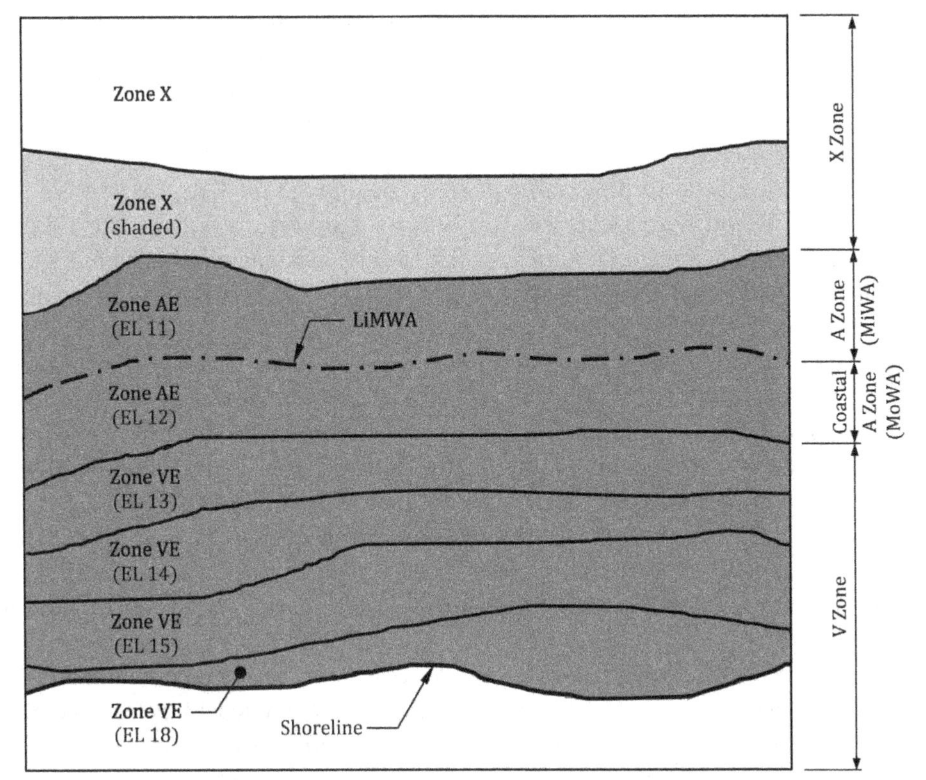

used to establish the elevation. FIRM panels can be obtained from Reference 7.2 and flood-related information needed to calculate flood loads can be acquired from Reference 7.3.

Floodways are channels of a river, creek, or other watercourse and adjacent land areas that must be reserved to discharge the base flood waters without cumulatively increasing the water surface elevation by more than a designated height. As such, floodways must be kept clear of encroachments (such as fill or buildings) so that the base flood can be discharged without increasing the water surface elevations by more than the designated height given in floodway data tables in FISs. A floodway schematic is given in Figure C1-2 of Reference 7.1.

Some local jurisdictions develop and subsequently adopt flood hazard maps that are more extensive than FEMA maps. In such cases, flood design and construction requirements must be satisfied in the areas delineated by the more extensive maps. Thus, a *design flood* is a flood associated with the greater of the area of a base flood based on a 1-percent or greater chance of flooding in any year or the area legally designated as a flood hazard area by a community.

7.2.3 Flood Hazard Zones

A flood hazard zone must be determined for any structure located in a flood hazard area. FEMA determines flood hazards at a given site based on the following:

1. Anticipated flood conditions, including the stillwater elevation, wave setup, wave runup and overtopping, and wave propagation during the base flood event (i.e., the flood level having a 1-percent chance of being equaled or exceeded in any given year).
2. Potential for storm-induced erosion during the base flood event.
3. Physical characteristics of the floodplain, including existing development and vegetation.
4. Topographic and bathymetric (underwater topography) information.

Flood hazards and water surface elevations are calculated by FEMA computer models and the results from these analyses are used to map flood hazard zones and BFEs.

FEMA flood hazard zones are given in Table 7.2. Coastal A Zones are not given by FEMA as a separate zone on FIRMs; a Coastal A Zone designation is used to facilitate application of load combinations in ASCE/SEI Chapter 2.

A shaded Zone X is an area of moderate flood hazard (areas between the limits of the 100-year and 500-year floods). On older FIRMs, Zone B corresponds to shaded Zone X and Zone C corresponds to unshaded Zone X.

Special flood hazard areas (high-risk areas) begin with the letter "A" or "V." A Zones are those areas within inland or coastal floodplains where high-velocity wave action is not expected during the base flood. In contrast, V Zones, which are also designated as coastal high hazard areas, are those areas within a coastal floodplain where high-velocity wave action from storms or seismic sources can occur during the base flood event.

A high-risk flood hazard area is an area where one or more of the following hazards may occur: alluvial fan flooding, flash floods, mudslides, ice jams, high-velocity flow, coastal wave heights greater than or equal to 1.5 ft (0.46 m) or erosion.

The concept of a Coastal A Zone is introduced in ASCE/SEI Chapter 5 and Reference 7.1 to facilitate application of load combinations in ASCE/SEI Chapter 2. IBC 1612.2 references ASCE/SEI Chapter 5 and Reference 7.1 for the design and construction of buildings and structures located in flood hazard areas. The requirements in these documents are covered below. The NFIP regulations do not differentiate between Coastal and Noncoastal A Zones.

According to ASCE/SEI 5.2, a Coastal A Zone is an area located within a special flood hazard area that is landward of a V Zone or landward of an open coast without mapped V Zones (such as the shorelines of the Great Lakes). Wave forces and erosion potential must be taken into consideration when designing a structure for the effects of flood loads in such zones.

To be classified as a Coastal A Zone, the principal source of flooding must be from astronomical tides, storm surges, seiches or tsunamis and not from riverine flooding. In addition, stillwater flood depths must be greater than or equal to 2.0 ft (0.61 m), and breaking wave heights must be greater than or equal to 1.5 ft (0.46 m) during the base flood conditions (see ASCE/SEI C5.2). Stillwater depth is the vertical distance between the ground and the stillwater elevation, which is the elevation that the surface of water would assume in the absence of waves. The stillwater elevation is referenced to the NAVD, the NGVD or other datum, and it is documented in FIRMs.

The principal sources of flooding in Noncoastal A Zones are runoff from rainfall, snowmelt, or a combination of both.

According to Section 4.1.1 of Reference 7.1, Coastal A Zones are located as follows:

- Landward of a V Zone or shoreline and seaward of the limit of moderate wave action (LiMWA) if the LiMWA is delineated on a FIRM
- Designated by the building official

Table 7.2 FEMA Flood Hazard Zones

Zone	Description
Moderate to Low-Risk Areas—X Zones	
X	These zones identify areas outside of the flood hazard area. • Shaded X Zone: areas subject to flooding having a 0.2% probability of being equaled or exceeded during any given year. • Unshaded X Zone: areas of minimal flood risk where the annual exceedance probability of flooding is less than 0.2%.
High-Risk Areas—A Zones	
A	• Detailed analyses are not performed for such areas. • No depths or BFEs are shown within these zones.
AE	• BFEs are provided. • AE Zones are used on new format FIRMs instead of A1 through A30 Zones.
A1 through A30	• These are known as numbered A Zones (e.g., A7 or A14). • Old-format FIRMs show a BFE for these zones.
A99	• Areas protected by a federal flood control system where construction has reached specified legal requirements. • No depths or BFEs are shown within these zones.
AR	Areas with a temporarily increased flood risk due to the building or restoration of a flood control system (such as a levee or a dam).
AO	• River or stream flood hazard areas and areas with a chance of shallow flooding, usually in the form of sheet flow, with an average depth ranging from 1 to 3 ft (0.30 to 0.91 m). • Average flood depths derived from detailed analyses are shown within these zones.
AH	• Areas of shallow flooding, usually in the form of a pond, with an average depth ranging from 1 to 3 ft (0.30 to 0.91 m). • BFEs are derived from detailed analyses and are shown at selected intervals within these zones.
High-Risk Areas—Coastal (V Zones)	
V	• Coastal areas with a 1% or greater chance of flooding and an additional hazard associated with storm waves. • No BFEs are shown within these zones.
VE, V1 through 30	• Coastal areas with a 1% or greater chance of flooding and an additional hazard associated with storm waves. • BFEs are derived from detailed analyses and are shown at selected intervals within these zones.

The LiMWA is defined as the line on a FIRM that indicates the inland limit of the 1.5-ft (0.46-m) breaking wave height during the base flood. FEMA indicates LiMWA lines on revised FIRMs prepared after December 2009. A schematic of a Coastal A Zone is given in Figure C1-1 of Reference 7.1. The definition of a Coastal A Zone in Reference 7.1 is the same as that in ASCE/SEI 7 and IBC 202.

It is recommended to check with the local building official for the most current information on flood hazard areas prior to designing a structure in a flood-prone area.

7.2.4 Design and Construction

According to IBC 1612.2, the design and construction of buildings and structures located in flood hazard areas must be in accordance with Chapter 5 of ASCE/SEI 7 and ASCE/SEI 24-14 (Reference 7.1). Section 1.6 of ASCE/SEI 24 requires that design flood loads and their combination with other loads must be determined by ASCE/SEI 7 (ASCE/SEI 24 references ASCE/SEI 7-10, which was the current edition of the standard at that time).

The provisions of ASCE/SEI 24 are intended to meet or exceed the requirements of the NFIP. Figures 1-1 and 1-2 in ASCE/SEI 24 illustrate the application of the standard and the application of the chapters in the standard, respectively.

The provisions in IBC Appendix G, Flood-resistant Construction, are intended to fulfill the floodplain management and administrative requirements of NFIP not included in the IBC. IBC appendix chapters are not mandatory unless they are specifically referenced in the adopting ordinance of the jurisdiction.

Other provisions related to construction in flood hazard areas worth noting are found in IBC Chapter 18. Requirements for grading and fill in flood hazard areas are given in IBC 1804.5. Raised floor buildings in flood hazard areas must have the finished grade elevation under the floor (such as at a crawlspace) to be equal to or higher than outside finished grade on at least one side (IBC 1805.1.2.1). The exception permits under-floor spaces in Group R-3 residential buildings to comply with the listed requirements instead.

7.2.5 Base Flood Elevation

The base flood elevation (BFE) is the height to which flood waters are anticipated to rise during a passage or occurrence of the base flood relative to the datum. In addition to showing the extent of the flood hazard zones, FIRMs and DFIRMs also give the BFE, which is rounded to the nearest whole foot. The BFE is typically located in parentheses below the flood hazard zone on the flood hazard map (see Figure 7.1). For riverine areas, the BFE is usually located adjacent to river cross-section locations.

Modern FIRMs contain BFEs based on the North American Vertical Datum of 1988 (NAVD 88). Historically, the most common vertical datum used by FEMA has been the National Geodetic Vertical Datum of 1929 (NGVD 29), and many existing FIRMs provide elevation values based on this old datum. Where the datums on two or more sources are different (e.g., the elevations on a FIRM are based on NAVD 88 and the elevations on an elevation certificate are based on NGVD 29), the elevations must be converted to the same datum before they are used subsequently. Differences between two datums vary from location to location and FEMA provides guidelines on how to calculate the conversions (see Reference 7.4).

7.2.6 Design Flood Elevation

The design flood elevation (DFE) is used in the determination of flood loads and minimum elevation requirements of the lowest floor of a building or structure (see Section 7.2.7 of this book).

The DFE is the elevation of the design flood including wave height (where applicable) and is equal to the following:

- For communities that have adopted minimum NFIP requirements: DFE = BFE
- For communities that have adopted requirements exceeding those from the NFIP: DFE > BFE

To account for uncertainties in the determination of flood elevations, local jurisdictions may require an additional depth of water above the BFE, which is defined as *freeboard*. Freeboard is essentially an additional factor of safety that provides an increased level of flood protection, which could reduce flood insurance premiums. Freeboard should not be included in the determination of the stillwater flood depth (see Section 7.2.8 of this book) but is used instead to raise the building to a level higher than the BFE level (see Section 7.2.7 of this book).

7.2.7 Minimum Elevation of Lowest Floor

IBC 1612.4 and ASCE/SEI 24 Sections 2.3 and 4.4 require the lowest floor in a building or structure be elevated to or above the minimum elevations identified in ASCE/SEI 24 Table 2-1 for A Zones and Table 4-1 for Coastal A Zones and V Zones; the minimum elevations are based on the flood design class, the BFE, the DFE, and the 500-year flood elevation (see Figure 7.4).

Figure 7.4 Minimum Elevation of the Lowest Floor in a Building or Structure

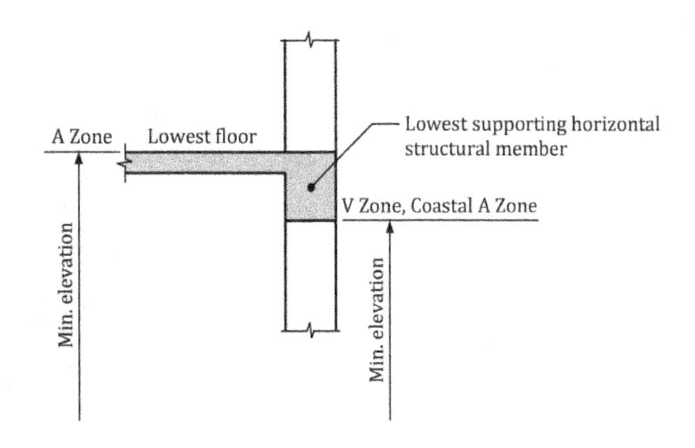

Flood Design Class	Minimum Elevation	
	A Zone	V Zone Coastal A Zone
1	DFE	DFE
2	Higher of {BFE + 1 ft DFE	Higher of {BFE + 1 ft DFE
3	Higher of {BFE + 1 ft DFE	Higher of {BFE + 2 ft DFE
4	Higher of {BFE + 2 ft DFE 500-yr FE	Higher of {BFE + 2 ft DFE 500-yr FE

1 ft = 0.3048 m
See ASCE/SEI 24 Table 1-1 for descriptions of Flood Design Class.
FE = flood elevation
See footnotes to ASCE/SEI 24 Table 2-1, which are applicable to A Zones.

Flood design classes are defined in ASCE/SEI 24 Table 1-1 and are similar to the risk categories in IBC Table 1604.5 and in ASCE/SEI Table 1.5-1.

For buildings located in an A Zone, the minimum elevation is to the top of the lowest floor whereas for buildings in Coastal A Zones or V Zones, the minimum elevation is to the bottom of the lowest supporting horizontal structural member of the lowest floor.

Enclosed areas used solely for parking of vehicles, building access, or storage are permitted to be below the DFE of elevated buildings in an A Zone provided the requirements in ASCE/SEI 24 Section 2.7 for such enclosures are satisfied (ASCE/SEI 24 Section 2.3). According to the exception in ASCE/SEI 24 Section 2.3, the lowest floor (including basements) in a nonresidential building or in nonresidential portions of a mixed-use building in an A Zone is permitted to be located below the minimum elevations in ASCE/SEI 24 Table 2-1 only if the dry floodproofing requirements in ASCE/SEI Section 6.2 are satisfied. There are no similar exceptions for buildings located in a Coastal A Zone or V Zone; all buildings must satisfy the elevation requirements in ASCE/SEI 24 Table 4-1 (ASCE/SEI 24 Section 4.4).

7.2.8 Stillwater Flood Elevation and Stillwater Flood Depth

Unless a local jurisdiction has adopted a more severe design flood, the SWEL in riverine and lake areas is equal to the BFE published in the FIS and shown on the FIRM (see Figure 7.5 for the case where DFE = BFE). In coastal areas, the SWEL is the average water level including wave setup (see Figure 7.6 for the case where DFE = BFE) and is published in the FIS (such values are not shown on FIRMs). The SWEL must be referenced to the same datum used in determining the BFE and the DFE.

Figure 7.5
Stillwater Flood Elevation and Stillwater Flood Depth for Riverine and Lake Areas

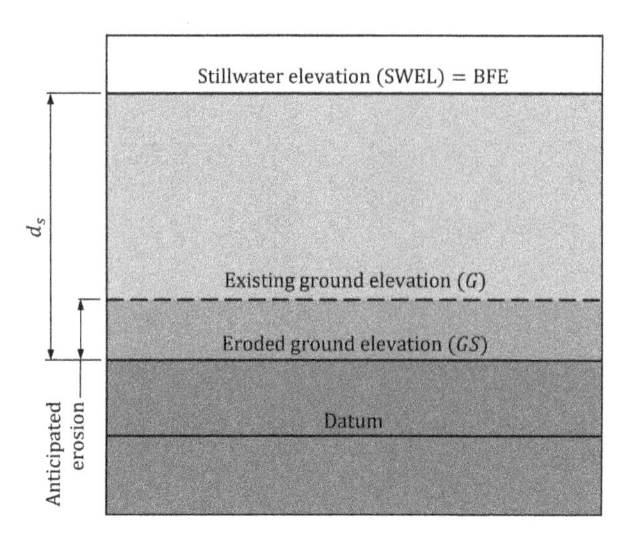

Figure 7.6
Stillwater Flood Elevation and Stillwater Flood Depth for Coastal Areas with Waves

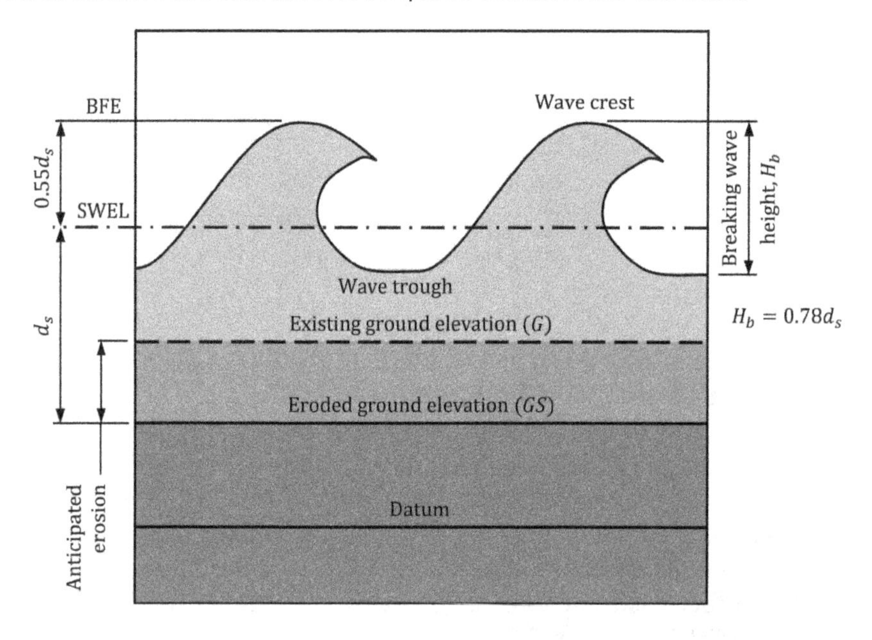

The stillwater flood depth, d_s, is defined as the vertical distance between the SWEL and the eroded ground level. In coastal areas, d_s must include the effects of wave setup, which is the increase in the stillwater surface near the shoreline due to the presence of breaking waves (see Section 7.2.9 of this book). Wave setup was typically not included in FIS reports and FIRMs prior to 1989. It is important to check the FIS report regarding wave setup: If wave setup is included in the BFE but not in the SWEL, wave setup must be added before calculating d_s, the design wave height, the design flood velocity, flood loads, and localized scour. If it has been determined that wave setup is included in the SWEL, d_s and the other parameters can be calculated based on that SWEL.

Floods may erode soil in both riverine and coastal areas. The existing ground elevation must be lowered if erosion has occurred in the vicinity of the site; erosion causes an increase in d_s, which in turn, increases the required design flood loads. Using historical information, contacting local geotechnical engineers, or contacting the local jurisdiction may provide some insight into the anticipated amount of erosion. Reference 7.5 provides guidance on estimating erosion in coastal areas.

7.2.9 Breaking Wave Height and Runup

Breaking wave height, H_b, is the vertical distance between the crest and the trough of a wave. In shallow waters, H_b depends on d_s, and is typically determined by ASCE/SEI Equation (5.4-2):

$$H_b = 0.78d_s \tag{7.1}$$

Where steep slopes exist immediately seaward of a building, wave heights can exceed those determined by ASCE/SEI Equation (5.4-2); in such cases, a reasonable alternative is to set $H_b = 1.0d_s$.

Because wave forms in shallow waters are distorted, the crest and trough of a wave are not equidistant from the SWEL. For NFIP mapping purposes, it is assumed that 70 percent of the wave height, H_b, is above the SWEL. Thus, the maximum elevation of a breaking wave crest above the SWEL is equal to $0.7 \times 0.78d_s = 0.55d_s$ (see Figure 7.6). Based on this assumption, the following relationship can be established:

$$\text{BFE} - GS = d_s + 0.55d_s = 1.55d_s$$

or

$$d_s = 0.65(\text{BFE} - GS) \tag{7.2}$$

ASCE/SEI Equation (5.4-3) is the same as Equation (7.2) except G is used in ASCE/SEI Equation (5.4-3) instead of GS where G is equal to the lowest eroded ground elevation.

ASCE/SEI Equation (5.4-3) is applicable in coastal flood hazard areas where the ground slopes up gradually from the shoreline and there are few obstructions seaward of a building, such as vegetation or other buildings. The BFE shown on the FIRM in such areas is approximately equal to $GS + 1.55d_s$. This case is illustrated in Figure 7.7, which also includes the associated flood hazard zones.

Figure 7.7 BFE and SWEL at Locations with a Gradually Sloping Shoreline

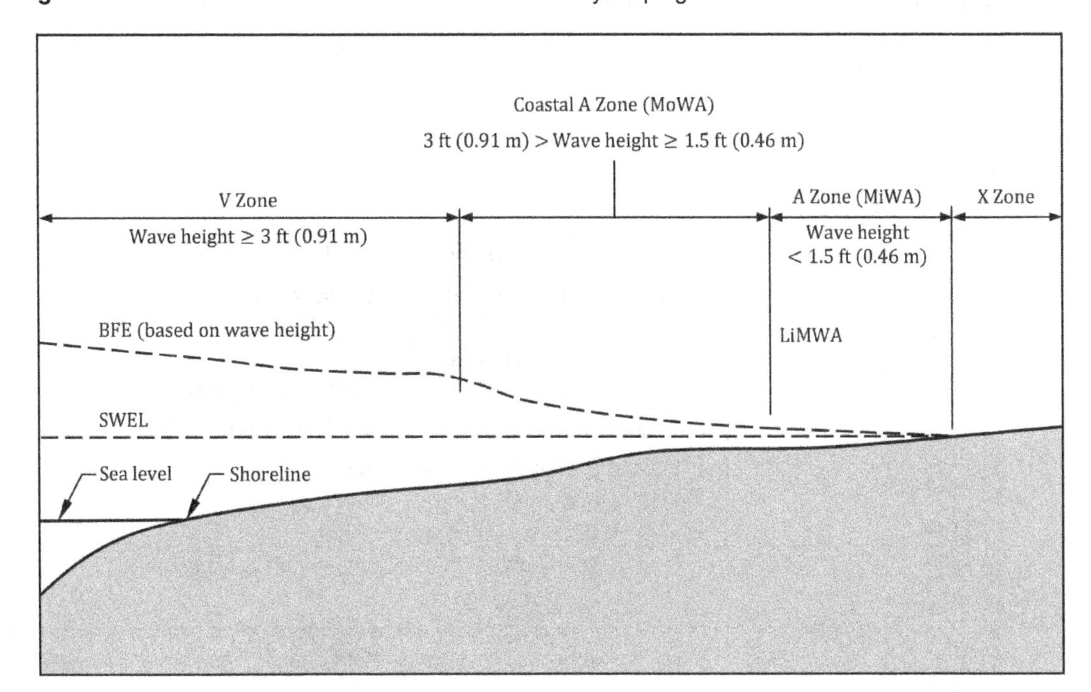

In areas with steeply sloping shorelines, water can rush up the surface, which could result in flood elevations higher than those based on breaking waves. This phenomenon is commonly referred to as wave runup. In coastal flood hazard areas where this can occur, the BFE shown on the FIRM is equal to the highest elevation reached by the water (see Figure 7.8).

Figure 7.8 BFE and SWEL at Locations with a Steeply Sloping Shoreline

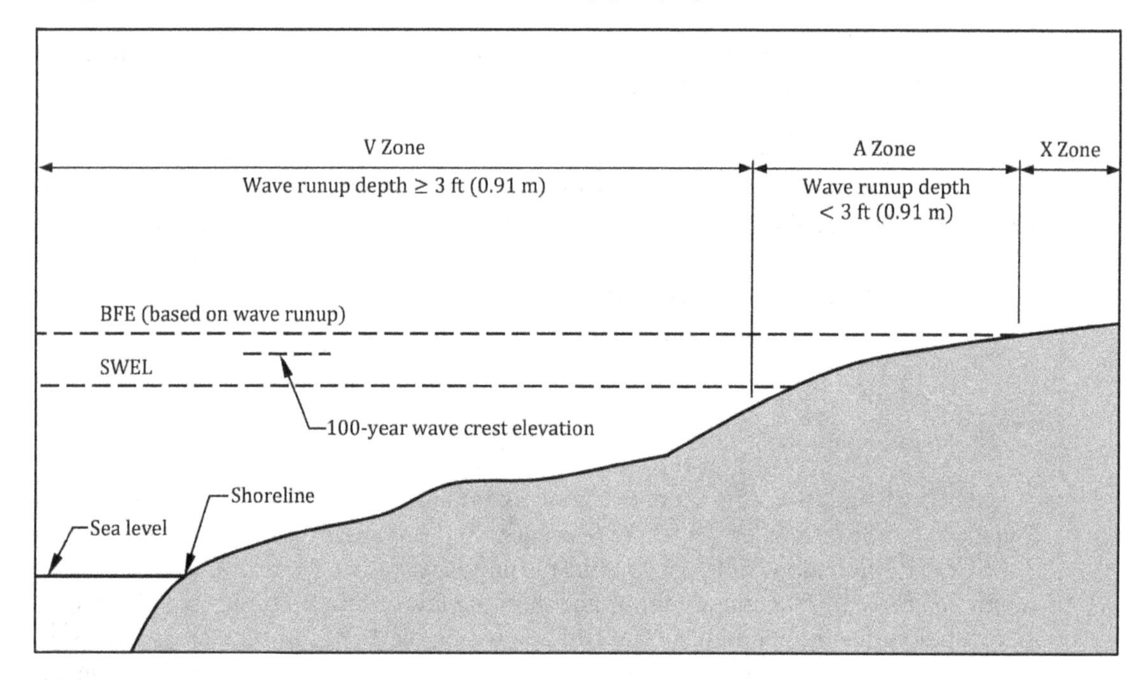

7.2.10 Design Flood Velocity

Design flood velocity, V, is used in the determination of hydrodynamic flood loads (see Section 7.2.12 of this book). Due to the inherent uncertainties of estimating design flood velocities, especially in coastal flood hazard areas, the lower and upper bound velocities in Table 7.3 can be used, which are based on d_s. Guidelines on when to use the lower and upper bound velocities are also given in the table [ASCE/SEI Equations (C5.4-1) and (C5.4-2)].

Table 7.3 Design Flood Velocity, *V*

Lower bound	$V = d_s/t$	To be used where one or more of the following conditions exist: • Where the site is distant from the flood source • In an A Zone • On flat or gently sloping terrain • Where the building is not affected by other buildings or obstructions
Upper bound	$V = (gd_s)^{0.5}$	To be used where one or more of the following conditions exist: • If the site is near the flood source • In a V Zone • In an AO Zone adjacent to a V Zone • In an A Zone subject to velocity flow and wave action • On steeply sloping terrain • Where the building is adjacent to other large buildings or obstructions that will confine or redirect floodwaters and increase local flood velocities

In the equations for design flood velocity, V is in ft/s (m/s), d_s is in ft (m), g = acceleration due to gravity = 32.2 ft/s^2 (9.81 m/s^2), and $t = 1$ s.

The design velocities in Table 7.3 are not applicable to tsunami events. For estimating flood velocities during tsunamis, see Section 7.3 of this book.

7.2.11 Hydrostatic Loads

Overview

Hydrostatic loads occur when stagnant or slow-moving water [velocity less than 5 ft/s (1.52 m/s); see ASCE/SEI C5.4.2] comes into contact with a foundation, building, or building component. The water can be above or below the ground surface. Such loads can be subdivided into lateral loads, vertical downward loads, and vertical upwards loads (uplift or buoyancy). In general, a hydrostatic load is equal to the water pressure times the surface area on which the pressure acts.

Lateral Hydrostatic Loads

For vertical components of a building above ground, the horizontal hydrostatic load, F_{sta}, is the resultant force of the hydrostatic water pressure, which is equal to zero at the surface of the water and increases linearly to $\gamma_s d_s$ at the stillwater depth. The horizontal hydrostatic load can be determined by the following equation:

$$F_{sta} = \frac{\gamma_w d_s^2 w}{2} \tag{7.3}$$

In this equation, γ_w is the unit weight of water, which is equal to 62.4 lb/ft^3 (9.80 kN/m^3) for fresh water and 64.0 lb/ft^3 (10.05 kN/m^3) for saltwater, and w is the width of the vertical building component in ft (m). The load F_{sta} acts perpendicular to the surface and is located $2d_s/3$ below the BFE. The hydrostatic loads on the above-ground foundation walls of a nonresidential building in an A Zone are illustrated in the upper part of Figure 7.9 where it is assumed the dry floodproofing requirements of ASCE/SEI 24 Section 6.2 are satisfied, which permits the lowest floor to be below the minimum elevation in ASCE/SEI 24 Table 2-1 (see Section 7.2.7 of this book). When calculating hydrostatic loads on surfaces exposed to free water, the design depth must be increased by 1.0 ft (0.30 m) [ASCE/SEI 5.4.2]. This is equivalent to increasing d_s by 1.0 ft (0.30 m) when calculating F_{sta} by Equation (7.3).

For vertical components of a building above and below ground (such as basement walls), the total lateral hydrostatic pressure consists of the pressure due to the standing floodwater and the pressure due to the saturated soil (see ASCE/SEI 3.2.1 and the lower part of Figure 7.9):

- Lateral hydrostatic load due to the water, which is based on the distance from the BFE to the top of the footing:

$$F_{sta} = \frac{\gamma_w (d_s + d_b + t)^2 w}{2}$$

- Lateral hydrostatic load due to the differential between the water and soil pressures, which is based on the depth of the saturated soil from the adjacent grade to the top of the footing:

$$F_{dif} = \frac{(\gamma_s - \gamma_w)(d_b + t)^2 w}{2}$$

In these equations, d_b is the height of the basement and t is the thickness of the basement slab. In the equation for F_{dif}, γ_s is the equivalent fluid weight of submerged soil and water, which is based on soil type. Values of γ_s are given in Table 7.4 (see Reference 7.6; it is always good practice to confirm γ_s for the site with the local jurisdiction or with other references that contain measured soil properties). The soil types and group symbols in Table 7.4 are from ASCE/SEI Table 3.2-1 and are defined in Table 7.5.

Figure 7.9 Hydrostatic Loads on Nonresidential Buildings Located in an A Zone Where the Dry Floodproofing Requirements of ASCE/SEI 24 Section 6.2 Are Satisfied

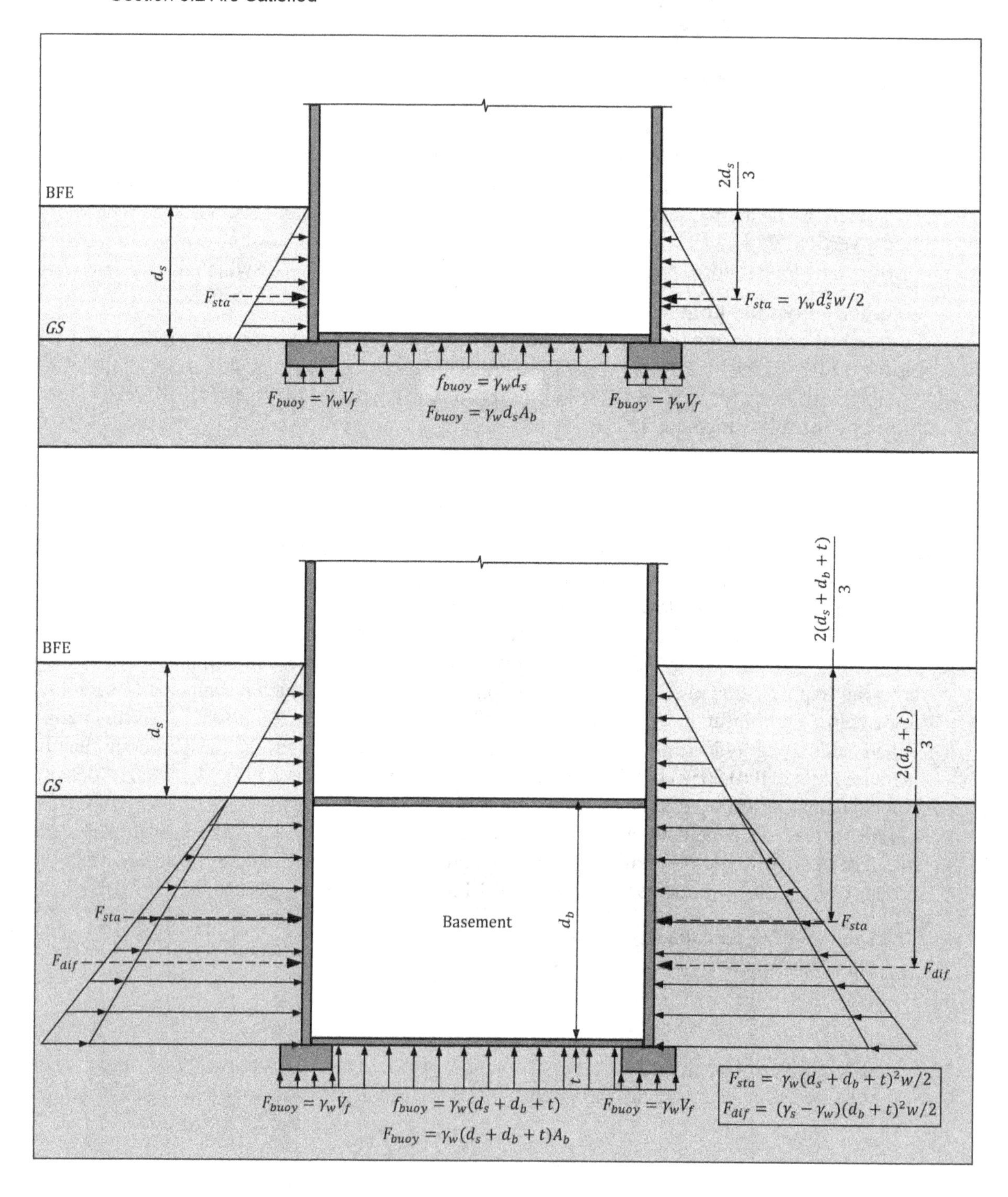

Vertical Hydrostatic Loads (Buoyancy Loads)

In addition to lateral hydrostatic loads, vertical structural elements above and below ground are subjected to upwards hydrostatic loads, which are commonly referred to as buoyancy loads. In general, the buoyancy load, F_{buoy}, is equal to the following:

$$F_{buoy} = \gamma_w V \tag{7.4}$$

Table 7.4 Equivalent Fluid Weight of Submerged Soil and Water, γ_s

Soil Type (Symbols)	γ_s, lb/ft³ (kN/m³)
Clean sand and gravel (GW, GP, SW, SP)	75 (11.78)
Dirty sand and gravel of restricted permeability (GM, GM-GP, SM, SM-SP)	77 (12.10)
Stiff residual silts and clays, silty fine sands, clayey sands, and gravels (CL, ML, CH, MH, SM, SC, GC)	82 (12.88)
Very soft to soft clay, silty clay, organic silt, and clay (CL, ML, OL, CH, MH, OH)	106 (16.65)
Medium to stiff clay deposited in chunks and protected from infiltration (CL, CH)	142 (22.31)

Table 7.5 Soil Type Definitions

Soil Type	Group Symbol	Description
Gravels	GW	Well-graded gravels and gravel mixtures
	GP	Poorly graded gravel-sand mixtures
	GM	Silty gravels, gravel-sand mixtures
	GC	Clayey gravels, gravel and clay mixtures
Sands	SW	Well-graded sands and gravelly sands
	SP	Poorly graded sands and gravelly sands
	SM	Silty sands, poorly graded sand-silt-mixtures
	SC	Clayey sands, poorly graded sand-clay mixtures
Fine Grain Silt and Clays	ML	Inorganic silts and clayey silts
	CL	Inorganic clays of low to medium plasticity
	OL	Organic silts and organic silty clays of low plasticity
	MH	Inorganic silts, micaceous or fine sands or silts, elastic silts
	CH	Inorganic clays of high plasticity, fine clays
	OH	Organic clays of medium to high plasticity

where V is the volume of floodwater displaced by the submerged object in ft³ (m³). For example, the buoyancy load on a column above ground and supporting a building or structure is equal to $F_{buoy} = \gamma_w d_s A_c$ where A_c is the area of the column. This load acts in the upwards direction and is perpendicular to A_c.

Hydrostatic uplift loads (buoyancy loads) are shown in Figure 7.9 on the lowest floor slab in each building assuming the soil below the slab becomes saturated by the floodwaters (see ASCE/SEI 3.2.2). The buoyancy pressure, f_{buoy}, on the underside of the slab in the upper part of Figure 7.9 is equal to $\gamma_w d_s$. The total buoyancy load on the slab is $F_{buoy} = \gamma_w d_s A_b$ where A_b is the plan area of the building. The volume of floodwater displaced by the submerged portion of the building is equal to $d_s A_b$. The buoyancy pressure on the underside of the slab in the lower part of Figure 7.9 is equal to $f_{buoy} = \gamma_w (d_s + d_b + t)$ and the total buoyancy load in this case is $F_{buoy} = \gamma_w (d_s + d_b + t) A_b$. Where the slab is isolated from the building structure (which is a typical slab-on-grade), the slab is subjected to buoyancy pressure and overall buoyancy (uplift) on the building need not be considered. However, if the slab is connected to the building structure and is designed to resist buoyancy pressure, uplift on the building must be considered.

Buoyancy loads due to the volume of water displaced by the portions of the foundation walls below the DFE (i.e., the submerged portions of the basement/foundation walls) and the wall footings must also be considered (see Figure 7.9). The buoyancy load on a footing is equal to $F_{buoy} = \gamma_w V_f$ where V_f is the volume of the footing. Similarly, the buoyancy load per unit length of a foundation or basement wall is equal to γ_w times the thickness of the wall times the depth of the submerged portion of the wall.

To prevent flotation of a building during a design flood event, the total buoyancy load, which can be relatively large for buildings that have dry floodproofed basements and structural slabs, is resisted by the weight of the building (including basement walls where applicable) and the weight of the foundations. Because floods can occur at any time, any transient loads, such as live loads, should not be included to resist uplift loads. Where the weights of the building and foundations are not sufficient, other structural measures (such as tension piles or additional mass) must be provided. Alternatively, vent openings in the foundation walls can be incorporated to alleviate the buoyancy loads.

Cases where hydrostatic loads do not need to be considered are given in Table 7.6 based on flood hazard area.

Table 7.6 Cases Where Hydrostatic Loads Do Not Need to Be Considered

Flood Hazard Area	Requirements	Notes
A Zones	Foundation walls and exterior enclosure walls below the DFE that do not meet the dry floodproofing requirements in ASCE/SEI 24 Section 6.2 must contain openings to allow for automatic entry and exit of floodwaters during design flood conditions (ASCE/SEI 24 Section 2.7.1)	For walls with flood openings, lateral hydrostatic loads are equal to zero because the external and internal lateral hydrostatic pressures are balanced.
V Zones and Coastal A Zones	• The bottom of the lowest horizontal structural member of the lowest floor must be elevated in accordance with the minimum requirements of ASCE/SEI 24 Table 4-1. • Buildings must be supported on open foundations (such as piles, posts, piers, or columns) free of obstructions that will not restrict or eliminate free passage of high-velocity floodwaters and waves (ASCE/SEI 24 Section 4.5.1).	Lateral hydrostatic loads are not applicable and buoyant forces are either zero or nominal.
	Breakaway walls and other similar nonbearing elements must be designed and constructed to fail under design flood or lesser conditions (ASCE/SEI 24 Section 4.6.1). Flood openings are required in breakaway walls (ASCE/SEI 24 Section 4.6.2).	Lateral and vertical hydrostatic loads need not be considered for breakaway walls located below the DFE.

7.2.12 Hydrodynamic Loads

Hydrodynamic loads are caused by floodwaters flowing past a fixed object, such as a building or a supporting element (e.g., a pile or a column). Like wind flowing around a building, the loads produced by moving water include an impact load on the upstream face of the building, drag forces along the sides of the building, and a negative force (suction) on the downstream face of the building. An approximation for the total hydrodynamic load, F_{dyn}, on a building or structural element can be determined by the following equation, which is applicable for all flow velocities, V (see Reference 7.5 and Figure 7.10 for a building with foundation walls and a building supported on piles):

$$F_{dyn} = \frac{1}{2} a \rho V^2 A \tag{7.5}$$

Figure 7.10 Hydrodynamic Loads on a Foundation Wall and on a Pile

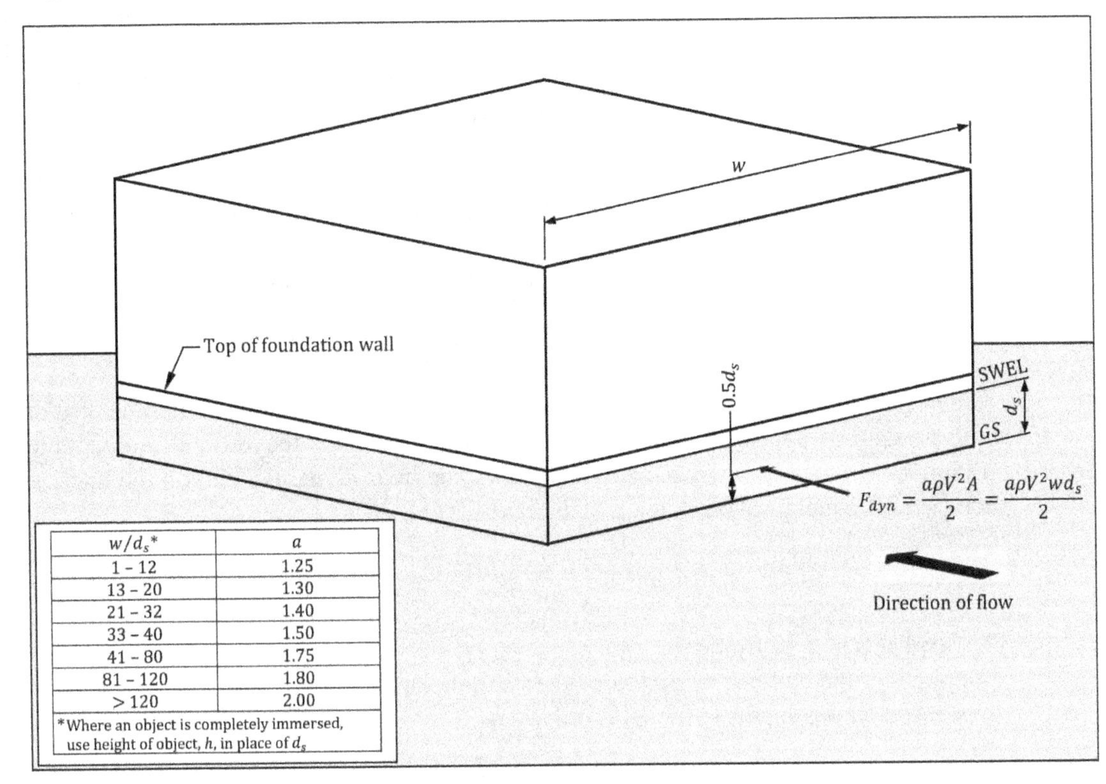

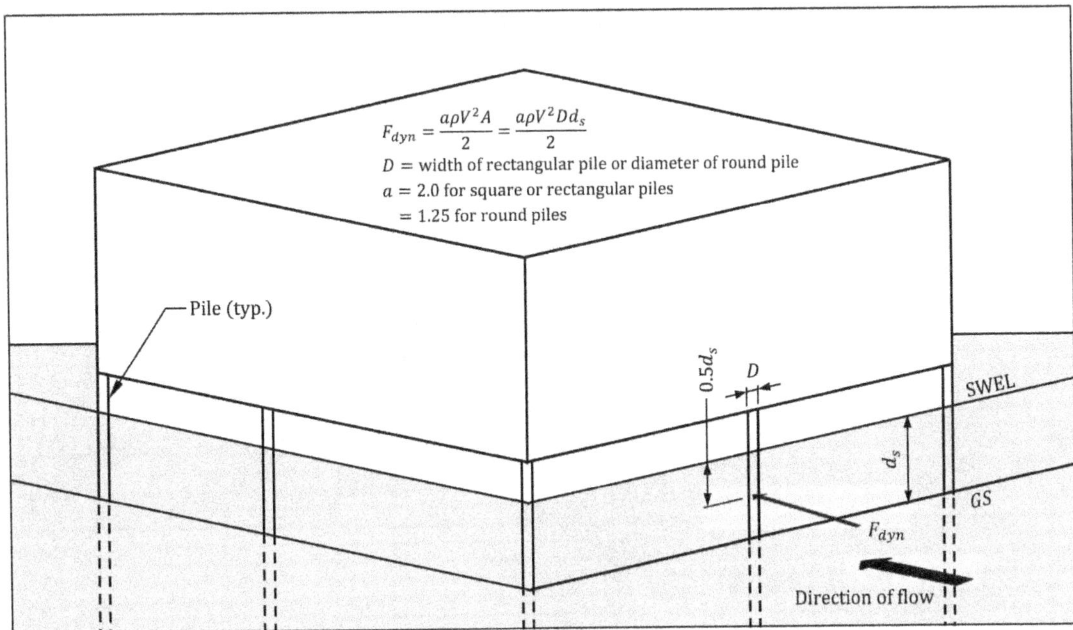

This force is assumed to act at the stillwater mid-depth (i.e., halfway between the SWEL and the eroded ground surface, *GS*; see Figure 7.10) on the upstream face of the building or structural element. The terms in Equation (7.5) are defined as follows:

- a = drag coefficient, which is a function of the shape of the object around which flow is directed. Recommended values of a are given in Figure 7.10 (see ASCE/SEI 5.4.3 and Reference 7.5).
- ρ = mass density of water. For fresh water, $\rho = 1.94$ slugs/ft^3 (1,000 kg/m^3) and for saltwater, $\rho = 2.0$ slugs/ft^3 (1,025 kg/m^3).

- V = design flood velocity, which is determined in accordance with Section 7.2.10 of this book.
- A = surface area of a building, structure, or structural element normal to the water flow. For the foundation wall in Figure 7.10, $A = wd_s$ where w is the width of the foundation wall perpendicular to the flow. For the pile, $A = Dd_s$ where D is the width of the rectangular pile perpendicular to the flow or the diameter of a round pile.

The dynamic effects of moving water are permitted to be converted into an equivalent hydrostatic load where $V \leq 10$ ft/s (3.05 m/s) [ASCE/SEI 5.4.3]; this is accomplished by increasing the depth of the floodwater above the flood level by an amount d_h, which is determined by ASCE/SEI Equation (5.4-1), on the upstream face and above the ground level only:

$$d_h = \frac{aV^2}{2g} \tag{7.6}$$

The equivalent surcharge depth, d_h, is added to the DFE design depth and the resultant hydrostatic pressure is applied to and uniformly distributed across the vertical projected area of the building or structure perpendicular to the flow. The hydrodynamic load on the upstream face in such cases is equal to the following (see Figure 7.11):

$$F_{dyn} = \gamma_w d_h d_s w = \gamma_w \left(\frac{aV^2}{2g}\right) d_s w = \frac{1}{2} a \left(\frac{\gamma_w}{g}\right) V^2 d_s w = \frac{1}{2} a \rho V^2 A \tag{7.7}$$

This load acts at $d_s/2$ from GS.

Surfaces of the building or structure other than the upstream face are subject to hydrostatic pressures based on the depths to the BFE only.

Figure 7.11 Hydrodynamic Load Where $V \leq 10$ ft/s (3.05 m/s)

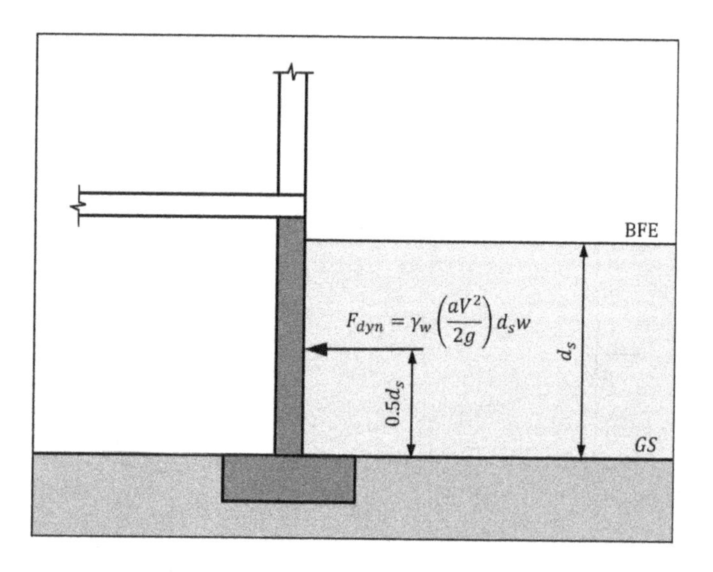

7.2.13 Wave Loads

Overview

Wave loads result from water waves propagating over the surface of the water and striking a building or structural element. These types of loads are typically separated into four categories: nonbreaking waves, breaking waves, broken waves, and uplift (see Table 7.7).

Wave loads must be included in the design of buildings and structural elements located in the following Zones:

- V Zones [wave heights greater than or equal to 3 ft (0.91 m)]
- Coastal A Zones [wave heights greater than or equal to 1.5 ft (0.46 m) and less than 3 ft (0.91 m)]
- A Zones subject to coastal waves [wave heights less than 1.5 ft (0.46 m)]

Table 7.7 Categories of Wave Loads

Category	Comments
Nonbreaking waves	The effects of nonbreaking waves can be determined using the procedures in ASCE/SEI 5.4.2 for hydrostatic loads on walls and in ASCE/SEI 5.4.3 for hydrodynamic loads on piles.
Breaking waves	These loads are caused by waves breaking on any portion of a building or structure. Although these loads are of short duration, they generally produce the largest magnitude of all the different types of wave loads.
Broken waves	The loads caused by broken waves are similar to hydrodynamic loads caused by flowing or surging water.
Uplift	Uplift effects are caused by wave runup striking any portion of a building or structure, deflection, or peaking against the underside of surfaces.

Because breaking wave loads are the largest of the four types, it is used as the design wave load. Three methods are given in ASCE/SEI 5.4.4 to determine breaking wave loads:

- Analytical procedures
- Advanced numerical modeling procedures
- Laboratory test procedures (physical modeling)

Regardless of the method used, an estimate of the breaking wave height, H_b, is needed to calculate breaking waves loads (see Section 7.2.9 of this book). Information on the FEMA model that can also be used to estimate wave heights and wave crest elevations can be found in Reference 7.7.

A summary of the breaking wave loads and their points of application in accordance with the analytical procedures in ASCE/SEI 5.4.4 is given in Figure 7.12.

Vertical Columns and Pilings

The net wave load, F_D, on vertical pilings and columns is applied to the element at the SWEL. The term C_D is the drag coefficient for breaking waves, which is equal to 1.75 for round piles or columns and 2.25 for square or rectangular piles or columns. The term D is the pile or column diameter for circular sections or 1.4 times the width of the pile or column for rectangular or square sections. The breaking wave height, H_b, is determined by Equation (7.1).

Breaking wave loads must be considered, for example, on vertical pilings and columns supporting buildings in V Zones where the lowest floor in the building must be elevated to or above the minimum elevations identified in ASCE/SEI 24 Table 4-1 (see Section 7.2.7 of this book).

Vertical Walls

For breaking wave loads on vertical walls, it is assumed in the analytical procedure the wall causes a reflected or standing wave to form against the waterward side of the wall and the crest of the reflected wave reaches a height of $1.2d_s$ above the SWEL. The resulting hydrostatic, dynamic, and total pressure distributions are given in Figure 7.12 where there is no water behind the wall and where the stillwater level is equal on both sides of the wall.

The case of a vertical wall with no water behind it corresponds to nonresidential buildings in an A Zone where the lowest floor is below the minimum elevation in ASCE/SEI 24 Table 2-1 and the dry floodproofing requirements of ASCE/SEI 24 Section 6.2 are satisfied for the space behind the wall (see ASCE/SEI 24 Section 2.3; under the NFIP, solid foundation walls are not permitted in V Zones). In this case, the maximum total pressure, P_{max}, occurs at the SWEL and is equal to the maximum dynamic pressure $[C_p \gamma_w d_s]$ plus the hydrostatic pressure $[\gamma_w (1.2d_s)]$ where C_p is the dynamic pressure coefficient, which is based on the risk category of the building (see ASCE/SEI Table 1.5-1 and ASCE/SEI Table 5.4-1). Risk category II buildings are assigned a value of C_p corresponding to a 1-percent probability of exceedance, which is consistent with wave analysis

Figure 7.12 Breaking Wave Loads in Accordance with ASCE/SEI 5.4.4

Structural Element	Breaking Wave Load
Vertical Pilings and Columns	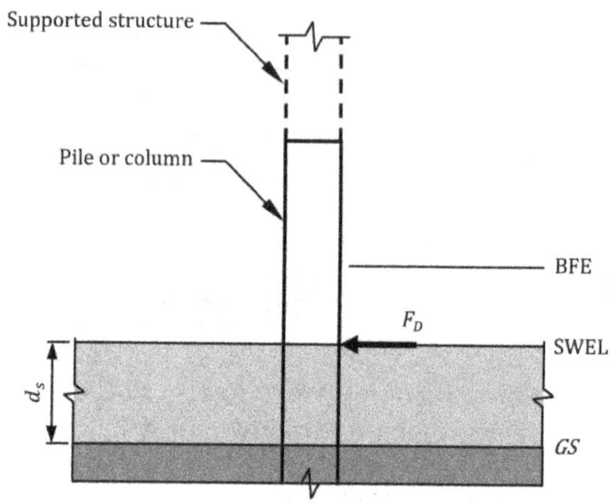 Supported structure Pile or column BFE F_D SWEL d_s GS $F_D = 0.5\gamma_w C_D D H_b^2$ C_D = 1.75 for round piles or columns = 2.25 for square piles or columns D = pile or column diameter for circular sections = 1.4 times the width of a square pile or column
Vertical Walls – Normally incident breaking wave with no water behind the wall	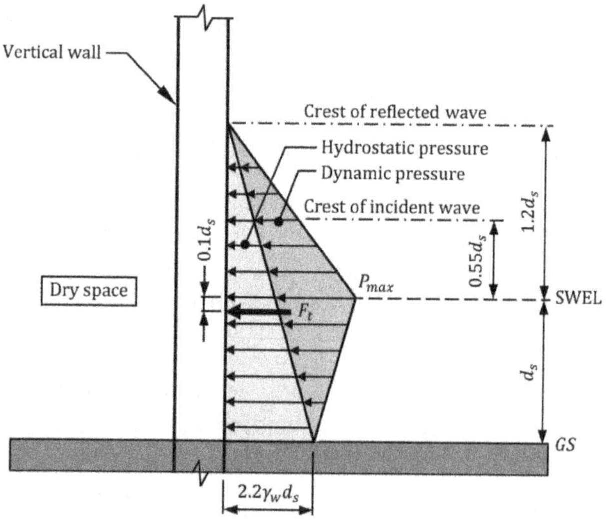 Vertical wall Crest of reflected wave Hydrostatic pressure Dynamic pressure Crest of incident wave Dry space $0.1d_s$ P_{max} F_t $0.55d_s$ $1.2d_s$ SWEL d_s GS $2.2\gamma_w d_s$ $P_{max} = C_p\gamma_w d_s + 1.2\gamma_w d_s$ $F_t = 1.1C_p\gamma_w d_s^2 + 2.4\gamma_w d_s^2$

Risk Category	C_p
I	1.6
II	2.8
III	3.2
IV	3.5

(continued)

Figure 7.12 Breaking Wave Loads in Accordance with ASCE/SEI 5.4.4 (*Continued*)

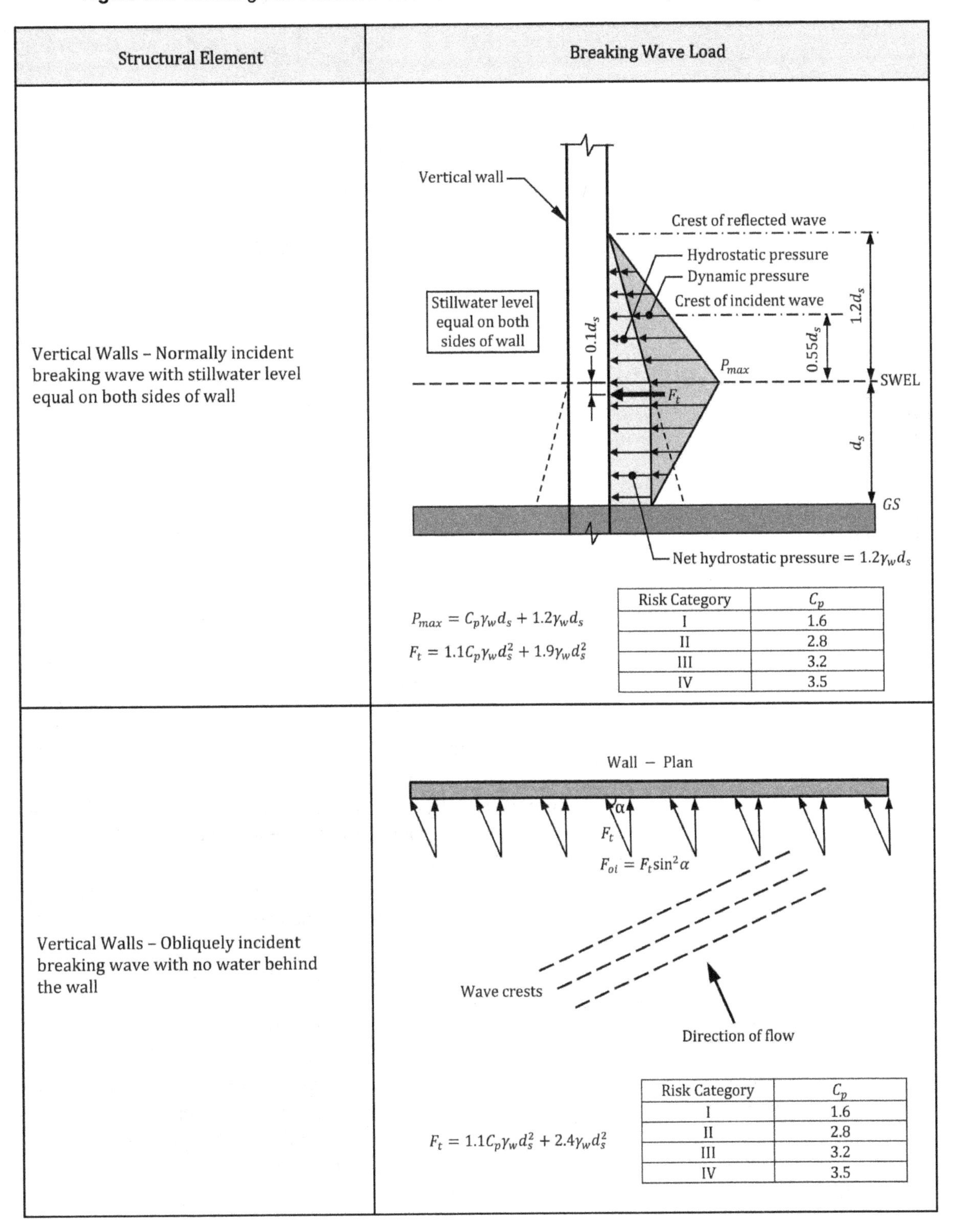

Risk Category	C_p
I	1.6
II	2.8
III	3.2
IV	3.5

Risk Category	C_p
I	1.6
II	2.8
III	3.2
IV	3.5

Figure 7.12 Breaking Wave Loads in Accordance with ASCE/SEI 5.4.4 (*Continued*)

Structural Element	Breaking Wave Load
Nonvertical Walls - No water behind the wall	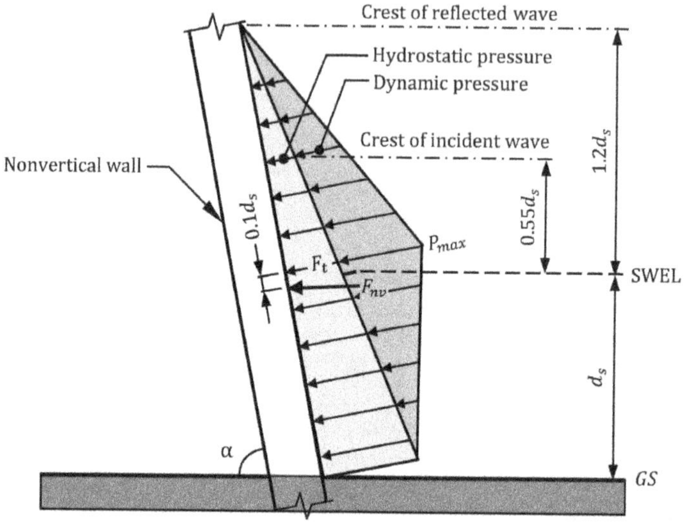 Crest of reflected wave Hydrostatic pressure Dynamic pressure Crest of incident wave Nonvertical wall $0.1d_s$ F_t F_{nv} P_{max} $0.55d_s$ $1.2d_s$ SWEL d_s α GS $F_{nv} = F_t \sin^2\alpha$ $P_{max} = C_p\gamma_w d_s + 1.2\gamma_w d_s$ $F_t = 1.1C_p\gamma_w d_s^2 + 2.4\gamma_w d_s^2$

Risk Category	C_p
I	1.6
II	2.8
III	3.2
IV	3.5

procedures used by FEMA in mapping coastal flood hazard areas and in establishing minimum floor elevations (see Section 7.2.7 of this book). The percent probability of exceedance for the other risk categories is given in ASCE/SEI C5.4.4.2.

The net breaking wave load per unit length of wall (i.e., the resultant force of the total pressure per unit length of wall), F_t, is equal to the resultant force of the dynamic pressure $[2.2C_p\gamma_w d_s^2(2.2-1.0)/(2\times1.2)=1.1C_p\gamma_w d_s^2]$ plus the resultant force of the hydrostatic pressure $[(2.2\gamma_w d_s)(2.2d_s)/2=2.4\gamma_w d_s^2]$. It is assumed F_t is applied to the wall a distance equal to $0.1d_s$ below the SWEL instead of at the centroid of the total pressure (Reference 7.5).

Water can occur behind a wall where a wave breaks against breakaway walls (which are required in V Zones and Coastal A Zones for enclosed areas below the DFE; see ASCE/SEI 24 Section 4.6.1) or walls equipped with openings that allow floodwaters to equalize on both sides of the wall. Based on the analytical procedure, the hydrostatic pressure equalized in this case is equal to $\gamma_w d_s$. Therefore, the net resultant force of the hydrostatic pressure over the height d_s of the wall is equal to $2.4\gamma_w d_s^2-(\gamma_w d_s^2/2)=1.9\gamma_w d_s^2$ (see Figure 7.12). The resultant force of the dynamic pressure is equal to $1.1C_p\gamma_w d_s^2$, which is the same as that for a wall with no water behind it.

When designing breakaway walls, a value of 1.0 should be used for C_p rather than the values given in ASCE/SEI Table 5.4-1 (Reference 7.5).

It is typically assumed the direction of wave approach is approximately perpendicular to the shoreline. For walls oriented perpendicular to the shoreline, the wave loads are maximum because the waves are normally incident to the wall; in such cases, the applicable equations in Figure 7.12 can be used to calculate the breaking wave loads. For walls not oriented perpendicular to the shoreline, the waves are not normally incident to the wall and the corresponding loads are smaller than those for walls perpendicular to the shoreline. The breaking wave load from obliquely incident waves is $F_{oi} = F_t \sin^2 \alpha$ where α is the horizontal angle between the direction of wave approach and the vertical surface of the wall [see ASCE/SEI Equation (5.4-9) and Figure 7.12].

Nonvertical Walls

Breaking wave loads on nonvertical walls can be determined by ASCE/SEI Equation (5.4-8):

$$F_{nv} = F_t \sin^2 \alpha \tag{7.8}$$

where α is the vertical angle between the nonvertical surface of the wall and the horizontal. The breaking wave pressures and loads illustrated in Figure 7.12 are for the case where no water is behind the wall.

7.2.14 Impact Loads

Overview

Impact loads occur where objects carried by moving water strike a building or structural element. Although the magnitude of such loads is difficult to predict, reasonable approximations can be made considering a number of different uncertainties.

Impact loads can be categorized as normal, special, and extreme (Reference 7.8). Information pertaining to these three categories is given in Table 7.8.

Table 7.8 Impact Loads

Impact Load Category	Comments
Normal	Normal impact loads result from the isolated impacts of commonly encountered objects. The size, shape, and weight of waterborne debris vary according to region.
Special	Special impact loads result from large objects such as broken up ice floats and accumulations of waterborne debris. The loads are caused by these objects either striking or resting against the building or structure.
Extreme	Extreme impact loads result from very large objects (such as boats, barges, or parts of collapsed buildings) striking a building or structure. Design for such extreme loads is usually not practical in most cases unless the probability of such an impact load during the design flood is high.

Normal Impact Loads

A rational method for calculating normal impact loads is given in ASCE/SEI C5.4.5. Flowchart 7.1 in Section 7.4 of this book can be used to determine the impact load, F [see ASCE/SEI Equation (C5.4-3)].

Information on the terms used to calculate F is given in Table 7.9. The term $\pi/2$ in ASCE/SEI Equation (C5.4-3) is a result of the half-sine form of the impulse load used in the derivation of this equation. It is presumed objects are at or near the water surface level when they strike a building. Thus, F is usually assumed to act at the stillwater flood level; in general, this load should be applied horizontally at the most critical location at or below the SWEL.

Special Impact Loads

The following equation can be used to determine special impact loads [see ASCE/SEI Equation (C5.4-4)]:

$$F = \frac{C_D \rho A V^2}{2} \tag{7.9}$$

Table 7.9 Terms Used to Calculate Normal Impact Loads, *F*

Term	ASCE/SEI Reference	Comments
Debris weight, W	C5.4.5	The debris impact weight must be selected considering local or regional conditions. A realistic weight for trees, logs, and other large wood-type debris, which are the most common forms of damaging debris, is 1,000 lb (4.5 kN). This is also a reasonable weight for small ice floes, boulders, and some man-made objects.
Velocity of object, V_b	C5.4.5	Debris velocity depends on the nature of the debris and the velocity of the floodwaters. Smaller pieces of debris usually travel at the velocity of the floodwaters; such items typically do not cause damage to buildings or structures. Large debris, which can cause damage, will likely travel at a velocity less than that of the floodwaters because it can drag at the bottom or can be slowed by prior collisions with nearby objects. For calculating debris loads, the velocity of the debris should be taken equal to the velocity of the floodwaters, V (see Section 7.2.10 of this book).
Importance coefficient, C_I	Table C5.4-1	Importance coefficients are based on a probability distribution of impact loads obtained from laboratory tests; this coefficient is larger for more important structures.
Orientation coefficient, C_O	C5.4.5	The orientation coefficient is used to reduce the impact load for other than head-on impact loads. Based on measurements taken during laboratory tests, an orientation factor of 0.8 has been adopted.
Depth coefficient, C_D	Table C5.4-2 Figure C5.4-1	The depth coefficient is used to account for reduced debris velocity due to debris dragging along the bottom in shallow water. Values of the coefficient are based on typical diameters of logs and trees and on anticipated diameters of root masses from drifting trees likely to be encountered in a flood hazard zone.
Blockage coefficient, C_B	Table C5.4-3 Figure C5.4-2	The blockage coefficient accounts for reduction of debris velocity due to screening and sheltering provided by trees or other structures within about 300 ft (91.4 m) upstream from a building or structure.
Maximum response ratio for impulsive loads, R_{max}	Table C5.4-4	The maximum response ratio modifies the impact load based on the natural period of the structure, T_n, and the impact duration, Δt. Stiff or rigid buildings and structural elements with natural periods similar to the impact duration will see an amplification of the impact load while more flexible buildings and structural elements with natural periods greater than approximately four times the impact duration will see a reduction of the impact load. The critical period is approximately 0.11 s. Buildings and structures with a natural period greater than 0.11 s will see a reduction in the impact load while those with a natural period less than 0.11 s will see an increase in the impact load. The natural period can be determined by any substantiated analysis. Recommendations on values of T_n to use for flood impact loads are given in ASCE/SEI C5.4.5 and in Reference 7.5.
Impact duration time, Δt	Sect. C5.4.5	Impact duration is the time it takes to reduce the velocity of the object to zero. The recommended value of Δt is 0.03 s.

In this equation, C_D is the drag coefficient, which is equal to 1.0; ρ is the mass density of water; A is the projected area of the debris accumulation into the flow, which is approximately the depth of the accumulation times the width of the accumulation perpendicular to the direction of the flow; and V is the velocity of the flow upstream of the debris accumulation. Like normal impact loads, this load is assumed to act at the stillwater flood level unless it is found that the critical location is below the SWEL.

7.2.15 Flood Load Combinations

Flood Loads to Use in the Determination of F_a

Overview

The horizontal flood loads that need to be considered in the determination of the design flood load, F_a, for buildings located in an A Zone where the flood source is from a river or lake are given in Table 7.10. The horizontal flood loads that need to be considered for buildings located

Table 7.10 Determination of F_a for Buildings in A Zones (Riverine or Lake Flood Source)

Foundation Type	Flood Loads, F_a		
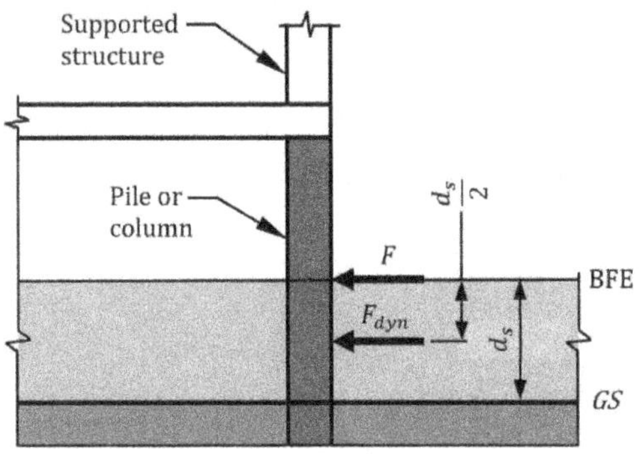 Supported structure; Pile or column; F; F_{dyn}; $d_s/2$; d_s; BFE; GS	Global Foundation Loads	Total Horizontal Load	$F_a = F + (n-1)F_{dyn}$
		Overturning Moment	$F_a = Fd_s + (n-1)F_{dyn}(d_s/2)$
	Loads on a Single Pile or Column	F_a consists of F and F_{dyn}, which are combined with gravity and other applicable load effects	
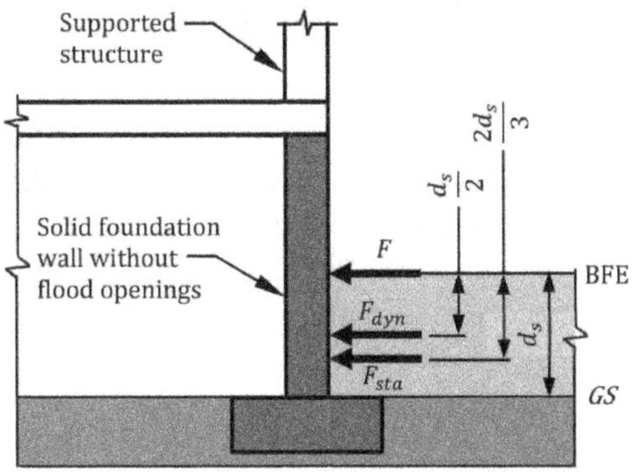 Supported structure; Solid foundation wall without flood openings; F; F_{dyn}; F_{sta}; $d_s/2$; $2d_s/3$; d_s; BFE; GS	Global Foundation Loads	Total Horizontal Load	$F_a = F + F_{dyn}$
		Overturning Moment	$F_a = Fd_s + F_{dyn}(d_s/2)$
	Loads on a Solid Wall	F_a consists of F, F_{dyn}, and F_{sta}, which are combined with gravity and other applicable load effects	
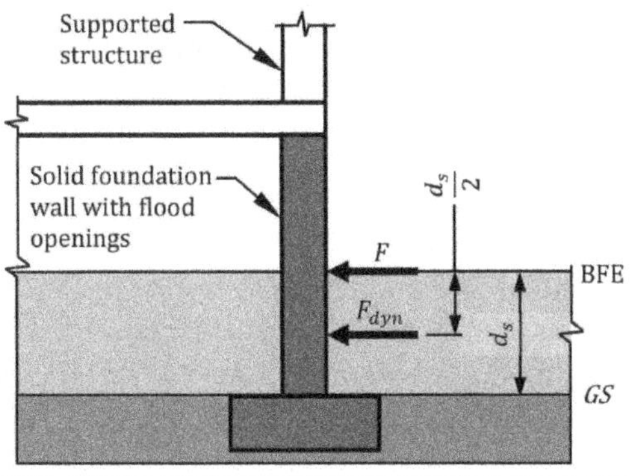 Supported structure; Solid foundation wall with flood openings; F; F_{dyn}; $d_s/2$; d_s; BFE; GS	Global Foundation Loads	Total Horizontal Load	$F_a = F + F_{dyn}$
		Overturning Moment	$F_a = Fd_s + F_{dyn}(d_s/2)$
	Loads on a Solid Wall	F_a consists of F and F_{dyn}, which are combined with gravity and other applicable load effects	

Table 7.11 Determination of F_a for Buildings in A Zones (Coastal Waves), Coastal A Zones, and V Zones

Foundation Type	Flood Loads, F_a		
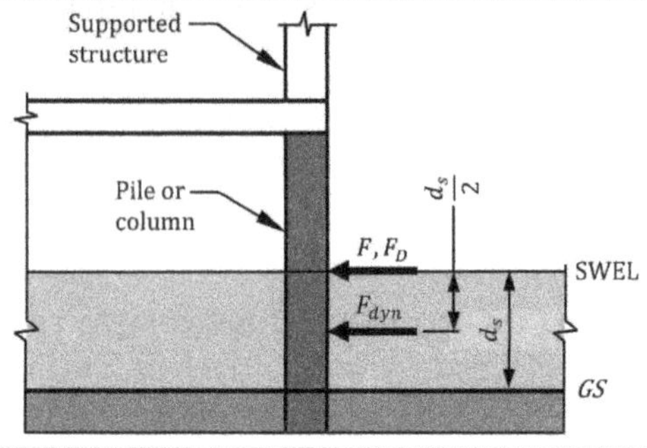 Supported structure Pile or column F, F_D F_{dyn} $\frac{d_s}{2}$ d_s SWEL GS	Global Foundation Loads	Total Horizontal Load	$F_a = F + (n-1)F_{dyn}$
		Overturning Moment	$F_a = Fd_s + (n-1)F_{dyn}(d_s/2)$
	Loads on a Single Pile or Column	F_a consists of F and the greater of F_{dyn} and F_D, which are combined with gravity and other applicable load effects	
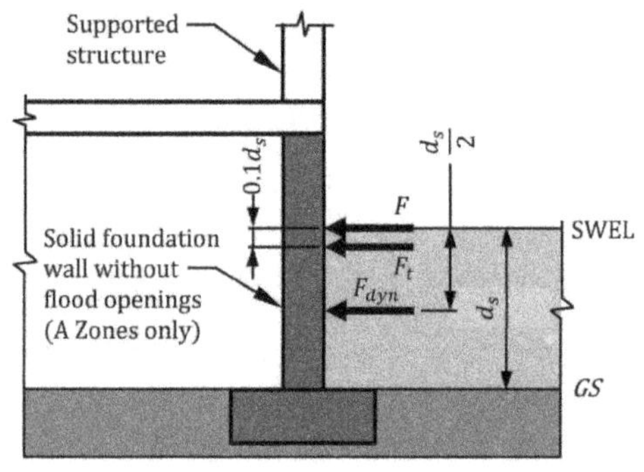 Supported structure Solid foundation wall without flood openings (A Zones only) $0.1d_s$ F F_t F_{dyn} $\frac{d_s}{2}$ d_s SWEL GS	Global Foundation Loads	Total Horizontal Load	$F_a + F +$ greater of $\begin{cases} F_{dyn} \\ F_t \end{cases}$
		Overturning Moment	$F_a = Fd_s + Fd_s +$ greater of $\begin{cases} F_{dyn}(d_s/2) \\ F_t(0.9d_s) \end{cases}$
	Loads on a Solid Wall	F_a consists of F, F_{dyn}, and F_{sta}, which are combined with gravity and other applicable load effects	

in A Zones subject to coastal waves, Coastal A Zones, and V Zones are given in Table 7.11. The effects of these loads are used in the load combinations in ASCE/SEI 2.3 and 2.4.

A Zones—Riverine or Lake Flood Source (Not Subjected to Coastal Wave Loads)

- Buildings Supported by Piles or Columns

  In an A Zone where the source of the floodwater is from a river or lake and where the building is not subjected to wave loads, it is reasonable to assume for the analysis of the entire (global) foundation of a building or structure supported by n piles or columns that one of the piles or columns will be subjected to the impact load, F, and the remaining piles or columns will be subjected to the hydrodynamic load, F_{dyn}. Therefore, the total horizontal load on the foundation due to flood loads in this case is equal to the following:

$$F_a = F + (n-1)F_{dyn} \tag{7.10}$$

  The total overturning moment about GS due to flood loads is equal to the following (see Table 7.10):

$$F_a = Fd_s + (n-1)F_{dyn}(d_s/2) \tag{7.11}$$

  Individual piles or columns in the front row (i.e., in the row experiencing the initial effects of the floodwaters) must be designed to resist the combined effects due to flood, gravity, and other applicable effects where F_a consists of F and F_{dyn}, which are applied to the pile or column at the locations indicated in Table 7.10.

- Buildings Supported by Solid Foundation Walls

  In the case of a solid foundation wall without flood openings where the dry floodproofing requirements of ASCE/SEI 24 Section 6.2 are satisfied, the total horizontal load on the foundation due to flood loads is equal to the following:

$$F_a = F + F_{dyn} \tag{7.12}$$

  The total overturning moment about *GS* is equal to the following:

$$F_a = Fd_s + F_{dyn}(d_s/2) \tag{7.13}$$

  When determining the global effects on the foundation in this case, it is evident the hydrostatic loads, F_{sta}, cancel out (see Table 7.10).

  Individual foundation walls must be designed to resist the combined effects due to flood, gravity, and other applicable effects where F_a consists of F, F_{dyn}, and F_{sta}, which are applied to the wall at the locations indicated in Table 7.10. For foundation walls that extend below *GS*, F_{dif} must also be considered.

  Equations (7.12) and (7.13) can also be used to determine global foundation effects on solid foundation walls with flood openings. Individual foundation walls in this case must be designed for the combined effects due to flood, gravity, and other applicable effects where F_a consists of F and F_{dyn}, which are applied to the wall at the locations indicated in Table 7.10. For foundation walls that extend below *GS*, F_{dif} must also be considered.

Breaking wave loads need not be considered in A Zones except at sites subject to coastal waves loads (see below).

The flood loads, F_a, determined above are used in the applicable load combinations given in Tables 7.12 through 7.15 of this book (see below). Load combinations that include vertical buoyant loads must also be considered where applicable.

A Zones (Subject to Coastal Wave Loads), Coastal A Zones, and V Zones

- Buildings Supported by Piles or Columns

  For a building or structure supported by *n* piles or columns, it is reasonable to assume that one of the piles or columns will be subjected to F and the remaining piles or columns will be subjected to F_{dyn}. Therefore, the total horizontal load on the foundation and the overturning moment about *GS* due to flood loads can be determined by Equations (7.10) and (7.11), respectively (see Table 7.11).

  Individual piles or columns in the front row must be designed to resist the combined effects due to flood, gravity, and other applicable effects where F_a consists of F and the greater of F_{dyn} and F_D, which are applied to the pile or column at the locations indicated in Table 7.11.

- Buildings Supported by Solid Foundation Walls

  Solid foundation walls are not permitted by the IBC or ASCE/SEI 24 in Coastal A Zones or V Zones. For buildings in A Zones subjected to coastal wave loads, solid foundation walls without flood openings that meet the dry floodproofing requirements of ASCE/SEI 24 Section 6.2 are permitted, and the total horizontal load on the foundation due to flood loads in this case is equal to the following:

$$F_a = F + \text{greater of} \begin{cases} F_{dyn} \\ F_t \end{cases} \tag{7.14}$$

The total overturning moment about *GS* is equal to the following:

$$F_a = Fd_s + \text{greater of} \begin{cases} F_{dyn}(d_s/2) \\ F_t(0.9d_s) \end{cases} \tag{7.15}$$

The individual foundation walls must be designed to resist the combined effects due to flood, gravity, and other applicable effects where F_a consists of F and the greater of F_{dyn} and F_t, which are applied to the wall at the locations indicated in Table 7.11 (F_{sta} need not be considered in the design of the foundation wall because the hydrostatic component of the flood load is included in F_t). For foundation walls that extend below *GS*, F_{dif} must also be considered.

The flood loads, F_a, determined above are used in the applicable load combinations given in Tables 7.12 through 7.15 of this book. Load combinations that include vertical buoyant loads must also be considered where applicable.

Strength Design and Allowable Stress Design Load Combinations

Strength design and allowable stress design load combinations pertaining to design flood loads, F_a, are given in ASCE/SEI 2.3.2 and 2.4.2, respectively. Strength design load combinations for structural elements above grade based on flood zones are given in Table 7.12 and allowable stress design load combinations are given in Table 7.13. The load factors on F_a are based on a statistical analysis of flood loads associated with hydrostatic pressures, hydrodynamic pressures due to steady overland flow, and hydrodynamic pressures due to waves, all of which are specified in ASCE/SEI 5.4. All other applicable load combinations in ASCE/SEI 2.3 and 2.4 must also be considered in the design of the structural elements.

Table 7.12 Strength Design Load Combinations Including Flood Loads for Structural Elements Above Grade

Flood Zone	ASCE/SEI Eq. No.	Load Combination
Noncoastal A	4b	$1.2D + 0.5W + 1.0F_a + L + (0.5L_r \text{ or } 0.3S \text{ or } 0.5R)$
	5b	$0.9D + 0.5W + 1.0F_a$
Coastal A and V	4b	$1.2D + 1.0W + 2.0F_a + L + (0.5L_r \text{ or } 0.3S \text{ or } 0.5R)$
	5b	$0.9D + 1.0W + 2.0F_a$

Table 7.13 Allowable Stress Design Load Combinations Including Flood Loads for Structural Elements Above Grade

Flood Zone	ASCE/SEI Eq. No.	Load Combination
Noncoastal A	5b	$D + 0.6W + 0.75F_a$
	6b	$D + 0.75L + 0.75(0.6W) + 0.75(L_r \text{ or } 0.7S \text{ or } R) + 0.75F_a$
	7b	$0.6D + 0.6W + 0.75F_a$
Coastal A and V	5b	$D + 0.6W + 1.5F_a$
	6b	$D + 0.75L + 0.75(0.6W) + 0.75(L_r \text{ or } 0.7S \text{ or } R) + 1.5F_a$
	7b	$0.6D + 0.6W + 1.5F_a$

Strength design load combinations for structural elements below grade subjected to flood-induced hydrostatic uplift or lateral hydrostatic loads, H, are given in Table 7.14. Because the variability in hydrostatic loads under flood conditions is small compared with the variability in hydrodynamic loads from overland flooding and from wave loads, a load factor of 1.6 on H is used in all flood zones instead of a load factor of 2.0, the latter of which is applicable to above-grade structural elements.

Table 7.14 Strength Design Load Combinations Including Flood-Induced Loads for Structural Elements Below Grade

ASCE/SEI Eq. No.	Load Combination
2a	$1.2D+1.6(L+H)+(0.5L_r$ or $0.3S$ or $0.5R)$
3a	$1.2D+(1.6L_r$ or $1.0S$ or $1.6R)+(L$ or $0.5W)+1.6H$
4a	$1.2D+1.0(W$ or $W_T)+L+(0.5L_r$ or $0.3S$ or $0.5R)+1.6H$
5a	$0.9D+1.0(W$ or $W_T)+1.6H$
6	$1.2D+E_v+E_h+L+0.15S+1.6H$
7	$0.9D-E_v+E_h+1.6H$

Allowable stress design load combinations for structural elements below grade subjected to flood-induced hydrostatic uplift or lateral hydrostatic loads, H, are given in Table 7.15.

Table 7.15 Allowable Stress Design Load Combinations Including Flood-Induced Loads for Structural Elements Below Grade

ASCE/SEI Eq. No.	Load Combination
2a	$D+L+H$
3a	$D+(L_r$ or $0.7S$ or $R)+H$
4a	$D+0.75L+0.75(L_r$ or $0.7S$ or $R)+H$
5a	$D+0.6(W$ or $W_T)+H$
6a	$D+0.75L+0.75[0.6(W$ or $W_T)]+0.75(L_r$ or $0.7S$ or $R)+H$
7a	$0.6D+0.6(W$ or $W_T)+H$
8	$D+0.7E_v+0.7E_v+H$
9	$D+0.525E_v+0.525E_h+0.75L+0.1S+H$
10	$0.6D-0.7E_v+0.7E_h+H$

7.3 Tsunami Loads

7.3.1 Overview

A tsunami is a series of waves with variable long periods that typically result from earthquake-induced uplift or subsidence of the seafloor (ASCE/SEI 6.2). Landslides near a coast or underwater volcanic eruptions can also generate tsunamis.

When a tsunamigenic earthquake occurs, the seafloor is uplifted and down-dropped, which pushes the column of water above the fault up and down. The potential energy that results from pushing the water above mean sea level is transferred into kinetic energy in the form of horizontal propagation of two oppositely traveling tsunami waves.

Amplification of the waves occurs as the waves travel outward from the source. The first part of the wave to reach the shoreline is a trough, which is followed by very strong and fastmoving tides. *Runup* occurs when the wave travels past the shoreline onto land and is a measure of the height of water onshore above a reference datum. The maximum horizontal inland extent of flooding due to a tsunami is defined as the *inundation limit* (see Figure 7.13, which is based on ASCE/SEI Figure 6.2-1).

Figure 7.13 Tsunami Parameters for a Topographic Flow Transect

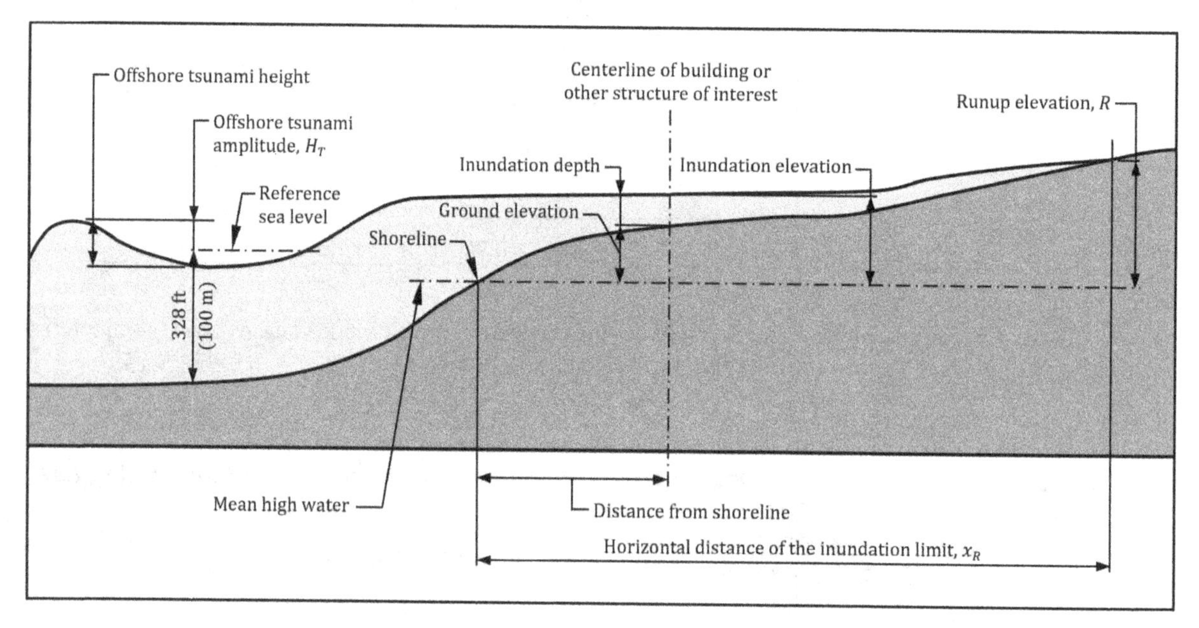

Runup elevation, *R*, is the vertical distance above ground elevation at the maximum tsunami inundation limit. This must include any relative change in sea level, which could be caused by ocean rise and/or ground subsidence due to the tsunami. Runup elevation is measured with respect to the North American Vertical Datum 1988 (NAVD 88) reference datum (ASCE/SEI 6.2). Effects of relative sea level change and seismic subsidence on tsunami inundation are given in ASCE/SEI Figure C6.1-3.

The *grade plane* is the horizontal reference plane at the site of the structure; it represents the average elevation of finished ground level adjoining the structure at all exterior walls. In cases where the finished ground level slopes away from the exterior walls, the grade plane is determined using the lowest points within the area between the structure and the property line. Where the property line is more than 6 ft (1.83 m) from the structure, the area to be used in determining the grade plane is that between the structure and 6 ft (1.83 m) from the structure. The depth of water at the grade plane of a structure, including any relative sea level change, is defined as the *inundation depth*.

Methods to calculate design tsunami loads and effects, F_{TSU}, are given in the following sections. Certain buildings and structures located in tsunami-prone regions must be designed and constructed to resist the effects of tsunamis in accordance with ASCE/SEI Chapter 6 (IBC 1615.1).

Tsunamis have the potential of generating very large hydrodynamic loads and it has been documented that one- and two-family dwellings and low-rise buildings of light-frame construction typically do not survive a design tsunami event. Therefore, it is not practical to provide tsunami design requirements for these types of structures (ASCE/SEI C6.1.1).

7.3.2 Terminology

Terminology related to tsunami loads is given in Table 7.16.

Table 7.16 Terminology Related to Tsunami Loads

Term	Definition
Bathymetric profile	A cross-section showing ocean depth plotted as a function of horizontal distance from a reference point (such as a coastline).
Closure ratio	Ratio of the area of enclosure, not including glazing and openings, that is inundated to the total projected vertical plane area of the inundated enclosure surface exposed to flow pressure.
Critical facility	• Buildings and structures providing services designated by federal, state, local, or tribal governments to be essential for the implementation of the response and management plan or for the continued service functioning of a community (such as, facilities for power, fuel, water, communications, public health, major transportation infrastructure, and essential government operations). • Critical facilities comprise all public and private facilities deemed by a community to be essential for the delivery of vital services, protection of special populations and the provision of other services of importance for that community.
Deadweight tonnage (DWT)	• A vessels displacement tonnage (DT) minus its lightship weight (LWT). • DWT is a classification used for the carrying capacity of a vessel, which is the sum of the weights of cargo, fuel, fresh water, ballast water, provisions, passengers, and crew. The weight of the vessel itself is not included in the DWT.
Design tsunami parameters	Tsunami parameters used for design, consisting of the inundation depths and flow velocities at the stages of inflow and outflow most critical to the structure and momentum flux.
Displacement tonnage (DT)	Total weight of a fully loaded vessel.
Froude number, F_r	A dimensionless number that is used to quantify the normalized tsunami flow velocity as a function of water depth.
Grade plane	• A horizontal reference plane at the site representing the average elevation of finished ground level adjoining the structure at all exterior walls. • Where the finished ground level slopes away from the exterior walls and the property line is less than or equal to 6 ft (1.83 m) from the structure, the grade plane is established by the lowest points within the area between the structure and the property line. • Where the finished ground slopes away from the exterior walls and the property line is more than 6 ft (1.83 m) from the structure, the grade plane is established by the lowest points within the area between the structure and points 6 ft (1.8 m) from the structure.
Inundation depth	The depth of the design tsunami water level, including relative sea level change, with respect to the grade plane at the structure.
Inundation elevation	The elevation of the design tsunami water surface, including relative sea level change, with respect to vertical datum at mean high water (MHW).
Inundation limit	• The maximum horizontal inland extent of flooding for the maximum considered tsunami (MCT), where the inundation depth above grade becomes zero. • The horizontal distance that is flooded relative to the shoreline defined where the elevation is at the MHW.
Lightship weight (LWT)	Weight of a vessel without cargo, fuel, fresh water, ballast water, provisions, passengers, and crew.

(continued)

Table 7.16 Terminology Related to Tsunami Loads (*Continued*)

Maximum considered tsunami (MCT)	A probabilistic tsunami having a 2% probability of being exceeded in a 50-year period (i.e., a 2,475-year mean recurrence interval).
Mean high water (MHW)	The average of all the high-water heights observed over the national tidal datum epoch. For stations with shorter series, comparison of simultaneous observations with a control tide station is made to derive the equivalent datum of the national tidal datum epoch.
Nearshore profile	Cross-sectional bathymetric profile from the shoreline to a water depth of 328 ft (100 m).
Nearshore tsunami amplitude	The MCT amplitude immediately off the coastline at 33 ft (10 m) of water depth.
Offshore tsunami amplitude	The MCT amplitude relative to the reference sea level, measured where the undisturbed water depth is 328 ft (100 m).
Offshore tsunami height	Waveform vertical dimension of the MCT from consecutive trough to crest, measured where the undisturbed water depth is 328 ft (100 m), after removing the tidal variation.
Reference sea level	The sea level datum used in site-specific inundation modeling, which is typically taken as the MHW.
Runup elevation	Ground elevation at the maximum tsunami inundation limit, including relative sea level change, with respect to the MHW datum.
Topographic transect	Profile of vertical elevation data versus horizontal distance along a cross-section of the terrain, in which the orientation of the cross-section is perpendicular or at some specified orientation angle to the shoreline.
Tsunami	A series of waves with variable long periods, typically resulting from earthquake-induced uplift or subsidence of the seafloor.
Tsunami amplitude	The absolute value of the difference between a particular peak or trough of the tsunami and the undisturbed sea level at the time.
Tsunami bore	A steep and turbulent broken wavefront generated on the front edge of a long-period tsunami waveform when shoaling over mild seabed slopes or abrupt seabed discontinuities such as fringing reefs or in a river estuary, in accordance with ASCE/SEI 6.6.4.
Tsunami bore height	The height of a broken tsunami surge above the water level in front of the bore or above the grade elevation if the bore arrives on nominally dry land.
Tsunami breakaway wall	A nonstructural wall designed to fail when subjected to tsunami loads.
Tsunami design zone (TDZ)	An area identified on the tsunami design zone map (TDZM) between the shoreline and the inundation limit, within which buildings and structures are analyzed and designed for inundation by the MCT.
Tsunami design zone map (TDZM)	• The map in ASCE/SEI Figure 6.1-1 designating the potential horizontal inundation limit of the MCT. • A state or local jurisdiction's probabilistic map produced in accordance with the requirements of ASCE/SEI 6.7.
Tsunami-prone region (TPR)	The coastal region in the United States addressed by ASCE/SEI Chapter 6 with quantified probability in the recognized literature of tsunami inundation hazard with a runup greater than 3 ft (0.91 m) caused by tsunamigenic earthquakes in accordance with the probabilistic tsunami hazard analysis (PTHA) method given in ASCE/SEI Chapter 6.
Tsunami risk category (TRC)	The risk category in ASCE/SEI 1.5, as modified for specific use related to ASCE/SEI Chapter 6 in accordance with ASCE/SEI 6.4.
Tsunami vertical evacuation refuge structure (TVERS)	A structure designated and designed to serve as a point of refuge to which a portion of the community's population can evacuate above a tsunami when high ground is not available.

7.3.3 General Requirements

The tsunami requirements in ASCE/SEI Chapter 6 are applicable to the following states: Alaska, California, Hawaii, Oregon, and Washington. These states have designated tsunami-prone regions (TPR), which are coastal regions where tsunamigenic earthquakes determined in accordance with the probabilistic tsunami hazard analysis method in ASCE/SEI Chapter 6 can cause a tsunami inundation hazard with a runup greater than 3 ft (0.91 m).

Certain structures in TPR must be analyzed and designed for inundation based on the maximum considered tsunami (MCT), which is defined as a probabilistic tsunami having a 2 percent probability of being exceed in a 50-year period (i.e., a 2,475-year mean recurrence interval). More information on the MCT can be found in ASCE/SEI C6.1.

The tsunami design zone map (TDZM) in ASCE/SEI Figure 6.1-1 designates the potential horizontal inundation limits of the MCT for the five states. The tsunami design zone (TDZ) is an area identified on the TDZM between the shoreline and the inundation limit.

The ASCE Tsunami Design Geodatabase contains maps that graphically depict the extent of inundation (i.e., the TDZ) due to the MCT (Reference 7.3). A site can be located by address, by latitude and longitude, or by clicking on the map. In addition to the TDZ (which is shaded light blue on the maps), offshore tsunami amplitudes, runup elevation points, and inundation depth points are given (which are designated by blue squares, red triangles, and red circles, respectively). Earthquake-induced regional ground subsidence contours, 3-ft (0.91-m) inundation depths, and bore points are indicated as well. Transects can be drawn and elevations of the MHW along the transect can be obtained in graphical or tabular form. A state or local jurisdiction may have adopted a TDZ and other data different from those in the Geodatabase, so it is good practice to check with the local authorities prior to performing the analysis.

In addition to the ASCE Geodatabase, ASCE 7-22 Tsunami Design Zone Maps for Selected Locations (https://doi.org/10.1061/9780784480748) contains TDZM in high-resolution PDF format for the 62 locations indicated by circles in ASCE/SEI Figures C6.1-1(a) to (i). For the purpose of identifying the TDZ, these downloadable PDF maps are considered equivalent to the results obtained from the ASCE Tsunami Design Geodatabase.

Sea level rise is not incorporated into the TDZM. The runup elevation and the inundation depth for a given MCT must be increased by at least the future projected relative sea level rise during the design life of a structure. A project life cycle of not less than 50 years should be considered (ASCE/SEI 6.5.3). Relative sea level change at a site may result from ground subsidence, long-term erosion, thermal expansion caused by warming of the ocean or increased melting of land-based ice. Sea level changes can be obtained from the NOAA Center for Operational Oceanographic Products and Services (https://tidesandcurrents.noaa.gov/). Once the potential increase in relative sea level has been determined, it must be added to the inundation depth results of the energy grade level analysis (EGLA).

The following buildings and other structures located within a TDZ must be designed for the MCT (ASCE/SEI 6.1.1):

- Tsunami risk category IV buildings and structures
- Tsunami risk category III buildings and structures with an inundation depth greater than 3 ft (0.91 m) at any location within the intended footprint of the structure
- Where required by a state or locally adopted building code statue to include design for tsunamis effects, tsunami risk category II buildings with a mean height above the grade plane greater than the height designated in the statue and with an inundation depth greater than 3 ft (0.91 m) at any location within the intended footprint of the building.

Tsunami risk categories are defined in ASCE/SEI 6.4 (see Section 7.3.4 of this book).

Two exceptions to these requirements are given in ASCE/SEI 6.1.1. In the first exception, tsunami loads and effects determined in accordance with ASCE/SEI Chapter 6 need not be applied to tsunami risk category II single-story buildings of any height without mezzanines or any occupiable roof level and not having any critical equipment or systems. The second exception permits the TDZ, inundation limits, and runup elevations to be determined by the site-specific procedures of ASCE/SEI 6.7 for coastal regions subject to tsunami inundation and not covered by ASCE/SEI Figure 6.1-1 or by the procedures in ASCE/SEI 6.5.1.1 for tsunami risk category II or II structures using ASCE/SEI Figure 6.7-1.

7.3.4 Tsunami Risk Categories

Tsunami risk categories are the same as the risk categories defined in ASCE/SEI 1.5 with the following modifications (ASCE/SEI 6.4; see Table 7.17):

1. Federal, state, local, or tribal governments are permitted to include critical facilities in tsunami risk category III, such as power-generating stations, water-treatment facilities for potable water, waste-water treatment facilities, and other public utility facilities not included in risk category IV. Additional information on this modification can be found in ASCE/SEI C6.4.
2. The following structures need not be included in tsunami risk category IV, and state, local, or tribal governments are permitted to designate them to tsunami risk category II or III:
   a. Fire stations, ambulance facilities and emergency vehicle garages;
   b. Earthquake or hurricane shelters;
   c. Emergency aircraft hangars; and
   d. Police stations without holding cells and not uniquely required for post-disaster emergency response as a critical facility.
3. Tsunami vertical evacuation refuge structures (TVERS) must be included in tsunami risk category IV.

Table 7.17 Risk Categories of Buildings and Other Structures (ASCE/SEI Table 1.5-1)

Risk Category	Use or Occupancy
I	Buildings and other structures that represent a low risk to humans.
II	All buildings and other structures except those listed in risk categories I, III, and IV.
III	• Buildings and other structures the failure of which could pose a substantial risk to human life. • Buildings and other structures with potential to cause a substantial economic impact and/or mass disruption of day-to-day civilian life in the event of failure.
IV	• Buildings and other structures designated as essential facilities. • Buildings and other structures, the failure of which could pose a substantial hazard to the community.

The essential facilities listed in item 2 above do not need to be included in tsunami risk category IV because these facilities are expected to be evacuated prior to the arrival of a tsunami. Additional information on these modifications can be found in ASCE/SEI C6.4.

A TVERS is a structure designated and designed to serve as a point of refuge where people can evacuate above a tsunami (ASCE/SEI 6.14). More information on such structures is given in ASCE/SEI C6.14.

7.3.5 Analysis of Design Inundation Depth and Flow Velocity

Permitted Analysis Procedures

The design inundation depth and flow velocity at a site are needed to calculate design tsunami loads. The applicable method to determine these parameters depends on the TRC assigned to the building or other structure and whether the inundation limit and runup elevation of the MCT can be determined by ASCE/SEI Figure 6.1-1 or not.

Tsunami Risk Category II and III Buildings and Other Structures

- Case 1: Inundation depth and runup elevation are given in ASCE/SEI Figure 6.1-1
  Where inundation depth and runup elevation are given in ASCE/SEI Figure 6.1-1 for a particular site, maximum inundation depths and flow velocities must be determined by the energy grade line analysis (EGLA) in ASCE/SEI 6.6. The site-specific probabilistic tsunami hazard analysis (PTHA) in ASCE/SEI 6.7 is permitted as an alternate to the EGLA. The flow velocities determined by the PTHA are subject to the limitations in ASCE/SEI 6.7.6.8.
- Case 2: No mapped inundation limit is given in ASCE/SEI Figure 6.1-1 and the offshore tsunami amplitude is given in ASCE/SEI Figure 6.7-1
  Where no mapped inundation limit is given in ASCE/SEI Figure 6.1-1 and the offshore tsunami amplitude (which is used to determine the runup elevation) is given in ASCE/SEI Figure 6.7-1, the R/H_T analysis method in ASCE/SEI 6.5.1.1 is permitted to be used for tsunami risk category II and III buildings and other structures.

Tsunami Risk Category IV Buildings and Other Structures

The EGLA and the site-specific PTHA must both be performed for tsunami risk category IV buildings and other structures. However, for structures other than tsunami vertical evacuation refuge structures (TVERS), a site-specific analysis need not be performed where the inundation depth determined by the EGLA method is less than 12 ft (3.66 m) at any point within the location of the structure.

Site-specific velocities less than those determined by the EGLA method are subject to the limitation in ASCE/SEI 6.7.6.8, while those greater than the velocities determined by the EGLA method must be used.

Required analysis methods to determine inundation depths and flow velocities are given in Tables 7.18 and 7.19 based on whether the runup elevation is given in ASCE/SEI Figure 6.1-1 or calculated from ASCE/SEI Figure 6.7-1, respectively.

Table 7.18 Inundation Depth and Flow Velocity Analysis Procedures Where Runup Is Given in the Tsunami Design Zone Maps of ASCE/SEI Figure 6.1-1

Analysis Procedure	Tsunami Risk Category (TRC)			
	II	III	IV: Excluding TVERS	IV: TVERS
R/H_T analysis (ASCE/SEI 6.5.1.1)	Not permitted	Not permitted	Not permitted	Not permitted
Energy grade line analysis (EGLA) [ASCE/SEI 6.6]	Required*	Required*	Required	Required
Site-specific probabilistic tsunami hazard analysis (PTHA) [ASCE/SEI 6.7]	Permitted*	Permitted*	Required**	Required

*The requirements of ASCE/SEI Chapter 6 do not apply to tsunami risk category II and III buildings and other structures where the MCT inundation depth is less than or equal to 3 ft (0.91 m) and where the MCT inundation depth must include a sea level rise component in accordance with ASCE/SEI 6.5.3 [ASCE/SEI 6.1.1].

**A site-specific PTHA must be performed where the inundation depth resulting from the EGLA is determined to be greater than or equal to 12 ft (3.66 m) at any point within the location of the tsunami risk category IV structure (other than TVERS) [ASCE/SEI 6.5.2].

Table 7.19 Inundation Depth and Flow Velocity Analysis Procedures Where Runup Is Calculated from ASCE/SEI Figure 6.7-1

Analysis Procedure	Tsunami Risk Category (TRC)			
	II	III	IV	
			Excluding TVERS	TVERS
R/H_T analysis (ASCE/SEI 6.5.1.1)	Required	Required	Not permitted	Not permitted
Energy grade line analysis (EGLA) [ASCE/SEI 6.6]	Required*	Required*	Required	Required
Site-specific probabilistic tsunami hazard analysis (PTHA) [ASCE/SEI 6.7]	Permitted*	Permitted*	Required	Required

*The requirements of ASCE/SEI Chapter 6 do not apply to tsunami risk category II and III buildings and other structures where the MCT inundation depth is less than or equal to 3 ft (0.91 m) and where the MCT inundation depth must include a sea level rise component in accordance with ASCE/SEI 6.5.3 [ASCE/SEI 6.1.1].

The R/H_T and EGLA methods are given in the following sections. Information on how to determine maximum inundation depths and flow velocities using a site-specific PTHA is given in ASCE/SEI C6.7.

Runup Evaluation for Areas Where No Map Values Are Given (R/H_T Analysis)

The runup elevation, R, can be determined using the procedure in ASCE/SEI 6.5.1.1 for tsunami risk category II and III buildings and other structures where no mapped inundation limit is shown in ASCE/SEI Figure 6.1-1 (see Table 7.19).

The ratio of tsunami runup elevation to offshore tsunami amplitude, R/H_T, is determined by ASCE/SEI Equations (6.5-2a) through (6.5-2e) as a function of the surf similarity parameter, ξ_{100} (the equations for R/H_T are plotted in ASCE/SEI Figure 6.5-1 and are given in Table 7.20). The offshore tsunami amplitude, H_T, is acquired from Reference 7.3 or the local jurisdiction and ξ_{100} is determined by ASCE/SEI Equation (6.5-1):

$$\xi_{100} = \frac{T_{\text{TSU}}}{\cot\Phi}\sqrt{\frac{g}{2\pi H_T}} \tag{7.16}$$

Table 7.20 Equations to Determine R/H_T

ASCE/SEI Eq. No.	ξ_{100}	R/H_T
6.5-2a	$\xi_{100} \leq 0.6$	1.5
6.5-2b	$0.6 < \xi_{100} \leq 6$	$2.5[\log_{10}(\xi_{100})] + 2.05$
6.5-2c	$6 < \xi_{100} \leq 20$	4.0
6.5-2d	$20 < \xi_{100} \leq 100$	$-2.15[\log_{10}(\xi_{100})] + 6.80$
6.5-2e	$\xi_{100} > 100$	2.5

In this equation, T_{TSU} is the predominant wave period or the time from the start of the first pulse to the end of the second pulse of the tsunami at the 328-ft (100-m) bathymetric depth, which is acquired from Reference 7.3 (the extent of which is given in ASCE/SEI Figure 6.7-1) or the local jurisdiction, Φ is the mean slope angle of the nearshore profile taken from the 328-ft (100-m) water depth to the MHW elevation along the axis of the topographic transect for the site and g is acceleration caused by gravity.

Runup elevation, R, for a site is equal to R/H_T times H_T; R can then be used in an EGLA to obtain inundation depths and flow velocities for tsunami risk category II and III buildings and other structures.

According to the exception in ASCE/SEI 6.5.1.1, the equations to determine R/H_T must not be used in the following cases:

- Where wave focusing, such as at headlands, are expected.
- In V-shaped bays.
- Where the on-land flow fields are expected to vary significantly in the direction parallel to the shoreline because of longshore variability of topography.

Flowchart 7.2 in Section 7.4 of this book can be used to determine to determine R/H_T.

7.3.6 Inundation Depths and Flow Velocities Based on Runup

Maximum Inundation Depth and Flow Velocities Based on Runup

Maximum inundation depths and flow velocities associated with the stages of tsunami flooding must be determined in accordance with ASCE/SEI 6.6.2.

According to ASCE/SEI 6.6.1, the calculated flow velocity must not be taken less than 10 ft/s (3.05 m/s) and need not be taken greater than the lesser of $1.5(gh_{max})^{1/2}$ and 50 ft/s (15.2 m/s) where h_{max} is the maximum inundation depth at the site. These minimum and maximum limits are based on data from past tsunamis of significant inundation depth. The upper limit of 50 ft/s (15.2 m/s) includes a 1.5 factor on the velocity observed during actual on-land flow.

In cases where the maximum topographic elevation along the transect between the shoreline and the inundation limit is greater than R, one of the following methods must be used to determine the inundation depth, h_i, and the tsunami flow velocity, u_i (ASCE/SEI 6.6.1):

1. The site specific PTHA of ASCE/SEI 6.7.6 subject to the minimum and maximum velocities noted above.
2. The EGLA method of ASCE/SEI 6.6.2 assuming a horizontal inundation limit distance, x_R (and corresponding runup elevation, R) that has at least 100 percent of the maximum topographic elevation along the topographic transect.
3. Where the site lies within a completely overwashed area where inundation depth points are given in Reference 7.3, the inundation elevation profiles must be determined using the EGLA with the modifications given in ASCE/SEI 6.6.1(3). This may occur, for example, where a tsunami flows over an island or peninsula into a second water body; in such cases, the horizontal distance of the inundation limit must be equal to the length of the overwashed land. Additional information can be found in ASCE/SEI C6.6.2.

Energy Grade Line Analysis (EGLA)

The EGLA method is required to determine inundation depths and velocities for the cases identified in Tables 7.18 and 7.19. The main assumption is that the flow velocity of the tsunami is maximum at the shoreline and zero at the inundation point (ASCE/SEI 6.6.2). Between the shoreline and the inundation point, the flow is decreased by friction from terrain roughness, which is accounted for by Manning's coefficient, n. Energy is accumulated based on the topographic (ground) slope and the equivalent slope due to surface friction. At any location, the total energy is a combination of potential energy (based on the depth of the water, h_i) and kinetic energy (based on the flow velocity, u_i). The Froude number, F_r, which is determined by ASCE/SEI Equation (6.6-3), provides the required relationship between h_i and u_i, so that the hydraulic head, $E_{g,i}$, can be calculated by ASCE/SEI Equation (6.6-1). Velocity is assumed to be a function of inundation depth, which, as noted previously, varies from the shoreline (maximum) to the inundation point (zero).

Analysis is performed incrementally across the ground transect at the site from the runup (where the hydraulic head, or energy, at the inundation limit, x_R, is zero) to the shoreline (see Figure 7.14). The transect must be broken into segments no longer than 100 ft (30.5 m) in length; this limit is intended to ensure accuracy of the hydraulic analysis. The ground elevation along the transect is represented by a series of one-dimensional, linear-sloped segments. Generally, more elevation points (i.e., smaller segment lengths) yield better results. The transect elevations, z_i,

Figure 7.14 Energy Grade Line Analysis (EGLA)

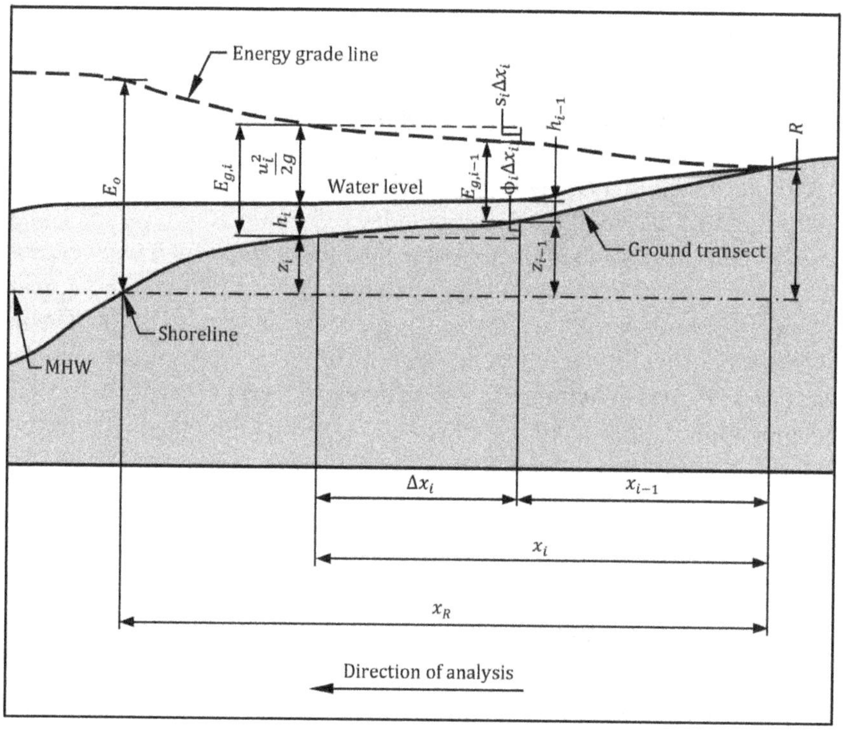

should be obtained from a topographic digital elevation model (DEM) of at least a 33-ft (10 m) resolution (ASCE/SEI C6.6.2).

The hydraulic head, $E_{g,i}$, at point i at the boundary of the segment is determined by ASCE/SEI Equation (6.6-1):

$$E_{g,i} = E_{g,i-1} + (\phi_i + s_i)\Delta x_i \tag{7.17}$$

where $E_{g,i-1}$ = hydraulic head at point $(i-1)$ of the segment

ϕ_i = average ground slope between points i and $(i-1) = \dfrac{\Delta z_i}{\Delta x_i} = \dfrac{z_{i-1} - z_i}{x_{i-1} - x_i}$

s_i = friction slope of the energy grade line between points i and $(i-1)$ determined by ASCE Equations (6.6-2) and (6.6-2.SI):

$$s_i = \frac{(u_i)^2}{\left(\frac{1.49}{n}\right)^2 h_i^{4/3}} = \frac{g(F_{ri})^2}{\left(\frac{1.49}{n}\right)^2 h_{i-1}^{1/3}}$$

In S.I.:

$$s_i = \frac{(u_i)^2}{\left(\frac{1.00}{n}\right)^2 h_i^{4/3}} = \frac{g(F_{ri})^2}{\left(\frac{1.00}{n}\right)^2 h_{i-1}^{1/3}}$$

u_i = maximum flow velocity at point i

n = Manning's coefficient of the terrain segment, obtained from ASCE/SEI Table 6.6-1 (see Table 7.21 and ASCE/SEI C6.6.3)

F_{ri} = Froude number at point i determined by ASCE/SEI Equation (6.6-3)

$$= \alpha\left(1 - \frac{x_i}{x_R}\right)^{0.5} = \frac{u_i}{(gh_i)^{0.5}}$$

α = Froude number coefficient = 1.0 in accordance with ASCE/SEI 6.6.2

x_R = distance from the shoreline to the runup point

Δx_i = increment of horizontal distance, which must be less than or equal to 100 ft (30.5 m) $= x_{i-1} - x_i$

Table 7.21 Manning's Roughness Coefficient, *n*, for EGLA

Description of Frictional Surface	*n*
Coastal water nearshore bottom friction	0.025 to 0.03
Open land or field	0.025
All other cases	0.03
Buildings of at least urban density	0.04

The analysis begins at the runup point, which is located a horizontal distance x_R from the shoreline where x_R is obtained from Reference 7.3 or an R/H_T analysis. At the runup point, the hydraulic head, E_R, is equal to zero. Although the inundation height, h_i, is equal to zero at this point, a small value (such as 0.1 ft or 0.03 m) is used in the analysis because at the next point along the transect, the friction slope, s_i, is determined by ASCE/SEI Equation (6.6-2) or (6.6-2.SI) using the inundation depth from the previous point; a value of zero for the inundation depth at the runup point would result in dividing by zero.

The analysis proceeds by calculating the inundation depth, h_i, and the flow velocity, u_i, at every point along the transect; it ends at the shoreline where the hydraulic head and flow velocity are maximum. At the site of the building or other structure, the inundation depth is h_{max} and the flow velocity is u_{max}, which are used to determine tsunami loads and effects on the structure.

The Froude number coefficient, α, is equal to 1.0 where tsunami bores need not be considered. Where tsunami bores must be considered in accordance with ASCE/SEI 6.6.4, tsunami loads on vertical structural components (ASCE/SEI 6.10.2.3) and the hydrodynamic loads for tsunami bore flow entrapped in structural wall-slab recesses (ASCE/SEI 6.10.3.3) must be determined based on inundation depths and flow velocities using $\alpha = 1.3$.

The flow velocity, u_i, must be increased by an amount equal to $\Delta u_i = u_i|\phi_{i,200}|^{0.25}$ [ASCE/SEI Equation (6.6-4)] at locations along the topographic transect where $\phi_{i,200}$, which is the topographic slope averaged over a distance of 200 ft (61.0 m) centered on the transect point x_i, is negative. Note that Δu_i need not be taken greater than 6 ft/s (1.83 m/s).

Flowchart 7.3 in Section 7.4 of this book can be used to determine the inundation depths, h_i, and the flow velocities, u_i, along the transect, including h_{max} and u_{max} at the site of a building or other structure, in accordance with the EGLA method in ASCE/SEI 6.6.2.

Figure 7.15 Directionality of Flow Requirements in ASCE/SEI 6.8.6

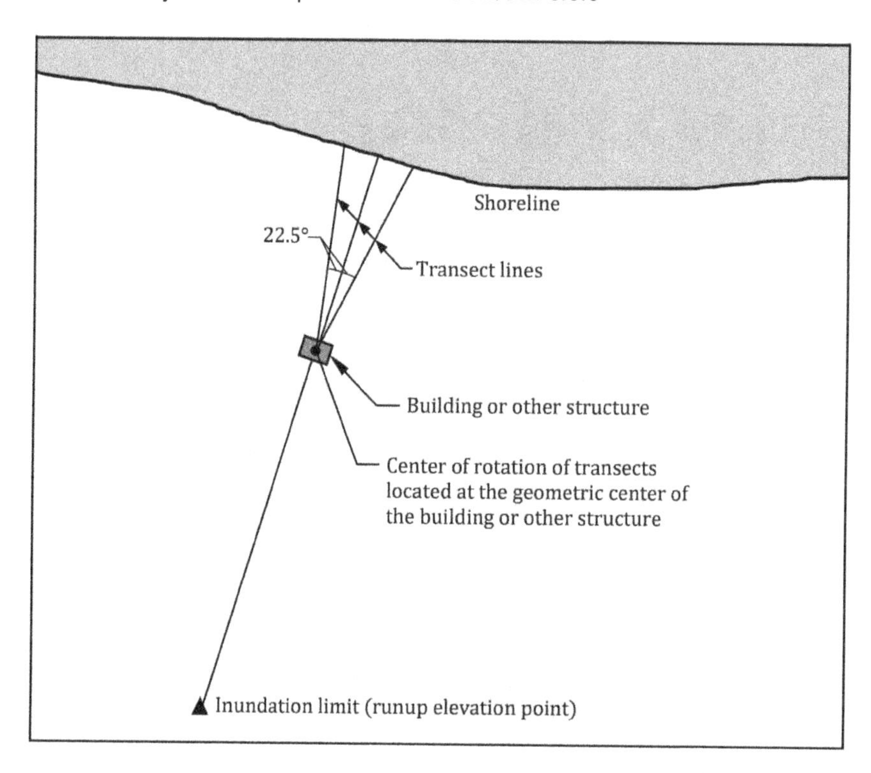

The directionality of flow requirements in ASCE/SEI 6.8.6 must be applied in the analysis. Both incoming and outgoing flow conditions must be considered for three flow directions (transects) averaged over 500 ft (152 m) to either side of the site (see Figure 7.15):

- Transect located perpendicular to the shoreline.
- Transect located 22.5-degrees clockwise from the transect that is perpendicular to the shoreline.
- Transect located 22.5-degrees counterclockwise from the transect that is perpendicular to the shoreline.

This requirement accounts for tsunamis that do not strike a building head on. The transect producing the maximum runup elevation may be used to perform the EGLA. Alternatively, an EGLA may be performed in each direction, and the one producing the maximum tsunami load effects on the structure should be used.

Tsunami bores must be considered where any of the following conditions exist (ASCE/SEI 6.6.4):

1. The prevailing nearshore bathymetric slope is 1/100 or milder.
2. Shallow fringing reefs or similar step discontinuities in nearshore bathymetric slope occur.
3. At locations where tsunami bores have historically been documented.
4. At locations where tsunami bores are described in recognized literature.
5. As determined by a site-specific inundation analysis.

The requirements in ASCE/SEI 6.10.2.3 for tsunami loads on vertical structural components and ASCE/SEI 6.10.3.3 for tsunami bore flow entrapped in structural wall-slab recesses are applicable where tsunami bores occur.

Flow velocities determined by the EGLA method must be adjusted for flow amplification in accordance with ASCE/SEI 6.8.5, as applicable (ASCE/SEI 6.6.5). The adjusted value need not exceed the maximum flow rate, which, as noted previously, is the lesser of $1.5(gh_{max})^{0.5}$ and 50 ft/s (15.2 m/s).

7.3.7 Inundation Depths and Flow Velocities Based on Site-Specific Probabilistic Tsunami Hazard Analysis

A site-specific PTHA is required to determine maximum inundation depths and velocities for the cases identified in Tables 7.18 and 7.19 (see Section 7.3.5 of this book).

In this method, Reference 7.3 is used to obtain the offshore tsunami amplitude and ASCE/SEI Figure 6.7-1 is used to obtain the dominant waveform period. These quantities are input to the inundation numerical model.

Detailed information on how to formulate such an analysis is given in ASCE/SEI C6.7 and in the references cited in that section.

7.3.8 Structural Design Procedures for Tsunami Effects

Performance Requirements

Structures, components, and foundations must be designed for the applicable performance criteria in ASCE/SEI 6.8.1 and 6.8.2 when subjected to the loads and effects of the MCT. The minimum performance criteria that must be satisfied based on the TRC are given in Table 7.22.

Additional information on performance criteria is given in ASCE/SEI C6.8.

Table 7.22 Performance Criteria Based on Tsunami Risk Category

Tsunami Risk Category	Performance Criteria
II and III	Collapse prevention structural performance criteria or better
• Critical facilities assigned to III • IV	• Operational nonstructural equipment necessary for essential functions and the elevation of the bottom of the lowest horizontal structural member at the level supporting such components and equipment must be above the inundation elevation of the MCT • Structural components and connections in occupiable levels and foundations must be designed in accordance with immediate occupancy structural performance criteria where the level of the occupiable levels equals or exceeds the MCT inundation elevation • Tsunami vertical evacuation refuge structures must also comply with the requirements in ASCE/SEI 6.14

Structural Performance Evaluation

Overview

Strength and stability of a structure must be evaluated to ensure it is capable of resisting the tsunami load cases in ASCE/SEI 6.8.3.1. The corresponding structural acceptance criteria are given in ASCE/SEI 6.8.3.4 for the lateral force-resisting system and in ASCE/SEI 6.8.3.5 for structural components.

Load Cases

The three load cases in ASCE/SEI 6.8.3.1 must be evaluated for any building or structure subjected to the MCT (see Figure 7.16). The hydrodynamic force, F_{dx}, is calculated by ASCE/SEI Equation (6.10-2) based on the conditions in each load case. These overall drag forces and other

Figure 7.16 Tsunami Load Cases

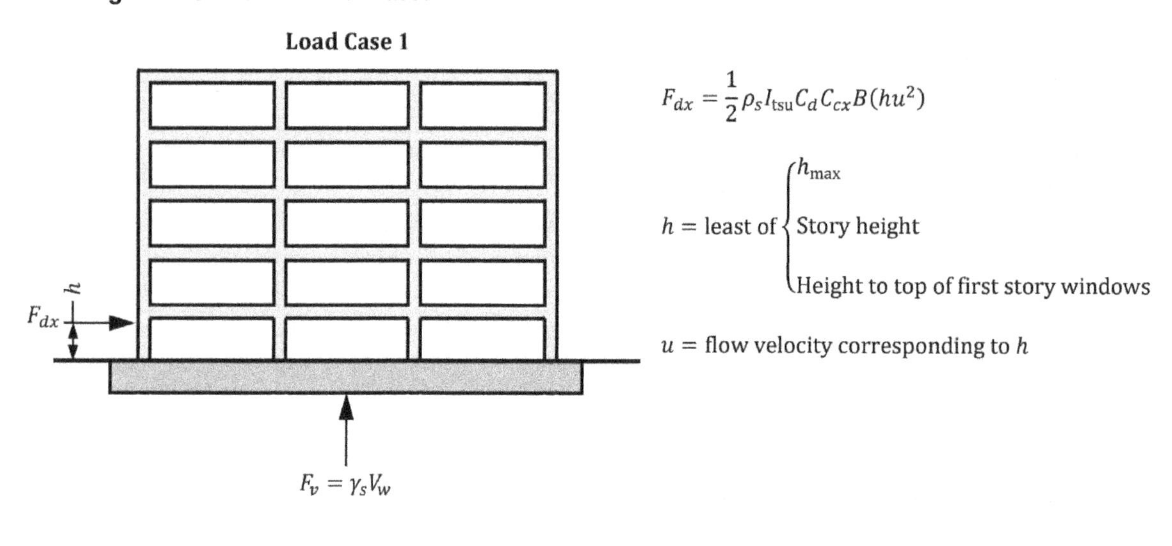

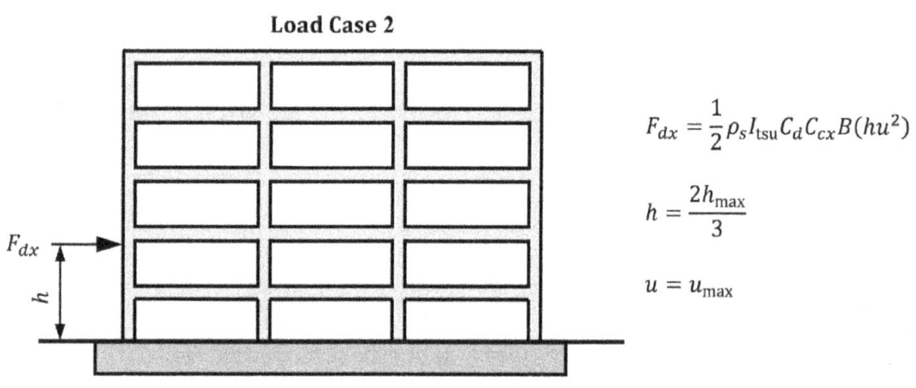

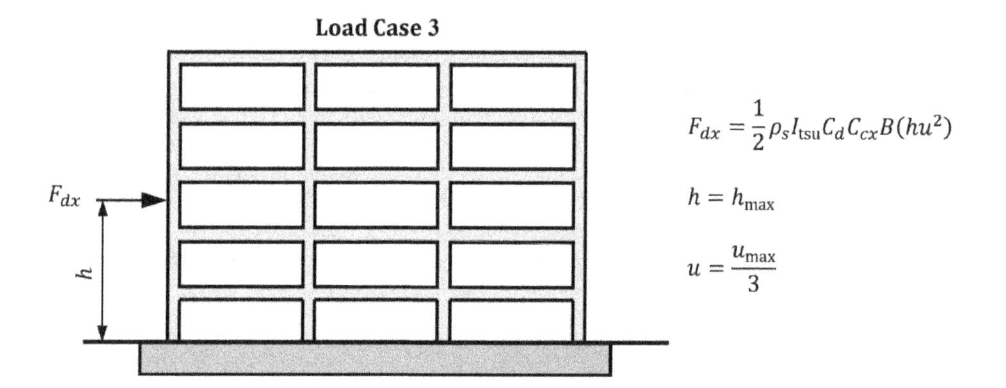

applicable forces are combined in accordance with the load combinations in ASCE/SEI 6.8.3.3 and are applied to the lateral force-resisting system of the building.

Load case 1 (buoyancy): In load case 1, the structure is subjected to the maximum buoyant force, F_v, and the hydrodynamic force, F_{dx}. The purpose of this load case is to check the overall stability of the structure and foundation anchorage for uplift. The uplift force is calculated in accordance with ASCE/SEI 6.9.1 and depends on the differential inundation depth exterior to the structure and the flooded depth within the structure. It need not be applied to open structures nor to structures where the soil properties or foundation and structural design prevent detrimental hydrostatic pressurization on the underside of the foundation and the lowest structural slab.

Load case 2 (maximum velocity): In load case 2, the structure is subjected to the maximum hydrodynamic force, F_{dx}. This force is determined using the maximum flow velocity, u_{max}, and two-thirds the maximum inundation depth, that is, $h = 2h_{max}/3$.

Load case 3 (maximum inundation depth): In load case 3, the structure is subjected to the hydrodynamic force, F_{dx}, based on the maximum inundation depth, h_{max}. The flow velocity is taken as one-third of the maximum flow velocity, that is, $u = u_{max}/3$.

Unless a site-specific tsunami analysis is performed in accordance with ASCE/SEI 6.7, the inundation depths and velocities defined for the load cases must be determined using the time-history curves in ASCE/SEI Fig. 6.8-1 (numerical values of normalized inundation depths and flow velocities from the curves are given in ASCE/SEI Table C6.8-2). In the top curve, normalized inundation depths, h/h_{max}, are plotted versus normalized time, t/T_{TSU}, where T_{TSU} is the predominant wave period obtained from Reference 7.3 for the site. Load cases 2 and 3 are identified in the curve for both incoming and receding directions of flow. In the bottom curve, normalized flow velocities, u/u_{max}, are plotted versus t/T_{TSU}. The intent of the curves is to capture all possible combinations of inundation depth and velocity flow during the time sequence of a tsunami inundation.

Where a site-specific inundation procedure is used, the local building official may approve a site-specific inundation and velocity time-history curve subject to the minimum values of 80 percent or 100 percent of the EGLA values.

Tsunami Importance Factors

Tsunami importance factors, I_{tsu}, are given in ASCE/SEI Table 6.8-1 (see Table 7.23). These importance factors must be applied to the tsunami hydrodynamic and impact loads in ASCE/SEI 6.10 and 6.11, respectively. Information on their derivation is given in ASCE/SEI C6.8.3.2.

Table 7.23 Tsunami Importance Factors, I_{tsu}

Tsunami Risk Category	I_{tsu}
II	1.0
III	1.25
• Tsunami risk category IV • Vertical evacuation refuges • Tsunami risk category III critical facilities	1.25
Tsunami risk category IV where the inundation depth is less than 12 ft (3.66 m) and a site-specific PTHA is not performed, in accordance with the exception in ASCE/SEI 6.5.2	1.5

Load Combinations

Principal tsunami forces and effects, F_{TSU}, for incoming and receding directions of flow must be combined with other applicable forces and effects using the following load combinations [see ASCE/SEI Equations (6.8-1a) and (6.8-1b)]:

$$0.9D + F_{TSU} + H_{TSU} \tag{7.18}$$

$$1.2D + F_{TSU} + 0.5L + 0.2S + H_{TSU} \tag{7.19}$$

In these equations, H_{TSU} is the load caused by the tsunami-induced lateral foundation pressures developed under submerged conditions. The load factor on H_{TSU} must be taken as 0.9 where the net effect of H_{TSU} counteracts the principal load effect. The resistance factors for foundation stability analysis, ϕ, are given in ASCE/SEI 6.12.1.

These load combinations are consistent with those given in ASCE/SEI 2.5 for extraordinary events.

Lateral Force-Resisting System Acceptance Criteria

For buildings designed and detailed in accordance with the requirements for seismic design category (SDC) A, B, or C, which is determined using the provisions in ASCE/SEI 11.6, the lateral force-resisting system must be designed to resist the effects due to the MCT, where applicable.

For structure assigned to SDC D, E, or F, the lateral force-resisting system is deemed adequate to resist the required tsunami force, F_{TSU}, where $0.75\Omega_0 E_h > F_{TSU}$ (ASCE/SEI 6.8.3.4). The overstrength factor, Ω_0, is given in ASCE/SEI Table 12.2-1 based on the seismic force-resisting system (SFRS) and E_h is the horizontal seismic load effect defined in ASCE/SEI 12.4.2.1. Structures typically resist the effects from the design earthquake prior to the effects from the design tsunami, and the overall capacity of the lateral force-resisting system is considered acceptable without additional analyses provided the maximum tsunami force acting on the structure is less than $0.75\Omega_0 E_h$.

Where $0.75\Omega_0 E_h \leq F_{TSU}$, one of the following two acceptance criteria must be satisfied for structures assigned to SDC D, E, or F:

1. The lateral force-resisting system at all levels below the maximum inundation depth, h_{max}, must be designed for a minimum seismic force, Q_E, equal to $F_{TSU}/(0.75\Omega_0)$.
2. The lateral force-resisting system must be explicitly analyzed to validate its resistance to the MCT.

For structures where the immediate occupancy structural performance objectives must be satisfied, an explicit analysis and evaluation for the MCT must be performed.

Structural Component Acceptance Criteria

Structural components must be designed for the load effects resulting from the overall tsunami forces on the structural system combined with any resultant actions caused by local tsunami pressures for a given flow direction. In particular, the design of structural components must comply with the acceptability criteria of ASCE/SEI 6.8.3.5.1, the performance-based criteria of ASCE/SEI 6.8.3.5.2, or the progressive collapse avoidance criteria of ASCE/SEI 6.8.3.5.3.

According to the acceptability criteria in ASCE/SEI 6.8.3.5.1, structural components must be designed for combined load effects, including the effects due to the MCT, using a linearly elastic, static analysis, and the load combinations in ASCE/SEI 6.8.3.3. The material resistance factors, ϕ, that must be used are those prescribed in the material-specific standards, like ACI 318 for reinforced concrete members.

Information on the proper application of the criteria of ASCE/SEI 6.8.3.5.2 and ASCE/SEI 6.8.3.5.3 is given in ASCE/SEI C6.8.3.5.2 and ASCE/SEI C6.8.3.5.3, respectively.

Minimum Fluid Density for Tsunami Loads

The specific weight density, γ_{sw}, and the mass density, ρ_{sw}, of seawater are to be taken as 64.0 lb/ft^3 (10 kN/m^2) and 2.0 slugs/ft^3 (1,025 kg/m^3), respectively (ASCE/SEI 6.8.4). For purposes of calculating tsunami hydrostatic loads, the minimum fluid specific weight density, γ_s, is determined by ASCE/SEI Equation (6.8-4):

$$\gamma_s = k_s \gamma_{sw} \tag{7.20}$$

The fluid density factor, k_s, is equal to 1.1, and accounts for suspended soils and debris objects of various sizes and materials that would be expected during a tsunami.

Similarly, the minimum fluid mass density, ρ_s, to be used for calculating tsunami hydrodynamic loads is determined by ASCE/SEI Equation (6.8-5):

$$\rho_s = k_s \rho_{sw} \tag{7.21}$$

Flow Velocity Amplification

The tsunami design zone maps are based on conditions without any discrete objects in the path of flow. Greater energy is required to reach the specified mapped runup when the EGLA is used with the Manning's roughness based on buildings of at least urban density. Such urban environments are expected to have a higher probability of generating flow amplifications effects because of numerous diffraction scenarios. However, as noted previously, the EGLA produces sufficiently conservative inundation depths and velocities. Therefore, for an urban environment, it follows that taking 100 percent of the inundation depths and flow velocities from an EGLA essentially accounts for any possible amplifications in the flow velocity. Some examples of flow diffraction effects, which are based on experimental studies performed in North America, are given in ASCE/SEI C6.8.5.

The effect of upstream obstructions on flow is considered where the obstructions are enclosed structures of concrete, masonry, or structural steel located within 500 ft (152 m) of the site and have the following additional characteristics:

1. Structures have plan width greater than 100 ft (30.5 m) or 50 percent of the width of the downstream structure, whichever is greater.
2. Structures exist within the sector between 10 and 55 degrees to either side of the flow vector aligned with the center third of the width of the downstream structure (see ASCE/SEI Figure C6.8-5).

Structures unable to resist tsunami hydrodynamic forces, such as buildings of light-frame construction, are excluded from this discussion because they are not able to redirect sustained flow to other buildings because such structures typically collapse during a design tsunami event.

The requirements of ASCE/SEI 6.8.5 do not permit shielding effects of neighboring structures to reduce flow velocity less than that found from an EGLA.

Directionality of Flow

Design of structures for tsunami effects must consider both incoming and outgoing flow conditions. Instead of assuming the flow is only perpendicular to the shoreline, the principal inflow direction is assumed to vary by ±22.5 degrees from the transect perpendicular to the shoreline averaged over 500 ft (152 m) to either side of the site (see Figure 7.15). The transect with the largest maximum runup elevation may be used to perform an EGLA, or an EGLA may be performed for each transect.

Including the 22.5-degree transects on either side of the transect that is perpendicular to the shoreline also accounts for variations in the load application to the structure, similar to what is required in wind and earthquake design.

Minimum Closure Ratio for Load Determination

Unless it is an open structure (i.e., a structure in which the portion within the inundation depth has a closure ratio less than or equal to 20 percent, the closure does not include tsunami breakaway walls and there are no interior partitions or content prevented from passing through and exiting the structure as unimpeded waterborne debris), tsunami loads on buildings must be calculated assuming a minimum closure ratio of 70 percent (ASCE/SEI 6.8.7). The closure ratio is the ratio of the area of the inundated enclosure (excluding glazing and openings) to the total projected vertical plane area of the inundated enclosure surface exposed to the flow.

This requirement accounts for the expected accumulation of waterborne debris trapped against the side of the structure and any internal blockage caused by building contents that cannot easily flow out of the structure. Based on observations of buildings subjected to destructive tsunamis, breakaway walls cannot be totally relied upon to relieve structural loading on a building because of the large amount of external debris caused by a tsunami.

Minimum Number of Tsunami Flow Cycles

Two tsunami inflow and outflow cycles must be considered in design: the first cycle is based on an inundation depth at 80 percent of the MCT and the second cycle is assumed to occur with the MCT inundation depth at the site (ASCE/SEI 6.8.8).

The purpose of this requirement is to account for any changes that may occur to the condition of a building and its foundation in each tsunami inflow and outflow cycle. Local scour effects caused by the first cycle must be determined in accordance with ASCE/SEI 6.12 based on an inundation depth equal to 80 percent of the MCT design level. The second tsunami cycle is at the MCT design level where the scour of the first cycle is combined with the load effects generated by the inflow of the second cycle.

Seismic Effects on Foundations

For buildings and other structures located on a site designated in ASCE/SEI Figure 6.7-3 as a site subject to a local subduction zone tsunami generated by an offshore subduction earthquake, the structure must be designed and detailed for the earthquake effects in accordance with the applicable material-specific standards. The foundation of the building or other structure must be designed to resist the earthquake effects according to ASCE/SEI Chapter 11 using the maximum considered earthquake geometric mean (MCE_G) peak ground acceleration given in ASCE/SEI 21.5.

Initial conditions for the design tsunami event must include any changes in the site surface and the in-situ soil properties caused by the design seismic event. The foundation design requirements of ASCE/SEI 6.12 must also be satisfied.

Physical Modeling of Tsunami Flow, Loads and Effects

Tsunami flow, loads and effects are permitted to be determined by a physical model instead of the prescriptive procedures in ASCE/SEI 6.8.5 (flow velocity amplification), 6.10 (hydrodynamic loads), 6.11 (impact loads), and 6.12 (foundation design) provided the eight criteria in ASCE/SEI 6.8.10 are satisfied. Detailed information on modeling requirements is given in ASCE/SEI C6.8.10.

7.3.9 Hydrostatic Loads

Buoyancy

Buoyancy must be evaluated for all inundated structural and designated nonstructural elements of a building or other structure in accordance with ASCE/SEI Equation (6.9-1):

$$F_v = \gamma_s V_w \tag{7.22}$$

where γ_s = minimum fluid weight density for design hydrostatic loads

$= k_s \gamma_{sw}$ [ASCE/SEI Equation (6.8-4)]

k_s = fluid density factor = 1.1

γ_{sw} = specific weight of seawater = 64.0 lb/ft^3 (10 kN/m^3)

V_w = displaced water volume

The buoyancy load on a building with a watertight exterior is illustrated in Figure 7.17 where h_{max} is the maximum inundation depth at the structure. This load is resisted by the weight of the building and the foundation (the weight of any deep foundations must not be included to resist uplift loads).

For purposes of discussion, a watertight exterior is assumed to be designed and detailed to resist all load effects from tsunamis, which means the exterior will not be breached. Unless designed for large missile wind-borne debris impact or blast loading, windows must be considered as openings

Figure 7.17 Buoyancy Load on a Building with a Watertight Exterior

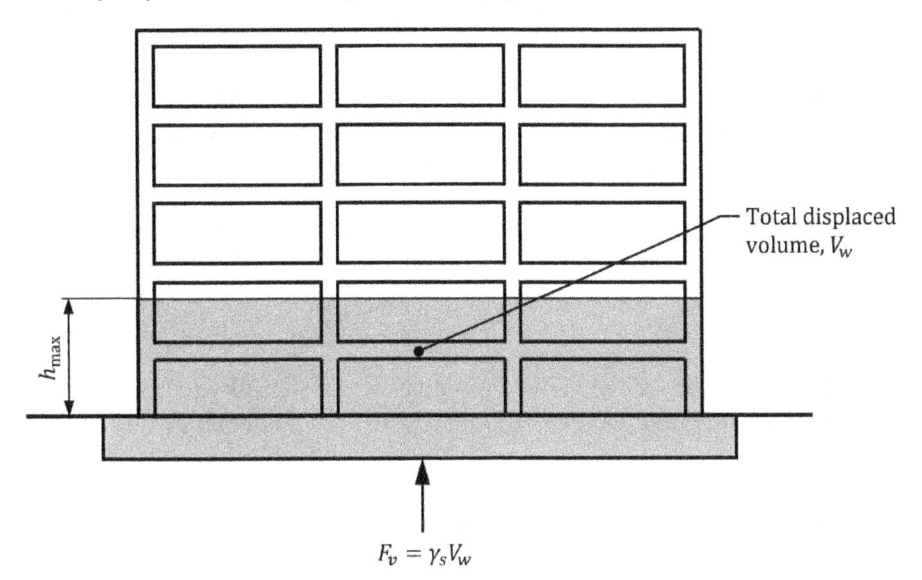

when the inundation depth reaches the top of the windows or the expected strength of the glazing (whichever is less), which means water can flow into enclosed spaces, thereby decreasing or eliminating buoyancy loads (ASCE/SEI 6.9.1).

Buoyancy pressures on slabs in buildings without and with a basement are illustrated in Figure 7.9 of this book where the buildings have a watertight exterior. In cases where the slab is isolated from the structure (which is a typical slab-on-grade), the slab is subjected to buoyancy pressure and overall buoyancy (uplift) on the structure need not be considered. However, if the slab is connected to the structure and is designed to resist buoyancy pressure, uplift on the structure must be considered.

Uplift caused by buoyancy must be considered for the following cases (ASCE/SEI 6.9.1):

- Enclosed spaces without tsunami breakaway walls where the walls have opening areas less than 25 percent of the inundated exterior wall area.
- Floors with trapped air below, including integral structural slabs and in enclosed spaces where the walls are not designed to break away during a tsunami event.

Load case 1 in ASCE/SEI 6.8.3.1 requires a minimum uplift condition be evaluated at an inundation depth of one story or the height of the top of the first story windows (see Section 7.3.8 of this book and Figure 7.16).

Net buoyancy may be avoided by considering one or more of the following:

- Preventing buildup of hydrostatic pressure beneath structural slabs
- Allowing water to enter the interior space
- Designing for pressure relief or structural yielding relief of the hydrostatic pressure
- Providing sufficient deadweight
- Providing sufficient anchorage

Unbalanced Lateral Hydrostatic Force

Lateral hydrostatic loads can occur where there is a difference in water level on either side of a member. Such loads can develop on walls with relatively long widths (water tends to equalize on both sides of walls with shorter widths) or on walls in the lower levels of a watertight building.

Unbalanced hydrostatic loads must be considered during load case 1 and load case 2 inflow cases defined in ASCE/SEI 6.8.3.1 on the following inundated structural walls with openings less than 10 percent of the wall area (ASCE/SEI 6.9.2):

- Walls longer than 30 ft (9.14 m) in width without adjacent tsunami breakaway walls
- Two- or three-sided perimeter structural wall configuration regardless of length where water is prevented from getting to the other side of the wall

Where flow does not overtop the structural wall, the lateral hydrostatic force on the wall, F_h, must be determined by ASCE/SEI Equation (6.9-2):

$$F_h = \frac{1}{2}\gamma_s b h_{max}^2 \tag{7.23}$$

where b is the width of the wall. The unbalanced lateral hydrostatic load on a building with exterior walls satisfying the criteria in ASCE/SEI 6.9.2 is illustrated in Figure 7.18.

Figure 7.18 Unbalanced Lateral Hydrostatic Load

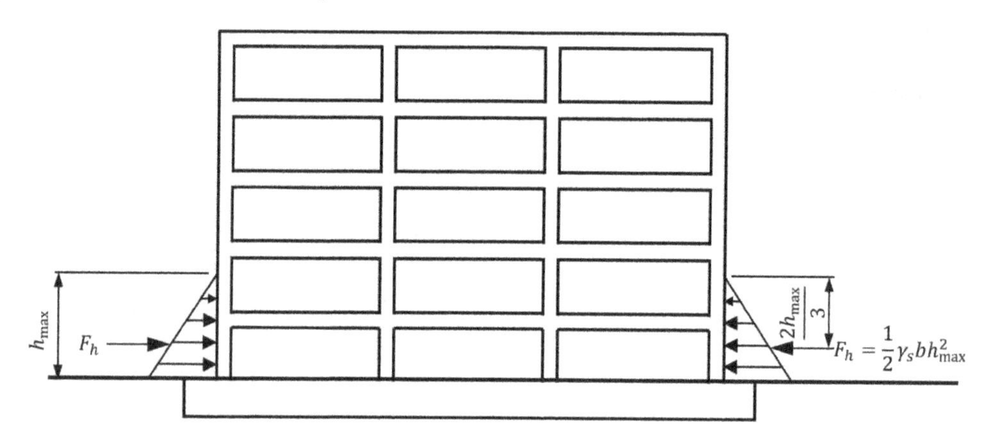

Where the flow overtops the wall, F_h must be determined by ASCE/SEI Equation (6.9-3):

$$F_h = 0.6\gamma_s(2h - h_w)bh_w \tag{7.24}$$

where h is the tsunami inundation depth above grade plane at the wall and h_w is the height of the wall.

Residual Water Surcharge Load on Floors and Walls

Water may not drain from elevated floors during tsunami drawdown on elevated slabs in buildings with perimeter structural components such as upturned beams, masonry or reinforced concrete walls, or parapets. This lack of drainage results in surcharge loads on the slab due to the weight of residual water.

Where water is unable to drain on floors located below h_{max} in buildings that are not watertight, the floors must be designed to resist the following residual water surcharge pressure, p_r, in addition to the effects due to dead loads [ASCE/SEI Equation (6.9-4)]:

$$p_r = \gamma_s h_r = \gamma_s(h_{max} - h_s) \tag{7.25}$$

where h_r is the residual water height in a building, which is equal to h_{max} minus the height of the top of the structural floor slab above the grade plane, h_s. Note that h_r need not exceed the height of the continuous portion of any perimeter structural element at the floor. This is illustrated in

Figure 7.19 for a continuous upturned reinforced concrete beam at the perimeter of the floor located below h_{max} in a building that is not watertight. The floor slab is subjected to p_r determined in accordance with Equation (7.25) where h_r is the depth of the upturned beam above the floor slab. The upturned beam is subjected to lateral hydrostatic pressure caused by the residual water, which is equal to zero at the top of the upturned beam and varies linearly over its depth with a maximum pressure of $\gamma_s h_r$ at the top of the floor slab (see Figure 7.19).

Figure 7.19 Determination of Residual Water Pressure, p_r

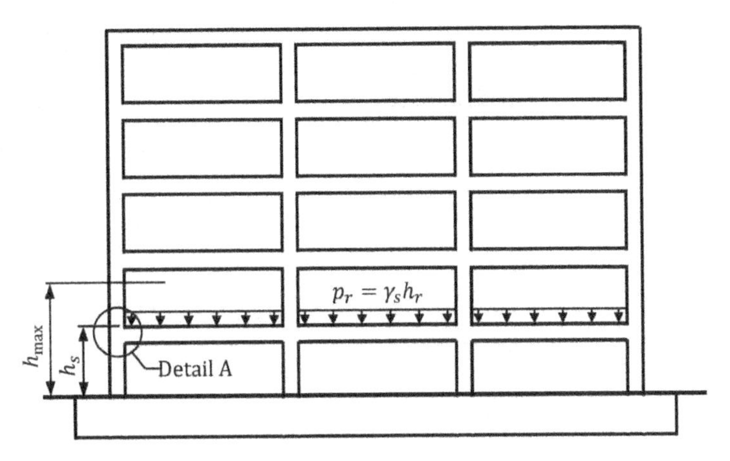

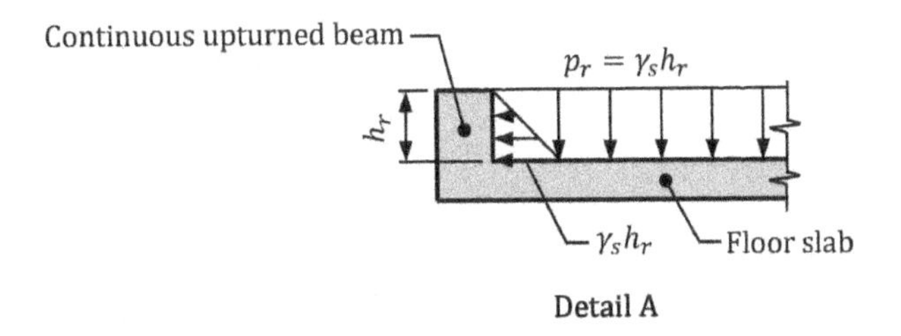

Structural walls that have the potential to retain water during drawdown must also be designed for residual water hydrostatic pressure.

Nonstructural elements above the perimeter structural elements are assumed to fail during tsunami inflow, which means they do not contribute to the retention of water on the slab during drawdown.

Hydrostatic Surcharge Pressure on Foundations

Different water levels may exist on opposite sides of a building, wall, or other structure during tsunami inundation and drawdown. The resulting differential in hydrostatic pressure on the foundation, p_s, that can occur is determined by ASCE/SEI Equation (6.9-5) [see Figure 7.20]:

$$p_s = \gamma_s h_{max} \tag{7.26}$$

7.3.10 Hydrodynamic Loads

Overview

Hydrodynamic loads, which are often referred to as drag forces, develop when water flows around an object in the flow path. The lateral force-resisting system and all structural components of a structure below the inundation level at a site must be designed for the hydrodynamic loads in accordance with the simplified method in ASCE/SEI 6.10.1 or the detailed method in ASCE/SEI 6.10.2 (ASCE/SEI 6.10).

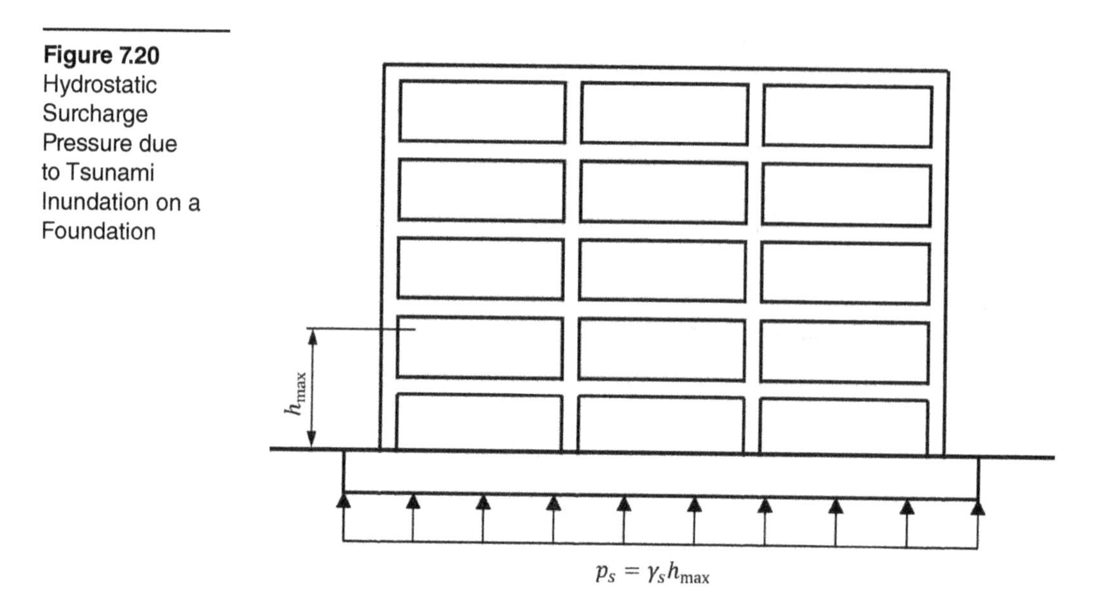

Figure 7.20 Hydrostatic Surcharge Pressure due to Tsunami Inundation on a Foundation

Simplified Equivalent Uniform Lateral Static Pressure

In lieu of performing a more detailed analysis, it is permitted to determine an equivalent uniform lateral static pressure, p_{uw}, which accounts for the combination of any unbalanced lateral hydrostatic and hydrodynamic loads caused by a tsunami [ASCE/SEI Equation (6.10-1)]:

$$p_{uw} = 1.25 I_{tsu} \gamma_s h_{max} \qquad (7.27)$$

where I_{tsu} is the importance factor for tsunami forces given in ASCE/SEI Table 6.8-1.

The uniform pressure is applied uniformly over a depth of $1.3h_{max}$ in each flow direction (see Figure 7.21) and conservatively accounts for the combination of unbalanced lateral hydrostatic and hydrodynamic loads caused by the tsunami (the method is based on the most conservative provisions in ASCE/SEI 6.10 for a rectangular building with no openings). The derivation of ASCE/SEI Equation (6.10-1) is given in ASCE/SEI C6.10.1.

Figure 7.21 Simplified Equivalent Uniform Lateral Static Pressure

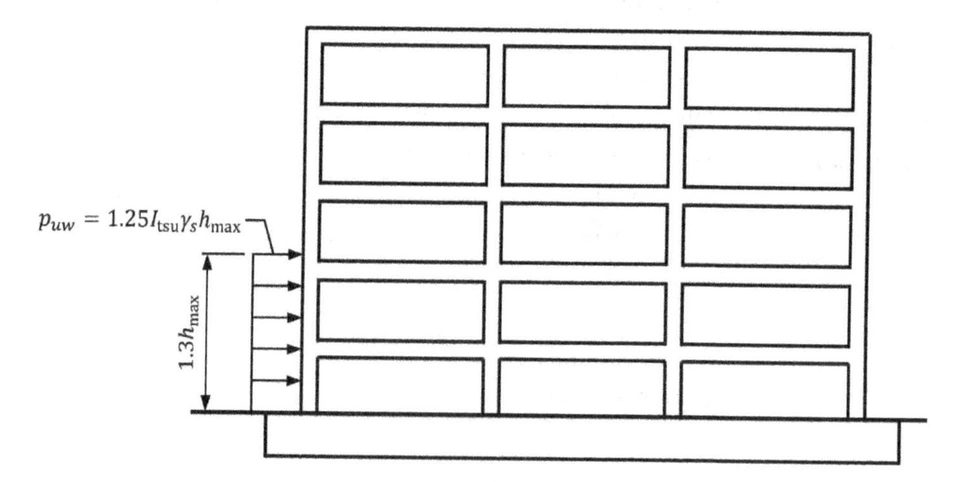

The lateral force-resisting system is evaluated for p_{uw} acting over the entire width of the building perpendicular to the flow direction for both incoming and outgoing flow, and all structural members below $1.3h_{max}$ must be evaluated for the effects of this pressure on their tributary width of projected area.

Detailed Hydrodynamic Lateral Forces

Overview

Drag forces on the entire building and on components must be considered for incoming and outgoing flow. Both of these cases are examined below.

Overall Drag Force on Buildings and Other Structures

The overall drag force on a building or other structure, F_{dx}, is determined by ASCE/SEI Equation (6.10-2) at each level caused by either incoming or outgoing flow for load case 2 defined in ASCE/SEI 6.8.3.1 (see Section 7.3.8 of this book):

$$F_{dx} = \frac{1}{2}\rho_s I_{tsu} C_d C_{cx} B(hu^2) \tag{7.28}$$

In this equation, h and u are the inundation depths and flow velocities, respectively, determined at the site for load case 2, that is, $h = 2h_{max}/3$ and $u = u_{max}$.

The drag force coefficient, C_d, is given in ASCE/SEI Table 6.10-1 for rectilinear structures based on the inundation depth, h, and the overall building width perpendicular to the flow, B (see Table 7.24). Linear interpolation is permitted to be used for intermediate values of B/h.

Table 7.24 Drag Coefficients for Rectilinear Structures

Width to Inundation Depth Ratio, B/h	Drag Coefficient, C_d
≤ 12	1.25
60	1.75
≥ 120	2.0

The closure coefficient, C_{cx}, which is determined by ASCE/SEI Equation (6.10-3), is the percentage of the total projected area of the inundated portion of the building that can block floating debris; that is, this coefficient accounts for debris accumulation during the tsunami event:

$$C_{cx} = \frac{\Sigma(A_{col} + A_{wall}) + 1.5A_{beam}}{Bh_{sx}} \tag{7.29}$$

The numerator in the equation represents the vertical projected area of inundated structural components (columns, walls, and beams, as applicable) and the denominator represents the vertical projected area of the submerged section of the building. Identified in Figure 7.22 are the projected areas of the beams (A_{beam}), columns (A_{col}), and walls (A_{wall}) for the building in the figure. The summation of A_{col} and A_{wall} is over all the columns and walls in the building, and A_{beam} is the combined vertical projected area of the slab facing the flow and the deepest beam laterally exposed to the flow. The equation for C_{cx} in the indicated direction of flow is given in Figure 7.22.

According to ASCE/SEI 6.8.7, C_{cx} must be at least 0.7 when determining F_{dx} by ASCE/SEI Equation (6.10-2).

The force F_{dx} is distributed to the inundated floors of a building based on tributary area. Flowchart 7.4 in Section 7.4 of this book can be used to determine F_{dx}.

Drag Force on Components

All structural components and wall assemblies of the building below the inundation depth are subject to the component drag force, F_d, which is determined by ASCE/SEI Equation (6.10-4):

$$F_d = \frac{1}{2}\rho_s I_{tsu} C_d b(h_e u^2) \tag{7.30}$$

This load must be applied as a pressure resultant on the projected inundated height, h_e, of all structural components and exterior wall assemblies below the inundation depth. The term b is the width of the component perpendicular to the flow.

Figure 7.22 Closure Coefficient, C_{cx}

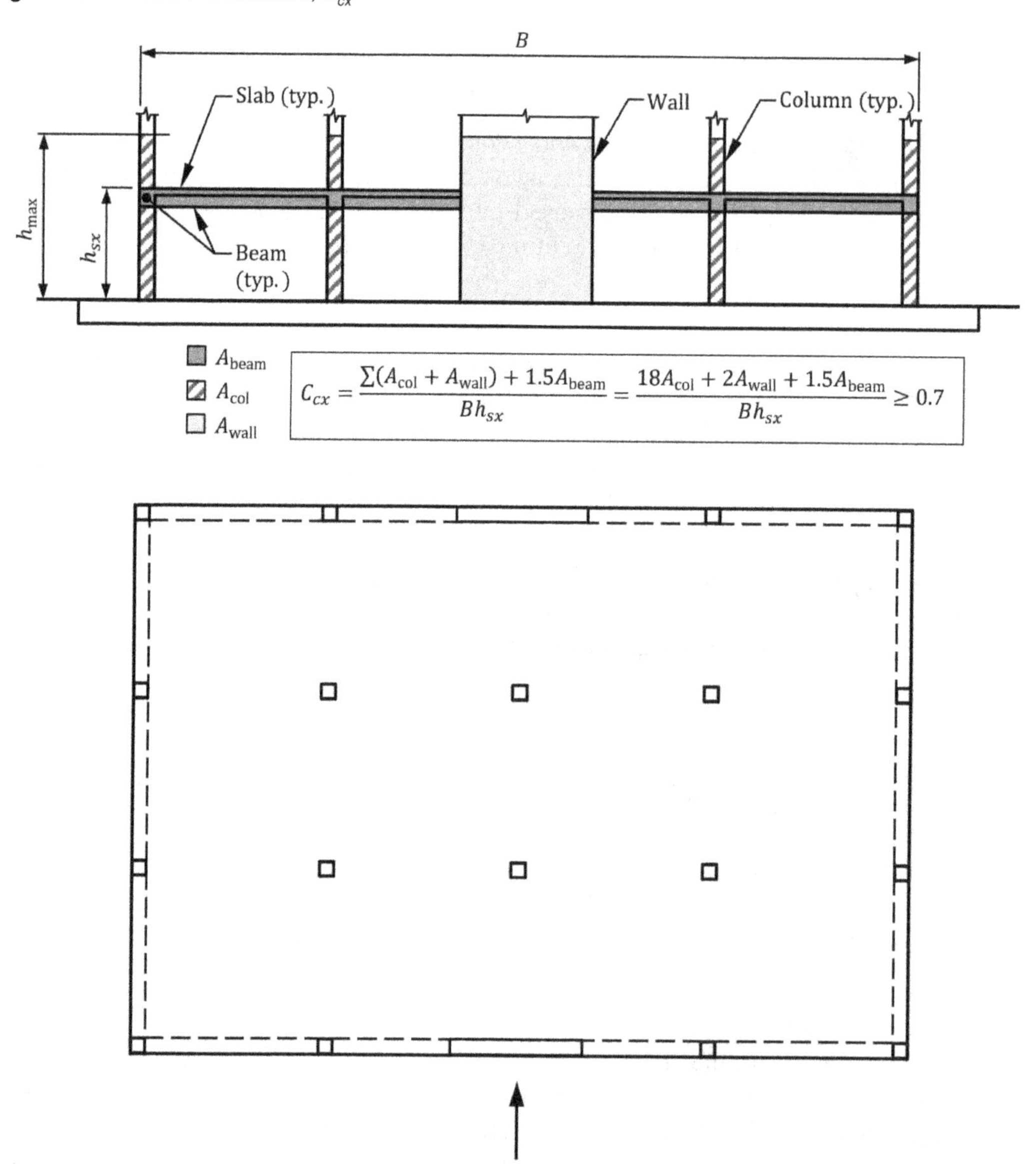

The drag coefficient, C_d, is determined in accordance with ASCE/SEI 6.10.2.2 (see Table 7.25)

All three load cases in ASCE/SEI 6.8.3.1 must be evaluated in the determination of F_d. Structural components that are part of the lateral force-resisting system are typically subjected to the effects due to the total drag force on the building, F_{dx}, and the local effects determined by ASCE/SEI Equation (6.10-4). According to ASCE/SEI 6.10.2.2, these effects are not additive.

Flowchart 7.5 in Section 7.4 of this book can be used to determine F_d.

Tsunami Loads on Vertical Structural Components

The hydrodynamic drag force, F_w, on a vertical component, such as a wall or column, where flow of a tsunami bore need not be considered is determined by ASCE/SEI Equation (6.10-5a), which is the same as ASCE/SEI Equation (6.10-4):

$$F_w = \frac{1}{2}\rho_s I_{tsu} C_d b(h_e u^2) \tag{7.31}$$

Table 7.25 Drag Coefficients for Structural Components

Structural Component	Drag Coefficient, C_d
Interior components other than vertical load-bearing components in storage warehouses and truck and bus garages	Obtain C_d from ASCE/SEI Table 6.10-2 (see Table 7.26) where b is the component width perpendicular to the flow
Interior vertical load-bearing components in storage warehouses	$C_d = 2.0$ where b is the greater of 9 ft (2.74 m) or the length of the structural wall
Interior vertical load-bearing components in truck and bus garages	$C_d = 2.0$ where b is the lesser of 40 ft (12.2 m) or the tributary width multiplied by the closure ratio value given in ASCE/SEI 6.8.7
Exterior components	$C_d = 2.0$ where b is the tributary width multiplied by the closure ratio value given in ASCE/SEI 6.8.7

Table 7.26 Drag Coefficients for Interior Structural Components*

Structural Element Section	Drag Coefficient, C_d
Round column or equilateral polygon with six sides or more	1.2
Rectangular column of at least 2:1 aspect ratio with the longer face oriented parallel to the flow	1.6
Triangular column pointing into the flow	1.6
Freestanding wall submerged in the flow	1.6
Square or rectangular column with the longer face oriented perpendicular to the flow	2.0
Triangular column pointing away from the flow	2.0
Wall or flat plate, normal to the flow	2.0
Diamond-shaped column, pointed into the flow (based on face width and not projected width)	2.5
Rectangular beam, normal to flow	2.0
I, L and channel shapes	2.0

*Applicable to interior components other than vertical load-bearing components in storage warehouses and in truck and bus garages (see Table 7.25).

ASCE/SEI Equation (6.10-5b) must be used to determine F_w where flow of a tsunami bore occurs with a Froude number at the site greater than 1.0 and where individual wall, wall pier or column components have a width-to-inundation-depth ratio of 3 or more:

$$F_w = \frac{3}{4}\rho_s I_{tsu} C_d b(h_e u^2)_{bore} \tag{7.32}$$

The force F_w determined by ASCE/SEI Equation (6.10-5b) is equal to 1.5 times the force F_w determined by ASCE/SEI Equation (6.10-5a).

The following two conditions must be considered when determining the maximum effects due to F_w (ASCE/SEI 6.10.2.3):

- Force F_w must be applied to all vertical structural components that are wider than 3 times the inundation depth corresponding to load case 2 during inflow as defined in ASCE/SEI 6.8.3.
- Force F_w must be determined based on an inundation depth equal to one-third of the structural wall width with the corresponding flow velocity determined using ASCE/SEI Figure 6.8-1.

A more detailed method to determine F_w is given in ASCE/SEI C6.10.2.3.

Hydrodynamic Load on Perforated Walls

The impulsive force determined in accordance with ASCE/SEI 6.10.2.3 for solid walls can be reduced for walls with openings that allow water to flow through (ASCE/SEI 6.10.2.4). For walls with openings that permit flow to pass between wall piers, the force on the elements of the perforated wall, F_{pw}, is determined by ASCE/SEI Equation (6.10-6) but must not be taken less than F_d determined by ASCE/SEI Equation (6.10-4):

$$F_{pw} = (0.4C_{cx} + 0.6)F_w \geq F_d = \frac{1}{2}\rho_s I_{\text{tsu}} C_d b(h_e u^2) \tag{7.33}$$

where the force F_w is determined by ASCE/SEI Equation (6.10-5a) or (6.10-5b), whichever is applicable.

Walls Angled to the Flow

For walls oriented at an angle of less than 90 degrees to the flow direction, the lateral load per unit width, $F_{w\theta}$, must be determined by ASCE/SEI Equation (6.10-7) [see Figure 7.23]:

$$F_{w\theta} = F_w \sin^2\theta \tag{7.34}$$

where θ is the angle between the wall and the direction of the flow and F_w is determined by ASCE/SEI Equations (6.10-5a) or (6.10-5b), whichever is applicable.

Figure 7.23 Lateral Hydrodynamic Load on a Wall Angled to the Flow

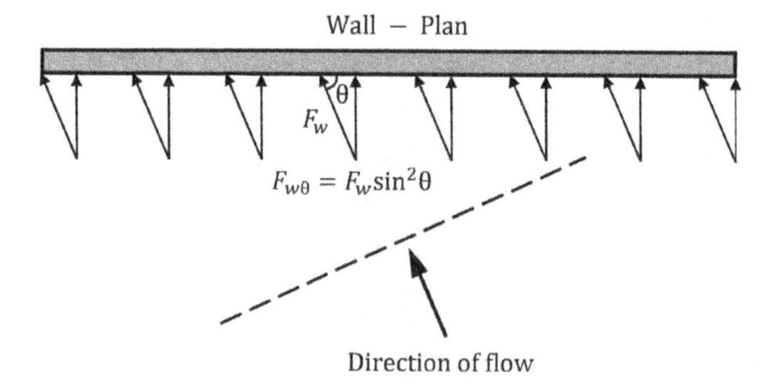

Loads on Above-Ground Horizontal Pipelines

Pipelines installed above ground in a tsunami inundation zone can be subjected to transient, highly turbulent incoming or outgoing flow loads. The pipelines must be designed to resist lateral loads and upward and downward vertical loads.

A summary of the tsunami loads on above-ground horizontal pipelines necessary for the functionality of tsunami risk category III and IV buildings and other structures is given in Table 7.27.

The terms in the equations for the tsunami loads are as follows:

- $C_{cp} = 1.5$
- C_r = pipe resistance coefficient given in ASCE/SEI Table 6.10-3 or ASCE/SEI Figure 6.10-1

Table 7.27 Tsunami Loads on Above-Ground Horizontal Pipelines*

Hydrodynamic Load Type	Tsunami Load	ASCE/SEI Equation No.
Horizontal	$F_{rp} = C_{cp} C_r \rho_s I_{\text{tsu}} D_p u^2$	6.10-8
Upward vertical	$F_{l+} = C_{cp} C_l^+ \rho_s I_{\text{tsu}} D_p u^2$	6.10-9
Downward vertical	$F_{l-} = C_{cp} C_l^- \rho_s I_{\text{tsu}} D_p u^2$	6.10-10

*The tsunami load equations in this table are applicable to above-ground horizontal pipelines necessary for the functionality of tsunami risk category III and IV buildings and other structures.

- C_l^+ = upward lift coefficient given in ASCE/SEI Table 6.10-4 or ASCE/SEI Figure 6.10-2
- C_l^- = downward lift coefficient given in ASCE/SEI Table 6.10-5 or ASCE/SEI Figure 6.10-3
- ρ_s = minimum fluid mass density for design hydrodynamic loads determined in accordance with ASCE/SEI 6.8.4
- I_{tsu} = tsunami importance factor
- D_p = pipe diameter
- u = tsunami flow velocity

The horizontal and vertical forces are illustrated in ASCE/SEI Figure C6.10-4.

Debris impact loads on above-ground horizontal pipelines must be determined in accordance with ASCE/SEI 6.11 (see Section 7.3.11 of this book).

Hydrodynamic Forces Associated with Slabs

Flow Stagnation Pressure

Enclosed spaces created by structural walls on three sides and structural slabs can become pressurized by flow entering the wall-enclosed space with no openings. Walls and slabs in such spaces must be subjected to the following pressure [ASCE/SEI Equation (6.10-11)]:

$$P_p = \frac{1}{2}\rho_s I_{tsu} u^2 \tag{7.35}$$

where u is the maximum free flow velocity at that location and load case in accordance with ASCE/SEI 6.8.3.1.

Hydrodynamic Surge Uplift at Elevated Horizontal Slabs

Uplift pressures can occur during tsunami inflow on submerged horizontal slabs with no flow obstructions (such as walls or columns) above or below the slab. A 20 lb/ft² (0.96 kN/m²) uplift pressure must be applied to the soffit of the slab in such cases (ASCE/SEI 6.10.3.2.1). This pressure is in addition to any hydrostatic buoyancy effects required by ASCE/SEI 6.9.1. Significantly larger uplift pressures occur where obstructions are present below the slab (see below).

Horizontal elevated slabs located over sloping ground are subjected to upward hydrodynamic pressure when the flow reaches the elevation of the slab soffit. Where the ground has an average slope, ϕ, greater than 10 degrees, the redirected uplift pressure, P_u, applied to the slab soffit is determined by ASCE/SEI Equation (6.10-12):

$$P_u = 1.5 I_{tsu} \rho_s u_v^2 \geq 20 \text{ lb/ft}^2 \text{ (0.96 kN/m}^2\text{)} \tag{7.36}$$

where $u_v = u \tan\phi$ is the vertical component of the flow velocity and u is the horizontal flow velocity corresponding to a water depth equal to or greater than the elevation to the soffit, h_{ss}.

Tsunami Bore Flow Entrapped in Structural Wall-Slab Recesses

Significant uplift pressure can occur below horizontal structural slabs where the flow below the slab is blocked by a wall (or other obstruction); in such cases, flow is diverted upward to the underside of the slab. Large pressures can occur on the wall and the slab soffit close to the face of the wall.

The wall and slab within a distance of h_s from the face of the wall must be subjected to an outward pressure, P_u, equal to 350 lb/ft² (16.8 kN/m²) where h_s is the height of the structural floor slab above the grade plane (ASCE/SEI 6.10.3.3.1). Between h_s and the length of the wall, ℓ_w, the upward pressure on the slab is equal to 175 lb/ft² (8.38 kN/m²). Beyond ℓ_w, the required upward pressure is 30 lb/ft² (1.44 kN/m²).

Uplift pressures specified in ASCE/SEI 6.10.3.3.1 may be reduced based on inundation depth (ASCE/SEI 6.10.3.3.2), wall openings (ASCE/SEI 6.10.3.3.3), slab openings (ASCE/SEI 6.10.3.3.4), and tsunami breakaway walls (ASCE/SEI 6.10.3.3.5).

A summary of the requirements in ASCE/SEI 6.10.3 pertaining to hydrodynamic pressures associated with slabs is given in Table 7.28.

Table 7.28 Hydrodynamic Pressures Associated with Slabs

Flow Stagnation Pressure	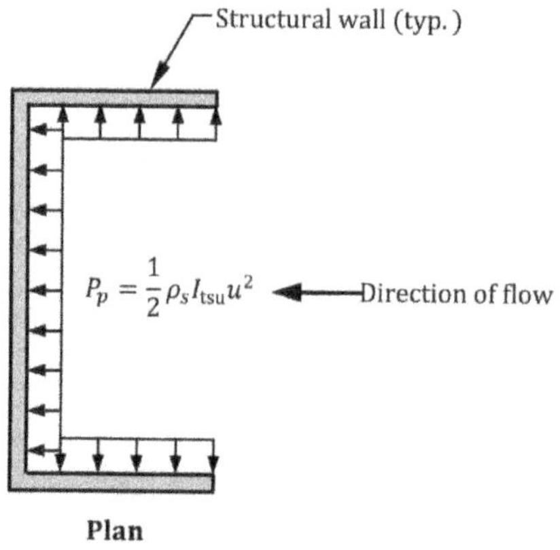 Structural wall (typ.) $P_p = \frac{1}{2}\rho_s I_{tsu} u^2$ ◄— Direction of flow **Plan**
Slabs Over Sloping Grade	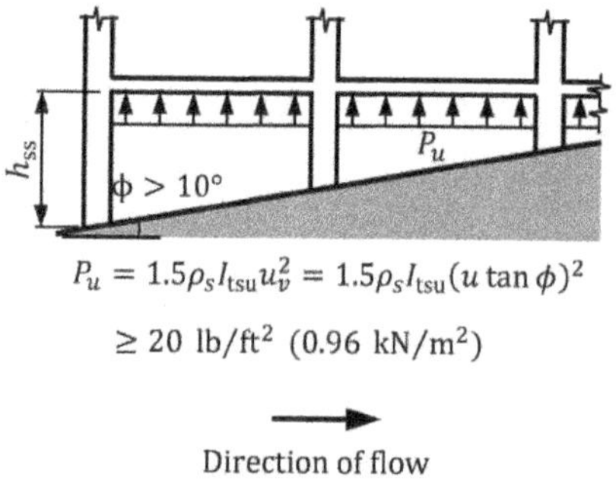 h_{ss} P_u $\phi > 10°$ $P_u = 1.5\rho_s I_{tsu} u_v^2 = 1.5\rho_s I_{tsu}(u \tan\phi)^2$ ≥ 20 lb/ft² (0.96 kN/m²) —► Direction of flow
Tsunami Bore Flow Entrapped In Structural Wall-Slab Recess	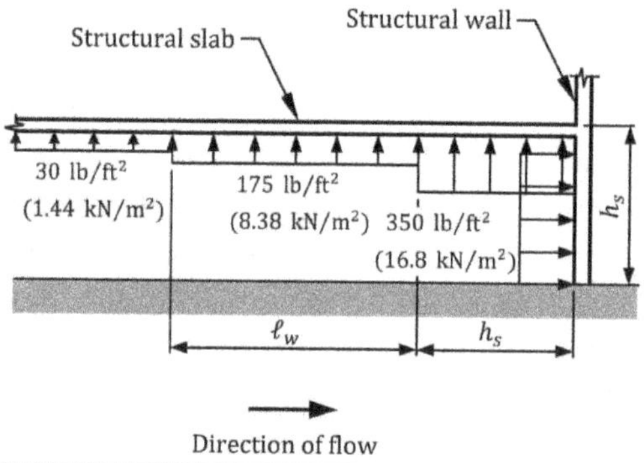 Structural slab; Structural wall 30 lb/ft² (1.44 kN/m²); 175 lb/ft² (8.38 kN/m²); 350 lb/ft² (16.8 kN/m²) h_s; ℓ_w; h_s —► Direction of flow

7.3.11 Debris Impact Loads

Overview

Large volumes of debris are typically transported by tsunamis, ranging in size from logs to ships. Requirements for debris impact loads are given in ASCE/SEI 6.11. All buildings and other structures located where the minimum inundation depth is 3 ft (0.91 m) or greater must be designed for debris impact loads due to the following:

- Floating wood logs and poles (ASE/SEI 6.11.2)
- Vehicles (ASCE/SEI 6.11.3)
- Tumbling boulders and concrete debris (ASCE/SEI 6.11.4)

For buildings or other structures located in proximity to a port or container yard, debris impact loads from shipping containers, ships, and barges may also be applicable (ASCE/SEI 6.11.5 through 6.11.7).

In general, debris impact loads must be applied at points critical for flexure and shear on all structural elements supporting gravity loads at the perimeter of the building, except as specified otherwise in the applicable provisions. Inundation depths and flow velocities corresponding to load cases 1, 2, and 3 in ASCE/SEI 6.8.3.1 must be used.

An alternative simplified debris impact static load method is given in ASCE/SEI 6.11.1, which is permitted to be used to evaluate debris impact loads instead of the methods noted above.

Debris impact loads are considered to be a separate load case, which means that they do not have to be applied simultaneously to all affected structural components (ASCE/SEI 6.11).

A summary of the conditions for which debris impact loads must be evaluated is given in Table 7.29 (see ASCE/SEI Table C6.11-1).

Table 7.29 Conditions for Which Design for Debris Impact Is Evaluated

Debris	Buildings and Other Structures*	Threshold Inundation Depth, ft (m)
Poles, logs, and passenger vehicles	All	3 (0.91)
Boulders and concrete debris	All	6 (1.83)
Shipping containers	All	3 (0.91)
Ships and/or barges**	• Tsunami risk category III critical facilities • Tsunami risk category IV	12 (3.66)

*"All" refers to all buildings and other structures located within a TDZ (ASCE/SEI 6.1.1).

**Applicable to tsunami risk category III critical facilities and tsunami risk category IV buildings and other structures located in the debris hazard impact region determined in accordance with ASCE/SEI 6.11.5.

A summary of the debris impact loads in ASCE/SEI 6.11.2 through 6.11.7 is given in Table 7.30.

Alternative Simplified Debris Impact Static Load

It is permitted to determine the debris impact load, F_i, by ASCE/SEI Equations (6.11-1) and (6.11-1.SI) instead of the detailed methods in ASCE/SEI 6.11.2 through 6.11.6:

$$F_i = 330C_o I_{tsu} \text{ (kips)} \tag{7.37}$$

$$F_i = 1{,}470C_o I_{tsu} \text{ (kN)} \tag{7.38}$$

In these equations, C_o is the orientation factor, which is equal to 0.65.

These equations, which are derived in ASCE/SEI C6.11.1, give conservative values of impact loads for poles, logs, vehicles, tumbling boulders, concrete debris, and shipping containers. The load is applied at points critical for flexure and shear on the perimeter structural members in the inundation depth corresponding to load case 3 in ASCE/SEI 6.8.3.1.

It is permitted to reduce the calculated value of F_i by 50 percent for buildings or other structures not located in an impact zone for shipping containers, ships, and barges, which is determined by the hazard assessment procedure in ASCE/SEI 6.11.5.

Wood Logs and Poles

It is assumed that a log or pole hits the structural element longitudinally rather than transversely. In lieu of a dynamic analysis, an equivalent elastic static analysis is permitted where the dynamic impact force is obtained by multiplying the design instantaneous debris impact force, F_i, by the dynamic response ratio for impulsive loads, R_{max}, given in ASCE/SEI Table 6.11-1. The dynamic response ratio is a function of the ratio of the impact duration, t_d, determined by ASCE/SEI Equation (6.11-4), to the natural period of the impacted structural element, T_n.

Table 7.30 Debris Impact Loads

Debris	Debris Impact Load	Notes
Wood logs and poles (ASCE/SEI 6.11.2)	• Nominal maximum instantaneous debris impact load: $F_{ni} = u_{max}\sqrt{km_d}$ • Design instantaneous debris impact load: $F_i = I_{tsu}C_oF_{ni}$ • Equivalent elastic static analysis: Impact force = $R_{max}F_i$	See Flowchart 7.6 for determination of impact load.
Vehicles (ASCE/SEI 6.11.3)	$F_i = 30I_{tsu}$ (kips) $F_i = 130I_{tsu}$ (kN)	F_i is applied to the vertical structural elements at any point greater than 3 ft (0.91 m) above grade up to the maximum depth.
Tumbling boulder and concrete debris (ASCE/SEI 6.11.4)	$F_i = 8I_{tsu}$ (kips) $F_i = 36I_{tsu}$ (kN)	F_i is applied to the vertical structural elements at 2 ft (0.61 m) above grade.
Shipping containers (ASCE/SEI 6.11.6)	• Nominal maximum instantaneous debris impact load: $F_{ni} = u_{max}\sqrt{km_d} \le 220$ kips (980 kN) • Design instantaneous debris impact load: $F_i = I_{tsu}C_oF_{ni}$ • Equivalent elastic static analysis: Impact force = $R_{max}F_i$	See Flowchart 7.7 for determination of impact load.
Ships and/or barges (ASCE/SEI 6.11.7)	• Nominal maximum instantaneous debris impact load: $F_{ni} = u_{max}\sqrt{km_d}$ • Design instantaneous debris impact load: $F_i = I_{tsu}C_oF_{ni}$	• See Flowchart 7.8 for determination of impact load. • Impact load is applied anywhere from the base of the structure up to 1.3 times the inundation depth plus the height to the deck of the vessel.

For walls, impact is assumed to act along the horizontal center of the wall. The natural period of the wall is permitted to be determined based on the fundamental period of an equivalent column with a width equal to one-half the vertical span of the wall.

Flowchart 7.6 in Section 7.4 of this book can be used to determine debris impact loads due to wood logs and poles in accordance with ASCE/SEI 6.11.2. Included in the flowchart are approximate equations to determine T_n for columns with various end conditions.

Impact by Vehicles

Requirements for debris impact loads due to vehicles are given in ASCE/SEI 6.11.3. Passenger vehicles easily float during a tsunami event. The impact load, F_i, in kips is equal to $30I_{tsu}$ ($130I_{tsu}$ in kN). The derivation of the 30-kip (130-kN) impact force is given in ASCE/SEI C6.11.3. This load is applied to vertical structural elements at any point greater than 3 ft (0.91 m) above grade up to the maximum inundation depth.

Impact by Submerged Tumbling Boulders and Concrete Debris

For submerged tumbling boulders and concrete debris, the impact load, F_i, in pounds is equal to $8{,}000 I_{tsu}$ ($36 I_{tsu}$ in kN) [ASCE/SEI 6.11.4]. The derivation of the 8,000-lb (36-kN) impact force is given in ASCE/SEI C6.11.4. It is applied to vertical structural elements at 2 ft (0.61 m) above grade.

Shipping Containers

Debris impact loads due to shipping containers must be considered for any building or other structure located within a debris impact hazard region, which can occur near a port or container yard. Information on how to determine such a region is given in ASCE/SEI 6.11.5 (see ASCE/SEI Figure 6.11-1).

The mass, m_d, used in determining the nominal maximum instantaneous debris impact force, F_{ni}, by ASCE/SEI Equation (6.11-2) is the mass of the empty shipping container, which is equal to W_d/g. The term W_d is the weight of the empty shipping container given in ASCE/SEI Table 6.11-2 for 20-ft (6.10-m) and 40-ft (12.2-m) standard shipping containers.

Design impact forces must be determined for both empty and loaded shipping containers. The impulse duration for elastic impact, t_d, which is used in determining the dynamic response ratio for impulsive loads, R_{max}, is calculated by ASCE/SEI Equation (6.11-4) for empty shipping containers. ASCE/SEI Equation (6.11-5) must be used to determine t_d for loaded shipping containers where $(m_d + m_{contents})$ is equal to the loaded shipping container weights in ASCE/SEI Table 6.11-2 divided by g.

Flowchart 7.7 in Section 7.4 of this book can be used to determine the debris impact loads due to shipping containers in accordance with ASCE/SEI 6.11.6.

Extraordinary Debris Impacts

Impact by ships and/or barges is considered an extraordinary event and must be considered for tsunami risk category III critical facilities and tsunami risk category IV buildings and other structures located in the debris hazard impact region of a port or harbor as defined in ASCE/SEI 6.11.5. Typical vessel sizes may be acquired from the local harbormaster or port authority or may be obtained in Appendix C of Reference 7.9.

Flowchart 7.8 in Section 7.4 of this book can be used to determine impact loads due to ships and barges in accordance with ASCE/SEI 6.11.7.

Alternative Methods of Response Analysis

In lieu of the prescriptive methods in ASCE/SEI 6.11, a dynamic analysis is permitted to be used to determine debris impact loads (ASCE/SEI 6.11.8). Information on how to perform such an analysis is given in ASCE/SEI C6.11.8.

7.3.12 Foundation Design

Requirements for the design of structure foundations are given in ASCE/SEI 6.12. Such elements must be designed for the effects due to the following:

- Lateral earth pressure determined in accordance with ASCE/SEI 3.2
- Hydrostatic forces determined in accordance with ASCE/SEI 6.9
- Hydrodynamic forces determined in accordance with ASCE/SEI 6.10
- Uplift and underseepage forces determined in accordance with ASCE/SEI 6.12.2.1

Foundations must provide the capacity to withstand uplift and overturning from tsunami hydrostatic, hydrodynamic, and debris impact loads applied to the building superstructure in accordance with the applicable requirements in ASCE/SEI Chapter 6. The effects of soil strength loss, general erosion and scour must be considered in accordance with the requirements in ASCE/SEI 6.12.

More information on the other requirements in ASCE/SEI 6.12 can be found in the corresponding ASCE/SEI commentary sections.

7.3.13 Other Requirements

Requirements for structural countermeasures for tsunami loading, TVERS, designated nonstructural components and systems, and nonbuilding tsunami risk category III and IV structures are given in ASCE/SEI 6.13, 6.14, 6.15, and 6.16, respectively. More information on these requirements can be found in the corresponding ASCE/SEI commentary sections.

7.4 Flowcharts

A summary of the flowcharts provided in this chapter is given in Table 7.31.

Table 7.31 Summary of Flowcharts Provided in Chapter 7

Flowchart	Title
7.1	Flood Impact Loads (ASCE/SEI 5.4.5)
7.2	Determination of R/H_T (ASCE/SEI 6.5.1.1)
7.3	Energy Grade Line Analysis (EGLA) (ASCE/SEI 6.6.2)
7.4	Overall Drag Force, F_{dx} (ASCE/SEI 6.10.2.1)
7.5	Component Drag Force, F_d (ASCE/SEI 6.10.2.2)
7.6	Debris Impact Load for Wood Logs and Poles (ASCE/SEI 6.11.2)
7.7	Debris Impact Load for Shipping Containers (ASCE/SEI 6.11.6)
7.8	Extraordinary Debris Impact Loads (ASCE/SEI 6.11.7)

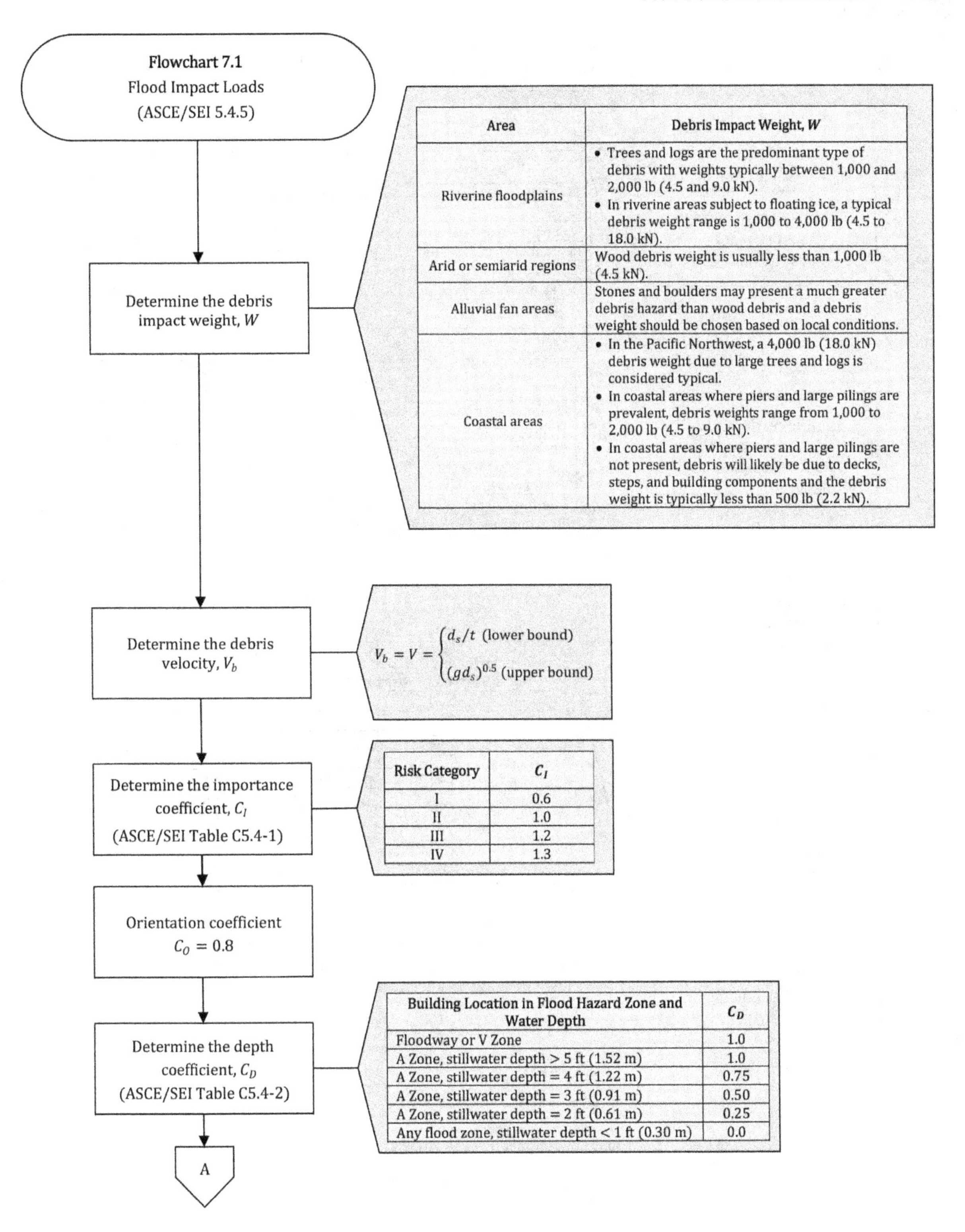
Flowchart 7.1
Flood Impact Loads
(ASCE/SEI 5.4.5)
Determine the debris impact weight, W
Determine the debris velocity, V_b
$V_b = V = \begin{cases} d_s/t \text{ (lower bound)} \\ (gd_s)^{0.5} \text{ (upper bound)} \end{cases}$
Determine the importance coefficient, C_I (ASCE/SEI Table C5.4-1)
Orientation coefficient $C_O = 0.8$
Determine the depth coefficient, C_D (ASCE/SEI Table C5.4-2)
A

Area	Debris Impact Weight, W
Riverine floodplains	• Trees and logs are the predominant type of debris with weights typically between 1,000 and 2,000 lb (4.5 and 9.0 kN). • In riverine areas subject to floating ice, a typical debris weight range is 1,000 to 4,000 lb (4.5 to 18.0 kN).
Arid or semiarid regions	Wood debris weight is usually less than 1,000 lb (4.5 kN).
Alluvial fan areas	Stones and boulders may present a much greater debris hazard than wood debris and a debris weight should be chosen based on local conditions.
Coastal areas	• In the Pacific Northwest, a 4,000 lb (18.0 kN) debris weight due to large trees and logs is considered typical. • In coastal areas where piers and large pilings are prevalent, debris weights range from 1,000 to 2,000 lb (4.5 to 9.0 kN). • In coastal areas where piers and large pilings are not present, debris will likely be due to decks, steps, and building components and the debris weight is typically less than 500 lb (2.2 kN).

Risk Category	C_I
I	0.6
II	1.0
III	1.2
IV	1.3

Building Location in Flood Hazard Zone and Water Depth	C_D
Floodway or V Zone	1.0
A Zone, stillwater depth > 5 ft (1.52 m)	1.0
A Zone, stillwater depth = 4 ft (1.22 m)	0.75
A Zone, stillwater depth = 3 ft (0.91 m)	0.50
A Zone, stillwater depth = 2 ft (0.61 m)	0.25
Any flood zone, stillwater depth < 1 ft (0.30 m)	0.0

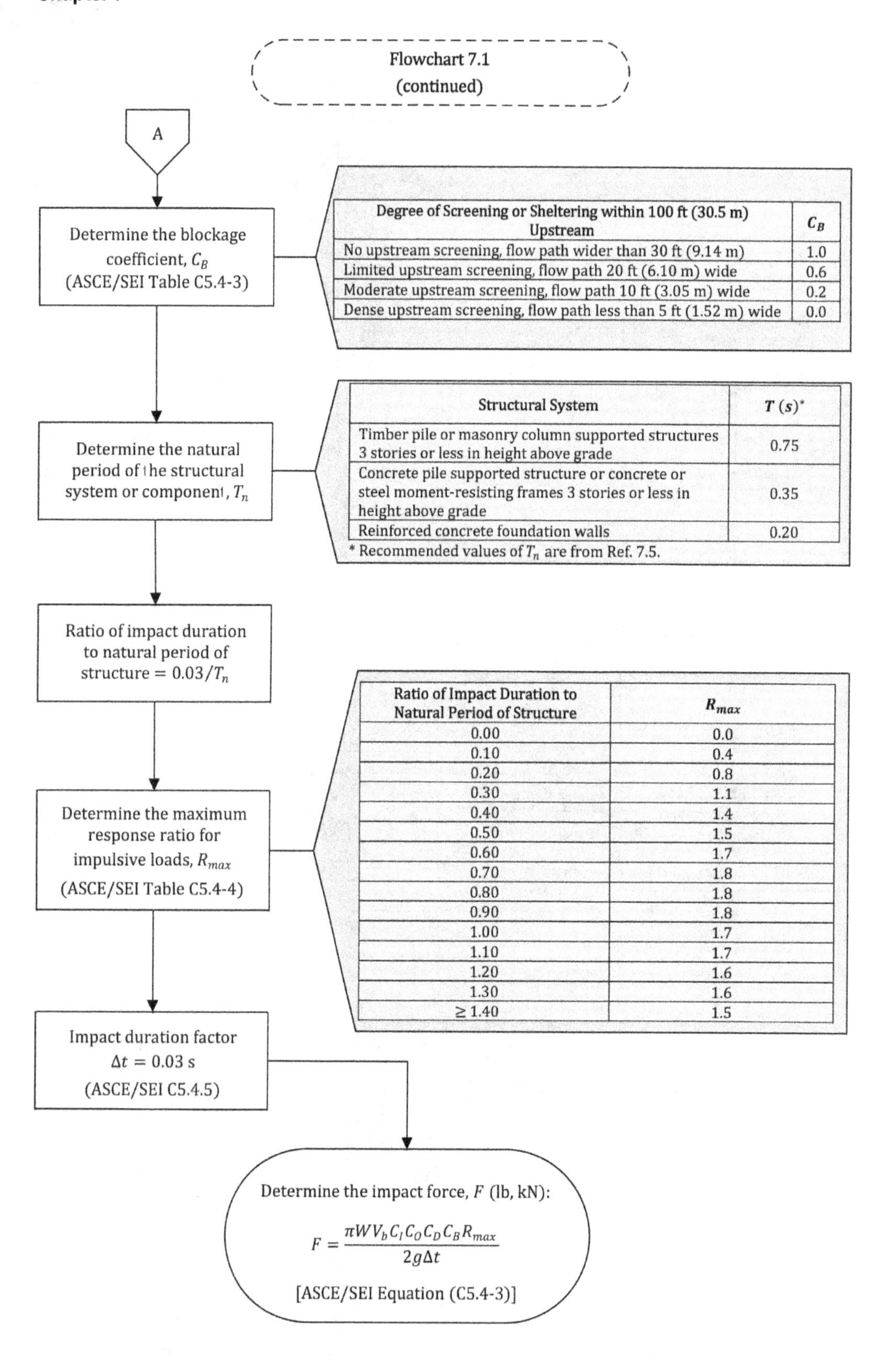

Flowchart 7.1
(continued)
A
Determine the blockage coefficient, C_B (ASCE/SEI Table C5.4-3)
Determine the natural period of the structural system or component, T_n
Ratio of impact duration to natural period of structure = $0.03/T_n$
Determine the maximum response ratio for impulsive loads, R_{max} (ASCE/SEI Table C5.4-4)
Impact duration factor $\Delta t = 0.03$ s (ASCE/SEI C5.4.5)
Determine the impact force, F (lb, kN):
$F = \frac{\pi W V_b C_I C_O C_D C_B R_{max}}{2g\Delta t}$
[ASCE/SEI Equation (C5.4-3)]

Degree of Screening or Sheltering within 100 ft (30.5 m) Upstream	C_B
No upstream screening, flow path wider than 30 ft (9.14 m)	1.0
Limited upstream screening, flow path 20 ft (6.10 m) wide	0.6
Moderate upstream screening, flow path 10 ft (3.05 m) wide	0.2
Dense upstream screening, flow path less than 5 ft (1.52 m) wide	0.0

Structural System	T (s)*
Timber pile or masonry column supported structures 3 stories or less in height above grade	0.75
Concrete pile supported structure or concrete or steel moment-resisting frames 3 stories or less in height above grade	0.35
Reinforced concrete foundation walls	0.20

* Recommended values of T_n are from Ref. 7.5.

Ratio of Impact Duration to Natural Period of Structure	R_{max}
0.00	0.0
0.10	0.4
0.20	0.8
0.30	1.1
0.40	1.4
0.50	1.5
0.60	1.7
0.70	1.8
0.80	1.8
0.90	1.8
1.00	1.7
1.10	1.7
1.20	1.6
1.30	1.6
≥ 1.40	1.5

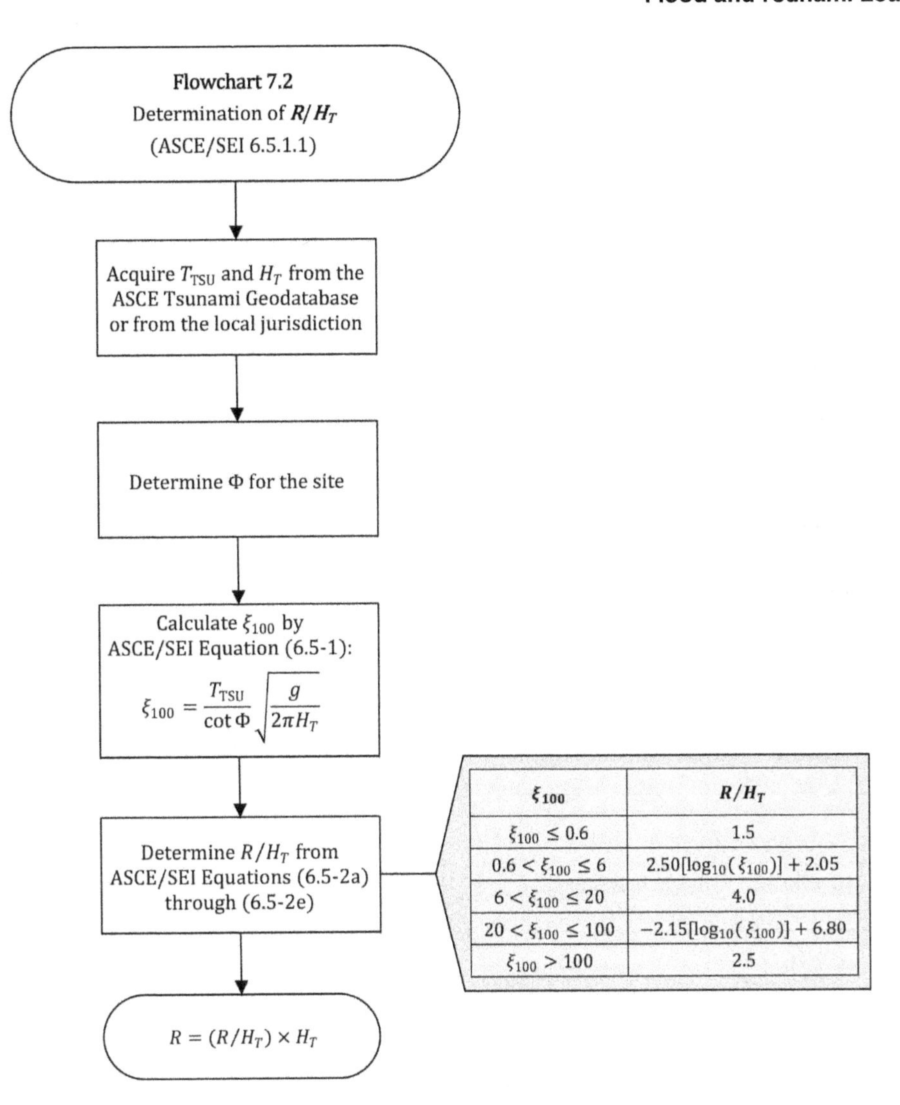

ξ_{100}	R/H_T
$\xi_{100} \le 0.6$	1.5
$0.6 < \xi_{100} \le 6$	$2.50[\log_{10}(\xi_{100})] + 2.05$
$6 < \xi_{100} \le 20$	4.0
$20 < \xi_{100} \le 100$	$-2.15[\log_{10}(\xi_{100})] + 6.80$
$\xi_{100} > 100$	2.5

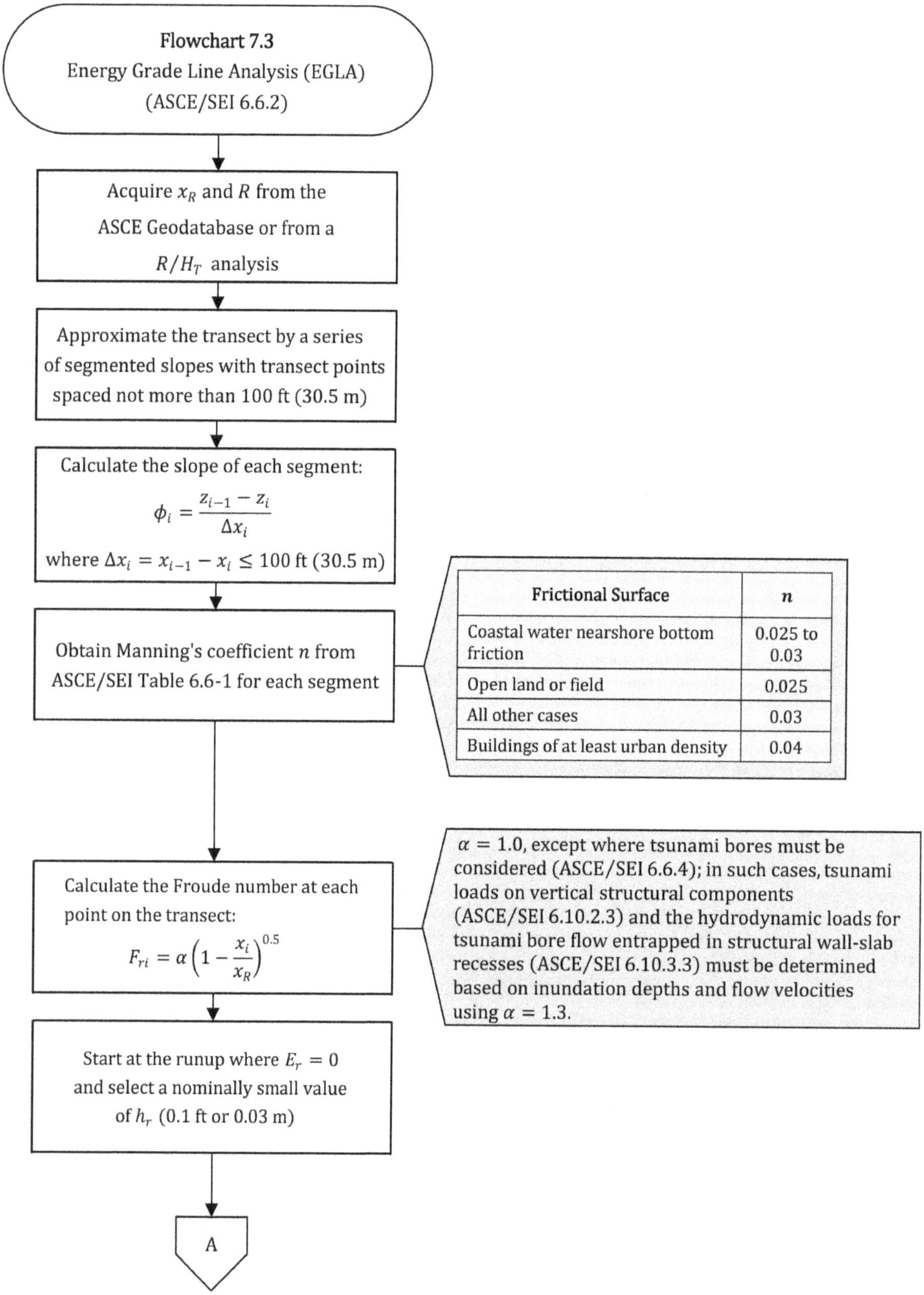

Flowchart 7.3
Energy Grade Line Analysis (EGLA)
(ASCE/SEI 6.6.2)

Acquire x_R and R from the ASCE Geodatabase or from a R/H_T analysis

Approximate the transect by a series of segmented slopes with transect points spaced not more than 100 ft (30.5 m)

Calculate the slope of each segment:

$$\phi_i = \frac{z_{i-1} - z_i}{\Delta x_i}$$

where $\Delta x_i = x_{i-1} - x_i \leq 100$ ft (30.5 m)

Obtain Manning's coefficient n from ASCE/SEI Table 6.6-1 for each segment

Frictional Surface	n
Coastal water nearshore bottom friction	0.025 to 0.03
Open land or field	0.025
All other cases	0.03
Buildings of at least urban density	0.04

Calculate the Froude number at each point on the transect:

$$F_{ri} = \alpha\left(1 - \frac{x_i}{x_R}\right)^{0.5}$$

$\alpha = 1.0$, except where tsunami bores must be considered (ASCE/SEI 6.6.4); in such cases, tsunami loads on vertical structural components (ASCE/SEI 6.10.2.3) and the hydrodynamic loads for tsunami bore flow entrapped in structural wall-slab recesses (ASCE/SEI 6.10.3.3) must be determined based on inundation depths and flow velocities using $\alpha = 1.3$.

Start at the runup where $E_r = 0$ and select a nominally small value of h_r (0.1 ft or 0.03 m)

A

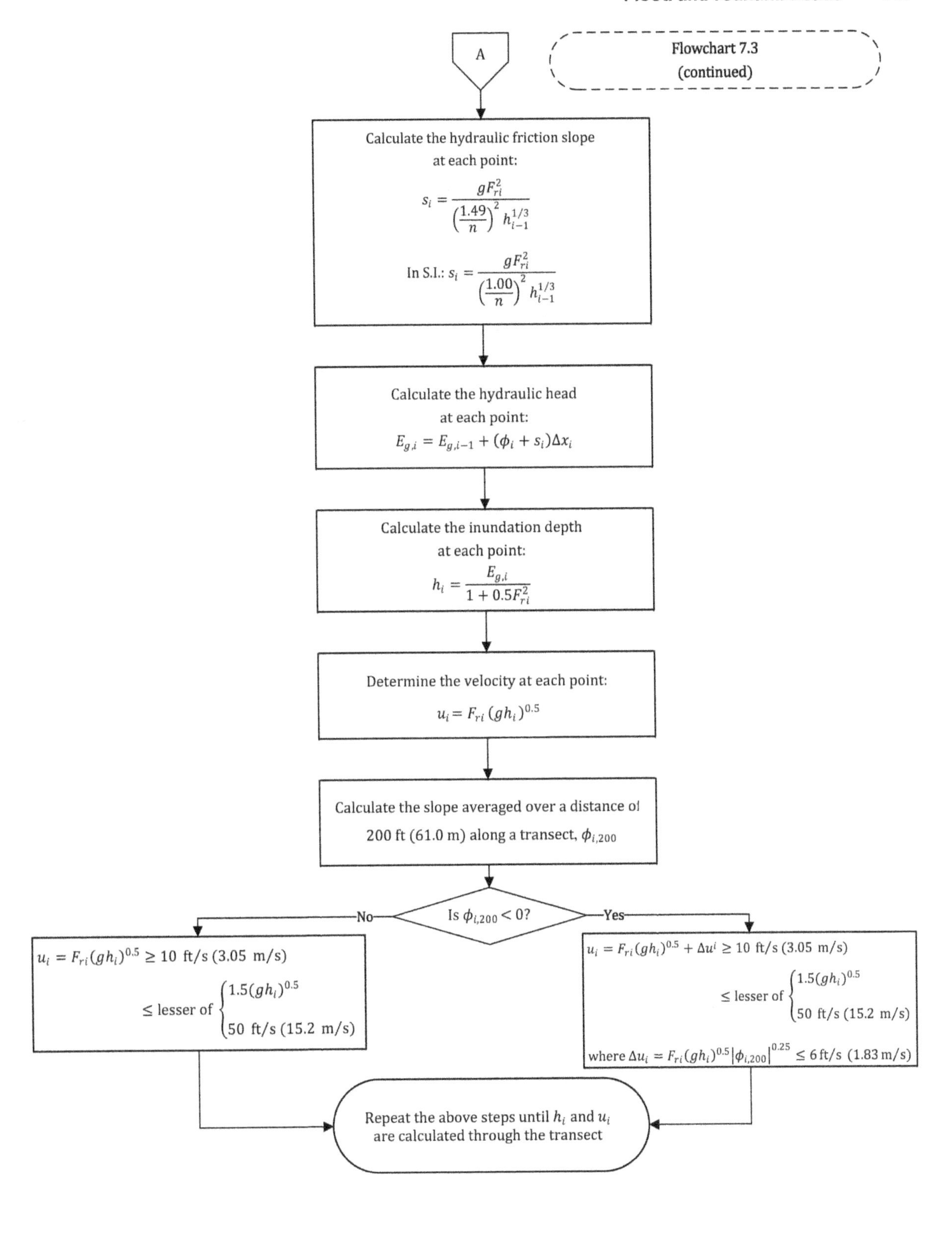

Flowchart 7.3
(continued)
A
Calculate the hydraulic friction slope
at each point:
$s_i = \dfrac{gF_{ri}^2}{\left(\frac{1.49}{n}\right)^2 h_{i-1}^{1/3}}$
In S.I.: $s_i = \dfrac{gF_{ri}^2}{\left(\frac{1.00}{n}\right)^2 h_{i-1}^{1/3}}$
Calculate the hydraulic head
at each point:
$E_{g,i} = E_{g,i-1} + (\phi_i + s_i)\Delta x_i$
Calculate the inundation depth
at each point:
$h_i = \dfrac{E_{g,i}}{1 + 0.5F_{ri}^2}$
Determine the velocity at each point:
$u_i = F_{ri}\,(gh_i)^{0.5}$
Calculate the slope averaged over a distance of
200 ft (61.0 m) along a transect, $\phi_{i,200}$
Is $\phi_{i,200} < 0$?
No
Yes
$u_i = F_{ri}(gh_i)^{0.5} \geq 10$ ft/s (3.05 m/s)
$\leq$ lesser of $\begin{cases} 1.5(gh_i)^{0.5} \\ 50 \text{ ft/s (15.2 m/s)} \end{cases}$
$u_i = F_{ri}(gh_i)^{0.5} + \Delta u^i \geq 10$ ft/s (3.05 m/s)
$\leq$ lesser of $\begin{cases} 1.5(gh_i)^{0.5} \\ 50 \text{ ft/s (15.2 m/s)} \end{cases}$
where $\Delta u_i = F_{ri}(gh_i)^{0.5} |\phi_{i,200}|^{0.25} \leq 6$ ft/s (1.83 m/s)
Repeat the above steps until h_i and u_i
are calculated through the transect

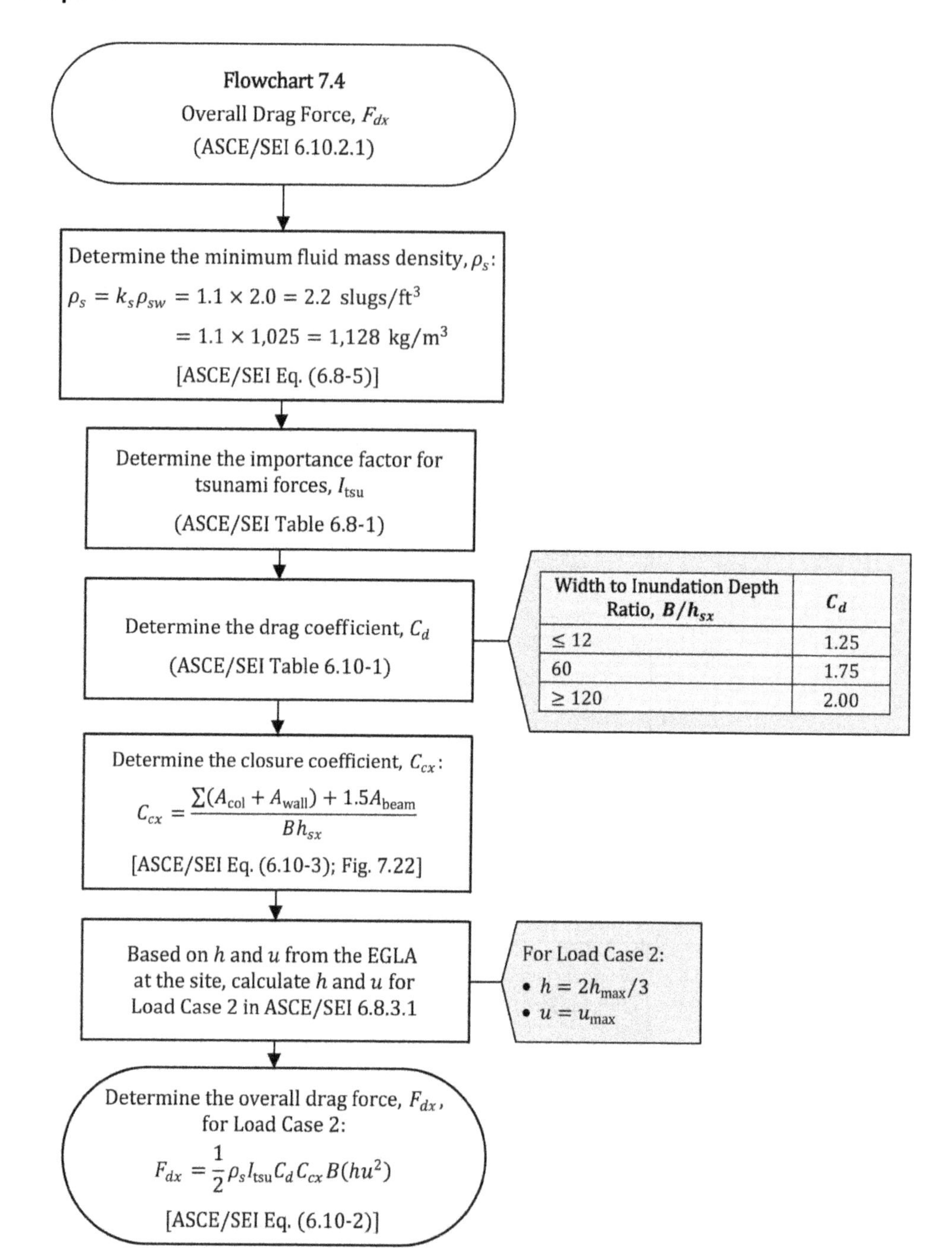

Flowchart 7.4
Overall Drag Force, F_{dx}
(ASCE/SEI 6.10.2.1)
Determine the minimum fluid mass density, ρ_s:
$\rho_s = k_s \rho_{sw} = 1.1 \times 2.0 = 2.2$ slugs/ft^3
$= 1.1 \times 1{,}025 = 1{,}128$ kg/m^3
[ASCE/SEI Eq. (6.8-5)]
Determine the importance factor for tsunami forces, I_{tsu}
(ASCE/SEI Table 6.8-1)
Determine the drag coefficient, C_d
(ASCE/SEI Table 6.10-1)
Width to Inundation Depth Ratio, B/h_{sx} | C_d
≤ 12 | 1.25
60 | 1.75
≥ 120 | 2.00
Determine the closure coefficient, C_{cx}:
$C_{cx} = \frac{\sum(A_{col} + A_{wall}) + 1.5A_{beam}}{Bh_{sx}}$
[ASCE/SEI Eq. (6.10-3); Fig. 7.22]
Based on h and u from the EGLA at the site, calculate h and u for Load Case 2 in ASCE/SEI 6.8.3.1
For Load Case 2:
• $h = 2h_{max}/3$
• $u = u_{max}$
Determine the overall drag force, F_{dx}, for Load Case 2:
$F_{dx} = \frac{1}{2}\rho_s I_{tsu} C_d C_{cx} B(hu^2)$
[ASCE/SEI Eq. (6.10-2)]

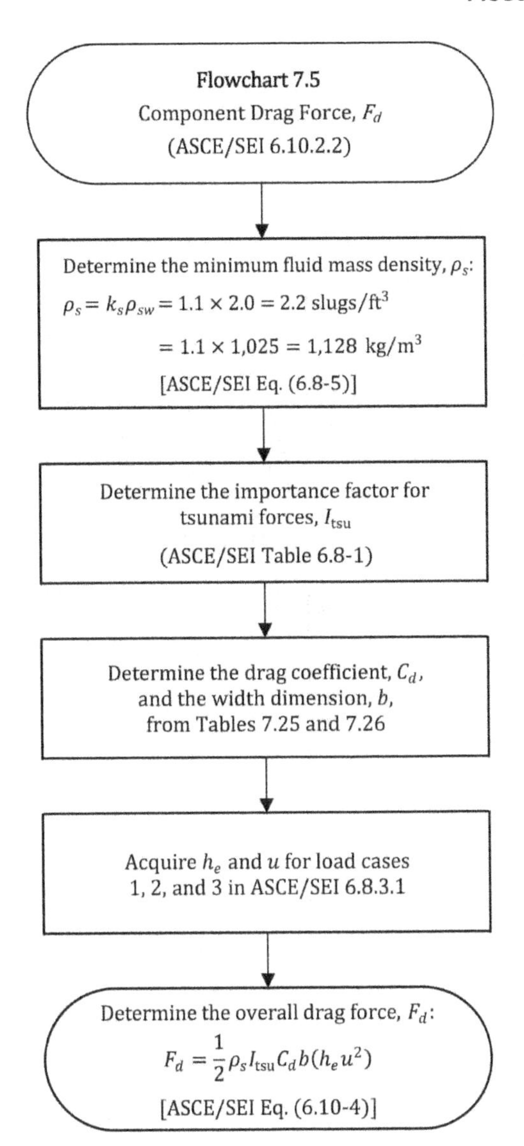

Flowchart 7.5
Component Drag Force, F_d
(ASCE/SEI 6.10.2.2)
Determine the minimum fluid mass density, ρ_s:
$\rho_s = k_s \rho_{sw} = 1.1 \times 2.0 = 2.2$ slugs/ft^3
$= 1.1 \times 1{,}025 = 1{,}128$ kg/m^3
[ASCE/SEI Eq. (6.8-5)]
Determine the importance factor for tsunami forces, I_{tsu}
(ASCE/SEI Table 6.8-1)
Determine the drag coefficient, C_d, and the width dimension, b, from Tables 7.25 and 7.26
Acquire h_e and u for load cases 1, 2, and 3 in ASCE/SEI 6.8.3.1
Determine the overall drag force, F_d:
$F_d = \frac{1}{2} \rho_s I_{tsu} C_d b (h_e u^2)$
[ASCE/SEI Eq. (6.10-4)]

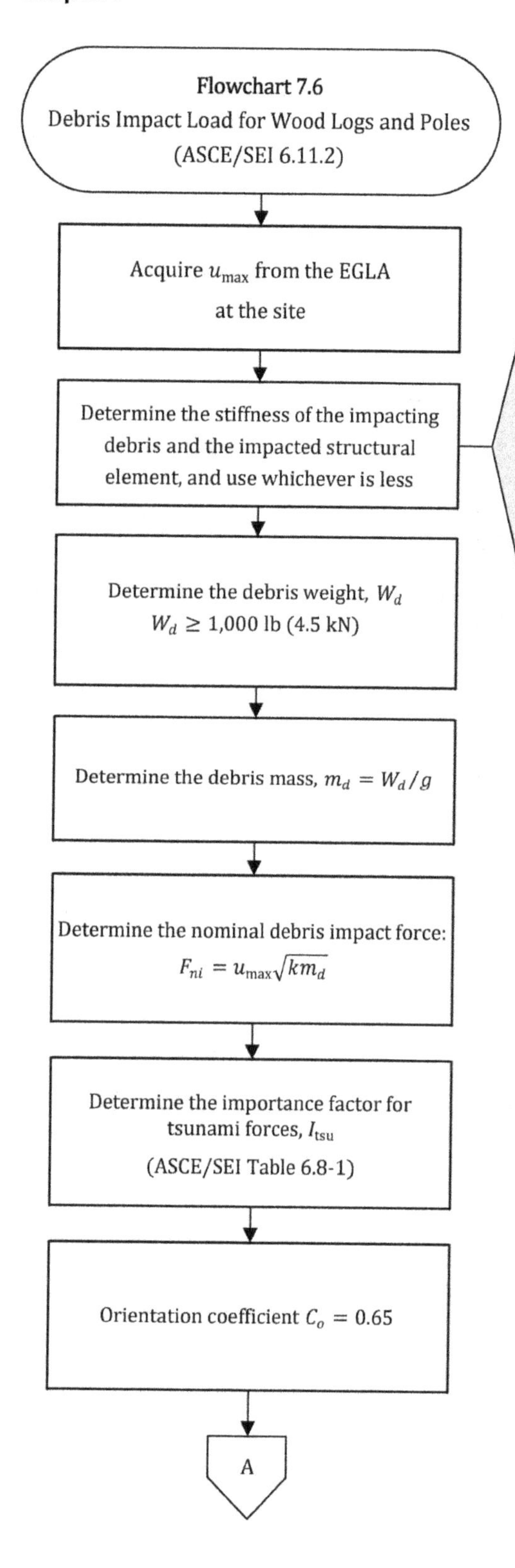
Flowchart 7.6
Debris Impact Load for Wood Logs and Poles
(ASCE/SEI 6.11.2)
Acquire u_{max} from the EGLA at the site
Determine the stiffness of the impacting debris and the impacted structural element, and use whichever is less
Determine the debris weight, W_d
$W_d \geq 1{,}000$ lb (4.5 kN)
Determine the debris mass, $m_d = W_d/g$
Determine the nominal debris impact force:
$F_{ni} = u_{max}\sqrt{km_d}$
Determine the importance factor for tsunami forces, I_{tsu}
(ASCE/SEI Table 6.8-1)
Orientation coefficient $C_o = 0.65$
A
• Stiffness of log or pole:
$k = EA/L \geq 350$ kip/in. (61,300 kN/m)
• Stiffness of column:
$k = c_1\, EI/L^3$
$c_1 = \begin{cases} 48 \text{ for simple connections at both ends} \\ (768/7) \text{ for simple and fixed connections} \\ 192 \text{ for fixed connections at both ends} \end{cases}$

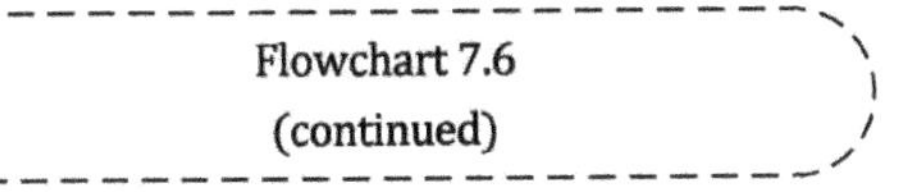

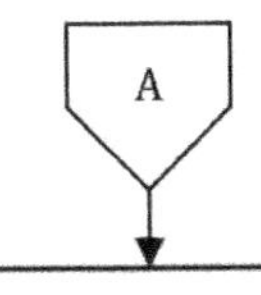

Determine the design debris impact force:
$F_i = I_{tsu} C_o F_{ni}$

Determine the impulse duration for elastic impact:
$t_d = \dfrac{2 m_d u_{max}}{F_{ni}}$

Determine the natural period of the impacted structural element, T_n

Impacted Structural Element		T_n
Column*	Simple connections at both ends	$2L^2/\pi\sqrt{EI/m}$
	Simple and fixed connections	$32L^2/25\pi\sqrt{EI/m}$
	Fixed connections at both ends	$8L^2/9\pi\sqrt{EI/m}$
Reinforced concrete foundation walls		0.20 s**

* The natural period equations for a column may also be used for a wall where the wall width is 0.5(vertical span of the wall).
** Recommended value of T_n is from Ref. 7.5.

Determine the ratio t_d/T_n

Determine the dynamic response ratio for impulsive loads, R_{max}
(ASCE/SEI Table 6.11-1)

t_d/T_n	R_{max}
0.00	0.0
0.10	0.4
0.20	0.8
0.30	1.1
0.40	1.4
0.50	1.5
0.60	1.7
0.70	1.8
0.80	1.8
0.90	1.8
1.00	1.7
1.10	1.7
1.20	1.6
1.30	1.6
≥1.40	1.5

Debris impact force for an equivalent elastic static analysis = $R_{max} F_i$

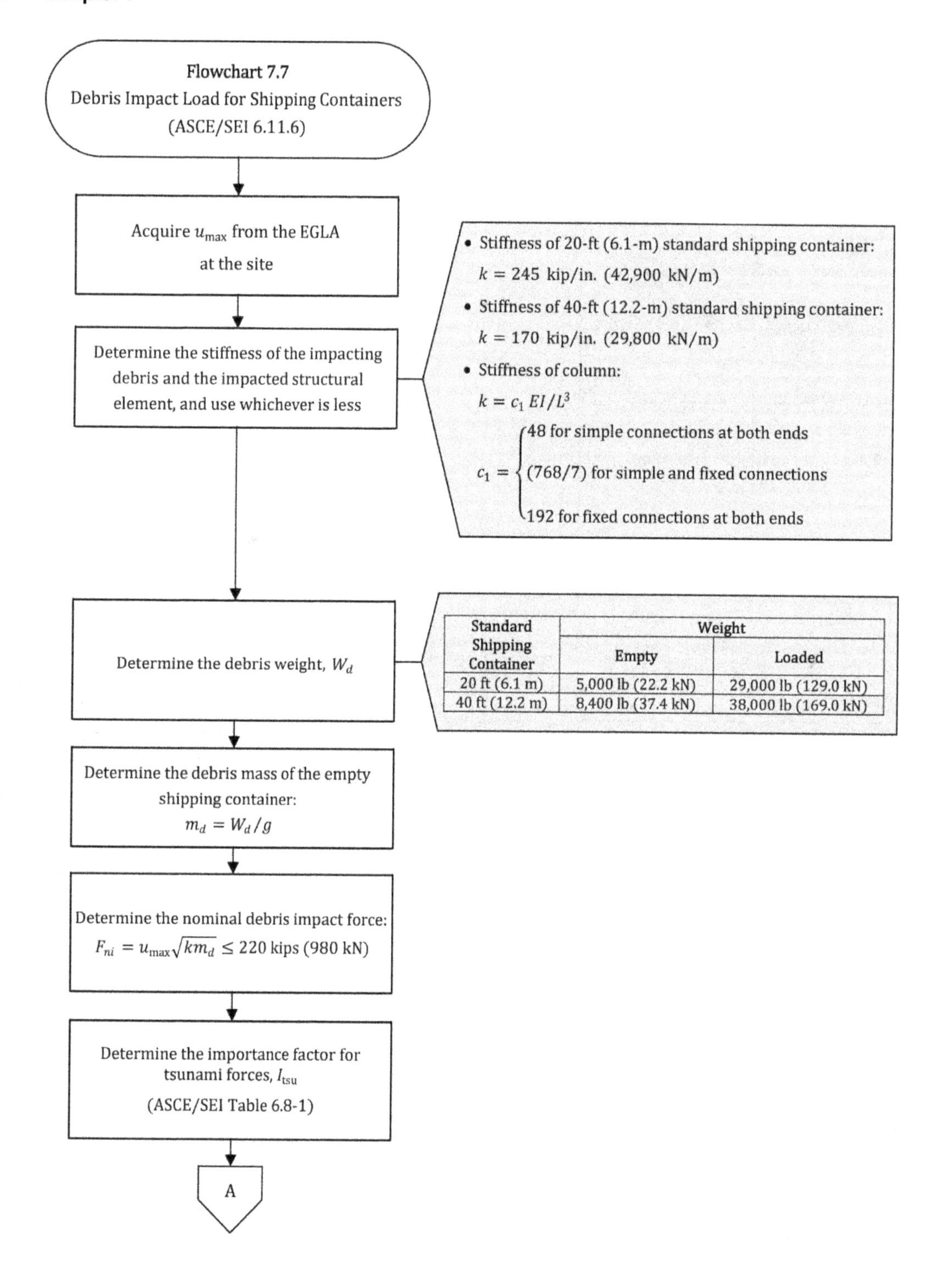

Flowchart 7.7
Debris Impact Load for Shipping Containers
(ASCE/SEI 6.11.6)
Acquire u_{max} from the EGLA at the site
Determine the stiffness of the impacting debris and the impacted structural element, and use whichever is less
• Stiffness of 20-ft (6.1-m) standard shipping container:
$k = 245$ kip/in. (42,900 kN/m)
• Stiffness of 40-ft (12.2-m) standard shipping container:
$k = 170$ kip/in. (29,800 kN/m)
• Stiffness of column:
$k = c_1 EI/L^3$
$c_1 =$ 48 for simple connections at both ends
(768/7) for simple and fixed connections
192 for fixed connections at both ends
Determine the debris weight, W_d
Standard Shipping Container
Weight
Empty
Loaded
20 ft (6.1 m)
5,000 lb (22.2 kN)
29,000 lb (129.0 kN)
40 ft (12.2 m)
8,400 lb (37.4 kN)
38,000 lb (169.0 kN)
Determine the debris mass of the empty shipping container:
$m_d = W_d/g$
Determine the nominal debris impact force:
$F_{ni} = u_{max}\sqrt{km_d} \leq 220$ kips (980 kN)
Determine the importance factor for tsunami forces, I_{tsu}
(ASCE/SEI Table 6.8-1)
A

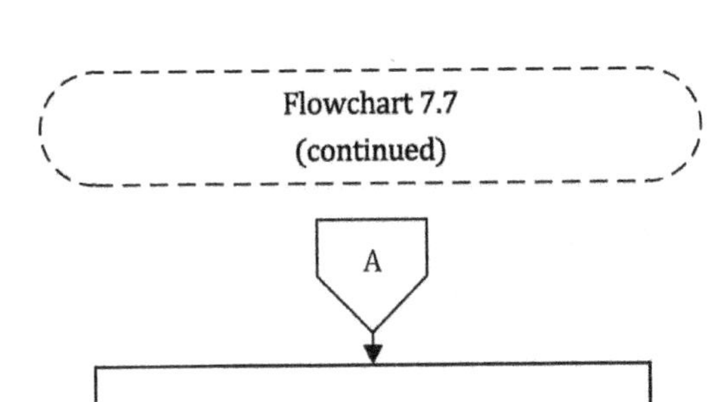

Orientation coefficient $C_o = 0.65$

Determine the design debris impact force:
$F_i = I_{tsu} C_o F_{ni}$

Determine the impulse duration for elastic impact, t_d

- For empty shipping containers:

$$t_d = \frac{2m_d u_{max}}{F_{ni}}$$

- For loaded shipping containers:

$$t_d = \frac{(m_d + m_{contents}) u_{max}}{F_{ni}}$$

$m_{contents}$ = max. rated content capacity/2 (see ASCE/SEI Table 6.11-2)

Determine the natural period of the impacted structural element, T_n

Impacted Structural Element		T_n
Column*	Simple connections at both ends	$2L^2/\pi\sqrt{EI/m}$
	Simple and fixed connections	$32L^2/25\pi\sqrt{EI/m}$
	Fixed connections at both ends	$8L^2/9\pi\sqrt{EI/m}$
Reinforced concrete foundation walls		0.20 s**

* The natural period equations for a column may also be used for a wall where the wall width is 0.5(vertical span of the wall).

** Recommended value of T_n is from Ref. 7.5.

Determine the ratio t_d/T_n

Determine the dynamic response ratio for impulsive loads, R_{max}

(ASCE/SEI Table 6.11-1)

t_d/T_n	R_{max}
0.00	0.0
0.10	0.4
0.20	0.8
0.30	1.1
0.40	1.4
0.50	1.5
0.60	1.7
0.70	1.8
0.80	1.8
0.90	1.8
1.00	1.7
1.10	1.7
1.20	1.6
1.30	1.6
≥ 1.40	1.5

Debris impact force for an equivalent elastic static analysis = $R_{max}F_i$

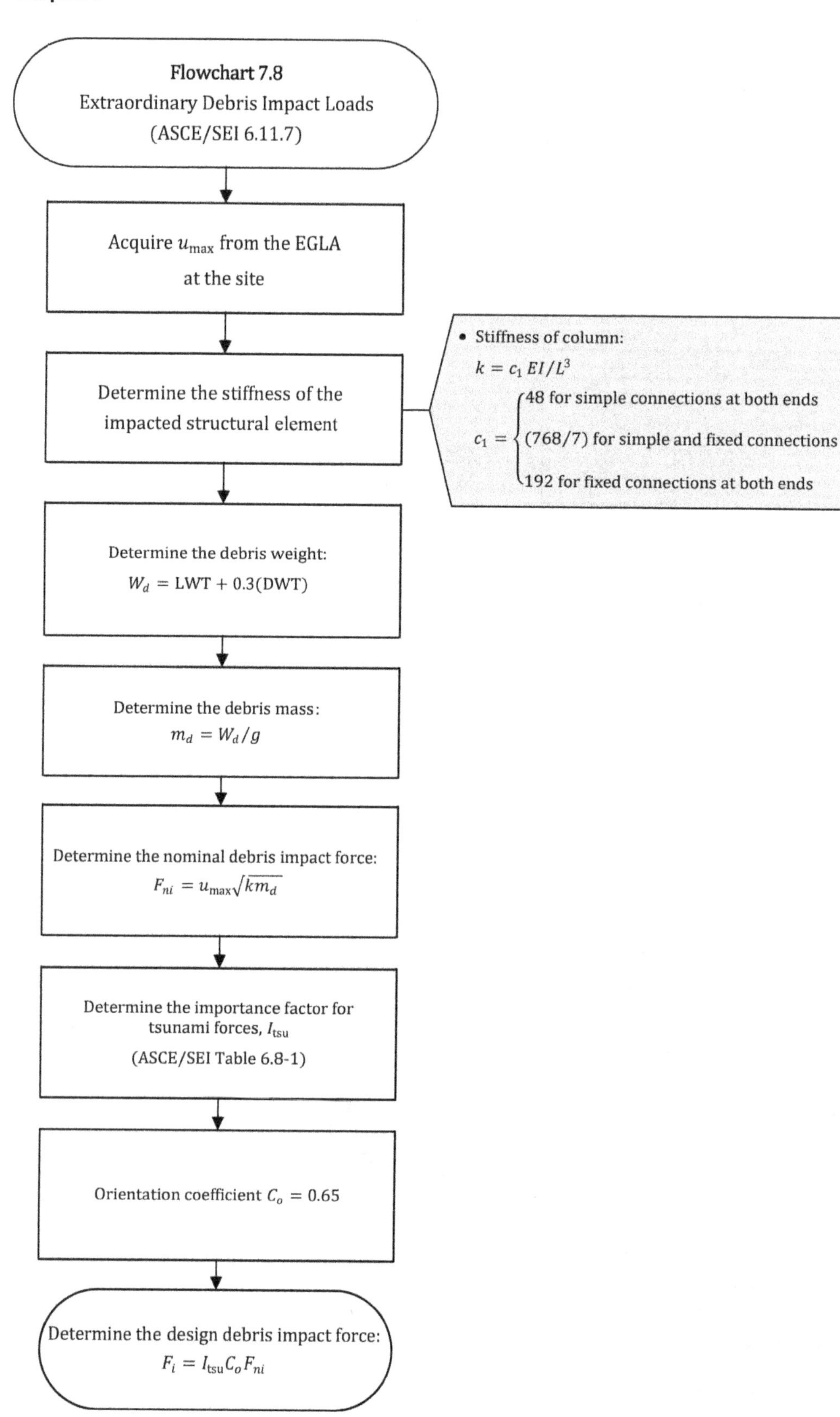
Flowchart 7.8
Extraordinary Debris Impact Loads
(ASCE/SEI 6.11.7)
Acquire u_{max} from the EGLA at the site
Determine the stiffness of the impacted structural element
• Stiffness of column:
$k = c_1 EI/L^3$
$c_1 = \begin{cases} 48 \text{ for simple connections at both ends} \\ (768/7) \text{ for simple and fixed connections} \\ 192 \text{ for fixed connections at both ends} \end{cases}$
Determine the debris weight:
$W_d = \text{LWT} + 0.3(\text{DWT})$
Determine the debris mass:
$m_d = W_d/g$
Determine the nominal debris impact force:
$F_{ni} = u_{max}\sqrt{km_d}$
Determine the importance factor for tsunami forces, I_{tsu}
(ASCE/SEI Table 6.8-1)
Orientation coefficient $C_o = 0.65$
Determine the design debris impact force:
$F_i = I_{tsu}C_oF_{ni}$

7.5 Examples

7.5.1 Example 7.1—Flood Loads on a Foundation Wall—AE Zone (River)

Determine the flood loads on the foundation wall and footing supporting the residential building in Figure 7.24. The location of the building is noted on the FIRM in Figure 7.25. Assume the following:

- Source of the floodwaters is from a river, which is fresh water.
- BFE = 209.0 ft NAVD (63.7 m NAVD) from the FIRM (see Figure 7.25).
- Dry floodproofing requirements of ASCE/SEI 24 Section 6.2 are met and no flood openings are provided in the foundation walls.
- The local jurisdiction has not adopted a more severe flood than that obtained from the FIRM.
- Floodwater debris hazards exist and are characterized as normal and special.
- The dimension of the building perpendicular to the floodwater flow is 40.0 ft (12.2 m).
- The thickness of the reinforced concrete foundation wall is 8 in. (203 mm).

Figure 7.24 Foundation Wall and Foundation in Example 7.1

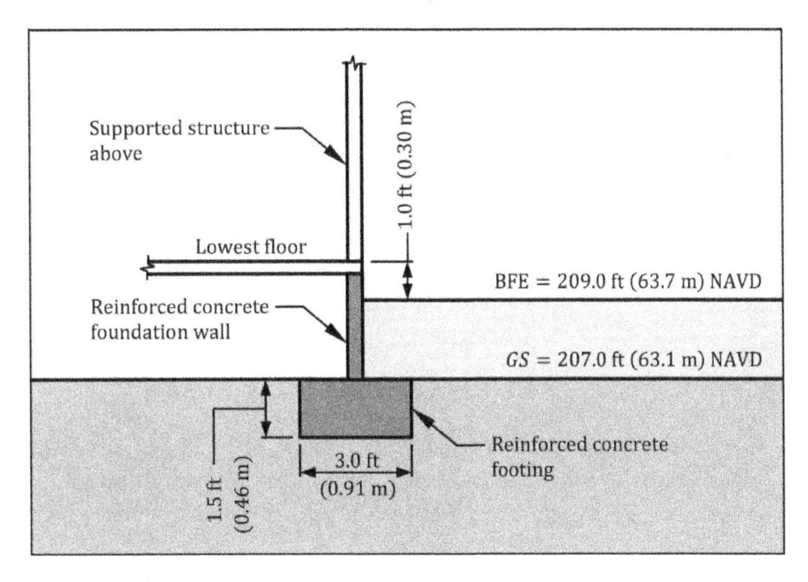

Figure 7.25
Location of the Residential Building in Example 7.1 (Source: FEMA)

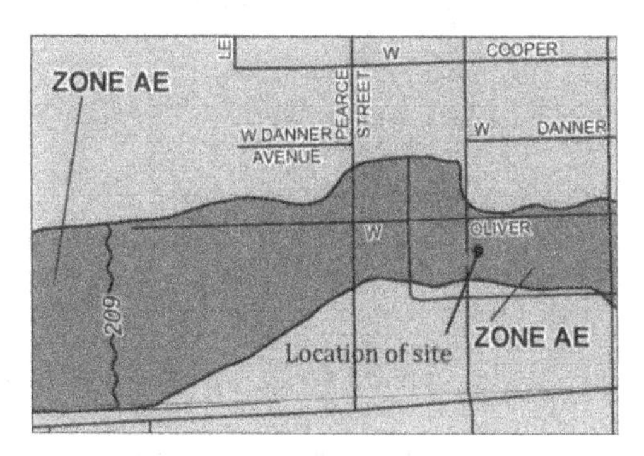

Solution

Step 1—Acquire the flood hazard zone

From the FIRM in Figure 7.25, the building is located in an AE Zone. From the design data, the source of the floodwaters is from a river.

Step 2—Acquire the base flood elevation (BFE)

From the FIRM in Figure 7.25, BFE = 209.0 ft NAVD (63.7 m NAVD).

Step 3—Determine the design flood elevation (DFE)

Because the local jurisdiction has not adopted a more severe flood than that obtained from the FIRM, DFE = BFE = 209.0 ft NAVD (63.7 m NAVD).

Step 4—Determine the minimum elevation of the lowest floor

Because the building has a residential occupancy, the flood design class is 2 (see Table 1-1 in ASCE/SEI 24 and Figure 7.4). Therefore, for an AE Zone, the minimum elevation of the top of the lowest floor is equal to the greater of the following:

- BFE + 1.0 = 209.0 + 1.0 = 210.0 ft NAVD (64.0 m NAVD)
- DFE = 209.0 ft NAVD (63.7 m NAVD)

Step 5—Determine the stillwater depth, d_s

For riverine areas (see Figure 7.5):

- SWEL = BFE = 209.0 ft (63.7 m)
- $d_s = \text{SWEL} - GS = 209.0 - 207.0 = 2.0$ ft (0.61 m)

Step 6—Determine the breaking wave height, H_b

Because the building is subjected to floodwaters from a river, wave loads are not applicable in this example and H_b need not be calculated.

Step 7—Determine the design flood velocity, V

Because the building is located in an AE Zone, the lower bound average water velocity is appropriate (see Table 7.3):

$$V = d_s/t = 2.0/1.0 = 2.0 \text{ ft/s}$$

In S.I.:

$$V = d_s/t = 0.61/1.0 = 0.61 \text{ m/s}$$

Step 8—Determine the hydrostatic loads, F_{sta} and F_{buoy}

The lateral hydrostatic load, F_{sta}, per linear foot of foundation wall is determined by Equation (7.3) with $w = 1.0$ ft (see Figure 7.9):

$$F_{sta} = \gamma_w d_s^2 w/2 = 62.4 \times (2.0+1.0)^2 \times 1.0/2 = 281 \text{ lb/linear ft of foundation wall}$$

where d_s is increased by 1.0 ft in accordance with ASCE/SEI 5.4.2 because the foundation wall above ground is exposed to free water (Note: d_s need not be increased for other than this hydrostatic load).

This load acts at $3.0/3 = 1.0$ ft above GS.

In S.I.:

$$F_{sta} = \gamma_w d_s^2/2 = 9.80 \times (0.61+0.30)^2 \times 1.0/2 = 4.10 \text{ kN/linear m of foundation wall}$$

This load acts at $0.91/3 = 0.30$ m above GS.

There is no buoyancy load on the lowest floor slab.

The buoyancy load on the foundation wall is equal to the following per linear foot of foundation wall:

$$F_{buoy} = \gamma_w V_w = \gamma_w(\text{wall thickness} \times d_s) = 62.4 \times (8.0/12) \times 2.0$$

$$= 83 \text{ lb/linear ft of foundation wall}$$

In S.I.:

$$F_{buoy} = 9.80 \times (203/1,000) \times 0.61 = 1.21 \text{ kN/linear m of foundation wall}$$

The buoyancy load on the wall footing per linear foot of foundation wall is equal to the following:

$$F_{buoy} = \gamma_w V_f = 62.4 \times 1.5 \times 3.0 = 281 \text{ lb/linear ft of foundation wall}$$

In S.I.:

$$F_{buoy} = 9.80 \times 0.46 \times 0.91 = 4.10 \text{ kN/linear m of foundation wall}$$

Step 9—Determine the hydrodynamic load, F_{dyn}

The hydrodynamic load, F_{dyn}, is determined by Equation (7.5) [see Figure 7.10]:

$$F_{dyn} = \frac{1}{2} a \rho V^2 A$$

For a building not totally immersed in floodwaters, use the ratio of the building width, w, to d_s to determine a:

$$\text{For } w/d_s = 40.0/2.0 = 20.0, a = 1.30.$$

Therefore, F_{dyn} per linear foot of foundation wall is equal to the following:

$$F_{dyn} = \frac{a \rho V^2 w d_s}{2} = \frac{1.30 \times 1.94 \times 2.0^2 \times 1.0 \times 2.0}{2} = 10 \text{ lb/linear ft of foundation wall}$$

This load acts at $d_s/2 = 2.0/2 = 1.0$ ft above GS.

Because $V < 10$ ft/s, F_{dyn} is permitted to be determined by Equation (7.7) where the dynamic effects of moving water are converted into an equivalent hydrostatic load (see Figure 7.11):

$$F_{dyn} = \gamma_w \left(\frac{aV^2}{2g} \right) d_s w = 62.4 \times \left(\frac{1.30 \times 2.0^2}{2 \times 32.2} \right) \times 2.0 \times 1.0$$

$$= 10 \text{ lb/linear ft of foundation wall}$$

In S.I.:

$$\text{For } w/d_s = 12.2/0.61 = 20.0, \ a = 1.30.$$

Therefore, F_{dyn} per linear meter of foundation wall is equal to the following:

$$F_{dyn} = \frac{a \rho V^2 w d_s}{2} = \frac{1.30 \times 1.0 \times 0.61^2 \times 1.0 \times 0.61}{2} = 0.15 \text{ kN/linear m of foundation wall}$$

This load acts at $d_s/2 = 0.61/2 = 0.30$ m above GS.

Because $V < 3.05$ m/s, F_{dyn} is permitted to be determined by Equation (7.7) where the dynamic effects of moving water are converted into an equivalent hydrostatic load (see Figure 7.11):

$$F_{dyn} = \gamma_w \left(\frac{aV^2}{2g} \right) d_s w = 9.80 \times \left(\frac{1.30 \times 0.61^2}{2 \times 9.81} \right) \times 0.61 \times 1.0$$

$$= 0.15 \text{ kN/linear m of foundation wall}$$

Step 10—Determine breaking wave loads.

Breaking wave loads need not be considered in this example because the flood source is from a river.

Step 11—Determine normal and special impact loads, F

Flowchart 7.1 is used to determine the impact loads.

Step 11a—Determine the debris impact weight, W

For riverine floodplains without floating ice, use an average $W = 1{,}500$ lb (6.7 kN) [see the table in Flowchart 7.1].

Step 11b—Determine the debris velocity, V_b

It is reasonable to assume $V_b = V = 2.0$ ft/s (0.61 m/s).

Step 11c—Determine the importance coefficient, C_I

For a risk category II building, $C_I = 1.0$ (ASCE/SEI Table C5.4-1).

Step 11d—Determine the orientation coefficient, C_O

From ASCE/SEI C5.4.5, $C_O = 0.8$.

Step 11e—Determine the depth coefficient, C_D

For $d_s = 2.0$ ft (0.61 m) in an AE Zone, $C_D = 0.25$ from ASCE/SEI Table C5.4-2.

Step 11f—Determine the blockage coefficient, C_B

Assuming there is no upstream screening and the flow path is wider than 30 ft (9.14 m), $C_B = 1.0$ from ASCE/SEI Table C5.4-3.

Step 11g—Determine the natural period, T_n, of the reinforced concrete foundation wall

From the table in Flowchart 7.1, approximate $T_n = 0.20$ s for reinforced concrete foundation walls.

Step 11h—Determine the ratio of the impact duration to the natural period of structure, $\Delta t / T_n$

Given the impact duration $\Delta t = 0.03$ s, $\Delta t / T_n = 0.03/0.20 = 0.15$.

Step 11i—Determine the maximum response ratio for impulsive loads, $R_{\max}$

For $\Delta t / T_n = 0.15$, $R_{\max} = 0.6$ by linear interpolation from ASCE/SEI Table C5.4-4.

Step 11j—Determine the normal impact load, F

The normal impact load, F, is determined by ASCE/SEI Equation (C5.4-3):

$$F = \frac{\pi W V_b C_I C_O C_D C_B R_{\max}}{2g\Delta t} = \frac{\pi \times 1{,}500 \times 2.0 \times 1.0 \times 0.8 \times 0.25 \times 1.0 \times 0.6}{2 \times 32.2 \times 0.03} = 585 \text{ lb}$$

In S.I.:

$$F = \frac{\pi W V_b C_I C_O C_D C_B R_{\max}}{2g\Delta t} = \frac{\pi \times 6.7 \times 0.61 \times 1.0 \times 0.8 \times 0.25 \times 1.0 \times 0.6}{2 \times 9.81 \times 0.03} = 2.6 \text{ kN}$$

This load acts at the SWEL = BFE, which is 2.0 ft (0.61 m) above the *GS*.

Step 11k—Determine the special impact load, F

Special impact loads are determined by Equation (7.9):

$$F = \frac{C_D \rho A V^2}{2} = \frac{1.0 \times 1.94 \times (2.0 \times 40.0) \times 2.0^2}{2} = 310 \text{ lb}$$

In S.I.:

$$F = \frac{C_D \rho A V^2}{2} = \frac{1.0 \times 1.0 \times (0.61 \times 12.2) \times 0.61^2}{2} = 1.4 \text{ kN}$$

This load acts at the SWEL = BFE, which is 2.0 ft (0.61 m) above *GS*.

Step 12—Determine the flood loads to use in the determination of F_a

- Global foundation loads

  For a solid foundation wall without openings that is dry floodproofed and located in an AE Zone where the source of the floodwaters is from a river, the total horizontal flood load on the foundation is equal to the following where the normal impact load is greater than the special impact load (see Table 7.10):

$$F_a = F + F_{dyn} = 585 + (10 \times 40.0) = 985 \text{ lb}$$

  In S.I.:

$$F_a = 2.6 + (0.15 \times 12.2) = 4.4 \text{ kN}$$

  The overturning moment about *GS* due to the flood loads is equal to the following:

$$F_a = Fd_s + F_{dyn}(d_s/2) = (585 \times 2.0) + (10 \times 40.0 \times 2.0/2) = 1{,}570 \text{ ft-lb}$$

  In S.I.:

$$F_a = (2.6 \times 0.61) + (0.15 \times 12.2 \times 0.61/2) = 2.14 \text{ kN-m}$$

- Foundation wall loads

  The flood loads on an individual foundation wall consist of F, F_{dyn}, and F_{sta} (see Figure 7.26). The load effects due to these loads are combined with the other applicable load effects using the appropriate load combinations in Section 7.2.15 of this book.

Figure 7.26 Flood Loads for the Residential Building in Example 7.1

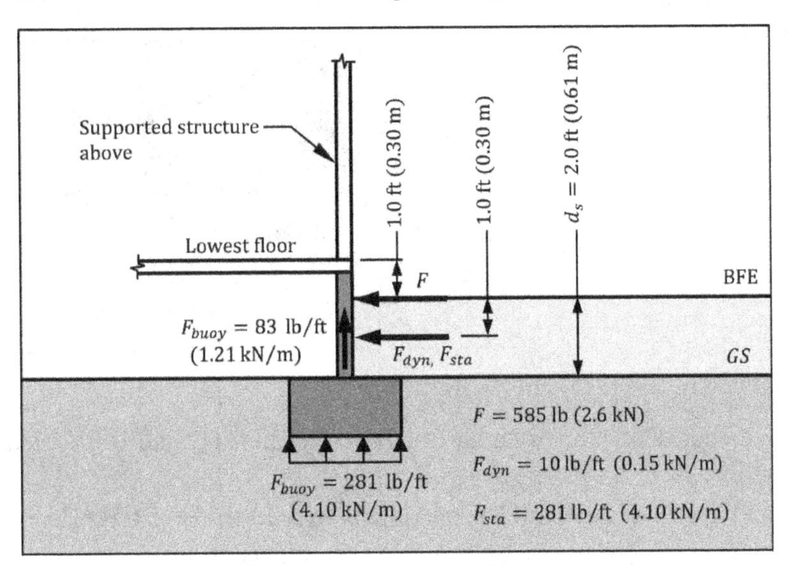

7.5.2 Example 7.2—Flood Loads on a Foundation Wall—AE Zone (Coastal Waters)

Determine the flood loads on the foundation wall and footing supporting the nonresidential building in Figure 7.27. The location of the building is noted on the FIRM in Figure 7.28. Assume the following:

- Source of the floodwaters is from coastal waters, which is saltwater.
- The ground slopes up gently from the shoreline and there are nominal obstructions between the shoreline and the site.

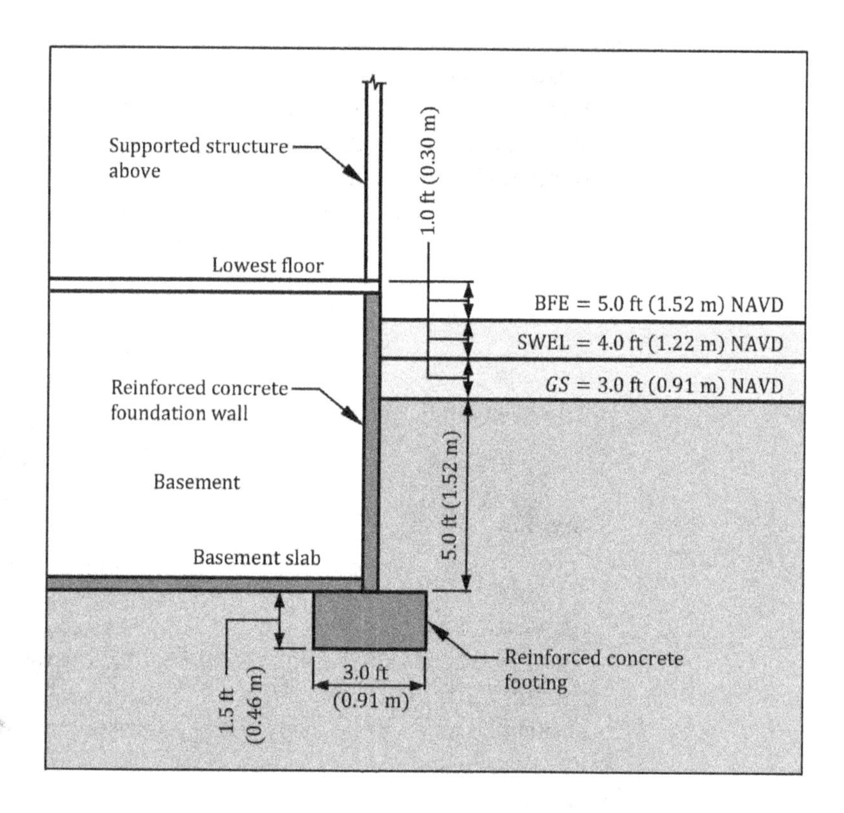

Figure 7.27 Nonresidential Building in Example 7.2

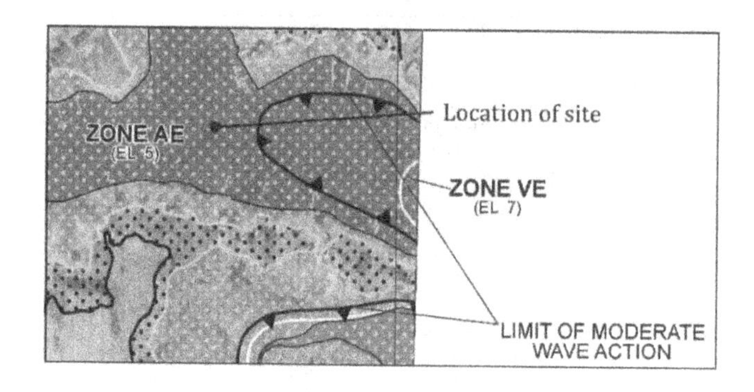

Figure 7.28 Location of the Nonresidential Building in Example 7.2 (Source: FEMA)

- The nonresidential building has a business occupancy and is oriented perpendicular to the shoreline.
- From the FIRM, the flood hazard zone is AE and BFE = 5.0 ft NAVD (1.52 m NAVD).
- From the FIS, SWEL = 4.0 ft NAVD (1.22 m NAVD).
- GS = 3.0 ft NAVD (0.91 m NAVD).
- Dry floodproofing requirements of ASCE/SEI 24 Section 6.2 are met and no flood openings are provided in the foundation walls.
- The local jurisdiction requires a 1.0 ft (0.30 m) freeboard.
- Floodwater debris hazards exist and are characterized as normal and special.
- The building plan dimensions are 80.0 ft by 80.0 ft (24.4 m by 24.4 m).
- The soil is a clayey sand (SC) with $\gamma_s = 82$ lb/ft^3 (12.88 kN/m^3).
- The thickness of the foundation wall is 10.0 in. (254 mm).

SOLUTION

Step 1—Acquire the flood hazard zone

From the FIRM, the building is located in an AE Zone and the source of the floodwaters is from coastal waters.

The site is not located in a Coastal A Zone because it is landward of the LiMWA (see Figure 7.28).

Step 2—Acquire the base flood elevation, BFE

From the FIRM, BFE = 5.0 ft NAVD (1.52 m NAVD).

Step 3—Determine the design flood elevation, DFE

Because the local jurisdiction has adopted a more severe flood than that obtained from the FIRM, DFE = BFE + freeboard = 5.0 +1.0 = 6.0 ft NAVD (1.83 m NAVD).

Step 4—Determine the minimum elevation of the lowest floor

Because the building has a business occupancy, the flood design class is 2 (see Table 1-1 in ASCE/SEI 24 and Figure 7.4). Therefore, for an AE Zone, the minimum elevation of the top of the lowest floor is equal to the greater of the following:

- BFE + 1.0 = 5.0 + 1.0 = 6.0 ft NAVD (1.83 m NAVD)
- DFE = 6.0 ft NAVD (1.83 m NAVD)

The basement in this nonresidential building is permitted to be located below the minimum elevation in ASCE/SEI 24 Table 2-1 because the dry floodproofing requirements in ASCE/SEI 24 Section 6.2 are satisfied.

Step 5—Determine the stillwater depth, d_s

For coastal areas with waves (see Figure 7.6):

$$d_s = \text{SWEL} - GS = 4.0 - 3.0 = 1.0 \text{ ft } (0.30 \text{ m})$$

Step 6—Determine the breaking wave height, H_b

For a gently sloping shoreline with nominal obstructions (see Figure 7.6):

$$H_b = 0.78 d_s = 0.78 \times 1.0 = 0.8 \text{ ft } (0.24 \text{ m})$$

Step 7—Determine the design flood velocity, V

Because the building is located in an AE Zone, the lower bound average water velocity is appropriate (see Table 7.3):

$$V = d_s / t = 1.0/1.0 = 1.0 \text{ ft/s}$$

In S.I.:

$$V = d_s / t = 0.30/1.0 = 0.30 \text{ m/s}$$

Step 8—Determine the hydrostatic loads, F_{sta}, F_{dif}, and F_{buoy}

The lateral hydrostatic load, F_{sta}, on the portion of the foundation wall above GS need not be determined because breaking wave loads must be considered, which include the hydrostatic component of the flood load (see Step 10).

Below GS, F_{sta} is equal to the following per linear foot of foundation wall with $w = 1.0$ ft (see Figure 7.9):

$$F_{sta} = \gamma_w (\text{distance from } GS \text{ to the top of the footing})^2 w/2$$

$$= 64.0 \times 5.0^2 \times 1.0/2 = 800 \text{ lb/ft}$$

This load acts at $5.0/3 = 1.7$ ft above the top of the footing.

In S.I.:

$$F_{sta} = 10.05 \times 1.52^2 \times 1.0/2 = 11.6 \text{ kN/m}$$

This load acts at $1.52/3 = 0.51$ m above the top of the footing.
The lateral load on the foundation wall below *GS* due the differential between the water and soil pressures, F_{dif}, is determined by the following equation with $w = 1.0$ ft (see Figure 7.9):

$$F_{dif} = (\gamma_s - \gamma_w)(\text{distance from } GS \text{ to the top of the footing})^2 w/2$$

$$= (82.0 - 64.0) \times 5.0^2 \times 1.0/2 = 225 \text{ lb/linear ft of foundation wall}$$

This load acts at $5.0/3 = 1.7$ ft above the top of the footing.
In S.I.:

$$F_{dif} = (12.88 - 10.05) \times 1.52^2 \times 1.0/2 = 3.27 \text{ kN/linear m of foundation wall}$$

This load acts at $1.52/3 = 0.51$ m above the top of the footing.
The buoyancy pressure on the basement slab is determined by the following equation (see Figure 7.9):

$$f_{buoy} = \gamma_w(\text{distance from the top of the flood water to the top of the footing})$$

$$= 64.0 \times (1.0 + 5.0) = 384 \text{ lb/ft}^2$$

With $A_b = 80.0 \times 80.0 = 6{,}400$ ft^2, the buoyancy load, F_{buoy}, is equal to the following:

$$F_{buoy} = 384 \times 6{,}400/1{,}000 = 2{,}458 \text{ kips}$$

In S.I.:

$$f_{buoy} = 10.05 \times (0.30 + 1.52) = 18.3 \text{ kN/m}^2$$

With $A_b = 24.4 \times 24.4 = 595.4$ m^2, the buoyancy load, F_{buoy}, is equal to the following:

$$F_{buoy} = 18.3 \times 595.4 = 10{,}896 \text{ kN}$$

The buoyancy load on the foundation wall is equal to the following per linear foot of foundation wall:

$$F_{buoy} = \gamma_w V_w = \gamma_w(\text{wall thickness} \times \text{submerged depth of the wall})$$

$$= 64.0 \times (10.0/12) \times (1.0 + 5.0) = 320 \text{ lb/linear ft of foundation wall}$$

In S.I.:

$$F_{buoy} = 10.05 \times (254/1{,}000) \times (0.30 + 1.52) = 4.65 \text{ kN/linear m of foundation wall}$$

The buoyancy load on the wall footing is equal to the following per linear foot of foundation wall:

$$F_{buoy} = \gamma_w V_f = 64.0 \times 1.5 \times 3.0 = 288 \text{ lb/linear ft of foundation wall}$$

In S.I.:

$$F_{buoy} = 10.05 \times 0.46 \times 0.91 = 4.21 \text{ kN/linear m of foundation wall}$$

Step 9—Determine the hydrodynamic load, F_{dyn}

The hydrodynamic load, F_{dyn}, is determined by Equation (7.5) [see Figure 7.10]:

$$F_{dyn} = \frac{1}{2} a \rho V^2 A$$

For a building not totally immersed in floodwaters, use the ratio of the building width, w, to d_s to determine a:

For $w/d_s = 80.0/1.0 = 80.0, a = 1.75$.

Therefore, F_{dyn} per linear foot of foundation wall is equal to the following:

$$F_{dyn} = \frac{a\rho V^2 w d_s}{2} = \frac{1.75 \times 1.99 \times 1.0^2 \times 1.0 \times 1.0}{2} = 1.7 \text{ lb/linear ft of foundation wall}$$

This load acts at $d_s/2 = 1.0/2 = 0.5$ ft above GS.

In S.I.:

$$F_{dyn} = \frac{a\rho V^2 w d_s}{2} = \frac{1.75 \times 1.026 \times 0.30^2 \times 1.0 \times 0.30}{2}$$

$$= 0.024 \text{ kN/linear m of foundation wall}$$

This load acts at $d_s/2 = 0.30/2 = 0.15$ m above GS.

Step 10—Determine breaking wave loads

Breaking wave loads occur on the vertical foundation wall. Because flood openings are not provided in the foundation wall to allow passage of the floodwaters, the breaking wave load, F_t, is determined by ASCE/SEI Equation (5.4-6), which is applicable where the space behind the vertical wall is dry (see Figure 7.12):

$$F_t = 1.1 C_p \gamma_w d_s^2 + 2.4 \gamma_w d_s^2$$

For a risk category II building, $C_p = 2.8$ (see ASCE/SEI Table 5.4-1).

Therefore, F_t per linear foot of foundation wall is equal to the following:

$$F_t = (1.1 \times 2.8 \times 64.0 \times 1.0^2) + (2.4 \times 64.0 \times 1.0^2) = 351 \text{ lb/linear ft of foundation wall}$$

This load acts at $0.1d_s = 0.1$ ft below the SWEL (for practical purposes, assume F_t acts at the SWEL, which is 1.0 ft above GS).

In S.I.:

$$F_t = (1.1 \times 2.8 \times 10.05 \times 0.30^2) + (2.4 \times 10.05 \times 0.30^2)$$

$$= 4.96 \text{ kN/linear m of foundation wall}$$

This load acts at $0.1d_s = 0.03$ m below the SWEL (for practical purposes, assume F_t acts at the SWEL, which is 0.30 m above GS).

Step 11—Determine normal and special impact loads, F

Flowchart 7.1 is used to determine the impact loads.

Step 11a—Determine the debris impact weight, W

For coastal areas where piers and large pilings are not present, use $W = 500$ lb (2.3 kN) [see the table in Flowchart 7.1].

Step 11b—Determine the debris velocity, V_b

It is reasonable to assume $V_b = V = 1.0$ ft/s (0.30 m/s).

Step 11c—Determine the importance coefficient, C_I

For a risk category II building, $C_I = 1.0$ (ASCE/SEI Table C5.4-1).

Step 11d—Determine the orientation coefficient, C_O

From ASCE/SEI C5.4.5, $C_O = 0.8$.

Step 11e—Determine the depth coefficient, C_D

A value of C_D is not given in ASCE/SEI Table C5.4-2 for $d_s = 1.0$ ft (0.30 m) in an AE Zone.

Use $C_D = 0.25$.

Step 11f—Determine the blockage coefficient, C_B

Assuming there is no upstream screening and the flow path is wider than 30 ft (9.14 m), $C_B = 1.0$ (ASCE/SEI Table C5.4-3).

Step 11g—Determine the natural period, approximate T_n, of the reinforced concrete foundation wall

From the table in Flowchart 7.1, $T_n = 0.20$ s for reinforced concrete foundation walls.

Step 11h—Determine the ratio of the impact duration to the natural period of structure, $\Delta t/T_n$

Given the impact duration $\Delta t = 0.03$ s, $\Delta t/T_n = 0.03/0.20 = 0.15$.

Step 11i—Determine the maximum response ratio for impulsive loads, R_{max}

For $\Delta t/T_n = 0.15$, $R_{max} = 0.6$ by linear interpolation from ASCE/SEI Table C5.4-4.

Step 11j—Determine the normal impact force, F

The normal impact load, F, is determined by ASCE/SEI Equation (C5.4-3):

$$F = \frac{\pi W V_b C_I C_O C_D C_B R_{max}}{2g\Delta t} = \frac{\pi \times 500 \times 1.0 \times 1.0 \times 0.8 \times 0.25 \times 1.0 \times 0.6}{2 \times 32.2 \times 0.03} = 98 \text{ lb}$$

In S.I.:

$$F = \frac{\pi W V_b C_I C_O C_D C_B R_{max}}{2g\Delta t} = \frac{\pi \times 2.2 \times 0.30 \times 1.0 \times 0.8 \times 0.25 \times 1.0 \times 0.6}{2 \times 9.81 \times 0.03} = 0.42 \text{ kN}$$

This load acts at the SWEL, which is 1.0 ft (0.30 m) above GS.

Special impact loads are determined by Equation (7.9):

$$F = \frac{C_D \rho A V^2}{2} = \frac{1.0 \times 1.99 \times (1.0 \times 80.0) \times 1.0^2}{2} = 80 \text{ lb}$$

In S.I.:

$$F = \frac{C_D \rho A V^2}{2} = \frac{1.0 \times 1.026 \times (0.30 \times 24.4) \times 0.30^2}{2} = 0.34 \text{ kN}$$

This load acts at the SWEL, which is 1.0 ft (0.30 m) above GS.

Step 12—Determine the flood loads to use in the determination of F_a

- Global foundation loads

  For a solid foundation wall without openings that is dry floodproofed and located in an AE Zone where the source of the floodwaters is from coastal waters, the total horizontal flood load on the foundation is equal to the following where the normal impact load is greater than the special impact load (see Table 7.11):

$$F_a = \text{greater of} \begin{cases} F + F_{dyn} = 98 + (1.7 \times 80.0) = 234 \text{ lb} \\ F + F_t = 98 + (351 \times 80.0) = 28{,}178 \text{ lb} \end{cases}$$

In S.I.:

$$F_a = \text{greater of} \begin{cases} F + F_{dyn} = 0.42 + (0.024 \times 24.4) = 1.0 \text{ kN} \\ F + F_t = 0.42 + (4.96 \times 24.4) = 121.4 \text{ kN} \end{cases}$$

The overturning moment about *GS* due to the flood loads is equal to the following assuming F_t acts at the SWEL:

$$F_a = \text{greater of} \begin{cases} Fd_s + F_{dyn}(d_s/2) = (98 \times 1.0) + (1.7 \times 80.0 \times 1.0/2) = 166 \text{ ft-lb} \\ Fd_s + F_t d_s = (98 \times 1.0) + (351 \times 80.0 \times 1.0) = 28{,}178 \text{ ft-lb} \end{cases}$$

In S.I.:

$$F_a = \text{greater of} \begin{cases} (0.42 \times 0.30) + (0.024 \times 24.4 \times 0.30/2) = 0.21 \text{ kN-m} \\ (0.42 \times 0.30) + (4.96 \times 24.4 \times 0.30) = 36.4 \text{ kN-m} \end{cases}$$

- Foundation wall loads
  The flood loads on an individual foundation wall consist of F, F_{dyn}, F_t, F_{sta}, and F_{dif} (see Figure 7.29). The load effects due to these loads are combined with the other applicable load effects using the appropriate load combinations in Section 7.2.15 of this book.

Figure 7.29 Flood Loads for the Nonresidential Building in Example 7.2

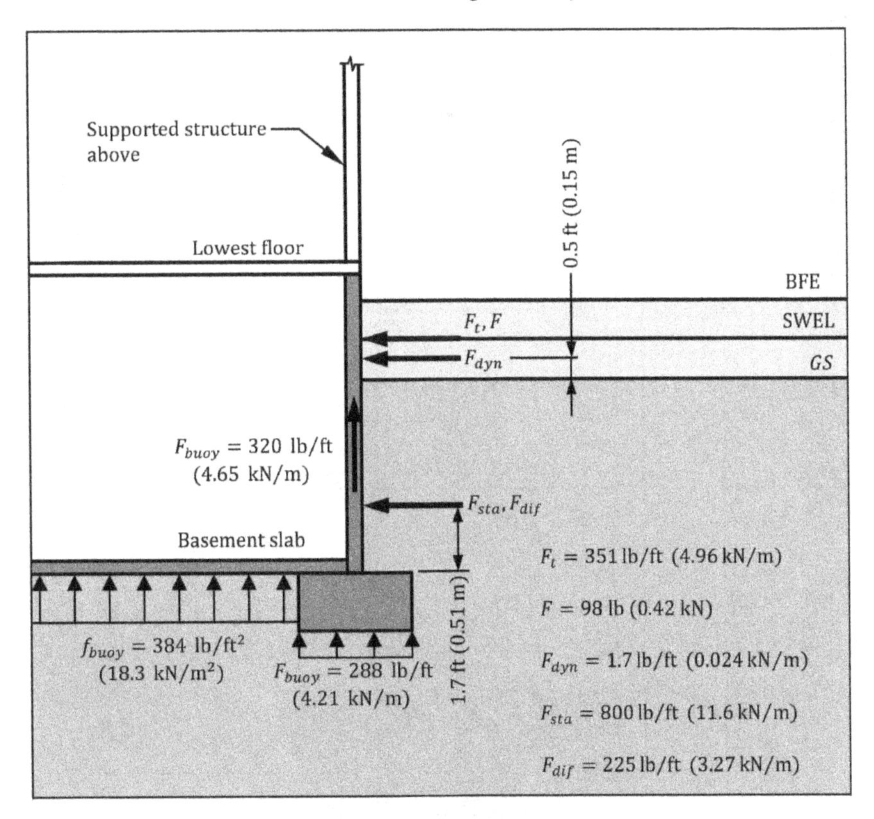

7.5.3 Example 7.3—Flood Loads on Columns—Coastal A Zone

Determine the flood loads, F_a, on the reinforced concrete columns and the reinforced concrete mat foundation supporting the nonresidential building depicted in Figure 7.30. The location of the building is noted on the FIRM in Figure 7.31. Assume the following:

- Source of the floodwaters is from coastal waters, which is saltwater.
- The ground slopes up gently from the shoreline and there are nominal obstructions between the shoreline and the site.

Figure 7.30 Nonresidential Building in Example 7.3

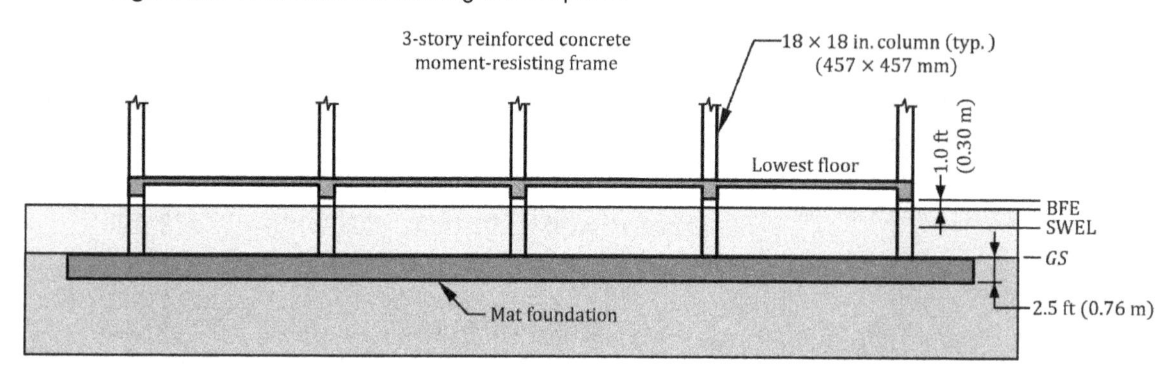

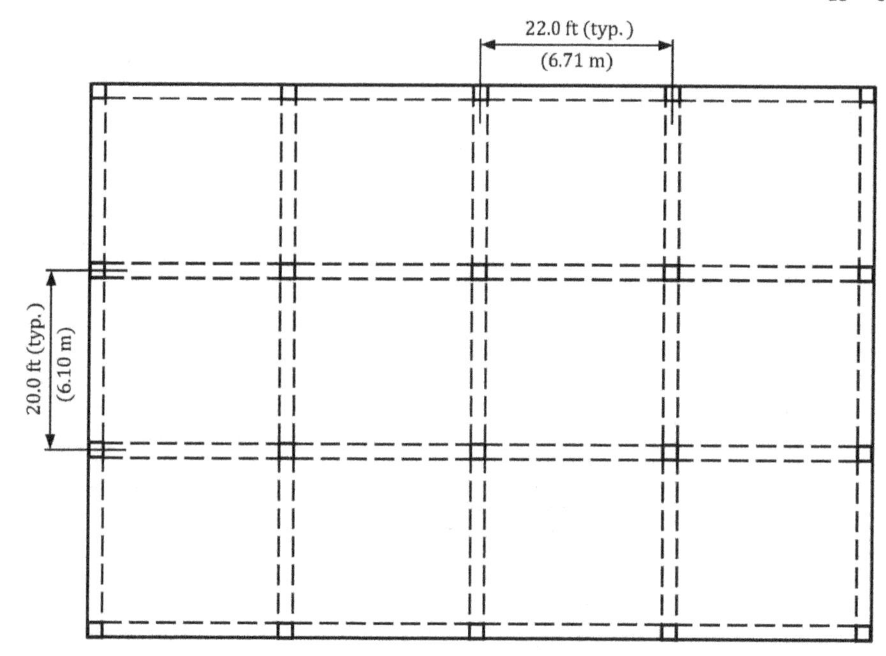

Figure 7.31 Location of the Nonresidential Building in Example 7.3

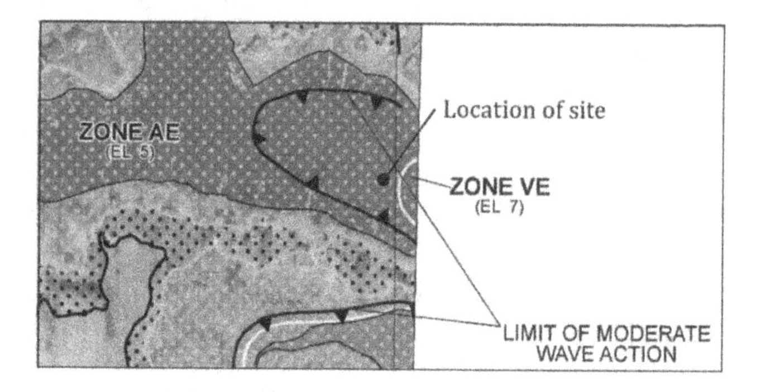

- The nonresidential building has a business occupancy and is oriented perpendicular to the shoreline.
- From the FIRM, the flood hazard zone is AE and BFE = 5.0 ft NAVD (1.52 m NAVD).
- From the FIS, SWEL = 3.0 ft NAVD (0.91 m NAVD).
- GS = 0.0 ft NAVD (0.0 m NAVD).
- The local jurisdiction requires a 1.0 ft (0.30 m) freeboard.
- Floodwater debris hazards exist and are characterized as normal and special.

Solution

Step 1—Acquire the flood hazard zone

From the design data, the building is located in an AE Zone and the source of the floodwaters is from coastal waters. Because the site is landward of the VE Zone and seaward of the LiMWA, it is designated a Coastal A Zone (see Figure 7.31).

Step 2—Acquire the base flood elevation, BFE

From the FIRM, BFE = 5.0 ft NAVD (1.52 m NAVD).

Step 3—Determine the design flood elevation, DFE

Because the local jurisdiction has adopted a more severe flood than that obtained from the FIRM, DFE = BFE + freeboard = 5.0 + 1.0 = 6.0 ft NAVD (1.83 m NAVD).

Step 4—Determine the minimum elevation of the lowest floor

Because the building has a business occupancy, the flood design class is 2 (see Table 1-1 in ASCE/SEI 24 and Figure 7.4). Therefore, for a Coastal A Zone, the minimum elevation of the bottom of the lowest supporting horizontal structural member is equal to the greater of the following:

- BFE + 1.0 = 5.0 + 1.0 = 6.0 ft NAVD (1.83 m NAVD)
- DFE = 6.0 ft NAVD (1.83 m NAVD)

Step 5—Determine the stillwater depth, d_s

For coastal areas with waves (Figure 7.6):

$$d_s = \text{SWEL} - GS = 3.0 - 0.0 = 3.0 \text{ ft } (0.91 \text{ m})$$

Step 6—Determine the breaking wave height, H_b

For a gently sloping shoreline with nominal obstructions (see Figure 7.6):

$$H_b = 0.78 d_s = 0.78 \times 3.0 = 2.3 \text{ ft } (0.71 \text{ m})$$

Step 7—Determine the design flood velocity, V

Because the building is located in a Coastal A Zone, the upper bound average water velocity is appropriate (see Table 7.3):

$$V = (g d_s)^{0.5} = (32.2 \times 3.0)^{0.5} = 9.8 \text{ ft/s}$$

In S.I.:

$$V = (g d_s)^{0.5} = (9.81 \times 0.91)^{0.5} = 2.99 \text{ m/s}$$

Step 8—Determine the hydrostatic loads, F_{sta} and F_{buoy}

Lateral hydrostatic loads on the columns are included in the hydrodynamic load, F_{dyn}.

There is no buoyancy load on the lowest floor slab.

The buoyancy pressure on the mat foundation is equal to the following:

$$f_{buoy} = \gamma_w (\text{thickness of the mat foundation}) = 64.0 \times 2.5 = 160 \text{ lb/ft}^2$$

In S.I.:

$$f_{buoy} = 10.05 \times 0.76 = 7.6 \text{ kN/m}^2$$

This buoyancy pressure is uniformly distributed over the entire base area of the mat foundation.
The buoyancy load on a column is equal to the following:

$$F_{buoy} = \gamma_w d_s A_c = 64.0 \times 3.0 \times (18.0/12)^2 = 432 \text{ lb}$$

In S.I.:

$$F_{buoy} = \gamma_w d_s A_c = 10.05 \times 0.91 \times (457/1{,}000)^2 = 1.9 \text{ kN}$$

Step 9—Determine the hydrodynamic load, F_{dyn}

Hydrodynamic loads occur on the columns and are determined by Equation (7.5) [see Figure 7.10]:

$$F_{dyn} = \frac{1}{2} a \rho V^2 A$$

From Figure 7.10, use $a = 2.0$ for rectangular columns.
Therefore, F_{dyn} on one of the columns is equal to the following:

$$F_{dyn} = \frac{a \rho V^2 D d_s}{2} = \frac{2.0 \times 1.99 \times 9.8^2 \times (18.0/12) \times 3.0}{2} = 860 \text{ lb}$$

This load acts at $d_s/2 = 3.0/2 = 1.5$ ft above GS.

In S.I.:

$$F_{dyn} = \frac{a \rho V^2 D d_s}{2} = \frac{2.0 \times 1.026 \times 2.99^2 \times (457/1{,}000) \times 0.91}{2} = 3.8 \text{ kN}$$

This load acts at $d_s/2 = 0.91/2 = 0.46$ m above GS.

Step 10—Determine breaking wave loads

Breaking wave loads, F_D, occur on the vertical columns and are determined by ASCE/SEI Equation (5.4-4):

$$F_D = 0.5 \gamma_w C_D D H_b^2$$

$C_D = 2.25$ for square columns (ASCE/SEI 5.4.4.1)
$D = 1.4 \times (18.0/12) = 2.1$ ft (ASCE/SEI 5.4.4.1)
Therefore, F_D on one of the columns is equal to the following:

$$F_D = 0.5 \times 64.0 \times 2.25 \times 2.1 \times 2.3^2 = 800 \text{ lb}$$

This load acts at the SWEL, which is 3.0 ft above GS.

In S.I.:

$$D = 1.4 \times (457/1{,}000) = 0.64 \text{ m}$$

$$F_D = 0.5 \times 10.05 \times 2.25 \times 0.64 \times 0.71^2 = 3.7 \text{ kN}$$

This load acts at the SWEL, which is 0.91 m above GS.

Step 11—Determine normal and special impact loads, F

Flowchart 7.1 is used to determine the impact loads.

Step 11a—Determine the debris impact weight, W

For coastal areas where piers and large pilings are present, use $W = 2{,}000$ lb (9.0 kN) [see the table in Flowchart 7.1].

Step 11b—Determine the debris velocity, V_b

It is reasonable to assume $V_b = V = 9.8$ ft/s (2.99 m/s).

Step 11c—Determine the importance coefficient, C_I

For a risk category II building, $C_I = 1.0$ (ASCE/SEI Table C5.4-1).

Step 11d—Determine the orientation coefficient, C_O

From ASCE/SEI C5.4.5, $C_O = 0.8$.

Step 11e—Determine the depth coefficient, C_D

For $d_s = 3.0$ ft (0.91 m) in a Coastal A Zone, $C_D = 0.50$ (ASCE/SEI Table C5.4-2).

Step 11f—Determine the blockage coefficient, C_B

Assuming there is no upstream screening and the flow path is wider than 30 ft (9.14 m), $C_B = 1.0$ (ASCE/SEI Table C5.4-3).

Step 11g—Determine the natural period, T_n, of the reinforced concrete moment-resisting frame

From the table in Flowchart 7.1, approximate $T_n = 0.35$ s for a reinforced concrete moment-resisting frame three stories in height.

Step 11h—Determine the ratio of the impact duration to the natural period of structure, $\Delta t / T_n$

Given the impact duration $\Delta t = 0.03$ s, $\Delta t / T_n = 0.03/0.35 = 0.09$.

Step 11i—Determine the maximum response ratio for impulsive loads, R_{max}

For $\Delta t / T_n = 0.09$, $R_{max} = 0.4$ (ASCE/SEI Table C5.4-4).

Step 11j—Determine the normal impact force, F

The normal impact load, F, is determined by ASCE/SEI Equation (C5.4-3):

$$F = \frac{\pi W V_b C_I C_O C_D C_B R_{max}}{2g\Delta t} = \frac{\pi \times 2{,}000 \times 9.8 \times 1.0 \times 0.8 \times 0.50 \times 1.0 \times 0.4}{2 \times 32.2 \times 0.03} = 5{,}099 \text{ lb}$$

In S.I.:

$$F = \frac{\pi W V_b C_I C_O C_D C_B R_{max}}{2g\Delta t} = \frac{\pi \times 9.0 \times 2.99 \times 1.0 \times 0.8 \times 0.50 \times 1.0 \times 0.4}{2 \times 9.81 \times 0.03} = 23.0 \text{ kN}$$

This load acts at the SWEL, which is 3.0 ft (0.91 m) above *GS*.

Special impact loads are determined by Equation (7.9):

$$F = \frac{C_D \rho A V^2}{2} = \frac{1.0 \times 1.99 \times [(18.0/12) \times 3.0] \times 9.8^2}{2} = 430 \text{ lb}$$

In S.I.:

$$F = \frac{C_D \rho A V^2}{2} = \frac{1.0 \times 1.026 \times (0.457 \times 0.91) \times 2.99^2}{2} = 1.9 \text{ kN}$$

This load acts at the SWEL, which is 3.0 ft (0.91 m) above *GS*.

Step 12—Determine the flood loads to use in the determination of F_a

- Global foundation loads

For the analysis of the entire (global) foundation, it is reasonable to assume that one of the 20 columns will be subjected to the impact load, F, and the remaining columns will be subjected to the hydrodynamic load, F_{dyn}. Therefore, the total horizontal flood load on the foundation is equal to the following where the normal impact load is greater than the special impact load:

$$F_a = F + (n-1)F_{dyn} = 5{,}099 + (19 \times 860) = 21{,}439 \text{ lb}$$

In S.I.:

$$F_a = 23.0 + (19 \times 3.8) = 95.2 \text{ kN}$$

The overturning moment about *GS* due to the flood loads is equal to the following:

$$F_a = Fd_s + (n-1)F_{dyn}(d_s/2)$$
$$= (5{,}099 \times 3.0) + (19 \times 860 \times 3.0/2) = 39{,}807 \text{ ft-lb}$$

In S.I.:

$$F_a = (23.0 \times 0.91) + (19 \times 3.8 \times 0.91/2) = 53.8 \text{ kN-m}$$

- Individual column loads
  The flood loads on an individual column in the front row consists of F and the greater of F_{dyn} and F_D (see Figure 7.32). The load effects due to these loads are combined with the other applicable loads using the appropriate load combinations in Section 7.2.15 of this book.

Figure 7.32 Flood Loads for the Nonresidential Building in Example 7.3

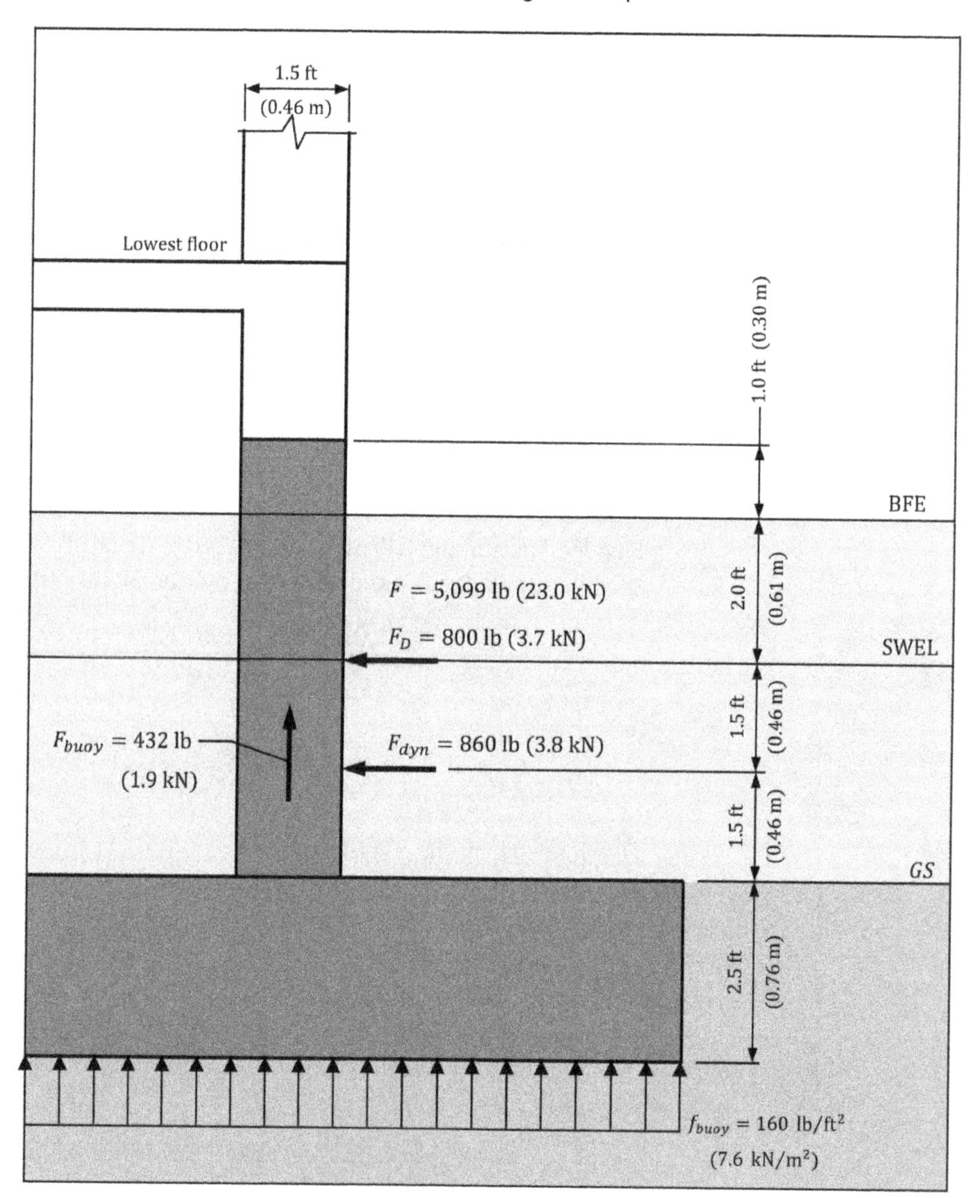

The design and construction of the mat foundation must satisfy the requirements of ASCE/SEI 24 Section 4.5. The top of the mat must be located below the *GS* and must extend to a depth sufficient to prevent flotation, collapse, or permanent lateral movement under the design load combinations (ASCE/SEI 24 Section 1.5.3).

7.5.4 Example 7.4—Flood Loads on Piles—VE Zone

Determine the flood loads, F_a, on the piles supporting the utility building depicted in Figure 7.33. The location of the building is noted on the FIRM in Figure 7.34. Assume the following:

- Source of the floodwaters is from coastal waters, which is saltwater.
- The ground slopes up gently from the shoreline and there are nominal obstructions between the shoreline and the site.
- The utility building must remain operational before and after a design flood event and is oriented perpendicular to the shoreline.
- From the FIRM, the flood hazard zone is VE and BFE = 7.0 ft NAVD (2.13 m NAVD).
- From the FIS, SWEL = 4.0 ft NAVD (1.22 m NAVD).
- *GS* = 0.0 ft NAVD (0.0 m NAVD).
- The local jurisdiction requires a 1.0 ft (0.30 m) freeboard.
- Floodwater debris hazards exist and are characterized as normal and special.

SOLUTION

Step 1—Acquire the flood hazard zone

From the design data, the building is located in a VE Zone.

Step 2—Acquire the base flood elevation, BFE

From the FIRM, BFE = 7.0 ft NAVD (2.13 m NAVD).

Step 3—Determine the design flood elevation, DFE

Because the local jurisdiction has adopted a more severe flood than that obtained from the FIRM, DFE = BFE + freeboard = 7.0 + 1.0 = 8.0 ft NAVD (2.44 m NAVD).

Step 4—Determine the minimum elevation of the lowest floor

Because the utility building must remain operational during and after a design flood event, the flood design class is 3 (see Table 1-1 in ASCE/SEI 24 and Figure 7.4). Therefore, for a VE Zone, the minimum elevation of the bottom of the lowest supporting horizontal structural member is equal to the greater of the following:

- BFE + 2.0 = 7.0 + 2.0 = 9.0 ft NAVD (2.74 m NAVD)
- DFE = 8.0 ft NAVD (2.44 m NAVD)

Step 5—Determine the stillwater depth, d_s

For coastal areas with waves (see Figure 7.6):

$$d_s = \text{SWEL} - GS = 4.0 - 0.0 = 4.0 \text{ ft } (1.22 \text{ m})$$

Step 6—Determine the breaking wave height, H_b

For a gently sloping shoreline with nominal obstructions (see Figure 7.6):

$$H_b = 0.78 d_s = 0.78 \times 4.0 = 3.1 \text{ ft } (0.95 \text{ m})$$

Step 7—Determine the design flood velocity, V

Because the building is located in a VE Zone, the upper bound average water velocity is appropriate:

$$V = (g d_s)^{0.5} = (32.2 \times 4.0)^{0.5} = 11.3 \text{ ft/s}$$

Figure 7.33 Utility Building in Example 7.4

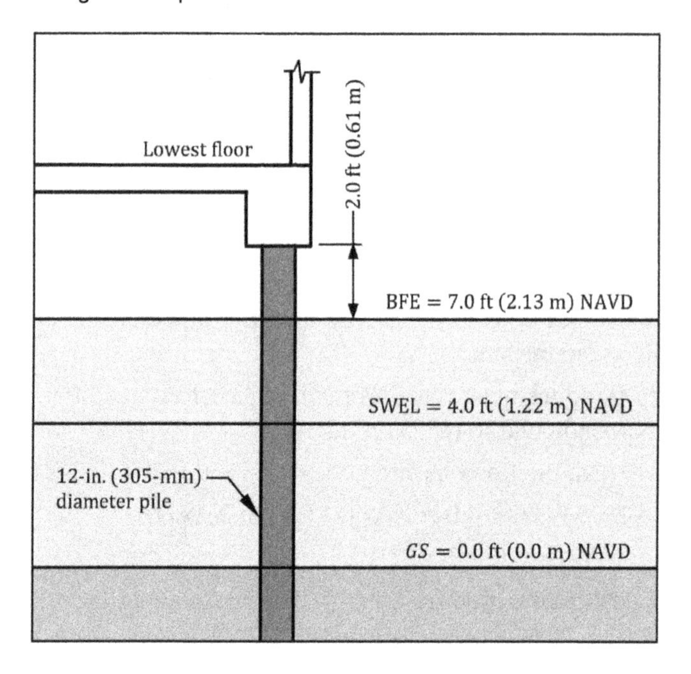

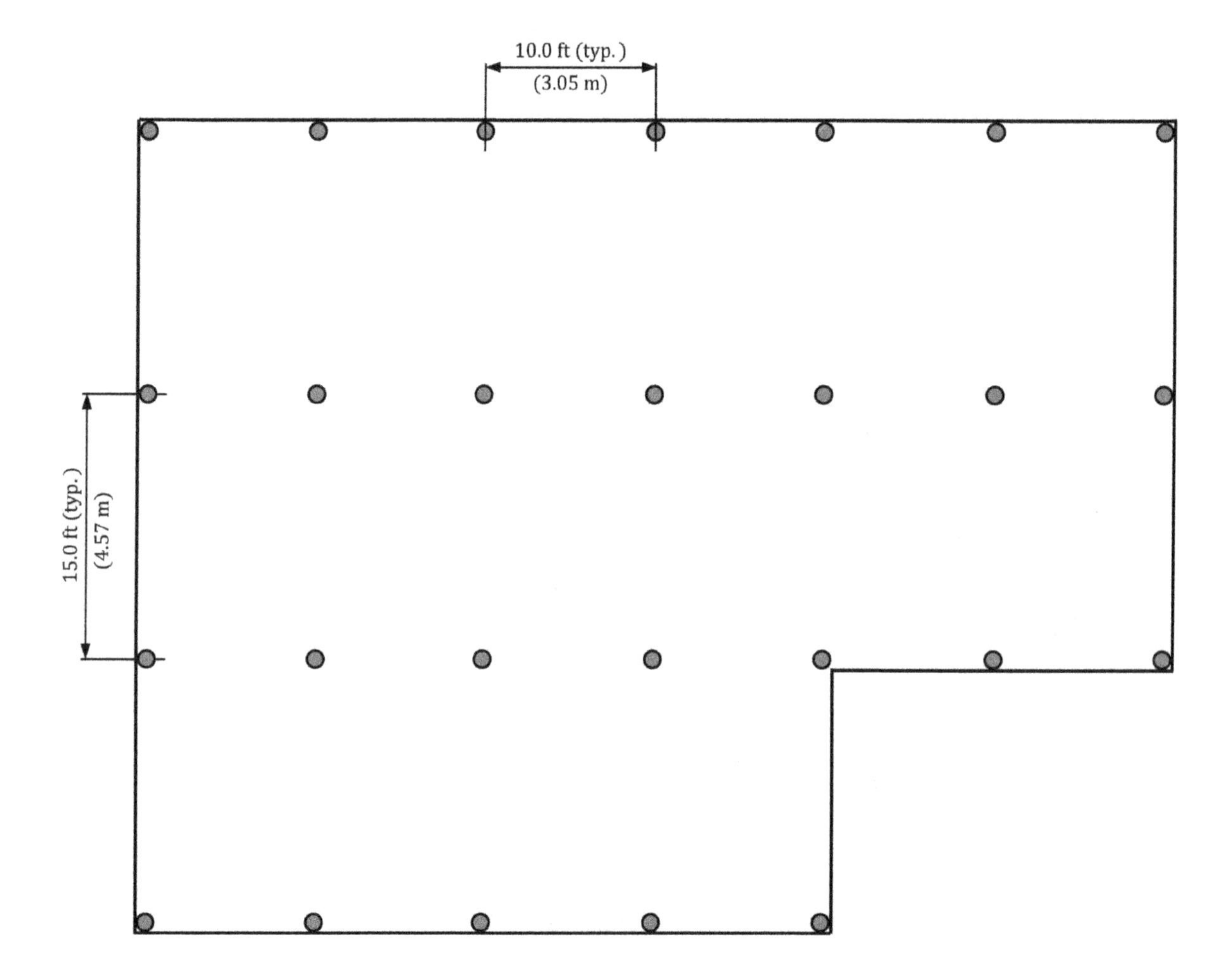

Figure 7.34 Location of the Utility Building in Example 7.4

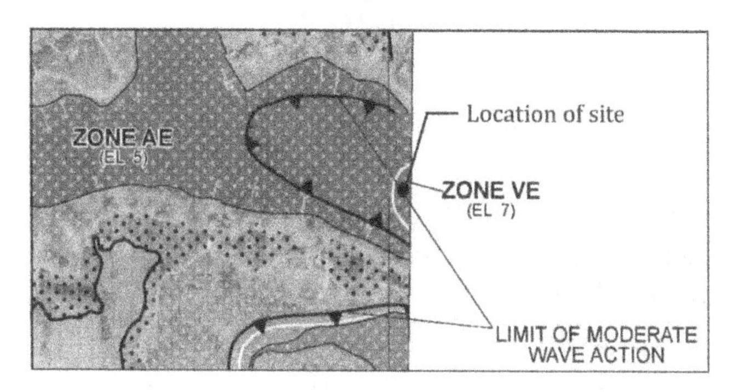

In S.I.:

$$V = (gd_s)^{0.5} = (9.81 \times 1.22)^{0.5} = 3.46 \text{ m/s}$$

Step 8—Determine the hydrostatic loads, F_{sta} and F_{buoy}

Lateral hydrostatic loads on the piles are included in the hydrodynamic load, F_{dyn}.

There is no buoyancy load on the lowest floor slab.

The buoyancy load on a pile is equal to the following:

$$F_{buoy} = \gamma_w d_s A_p = 64.0 \times 4.0 \times \pi \times (12.0/12)^2/4 = 201 \text{ lb}$$

In S.I.:

$$F_{buoy} = \gamma_w d_s A_c = 10.05 \times 1.22 \times \pi \times (305/1{,}000)^2/4 = 0.90 \text{ kN}$$

Step 9—Determine the hydrodynamic load, F_{dyn}

Hydrodynamic loads occur on the columns and are determined by Equation (7.5) [see Figure 7.10]:

$$F_{dyn} = \frac{1}{2} a\rho V^2 A$$

From Figure 7.10, use $a = 1.25$ for round piles.

Therefore, F_{dyn} on one of the piles is equal to the following:

$$F_{dyn} = \frac{a\rho V^2 D d_s}{2} = \frac{1.25 \times 1.99 \times 11.3^2 \times (12.0/12) \times 4.0}{2} = 635 \text{ lb}$$

This load acts at $d_s/2 = 4.0/2 = 2.0$ ft above GS.

In S.I.:

$$F_{dyn} = \frac{a\rho V^2 D d_s}{2} = \frac{1.25 \times 1.026 \times 3.46^2 \times (305/1{,}000) \times 1.22}{2} = 2.9 \text{ kN}$$

This load acts at $d_s/2 = 1.22/2 = 0.61$ m above GS.

Step 10—Determine breaking wave loads

Breaking wave loads, F_D, occur on the vertical piles and are determined by ASCE/SEI Equation (5.4-4):

$$F_D = 0.5\gamma_w C_D D H_b^2$$

$C_D = 1.75$ for round piles (ASCE/SEI 5.4.4.1).

$D = 1.0$ ft (ASCE/SEI 5.4.4.1)

Therefore, F_D on one of the piles is equal to the following:

$$F_D = 0.5 \times 64.0 \times 1.75 \times 1.0 \times 3.1^2 = 538 \text{ lb}$$

This load acts at the SWEL, which is 4.0 ft above *GS*.

In S.I.:

$$D = 0.3 \text{ m}$$

$$F_D = 0.5 \times 10.05 \times 1.75 \times (305/1{,}000) \times 0.95^2 = 2.4 \text{ kN}$$

This load acts at the SWEL, which is 1.22 m above *GS*.

Step 11—Determine normal and special impact loads, F

Flowchart 7.1 is used to determine the impact loads.

Step 11a—Determine the debris impact weight, W

For coastal areas where piers and large pilings are present, use $W = 2{,}000$ lb (9.0 kN) [see the table in Flowchart 7.1].

Step 11b—Determine the debris velocity, V_b

It is reasonable to assume $V_b = V = 11.3$ ft/s (3.46 m/s).

Step 11c—Determine the importance coefficient, C_I

For a risk category III building, $C_I = 1.2$ (ASCE/SEI Table C5.4-1).

Step 11d—Determine the orientation coefficient, C_O

From ASCE/SEI C5.4.5, $C_O = 0.8$.

Step 11e—Determine the depth coefficient, C_D

For a VE Zone, $C_D = 1.0$ (ASCE/SEI Table C5.4-2).

Step 11f—Determine the blockage coefficient, C_B

Assuming there is no upstream screening and the flow path is wider than 30 ft (9.14 m), $C_B = 1.0$ (ASCE/SEI Table C5.4-3).

Step 11g—Determine the natural period, T_n, of the utility building.

From a dynamic analysis, $T_n = 0.30$ s.

Step 11h—Determine the ratio of the impact duration to the natural period of structure, $\Delta t / T_n$

Given the impact duration $\Delta t = 0.03$ s, $\Delta t / T_n = 0.03/0.30 = 0.10$.

Step 11i—Determine the maximum response ratio for impulsive loads, R_{max}

For $\Delta t / T_n = 0.10$, $R_{max} = 0.4$ (ASCE/SEI Table C5.4-4).

Step 11j—Determine the normal impact force, F

The normal impact load, F, is determined by ASCE/SEI Equation (C5.4-3):

$$F = \frac{\pi W V_b C_I C_O C_D C_B R_{max}}{2g\Delta t} = \frac{\pi \times 2{,}000 \times 11.3 \times 1.2 \times 0.8 \times 1.0 \times 1.0 \times 0.4}{2 \times 32.2 \times 0.03} = 14{,}112 \text{ lb}$$

In S.I.:

$$F = \frac{\pi W V_b C_I C_O C_D C_B R_{max}}{2g\Delta t} = \frac{\pi \times 9.0 \times 3.46 \times 1.2 \times 0.8 \times 1.0 \times 1.0 \times 0.4}{2 \times 9.81 \times 0.03} = 63.8 \text{ kN}$$

This load acts at the SWEL, which is 4.0 ft (1.22 m) above *GS*.

Special impact loads are determined by Equation (7.9):

$$F = \frac{C_D \rho A V^2}{2} = \frac{1.0 \times 1.99 \times [(12.0/12) \times 4.0] \times 11.3^2}{2} = 508 \text{ lb}$$

In S.I.:

$$F = \frac{C_D \rho A V^2}{2} = \frac{1.0 \times 1.026 \times (0.305 \times 1.22) \times 3.46^2}{2} = 2.3 \text{ kN}$$

This load acts at the SWEL, which is 4.0 ft (1.22 m) above *GS*.

Step 12—Determine the flood loads to use in the determination of F_a

- Global foundation loads

  For the analysis of the entire (global) foundation, it is reasonable to assume that one of the 26 piles will be subjected to the impact load, F, and the remaining piles will be subjected to the hydrodynamic load, F_{dyn}. Therefore, the total horizontal flood load on the foundation is equal to the following where the normal impact load is greater than the special impact load:

$$F_a = F + (n-1)F_{dyn} = 14{,}112 + (25 \times 635) = 29{,}987 \text{ lb}$$

In S.I.:

$$F_a = 63.8 + (25 \times 2.9) = 136.3 \text{ kN}$$

The overturning moment about *GS* due to the flood loads is equal to the following:

$$F_a = Fd_s + (n-1)F_{dyn}(d_s/2)$$
$$= (14{,}112 \times 4.0) + (25 \times 635 \times 4.0/2) = 88{,}198 \text{ ft-lb}$$

In S.I.:

$$F_a = (63.8 \times 1.22) + (25 \times 2.9 \times 1.22/2) = 122.1 \text{ kN-m}$$

- Individual pile loads

  The flood loads on an individual pile in the front row consists of F and the greater of F_{dyn} and F_D (see Figure 7.35). The load effects due to these loads are combined with the other applicable loads using the appropriate load combinations in Section 7.2.15 of this book.

Figure 7.35 Flood Loads for the Utility Building in Example 7.4

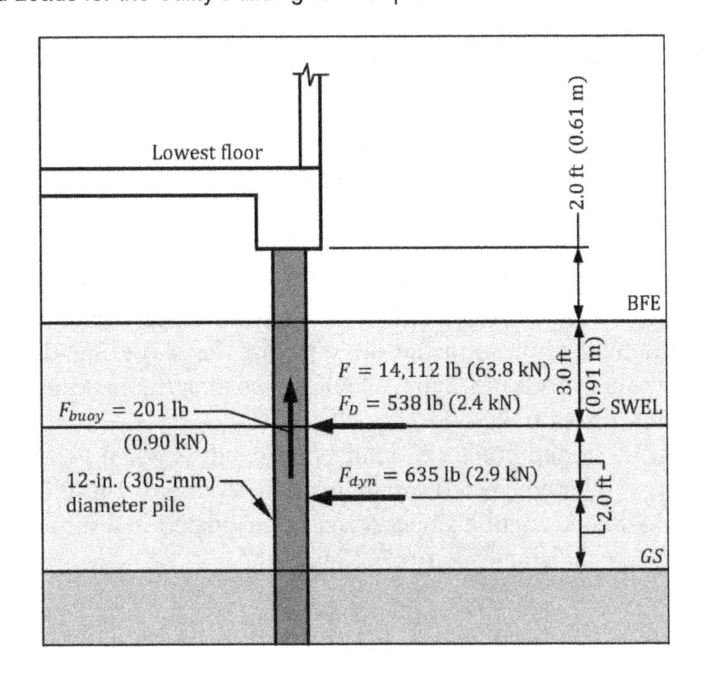

7.5.5 Example 7.5—Tsunami Loads on an Essential Facility

Determine the tsunami loads, F_{TSU}, on the six-story essential facility in Figure 7.36. Assume the following:

- Building site: latitude: 36.59989°, longitude: −121.89101°
- The building is located 820 ft (249.9 m) from the shoreline as shown in Figure 7.37.
- The ground transect perpendicular to the shoreline is given in Figure 7.37.
- Runup elevation = 29.6 ft (9.0 m) located 1,570 ft (478.5 m) from the shoreline (obtained from Reference 7.3).
- No tsunami bores are expected within this area.
- The site is located in an impact zone for shipping containers in accordance with ASCE/SEI 6.11.5.
- Windows are not designed for large impact loads or blast loads and span from the top of the slab to the underside of the beams.
- The tops of the beams are flush with the top of the slab.
- The structural members are constructed of reinforced concrete with a unit weight equal to 150 lb/ft^3 (23.6 kN/m^3) and a modulus of elasticity, E, equal to 3,605,000 lb/in.^{2} (24,900 MPa).

SOLUTION

Step 1—Determine the tsunami risk category

From the design data, the building is an essential facility, so the tsunami risk category is IV (see Section 7.3.4 of this book).

Step 2—Determine the permitted analysis procedure

Because the runup elevation is obtained from Reference 7.3, the permitted analysis procedures are given in Table 7.19.

For a tsunami risk category IV building, the EGLA method and a site-specific PTHA must both be performed. It is determined in step 4 below from an EGLA that $h_{max} = 6.51$ ft (1.99 m) at the site, which is less than 12 ft (3.66 m), so a site-specific PTHA need not be performed (see the exception in ASCE/SEI 6.5.2).

Step 3—Perform an R/H_T analysis

An R/H_T analysis is not required because the runup elevation is obtained from Reference 7.3 for the building in this example.

Step 4—Perform the EGLA

Flowchart 7.3 is used to determine the inundation depths and flow velocities along the ground transect in Figure 7.37 and the information provided in the design data.

The analysis begins at the inundation point where $x_R = 1{,}570$ ft (478.5 m) and continues stepwise in the direction of the shoreline. The interval chosen in this example is 50 ft (15.2 m), which is less than the maximum limit of 100 ft (30.5 m) in ASCE/SEI 6.6.2.

At the runup point, the inundation depth is equal to zero, as are the velocity and energy (hydraulic head). An inundation depth of 0.1 ft (0.03 m) is used at this point so the hydraulic friction slope, s, can be calculated in a subsequent step (otherwise there would be division by zero).

Figure 7.36 Six-Story Essential Facility in Example 7.5

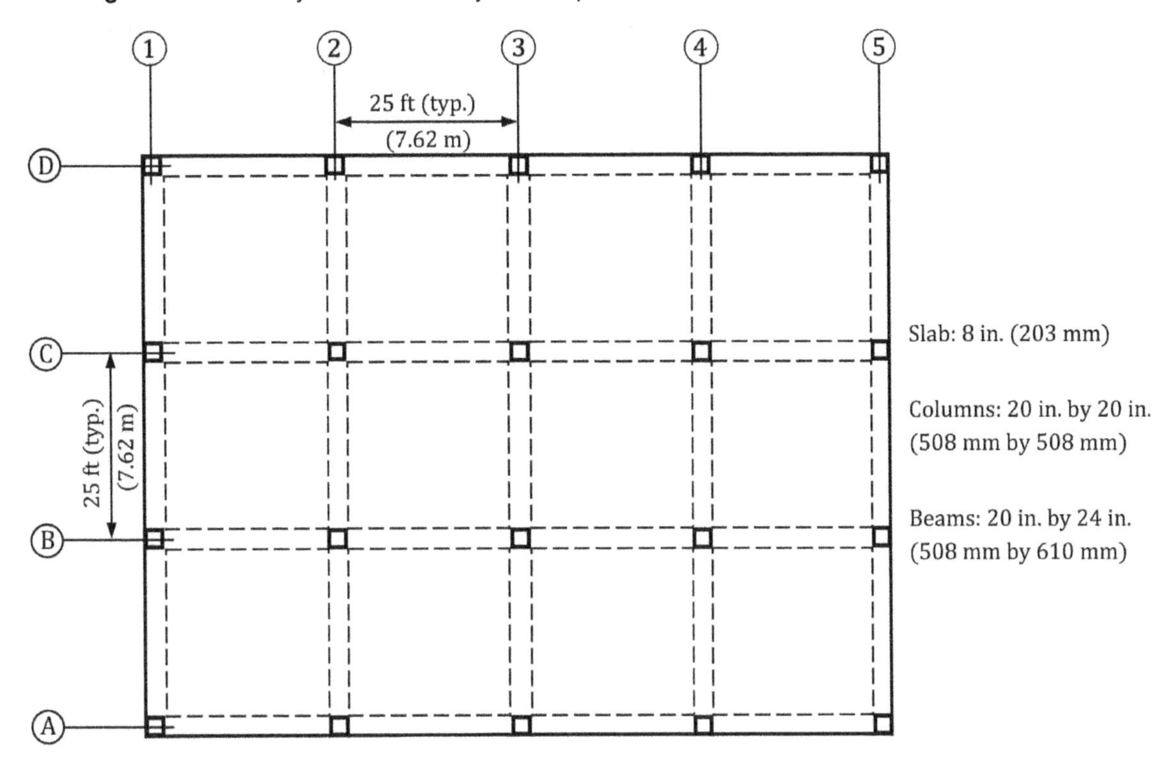

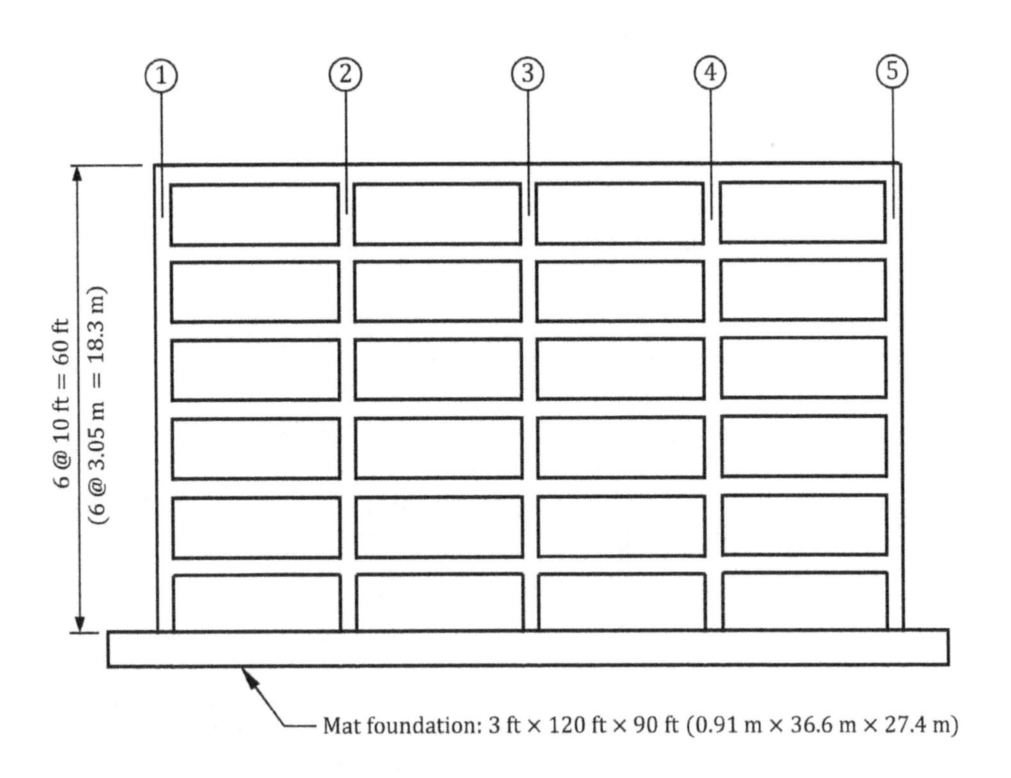

Figure 7.37 Site Plan and Ground Transect for the Essential Facility in Example 7.5

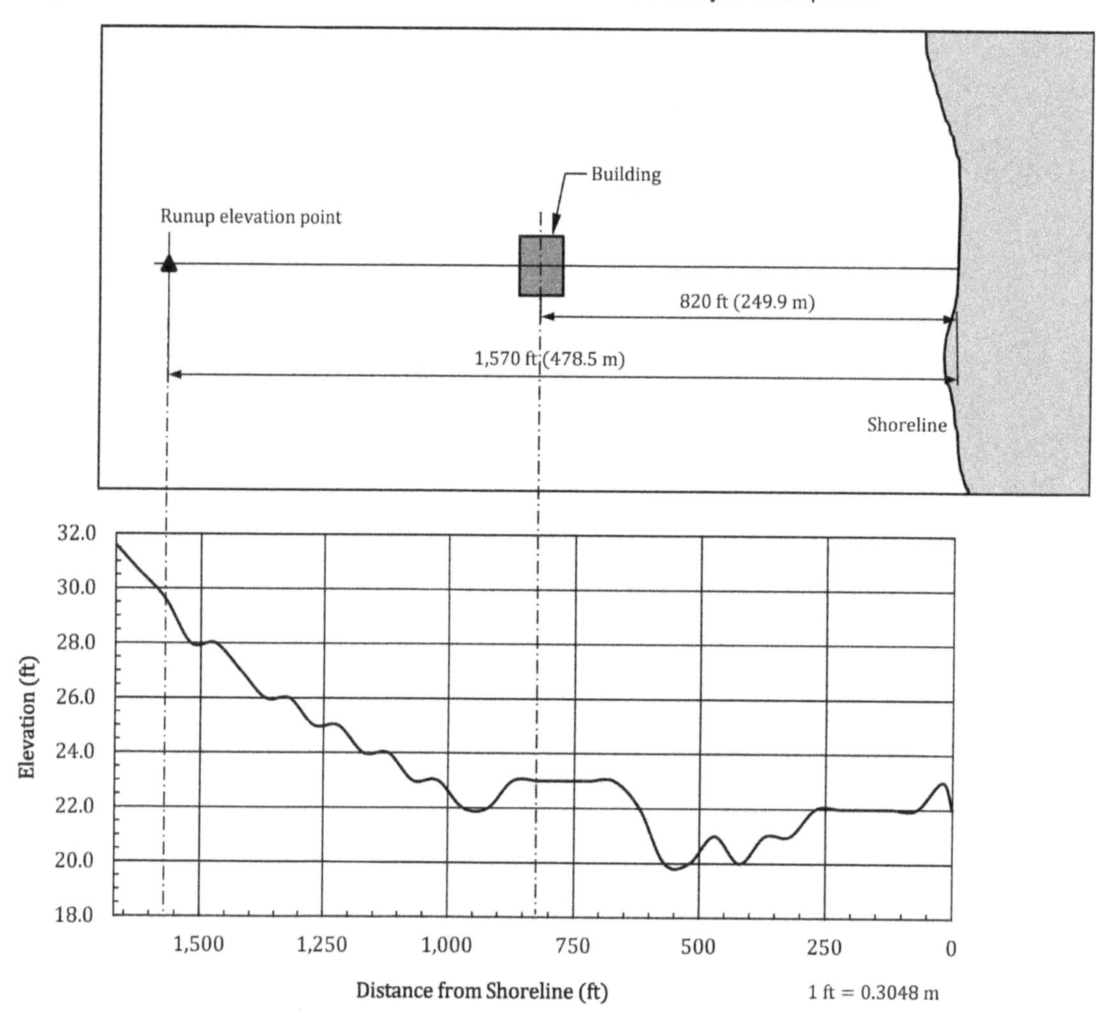

Inundation depths and flow velocities along the transect are given in Table 7.32.

Sample calculations at the centerline of the building location, which is point 15 in Table 7.32, are given below using Flowchart 7.3.

Step 4a—Acquire x_R and R

$R = 29.6$ ft (9.0 m) is obtained from Reference 7.3.

R occurs at $x_R = 1{,}570$ ft (478.5 m) from the shoreline.

Step 4b—Approximate the transect by a series of segmented slopes with transect points spaced not more than 100 ft (30.5 m).

The ground transect at this location is given in Figure 7.37 with a maximum transect point spacing of 50 ft (15.2 m).

Step 4c—Calculate the slope within the segment, ϕ_{15}

$$\phi_{15} = \frac{z_{14} - z_{15}}{x_{14} - x_{15}} = \frac{23.0 - 23.0}{870.0 - 820.0} = 0.0000$$

Step 4d—Obtain Manning's coefficient, n, from ASCE/SEI Table 6.6-1 for the segment

Assume the frictional surface for this segment (as well as all the other segments in the transect) does not fall under any specific description in ASCE/SEI Table 6.6-1. Therefore, $n = 0.03$.

Table 7.32 EGLA Calculations for the Building in Example 7.5

Point	Distance from Shoreline, x_i (ft)	Transect Elevation, z_i (ft)	Topographic Slope, ϕ_i	Froude Number, F_r	Friction Slope, s_i	Topographic Slope, $\phi_{i,200}$	Energy Head, E_g (ft)	Inundation Depth, h_i (ft)	Flow Velocity, u_i (ft/s)
Runup	1,570.0	29.6		0.000	0.000000		0.00	0.10	0.00
1	1,520.0	28.0	0.0320	0.178	0.000896	0.0180	1.64	1.62	1.29
2	1,470.0	28.0	0.0000	0.252	0.000708	0.0180	1.68	1.63	1.83
3	1,420.0	27.0	0.0200	0.309	0.001060	0.0100	2.73	2.61	2.83
4	1,370.0	26.0	0.0200	0.357	0.001208	0.0150	3.79	3.57	3.82
5	1,320.0	26.0	0.0000	0.399	0.001360	0.0100	3.86	3.58	4.28
6	1,270.0	25.0	0.0200	0.437	0.001631	0.0100	4.94	4.51	5.27
7	1,220.0	25.0	0.0000	0.472	0.001761	0.0100	5.03	4.53	5.70
8	1,170.0	24.0	0.0200	0.505	0.002010	0.0100	6.13	5.44	6.68
9	1,120.0	24.0	0.0000	0.535	0.002127	0.0100	6.24	5.46	7.10
10	1,070.0	23.0	0.0200	0.564	0.002361	0.0100	7.36	6.35	8.07
11	1,020.0	23.0	0.0000	0.592	0.002470	0.0100	7.48	6.36	8.47
12	970.0	22.0	0.0200	0.618	0.002692	0.0000	8.61	7.23	9.43
13	920.0	22.0	0.0000	0.643	0.002795	0.0000	8.75	7.25	9.83
14	870.0	23.0	−0.0200	0.668	0.003007	−0.0050	7.90	6.46	12.19
15	820.0	23.0	0.0000	0.691	0.003348	−0.0050	8.07	6.51	12.67
16	770.0	23.0	0.0000	0.714	0.003561	0.0000	8.25	6.57	10.39
17	720.0	23.0	0.0000	0.736	0.003772	0.0050	8.44	6.64	10.76
18	670.0	23.0	0.0000	0.757	0.003981	0.0150	8.64	6.71	11.13
19	620.0	22.0	0.0200	0.778	0.004187	0.0150	9.85	7.56	12.14

(continued)

Table 7.32 EGLA Calculations for the Building in Example 7.5 (*Continued*)

Point	Distance from Shoreline, x_i (ft)	Transect Elevation, z_i (ft)	Topographic Slope, ϕ_i	Froude Number, F_r	Friction Slope, s_i	Topographic Slope, $\phi_{i,200}$	Energy Head, E_g (ft)	Inundation Depth, h_i (ft)	Flow Velocity, u_i (ft/s)
20	570.0	20.0	0.0400	0.798	0.004236	0.0100	12.06	9.15	13.70
21	520.0	20.0	0.0000	0.818	0.004175	0.0100	12.27	9.19	14.07
22	470.0	21.0	−0.0200	0.837	0.004366	−0.0050	11.49	8.51	17.54
23	420.0	20.0	0.0200	0.856	0.004684	−0.0050	12.72	9.31	18.76
24	370.0	21.0	−0.0200	0.874	0.004743	−0.0050	11.96	8.65	18.47
25	320.0	21.0	0.0000	0.892	0.005063	−0.0100	12.21	8.73	19.70
26	270.0	22.0	−0.0200	0.910	0.005249	−0.0050	11.47	8.11	18.62
27	220.0	22.0	0.0000	0.927	0.005586	−0.0050	11.75	8.22	19.10
28	170.0	22.0	0.0000	0.944	0.005768	0.0000	12.04	8.33	15.46
29	120.0	22.0	0.0000	0.961	0.005948	−0.0050	12.34	8.44	20.06
30	70.0	22.0	0.0000	0.977	0.006125	0.0000	12.64	8.56	16.22
31	20.0	23.0	−0.0200	0.994	0.006301		11.96	8.01	15.95
32	0.0	22.0	0.0500	1.000	0.006525		13.09	8.73	16.76

Step 4e—Calculate the Froude number, F_{r15}

The Froude number is determined by ASCE/SEI Equation (6.6-3):

$$F_{r15} = \alpha\left(1 - \frac{x_{15}}{x_R}\right)^{0.5} = 1.0 \times \left(1 - \frac{820.0}{1{,}570.0}\right)^{0.5} = 0.6912$$

Step 4f—Calculate the hydraulic friction slope, s_{15}

The hydraulic friction slope, s_{15}, is determined by ASCE/SEI Equations (6.6-2) and (6.6-2.SI):

$$s_{15} = \frac{gF_{r15}^2}{\left(\frac{1.49}{n}\right)^2 h_{14}^{1/3}} = \frac{32.2 \times 0.6912^2}{\left(\frac{1.49}{0.03}\right)^2 \times (6.463)^{1/3}} = 0.003348$$

In S. I.: $$s_{15} = \frac{gF_{r15}^2}{\left(\frac{1.00}{n}\right)^2 h_{14}^{1/3}} = \frac{9.81 \times 0.6912^2}{\left(\frac{1.00}{0.03}\right)^2 \times (1.97)^{1/3}} = 0.003365$$

Step 4g—Calculate the hydraulic head, $E_{g,15}$

$$E_{g,15} = E_{g,14} + (\phi_{15} + s_{15})(x_{14} - x_{15})$$

$$= 7.90 + [(0.0000 + 0.003348) \times 50] = 8.07 \text{ ft}$$

In S.I.: $$E_{g,15} = 2.41 + [(0.0000 + 0.003365) \times 15.2] = 2.46 \text{ m}$$

Step 4h—Calculate the inundation depth, h_{15}

$$h_{15} = \frac{E_{g,15}}{1 + 0.5F_{r15}^2} = \frac{8.07}{1 + (0.5 \times 0.6912^2)} = 6.51 \text{ ft}$$

In S.I.: $$h_{15} = \frac{E_{g,15}}{1 + 0.5F_{r15}^2} = \frac{2.46}{1 + (0.5 \times 0.6912^2)} = 1.99 \text{ m}$$

Step 4i—Calculate the flow velocity, u_{15}

$$u_{15} = F_{r,15}(gh_{15})^{0.5} = 0.6912 \times (32.2 \times 6.51)^{0.5} = 10.01 \text{ ft/s}$$

In S.I.: $$u_{15} = F_{r,15}(gh_{15})^{0.5} = 0.6912 \times (9.81 \times 1.99)^{0.5} = 3.05 \text{ m/s}$$

Step 4j—Calculate the slope averaged over a distance of 200 ft (61.0 m) along the transect, $\phi_{15,200}$

$$\phi_{15,200} = \frac{z_{13} - z_{17}}{x_{13} - x_{17}} = \frac{22.0 - 23.0}{920.0 - 720.0} = -0.0050$$

Step 4k—Determine the increase in the flow velocity, Δu_{15}, where applicable

Because $\phi_{15,200} < 0$, Δu_{15} must be determined by ASCE/SEI Equation (6.6-4):

$$\Delta u_{15} = u_{15}\left|\phi_{15,200}\right|^{0.25} = 10.01 \times (0.0050)^{0.25} = 2.66 \text{ ft/s} < 6.00 \text{ ft/s}$$

In S.I.: $$\Delta u_{15} = 3.05 \times (0.0050)^{0.25} = 0.81 \text{ m/s} < 1.83 \text{ m/s}$$

Step 4l—Determine adjusted flow velocity, u_i

Adjusted flow velocity $= u_{15} + \Delta u_{15}$

$$= 10.01 + 2.66 = 12.67 \text{ ft/s}$$

$$> 10.00 \text{ ft/s}$$

$$< \text{lesser of} \begin{cases} 1.5(gh_{15})^{0.5} = 21.72 \text{ ft/s} \\ 50.00 \text{ ft/s} \end{cases}$$

In S.I.:
Adjusted flow velocity $= 3.05 + 0.81 = 3.86$ m/s

$$\geq 3.05 \text{ ft/s}$$

$$< \text{lesser of} \begin{cases} 1.5(gh_{15})^{0.5} = 6.63 \text{ m/s} \\ 15.24 \text{ m/s} \end{cases}$$

Plots of elevation, inundation depth and flow velocity (including adjusted flow velocity, where required) along the transect are given in Figure 7.38.

The effects of sea level change in accordance with ASCE/SEI 6.5.3 are not considered in this example but must be included wherever appropriate. In addition, it is assumed the flow velocity amplification requirements in ASCE/SEI 6.8.5 are not applicable.

Similar analyses must be performed for transects located 22.5 degrees on either side of the transect in this example (see ASCE/SEI 6.8.6.1 and Figure 7.15).

Step 5—Determine the hydrostatic loads

Step 5a—Determine the buoyancy loads

Buoyancy loads are considered in load case 1 where the inundation depth is equal to the least of the following (ASCE/SEI 6.8.3.1):

$$h_{sx} = \text{least of} \begin{cases} h_{max} = 6.51 \text{ ft (1.99 m)} \\ \text{Story height} = 10.0 \text{ ft (3.05 m)} \\ \text{Height to top of first story window} = 10.0 - 2.0 = 8.0 \text{ ft (2.44 m)} \end{cases}$$

The buoyancy load, F_v, is determined by ASCE/SEI Equation (6.9-1):

$$F_v = \gamma_s V_w$$

The volume of displaced water, V_w, in this case consists of the volume of the inundated building plus the volume of the mat foundation, which is connected to the building columns:

$$V_w = (6.51 \times 101.7 \times 76.7) + (3.0 \times 120.0 \times 90.0) = 83{,}181 \text{ ft}^3$$

$$F_v = (1.1 \times 64.0) \times 83{,}181/1{,}000 = 5{,}856 \text{ kips}$$

In S.I.:

$$V_w = (1.99 \times 31.0 \times 23.4) + (0.91 \times 36.6 \times 27.4) = 2{,}356 \text{ m}^3$$

$$F_v = (1.1 \times 10.05) \times 2{,}356 = 26{,}046 \text{ kN}$$

Step 5b—Unbalanced lateral hydrostatic loads

This load is not applicable in this example because no inundated exterior walls are present.

Step 5c—Residual water surcharge load on floors and walls

Because $h_{max} = 6.51$ ft (1.99 m) is less than the first story height, there are no residual water surcharge loads on the floors.

Step 5d—Hydrostatic surcharge pressure on foundation

Hydrostatic surcharge pressure on the foundation, p_s, is determined by ASCE/SEI Equation (6.9-5):

$$p_s = \gamma_s h_{max} = (1.1 \times 64.0) \times 6.51 = 458 \text{ lb/ft}^2$$

In S.I.:

$$p_s = \gamma_s h_{max} = (1.1 \times 10.05) \times 1.99 = 22.0 \text{ kN/m}^2$$

Figure 7.38 Elevation, Inundation Depth, and Flow Velocity along the Transect in Example 7.5

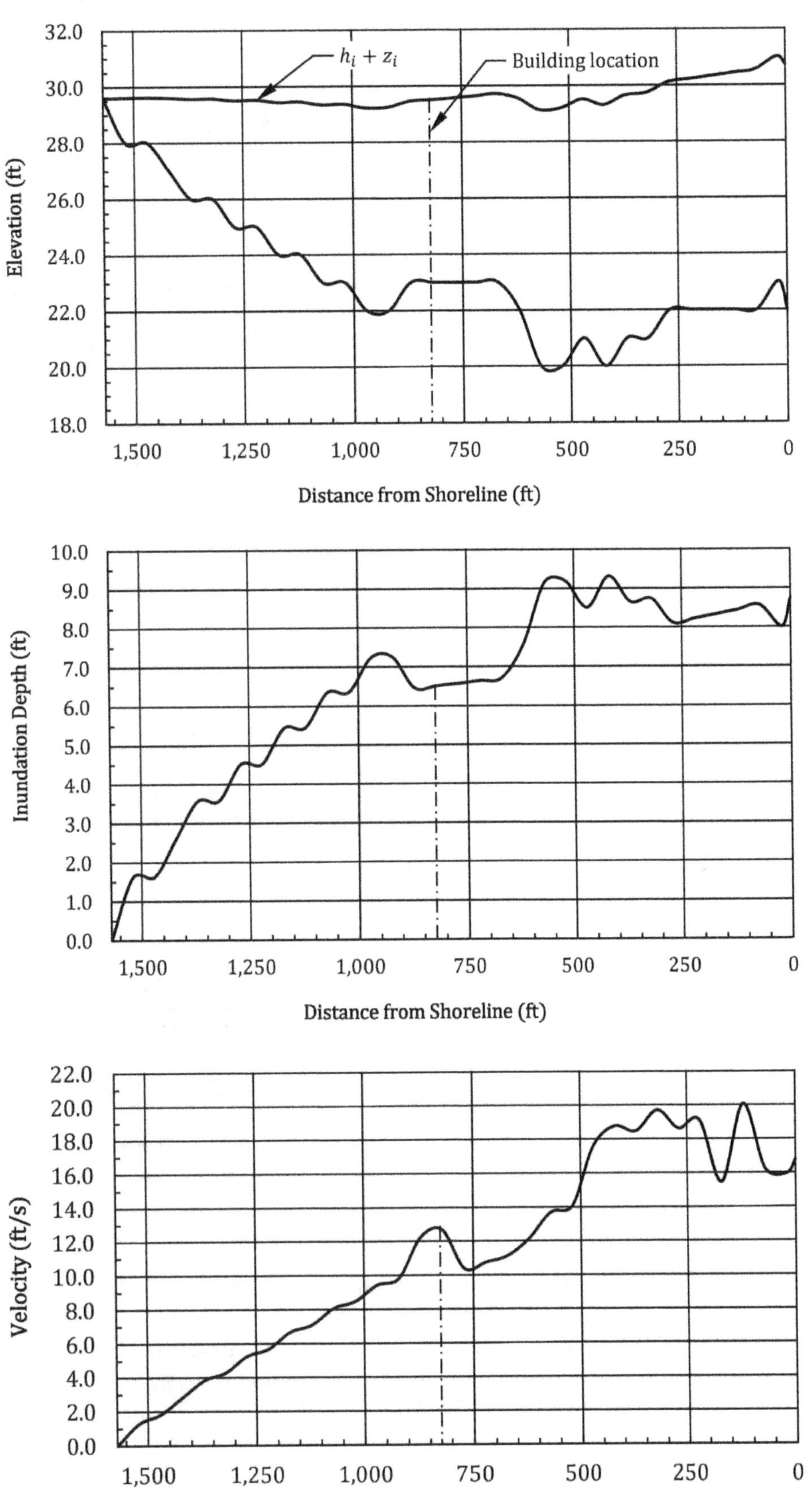

Step 6—Determine the hydrodynamic loads

- Option 1: Simplified uniform lateral static pressure
  The simplified uniform lateral static pressure, p_{sw} is determined by ASCE/SEI Equation (6.10-1):

$$p_{uw} = 1.25 I_{tsu} \gamma_s h_{max} = 1.25 \times 1.50 \times (1.1 \times 64.0) \times 6.51 = 859 \text{ lb/ft}^2$$

where $I_{tsu} = 1.50$ in accordance with ASCE/SEI Table 6.8-1 and the exception in ASCE/SEI 6.5.2.

This pressure is applied over the width of the building and over a height of $1.3h_{max} = 8.5$ ft.

In S.I.:

$$p_{uw} = 1.25 I_{tsu} \gamma_s h_{max} = 1.25 \times 1.50 \times (1.1 \times 10.05) \times 1.99 = 41.3 \text{ kN/m}^2$$

This pressure is applied over the width of the building and over a height of $1.3h_{max} = 2.59$ m.

$$\text{Total force} = 859 \times 101.7 \times 8.5/1{,}000 = 743 \text{ kips}$$

In S.I.: $\text{Total force} = 41.3 \times 31.0 \times 2.59 = 3{,}316 \text{ kN}$

The simplified uniform lateral static pressure is given in Figure 7.39.

Figure 7.39 Simplified Uniform Lateral Static Hydrodynamic Pressure on the Essential Facility in Example 7.5

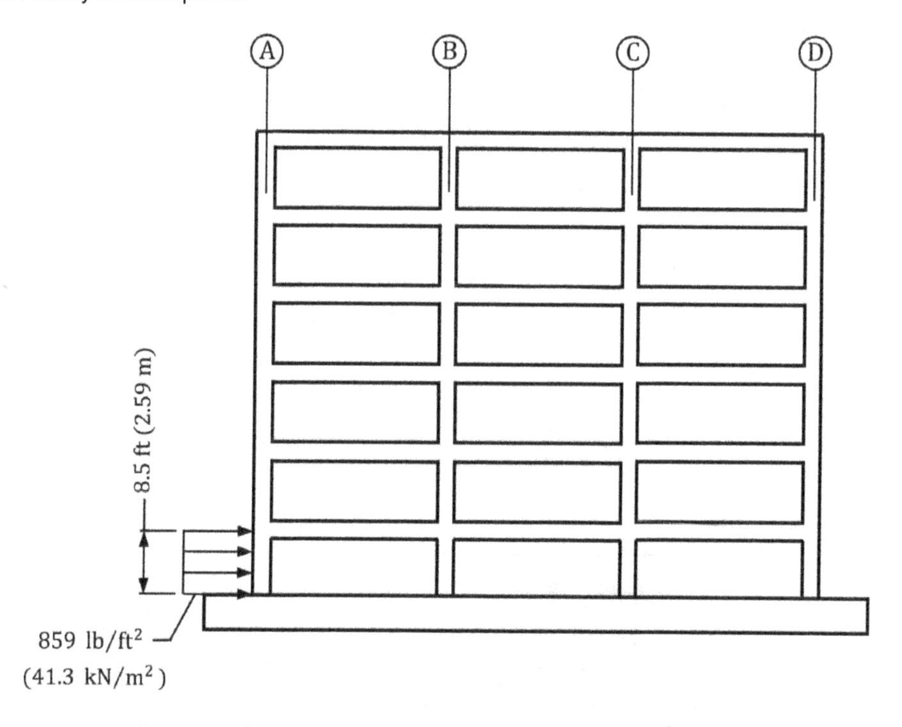

- Option 2—Detailed hydrodynamic lateral forces
  1. Overall drag force on the building
     Flowchart 7.4 is used to determine F_{dx} for load cases 1, 2, and 3.
     a. Determine the minimum fluid mass density, ρ_s:
        From ASCE/SEI 6.8.4, $\rho_s = k_s \rho_{sw} = 1.1 \times 2.0 = 2.2$ slugs/ft³
        In S.I.: $\rho_s = 1.1 \times 1{,}025 = 1{,}128$ kg/m³

b. Determine the importance factor, I_{tsu}
The importance factor, I_{tsu}, is determined from ASCE/SEI Table 6.8-1.
Because the inundation depth at the site resulting from the EGLA for this TRC IV building has been determined to be less than 12 ft (3.66 m) and a PTHA has not been performed, $I_{tsu} = 1.50$.

c. Determine the drag coefficient, C_d
Values of C_d are determined from ASCE/SEI Table 6.10-1 for load cases 1, 2, and 3 based on the width to inundation depth ratio, B/h.
Load case 1:

$$h_{sx} = \text{lesser of} \begin{cases} h_{max} = 6.51 \text{ ft } (1.99 \text{ m}) \\ \text{Story height} = 10.0 \text{ ft } (3.05 \text{ m}) \\ \text{Height to top of first story window} = 10.0 - 2.0 = 8.0 \text{ ft } (2.44 \text{ m}) \end{cases}$$

$$B/h_{sx} = 101.7/6.51 = 15.6$$

$C_d = 1.29$ by linear interpolation from ASCE/SEI Table 6.10-1
Load case 2:

$$h_{sx} = 2h_{max}/3 = 2 \times 6.51/3 = 4.34 \text{ ft } (1.33 \text{ m})$$
$$B/h_{sx} = 101.7/4.34 = 23.4$$

$C_d = 1.37$ by linear interpolation from ASCE/SEI Table 6.10-1
Load case 3:

$$h_{sx} = h_{max} = 6.51 \text{ ft } (1.99 \text{ m})$$
$$B/h_{sx} = 101.7/6.51 = 15.6$$

$C_d = 1.29$ by linear interpolation from ASCE/SEI Table 6.10-1

d. Determine the closure coefficient, C_{cx}
The closure coefficient, C_{cx}, is determined by ASCE/SEI Equation (6.10-3) for load cases 1, 2, and 3.
Load case 1:
$C_{cx} = 1.0$ because it is assumed the exterior walls have not failed
Load case 2:

$$h_{sx} = 2h_{max}/3 = 4.34 \text{ ft } (1.33 \text{ m})$$
$$A_{col} = (20.0/12) \times 4.34 = 7.23 \text{ ft}^2 \ (0.67 \text{ m}^2)$$

$A_{wall} = 0$ (there are no walls)
$A_{beam} = 0$ (the beams are above the inundation depth)

$$C_{cx} = \frac{\sum(A_{col} + A_{wall}) + 1.5A_{beam}}{Bh_{sx}} = \frac{20 \times 7.23}{101.7 \times 4.34} = 0.33 < 0.7, \text{ use } 0.7 \text{ (ASCE/SEI 6.8.7)}$$

In S.I.:
$$C_{cx} = \frac{20 \times 0.67}{31.0 \times 1.32} = 0.33 < 0.7, \text{ use } 0.7$$

Load case 3:

$$h_{sx} = h_{max} = 6.51 \text{ ft } (1.99 \text{ m})$$

$$C_{cx} = \frac{\sum(A_{col} + A_{wall}) + 1.5A_{beam}}{Bh_{sx}} = \frac{20 \times 7.23}{101.7 \times 6.51} = 0.22 < 0.7, \text{ use } 0.7 \text{ (ASCE/SEI 6.8.7)}$$

In S.I.:
$$C_{cx} = \frac{20 \times 0.67}{31.0 \times 1.99} = 0.22 < 0.7, \text{ use } 0.7$$

e. Determine the overall drag force, F_{dx}

The overall drag force, F_{dx}, is determined by ASCE/SEI Equation (6.10-2) for load cases 1, 2, and 3:

$$F_{dx} = \frac{1}{2}\rho_s I_{tsu} C_d C_{cx} B(h_{sx} u^2)$$

Load case 1:

$h_{sx} = 6.51$ ft (1.99 m) and u is the minimum flow velocity of 10.0 ft/s (3.05 m/s)

$$F_{dx} = \frac{1}{2} \times 2.2 \times 1.5 \times 1.29 \times 1.0 \times 101.7 \times 6.51 \times 10.0^2 / 1{,}000 = 141 \text{ kips}$$

In S.I.:

$$F_{dx} = \frac{1}{2} \times (1{,}128/1{,}000) \times 1.5 \times 1.29 \times 1.0 \times 31.0 \times 1.99 \times 3.05^2 = 626 \text{ kN}$$

The hydrodynamic and buoyancy loads in load case 1 are given in Figure 7.40.

Load case 2:

$h_{sx} = 2h_{max}/3 = 4.34$ ft (1.33 m) and u is the maximum flow velocity of 12.67 ft/s (3.86 m/s)

$$F_{dx} = \frac{1}{2} \times 2.2 \times 1.5 \times 1.37 \times 0.7 \times 101.7 \times 4.34 \times 12.67^2 / 1{,}000 = 112 \text{ kips}$$

Figure 7.40 Load Case 1 for the Essential Facility in Example 7.5

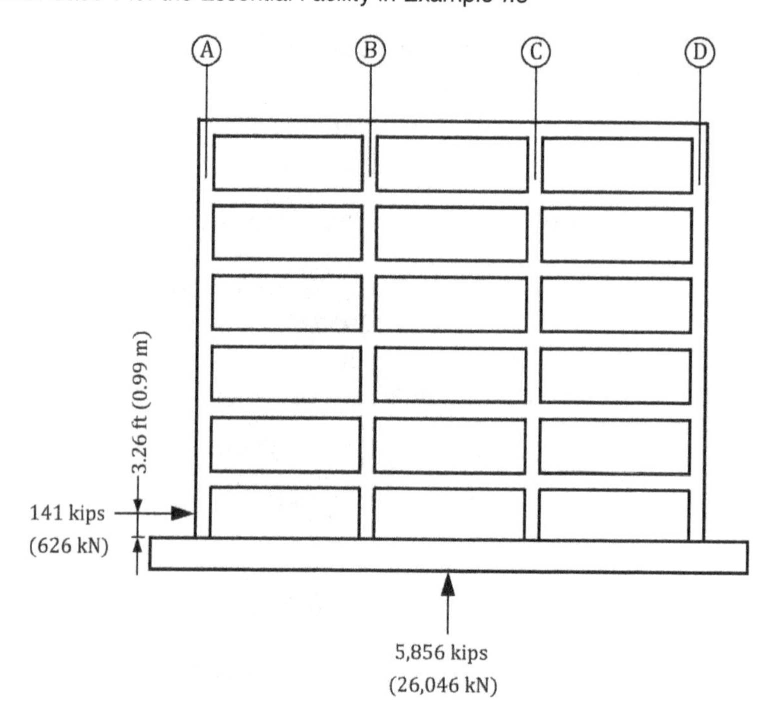

In S.I.:

$$F_{dx} = \frac{1}{2} \times (1{,}128/1{,}000) \times 1.5 \times 1.37 \times 0.7 \times 31.0 \times 1.33 \times 3.86^2 = 498 \text{ kN}$$

This force is applied to the building at $(4.34/2) = 2.17$ ft (0.66 m) above the ground level, as shown in Figure 7.41.

Figure 7.41 Load Case 2 for the Essential Facility in Example 7.5

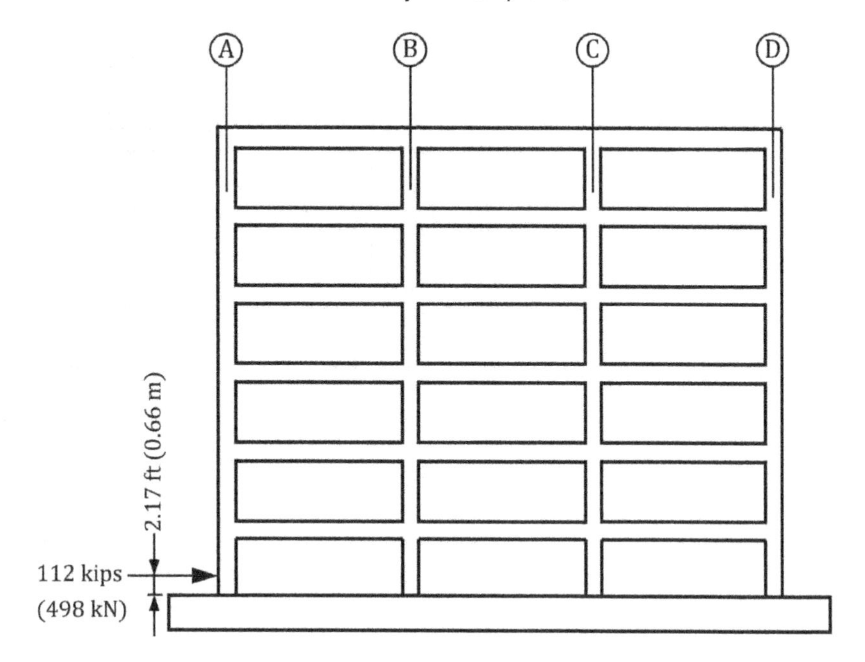

Load case 3:
$h = h_{max} = 6.51$ ft (1.99 m) and $u = u_{max}/3 = 12.67/3 = 4.22$ ft/s (1.29 m/s) < 10.0 ft/s (3.05 m/s)

$$F_{dx} = \frac{1}{2} \times 2.2 \times 1.5 \times 1.29 \times 0.7 \times 101.7 \times 6.51 \times 10.0^2 / 1{,}000 = 99 \text{ kips}$$

In S.I.:

$$F_{dx} = \frac{1}{2} \times (1{,}128/1{,}000) \times 1.5 \times 1.29 \times 0.7 \times 31.0 \times 1.99 \times 3.05^2 = 438 \text{ kN}$$

This force is applied at $(6.51/2) = 3.26$ ft (0.99 m) above the base of the column (see Figure 7.42).

2. Drag force on components
   Flowchart 7.5 is used to determine F_d for the interior and exterior columns.
   a. Determine the minimum fluid mass density, ρ_s:
      From ASCE/SEI 6.8.4, $\rho_s = k_s \rho_{sw} = 1.1 \times 2.0 = 2.2$ slugs/ft^3
      In S.I.: $\rho_s = 1.1 \times 1{,}025 = 1{,}128$ kg/m^3
   b. Determine the importance factor, I_{tsu}
      The importance factor, I_{tsu}, is determined from ASCE/SEI Table 6.8-1.
      Because the inundation depth at the site resulting from the EGLA for this TRC IV building has been determined to be less than 12 ft (3.66 m) and a PTHA has not been performed, $I_{tsu} = 1.50$.
   c. Determine the drag coefficient, C_d, and the component width, b
      For interior square columns, $C_d = 2.0$ from ASCE/SEI Table 6.10-2 and $b = 20/12 = 1.7$ft (0.51 m).
      For exterior columns, $C_d = 2.0$ and $b = C_{cx} \times$ tributary width $= 0.7 \times 25.0 =$ 17.5 ft (5.33 m). (ASCE/SEI 6.8.7)
   d. Determine h_e and u
      Load case 2 governs where inundated height $h_e = 2h_{max}/3 = 2 \times 6.51/3 =$ 4.34 ft (1.33 m) and $u = u_{max} = 12.67$ ft/s (3.86 m/s).

Figure 7.42 Load Case 3 for the Essential Facility in Example 7.5

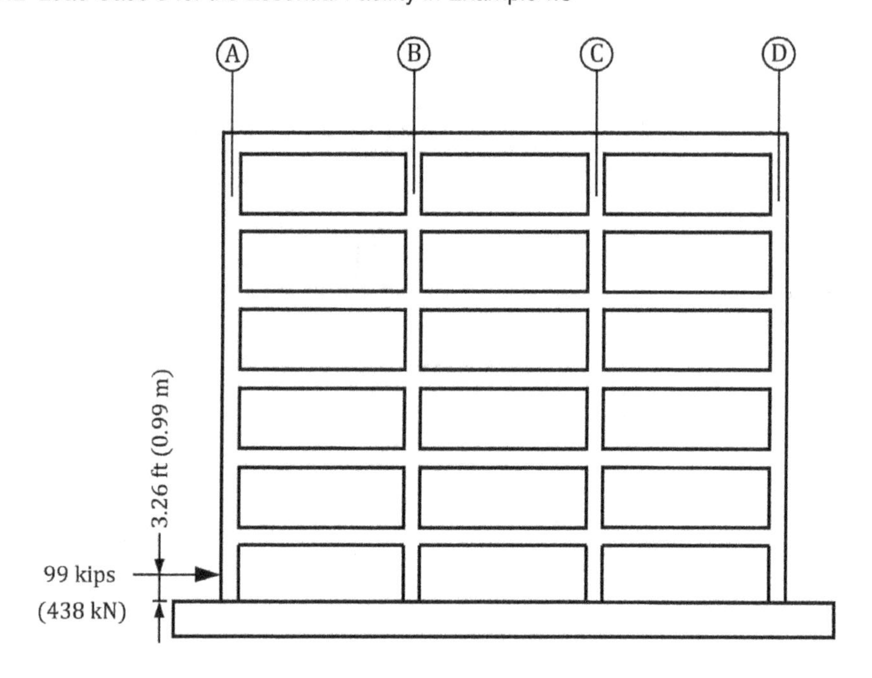

e. Determine the component drag force, F_x

The component drag force, F_x, is determined by ASCE/SEI Equation (6.10-4):

$$F_d = \frac{1}{2}\rho_s I_{tsu} C_d b(h_e u^2)$$

For interior columns:

$$F_d = \frac{1}{2}\times 2.2\times 1.5\times 2.0\times 1.7\times 4.34\times 12.67^2/1{,}000 = 4 \text{ kips}$$

This load is applied as an equivalent uniformly distributed lateral load of $4/4.34 = 0.9$ kips/ft over the inundated portion of the column (see Figure 7.43).

In S.I.:

$$F_d = \frac{1}{2}\times(1{,}128/1{,}000)\times 1.5\times 2.0\times 0.51\times 1.33\times 3.86^2 = 17 \text{ kN}$$

Figure 7.43 Drag Force on an Interior Column in the Essential Facility in Example 7.5

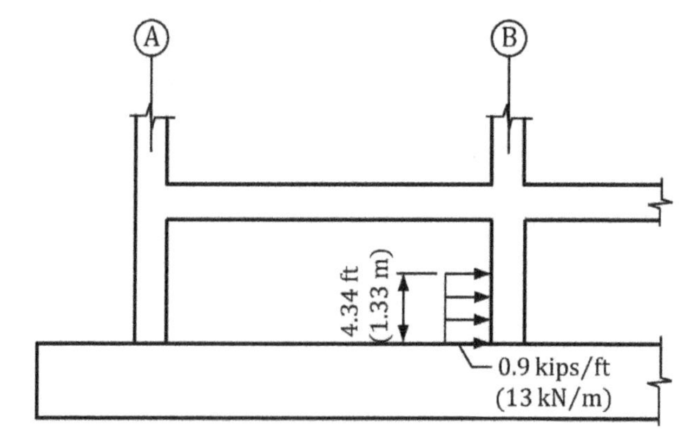

This load is applied as an equivalent uniformly distributed lateral load of $17/1.33 = 13$ kN/m over the inundated portion of the column (see Figure 7.43).
For exterior columns:

$$F_d = \frac{1}{2} \times 2.2 \times 1.5 \times 2.0 \times 17.5 \times 4.34 \times 12.67^2 / 1,000 = 40 \text{ kips}$$

This load is applied as an equivalent uniformly distributed lateral load of $40/4.34 = 9$ kips/ft over the inundated portion of the column (see Figure 7.44).

Figure 7.44 Drag Force on an Exterior Column in the Essential Facility in Example 7.5

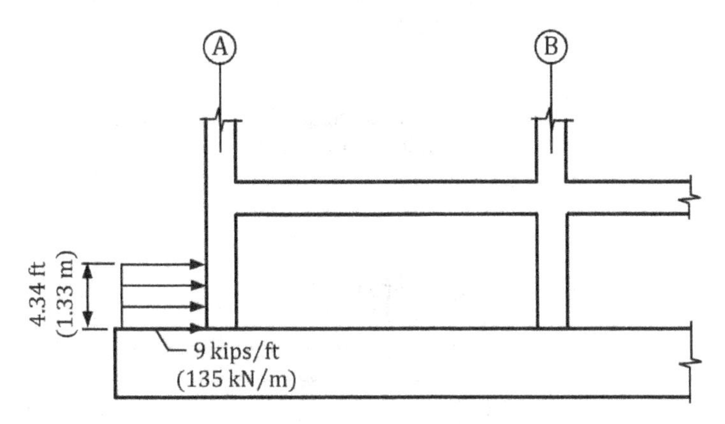

In S.I.:

$$F_d = \frac{1}{2} \times (1,128/1,000) \times 1.5 \times 2.0 \times 5.33 \times 1.33 \times 3.86^2 = 179 \text{ kN}$$

This load is applied as an equivalent uniformly distributed lateral load of $179/1.33 = 135$ kN/m over the inundated portion of the column (see Figure 7.44).

3. Tsunami loads on vertical structural components
   These loads are not applicable in this example based on the limitations in ASCE/SEI 6.10.2.3.
4. Hydrodynamic load on perforated walls
   This load is not applicable in this example because there are no perforated walls.
5. Walls angled to the flow
   This load is not applicable in this example because there are no walls angled to the flow.
6. Loads on above-ground horizontal pipelines
   This load is not applicable in this example because there are no above-ground horizontal pipelines.
7. Hydrodynamic pressures associated with slabs
   The loads associated with the following are not applicable in this example: (1) spaces in buildings subjected to flow stagnation pressurization, (2) hydrodynamic surge uplift at horizontal slabs, and (3) tsunami pore flow entrapped in structural wall-slab recess.

Step 7—Determine the debris impact loads

Because the inundation depth exceeds 3 ft (0.91 m) at the site, the exterior columns below the flow depth are subjected to debris impact loads in accordance with ASCE/SEI 6.11.

Step 7a—Option 1: Alternative simplified debris impact static load

The maximum debris impact static load, F_i, is determined by ASCE/SEI Equations (6.11-1) and (6.11-1.SI):

$$F_i = 330C_o I_{tsu} = 330 \times 0.65 \times 1.5 = 322 \text{ kips}$$

In S.I.: $F_i = 1{,}470C_o I_{tsu} = 1{,}470 \times 0.65 \times 1.5 = 1{,}433$ kN

Because the site is located in an impact zone for shipping containers, it is not permitted to reduce the calculated value of the impact force by 50 percent.

The impact force is applied to an exterior column at points along the height of the column critical for flexure and shear. For example, the critical section for flexure is located 4 ft (1.22 m) from the base of the column, and F_i is applied at this section for the flexural design of the column (see Figure 7.45).

Figure 7.45 Alternative Simplified Debris Impact Static Load on an Exterior Column in the Essential Facility in Example 7.5

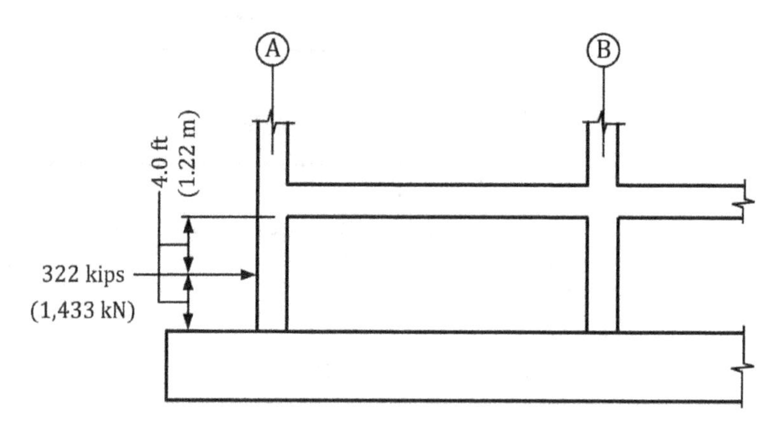

Step 7b—Option 2: Detailed debris impact loads

1. Wood logs and poles

   Flowchart 7.6 is used to determine the impact force due to wood logs and poles.

   a. Acquire the maximum flow velocity, u_{max}

      From Table 7.32, $u_{max} = 12.67$ ft/s (3.86 m/s) at the site.

   b. Determine the stiffness of the impacting debris and the impacted structural element and use whichever is less

      Minimum log stiffness = 350 kip/in. (61,300 kN/m) [ASCE/SEI 6.11.2]

      Assuming the column has simple connections at both ends, the stiffness of the column is determined as follows (see Flowchart 7.6):

$$k = \frac{48EI}{L^3} = \frac{48 \times (3{,}605{,}000/1{,}000) \times \left(\frac{1}{12} \times 20^4\right)}{(8.0 \times 12)^3} = 2{,}608 \text{ kip/in.} > 350 \text{ kip/in.}$$

In S.I.:

$$k = \frac{48 \times (24{,}900 \times 10^3) \times \left[\frac{1}{12} \times \left(\frac{508}{1{,}000}\right)^4\right]}{(2.44)^3} = 456{,}609 \text{ kN/m} > 61{,}300 \text{ kN/m}$$

Therefore, use $k = 350$ kip/in. (61,300 kN/m).

c. Determine the debris weight, W_d
Assume the minimum debris weight $W_d = 1{,}000$ lb (4.5 kN) [ASCE/SEI 6.11.2].

d. Determine the debris mass, m_d

$$m_d = W_d/g = 1{,}000/32.2 = 31.1 \text{ slugs}$$

In S.I.: $$m_d = 4.5\times10^3/9.81 = 459 \text{ kg}$$

e. Determine the nominal debris impact force, F_{ni}
The nominal debris impact force, F_{ni}, is determined by ASCE/SEI Equation (6.11-2):

$$F_{ni} = u_{\max}\sqrt{km_d} = 12.67\times\sqrt{(350\times12)\times(31.1/1{,}000)} = 145 \text{ kips}$$

In S.I.: $$F_{ni} = 3.86\times\sqrt{(61{,}300/1{,}000)\times459} = 648 \text{ kN}$$

f. Determine the importance factor, I_{tsu}
The importance factor, I_{tsu}, is determined from ASCE/SEI Table 6.8-1.
Because the inundation depth at the site resulting from the EGLA for this TRC IV building has been determined to be less than 12 ft (3.66 m) and a PTHA has not been performed, $I_{\text{tsu}} = 1.50$.

g. Determine the orientation coefficient, C_o
The orientation coefficient, C_o, is equal to 0.65 for logs and poles (ASCE/SEI 6.11.2).

h. Determine the design debris impact force, F_i
The design debris impact force, F_i, is determined by ASCE/SEI Equation (6.11-3):

$$F_i = I_{\text{tsu}}C_oF_{ni} = 1.50\times0.65\times145 = 141 \text{ kips}$$

In S.I.: $F_i = 1.50\times0.65\times648 = 632$ kN

i. Determine the impulse duration for elastic impact, t_d
The impulse duration for elastic impact, t_d, is determined by ASCE/SEI Equation (6.11-4):

$$t_d = \frac{2m_du_{\max}}{F_{ni}} = \frac{2\times31.1\times12.67}{145\times1{,}000} = 0.0054 \text{ s}$$

In S.I.: $$t_d = \frac{2m_du_{\max}}{F_{ni}} = \frac{2\times459\times3.86}{648\times1{,}000} = 0.0055 \text{ s}$$

j. Determine the natural period of the impacted structural element, T_n
Assuming the column has simple connections at both ends, the natural period of the column is determined as follows (see Flowchart 7.6):

$$\text{Mass of column per unit length } m = \frac{20\times20}{144}\times\frac{150}{32.2} = 12.9 \text{ slugs/ft}$$

$$T_n = \frac{2L^2}{\pi\sqrt{\dfrac{EI}{m}}} = \frac{2\times8.0^2}{\pi\sqrt{\dfrac{(3{,}605{,}000\times144)\times\left[\dfrac{1}{12}\times\left(\dfrac{20}{12}\right)^4\right]}{12.9}}} = 0.0080 \text{ s}$$

In S.I.:

$$m = \left(\frac{508}{1{,}000}\right)^2 \times \frac{(23.6\times10^3)}{9.81} = 621 \text{ kg/m}$$

$$T_n = \frac{2\times2.44^2}{\pi\sqrt{\dfrac{(29{,}600\times10^6)\times\left[\dfrac{1}{12}\times\left(\dfrac{508}{1{,}000}\right)^4\right]}{621}}} = 0.0074 \text{ s}$$

k. Determine t_d/T_n

$$t_d/T_n = 0.0054/0.0080 = 0.68$$

In S.I.: $t_d/T_n = 0.0055/0.0074 = 0.74$

l. Determine the dynamic response ratio for impulsive loads, R_{max}
The dynamic response ratio for impulsive loads, R_{max}, is determined from ASCE/SEI Table 6.11-1:
For $t_d/T_n = 0.68$, $R_{max} = 1.78$ by linear interpolation
In S.I.: For $t_d/T_n = 0.74$, $R_{max} = 1.80$

m. Determine the debris dynamic impact force
Debris dynamic impact force $= R_{max}F_i = 1.78\times141 = 251$ kips
In S.I.: Debris dynamic impact force $= 1.80\times632 = 1{,}138$ kN

The dynamic impact force is applied anywhere from 3 ft (0.91 m) up to $h_{max} = 6.51$ ft (1.99 m) from the base of the column. Illustrated in Figure 7.46 is the dynamic impact force applied at 4 ft (1.22 m) from the base of the column, which is the critical section for flexure.

Figure 7.46 Dynamic Impact Force Due to Wood Logs and Poles for the Essential Facility in Example 7.5

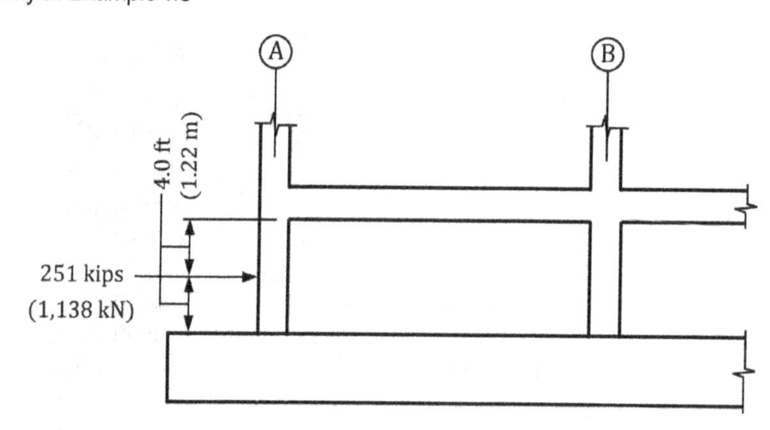

2. Impact by Vehicles
The force due to impact by floating vehicles, F_i, is determined in accordance with ASCE/SEI 6.11.3:

$$F_i = 30I_{tsu} = 30\times1.50 = 45 \text{ kips}$$

In S.I.: $F_i = 130 I_{tsu} = 130 \times 1.50 = 195$ kN
This impact force is applied anywhere from 3 ft (0.91 m) up to $h_{max} = 6.51$ ft (1.99 m) from the base of the column.

3. Tumbling boulders and concrete debris
This impact force must be considered because $h_{max} = 6.51$ ft (1.99 m) 6.0 ft (1.83 m).

$$F_i = 8{,}000 I_{tsu} = 8{,}000 \times 1.50/1{,}000 = 12 \text{ kips}$$

In S.I.: $F_i = 36 I_{tsu} = 36 \times 1.50 = 54$ kN

This impact force is applied 2 ft (0.61 m) above the base of the column.

4. Shipping containers
Flowchart 7.7 is used to determine the impact load due to shipping containers.
   a. Acquire u_{max} from the EGLA
   From Table 7.32, $u_{max} = 12.67$ ft/s (3.86 m/s)
   b. Determine the stiffness of the impacting debris and the impacted structural element and use whichever is less
   For a 20-ft (6.10-m) shipping container, stiffness $= 245$ kip/in. (42,900 kN/m) [ASCE/SEI Table 6.11-2 and Flowchart 7.7].
   Assuming the column has simple connections at both ends, the stiffness of the column is determined as follows (see Flowchart 7.7):

$$k = \frac{48EI}{L^3} = \frac{48 \times (3{,}605{,}000/1{,}000) \times \left(\frac{1}{12} \times 20^4\right)}{(8.0 \times 12)^3} = 2{,}608 \text{ kip/in.} > 245 \text{ kip/in.}$$

   In S.I.:

$$k = \frac{48 \times (24{,}900 \times 10^3) \times \left[\frac{1}{12} \times \left(\frac{508}{1{,}000}\right)^4\right]}{(2.44)^3} = 456{,}609 \text{ kN/m} > 42{,}900 \text{ kN/m}$$

   Therefore, use $k = 245$ kip/in. (42,900 kN/m).
   c. Determine the debris mass of the empty shipping container
   From ASCE/SEI Table 6.11-2:
   Empty debris weight $W_d = 5{,}000$ lb (22.2 kN)
   Loaded debris weight $W_d = 29{,}000$ lb (129.0 kN)
   Debris mass of the empty shipping container:

$$m_d = W_d / g = 5{,}000/32.2 = 155.3 \text{ slugs}$$

   In S.I.: $m_d = W_d / g = 22.2 \times 1{,}000/9.81 = 2{,}263$ kg

   d. Determine the nominal debris impact force, F_{ni}:
   The nominal debris impact force, F_{ni}, is determined by ASCE/SEI Equation (6.11-2):

$$F_{ni} = u_{max}\sqrt{km_d} = 12.67 \times \sqrt{(245 \times 12) \times (155.3/1{,}000)} = 271 \text{ kips} > 220 \text{ kips}$$

   In S.I.: $F_{ni} = 3.86 \times \sqrt{(42{,}900/1{,}000) \times 2{,}263} = 1{,}203 \text{ kN} > 980 \text{ kN}$

   Use $F_{ni} = 220$ kips (980 kN) in accordance with ASCE/SEI 6.11.6.

e. Determine the importance factor, I_{tsu}
The importance factor, I_{tsu}, is determined from ASCE/SEI Table 6.8-1.
Because the inundation depth at the site resulting from the EGLA for this TRC IV building has been determined to be less than 12 ft (3.66 m) and a PTHA has not been performed, $I_{tsu} = 1.50$.

f. Determine the orientation coefficient, C_o
The orientation coefficient, C_o, is equal to 0.65 for shipping containers (ASCE/SEI 6.11.6).

g. Determine the design debris impact force, F_i
The design debris impact force, F_i, is determined by ASCE/SEI Equation (6.11-3):

$$F_i = I_{tsu} C_o F_{ni} = 1.50 \times 0.65 \times 220 = 215 \text{ kips}$$

In S.I.: $$F_i = 1.50 \times 0.65 \times 980 = 956 \text{ kN}$$

h. Determine the impulse duration for elastic impact, t_d
For empty shipping containers, t_d is determined by ASCE/SEI Equation (6.11-4):

$$t_d = \frac{2m_d u_{max}}{F_{ni}} = \frac{2 \times 155.3 \times 12.67}{220 \times 1{,}000} = 0.0179 \text{ s}$$

In S.I.: $$t_d = \frac{2 \times 2{,}263 \times 3.86}{980 \times 1{,}000} = 0.0178 \text{ s}$$

For loaded shipping containers, t_d is determined by ASCE/SEI Equation (6.11-5):

$$t_d = \frac{(m_d + m_{contents}) u_{max}}{F_{ni}} = \frac{(29{,}000/32.2) \times 12.67}{220 \times 1{,}000} = 0.0519 \text{ s}$$

In S.I.: $$t_d = \frac{13{,}150 \times 3.86}{980 \times 1{,}000} = 0.0518 \text{ s}$$

where minimum $m_d + m_{contents}$ is given in ASCE/SEI Table 6.11-2

i. Determine the natural period of the impacted structural element, T_n
Assuming the column has simple connections at both ends, the natural period of the column is determined above for wood logs and poles as 0.0080 s (the natural period is equal to 0.0074 s when S.I. units are used; see the calculations under the wood logs and poles section of this example).

j. Determine t_d / T_n
For empty shipping containers: $t_d / T_n = 0.0179/0.0080 = 2.24$
For loaded shipping containers: $t_d / T_n = 0.0519/0.0080 = 6.49$

k. Determine the dynamic response ratio for impulsive loads, R_{max}
The dynamic response ratio for impulsive loads, R_{max}, is determined from ASCE/SEI Table 6.11-1:
For empty shipping containers, $R_{max} = 1.5$ for $t_d / T_n = 2.24$
For loaded shipping containers, $R_{max} = 1.5$ for $t_d / T_n = 6.49$

l. Determine the debris impact force
For both empty and loaded shipping containers:

$$\text{Debris dynamic impact force} = R_{max} F_i = 1.5 \times 215 = 323 \text{ kips}$$

In S.I.: Debris dynamic impact force $= 1.5 \times 956 = 1{,}434$ kN

In this example, the dynamic impact forces of the empty and fully load shipping containers are equal.

The dynamic impact force is applied anywhere from 3 ft (0.91 m) up to $h_{max} = 6.51$ ft (1.99 m) from the base of the column. Illustrated in Figure 7.47 is the dynamic impact force applied at 4 ft (1.22 m) from the base of the column, which is the critical section for flexure.

Figure 7.47 Dynamic Impact Force due to Shipping Containers on the Essential Facility in Example 7.5

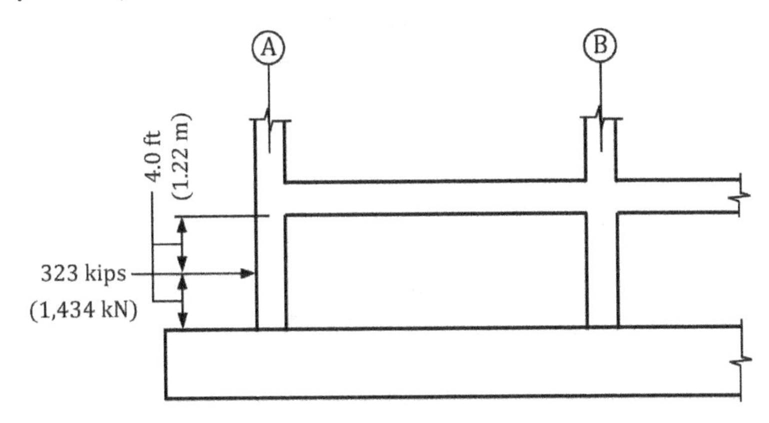

Step 8—Determine the tsunami load combinations

Principal tsunami forces and effects on the lateral force-resisting system and the individual components must be combined with other specified loads using the load combinations in ASCE/SEI 6.8.3.3. As noted previously, the debris impact loads need not be combined with other tsunami-related loads determined in accordance with ASCE/SEI Chapter 6.

7.6 References

7.1. American Society of Civil Engineers (ASCE). 2014. *Flood Resistant Design and Construction*, ASCE/SEI 24-14, Reston, VA.

7.2. Federal Emergency Management Agency (FEMA). FEMA Flood Map Service Center. https://msc.fema.gov/portal/home.

7.3. American Society of Civil Engineers (ASCE). 2022. ASCE Tsunami Hazard Tool. https://asce7tsunami.online/.

7.4. Federal Emergency Management Agency (FEMA). 2014. *Guidance for Flood Risk Analysis and Mapping—Vertical Datum Conversions*, Washington, D.C.

7.5. Federal Emergency Management Agency (FEMA). 2011. *Coastal Construction Manual: Principles and Practices of Planning, Siting, Designing, Constructing, and Maintaining Residential Buildings in Coastal Areas,* 4th ed, FEMA P-55, Washington, D.C.

7.6. Federal Emergency Management Agency (FEMA). 2012. *Engineering Principles and Practices of Retrofitting Floodprone Residential Structures*, 3rd ed, FEMA P-259, Washington, D.C.

7.7. Federal Emergency Management Agency (FEMA). Wave Height Analysis for Flood Insurance Studies (WHAFIS), Version 4.0, https://www.fema.gov/wave-height-analysis-flood-insurance-studies-version-40, Washington, D.C.

7.8. US Army Corps of Engineers (USACE). 1995. Flood Proofing Regulations, Washington, D.C.

7.9. PIANC. 2014. *Harbor Approach Channels: Design Guidelines*, Appendix C, Typical Ship Dimensions, Report No. 121-2014, Brussels, Belgium.

7.7 Problems

7.1. A 15-ft-tall (4.57-m-tall), vertical reinforced concrete wall is located in a V Zone that supports the base of a risk category III structure. Determine the design flood loads on the wall assuming the following:

- Design stillwater elevation, $d_s = 6$ ft (1.83 m)
- Base flood elevation (BFE) = 10 ft (3.05 m)

The local jurisdiction has adopted a freeboard of 2 ft (0.61 m) above the BFE. Assume the wall length is 60 ft (18.3 m).

7.2. For the reinforced concrete wall in Problem 7.1, determine the design flood loads on the wall assuming the wall is nonvertical where the longitudinal axis of the wall is 5 degrees from vertical.

7.3. For the reinforced concrete wall in Problem 7.1, determine the design flood loads on the wall assuming the wall is vertical and the waves are obliquely incident to the wall where the horizontal angle between the direction of the waves and the vertical surface is 30 degrees.

7.4. A 125 ft by 200 ft (38.1 m by 61.0 m) commercial building is supported by 81 circular reinforced concrete piles that have a diameter of 12 in. (305 mm) each. The site has been categorized as Zone AE with a 100-year stillwater elevation of 10.5 ft (3.20 m) NGVD. The lowest ground elevation at the site is 6.0 ft (1.83 m) NGVD and a geotechnical report has indicated that 8 in. (203 mm) of erosion is expected to occur during the design flood. The local jurisdiction has not mandated any additional freeboard. Determine the design flood loads on an individual pile.

7.5. Given the design data in Problem 7.4, determine the design floods on an individual pile assuming the building is designated an emergency shelter.

7.6. Determine the tsunami loads, F_{TSU}, on the six-story residential building in Figure 7.48. Assume the following:

- Building site: latitude = 19.71717°, longitude = 155.06650°
- The building is located 4,287 ft (1,307 m) from the shoreline as shown in Figure 7.49.
- The ground transect perpendicular to the shoreline is given in Figure 7.49 with the transect elevations given in Table 7.33. The transect elevations located inland from the runup location are the same as that at the runup elevation.
- Runup elevation is not given in Reference 7.3.
- The mean slope angle of the nearshore profile, Φ, is equal to 15 degrees.
- Offshore tsunami amplitude, H_T, from Reference 7.3 is equal to 28.0 ft (8.5 m).
- Predominant wave period, T_{TSU}, from Reference 7.3 is equal to 16 minutes.
- Tsunami bores are expected within this area.

Figure 7.48 Six-Story Residential Building in Problem 7.6

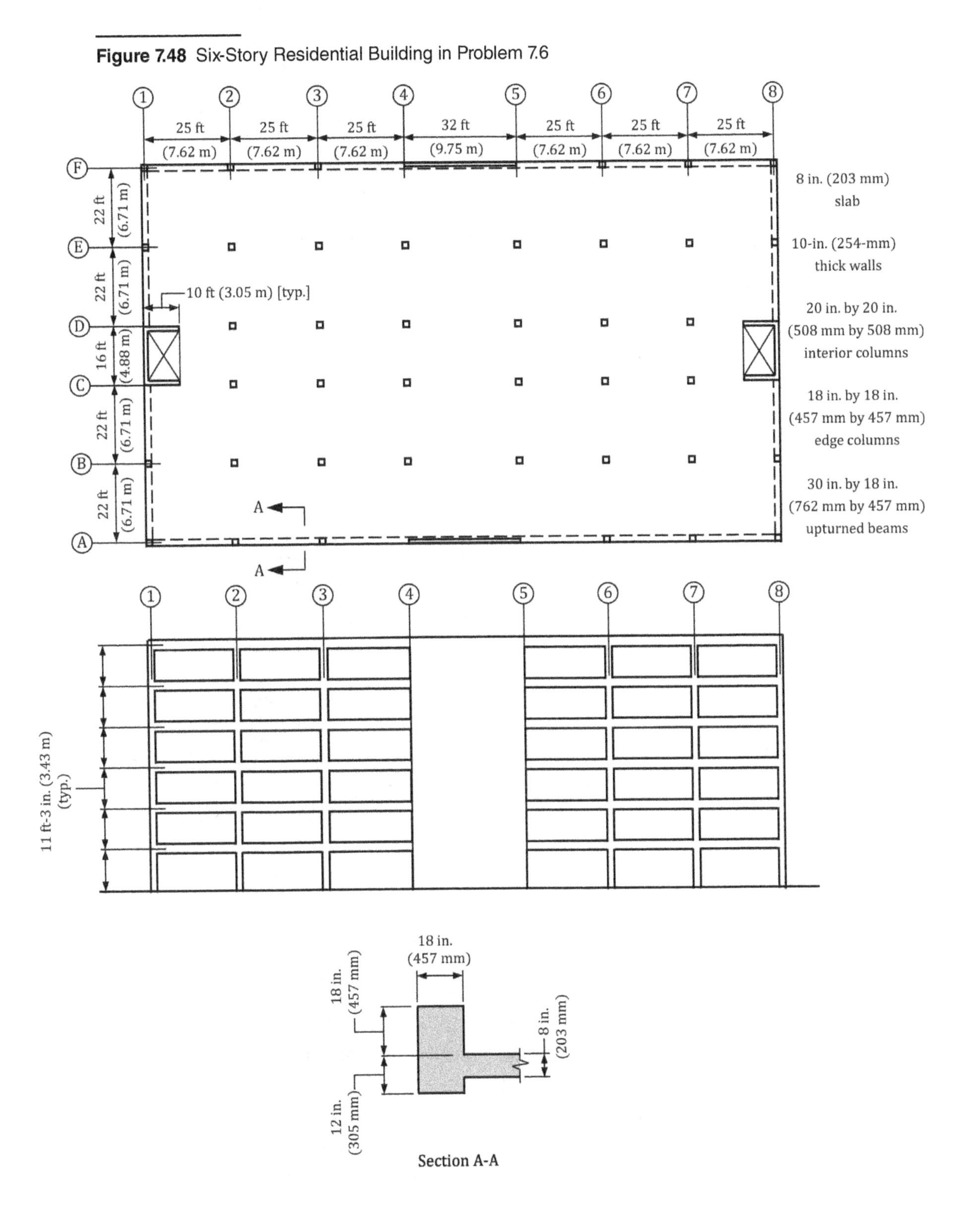

Figure 7.49 Site Plan and Ground Transect for the Residential Building in Problem 7.6

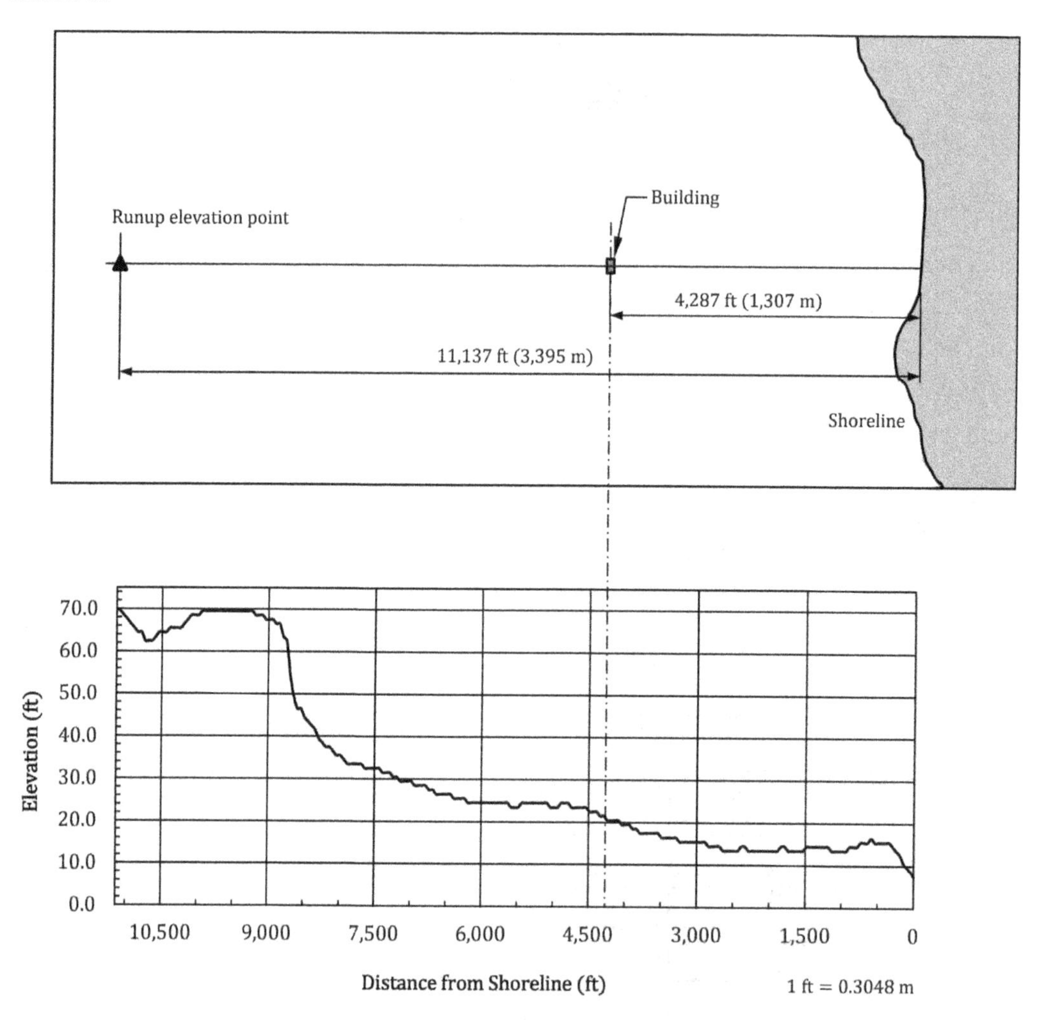

- The site is not located in an impact zone for shipping containers in accordance with ASCE/SEI 6.11.5.
- Windows are not designed for large impact loads or blast loads and span from the top of the slab to the underside of the beams.
- The building is supported by pile foundations beneath the columns and walls.
- The slab-on-grade is isolated from the columns and walls.
- The perimeter beams are upturned as shown in detail A-A in Figure 7.48.
- The structural members are constructed of reinforced concrete with a unit weight equal to 150 lb/ft^3 (23.6 kN/m^3) and a modulus of elasticity, E, equal to 3,605,000 lb/in.^{2} (24,900 MPa).

Table 7.33 Transect Elevations for the Residential Building in Problem 7.6

Point	Distance from Shoreline, x_i (ft)	Transect Elevation, z_i (ft)
1	11,087.0	69.5
2	11,037.0	68.5
3	10,987.0	67.5
4	10,937.0	66.5
5	10,887.0	65.5
6	10,837.0	64.5
7	10,787.0	64.5
8	10,737.0	62.5
9	10,687.0	62.5
10	10,637.0	62.5
11	10,587.0	63.5
12	10,537.0	64.5
13	10,487.0	64.5
14	10,437.0	64.5
15	10,387.0	65.5
16	10,337.0	65.5
17	10,287.0	65.5
18	10,237.0	65.5
19	10,187.0	66.5
20	10,137.0	67.5
21	10,087.0	68.5
22	10,037.0	68.5
23	9,987.0	68.5
24	9,937.0	69.5
25	9,887.0	69.5
26	9,837.0	69.5
27	9,787.0	69.5
28	9,737.0	69.5
29	9,687.0	69.5
30	9,637.0	69.5
31	9,587.0	69.5
32	9,537.0	69.5
33	9,487.0	69.5
34	9,437.0	69.5
35	9,387.0	69.5
36	9,337.0	69.5
37	9,287.0	69.5
38	9,237.0	69.5
39	9,187.0	68.5
40	9,137.0	68.5
41	9,087.0	68.5
42	9,037.0	67.5
43	8,987.0	67.5
44	8,937.0	67.5
45	8,887.0	66.5
46	8,837.0	66.5
47	8,787.0	63.5
48	8,737.0	62.5
49	8,687.0	54.5
50	8,637.0	49.5
51	8,587.0	46.5
52	8,537.0	46.5
53	8,487.0	44.5
54	8,437.0	43.5
55	8,387.0	42.5
56	8,337.0	41.5
57	8,287.0	39.5
58	8,237.0	38.5
59	8,187.0	37.5
60	8,137.0	37.5
61	8,087.0	36.5
62	8,037.0	35.5
63	7,987.0	35.5
64	7,937.0	34.5
65	7,887.0	33.5
66	7,837.0	33.5
67	7,787.0	33.5
68	7,737.0	33.5
69	7,687.0	33.5
70	7,637.0	32.5
71	7,587.0	32.5
72	7,537.0	32.5
73	7,487.0	32.5
74	7,437.0	32.5
75	7,387.0	31.5
76	7,337.0	31.5
77	7,287.0	31.5
78	7,237.0	30.5
79	7,187.0	30.5
80	7,137.0	29.5
81	7,087.0	29.5
82	7,037.0	29.5
83	6,987.0	29.5
84	6,937.0	28.5
85	6,887.0	28.5
86	6,837.0	28.5
87	6,787.0	28.5
88	6,737.0	27.5
89	6,687.0	27.5
90	6,637.0	26.5
91	6,587.0	26.5
92	6,537.0	26.5
93	6,487.0	26.5
94	6,437.0	26.5
95	6,387.0	25.5
96	6,337.0	25.5
97	6,287.0	25.5
98	6,237.0	25.5
99	6,187.0	24.5
100	6,137.0	24.5
101	6,087.0	24.5
102	6,037.0	24.5
103	5,987.0	24.5
104	5,937.0	24.5
105	5,887.0	24.5
106	5,837.0	24.5
107	5,787.0	24.5
108	5,737.0	24.5
109	5,687.0	24.5
110	5,637.0	24.5
111	5,587.0	23.5
112	5,537.0	23.5
113	5,487.0	23.5
114	5,437.0	24.5

(*continued*)

Table 7.33 Transect Elevations for the Residential Building in Problem 7.6 (*Continued*)

Point	Distance from Shoreline, x_i (ft)	Transect Elevation, z_i (ft)
115	5,387.0	24.5
116	5,337.0	24.5
117	5,287.0	24.5
118	5,237.0	24.5
119	5,187.0	24.5
120	5,137.0	24.5
121	5,087.0	24.5
122	5,037.0	23.5
123	4,987.0	23.5
124	4,937.0	23.5
125	4,887.0	24.5
126	4,837.0	24.5
127	4,787.0	24.5
128	4,737.0	23.5
129	4,687.0	23.5
130	4,637.0	23.5
131	4,587.0	23.5
132	4,537.0	23.5
133	4,487.0	22.5
134	4,437.0	22.5
135	4,387.0	22.5
136	4,337.0	21.5
137	4,287.0	21.5
138	4,237.0	20.5
139	4,187.0	20.5
140	4,137.0	20.5
141	4,087.0	20.5
142	4,037.0	19.5
143	3,987.0	19.5
144	3,937.0	19.5
145	3,887.0	18.5
146	3,837.0	18.5
147	3,787.0	17.5
148	3,737.0	17.5
149	3,687.0	17.5
150	3,637.0	17.5
151	3,587.0	17.5
152	3,537.0	17.5
153	3,487.0	16.5
154	3,437.0	16.5
155	3,387.0	16.5
156	3,337.0	16.5
157	3,287.0	16.5
158	3,237.0	15.5
159	3,187.0	15.5
160	3,137.0	15.5
161	3,087.0	15.5
162	3,037.0	15.5
163	2,987.0	15.5
164	2,937.0	15.5
165	2,887.0	15.5
166	2,837.0	14.5
167	2,787.0	14.5
168	2,737.0	14.5
169	2,687.0	14.5
170	2,637.0	13.5
171	2,587.0	13.5
172	2,537.0	13.5
173	2,487.0	13.5
174	2,437.0	13.5
175	2,387.0	14.5
176	2,337.0	14.5
177	2,287.0	13.5
178	2,237.0	13.5
179	2,187.0	13.5
180	2,137.0	13.5
181	2,087.0	13.5
182	2,037.0	13.5
183	1,987.0	13.5
184	1,937.0	13.5
185	1,887.0	13.5
186	1,837.0	14.5
187	1,787.0	14.5
188	1,737.0	13.5
189	1,687.0	13.5
190	1,637.0	13.5
191	1,587.0	13.5
192	1,537.0	13.5
193	1,487.0	14.5
194	1,437.0	14.5
195	1,387.0	14.5
196	1,337.0	14.5
197	1,287.0	14.5
198	1,237.0	14.5
199	1,187.0	14.5
200	1,137.0	13.5
201	1,087.0	13.5
202	1,037.0	13.5
203	987.0	13.5
204	937.0	13.5
205	887.0	14.5
206	837.0	14.5
207	787.0	14.5
208	737.0	15.5
209	687.0	15.5
210	637.0	15.5
211	587.0	16.5
212	537.0	15.5
213	487.0	15.5
214	437.0	15.5
215	387.0	15.5
216	337.0	15.5
217	287.0	14.5
218	237.0	13.5
219	187.0	12.5
220	137.0	10.5
221	87.0	9.5
222	37.0	8.5
223	0.0	7.5

1 ft = 0.3048 m.

CHAPTER 8

Load Paths

8.1 Introduction

To properly ensure the strength and stability of a building as a whole and all of its members, it is important to understand the paths that loads take through a structure. A continuous load path is required from the top of a building or structure to the foundation and into the ground. Structural systems must be capable of providing a clearly defined and uninterrupted load path for the effects due to gravity, uplift, and lateral loads.

Section 6.6 of Reference 8.1 provides the following analogy on a continuous load path:

> A continuous load path can be thought of as a "chain" running through a building. The "links" of the chain are structural members, connections between members, and any fasteners used in the connections (e.g., nails, screws, bolts, welds or reinforcing steel). To be effective, each "link" in the continuous load path must be strong enough to transfer loads without permanently deforming or breaking. Because all applied loads (e.g., gravity, dead, live, uplift, lateral) must be transferred into the ground, the load path must continue unbroken from the uppermost building element through the foundation and into the ground.

A continuous load path is essential for a building or structure to perform as intended when subjected to the code-prescribed loads; this helps ensure the life safety of its occupants. The general requirements for providing a complete load path that is capable of transferring all loads from their point of origin to the resisting elements are found in IBC 1604.4 and ASCE/SEI 1.4.1. Specific requirements on continuous load path and interconnection for building structures subject to seismic loads are given in ASCE/SEI 12.1.3.

The path that loads take through a structure includes the structural members, the connections between the members, and the interface between the foundation and the soil. Each element in the path must be designed and detailed to resist the applicable combinations of gravity, uplift, and lateral load effects. The level of detailing needed to ensure a complete load path depends to some extent on the material and type of structural system involved. For example, the load path in light wood frame and cold-formed steel (CFS) structures consists of many individual elements; as such, considerable detailing is often required to guarantee the lateral forces are transferred from the out-of-plane walls, through the diaphragm and collectors and into the supporting shear walls and foundation. In addition, the structure as a whole must be capable of resisting the effects of overturning and sliding. Because light-frame wood and CFS structures often do not have the large dead loads that are present in structures built with concrete, masonry, and steel, the uplift load path is important for building performance.

The following sections cover typical load paths for gravity, uplift, and lateral loads. Even though almost every building or structure must contain a gravity, uplift, and lateral force-resisting system (LFRS), the systems are usually interconnected; that is, some or even all the members in a structure can be part of these systems.

8.2 Load Paths for Gravity Loads

The path that gravity loads travel through a conventional building or structure (i.e., a building or structure consisting of floor and roof slabs, beams, columns, walls, and foundations) is usually straightforward and depends on the type of structural members present in the structure. In conventional building construction, a majority of the gravity loads (superimposed dead loads, live loads, rain loads, and snow loads) is initially supported by a floor or roof deck (diaphragm), which is usually a concrete slab, a steel metal deck, wood sheathing or any combination thereof. The loads are then transferred to the structural members supporting the floor or roof deck. Individual structural members are designed to support their self-weight and other applicable gravity loads, which in turn are transferred to their supporting members. Ultimately, the gravity loads are transmitted to the soil supporting the foundation system.

Crane loads and elevator loads are examples of gravity loads not transmitted first to a floor or roof deck. In such cases, the loads are initially transferred to beams. Like conventional structures, the gravity loads are eventually transferred to the foundations and the adjoining soil.

Gravity load paths in structures with conventional framing are illustrated in Figures 8.1 through 8.4. These figures are not meant to be comprehensive; their purpose is to exemplify the common paths for gravity loads in such structures.

Figure 8.1 Gravity Load Path—Conventional Structure with Slab, Walls, and Footings

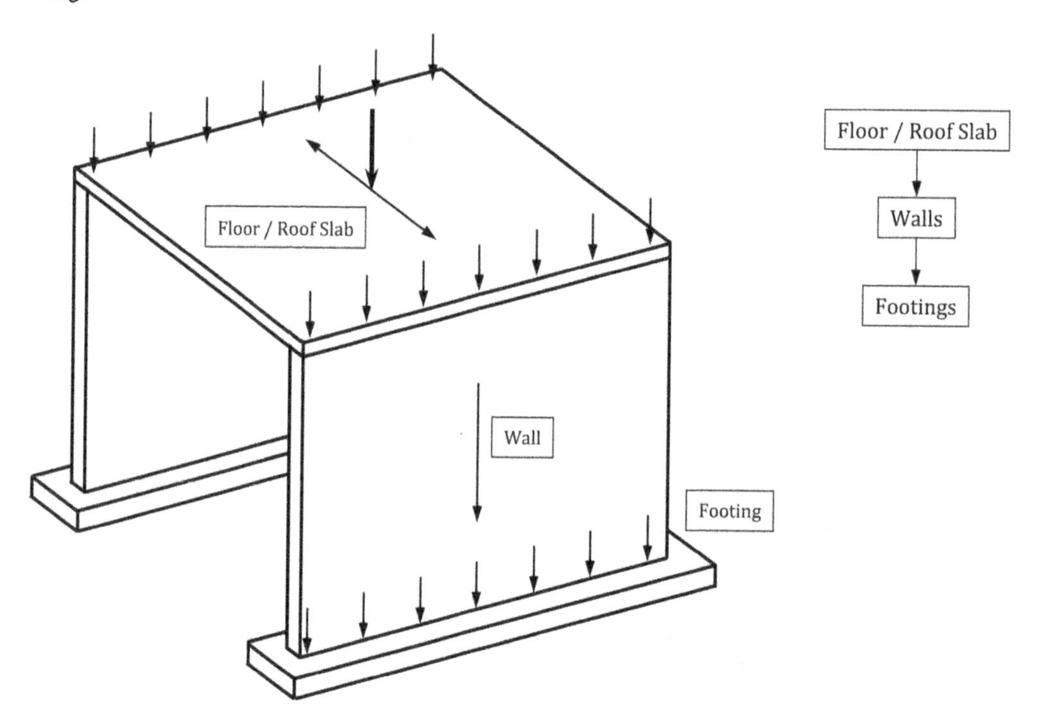

A generic structure consisting of a roof or floor slab supported by walls and spread footings is shown in Figure 8.1. Gravity loads are supported by the floor or roof slab and are transferred to the walls, which in turn transfer them to the footings. The ground beneath the footings ultimately supports these loads and the footings must be properly sized based on the properties of the soil.

The structure depicted in Figure 8.2 is similar to the one in Figure 8.1 except the floor or roof slab is supported directly on beams. The gravity loads from the slab are transferred to the beams,

Figure 8.2
Gravity Load Path—Conventional Structure with Slab, Beams, Walls, and Footings

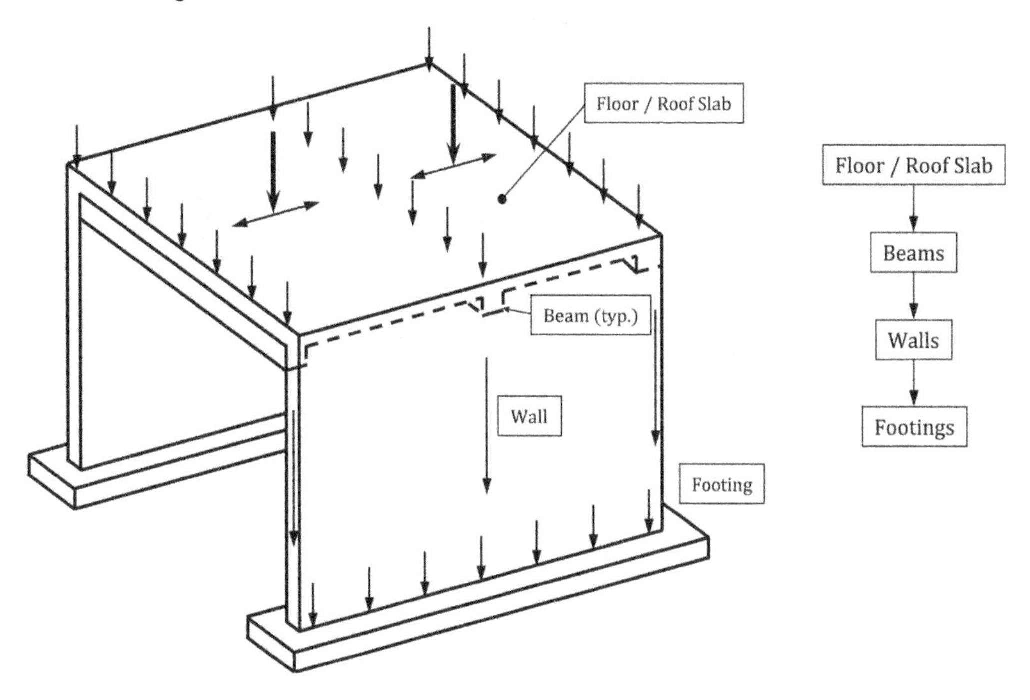

which subsequently transfer them to the walls. Like the structure in Figure 8.1, the walls transfer the loads to the footings and into the ground.

Beams are often supported directly by columns in conventional structures (see Figure 8.3). The beams are supported by walls on one end and by individual columns at the other end.

Figure 8.3 Gravity Load Path—Conventional Structure with Slab, Beams, Columns, Walls, and Footings

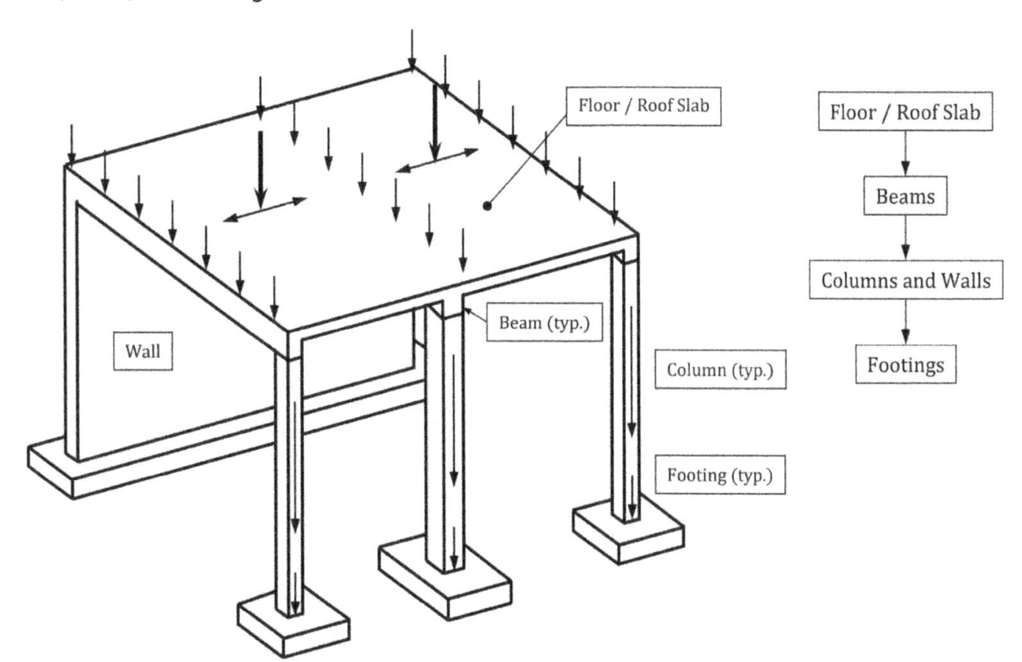

Another common type of conventional framing utilizes a slab, beams, girders, and columns (see Figure 8.4; the footings below two of the columns are not shown for clarity). The gravity loads from the slab are transferred to the beams, which are supported at both ends by girders. The girders in turn transfer the loads to the columns (or walls), which are supported by footings.

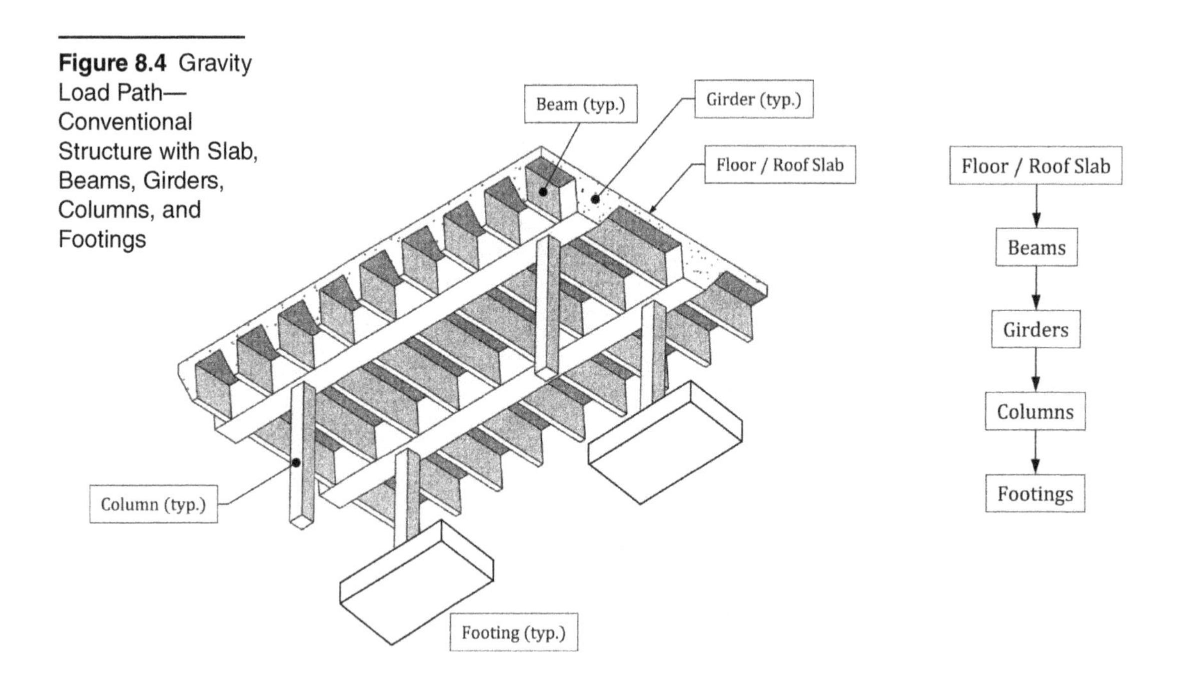

Figure 8.4 Gravity Load Path—Conventional Structure with Slab, Beams, Girders, Columns, and Footings

Although the structures in Figures 8.1 through 8.4 consist of only one story and one bay, the same load paths described above are applicable to multistory and multi-bay structures. In a multistory building, the gravity loads accumulate over the height of the structure from the top to the bottom. The walls and/or columns in the lowest level must be designed to support the sum of the applicable gravity loads from the roof through the first level.

In certain situations, there may be a need to discontinue one or more columns in a conventionally framed structure. One common example of this occurs in office buildings where a column-free lobby is desired. In such cases, a transfer beam (or girder) is used to transfer the gravity loads from the column(s) above to one or more columns below (see Figure 8.5). The transfer beam and all of the connections to the columns must be designed to support the gravity loads from above.

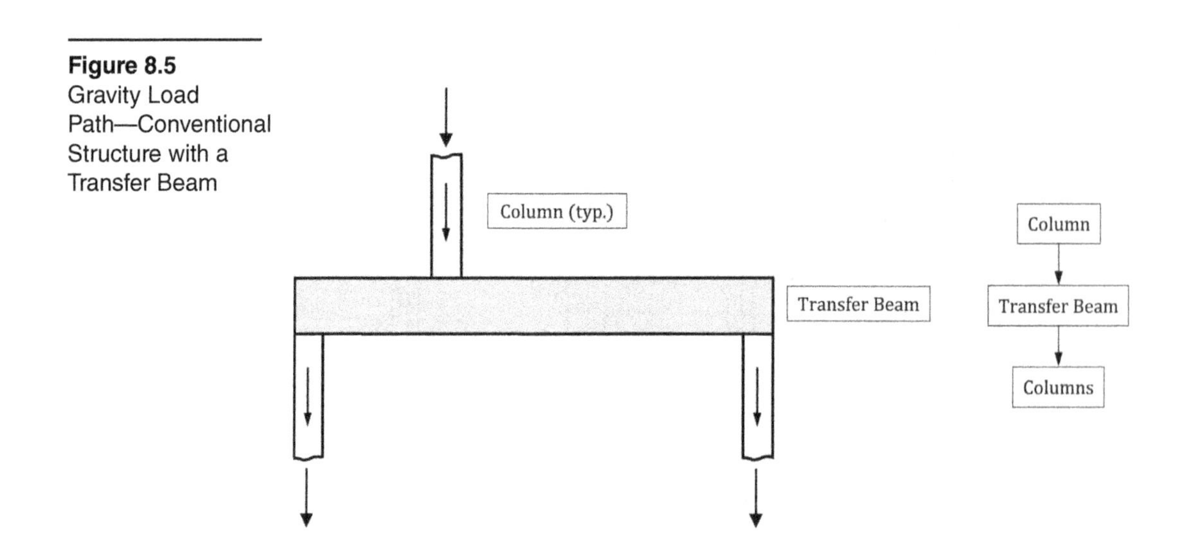

Figure 8.5 Gravity Load Path—Conventional Structure with a Transfer Beam

As noted previously, the connections between the structural members in any structural system must be adequately designed and detailed to ensure a complete load path. For example, the interfaces between the slab and the walls and the walls and the footings in the structure depicted in Figure 8.1 must be designed and detailed for the applicable gravity load effects obtained from the structural analysis at the respective locations. The same is true for all of the other connections in the systems shown in the other figures.

In nonconventional buildings or structures, the gravity load paths may not be as apparent as those described above for conventional ones. For example, in a cable-stayed roof, the gravity loads are transferred from the roof structure to cables, which are connected to masts (see Figure 8.6). The masts subsequently transmit the loads to columns, walls, or trusses, for example, which are supported by a foundation system. Tension and shell structures are other types of nonconventional systems where the paths for gravity loads may not be as obvious as those in more conventional framing systems.

Figure 8.6 Gravity Load Path—Nonconventional Structure

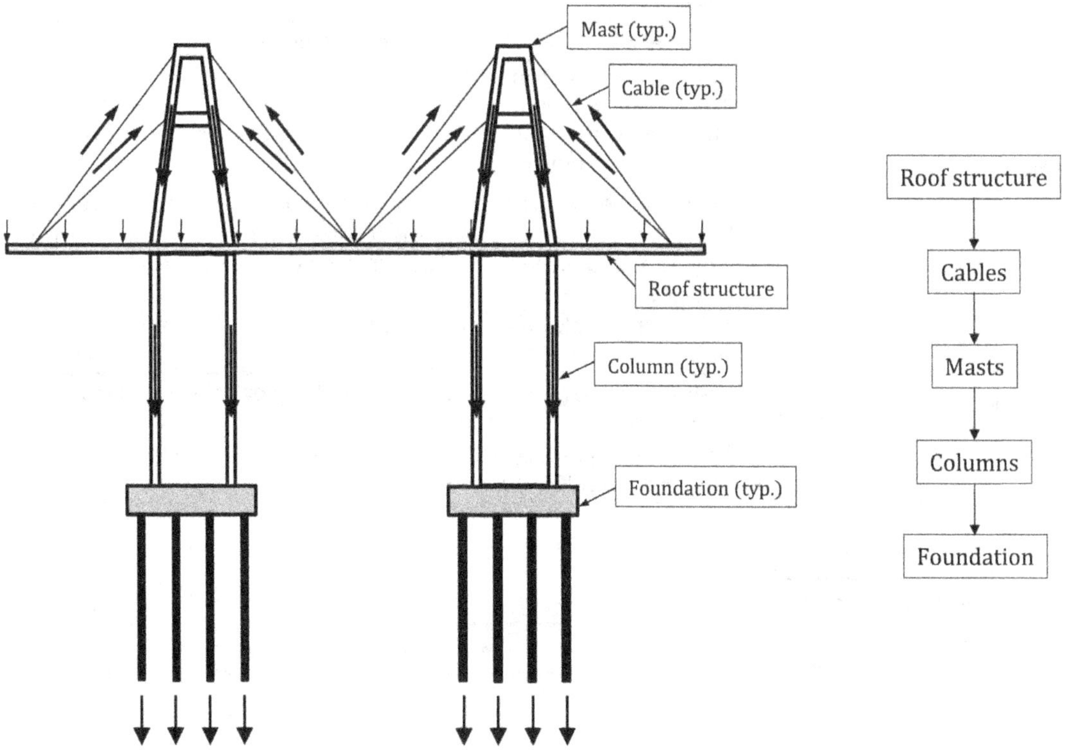

8.3 Load Paths for Lateral and Uplift Loads

8.3.1 Overview

A continuous load path for lateral and uplift loads must be provided for any building or structure. Like in the case of gravity loads, the load path for lateral and uplift loads begins at the point of application of the loads. The lateral and uplift loads travel through the structural members (and connections) of the LFRS and eventually find their way into the soil surrounding the foundations.

The types of lateral and uplift loads that are commonly encountered in the design of building structures are caused by wind, earthquakes, and soil pressure. As discussed in Chapter 5 of this book, the effects of wind must be considered in the design of any above-ground structure; the effects from earthquakes and soil pressure are considered only where applicable. The focus of this section is on load paths for wind and earthquakes.

The paths that lateral loads travel through a building or structure depend on the type of LFRS. A LFRS that resists wind load is referred to as the main wind force-resisting system (MWFRS) as defined in ASCE/SEI 26.2, and a LFRS that resists seismic load is denoted the seismic force-resisting system (SFRS) as defined in ASCE/SEI 11.2. In either case, the LFRS is the assemblage of structural elements specifically designed and detailed to resist the effects from wind or seismic loads. Descriptions of common LFRSs are given in Table 8.1. Information on the specific load paths for wind and seismic loads is covered in the next sections.

Table 8.1 Lateral Force Resisting Systems

System	Description	Figure No.
Bearing wall	• Bearing walls support all or a major portion of the vertical loads. • Some or all walls also provide resistance to the lateral loads.	8.7
Building frame	• An essentially complete space frame supports the vertical loads. • Shear walls or braced frames provide resistance to the lateral loads.	8.8
Moment-resisting frame	• An essentially complete space frame supports the vertical loads. • Lateral loads are resisted primarily by flexural action of the frame members through the joints. • The entire space frame or selected portions of the frame may be designated as the LFRS.	8.9
Dual	• An essentially complete space frame supports the vertical loads. • Moment-resisting frames and shear walls or braced frames provide resistance to lateral loads in accordance with their rigidities. • This system is also referred to as shear wall-frame interactive system.	8.10
Cantilever column	• Vertical and lateral loads are resisted entirely by columns acting as cantilevers from their base. • This system is usually used in one-story buildings or in the top story of a multistory building.	—

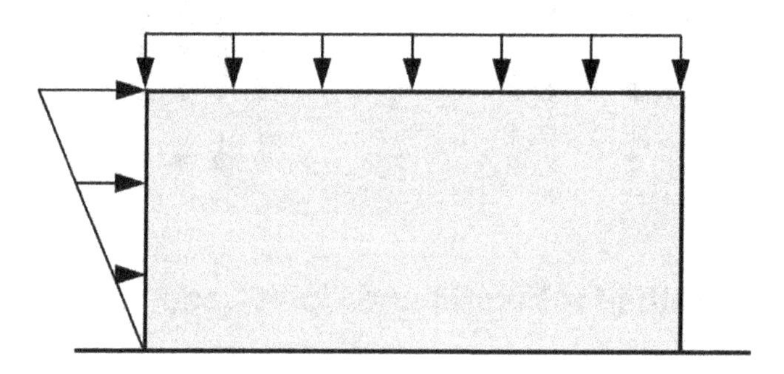

Figure 8.7
Bearing Wall System

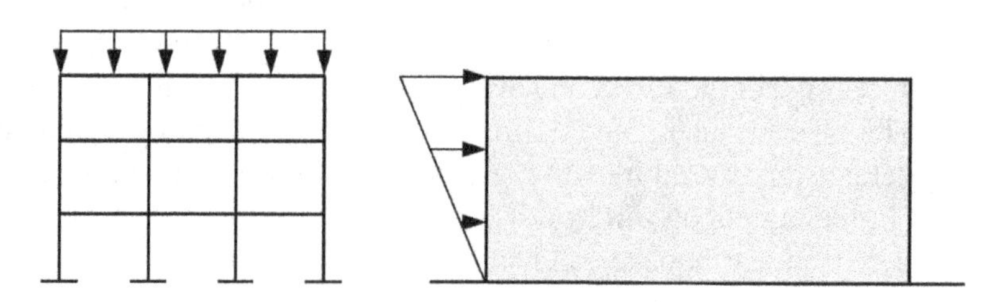

Figure 8.8
Building Frame System

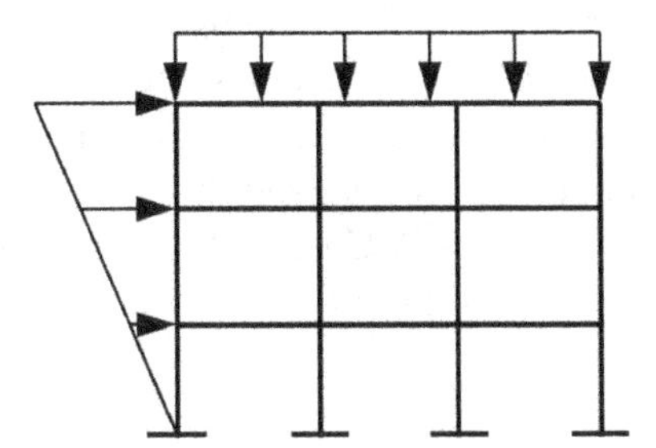

Figure 8.9 Moment-Resisting Frame System

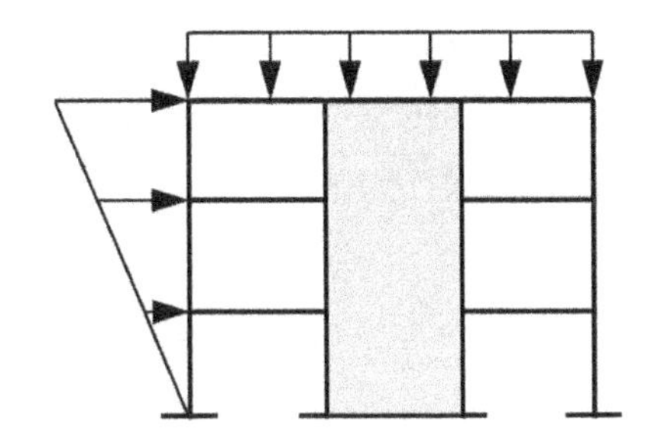

Figure 8.10 Dual-Frame System

8.3.2 Load Paths for Wind Loads

Overview

The continuous path for wind loads through a one-story building with perimeter walls is illustrated in Figure 8.11. The load path begins at the windward wall: this wall receives the wind pressures and transfers the resulting loads to the diaphragm, which laterally supports the wall at the roof. The wind loads are then transferred from the diaphragm to the walls parallel to the direction of the wind. These walls, in turn, transfer the wind loads to the spread footing foundation, which is supported by the soil beneath the footing.

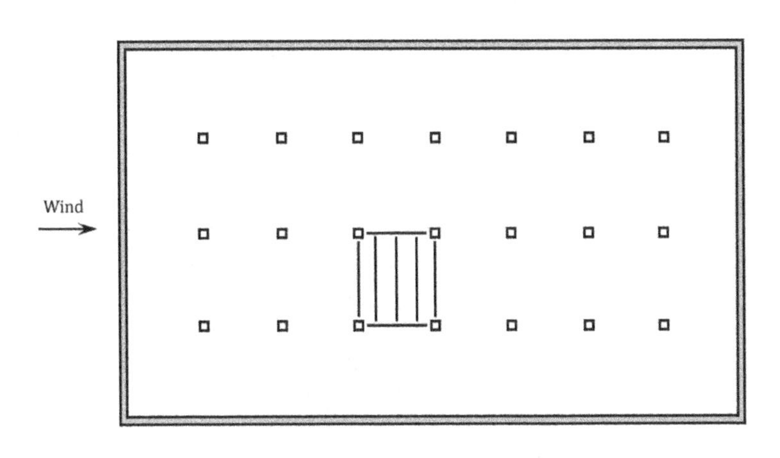

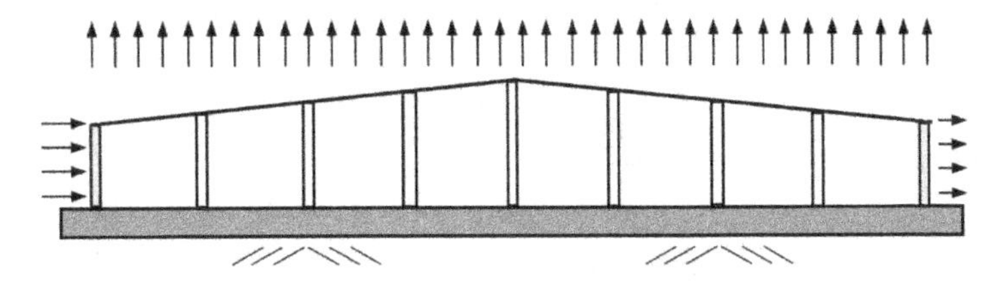

Figure 8.11 Load Path for Wind Loads in a One-Story Building with Perimeter Walls

The same sequence of events would occur for a multistory building (see Figure 8.12) or if moment frames or any other type of LFRS were used instead of or in conjunction with the perimeter walls.

The windward wall (and/or windows and doors) initially receives wind pressure and is generally perpendicular to the direction of wind. This wall is classified as components and cladding (C&C) for wind in this direction, and the wind pressure on it is determined differently than that on the MWFRS, as shown in Chapter 5 of this book. The leeward wall is similarly loaded. For wind in the perpendicular direction, this wall would be part of the MWFRS of the building.

Figure 8.12 Load Path for Wind Loads in a Multistory Building with Perimeter Walls

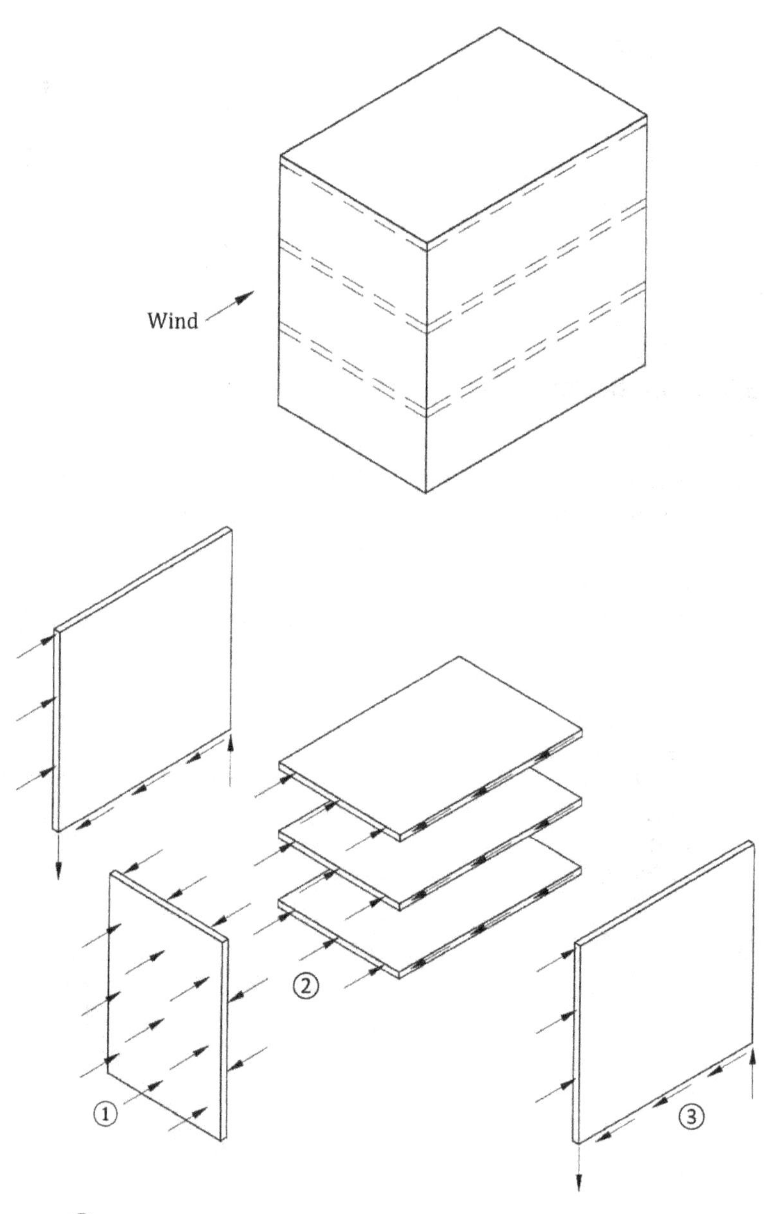

① Wind pressure on windward wall.

② Diaphragms receive wind loads from the windward wall and distribute them to the walls parallel to the wind pressure.

③ Walls parallel to the wind pressure receive the wind loads from the diaphragms and transfer them to the foundations.

The uplift loads on the roof due to wind are resisted by the walls and columns and must be properly combined with the applicable gravity load effects. If the wind load effects are large enough, a net upward load could occur on the vertical members and the foundations.

A similar scenario occurs for a wood-frame building (see Figure 8.13). Wind pressure on the upper half of the windward wall and the vertical projection of the windward roof (for pitched roofs) are transferred to the roof diaphragm. Similar loads are generated on the leeward roof and wall. These loads are transferred to the shear walls parallel to the direction of wind by boundary elements and collectors. The shear walls transfer the lateral loads to the reinforced concrete foundation walls by fasteners. Reinforcement in the foundation walls transmits the loads to the concrete footings, which then transfers them to the surrounding soil. For clarity, only the wind pressure on the windward wall is depicted in Figure 8.13.

Figure 8.13 Load Path for Wind Loads in a Wood-Frame Building

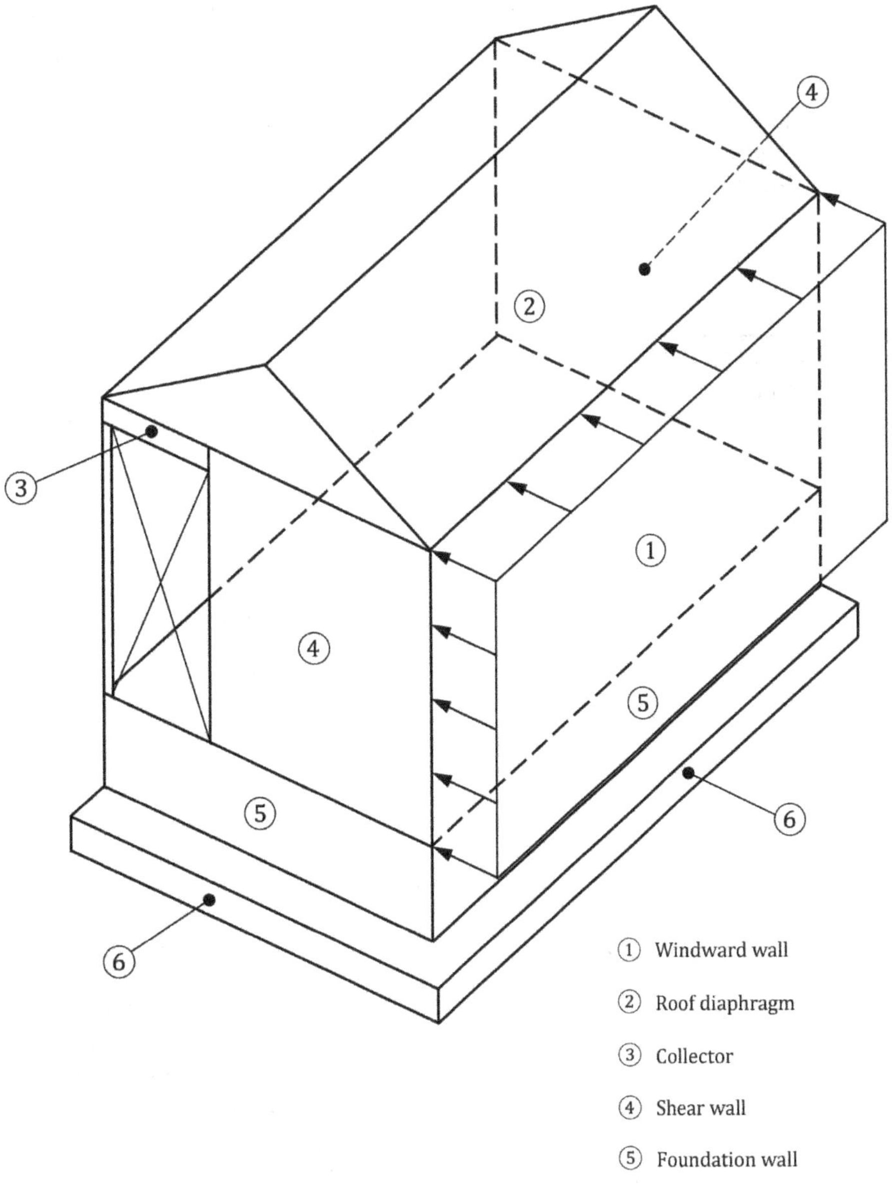

The following sections cover how wind loads propagate through diaphragms and collectors, both of which are essential elements in the lateral load path for any building or structure.

Diaphragms

Overview

Diaphragms are defined in IBC 202 as a horizontal or sloped system that transmits lateral forces to the elements of the LFRS. Horizontal bracing systems are included in this definition. In addition to providing global stability against the effects from lateral loads, diaphragms are commonly the initial point in the load path for gravity loads (see Section 8.2 of this book). Overall behavior of diaphragms subjected to wind loads is discussed below. Additional information can be found in Reference 8.2.

Typical diaphragms in building structures include wood sheathing, corrugated metal deck, concrete fill over corrugated metal deck, concrete topping slab over precast concrete planks, and cast-in-place concrete slab.

Diaphragm Behavior

As discussed above, wind loads from both the windward and leeward walls and roof are transferred to diaphragms, which essentially act as beams spanning between the walls oriented parallel to the direction of wind. For multistory buildings, wind load on the upper half of the wall below the diaphragm and the lower half of the wall above the diaphragm transfer into the diaphragm. Wind load propagation at the roof level of the one-story bearing wall building in Figure 8.11 for wind loads in the direction perpendicular to that shown in Figure 8.11 is illustrated in Figure 8.14 where the windward and leeward loads are combined and applied to the windward wall. Note that low-pitched roofs do not generate significant lateral loads into the diaphragm.

Figure 8.14 Wind Load Propagation from a Diaphragm to the MWFRS

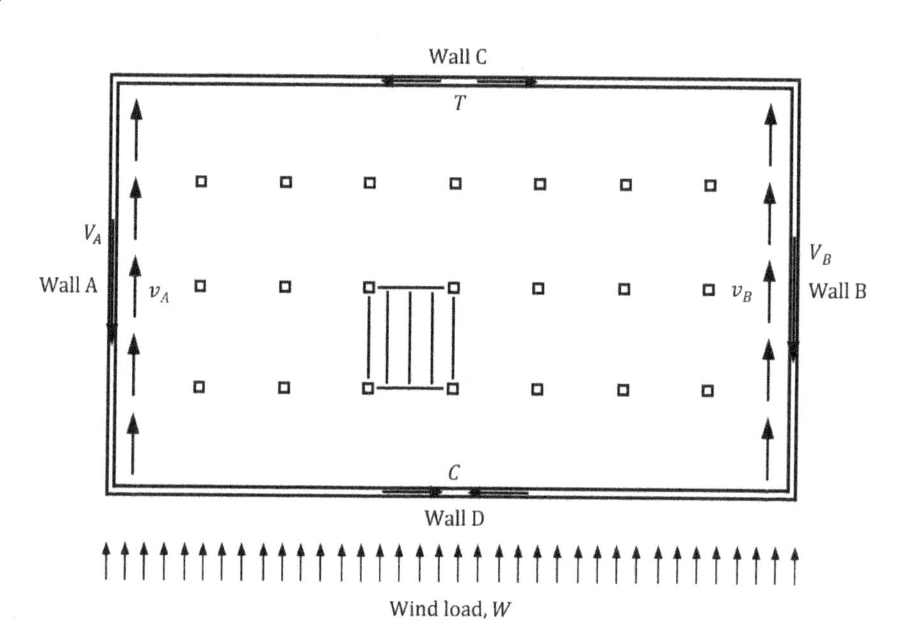

Wall D is the windward wall, which receives the wind load, W. The roof diaphragm behaves essentially as a deep flexural member spanning between Walls A and B, which are oriented parallel to the direction of the wind; the diaphragm laterally supports Wall D (and Wall C) and receives the combined wind load from this wall along its windward edge. The web of the beam transfers the uniform shear forces v_A and v_B along the length of the diaphragm to the supports at Walls A and B. The sum of the forces V_A and V_B at Walls A and B, respectively, is equal to the total wind load, W, at this level.

Chords behave as flange elements at the edges of the diaphragm perpendicular to the wind loads. These elements must be designed to resist the axial tension and compression forces, T and C, respectively, which result from flexural behavior of the diaphragm (see Figure 8.14).

Relatively large openings in a diaphragm can have a dramatic effect on the flow of loads through the diaphragm to the MWFRS. Typical elevator and stair openings in a diaphragm will usually not have a major impact on the overall behavior and load distribution. Larger openings can have a significant influence on overall behavior and a more refined analysis is generally warranted to ensure a complete load path is achieved around the opening. Regardless of the opening size, secondary chord forces occur due to local bending of the diaphragm segments on either side of the opening; these chord forces need to be accounted for in the design.

Diaphragm Flexibility

The manner in which wind loads are transferred from the diaphragm to the elements of the MWFRS depends on the stiffness of the diaphragm in relation to the vertical-resisting elements. Diaphragms can usually be idealized as either flexible or rigid. Diaphragms constructed of wood structural panels can be classified as flexible, rigid, or semi-rigid while those constructed of untopped metal deck, concrete-filled metal deck, and concrete slabs can be idealized as rigid provided the span-to-depth ratio of the diaphragm is less than or equal to 2.0 for consideration of wind loading. It is possible for a diaphragm to be idealized differently depending on the direction of analysis.

It is commonly assumed that flexible diaphragms transfer the wind loads to the elements of the MWFRS by a lateral tributary area basis without any participation of the walls perpendicular to the direction of analysis. Thus, assuming the building in Figures 8.11 and 8.14 has a flexible roof diaphragm, Walls A and B would each carry one-half of the total wind load, W, at the roof level.

In the case of rigid diaphragms, wind loads are transferred based on the relative stiffness of the members in the MWFRS considering all such elements at that level in both directions: the stiffer the element, the greater the load it must resist. Torsional effects must also be considered in rigid diaphragms. Torsion occurs where the center of rigidity of the elements of the MWFRS and the centroid of the wind pressure are at different locations. Elements of the MWFRS perpendicular to the wind load also resist the effects from torsion. In any situation, a finite element model of the system can be utilized to determine the loads in the diaphragm and in the elements of the MWFRS.

Load Paths for Wind

The connections between the diaphragm and the elements of the MWFRS must be properly designed and detailed to ensure a continuous load path. The connection details depend on the material type.

The load path for the building illustrated in Figures 8.11 and 8.14 is given in Figure 8.15 where the diaphragm is a corrugated metal deck and the walls of the MWFRS are reinforced concrete masonry units. The load path from the diaphragm to the foundation is as follows:

1. The shear force from the diaphragm (v_A or v_B) is transferred to a continuous angle by a welded connection between these elements.
2. The angle transfers the load to the top chord of the joists, which is welded to the base plate that transfers the horizontal and vertical loads to the bond beam through the headed shear studs.
3. Vertical reinforcement in the wall transfers the loads through the load-bearing masonry wall, which is the MWFRS in this direction.
4. At the base of the wall, the vertical reinforcement is spliced to reinforcement in the reinforced concrete footing that transfers the loads to the soil.

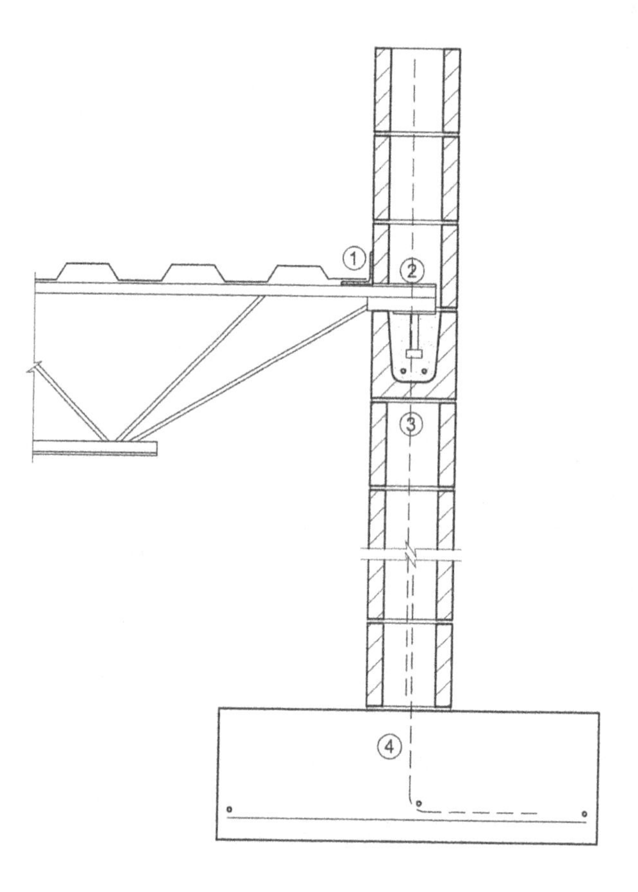

Figure 8.15 Lateral Load Path from the Roof to the Foundation of a One-Story Building with Perimeter Walls—MWFRS

The connections must be designed for the appropriate load effect combinations to ensure a continuous load path through the system. Also, the individual deck sheets must be adequately fastened together at the seams; the design of the seam fasteners is important to ensure that the shear in the diaphragm is transferred as intended. A similar load path would be applicable if reinforced concrete walls were utilized instead of reinforced masonry walls.

A typical section through Wall D of the one-story building in Figure 8.14 is illustrated in Figure 8.16. The load path from the diaphragm to the masonry wall is as follows:

1. The chord forces that develop along the edges of the diaphragm are transferred to a continuous angle via a welded connection between these elements.
2. The angle transfers the load to the masonry bond beam by a connector that is spaced along the length of the wall.

Once the load is transferred into the wall, it follows the same path to the foundation as described in items 3 and 4 above except that in this case, the wall is not load-bearing and is not part of the MWFRS for wind in that direction.

The continuous angle must be designed to resist the combined loads from gravity and the axial tension or compression load from wind. Similarly, the connectors must be designed for the combined effects due to gravity loads, the shear force from the diaphragm due to wind, and the out-of-plane tension force due to negative wind pressure on the wall. For wind in the perpendicular direction, this assemblage must resist the diaphragm shear forces transferred into Walls C and D, which are part of the MWFRS in that direction. In general, all members and connections must be designed to resist the critical combined effects produced by wind in either direction.

Figure 8.16 Lateral Load Path from the Roof to the Foundation of a One-Story Building with Perimeter Walls—Chords

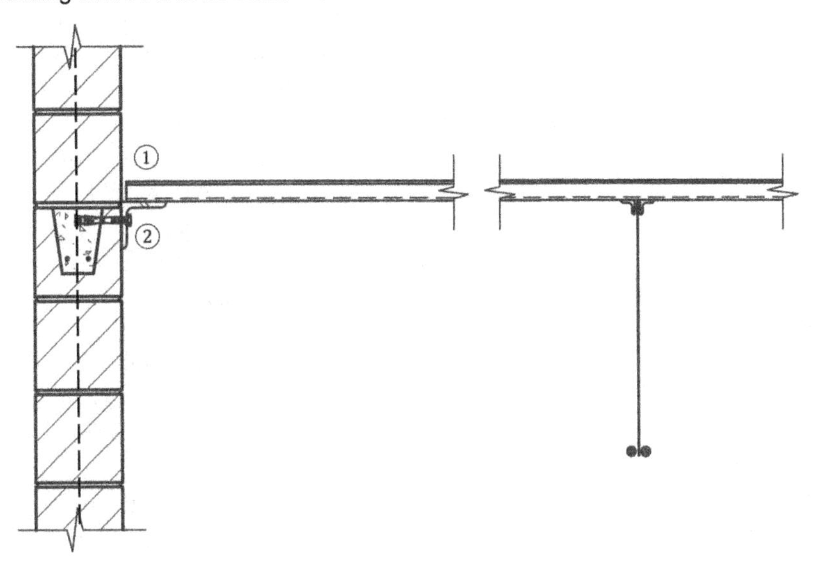

The load paths described above are basically the same for bearing wall buildings with diaphragms constructed of concrete-filled metal deck, concrete topping slab over precast concrete planks, and cast-in-place concrete slab. In these cases, dowel bars are regularly spaced along the length of the masonry or concrete walls and are used to transfer the shear forces from the diaphragm into the wall. Dowel bars for a cast-in-place reinforced concrete system consisting of a slab and wall are illustrated in Figure 8.17.

Figure 8.17 Lateral Load Path at the Interface between a Reinforced Concrete Diaphragm and Wall

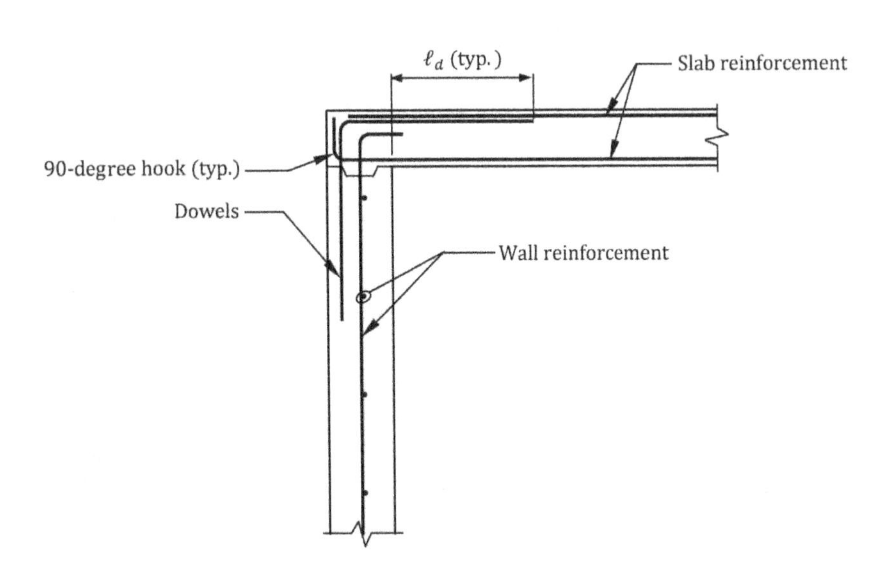

Regardless of the type of MWFRS or the number of stories in a building, the load path begins at the windward wall, continues through the diaphragm to the members of the MWFRS and to the foundation, and ends in the soil surrounding the foundation. All the members and connections along this path must be properly designed for the most critical combined load effects due to gravity and wind loads.

For moment-resisting frames, the diaphragm must be adequately connected to the beams and columns that form the MWFRS. In a cast-in-place reinforced concrete system, continuity is achieved by providing dowel bars between the diaphragm (reinforced concrete slab) and the beams that support it. Reinforcement that connects the beams to the columns and the columns to the foundations is provided, which completes the load path. Because cast-in-place concrete systems are inherently redundant, a continuous load path for wind loads is achieved rather easily.

In the case of a concrete-filled composite steel deck supported by structural steel beams, the concrete diaphragm is usually connected to the steel beams by steel headed stud connectors that are spaced along the length of the beams. These stud connectors, which provide the composite action between the steel beams and the concrete in the metal deck, must be designed for the combined shear flow effects due to gravity and wind loads.

The load path in light-frame wood construction begins at the windward wall, as in all other types of construction, but follows a much more complex path to the soil surrounding the foundation. In general, wind loads are transferred from the exterior out-of-plane walls to diaphragms, which subsequently transfer them to shear walls or frames and then to foundations. The connections between the various elements in the load path typically consist of nails, staples, bolts, straps, and other types of steel hardware. Many of these connections are designed and detailed to transmit the effects from lateral, shear, or uplift loads (i.e., they are not designed to support gravity loads).

Load transfer in a light-frame wood diaphragm begins at the sheathing, which is fastened to the top surface of a wood-framed floor or roof usually by nails or staples. The diaphragm itself consists of the sheathing, boundary members, and intermediate framing that act as stiffeners. The sheathing provides the required shear capacity to transmit the wind loads to the members of the MWFRS, which typically consist of wood shear walls or frame members. Shear capacity of the diaphragm depends on the panel thickness of the sheathing, size of framing members, the size and spacing of the fasteners that attach the edges of the sheathing to the framing members below, the sheathing layout pattern and the presence of wood blocking below the edges of the sheathing.

Transfer of the chord forces at the ends of the diaphragm must also be considered. In a typical wood-frame building with stud walls, the top plates are usually designed as the chord member. Unless the dimensions of a building are very small, the top plate members are not continuous and, thus, must be spliced together. At least two plates are typically provided so that the splice in one plate can be staggered with respect to the splice in the other plate(s). In this way, a continuous chord is provided with at least one member being effective at any given point along the chord. Nails can be used to connect the members when chord forces are relatively small. Screws, bolts, or steel straps are required for relatively large chord forces; in such cases, more than two plates may be necessary.

A typical detail where masonry walls are used in conjunction with horizontal wood framing is given in Figure 8.18 (a similar detail would be applicable to a cast-in-place, reinforced concrete wall). It is usually assumed that the chord forces are resisted by the horizontal reinforcement in the wall at or near the level of the diaphragm. Although it is possible to design the wood ledger to function as the chord member, it is preferable to use the horizontal reinforcement in the wall because the wall is much stiffer than the ledger. This detail shows in-plane shear transfer and not out-of-plane anchorage of the masonry wall to the diaphragm to resist seismic forces (see Section 8.3.3 for more information on load paths for seismic forces).

Collectors

Overview

A collector is defined in IBC 202 as a horizontal diaphragm element that is parallel and in line with the applied force that collects and transfers diaphragm shear forces to the vertical elements of the LFRS and/or distributes forces within the diaphragm.

Figure 8.18 Transfer of Chord Forces in a Wood-Frame Building with Masonry Walls

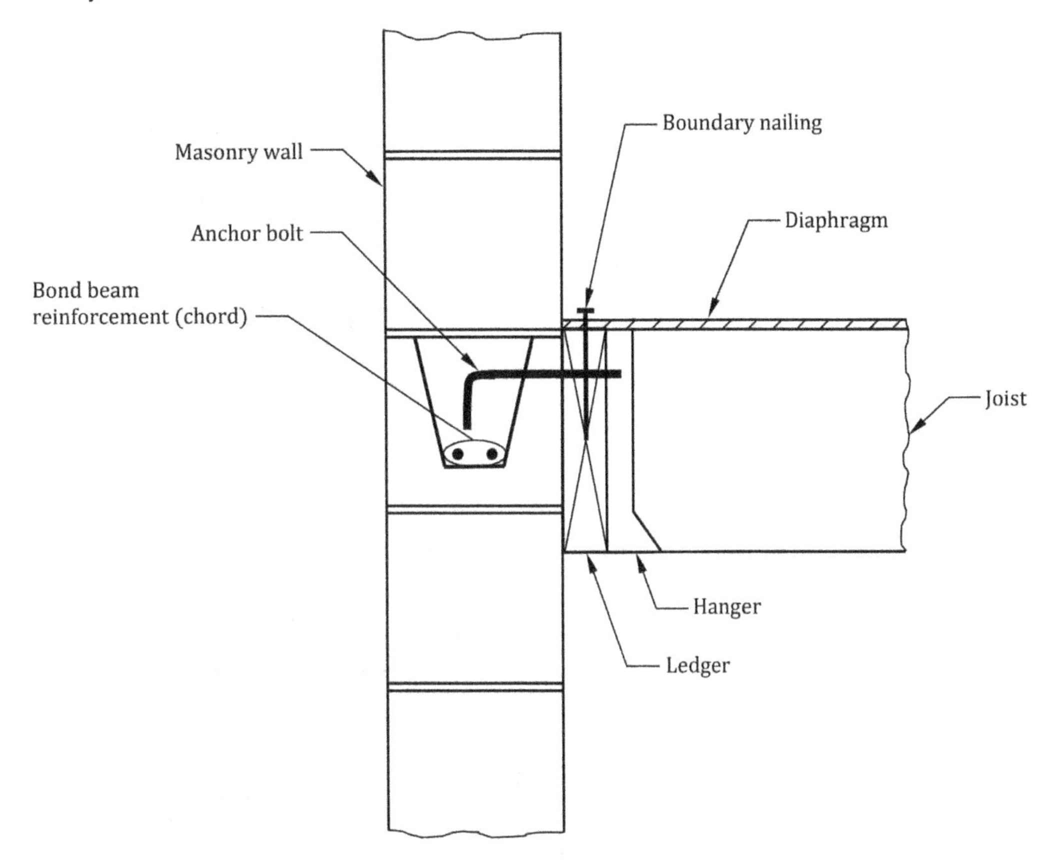

Collectors are used in situations where, in the case of wind loads, the MWFRS does not extend the full length of the building in the direction of analysis (see Figure 8.19 for the case of collector beams with shear walls). As noted in the definition above, the main purpose of a collector is to collect the shear force from the diaphragm over the length where there are no elements of the MWFRS and to transfer this force to the MWFRS.

Load Paths for Wind

The load paths described previously for wind loads can be modified accordingly in cases where collectors are present. In general, the shear forces from the diaphragm must be transferred to the collectors, which in turn, transfer them to the vertical elements of the MWFRS. The general paths before and after this link are the same as outlined above.

Collector forces are obtained from equilibrium. The forces along the line of the wall and collector beams in Figure 8.19 are illustrated in Figure 8.20. Assuming the calculated shear force in the wall due to wind is equal to V_W, the unit shear force per length in the diaphragm is equal to $V_W/(\ell_1+\ell_2+\ell_3)$ and the unit shear force per length in the wall is equal to V_W/ℓ_2. The net shear forces per unit length are determined by subtracting the unit shear force in the diaphragm from that in the wall. The force at any point along the length of a collector is equal to the net shear force in that segment times the length to that point. For example, at point B, the force in the collector is equal to $v_{AB,\text{net}}\ell_1$, and at point C it is $v_{CD,\text{net}}\ell_3$ or, equivalently, $v_{BC,\text{net}}\ell_2 - v_{AB,\text{net}}\ell_1$.

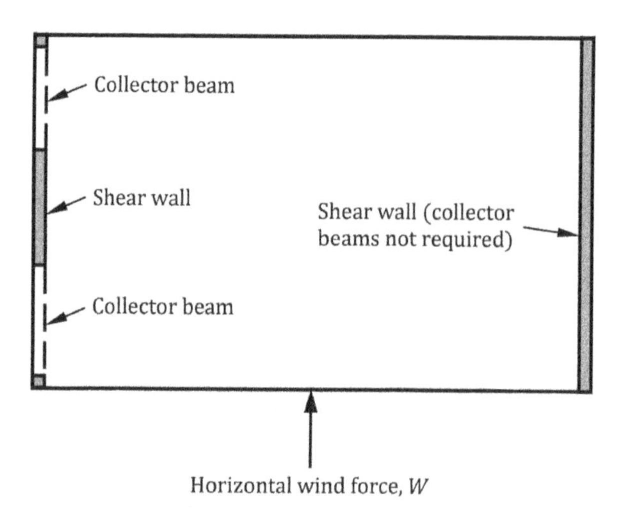

Figure 8.19 Collector Beams and Shear Walls

Figure 8.20 Unit Shear Forces, Net Shear Forces, and Collector Forces in a Diaphragm

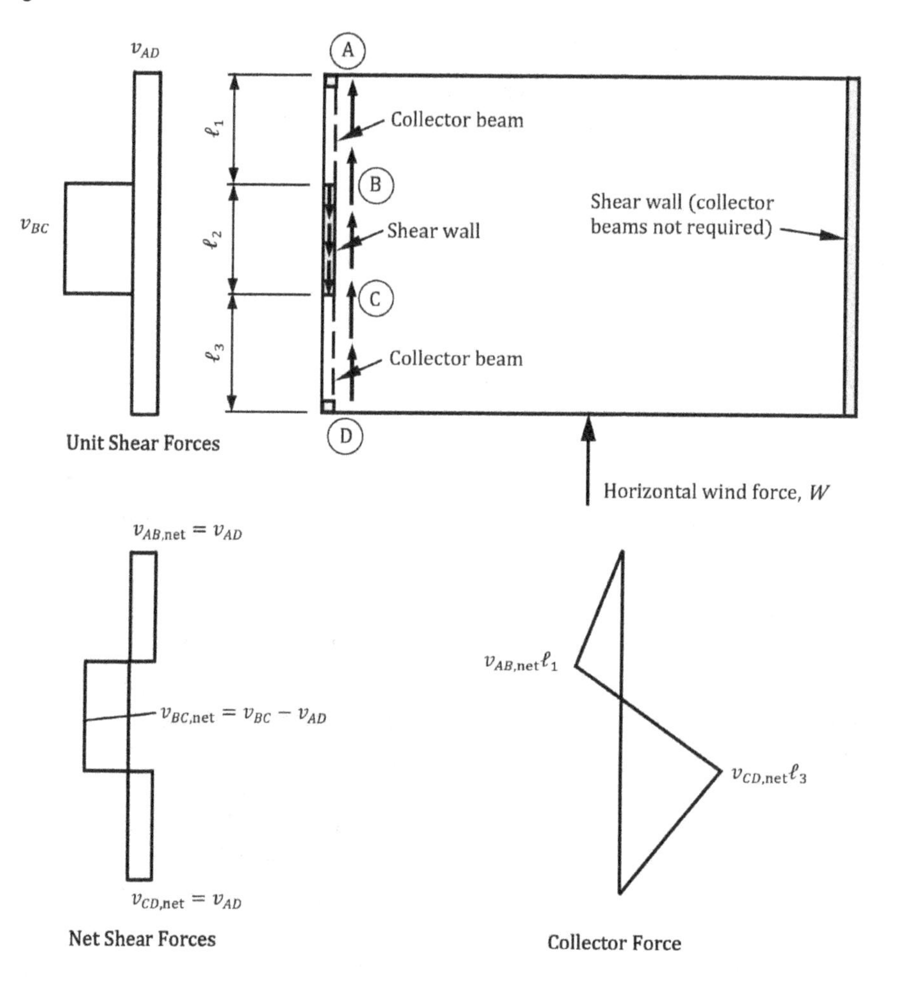

The collector beams must be designed for the applicable gravity and wind load combinations where the axial force in the collector due to wind load effects must be tension and compression. Connections between the diaphragm and the collector beams and between the collector beams and the elements of the MWFRS must also be properly designed and detailed to ensure a continuous load path.

In cast-in-place concrete systems that utilize shear walls, reinforced concrete beams, or a portion of the slab adjacent to the walls are typically designated as collectors. The reinforcement in the collectors must be adequately developed into the shear walls to guarantee continuity in this link of the wind load path.

In other types of construction, it is common to use structural steel members as collectors. A roof system consisting of a wood diaphragm supported on open-web joists, which in turn, are supported by steel columns and precast concrete walls is illustrated in Figure 8.21. The precast walls are the MWFRS, and steel collector beams are utilized at the locations shown in the figure to transfer the loads into the walls.

Figure 8.21 Structural Steel Collector Beams

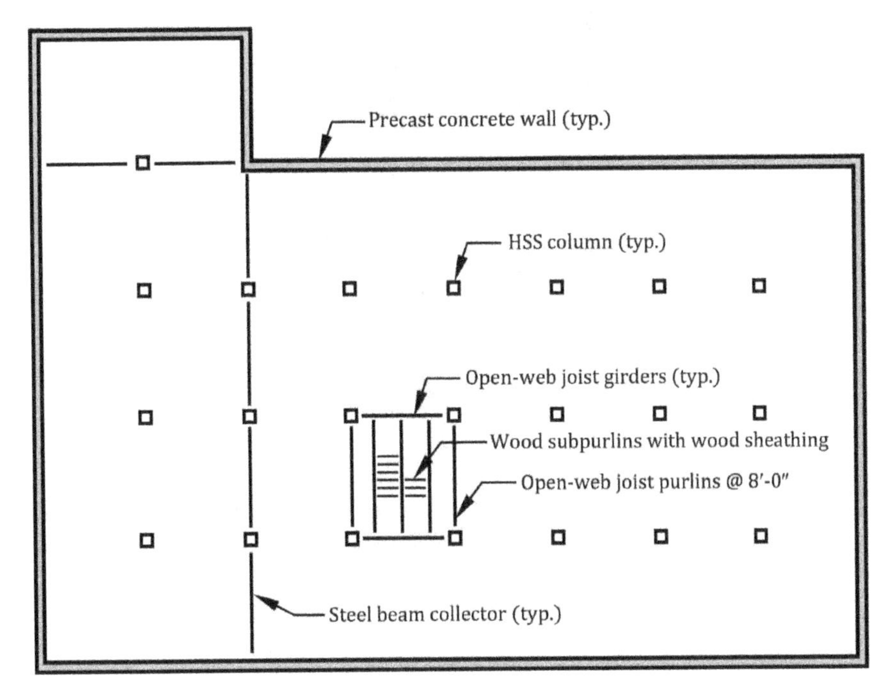

Steel beams are also used as collectors in composite construction, that is, floor or roof systems that utilize a concrete-filled composite metal deck supported by structural steel beams.

In wood-frame construction, the collector is usually the double top plate of the stud wall. In cases where the top plates are not continuous over openings, a beam or header is often used as the collector over the opening. Where reinforced concrete or masonry walls are present, a wood ledger bolted to the wall typically takes on the role of the collector (see Figure 8.22). At interior shear walls, the collector is most often a wood beam located in the plane of the floor or roof.

Illustrated in Figure 8.23 is a wood shear wall with openings on both sides. In this case, the shear forces are transferred from the diaphragm to the blocking between the roof joists and then into the collectors (double top plate on one side of the shear wall and the beam on the other side) and shear wall. These forces are then transferred from the shear wall into the reinforced concrete

Figure 8.22 Wood Ledger Collector

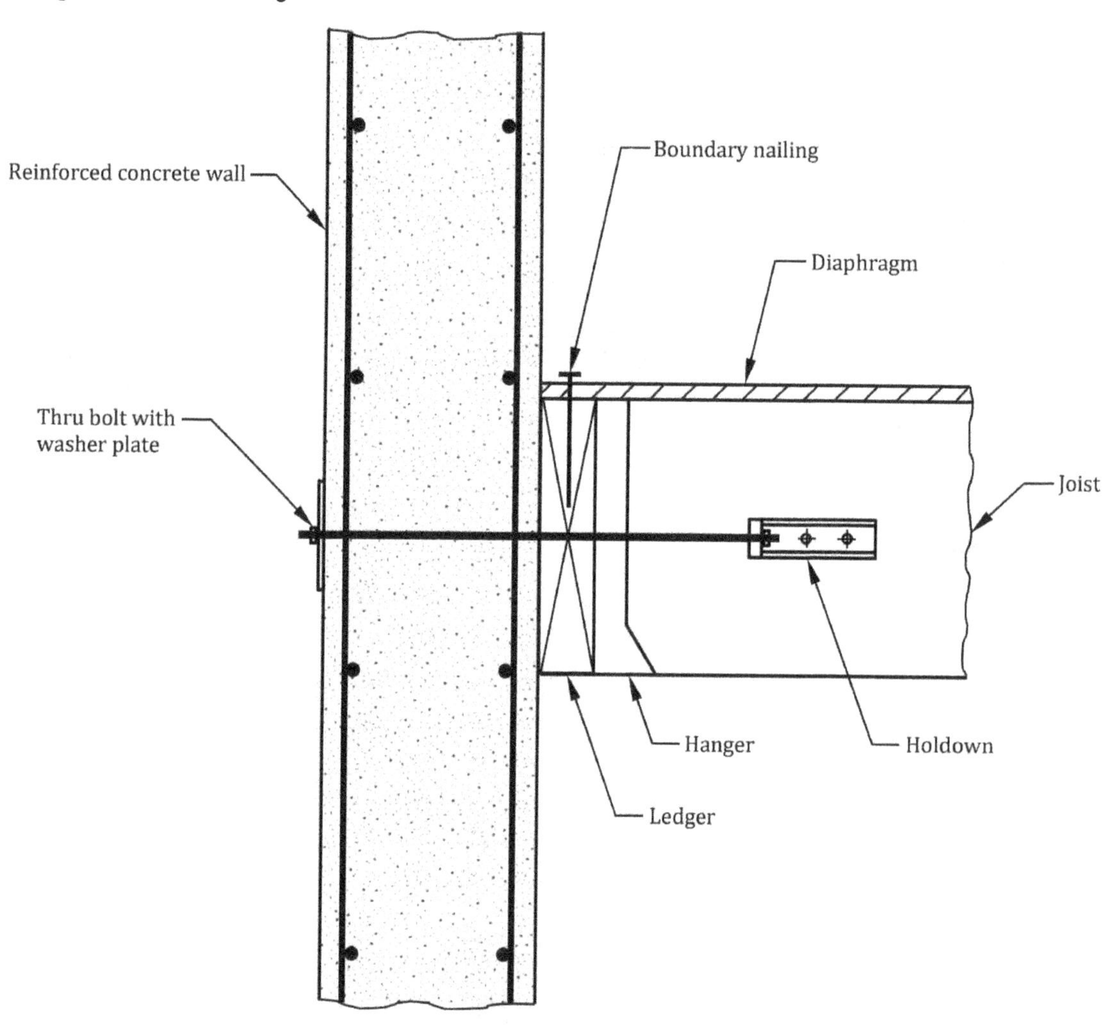

footing through anchor bolts and from the footing into the surrounding soil. The connections and members in this load path must be properly designed for the applicable combination of gravity and lateral forces. In lieu of a load path with blocking, one with a continuous rim joist acting as a collector can be used.

In-depth discussions on load paths for a variety of situations can be found in Reference 8.3.

8.3.3 Load Paths for Earthquake Loads

Overview

Unlike wind loads, the magnitude of which are proportional to the surface area of a building, earthquake loads are generated by the inertia of the building mass that counteracts ground motion. Although the loads are generated differently, the propagation of the loads through the building and into the ground is assumed to be essentially the same under particular conditions.

It is shown in Chapter 6 of this book that for certain types of buildings, the effects of complex seismic ground motion can be adequately represented by applying a set of static forces over the height of the building. These seismic loads, which are assumed to act at the center of mass at each level in the building or structure, are distributed to the elements of the SFRS in essentially the same way wind loads are distributed by considering the relative stiffness of the diaphragm.

Figure 8.23 Collectors in Wood-Frame Construction

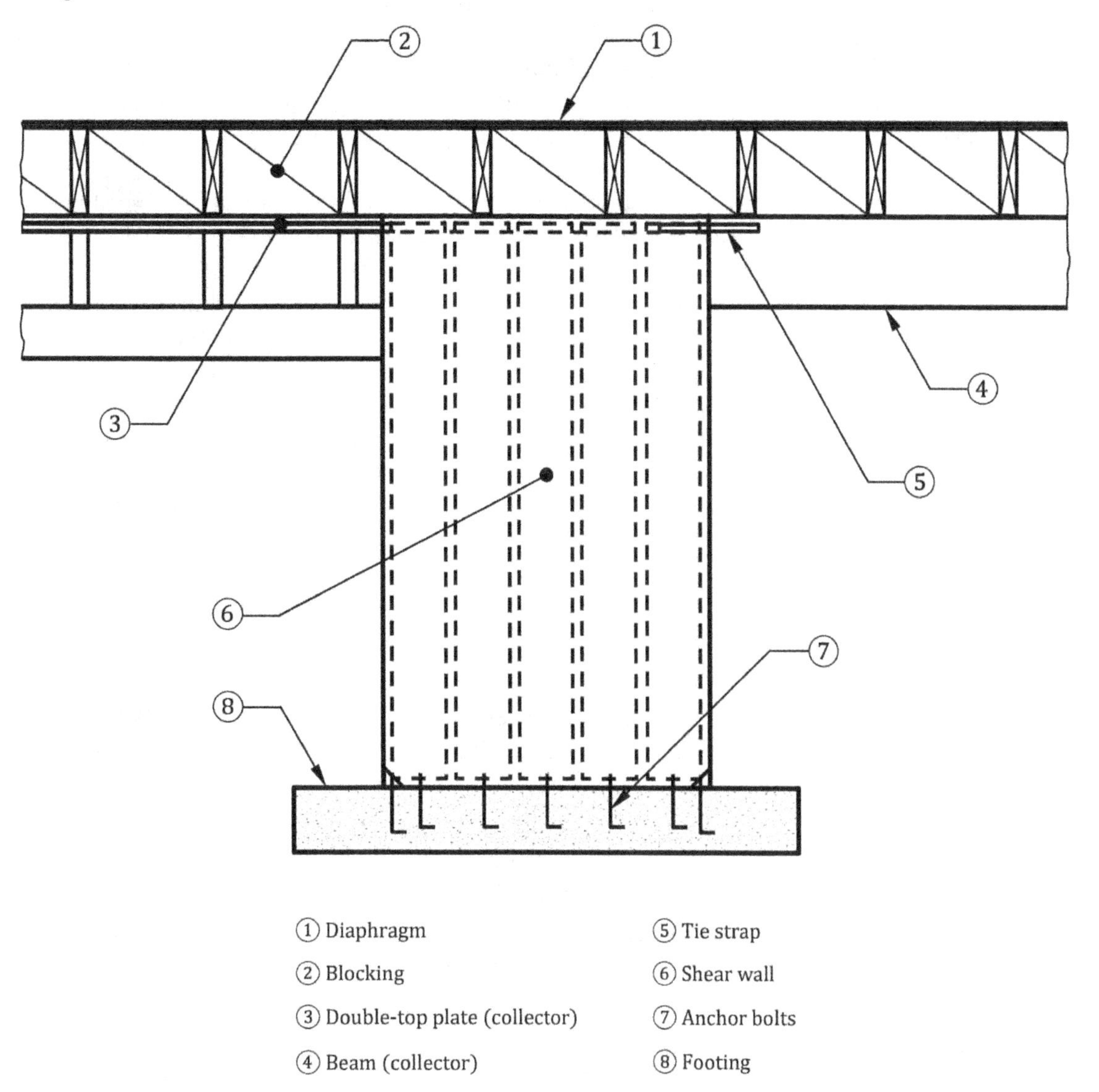

Specific requirements are given in ASCE/SEI 12.1.3 for a continuous load path and interconnection of members in a structure subjected to seismic loads. In particular, all smaller portions of a structure must be tied to the remainder of the structure with elements having a design strength capable of transmitting the greater of $0.133S_{DS}$ times the weight of the smaller portion or 5 percent of the portion's weight where S_{DS} is the short-period design spectral response acceleration parameter. The purpose of this requirement is to help ensure that a minimum amount of continuity is provided in a structure. In structures without redundant components to resist the seismic effects, the members in the load path must be designed to resist the seismic effects so that the integrity of the structure is preserved. In structures with a higher degree of redundancy, one or more of the elements in the load path may fail, but other members are available to resist the seismic effects without loss of overall integrity.

A similar requirement applicable to connections to supporting elements is given in ASCE/SEI 12.1.4: every connection between each beam, girder, or truss to its supporting elements, including diaphragms, must have a minimum design strength of 5 percent of the dead load plus live load reaction on that member. This includes the connections to the foundations.

Additional general information on seismic load paths can be found in Reference 8.4.

Diaphragms and Collectors

Diaphragms and collectors subjected to seismic loads perform essentially the same roles as those subjected to wind loads (see Section 8.3.2 of this book).

Information on how to calculate seismic diaphragm design forces at roof and floor levels is given in Chapter 6 of this book. Special design and detailing requirements must be satisfied for collector elements and their connections in buildings assigned to SDC C and above; similar load combinations are not applicable in the case of wind loads.

In the case of cast-in-place concrete construction, reinforcing bars are designed and detailed to transfer the seismic forces from the diaphragms and collectors into the vertical members of the SFRS (such as walls and columns). Such reinforcing bars must be adequately developed to ensure that load transfer occurs during the design earthquake. Additional information on cast-in-place concrete diaphragms, chords, and collectors can be found in Reference 8.5.

The load paths described in the previous section for composite steel deck and concrete-filled diaphragms subjected to wind loads are basically the same in the case of seismic loads.

In the case of wood-frame structures, the load path described in Section 8.3.2 for wind loads is essentially the same for seismic loads. The lateral load paths for wood structures commonly entail many more members and connections compared to those for other types of construction.

8.4 References

8.1. Federal Emergency Management Agency. 2008. *Design and Construction Guidance for Community Safe Rooms*, FEMA 361. Washington, DC.

8.2. National Council of Structural Engineering Associations (NCSEA). 2009. *Guide to the Design of Diaphragms, Chords and Collectors Based on the 2006 IBC and ASCE/SEI 7-05*. International Code Council, Washington, DC.

8.3. Malone, R.T. and Rice, R.W. 2012. *The Analysis of Irregular Shaped Structures: Diaphragms and Shear Walls*. McGraw Hill: New York, NY.

8.4. Applied Technology Council and Structural Engineers Association of California (ATC/SEAOC). 1999. *Briefing Paper 1 – Building Safety and Earthquakes, Part D: The Seismic Load Path*, ATC/SEAOC Training Curriculum: The Path to Quality Seismic Design and Construction, ATC-48. Redwood City, CA.

8.5. Moehle, J.P., Hooper, J.D., Kelly, D.J., and Meyer, T.R. 2016. "Seismic Design of Cast-in-place Concrete Diaphragms, Chords and Collectors: A Guide for Practicing Engineers," Second Edition, *NEHRP Seismic Design Technical Brief No. 3*, GCR 16-917-42, produced by the Applied Technology Council for the National Institute of Standards and Technology, Gaithersburg, MD.

Index

Note: Page numbers followed by *f* denote figures; by *t*, tables.

T

U

V

W

X

NCSEA
National Council of Structural Engineers Associations

Discover the Power of
Progress with NCSEA
Your ultimate resource fo
elevated education and
unparalleled standards in
structural engineering.

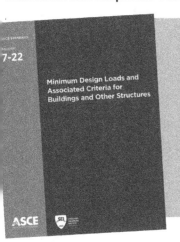